Probability Theory and Stochastic Modelling

Volume 84

AF822801

The **Probability Theory and Stochastic Modelling** series is a merger and continuation of Springer's two well established series Stochastic Modelling and Applied Probability and Probability and Its Applications series. It publishes research monographs that make a significant contribution to probability theory or an applications domain in which advanced probability methods are fundamental. Books in this series are expected to follow rigorous mathematical standards, while also displaying the expository quality necessary to make them useful and accessible to advanced students as well as researchers. The series covers all aspects of modern probability theory including

- Gaussian processes
- Markov processes
- Random fields, point processes and random sets
- Random matrices
- Statistical mechanics and random media
- Stochastic analysis

as well as applications that include (but are not restricted to):

- Branching processes and other models of population growth
- Communications and processing networks
- Computational methods in probability and stochastic processes, including simulation
- Genetics and other stochastic models in biology and the life sciences
- Information theory, signal processing, and image synthesis
- Mathematical economics and finance
- Statistical methods (e.g. empirical processes, MCMC)
- Statistics for stochastic processes
- Stochastic control
- Stochastic models in operations research and stochastic optimization
- Stochastic models in the physical sciences

More information about this series at http://www.springer.com/series/13205

René Carmona • François Delarue

Probabilistic Theory of Mean Field Games with Applications II

Mean Field Games with Common Noise and Master Equations

Springer

René Carmona
ORFE Department
Program in Applied and
Computational Mathematics
Princeton University
Princeton, NJ, USA

François Delarue
Institut Universitaire de France
& Laboratoire J.A. Dieudonné
Université Nice Sophia Antipolis
Nice, France

ISSN 2199-3130 ISSN 2199-3149 (electronic)
Probability Theory and Stochastic Modelling
ISBN 978-3-030-13259-0 ISBN 978-3-319-56436-4 (eBook)
DOI 10.1007/978-3-319-56436-4

Printed on acid-free paper

This Springer imprint is published by Springer Nature
The registered company is Springer International Publishing AG
The registered company address is: Gewerbestrasse 11, 6330 Cham, Switzerland

Foreword

Since its inception about a decade ago, the theory of Mean Field Games has rapidly developed into one of the most significant and exciting sources of progress in the study of the dynamical and equilibrium behavior of large systems. The introduction of ideas from statistical physics to identify approximate equilibria for sizeable dynamic games created a new wave of interest in the study of large populations of competitive individuals with "mean field" interactions. This two-volume book grew out of series of lectures and short courses given by the authors over the last few years on the mathematical theory of Mean Field Games and their applications in social sciences, economics, engineering, and finance. While this is indeed the object of the book, by taste, background, and expertise, we chose to focus on the probabilistic approach to these game models.

In a trailblazing contribution, Lasry and Lions proposed in 2006 a methodology to produce approximate Nash equilibria for stochastic differential games with symmetric interactions and a large number of players. In their models, a given player *feels* the presence and the behavior of the other players through the empirical distribution of their private states. This type of interaction was studied in the statistical physics literature under the name of *mean field* interaction, hence the terminology Mean Field Game coined by Lasry and Lions. The theory of these new game models was developed in lectures given by Pierre-Louis Lions at the Collège de France which were video-taped and made available on the internet. Simultaneously, Caines, Huang, and Malhamé proposed a similar approach to large games under the name of Nash Certainty Equivalence principle. This terminology fell from grace and the standard reference to these game models is now Mean Field Games.

While slow to pick up momentum, the subject has seen a renewed wave of interest over the last seven years. The mean field game paradigm has evolved from its seminal principles into a full-fledged field attracting theoretically inclined investigators as well as applied mathematicians, engineers, and social scientists. The number of lectures, workshops, conferences, and publications devoted to the subject has grown exponentially, and we thought it was time to provide the applied mathematics community interested in the subject with a textbook presenting the state of the art, as we see it. Because of our personal taste, we chose to focus on what we like to call the probabilistic approach to mean field games. While a significant portion of the text is based on original research by the authors, great care was taken to include models and results contributed by others, whether or not they were aware of the fact they were working with mean field games. So the book should feel and read like a textbook, not a research monograph.

Most of the material and examples found in the text appear for the first time in book form. In fact, a good part of the presentation is original, and the lion's share of the arguments used in the text have been designed especially for the purpose of the book. Our concern for pedagogy justifies (or at least explains) why we chose to divide the material in two volumes and present mean field games without a common noise first. We ease the introduction of the technicalities needed to treat models with a common noise in a crescendo of sophistication in the complexity of the models. Also, we included at the end of each volume four extensive indexes (author index, notation index, subject index, and assumption index) to make navigation throughout the book seamless.

Acknowledgments

First and foremost, we want to thank our wives Debbie and Mélanie for their understanding and unwavering support. The intensity of the research collaboration which led to this two-volume book increased dramatically over the years, invading our academic lives as well as our social lives, pushing us to the brink of sanity at times. We shall never be able to thank them enough for their patience and tolerance. This book project would not have been possible without them: our gratitude is limitless.

Next we would like to thank Pierre-Louis Lions, Jean-Michel Lasry, Peter Caines, Minyi Huang, and Roland Malhamé for their incredible insight in introducing the concept of mean field games. Working independently on both sides of the pond, their original contributions broke the grounds for an entirely new and fertile field of research. Next in line is Pierre Cardaliaguet, not only for numerous private conversations on game theory but also for the invaluable service provided by the notes he wrote from Pierre-Louis Lions' lectures at the Collège de France. Although they were never published in printed form, these notes had a tremendous impact on the mathematical community trying to learn about the subject, especially at a time when writings on mean field games were few and far between.

We also express our gratitude to the organizers of the 2013 and 2015 conferences on mean field games in Padova and Paris: Yves Achdou, Pierre Cardaliaguet, Italo Capuzzo-Dolcetta, Paolo Dai Pra, and Jean-Michel Lasry.

While we like to cast ourselves as proponents of the probabilistic approach to mean field games, it is fair to say that we were far from being the only ones following this path. In fact, some of our papers were posted essentially at the same time as papers of Bensoussan, Frehse, and Yam, addressing similar questions, with the same type of methods. We benefitted greatly from this stimulating and healthy competition.

We also thank our coauthors, especially Jean-François Chasagneux, Dan Crisan, Jean-Pierre Fouque, Daniel Lacker, Peiqi Wang, and Geoffrey Zhu. We used our joint works as the basis for parts of the text which they will recognize easily.

Also, we would like to express our gratitude to the many colleagues and students who gracefully tolerated our relentless promotion of this emerging field of research through courses, seminar, and lecture series. In particular, we would like to thank Jean-François Chasagneux, Rama Cont, Dan Crisan, Romuald Elie, Josselin Garnier, Marcel Nutz, Huyen Pham, and Nizar Touzi for giving us the opportunity to do just that.

Finally, we would like to thank Paul-Eric Chaudru de Raynal, Mark Cerenzia, Christy Graves, Dan Lacker, Victor Marx, and Peiqi Wang for their help in proofreading an earlier version of the manuscript.

The work of R. Carmona was partially supported by NSF DMS-1211928 and DMS-1716673, and by ARO W911NF-17-1-0578. The work of F. Delarue was partially supported by the Institut Universitaire de France and by the Agence Nationale de la Recherche (project MFG, ANR-16-CE40-0015-01).

Princeton, NJ, USA — René Carmona
Nice, France — François Delarue
July 29, 2016

Preface of Volume II

While the first volume of the book only addressed mean field games where the sources of random shocks were idiosyncratic to the individual players, this second volume tackles the analysis of mean field games in which the players are subject to a common source of random shocks. We call these models *games with a common noise*. General solvability results, as well as uniqueness conditions, are established in Part I. Part II is devoted to the study of the so-called master equation. In the final chapter of this part, we connect the various asymptotic models investigated in the book, starting from games with finitely many players. Even for models without a common noise, this type of analysis heavily relies on the tools developed throughout this second volume. Like in the case of Volume I, Volume II ends with an epilogue. There, we discuss several extensions which fit naturally the framework of models with a common noise. They include games with major and minor players, and games of timing.

The reader will find on page xiii below a diagram summarizing the interconnections between the different chapters and parts of the book.

Chapter 1 is the cornerstone of the second volume. It must be seen as a preparation for the analysis of mean field games with a common noise. As equilibria become random in the presence of a common noise, we need to revisit the tools introduced and developed in the first volume for the solution of optimal stochastic control problems and establish a similar technology for stochastic dynamics and cost depending on an additional source of randomness. To that effect, we provide a general introduction to forward-backward systems in random environment and possibly non-Brownian filtrations. A key point in our analysis is to allow the random environment not to be adapted to the Brownian motions driving the controlled dynamics. This forces us to impose *compatibility* conditions constraining the correlation between the noise carrying the random environment and the general filtration to which the solution of the forward-backward system is required to be adapted. Although these *compatibility* conditions are rather difficult to handle, they turn out to be absolutely crucial as they play a fundamental role throughout the text, in particular for the notion of weak equilibria defined in the subsequent Chapter 2. As far as we know, this material does not exist in book form.

The notion of solution for mean field games with a common noise ends up being more subtle than one could expect. In contrast with the case addressed in Volume I, the fixed point problem underpinning the definition of an equilibrium cannot be tackled by means of Schauder's fixed point theorem. In order to account for the dependence of the equilibria upon the realization of the common noise, it is indeed necessary to enlarge the space in which the fixed point has to be sought. Unfortunately, proceeding in this way increases dramatically the complexity of the problem, as the new space of solutions becomes so big that it becomes very difficult to identify tractable compactness criteria to use with Schauder's theorem. The purposes of Chapters 2 and 3 is precisely to overcome this issue. The first step of our strategy is to discretize the realization of the common noise entering the definition of an equilibrium in such a way that the equilibrium can take at most a finite number of outcomes as the realization of the common noise varies. This makes the size of the space of solutions much more reasonable. The second step is to pass to the limit along discretized solutions. The success of this approach comes at the price of weakening the notion of equilibrium in the sense that the equilibrium is only adapted to a filtration which is larger than the filtration generated by the common noise. This notion of weak solutions is the rationale for the Compatibility Condition introduced in Chapter 1. The concepts of weak and strong solutions are explained in Chapter 2. In analogy with strong solutions of stochastic differential equations, strong equilibria are required to be adapted to the filtration generated by the common noise. Naturally, we establish a form of Yamada-Watanabe theory for Nash equilibria. Quite expectedly, it says that weak solutions are strong provided that uniqueness holds in the strong sense. In analogy with the theory developed in the first volume, we prove that strong and weak equilibria may be represented by means of forward-backward systems of the conditional McKean-Vlasov type. Due to the presence of the common noise, conditioning appears in the formulation. To wit, we first develop, in the first section of the chapter, a theory of *conditional propagation of chaos* for particle systems with mean field interaction driven by a common noise.

The construction of weak solutions is addressed in Chapter 3. There, we implement the aforementioned discretization approach. It consists in forcing the total number of realizations of an equilibrium to be finite. The passage to the limit along discretized solutions is achieved in the weak sense: we consider the asymptotic behavior of the joint law of the equilibria and the forward and backward processes characterizing the best response of the representative agent under the discretized environment. While tightness of the forward component is investigated for the standard uniform topology on the space of continuous paths, tightness of the backward component is established for the Meyer-Zheng topology on the space of right-continuous paths with left limits. The results of the Meyer-Zheng topology needed for this purpose are recalled in the first section of Chapter 3. In the last sections, we address the question of uniqueness. Like for mean field games without a common noise, mean field games with a common noise are shown to have at most one solution under the Lasry-Lions monotonicity condition. Furthermore, we prove that the common noise may restore uniqueness in some cases for which the game without a common noise has several solutions.

The next two chapters are dedicated to another major aspect of mean field games: the *master equation*. The master equation was introduced by Lions in his lectures at the *Collège de France*. It is a special partial differential equation over the enlarged state space made of the physical Euclidean space times the space of probability measures on the physical space. In order to define it properly, we appeal to the differential calculus on the Wasserstein space introduced in Chapter 5 of the first volume. For mean field games without a common noise, the master equation is a first-order equation in the measure argument. It becomes of the second order with respect to the measure argument when the mean field game includes a common noise. We call *master field* the solution of the master equation. This terminology is inspired by the notion of *decoupling field* introduced in the first volume to connect the forward and backward components of a forward-backward system. In full analogy, the master field is a function that makes the connection between the forward and backward components of a forward-backward system of the McKean-Vlasov type. The fact that it is defined over the enlarged state space is fully consistent with the fact that the solution of a (conditional) McKean-Vlasov stochastic differential equation can only be a Markov process on the enlarged state space. In the first two sections of Chapter 4, we show that the master field is well defined provided that there exists a unique equilibrium for any initial state of the population. In such a case, we prove that it satisfies a dynamic programming principle. By adapting the chain rule for functions of probability measures proven in Chapter 5 of the first volume, we show that the master field is a viscosity solution of the master equation. We then derive explicitly the master equation for some of the examples introduced in Chapter 1 of the first volume.

The purpose of Chapter 5 is to prove that the master equation has a classical solution provided that the coefficients are smooth enough in all the directions of the enlarged state space and satisfy the Lasry-Lions monotonicity condition. The proof comprises two main steps. The first one is to prove that the result holds when the time horizon is small enough. In the second step, we show that the small time construction can be iterated when the monotonicity condition is in force. The key point in both steps is to prove that the master field is smooth enough. To do so, we view the forward-backward system of the conditional McKean-Vlasov type which characterizes the equilibrium as a system of random characteristics for the master equation. We then make intensive use of Lions's approach to the differential calculus on the Wasserstein space, which we called L-differential calculus in Volume I. We establish the regularity of the master field by proving that the forward-backward system responds smoothly to perturbations of the initial condition in the L^2 space.

The last chapter of Part II is devoted to approximation and convergence problems. We first prove that solutions of a mean field game may be used to construct approximate Nash equilibria for the corresponding games with finitely many players. The proof is given for games with and without common noise. To do so, we appeal to some of the results obtained in Chapters 3 and 4 of the first volume. We also prove an analog result for mean field control problems studied in Chapter 6 of the first volume but for a different notion of equilibrium: we show that solutions of the limiting problem induce an almost optimal strategy for a central planner optimizing the common reward of a collectivity of N players over exchangeable control profiles. In the last two sections of Chapter 6, we address the converse question, which is known as the convergence problem for mean field games. We prove, under suitable conditions, that the Nash equilibria of the N-player games converge to solutions of the corresponding mean field game. We give two proofs, depending on the nature of the equilibria, whether they are computed over strategies in open or closed loop forms. The convergence problem for equilibria over open loop strategies is tackled by a compactness argument which is very similar to that used to construct weak solutions in Chapter 3. In particular, it does not require the limiting mean field game to be uniquely solvable. This is in contrast with the approach we use to tackle the same problem for equilibria over closed loop strategies for which we require the master equation to have a classical solution and, as a by-product, the mean field game to have a unique solution. The proof appears as a variation over the so-called *four-step-scheme* for forward-backward systems as it consists in expanding the master equation along the N-equilibrium and in comparing the resulting process with the N-equilibrium value process.

As for the first volume, the final chapter leverages the technology developed in the second volume to revisit some of the examples introduced in the introductory Chapter 1 of the first volume, and complete their mathematical analysis. We use some of the tools introduced for the analysis of mean field games with a common noise to study important game models which are not amenable to the theory covered by the first volume. These models include extensions to games with major and minor players, and games of timing. We believe that these mean field game models have a great potential for the quantitative analysis of very important practical applications, and we show how the technology developed in the second volume of the book can be brought to bear on their solutions.

Princeton, NJ, USA
Nice, France
December 29, 2016

René Carmona
François Delarue

Organization of the Book: Volume II Organigram

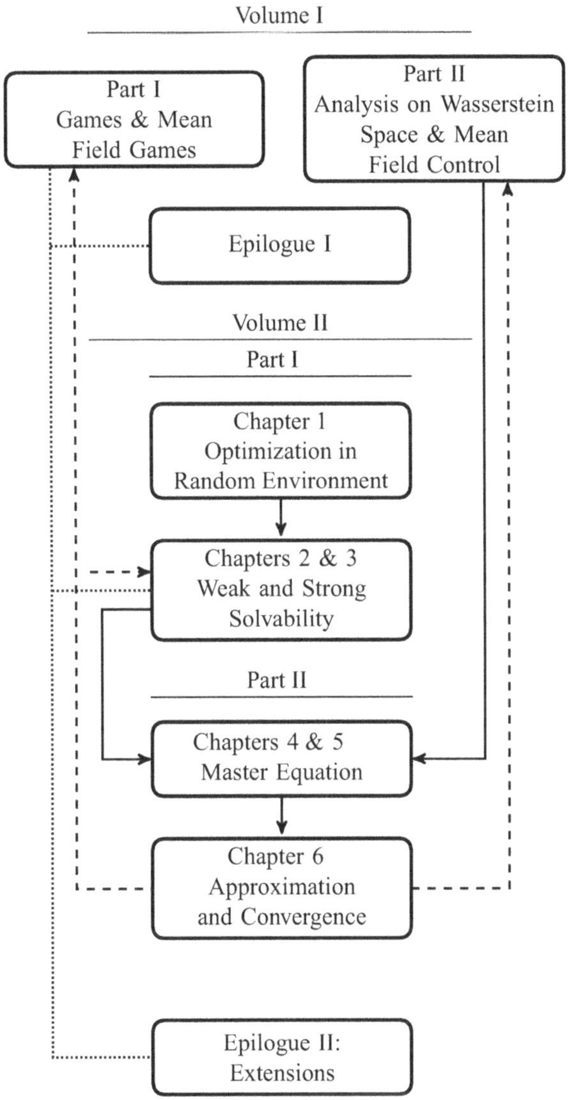

Thick lines indicate the logical order of the chapters. The dotted lines between Part I, Epilogue I, Chapters 2 and 3, and Epilogue II connect the various types of mean field games studied in the book. Finally, the dashed lines starting from Part II point toward the games and the optimization problems for which we can solve approximately the finite-player versions or for which the finite-player equilibria are shown to converge.

References to the first volume appear in the text in the following forms: Chapter (Vol I)-X, Section (Vol I)-$X.x$, Theorem (Vol I)-$X.x$, Proposition (Vol I)-$X.x$, Equation (Vol I)-($X.x$), ..., where X denotes the corresponding chapter in the first volume and x the corresponding label.

Contents

Volume II

Volume I

Part I The Probabilistic Approach to Mean Field Games

Part I

MFGs with a Common Noise

1 Optimization in a Random Environment

Abstract

This chapter is a preparation for the analysis of mean field games with a common noise, to which we dedicate the entire first half of this second volume. By necessity, we revisit the basic tools introduced in Chapters (Vol I)-3 and (Vol I)-4 for mean field games without common noise, and in particular, the theory of forward-backward stochastic differential equations and its connection with optimal stochastic control. Our goal is to investigate optimal stochastic control problems based on stochastic dynamics and cost functionals depending on an additional random environment. To that effect, we provide a general discussion of forward-backward systems in a random environment. In the framework of mean field games, this random environment will account for the random state of the population in equilibrium given the (random) realization of the systemic noise source common to all the players.

1.1 FBSDEs in a Random Environment

While the first models of mean field games which we studied in Chapters (Vol I)-3 and (Vol I)-4 led to optimal control problems which can be solved by means of standard BSDEs and FBSDEs, more sophisticated game models, including models with a common noise and games with major and minor players, require the manipulation of FBSDEs with coefficients depending upon an extra source of randomness. The purpose of this section is to revisit our discussion of FBSDEs to accommodate coefficients depending upon a random environment, say a process $\boldsymbol{\mu} = (\mu_t)_{0 \le t \le T}$ taking values in an auxiliary Polish space. To be more specific, we shall consider systems of the form:

R. Carmona, F. Delarue, *Probabilistic Theory of Mean Field Games with Applications II*, Probability Theory and Stochastic Modelling 84,
DOI 10.1007/978-3-319-56436-4_1

$$\begin{cases} dX_t = B(t, X_t, \mu_t, Y_t, Z_t, Z_t^0)dt \\ \qquad\qquad + \Sigma(t, X_t, \mu_t)dW_t + \Sigma^0(t, X_t, \mu_t)dW_t^0, \\ dY_t = -F(t, X_t, \mu_t, Y_t, Z_t, Z_t^0)dt + Z_t dW_t + Z_t^0 dW_t^0, \end{cases} \tag{1.1}$$

with a terminal condition $Y_T = G(X_T, \mu_T)$ for some finite time horizon $T > 0$. Notice the similarity with (3.17) in Chapter (Vol I)-3. However, the forward dynamics are now subject to an additional Brownian motion $\boldsymbol{W}^0 = (W_t^0)_{0 \le t \le T}$ which will account for the common noise when dealing with mean field games. In the next paragraph, we make clear the correlation between the three sources of noise $\boldsymbol{W} = (W_t)_{0 \le t \le T}$, $\boldsymbol{W}^0 = (W_t^0)_{0 \le t \le T}$ and $\boldsymbol{\mu} = (\mu_t)_{0 \le t \le T}$. Essentially, we shall assume that $\boldsymbol{W}$ is independent of $(\boldsymbol{W}^0, \boldsymbol{\mu})$.

Part of the material presented in this section is needed in Chapters 2 and 3 for the analysis of mean field games with a common noise. Since this material is rather technical, and since the measure theoretic and probabilistic issues it resolves may be viewed as esoteric, some of the detailed arguments can be skipped in a first reading.

Readers familiar with the classical theory of stochastic differential equations will recognize standard arguments aimed at defining and comparing strong and weak solutions of these equations. A good part of this chapter is devoted to extend these arguments to FBSDEs in random environment. Theorem 1.33, which essentially says that strong uniqueness implies uniqueness in law, is a case in point. Its proof, like several others in this chapter, relies heavily on the notion of regular conditional probability whose definition and existence we recall in the form of a theorem for the convenience of the reader.

Theorem 1.1 *Let* $\mathbf{Q}$ *be a probability measure on a Polish space* $\mathcal{S}$ *equipped with its Borel* σ*-field* $\mathcal{B}(\mathcal{S})$ *and let* $\mathcal{G} \subset \mathcal{B}(\mathcal{S})$ *be any sub-*σ*-field. There exists a family* $(Q(\omega, D))_{\omega \in \mathcal{S}, D \in \mathcal{B}(\mathcal{S})}$*, called the regular conditional probability of* $\mathbf{Q}$ *given* $\mathcal{G}$*, such that:*

1. *For any* $\omega \in \mathcal{S}$*, the mapping* $\mathcal{B}(\mathcal{S}) \ni D \mapsto Q(\omega, D)$ *is a probability measure on* $(\mathcal{S}, \mathcal{B}(\mathcal{S}))$*;*
2. *For any* $D \in \mathcal{B}(\mathcal{S})$*, the mapping* $\mathcal{S} \ni \omega \mapsto Q(\omega, D)$ *is measurable from the space* $(\mathcal{S}, \mathcal{G})$ *into the space* $([0, 1], \mathcal{B}([0, 1]))$*;*
3. *For any* $D \in \mathcal{B}(\mathcal{S})$*, for* $\mathbf{Q}$*-almost every* $\omega \in \mathcal{S}$*,*

$$Q(\omega, D) = \mathbf{E}^{\mathbf{Q}}\big[\mathbf{1}_D | \mathcal{G}\big](\omega).$$

If $\mathcal{G}$ *is generated by a countable* π*-system, the family* $(Q(\omega, D))_{\omega \in \Omega, D \in \mathcal{B}(\mathcal{S})}$ *is said to be proper as it satisfies:*

4. *For* $\mathbf{Q}$*-almost every* $\omega \in \Omega$*, for all* $D \in \mathcal{G}$*,* $Q(\omega, D) = \mathbf{1}_D(\omega)$*.*

When $(\Xi, \mathcal{H})$ *is another measurable space and* $\mathbf{P}$ *is a probability measure on the product space* $(\Xi \times \mathcal{S}, \mathcal{H} \otimes \mathcal{B}(\mathcal{S}))$*, there exists a family of kernels* $(q(x, D))_{x\in\Xi, D\in\mathcal{B}(\mathcal{S})}$*, called regular conditional probability of the second marginal of* $\mathbf{P}$ *given the first one, such that:*

1. *For any* $x \in \Xi$*, the mapping* $\mathcal{B}(\mathcal{S}) \ni D \mapsto q(x, D)$ *is a probability measure on* $(\mathcal{S}, \mathcal{B}(\mathcal{S}))$*,*
2. *For any* $D \in \mathcal{B}(\mathcal{S})$*, the mapping* $\Xi \ni x \mapsto q(x, D)$ *is measurable from* $(\Xi, \mathcal{H})$ *into* $([0, 1], \mathcal{B}([0, 1]))$.
3. *For any* $(C, D) \in \mathcal{H} \times \mathcal{B}(\mathcal{S})$*,*

$$\mathbf{P}\big(C \times D\big) = \int_C q(x, D) d\lambda(x),$$

where λ *is the marginal distribution of* $\mathbf{P}$ *on* $(\Xi, \mathcal{H})$.

Since $\mathcal{B}(\mathcal{S})$ is generated by a countable π-system, note that, in both cases, the conditional probability is almost surely unique: In the first case, any other family $(Q'(\omega, D))_{\omega\in\Omega, D\in\mathcal{B}(\mathcal{S})}$ fulfilling the same conditions as $(Q(\omega, D))_{\omega\in\mathcal{S}, D\in\mathcal{B}(\mathcal{S})}$ satisfies, for $\mathbf{Q}$ almost every $\omega \in \mathcal{S}$ and for all $D \in \mathcal{B}(\mathcal{S})$, $Q(\omega, D) = Q'(\omega, D)$; Similarly, in the second case, any other family $(q'(x, D))_{x\in\Xi, D\in\mathcal{B}(\mathcal{S})}$ fulfilling the same conditions as $(q(x, D))_{x\in\Xi, D\in\mathcal{B}(\mathcal{S})}$ satisfies, for λ almost every $x \in \Xi$ and for all $D \in \mathcal{B}(\mathcal{S})$, $q(x, D) = q'(x, D)$.

1.1.1 Immersion and Compatibility

For the remainder of the book, we shall have to deal with several filtrations on the same probability space. This should not be a surprise since we shall have to disentangle the effects of the idiosyncratic sources of noise from the common noise affecting the whole system. This subsection gathers the most important theoretical definitions and properties we shall need in the sequel.

Throughout the subsection, T is a fixed finite time horizon.

Definition 1.2 *If* $\mathbb{F}$ *and* $\mathbb{G}$ *are two filtrations on the probability space* $(\Omega, \mathcal{F}, \mathbb{P})$ *such that* $\mathbb{G} \subset \mathbb{F}$ *by which we mean* $\mathcal{G}_t \subset \mathcal{F}_t$ *for all* $t \in [0, T]$*, we say that* $\mathbb{G}$ *is immersed in* $\mathbb{F}$ *if every square integrable* $\mathbb{G}$*-martingale is an* $\mathbb{F}$*-martingale.*

Of course, we stress the fact that the immersion property depends on the underlying probability $\mathbb{P}$, so that $\mathbb{G}$ should be said to be immersed in $\mathbb{F}$ under $\mathbb{P}$. Most of the time, we shall omit to specify $\mathbb{P}$, since the probability measure should be clear from the context.

The immersion property is important and is often called the (H)-hypothesis, see the citations in the Notes & Complements at the end of the chapter. It can be characterized in a few convenient ways.

Proposition 1.3 *If $\mathbb{F}$ and $\mathbb{G}$ are two filtrations on $(\Omega, \mathcal{F}, \mathbb{P})$ such that $\mathbb{G} \subset \mathbb{F}$, $\mathbb{G}$ is immersed in $\mathbb{F}$ if and only if any of the three following properties holds:*

(H1) *For every $t \in [0, T]$, $\mathcal{G}_T$ and $\mathcal{F}_t$ are conditionally independent given $\mathcal{G}_t$.*
(H2) *For every $t \in [0, T]$ and any $\zeta \in L^1(\mathcal{F}_t)$, $\mathbb{E}[\zeta|\mathcal{G}_T] = \mathbb{E}[\zeta|\mathcal{G}_t]$.*
(H3) *For every $t \in [0, T]$ and any $\xi \in L^1(\mathcal{G}_T)$, $\mathbb{E}[\xi|\mathcal{F}_t] = \mathbb{E}[\xi|\mathcal{G}_t]$.*

Proof. Clearly (H3) $\Rightarrow$ (H), so we only prove (H) $\Rightarrow$ (H1), (H1) $\Rightarrow$ (H2), and (H2) $\Rightarrow$ (H3).
(H) $\Rightarrow$ (H1). Let $\xi \in L^2(\mathcal{G}_T)$ and let us assume that (H) holds. This implies that the process $\boldsymbol{M}$ defined by $M_t = \mathbb{E}[\xi|\mathcal{G}_t]$ is an $\mathbb{F}$-martingale and $M_T = \xi$, and consequently, $M_t = \mathbb{E}[\xi|\mathcal{F}_t]$. It follows that for any $\zeta \in L^2(\mathcal{F}_t)$, we have:

$$\mathbb{E}[\xi\zeta|\mathcal{G}_t] = \mathbb{E}[\zeta\mathbb{E}[\xi|\mathcal{F}_t]|\mathcal{G}_t] = \mathbb{E}[\zeta\mathbb{E}[\xi|\mathcal{G}_t]|\mathcal{G}_t] = \mathbb{E}[\zeta|\mathcal{G}_t]\mathbb{E}[\xi|\mathcal{G}_t],$$

which is exactly (H1).
(H1) $\Rightarrow$ (H2). Let $\xi \in L^2(\mathcal{G}_T)$ and $\zeta \in L^2(\mathcal{F}_t)$. Assuming (H1), we have:

$$\mathbb{E}[\xi\mathbb{E}[\zeta|\mathcal{G}_t]] = \mathbb{E}\big[\mathbb{E}[\xi|\mathcal{G}_t]\mathbb{E}[\zeta|\mathcal{G}_t]\big] = \mathbb{E}\big[\mathbb{E}[\xi\zeta|\mathcal{G}_t]\big] = \mathbb{E}[\xi\zeta],$$

which is exactly (H2) modulo the integrability conditions which can be checked by simple density arguments.
(H2) $\Rightarrow$ (H3). Let $\xi \in L^2(\mathcal{G}_T)$ and $\zeta \in L^2(\mathcal{F}_t)$ and let us assume that (H2) holds. Then:

$$\mathbb{E}[\zeta\mathbb{E}[\xi|\mathcal{G}_t]] = \mathbb{E}[\xi\mathbb{E}[\zeta|\mathcal{G}_t]] = \mathbb{E}[\xi\zeta],$$

which is exactly (H3) modulo the integrability conditions, and as before, the general case is easily obtained by approximation. □

Example. A trivial example is given by $\mathbb{F} = \mathbb{G} \vee \mathbb{H}$ where $\mathbb{G}$ and $\mathbb{H}$ are two filtrations such that $\mathcal{G}_T$ and $\mathcal{H}_T$ are independent.

Remark 1.4 *If a process $\mathbf{W} = (W_t)_{0 \le t \le T}$ is a $\mathbb{G}$-Wiener process and if $\mathbb{G}$ is immersed in $\mathbb{F}$ (i.e., hypothesis* (H) *holds), then $\mathbf{W}$ is also an $\mathbb{F}$-martingale, and since its increasing process (square bracket) does not depend upon the filtration, it is still an $\mathbb{F}$-Wiener process.*

Remark 1.5 *If $\mathbb{G}$ is immersed in $\mathbb{F}$, then for any $t \in [0, T]$, $\mathcal{G}_t = \mathcal{F}_t \cap \mathcal{G}_T$ up to negligible events, which means that $\mathcal{G}_t \subset \mathcal{F}_t \cap \mathcal{G}_T$ and that, for every event in $\mathcal{F}_t \cap \mathcal{G}_T$, we can find another event in $\mathcal{G}_t$ such that both have a symmetric difference of zero probability. Moreover, if τ is an $\mathbb{F}$-stopping time which is $\mathcal{G}_T$-measurable, then τ is also a $\mathbb{G}$-stopping time if $\mathbb{G}$ is complete and immersion holds. Indeed, if $A \in \mathcal{F}_t \cap \mathcal{G}_T$, then $\mathbf{1}_A = \mathbb{E}[\mathbf{1}_A|\mathcal{G}_T] = \mathbb{E}[\mathbf{1}_A|\mathcal{G}_t]$; the claim follows by choosing $A = \{\tau \le t\}$.*

We now specialize the immersion property (H) to the particular case when the smaller filtration $\mathbb{G}$ is generated by an $\mathbb{F}$-adapted process on a probability space $(\Omega, \mathcal{F}, \mathbb{P})$.

Definition 1.6 *An $\mathbb{F}$-adapted càd-làg process $\boldsymbol{\theta} = (\theta_t)_{0 \le t \le T}$, on $(\Omega, \mathcal{F}, \mathbb{P})$ with values in a Polish space, is said to be compatible with $\mathbb{F}$ (under $\mathbb{P}$) if its filtration $\mathbb{F}^{\boldsymbol{\theta}}$ is immersed in $\mathbb{F}$.*

Here and throughout the book, *càd-làg* stands for right continuous with left limits (*continue à droite* with *limites à gauche* in French). The filtration $\mathbb{F}^{\boldsymbol{\theta}} = (\mathcal{F}_t^{\boldsymbol{\theta}})_{0 \le t \le T}$ denotes the smallest right-continuous filtration on $(\Omega, \mathcal{F}, \mathbb{P})$ which contains $\mathbb{P}$-null events and which renders $\boldsymbol{\theta}$ adapted, namely $\mathcal{F}_t^{\boldsymbol{\theta}}$ is the completion of $\cap_{s \in (t,T]} \mathcal{F}_s^{\mathrm{nat},\boldsymbol{\theta}}$, where $\mathbb{F}^{\mathrm{nat},\boldsymbol{\theta}} = (\mathcal{F}_s^{\mathrm{nat},\boldsymbol{\theta}})_{0 \le s \le T}$ is the natural filtration generated by $\boldsymbol{\theta}$:

$$\mathcal{F}_t^{\boldsymbol{\theta}} = \sigma\{\mathcal{N}\} \vee \Big(\bigcap_{s \in (t,T]} \mathcal{F}_s^{\mathrm{nat},\boldsymbol{\theta}} \Big) = \bigcap_{s \in (t,T]} \Big(\sigma\{\mathcal{N}\} \vee \mathcal{F}_s^{\mathrm{nat},\boldsymbol{\theta}} \Big),$$

where $\mathcal{N} = \{B \subset \Omega : \exists C \in \mathcal{F},\ B \subset C,\ \mathbb{P}(C) = 0\}$.

Notice that because of (H1) of Proposition 1.3 above, $\boldsymbol{\theta}$ is compatible with $\mathbb{F}$ if and only if for any $t \in [0, T]$, $\mathcal{F}_T^{\boldsymbol{\theta}}$ and $\mathcal{F}_t$ are conditionally independent under $\mathbb{P}$ given the σ-field $\mathcal{F}_t^{\boldsymbol{\theta}}$.

Lemma 1.7 *Let $\boldsymbol{\theta}$ be an $\mathbb{F}$-adapted càd-làg process with values in a Polish space. Then, $\boldsymbol{\theta}$ is compatible with $\mathbb{F}$ if for any $t \in [0, T]$, the σ-fields $\mathcal{F}_t$ and $\mathcal{F}_T^{\mathrm{nat},\boldsymbol{\theta}}$ are conditionally independent given $\mathcal{F}_t^{\mathrm{nat},\boldsymbol{\theta}}$.*

Proof. For any $D \in \mathcal{F}_t$ and any $E \in \mathcal{F}_T^{\mathrm{nat},\boldsymbol{\theta}}$, for $\varepsilon > 0$ such that $t + \varepsilon \le T$, we have:

$$\mathbb{P}\big[D \cap E | \mathcal{F}_{t+\varepsilon}^{\mathrm{nat},\boldsymbol{\theta}}\big] = \mathbb{P}\big[D | \mathcal{F}_{t+\varepsilon}^{\mathrm{nat},\boldsymbol{\theta}}\big] \mathbb{P}\big[E | \mathcal{F}_{t+\varepsilon}^{\mathrm{nat},\boldsymbol{\theta}}\big].$$

Letting ε tend to 0, we get:

$$\mathbb{P}\big[D \cap E | \mathcal{F}_{t+}^{\mathrm{nat},\boldsymbol{\theta}}\big] = \mathbb{P}\big[D | \mathcal{F}_{t+}^{\mathrm{nat},\boldsymbol{\theta}}\big] \mathbb{P}\big[E | \mathcal{F}_{t+}^{\mathrm{nat},\boldsymbol{\theta}}\big],$$

where $\mathcal{F}_{t+}^{\mathrm{nat},\boldsymbol{\theta}} = \cap_{\varepsilon \in (0, T-t]} \mathcal{F}_{t+\varepsilon}^{\mathrm{nat},\boldsymbol{\theta}}$.

Recalling that, for any $F \in \mathcal{F}_t^{\theta}$, there exists $G \in \mathcal{F}_{t+}^{\mathrm{nat},\boldsymbol{\theta}}$ such that the symmetric set difference $F \Delta G$ between F and G is of zero measure under $\mathbb{P}$, we deduce that the left-hand side is $\mathbb{P}$ almost-surely equal to $\mathbb{P}(D \cap E | \mathcal{F}_t^{\boldsymbol{\theta}})$. Proceeding similarly with the two terms in the right-hand side, we complete the argument. □

Remark 1.8 *What Lemma 1.7 says is that compatibility is not sensitive to completeness of filtrations. More generally, $\mathbb{G}$ is immersed in $\mathbb{F}$ if and only if either $\mathbb{G}$ or its completion is immersed in the completion of $\mathbb{F}$.*

The following immediate consequence of the definition of compatibility and Proposition 1.3 will be crucial for the analysis of FBSDEs of the same type as (1.1). We state it as a lemma for the sake of later reference.

Lemma 1.9 *Let $\boldsymbol{\theta}$ be an $\mathbb{F}$-adapted càd-làg process on the space $(\Omega, \mathcal{F}, \mathbb{P})$ with values in a Polish space, and let us assume that $\boldsymbol{\theta}$ is $\mathbb{F}$ compatible. Then if $\ell \geq 1$, any $\mathbb{R}^\ell$-valued martingale $\boldsymbol{M} = (M_t)_{0 \leq t \leq T}$ with respect to the filtration $\mathbb{F}^{\boldsymbol{\theta}}$ is also a martingale with respect to the filtration $\mathbb{F}$.*

This chapter is not the only place where the notion of compatibility will be needed. Indeed, we shall use it in Chapter 7 for the study of weak equilibria for mean field games of timing. There we shall see that compatibility properties will provide tools to approximate compatible (randomized) stopping times with nonrandomized stopping times.

In anticipation of the set-ups in which we shall use compatibility, we move to a somewhat special setting for the following discussion. We consider a probability space with a product structure $\Omega = \Omega^1 \times \Omega^2$. Suppose $\mathbb{F}^1$ is a filtration on Ω^1 and $\mathbb{F}^2$ a filtration on Ω^2. Without mentioning it explicitly, we canonically extend the filtration $\mathbb{F}^1$ (resp. $\mathbb{F}^2$) from Ω^1 (resp. Ω^2) to $\Omega = \Omega^1 \times \Omega^2$ by replacing $\mathcal{F}^1_t$ (resp. $\mathcal{F}^2_t$) by $\mathcal{F}^1_t \otimes \{\emptyset, \Omega^2\}$ (resp. $\{\emptyset, \Omega^1\} \otimes \mathcal{F}^2_t$) for each $t \in [0, T]$.

Given a probability measure $\mathbb{P}$ on $(\Omega, \mathcal{F})$ where we set $\mathcal{F} = \mathcal{F}^1_T \otimes \mathcal{F}^2_T$, we consider the problem of the immersion of the extension $\mathbb{F}^1$ into $\mathbb{F}$ where $\mathbb{F}$ is the product filtration defined by $\mathcal{F}_t = \mathcal{F}^1_t \otimes \mathcal{F}^2_t$ for $t \in [0, T]$. Even though we shall not need them, similar results could be stated and proved for the immersion of the extension of $\mathbb{F}^2$ into $\mathbb{F}$. In terms of the definition of compatibility given above, this means that we are interested in finding out when the first projection from Ω onto Ω^1 is compatible with $\mathbb{F}$. The following proposition provides a clean characterization of the immersion property, or equivalently of this compatibility condition, in terms of a measurability property of the regular version of the conditional probability of $\mathbb{P}$ with respect to its first marginal projection.

Proposition 1.10 *If $\mathbb{P}$ is a probability measure on $(\Omega^1 \times \Omega^2, \mathcal{F}^1_T \otimes \mathcal{F}^2_T)$ of the form*

$$\mathbb{P}(d\omega^1, d\omega^2) = \mathbb{P}^1(d\omega^1) P^2(\omega^1, d\omega^2), \tag{1.2}$$

for some probability $\mathbb{P}^1$ on $(\Omega^1, \mathcal{F}^1_T)$ and a kernel P^2 from $(\Omega^1, \mathcal{F}^1_T)$ to $(\Omega^2, \mathcal{F}^2_T)$, then the following are equivalent:

(i) For each $t \in [0, T)$ and $A \in \mathcal{F}^2_t$, the map $\omega^1 \mapsto P^2(\omega^1, A)$ is measurable with respect to the completion of $\mathcal{F}^1_t$.
(ii) Every martingale $\boldsymbol{M}$ on $(\Omega^1, \mathbb{F}^1, \mathbb{P}^1)$, extended to $\Omega^1 \times \Omega^2$ by $M_t(\omega^1, \omega^2) = M_t(\omega^1)$, remains a martingale on $(\Omega^1 \times \Omega^2, \mathbb{F}^1 \otimes \mathbb{F}^2, \mathbb{P})$.

We refer to Theorem 1.1 for the notion of *kernel*.

Proof. (*i*) $\Rightarrow$ (*ii*). Let us assume that for every $t \in [0,T]$, $P^2(\cdot,A)$ is $\mathcal{F}_t^1$-measurable for every $A \in \mathcal{F}_t^2$, and let $\boldsymbol{M} = (M_t)_{0 \le t \le T}$ be a martingale on $(\Omega^1, \mathbb{F}^1, \mathbb{P}^1)$. Then, if $0 \le t \le s \le T$, $C \in \mathcal{F}_t^1$ and $A \in \mathcal{F}_t^2$, we get:

$$\mathbb{E}\big[M_s \mathbf{1}_{C\times A}\big] = \mathbb{E}^1\big[M_s \mathbf{1}_C P^2(\cdot,A)\big] = \mathbb{E}^1\big[M_t \mathbf{1}_C P^2(\cdot,A)\big] = \mathbb{E}\big[M_t \mathbf{1}_{C\times A}\big].$$

Thus, $M_t = \mathbb{E}[M_s | \mathcal{F}_t^1 \otimes \mathcal{F}_t^2]$, which proves (*ii*).
(*ii*) $\Rightarrow$ (*i*). Conversely, if $t \in [0,T)$ and $A \in \mathcal{F}_t^2$, for any $C \in \mathcal{F}_T^1$ we have $\mathbb{P}^1[C|\mathcal{F}_t^1] = \mathbb{P}[C \times \Omega^2 | \mathcal{F}_t]$ by hypothesis. So:

$$\mathbb{E}^1[\mathbf{1}_C P^2(\cdot,A)] = \mathbb{E}[\mathbf{1}_C \mathbf{1}_A] = \mathbb{E}\big[\mathbf{1}_A \mathbb{P}^1[C|\mathcal{F}_t^1]\big] = \mathbb{E}\big[\mathbb{P}^1[C|\mathcal{F}_t^1] P^2(\cdot,A)\big],$$

which implies that $P^2(\cdot,A)$ is measurable with respect to the completion of $\mathcal{F}_t^1$, proving (i). □

In the following, our choices of Ω^1 and Ω^2 are always Polish spaces, and the filtrations are always such that $\mathcal{F}_T^1$ and $\mathcal{F}_T^2$ are the respective Borel σ-fields. In these cases, Theorem 1.1 says that every probability measure $\mathbb{P}$ on $\Omega = \Omega^1 \times \Omega^2$ admits a disintegration of the form (1.2). Notice that Theorem 1.1 only requires that Ω^2 is a Polish space.

1.1.2 Compatible Probabilistic Set-Up

Thanks to our preliminary discussion on compatibility, we are now in position to provide a careful description of the probabilistic structure used to investigate forward-backward systems of the type (1.1).

We shall work on a general probabilistic set-up, based on a complete probability space $(\Omega, \mathcal{F}, \mathbb{P})$ equipped with a complete and right-continuous filtration $\mathbb{F}$, and with a $2d$-dimensional $\mathbb{F}$-Wiener process $(\boldsymbol{W}^0, \boldsymbol{W}) = (W_t^0, W_t)_{t\ge 0}$, $\boldsymbol{W}^0 = (W_t^0)_{t\ge 0}$ and $\boldsymbol{W} = (W_t)_{t\ge 0}$ being both of dimension d. As suggested by the form of the FBSDE (1.1), we require only two d-dimensional Wiener processes. This is consistent with the fact that we shall apply the results of this section to the analysis of mean field games in which a generic player is interacting (and competing) with a continuum of players subject to a common systemic noise of dimension d. By convention, the index 0 always refers to the common noise and we reserve the index 1, or no index at all, to the idiosyncratic noise. Indeed, in order to alleviate notation, we shall often drop the index 1, and use the notation $\boldsymbol{W}$ rather than $\boldsymbol{W}^1$.

Throughout the section, the common random environment manifests in the form of an input $\boldsymbol{\mu} = (\mu_t)_{t\ge 0}$ which is an $\mathbb{F}$-progressively measurable, right continuous with left limits process (or *càd-làg* process for short) with values in an auxiliary metric space $(\mathcal{X}, d)$ which we will assume to be a Polish space (meaning that it is complete and separable). Importantly, the pair process $(\boldsymbol{W}^0, \boldsymbol{\mu})$ is required to be independent of $\boldsymbol{W}$. As demonstrated by the analysis of mean field games without common noise performed in Chapters (Vol I)-3 and (Vol I)-4, see for instance Remark (Vol I)-3.7, the typical example we should keep in mind for $\mathcal{X}$ is $\mathcal{P}_2(\mathbb{R}^d)$, the space of probability measures on $\mathbb{R}^d$ with finite second moments, endowed with the 2-Wasserstein distance (see Chapter (Vol I)-5). The measure μ_t should be thought of as the statistical distribution at time t of the state of a population subject to some random forcing under the action of $\boldsymbol{W}^0$. In this respect, independence between $(\boldsymbol{W}^0, \boldsymbol{\mu})$ and $\boldsymbol{W}$ accounts for the fact that the stochastic flow of random measures

$$\boldsymbol{\mu} = \Big(\mu_t : \Omega \ni \omega \mapsto \mu_t(\omega) \in \mathcal{P}_2(\mathbb{R}^d)\Big)_{t\geq 0}$$

is not subject to $\boldsymbol{W}$.

In the context of mean field games without a common noise, the common noise $\boldsymbol{W}^0$ is not present, and as explained in Chapter (Vol I)-3, $\boldsymbol{\mu} = (\mu_t)_{0\leq t\leq T}$ is deterministic. It is understood as a candidate for the flow of marginal distributions of the state of a generic player in equilibrium. For mean field games with a common noise, $\boldsymbol{\mu}(\omega^0) = (\mu_t(\omega^0))_{0\leq t\leq T}$ must be understood as a candidate for, still in equilibrium, the flow of conditional marginal distributions of the state of a generic player, given the realization $\boldsymbol{W}^0(\omega^0)$ of the common noise. For the time being, $\boldsymbol{\mu} = (\mu_t)_{t\geq 0}$ is used as a random *input* only. See Remark 1.11 below for the case of deterministic input $\boldsymbol{\mu}$. With d denoting the distance of $\mathcal{X}$ as well, we assume that for a fixed element $0_{\mathcal{X}} \in \mathcal{X}$:

$$\mathbb{E}[\sup_{0\leq t\leq T} d(0_{\mathcal{X}}, \mu_t)^2] < \infty, \tag{1.3}$$

condition which is independent of the particular choice of the point $0_{\mathcal{X}} \in \mathcal{X}$. In our probabilistic set-up, we also include an initial condition $X_0 \in L^2(\Omega, \mathcal{F}_0, \mathbb{P}; \mathbb{R}^d)$, which will be used to model the initial condition of one typical player in the population when we deal with mean field games. The triplet $(X_0, \boldsymbol{W}^0, \boldsymbol{\mu})$ is assumed to be independent of $\boldsymbol{W}$.

In the FBSDE (1.1), $\boldsymbol{X} = (X_t)_{0\leq t\leq T}$, $\boldsymbol{Y} = (Y_t)_{0\leq t\leq T}$, $\boldsymbol{Z} = (Z_t)_{0\leq t\leq T}$ and $\boldsymbol{Z}^0 = (Z_t^0)_{0\leq t\leq T}$ are $\mathbb{F}$-progressively measurable processes defined on $(\Omega, \mathcal{F}, \mathbb{P})$ with values in $\mathbb{R}^d$, $\mathbb{R}^m$, $\mathbb{R}^{m\times d}$ and $\mathbb{R}^{m\times d}$ respectively. The coefficients B and F, (resp. Σ and Σ^0, resp. G) are measurable mappings from $[0,T]\times\mathbb{R}^d\times\mathcal{X}\times\mathbb{R}^m\times\mathbb{R}^{m\times d}\times\mathbb{R}^{m\times d}$ (resp. $[0,T]\times\mathbb{R}^d\times\mathcal{X}$, resp. $\mathbb{R}^d\times\mathcal{X}$) with values in $\mathbb{R}^d$ (resp. $\mathbb{R}^{d\times d}$, resp. $\mathbb{R}^m$). To alleviate the notation, we often identify $\mathbb{R}^{m\times d}\times\mathbb{R}^{m\times d}$ with $\mathbb{R}^{2(m\times d)}$.

Throughout the section, we assume:

Assumption (FBSDE in Random Environment). There exists a constant $C \geq 0$ such that, for all $(t, x, \mu, y, z, z^0) \in [0,T]\times\mathbb{R}^d\times\mathcal{X}\times\mathbb{R}^m\times(\mathbb{R}^{m\times d})^2$,

$$\big|(\Sigma, \Sigma^0)(t,x,\mu)\big| \leq C\big(1 + |x| + d(0_{\mathcal{X}}, \mu)\big),$$

$$\big|(F,B,G)(t,x,\mu,y,z,z^0)\big| \leq C\big[1 + |x|^2 + \big(d(0_{\mathcal{X}},\mu)\big)^2 + |y|^2 + |z|^2 + |z^0|^2\big].$$

Once again, notice that $0_{\mathcal{X}}$ could be replaced by any other point of $\mathcal{X}$. We restrict ourselves to quite a mild type of FBSDEs for simplicity only. In particular, the diffusion coefficients of (1.1) do not contain backward terms. Some of the results we provide below could be generalized to a more general framework, but we refrain from doing so because there is no real need for a higher level of generality given the nature of the applications we have in mind.

Inspired by the preliminary discussion on compatibility, we shall require the following assumption throughout the analysis:

Assumption (Compatibility Condition). The process $(X_0, \boldsymbol{W}^0, \boldsymbol{\mu}, \boldsymbol{W})$ is compatible with $\mathbb{F}$.

From a practical point of view, compatibility of $(X_0, \boldsymbol{W}^0, \boldsymbol{\mu}, \boldsymbol{W})$ with $\mathbb{F}$ means that, given the observations of the initial condition X_0, of the realizations of the noises $\boldsymbol{W}^0$ and $\boldsymbol{W}$ and of the environment $\boldsymbol{\mu}$ up until time t, the observation of other events in the σ-field $\mathcal{F}_t$ does not supply any additional information on the joint behavior of the three processes $\boldsymbol{W}^0$, $\boldsymbol{\mu}$ and $\boldsymbol{W}$ in the future after t. See Definition 1.6.

Remark 1.11 *When $\boldsymbol{\mu}$ is deterministic,* (1.3) *is irrelevant since $\boldsymbol{\mu} = (\mu_t)_{0 \leq t \leq T}$ is taken as a right-continuous with left limits function from $[0, T]$ into $\mathcal{X}$. The compatibility condition is automatically satisfied since $(\boldsymbol{W}^0, \boldsymbol{W})$ is an $\mathbb{F}$-Brownian motion under $\mathbb{P}^0$: All the increments of the form $(W_s^0 - W_t^0, W_s - W_t)_{t \leq s \leq T}$ are independent of $\mathcal{F}_t$.*

If in addition the coefficients B and F are independent of z^0 and $\boldsymbol{W}^0 \equiv 0$ (which fits with the above prescriptions if we accept 0 as a degenerate Brownian motion), then the two stochastic integrals driven by $\boldsymbol{W}^0$ in (1.1) *disappear. We then recover the system (Vol I)-(3.17) and our framework is consistent with Chapters (Vol I)-3 and (Vol I)-4.*

To wit, Remark 1.11 says that the **compatibility condition is useless when $\boldsymbol{\mu}$ is deterministic**.

Remark 1.12 *For the same reasons as in Remark 1.11, the compatibility condition is always satisfied whenever $\boldsymbol{\mu}$ is adapted to the filtration generated by X_0 and $\boldsymbol{W}^0$.*

The following definition is prompted by the above discussion.

Definition 1.13 *A complete probability space $(\Omega, \mathcal{F}, \mathbb{P})$ equipped with a complete and right-continuous filtration $\mathbb{F} = (\mathcal{F}_t)_{0 \leq t \leq T}$ and with a tuple $(X_0, \boldsymbol{W}^0, \boldsymbol{\mu}, \boldsymbol{W})$ is* admissible *if*

1. $X_0 \in L^2(\Omega, \mathcal{F}_0, \mathbb{P}; \mathbb{R}^d)$,
2. $(\mathbf{W}^0, \mathbf{W})$ *is a 2d-Brownian motion with respect to* $\mathbb{F}$ *under* $\mathbb{P}$,
3. $(X_0, \mathbf{W}^0, \boldsymbol{\mu})$ *is independent of* $\mathbf{W}$ *under* $\mathbb{P}$,
4. $(X_0, \mathbf{W}^0, \boldsymbol{\mu}, \mathbf{W})$ *and* $\mathbb{F}$ *are compatible under* $\mathbb{P}$.

More on the Compatibility Condition

We now collect several useful properties in the form of two lemmas.

Lemma 1.14 *The process* $(X_0, \mathbf{W}^0, \boldsymbol{\mu}, \mathbf{W})$ *and the filtration* $\mathbb{F}$ *are compatible if* X_0 *is an* $\mathcal{F}_0$*-measurable initial condition and the process* $(\mathbf{W}^0, \boldsymbol{\mu}, \mathbf{W})$ *is compatible with the filtration* $\mathbb{F}$.

Proof. The proof relies on the following observation. For any $t \in [0, T]$ and any three events $B_0 \in \sigma\{X_0\}$, $C_t \in \mathcal{F}_t^{(\mathbf{W}^0,\boldsymbol{\mu},\mathbf{W})}$ and $C_T \in \mathcal{F}_T^{(\mathbf{W}^0,\boldsymbol{\mu},\mathbf{W})}$, compatibility of $(\mathbf{W}^0, \boldsymbol{\mu}, \mathbf{W})$ with $\mathbb{F}$ says that:

$$\begin{aligned}\mathbb{P}(B_0 \cap C_t \cap C_T) &= \mathbb{E}\Big[\mathbf{1}_{C_t}\mathbb{P}\big(B_0|\mathcal{F}_t^{(\mathbf{W}^0,\boldsymbol{\mu},\mathbf{W})}\big)\mathbb{P}\big(C_T|\mathcal{F}_t^{(\mathbf{W}^0,\boldsymbol{\mu},\mathbf{W})}\big)\Big] \\ &= \mathbb{E}\Big[\mathbf{1}_{B_0 \cap C_t}\mathbb{P}\big(C_T|\mathcal{F}_t^{(\mathbf{W}^0,\boldsymbol{\mu},\mathbf{W})}\big)\Big],\end{aligned}$$

from which we deduce (since $\mathcal{F}_t^{(X_0,\mathbf{W}^0,\boldsymbol{\mu},\mathbf{W})}$ is generated by $\sigma\{X_0\}$ and $\mathcal{F}_t^{(\mathbf{W}^0,\boldsymbol{\mu},\mathbf{W})}$) that, $\mathbb{P}$ almost surely,

$$\mathbb{P}\Big(C_T|\mathcal{F}_t^{(X_0,\mathbf{W}^0,\boldsymbol{\mu},\mathbf{W})}\Big) = \mathbb{P}\big(C_T|\mathcal{F}_t^{(\mathbf{W}^0,\boldsymbol{\mu},\mathbf{W})}\big). \tag{1.4}$$

Therefore, for an additional $D_t \in \mathcal{F}_t$,

$$\begin{aligned}&\mathbb{P}\Big(\big(B_0 \cap C_t\big) \cap C_T \cap D_T\Big) \\ &= \mathbb{P}\Big(C_t \cap \big(B_0 \cap D_T\big) \cap C_T\Big) \\ &= \mathbb{E}\Big[\mathbf{1}_{C_t}\mathbb{P}\big(B_0 \cap D_T|\mathcal{F}_t^{(\mathbf{W}^0,\boldsymbol{\mu},\mathbf{W})}\big)\mathbb{P}\big(C_T|\mathcal{F}_t^{(\mathbf{W}^0,\boldsymbol{\mu},\mathbf{W})}\big)\Big] \\ &= \mathbb{E}\Big[\mathbf{1}_{C_t}\mathbf{1}_{B_0 \cap D_T}\mathbb{P}\big(C_T|\mathcal{F}_t^{(\mathbf{W}^0,\boldsymbol{\mu},\mathbf{W})}\big)\Big] \\ &= \mathbb{E}\Big[\mathbf{1}_{C_t}\mathbf{1}_{B_0 \cap D_T}\mathbb{P}\big(C_T|\mathcal{F}_t^{(X_0,\mathbf{W}^0,\boldsymbol{\mu},\mathbf{W})}\big)\Big] \\ &= \mathbb{E}\Big[\mathbf{1}_{B_0 \cap C_t}\mathbb{P}\big(D_T|\mathcal{F}_t^{(X_0,\mathbf{W}^0,\boldsymbol{\mu},\mathbf{W})}\big)\mathbb{P}\big(C_T|\mathcal{F}_t^{(X_0,\mathbf{W}^0,\boldsymbol{\mu},\mathbf{W})}\big)\Big],\end{aligned}$$

where we used, once again, the compatibility condition in order to pass from the first to the second line, and (1.4) to get the fourth equality. This completes the proof. □

Lemma 1.15 *Under the conditions of Definition 1.13, the filtration $\mathbb{F}$ is compatible with $(X_0, \boldsymbol{W}^0, \boldsymbol{\mu})$ under $\mathbb{P}$.*

Proof. For a given $t \in [0, T]$, we consider $C_t^0 \in \mathcal{F}_t^{(X_0,\boldsymbol{W}^0,\boldsymbol{\mu})}$, $D_t \in \mathcal{F}_t$ and $C_T^0 \in \mathcal{F}_T^{(X_0,\boldsymbol{W}^0,\boldsymbol{\mu})}$. Then, by the compatibility property in Definition 1.13,

$$\begin{aligned}\mathbb{P}\Big(C_t^0 \cap D_t \cap C_T^0\Big) &= \mathbb{E}\Big[\mathbf{1}_{C_t^0}\mathbb{P}\big(C_T^0 \,|\, \mathcal{F}_t^{(X_0,\boldsymbol{W}^0,\boldsymbol{\mu},\boldsymbol{W})}\big)\mathbb{P}\big(D_t \,|\, \mathcal{F}_t^{(X_0,\boldsymbol{W}^0,\boldsymbol{\mu},\boldsymbol{W})}\big)\Big] \\ &= \mathbb{E}\Big[\mathbf{1}_{C_t^0 \cap D_t}\mathbb{P}\big(C_T^0 \,|\, \mathcal{F}_t^{(X_0,\boldsymbol{W}^0,\boldsymbol{\mu},\boldsymbol{W})}\big)\Big].\end{aligned}$$

Since $(X_0, \boldsymbol{W}^0, \boldsymbol{\mu})$ and $\boldsymbol{W}$ are independent, we have:

$$\mathbb{P}\big(C_T^0 \,|\, \mathcal{F}_t^{(X_0,\boldsymbol{W}^0,\boldsymbol{\mu},\boldsymbol{W})}\big) = \mathbb{P}\big(C_T^0 \,|\, \mathcal{F}_t^{(X_0,\boldsymbol{W}^0,\boldsymbol{\mu})}\big).$$

Therefore,

$$\mathbb{P}\Big(C_t^0 \cap D_t \cap C_T^0\Big) = \mathbb{E}\Big[\mathbf{1}_{C_t^0}\mathbb{P}\big(D_t \,|\, \mathcal{F}_t^{(X_0,\boldsymbol{W}^0,\boldsymbol{\mu})}\big)\mathbb{P}\big(C_T^0 \,|\, \mathcal{F}_t^{(X_0,\boldsymbol{W}^0,\boldsymbol{\mu})}\big)\Big],$$

which completes the proof. □

1.1.3 Kunita-Watanabe Decomposition and Definition of a Solution

When compared with the forward-backward stochastic differential equations encountered in Volume I, the system (1.1) exhibits two new additional features. First, the system is driven by two noise terms instead of one. This does not make much difference except for the fact that another stochastic integral is needed in order to keep track of the randomness generated by the common noise $\boldsymbol{W}^0$. Second, although it is independent of $\boldsymbol{W}$, the random environment $\boldsymbol{\mu}$ may not be adapted to the filtration $\mathbb{F}^{\boldsymbol{W}^0}$ generated by $\boldsymbol{W}^0$, and in this case, the martingale term given by a stochastic integral with respect to $\boldsymbol{W}^0$ may not suffice to account for the randomness of the environment $\boldsymbol{\mu}$. This remark is very important since as we shall see in Chapter 2, solutions to mean field games with a common noise may not be adapted to the filtration of the common noise, and as a consequence, may contain additional randomness.

In order to overcome the fact that the random environment may not be adapted to $\mathbb{F}^{\boldsymbol{W}^0}$, an extra term needs to be added to the martingale terms already appearing in (1.1). The important thing to keep in mind is that this extra term can only be a martingale, and we cannot assume that it can be represented as a stochastic integral with respect to the Wiener processes providing the sources of noise. In any case, such a decomposition is known as the *Kunita-Watanabe decomposition*. It has already been used in the theory of backward SDEs. See the section Notes & Complements at the end of the chapter for references.

New Formulation of the Forward-Backward System

According to this decomposition, the backward component in (1.1) needs to be rewritten, and the appropriate form of the FBSDE becomes:

$$\begin{cases} dX_t = B(t, X_t, \mu_t, Y_t, Z_t, Z_t^0)dt + \Sigma(t, X_t, \mu_t)dW_t \\ \qquad +\Sigma^0(t, X_t, \mu_t)dW_t^0, \\ dY_t = -F(t, X_t, \mu_t, Y_t, Z_t, Z_t^0)dt \\ \qquad +Z_t dW_t + Z_t^0 dW_t^0 + dM_t, \end{cases} \tag{1.5}$$

with the same terminal condition $Y_T = G(X_T, \mu_T)$, where $\boldsymbol{M} = (M_t)_{0\le t\le T}$ is a càd-làg $\mathbb{F}$-martingale with values in $\mathbb{R}^m$ starting from 0, and orthogonal to $\boldsymbol{W}$ and $\boldsymbol{W}^0$ in the sense that, for any coordinate $j \in \{1, \cdots, m\}$,

$$\mathbb{E}\left[\left(M_T\right)_j \int_0^T \phi_s \cdot dW_s\right] = 0, \quad \text{and} \quad \mathbb{E}\left[\left(M_T\right)_j \int_0^T \phi_s \cdot dW_s^0\right] = 0, \tag{1.6}$$

for any square-integrable $\mathbb{F}$-progressively measurable processes $\boldsymbol{\phi} = (\phi_t)_{0\le t\le T}$ with values in $\mathbb{R}^d$. By stochastic integration by parts, it can be checked that the orthogonality condition is equivalent to $[\boldsymbol{M}, \boldsymbol{W}]_\cdot \equiv 0$ and $[\boldsymbol{M}, \boldsymbol{W}^0]_\cdot \equiv 0$, where $[\cdot,\cdot]$ denotes the quadratic covariation (or bracket) of two martingales. In the present situation, this means that $[\boldsymbol{M}, \boldsymbol{W}]$ is the unique $\mathbb{R}^{m\times d}$-valued continuous process with bounded variation such that $(M_t \otimes W_t - [\boldsymbol{M}, \boldsymbol{W}]_t)_{0\le t\le T}$ is a local $\mathbb{F}$-martingale where $(M_t \otimes W_t)_{i,j} = (M_t)_i (W_t)_j$. As any square-integrable function of $\boldsymbol{W}$ may be written as a stochastic integral with respect to $\boldsymbol{W}$, the orthogonality condition implies that M_T, and thus any M_t with $t \in [0, T]$, are orthogonal to any random variable in $L^2(\Omega, \sigma\{W_s; 0 \le s \le T\}, \mathbb{P})$. Obviously, the same holds with $\boldsymbol{W}$ replaced by $\boldsymbol{W}^0$.

The intuitive reason for such a decomposition is as follows. Given a general random variable $\xi \in L^2(\Omega, \mathcal{F}, \mathbb{P}; \mathbb{R})$, we claim that ξ may be written as:

$$\xi = \mathbb{E}[\xi|\mathcal{F}_0] + \int_0^T (Z_t \cdot dW_t + Z_t^0 \cdot dW_t^0) + M_T, \tag{1.7}$$

where both $(Z_t)_{0\le t\le T}$ and $(Z_t^0)_{0\le t\le T}$ are square integrable d-dimensional $\mathbb{F}$-progressively measurable processes and M_T is a one-dimensional random variable which has the property (1.6). To prove (1.7), it suffices to notice from the Kunita-Watanabe inequality that the brackets of $(\mathbb{E}[\xi|\mathcal{F}_t])_{0\le t\le T}$ with $\boldsymbol{W}$ and $\boldsymbol{W}^0$ are absolutely continuous with respect to the Lebesgue measure dt and consequently, may be represented as integrals with respect to the Lebesgue measure dt, the integrands providing the processes $(Z_t)_{0\le t\le T}$ and $(Z_t^0)_{0\le t\le T}$. Then, defining:

$$M_T = \xi - \mathbb{E}[\xi|\mathcal{F}_0] - \int_0^T (Z_t \cdot dW_t + Z_t^0 \cdot dW_t^0),$$

it is easy to check (1.6).

Remark 1.16 *We emphasize that the martingale* $\boldsymbol{M} = (M_t)_{0 \le t \le T}$ *may not be continuous. For that reason,* $\boldsymbol{Y} = (Y_t)_{0 \le t \le T}$ *may be discontinuous as well.*

If $\boldsymbol{\mu}$ *is deterministic,* B *and* F *are independent of* z^0*, and* $\boldsymbol{W}^0 \equiv 0$*, which is the framework investigated in Chapters (Vol I)-3 and (Vol I)-4, then* $\boldsymbol{M} \equiv 0$*. In this case,* $\boldsymbol{Y} = (Y_t)_{0 \le t \le T}$ *has continuous paths. More generally, when* $\boldsymbol{\mu}$ *is adapted to the filtration generated by* $\boldsymbol{W}^0$*,* $\boldsymbol{M}$ *is also equal to 0 and* $\boldsymbol{Y} = (Y_t)_{0 \le t \le T}$ *has continuous sample paths as well.*

Definition of a Solution

We now have all the necessary ingredients to define the appropriate notion of solution.

Definition 1.17 *Given a probabilistic set-up* $(\Omega, \mathcal{F}, \mathbb{F}, \mathbb{P})$ *and an admissible four-tuple* $(X_0, \boldsymbol{W}^0, \boldsymbol{\mu}, \boldsymbol{W})$*, an* $\mathbb{F}$*-progressively measurable process*

$$\big(X_t, Y_t, Z_t, Z_t^0, M_t\big)_{0 \le t \le T}$$

with values in $\mathbb{R}^d \times \mathbb{R}^m \times (\mathbb{R}^{m \times d})^2 \times \mathbb{R}^m$ *is said to be a solution of the forward-backward system* (1.5) *with coefficients satisfying assumption* **FBSDE in Random Environment***, if* $(X_t)_{0 \le t \le T}$ *has continuous paths,* $(Y_t, M_t)_{0 \le t \le T}$ *has càd-làg paths,* $(M_t)_{0 \le t \le T}$ *is an* $\mathbb{F}$*-martingale with* $M_0 = 0$ *and zero bracket with* $(\boldsymbol{W}^0, \boldsymbol{W})$*,* $(\boldsymbol{X}, \boldsymbol{Y}, \boldsymbol{M}, \boldsymbol{Z}, \boldsymbol{Z}^0)$ *satisfies:*

$$\mathbb{E}\Big[\sup_{0 \le t \le T} \big(|X_t|^2 + |Y_t|^2 + |M_t|^2\big) + \int_0^T \big(|Z_t|^2 + |Z_t^0|^2\big) dt\Big] < \infty,$$

and (1.5) *is true with* $\mathbb{P}$*-probability* 1.

We call X_0 *the initial condition of the equation, and we say that the forward-backward system* (1.5) *is* ***strongly solvable*** *if a solution exists on any probabilistic set-up as above.*

We stress the fact that *a priori*, the notion of solution heavily depends upon the choice of the filtration $\mathbb{F}$. Indeed the martingale property of $\boldsymbol{M}$ in the definition of a solution is somehow predicated on the structure of $\mathbb{F}$. In this respect, the compatibility condition between $\mathbb{F}$ and $(X_0, \boldsymbol{W}^0, \boldsymbol{\mu}, \boldsymbol{W})$ is a way to select solutions which are somehow *physically meaningful*. Without the compatibility property, solutions could anticipate the future of $\boldsymbol{\mu}$. For instance, this is the case when $\mathcal{F}_0$ contains the σ-field generated by $\boldsymbol{\mu}$.

Another important remark is that, under the assumptions and notation of Definition 1.17, we can choose a somewhat canonical version of $(Z_t, Z_t^0)_{0 \le t \le T}$ by observing that:

$$(Z_t, Z_t^0) = \lim_{n \to \infty} n \int_{(t-1/n)_+}^{t} (Z_s, Z_s^0) ds, \quad \text{Leb}_1 \otimes \mathbb{P} \quad a.e.,$$

where Leb_1 denotes the one-dimensional Lebesgue measure. In particular, the 5-tuple $(X_t, Y_t, \tilde{Z}_t, \tilde{Z}_t^0, M_t)_{0\le t\le T}$ with

$$(\tilde{Z}_t, \tilde{Z}_t^0) = \begin{cases} \lim_{n\to\infty} n \int_{(t-1/n)_+}^{t} (Z_s, Z_s^0) ds \text{ if the limit exists,} \\ 0 \qquad\qquad\qquad\qquad\qquad \text{otherwise,} \end{cases} \quad t \in [0,T],$$

is a progressively measurable solution with respect to the right-continuous and complete augmentation $\mathbb{G} = (\mathcal{G}_t)_{0\le t\le T}$ of the filtration generated by $(\boldsymbol{W}^0, \boldsymbol{\mu}, \boldsymbol{W}, \boldsymbol{X}, \boldsymbol{Y}, \int_0^{\cdot} Z_s ds, \int_0^{\cdot} Z_s^0 ds, \boldsymbol{M})$. Indeed, it is easily checked that $(\boldsymbol{W}^0, \boldsymbol{W})$ remains a $\mathbb{G}$-Brownian motion, and that $\boldsymbol{M}$ remains a $\mathbb{G}$-martingale of zero covariation with $(\boldsymbol{W}^0, \boldsymbol{W})$. Moreover, since the filtration $\mathbb{G}$ is included in $\mathbb{F}$, it is compatible with $(X_0, \boldsymbol{W}^0, \boldsymbol{\mu}, \boldsymbol{W})$ in the sense that, for any $t \in [0,T]$, $\mathcal{G}_t$ and $\mathcal{F}_T^{(X_0,\boldsymbol{W}^0,\boldsymbol{\mu},\boldsymbol{W})}$ are conditionally independent under $\mathbb{P}$ given $\mathcal{F}_t^{(X_0,\boldsymbol{W}^0,\boldsymbol{\mu},\boldsymbol{W})}$.

1.2 Strong Versus Weak Solution of an FBSDE

1.2.1 Notions of Uniqueness

We shall not discuss general solvability results for forward-backward stochastic differential equations of type (1.5) at this stage. When $\boldsymbol{\mu}$ is random, specific theorems of existence and uniqueness will be given later in the text when we make the connection with optimization in a random environment, see Subsection 1.4. When $\boldsymbol{\mu}$ is deterministic, we already accounted for solvability results in Chapters (Vol I)-3 and (Vol I)-4.

For the time being, we stress an important feature of the concept of uniqueness in law for solutions of an FBSDE, whether or not it is in random environment. We already appealed to this uniqueness result in our study of mean field games without common noise (see for instance Remark (Vol I)-4.6) and we shall use it again in the sequel. The main underpinning is the realization that the Yamada-Watanabe theorem still holds for forward-backward SDEs. This fact has already been pointed out by several authors, but as it is at the core of some of the arguments we use below, we feel confident that a self-contained version of this result will enlighten the lengthy derivations which follow. But first, we specify what we mean by strong uniqueness and uniqueness in law.

1. **Strong uniqueness** is the standard notion of pathwise uniqueness.

Definition 1.18 *Under assumption* **FBSDE in Random Environment**, *we say that uniqueness holds for the forward-backward system* (1.5) *on an admissible set-up* $(\Omega, \mathcal{F}, \mathbb{F}, \mathbb{P})$ *for* $(X_0, \boldsymbol{W}^0, \boldsymbol{\mu}, \boldsymbol{W})$ *if, for any two* $\mathbb{F}$*-progressively measurable five-tuples*

$$(\boldsymbol{X}, \boldsymbol{Y}, \boldsymbol{Z}, \boldsymbol{Z}^0, \boldsymbol{M}) = (X_t, Y_t, Z_t, Z_t^0, M_t)_{0\le t\le T}$$

and

$$(\boldsymbol{X}',\boldsymbol{Y}',\boldsymbol{Z}',\boldsymbol{Z}^{0\prime},\boldsymbol{M}') = (X'_t, Y'_t, Z'_t, Z^{0\prime}_t, M'_t)_{0\leq t\leq T}$$

satisfying the FBSDE (1.5) *with the same initial condition* X_0 *(up to an exceptional event), it holds that:*

$$\mathbb{E}\Big[\sup_{0\leq t\leq T}\big(|X_t - X'_t|^2 + |Y_t - Y'_t|^2 + |M_t - M_{t'}|^2\big) + \int_0^T \big(|Z_t - Z'_t|^2 + |Z^0_t - Z^{0\prime}_t|^2\big)dt\Big] = 0.$$

We say that **strong uniqueness** *holds if uniqueness of the forward-backward system* (1.5) *holds on any admissible set-up. Sometimes, we shall specialize the definition by saying that strong uniqueness holds but only for a prescribed value of* $\mathcal{L}(X_0, \boldsymbol{W}^0, \boldsymbol{\mu}, \boldsymbol{W})$.

Recalling from Remark 1.16 that $\boldsymbol{M} \equiv \boldsymbol{M}' \equiv 0$ when $\boldsymbol{\mu}$ is deterministic and $\boldsymbol{W}^0 \equiv 0$, we see that the above definition is consistent with the notion of uniqueness discussed in Remark (Vol I)-4.3 for standard FBSDEs when B and F do not depend on the variable z^0.

Definition 1.19 *Whenever the forward-backward system* (1.5) *is strongly solvable and satisfies the strong uniqueness property, we say that the system is strongly uniquely solvable.*

Example 1.20. *As an important example, although for a somewhat different class of equations, notice that, on a filtered complete probability space* $(\Omega, \mathcal{F}, \mathbb{F}, \mathbb{P})$ *equipped with two independent Brownian motions* $\boldsymbol{W}^0$ *and* $\boldsymbol{W}$*, any BSDE of the form:*

$$Y_t = \xi + \int_t^T F(s, Y_s, Z_s, Z^0_s)ds - \int_t^T \big(Z_s dW_s + Z^0_s dW^0_s\big) - \big(M_T - M_t\big), \quad t \in [0,T],$$

where $\xi \in L^2(\Omega, \mathcal{F}_T, \mathbb{P}; \mathbb{R}^m)$*, and* $F : [0,T] \times \Omega \times \mathbb{R}^m \times (\mathbb{R}^{m\times d})^2 \to \mathbb{R}^m$ *satisfies:*

1. *For any* $t \in [0,T]$ *and* $\omega \in \Omega$*, the mapping* $\mathbb{R}^m \times (\mathbb{R}^{m\times d})^2 \ni (y,z,z^0) \mapsto F(t,\omega,y,z,z^0)$ *is L-Lipschitz continuous, for some* $L \geq 0$*;*
2. *For any* $(y,z,z^0) \in \mathbb{R}^m \times (\mathbb{R}^{m\times d})^2$*, the process* $[0,T] \times \Omega \ni (t,\omega) \mapsto F(t,\omega,y,z,z^0)$ *is* $\mathbb{F}$*-progressively measurable and is square-integrable under* $\mathrm{Leb}_1 \otimes \mathbb{P}$*;*

must have a unique solution $(\boldsymbol{Y}, \boldsymbol{Z}, \boldsymbol{Z}^0, \boldsymbol{M}) = (Y_t, Z_t, Z_t^0, M_t)_{0\le t\le T}$ *which is* $\mathbb{F}$*-progressively measurable,* $(\boldsymbol{Y}, \boldsymbol{M})$ *having càd-làg paths,* $\boldsymbol{M} = (M_t)_{0\le t\le T}$ *being an* $\mathbb{F}$*-martingale with* $M_0 = 0$ *and of zero cross variation with* $(\boldsymbol{W}^0, \boldsymbol{W})$*, and such that:*

$$\mathbb{E}\Big[\sup_{0\le t\le T}\big(|Y_t|^2 + |M_t|^2\big) + \int_0^T \big(|Z_t|^2 + |Z_t^0|^2\big)dt\Big] < \infty.$$

Proof. Existence and uniqueness follow from a standard application of Picard's fixed point argument. We provide the proof for completeness.

First Step. Recalling that $\mathbb{H}^{2,n}$ denotes the space of $\mathbb{F}$-progressively measurable processes with values in $\mathbb{R}^n$ that are square-integrable under $\mathrm{Leb}_1 \otimes \mathbb{P}$, we construct a mapping Φ from $\Xi = \mathbb{H}^{2,m} \times \mathbb{H}^{2,m\times d} \times \mathbb{H}^{2,m\times d}$ into itself by mapping $(\boldsymbol{Y}, \boldsymbol{Z}, \boldsymbol{Z}^0)$ onto $(\boldsymbol{Y}', \boldsymbol{Z}', \boldsymbol{Z}^{0\prime})$ as given by implementing (1.7) with ξ replaced by

$$\xi + \int_0^T F(s, Y_s, Z_s, Z_s^0)ds,$$

which is square integrable under the standing assumption. Namely, we write:

$$\xi + \int_0^T F(s, Y_s, Z_s, Z_s^0)ds = \mathbb{E}\Big[\xi + \int_0^T F(s, Y_s, Z_s, Z_s^0)ds \,\big|\, \mathcal{F}_0\Big] + \int_0^T \big(\bar{Z}_t dW_t + \bar{Z}_t^{0\prime} dW_t^0\big) + \bar{M}_T,$$

where $\bar{\boldsymbol{Z}}$ and $\bar{\boldsymbol{Z}}^0$ belong to $\mathbb{H}^{2,(m\times d)}$ and $(\bar{M}_t)_{0\le t\le T}$ is a square-integrable martingale with respect to $\mathbb{F}$ with values in $\mathbb{R}^m$, starting from 0 and of zero cross variation with $(\boldsymbol{W}^0, \boldsymbol{W})$.

Letting

$$\bar{Y}_t = \mathbb{E}\Big[\xi + \int_t^T F(s, Y_s, Z_s, Z_s^0)ds \,\big|\, \mathcal{F}_t\Big],$$

we deduce that $(\bar{\boldsymbol{Y}}, \bar{\boldsymbol{Z}}, \bar{\boldsymbol{Z}}^0)$ solves:

$$\bar{Y}_t + \int_0^t F(s, Y_s, Z_s, Z_s^0)ds = \bar{Y}_0 + \int_0^t \big(\bar{Z}_s dW_s + \bar{Z}_s^{0\prime} dW_s\big) + \bar{M}_t,$$

from which we get, by writing the difference $\bar{Y}_T - \bar{Y}_t$:

$$\bar{Y}_t = \xi + \int_t^T F(s, Y_s, Z_s, Z_s^0)ds - \int_t^T \big(\bar{Z}_s dW_s + \bar{Z}_s^0 dW_s^0\big) - (\bar{M}_T - \bar{M}_t), \quad t \in [0, T].$$

By construction, we have:

$$\mathbb{E}\int_0^T \big(|\bar{Z}_t|^2 + |\bar{Z}_t^0|^2\big)dt + \mathbb{E}\big[\sup_{0\le t\le T}|\bar{M}_t|^2\big] < \infty.$$

And then, by Doob's maximal inequality,

$$\mathbb{E}\big[\sup_{0\le t\le T} |\bar{Y}_t|^2\big] < \infty.$$

Second Step. For two tuples $(\boldsymbol{Y}, \boldsymbol{Z}, \boldsymbol{Z}^0)$ and $(\boldsymbol{Y}', \boldsymbol{Z}', \boldsymbol{Z}^{0\prime})$ in $\mathbb{H}^{2,m} \times \mathbb{H}^{2,(m\times d)} \times \mathbb{H}^{2,(m\times d)}$, we call $(\bar{\boldsymbol{Y}}, \bar{\boldsymbol{Z}}, \bar{\boldsymbol{Z}}^0)$ and $(\bar{\boldsymbol{Y}}', \bar{\boldsymbol{Z}}', \bar{\boldsymbol{Z}}^{0\prime})$ the respective images by Φ, and $\bar{\boldsymbol{M}}$ and $\bar{\boldsymbol{M}}'$ the corresponding martingale parts following from the Kunita-Watanabe decomposition.

We then compute $(|\bar{Y}_t - \bar{Y}'_t|^2)_{0\le t\le T}$ by integration by parts. In comparison with the standard case when the filtration is Brownian, we must pay special attention to the fact that the processes $\bar{\boldsymbol{Y}}$ and $\bar{\boldsymbol{Y}}'$ may be discontinuous. We get:

$$\begin{aligned}
|\bar{Y}_t - \bar{Y}'_t|^2 &= 2\int_t^T (\bar{Y}_s - \bar{Y}'_s)\cdot\Big(F(s, Y_s, Z_s, Z^0_s) - F(s, Y'_s, Z'_s, Z^{0\prime}_s)\Big)ds \\
&\quad - 2\int_{(t,T]} (\bar{Y}_{s-} - \bar{Y}'_{s-})\cdot\Big[(\bar{Z}_s - \bar{Z}'_s)dW_s + (\bar{Z}^0_s - \bar{Z}^{0\prime}_s)dW^0_s + d(\bar{M}_s - \bar{M}'_s)\Big] \\
&\quad - \mathrm{trace}\bigg(\Big[\int_0^\cdot (\bar{Z}_s - \bar{Z}'_s)dW_s + \int_0^\cdot (\bar{Z}^0_s - \bar{Z}^{0\prime}_s)dW^0_s + M.\Big]_T \\
&\quad - \Big[\int_0^\cdot (\bar{Z}_s - \bar{Z}'_s)dW_s + \int_0^\cdot (\bar{Z}^0_s - \bar{Z}^{0\prime}_s)dW^0_s + M.\Big]_t\bigg),
\end{aligned} \tag{1.8}$$

where as before, for an m-dimensional $\mathbb{F}$-martingale $\boldsymbol{N} = (N_t)_{0\le t\le T}$, $([N.]_t)_{0\le t\le T}$ denotes the quadratic variation of $\boldsymbol{N}$, regarded as a process with values in $\mathbb{R}^{m\times m}$.

Notice by orthogonality that the bracket terms in the last two lines satisfy:

$$\begin{aligned}
&\Big[\int_0^\cdot (\bar{Z}_s - \bar{Z}'_s)dW_s + \int_0^\cdot (\bar{Z}^0_s - \bar{Z}^{0\prime}_s)dW^0_s + M.\Big]_t \\
&\qquad = \int_0^t |\bar{Z}_s - \bar{Z}'_s|^2 ds + \int_0^t |\bar{Z}^0_s - \bar{Z}^{0\prime}_s|^2 ds + \big[\bar{M} - \bar{M}'\big]_t,
\end{aligned}$$

and similarly at time T.

Observing also that:

$$\begin{aligned}
&\mathbb{E}\bigg[\Big(\int_0^T |\bar{Y}_{s-} - \bar{Y}'_{s-}|^2 d\Big(\mathrm{trace}\big[\bar{M}. - \bar{M}'.\big]_s\Big)\Big)^{1/2}\bigg] \\
&\qquad \le \mathbb{E}\big[\sup_{0\le s\le T} |\bar{Y}_s - \bar{Y}'_s|^2\big]^{1/2}\mathbb{E}\big[\mathrm{trace}\big([\bar{M} - \bar{M}']_T\big)\big]^{1/2},
\end{aligned}$$

we deduce that the martingale in the second line appearing in the right-hand side of (1.8) is the increment of a true martingale. Therefore, by taking the expectation in (1.8), we get:

$$\begin{aligned}
&\mathbb{E}[|\bar{Y}_t - \bar{Y}'_t|^2] + \mathbb{E}\int_t^T \big(|\bar{Z}_s - \bar{Z}'_s|^2 + |\bar{Z}^0_s - \bar{Z}^{0\prime}_s|^2\big)ds \\
&\qquad \le 2L\mathbb{E}\int_t^T |\bar{Y}_s - \bar{Y}'_s|\big(|Y_s - Y'_s| + |Z_s - Z'_s| + |Z^0_s - Z^{0\prime}_s|\big)ds.
\end{aligned}$$

Then, it is completely standard to deduce that there exists a constant C such that, for $t \in [0, T]$:

$$\mathbb{E}[|\bar{Y}_t - \bar{Y}'_t|^2] + \mathbb{E}\int_t^T (|\bar{Z}_s - \bar{Z}'_s|^2 + |\bar{Z}^0_s - \bar{Z}^{0\prime}_s|^2)ds \tag{1.9}$$
$$\leq C\mathbb{E}\Big[\int_t^T |\bar{Y}_s - \bar{Y}'_s|^2 ds\Big] + \frac{1}{2}\mathbb{E}\int_t^T (|Y_s - Y'_s|^2 + |Z_s - Z'_s|^2 + |Z^0_s - Z^{0\prime}_s|^2)ds.$$

By multiplying by $e^{\alpha t}$ and integrating with respect to t, we obtain:

$$\int_0^T e^{\alpha t}\mathbb{E}[|\bar{Y}_t - \bar{Y}'_t|^2]dt + \mathbb{E}\int_0^T e^{\alpha s}(|\bar{Z}_s - \bar{Z}'_s|^2 + |\bar{Z}^0_s - \bar{Z}^{0\prime}_s|^2)\Big(\int_0^s e^{\alpha(t-s)}dt\Big)ds.$$
$$\leq \frac{C}{\alpha}\int_0^T e^{\alpha s}\mathbb{E}[|\bar{Y}_s - \bar{Y}'_s|^2]ds$$
$$+ \frac{1}{2}\mathbb{E}\int_0^T e^{\alpha s}(|Y_s - Y'_s|^2 + |Z_s - Z'_s|^2 + |Z^0_s - Z^{0\prime}_s|^2)\Big(\int_0^s e^{\alpha(t-s)}dt\Big)ds.$$

Now, multiplying (1.9) by $\varepsilon > 0$ and summing with the inequality right above, we deduce that:

$$\int_0^T e^{\alpha t}\mathbb{E}[|\bar{Y}_t - \bar{Y}'_t|^2]dt + \mathbb{E}\int_0^T \theta^\varepsilon(s)(|\bar{Z}_s - \bar{Z}'_s|^2 + |\bar{Z}^0_s - \bar{Z}^{0\prime}_s|^2)ds$$
$$\leq C(\frac{1}{\alpha} + \varepsilon)\int_0^T e^{\alpha s}\mathbb{E}[|\bar{Y}_s - \bar{Y}'_s|^2]ds$$
$$+ \frac{1}{2}\mathbb{E}\int_0^T \theta^\varepsilon(s)(|Y_s - Y'_s|^2 + |Z_s - Z'_s|^2 + |Z^0_s - Z^{0\prime}_s|^2)ds,$$

with:

$$\theta^\varepsilon(s) = \varepsilon + e^{\alpha s}\int_0^s e^{\alpha(t-s)}dt = \varepsilon + \frac{1}{\alpha}(e^{\alpha s} - 1).$$

Choose $\alpha \geq 1$ large enough and $\varepsilon < 1/\alpha < 1$ small enough so that $C(1/\alpha + \varepsilon) < 1/4$ and deduce that:

$$\int_0^T e^{\alpha t}\mathbb{E}[|\bar{Y}_t - \bar{Y}'_t|^2]dt + \mathbb{E}\int_0^T \theta^\varepsilon(s)(|\bar{Z}_s - \bar{Z}'_s|^2 + |\bar{Z}^0_s - \bar{Z}^{0\prime}_s|^2)ds$$
$$\leq \frac{2}{3}\mathbb{E}\int_0^T \theta^\varepsilon(s)(|Y_s - Y'_s|^2 + |Z_s - Z'_s|^2 + |Z^0_s - Z^{0\prime}_s|^2)ds$$
$$\leq \frac{2}{3}\mathbb{E}\int_0^T e^{\alpha s}|Y_s - Y'_s|^2 ds + \frac{2}{3}\mathbb{E}\int_0^T \theta^\varepsilon(s)(|Z_s - Z'_s|^2 + |Z^0_s - Z^{0\prime}_s|^2)ds,$$

where we have used the fact that $\theta^\varepsilon(s) \leq e^{\alpha s}$ for the prescribed values of α and ε.

Third Step. Equipping $\mathbb{H}^{2,m} \times \mathbb{H}^{2,m\times d} \times \mathbb{H}^{2,m\times d}$ with the norm

$$|||(\boldsymbol{Y}, \boldsymbol{Z}, \boldsymbol{Z}^0)||| = \mathbb{E}\int_0^T e^{\alpha s}|Y_s|^2 ds + \mathbb{E}\int_0^T \theta^{\varepsilon}(s)\big(|Z_s|^2 + |Z_s^0|^2\big)ds,$$

under which it is complete, we conclude by Picard's fixed point theorem. □

2. **Uniqueness in law** concerns uniqueness of the distribution of the solution. A nice feature of this concept of uniqueness is that it deals with solutions defined on possibly different probability spaces. As one can easily guess, uniqueness in law says that, whatever the probabilistic set-up, the distribution of the solution is the same provided that the distribution of the inputs is fixed. In our framework, a challenging question is to determine the spaces on which the distributions of the inputs and the solutions should be considered.

The inputs consist of the initial condition X_0 constructed as an $\mathbb{R}^d$-valued square-integrable $\mathcal{F}_0$-measurable random variable, the random shock processes $\boldsymbol{W}$ and $\boldsymbol{W}^0$, and the environment $\boldsymbol{\mu}$. Since $(X_0, \boldsymbol{W}^0, \boldsymbol{\mu})$ is independent of $\boldsymbol{W}$, and the law of $\boldsymbol{W}$ is fixed (recall that it is the standard Wiener measure), specifying the law of the input may be restricted to the distribution of $(X_0, \boldsymbol{W}^0, \boldsymbol{\mu})$ on the space $\mathbb{R}^d \times \mathcal{C}([0,T];\mathbb{R}^d) \times \mathcal{D}([0,T];\mathcal{X})$. Recall that $\mathcal{C}([0,T];\mathbb{R}^d)$ is the space of continuous functions from $[0,T]$ to $\mathbb{R}^d$ and that $\mathcal{D}([0,T];\mathcal{X})$ is the space of càd-làg functions from $[0,T]$ to $\mathcal{X}$ which are left-continuous at T. Here, we endow $\mathcal{C}([0,T];\mathbb{R}^d)$v and $\mathcal{D}([0,T];\mathcal{X})$ with the topology of the uniform convergence and the Skorohod topology (as given by the so-called $J1$ metric) respectively, and with the corresponding Borel σ-fields. See the Notes & Complements at the end of the chapter for references. The product space $\mathbb{R}^d \times \mathcal{C}([0,T];\mathbb{R}^d) \times \mathcal{D}([0,T];\mathcal{X})$ is equipped with the product topology and the product σ-field.

In this context, a solution consists of a four-tuple $(X_t, Y_t, (Z_t, Z_t^0), M_t)_{0\le t\le T}$ defined on some probabilistic set-up as in Subsection 1.1.1. In order to make things slightly more regular, we shall focus on $(X_t, Y_t, \int_0^t (Z_s, Z_s^0)ds, M_t)_{0\le t\le T}$, which can be seen as a process with trajectories in $\mathcal{C}([0,T];\mathbb{R}^d) \times \mathcal{D}([0,T];\mathbb{R}^m) \times \mathcal{C}([0,T];\mathbb{R}^{2(m\times d)}) \times \mathcal{D}([0,T];\mathbb{R}^m)$. The reason is the same as above. By Lebesgue differentiation theorem, we may recover a version of $(\mathbf{Z}, \mathbf{Z}^0) = (Z_t, Z_t^0)_{0\le t\le T}$ from the equality:

$$(Z_t, Z_t^0) = \lim_{n\to\infty} n\int_{(t-1/n)_+}^t (Z_s, Z_s^0)ds, \quad \mathrm{Leb}_1 \otimes \mathbb{P} \quad a.e..$$

Accordingly, the joint distribution of the input and the output is given by the distribution of the process:

$$\left(W_t^0, \mu_t, W_t, X_t, Y_t, \int_0^t (Z_s, Z_s^0)ds, M_t\right)_{0\le t\le T}$$

on

$$\Omega_{\text{total}} = \mathcal{C}([0,T];\mathbb{R}^d) \times \mathcal{D}([0,T];\mathcal{X}) \times \mathcal{C}([0,T];\mathbb{R}^d) \times \mathcal{C}([0,T];\mathbb{R}^d) \\ \times \mathcal{D}([0,T];\mathbb{R}^m) \times \mathcal{C}([0,T];\mathbb{R}^{2(m\times d)}) \times \mathcal{D}([0,T];\mathbb{R}^m).$$

Observe that the initial condition is encoded in the process $\boldsymbol{X}$.

Definition 1.21 *Under assumption* **FBSDE in Random Environment***, we say that weak uniqueness holds for* (1.5) *if for any two admissible set-ups* $(\Omega,\mathcal{F},\mathbb{F},\mathbb{P})$ *and* $(\Omega',\mathcal{F}',\mathbb{F}',\mathbb{P}')$ *with inputs* $(X_0,\boldsymbol{W}^0,\boldsymbol{\mu},\boldsymbol{W})$ *and* $(X_0',\boldsymbol{W}^{0\prime},\boldsymbol{\mu}',\boldsymbol{W}')$*,* $(X_0,\boldsymbol{W}^0,\boldsymbol{\mu})$ *and* $(X_0',\boldsymbol{W}^{0\prime},\boldsymbol{\mu}')$ *having the same law on* $\mathbb{R}^d \times \mathcal{C}([0,T];\mathbb{R}^d) \times \mathcal{D}([0,T];\mathcal{X})$*, any two solutions on* Ω *and* Ω' *respectively:*

$$\Big(W_t^0,\mu_t,W_t,X_t,Y_t,\int_0^t (Z_s,Z_s^0)ds,M_t\Big)_{0\le t\le T} \quad \textit{and}$$
$$\Big(W_t^{0\prime},\mu_t',W_t',X_t',Y_t',\int_0^t (Z_s',Z_s^{0\prime})ds,M_t'\Big)_{0\le t\le T},$$

have the same distribution on Ω_{total}.

Remark 1.22 *Instead of handling the processes* $\mathbf{Z}$ *and* $\mathbf{Z}^0$ *through the integrals* $\int_0^\cdot Z_s ds$ *and* $\int_0^\cdot Z_s ds$ *and regarding the latter as elements of the space* $\mathcal{C}([0,T];\mathbb{R}^{m\times d})$ *equipped with the uniform topology, we could directly see* $\mathbf{Z}$ *and* $\mathbf{Z}^0$ *as random variables with values in the space of Borel functions on* $[0,T]$ *equipped with a suitable topology. This is exactly what we shall do in Chapter 3 when solving mean field games with a common noise. However, at this stage of our presentation, it suffices to work with the integrals of* $\mathbf{Z}$ *and* $\mathbf{Z}^0$*; this is much easier as the space* $\mathcal{C}([0,T];\mathbb{R}^{2(m\times d)})$ *is of a familiar use in the theory of stochastic processes.*

1.2.2 Canonical Spaces

As it is usually done for weak solutions of SDEs, it makes sense to distinguish one specific canonical probability space among others, and to transfer solutions from generic set-ups to this particular canonical space. Here, what we mean by canonical set-up is the canonical space carrying the various data, namely the inputs and the outputs. It reads $\Omega_{\text{canon}} = \mathbb{R}^d \times \Omega_{\text{total}} = \Omega_{\text{input}} \times \Omega_{\text{output}}$, with

$$\Omega_{\text{input}} = \mathbb{R}^d \times \mathcal{C}([0,T];\mathbb{R}^d) \times \mathcal{D}([0,T];\mathcal{X}) \times \mathcal{C}([0,T];\mathbb{R}^d),$$
$$\Omega_{\text{output}} = \mathcal{C}([0,T];\mathbb{R}^d) \times \mathcal{D}([0,T];\mathbb{R}^m) \times \mathcal{C}([0,T];\mathbb{R}^{2(m\times d)}) \times \mathcal{D}([0,T];\mathbb{R}^m).$$

On Ω_{input}, the first coordinate carries the initial condition of the forward-backward equation, the second one the common source of noise (that is $\boldsymbol{W}^0$), the third one the environment (that is $\boldsymbol{\mu}$) and the last one the idiosyncratic noise (that is $\boldsymbol{W}$). On Ω_{output}, the first coordinate carries the forward component of the solution, the second one the backward component of the solution, the third one the integral of the martingale representation term, and the last one the martingale part deriving from the Kunita-Watanabe decomposition.

The canonical process on Ω_{input} will be denoted by $(\xi, \boldsymbol{w}^0, \boldsymbol{\nu}, \boldsymbol{w})$ while it will be denoted by $(\boldsymbol{x}, \boldsymbol{y}, (\boldsymbol{\zeta}, \boldsymbol{\zeta}^0), \boldsymbol{m})$ on Ω_{output}. Note that we change the notation for the canonical process from the Greek letter μ to ν because we do not have fonts to switch from upper case to lower case μ. For this reason, we shall use the nearby Greek letter ν to represent the input μ on the canonical space, hoping that this will not be the source of confusion.

With a slight abuse of notation, we shall extend the two processes $(\xi, \boldsymbol{w}^0, \boldsymbol{\nu}, \boldsymbol{w})$ and $(\boldsymbol{x}, \boldsymbol{y}, (\boldsymbol{\zeta}, \boldsymbol{\zeta}^0), \boldsymbol{m})$ to the entire Ω_{canon} so that the canonical process on Ω_{canon} will be also denoted by $(\xi, \boldsymbol{w}^0, \boldsymbol{\nu}, \boldsymbol{w}, \boldsymbol{x}, \boldsymbol{y}, (\boldsymbol{\zeta}, \boldsymbol{\zeta}^0), \boldsymbol{m})$.

Following the prescription in Definition 1.13, we shall equip Ω_{canon} with the completion $\mathcal{F}_{\text{canon}}$ of the Borel σ-field under a probability measure $\mathbb{Q}$ that is required to satisfy:

Definition 1.23 *A pair* $(\mathcal{F}_{\text{canon}}, \mathbb{Q})$ *is a said to be* admissible *if there exists a probability measure* $\mathbb{Q}^{\mathcal{B}}$ *on* Ω_{canon} *equipped with its Borel* σ*-field* $\mathcal{B}(\Omega_{\text{canon}})$ *such that* $(\Omega_{\text{canon}}, \mathcal{F}_{\text{canon}}, \mathbb{Q})$ *is the completion of* $(\Omega_{\text{canon}}, \mathcal{B}(\Omega_{\text{canon}}), \mathbb{Q}^{\mathcal{B}})$ *and satisfies:*

1. *under* $\mathbb{Q}$, $(\xi, \boldsymbol{w}^0, \boldsymbol{\nu})$ *and* $\boldsymbol{w}$ *are independent;*
2. *under* $\mathbb{Q}$, *the process* $(\boldsymbol{w}^0, \boldsymbol{w})$ *is a 2d-dimensional Brownian motion with respect to the filtration* $\mathbb{G} = (\mathcal{G}_t)_{0 \le t \le T}$, *defined as the complete and right-continuous augmentation under* $\mathbb{Q}$ *of the canonical filtration* $\mathbb{G}^{\text{nat}} = (\mathcal{G}_t^{\text{nat}})_{0 \le t \le T}$ *on* Ω_{canon};
3. *the process* $(\xi, \boldsymbol{w}^0, \boldsymbol{\nu}, \boldsymbol{w})$ *and the filtration* $\mathbb{G}$ *are compatible under* $\mathbb{Q}$.

Remark 1.24 *When the input* $\boldsymbol{\mu}$ *is deterministic, the coefficients are independent of* z^0 *and* $\boldsymbol{W}^0$ *is not present,* Ω_{input} *may be taken to be* $\mathbb{R}^d \times \mathcal{C}([0, T]; \mathbb{R}^d)$ *and* Ω_{output} *as* $\mathcal{C}([0, T]; \mathbb{R}^d) \times \mathcal{C}([0, T]; \mathbb{R}^m) \times \mathcal{C}([0, T]; \mathbb{R}^{m \times d})$. *In particular, Definition 1.23 of an admissible probability becomes somewhat irrelevant.*

Remark 1.25 *Sometimes, we merely say that* $\mathbb{Q}$ *is admissible instead of* $(\mathcal{F}_{\text{canon}}, \mathbb{Q})$. *The* σ*-field* $\mathcal{F}_{\text{canon}}$ *is then automatically understood as the completion of the Borel* σ*-field* $\mathcal{B}(\Omega_{\text{canon}})$ *under* $\mathbb{Q}^{\mathcal{B}}$. *Observe in particular that* $\mathbb{Q}$ *is in one-to-one correspondence with* $\mathbb{Q}^{\mathcal{B}}$.

We shall use the fact that conditions 1, 2, and 3 in Definition 1.23 can be formulated under $\mathbb{Q}^{\mathcal{B}}$.

Proposition 1.26 $(\mathcal{F}_{\text{canon}}, \mathbb{Q})$ *is admissible if and only if there exists a probability measure* $\mathbb{Q}^{\mathcal{B}}$ *on* $(\Omega_{\text{canon}}, \mathcal{B}(\Omega_{\text{canon}}))$ *such that* $(\Omega_{\text{canon}}, \mathcal{F}_{\text{canon}}, \mathbb{Q})$ *is the completion of* $(\Omega_{\text{canon}}, \mathcal{B}(\Omega_{\text{canon}}), \mathbb{Q}^{\mathcal{B}})$ *and*

1. *under* $\mathbb{Q}^{\mathcal{B}}$, $(\xi, \boldsymbol{w}^0, \boldsymbol{\nu})$ *and* $\boldsymbol{w}$ *are independent;*
2. *under* $\mathbb{Q}^{\mathcal{B}}$, *the process* $(\boldsymbol{w}^0, \boldsymbol{w})$ *is a 2d-Brownian motion with respect to the right-continuous augmentation* $\mathbb{G}^{\text{nat}}_{\cdot+} = (\mathcal{G}^{\text{nat}}_{t+} = \cap_{s\in(t,T]}\mathcal{G}^{\text{nat}}_s)_{0\le t\le T}$ *of* $\mathbb{G}^{\text{nat}}$*;*
3. *for each* $t \in [0, T]$, *the* σ*-fields* $\mathcal{G}^{\text{nat}}_{t+}$ *and* $\mathcal{F}^{\text{nat},(\xi,\boldsymbol{w}^0,\boldsymbol{\nu},\boldsymbol{w})}_T$ *are conditionally independent for* $\mathbb{Q}^{\mathcal{B}}$ *given the* σ*-field* $\mathcal{F}^{\text{nat},(\xi,\boldsymbol{w}^0,\boldsymbol{\nu},\boldsymbol{w})}_{t+}$.

Proof. The equivalence regarding the first two conditions is easily established. As for the compatibility conditions, the result follows from the following two facts. a) for any $D \in \mathcal{F}^{(\xi,\boldsymbol{w}^0,\boldsymbol{\nu},\boldsymbol{w})}_t$ (respectively $\mathcal{F}^{(\xi,\boldsymbol{w}^0,\boldsymbol{\nu},\boldsymbol{w})}_T$), there exists $E \in \mathcal{F}^{\text{nat},(\xi,\boldsymbol{w}^0,\boldsymbol{\nu},\boldsymbol{w})}_{t+}$ (respectively $E \in \mathcal{F}^{\text{nat},(\xi,\boldsymbol{w}^0,\boldsymbol{\nu},\boldsymbol{w})}_T$) such that the symmetric difference $D\Delta E$ has zero measure under $\mathbb{Q}$. b) for any $D \in \mathcal{B}(\Omega_{\text{canon}})$, $\mathbb{Q}(D|\mathcal{F}^{(\xi,\boldsymbol{w}^0,\boldsymbol{\nu},\boldsymbol{w})}_t)$ and $\mathbb{Q}^{\mathcal{B}}(D|\mathcal{F}^{\text{nat},(\xi,\boldsymbol{w}^0,\boldsymbol{\nu},\boldsymbol{w})}_{t+})$ are $\mathbb{Q}$ almost surely equal. □

Transfer of Solutions to the Canonical Space

Fortunately, any solution of (1.5) may be transferred to the canonical space.

Lemma 1.27 *If* $(X_t, Y_t, Z_t, Z^0_t, M_t)_{0\le t\le T}$ *is a solution of the FBSDE* (1.5) *on some probabilistic set-up* $(\Omega, \mathcal{F}, \mathbb{F}, \mathbb{P})$ *equipped with* $(X_0, W^0_t, \mu_t, W_t)_{0\le t\le T}$ *and if we define* $(\Omega_{\text{canon}}, \mathcal{F}_{\text{canon}}, \mathbb{Q})$ *as the completion induced by the distribution of*

$$\left(X_0, W^0_t, \mu_t, W_t, X_t, Y_t, \int_0^t (Z_s, Z^0_s)ds, M_t\right)_{0\le t\le T}$$

on Ω_{canon} *equipped with its Borel* σ*-field, and if we define* $(z_t, z^0_t)(\omega)$ *by:*

$$(z_t, z^0_t)(\omega) = \begin{cases} \lim_{n\to\infty} n\big((\zeta, \zeta^0)_t(\omega) - (\zeta, \zeta^0)_{(t-1/n)_+}(\omega)\big) & \textit{if the limit exists,} \\ 0 \quad \textit{otherwise,} \end{cases}$$

for any $t \in [0, T]$ *and* $\omega \in \Omega_{\text{canon}}$, *then the pair* $(\mathcal{F}_{\text{canon}}, \mathbb{Q})$ *is admissible and the process* $(x_t, y_t, z_t, z^0_t, m_t)_{0\le t\le T}$ *is, under* $\mathbb{Q}$, *a solution of the FBSDE* (1.5) *on* Ω_{canon} *for the complete and right-continuous augmentation* $\mathbb{G}$ *of the canonical filtration.*

Proof.

First Step. We first check that $(\mathcal{F}_{\text{canon}}, \mathbb{Q})$ is admissible. Recall that $\mathbb{G} = (\mathcal{G}_t)_{0\le t\le T}$ denotes the complete and right-continuous augmentation under $\mathbb{Q}$ of the canonical filtration on Ω_{canon}. Using the fact that $(M_t)_{0\le t\le T}$ is an $\mathbb{F}$-martingale on $(\Omega, \mathcal{F}, \mathbb{F}, \mathbb{P})$, it is plain to check that $(m_t)_{0\le t\le T}$ is a $\mathbb{G}$-square-integrable martingale under $\mathbb{Q}$. Indeed, for any integer $n \ge 1$ and any bounded and measurable function ψ from $(\mathbb{R}^d \times \mathbb{R}^d \times \mathcal{X} \times \mathbb{R}^d \times \mathbb{R}^d \times \mathbb{R}^m \times \mathbb{R}^{2(m\times d)} \times \mathbb{R}^m)^n$ into $\mathbb{R}$ and for all $0 \le t < s \le T$ and $0 \le t_0 < t_1 < \cdots < t_n \le t$, it holds:

$$\begin{aligned} \mathbb{E}\big[\psi(\Theta_{t_1}, \cdots, \Theta_{t_n})M_s\big] &= \mathbb{E}^{\mathbb{Q}}\big[\psi(\theta_{t_1}, \cdots, \theta_{t_n})m_s\big], \\ \mathbb{E}\big[\psi(\Theta_{t_1}, \cdots, \Theta_{t_n})M_t\big] &= \mathbb{E}^{\mathbb{Q}}\big[\psi(\theta_{t_1}, \cdots, \theta_{t_n})m_t\big], \end{aligned} \tag{1.10}$$

where we have let:

$$\Theta_t = \Big(X_0, W_t^0, \mu_t, W_t, X_t, Y_t, \int_0^t (Z_s, Z_s^0) ds, M_t\Big),$$

$$\theta_t = \Big(\xi, w_t^0, \nu_t, w_t, x_t, y_t, \int_0^t (z_s, z_s^0) ds, m_t\Big),$$

for $t \in [0, T]$. Since the two left-hand sides in (1.10) are equal, we deduce that the right-hand sides are also equal, from which we deduce that $\boldsymbol{m} = (m_t)_{0 \le t \le T}$ is a martingale with respect to the canonical filtration. By right-continuity, we deduce that $\boldsymbol{m}$ is also a martingale with respect to $\mathbb{G}$.

It is also clear that, under $\mathbb{Q}$, $(w_t^0)_{0 \le t \le T}$ and $(w_t)_{0 \le t \le T}$ are independent d-dimensional Wiener processes with respect to $\mathbb{G}$. Moreover, $(\xi, w_t^0, \nu_t)_{0 \le t \le T}$ is independent of $(w_t)_{0 \le t \le T}$ so that the four-tuple $(\xi, w_t^0, \nu_t, w_t)_{0 \le t \le T}$ together with the filtration $\mathbb{G}$ satisfy the first two prescriptions in Definition 1.23.

In order to prove the compatibility condition, we shall check (H2) in Proposition 1.3. In order to do so, we use an argument inspired by the proof of Lemma 1.7. We denote by $\mathbb{G}^{\text{nat}} = (\mathcal{G}_t^{\text{nat}})_{0 \le t \le T}$ (respectively $\mathbb{G}^{\text{nat,input}} = (\mathcal{G}_t^{\text{nat,input}})_{0 \le t \le T}$) the canonical filtration on Ω_{canon} (respectively Ω_{input}). We do not require that it is augmented nor right continuous. For any $t \in [0, T)$, any $\varepsilon > 0$ such that $t + \varepsilon \le T$, and any $\mathcal{G}_t^{\text{nat}}$ measurable function ψ_t from Ω_{canon} into $\mathbb{R}$, we can find two bounded functions $\phi_{t+\varepsilon}$ and ϕ_T from Ω_{input} into $\mathbb{R}$, that are respectively $\mathcal{G}_{t+\varepsilon}^{\text{nat}}$ and $\mathcal{G}_T^{\text{nat}}$ measurable, such that, $\mathbb{Q}$ almost surely:

$$\mathbb{E}^{\mathbb{Q}}\big[\psi_t(\xi, \boldsymbol{w}^0, \boldsymbol{\nu}, \boldsymbol{w}, \boldsymbol{x}, \boldsymbol{y}, (\boldsymbol{\zeta}, \boldsymbol{\zeta}^0), \boldsymbol{m}) \,|\, \mathcal{G}_{t+\varepsilon}^{\text{nat,input}}\big] = \phi_{t+\varepsilon}(\xi, \boldsymbol{w}^0, \boldsymbol{\nu}, \boldsymbol{w}),$$

$$\mathbb{E}^{\mathbb{Q}}\big[\psi_t(\xi, \boldsymbol{w}^0, \boldsymbol{\nu}, \boldsymbol{w}, \boldsymbol{x}, \boldsymbol{y}, (\boldsymbol{\zeta}, \boldsymbol{\zeta}^0), \boldsymbol{m}) \,|\, \mathcal{G}_{T}^{\text{nat,input}}\big] = \phi_{T}(\xi, \boldsymbol{w}^0, \boldsymbol{\nu}, \boldsymbol{w}).$$

Observe that the left-hand side in the first line converges almost surely as ε tends to 0 towards the conditional expectation with respect to $\mathcal{G}_{t+}^{\text{nat,input}} = \cap_{\varepsilon > 0} \mathcal{G}_{t+\varepsilon}^{\text{nat,input}}$. Therefore, the right-hand side converges almost surely as well. Going back to the original space Ω, we also have:

$$\mathbb{E}\Big[\psi_t\Big(X_0, \boldsymbol{W}^0, \boldsymbol{\mu}, \boldsymbol{W}, \boldsymbol{X}, \boldsymbol{Y}, \int_0^{\cdot} (Z_s, Z_s^0) ds, \boldsymbol{M}\Big) \,|\, \mathcal{F}_t^{(X_0, \boldsymbol{W}^0, \boldsymbol{\mu}, \boldsymbol{W})}\Big] = \lim_{\varepsilon \to 0} \phi_{t+\varepsilon}(X_0, \boldsymbol{W}^0, \boldsymbol{\mu}, \boldsymbol{W}),$$

$$\mathbb{E}\Big[\psi_t\Big(X_0, \boldsymbol{W}^0, \boldsymbol{\mu}, \boldsymbol{W}, \boldsymbol{X}, \boldsymbol{Y}, \int_0^{\cdot} (Z_s, Z_s^0) ds, \boldsymbol{M}\Big) \,|\, \mathcal{F}_T^{(X_0, \boldsymbol{W}^0, \boldsymbol{\mu}, \boldsymbol{W})}\Big] = \phi_T(X_0, \boldsymbol{W}^0, \boldsymbol{\mu}, \boldsymbol{W}).$$

By compatibility of the original probabilistic set-up, we have:

$$\mathbb{P}\Big[\lim_{\varepsilon \to 0} \phi_{t+\varepsilon}(X_0, \boldsymbol{W}^0, \boldsymbol{\mu}, \boldsymbol{W}) = \phi_T(X_0, \boldsymbol{W}^0, \boldsymbol{\mu}, \boldsymbol{W})\Big] = 1,$$

which gives on the canonical space:

$$\mathbb{Q}\Big[\lim_{\varepsilon \to 0} \phi_{t+\varepsilon}(\xi, \boldsymbol{w}^0, \boldsymbol{\nu}, \boldsymbol{w}) = \phi_T(\xi, \boldsymbol{w}^0, \boldsymbol{\nu}, \boldsymbol{w})\Big] = 1.$$

Therefore, with probability 1 under $\mathbb{Q}$:

$$\begin{aligned}
&\mathbb{E}^{\mathbb{Q}}\big[\psi_t(\xi, \boldsymbol{w}^0, \boldsymbol{\nu}, \boldsymbol{w}, \boldsymbol{x}, \boldsymbol{y}, (\boldsymbol{\zeta}, \boldsymbol{\zeta}^0), \boldsymbol{m}) \,|\, \mathcal{G}_{t+\varepsilon}^{\text{nat,input}}\big] \\
&= \mathbb{E}^{\mathbb{Q}}\big[\psi_t(\xi, \boldsymbol{w}^0, \boldsymbol{\nu}, \boldsymbol{w}, \boldsymbol{x}, \boldsymbol{y}, (\boldsymbol{\zeta}, \boldsymbol{\zeta}^0), \boldsymbol{m}) \,|\, \mathcal{G}_{T}^{\text{nat,input}}\big].
\end{aligned}$$

Let us now consider an event $D \in \mathcal{G}_t$. We know that $\mathbf{1}_D$ is almost surely equal to a random variable of the form $\psi_{t+\delta}(\xi, \boldsymbol{w}^0, \boldsymbol{v}, \boldsymbol{w}, \boldsymbol{x}, \boldsymbol{y}, (\boldsymbol{\zeta}, \boldsymbol{\zeta}^0), \boldsymbol{m})$, with δ as small as we want. Therefore, the above identity, with t replaced by $t + \delta$, says that:

$$\mathbb{Q}(D|\mathcal{G}_{(t+\delta)+}^{\text{nat,input}}) = \mathbb{Q}(D|\mathcal{G}_T^{\text{nat,input}}).$$

Letting δ tend to 0, we complete the first step of the proof by noticing that $\mathbb{Q}(D|\mathcal{G}_{t+}^{\text{nat,input}})$ is almost surely equal to $\mathbb{Q}(D|\mathcal{G}_t^{\text{input}})$, and similarly for $\mathbb{Q}(D|\mathcal{G}_T^{\text{nat,input}})$.

Second Step. We now prove that the FBSDE (1.5) is satisfied on Ω_{canon}. By Cauchy-Schwarz's inequality, $\mathbb{P}$ almost surely we have:

$$\int_O (|Z_t| + |Z_t^0|)dt \leq |O|^{1/2} \bigg(\int_0^T (|Z_t|^2 + |Z_t^0|^2)dt \bigg)^{1/2},$$

for every Borel subset $O \subset [0, T]$. Using the fact that:

$$\lim_{\varepsilon \searrow 0} \mathbb{P}\bigg[\int_0^T (|Z_t|^2 + |Z_t^0|^2)dt \leq \varepsilon^{-2} \bigg] = 1,$$

the following limit:

$$\lim_{\varepsilon \searrow 0} \mathbb{P}\bigg[\forall n \geq 1,\ \forall 0 = t_0 < \cdots < t_n = T,$$
$$\sum_{i=1}^n \bigg| \int_{t_{i-1}}^{t_i} (Z_s, Z_s^0)ds \bigg| \leq \varepsilon^{-1} \Big(\sum_{i=1}^n (t_i - t_{i-1}) \Big)^{1/2} \bigg]$$

is equal to 1, and by definition of the distribution $\mathbb{Q}$, the limit:

$$\lim_{\varepsilon \searrow 0} \mathbb{Q}\bigg[\forall n \geq 1,\ \forall 0 = t_0 < \cdots < t_n = T,$$
$$\sum_{i=1}^n \big| (\zeta, \zeta^0)_{t_i} - (\zeta, \zeta^0)_{t_{i-1}} \big| \leq \varepsilon^{-1} \Big(\sum_{i=1}^n (t_i - t_{i-1}) \Big)^{1/2} \bigg]$$

is also equal to 1. We deduce that $\mathbb{Q}$-almost surely, the path of $((\zeta_t, \zeta_t^0) = (\zeta, \zeta^0)_t)_{0 \leq t \leq T}$ is absolutely continuous with respect to the Lebesgue measure, proving that the set of pairs (t, ω) such that the limit $\lim_{n \to \infty} n((\zeta, \zeta^0)_t - (\zeta, \zeta^0)_{(t-1/n)+})$ does not exist has zero $\text{Leb}_1 \otimes \mathbb{Q}$-measure. Moreover, for almost every realization under $\mathbb{Q}$, we have:

$$(\zeta, \zeta^0)_t = (\zeta_t, \zeta_t^0) = \int_0^t (z_s, z_s^0)ds, \quad t \in [0, T],$$

with $(z_t, z_t^0)_{0 \leq t \leq T}$ as in the statement.

Notice also that $(z_t, z_t^0)_{0\le t\le T}$ is $\mathbb{G}$-progressively measurable. It satisfies:

$$\begin{aligned}
\mathbb{E}^{\mathbb{Q}} \int_0^T \big(|z_t|^2 + |z_t^0|^2\big)dt &\le \liminf_{n\to\infty} \mathbb{E}^{\mathbb{Q}} \int_0^T \big|n\big((\zeta,\zeta^0)_t - (\zeta,\zeta^0)_{(t-1/n)+}\big)\big|^2 dt \\
&= \liminf_{n\to\infty} n^2 \mathbb{E} \int_0^T \left| \int_{(t-1/n)+}^t (Z_s, Z_s^0) ds \right|^2 dt \\
&\le \liminf_{n\to\infty} n \mathbb{E} \int_0^T \int_{(t-1/n)+}^t \big(|Z_s|^2 + |Z_s^0|^2\big) ds dt \\
&\le \mathbb{E} \int_0^T \big(|Z_s|^2 + |Z_s^0|^2\big) ds < \infty,
\end{aligned}$$

the last claim following from Fubini's theorem.

Third Step. We now approximate the drift coefficient B. We construct a sequence $(B_\ell)_{\ell\ge1}$ of bounded measurable functions on $\Xi = [0,T] \times \mathbb{R}^d \times \mathcal{X} \times \mathbb{R}^m \times (\mathbb{R}^{m\times d})^2$, each B_ℓ being continuous in $(x, y, z, z^0) \in \mathbb{R}^d \times \mathbb{R}^m \times (\mathbb{R}^{m\times d})^2$ when $(s, \mu) \in [0,T] \times \mathcal{X}$ is kept fixed, and such that:

$$\lim_{\ell\to\infty} \mathbb{E} \int_0^T \big|(B - B_\ell)\big(s, X_s, \mu_s, Y_s, Z_s, Z_s^0\big)\big| ds = 0. \tag{1.11}$$

In order to do so, we let Λ be the image on Ξ (equipped with its Borel σ-field) of the product measure $\mathrm{Leb}_1 \otimes \mathbb{P}$ by the mapping

$$[0,T] \times \Omega \ni (s,\omega) \mapsto \big(s, X_s(\omega), \mu_s(\omega), Y_s(\omega), Z_s(\omega), Z_s^0(\omega)\big) \in \Xi,$$

and we approximate B in $L^1(\Xi, \Lambda; \mathbb{R})$ by a sequence of finite linear combinations of indicator functions of the form $(s, x, \mu, y, z, z^0) \mapsto \mathbf{1}_D(s,\mu)\mathbf{1}_E(x, y, z, z^0)$, with $D \in \mathcal{B}([0,T] \times \mathcal{X})$ and $E \in \mathcal{B}(\mathbb{R}^d \times \mathbb{R}^m \times (\mathbb{R}^{m\times d})^2)$, see [57, Theorems 11.4 and 19.2]. In this way:

$$\lim_{\ell\to\infty} \int_\Xi \left| B(s,x,\mu,y,z,z^0) - \sum_{i=1}^{\ell} \alpha_i^\ell \mathbf{1}_{D_i^\ell}(s,\mu) \mathbf{1}_{E_i^\ell}(x,y,z,z^0) \right| d\Lambda(s,x,\mu,y,z,z^0) = 0,$$

where, for any $\ell \ge 1$ and any $i \in \{1,\cdots,\ell\}$, $\alpha_i^\ell \in \mathbb{R}$, $D_i^\ell \in \mathcal{B}([0,T] \times \mathcal{X})$ and $E_i^\ell \in \mathcal{B}(\mathbb{R}^d \times \mathbb{R}^m \times (\mathbb{R}^{m\times d})^2)$. Then, denoting by λ the image on $\mathbb{R}^d \times \mathbb{R}^m \times (\mathbb{R}^{m\times d})^2$ of Λ by the mapping $\Xi \ni (s, x, \mu, y, z, z^0) \mapsto (x, y, z, z^0)$, we replace for each $\ell \ge 1$ and $i \in \{1,\cdots,\ell\}$, the indicator function $\mathbf{1}_{E_i^\ell}$ by a bounded continuous function f_i^ℓ such that:

$$\int_{\mathbb{R}^d\times\mathbb{R}^m\times(\mathbb{R}^{m\times d})^2} \big|\mathbf{1}_{E_i^\ell}(x,y,z,z^0) - f_i^\ell(x,y,z,z^0)\big| d\lambda(x,y,z,z^0)$$

is as small as desired. This gives the desired approximation B_ℓ.

Fourth Step. Now, for a bounded nondecreasing smooth function $\vartheta : \mathbb{R}_+ \to \mathbb{R}$ satisfying $\vartheta(x) = x$ whenever $x \in [0,1]$, it is plain to check that, for any integers $\ell \ge 1$ and $p > n$,

$$
\begin{aligned}
&\mathbb{E}\Bigg[\vartheta\Bigg(\sup_{0\le t\le T}\bigg|X_t-\int_0^t B_\ell\bigg(s,X_s,\mu_s,Y_s,n\int_{(s-1/n)+}^{s}(Z_r,Z_r^0)dr\bigg)ds \\
&\qquad -\int_0^t\bigg(n\int_{(\lfloor ps\rfloor/p-1/n)+}^{\lfloor ps\rfloor/p}\Sigma(r,X_r,\mu_r)dr\bigg)dW_s \\
&\qquad -\int_0^t\bigg(n\int_{(\lfloor ps\rfloor/p-1/n)+}^{\lfloor ps\rfloor/p}\Sigma^0(r,X_r,\mu_r)dr\bigg)dW_s^0\bigg|\Bigg)\Bigg] \\
&=\mathbb{E}^{\mathbb{Q}}\Bigg[\vartheta\Bigg(\sup_{0\le t\le T}\bigg|x_t-\int_0^t B_\ell\Big(s,x_s,\nu_s,y_s,n\big((\zeta,\zeta^0)_s-(\zeta,\zeta^0)_{(s-1/n)+}\big)\Big)ds \\
&\qquad -\int_0^t\bigg(n\int_{(\lfloor ps\rfloor/p-1/n)+}^{\lfloor ps\rfloor/p}\Sigma(r,x_r,\nu_r)dr\bigg)dw_s \\
&\qquad -\int_0^t\bigg(n\int_{(\lfloor ps\rfloor/p-1/n)+}^{\lfloor ps\rfloor/p}\Sigma^0(r,x_r,\nu_r)dr\bigg)dw_s^0\bigg|\Bigg)\Bigg],
\end{aligned}
\tag{1.12}
$$

where we used the fact that, in that case, the stochastic integrals reduce to finite Riemann sums.

Our goal is now to let p tend first to ∞, then $n\to\infty$ and finally $\ell\to\infty$. Recall that, for a square-integrable $\mathbb{F}$-progressively measurable real-valued process $\boldsymbol{\phi}=(\phi_t)_{0\le t\le T}$ on $(\Omega,\mathcal{F},\mathbb{P})$,

$$
\lim_{n\to\infty}\mathbb{E}\int_0^T\bigg|\phi_s-n\int_{(s-1/n)+}^{s}\phi_r dr\bigg|^2 ds=0,
$$

which follows from the fact that:

$$
\begin{aligned}
&\mathbb{E}\int_0^T\bigg|\phi_s-n\int_{(s-1/n)+}^{s}\phi_r dr\bigg|^2 ds \\
&=\mathbb{E}\int_0^T|\phi_s|^2ds+\mathbb{E}\int_0^T\bigg|n\int_{(s-1/n)+}^{s}\phi_r dr\bigg|^2 ds-2\mathbb{E}\int_0^T\phi_s\bigg(n\int_{(s-1/n)+}^{s}\phi_r dr\bigg)ds \\
&\le 2\mathbb{E}\int_0^T|\phi_s|^2ds-2\mathbb{E}\int_0^T\phi_s\bigg(n\int_{(s-1/n)+}^{s}\phi_r dr\bigg)ds,
\end{aligned}
$$

and a standard uniform integrability argument. We deduce that:

$$
\begin{aligned}
&\lim_{n\to\infty}\lim_{p\to\infty}\mathbb{E}\bigg[\int_0^T\bigg|\Sigma(s,X_s,\mu_s)-\bigg(n\int_{(\lfloor ps\rfloor/p-1/n)+}^{\lfloor ps\rfloor/p}\Sigma(r,X_r,\mu_r)dr\bigg)\bigg|^2 ds\bigg]=0, \\
&\lim_{n\to\infty}\lim_{p\to\infty}\mathbb{E}\bigg[\int_0^T\bigg|\Sigma^0(s,X_s,\mu_s)-\bigg(n\int_{(\lfloor ps\rfloor/p-1/n)+}^{\lfloor ps\rfloor/p}\Sigma^0(r,X_r,\mu_r)dr\bigg)\bigg|^2 ds\bigg]=0.
\end{aligned}
$$

Similarly,

$$\lim_{n\to\infty}\lim_{p\to\infty}\mathbb{E}^{\mathbb{Q}}\biggl[\int_0^T\biggl|\Sigma\bigl(s,x_s,\nu_s\bigr)-\biggl(n\int_{(\lfloor ps\rfloor/p-1/n)+}^{\lfloor ps\rfloor/p}\Sigma\bigl(r,x_r,\nu_r\bigr)dr\biggr)\biggr|^2ds\biggr]=0,$$

$$\lim_{n\to\infty}\lim_{p\to\infty}\mathbb{E}^{\mathbb{Q}}\biggl[\int_0^T\biggl|\Sigma^0\bigl(s,x_s,\nu_s\bigr)-\biggl(n\int_{(\lfloor ps\rfloor/p-1/n)+}^{\lfloor ps\rfloor/p}\Sigma^0\bigl(r,x_r,\nu_r\bigr)dr\biggr)\biggr|^2ds\biggr]=0.$$

Therefore, when we let p tend first to ∞, then n to ∞ and finally ℓ to ∞, the left-hand side of (1.12) tends to 0. Here we use (1.11) together with the fact that each B_ℓ is bounded and continuous in (x,y,z,z^0); also we use the fact that $\boldsymbol{X}$ satisfies the forward SDE in (1.5). This shows that the right-hand side must also tend to 0. Therefore, provided that the process $(x_s,\nu_s,y_s,z_s,z_s^0)_{0\le s\le T}$ satisfies the analogue of (1.11), $(x_t,y_t,z_t,z_t^0)_{0\le t\le T}$ satisfies the forward equation in (1.5). In order to prove that $(x_s,\nu_s,y_s,z_s,z_s^0)_{0\le s\le T}$ satisfies the analogue of (1.11), it suffices to prove that, for almost every $s\in[0,T]$, $(x_s,\nu_s,y_s,z_s,z_s^0)$ and $(X_s,\mu_s,Y_s,Z_s,Z_s^0)$ have the same distribution, which is easily checked by noticing that, for any bounded function ψ from $\mathbb{R}^d\times\mathcal{X}\times\mathbb{R}^m\times\mathbb{R}^{2(m\times d)}$ into $\mathbb{R}$ which is continuous in the last argument and for any $n\ge 1$, it holds:

$$\begin{aligned}\forall t\in[0,T],\quad &\mathbb{E}\int_0^t\psi\biggl(X_s,\mu_s,Y_s,n\int_{(s-1/n)+}^s(Z_r,Z_r^0)dr\biggr)ds\\&=\mathbb{E}^{\mathbb{Q}}\int_0^t\psi\Bigl(x_s,\nu_s,y_s,n\bigl((\zeta,\zeta^0)_s-(\zeta,\zeta^0)_{(s-1/n)+}\bigr)\Bigr)ds,\end{aligned}$$

and then by letting n tend to ∞.

Last Step. We use the same argument for the backward equation. For a sequence $(F_\ell)_{\ell\ge1}$ of bounded measurable functions on Ξ, each F_ℓ being continuous in (x,y,z,z^0) when (s,μ) is kept fixed, such that:

$$\lim_{\ell\to\infty}\mathbb{E}\int_0^T\bigl|(F-F_\ell)\bigl(s,X_s,\mu_s,Y_s,Z_s,Z_s^0\bigr)\bigr|ds=0,$$

we have for any $\ell\ge1$ and $p>n$:

$$\begin{aligned}\mathbb{E}\biggl[\vartheta\biggl(\sup_{0\le t\le T}\biggl|&Y_t-G(X_T,\mu_T)\\&-\int_t^TF_\ell\biggl(s,X_s,\mu_s,Y_s,n\int_{(s-1/n)+}^s(Z_r,Z_r^0)dr\biggr)ds\\&-\int_t^T\biggl(\biggl(n\int_{(\lfloor ps\rfloor/p-1/n)+}^{\lfloor ps\rfloor/p}Z_rdr\biggr)dW_s+\biggl(n\int_{(\lfloor ps\rfloor/p-1/n)+}^{\lfloor ps\rfloor/p}Z_r^0dr\biggr)dW_s^0\\&-(M_T-M_t)\biggr|\biggr)\biggr]\end{aligned}$$

$$= \mathbb{E}^{\mathbb{Q}}\Bigg[\vartheta\Bigg(\sup_{0\le t\le T}\Bigg|y_t - G(x_T, \nu_T)$$

$$- \int_t^T F_\ell\Big(s, x_s, \nu_s, y_s, n\big((\zeta,\zeta^0)_s - (\zeta,\zeta^0)_{(s-1/n)_+}\big)\Big)ds$$

$$- \int_t^T \Big(n\big(\zeta_{\lfloor ps\rfloor/p} - \zeta_{(\lfloor ps\rfloor/p-1/n)_+}\big)dw_s + n\big(\zeta^0_{\lfloor ps\rfloor/p} - \zeta^0_{(\lfloor ps\rfloor/p-1/n)_+}\big)dw^0_s\Big)$$

$$- (m_T - m_t)\Bigg|\Bigg)\Bigg].$$

Again, when we let p tend first to ∞, then n to ∞ and finally ℓ to ∞, the left-hand side tends to 0 since $(X_t, Y_t, Z_t, Z^0_t, M_t)_{0\le t\le T}$ solves the backward equation in (1.5). This proves that the right-hand side also tends to 0. Proceeding as for the forward equation, we deduce that $(x_t, y_t, z_t, z^0_t, m_t)_{0\le t\le T}$ satisfies the backward equation in (1.5). □

1.2.3 Yamada-Watanabe Theorem for FBSDEs

We now address the question of weak uniqueness, the main objective being to introduce and prove an appropriate version for FBSDEs of the classical Yamada-Watanabe theorem.

Measurability of the Law of the Output with Respect to the Law of the Input

We first prove a technical result which asserts that the distribution of the solution is a measurable function of the distribution of the input whenever weak uniqueness holds. As a preliminary, we introduce a version of Definition 1.23 with main focus on the distribution of the input.

Definition 1.28 *A pair $(\mathcal{F}_{\text{input}}, \mathbb{Q}_{\text{input}})$ is a said to be* an admissible input *if there exists a probability measure $\mathbb{Q}^{\mathcal{B}}_{\text{input}}$ on Ω_{input} equipped with its Borel σ-field $\mathcal{B}(\Omega_{\text{input}})$ such that $(\Omega_{\text{input}}, \mathcal{F}_{\text{input}}, \mathbb{Q}_{\text{input}})$ is the completion of $(\Omega_{\text{input}}, \mathcal{B}(\Omega_{\text{input}}), \mathbb{Q}^{\mathcal{B}}_{\text{input}})$ and*

1. *under $\mathbb{Q}_{\text{input}}$, $(\xi, \boldsymbol{w}^0, \boldsymbol{\nu})$ and $\boldsymbol{w}$ are independent,*
2. *the process $(\boldsymbol{w}^0, \boldsymbol{w})$ is a 2d-dimensional Brownian motion with respect to $\mathbb{G}^{\text{input}} = (\mathcal{G}^{\text{input}}_t)_{0\le t\le T}$, where $\mathbb{G}^{\text{input}} = (\mathcal{G}^{\text{input}}_t)_{0\le t\le T}$ is the complete and right-continuous augmentation under $\mathbb{Q}_{\text{input}}$ of the canonical filtration $\mathbb{G}^{\text{nat,input}}$ on Ω_{input}.*

Notice that, in contrast with Definition 1.23, there is no need to require compatibility since the process $(\xi, \boldsymbol{w}^0, \boldsymbol{\nu}, \boldsymbol{w})$ and the filtration $\mathbb{G}^{\text{input}}$ are necessarily compatible.

Remark 1.29 *Following Remark 1.25, we shall sometimes say that $\mathbb{Q}_{\text{input}}$ (instead of $(\mathcal{F}_{\text{input}}, \mathbb{Q}_{\text{input}})$) is an admissible input. The σ-field $\mathcal{F}_{\text{input}}$ is then automatically* *understood as the completion of the Borel* σ*-field under* $\mathbb{Q}_{\text{input}}$. *Also, note that* $\mathbb{Q}_{\text{input}}$ *is in one-to-one correspondence with* $\mathbb{Q}^{\mathcal{B}}_{\text{input}}$.

The proof of the following result is identical to the proof of Proposition 1.26.

Proposition 1.30 *A pair* $(\mathcal{F}_{\text{input}}, \mathbb{Q}_{\text{input}})$ *is an admissible input if and only if there exists a probability measure* $\mathbb{Q}^{\mathcal{B}}_{\text{input}}$ *on* $(\Omega_{\text{input}}, \mathcal{B}(\Omega_{\text{input}}))$ *such that* $(\Omega_{\text{input}}, \mathcal{F}_{\text{input}}, \mathbb{Q}_{\text{input}})$ *is the completion of* $(\Omega_{\text{input}}, \mathcal{B}(\Omega_{\text{input}}), \mathbb{Q}^{\mathcal{B}}_{\text{input}})$ *and*

1. *under* $\mathbb{Q}^{\mathcal{B}}$, $(\xi, \boldsymbol{w}^0, \boldsymbol{\nu})$ *and* $\boldsymbol{w}$ *are independent,*
2. *under* $\mathbb{Q}^{\mathcal{B}}$, *the process* $(\boldsymbol{w}^0, \boldsymbol{w})$ *is a 2d-dimensional Brownian motion with respect to the right-continuous augmentation* $\mathbb{G}^{\text{nat,input}}_{\cdot+}$ *of* $\mathbb{G}^{\text{nat,input}}$.

We will use the following result:

Proposition 1.31 *Denote by* $\mathscr{P} \subset \mathcal{P}_2(\Omega_{\text{input}})$ *the subset of* $\mathcal{P}_2(\Omega_{\text{input}})$ *of the probability measures on* $(\Omega_{\text{input}}, \mathcal{B}(\Omega_{\text{input}}))$ *whose completions are admissible inputs, and assume that for any admissible input* $(\mathcal{F}_{\text{input}}, \mathbb{Q}_{\text{input}})$, *there exists a unique admissible pair* $(\mathcal{F}_{\text{canon}}, \mathbb{Q})$ *such that:*

1. *The image of* $\mathbb{Q}^{\mathcal{B}}$ *by the canonical projection mapping* $\Omega_{\text{canon}} \to \Omega_{\text{input}}$ *coincides with* $\mathbb{Q}^{\mathcal{B}}_{\text{input}}$ *on the Borel* σ*-field of* $\mathcal{B}(\Omega_{\text{input}})$;
2. *Under* $\mathbb{Q}$, *the process* $(x_t, y_t, z_t, m_t)_{0 \le t \le T}$ *is a solution of the FBSDE* (1.5) *on* Ω_{canon} *for the complete and right-continuous augmentation* $\mathbb{G}$ *of the canonical filtration.*

Then, the mapping:

$$\mathscr{P} \ni \mathbb{Q}^{\mathcal{B}}_{\text{input}} \mapsto \mathbb{Q}^{\mathcal{B}} \in \mathcal{P}_2\big(\Omega_{\text{canon}}\big)$$

is measurable when the spaces are equipped with their respective Borel σ*-fields.*

Recall that $\mathbb{Q}^{\mathcal{B}}_{\text{input}}$ and $\mathbb{Q}^{\mathcal{B}}$ are probability measures on the Borel σ-fields of Ω_{input} and Ω_{canon} respectively, whose completions are $\mathbb{Q}_{\text{input}}$ and $\mathbb{Q}$. The proof uses the so-called Souslin-Kuratowski theorem whose statement we give below for the sake of completeness. See the Notes & Complements at the end of the chapter for references.

Proposition 1.32 *Let* $\mathcal{X}$ *and* $\mathcal{Y}$ *be two Polish spaces,* $\mathcal{S}$ *a Borel subset of* $\mathcal{X}$, *and* ℓ *a one-to-one Borel mapping from* $\mathcal{S}$ *to* $\mathcal{Y}$. *Then* ℓ^{-1} *is Borel.*

Proof of Proposition 1.31. As before, we denote by $\mathbb{G}^{\text{nat}} = (\mathcal{G}^{\text{nat}}_t)_{0 \le t \le T}$ the canonical filtration on Ω_{canon} and by $\mathbb{G}^{\text{nat}}_{\cdot+}$ its right-continuous augmentation. We consider the subset $\mathcal{S}^0 \subset \mathcal{P}_2(\Omega_{\text{canon}})$ formed by the probability measures $\mathbb{Q}^{\mathcal{B}}$ on $(\Omega_{\text{canon}}, \mathcal{B}(\Omega_{\text{canon}}))$ satisfying the three items in the statement of Proposition 1.26 and under which the process $\boldsymbol{m}$ is a square-integrable martingale with respect to $\mathbb{G}^{\mathrm{nat}}_{\cdot +}$, of zero-cross covariation with $(\boldsymbol{w}^0, \boldsymbol{w})$, and starting from $m_0 = 0$ at time 0. Denoting by $\mathbb{Q}$ the completion of $\mathbb{Q}^{\mathcal{B}}$, we observe that $\boldsymbol{m}$ has these properties if and only if under $\mathbb{Q}$, $\boldsymbol{m}$ is a square-integrable martingale with respect to $\mathbb{G}$ (the completion of $\mathbb{G}^{\mathrm{nat}}_{\cdot +}$), of zero-cross covariation with $(\boldsymbol{w}^0, \boldsymbol{w})$, and starting from $m_0 = 0$ at time 0.

First Step. First, we claim that $\mathcal{S}^0$ is a Borel subset of $\mathcal{P}_2(\Omega_{\mathrm{canon}})$ equipped with the Borel σ-field associated with the Wasserstein distance.

The set $\mathcal{S}^0$ is characterized by four conditions. The first three correspond to the three items in the statement of Proposition 1.26. The last one follows from the constraint that $\boldsymbol{m}$ has to be a martingale with respect to $\mathbb{G}^{\mathrm{nat}}_{\cdot +}$, starting from 0 and of zero-cross covariation with respect to $(\boldsymbol{w}, \boldsymbol{w}^0)$. We shall check that each of them defines a measurable subset of $\mathcal{P}_2(\Omega_{\mathrm{canon}})$.

We start with the first two items in Proposition 1.26. We claim that the set of probability measures $\mathbb{Q}^{\mathcal{B}}$ on $(\Omega_{\mathrm{canon}}, \mathcal{B}(\Omega_{\mathrm{canon}}))$ under which $(\xi, \boldsymbol{w}^0, \boldsymbol{\nu})$ and $\boldsymbol{w}$ are independent and $(\boldsymbol{w}^0, \boldsymbol{w})$ is a $\mathbb{G}^{\mathrm{nat}}_{\cdot +}$-Brownian motion of dimension $2d$ is a closed subset of $\mathcal{P}_2(\Omega_{\mathrm{canon}})$. The reason is that according to Theorem (Vol I)-5.5, convergence in the Wasserstein space is equivalent to convergence in law plus uniform square integrability. Clearly, the first two items in Proposition 1.26 are stable under convergence in law, and thus under convergence in the Wasserstein space as well. This shows the required closure property.

We proceed in a similar way with the fourth condition, namely the martingale property of $\boldsymbol{m}$. To do so, we invoke two well-known properties of the $J1$ Skorohod topology:

1. From [57, Chapter 3, Section 12, Theorem 12.5], the functional which maps a path from $\mathcal{D}([0,T]; \mathcal{X})$ onto its initial condition at time 0, or its terminal condition at time T, is continuous for the $J1$ topology. This shows that the condition $m_0 = 0$ is stable under convergence in the Wasserstein space.
2. From [15, Chapter 1, Proposition 5.1], the martingale property is stable under convergence in law with respect to $J1$ provided uniform integrability holds, which proves that the martingale property of $\boldsymbol{m}$ is stable under convergence in the Wasserstein space. Similarly, the orthogonality property between $\boldsymbol{m}$ and $(\boldsymbol{w}^0, \boldsymbol{w})$ is preserved.

It is somehow more difficult to prove that the compatibility condition in the third item of Proposition 1.26 defines a Borel subset of $\mathcal{P}_2(\Omega_{\mathrm{canon}})$. The crucial fact which we use is the following:

3. For any $t \in [0,T]$, the σ-field $\mathcal{G}^{\mathrm{nat}}_t$ is generated by a countable field $\mathcal{C}^{\mathrm{canon}}_t$ that may be chosen independently of $\mathbb{Q}$ in such a way that rational linear combinations of indicator functions of events in $\mathcal{C}^{\mathrm{canon}}_t$ are dense in $L^2(\Omega_{\mathrm{canon}}, \mathcal{G}^{\mathrm{nat}}_t, \mathbb{Q}^{\mathcal{B}})$. We denote by $\mathcal{E}^{\mathrm{canon}}_t$ the countable collection of such linear combinations. See for example [57, Theorem 19.2]. Similarly with a slight abuse of notation, we can regard $\mathcal{G}^{\mathrm{nat,input}}_t$ as a sub-σ-field of the Borel σ-field on Ω_{canon} (and not Ω_{input}), and check that it is generated by a countable field $\mathcal{C}^{\mathrm{input}}_t$ that may be chosen independently of $\mathbb{Q}$ in such a way that rational linear combinations of indicator functions of events in $\mathcal{C}^{\mathrm{input}}_t$ are dense in $L^2(\Omega_{\mathrm{canon}}, \mathcal{G}^{\mathrm{nat,input}}_t, \mathbb{Q}^{\mathcal{B}})$. We denote by $\mathcal{E}^{\mathrm{input}}_t$ the countable collection of such linear combinations.

We claim that compatibility reads as follows: $\mathbb{Q}^{\mathcal{B}}$ satisfies item 3 in Proposition 1.26 if and only if, for any t in a countable dense subset of $[0,T)$, any rational $\varepsilon \in (0,1]$, any

$D \in \mathcal{C}_t^{\text{canon}}$, there exist a rational $\delta > 0$ with $t + \delta \le T$, together with a $[0,1]$-valued function $\psi^{\varepsilon,\delta}$ in $\mathcal{E}_{t+\delta}^{\text{input}}$, such that, for all $[0,1]$-valued functions ψ in $\mathcal{E}_T^{\text{input}}$, we have:

$$\int_{\Omega_{\text{canon}}} \left|\mathbf{1}_D - \psi^{\varepsilon,\delta}\right|^2 d\mathbb{Q}^{\mathcal{B}} \le \int_{\Omega_{\text{canon}}} \left|\mathbf{1}_D - \psi\right|^2 d\mathbb{Q}^{\mathcal{B}} + \varepsilon. \tag{1.13}$$

If this is the case, for prescribed values of t, ε, δ, D, $\psi^{\varepsilon,\delta}$, ψ, the condition above defines a Borel subset of $\mathcal{P}_2(\Omega_{\text{canon}})$. This follows from Proposition (Vol I)-5.7. By intersection and union over countable sets of indices, we deduce that the compatibility condition defines a Borel subset of $\mathcal{P}_2(\Omega_{\text{canon}})$, provided that the way we characterized compatibility is indeed correct.

In order to check that the characterization is correct, we first fix $t \in [0,T)$ and $D \in \mathcal{C}_t^{\text{canon}}$. We claim that the following two assertions are equivalent:

(i.) for any rational $\varepsilon \in (0,1]$, there exist a rational $\delta > 0$, with $t + \delta \le T$, together with a $[0,1]$-valued function $\psi^{\varepsilon,\delta}$ in $\mathcal{E}_{t+\delta}^{\text{input}}$ such that, for all $[0,1]$-valued functions ψ in $\mathcal{E}_T^{\text{input}}$, (1.13) holds;

(ii.) $\mathbb{Q}^{\mathcal{B}}$ almost surely, we have:

$$\mathbb{Q}^{\mathcal{B}}\big(D|\mathcal{G}_{t+}^{\text{nat,input}}\big) = \mathbb{Q}^{\mathcal{B}}\big(D|\mathcal{G}_T^{\text{nat,input}}\big). \tag{1.14}$$

The proof of the equivalence is as follows. If (1.14) holds, then,

$$\lim_{\delta \to 0} \mathbb{E}^{\mathbb{Q}^{\mathcal{B}}}\Big[\big|\mathbb{Q}^{\mathcal{B}}\big(D|\,\mathcal{G}_{t+\delta}^{\text{nat,input}}\big) - \mathbb{Q}^{\mathcal{B}}\big(D|\mathcal{G}_T^{\text{nat,input}}\big)\big|^2\Big] = 0.$$

Therefore, for δ small enough, the left-hand side is less than ε. Now, we can approximate $\mathbb{Q}^{\mathcal{B}}(D|\mathcal{G}_{t+\delta}^{\text{nat,input}})$ by an element $\psi^{\varepsilon,\delta}$ in $\mathcal{E}_{t+\delta}^{\text{input}}$ up to ε in L^2-norm. Up to a multiplicative constant in front of ε, we get (1.13) by using the characterization of the conditional expectation as an orthogonal projection in L^2. Conversely, if (i.) holds, then for any ε, we can find δ small enough such that:

$$\mathbb{E}^{\mathbb{Q}^{\mathcal{B}}}\Big[\big|\mathbb{Q}^{\mathcal{B}}\big(D|\mathcal{G}_T^{\text{nat,input}}\big) - \psi^{\varepsilon,\delta}\big|^2\Big] \le \varepsilon,$$
$$\mathbb{E}^{\mathbb{Q}^{\mathcal{B}}}\Big[\big|\mathbb{Q}^{\mathcal{B}}\big(D|\mathcal{G}_{t+\delta}^{\text{nat,input}}\big) - \psi^{\varepsilon,\delta}\big|^2\Big] \le \varepsilon,$$

from which we get (1.14).

Now that (i.) and (ii.) are known to be equivalent, the necessary part of our characterization of compatibility through (1.13) follows from (H2) in Proposition 1.3. In order to prove that the sufficient part is also true, we use the fact that in our construction of $\mathcal{C}_t^{\text{canon}}$, we can assume that, using again [57, Theorems 11.4 and 19.2], for any $\varepsilon > 0$, $D \in \mathcal{G}_t^{\text{nat}}$, there exist $D^{\varepsilon} \in \mathcal{C}_t^{\text{canon}}$ such that:

$$\mathbb{E}^{\mathbb{Q}^{\mathcal{B}}}\big[|\mathbf{1}_D - \mathbf{1}_{D^{\varepsilon}}|^2\big] \le \varepsilon.$$

Now, by a standard approximation argument and our characterization, (1.14) holds for any $D \in \mathcal{G}_t^{\text{nat}}$. This is true for any t in a dense countable subset of $[0,T)$. Actually, it is obviously true for $t = T$ as well. By right-continuity of the filtration $\mathbb{G}$, this is true for any $t \in [0,T]$. In order to complete the proof, we have to replace the condition $D \in \mathcal{G}_t^{\mathrm{nat}}$ by $D \in \mathcal{G}_{t+}^{\mathrm{nat}}$. This can be done by following the argument of the first step of the proof of Lemma 1.7.

Second Step. Consider now the set $\mathcal{S} \subset \mathcal{S}^0$ of the probability measures $\mathbb{Q}^{\mathcal{B}} \in \mathcal{S}^0$ such that, under the completion $\mathbb{Q}$ of $\mathbb{Q}^{\mathcal{B}}$, the canonical process $(\xi, \boldsymbol{w}^0, \boldsymbol{\nu}, \boldsymbol{w}, \boldsymbol{x}, \boldsymbol{y}, (\boldsymbol{\zeta}, \boldsymbol{\zeta}^0), \boldsymbol{m})$ satisfies the FBSDE (1.5) on the canonical space Ω_{canon} equipped with the complete and right-continuous filtration $\mathbb{G}$.

We use the approximating sequence $(B_\ell)_{\ell \geq 1}$ and the bounded and nondecreasing smooth function $\vartheta : \mathbb{R} \to \mathbb{R}$ satisfying $\vartheta(x) = x$ for any $x \in [0, 1]$ which we introduced in the proof of Lemma 1.27. For any integers $\ell, n, p \geq 1$, with $p > n$, and any $\mathbb{Q}^{\mathcal{B}} \in \mathcal{P}_2(\Omega_{\mathrm{canon}})$, we write:

$$\begin{aligned}\Theta_{n,p}^{\ell}(\mathbb{Q}^{\mathcal{B}}) = \mathbb{E}^{\mathbb{Q}^{\mathcal{B}}}\biggl[\vartheta\biggl(\sup_{0 \leq t \leq T}\biggl| x_t - \int_0^t B_\ell\Bigl(s, x_s, \nu_s, y_s, n\bigl((\zeta, \zeta^0)_s - (\zeta, \zeta^0)_{(s-1/n)+}\bigr)\Bigr)ds \\ - \int_0^t \biggl(n\int_{(\lfloor ps \rfloor p - 1/n)+}^{\lfloor ps \rfloor/p} \Sigma(r, x_r, \nu_r)dr\biggr)dw_s \\ - \int_0^t \biggl(n\int_{(\lfloor ps \rfloor/p - 1/n)+}^{\lfloor ps \rfloor/p} \Sigma^0(r, x_r, \nu_r)dr\biggr)dw_s^0\biggr|\biggr)\biggr],\end{aligned}$$

which makes sense even if $\mathbb{Q}^{\mathcal{B}} \notin \mathcal{S}^0$, since the above stochastic integrals can be interpreted as finite Riemann sums. It is clear that, for any $\ell, n, p \geq 1$, with $p > n$, there exists a measurable and bounded mapping $\Psi_{n,p}^{\ell} : \Omega_{\mathrm{canon}} \to \mathbb{R}$, such that:

$$\Theta_{n,p}^{\ell}(\mathbb{Q}^{\mathcal{B}}) = \int_{\Omega_{\mathrm{canon}}} \Psi_{n,p}^{\ell} d\mathbb{Q}^{\mathcal{B}}.$$

Thanks to Proposition (Vol I)-5.7, this proves that $\Theta_{n,p}^{\ell}$ is a measurable mapping from $\mathcal{P}^2(\Omega_{\mathrm{canon}})$ into $\mathbb{R}$.

Proceeding in the same way with the backward equation, we may assume without any loss of generality that $\mathbb{Q}^{\mathcal{B}} \in \mathcal{S}$ if and only if $\mathbb{Q}^{\mathcal{B}} \in \mathcal{S}^0$ and:

$$\limsup_{\ell \to \infty} \limsup_{n \to \infty} \limsup_{p \to \infty} \Theta_{n,p}^{\ell}(\mathbb{Q}^{\mathcal{B}}) = 0,$$

from which we deduce that $\mathcal{S}$ is a Borel subset of $\mathcal{P}^2(\Omega_{\mathrm{canon}})$.

We finally notice that the mapping $\mathcal{S} \ni \mathbb{Q}^{\mathcal{B}} \mapsto \mathbb{Q}^{\mathcal{B}} \circ (\xi, \boldsymbol{w}^0, \boldsymbol{\nu}, \boldsymbol{w})^{-1} \in \mathcal{P}_2(\Omega_{\mathrm{input}})$ is measurable as it is the restriction of a continuous function defined on the whole $\mathcal{P}_2(\Omega_{\mathrm{canon}})$. By weak uniqueness, it is obviously one-to-one. We complete the proof using the Souslin-Kuratowski theorem recalled as Proposition 1.32. □

Main Statement

We now state and prove the main result of this section.

Theorem 1.33 *Assume that, on an admissible probabilistic set-up* $(\Omega, \mathcal{F}, \mathbb{F}, \mathbb{P})$ *with input* $(X_0, W_t^0, \mu_t, W_t)_{0 \leq t \leq T}$, *the FBSDE* (1.5) *has a solution which we denote* *by* $(X_t, Y_t, \int_0^t (Z_s, Z_s^0)ds, M_t)_{0\le t\le T}$, *and that strong uniqueness holds for a given choice of the joint distribution* $\mathcal{L}(X_0, \mathbf{W}^0, \boldsymbol{\mu})$ *of* $(X_0, \mathbf{W}^0, \boldsymbol{\mu})$ *on* $\mathbb{R}^d \times \mathcal{C}([0,T];\mathbb{R}^d) \times \mathcal{D}([0,T];\mathcal{X})$.

Then, the law of $(X_0, W_t^0, \mu_t, W_t, X_t, Y_t, \int_0^t (Z_s, Z_s^0)ds, M_t)_{0\le t\le T}$ *on the space* Ω_{canon} *only depends upon* $\mathcal{L}(X_0, \mathbf{W}^0, \boldsymbol{\mu})$. *Moreover, there exists a measurable function* Φ *from* Ω_{input} *into* Ω_{output}, *only depending on the joint distribution of* $(X_0, \mathbf{W}^0, \boldsymbol{\mu})$, *such that,* $\mathbb{P}$ *almost surely,*

$$\left(X_t, Y_t, \int_0^t (Z_s, Z_s^0)ds, M_t\right)_{0\le t\le T} = \Phi\big(X_0, \mathbf{W}^0, \boldsymbol{\mu}, \mathbf{W}\big).$$

Furthermore, if $(\Omega', \mathcal{F}', \mathbb{F}', \mathbb{P}')$ *is another (possibly different) admissible set-up with input* $(X_0', W_t^{0\prime}, \mu_t', W_t')_{0\le t\le T}$ *such that* $\mathcal{L}(X_0', \mathbf{W}^{0\prime}, \boldsymbol{\mu}') = \mathcal{L}(X_0, \mathbf{W}^0, \boldsymbol{\mu})$, *the process* $(X_t', Y_t', Z_t', Z_t^{0\prime}, M_t')_{0\le t\le T}$ *defined by:*

$$\left(X_t', Y_t', (\mathcal{Z}', \mathcal{Z}^{0\prime})_t, M_t'\right)_{0\le t\le T} = \Phi\big(X_0', \mathbf{W}^{0\prime}, \boldsymbol{\mu}', \mathbf{W}'\big),$$

and

$$\begin{aligned}
&\big(Z_t', Z_t^{0\prime}\big)(\omega') \\
&= \begin{cases} \lim_{n\to\infty} n\big((\mathcal{Z}', \mathcal{Z}^{0\prime})_t(\omega') - (\mathcal{Z}', \mathcal{Z}^{0\prime})_{(t-1/n)_+}(\omega')\big) & \text{if the limit exists,} \\ 0 \quad \text{otherwise,} \end{cases}
\end{aligned}$$

for $(t, \omega') \in [0,T] \times \Omega'$, *is also a solution of* (1.5).

Remark 1.34 *Importantly, in the proof below, the functional* Φ *is shown to be progressively measurable. Namely, writing* Φ *in the form* $\Phi = (\Phi_t = e_t \circ \Phi)_{0\le t\le T}$, *where* e_t *is the evaluation mapping* $(\boldsymbol{x}, \boldsymbol{y}, \boldsymbol{\zeta}, \boldsymbol{\zeta}^0, \boldsymbol{m}) \mapsto (x_t, y_t, \zeta_t, \zeta_t^0, m_t)$, *the process* $(\Phi_t)_{0\le t\le T}$ *is progressively measurable with respect to the complete and right-continuous augmentation of the canonical filtration.*

In particular, Theorem 1.33 says that whenever the forward-backward system (1.5) *satisfies the strong uniqueness property, any solution constructed on some admissible set-up* $(\Omega, \mathcal{F}, \mathbb{F}, \mathbb{P})$ *with input* $(X_0, \mathbf{W}^0, \boldsymbol{\mu}, \mathbf{W})$ *and with* X_0 *as initial condition is* $\mathbb{F}^{(X_0, \mathbf{W}^0, \boldsymbol{\mu}, \mathbf{W})}$*-progressively measurable.*

Remark 1.35 *When the coefficients do not depend on the variable* z^0, $\mathbf{W}^0 \equiv 0$ *and* $\boldsymbol{\mu}$ *is deterministic,* $(M_t)_{0\le t\le T}$ *may be ignored and the law of the process* $(X_0, W_t, X_t, Y_t, \int_0^t Z_s ds)_{0\le t\le T}$ *on* $\mathbb{R}^d \times \mathcal{C}([0,T];\mathbb{R}^d) \times \mathcal{C}([0,T];\mathbb{R}^d) \times \mathcal{C}([0,T];\mathbb{R}^m) \times \mathcal{C}([0,T];\mathbb{R}^{m\times d})$ *only depends upon the law of* X_0. *In particular, the function* Φ *may be constructed on* $\mathbb{R}^d \times \mathcal{C}([0,T];\mathbb{R}^d)$ *and the solution* $(X_0, W_t, X_t, Y_t, \int_0^t Z_s ds)_{0\le t\le T}$ *is,* $\mathbb{P}$*-almost surely, equal to* $\Phi(X_0, \mathbf{W})$. *The reconstruction of a solution on any other set-up is obtained accordingly.*

Proof of Theorem 1.33.

First Step. Let $(X_t, Y_t, Z_t, Z_t^0, M_t)_{0\le t\le T}$ and $(X_t', Y_t', Z_t', Z_t^{0\prime}, M_t')_{0\le t\le T}$ be two solutions of the FBSDE (1.1) associated with two set-ups $(\Omega, \mathcal{F}, \mathbb{F}, \mathbb{P})$ and $(\Omega', \mathcal{F}', \mathbb{F}', \mathbb{P}')$ equipped with two identically distributed processes $(X_0, W_t^0, \mu_t, W_t)_{0\le t\le T}$ and $(X_0', W_t^{0\prime}, \mu_t', W_t')_{0\le t\le T}$. We denote by $\mathbb{Q}^{\mathcal{B}}$ and $\mathbb{Q}^{\mathcal{B},\prime}$ the distributions of the processes:

$$\Big(X_0, W_t^0, \mu_t, W_t, X_t, Y_t, \int_0^t (Z_s, Z_s^0) ds, M_t\Big)_{0\le t\le T}$$

and

$$\Big(X_0', W_t^{0\prime}, \mu_t', W_t', X_t', Y_t', \int_0^t (Z_s', Z_s^{0\prime}) ds, M_t'\Big)_{0\le t\le T}$$

on $\Omega_{\mathrm{canon}} = \Omega_{\mathrm{input}} \times \Omega_{\mathrm{output}}$ equipped with its Borel σ-field, and by $\mathbb{Q}^{\mathcal{B}}_{\mathrm{input}}$ the (common) distribution of the processes:

$$\Big(X_0, W_t^0, \mu_t, W_t\Big)_{0\le t\le T} \quad \text{and} \quad \Big(X_0', W_t^{0\prime}, \mu_t', W_t'\Big)_{0\le t\le T}$$

on Ω_{input} equipped with its Borel σ-field.

On the canonical space Ω_{canon}, we denote by q (respectively q') the regular conditional probability of the second marginal of $\mathbb{Q}^{\mathcal{B}}$ (respectively $\mathbb{Q}^{\mathcal{B},\prime}$) on Ω_{output} given the first one on Ω_{input}. On the extended space $\bar{\Omega} = \Omega_{\mathrm{input}} \times \Omega_{\mathrm{output}} \times \Omega_{\mathrm{output}}$ equipped with the product Borel σ-field, we define the probability measure $\bar{\mathbb{Q}}^{\mathcal{B}}$ by setting:

$$\bar{\mathbb{Q}}^{\mathcal{B}}\big(C \times D \times D'\big) = \int_C q\big(\omega_{\mathrm{input}}, D\big) q'\big(\omega_{\mathrm{input}}, D'\big) d\mathbb{Q}^{\mathcal{B}}_{\mathrm{input}}(\omega_{\mathrm{input}}),$$

where C is a Borel subset of Ω_{input} and D and D' are Borel subsets of Ω_{output}. Notice that:

$$\bar{\mathbb{Q}}^{\mathcal{B}}\big(C \times D \times \Omega_{\mathrm{output}}\big) = \int_C q\big(\omega_{\mathrm{input}}, D\big) d\mathbb{Q}^{\mathcal{B}}_{\mathrm{input}}(\omega_{\mathrm{input}}) = \mathbb{Q}^{\mathcal{B}}(C \times D),$$

and similarly, $\bar{\mathbb{Q}}^{\mathcal{B}}(C \times \Omega_{\mathrm{output}} \times D') = \mathbb{Q}^{\mathcal{B},\prime}(C \times D')$. In particular, if we denote by $(\xi, \boldsymbol{w}^0, \boldsymbol{\nu}, \boldsymbol{w}, \boldsymbol{x}, \boldsymbol{y}, \boldsymbol{\zeta}, \boldsymbol{m}, \boldsymbol{x}', \boldsymbol{y}', \boldsymbol{\zeta}', \boldsymbol{m}')$ the canonical process on $\bar{\Omega}$, then, under $\bar{\mathbb{Q}}^{\mathcal{B}}$, the process $(\xi, \boldsymbol{w}^0, \boldsymbol{\nu}, \boldsymbol{w}, \boldsymbol{x}, \boldsymbol{y}, \boldsymbol{\zeta}, \boldsymbol{m})$ has distribution $\mathbb{Q}^{\mathcal{B}}$ and $(\xi, \boldsymbol{w}^0, \boldsymbol{\nu}, \boldsymbol{w}, \boldsymbol{x}', \boldsymbol{y}', \boldsymbol{\zeta}', \boldsymbol{m}')$ has distribution $\mathbb{Q}^{\mathcal{B},\prime}$.

We now use the proof and the result of Lemma 1.27. For any $t \in [0, T]$ and $\omega \in \bar{\Omega}$ we set:

$$(z_t, z_t^0)(\omega) = \begin{cases} \lim_{n\to\infty} n\big((\zeta, \zeta^0)_t(\omega) - (\zeta, \zeta^0)_{t-1/n}(\omega)\big) & \text{if the limit exists} \\ 0 \quad \text{otherwise} \end{cases},$$

$$(z_t', z_t^{0\prime})(\omega) = \begin{cases} \lim_{n\to\infty} n\big((\zeta', \zeta^{0\prime})_t(\omega) - (\zeta', \zeta^{0\prime})_{t-1/n}(\omega)\big) & \text{if the limit exists} \\ 0 \quad \text{otherwise} \end{cases}.$$

On Ω_{canon} equipped with the completion $\mathbb{Q}$ of $\mathbb{Q}^{\mathcal{B}}$ and endowed with the complete and right-continuous augmentation $\mathbb{G}$ of the canonical filtration (respectively equipped with the completion $\mathbb{Q}'$ of $\mathbb{Q}^{\mathcal{B},\prime}$ and endowed with the complete and right-continuous augmentation $\mathbb{G}'$ of the canonical filtration), the canonical process

$$(\xi, \boldsymbol{w}^0, \boldsymbol{v}, \boldsymbol{w}, \boldsymbol{x}, \boldsymbol{y}, \boldsymbol{z}, \boldsymbol{z}^0, \boldsymbol{m})$$

satisfies the FBSDE (1.5). Proceeding in the same way as in the proof of Lemma 1.27, we prove that, if $\bar{\Omega}$ is equipped with the completion $\bar{\mathbb{Q}}$ of $\bar{\mathbb{Q}}^{\mathcal{B}}$ and endowed with the complete and right-continuous augmentation under $\bar{\mathbb{Q}}$ of the filtration generated by

$$(\xi, \boldsymbol{w}^0, \boldsymbol{v}, \boldsymbol{w}, \boldsymbol{x}, \boldsymbol{y}, (\boldsymbol{\zeta}, \boldsymbol{\zeta}^0), \boldsymbol{m}) \quad \text{(respectively } (\xi, \boldsymbol{w}^0, \boldsymbol{v}, \boldsymbol{w}, \boldsymbol{x}', \boldsymbol{y}', (\boldsymbol{\zeta}', \boldsymbol{\zeta}^{0\prime}), \boldsymbol{m}')),$$

the process

$$(\xi, \boldsymbol{w}^0, \boldsymbol{v}, \boldsymbol{w}, \boldsymbol{x}, \boldsymbol{y}, \boldsymbol{z}, \boldsymbol{z}^0, \boldsymbol{m}) \quad \text{(respectively } (\xi, \boldsymbol{w}^0, \boldsymbol{v}, \boldsymbol{w}, \boldsymbol{x}', \boldsymbol{y}', \boldsymbol{z}', \boldsymbol{z}^{0\prime}, \boldsymbol{m}'))$$

satisfies the FBSDE (1.5). The difficulty is that this does not suffice to compare the two solutions since they are not defined on the same probabilistic set-up since the underlying filtrations differ.

Second Step. Denoting by $\bar{\mathbb{G}} = (\bar{\mathcal{G}}_t)_{0 \le t \le T}$ the complete and right-continuous augmentation under $\bar{\mathbb{Q}}$ of the canonical filtration on $\bar{\Omega}$, we must prove that $(\bar{\Omega}, \bar{\mathbb{G}}, \bar{\mathbb{Q}})$, equipped with $(\xi, \boldsymbol{w}^0, \boldsymbol{v}, \boldsymbol{w})$ forms an admissible probabilistic set-up as defined in Subsection 1.1.1. Then, we must show that on this set-up, both processes $(\xi, \boldsymbol{w}^0, \boldsymbol{v}, \boldsymbol{w}, \boldsymbol{x}, \boldsymbol{y}, \boldsymbol{z}, \boldsymbol{z}^0, \boldsymbol{m})$ and $(\xi, \boldsymbol{w}^0, \boldsymbol{v}, \boldsymbol{w}, \boldsymbol{x}', \boldsymbol{y}', \boldsymbol{z}', \boldsymbol{z}^{0\prime}, \boldsymbol{m}')$ solve the FBSDE (1.5).

In order to show that $(\bar{\Omega}, \bar{\mathbb{G}}, \bar{\mathbb{Q}})$, equipped with $(\xi, \boldsymbol{w}^0, \boldsymbol{v}, \boldsymbol{w})$ indeed forms an admissible probabilistic set-up, we must check that, under $\bar{\mathbb{Q}}$, $(\boldsymbol{w}^0, \boldsymbol{w})$ is a $2d$-dimensional $\bar{\mathbb{G}}$-Brownian motion and that $(\xi, \boldsymbol{w}^0, \boldsymbol{v}, \boldsymbol{w})$ is compatible with the filtration $\bar{\mathbb{G}}$. The argument is deferred to the third step below.

For now, we claim that, in order to show that $(\xi, \boldsymbol{w}^0, \boldsymbol{v}, \boldsymbol{w}, \boldsymbol{x}, \boldsymbol{y}, \boldsymbol{z}, \boldsymbol{z}^0, \boldsymbol{m})$ (the argument being similar for $(\xi, \boldsymbol{w}^0, \boldsymbol{v}, \boldsymbol{w}, \boldsymbol{x}', \boldsymbol{y}', \boldsymbol{z}', \boldsymbol{z}^{0\prime}, \boldsymbol{m}')$) solves (1.5) on $(\bar{\Omega}, \bar{\mathbb{G}}, \bar{\mathbb{Q}})$, it suffices to prove that $(\xi, \boldsymbol{w}^0, \boldsymbol{v}, \boldsymbol{w}, \boldsymbol{x}, \boldsymbol{y}, (\boldsymbol{\zeta}, \boldsymbol{\zeta}^0), \boldsymbol{m})$ (respectively $(\xi, \boldsymbol{w}^0, \boldsymbol{v}, \boldsymbol{w}, \boldsymbol{x}', \boldsymbol{y}', (\boldsymbol{\zeta}', \boldsymbol{\zeta}^{0\prime}), \boldsymbol{m}')$) is compatible with the filtration $\bar{\mathbb{G}}$. Indeed, we know from Lemma 1.9 that, whenever compatibility holds, any martingale with respect to the augmentation of the filtration generated by $(\xi, \boldsymbol{w}^0, \boldsymbol{v}, \boldsymbol{w}, \boldsymbol{x}, \boldsymbol{y}, (\boldsymbol{\zeta}, \boldsymbol{\zeta}^0), \boldsymbol{m})$ is also a martingale with respect to $\bar{\mathbb{G}}$. Since $\boldsymbol{m}$ is already known to be a martingale with respect to the augmentation of the filtration generated by $(\xi, \boldsymbol{w}^0, \boldsymbol{v}, \boldsymbol{w}, \boldsymbol{x}, \boldsymbol{y}, (\boldsymbol{\zeta}, \boldsymbol{\zeta}^0), \boldsymbol{m})$, it is also a martingale with respect to $\bar{\mathbb{G}}$ whenever compatibility holds. Similarly, since $\boldsymbol{m} \otimes (\boldsymbol{w}^0, \boldsymbol{w})$ is a martingale with respect to the augmentation of the filtration generated by $(\xi, \boldsymbol{w}^0, \boldsymbol{v}, \boldsymbol{w}, \boldsymbol{x}, \boldsymbol{y}, (\boldsymbol{\zeta}, \boldsymbol{\zeta}^0), \boldsymbol{m})$, it is also a martingale with respect to $\bar{\mathbb{G}}$ whenever compatibility holds, namely the bracket of $\boldsymbol{m}$ and $(\boldsymbol{w}^0, \boldsymbol{w})$ is zero on the probabilistic set-up $(\bar{\Omega}, \bar{\mathbb{G}}, \bar{\mathbb{Q}})$. Finally, it is indeed true that $(\xi, \boldsymbol{w}^0, \boldsymbol{v}, \boldsymbol{w}, \boldsymbol{x}, \boldsymbol{y}, \boldsymbol{z}, \boldsymbol{z}^0, \boldsymbol{m})$ is a solution of (1.5) whenever compatibility holds. Again, the proof of the compatibility is deferred to the third step of the proof below.

Assuming momentarily that

$$(\xi, \boldsymbol{w}^0, \boldsymbol{v}, \boldsymbol{w}, \boldsymbol{x}, \boldsymbol{y}, \boldsymbol{z}, \boldsymbol{z}^0, \boldsymbol{m}) \text{ and } (\xi, \boldsymbol{w}^0, \boldsymbol{v}, \boldsymbol{w}, \boldsymbol{x}', \boldsymbol{y}', \boldsymbol{z}', \boldsymbol{z}^{0\prime}, \boldsymbol{m}')$$

both solve (1.5) on the admissible set-up $(\bar{\Omega}, \bar{\mathbb{G}}, \bar{\mathbb{Q}})$, it makes sense to compare them pathwise. If strong uniqueness holds, they must match $\bar{\mathbb{P}}$ almost surely, proving that they have the same law.

It then remains to construct the mapping Φ. By strong uniqueness, we know that, for $\mathbb{Q}^{\mathcal{B}}_{\text{input}}$-almost every $\omega_{\text{input}} \in \Omega_{\text{input}}$, we have:

$$\big(q(\omega_{\text{input}}, \cdot) \otimes q'(\omega_{\text{input}}, \cdot)\big)\big[\forall t \in [0,T],\ \big(x_t, y_t, (\zeta, \zeta^0)_t, m_t\big) = \big(x'_t, y'_t, (\zeta', \zeta^{0\prime})_t, m'_t\big)\big] = 1.$$

Since $(\boldsymbol{x}, \boldsymbol{y}, (\boldsymbol{\zeta}, \boldsymbol{\zeta}^0), \boldsymbol{m})$ and $(\boldsymbol{x}', \boldsymbol{y}', (\boldsymbol{\zeta}', \boldsymbol{\zeta}^{0\prime}), \boldsymbol{m}')$ are independent under the probability measure $q(\omega_{\text{input}}, \cdot) \otimes q'(\omega_{\text{input}}, \cdot)$, the only way they can be equal with probability 1 is if they are constant, in other words if $q(\omega_{\text{input}}, \cdot)$ almost surely, for all $t \in [0,T]$,

$$(x_t, y_t, \zeta_t, \zeta^0_t, m_t) = \mathbb{E}^{q(\omega_{\text{input}}, \cdot)}\big[(x_t, y_t, \zeta_t, \zeta^0_t, m_t)\big].$$

Letting

$$\Phi(\omega_{\text{input}}) = \Big(\mathbb{E}^{q(\omega_{\text{input}}, \cdot)}\big[(x_t, y_t, (\zeta, \zeta^0)_t, m_t)\big]\mathbf{1}_{\{\mathbb{E}^{q(\omega_{\text{input}}, \cdot)}[\sup_{0 \le s \le T} |(x_s, y_s, (\zeta, \zeta^0)_s, m_s)|] < \infty\}}\Big)_{0 \le t \le T}, \tag{1.15}$$

we complete the proof of the representation formula. By dominated convergence, Φ takes values in the set Ω_{output}.

Third Step. We now check all the compatibility conditions. We use the same notations as in Definitions 1.23 and 1.28. We denote by $\mathbb{G}^{\text{nat,input}} = (\mathcal{G}^{\text{nat,input}}_t)_{0 \le t \le T}$ the natural filtration generated by the input process $(\xi, \boldsymbol{w}^0, \boldsymbol{v}, \boldsymbol{w})$. With a slight abuse of notation, we regard $\mathbb{G}^{\text{nat,input}}$ as a filtration on Ω_{input} and on $\bar{\Omega}$ as well. Similarly, we denote by $\mathbb{G}^{\text{nat,output}} = (\mathcal{G}^{\text{nat,output}}_t)_{0 \le t \le T}$ (resp. $\mathbb{G}^{\text{nat,canon}} = (\mathcal{G}^{\text{nat,canon}}_t)_{0 \le t \le T}$) the natural filtration generated by the output process $(\boldsymbol{x}, \boldsymbol{y}, (\boldsymbol{\zeta}, \boldsymbol{\zeta}^0), \boldsymbol{m})$ (respectively the process $(\xi, \boldsymbol{w}^0, \boldsymbol{v}, \boldsymbol{w}, \boldsymbol{x}, \boldsymbol{y}, (\boldsymbol{\zeta}, \boldsymbol{\zeta}^0), \boldsymbol{m})$) on Ω_{output} (respectively Ω_{canon}) and again, we regard them as filtrations on Ω_{output} (respectively Ω_{canon}) and $\bar{\Omega}$. Finally, we denote by $\mathbb{G}^{\text{nat}} = (\mathcal{G}^{\text{nat}}_t)_{0 \le t \le T}$ the canonical filtration on $\bar{\Omega}$. In all cases, we denote the corresponding complete and right-continuous augmentation on $\bar{\Omega}$ under $\bar{\mathbb{Q}}$ by dropping the label 'nat' in the notation $\mathbb{G}^{\text{nat},\cdots}$.

Generally speaking, the aforementioned compatibility properties follow from the following measurability result whose proof is deferred to the last step below: For any $t \in [0,T]$ and any $D \in \mathcal{G}^{\text{nat,output}}_t$ seen as a Borel subset of Ω_{output}, the random variables $\Omega_{\text{input}} \ni \omega_{\text{input}} \mapsto q(\omega_{\text{input}}, D)$ and $\Omega_{\text{input}} \ni \omega_{\text{input}} \mapsto q'(\omega_{\text{input}}, D)$ are measurable with respect to the completion under $\mathbb{Q}_{\text{input}}$ of the σ-field $\mathcal{G}^{\text{nat,input}}_{t+}$ on Ω_{input}, which is the completion of $\mathbb{Q}^{\mathcal{B}}_{\text{input}}$. In particular, the random variables $\bar{\Omega} \ni (\omega_{\text{input}}, \omega_{\text{output}}, \omega'_{\text{output}}) \mapsto q(\omega_{\text{input}}, D)$ and $\bar{\Omega} \ni (\omega_{\text{input}}, \omega_{\text{output}}, \omega'_{\text{output}}) \mapsto q'(\omega_{\text{input}}, D)$ are measurable with respect to $\mathcal{G}^{\text{input}}_t$, which is a σ-field on $\bar{\Omega}$.

We first check that $(\boldsymbol{w}^0, \boldsymbol{w})$ is a $2d$-dimensional $\bar{\mathbb{G}}$-Brownian motion. To start with, notice that $(\boldsymbol{w}^0, \boldsymbol{w})$ is a $2d$-dimensional $\mathbb{G}^{\text{input}}$-Brownian motion on Ω_{input} for $\mathbb{Q}_{\text{input}}$. Consider now $C \in \mathcal{G}^{\text{nat,input}}_t$, $C' \in \sigma\{(w^0_s - w^0_t, w_s - w_t), t \le s \le T\}$, $D \in \mathcal{G}^{\text{nat,output}}_t$ and $D' \in \mathcal{G}^{\text{nat,output}}_t$. Then, identifying C and C' with Borel subsets of Ω_{input} and D and D' with Borel subsets of Ω_{output}, we have:

$$\bar{\mathbb{Q}}((C \cap C') \times D \times D') = \int_{C \cap C'} q(\omega_{\text{input}}, D) q'(\omega_{\text{input}}, D') d\mathbb{Q}_{\text{input}}(\omega_{\text{input}})$$

$$= \mathbb{Q}_{\text{input}}(C') \int_C q(\omega_{\text{input}}, D) q'(\omega_{\text{input}}, D') d\mathbb{Q}_{\text{input}}(\omega_{\text{input}}),$$

from which we easily deduce that $(w_s^0 - w_t^0, w_s - w_t)_{t \le s \le T}$ is independent of $\mathcal{G}_t$. Above, the passage from the first to the second line follows from the aforementioned measurability properties of the kernels q and q', and from the fact that $(\boldsymbol{w}^0, \boldsymbol{w})$ is a $2d$-dimensional $\mathbb{G}^{\text{input}}$-Brownian motion for $\mathbb{Q}_{\text{input}}$.

We now prove that $(\xi, \boldsymbol{w}^0, \boldsymbol{v}, \boldsymbol{w}, \boldsymbol{x}, \boldsymbol{y}, (\boldsymbol{\zeta}, \boldsymbol{\zeta}^0), \boldsymbol{m})$ is compatible with the filtration $\mathbb{G}$, the argument being the same for $(\xi, \boldsymbol{w}^0, \boldsymbol{v}, \boldsymbol{w}, \boldsymbol{x}', \boldsymbol{y}', (\boldsymbol{\zeta}', \boldsymbol{\zeta}^{0\prime}), \boldsymbol{m}')$. Given $C \in \mathcal{G}_T^{\text{nat,input}}$, and $D, D' \in \mathcal{G}_T^{\text{nat,output}}$ (and again identifying C with a Borel subset of Ω_{input} and D and D' with Borel subsets of Ω_{output}), it holds:

$$\bar{\mathbb{Q}}(C \times D \times D') = \int_C q(\omega_{\text{input}}, D) q'(\omega_{\text{input}}, D') d\mathbb{Q}_{\text{input}}(\omega_{\text{input}})$$

$$= \int_{C \times D} q'(\omega_{\text{input}}, D') d\mathbb{Q}(\omega_{\text{input}}, \omega_{\text{output}}).$$

We now recall that $q'(\cdot, D')$ is $\mathcal{G}_t^{\text{input}}$-measurable when $D' \in \mathcal{G}_t^{\text{nat,output}}$ for some $t \in [0, T]$. We thus recover the statement of Proposition 1.10 with $\mathbb{F}^1 = \mathbb{G}^{\text{canon}}$ and $\mathbb{F}^2 = \mathbb{G}^{\text{nat,output}}$. We deduce that for any $t \in [0, T]$, $\mathcal{G}_t^{\text{nat}}$ and $\mathcal{G}_T^{\text{canon}}$ are conditionally independent given $\mathcal{G}_t^{\text{canon}}$, when the three σ-fields are regarded as σ-fields on $\bar{\Omega}$. Proceeding as in the proof of Lemma 1.7, we deduce that for any $t \in [0, T]$, $\mathcal{G}_t$ and $\mathcal{G}_T^{\text{canon}}$ are conditionally independent given $\mathcal{G}_t^{\text{canon}}$.

We check that $\mathcal{G}_t$ and $\mathcal{G}_T^{\text{input}}$ are conditionally independent given $\mathcal{G}_t^{\text{input}}$ in the same way. With the same notations as above, it suffices to use the first of the two lines above:

$$\bar{\mathbb{Q}}(C \times D \times D') = \int_C q(\omega_{\text{input}}, D) q'(\omega_{\text{input}}, D') d\mathbb{Q}_{\text{input}}(\omega_{\text{input}}),$$

and then to invoke Proposition 1.10 with $\mathbb{F}^1 = \mathbb{G}^{\text{input}}$ and $\mathbb{F}^2 = \mathbb{G}^{\text{nat,output}} \otimes \mathbb{G}^{\text{nat,output}}$.

Fourth Step. It remains to prove that, for any $t \in [0, T]$ and any $D \in \mathcal{G}_t^{\text{nat,output}}$ seen as a Borel subset of Ω_{output}, the random variable $\Omega_{\text{input}} \ni \omega_{\text{input}} \mapsto q(\omega_{\text{input}}, D)$ is measurable with respect to the completion $\tilde{\mathcal{G}}_t^{\text{input}}$ for $\mathbb{Q}_{\text{input}}$ of the σ-field $\mathcal{G}_{t+}^{\text{nat,input}}$ on Ω_{input}. The proof relies on the crucial fact that the original solutions are built on admissible (and thus compatible) set-ups.

Given $C \in \mathcal{G}_T^{\text{nat,input}}$, seen as a Borel subset of Ω_{input}, we know from the proof of Lemma 1.27 that $C \times \Omega_{\text{output}}$ and $\Omega_{\text{input}} \times D$ are conditionally independent for $\mathbb{Q}^{\mathcal{B}}$ given the σ-field $\mathcal{G}_{t+}^{\text{nat,input}}$ when the σ-field $\mathcal{G}_{t+}^{\text{nat,input}}$ is understood as a σ-field on Ω_{canon}, namely

$$\mathbb{Q}^{\mathcal{B}}(C \times D) = \mathbb{E}^{\mathbb{Q}^{\mathcal{B}}}\big[\mathbb{Q}^{\mathcal{B}}(C \times \Omega_{\text{output}} | \mathcal{G}_{t+}^{\text{nat,input}}) \mathbb{Q}^{\mathcal{B}}(\Omega_{\text{input}} \times D | \mathcal{G}_{t+}^{\text{nat,input}})\big]$$

$$= \mathbb{E}^{\mathbb{Q}^{\mathcal{B}}}\big[\mathbf{1}_{C \times \Omega_{\text{output}}} \mathbb{Q}^{\mathcal{B}}(\Omega_{\text{input}} \times D | \mathcal{G}_{t+}^{\text{nat,input}})\big].$$

Observe that we may regard $\mathbb{Q}^{\mathcal{B}}(\Omega_{\text{input}} \times D|\mathcal{G}_{t+}^{\text{nat,input}})$ as a random variable $\theta : \Omega_{\text{input}} \to \mathbb{R}$, which is $\mathcal{G}_{t+}^{\text{nat,input}}$-measurable (the σ-field being now regarded as a σ-field on Ω_{input}). We then have $\mathbb{Q}^{\mathcal{B}}(C \times D) = \mathbb{E}^{\mathbb{Q}_{\text{input}}^{\mathcal{B}}}[\mathbf{1}_C \theta]$.

Now, we can also write:

$$\mathbb{Q}^{\mathcal{B}}(C \times D) = \int_C q(\omega_{\text{input}}, D) d\mathbb{Q}_{\text{input}}^{\mathcal{B}}(\omega_{\text{input}}) = \mathbb{E}^{\mathbb{Q}_{\text{input}}^{\mathcal{B}}}\big[\mathbf{1}_C q(\cdot, D)\big],$$

where $q(\cdot, D)$ is understood as a $\mathcal{G}_T^{\text{nat,input}}$-measurable random variable from Ω_{input} to $\mathbb{R}$. We deduce that, $\mathbb{Q}_{\text{input}}^{\mathcal{B}}$ almost surely, $q(\cdot, D) = \theta$, which completes the proof of the measurability property of the kernel q.

Last Step. It only remains to prove that we can construct a solution on any admissible probability set-up $(\Omega', \mathcal{F}', \mathbb{F}', \mathbb{P}')$ with input $(X_0', \boldsymbol{W}^{0\prime}, \boldsymbol{\mu}, \boldsymbol{W})$ by setting:

$$\big(X_t', Y_t', (\mathcal{Z}', \mathcal{Z}^{0\prime})_t, M_t'\big)_{0 \le t \le T} = \Phi\big(X_0', \boldsymbol{W}^{0\prime}, \boldsymbol{\mu}, \boldsymbol{W}\big),$$

and defining Z_t and Z_t^0 accordingly. The important thing is that, as we just showed, for any $t \in [0, T]$ and any $D \in \mathcal{G}_t^{\text{nat,output}}$ seen as a Borel subset of Ω_{output}, the random variable $\Omega_{\text{input}} \ni \omega_{\text{input}} \mapsto q(\omega_{\text{input}}, D)$ is measurable with respect to the completion $\tilde{\mathcal{G}}_t^{\text{input}}$ of the σ-field $\mathcal{G}_{t+}^{\text{nat,input}}$ on Ω_{input} under $\mathbb{Q}_{\text{input}}$ which is the completion of $\mathbb{Q}_{\text{input}}^{\mathcal{B}}$. In particular, if we write $\Phi(\omega_{\text{input}})$ in (1.15) in the form $\Phi(\omega_{\text{input}}) = (\Phi_t(\omega_{\text{input}}))_{0 \le t \le T}$, with

$$\Phi_t(\omega_{\text{input}}) = \mathbb{E}^{q(\omega_{\text{input}}, \cdot)}\big[(x_t, y_t, (\zeta, \zeta^0)_t, m_t)\big] \\ \times \mathbf{1}_{\{\mathbb{E}^{q(\omega_{\text{input}}, \cdot)}[\sup_{0 \le s \le T} |(x_s, y_s, (\zeta, \zeta^0)_s, m_s)|] < \infty\}},$$

then $\Phi_t : \Omega_{\text{input}} \to \mathbb{R}^d \times \mathbb{R}^m \times \mathbb{R}^{d \times m} \times \mathbb{R}^m$ is $\tilde{\mathcal{G}}_t^{\text{input}}$-measurable. Notice indeed that the event:

$$\{\omega_{\text{input}} \in \Omega_{\text{input}} : \mathbb{E}^{q(\omega_{\text{input}}, \cdot)}[\sup_{0 \le s \le T} |(x_s, y_s, \zeta_s, m_s)|] < \infty\}$$

is of full measure under $\mathbb{Q}_{\text{input}}$ and thus belongs to $\tilde{\mathcal{G}}_t^{\text{input}}$ since the latter is complete.

Therefore, the process $(X_t', Y_t', (\mathcal{Z}', \mathcal{Z}^{0\prime})_t, M_t')_{0 \le t \le T}$ is $\mathbb{F}$-adapted and thus $\mathbb{F}$-progressively measurable since it has right-continuous paths. Moreover, by construction, the process $(X_0', W_t^{0\prime}, \mu_t', W_t', X_t', Y_t', (\mathcal{Z}', \mathcal{Z}^{0\prime})_t, M_t')_{0 \le t \le T}$ has exactly the same distribution as the process $(X_0, W_t^0, \mu_t, W_t, X_t, Y_t, \int_0^t (Z_s, Z_s^0) ds, M_t)_{0 \le t \le T}$ on Ω_{canon}. In particular, the process $(M_t')_{0 \le t \le T}$ is a martingale with respect to the canonical filtration of

$$(X_0', W_t^{0\prime}, \mu_t', W_t', X_t', Y_t', (\mathcal{Z}', \mathcal{Z}^{0\prime})_t, M_t')_{0 \le t \le T}$$

and thus with respect to the canonical filtration generated by $(X_0', W_t^{0\prime}, \mu_t', W_t')_{0 \le t \le T}$ and also with respect to its complete and right-continuous augmentation under $\mathbb{Q}'$. Since $(X_0', W_t^{0\prime}, \mu_t', W_t')_{0 \le t \le T}$ is assumed to be compatible with $\mathbb{F}$, $(M_t')_{0 \le t \le T}$ is also a martingale with respect to $\mathbb{F}$. See Lemma 1.9. By the same argument, $((W_t^{0\prime}, W_t') \otimes M_t')_{0 \le t \le T}$ is a martingale with respect to $\mathbb{F}$, and thus the bracket of $\boldsymbol{M}'$ and $(\boldsymbol{W}^{0\prime}, \boldsymbol{W}')$ is 0. The proof will be complete if we can prove that $(X_0', W_t^{0\prime}, \mu_t', W_t', X_t', Y_t', Z_t', Z_t^{0\prime}, M_t')_{0 \le t \le T}$

satisfies the two equations forming the FBSDE (1.5), where $(Z'_t, Z_t^{0\prime})(\omega')$ is given as $\lim_{n\to\infty} n\big((\mathcal{Z}', \mathcal{Z}^{0\prime})_t(\omega') - (\mathcal{Z}', \mathcal{Z}^{0\prime})_{(t-1/n)+}(\omega')\big)$ (whenever the limit exists). This follows from the same argument as in the proof of Lemma 1.27. □

Compatible Product Probabilistic Set-Up

We learned from Theorem 1.33 that, provided that the FBSDE is strongly uniquely solvable, the choice of the probabilistic set-up for the search of a solution does not really matter if one is only interested in the law of the solution.

In order to disentangle the different sources of noise (we shall come back to this question in Chapter 2), it may be convenient to use a probabilistic set-up $(\Omega, \mathcal{F}, \mathbb{P})$ based on the product of two complete probability spaces:

$$(\Omega^0, \mathcal{F}^0, \mathbb{P}^0) \quad \text{and} \quad (\Omega^1, \mathcal{F}^1, \mathbb{P}^1),$$

endowed with two complete and right-continuous filtrations $\mathbb{F}^0 = (\mathcal{F}_t^0)_{0\le t\le T}$ and $\mathbb{F}^1 = (\mathcal{F}_t^1)_{0\le t\le T}$ and two d-dimensional Wiener processes $\boldsymbol{W}^0 = (W_t^0)_{0\le t\le T}$ and $\boldsymbol{W} = (W_t)_{0\le t\le T}$ for the filtrations $\mathbb{F}^0$ and $\mathbb{F}^1$ respectively, and to work with the product structure:

$$\Omega = \Omega^0 \times \Omega^1, \quad \mathcal{F}, \quad \mathbb{F} = \big(\mathcal{F}_t\big)_{0\le t\le T}, \quad \mathbb{P} = \mathbb{P}^0 \otimes \mathbb{P}^1, \tag{1.16}$$

where $(\mathcal{F}, \mathbb{P})$ is the completion of $(\mathcal{F}^0 \otimes \mathcal{F}^1, \mathbb{P}^0 \otimes \mathbb{P}^1)$, and $\mathbb{F}$ is the complete and right continuous augmentation of $(\mathcal{F}_t^0 \otimes \mathcal{F}_t^1)_{0\le t\le T}$. Generic elements of Ω will be denoted by $\omega = (\omega^0, \omega^1)$, with $\omega^0 \in \Omega^0$ and $\omega^1 \in \Omega^1$.

In this set-up, we require the input $\boldsymbol{\mu} = (\mu_t)_{0\le t\le T}$ to be defined on Ω^0, and to be an $\mathbb{F}^0$-progressively measurable, right continuous with left limits (càdlàg) process with values in a Polish space $(\mathcal{X}, d)$. The stochastic flow $\boldsymbol{\mu}$ then reads:

$$\boldsymbol{\mu} = \Big(\mu_t : \Omega^0 \ni \omega^0 \mapsto \mu_t(\omega^0) \in \mathcal{X}\Big)_{0\le t\le T}.$$

In what follows, we often write $\mu_t(\omega)$ instead of $\mu_t(\omega^0)$, μ_t being identified with its natural extension to Ω. Similarly, the initial condition is required to be a random variable $X_0 \in L^2(\Omega^0, \mathcal{F}_0^0, \mathbb{P}^0; \mathbb{R}^d)$.

In this framework, compatibility of $(X_0, \boldsymbol{W}^0, \boldsymbol{\mu}, \boldsymbol{W})$ with $\mathbb{F}$ holds if the filtration $\mathbb{F}^0$ on $(\Omega^0, \mathcal{F}^0, \mathbb{P}^0)$, is compatible with the process $(X_0, \boldsymbol{W}^0, \boldsymbol{\mu})$.

This follows from the fact that under $\mathbb{P}$, the σ-fields $(\mathcal{F}_t^0 \otimes \{\emptyset, \Omega^1\})_{0\le t\le T}$ and $(\{\emptyset, \Omega^0\} \otimes \mathcal{F}_t^1)_{0\le t\le T}$ are independent. A proof can easily be given by following the same steps as in the proof of Lemma 1.7.

Obviously, a simple way to guarantee this product structure is to choose canonical spaces for Ω^0 and Ω^1. Below, we use the specific notation $\bar{\Omega}^0 = \mathbb{R}^d \times \mathcal{C}([0,T];\mathbb{R}^d) \times \mathcal{D}([0,T];\mathcal{X})$ and $\bar{\Omega}^1 = \mathcal{C}([0,T];\mathbb{R}^d)$ for these canonical spaces. We equip them with their respective Borel σ-fields, and with the probability measures $\bar{\mathbb{P}}^0$ and $\bar{\mathbb{P}}^1$, where $\bar{\mathbb{P}}^0$ is the law of the input $(\xi, \boldsymbol{W}^0, \boldsymbol{\mu})$, and $\bar{\mathbb{P}}^1$ is the d-dimensional Wiener probability measure. We then call $\bar{\mathcal{F}}^0$ (respectively $\bar{\mathcal{F}}^1$) the completion of the Borel σ-field of $\bar{\Omega}^0$ (respectively $\bar{\Omega}^1$), and $\bar{\mathbb{F}}^0$ (respectively $\bar{\mathbb{F}}^1$) the complete and right-continuous augmentation of the canonical filtration. We still denote by $\bar{\mathbb{P}}^0$ and $\bar{\mathbb{P}}^1$ the extensions of the two probability measures to the σ-fields $\bar{\mathcal{F}}^0$ and $\bar{\mathcal{F}}^1$ respectively. On the product space $\Omega_{\text{input}} = \bar{\Omega}^0 \times \bar{\Omega}^1$, we let $\bar{\mathbb{P}}$ be the completion of $\bar{\mathbb{P}}^0 \otimes \bar{\mathbb{P}}^1$ to the σ-field $\bar{\mathcal{F}}$, obtained as the completion of the Borel σ-field (which is also the completion of $\bar{\mathcal{F}}^0 \otimes \bar{\mathcal{F}}^1$). We then let $\mathbb{F}$ be the complete and right-continuous augmentation of the canonical filtration on $\bar{\Omega}^0 \times \bar{\Omega}^1$, the canonical filtration being also given by the product of the canonical filtrations on $\bar{\Omega}^0$ and $\bar{\Omega}^1$. As above, the canonical processes are denoted by $(\xi, \boldsymbol{\nu}, \boldsymbol{w}^0) = (\xi, (w^0_s)_{0\le s\le T}, (\nu_s)_{0\le s\le T})$ on $\bar{\Omega}^0$ and $\boldsymbol{w} = (w_s)_{0\le s\le T}$ on $\bar{\Omega}^1$.

Whenever the FBSDE (1.5) is strongly uniquely solvable, we shall denote by $\mathcal{L}(\mathbf{FBSDE}(\mathbb{P}^0))$ the law of the solution of the FBSDE (1.5) when the input $(X_0, \boldsymbol{\mu}, \boldsymbol{W}^0)$ has $\mathbb{P}^0$ as distribution, irrespective of the probabilistic set-up.

1.3 Initial Information, Small Time Solvability, and Decoupling Field

1.3.1 Disintegration of FBSDEs

Initial Information

A common practice in the study of SDEs, FBSDEs, and even stochastic games, is the analysis of the problem on a sub time interval $[t, T]$ for all times $t \in [0, T]$. Although quite natural, the procedure requires a modicum of care, since one must decide how to treat the information contained in the observations up until t. Two rigorous points of view are conceivable:

Admissible set-up with initial information. The first one consists in handling the equation at time t as a new equation, like the one set at time 0, but with an additional initial information set, larger than that enclosed in the simple observation of X_0 and μ_0. This requires a slight modification of the notion of compatible probabilistic set-up used so far in solving the equation. Indeed, when the initial information set is larger than the σ-field $\sigma\{X_0, \mu_0\}$, the basic compatibility condition described in Subsection 1.1.1 is no longer sufficient. Compatibility should now require that, for any $t \in [0, T]$, the σ-fields

$$\mathcal{F}_T^{\mathcal{G},\boldsymbol{W}^0,\boldsymbol{\mu},\boldsymbol{W}} \quad \text{and} \quad \mathcal{F}_t,$$

are conditionally independent given $\mathcal{F}_t^{\mathcal{G},\boldsymbol{W}^0,\boldsymbol{\mu},\boldsymbol{W}}$, where $\mathcal{G}$ is a σ-field describing the initial information (in addition to that enclosed in μ_0) and $\mathbb{F}^{\mathcal{G},\boldsymbol{W}^0,\boldsymbol{\mu},\boldsymbol{W}} = (\mathcal{F}_t^{\mathcal{G},\boldsymbol{W}^0,\boldsymbol{\mu},\boldsymbol{W}})_{0\le t\le T}$ is the complete and right-continuous augmentation under $\mathbb{P}$ of the filtration generated by $\mathcal{G}$ and $(\boldsymbol{W}^0, \boldsymbol{\mu}, \boldsymbol{W})$. We then say that the process $(\boldsymbol{W}^0, \boldsymbol{\mu}, \boldsymbol{W})$ together with the initial σ-field $\mathcal{G}$ are compatible with the filtration $\mathbb{F}$. So in analogy with Definition 1.13, we define the new concept of compatibility as follows.

Definition 1.36 *A complete probability space $(\Omega, \mathcal{F}, \mathbb{P})$ equipped with a complete and right-continuous filtration $\mathbb{F} = (\mathcal{F}_t)_{0\le t\le T}$ and a tuple $(\mathcal{G}, \boldsymbol{W}^0, \boldsymbol{\mu}, \boldsymbol{W})$ for a sub-σ-algebra $\mathcal{G}$ is said to be admissible if:*

1. $\mathcal{G} \subset \mathcal{F}_0$;
2. $(\boldsymbol{W}^0, \boldsymbol{W})$ *is a 2d-dimensional Brownian motion with respect to $\mathbb{F}$ (under $\mathbb{P}$);*
3. $(\mathcal{G}, \boldsymbol{W}^0, \boldsymbol{\mu})$ *is independent of $\boldsymbol{W}$ (under $\mathbb{P}$);*
4. $(\mathcal{G}, \boldsymbol{W}^0, \boldsymbol{\mu}, \boldsymbol{W})$ *and $\mathbb{F}$ are compatible (under $\mathbb{P}$).*

The σ-field $\mathcal{G}$ is then referred to as the initial information. *Whenever $\mathcal{G}$ is trivial, we say that there is* no initial information.

The following lemma provides some consistency in the use of the initial information.

Lemma 1.37 *If $(\mathcal{G}, \boldsymbol{W}^0, \boldsymbol{\mu}, \boldsymbol{W})$ is admissible and $\mathcal{G}'$ is another σ-field with $\mathcal{G} \subset \mathcal{G}' \subset \mathcal{F}_0$, then $(\mathcal{G}', \boldsymbol{W}^0, \boldsymbol{\mu}, \boldsymbol{W})$ is also admissible.*

Proof.

First Step. The first step is to check 3 in Definition 1.36. The proof is a variation of Blumenthal's zero-one law, and is based on the fact that $\boldsymbol{W}$ is independent of $\mathcal{F}_0^{\mathcal{G},\boldsymbol{W}^0,\boldsymbol{\mu},\boldsymbol{W}}$.

For $B_0 \in \mathcal{F}_0$, $C^0 \in \sigma\{\boldsymbol{W}^0, \boldsymbol{\mu}\}$ and $C \in \sigma\{\boldsymbol{W}\}$, it holds:

$$\mathbb{P}(B_0 \cap C^0 \cap C) = \mathbb{E}\big[\mathbb{P}(B_0|\mathcal{F}_0^{\mathcal{G},\boldsymbol{W}^0,\boldsymbol{\mu},\boldsymbol{W}})\mathbf{1}_{C^0\cap C}\big], \tag{1.17}$$

which follows from the compatibility property.

Assume that, for some $t > 0$, $C \in \sigma\{W_s - W_t; s \in [t, T]\}$. Then, for any $C_t^0 \in \mathcal{G} \vee \sigma\{W_s^0, \mu_s; s \le t\}$ and $C_t \in \sigma\{W_s; s \le t\}$, $\mathbb{P}(C_t^0 \cap C_t \cap C^0 \cap C) = \mathbb{P}(C_t^0 \cap C^0)\mathbb{P}(C_t)\mathbb{P}(C)$ since $(\mathcal{G}, \boldsymbol{W}^0, \boldsymbol{\mu})$ is independent of $\boldsymbol{W}$. Also,

$$\mathbb{P}(C_t^0 \cap C_t \cap C^0 \cap C) = \mathbb{P}(C_t^0 \cap C^0 \cap C_t)\mathbb{P}(C),$$

from which we deduce that $\mathcal{F}_0^{(\mathcal{G},\boldsymbol{W}^0,\boldsymbol{\mu},\boldsymbol{W})} \vee \sigma\{\boldsymbol{W}^0, \boldsymbol{\mu}\}$ is independent of $\sigma\{W_s - W_t; s \in [t, T]\}$. And then, $\mathcal{F}_0^{(\mathcal{G},\boldsymbol{W}^0,\boldsymbol{\mu},\boldsymbol{W})} \vee \sigma\{\boldsymbol{W}^0, \boldsymbol{\mu}\}$ is independent of $\sigma\{\boldsymbol{W}\}$. Therefore, (1.17) yields:

$$\begin{aligned}\mathbb{P}(B_0 \cap C^0 \cap C) &= \mathbb{E}\big[\mathbb{P}(B_0|\mathcal{F}_0^{\mathcal{G},\boldsymbol{W}^0,\boldsymbol{\mu},\boldsymbol{W}})\mathbf{1}_{C^0}\big]\mathbb{P}(C)\\ &= \mathbb{P}(B_0 \cap C^0)\mathbb{P}(C),\end{aligned}$$

which shows that $\boldsymbol{W}$ is independent of $(\mathcal{F}_0, \boldsymbol{W}^0, \boldsymbol{\mu})$. In particular, $\boldsymbol{W}$ is independent of $(\mathcal{G}', \boldsymbol{W}^0, \boldsymbol{\mu})$.

Second Step. We now check the compatibility property. Given $t \in [0, T]$, we consider $B_0 \in \mathcal{G}'$, $C_t^0 \in \mathcal{F}_t^{\mathrm{nat},(\boldsymbol{W}^0,\boldsymbol{\mu},\boldsymbol{W})}$ and $C_T^0 \in \mathcal{F}_T^{(\boldsymbol{W}^0,\boldsymbol{\mu},\boldsymbol{W})}$. Then, by compatibility of $(\mathcal{G}, \boldsymbol{W}^0, \boldsymbol{\mu}, \boldsymbol{W})$ with $\mathbb{F}$, we get:

$$\begin{aligned}\mathbb{P}(B_0 \cap C_t^0 \cap C_T^0) &= \mathbb{E}\big[\mathbb{P}(B_0|\mathcal{F}_0^{(\mathcal{G},\mathbf{W}^0,\boldsymbol{\mu},\mathbf{W})})\mathbf{1}_{C_t^0\cap C_T^0}\big] \\ &= \mathbb{E}\big[\mathbb{P}(B_0|\mathcal{F}_0^{(\mathcal{G},\mathbf{W}^0,\boldsymbol{\mu},\mathbf{W})})\mathbf{1}_{C_t^0}\mathbb{P}(C_T^0|\mathcal{F}_t^{(\mathcal{G},\mathbf{W}^0,\boldsymbol{\mu},\mathbf{W})})\big] \\ &= \mathbb{E}\big[\mathbf{1}_{B_0}\mathbf{1}_{C_t^0}\mathbb{P}(C_T^0|\mathcal{F}_t^{(\mathcal{G},\mathbf{W}^0,\boldsymbol{\mu},\mathbf{W})})\big],\end{aligned}$$

which shows that:

$$\mathbb{P}\big(C_T^0|\mathcal{F}_t^{\mathrm{nat},(\mathcal{G}',\mathbf{W}^0,\boldsymbol{\mu},\mathbf{W})}\big) = \mathbb{E}\Big[\mathbb{P}\big(C_T^0|\mathcal{F}_t^{(\mathcal{G},\mathbf{W}^0,\boldsymbol{\mu},\mathbf{W})}\big)|\mathcal{F}_t^{\mathrm{nat},(\mathcal{G}',\mathbf{W}^0,\boldsymbol{\mu},\mathbf{W})}\Big].$$

Replace t by $t+\delta$ for $\delta > 0$ and write:

$$\begin{aligned}&\mathbb{P}\big(C_T^0|\mathcal{F}_{t+\delta}^{\mathrm{nat},(\mathcal{G}',\mathbf{W}^0,\boldsymbol{\mu},\mathbf{W})}\big) \\ &= \mathbb{E}\Big[\mathbb{P}\big(C_T^0|\mathcal{F}_{t+\delta}^{(\mathcal{G},\mathbf{W}^0,\boldsymbol{\mu},\mathbf{W})}\big) - \mathbb{P}\big(C_T^0|\mathcal{F}_t^{(\mathcal{G},\mathbf{W}^0,\boldsymbol{\mu},\mathbf{W})}\big)|\mathcal{F}_{t+\delta}^{\mathrm{nat},(\mathcal{G}',\mathbf{W}^0,\boldsymbol{\mu},\mathbf{W})}\Big] \\ &\quad + \mathbb{E}\Big[\mathbb{P}\big(C_T^0|\mathcal{F}_t^{(\mathcal{G},\mathbf{W}^0,\boldsymbol{\mu},\mathbf{W})}\big)|\mathcal{F}_{t+\delta}^{\mathrm{nat},(\mathcal{G}',\mathbf{W}^0,\boldsymbol{\mu},\mathbf{W})}\Big].\end{aligned}$$

The first term in the right-hand side tends to 0 in L^1-norm as δ tends to 0. The second term is (almost surely) equal to $\mathbb{P}(C_T^0|\mathcal{F}_t^{(\mathcal{G},\mathbf{W}^0,\boldsymbol{\mu},\mathbf{W})})$. Therefore, letting δ tend to 0, we obtain:

$$\mathbb{P}\big(C_T^0|\mathcal{F}_t^{(\mathcal{G}',\mathbf{W}^0,\boldsymbol{\mu},\mathbf{W})}\big) = \mathbb{P}\big(C_T^0|\mathcal{F}_t^{(\mathcal{G},\mathbf{W}^0,\boldsymbol{\mu},\mathbf{W})}\big).$$

Also,

$$\mathbb{P}\big(B_0 \cap C_T^0|\mathcal{F}_t^{(\mathcal{G}',\mathbf{W}^0,\boldsymbol{\mu},\mathbf{W})}\big) = \mathbf{1}_{B_0}\mathbb{P}\big(C_T^0|\mathcal{F}_t^{(\mathcal{G},\mathbf{W}^0,\boldsymbol{\mu},\mathbf{W})}\big).$$

Now, by compatibility of $(\mathcal{G}, \mathbf{W}^0, \boldsymbol{\mu}, \mathbf{W})$ and by (H3) in Proposition 1.3, we know that:

$$\mathbb{P}\big(B_0 \cap C_T^0|\mathcal{F}_t\big) = \mathbf{1}_{B_0}\mathbb{P}\big(C_T^0|\mathcal{F}_t\big) = \mathbf{1}_{B_0}\mathbb{P}\big(C_T^0|\mathcal{F}_t^{(\mathcal{G},\mathbf{W}^0,\boldsymbol{\mu},\mathbf{W})}\big),$$

so that:

$$\mathbb{P}\big(B_0 \cap C_T^0|\mathcal{F}_t^{(\mathcal{G}',\mathbf{W}^0,\boldsymbol{\mu},\mathbf{W})}\big) = \mathbb{P}\big(B_0 \cap C_T^0|\mathcal{F}_t\big),$$

from which compatibility follows since events of the form $B_0 \cap C_T^0$ generate the σ-field $\mathcal{F}_T^{(\mathcal{G}',\mathbf{W}^0,\boldsymbol{\mu},\mathbf{W})}$. □

Despite the fact that it offers a natural generalization of Definition 1.13, using this notion of admissibility may not be the best way to proceed. Indeed, it would require to revisit *all* the results of the previous subsections. For instance, we could redefine the notion of strong uniqueness in Definition 1.18 by requiring that uniqueness holds for any arbitrary admissible set-up equipped with any initial information $\mathcal{G}$. In this regard, the following observation could make our life easier. Whenever, notice that this is what happens in practice, the initial σ-field $\mathcal{G}$ is generated by a random variable θ_0 with values in a Polish space $\mathcal{S}$, we may easily adapt all the results above by letting θ_0 play the role of the initial condition X_0. Namely, if strong uniqueness holds, then the laws of the solutions constructed on possibly different spaces have to be the same provided that the initial triples $(\theta_0, \boldsymbol{W}^0, \boldsymbol{\mu})$ have the same distributions, and furthermore, they have to be functions of $(\theta_0, \boldsymbol{W}^0, \boldsymbol{\mu}, \boldsymbol{W})$. In other words, it would suffice to use $\mathcal{S}$ (equipped with its Borel σ-field) instead of $\mathbb{R}^d$ as the canonical space carrying the initial condition.

Remark 1.38 *Observe from Lemma 1.37 that any solution* $(\boldsymbol{X}, \boldsymbol{Y}, \boldsymbol{Z}, \boldsymbol{Z}^0, \boldsymbol{M})$ *to the FBSDE* (1.5) *constructed on some admissible set-up equipped with some tuple* $(\mathcal{G}, \boldsymbol{W}^0, \boldsymbol{\mu}, \boldsymbol{W})$ *is also a solution on any admissible set-up equipped with some tuple* $(\mathcal{G}', \boldsymbol{W}^0, \boldsymbol{\mu}, \boldsymbol{W})$ *for a larger initial information in the sense that* $\mathcal{G}'$ *is another* σ*-field with* $\mathcal{G} \subset \mathcal{G}' \subset \mathcal{F}_0$*. In particular, if uniqueness holds on the set-up equipped with* $(\mathcal{G}', \boldsymbol{W}^0, \boldsymbol{\mu}, \boldsymbol{W})$*, then it also holds on the set-up equipped with* $(\mathcal{G}, \boldsymbol{W}^0, \boldsymbol{\mu}, \boldsymbol{W})$.

Conditioning on the Initial Information. A different strategy is based on a conditioning argument. Instead of allowing for solutions initialized with additional information, it suffices to change the probability in such a way that this additional information becomes deterministic. In other words, the strategy is to work with the conditional probability given the additional information $\mathcal{G}$. In such a way, there is no need to revisit the previous results. However, there is still a heavy price to pay, essentially due to the technicalities inherent to the construction and the use of conditional probabilities.

We shall use both points of view, switching from one to the other depending upon the context. This being said, the reader should understand that a modicum of care has to be taken regarding the structure of the probabilistic set-up when additional initial information is available and used. As mentioned earlier, the issue is not present when $\boldsymbol{\mu}$ is deterministic, which is the standard framework for FBSDEs with deterministic coefficients discussed in Chapters (Vol I)-3 and (Vol I)-4. Indeed, in that case, the Markov structure of the Brownian motion can easily accommodate any initial information as long as it is independent of the future increments of the Brownian motion.

If the reader is still unsure of the need to revisit the notion of solution under an enlarged initial σ-field, a convincing argument is to come back to the original motivation. Indeed, let us consider a solution $(\boldsymbol{X}, \boldsymbol{Y}, \boldsymbol{Z}, \boldsymbol{Z}^0, \boldsymbol{M})$ to (1.5) constructed on some set-up equipped with an input $(X_0, \boldsymbol{W}^0, \boldsymbol{\mu}, \boldsymbol{W})$ and a compatible filtration $\mathbb{F}$, and for $t \in (0, T)$, let us address the following question: "*In which sense does* $(X_s, Y_s, Z_s, Z_s^0, M_s)_{t \le s \le T}$ *solve the FBSDE* (1.5) *on* $[t, T]$*?*" The answer is given by the following simple observation: the filtration used to define the solution must remain the original filtration $\mathbb{F}$. However, it is unlikely that the process $(X_t, W_s^0 - W_t^0, \mu_s, W_s - W_t)_{t \le s \le T}$ which drives the equation (1.5) is compatible with $\mathbb{F}$. Indeed, $\mathbb{F}$ incorporates the past of $\boldsymbol{\mu}$ before t, whereas the filtration generated by $(X_t, W_s^0 - W_t^0, \mu_s, W_s - W_t)_{t \le s \le T}$ does not. The fact that $\mathbb{F}$ is not compatible with

$(X_t, W_s^0 - W_t^0, \mu_s, W_s - W_t)_{t \le s \le T}$ says that, if we manage to solve (1.5) on $[t, T]$ with respect to the complete and right-continuous augmentation of the filtration generated by $(X_t, W_s^0 - W_t^0, \mu_s, W_s - W_t)_{t \le s \le T}$, then the solution may not have the same law as $(X_s, Y_s, Z_s, Z_s^0, M_s)_{t \le s \le T}$, even if strong uniqueness holds! In order to guarantee uniqueness in law whenever strong uniqueness holds, we must solve (1.5) on $[t, T]$ with respect to the filtration generated by the σ-field $\sigma\{X^0, W_s^0, \mu_s, W_s;\ s \le t\}$ and by the process $(W_s^0 - W_t^0, \mu_s, W_s - W_t)_{t \le s \le T}$, which is obviously compatible with $\mathbb{F}$. Below, we shall say that we solve (1.5) with $\sigma\{X^0, W_s^0, \mu_s, W_s; s \le t\}$ as *initial information*, which suggests the following version of Definition 1.36.

Definition 1.39 *For any $t \in [0, T]$, any complete probabilistic set-up $(\Omega, \mathcal{F}, \mathbb{P})$ with a complete and right-continuous filtration $\mathbb{F} = (\mathcal{F}_s)_{t \le s \le T}$ and a tuple $(\mathcal{G}, (W_s^0, \mu_s, W_s)_{t \le s \le T})$ for a sub-σ-algebra $\mathcal{G}$, is said to be a* t-initialized admissible set-up *if:*

1. $\mathcal{G} \subset \mathcal{F}_t$,
2. $(W_s^0, W_s)_{t \le s \le T}$ *is a 2d-dimensional Brownian motion starting from* 0 *at time t with respect to* $\mathbb{F}$ *(under* $\mathbb{P}$*);*
3. $(\mathcal{G}, (W_s^0, \mu_s)_{t \le s \le T})$ *is independent of* $(W_s)_{t \le s \le T}$ *(under* $\mathbb{P}$*);*
4. $(\mathcal{G}, (W_s^0, \mu_s, W_s)_{t \le s \le T})$ *and* $\mathbb{F}$ *are compatible (under* $\mathbb{P}$*).*

The σ-field $\mathcal{G}$ is then referred to as the initial information at time t.

Disintegration of FBSDEs

As we already alluded to, there is another point of view for letting a solution, originally initialized at time 0, restart at a later time $t > 0$. The following lemma shows that such a procedure may be understood in terms of conditioning arguments. For these arguments to be more transparent, it is convenient to work on the canonical set-up $\Omega_{\text{canon}} = \Omega_{\text{input}} \times \Omega_{\text{output}}$ defined in Subsection 1.2.2 so we can use regular conditional probabilities as introduced in Theorem 1.1.

Given the canonical set-up $\Omega_{\text{canon}} = \Omega_{\text{input}} \times \Omega_{\text{output}}$ endowed with some admissible probability measure $\mathbb{Q}$ as in Definition 1.23, we denote by $\mathbb{G}^{\text{nat}} = (\mathcal{G}_t^{\text{nat}})_{0 \le t \le T}$ the canonical filtration. Notice that this filtration has not been augmented. Then, for any $t \in [0, T]$ we can find a family of probability measures $(\mathbb{Q}_\omega^t)_{\omega \in \Omega}$ such that, for any Borel subset $D \subset \Omega_{\text{canon}}$, the mapping $\Omega \ni \omega \mapsto \mathbb{Q}_\omega^t(D) \in [0, 1]$ is measurable, both spaces being endowed with Borel σ-fields, and $\mathbb{Q}_\omega^t(D)$ is a version of the conditional expectation of $\mathbf{1}_D$ given $\mathcal{G}_t^{\text{nat}}$ under $\mathbb{Q}$.

Lemma 1.40 *Consider the FBSDE* (1.5), *assume that it has a solution on some admissible probabilistic set-up $(\Omega, \mathcal{F}, \mathbb{F}, \mathbb{P})$ equipped with a tuple $(X_0, \mathbf{W}^0, \boldsymbol{\mu}, \mathbf{W})$, transfer this solution onto the canonical set-up as we did in Lemma 1.27, and denote by $\mathbb{Q}^{\mathcal{B}}$ the resulting probability law on Ω_{canon} equipped with its Borel σ-field. Then, for any $t \in [0, T]$ and $\mathbb{Q}^{\mathcal{B}}$-almost every $\omega \in \Omega_{\text{canon}}$, under the completion of the regular conditional probability $\mathbb{Q}_\omega^t$ of $\mathbb{Q}^{\mathcal{B}}$ given $\mathcal{G}_t^{\text{nat}}$, $(x_s, y_s, z_s, z_s^0, m_s - m_t)_{t \le s \le T}$ is a solution of* (1.5) *on $\Omega_{\text{canon}}^t = \Omega_{\text{input}}^t \times \Omega_{\text{output}}^t$ equipped with the completion of* *the Borel σ-field and the complete and right-continuous augmentation $\mathbb{G}^t_\omega$ of the canonical filtration $\mathbb{G}^{t,\mathrm{nat}} = (\mathcal{G}^{t,\mathrm{nat}}_s)_{t \le s \le T}$, with:*

$$\mathcal{G}^{t,\mathrm{nat}}_s = \sigma\{w^0_r - w^0_t, \mu_r, w_r - w_t, x_r, y_r, (\zeta, \zeta^0)_r, m_r - m_t; r \in [t,s]\}$$

$$\Omega^t_{\mathrm{input}} = \mathbb{R}^d \times \mathcal{C}([t,T];\mathbb{R}^d) \times \mathcal{D}([t,T];\mathcal{X}) \times \mathcal{C}([t,T];\mathbb{R}^d),$$

$$\Omega^t_{\mathrm{output}} = \mathcal{C}([t,T];\mathbb{R}^d) \times \mathcal{D}([t,T];\mathbb{R}^m) \times \mathcal{C}([t,T];\mathbb{R}^{2(m\times d)}) \times \mathcal{D}([t,T];\mathbb{R}^m),$$

the input $(x_t, w^0_s - w^0_t, \nu_s, w_s - w_t)_{t \le s \le T}$ having the distribution:

$$\delta_{x_t(\omega)} \otimes \Big(\mathbb{Q}^t_\omega \circ \big((w^0_s - w^0_t, \nu_s)_{t \le s \le T}\big)^{-1}\Big) \otimes \mathcal{W}^t_d,$$

where $\mathcal{W}^t_d$ is the d-dimensional Wiener measure on $\mathcal{C}([t,T];\mathbb{R}^d)$.

Remark 1.41 *In the statement, it is implicitly understood that the probabilistic set-up formed by the filtered probability space supporting the input $(x_t, w^0_s - w^0_t, \nu_s, w_s - w_t)_{t \le s \le T}$ is admissible. Here it is t-initialized and has no initial information.*

Remark 1.42 *As we just explained, the random variable X_0 could be replaced by a more general random variable θ_0 taking values in a Polish space $\mathcal{S}$ such that $\sigma\{X_0\} \subset \sigma\{\theta_0\} \subset \mathcal{F}_0$ and $(\theta_0, \boldsymbol{W}^0, \boldsymbol{\mu}, \boldsymbol{W})$ is compatible with $\mathbb{F}$.*

The interpretation of Lemma 1.40 is quite clear. When conditioning on the past before t, the solution of the FBSDE on the whole $[0,T]$ generates a solution on $[t,T]$ with the current position at time t as initial condition and with the conditional law of $(\boldsymbol{W}^0, \boldsymbol{\mu})$ given the past as input. When $\boldsymbol{\mu}$ is deterministic, $(\boldsymbol{W}^0, \boldsymbol{\mu})$ may be ignored and only the current position at time t matters for determining the input.

Proof. We make use of the same notations as in Lemma 1.27. In particular, the process $(x_s, y_s, (\zeta, \zeta^0)_s, m_s)_{0 \le s \le T}$ denotes the canonical process on Ω_{output}. Also, $\mathbb{Q}$ denotes the completion of $\mathbb{Q}^{\mathcal{B}}$ and $\mathbb{G}$ the complete and right-continuous augmentation of the canonical filtration on Ω_{canon} under $\mathbb{Q}$. Throughout the proof, we always use $\mathbb{Q}$ instead of $\mathbb{Q}^{\mathcal{B}}$, even when working with Borel subsets of Ω_{canon}.

First Step. We first notice that, for $\mathbb{Q}$-almost every $\omega \in \Omega_{\mathrm{canon}}$, x_t is almost surely equal to $x_t(\omega)$ under $\mathbb{Q}^t_\omega$. We then check that, for $\mathbb{Q}$-almost every $\omega \in \Omega_{\mathrm{canon}}$, under $\mathbb{Q}^t_\omega$, the process $(w^0_s - w^0_t, w_s - w_t)_{t \le s \le T}$ is a Wiener process with respect to the filtration $\mathbb{G}^{t,\mathrm{nat}}$. Obviously, if so, the property is also true under $\mathbb{G}^t_\omega$. The proof is quite standard. Given an event E in $\mathcal{G}^{\mathrm{nat}}_t$ and an event F in $\mathcal{G}^{t,\mathrm{nat}}_s$ for some $s \in [t,T]$, we know that for two Borel subsets C^0 and C of $\mathcal{C}([s,T];\mathbb{R}^d)$:

$$\begin{aligned}&\mathbb{Q}\Big(E \cap F \cap \big\{(w^0_r - w^0_s)_{s \le r \le T} \in C^0, (w_r - w_s)_{s \le r \le T} \in C\big\}\Big)\\&\quad = \mathbb{Q}(E \cap F)\mathbb{Q}\big[(w^0_r - w^0_s)_{s \le r \le T} \in C^0, (w_r - w_s)_{s \le r \le T} \in C\big],\end{aligned}$$

so that:

$$\int_E \mathbb{Q}^t_\omega\Big(F \cap \{(w^0_r - w^0_s)_{s\le r\le T} \in C^0, (w_r - w_s)_{s\le r\le T} \in C\}\Big) d\mathbb{Q}(\omega)$$
$$= \Big(\int_E \mathbb{Q}^t_\omega(F) d\mathbb{Q}(\omega)\Big)\mathbb{Q}\big[(w^0_r - w^0_s)_{s\le r\le T} \in C^0, (w_r - w_s)_{s\le r\le T} \in C\big],$$

which implies, since E is arbitrary, that for $\mathbb{Q}$-almost every ω, we have:

$$\mathbb{Q}^t_\omega\Big(F \cap \{(w^0_r - w^0_s)_{s\le r\le T} \in C^0, (w_r - w_s)_{s\le r\le T} \in C\}\Big)$$
$$= \mathbb{Q}^t_\omega(F)\mathbb{Q}\big[(w^0_r - w^0_s)_{s\le r\le T} \in C^0, (w_r - w_s)_{s\le r\le T} \in C\big]$$

which is enough to conclude by choosing s in a dense countable subset of $[t, T]$ and F, C and C^0 in countable generating π-systems. Recall that we are working on Polish spaces equipped with their Borel σ-fields.

In the same way, we now prove that for $\mathbb{Q}$-almost every ω, $(w^0_s - w^0_t, v_s)_{t\le s\le T}$ and $(w_s - w_t)_{t\le s\le T}$ are independent under $\mathbb{Q}^t_\omega$ and that $(m_s - m_t)_{t\le s\le T}$ is a $\mathbb{G}^{t,\mathrm{nat}}$ square-integrable martingale of zero cross variation with $(w^0_s - w^0_t, w_s - w_t)_{t\le s\le T}$. As above, the property will remain true under $\mathbb{G}^t_\omega$. Taking as before E in $\mathcal{G}^{\mathrm{nat}}_t$ and choosing now C and C^0 as Borel subsets of $\mathcal{C}([t, T]; \mathbb{R}^d)$ and D as a Borel subset of $\mathcal{D}([t, T]; \mathcal{X})$, we indeed have from the compatibility condition (recall item 3 in Proposition 1.26):

$$\mathbb{Q}\Big(E \cap \{(w^0_s - w^0_t)_{t\le s\le T} \in C^0, (v_s)_{t\le s\le T} \in D, (w_s - w_t)_{t\le s\le T} \in C\}\Big)$$
$$= \mathbb{E}^{\mathbb{Q}}\Big[\mathbb{Q}(E|\mathcal{G}^{\mathrm{nat,input}}_{t+})\mathbf{1}_{\{(w^0_s - w^0_t)_{t\le s\le T} \in C^0, (v_s)_{t\le s\le T} \in D, (w_s - w_t)_{t\le s\le T} \in C\}}\Big],$$

where $\mathcal{G}^{\mathrm{nat,input}}_{t+} = \cap_{s\in(t,T]}\mathcal{G}^{\mathrm{nat,input}}_s$, with $\mathbb{G}^{\mathrm{nat,input}} = (\mathcal{G}^{\mathrm{nat,input}}_s)_{0\le s\le T}$ being the canonical filtration generated by $(\xi, \boldsymbol{w}^0, \boldsymbol{v}, \boldsymbol{w})$, now regarded as a filtration on Ω_{canon}. Since $(\xi, \boldsymbol{w}^0, \boldsymbol{v})$ is independent of $\boldsymbol{w}$ under $\mathbb{Q}$, it is easily seen that under $\mathbb{Q}$, $\mathcal{G}^{\mathrm{nat,input}}_{t+} \vee \sigma\{w^0_s - w^0_t, v_s; t \le s \le T\}$ is independent of $\sigma\{w_s - w_t; t \le s \le T\}$. We deduce that:

$$\mathbb{Q}\Big(E \cap \{(w^0_s - w^0_t)_{t\le s\le T} \in C^0, (v_s)_{t\le s\le T} \in D, (w_s - w_t)_{t\le s\le T} \in C\}\Big)$$
$$= \mathbb{Q}\Big(E \cap \{(w^0_s - w^0_t)_{t\le s\le T} \in C^0, (v_s)_{t\le s\le T} \in D\}\Big)\mathbb{Q}\big[(w_s - w_t)_{s\le r\le T} \in C\big],$$

so that, again because E is arbitrary, for $\mathbb{Q}$-almost every ω, we have:

$$\mathbb{Q}^t_\omega\Big(\{(w^0_s - w^0_t)_{t\le s\le T} \in C^0, (v_s)_{t\le s\le T} \in D, (w_s - w_t)_{t\le s\le T} \in C\}\Big)$$
$$= \mathbb{Q}^t_\omega\Big(\{(w^0_s - w^0_t)_{t\le s\le T} \in C^0, (v_s)_{t\le s\le T} \in D\}\Big)\mathbb{Q}\big[(w_s - w_t)_{s\le r\le T} \in C\big],$$

which again is enough to conclude that, for $\mathbb{Q}$-almost every ω, $(w^0_s - w^0_t, v_s)_{t\le s\le T}$ and $(w_s - w_t)_{t\le s\le T}$ are independent under $\mathbb{Q}^t_\omega$.

The martingale property of $(m_s - m_t)_{t \le s \le T}$ under $\mathbb{Q}_\omega^t$ is quite obvious. Indeed, for E and F as above, namely E is in $\mathcal{G}_t^{\text{nat}}$ and, for some $s \in [t, T]$, F is in $\mathcal{G}_s^{t,\text{nat}}$, the martingale property of $(m_r - m_t)_{t \le r \le T}$ under $\mathbb{Q}$ says that for two Borel subsets C^0 and C of $\mathcal{C}([s, T]; \mathbb{R}^d)$ and for $r \in [s, T]$, we have:

$$\begin{aligned}\mathbb{E}^{\mathbb{Q}}\big[\mathbf{1}_E \mathbb{E}^{\mathbb{Q}_\omega^t}\big(\mathbf{1}_F[m_r - m_t]\big)\big] = \mathbb{E}^{\mathbb{Q}}\big[\mathbf{1}_E \mathbf{1}_F[m_r - m_t]\big] = \mathbb{E}^{\mathbb{Q}}\big[\mathbf{1}_E \mathbf{1}_F[m_s - m_t]\big] \\ = \mathbb{E}^{\mathbb{Q}}\big[\mathbf{1}_E \mathbb{E}^{\mathbb{Q}_\omega^t}\big(\mathbf{1}_F[m_s - m_t]\big)\big],\end{aligned}$$

from which we get that, for almost every ω under $\mathbb{Q}$, we must have:

$$\mathbb{E}^{\mathbb{Q}_\omega^t}[\mathbf{1}_F(m_r - m_t)] = \mathbb{E}^{\mathbb{Q}_\omega^t}[\mathbf{1}_F(m_s - m_t)].$$

By choosing r and s in a dense countable subset of $[t, T]$ and F in a countable generating π-system, we deduce that $(m_s - m_t)_{t \le s \le T}$ is a $\mathbb{G}^{t,\text{nat}}$- martingale under $\mathbb{Q}_\omega^t$. The square-integrability property of the martingale is also easily proved. Furthermore, one proves in the same way that $((m_s - m_t) \otimes (w_s^0 - w_t^0, w_s - w_t))_{t \le s \le T}$ is a $\mathbb{G}^{t,\text{nat}}$-martingale under $\mathbb{Q}_\omega^t$, the tensor product acting on elements of $\mathbb{R}^m \times \mathbb{R}^{2d}$, implying that the bracket between $(m_s - m_t)_{t \le s \le T}$ and $(w_s^0 - w_t^0, w_s - w_t)_{t \le s \le T}$ is zero.

Second Step. We now check compatibility. We call $\mathbb{G}^{t,\text{nat},\text{input}} = (\mathcal{G}_s^{t,\text{nat},\text{input}})_{t \le s \le T}$ the filtration generated by $(w_s^0 - w_t^0, \nu_s, w_s - w_t)_{t \le s \le T}$. For a given $s \in [t, T]$, we consider $E \in \mathcal{G}_t^{\text{nat}}$, $C \in \mathcal{G}_s^{t,\text{nat},\text{input}}$, $D \in \mathcal{G}_{s+}^{t,\text{nat}}$, $F \in \mathcal{G}_T^{t,\text{nat},\text{input}}$. By compatibility of $(\xi, \boldsymbol{w}^0, \boldsymbol{\mu}, \boldsymbol{w})$ and $\mathbb{G}$ under $\mathbb{Q}$, see Lemma 1.27, we have:

$$\mathbb{E}^{\mathbb{Q}}\big[\mathbf{1}_E \mathbf{1}_C \mathbf{1}_D \mathbf{1}_F\big] = \mathbb{E}^{\mathbb{Q}}\big[\mathbf{1}_{E \cap D} \mathbf{1}_C \mathbb{Q}(F | \mathcal{G}_s^{\text{input}})\big],$$

where $\mathcal{G}_s^{\text{input}}$ denotes the completion of $\mathcal{G}_{s+}^{\text{nat},\text{input}}$ under $\mathbb{Q}$. In particular, we also have:

$$\begin{aligned}\mathbb{E}^{\mathbb{Q}}\big[\mathbf{1}_E \mathbf{1}_C \mathbf{1}_D \mathbf{1}_F\big] &= \lim_{\varepsilon \searrow 0} \mathbb{E}^{\mathbb{Q}}\big[\mathbf{1}_{E \cap D} \mathbf{1}_C \mathbb{Q}(F | \mathcal{G}_{s+\varepsilon}^{\text{nat},\text{input}})\big] \\ &= \mathbb{E}^{\mathbb{Q}}\big[\mathbf{1}_{E \cap D} \mathbf{1}_C \liminf_{\varepsilon \searrow 0} \mathbb{Q}(F | \mathcal{G}_{s+\varepsilon}^{\text{nat},\text{input}})\big].\end{aligned}$$

Therefore, for almost every $\omega \in \Omega$ under $\mathbb{Q}$,

$$\mathbb{Q}_\omega^t\big(C \cap D \cap F\big) = \mathbb{E}^{\mathbb{Q}_\omega^t}\big[\mathbf{1}_C \mathbf{1}_D \liminf_{\varepsilon \searrow 0} \mathbb{Q}(F | \mathcal{G}_{s+\varepsilon}^{\text{nat},\text{input}})\big]. \tag{1.18}$$

Notice that, up a to a null event under $\mathbb{Q}$, the right-hand side is independent of the version used for the random variable $\liminf_{\varepsilon \searrow 0} \mathbb{Q}(F | \mathcal{G}_{s+\varepsilon}^{\text{nat},\text{input}})$. Now, for $B \in \mathcal{G}_s^{\text{nat},\text{input}}$ and $C' \in \mathcal{G}_t^{\text{nat},\text{input}}$, and for $\mathbb{Q}$-almost every $\omega \in \Omega$, we have:

$$\begin{aligned}\mathbb{Q}_\omega^t\big(C \cap C' \cap B\big) = \mathbf{1}_{C'} \mathbb{Q}_\omega^t\big(C \cap B\big) = \mathbf{1}_{C'} \mathbb{E}^{\mathbb{Q}_\omega^t}\big[\mathbf{1}_C \mathbb{Q}_\omega^t(B | \mathcal{G}_s^{t,\text{nat},\text{input}})\big] \\ = \mathbb{E}^{\mathbb{Q}_\omega^t}\big[\mathbf{1}_{C \cap C'} \mathbb{Q}_\omega^t(B | \mathcal{G}_s^{t,\text{nat},\text{input}})\big].\end{aligned}$$

Choosing C and C' in countable generating π-systems, we deduce that, for $\mathbb{Q}$-almost every ω, $\mathbb{Q}_\omega^t$ almost surely, $\mathbf{1}_B = \mathbb{Q}_\omega^t(B|\mathcal{G}_s^{t,\mathrm{nat,input}})$. In particular $B \in \mathcal{G}_{\omega,s}^{t,\mathrm{input}}$. Therefore, returning to (1.18), we deduce that $\liminf_{\varepsilon\searrow 0}\mathbb{Q}(F|\mathcal{G}_{s+\varepsilon}^{\mathrm{nat,input}})$ is measurable with respect to the completion $\mathcal{G}_{\omega,s}^{t,\mathrm{input}}$ of $\mathcal{G}_{s+}^{t,\mathrm{nat,input}}$ under $\mathbb{Q}_\omega^t$. Choosing $D \in \mathcal{G}_{s+}^{t,\mathrm{nat,input}}$ in (1.18), we deduce that $\liminf_{\varepsilon\searrow 0}\mathbb{Q}(F|\mathcal{G}_{s+\varepsilon}^{\mathrm{nat,input}})$ matches the conditional probability of F given $\mathcal{G}_{\omega,s}^{t,\mathrm{input}}$.

We obtain:

$$\mathbb{Q}_\omega^t\big(C\cap D\cap F\big) = \mathbb{E}^{\mathbb{Q}_\omega^t}\big[\mathbf{1}_C\mathbb{Q}_\omega^t(D|\mathcal{G}_{\omega,s}^{t,\mathrm{input}})\mathbb{Q}_\omega^t(F|\mathcal{G}_{\omega,s}^{t,\mathrm{input}})\big].$$

The above is true for $s \in [t,T]$, for C, D, and F as above and for $\mathbb{Q}$-almost every ω. In order to prove it $\mathbb{Q}$-almost surely, for any $s \in [t,T]$ and for all C, D, and F, we first restrict ourselves to the case when D belongs to $\mathcal{G}_s^{t,\mathrm{nat}}$. Then, the above is true $\mathbb{Q}$-almost surely, for all s in a dense countable subset of $[t,T]$ and for all C, D, and F in countable generating π-systems. Therefore, it is true, $\mathbb{Q}$-almost surely, for all s in a dense countable subset of $[t,T]$ and for all $C \in \mathcal{G}_s^{t,\mathrm{nat,input}}$, $D \in \mathcal{G}_s^{t,\mathrm{nat}}$ and F in $\mathcal{G}_T^{t,\mathrm{nat,input}}$. The end of the proof is standard. For $\mathbb{Q}$-almost every ω and for any $s \in [t,T]$, we can find a decreasing sequence $(s_n)_{n\geq 1}$, converging to s, such that for $C \in \mathcal{G}_{s+}^{t,\mathrm{nat,input}}$, $D \in \mathcal{G}_{s+}^{t,\mathrm{nat}}$ and F in $\mathcal{G}_T^{t,\mathrm{nat,input}}$, we have:

$$\mathbb{Q}_\omega^t\big(C\cap D\cap F\big) = \mathbb{E}^{\mathbb{Q}_\omega^t}\big[\mathbf{1}_C\mathbb{Q}_\omega^t(D|\mathcal{G}_{\omega,s_n}^{t,\mathrm{input}})\mathbb{Q}_\omega^t(F|\mathcal{G}_{\omega,s_n}^{t,\mathrm{input}})\big].$$

Letting n tend to ∞, we deduce that the compatibility holds for $\mathbb{Q}$-almost every $\omega \in \Omega$.

Recalling that x_t is almost surely constant under $\mathbb{Q}_\omega^t$, the above computations show that, for $\mathbb{Q}$-almost every $\omega \in \Omega_{\mathrm{canon}}$, the variables:

$$\big(x_t, (w_s^0 - w_t^0)_{t\leq s\leq T}, (\nu_s)_{t\leq s\leq T}, (w_s - w_t)_{t\leq s\leq T}\big)$$

form an admissible probabilistic set-up on $\Omega_{\mathrm{input}} \times \Omega_{\mathrm{output}}$ equipped with $\mathbb{Q}_\omega^t$ and the completed filtration $\mathbb{G}_\omega^t$ of $\mathbb{G}^{t,\mathrm{nat}}$.

Third Step. It then remains to prove that, on such a set-up, the process $(x_s, y_s, z_s, z_s^0, m_s - m_t)_{t\leq s\leq T}$ solves the FBSDE (1.5). As in the proof of Lemma 1.27, we shall make use of the process $(\zeta_s, \zeta_s^0)_{t\leq s\leq T}$. Following the second step in the proof of Lemma 1.27, it is clear that, for $\mathbb{Q}$-almost every ω, almost every trajectory of $(\zeta_s)_{t\leq s\leq T}$ under $\mathbb{Q}_\omega^t$ is absolutely continuous. Moreover, since:

$$\mathbb{Q}\Big(\mathrm{Leb}_1\big\{s \in [t,T] : \lim_{n\to\infty} n\big((\zeta,\zeta^0)_s - (\zeta,\zeta^0)_{(s-1/n)+}\big) = (z_s, z_s^0)\big\}\Big) = 1,$$

we deduce that for $\mathbb{Q}$-almost every ω, almost surely under $\mathbb{Q}_\omega^t$, for almost every $s \in [t,T]$, we have:

$$\lim_{n\to\infty} n\big((\zeta,\zeta^0)_s - (\zeta,\zeta^0)_{(s-1/n)+}\big) = (z_s, z_s^0).$$

Now, as in the proof of Lemma 1.27, we consider a sequence of bounded measurable functions $(B_\ell : [0,T]\times\mathbb{R}^d\times\mathcal{X}\times\mathbb{R}^m\times\mathbb{R}^{2(m\times d)} \ni (s,x,\mu,y,z,z^0) \mapsto B_\ell(s,x,\mu,y,z,z^0) \in \mathbb{R}^d)_{\ell\geq 1}$, each B_ℓ being continuous in (x,y,z,z^0) when (s,μ) is fixed, such that:

$$\lim_{\ell\to\infty} \mathbb{E}^{\mathbb{Q}} \int_0^T \big|(B - B_\ell)\big(s, x_s, \nu_s, y_s, z_s, z_s^0\big)\big| ds = 0.$$

Using the same argument as in the proof of Lemma 1.27, we find that, for a bounded and nondecreasing smooth function $\vartheta : \mathbb{R} \to \mathbb{R}$ equal to the identity on $[0, 1]$, we have:

$$\begin{aligned}\lim_{\ell\to\infty} \lim_{n\to\infty} \lim_{p\to\infty} \mathbb{E}^{\mathbb{Q}}\bigg[\vartheta\bigg(\sup_{t\le s\le T}\bigg| x_t - \int_t^s \bigg(n\int_{(\lfloor pr\rfloor/p-1/n)+}^{\lfloor pr\rfloor/p} \Sigma\big(u, x_u, \nu_u\big)du\bigg)dw_r \\ - \int_t^s B_\ell\Big(r, x_r, \nu_r, y_r, n\big((\zeta, \zeta^0)_r - (\zeta, \zeta^0)_{(r-1/n)+}\big)\Big)dr \\ - \int_t^s \bigg(n\int_{(\lfloor pr\rfloor/p-1/n)+}^{\lfloor pr\rfloor/p} \Sigma^0\big(u, x_u, \nu_u\big)du\bigg)dw_r^0\bigg|\bigg)\bigg] = 0,\end{aligned}$$

proving that, for $\varepsilon > 0$,

$$\begin{aligned}\lim_{\ell\to\infty} \limsup_{n\to\infty} \limsup_{p\to\infty} \mathbb{Q}\bigg(\mathbb{E}^{\mathbb{Q}_\omega^t}\bigg[\vartheta\bigg(\sup_{t\le s\le T}\bigg| x_s - \int_t^s \bigg(n\int_{(\lfloor pr\rfloor/p-1/n)+}^{\lfloor pr\rfloor/p} \Sigma\big(u, x_u, \nu_u\big)du\bigg)dw_r \\ - \int_t^s B_\ell\Big(r, x_r, \nu_r, y_r, n\big((\zeta, \zeta^0)_r - (\zeta, \zeta^0)_{(r-1/n)+}\big)\Big)dr \\ - \int_t^s \bigg(n\int_{(\lfloor pr\rfloor/p-1/n)+}^{\lfloor pr\rfloor/p} \Sigma^0\big(u, x_u, \nu_u\big)du\bigg)dw_r^0\bigg|\bigg)\bigg] > \varepsilon\bigg) = 0.\end{aligned}$$

Now, for $\mathbb{Q}$-almost every ω, we have:

$$\begin{aligned}\lim_{\ell\to\infty} \lim_{n\to\infty} \lim_{p\to\infty} \mathbb{E}^{\mathbb{Q}_\omega^t}\bigg[\vartheta\bigg(\sup_{t\le s\le T}\bigg| x_s - \int_t^s \bigg(n\int_{(\lfloor pr\rfloor/p-1/n)+}^{\lfloor pr\rfloor/p} \Sigma\big(u, x_u, \nu_u\big)du\bigg)dw_r \\ - \int_t^s B_\ell\Big(r, x_r, \nu_r, y_r, n\big((\zeta, \zeta^0)_r - (\zeta, \zeta^0)_{(r-1/n)+}\big)\Big)dr \\ - \int_t^s \bigg(n\int_{(\lfloor pr\rfloor/p-1/n)+}^{\lfloor pr\rfloor/p} \Sigma^0\big(u, x_u, \nu_u\big)du\bigg)dw_r^0\bigg|\bigg)\bigg] \\ = \mathbb{E}^{\mathbb{Q}_\omega^t}\bigg[\vartheta\bigg(\sup_{t\le s\le T}\bigg| x_s - \int_t^s B\big(r, x_r, \nu_r, y_r, z_r, z_r^0\big)dr \\ - \int_t^s \Sigma\big(r, x_r, \nu_r\big)dw_r - \int_t^s \Sigma^0\big(r, x_r, \nu_r\big)dw_r^0\bigg|\bigg)\bigg].\end{aligned}$$

which shows that under $\mathbb{Q}_\omega^t$, the forward equation in (1.5) holds on Ω_{canon}^t equipped with the filtration $\mathbb{G}_\omega^t$. We handle the backward equation in the same way as in the proof of Lemma 1.27. This completes the proof. □

Lemma 1.40 may be reformulated when strong uniqueness holds. In this case, we know that solutions can be constructed directly on the canonical space Ω_{input}

carrying the initial condition, the two noise processes and the random environment process, and the conditioning argument can be directly implemented on Ω_{input}.

Lemma 1.43 *Let us assume that the FBSDE* (1.5) *satisfies the strong uniqueness property on set-ups initialized with an input* $(X_0, \mathbf{W}^0, \boldsymbol{\mu}, \mathbf{W})$ *of a given prescribed law, that it has a solution on such a set-up, and let us transfer this solution onto the set-up* Ω_{input} *equipped with the law* $\mathbb{Q}^{\mathcal{B}}_{\text{input}}$ *of the input* $(X_0, \mathbf{W}^0, \boldsymbol{\mu}, \mathbf{W})$ *on* Ω_{input} *equipped with the Borel* σ*-field. Then, for any* $t \in [0, T]$, *we consider the regular conditional probability of* $\mathbb{Q}^{\mathcal{B}}_{\text{input}}$ *given* $\mathcal{G}^{\text{nat,input}}_t$, *and for each realization* $\omega \in \Omega_{\text{input}}$, *we denote by* $\mathbb{Q}^t_{\omega,\text{input}}$ *its completion. Similarly, we denote by* $\tilde{\mathbb{G}}^{t,\text{input}}_{\omega} = (\tilde{\mathcal{G}}^{t,\text{input}}_{\omega,s})_{t\le s\le T}$ *the completion of the filtration* $\mathbb{G}^{t,\text{nat,input}}_{\cdot +} = (\mathcal{G}^{t,\text{nat,input}}_{s+})_{t\le s\le T}$, *where* $\mathbb{G}^{t,\text{nat,input}} = (\mathcal{G}^{t,\text{nat,input}}_s)_{t\le s\le T}$ *is the filtration generated by* $(w^0_s - w^0_t, \nu_s, w_s - w_t)_{t\le s\le T}$. *Now, if we set:*

$$\big(\boldsymbol{x}, \boldsymbol{y}, (\boldsymbol{\zeta}, \boldsymbol{\zeta}^0), \boldsymbol{m}\big) = \Phi(\xi, \boldsymbol{w}^0, \boldsymbol{\mu}, \boldsymbol{w}),$$

with Φ *as in the statement of Theorem 1.33 and:*

$$(z_t, z^0_t)(\omega) = \begin{cases} \lim_{n\to\infty} n\big((\zeta, \zeta^0)_t(\omega) - (\zeta, \zeta^0)_{(t-1/n)_+}(\omega)\big) & \text{if the limit exists,} \\ 0 \quad \text{otherwise,} \end{cases}$$

then, for $\mathbb{Q}^{\mathcal{B}}_{\text{input}}$*-almost every* $\omega \in \Omega_{\text{input}}$, *the process* $(x_s, y_s, z_s, z^0_s, m_s - m_t)_{t\le s\le T}$ *solves* (1.5) *on the space* Ω^t_{input} *equipped with the probability* $\mathbb{Q}^t_{\omega,\text{input}}$ *and with the filtration* $\tilde{\mathbb{G}}^{t,\text{input}}_{\omega}$. *The input reads* $(x_t, w^0_s - w^0_t, \nu_s, w_s - w_t)_{t\le s\le T}$ *and has the distribution:*

$$\delta_{x_t(\omega)} \otimes \Big(\mathbb{Q}^t_{\omega,\text{input}} \circ \big((w^0_s - w^0_t, \nu_s)_{t\le s\le T}\big)^{-1}\Big) \otimes \mathcal{W}^t_d,$$

where $\mathcal{W}^t_d$ *is the d-dimensional Wiener measure on* $\mathcal{C}([t, T]; \mathbb{R}^d)$.

Remark 1.44 *Notice that in order to lighten the notation, we used the letter* ω *to denote a generic element of* Ω_{input}, *instead of the symbol* ω_{input} *which we used previously.*

Moreover, in analogy with Remark 1.41, it is implicitly understood that the probabilistic set-up formed by the filtered probability space and the input $(x_t, w^0_s - w^0_t, \nu_s, w_s - w_t)_{t\le s\le T}$ *is admissible. Finally, notice also that a similar version of Remark 1.42 holds.*

Proof. Throughout the proof, we denote by $\mathbb{Q}_{\text{input}}$ the completion of $\mathbb{Q}^{\mathcal{B}}_{\text{input}}$.

First Step. Following the steps of the proof of Lemma 1.40, we check that, for $\mathbb{Q}_{\text{input}}$-almost every $\omega \in \Omega_{\text{input}}$, $(w^0_s - w^0_t, w_s - w_t)_{t\le s\le T}$ is a $2d$-dimensional $\tilde{\mathbb{G}}^{t,\text{input}}_{\omega}$-Brownian motion for

$\mathbb{Q}^t_{\omega,\mathrm{input}}$, that $(w^0_s - w^0_t, v_s)_{t\le s\le T}$ and $(w_s - w_t)_{t\le s\le T}$ are independent, and that $\tilde{\mathbb{G}}^{t,\mathrm{input}}_\omega$ and $(w^0_s - w^0_t, v_s, w_s - w_t)_{t\le s\le T}$ are compatible. We shall prove in the fourth step below that, under $\mathbb{Q}^t_{\omega,\mathrm{input}}$, x_t is almost surely equal to $x_t(\omega)$, which shows that the probabilistic set-up formed by $(\Omega^t_{\mathrm{input}}, \tilde{\mathbb{G}}^{t,\mathrm{input}}_\omega, \mathbb{Q}^t_{\omega,\mathrm{input}})$ and $(x_t, w^0_s - w^0_t, v_s, w_s - w_t)_{t\le s\le T}$ is admissible and has the right law.

Second Step. We now check that, for $\mathbb{Q}_{\mathrm{input}}$-almost every $\omega \in \Omega_{\mathrm{input}}$, the process

$$(x_s, y_s, (\zeta,\zeta^0)_s, m_s - m_t)_{t\le s\le T}$$

is adapted to $\tilde{\mathbb{G}}^{t,\mathrm{input}}_\omega$. By construction, we know that, for each $s \in [t,T]$, $(x_s, y_s, (\zeta,\zeta^0)_s, m_s)$ is $\mathcal{G}^{\mathrm{input}}_s$-measurable. In particular, there exists a $\mathcal{G}^{\mathrm{nat,input}}_{s+}$-measurable random variable, denoted by $(\tilde{x}_s, \tilde{y}_s, (\tilde{\zeta},\tilde{\zeta}^0)_s, \tilde{m}_s)$, which is $\mathbb{Q}_{\mathrm{input}}$-almost surely equal to $(x_s, y_s, (\zeta,\zeta^0)_s, m_s)$.

In the third step below, we prove that given $s \in [t,T]$, for $\mathbb{Q}_{\mathrm{input}}$-almost every $\omega \in \Omega_{\mathrm{input}}$, $\mathcal{G}^{\mathrm{nat,input}}_{s+} \subset \tilde{\mathcal{G}}^{t,\mathrm{input}}_{\omega,s}$. Assuming momentarily this result, this shows that, for $\mathbb{Q}_{\mathrm{input}}$-almost every $\omega \in \Omega_{\mathrm{input}}$, $(\tilde{x}_s, \tilde{y}_s, (\tilde{\zeta},\tilde{\zeta}^0)_s, \tilde{m}_s - \tilde{m}_t)$ is $\tilde{\mathcal{G}}^{t,\mathrm{input}}_{\omega,s}$-measurable.

Now, we use the fact that for every $\mathbb{Q}$-null subset $N \subset \Omega_{\mathrm{input}}$, N belongs to $\tilde{\mathcal{G}}^{t,\mathrm{input}}_{\omega,s}$ for $\mathbb{Q}_{\mathrm{input}}$-almost every $\omega \in \Omega_{\mathrm{input}}$. This follows from the fact that, for such an N, there exists a Borel subset N' containing N such that $0 = \mathbb{Q}_{\mathrm{input}}(N') = \int_{\Omega_{\mathrm{input}}} \mathbb{Q}^t_{\omega,\mathrm{input}}(N') d\mathbb{Q}_{\mathrm{input}}(\omega)$. We use this fact in the following way. Notice that the difference $(x_s - \tilde{x}_s, y_s - \tilde{y}_s, (\zeta,\zeta^0)_s - (\tilde{\zeta},\tilde{\zeta}^0)_s, m_s - \tilde{m}_s - (m_t - \tilde{m}_t))$ is $\mathbb{Q}_{\mathrm{input}}$-almost surely equal to 0. Therefore, for any Borel subset $B \subset \mathbb{R}^d \times \mathbb{R}^m \times \mathbb{R}^{2(m\times d)} \times \mathbb{R}^m$,

$$\mathbb{Q}_{\mathrm{input}}\Big[\big(x_s - \tilde{x}_s, y_s - \tilde{y}_s, (\zeta,\zeta^0)_s - (\tilde{\zeta},\tilde{\zeta}^0)_s, m_s - \tilde{m}_s - (m_t - \tilde{m}_t)\big) \in B\Big] \in \{0,1\}.$$

We deduce that, for $\mathbb{Q}_{\mathrm{input}}$-almost every ω, for any B in a countable generating class of $\mathcal{B}(\mathbb{R}^d \times \mathbb{R}^m \times \mathbb{R}^{2(m\times d)} \times \mathbb{R}^m)$, we have:

$$\big(x_s - \tilde{x}_s, y_s - \tilde{y}_s, (\zeta,\zeta^0)_s - (\tilde{\zeta},\tilde{\zeta}^0)_s, m_s - \tilde{m}_s - (m_t - \tilde{m}_t)\big)^{-1}(B) \in \tilde{\mathcal{G}}^{t,\mathrm{input}}_{\omega,s}.$$

Therefore, $(x_s - \tilde{x}_s, y_s - \tilde{y}_s, (\zeta,\zeta^0)_s - (\tilde{\zeta},\tilde{\zeta}^0)_s, m_s - \tilde{m}_s - (m_t - \tilde{m}_t))$ is $\tilde{\mathcal{G}}^{t,\mathrm{input}}_{\omega,s}$ -measurable, and almost surely equal to 0 for $\mathbb{Q}^t_{\omega,\mathrm{input}}$. So for any $s \in [t,T]$ and for $\mathbb{Q}_{\mathrm{input}}$-almost every ω, $(x_s, y_s, (\zeta,\zeta^0)_s, m_s - m_t)$ is $\tilde{\mathcal{G}}^{t,\mathrm{input}}_{\omega,s}$-measurable and is $\mathbb{Q}^t_{\omega,\mathrm{input}}$-almost surely equal to $(\tilde{x}_s, \tilde{y}_s, (\tilde{\zeta},\tilde{\zeta}^0)_s, \tilde{m}_s - \tilde{m}_t)$. Choosing s in a dense countable subset of $[t,T]$, and using the right-continuity of the process and the filtration, we deduce that for $\mathbb{Q}_{\mathrm{input}}$-almost every ω, for all $s \in [t,T]$, $(x_s, y_s, (\zeta,\zeta^0)_s, m_s - m_t)$ is $\tilde{\mathcal{G}}^{t,\mathrm{input}}_{\omega,s}$-measurable.

Third Step. We now check that, for all $s \in [t,T]$ and for $\mathbb{Q}_{\mathrm{input}}$-almost every ω, $\mathcal{G}^{\mathrm{nat,input}}_{s+} \subset \tilde{\mathcal{G}}^{t,\mathrm{input}}_{\omega,s}$. In order to do so, we proceed as follows. Consider $C \in \mathcal{G}^{\mathrm{nat,input}}_t$ and $D \in \mathcal{G}^{t,\mathrm{nat,input}}_s$. By definition, $D \in \tilde{\mathcal{G}}^{t,\mathrm{input}}_{\omega,s}$. Moreover, for $\mathbb{Q}_{\mathrm{input}}$-almost every ω, we have $\mathbb{Q}^t_{\omega,\mathrm{input}}(C) \in \{0,1\}$ (according to $\omega \in C$ or not) and thus $C \in \tilde{\mathcal{G}}^{t,\mathrm{input}}_{\omega,s}$. We deduce that, for $\mathbb{Q}_{\mathrm{input}}$-almost every $\omega \in \Omega_{\mathrm{input}}$, for all C and D in countable generating classes of $\mathcal{G}^{\mathrm{nat,input}}_t$ and $\mathcal{G}^{t,\mathrm{nat,input}}_s$, $C \cap D$ belongs to $\tilde{\mathcal{G}}^{t,\mathrm{input}}_s$. Therefore, for $\mathbb{Q}_{\mathrm{input}}$-almost every $\omega \in \Omega_{\mathrm{input}}$, $\mathcal{G}^{\mathrm{nat,input}}_s \subset \tilde{\mathcal{G}}^{t,\mathrm{input}}_{\omega,s}$ and then $\mathcal{G}^{\mathrm{nat,input}}_{s+} \subset \tilde{\mathcal{G}}^{t,\mathrm{input}}_{\omega,s}$.

Fourth Step. By right-continuity of the trajectories of $(x_s, y_s, (\zeta, \zeta^0)_s, m_s - m_t)_{t \le s \le T}$, we deduce from the second step that, for $\mathbb{Q}_{\text{input}}$-almost every $\omega \in \Omega_{\text{input}}$, $(x_s, y_s, (\zeta, \zeta^0)_s, m_s - m_t)_{t \le s \le T}$ is $\tilde{\mathbb{G}}_\omega^{t,\text{input}}$-progressively measurable. We now prove that, for $\mathbb{Q}_{\text{input}}$-almost every $\omega \in \Omega_{\text{input}}$, x_t is $\mathbb{Q}^t_{\omega,\text{input}}$-almost surely equal to $x_t(\omega)$. To do so, we claim that the random variable x_t is $\mathbb{Q}_{\text{input}}$-almost surely equal to a $\mathcal{G}_t^{\text{nat,input}}$-measurable random variable $\tilde{x}_t$. Indeed, the process $(x_s)_{0 \le s \le T}$ has continuous paths and each x_s is $\mathbb{Q}_{\text{input}}$-almost surely equal to a $\mathcal{G}_t^{\text{nat,input}}$-measurable random variable. Since $\tilde{x}_t$ is $\mathcal{G}_t^{\text{nat,input}}$-measurable, we deduce that for $\mathbb{Q}_{\text{input}}$ almost every ω, $\tilde{x}_t$ is almost surely equal to $\tilde{x}_t(\omega)$ under $\mathbb{Q}^t_{\omega,\text{input}}$. Arguing as in the second step of the proof, we deduce that, for $\mathbb{Q}_{\text{input}}$ almost every ω, x_t is almost surely equal to $\tilde{x}_t$ under $\mathbb{Q}^t_{\omega,\text{input}}$ and is thus almost surely equal to $\tilde{x}_t(\omega)$ under $\mathbb{Q}^t_{\omega,\text{input}}$.

Now that we have the progressive-measurability property, the fact that for $\mathbb{Q}_{\text{input}}$-almost every $\omega \in \Omega_{\text{input}}$, $(m_s - m_t)_{t \le s \le T}$ is a $\tilde{\mathbb{G}}_\omega^{t,\text{input}}$-martingale may be shown as in the proof of Lemma 1.40. The fact that $(x_s, y_s, (\zeta, \zeta^0)_s, m_s - m_t)_{t \le s \le T}$ solves the FBSDE (1.5) may be also shown as in the proof of Lemma 1.40. □

1.3.2 Small Time Solvability and Decoupling Field

One of the main motivation underpinning the restarting procedure discussed above is the fact that Cauchy-Lipschitz theory for FBSDEs holds in small time only, and that we shall need sharp tools if we want to piece together solutions on small intervals in order to construct solutions on large intervals with prescribed lengths. We shall use the following assumption on the coefficients:

Assumption (Lipschitz FBSDE in Random Environment). There exist two nonnegative constants L and Λ such that:

(A1) For any $t \in [0, T]$ and any $\mu \in \mathcal{X}$,

$$|B(t, 0, \mu, 0, 0, 0)| + |\Sigma(t, 0, \mu)| + |\Sigma^0(t, 0, \mu)| + |F(t, 0, \mu, 0, 0, 0)| + |G(0, \mu)| \le \Lambda\big(1 + d(0_{\mathcal{X}}, \mu)\big),$$

for some $0_{\mathcal{X}} \in \mathcal{X}$.

(A2) For each $t \in [0, T]$ and each $\mu \in \mathcal{X}$, the functions $B(t, \cdot, \mu, \cdot, \cdot, \cdot)$, $F(t, \cdot, \mu, \cdot, \cdot, \cdot)$, $\Sigma(t, \cdot, \mu)$, $\Sigma^0(t, \cdot, \mu)$ and G are L-Lipschitz continuous on their own domain.

Then, on any admissible set-up $(\Omega, \mathcal{F}, \mathbb{F}, \mathbb{P})$ equipped with some $(\mathcal{G}, \boldsymbol{W}^0, \boldsymbol{\mu}, \boldsymbol{W})$, the FBSDE (1.5) with $X_0 \in L^2(\Omega, \mathcal{G}, \mathbb{P}; \mathbb{R}^d)$ as initial condition has a unique solution when the time horizon T is less than some $c > 0$, c only depending on the Lipschitz constant of the coefficients. We refer to Theorem 1.45 right below for a precise statement.

As we see repeatedly throughout the book, it is sometimes possible to iterate the small time solvability result. As it might be easily guessed, the iteration argument consists in restarting the forward process along a sequence of initial times with a sufficiently small step. We already used this argument in Chapter (Vol I)-4 for handling FBSDEs with a deterministic environment $\boldsymbol{\mu}$. The argument in the case when $\boldsymbol{\mu}$ is random is made clear at the end of the subsection.

Small Time Solvability

We start with a stability result.

Theorem 1.45 *Consider a filtered probability space $(\Omega, \mathcal{F}, \mathbb{F}, \mathbb{P})$ together with two sub-σ-algebras $\mathcal{G}$ and $\mathcal{G}'$ and a tuple $(\mathbf{W}^0, \boldsymbol{\mu}, \boldsymbol{\mu}', \mathbf{W})$ such that $(\Omega, \mathcal{F}, \mathbb{F}, \mathbb{P})$ equipped with $(\mathcal{G}, \mathbf{W}^0, \boldsymbol{\mu}, \mathbf{W})$ and $(\Omega, \mathcal{F}, \mathbb{F}, \mathbb{P})$ equipped with $(\mathcal{G}', \mathbf{W}^0, \boldsymbol{\mu}', \mathbf{W})$ form admissible set-ups.*

For a tuple of coefficients $(B, \Sigma, \Sigma^0, F, G)$ satisfying assumption **Lipschitz FBSDE in Random Environment**, *there exists a constant $c > 0$, only depending on the Lipschitz constant of the coefficients, such that, for $T \leq c$ and any $X_0 \in L^2(\Omega, \mathcal{G}, \mathbb{P}; \mathbb{R}^d)$, the FBSDE* (1.5) *has a unique solution $(\boldsymbol{X}, \boldsymbol{Y}, \boldsymbol{Z}, \boldsymbol{Z}^0, \boldsymbol{M})$ on $(\Omega, \mathcal{F}, \mathbb{F}, \mathbb{P})$ equipped with $(\mathcal{G}, \mathbf{W}^0, \boldsymbol{\mu}, \mathbf{W})$.*

In addition to $(B, \Sigma, \Sigma^0, F, G)$, consider coefficients $(B', \Sigma', \Sigma^{0\prime}, F', G')$, also satisfying assumption **Lipschitz FBSDE in Random Environment** *with the same constants. Then, there exist a constant $\Gamma \geq 0$, only depending on the Lipschitz constant of the coefficients, such that, for $T \leq c$, for any $X_0 \in L^2(\Omega, \mathcal{G}, \mathbb{P}; \mathbb{R}^d)$ and $X_0' \in L^2(\Omega, \mathcal{G}', \mathbb{P}; \mathbb{R}^d)$, the unique solutions $(\boldsymbol{X}, \boldsymbol{Y}, \boldsymbol{Z}, \boldsymbol{Z}^0, \boldsymbol{M})$ and $(\boldsymbol{X}', \boldsymbol{Y}', \boldsymbol{Z}', \boldsymbol{Z}^{0\prime}, \boldsymbol{M}')$ of the corresponding FBSDEs* (1.5), *with X_0 and X_0' as respective initial conditions and with $(\mathcal{G}, \mathbf{W}^0, \boldsymbol{\mu}, \mathbf{W})$ and $(\mathcal{G}', \mathbf{W}^0, \boldsymbol{\mu}', \mathbf{W})$ as respective inputs, satisfy:*

$$
\begin{aligned}
&\mathbb{E}\biggl[\sup_{0\leq t\leq T}\bigl(|X_t - X_t'|^2 + |Y_t - Y_t'|^2 + |M_t - M_t'|^2\bigr) \\
&\qquad + \int_0^T \bigl(|Z_t - Z_t'|^2 + |Z_t^0 - Z_t^{0\prime}|^2\bigr)dt \,\Big|\, \mathcal{F}_0\biggr] \\
&\leq \Gamma \mathbb{E}\biggl[|X_0 - X_0'|^2 + \bigl|G(X_T, \mu_T) - G'(X_T, \mu_T')\bigr|^2 \\
&\qquad + \int_0^T \bigl|(B, F, \Sigma, \Sigma^0)\bigl(t, X_t, \mu_t, Y_t, Z_t, Z_t^0\bigr) \\
&\qquad\qquad - (B', F', \Sigma', \Sigma^{0\prime})\bigl(t, X_t, \mu_t', Y_t, Z_t, Z_t^0\bigr)\bigr|^2 dt \,\Big|\, \mathcal{F}_0\biggr].
\end{aligned}
\tag{1.19}
$$

Proof. The proof is pretty standard and is similar to the proof given in Section (Vol I)-4.2 for the short time solvability of FBSDEs of the McKean-Vlasov type. We provide it for the sake of completeness.

First Step. We start with the existence and uniqueness, for a given initial condition $X_0 \in L^2(\Omega, \mathcal{G}, \mathbb{P}; \mathbb{R}^d)$. Throughout the proof, we use the notation $\mathbb{S}^{2,d}$ for the space of $\mathbb{F}$-progressively measurable continuous processes whose supremum norm on $[0, T]$ is square-integrable.

For an element $\boldsymbol{X} = (X_t)_{0 \le t \le T} \in \mathbb{S}^{2,d}$ starting from X_0, we call $(\boldsymbol{Y}, \boldsymbol{Z}, \boldsymbol{Z}^0, \boldsymbol{M}) = (Y_t, Z_t, Z_t^0, M_t)_{0 \le t \le T}$ the solution of the BSDE:

$$dY_t = -F(t, X_t, \mu_t, Y_t, Z_t, Z_t^0)dt + Z_t dW_t + Z_t^0 dW_t^0 + dM_t, \quad t \in [0, T], \tag{1.20}$$

with the terminal condition $Y_T = G(X_T, \mu_T)$. The tuple $(\boldsymbol{Y}, \boldsymbol{Z}, \boldsymbol{Z}^0, \boldsymbol{M})$ is progressively measurable with respect to $\mathbb{F}$. Its existence and uniqueness are guaranteed by the result of Example 1.20.

With this $(\boldsymbol{Y}, \boldsymbol{Z}) \in \mathbb{H}^{2,m} \times \mathbb{H}^{2,m \times d}$, we associate the solution $\bar{\boldsymbol{X}} = (\bar{X}_t)_{0 \le t \le T}$ of the SDE:

$$\begin{aligned} d\bar{X}_t &= B(t, \bar{X}_t, \mu_t, Y_t, Z_t, Z_t^0)dt \\ &\quad + \Sigma(t, \bar{X}_t, \mu_t)dW_t + \Sigma^0(t, \bar{X}_t, \mu_t)dW_t^0, \quad t \in [0, T], \end{aligned}$$

with X_0 as initial condition. Obviously, $\bar{\boldsymbol{X}}$ is $\mathbb{F}$-progressively measurable. In this way, we created a map:

$$\Phi : \mathbb{S}^{2,d} \ni \boldsymbol{X} \mapsto \bar{\boldsymbol{X}} \in \mathbb{S}^{2,d}.$$

Our goal is now to prove that Φ is a contraction when T is small enough.

Given two inputs $\boldsymbol{X}^1$ and $\boldsymbol{X}^2$ in $\mathbb{S}^{2,d}$, with X_0 as common initial condition, we denote by $(\boldsymbol{Y}^1, \boldsymbol{Z}^1, \boldsymbol{Z}^{0,1}, \boldsymbol{M}^1)$ and $(\boldsymbol{Y}^2, \boldsymbol{Z}^2, \boldsymbol{Z}^{0,2}, \boldsymbol{M}^2)$ the solutions of the BSDE (1.20) when $\boldsymbol{X}$ is replaced by $\boldsymbol{X}^1$ and $\boldsymbol{X}^2$ respectively. Moreover, we let $\bar{\boldsymbol{X}}^1 = \Phi(\boldsymbol{X}^1)$ and $\bar{\boldsymbol{X}}^2 = \Phi(\boldsymbol{X}^2)$. Then, following the proof of Example 1.20, we can find a constant $C \ge 0$, depending on L in **MKV FBSDE in Small Time** such that, for $T \le 1$:

$$\mathbb{E}\Big[\int_0^T \big(|Y_t^1 - Y_t^2|^2 + |Z_t^1 - Z_t^2|^2 + |Z_t^{0,1} - Z_t^{0,2}|^2\big)dt\Big] \le C\mathbb{E}\Big[\sup_{0 \le t \le T} |X_t^1 - X_t^2|^2\Big].$$

Also, it is well checked that for a possibly new value of the constant C:

$$\mathbb{E}\Big[\sup_{0 \le t \le T} |\bar{X}_t^1 - \bar{X}_t^2|^2\Big] \le CT\mathbb{E}\int_0^T \big(|Y_t^1 - Y_t^2|^2 + |Z_t^1 - Z_t^2|^2 + |Z_t^{0,1} - Z_t^{0,2}|^2\big)dt,$$

so that, increasing the constant C if needed, we get:

$$\mathbb{E}\Big[\sup_{0 \le t \le T} |\bar{X}_t^1 - \bar{X}_t^2|^2\Big] \le CT\mathbb{E}\Big[\sup_{0 \le t \le T} |X_t^1 - X_t^2|^2\Big],$$

which proves that Φ is a contraction when T is small enough.

Second Step. The proof of (1.19) is pretty similar. Repeating the analysis of Example 1.20, but working with the conditional expectation with respect to $\mathcal{F}_0$ instead of the expectation itself, we first check that:

$$
\begin{aligned}
&\mathbb{E}\bigg[\int_0^T \big(|Y_t - Y_t'|^2 + |Z_t - Z_t'|^2 + |Z_t^0 - Z_t^{0\prime}|^2\big)dt \,\Big|\, \mathcal{F}_0\bigg] \\
&\leq \Gamma \mathbb{E}\bigg[\sup_{0 \leq t \leq T} |X_t - X_t'|^2 + \big|G(X_T, \mu_T) - G'(X_T, \mu_T')\big|^2 \\
&\quad + \int_0^T \big|F\big(t, X_t, \mu_t, Y_t, Z_t, Z_t^0\big) - F'\big(t, X_t, \mu_t', Y_t, Z_t, Z_t^0\big)\big|^2 dt \,\Big|\, \mathcal{F}_0\bigg],
\end{aligned}
\tag{1.21}
$$

for a constant Γ only depending on L.

As above, it is pretty standard to compare the forward processes $\boldsymbol{X}$ and $\boldsymbol{X}'$. We get:

$$
\begin{aligned}
&\mathbb{E}\Big[\sup_{0 \leq t \leq T} |X_t - X_t'|^2 \,|\, \mathcal{F}_0\Big] \\
&\leq \Gamma \mathbb{E}\bigg[|X_0 - X_0'|^2 + T\int_0^T \big(|Y_t - Y_t'|^2 + |Z_t - Z_t'|^2 + |Z_t^0 - Z_t^{0\prime}|^2\big)dt \\
&\quad + \int_0^T \big|(B, \Sigma, \Sigma^0)\big(t, X_t, \mu_t, Y_t, Z_t, Z_t^0\big) \\
&\qquad - (B', \Sigma', \Sigma^{0\prime})\big(t, X_t, \mu_t', Y_t, Z_t, Z_t^0\big)\big|^2 dt \,\Big|\, \mathcal{F}_0\bigg].
\end{aligned}
\tag{1.22}
$$

Collecting (1.21) and (1.22), we deduce that, for T small enough:

$$
\begin{aligned}
&\mathbb{E}\bigg[\sup_{0 \leq t \leq T} |X_t - X_t'|^2 + \int_0^T \big(|Y_t - Y_t'|^2 + |Z_t - Z_t'|^2 + |Z_t^0 - Z_t^{0\prime}|^2\big)dt \,\Big|\, \mathcal{F}_0\bigg] \\
&\leq \Gamma \mathbb{E}\bigg[|X_0 - X_0'|^2 + \big|G(X_T, \mu_T) - G'(X_T, \mu_T')\big|^2 \\
&\quad + \int_0^T \big|(B, F, \Sigma, \Sigma^0)\big(t, X_t, \mu_t, Y_t, Z_t, Z_t^0\big) \\
&\qquad - \big(B', F', \Sigma', \Sigma^{0\prime}\big)\big(t, X_t, \mu_t', Y_t, Z_t, Z_t^0\big)\big|^2 dt \,\Big|\, \mathcal{F}_0\bigg].
\end{aligned}
\tag{1.23}
$$

Returning to (1.8) in the proof of Example 1.20, we easily deduce that $\mathbb{E}[\mathrm{trace}([M - M']_T)|\mathcal{F}_0]$ is also bounded by the right-hand side of the above inequality. By conditional Doob's inequality, we deduce that the same holds true for $\mathbb{E}[\sup_{0 \leq t \leq T} |M_t - M_t'|^2 | \mathcal{F}_0]$.

Now, by Burkholder-Davis-Gundy inequality for *càd-làg* martingales, the bound also holds true for $\mathbb{E}[\sup_{0 \leq t \leq T} |Y_t - Y_t'|^2 | \mathcal{F}_0]$. □

Decoupling Field

As already discussed in the frameworks of standard FBSDEs in Chapter (Vol I)-4 and of McKean-Vlasov FBSDEs in Chapter (Vol I)-6, a challenging question concerns the possible extension of the short time solvability result to time intervals with arbitrary lengths. The standard procedure presented in Chapter (Vol I)-4 relies on an induction argument based on the properties of the so-called *decoupling field*.

Basically, the decoupling field permits to reproduce, at any time $t \in [0, T]$, the role played by the terminal condition G at time T. The following result captures the flavor of this fact.

Proposition 1.46 *Under assumption* **FBSDE in Random Environment**, *assume that for some $t \in [0, T]$, the FBSDE* (1.5) *satisfies the strong uniqueness property on any t-initialized admissible set-up with an input $(\theta_0, (W_s^0, \mu_s, W_s)_{t \le s \le T})$, the random variable θ_0 taking values in a prescribed auxiliary Polish space $\mathcal{S}$ and the law of $(\theta_0, \boldsymbol{W}^0, \boldsymbol{\mu})$ being also prescribed. Assume also that, on any t-initialized admissible set-up $(\Omega, \mathcal{F}, \mathbb{F}, \mathbb{P})$ equipped with some $(\boldsymbol{W}^0, \boldsymbol{\mu}, \boldsymbol{W})$ and no initial information, for any initial condition $x \in \mathbb{R}^d$ at time t,* (1.5) *has a (unique) solution $(\boldsymbol{X}^{t,x}, \boldsymbol{Y}^{t,x}, \boldsymbol{Z}^{t,x}, \boldsymbol{Z}^{0,t,x}, \boldsymbol{M}^{t,x})$, and there exists a constant $\Gamma \ge 0$ such that, for any $x, x' \in \mathbb{R}^d$,*

$$\mathbb{P}\big[|Y_t^{t,x} - Y_t^{t,x'}| \le \Gamma |x - x'|\big] = 1. \tag{1.24}$$

Then, letting $\bar{\Omega}^{0,t} = \mathcal{C}([t, T]; \mathbb{R}^d) \times \mathcal{D}([t, T]; \mathcal{X})$ and denoting by $(w_s^0, \nu_s)_{t \le s \le T}$ the canonical process on $\bar{\Omega}^{0,t}$, there exists a mapping $U_t : \mathbb{R}^d \times \mathcal{P}_2(\bar{\Omega}^{0,t}) \times \bar{\Omega}^{0,t} \to \mathbb{R}^m$, measurable with respect to $\mathcal{B}(\mathbb{R}^d) \otimes \mathcal{B}(\mathcal{P}_2(\bar{\Omega}^{0,t})) \otimes \cap_{\varepsilon > 0} \sigma\{w_s^0, \nu_s;\ t \le s \le t + \varepsilon\}$, such that, for any solution $(X_s, Y_s, Z_s, Z_s^0, M_s)_{t \le s \le T}$ of the FBSDE (1.5) *constructed on a t-initialized set-up $(\Omega, \mathcal{F}, \mathbb{F}, \mathbb{P})$ with some $(\theta_0, W_s^0, \mu_s, W_s)_{t \le s \le T}$, θ_0 taking values in $\mathcal{S}$ and $(\theta_0, \boldsymbol{W}^0, \boldsymbol{\mu})$ following the prescribed distribution, it holds that:*

$$\mathbb{P}\Big[Y_t = U_t\Big(X_t, \mathcal{L}\big((W_s^0 - W_t^0, \mu_s)_{t \le s \le T} | \theta_0, \mu_t\big), (W_s^0 - W_t^0, \mu_s)_{t \le s \le T}\Big)\Big] = 1,$$

where $\mathcal{L}((W_s^0 - W_t^0, \mu_s)_{t \le s \le T} | \theta_0, \mu_t)$ denotes the conditional law of $(W_s^0 - W_t^0, \mu_s)_{t \le s \le T}$ given (θ_0, μ_t). Moreover, U_t is Γ-Lipschitz continuous in the space variable $x \in \mathbb{R}^d$.

Remark 1.47 *The random variable $U_t(x, \bar{\mathbb{P}}^0, \cdot)$ is constructed for $x \in \mathbb{R}^d$ and $\bar{\mathbb{P}}^0$ a probability measure on $\bar{\Omega}^{0,t}$. As shown by the proof, it suffices to construct it for $\bar{\mathbb{P}}^0$ a probability measure under which $(w_s^0)_{t \le s \le T}$ is a d-dimensional Brownian motion starting from 0 at time t with respect to the natural filtration generated by $(w_s^0, \nu_s)_{t \le s \le T}$. Otherwise, the random variable $U_t(x, \bar{\mathbb{P}}^0, \cdot)$ is useless and may be defined in an arbitrary way.*

Proof. Given $t \in [0, T]$, we consider a probability measure $\bar{\mathbb{P}}^0$ on $\bar{\Omega}^{0,t} = \mathcal{C}([t, T]; \mathbb{R}^d) \times \mathcal{D}([t, T]; \mathcal{X})$ equipped with its Borel σ-field and the canonical process $(w_s^0, \nu_s)_{t \le s \le T}$, such that $(w_s^0)_{t \le s \le T}$ is a d-dimensional Brownian motion starting from 0 with respect to the natural filtration generated by $(w_s^0, \nu_s)_{t \le s \le T}$. With a slight abuse of notation, we still denote by $\bar{\mathbb{P}}^0$ the extension of $\bar{\mathbb{P}}^0$ to the completion of the Borel σ-field. The complete and right-continuous augmentation of the canonical filtration is denoted by $\bar{\mathbb{F}}^{0,t}$, imitating the notations used in (1.16). We then construct an admissible set-up of the product form by following the procedure described at the end of Subsection 1.2.3, see for instance page 41. We let $\bar{\Omega}^{1,t} = \mathcal{C}([t,T];\mathbb{R}^d)$ and we equip it with the completion of the Wiener measure $\bar{\mathbb{P}}^1 = \mathcal{W}_d^t$ and with the complete and right-continuous filtration $\bar{\mathbb{F}}^{1,t} = (\bar{\mathcal{F}}_s^{1,t})_{t \le s \le T}$ generated by the canonical process. On the product space $\bar{\Omega}^t = \bar{\Omega}^{0,t} \times \bar{\Omega}^{1,t}$, endowed with the completion $\bar{\mathbb{P}}$ of the product measure $\bar{\mathbb{P}}^0 \otimes \bar{\mathbb{P}}^1$, we consider the filtration $\bar{\mathbb{F}}^t = (\bar{\mathcal{F}}_s^t = (\cap_{\varepsilon>0}\bar{\mathcal{F}}_{s+\varepsilon}^{0,t} \otimes \bar{\mathcal{F}}_{s+\varepsilon}^{1,t}) \vee \mathcal{N})_{t \le s \le T}$, where $\mathcal{N}$ denotes the collection of $\bar{\mathbb{P}}$ null sets.

First Step. We first notice that, for any event $C \in \bar{\mathcal{F}}_t^t$, there exists an event $C^0 \in \bar{\mathcal{F}}_t^{0,t}$ such that the symmetric difference between C and $C^0 \times \bar{\Omega}^{1,t}$ has zero probability under $\bar{\mathbb{P}}$. Consider indeed such an event C and assume without any loss of generality that it belongs to $\cap_{\varepsilon>0}\bar{\mathcal{F}}_{t+\varepsilon}^{0,t} \otimes \bar{\mathcal{F}}_{t+\varepsilon}^{1,t}$. As in the proof of Blumenthal's zero-one law, observe that C is independent of any event of the form $\bar{\Omega}^{0,t} \times D$, with $D \in \bar{\mathcal{F}}_T^{1,t}$. More generally, for any $C^0 \in \bar{\mathcal{F}}_t^{0,t}$, $C \cap (C^0 \times \bar{\Omega}^{1,t})$ is independent of any event of the same form $\bar{\Omega}^{0,t} \times D$, so that:

$$\bar{\mathbb{P}}\Big((C^0 \times D) \cap C\Big) = \bar{\mathbb{P}}\Big((C^0 \times \bar{\Omega}^{1,t}) \cap C\Big)\bar{\mathbb{P}}(\bar{\Omega}^0 \times D).$$

Then,

$$\int_{\bar{\Omega}^{0,t}} \big(\mathbb{E}^{\bar{\mathbb{P}}^1}\big[\mathbf{1}_C(\omega^0,\cdot)\mathbf{1}_D\big] - \mathbb{E}^{\bar{\mathbb{P}}^1}\big[\mathbf{1}_C(\omega^0,\cdot)\big]\mathbb{E}^{\bar{\mathbb{P}}^1}\big[\mathbf{1}_D\big]\big)\mathbf{1}_{C^0}(\omega^0)d\bar{\mathbb{P}}^0(\omega^0) = 0.$$

By Fubini's theorem, the integrand is $\bar{\mathcal{F}}_t^{0,t}$-measurable in ω^0. Therefore, for $\bar{\mathbb{P}}^0$-almost every ω^0, we have that $C_{\omega^0} = \{\omega^1 \in \bar{\Omega}^{1,t} : (\omega^0,\omega^1) \in C\}$ is independent of D. By choosing D in a countable generating π-system of the Borel σ-field on $\bar{\Omega}^{1,t}$, we deduce that, for $\bar{\mathbb{P}}^0$-almost every ω^0, C_{ω^0} is independent of $\bar{\mathcal{F}}_T^{1,t}$ (and thus of itself), that is:

$$\mathbb{E}^{\bar{\mathbb{P}}^1}\big[\mathbf{1}_C(\omega^0,\cdot)\big] = \bar{\mathbb{P}}^1\big(C_{\omega^0}\big) = 0 \text{ or } 1.$$

Now, let us define $C^0 = \{\omega^0 : \mathbb{E}^{\bar{\mathbb{P}}^1}[\mathbf{1}_C(\omega^0,\cdot)] = 1\}$. Clearly $C^0 \in \bar{\mathcal{F}}_t^{0,t}$ and

$$\begin{aligned}\bar{\mathbb{P}}(C) = \int_{\bar{\Omega}^{0,t}} \mathbb{E}^{\bar{\mathbb{P}}^1}\big[\mathbf{1}_C(\omega^0,\cdot)\big]d\bar{\mathbb{P}}^0(\omega^0) &= \int_{\bar{\Omega}^{0,t}} \mathbf{1}_{C^0}(\omega^0)\mathbb{E}^{\bar{\mathbb{P}}^1}\big[\mathbf{1}_C(\omega^0,\cdot)\big]d\bar{\mathbb{P}}^0(\omega^0) \\ &= \bar{\mathbb{P}}\big(C \cap (C^0 \times \bar{\Omega}^{1,t})\big),\end{aligned}$$

which is also equal to $\bar{\mathbb{P}}^0(C^0) = \bar{\mathbb{P}}(C^0 \times \bar{\Omega}^{1,t})$. Finally, we have:

$$\bar{\mathbb{P}}(C) = \bar{\mathbb{P}}\big(C^0 \times \bar{\Omega}^{1,t}\big) = \bar{\mathbb{P}}\big(C \cap (C^0 \times \bar{\Omega}^{1,t})\big),$$

which is the desired result.

Second Step. As an application, consider an $\bar{\mathcal{F}}_t$-measurable random variable X (with values in $\mathbb{R}$) and call $\tilde{X}$ another random variable, measurable with respect to $\cap_{\varepsilon>0}\bar{\mathcal{F}}_{t+\varepsilon}^{0,t} \otimes \bar{\mathcal{F}}_{t+\varepsilon}^{1,t}$, such that $\bar{\mathbb{P}}(X = \tilde{X}) = 1$. For such an $\tilde{X}$, let:

$$Y(\omega^0) = \int_{\bar{\Omega}^t} \tilde{X}(\omega^0,\omega^1)d\bar{\mathbb{P}}^1(\omega^1), \quad \omega^0 \in \bar{\Omega}^{0,t},$$

which is an $\bar{\mathcal{F}}_t^{0,t}$-measurable random variable. Let us now choose $C = \{(\omega^0, \omega^1) : Y(\omega^0) \le \tilde{X}(\omega^0, \omega^1)\} \in \bar{\mathcal{F}}_t^t$. Then, with the same notation as above, we have, for $\bar{\mathbb{P}}^0$-almost every ω^0, $\bar{\mathbb{P}}^1(C_{\omega^0}) \in \{0, 1\}$. Since:

$$
\begin{aligned}
0 &= \mathbb{E}^{\bar{\mathbb{P}}^1}\big[\big(\tilde{X}(\omega^0, \cdot) - Y(\omega^0)\big)\big] \\
&= \mathbb{E}^{\bar{\mathbb{P}}^1}\big[\big(\tilde{X}(\omega^0, \cdot) - Y(\omega^0)\big)\mathbf{1}_{C_{\omega^0}}\big] + \mathbb{E}^{\bar{\mathbb{P}}^1}\big[\big(\tilde{X}(\omega^0, \cdot) - Y(\omega^0)\big)\big(1 - \mathbf{1}_{C_{\omega^0}}\big)\big],
\end{aligned}
$$

we must have $\bar{\mathbb{P}}^1(C_{\omega^0}) = 1$ as otherwise we would have $0 < 0$. Regarding Y as a random variable constructed on $\bar{\Omega}^t$, we deduce that $\bar{\mathbb{P}}(Y \le \tilde{X}) = 1$ and by symmetry, that $\bar{\mathbb{P}}(Y = \tilde{X}) = 1$. In the end, we have $\bar{\mathbb{P}}(Y = X) = 1$.

Here is a typical example for Y. On the set-up formed by $(\bar{\Omega}, \bar{\mathbb{F}}, \bar{\mathbb{P}})$, the process $(w_s^0, \nu_s, w_s)_{t \le s \le T}$ is compatible with the filtration $(\bar{\mathcal{F}}_s^t)_{t \le s \le T}$ for the probability $\bar{\mathbb{P}}$. The proof is obvious since $(\bar{\mathcal{F}}_s^t)_{t \le s \le T}$ is precisely the complete and right-continuous augmentation of the filtration generated by $(w_s^0, \nu_s, w_s)_{t \le s \le T}$. Therefore, we can use this set-up in order to solve the FBSDE (1.5) with some $x \in \mathbb{R}^d$ as initial condition and with no initial information. We then call $y_t^{t,x}$ the initial value of the backward process. By construction, it is an $\bar{\mathcal{F}}_t^t$-measurable random variable and, by what we just proved, it is almost surely equal to a $\cap_{\varepsilon > 0} \bar{\mathcal{F}}_{t+\varepsilon}^{0,t}$-measurable random variable.

Third Step. We now construct U_t. As a starter, we recall that we can find a measurable mapping L from $\mathbb{R}^d \times \mathcal{X}$ into $[0, 1]$ such that $\mathcal{B}(\mathbb{R}^d \times \mathcal{X}) = \{L^{-1}(B); B \in \mathcal{B}([0, 1])\}$. See Theorem 6.5.5 in [64]. For any $n \ge 0$, we call $(A_{k,n})_{0 \le k < 2^n}$ the dyadic partition of $[0, 1]$. Then $(L^{-1}(A_{k,n}))_{0 \le k < 2^n}$ is a partition of $\mathbb{R}^d \times \mathcal{X}$ and, if we call $\mathcal{H}^n$ the σ-field generated by $(L^{-1}(A_{k,n}))_{0 \le k < 2^n}$, then, $\mathcal{H}^n \subset \mathcal{H}^{n+1}$ and $\bigvee_{n \ge 1} \mathcal{H}^n = \mathcal{B}(\mathbb{R}^d \times \mathcal{X})$.

Now, for a new integer $p \ge 1$ and for any tuple $\boldsymbol{k} = (k_1, \cdots, k_N) \in \{0, \cdots, N-1\}^N$, with $N = 2^n$, we let $C_{\boldsymbol{k},n}^p = \{(w_{t+(T-t)\ell/(pN)}^0, \nu_{t+(T-t)\ell/(pN)}) \in L^{-1}(A_{k_\ell,n}),\ \ell = 0, \cdots, N\} \subset \bar{\Omega}^{0,t}$. If we call $\mathcal{G}^{p,n}$ the σ-field generated by $(C_{\boldsymbol{k},n}^p)_{\boldsymbol{k} \in \{0, \cdots, N-1\}^N}$, then $\mathcal{G}^{p,n} \subset \mathcal{G}^{p,n+1}$ and $\bigvee_{n \ge 1} \mathcal{G}^{p,n+1} = \sigma\{w_s^0, \nu_s;\ t \le s \le t + (T-t)/p\}$. We then let:

$$
\begin{aligned}
&U_t^{n,p}\big(x, \bar{\mathbb{P}}^0, (w_s^0, \nu_s)_{t \le s \le T}\big) \\
&\quad = \sum_{\boldsymbol{k} \in \{0, \cdots, N-1\}^N} \frac{1}{\bar{\mathbb{P}}^0(C_{\boldsymbol{k},n}^p)} \bar{\mathbb{E}}\big[\mathbf{1}_{C_{\boldsymbol{k},n}^p \times \bar{\Omega}^{1,t}} y_t^{t,x}\big] \mathbf{1}_{\{\bar{\mathbb{P}}^0(C_{\boldsymbol{k},n}^p) > 0\}} \mathbf{1}_{C_{\boldsymbol{k},n}^p}\big((w_s^0, \nu_s)_{t \le s \le T}\big).
\end{aligned}
$$

We notice that the map $U_t^{n,p}$ is jointly measurable with respect to $\mathcal{B}(\mathbb{R}^d) \times \mathcal{B}(\mathcal{P}_2(\bar{\Omega}^{0,t})) \times \sigma\{(w_s^0, \nu_s);\ t \le s \le t + (T-t)/p\}$, which follows from the fact that the mapping which takes $(x, \bar{\mathbb{P}}^0)$ to the law of the solution under the input $\delta_x \otimes \bar{\mathbb{P}}^0 \otimes \mathcal{W}_d^t$ is measurable. See Proposition 1.31. Moreover, $\bar{\mathbb{P}}^0$-almost surely,

$$
U_t^{n,p}(x, \bar{\mathbb{P}}^0, \cdot) = \bar{\mathbb{E}}\big[y_t^{t,x} | \mathcal{G}^{p,n}\big].
$$

We then let:

$$
\begin{aligned}
&U_t^p\big(x, \bar{\mathbb{P}}^0, (w_s^0, \nu_s)_{t \le s \le T}\big) \\
&\quad = \begin{cases} \lim_{n \to \infty} U_t^{n,p}\big(x, \bar{\mathbb{P}}^0, (w_s^0, \nu_s)_{t \le s \le T}\big) & \text{whenever the limit exists,} \\ 0 & \text{otherwise.} \end{cases}
\end{aligned}
$$

Then, U_t^p is also jointly measurable with respect to $\mathcal{B}(\mathbb{R}^d) \times \mathcal{B}(\mathcal{P}_2(\bar{\Omega}^{0,t})) \times \sigma\{(w_s^0, \nu_s);\ t \le s \le t+(T-t)/p\}$. Moreover, $\bar{\mathbb{P}}^0$-almost surely, $U_t^{n,p}(x, \bar{\mathbb{P}}^0, \cdot)$ converges to the random variable $\bar{\mathbb{E}}[y_t^{t,x}|\sigma\{w_s^0, \nu_s;\ t \le s \le t + (T-t)/p)\}]$, which is almost surely equal to $\tilde{y}_t^{t,x}$ and thus $y_t^{t,x}$, where $\tilde{y}_t^{t,x}$ is a version of $y_t^{t,x}$ which is measurable with respect to $\cap_{\varepsilon>0}\sigma\{w_s^0, \nu_s; t \le s \le t+\varepsilon\}$.

Next we set:

$$
U_t\big(x, \bar{\mathbb{P}}^0, (w_s^0, \nu_s)_{t \le s \le T}\big) = \begin{cases} \lim_{p\to\infty} U_t^p\big(x, \bar{\mathbb{P}}^0, (w_s^0, \nu_s)_{t\le s\le T}\big) & \text{whenever the limit exists,} \\ 0 & \text{otherwise.} \end{cases}
$$

U_t is jointly measurable with respect to $\mathcal{B}(\mathbb{R}^d) \times \mathcal{B}(\mathcal{P}_2(\bar{\Omega}^{0,t})) \times \cap_{p\ge 1}\sigma\{(w_s^0, \nu_s);\ t \le s \le t + (T-t)/p\}$ and, $\bar{\mathbb{P}}^0$-almost surely, $U_t(x, \bar{\mathbb{P}}^0, \cdot)$ is equal to $y_t^{t,x}$.

By the stability property (1.19), we easily deduce that, for any two different $x, x' \in \mathbb{R}^d$, $\bar{\mathbb{P}}^0$-almost surely, $|U_t(x, \bar{\mathbb{P}}^0, \cdot) - U_t(x', \bar{\mathbb{P}}^0, \cdot)| \le C|x - x'|$, for a universal constant C. In particular, under $\bar{\mathbb{P}}^0$, we can choose a modification of each $U_t(x, \bar{\mathbb{P}}^0, \cdot)$ for $x \in \mathbb{R}^d$, such that the mapping $U_t(\cdot, \bar{\mathbb{P}}^0, \cdot)$ is C-Lipschitz continuous in x.

Fourth Step. It remains to check that the decoupling field has the required representation property. This is a consequence of the disintegration Lemma 1.43 and the uniqueness in law property. Indeed, consider a solution $(X_s, Y_s, Z_s, Z_s^0, M_s)_{t\le s\le T}$ defined on an arbitrary set-up equipped with some $(W_s^0, \mu_s, W_s)_{t\le s\le T}$ and with some $\sigma\{\theta_0\}$ as initial information, where θ_0 is a random variable taking values in an auxiliary Polish space $\mathcal{S}$.

With the same notation as in the statement of Lemma 1.43, this solution can be also transferred into a solution $(\boldsymbol{x}, \boldsymbol{y}, \boldsymbol{z}, \boldsymbol{z}^0, \boldsymbol{m})$ on the space $\Omega_{\text{input}} = \mathcal{S} \times \mathcal{C}([t,T];\mathbb{R}^d) \times \mathcal{D}([t,T];\mathcal{X}) \times \mathcal{C}([t,T];\mathbb{R}^d)$ equipped with the law $\mathbb{Q}_{\text{input}}$ of $(\theta_0, W_s^0, \mu_s, W_s)_{t\le s\le T}$. For $\mathbb{Q}_{\text{input}}$-almost every $\omega \in \Omega_{\text{input}}$, we know from Lemma 1.43 that, under the regular conditional probability $\mathbb{Q}_{\omega,\text{input}}$ of $\mathbb{Q}_{\text{input}}$ given $\sigma\{\theta_0, \nu_t\}$, the process $(\boldsymbol{x}, \boldsymbol{y}, \boldsymbol{z}, \boldsymbol{z}^0, \boldsymbol{m})$ is a solution of (1.40) with $x_t(\omega)$ as initial (deterministic) condition and no initial information, the input having $\delta_{x_t(\omega)} \otimes \mathbb{Q}^0_{\omega,\text{input}} \otimes \mathcal{W}_d^t$ as distribution, where $\mathbb{Q}^0_{\omega,\text{input}}$ denotes the law of $(w_s^0, \nu_s)_{t\le s\le T}$ under $\mathbb{Q}_{\omega,\text{input}}$. Here, with a slight abuse of notation, θ_0 also denotes the canonical mapping $\Omega_{\text{input}} \to \mathcal{S}$. By the third step, we deduce that $\mathbb{Q}_{\omega,\text{input}}[y_t = U(x_t, \mathbb{Q}^0_{\omega,\text{input}}, (w_s^0 - w_t^0, \nu_s)_{t\le s\le T})] = 1$, where we used the fact that $\mathbb{Q}_{\omega,\text{input}}[w_t^0 = 0] = 1$. Using once again the fact that $\mathbb{Q}_{\text{input}}[w_t^0 = 0] = 1$, we observe that $\Omega_{\text{input}} \ni \omega \mapsto \mathbb{Q}^0_{\omega,\text{input}}$ coincides with the law of $(w_s^0 - w_t^0, \nu_s)_{t\le s\le T}$ under $\mathbb{Q}_{\omega,\text{input}}$. Therefore, we must have, with probability 1 on the original set-up, that $Y_t = U_t(X_t, \mathcal{L}((W_s^0 - W_t^0, \mu_s)_{t\le s\le T}|\theta_0, \mu_t), (W_s^0 - W_t^0, \mu_s)_{t\le s\le T})$. □

Remark 1.48 *The reader must keep in mind the fact that Y_0 is not necessarily deterministic, even when X_0 is deterministic. Of course, it is deterministic when the σ-field $\mathcal{F}_0$ is almost surely trivial, which is the case when $\mathbb{F}$ is generated by $\boldsymbol{W}^0$ and $\boldsymbol{W}$. However, it is easy to construct examples for which $\boldsymbol{\mu}$ is not adapted to the filtration generated by $\boldsymbol{W}^0$ and $\boldsymbol{W}$, and for which Y_0 is random.*

For instance, if $d = 1$ and $\boldsymbol{X}$ has the dynamics:

$$
dX_t = \text{sign}\big(\mu_t\big)dt, \quad t \in [0,1]\ ; \quad X_0 = 0,
$$

where $\mu_t = t\mu_1$ and μ_1 is a symmetric Bernoulli random variable with values in $\{-1, 1\}$, $(\mu_t)_{0\le t\le 1}$ is a continuous process and $X_1 = \mu_1$.

Now, constructing μ_1 on the canonical space $\Omega = \{-1, 1\}$ equipped with $\mathbb{P} = \frac{1}{2}\delta_{-1} + \frac{1}{2}\delta_1$, the right continuous filtration generated by $(\mu_t)_{0\le t\le 1}$ is equal to $(\mathcal{F}_t = \sigma\{\mu_1\})_{0\le t\le 1}$. In particular, for a given Borel-measurable function $g : \mathbb{R} \to \mathbb{R}$, it holds that $\mathbb{E}[g(X_1)|\mathcal{F}_0] = g(\mu_1)$, which is obviously random. Since $\mathbb{E}[g(X_1)|\mathcal{F}_0]$ should be interpreted as Y_0, this is an example for which Y_0 is not deterministic.

Remark 1.49 *Whenever $(\mu_s)_{t\le s\le T}$ is adapted to the filtration generated by θ_0 and $(W_s^0)_{t\le s\le T}$, $U_t(x, \mathcal{L}((W_s^0, \mu_s)_{t\le s\le T}|\theta_0, \mu_t), (W_s^0, \mu_s)_{t\le s\le T})$ in the representation formula for Y_t should merely write $U_t(x, \mathcal{L}((W_s^0, \mu_s)_{t\le s\le T}|\theta_0), \theta_0)$. The reason is that the σ-field $\cap_{\varepsilon>0}\sigma\{w_s^0, \nu_s;\ t \le s \le t+\varepsilon\}$ is included in $\cap_{\varepsilon>0}\sigma\{w_s^0, \theta_0;\ t \le s \le t+\varepsilon\}$, which is almost surely equal to $\sigma\{\theta_0\}$ by Blumenthal's zero-one law.*

In practice, we shall use Proposition 1.46 in the following form.

Proposition 1.50 *Assume that $(X_s, Y_s, Z_s, Z_s^0, M_s)_{0\le s\le T}$ is a solution of (1.5) on $[0, T]$ constructed on some 0-initialized admissible set-up $(\Omega, \mathcal{F}, \mathbb{F}, \mathbb{P})$ with $(X_0, \boldsymbol{W}^0, \boldsymbol{\mu}, \boldsymbol{W})$ as input, X_0 denoting the initial condition. Then, for any time t satisfying the required conditions for any choice of $\mathcal{S}$ in the statement of Proposition 1.46, it holds that:*

$$\mathbb{P}\Big[Y_t = U_t\Big(X_t, \mathcal{L}\Big(\big(W_s^0 - W_t^0, \mu_s\big)_{t\le s\le T} \,|\, \mathcal{F}_t^{\mathrm{nat},(X_0,\boldsymbol{W}^0,\boldsymbol{\mu})}\Big),\ (W_s^0 - W_t^0, \mu_s)_{t\le s\le T}\Big)\Big] = 1.$$

Proof. We may regard $(X_s, Y_s, Z_s, Z_s^0, M_s)_{0\le s\le T}$ as a solution of (1.5) on the interval $[t, T]$ for the t-initialized admissible set-up $(\Omega, \mathcal{F}, (\mathcal{F}_s)_{t\le s\le T}, \mathbb{P})$ with $(\mathcal{G} = \sigma\{X_t\} \vee \mathcal{F}_t^{\mathrm{nat},(X_0,\boldsymbol{W},\boldsymbol{\mu},\boldsymbol{W})}, (W_s^0 - W_t^0, \mu_s, W_s - W_t)_{t\le s\le T})$ as input. Admissibility of the set-up follows from Lemma 1.37 and from the fact that the set-up equipped with $(\mathcal{F}_t^{\mathrm{nat},(X_0,\boldsymbol{W},\boldsymbol{\mu},\boldsymbol{W})}, (W_s^0 - W_t^0, \mu_s, W_s - W_t)_{t\le s\le T})$ is admissible.

Observe that, here, the initial information is generated by the random variable

$$\theta_0 = (X_0, X_t, (W_s^0, \mu_s, W_s)_{0\le s\le t}).$$

Now, Proposition 1.46 implies:

$$\mathbb{P}\Big[Y_t = U_t\Big(X_t, \mathcal{L}\Big(\big(W_s^0 - W_t^0, \mu_s\big)_{t\le s\le T} \,|\, \sigma\{X_t\} \vee \mathcal{F}_t^{\mathrm{nat},(X_0,\boldsymbol{W}^0,\boldsymbol{\mu},\boldsymbol{W})}\Big),\ (W_s^0 - W_t^0, \mu_s)_{t\le s\le T}\Big)\Big] = 1,$$

which is not exactly the required identity because the σ-field in the conditional law is not the right one. In order to complete the proof, we notice that by continuity of $\boldsymbol{X}$, $\sigma\{X_t\} \subset \bigvee_{s<t}\mathcal{F}_s$.

Also, we observe by compatibility that for any $s < t$, $C_s \in \mathcal{F}_s$, $B_t \in \mathcal{F}_t^{\mathrm{nat},(X_0,\boldsymbol{W}^0,\boldsymbol{\mu},\boldsymbol{W})}$ and $B_T \in \mathcal{F}_T^{\mathrm{nat},(\boldsymbol{W}^0,\boldsymbol{\mu})}$:

$$\begin{aligned}\mathbb{P}\big(C_s \cap B_t \cap B_T\big) &= \mathbb{E}\Big[\mathbb{P}\big(C_s|\mathcal{F}_s^{(X_0,\boldsymbol{W}^0,\boldsymbol{\mu},\boldsymbol{W})}\big)\mathbf{1}_{B_t \cap B_T}\Big] \\ &= \mathbb{E}\Big[\mathbb{P}\big(C_s|\mathcal{F}_{s+}^{\mathrm{nat},(X_0,\boldsymbol{W}^0,\boldsymbol{\mu},\boldsymbol{W})}\big)\mathbf{1}_{B_t \cap B_T}\Big],\end{aligned}$$

so that:

$$\mathbb{P}\big(C_s \cap B_T|\mathcal{F}_t^{\mathrm{nat},(X_0,\boldsymbol{W}^0,\boldsymbol{\mu},\boldsymbol{W})}\big) = \mathbb{P}\big(C_s|\mathcal{F}_{s+}^{\mathrm{nat},(X_0,\boldsymbol{W}^0,\boldsymbol{\mu},\boldsymbol{W})}\big)\mathbb{P}\big(B_T|\mathcal{F}_t^{\mathrm{nat},(X_0,\boldsymbol{W}^0,\boldsymbol{\mu},\boldsymbol{W})}\big),$$

where we use the fact that, for $s < t$, $\mathcal{F}_{s+}^{\mathrm{nat},(X_0,\boldsymbol{W}^0,\boldsymbol{\mu},\boldsymbol{W})} \subset \mathcal{F}_t^{\mathrm{nat},(X_0,\boldsymbol{W}^0,\boldsymbol{\mu},\boldsymbol{W})}$. In particular,

$$\mathbb{P}\big(C_s|\mathcal{F}_t^{\mathrm{nat},(X_0,\boldsymbol{W}^0,\boldsymbol{\mu},\boldsymbol{W})}\big) = \mathbb{P}\big(C_s|\mathcal{F}_{s+}^{\mathrm{nat},(X_0,\boldsymbol{W}^0,\boldsymbol{\mu},\boldsymbol{W})}\big),$$

and then:

$$\mathbb{P}\big(C_s \cap B_T|\mathcal{F}_t^{\mathrm{nat},(X_0,\boldsymbol{W}^0,\boldsymbol{\mu},\boldsymbol{W})}\big) = \mathbb{P}\big(C_s|\mathcal{F}_t^{\mathrm{nat},(X_0,\boldsymbol{W}^0,\boldsymbol{\mu},\boldsymbol{W})}\big)\mathbb{P}\big(B_T|\mathcal{F}_t^{\mathrm{nat},(X_0,\boldsymbol{W}^0,\boldsymbol{\mu},\boldsymbol{W})}\big),$$

which shows that $\mathcal{F}_s$ and $\mathcal{F}_T^{\mathrm{nat},(\boldsymbol{W}^0,\boldsymbol{\mu})}$ are conditionally independent given $\mathcal{F}_t^{\mathrm{nat},(X_0,\boldsymbol{W}^0,\boldsymbol{\mu},\boldsymbol{W})}$. In particular, $\bigvee_{s<t}\mathcal{F}_s$ and $\mathcal{F}_T^{\mathrm{nat},(\boldsymbol{W}^0,\boldsymbol{\mu})}$ are conditionally independent on $\mathcal{F}_t^{\mathrm{nat},(X_0,\boldsymbol{W}^0,\boldsymbol{\mu},\boldsymbol{W})}$.

This shows that the conditional law of $(\boldsymbol{W}^0, \boldsymbol{\mu})$ given $\sigma\{X_t\} \vee \mathcal{F}_t^{\mathrm{nat},(X_0,\boldsymbol{W}^0,\boldsymbol{\mu},\boldsymbol{W})}$ is the same as the conditional law of $(\boldsymbol{W}^0, \boldsymbol{\mu})$ given $\mathcal{F}_t^{\mathrm{nat},(X_0,\boldsymbol{W}^0,\boldsymbol{\mu},\boldsymbol{W})}$. By independence of $(X_0, \boldsymbol{W}^0, \boldsymbol{\mu})$ and $\boldsymbol{W}$, this is also the conditional law of $(\boldsymbol{W}^0, \boldsymbol{\mu})$ given $\mathcal{F}_t^{\mathrm{nat},(X_0,\boldsymbol{W}^0,\boldsymbol{\mu})}$. □

1.3.3 Induction Procedure

Throughout this subsection, we assume that assumption **Lipschitz FBSDE in Random Environment** is in force.

General Mechanism

As we already alluded to, one generic way for solving an FBSDE of the type (1.5) is to iterate the small time solvability Theorem 1.45. The iterative procedure consists in constructing the decoupling field by means of a backward induction.

The main steps of the induction are as follows. Assuming that T is arbitrarily fixed, we are given an admissible probabilistic set-up $(\Omega, \mathcal{F}, \mathbb{F}, \mathbb{P})$ equipped with an input process of the form $(X_0, \boldsymbol{W}^0, \boldsymbol{\mu}, \boldsymbol{W})$. From Theorem 1.45, we know that we can find a constant $\delta > 0$, only depending upon the Lipschitz constant of the coefficients, such that, for any square-integrable and $\mathcal{F}_{T-\delta}^{\mathrm{nat},(X_0,\boldsymbol{W}^0,\boldsymbol{\mu},\boldsymbol{W})}$-measurable initial condition ξ, (1.5), regarded as an FBSDE defined on the $(T-\delta)$-initialized set-up $(\Omega, \mathcal{F}, (\mathcal{F}_s)_{T-\delta \le s \le T}, \mathbb{P})$ with

$$\big(\mathcal{F}_{T-\delta}^{\mathrm{nat},(X_0,\boldsymbol{W}^0,\boldsymbol{\mu},\boldsymbol{W})}, (W_s^0 - W_{T-\delta}^0, \mu_s, W_s - W_{T-\delta})_{T-\delta \le s \le T}\big)$$

as input, has a unique solution

$$(X_s^{T-\delta,\xi,(1)}, Y_s^{T-\delta,\xi,(1)}, Z_s^{T-\delta,\xi,(1)}, Z_s^{0,T-\delta,\xi,(1)}, M_s^{T-\delta,\xi,(1)})_{T-\delta\le s\le T}$$

with ξ as initial condition. Indeed, the filtration $(\mathcal{F}_s)_{T-\delta\le s\le T}$ is compatible with the filtration generated by the process $(W_s^0 - W_{T-\delta}^0, \mu_s, W_s - W_{T-\delta})_{T-\delta\le s\le T}$ and the σ-field $\mathcal{F}_{T-\delta}^{\mathrm{nat},(X_0,\boldsymbol{W}^0,\boldsymbol{\mu},\boldsymbol{W})}$.

By Propositions 1.46 and 1.50, there exists a mapping

$$U_{T-\delta} : \mathbb{R}^d \times \mathcal{P}_2(\bar{\Omega}^{0,T-\delta}) \otimes \bar{\Omega}^{0,T-\delta} \to \mathbb{R}^m, \tag{1.25}$$

measurable with respect to the σ-field $\mathcal{B}(\mathbb{R}^d) \otimes \mathcal{B}(\mathcal{P}_2(\bar{\Omega}^{0,T-\delta})) \otimes \cap_{\varepsilon>0}\sigma\{w_s^0, \nu_s;\ T-\delta \le s \le T-\delta+\varepsilon\}$, Lipschitz continuous in $x \in \mathbb{R}^d$, uniformly in the other variables, such that, with probability one,

$$Y_{T-\delta}^{T-\delta,\xi,(1)} = U_{T-\delta}\Big(\xi, \mathcal{L}\big((W_s^0 - W_{T-\delta}^0, \mu_s)_{T-\delta\le s\le T} \,|\, \mathcal{F}_{T-\delta}^{\mathrm{nat},(X_0,\boldsymbol{W}^0,\boldsymbol{\mu})}\big),$$
$$(W_s^0 - W_{T-\delta}^0, \mu_s)_{T-\delta\le s\le T}\Big).$$

In particular, if we define $V_{T-\delta} : \mathbb{R}^d \times \Omega \to \mathbb{R}^m$ by:

$$V_{T-\delta}(x,\cdot) = U_{T-\delta}\Big(x, \mathcal{L}\big((W_s^0 - W_{T-\delta}^0, \mu_s)_{T-\delta\le s\le T} \,|\, \mathcal{F}_{T-\delta}^{\mathrm{nat},(X_0,\boldsymbol{W}^0,\boldsymbol{\mu})}\big),$$
$$(W_s^0 - W_{T-\delta}^0, \mu_s)_{T-\delta\le s\le T}\Big),$$

for $x \in \mathbb{R}^d$, then, by a standard composition argument, the mapping $V_{T-\delta}$ is $\mathcal{B}(\mathbb{R}^d)\otimes \mathcal{F}_{(T-\delta)+}^{\mathrm{nat},(X_0,\boldsymbol{W}^0,\boldsymbol{\mu})}$-measurable.

Below, we denote by $\Gamma_{T-\delta}$ the Lipschitz constant of $U_{T-\delta}$ (and thus of $V_{T-\delta}$) in x. Keep in mind that we shall not always specify the dependence upon ω in the random field $V_{T-\delta}$. As a representation formula for $Y_{T-\delta}^{T-\delta,\xi,(1)}$, we shall just write $Y_{T-\delta}^{T-\delta,\xi,(1)} = V_{T-\delta}(\xi)$.

Importantly, we claim that this representation formula remains true, with the same random field $V_{T-\delta}$, when ξ is assumed to be $\bigvee_{s<T-\delta}\mathcal{F}_s$-measurable, or more generally when ξ is independent of $\sigma\{W_s^0 - W_{T-\delta}^0, \mu_s;\ T-\delta \le s \le T\}$ conditional on $\mathcal{F}_{T-\delta}^{\mathrm{nat},(X_0,\boldsymbol{W}^0,\boldsymbol{\mu},\boldsymbol{W})}$. In that case, we must incorporate ξ as part of the initial information, so we can apply Proposition 1.46, with $\theta_0 = (X_0, \xi, (W_s^0, \mu_s, W_s)_{0\le s\le T-\delta})$ as a random variable generating the initial information. As in the proof of Proposition 1.50, the compatibility of the process $(\theta_0, (W_s^0, \mu_s, W_s)_{T-\delta\le s\le T})$ with the filtration $(\mathcal{F}_s)_{T-\delta\le s\le T}$, and more generally the admissibility of the set-up, are consequences of Lemma 1.37. What we have to prove is that this new definition of θ_0 does not change the decoupling field. Arguing as in the proof of Proposition 1.50, we can prove that the conditional law of $(W_s^0 - W_{T-\delta}^0, \mu_s)_{T-\delta\le s\le T}$ given θ_0 is the same as the conditional law of $(W_s^0 - W_{T-\delta}^0, \mu_s)_{T-\delta\le s\le T}$ given $(X_0, W_s^0, \mu_s)_{0\le s\le T-\delta}$.

The goal is now to iterate the argument and solve the FBSDE (1.5) with $T-\delta$ as terminal time, $Y_{T-\delta} = V_{T-\delta}(X_{T-\delta})$ as terminal condition at time $T-\delta$ and $(\mathcal{F}_s)_{0\le s\le T-\delta}$ as underlying filtration. This is possible in small time, that is on an interval $[T-(\delta+\delta'), T-\delta]$, for some $\delta' > 0$ depending upon L (the Lipschitz constant of the coefficients) and $\Gamma_{T-\delta}$. This follows again from Theorem 1.45 but with the slight difference that the terminal condition $V_{T-\delta}$ now depends on the whole trajectory $(X_0, \boldsymbol{W}^0, \boldsymbol{\mu})$ up until time $(T-\delta)+$, and not only on $\mu_{T-\delta}$. Intuitively, what happens is that the environment at the final time $T-\delta$ is the whole trajectory $(X_0, \boldsymbol{W}^0, \boldsymbol{\mu})$. The reader may check that Theorem 1.45 still applies in this more general setting. Notice that things would have been obvious had the terminal condition depended on the whole path $(X_0, \boldsymbol{W}^0, \boldsymbol{\mu})$ up until time $(T-\delta)$ only. One way is to see the whole path $(X_0, \boldsymbol{W}^0, \boldsymbol{\mu})$ as the environment at the terminal time, and to rewrite the equation on $[T-(\delta+\delta'), T]$ instead of $[T-(\delta+\delta'), T-\delta]$, with $B(s,\cdot)$ replaced by $B(s,\cdot)\mathbf{1}_{[T-(\delta+\delta'),T-\delta')}(s)$ and the same for F, Σ and Σ^0. The equation is then solved on $[T-(\delta+\delta'), T]$ and not on $[T-(\delta+\delta'), T-\delta]$. Since the coefficients are null between $T-\delta$ and T, there is no difficulty for applying Theorem 1.45 when δ' is small enough.

Given a square-integrable and $\mathcal{F}^{\mathrm{nat},(X_0,\boldsymbol{W}^0,\boldsymbol{\mu},\boldsymbol{W})}_{T-(\delta+\delta')}$-measurable random variable ξ, (1.5) has a unique solution, with ξ as initial condition, $\mathcal{F}^{\mathrm{nat},(X_0,\boldsymbol{W}^0,\boldsymbol{\mu},\boldsymbol{W})}_{T-(\delta+\delta')}$ as initial information and

$$Y_T = V_{T-\delta}\big(X_{T-\delta}\big)$$

as terminal condition at time T, on the $(T-(\delta+\delta'))$-initialized set-up

$$\big(\Omega, \mathcal{F}, (\mathcal{F}_s)_{T-(\delta+\delta')\le s\le T}, \mathbb{P}\big)$$

equipped with

$$\big(\mathcal{F}^{\mathrm{nat},(X_0,\boldsymbol{W}^0,\boldsymbol{\mu},\boldsymbol{W})}_{T-(\delta+\delta')}, (W^0_s - W^0_{T-(\delta+\delta')}, \mu_s, W_s - W_{T-(\delta+\delta')}\big)_{T-(\delta+\delta')\le s\le T}\big).$$

Taking the conditional expectation given $\mathcal{F}_{T-\delta}$ in the backward equation of (1.5), we get, as required, $Y_{T-\delta} = V_{T-\delta}(X_{T-\delta})$, since the coefficients in the backward equation are null between $T-\delta$ and T. The solution is denoted by:

$$\Big(X_s^{T-(\delta+\delta'),\xi,(2)}, Y_s^{T-(\delta+\delta'),\xi,(2)}, Z_s^{T-(\delta+\delta'),\xi,(2)},$$
$$Z_s^{0,T-(\delta+\delta'),\xi,(2)}, M_s^{T-(\delta+\delta'),\xi,(2)}\Big)_{T-(\delta+\delta')\le s\le T}.$$

Notice that the process $(Y_s^{T-(\delta+\delta'),\xi,(2)}, M_s^{T-(\delta+\delta'),\xi,(2)})_{T-(\delta+\delta')\le s\le T}$ has trajectories in $\mathcal{D}([T-(\delta+\delta'), T]; \mathbb{R}^m \times \mathbb{R}^m)$, but the restriction of the trajectories to $[T-(\delta+\delta'), T-\delta]$ are not in $\mathcal{D}([T-(\delta+\delta'), T-\delta]; \mathbb{R}^m \times \mathbb{R}^m)$. Indeed we usually assume that paths in $\mathcal{D}([a,b]; \mathbb{R}^m \times \mathbb{R}^m)$ are left-continuous at terminal time b.

We now glue the solution constructed on $[T-(\delta+\delta'), T-\delta]$ with the solution previously constructed on $[T-\delta, T]$. At time $T-\delta$, we consider $X_{T-\delta}^{T-(\delta+\delta'),\xi,(2)}$ as initial condition for the problem set over $[T-\delta, T]$ with $\sigma\{X_{T-\delta}^{T-(\delta+\delta'),\xi,(2)}\} \vee \mathcal{F}_{T-\delta}^{\mathrm{nat},(X_0,\boldsymbol{W}^0,\boldsymbol{\mu},\boldsymbol{W})}$ as initial information. The solution reads:

$$\Big(X_s^{T-\delta,X_{T-\delta}^{T-(\delta+\delta'),\xi,(2)},(1)}, Y_s^{T-\delta,X_{T-\delta}^{T-(\delta+\delta'),\xi,(2)},(1)}, Z_s^{T-\delta,X_{T-\delta}^{T-(\delta+\delta'),\xi,(2)},(1)},$$
$$Z_s^{0,T-\delta,X_{T-\delta}^{T-(\delta+\delta'),\xi,(2)},(1)}, M_s^{T-\delta,X_{T-\delta}^{T-(\delta+\delta'),\xi,(2)},(1)}\Big)_{T-\delta\le s\le T}.$$

By continuity of the paths of $\boldsymbol{X}$, $X_{T-\delta}^{T-(\delta+\delta'),\xi,(2)}$ is $\bigvee_{s<T-\delta}\mathcal{F}_s$-measurable and by definition of the decoupling field at time $T-\delta$, we must have:

$$\begin{aligned} Y_{T-\delta}^{T-\delta,X_{T-\delta}^{T-(\delta+\delta'),\xi,(2)},(1)} &= \mathcal{V}_{T-\delta}\Big(X_{T-\delta}^{T-\delta,X_{T-\delta}^{T-(\delta+\delta'),\xi,(2)},(1)}\Big) \\ &= \mathcal{V}_{T-\delta}\Big(X_{T-\delta}^{T-(\delta+\delta'),\xi,(2)}\Big) \\ &= Y_{T-\delta}^{T-(\delta+\delta'),\xi,(2)}. \end{aligned}$$

Therefore, letting

$$X_s^{T-(\delta+\delta'),\xi} = \begin{cases} X_s^{T-(\delta+\delta'),\xi,(2)} & \text{if } s\in[T-(\delta+\delta'), T-\delta], \\ X_s^{T-\delta,X_{T-\delta}^{T-(\delta+\delta'),\xi,(2)},(1)} & \text{if } s\in(T-\delta, T], \end{cases}$$

with similar definitions for $\boldsymbol{Y}, \boldsymbol{Z}$, and $\boldsymbol{Z}^0$ with $\boldsymbol{X}$ replaced by $\boldsymbol{Y}, \boldsymbol{Z}$ and $\boldsymbol{Z}^0$ respectively, if we set:

$$M_s^{T-(\delta+\delta'),\xi} = \begin{cases} M_s^{T-(\delta+\delta'),\xi,(2)} & \text{if } s\in[T-(\delta+\delta'), T-\delta], \\ M_s^{T-\delta,X_{T-\delta}^{T-(\delta+\delta'),\xi,(2)},(1)} + M_{T-\delta}^{T-(\delta+\delta'),\xi,(2)} & \text{if } s\in(T-\delta, T], \end{cases}$$

then we get that

$$(X_s^{T-(\delta+\delta'),\xi}, Y_s^{T-(\delta+\delta'),\xi}, Z_s^{T-(\delta+\delta'),\xi},$$
$$Z_s^{0,T-(\delta+\delta'),\xi}, M_s^{T-(\delta+\delta'),\xi})_{T-(\delta+\delta')\le s\le T}$$

is a solution of (1.5) on $[T-(\delta+\delta'), T]$, with $X_{T-(\delta+\delta')} = \xi$ as initial condition. As already explained, the argument extends to the case when ξ is $\bigvee_{s<T-(\delta+\delta')}\mathcal{F}_s$-measurable choosing $(X_0, \xi, (W_s^0, \mu_s, W_s)_{0\le s\le T-(\delta+\delta')})$ as initial information.

We emphasize that the solution we constructed with ξ as initial condition at time $T-(\delta+\delta')$ and with the corresponding initial information must be unique. Indeed, any other solution $(\boldsymbol{X}', \boldsymbol{Y}', \boldsymbol{Z}', \boldsymbol{Z}^{0\prime}, \boldsymbol{M}')$ must also satisfy $Y'_{T-\delta} = \mathcal{V}_{T-\delta}(X'_{T-\delta})$.

Uniqueness is then proved in two steps: first on $[T-(\delta+\delta'), T-\delta]$ and then on $[T-\delta, T]$. We refer to Subsection (Vol I)-4.1.2 for a complete account in the standard case without random environment.

So, we proved that, for any square-integrable and $\mathcal{F}_{T-(\delta+\delta')}^{\mathrm{nat},(X_0,\boldsymbol{W}^0,\boldsymbol{\mu},\boldsymbol{W})}$-measurable random variable ξ, the FBSDE (1.5) has a unique solution on $[T-(\delta+\delta'), T]$ with ξ as initial condition at time $T-(\delta+\delta')$, $\mathcal{F}_{T-(\delta+\delta')}^{\mathrm{nat},(X_0,\boldsymbol{W}^0,\boldsymbol{\mu},\boldsymbol{W})}$ as initial information and

$$Y_T = G\big(X_T, \mu_T\big)$$

as terminal condition at time T, on the $(T-(\delta+\delta'))$-initialized set-up $(\Omega, \mathcal{F}, (\mathcal{F}_s)_{T-(\delta+\delta')\le s\le T}, \mathbb{P})$ equipped with

$$\big(\mathcal{F}_{T-(\delta+\delta')}^{\mathrm{nat},(X_0,\boldsymbol{W}^0,\boldsymbol{\mu},\boldsymbol{W})}, (W_s^0 - W_{T-(\delta+\delta')}^0, \mu_s, W_s - W_{T-(\delta+\delta')})\big)_{T-(\delta+\delta')\le s\le T}.$$

Also, the same argument shows that, for any $x \in \mathbb{R}^d$, the FBSDE (1.5) has a unique solution on $[T-(\delta+\delta'), T]$, with x as initial condition on any $T-(\delta+\delta')$-initialized admissible set-up with no initial information. By combining the two stability estimates in Theorem 1.45 on the intervals $[T-(\delta+\delta'), T-\delta]$ and $[T-(\delta+\delta'), T]$, we can even prove that there exists a constant $\Gamma^{(2)}$ such that, for any two initial conditions $x, x' \in \mathbb{R}^d$, the corresponding solutions $(\boldsymbol{X}, \boldsymbol{Y}, \boldsymbol{Z}, \boldsymbol{Z}^0, \boldsymbol{M})$ and $(\boldsymbol{X}', \boldsymbol{Y}', \boldsymbol{Z}', \boldsymbol{Z}^{0\prime}, \boldsymbol{M}')$ on the same $T-(\delta+\delta')$-initialized set-up with no initial information satisfy $|Y_{T-(\delta+\delta')} - Y'_{T-(\delta+\delta')}| \le \Gamma^{(2)}|x-x'|$ with probability 1. The proof is deferred to the next paragraph, where we provide a more general stability property, see the statement of Theorem 1.53. Hence, by Proposition 1.46, we can construct the analogue $U_{T-(\delta+\delta')}$ of $U_{T-\delta}$ in 1.25 but at time $T-(\delta+\delta')$ instead of $T-\delta$. This permits to repeat the argument on a third interval of the form $[T-(\delta+\delta'+\delta''), T-(\delta+\delta')]$.

In order to iterate the argument, we need to control the Lipschitz constant of the decoupling field in x, namely the Lipschitz constant of U_t in Proposition 1.46, uniformly in the other variables, and in time. Indeed, if the Lipschitz constant blows up in finite time, the intervals on which existence and uniqueness hold will become smaller and smaller, and their concatenation may not cover the entire time interval $[0, T]$. Again, we refer to Subsection (Vol I)-4.1.2 for a complete account in the standard case without random environment.

Mimicking Proposition (Vol I)-4.8 and Lemma (Vol I)-4.9, we introduce the following assumption:

Assumption (Iteration in Random Environment). There exists a constant $\Gamma_0 \ge 0$, such that, if $t \in [0, T]$, $(\Omega, \mathcal{F}, (\mathcal{F}_s)_{t\le s\le T}, \mathbb{P})$ is a t-initialized admissible set-up equipped with some $(W_s^0, \mu_s, W_s)_{t\le s\le T}$ and no initial infor-

(continued)

mation, and $x, x' \in \mathbb{R}^d$, then for any two solutions $(X_s, Y_s, Z_s, Z_s^0, M_s)_{t \le s \le T}$ and $(X'_s, Y'_s, Z'_s, Z_s^{0\prime}, M'_s)_{t \le s \le T}$ of the FBSDE (1.5) with x and x' as respective initial condition at time t, it holds that:

$$\mathbb{P}\big[|Y_t - Y'_t| \le \Gamma_0 |x - x'|\big] = 1.$$

Remark 1.51 *Alternatively, assumption* **Iteration in Random Environment** *may be written by replacing "for any two solutions" by "there exist two solutions." The reason is that existence and uniqueness of solutions are proved simultaneously along the induction procedure introduced right above.*

Proposition 1.52 *If assumptions* **Lipschitz FBSDE in Random Environment** *and* **Iteration in Random Environment** *hold true, then, for any $t \in [0, T]$, for any t-initialized admissible set-up $(\Omega, \mathcal{F}, (\mathcal{F}_s)_{t \le s \le T}, \mathbb{P})$ equipped with some $(\mathcal{G} = \sigma\{\theta_0\}, (W_s^0, \mu_s, W_s)_{t \le s \le T})$ for some random variable θ_0 with values in an auxiliary Polish space $\mathcal{S}$, and for any $\xi \in L^2(\Omega, \mathcal{G}, \mathbb{P}; \mathbb{R}^d)$, the FBSDE set on $[t, T]$ with ξ as initial condition has a unique solution. Moreover, the decoupling field given by Proposition 1.46 is Γ_0-Lipschitz.*

In particular, on a standard (0-initialized) admissible set-up $(\Omega, \mathcal{F}, \mathbb{F}, \mathbb{P})$ equipped with some $(X_0, \boldsymbol{W}^0, \boldsymbol{\mu}, \boldsymbol{W})$, the FBSDE set on $[0, T]$ with X_0 as initial condition and initial information, is uniquely solvable. The solution is progressively measurable with respect to the (augmentation) of the filtration generated by the process $(X_0, \boldsymbol{W}^0, \boldsymbol{\mu}, \boldsymbol{W})$ and it satisfies the representation formula in Proposition 1.50.

Proof. Since the proof is similar to the proof of Proposition (Vol I)-4.8, we only sketch its main steps. The reader may also want to consult the proof of Theorem 1.53 below on the stability of solution, which somehow covers Proposition 1.52.

The existence and uniqueness of a solution are proved by using the induction principle explained above. At each step of the induction, we know from assumption **Iteration in Random Environment** that U_t is Γ_0-Lipschitz continuous in x. We use the fact that, for each $\bar{\mathbb{P}}^0$ on $\bar{\Omega}^{0,t} = \mathcal{C}([t, T]; \mathbb{R}^d) \times \mathcal{D}([t, T]; \mathcal{X})$ under which the process $(w_s^0)_{t \le s \le T}$ is a d-dimensional Brownian motion starting from 0 with respect to the natural filtration generated by the canonical process $(w_s^0, \nu_s)_{t \le s \le T}$, we can construct a t-initialized admissible set-up with no initial information under which $(\boldsymbol{W}^0, \boldsymbol{\mu})$ has $\bar{\mathbb{P}}^0$ as distribution, see the proof of Proposition 1.46. This suffices to construct the decoupling field at all the probability measures $\bar{\mathbb{P}}^0$ that are useful, see Remark 1.47. The fact that $\mathcal{G}$ is generated by θ_0 permits to identify the initial information at each step of the induction with a σ-algebra generated by a random variable taking values in a Polish space. □

Stability Property

Quite remarkably, assumptions **Lipschitz FBSDE** and **Iteration in Random Environment** guarantee not only the existence and uniqueness of solutions, but also stability of these solutions provided that they are defined on the same probabilistic set-up with a *common input*. Here stability is understood as an estimate for the difference of two solutions, defined on the same space $(\Omega, \mathcal{F}, \mathbb{F}, \mathbb{P})$ but with respect to two different initial conditions X_0 and X'_0, two different environments $\boldsymbol{\mu}$ and $\boldsymbol{\mu}'$ and two sets of coefficients $(B, \Sigma, \Sigma^0, F, G)$ and $(B', \Sigma', \Sigma^{0\prime}, F', G')$, exactly as in the short time stability result of Theorem 1.45. Requiring both solutions to be defined on the same probabilistic set-up with a *common input* is to ask the initial conditions X_0 and X'_0 to be measurable with respect to a common information and the environments $\boldsymbol{\mu}$ and $\boldsymbol{\mu}'$ to derive from a common *super-environment*. Together with the noises $\boldsymbol{W}^0$ and $\boldsymbol{W}$, the common initial information and common *super-environment* form what we call a *common input*.

The common initial information will be given in the form of a σ-field $\mathcal{G} = \sigma\{\theta_0\}$ generated by a random variable θ_0 taking values in an auxiliary Polish space $\mathcal{S}$. As for the super-environment, we shall consider a process $\underline{\mathfrak{M}} = (\mathfrak{M}_t)_{0 \le t \le T}$ with realizations in $\mathcal{D}([0,T]; \mathcal{Y})$ for another Polish space $\mathcal{Y}$ such that $(\Omega, \mathcal{F}, \mathbb{F}, \mathbb{P})$ equipped with $(\mathcal{G} = \sigma\{\theta_0\}, \boldsymbol{W}^0, \underline{\mathfrak{M}}, \boldsymbol{W})$ is admissible. Processes $\boldsymbol{\mu}$ and $\boldsymbol{\mu}'$ will be regarded as *sub-environments*. We shall consider two continuous mappings ψ and ψ' from $\mathcal{Y}$ to $\mathcal{X}$ such that $\boldsymbol{\mu} = (\mu_t = \psi(\mathfrak{M}_t))_{0 \le t \le T}$ and $\boldsymbol{\mu}' = (\mu'_t = \psi'(\mathfrak{M}_t))_{0 \le t \le T}$. We require ψ and ψ' to be at most of linear growth in the following sense: denoting by $d_{\mathcal{Y}}$ the distance on $\mathcal{Y}$ and $d_{\mathcal{X}}$ the distance on $\mathcal{X}$, we impose that, for some $C \ge 0$, for all $\mathfrak{m} \in \mathcal{Y}$,

$$d_{\mathcal{X}}\big(0_{\mathcal{X}}, \psi(\mathfrak{m})\big) \le C d_{\mathcal{Y}}\big(0_{\mathcal{Y}}, \mathfrak{m}\big), \quad d_{\mathcal{X}}\big(0_{\mathcal{X}}, \psi'(\mathfrak{m})\big) \le C d_{\mathcal{Y}}\big(0_{\mathcal{Y}}, \mathfrak{m}\big),$$

where $0_{\mathcal{X}}$ and $0_{\mathcal{Y}}$ are arbitrary points in $\mathcal{X}$ and $\mathcal{Y}$. Since $\mathbb{E}[\sup_{0 \le t \le T} d_{\mathcal{Y}}(0_{\mathcal{Y}}, \mathfrak{M}_t)^2]$ is finite, we also have that $\mathbb{E}[\sup_{0 \le t \le T} d_{\mathcal{X}}(0_{\mathcal{X}}, \mu_t)^2]$ and $\mathbb{E}[\sup_{0 \le t \le T} d_{\mathcal{X}}(0_{\mathcal{X}}, \mu'_t)^2]$ are finite.

With this description of the *common input*, we may regard the FBSDEs (1.5), driven by the environments $\boldsymbol{\mu}$ and $\boldsymbol{\mu}'$ and by the sets of coefficients $(B, \Sigma, \Sigma^0, F, G)$ and $(B', \Sigma', \Sigma^{0\prime}, F', G')$ respectively, as FBSDEs driven by the same environment $\underline{\mathfrak{M}}$ and by the sets of coefficients defined by the formula:

$$D\big(\cdot, \cdot, \psi(\cdot), \cdot, \cdot, \cdot\big)(t, x, \mathfrak{m}, y, z, z^0) = D(t, x, \mu, y, z, z^0)$$

for $D \in \{B, \Sigma, \Sigma^0, F, G\}$, $(t, x, \mathfrak{m}, y, (z, z^0)) \in [0,T] \times \mathbb{R}^d \times \mathcal{Y} \times \mathbb{R}^m \times \mathbb{R}^{2(m \times d)}$ and with $\mu = \psi(\mathfrak{m})$, and similarly for the other FBSDE driven by the coefficients $(B', \Sigma', \Sigma^{0\prime}, F', G')$. Below, we shall use the notation $(B, \Sigma, \Sigma^0, F, G) \circ \psi$ and $(B', \Sigma', \Sigma^{0\prime}, F', G') \circ \psi'$ to denote these new sets of coefficients.

It is clear that if the original set of coefficients $(B, \Sigma, \Sigma^0, F, G)$ satisfies assumption **Lipschitz FBSDE in Random Environment**, the same is true for the new set of coefficients $(B, \Sigma, \Sigma^0, F, G) \circ \psi$, and similarly for the coefficients labeled with a 'prime'. Also, it makes perfect sense to require the new set of coefficients $(B, \Sigma, \Sigma^0, F, G) \circ \psi$ to satisfy assumption **Iteration in Random Environment**, and similarly for the coefficients labeled with a "prime."

Theorem 1.53 *Assume that coefficients* $(B, \Sigma, \Sigma^0, F, G)$ *and* $(B', \Sigma', \Sigma^{0\prime}, F', G')$ *satisfy assumption* **Lipschitz FBSDE in Random Environment** *with the same constant L, and that the new ones* $(B, \Sigma, \Sigma^0, F, G) \circ \psi$ *and* $(B', \Sigma', \Sigma^{0\prime}, F', G') \circ \psi'$ *satisfy assumption* **Iteration in Random Environment** *with the same constant* Γ_0. *On a complete filtered probability space* $(\Omega, \mathcal{F}, \mathbb{F}, \mathbb{P})$, *consider a random variable* θ_0 *with values in the auxiliary Polish space* $\mathcal{S}$ *together with a super-environment* $\underline{\mathfrak{M}} = (\mathfrak{M}_t)_{0 \le t \le T}$ *with realizations in* $\mathcal{D}([0, T]; \mathcal{Y})$ *such that* $(\Omega, \mathcal{F}, \mathbb{F}, \mathbb{P})$ *equipped with* $(\mathcal{G} = \sigma\{\theta_0\}, \boldsymbol{W}^0, \underline{\mathfrak{M}}, \boldsymbol{W})$ *is admissible.*

Then, there exists a constant $\Gamma \ge 0$, *only depending on T, L and* Γ_0, *such that for any initial conditions* $X_0, X_0' \in L^2(\Omega, \mathcal{G}, \mathbb{P}; \mathbb{R}^d)$, *the unique solutions to the FBSDEs* (1.5), *when driven by the new coefficients and regarded with the input* $\underline{\mathfrak{M}}$, *satisfy the stability estimate* (1.19).

Proof. Throughout the proof, we consider the admissible set-up $(\Omega, \mathcal{F}, (\mathcal{F}_t)_{0 \le t \le T}, \mathbb{P})$ equipped with some $(\mathcal{G} = \sigma\{\theta_0\}, \boldsymbol{W}^0, \underline{\mathfrak{M}}, \boldsymbol{W})$, as given in the statement.

First Step. With each of the two sets of coefficients and the common super-environment, we associate a decoupling field along the lines of Proposition 1.52. The two decoupling fields are denoted by $(U_t)_{0 \le t \le T}$ and $(U_t')_{0 \le t \le T}$ respectively. For any $t \in [0, T]$, U_t and U_t' are mappings from $\mathbb{R}^d \times \mathcal{P}_2(\mathcal{C}([t, T]; \mathbb{R}^d) \times \mathcal{D}([t, T]; \mathcal{Y})) \times (\mathcal{C}([t, T]; \mathbb{R}^d) \times \mathcal{D}([t, T]; \mathcal{Y}))$ into $\mathbb{R}^m$. Both are constructed with respect to the super-environment $\underline{\mathfrak{M}}$, in the sense that $\underline{\mathfrak{M}}$ here plays the role played by $\boldsymbol{\mu}$ in the statement of Proposition 1.46. The decoupling field $(U_t)_{0 \le t \le T}$ is associated with the coefficients $(B, \Sigma, \Sigma^0, F, G) \circ \psi$, while $(U_t')_{0 \le t \le T}$ is associated with the coefficients $(B', \Sigma', \Sigma^{0\prime}, F', G') \circ \psi'$. In order to fit the framework of Propositions 1.46 and 1.52, we reformulate each problem in the following way: $(U_t)_{0 \le t \le T}$ is associated with the super-environment $\underline{\mathfrak{M}}$ and with coefficients $(B, \Sigma, \Sigma^0, F, G) \circ \psi$, and similarly for $(U_t')_{0 \le t \le T}$.

Using the notation used in the introductory description of the induction procedure under the heading *general mechanism* at the very beginning of this section, we let for any $t \in [0, T]$, $V_t : \mathbb{R}^d \times \Omega \to \mathbb{R}^m$ be the random field defined by:

$$V_t(x, \cdot) = U_t\Big(x, \mathcal{L}\big((W_s^0 - W_t^0, \mathfrak{M}_s)_{t \le s \le T} \,|\, \mathcal{F}_t^{\mathrm{nat},(\theta_0, \boldsymbol{W}^0, \underline{\mathfrak{M}})}\big), (W_s^0 - W_t^0, \mathfrak{M}_s)_{t \le s \le T}\Big),$$

for $x \in \mathbb{R}^d$, with a similar definition for V_t'.

Second Step. We proceed by induction. Given the constant Γ_0 in assumption **Iteration in Random Environment**, we consider the same constants c and Γ as those provided by the statement of Theorem 1.45 when applied with coefficients that satisfy assumption **Lipschitz FBSDE in Random Environment** with $\max(L, \Gamma_0)$ as Lipschitz constant.

For $t \in [0, T]$ we denote by $\mathscr{H}_t$ the following property: for any two random variables ξ and ξ' in $L^2(\Omega, \bigvee_{s<t} \mathcal{F}_s, \mathbb{P}; \mathbb{R}^d)$, the solutions $(X_s, Y_s, Z_s, Z_s^0, M_s)_{t \le s \le T}$ and $(X_s', Y_s', Z_s', Z_s^{0\prime}, M_s')_{t \le s \le T}$ to (1.5) with ξ and ξ' as respective initial conditions, with $(\mu_s)_{t \le s \le T}$ and $(\mu_s')_{t \le s \le T}$ as respective sub-environments and with $(B, \Sigma, \Sigma^0, F, G)$ and $(B', \Sigma', \Sigma^{0\prime}, F', G')$ as respective coefficients, satisfy:

$$
\begin{aligned}
&\mathbb{E}\bigg[\sup_{t\le s\le T}\big(|X_s-X'_s|^2+|Y_s-Y'_s|^2+|M_s-M'_s|^2\big)\\
&\qquad+\int_t^T\big(|Z_s-Z'_s|^2+|Z^0_s-Z^{0\prime}_s|^2\big)ds\,\Big|\,\mathcal{F}_t\bigg]\\
&\le \Gamma(t)\,\mathbb{E}\bigg[|\xi-\xi'|^2+\big|G(X_T,\mu_T)-G'(X_T,\mu'_T)\big|^2\\
&\qquad+\int_t^T\big|(B,F,\Sigma,\Sigma^0)\big(s,X_s,\mu_s,Y_s,Z_s,Z^0_s\big)\\
&\qquad\qquad-(B',F',\Sigma',\Sigma^{0\prime})\big(s,X_s,\mu'_s,Y_s,Z_s,Z^0_s\big)\big|^2ds\,\Big|\,\mathcal{F}_t\bigg],
\end{aligned}
\tag{1.26}
$$

for some constant $\Gamma(t)$ only depending upon Γ_0, L, T and t, both FBSDEs being regarded on the t-initialized set-up $(\Omega,\mathcal{F},(\mathcal{F}_s)_{t\le s\le T},\mathbb{P})$ with input $(\mathcal{F}_t^{\mathrm{nat},(\theta_0,\xi,\xi',W^0,\mathfrak{M},W)},(W^0_s-W^0_t,\mathfrak{M}_s,W_s-W_t)_{t\le s\le T})$. Recall from Lemma 1.37 that such a set-up is admissible and from the proof of Proposition 1.50 that the additional presences of ξ and ξ' does not affect the form of the decoupling fields.

Below, we make use of (1.26) in order to compare the decoupling fields U_t and U'_t at times $t=ck$ for integers k.

As often in the book, we use the following notation: for a given $t\in[0,T]$, on the same t-initialized set-up $(\Omega,\mathcal{F},(\mathcal{F}_s)_{t\le s\le T},\mathbb{P})$ with $(\theta_0,\xi,(W^0_s,\mathfrak{M}_s,W_s)_{0\le s\le t})$ as initial information, we denote by $(X^{t,\xi}_s,Y^{t,\xi}_s,Z^{t,\xi}_s,Z^{0,t,\xi}_s,M^{t,\xi}_s)_{t\le s\le T}$ and $(X^{t,\xi,\prime}_s,Y^{t,\xi,\prime}_s,Z^{t,\xi,\prime}_s,Z^{0,t,\xi,\prime}_s,M^{t,\xi,\prime}_s)_{t\le s\le T}$ the two solutions corresponding to the two sets of coefficients and to the same initial condition ξ. The random variable $Y^{t,\xi}_t-Y^{t,\xi,\prime}_t$ being $\mathcal{F}_t$-measurable, (1.26) implies:

$$
\begin{aligned}
|Y^{t,\xi}_t-Y^{t,\xi,\prime}_t|^2 &\le \Gamma(t)\,\mathbb{E}\bigg[\big|G(X^{t,\xi}_T,\mu_T)-G'(X^{t,\xi}_T,\mu'_T)\big|^2\\
&\qquad+\int_t^T\big|(B,F,\Sigma,\Sigma^0)\big(s,X^{t,\xi}_s,\mu_s,Y^{t,\xi}_s,Z^{t,\xi}_s,Z^{0,t,\xi}_s\big)\\
&\qquad\qquad-(B',F',\Sigma',\Sigma^{0\prime})\big(s,X^{t,\xi}_s,\mu'_s,Y^{t,\xi}_s,Z^{t,\xi}_s,Z^{0,t,\xi}_s\big)\big|^2ds\,\Big|\,\mathcal{F}_t\bigg],
\end{aligned}
$$

from which we get:

$$
\begin{aligned}
|V_t(\xi)-V'_t(\xi)|^2 &\le \Gamma(t)\,\mathbb{E}\bigg[\big|G(X^{t,\xi}_T,\mu_T)-G'(X^{t,\xi}_T,\mu'_T)\big|^2\\
&\qquad+\int_t^T\big|(B,F,\Sigma,\Sigma^0)\big(s,X^{t,\xi}_s,\mu_s,Y^{t,\xi}_s,Z^{t,\xi}_s,Z^{0,t,\xi}_s\big)\\
&\qquad\qquad-(B',F',\Sigma',\Sigma^{0\prime})\big(s,X^{t,\xi}_s,\mu'_s,Y^{t,\xi}_s,Z^{t,\xi}_s,Z^{0,t,\xi}_s\big)\big|^2ds\,\Big|\,\mathcal{F}_t\bigg].
\end{aligned}
\tag{1.27}
$$

Third Step. We now prove the desired stability estimate by backward induction. Without any loss of generality, we assume that there exists an integer $N\ge 1$ such that $T=cN$, and we prove $\mathscr{H}_{kc}$ for $k=N-1,\cdots,1,0$ by induction. When $k=N-1$, the property

$\mathscr{H}_{(N-1)c}$ follows from Theorem 1.45, applied on the time interval $[T-c,T]$ with the input $(\theta_0,\xi,\xi',(W_s^0,\mathfrak{M}_s,W_s)_{0\le s\le T-c})$. Next, we prove that for any $k\in\{0,\cdots,N-2\}$, $\mathscr{H}_{(k+1)c}$ implies $\mathscr{H}_{kc}$. Given $k\in\{0,\cdots,N-2\}$ for which $\mathscr{H}_{(k+1)c}$ holds, we set $t=kc$, $S=(k+1)c$, and we denote by $(X_s,Y_s,Z_s,Z_s^0,M_s)_{t\le s\le T}$ and $(X_s',Y_s',Z_s',Z_s^{0\prime},M_s')_{t\le s\le T}$ the solutions corresponding to two initial conditions ξ and ξ' at time t. Our goal is to make use of (1.27) in order to prove $\mathscr{H}_{kc}$. By Proposition 1.50, we know that $\mathbb{P}$-almost surely:

$$Y_S=V_S(X_S),\quad Y_S'=V_S'(X_S'),$$

with

$$V_S(x,\cdot)=U_S\Big(x,\mathcal{L}((W_s^0-W_S^0,\mathfrak{M}_s)_{S\le s\le T}\,|\,\mathcal{F}_S^{\mathrm{nat},(\theta_0,\mathbf{W}^0,\mathfrak{M})}),(W_s^0-W_S^0,\mu_s)_{S\le s\le T}\Big),$$

where U_S is associated with the coefficients (B,F,Σ,Σ^0,G) through Proposition 1.50, and similarly for V_S'.

Then, we regard $(X_{s\wedge S},Y_{s\wedge S},Z_s\mathbf{1}_{s<S},Z_s^0\mathbf{1}_{s<S},M_{s\wedge S})_{t\le s\le T}$ as the solution of the FBSDE (1.5) with coefficients:

$$\Big(\mathbf{1}_{[t,S]}B(s,\cdot,\cdot,\cdot,\cdot,\cdot),\mathbf{1}_{[t,S]}\Sigma(s,\cdot,\cdot),\mathbf{1}_{[t,S]}\Sigma^0(s,\cdot,\cdot),\mathbf{1}_{[t,S]}F(s,\cdot,\cdot,\cdot,\cdot,\cdot)\Big),$$

instead of (B,Σ,Σ^0,F) and with the random field $V_S(\cdot)$ instead of $G(\cdot,\mu_T)$ as terminal condition, but with the same difference as above regarding the fact that the randomness in the random field V_S has a more intricate structure. Proceeding in the same way with the process $(X'_{s\wedge S},Y'_{s\wedge S},Z_s'\mathbf{1}_{s<S},Z_s^{0\prime}\mathbf{1}_{s<S},M'_{s\wedge S})_{t\le s\le T}$, Theorem 1.45, applied with the tuple $(\theta_0,\xi,\xi',(W_s^0,\mathfrak{M}_s,W_s)_{0\le s\le t})$ as input at time t, shows that, for Γ only depending on L and Γ_0,

$$\begin{aligned}
&\mathbb{E}\bigg[\sup_{t\le s\le S}\big(|X_s-X_s'|^2+|Y_s-Y_s'|^2+|M_s-M_s'|^2\big)\\
&\qquad+\int_t^S\big(|Z_s-Z_s'|^2+|Z_s^0-Z_s^{0\prime}|^2\big)ds\,\Big|\,\mathcal{F}_t\bigg]\\
&\le\Gamma\mathbb{E}\bigg[|\xi-\xi'|^2+\big|V_S(X_S)-V_S'(X_S)\big|^2\\
&\qquad+\int_t^S\big|(B,F,\Sigma,\Sigma^0)\big(s,X_s,\mu_s,Y_s,Z_s,Z_s^0\big)\\
&\qquad\qquad-(B',F',\Sigma',\Sigma^{0\prime})\big(s,X_s,\mu_s',Y_s,Z_s,Z_s^0\big)\big|^2ds\,\Big|\,\mathcal{F}_t\bigg].
\end{aligned}$$

We now use (1.27) with $\xi=X_S$ as $\bigvee_{s\le S}\mathcal{F}_s$-measurable initial condition and with $t=S$. We also use the fact that $(X_s^{S,X_S},Y_s^{S,X_S},Z_s^{S,X_S},Z_s^{0,S,X_S},M_s^{S,X_S})_{S\le s\le T}$ matches $(X_s,Y_s,Z_s,Z_s^0,M_s)_{S\le s\le T}$. We deduce that there exists a constant $\Gamma^{(k)}$, only depending on Γ and $\Gamma((k+1)c)$, such that:

$$\begin{aligned}
&\mathbb{E}\bigg[\sup_{t\le s\le S}\big(|X_s-X_s'|^2+|Y_s-Y_s'|^2+|M_s-M_s'|^2\big)\\
&\qquad+\int_t^S\big(|Z_s-Z_s'|^2+|Z_s^0-Z_s^{0\prime}|^2\big)ds\,\Big|\,\mathcal{F}_t\bigg]\\
&\le\Gamma^{(k)}\mathbb{E}\bigg[|\xi-\xi'|^2+\big|G(X_T,\mu_T)-G'(X_T,\mu_T')\big|^2
\end{aligned}\tag{1.28}$$

$$+ \int_t^T \big|\big(B, F, \Sigma, \Sigma^0\big)\big(s, X_s, \mu_s, Y_s, Z_s, Z_s^0\big)$$
$$- \big(B', F', \Sigma', \Sigma^{0\prime}\big)\big(s, X_s, \mu'_s, Y_s, Z_s, Z_s^0\big)\big|^2 ds \,\Big|\, \mathcal{F}_t\Big].$$

In order to complete the proof of $\mathscr{H}_{kc}$, it suffices to use $\mathscr{H}_{(k+1)c}$ with $\xi = X_S$ and $\xi' = X'_S$ as initial conditions, recalling that $S = (k+1)c$. Indeed, we get:

$$\mathbb{E}\Big[\sup_{S \le s \le T} \big(|X_s - X'_s|^2 + |Y_s - Y'_s|^2 + |M_s - M'_s|^2\big)$$
$$+ \int_S^T \big(|Z_s - Z'_s|^2 + |Z_s^0 - Z_s^{0\prime}|^2\big) ds \,\Big|\, \mathcal{F}_S\Big]$$
$$\le \Gamma\big((k+1)c\big)\, \mathbb{E}\Big[|X_S - X'_S|^2 + \big|G(X_T, \mu_T) - G'(X_T, \mu'_T)\big|^2 \tag{1.29}$$
$$+ \int_S^T \big|\big(B, F, \Sigma, \Sigma^0\big)\big(s, X_s, \mu_s, Y_s, Z_s, Z_s^0\big)$$
$$- \big(B', F', \Sigma', \Sigma^{0\prime}\big)\big(s, X_s, \mu'_s, Y_s, Z_s, Z_s^0\big)\big|^2 ds \,\Big|\, \mathcal{F}_S\Big].$$

Taking conditional expectations with respect to $\mathcal{F}_t$, and then using (1.28) to estimate $|X_S - X'_S|^2$, we obtain the same inequality as above, but with $\Gamma((k+1)c)$ replaced by a new value of $\Gamma^{(k)}$ and $|X_S - X'_S|^2$ replaced by $|\xi - \xi'|^2$. We complete the proof by adding the resulting inequality to (1.28). □

1.4 Optimization with Random Coefficients

Throughout this section, we focus on optimal stochastic control problems depending upon a random environment. As we elucidate in Chapter 2, the rationale for doing so is that, in mean field games with a common noise, the mean field interaction manifests under the form of a theoretical flow of conditional marginal measures given the realization of the systemic noise.

1.4.1 Optimization Problem

To address stochastic control problems in a random environment, we use the framework introduced in the previous sections to discuss forward-backward SDEs in a random environment. Namely, we consider a complete probability space $(\Omega, \mathcal{F}, \mathbb{P})$ equipped with a complete and right-continuous filtration $\mathbb{F} = (\mathcal{F}_t)_{0 \le t \le T}$, a $2d$-dimensional Brownian motion $(\boldsymbol{W}^0, \boldsymbol{W})$ with respect to the filtration $\mathbb{F}$ (both $\boldsymbol{W}^0$ and $\boldsymbol{W}$ being of dimension d), an initial condition $X_0 \in L^2(\Omega, \mathcal{F}_0, \mathbb{P}; \mathbb{R}^d)$ and a random process $\boldsymbol{\mu} = (\mu_t)_{0 \le t \le T}$, called *environment*, with values in a Polish metric space $(\mathcal{X}, d)$ and with *càd-làg* paths satisfying $\mathbb{E}[\sup_{0 \le t \le T} d(0_{\mathcal{X}}, \mu_t)^2] < \infty$, where $0_{\mathcal{X}}$ is some arbitrary point in $\mathcal{X}$. As we shall see next, a typical example for $\mathcal{X}$ is $\mathcal{X} = \mathcal{P}_2(\mathbb{R}^d)$, equipped with the 2-Wasserstein distance.

We assume that $(\Omega, \mathcal{F}, \mathbb{F}, \mathbb{P})$ equipped with $(X_0, \boldsymbol{W}^0, \boldsymbol{\mu}, \boldsymbol{W})$ is admissible.

Description of the Dynamics

Given this set-up, we consider controlled dynamics of the type:

$$dX_t = b\big(t, X_t, \mu_t, \alpha_t\big)dt + \sigma\big(t, X_t, \mu_t, \alpha_t\big)dW_t + \sigma^0\big(t, X_t, \mu_t, \alpha_t\big)dW_t^0, \tag{1.30}$$

for $t \in [0, T]$. Throughout the section, the control process $\boldsymbol{\alpha} = (\alpha_t)_{0 \le t \le T}$ is taken in the set $\mathbb{A}$ of $\mathrm{Leb}_1 \otimes \mathbb{P}$-square-integrable $\mathbb{F}$-progressively measurable processes with values in a closed convex subset A of a Euclidean space $\mathbb{R}^k$ for some integer $k \ge 1$. We stress the fact that the compatibility condition enclosed in the admissibility property of the set-up plays a key role: it ensures that, conditioned on the observation of the input $(X_0, \boldsymbol{W}^0, \boldsymbol{\mu}, \boldsymbol{W})$ up until some time $t \in [0, T]$, the randomness enclosed in any control process $\boldsymbol{\alpha}$ up to time t is independent of the process $(X_0, \boldsymbol{W}^0, \boldsymbol{\mu}, \boldsymbol{W})$ up until time T. In particular, the control process $\boldsymbol{\alpha}$ does not anticipate the future of the input. We recall that the compatibility constraint is automatically satisfied when $\boldsymbol{\mu}$ is assumed to be adapted with respect to the complete and right-continuous augmentation of the filtration generated by $(X_0, \boldsymbol{W}^0)$.

The variable X_t stands for the state variable in $\mathbb{R}^d$. The coefficients b, σ, and σ^0 are deterministic continuous functions from $[0, T] \times \mathbb{R}^d \times \mathcal{X} \times A$ into $\mathbb{R}^d$ and $\mathbb{R}^{d \times d}$ respectively.

Regularity conditions will be imposed later on to guarantee the solvability of the state equation (1.30). We shall use the parameterized family of generators $(\mathscr{L}_{t,\mu,\alpha})_{t \ge 0, \mu \in \mathcal{X}, \alpha \in A}$ defined by:

$$\mathscr{L}_{t,\mu,\alpha}\phi(x) = b(t, x, \mu, \alpha) \cdot \partial_x \phi(x) + \frac{1}{2}\mathrm{trace}\Big[\big(\sigma\sigma^\dagger + \sigma^0(\sigma^0)^\dagger\big)(t, x, \mu, \alpha)\partial^2_{xx}\phi(x)\Big],$$

where $a^\dagger$ denotes the transpose of the matrix a. When σ and σ^0 do not depend upon α, which is the typical case that we consider below, we denote by $\mathscr{L}^0_{t,\mu}$ the second-order part of $\mathscr{L}_{t,\mu,\alpha}$:

$$\mathscr{L}^0_{t,\mu}\phi(x) = \frac{1}{2}\mathrm{trace}\Big[\big(\sigma\sigma^\dagger + \sigma^0(\sigma^0)^\dagger\big)(t, x, \mu)\partial^2_{xx}\phi(x)\Big].$$

Notations. As in Volume I, gradients of scalar valued functions will be regarded, when needed, as column vectors. Differently, derivatives of vector valued functions will be regarded as matrices, the number of lines being given by the dimension of the arrival vector space and the number of columns being given by the number of directions in the differentiation. Hence, if $v : \mathbb{R}^d \to \mathbb{R}$, $\partial_x v = (\partial_{x_i} v(x))_{1 \le i \le d}$ is regarded as d-dimensional column vector, while, if $v = (v^1, \cdots, v^n) : \mathbb{R}^d \to \mathbb{R}^n$, $n \ge 2$, $\partial_x v = (\partial_{x_j} v^i(x))_{1 \le i \le n, 1 \le j \le d}$ is regarded as a matrix of dimension $n \times d$.

Cost Functionals

Given input and control processes $\boldsymbol{\mu}$ and $\boldsymbol{\alpha}$ as above, we consider, for some maturity time T, the cost functional:

$$J^{\boldsymbol{\mu}}(\boldsymbol{\alpha}) = \mathbb{E}\left[\int_0^T f(s, X_s, \mu_s, \alpha_s)ds + g(X_T, \mu_T)\right], \tag{1.31}$$

where the terminal cost g is a continuous function from $\mathbb{R}^d \times \mathcal{X}$ into $\mathbb{R}$, and f is a continuous function from $[0, T] \times \mathbb{R}^d \times \mathcal{X} \times A$ into $\mathbb{R}$. Recall that in most applications of interest the input space $\mathcal{X}$ will be the space $\mathcal{P}_2(\mathbb{R}^d)$ of probability measures of order 2 on $\mathbb{R}^d$. The initial value X_0 of $\boldsymbol{X}$ being prescribed, the objective is to minimize $J^{\boldsymbol{\mu}}(\boldsymbol{\alpha})$ over $\boldsymbol{\alpha} \in \mathbb{A}$.

As in the previous chapters, we restrict our analysis to the case when σ and σ^0 do not depend upon the control. Following (Vol I)-(3.5), it allows us to use the reduced Hamiltonian:

$$H^{(r)}\big(t, x, \mu, y, \alpha\big) = b(t, x, \mu, \alpha) \cdot y + f(t, x, \mu, \alpha), \tag{1.32}$$

for $t \in [0, T]$, $x \in \mathbb{R}^d$, $\mu \in \mathcal{X}$, $y \in \mathbb{R}^d$, and $\alpha \in A$. A minimizer of this Hamiltonian, if and when it exists, then reads:

$$\hat{\alpha}\big(t, x, \mu, y\big) \in \text{argmin}_{\alpha \in A} H^{(r)}\big(t, x, \mu, y, \alpha\big). \tag{1.33}$$

Remark 1.54 *The reduced Hamiltonian $H^{(r)}$ defined above in* (1.32) *is sufficient as long as we look for a minimizer $\hat{\alpha}$. However, the full Hamiltonian may be needed for other computations, for example in writing forward-backward systems attempting to solve the optimization problem or, even though we shall not make much use of this fact, when σ and σ^0 are controlled. To wit, the full Hamiltonian H should involve an additional adjoint variable, to account for the dual of the volatility. But because of the presence of two sources of noise, this additional adjoint variable must be split into z and z^0 in order to emphasize the origin of the noise. Precisely, the full Hamiltonian should read:*

$$\begin{aligned} H\big(t, x, \mu, y, z, z^0, \alpha\big) &= b(t, x, \mu, \alpha) \cdot y + f(t, x, \mu, \alpha) \\ &\quad + \text{trace}\big[\sigma(t, x, \mu, \alpha) z^\dagger + \sigma^0(t, x, \mu, \alpha)(z^0)^\dagger\big], \end{aligned} \tag{1.34}$$

for $t \in [0, T]$, $x \in \mathbb{R}^d$, $\mu \in \mathcal{X}$, $\alpha \in A$, $y \in \mathbb{R}^d$, and $z, z^0 \in \mathbb{R}^{d \times d}$.

Standing Assumptions

In complete analogy with our approach of Chapter (Vol I)-3, we assume:

Assumption (Optimization in Random Environment). The coefficients $b : [0,T]\times\mathbb{R}^d\times\mathcal{X}\times A \to \mathbb{R}^d$, $\sigma : [0,T]\times\mathbb{R}^d\times\mathcal{X} \to \mathbb{R}^{d\times d}$, $\sigma^0 : [0,T]\times\mathbb{R}^d\times\mathcal{X} \to \mathbb{R}^{d\times d}$, $f : [0,T]\times\mathbb{R}^d\times\mathcal{X}\times A \to \mathbb{R}$ and $g : \mathbb{R}^d\times\mathcal{X} \to \mathbb{R}$ are Borel-measurable and satisfy:

(A1) b, σ and σ^0 are Lipschitz continuous in x, uniformly in $t \in [0,T]$, $\mu \in \mathcal{X}$ and $\alpha \in A$.

(A2) b, σ and σ^0 are at most of linear growth in (x,μ,α), uniformly in $t \in [0,T]$, *i.e.* there exists a constant $C \geq 0$ such that, for all $t \in [0,T]$, $x \in \mathbb{R}^d$, $\mu \in \mathcal{X}$ and $\alpha \in A$,

$$|b(t,x,\mu,\alpha)| \leq C\big(1 + |x| + d(0_{\mathcal{X}},\mu) + |\alpha|\big),$$

$$|(\sigma,\sigma^0)(t,x,\mu)| \leq C\big(1 + |x| + d(0_{\mathcal{X}},\mu)\big),$$

where $0_{\mathcal{X}}$ is some arbitrary point in $\mathcal{X}$.

Regarding the cost functions f and g, we require:

(A3) f and g are at most of quadratic growth in (x,μ,α), uniformly in $t \in [0,T]$, i.e. there exists a constant $C \geq 0$ such that, for all $t \in [0,T]$, $x \in \mathbb{R}^d$, $\mu \in \mathcal{X}$ and $\alpha \in A$,

$$|f(t,x,\mu,\alpha)| \leq C\big(1 + |x|^2 + d(0_{\mathcal{X}},\mu)^2 + |\alpha|^2\big),$$

$$|g(x,\mu)| \leq C\big(1 + |x|^2 + d(0_{\mathcal{X}},\mu)^2\big).$$

It is then clear that, under assumption **Optimization in a Random Environment**, for any input $\boldsymbol{\mu} = (\mu_t)_{0\leq t\leq T}$ as above satisfying:

$$\mathbb{E}[\sup_{0\leq t\leq T} d(0_{\mathcal{X}},\mu_t)^2] < \infty,$$

and any square-integrable $\mathbb{F}$-progressively measurable process $\boldsymbol{\alpha} = (\alpha_t)_{0\leq t\leq T}$ with values in A, the SDE (1.30) is uniquely solvable and the cost $J^{\boldsymbol{\mu}}(\boldsymbol{\alpha})$ is well defined.

1.4.2 Value Function and Stochastic HJB Equation

We start with a preliminary discussion of the case when X_0 is deterministic and $\boldsymbol{\mu}$ is $\mathbb{F}^{W^0} = (\mathcal{F}_t^{W^0})_{0\le t\le T}$-adapted. In that case, it is natural to introduce the following random value function:

$$U^{\mu}(t,x) = \operatorname*{essinf}_{(\alpha_s)_{t\le s\le T}} \mathbb{E}\bigg[\int_t^T f(s,X_s,\mu_s,\alpha_s)ds + g(X_T,\mu_T) \,\big|\, \mathcal{F}_t^{W^0}\bigg], \tag{1.35}$$

the essential infimum being taken over square-integrable controls $\boldsymbol{\alpha} = (\alpha_s)_{t\le s\le T}$ which are progressively measurable with respect to the completion of the filtration $(\mathcal{F}_s^{W^0} \vee \sigma\{W_u - W_t;\ t\le u\le s\})_{t\le s\le T}$, and the state process $(X_s)_{t\le s\le T}$ satisfying the SDE (1.30) with $X_t = x$ as initial condition. The use of an essential infimum instead of an infimum is dictated by the fact that the conditional expectations in the right-hand side are random variables. It guarantees that the infimum over a continuum of random variables is still a random variable. The conditional expectation with respect to $\mathcal{F}_t^{W^0}$ accounts for the fact that the environment has been observed up until time t.

The value function U^{μ} being a random field, it cannot solve a standard HJB equation. However, as the following verification result shows, it can still be associated with a backward stochastic PDE.

Proposition 1.55 *Under assumption* **Optimization in a Random Environment**, *let us assume that the Hamiltonian H has a minimizer $\hat{\alpha}$ as in* (1.33), *that there exist an $\mathbb{F}^{W^0}$-adapted input $\boldsymbol{\mu} = (\mu_t)_{0\le t\le T}$, with $\mathbb{E}[\sup_{0\le t\le T} d(0_{\mathcal{X}},\mu_t)^2] < \infty$, and $\mathbb{F}^{W^0}$-progressively measurable processes $\boldsymbol{U} = (U_t)_{0\le t\le T}$ and $\boldsymbol{V} = (V_t)_{0\le t\le T}$ defined on Ω and with values in $\mathcal{C}^2(\mathbb{R}^d;\mathbb{R})$ and $\mathcal{C}^1(\mathbb{R}^d;\mathbb{R}^d)$ respectively, both spaces being equipped with the topology of uniform convergence on compact subsets, such that $\boldsymbol{U}$ has continuous sample paths and $(\boldsymbol{U},\boldsymbol{V})$ satisfies:*

$$\mathbb{E}\Big[\sup_{0\le t\le T}\sup_{x\in\mathbb{R}^d}\Big(|U(t,x)|^2 + |\partial_x U(t,x)|^2 + |\partial_{xx}^2 U(t,x)|^2\Big)\Big] < \infty,$$

$$\mathbb{E}\int_0^T \sup_{x\in\mathbb{R}^d}\big(|V(t,x)|^2 + |\partial_x V(t,x)|^2\big)dt < \infty,$$

and, $\mathbb{P}$-almost surely:

$$\begin{aligned}U(t,x) = g(x,\mu_T) + \int_t^T \Big\{&\mathscr{L}^0_{s,\mu_s}U(s,x)\\ &+ b\Big(s,x,\mu_s,\hat{\alpha}\big(s,x,\mu_s,\partial_x U(s,x)\big)\Big)\cdot\partial_x U(s,x)\\ &+ f\Big(s,x,\mu_s,\hat{\alpha}\big(s,x,\mu_s,\partial_x U(s,x)\big)\Big)\\ &+ \operatorname{trace}\big[\sigma^0(s,x,\mu_s)\partial_x V(s,x)\big]\Big\}ds - \int_t^T V(s,x)\cdot dW_s^0,\end{aligned} \tag{1.36}$$

for $(t,x) \in [0,T] \times \mathbb{R}^d$, *where we write* $U(t,x)$ *for* $U_t(x)$ *and* $V(t,x)$ *for* $V_t(x)$ *(and similarly for the derivatives) in order to make more explicit the connection with the theory of PDEs.*

Finally, we also assume that the minimizer $\hat{\alpha}$ *of the reduced Hamiltonian is such that, for any initial condition* $(t,x) \in [0,T] \times \mathbb{R}^d$, *the state dynamics SDE* (1.30) *with* $X_t = x$ *and* $(\alpha_s = \hat{\alpha}(s, X_s, \mu_s, \partial_x U(s, X_s)))_{t \le s \le T}$ *has a unique strong solution which satisfies* $\mathbb{E}\int_t^T |\alpha_s|^2 ds < \infty$. *Then,*

$$\mathbb{P}\Big[U(t,x) = U^{\mu}(t,x)\Big] = 1, \quad (t,x) \in [0,T] \times \mathbb{R}^d.$$

Proof. Let $t \in [0,T]$ be fixed, $\boldsymbol{\beta} = (\beta_s)_{t \le s \le T}$ be an admissible control over the interval $[t,T]$ as defined above, and let us denote by $\boldsymbol{X}^{\boldsymbol{\beta}}$ the corresponding controlled state, namely the solution of the state equation:

$$dX_s^{\boldsymbol{\beta}} = b(s, X_s^{\boldsymbol{\beta}}, \mu_s, \beta_s)ds + \sigma(s, X_s^{\boldsymbol{\beta}}, \mu_s)dW_s + \sigma^0(s, X_s^{\boldsymbol{\beta}}, \mu_s)dW_s^0, \quad s \in [t,T],$$

with $X_t = x$. Applying Itô-Wentzell's formula (see the Notes & Complements for references) to $(U(s, X_s^{\boldsymbol{\beta}}))_{t \le s' T}$ over the interval $[t,T]$ and using (1.36), we get:

$$\begin{aligned}
&U(T, X_T^{\boldsymbol{\beta}}) \\
&= U(t,x) + \int_t^T d_s\big[U(s, X_s^{\boldsymbol{\beta}})\big] \\
&= U(t,x) + \int_t^T d_s U(s,x)\big|_{x = X_s^{\boldsymbol{\beta}}} + \partial_x U(s, X_s^{\boldsymbol{\beta}}) \cdot dX_s^{\boldsymbol{\beta}} \\
&\quad + \text{trace}\bigg[\frac{1}{2}\big[\sigma\sigma^\dagger + \sigma^0(\sigma^0)^\dagger\big](s, X_s^{\boldsymbol{\beta}}, \mu_s)\partial_{xx}^2 U(s, X_s^{\boldsymbol{\beta}}) + \sigma^0(s, X_s^{\boldsymbol{\beta}}, \mu_s)\partial_x V(s, X_s^{\boldsymbol{\beta}})\bigg]ds \\
&= U(t,x) + \int_t^T \Big[- b\Big(s, X_s^{\boldsymbol{\beta}}, \mu_s, \hat{\alpha}\big(s, X_s^{\boldsymbol{\beta}}, \mu_s, \partial_x U(s, X_s^{\boldsymbol{\beta}})\big)\Big) \cdot \partial_x U(s, X_s^{\boldsymbol{\beta}}) \\
&\qquad\qquad - f\Big(s, X_s^{\boldsymbol{\beta}}, \mu_s, \hat{\alpha}\big(s, X_s^{\boldsymbol{\beta}}, \mu_s, \partial_x U(s, X_s^{\boldsymbol{\beta}})\big)\Big)\Big]ds \\
&\quad + \int_t^T V(s, X_s^{\boldsymbol{\beta}}) \cdot dW_s^0 \\
&\quad + \int_t^T \partial_x U(s, X_s^{\boldsymbol{\beta}}) \cdot \big[b(s, X_s^{\boldsymbol{\beta}}, \mu_s, \beta_s)ds + \sigma(s, X_s^{\boldsymbol{\beta}}, \mu_s)dW_s + \sigma^0(s, X_s^{\boldsymbol{\beta}}, \mu_s)dW_s^0\big] \\
&\ge U(t,x) - \int_t^T f(s, X_s^{\boldsymbol{\beta}}, \mu_s, \beta_s)ds \\
&\quad + \int_t^T V(s, X_s^{\boldsymbol{\beta}}) \cdot dW_s^0 + \partial_x U(s, X_s^{\boldsymbol{\beta}}) \cdot \big[\sigma(s, X_s^{\boldsymbol{\beta}}, \mu_s)dW_s + \sigma^0(s, X_s^{\boldsymbol{\beta}}, \mu_s)dW_s^0\big],
\end{aligned}$$

with equality when $\beta_s = \hat{\alpha}(s, X_s^{\beta}, \mu_s, \partial_x U(s, X_s^{\beta}))$, where we used the fact that $\hat{\alpha}$ minimizes the Hamiltonian to pass from the equality to the inequality. So taking conditional expectations of both sides, we get:

$$\mathbb{E}\big[U(T, X_T^{\beta}) \,|\, \mathcal{F}_t^{W^0}\big] \geq U(t,x) - \mathbb{E}\bigg[\int_t^T f(s, X_s^{\beta}, \mu_s, \beta_s) ds \,\Big|\, \mathcal{F}_t^{W^0}\bigg],$$

which gives:

$$U(t,x) \leq \mathbb{E}\bigg[\int_t^T f(s, X_s^{\beta}, \mu_s, \beta_s) ds + g(X_T^{\beta}, \mu_T) \,\Big|\, \mathcal{F}_t^{W^0}\bigg],$$

and $U(t,x) \leq U^{\mu}(t,x)$ since the choice of $\boldsymbol{\beta}$ was arbitrary. We conclude because the inequality becomes an equality if we use the control $\hat{\boldsymbol{\alpha}}$. □

We can reformulate (1.36) as a stochastic HJB equation with terminal condition $U(T,x) = g(x, \mu_T)$. Indeed, we can rewrite (1.36) using the fact that $\hat{\alpha}(t,x,\mu,y)$ is the minimizer of the reduced Hamiltonian $H^{(r)}(t,x,\mu,y,\alpha) = b(t,x,\mu,\alpha) \cdot y + f(t,x,\mu,\alpha)$, fact which we used in the proof. Doing so, (1.36) appears as a backward stochastic HJB equation:

$$\begin{aligned} d_t U(t,x) + \bigg[&\mathscr{L}^0_{t,\mu_t} U(t,x) + \inf_{\alpha}\big[b(t,x,\mu_t,\alpha) \cdot \partial_x U(t,x) + f(t,x,\mu_t,\alpha)\big] \\ &+ \mathrm{trace}\big[\sigma^0(t,x,\mu)\partial_x V(t,x)\big]\bigg] dt - V(t,x) \cdot dW_t^0 = 0. \end{aligned}$$

The backward term is specifically designed in order to make possible the progressive-measurability of the value function U.

Although it may be very difficult to check the strong solvability of (1.36) (see however some references in the Notes & Complements below), its form gives some insight into the nature of the optimal control. As in the classical case, it is of a somewhat feedback form $\hat{\alpha}(X_t, \mu_t, \partial_x U^{\mu}(t, X_t))$. However , what we would like to call the optimal feedback function giving this optimal control, is now a *random field*, namely $[0,T] \times \mathbb{R}^d \ni (t,x) \mapsto \hat{\alpha}(x, \mu_t, \partial_x U^{\mu}(t,x))$.

1.4.3 Revisiting the Connection Between the HJB Equations and BSDEs

We now provide an interpretation of the optimization problem in a random environment by means of forward-backward SDEs. In full analogy with the discussion of Chapters (Vol I)-3 and (Vol I)-4, two options are available:

1. Representing the value function of the optimization problem as the solution of a backward stochastic differential equation;
2. Making use of a stochastic version of the Pontryagin maximum principle.

Since the random environment contains an additional source of randomness, the martingale component of the backward equation may not be a stochastic integral, and it is mandatory to make use of the representation (1.5) based on the Kunita-Watanabe decomposition.

In this paragraph, we discuss the representation of the value function of the optimization problem (1.30)–(1.31) when, as already indicated in Subsection 1.4.1, the volatilities σ and σ^0 do not depend upon the control α and in addition, the matrix σ is invertible. In such a case, we use the minimizer of the reduced Hamiltonian as given in (1.33). Following the discussion of Subsection (Vol I)-4, we introduce the FBSDE:

$$\begin{cases} dX_t = b\big(t, X_t, \mu_t, \hat{\alpha}(t, X_t, \mu_t, \sigma(t, X_t, \mu_t)^{-1\dagger} Z_t)\big)dt \\ \qquad + \sigma(t, X_t, \mu_t)dW_t + \sigma^0(t, X_t, \mu_t)dW_t^0, \\ dY_t = -f\big(t, X_t, \mu_t, \hat{\alpha}(t, X_t, \mu_t, \sigma(t, X_t, \mu_t)^{-1\dagger} Z_t)\big)dt \\ \qquad + Z_t \cdot dW_t + Z_t^0 \cdot dW_t^0 + dM_t, \quad t \in [0, T], \\ Y_T = g(X_T, \mu_T), \end{cases} \tag{1.37}$$

$\boldsymbol{M} = (M_t)_{0 \le t \le T}$ being a square-integrable *càd-làg* martingale with respect to the filtration $\mathbb{F}$, with $M_0 = 0$ as initial condition, the bracket of $\boldsymbol{M}$ with $(\boldsymbol{W}^0, \boldsymbol{W})$ being zero. The initial condition is given by X_0 in the set-up.

The above formulation requires the existence of a minimizer $\hat{\alpha}$. One way to guarantee existence of a minimizer is to assume that the reduced Hamiltonian $H^{(r)}(t, x, \mu, y, \alpha) = b(t, x, \mu, \alpha) \cdot y + f(t, x, \mu, \alpha)$ is strictly convex in α, which requires, as already explained in Chapters (Vol I)-3 and (Vol I)-4, to assume that b is linear in α. Although rather restrictive, this convexity assumption has several advantages: the minimizer is uniquely defined and, as discussed in Chapter (Vol I)-3, inherits the regularity properties of the coefficients. Here is a precise set of assumptions under which we can repeat the proof of Lemma (Vol I)-3.3:

Assumption (Hamiltonian Minimization in Random Environment). Assume that there exist two positive constants λ and L such that the coefficients b and f satisfy:

(A1) The drift b is an affine function of α in the sense that it is of the form

$$b(t, x, \mu, \alpha) = b_1(t, x, \mu) + b_2(t)\alpha,$$

where the mapping $[0, T] \ni t \mapsto b_2(t) \in \mathbb{R}^{d \times k}$ is measurable and bounded, and the mapping $[0, T] \times \mathbb{R}^d \times \mathcal{X} \ni (t, x, \mu) \mapsto b_1(t, x, \mu) \in \mathbb{R}^d$ is measurable and bounded on bounded subsets of $[0, T] \times \mathbb{R}^d \times \mathcal{X}$.

(continued)

(A2) For any $t \in [0, T]$, $x \in \mathbb{R}^d$ and $\mu \in \mathcal{X}$, the function $A \ni \alpha \mapsto f(t, x, \mu, \alpha) \in \mathbb{R}$ is once continuously differentiable, the derivative $\partial_\alpha f$ being L-Lipschitz-continuous in x. Moreover, it satisfies the λ-convexity assumption:

$$f(t, x, \mu, \alpha') - f(t, x, \mu, \alpha) - (\alpha' - \alpha) \cdot \partial_\alpha f(t, x, \mu, \alpha) \geq \lambda |\alpha' - \alpha|^2.$$

(A3) For all $t \in [0, T]$, $x \in \mathbb{R}^d$, $\mu \in \mathcal{X}$ and $\alpha \in A$,

$$|\partial_\alpha f(t, x, \mu, \alpha)| \leq L\big(1 + |x| + d(0_{\mathcal{X}}, \mu) + |\alpha|\big),$$

for some arbitrary point $0_{\mathcal{X}} \in \mathcal{X}$.

Then, repeating mutatis mutandis the proof of Lemma (Vol I)-3.3, we prove the following result:

Lemma 1.56 *If we assume that assumption* **Hamiltonian Minimization in Random Environment** *holds, then for all $(t, x, \mu, y) \in [0, T] \times \mathbb{R}^d \times \mathcal{X} \times A$, there exists a unique minimizer $\hat{\alpha}(t, x, \mu, y)$ of $A \ni \alpha \mapsto H^{(r)}(t, x, \mu, y, \alpha)$. Moreover, the function $[0, T] \times \mathbb{R}^d \times \mathcal{X} \times A \ni (t, x, \mu, y) \mapsto \hat{\alpha}(t, x, \mu, y)$ is measurable, locally bounded, and Lipschitz-continuous with respect to (x, y), uniformly in $(t, \mu) \in [0, T] \times \mathcal{X}$, the Lipschitz constant depending only upon λ, the supremum norm of b_2 and the Lipschitz constant of $\partial_\alpha f$ in x. Moreover, there exists a constant $C > 0$ such that:*

$$\begin{aligned} &\forall t \in [0, T],\ x, y \in \mathbb{R}^d,\ \mu \in \mathcal{X}, \\ &\qquad \big|\hat{\alpha}(t, x, \mu, y)\big| \leq C\big(1 + d(0_{\mathcal{X}}, \mu) + |x| + |y|\big). \end{aligned} \tag{1.38}$$

Main Statement

Our goal is to prove that the forward-backward system (1.37) is the right one, and is indeed solvable under the following assumption which is reminiscent of assumption **MFG Solvability HJB** in Subsection (Vol I)-4.4, except for the fact that we ask less regularity in the variable μ since we do not address the MFG problem yet.

Assumption (HJB in Random Environment). On top of assumption **Optimization in Random Environment**, assume that there exists a constant $L \geq 0$ such that the coefficients b, σ, σ^0 and f satisfy:

(continued)

(A1) For all $(t, x, \mu, \alpha) \in [0, T] \times \mathbb{R}^d \times \mathcal{X} \times A$,

$$|b(t,x,\mu,\alpha)| \leq L\big(1+|\alpha|\big), \quad |f(t,x,\mu,\alpha)| \leq L\big(1+|\alpha|^2\big),$$
$$|(\sigma,\sigma^{-1},\sigma^0)(t,x,\mu)| \leq L, \quad |g(x,\mu)| \leq L,$$

which implies in particular that σ is invertible.

(A2) For all $(t, x, x', \mu, \alpha', \alpha) \in [0, T] \times \mathbb{R}^d \times \mathbb{R}^d \times \mathcal{X} \times A \times A$,

$$\begin{aligned}
&|(b,f)(t,x',\mu,\alpha) - (b,f)(t,x,\mu,\alpha)| \\
&\qquad + |(\sigma,\sigma^0)(t,x',\mu) - (\sigma,\sigma^0)(t,x,\mu)| + |g(x') - g(x)| \leq L|x'-x|, \\
&|b(t,x,\mu,\alpha') - b(t,x,\mu,\alpha)| \leq L|\alpha'-\alpha|, \\
&|f(t,x,\mu,\alpha') - f(t,x,\mu,\alpha)| \leq L\big(1+\max(|\alpha|,|\alpha'|)\big)|\alpha'-\alpha|.
\end{aligned}$$

(A3) There exists a minimizer $\hat{\alpha}(t, x, \mu, y) \in \text{argmin}_\alpha H^{(r)}(t, x, \mu, y, \alpha)$ which satisfies:

$$\begin{aligned}
&|\hat{\alpha}(t,x,\mu,y)| \leq L\big(1+|y|\big), \\
&|\hat{\alpha}(t,x',\mu,y') - \hat{\alpha}(t,x,\mu,y)| \leq L\big(|x'-x| + |y'-y|\big),
\end{aligned}$$

for all $(t, x, x', \mu, y, y') \in [0, T] \times \mathbb{R}^d \times \mathbb{R}^d \times \mathcal{X} \times \mathbb{R}^d \times \mathbb{R}^d$.

Theorem 1.57 *Under assumption* **HJB in Random Environment**, *consider an admissible set-up* $(\Omega, \mathcal{F}, \mathbb{F}, \mathbb{P})$ *for an input* $(X_0, \boldsymbol{W}^0, \boldsymbol{\mu}, \boldsymbol{W})$. *Then, on this set-up, the FBSDE* (1.37) *with* X_0 *as initial condition has a unique solution* $(\boldsymbol{X}, \boldsymbol{Y}, \boldsymbol{Z}, \boldsymbol{Z}^0, \boldsymbol{M})$ *such that, for some* $R \geq 0$, $|\sigma(t, X_t, \mu_t)^{-1\dagger} Z_t| \leq R\,\text{Leb}_1 \otimes \mathbb{P}$*-almost everywhere. The process:*

$$\hat{\boldsymbol{\alpha}} = \Big(\hat{\alpha}_t = \hat{\alpha}\big(t, X_t, \mu_t, \sigma(t, X_t, \mu_t)^{-1\dagger} Z_t\big)\Big)_{0 \leq t \leq T},$$

is a minimizer of $J^{\boldsymbol{\mu}}$, *and if for any* $(t, x, \mu, y) \in [0, T] \times \mathbb{R}^d \times \mathcal{X} \times \mathbb{R}^d$, $\hat{\alpha}(t, x, \mu, y)$ *is a strict minimizer of the Hamiltonian* $H^{(r)}(t, x, \mu, y, \cdot)$, *then* $\hat{\boldsymbol{\alpha}}$ *is the unique minimizer of* $J^{\boldsymbol{\mu}}$ *in* (1.31).

The solution $(\boldsymbol{X}, \boldsymbol{Y}, \boldsymbol{Z}, \boldsymbol{Z}^0, \boldsymbol{M})$ *is given as the solution of a strongly uniquely solvable FBSDE with Lipschitz coefficients, whose construction is independent of the underlying probabilistic set-up and in particular, of the initial condition. This Lipschitz FBSDE satisfies the assumptions of Proposition 1.52 and thus admits a decoupling field which is uniformly bounded in all the variables, and which is Lipschitz-continuous (in x), uniformly in time* $t \in [0, T]$ *and the other variables, the bound and the Lipschitz constant depending only on the constants L and T appearing in the assumption. As a by product,* $\boldsymbol{Y}$ *may be represented in terms of* $\boldsymbol{X}$ *through this decoupling field.*

The proof of Theorem 1.57 given below is reminiscent of the proof of Theorem (Vol I)-4.45. The difficult point is that, under the standing assumptions on the coefficients (quadratic growth of f in α, and linear growths of b in α and of $\hat{\alpha}$ in z), the FBSDE (1.37) is quadratic in Z. As a result, its analysis requires a modicum of care. This is all the more true that the equation is rendered nonstandard by the presence of the discontinuous martingale $\boldsymbol{M}$. Below, we provide a tailored-made argument based upon a specific truncation procedure of the coefficients.

An Intermediate Result

The truncation argument is a variation on the proof of Proposition (Vol I)-4.51. The rationale for the introduction of the cut-off functions is quite clear: we want to recover the Lipschitz setting that we accounted for in Subsection 1.3.

Proposition 1.58 *Under assumption* **HJB in Random Environment** *with* $L \geq 1$ *therein, consider a set-up* $(\Omega, \mathcal{F}, \mathbb{F}, \mathbb{P})$, *admissible for an input* $(X_0, \boldsymbol{W}^0, \boldsymbol{\mu}, \boldsymbol{W})$, *and assume that the FBSDE* (1.37) *has a solution* $(\boldsymbol{X}, \boldsymbol{Y}, \boldsymbol{Z}, \boldsymbol{Z}^0, \boldsymbol{M})$ *such that, for some* $R \geq 1$, $|\sigma(t, X_t, \mu_t)^{-1\dagger} Z_t| \leq R\, \mathrm{Leb}_1 \otimes \mathbb{P}$*-almost everywhere.*

Assume also that for any 1-Lipschitz-continuous cut-off functions $\phi : A \to A$ *and* $\psi : \mathbb{R}^d \to [0,1]$, ϕ *being bounded and satisfying* $\phi(\alpha) = \alpha$ *for* $|\alpha| \leq L(1+R)$, *and* ψ *satisfying* $|\psi(z)| \leq (4|z-z'|)/\min(|z|,|z'|)$, $\psi(z) = 1$ *for* $|z| \leq LR$ *and* $\psi(z) = 0$ *for* $|z| \geq 2LR$, *with L as in the assumption, the FBSDE:*

$$\begin{cases} dX'_t = \psi(Z'_t) b\big(t, X'_t, \mu_t, \phi\big(\hat{\alpha}(t, X'_t, \mu_t, \sigma(t, X'_t, \mu_t)^{-1\dagger} Z'_t)\big)\big)dt \\ \qquad + \sigma(t, X'_t, \mu_t) dW_t + \sigma^0(t, X'_t, \mu_t) dW^0_t, \\ dY'_t = -\psi(Z'_t) f\big(t, X'_t, \mu_t, \phi\big(\hat{\alpha}(t, X'_t, \mu_t, \sigma(t, X'_t, \mu_t)^{-1\dagger} Z'_t)\big)\big)dt \\ \qquad + Z'_t \cdot dW_t + Z^{0\prime}_t \cdot dW^0_t + dM'_t, \quad t \in [0,T], \\ Y'_T = g(X'_T, \mu_T), \end{cases} \tag{1.39}$$

satisfies the strong uniqueness property and has $(X_t, Y_t, Z_t, Z^0_t, M_t)_{0 \leq t \leq T}$ *as solution when initialized with* X_0 *on the set-up* $(\Omega, \mathcal{F}, \mathbb{F}, \mathbb{P})$ *with the same input* $(X_0, \boldsymbol{W}^0, \boldsymbol{\mu}, \boldsymbol{W})$.

Then $\hat{\boldsymbol{\alpha}} = (\hat{\alpha}_t = \hat{\alpha}(t, X_t, \mu_t, \sigma(t, X_t, \mu_t)^{-1\dagger} Z_t))_{0 \leq t \leq T}$ *is a minimizer of* $J^{\boldsymbol{\mu}}$. *If, for any* $(t, x, \mu, y) \in [0,T] \times \mathbb{R}^d \times \mathcal{X} \times \mathbb{R}^d$, $\hat{\alpha}(t, x, \mu, y)$ *is a strict minimizer of the Hamiltonian* $H^{(r)}(t, x, \mu, y, \cdot)$, *then* $\hat{\boldsymbol{\alpha}}$ *is the unique minimizer of* $J^{\boldsymbol{\mu}}$.

Proof.

First Step. We start with a preliminary remark. Since the FBSDE (1.39) is uniquely solvable, we know from the Yamada-Watanabe Theorem 1.33 for FBSDEs that there exists a measurable function Φ from $\mathbb{R}^d \times \mathcal{C}([0,T];\mathbb{R}^d) \times \mathcal{D}([0,T];\mathbb{R}^d) \times \mathcal{C}([0,T];\mathbb{R}^d)$ into

$\mathcal{C}([0,T];\mathbb{R}^d) \times \mathcal{D}([0,T];\mathbb{R}) \times \mathcal{C}([0,T];\mathbb{R}^d) \times \mathcal{C}([0,T];\mathbb{R}^d) \times \mathcal{D}([0,T];\mathbb{R})$ such that, $\mathbb{P}$-almost surely:

$$\Big(\boldsymbol{X}, \boldsymbol{Y}, \int_0^\cdot Z_s ds, \int_0^\cdot Z_s^0 ds, \boldsymbol{M}\Big) = \Phi\big(X_0, \boldsymbol{W}^0, \boldsymbol{\mu}, \boldsymbol{W}\big).$$

Therefore, if we can find a new probability measure $\mathbb{P}'$ on $(\Omega, \mathcal{F})$, equivalent to $\mathbb{P}$ so that $(\Omega, \mathcal{F}, \mathbb{F}, \mathbb{P}')$ is a complete probability space equipped with a complete and right-continuous filtration, and a new d-dimensional Wiener process $\boldsymbol{W}'$ such that the tuple $(X_0, \boldsymbol{W}^0, \boldsymbol{\mu}, \boldsymbol{W}')$ is admissible under $\mathbb{P}'$, and in particular is compatible with $(\Omega, \mathcal{F}, \mathbb{F}, \mathbb{P}')$, then for any solution $(\boldsymbol{X}', \boldsymbol{Y}', \boldsymbol{Z}', \boldsymbol{Z}^{0\prime}, \boldsymbol{M}')$ of (1.39), constructed on $(\Omega, \mathcal{F}, \mathbb{F}, \mathbb{P}')$ for the input process $(X_0, \boldsymbol{W}^0, \boldsymbol{\mu}, \boldsymbol{W}')$, and with X_0 as initial condition, it must hold $\mathbb{P}'$ almost surely that $(\boldsymbol{X}', \boldsymbol{Y}', \int_0^\cdot Z_s' ds, \int_0^\cdot Z_s^{0\prime} ds, \boldsymbol{M}') = \Phi(X_0, \boldsymbol{W}^0, \boldsymbol{\mu}, \boldsymbol{W}')$. Hence, $|\sigma(t, X_t', \mu_t)^{-1\dagger} Z_t'| \leq R$ $dt \otimes \mathbb{P}'$-almost everywhere. Moreover $(\boldsymbol{X}', \boldsymbol{\mu}, \int_0^\cdot \hat{\boldsymbol{\alpha}}_s' ds)$ and $(\boldsymbol{X}, \boldsymbol{\mu}, \int_0^\cdot \hat{\boldsymbol{\alpha}}_s ds)$ have the same distributions, where:

$$\hat{\alpha}_t = \hat{\alpha}\big(t, X_t, \mu_t, \sigma(t, X_t, \mu_t)^{-1\dagger} Z_t\big), \quad \hat{\alpha}_t' = \hat{\alpha}'\big(t, X_t', \mu_t, \sigma(t, X_t', \mu_t)^{-1\dagger} Z_t'\big).$$

Second Step. We now return to the control problem. Given another controlled path $(X^{\boldsymbol{\beta}}, \boldsymbol{\beta})$ on the original set-up $(\Omega, \mathcal{F}, \mathbb{F}, \mathbb{P})$ with input $(X_0, \boldsymbol{W}^0, \boldsymbol{\mu}, \boldsymbol{W})$, the control $\boldsymbol{\beta}$ being bounded by some deterministic constant, we consider, still on $(\Omega, \mathcal{F}, \mathbb{F}, \mathbb{P})$ equipped with $(X_0, \boldsymbol{W}^0, \boldsymbol{\mu}, \boldsymbol{W})$, the BSDE (with the same ϕ and ψ as in the statement):

$$\begin{aligned} dY_t^{\beta} &= \psi(Z_t^{\beta}) f\big(t, X_t^{\beta}, \mu_t, \phi(\hat{\alpha}_t^{\beta})\big) dt \\ &\quad + Z_t^{\beta} \cdot \Big[\big(\sigma^{-1} b\big)\big(t, X_t^{\beta}, \mu_t, \beta_t\big) - \psi(Z_t^{\beta})\big(\sigma^{-1} b\big)\big(t, X_t^{\beta}, \mu_t, \phi(\hat{\alpha}_t^{\beta})\big)\Big] dt \\ &\quad + Z_t^{\beta} \cdot dW_t + Z_t^{0,\beta} \cdot dW_t^0 + dM_t^{\beta}, \end{aligned} \tag{1.40}$$

with $\hat{\alpha}_t^{\beta} = \hat{\alpha}(t, X_t^{\beta}, \mu_t, \sigma(t, X_t^{\beta}, \mu_t)^{-1\dagger} Z_t^{\beta})$ and $Y_T^{\beta} = g(X_T^{\beta}, \mu_T)$. Here, $(M_t^{\beta})_{0 \leq t \leq T}$ denotes an $\mathbb{F}$-square integrable martingale such that $[\boldsymbol{M}^{\beta}, \boldsymbol{W}]_\cdot \equiv 0$ and $[\boldsymbol{M}^{\beta}, \boldsymbol{W}^0]_\cdot \equiv 0$ and with 0 as initial condition.

Thanks to the cut-off functions ϕ and ψ, the system (1.40) has Lipschitz continuous coefficients in z. Although the standard theory does not apply directly because of the presence of the additional martingale $(M_t^{\beta})_{0 \leq t \leq T}$, it may be easily extended to the present situation and, as a result, (1.40) has a unique solution, see Example 1.20. We now let $(\mathcal{E}_t^{\beta})_{0 \leq t \leq T}$ be the Doléans exponential of the stochastic integral:

$$\Big(- \int_0^t \big[(\sigma^{-1} b)(s, X_s, \mu_s, \beta_s) - \psi(Z_s^{\beta})(\sigma^{-1} b)\big(s, X_s, \mu_s, \phi(\hat{\alpha}_s^{\beta})\big)\big] \cdot dW_s \Big)_{0 \leq t \leq T}.$$

Since the integrand is bounded, $(\mathcal{E}_t^{\beta})_{0 \leq t \leq T}$ is a true martingale, and we can define the probability measure $\mathbb{P}^{\boldsymbol{\beta}} = \mathcal{E}_T^{\beta} \cdot \mathbb{P}$. Under $\mathbb{P}^{\boldsymbol{\beta}}$, the process:

$$\Big(W_t^{\beta} = W_t + \int_0^t \big[(\sigma^{-1} b)(s, X_s, \mu_s, \beta_s) - \psi(Z_s^{\beta})(\sigma^{-1} b)\big(s, X_s, \mu_s, \phi(\hat{\alpha}_s^{\beta})\big)\big] ds \Big)_{0 \leq t \leq T},$$

is a d-dimensional Brownian motion. We show at the end of the proof that $(X_0, \boldsymbol{W}^0, \boldsymbol{\mu}, \boldsymbol{W}^{\boldsymbol{\beta}})$ is admissible and compatible with $(\Omega, \mathcal{F}, \mathbb{F}, \mathbb{P}^{\boldsymbol{\beta}})$, and that $(\boldsymbol{X}^{\boldsymbol{\beta}}, \boldsymbol{Y}^{\boldsymbol{\beta}}, \boldsymbol{Z}^{\boldsymbol{\beta}}, \boldsymbol{Z}^{0,\boldsymbol{\beta}}, \boldsymbol{M}^{\boldsymbol{\beta}})$ is a solution of the FBSDE (1.39) on $(\Omega, \mathcal{F}, \mathbb{F}, \mathbb{P}^{\boldsymbol{\beta}})$ equipped with $(X_0, \boldsymbol{W}^0, \boldsymbol{\mu}, \boldsymbol{W}^{\boldsymbol{\beta}})$. Therefore, taking these facts for granted momentarily, we infer that the law of $(\boldsymbol{X}^{\boldsymbol{\beta}}, \boldsymbol{\mu}, \hat{\boldsymbol{\alpha}}^{\boldsymbol{\beta}})$ under $\mathbb{P}^{\boldsymbol{\beta}}$ is the same as the law of $(\boldsymbol{X}, \boldsymbol{\mu}, \hat{\boldsymbol{\alpha}})$ under $\mathbb{P}$, which proves in particular that:

$$J^{\mu}(\hat{\boldsymbol{\alpha}}) = \mathbb{E}^{\mathbb{P}^{\boldsymbol{\beta}}}\left[\int_0^T f(t, X_t^{\beta}, \mu_t, \hat{\alpha}_t^{\beta})dt + g(X_T^{\beta}, \mu_T)\right].$$

Also, $(\sigma(t, X_t^{\beta}, \mu_t)^{-1\dagger} Z_t^{\beta})_{0 \le t \le T}$ is bounded by R $\text{Leb}_1 \otimes \mathbb{P}^{\boldsymbol{\beta}}$ almost everywhere, and thus $\text{Leb}_1 \otimes \mathbb{P}$-almost everywhere, which shows that, in (1.40), $\psi(Z_t^{\beta})$ is equal to 1 and $\phi(\hat{\alpha}_t^{\beta})$ is equal to $\hat{\alpha}_t^{\beta}$. In particular, $\mathbb{E}^{\mathbb{P}^{\boldsymbol{\beta}}}[Y_0^{\beta}]$ is equal to the right-hand side above and thus to $J^{\mu}(\hat{\boldsymbol{\alpha}})$. Since:

$$\mathbb{E}^{\mathbb{P}^{\boldsymbol{\beta}}}[Y_0^{\beta}] = \mathbb{E}^{\mathbb{P}}\big[\mathcal{E}_T^{\beta} Y_0^{\beta}\big] = \mathbb{E}^{\mathbb{P}}\big[\mathcal{E}_0^{\beta} Y_0^{\beta}\big] = \mathbb{E}^{\mathbb{P}}\big[Y_0^{\beta}\big],$$

we have:

$$\begin{aligned} J^{\mu}(\hat{\boldsymbol{\alpha}}) - J^{\mu}(\boldsymbol{\beta}) &= \mathbb{E}^{\mathbb{P}}\big[Y_0^{\beta}\big] - J^{\mu}(\boldsymbol{\beta}) \\ &= \mathbb{E}^{\mathbb{P}}\bigg[\int_0^T \big[H^{(r)}\big(t, X_t^{\beta}, \mu_t, \sigma(t, X_t^{\beta}, \mu_t)^{-1\dagger} Z_t^{\beta}, \hat{\alpha}_t^{\beta}\big) \\ &\qquad - H^{(r)}\big(t, X_t^{\beta}, \mu_t, \sigma(t, X_t^{\beta}, \mu_t)^{-1\dagger} Z_t^{\beta}, \beta_t\big)\big]dt\bigg], \end{aligned} \tag{1.41}$$

so that $J^{\mu}(\hat{\boldsymbol{\alpha}}) \le J^{\mu}(\boldsymbol{\beta})$.

For a generic $\boldsymbol{\beta}$ satisfying $\mathbb{E}\int_0^T |\beta_t|^2 dt < \infty$, we can apply the previous inequality with $\boldsymbol{\beta}$ replaced by $\boldsymbol{\beta}^n = (\beta_t \mathbf{1}_{|\beta_t| \le n})_{0 \le t \le T}$. Using the continuity and growth assumptions on the coefficients, it is easy to prove that $J^{\mu}(\boldsymbol{\beta}^n)$ converges to $J^{\mu}(\boldsymbol{\beta})$ as n tends to ∞, from which we deduce that $\hat{\boldsymbol{\alpha}}$ is a control minimizing the cost.

Third Step. If $\hat{\alpha}(t, x, \mu, y)$ is a strict minimizer of $H^{(r)}(t, x, \mu, y, \cdot)$, then for any bounded control $\boldsymbol{\beta}$, $J^{\mu}(\boldsymbol{\beta}) = J^{\mu}(\hat{\boldsymbol{\alpha}})$ if and only if $\boldsymbol{\beta} = \hat{\boldsymbol{\alpha}}^{\beta}$ $\text{Leb}_1 \otimes \mathbb{P}$-almost everywhere. Using the fact that $|\hat{\boldsymbol{\alpha}}^{\beta}| \le L(1 + R)$, we deduce that $(\boldsymbol{X}^{\boldsymbol{\beta}}, \boldsymbol{Y}^{\boldsymbol{\beta}}, \boldsymbol{Z}^{\boldsymbol{\beta}}, \boldsymbol{Z}^{0,\boldsymbol{\beta}}, \boldsymbol{M}^{\boldsymbol{\beta}})$ satisfies the FBSDE (1.39) under $\mathbb{P}$, and by uniqueness, $\boldsymbol{X}^{\boldsymbol{\beta}} = \boldsymbol{X}$ and $\boldsymbol{\beta} = \hat{\boldsymbol{\alpha}}^{\beta} = \hat{\boldsymbol{\alpha}}$. If $\boldsymbol{\beta}$ is not bounded, we can use the same approximating sequence $(\boldsymbol{\beta}^n)_{n \ge 0}$ as above, and since $\boldsymbol{X}^{\boldsymbol{\beta}^n}$ converges to $\boldsymbol{X}^{\boldsymbol{\beta}}$ for the norm $\mathbb{E}^{\mathbb{P}}[\sup_{0 \le t \le T} |\cdot|^2]^{1/2}$, we have from (1.41):

$$\begin{aligned} J^{\mu}(\boldsymbol{\beta}) - J^{\mu}(\hat{\boldsymbol{\alpha}}) &= \lim_{n \to \infty} J^{\mu}(\boldsymbol{\beta}^n) - J^{\mu}(\hat{\boldsymbol{\alpha}}) \\ &\ge \mathbb{E}^{\mathbb{P}} \int_0^T \liminf_{n \to \infty} \Big[\Big(H^{(r)}\big(t, X_t^{\beta^n}, \mu_t, \sigma(t, X_t^{\beta^n}, \mu_t)^{-1\dagger} Z_t^{\beta^n}, \beta_t^n\big) \\ &\qquad - H^{(r)}\big(t, X_t^{\beta^n}, \mu_t, \sigma(t, X_t^{\beta^n}, \mu_t)^{-1\dagger} Z_t^{\beta^n}, \hat{\alpha}_t^{\beta^n}\big)\Big)\Big]dt. \end{aligned}$$

Again, if $\boldsymbol{\beta}$ is not bounded, we can find $R' > L(R+1)+1$ (for the same L as in the statement) such that $\mathbb{E}^{\mathbb{P}} \int_0^T \mathbf{1}_{L(R+1)+1 \le |\beta_t| \le R'} dt \neq 0$. Given $(t, x, \mu) \in [0, T] \times \mathbb{R}^d \times \mathcal{X}$, we then let:

$$\Theta(t, x, \mu) = \{(y, \beta) \in \mathbb{R}^d \times A : |y| \le R, |\beta| \le R', |\beta - \hat{\alpha}(t, x, \mu, y)| \ge 1\}.$$

Then,

$$\begin{aligned} J^{\mu}(\boldsymbol{\beta}) - J^{\mu}(\hat{\boldsymbol{\alpha}}) \ge \mathbb{E}^{\mathbb{P}} \int_0^T \liminf_{n \to \infty} \Big[\inf_{(y,\beta) \in \Theta(t, X_t^{\boldsymbol{\beta}^n}, \mu_t)} \Big(H^{(r)}(t, X_t^{\boldsymbol{\beta}^n}, \mu_t, y, \beta) \\ - H^{(r)}(t, X_t^{\boldsymbol{\beta}^n}, \mu_t, y, \hat{\alpha}(t, X_t^{\boldsymbol{\beta}^n}, \mu_t, y)) \Big) \mathbf{1}_{L(R+1)+1 \le |\beta_t^n| \le R'} \Big] dt. \end{aligned}$$

By continuity of $H^{(r)}(t, \cdot, \mu_t, \cdot, \cdot)$ and $\hat{\alpha}(t, \cdot, \mu_t, \cdot)$ and by compactness of $\Theta(t, x, \mu)$ for each (t, x, μ), it is plain to deduce that:

$$\begin{aligned} J^{\mu}(\boldsymbol{\beta}) - J^{\mu}(\hat{\boldsymbol{\alpha}}) \ge \mathbb{E}^{\mathbb{P}} \int_0^T \Big[\inf_{(y,\beta) \in \Theta(t, X_t^{\boldsymbol{\beta}}, \mu_t)} \Big(H^{(r)}(t, X_t^{\boldsymbol{\beta}}, \mu_t, y, \beta) \\ - H^{(r)}(t, X_t^{\boldsymbol{\beta}}, \mu_t, y, \hat{\alpha}(t, X_t^{\boldsymbol{\beta}}, \mu_t, y)) \Big) \mathbf{1}_{L(R+1)+1 \le |\beta_t^n| \le R'} \Big] dt, \end{aligned}$$

which cannot be zero by definition of $\Theta(t, X_t^{\boldsymbol{\beta}}, \mu_t)$. This proves that $\boldsymbol{X}$ is the unique minimizing path.

Fourth Step. We prove that, if the control $\boldsymbol{\beta}$ is bounded, the set-up formed by $(\Omega, \mathcal{F}, \mathbb{F}, \mathbb{P}^{\boldsymbol{\beta}})$ and $(X_0, \boldsymbol{W}^0, \boldsymbol{\mu}, \boldsymbol{W}^{\boldsymbol{\beta}})$ is admissible. It is completely standard to check that $(\boldsymbol{W}^0, \boldsymbol{W}^{\boldsymbol{\beta}})$ is a $2d$-dimensional Brownian motion process with respect to $\mathbb{F}$. In order to prove the independence of $\boldsymbol{W}^{\boldsymbol{\beta}}$ and $(X_0, \boldsymbol{W}^0, \boldsymbol{\mu})$, we proceed as follows. First, we claim that for any square-integrable $\mathbb{F}$-progressively measurable process $(\vartheta_t)_{0 \le t \le T}$ with values in $\mathbb{R}^d$, for any $t \in [0, T]$, we have:

$$\mathbb{E}^{\mathbb{P}}\Big[\int_t^T \vartheta_s \cdot dW_s \,|\, \mathcal{F}_t^{\boldsymbol{W}} \vee \mathcal{F}_T^{(X_0, \boldsymbol{W}^0, \boldsymbol{\mu})} \Big] = 0. \tag{1.42}$$

Indeed, by approximating $\boldsymbol{\vartheta}$ by a sequence of simple processes, it suffices to check that, for any $t \le s \le T$, for any random variable $\vartheta_t \in L^2(\Omega, \mathcal{F}_t, \mathbb{P}; \mathbb{R}^d)$,

$$\mathbb{E}^{\mathbb{P}}\Big[\vartheta_t \cdot (W_s - W_t) \,|\, \mathcal{F}_t^{\boldsymbol{W}} \vee \mathcal{F}_T^{(X_0, \boldsymbol{W}^0, \boldsymbol{\mu})} \Big] = 0.$$

By the compatibility condition, observe that for any events $C \in \mathcal{F}_t^{(X_0, \boldsymbol{W}^0, \boldsymbol{\mu}, \boldsymbol{W})}$ and $D \in \mathcal{F}_T^{(X_0, \boldsymbol{W}^0, \boldsymbol{\mu})}$:

$$\mathbb{E}^{\mathbb{P}}\big[\mathbf{1}_C \mathbf{1}_D \vartheta_t \cdot (W_s - W_t) \big] = \mathbb{E}^{\mathbb{P}}\big[\mathbf{1}_C \mathbf{1}_D \mathbb{E}^{\mathbb{P}}[\vartheta_t \,|\, \mathcal{F}_t^{(X_0, \boldsymbol{W}^0, \boldsymbol{\mu}, \boldsymbol{W})}] \cdot (W_s - W_t) \big].$$

Now, using the fact that $\boldsymbol{W}$ and $(X_0, \boldsymbol{W}^0, \boldsymbol{\mu})$ are independent, we can check that $W_s - W_t$ is orthogonal to $\mathcal{F}_t^{(X_0, \boldsymbol{W}^0, \boldsymbol{\mu}, \boldsymbol{W})} \vee \mathcal{F}_T^{(X_0, \boldsymbol{W}^0, \boldsymbol{\mu})}$, from which we deduce that the right-hand side above is zero. This completes the proof of (1.42).

We now make use of (1.42) to prove that $\boldsymbol{W}^{\beta}$ and $(X_0, \boldsymbol{W}^0, \boldsymbol{\mu})$ are independent under $\mathbb{P}^{\beta}$. Indeed, given a bounded and measurable function Ψ on $\mathbb{R}^d \times \mathcal{C}([0,T];\mathbb{R}^d) \times \mathcal{D}([0,T];\mathcal{X})$ and a function $f \in L^2([0,T];\mathbb{R}^d)$, we have:

$$\begin{aligned}
&\mathbb{E}^{\mathbb{P}^{\beta}}\bigg[\Psi(X_0, \boldsymbol{W}^0, \boldsymbol{\mu}) \exp\bigg(\int_0^T f_t \cdot dW_t^{\beta} - \frac{1}{2}\int_0^T |f_t|^2 dt\bigg)\bigg] \\
&= \mathbb{E}^{\mathbb{P}}\bigg[\mathcal{E}_T^{\beta}\Psi(X_0, \boldsymbol{W}^0, \boldsymbol{\mu}) \exp\bigg(\int_0^T f_t \cdot dW_t - \frac{1}{2}\int_0^T |f_t|^2 dt + \int_0^T u_t \cdot f_t dt\bigg)\bigg],
\end{aligned}$$

where we set $u_t = (\sigma^{-1}b)(t, X_t^{\beta}, \mu_t, \beta_t) - \psi(Z_t^{\beta})(\sigma^{-1}b)(t, X_t^{\beta}, \mu_t, \hat{\alpha}_t^{\beta})$, which is bounded. Therefore,

$$\begin{aligned}
&\mathbb{E}^{\mathbb{P}^{\beta}}\bigg[\Psi(X_0, \boldsymbol{W}^0, \boldsymbol{\mu}) \exp\bigg(\int_0^T f_t \cdot dW_t^{\beta} - \frac{1}{2}\int_0^T |f_t|^2 dt\bigg)\bigg] \\
&= \mathbb{E}^{\mathbb{P}}\bigg[\Psi(X_0, \boldsymbol{W}^0, \boldsymbol{\mu}) \\
&\qquad \mathbb{E}^{\mathbb{P}}\bigg[\mathcal{E}_T^{\beta} \exp\bigg(\int_0^T f_t \cdot dW_t - \frac{1}{2}\int_0^T |f_t|^2 dt + \int_0^T u_t \cdot f_t dt\bigg) \,\Big|\, \mathcal{F}_T^{(X_0, \boldsymbol{W}^0, \boldsymbol{\mu})}\bigg]\bigg] \\
&= \mathbb{E}^{\mathbb{P}}\bigg[\Psi(X_0, \boldsymbol{W}^0, \boldsymbol{\mu}) \\
&\qquad \mathbb{E}^{\mathbb{P}}\bigg[\exp\bigg(\int_0^T (f_t - u_t) \cdot dW_t - \frac{1}{2}\int_0^T |f_t - u_t|^2 dt\bigg) \,\Big|\, \mathcal{F}_T^{(X_0, \boldsymbol{W}^0, \boldsymbol{\mu})}\bigg]\bigg] \\
&= \mathbb{E}^{\mathbb{P}}\big[\Psi(X_0, \boldsymbol{W}^0, \boldsymbol{\mu})\big],
\end{aligned}$$

the passage from the third to the fourth lines being proved by expanding the exponential as a stochastic integral and then applying (1.42) with $t = 0$. Using the fact that the collection of random variables of the form $\exp(\int_0^T f_s \cdot dW_s^{\beta} - \frac{1}{2}\int_0^T |f_s|^2 ds)$, with $f \in L^2([0,T];\mathbb{R}^d)$, is a total subset of $L^2(\Omega, \sigma\{W_s^{\beta};\ s \le T\}, \mathbb{P}^{\beta}; \mathbb{R})$, we complete the proof of the independence property.

We check in a similar way that $\mathbb{F}$ and $(X_0, \boldsymbol{W}^0, \boldsymbol{\mu}, \boldsymbol{W}^{\beta})$ are compatible under $\mathbb{P}^{\beta}$. To do so, we make use of the following analogue of (1.42):

$$\mathbb{E}^{\mathbb{P}}\bigg[\int_t^T \vartheta_s \cdot dW_s \,\Big|\, \mathcal{F}_t \vee \mathcal{F}_T^{(X_0, \boldsymbol{W}^0, \boldsymbol{\mu})}\bigg] = 0, \tag{1.43}$$

for a square-integrable $\mathbb{F}$-progressively measurable process $(\vartheta_s)_{0 \le s \le T}$ with values in $\mathbb{R}^d$. In order to prove (1.43), it suffices to show that, for any $t \le s \le T$,

$$\mathbb{E}^{\mathbb{P}}\big[W_s - W_t | \mathcal{F}_t \vee \mathcal{F}_T^{(X_0, \boldsymbol{W}^0, \boldsymbol{\mu})}\big] = 0.$$

By the compatibility property, the above is true if

$$\mathbb{E}^{\mathbb{P}}\big[W_s - W_t | \mathcal{F}_t^{(X_0, \boldsymbol{W}^0, \boldsymbol{\mu}, \boldsymbol{W})} \vee \mathcal{F}_T^{(X_0, \boldsymbol{W}^0, \boldsymbol{\mu})}\big] = 0,$$

but the latter is indeed true as we already noticed in the proof of (1.42).

Now that (1.43) has been proved, for $t \in [0, T]$, $C \in \mathcal{F}_t^{(X_0, \boldsymbol{W}^0, \boldsymbol{\mu})}$, $D \in \mathcal{F}_T^{(X_0, \boldsymbol{W}, \boldsymbol{\mu})}$, $E \in \mathcal{F}_t$ and $f \in L^2([0, T]; \mathbb{R}^d)$, we have:

$$\begin{aligned}
&\mathbb{E}^{\mathbb{P}^{\beta}}\Big[\mathbf{1}_C \mathbf{1}_D \mathbf{1}_E \exp\Big(\int_0^T f_s \cdot dW_s^{\beta} - \frac{1}{2}\int_0^T |f_s|^2 ds\Big)\Big] \\
&= \mathbb{E}^{\mathbb{P}}\Big[\mathbf{1}_C \mathbf{1}_D \mathbf{1}_E \exp\Big(\int_0^T (f_s - u_s) \cdot dW_s - \frac{1}{2}\int_0^T |f_s - u_s|^2 ds\Big)\Big] \\
&= \mathbb{E}^{\mathbb{P}}\Big[\mathbf{1}_C \mathbf{1}_D \mathbf{1}_E \exp\Big(\int_0^t (f_s - u_s) \cdot dW_s - \frac{1}{2}\int_0^t |f_s - u_s|^2 ds\Big) \\
&\qquad \times \mathbb{E}^{\mathbb{P}}\Big[\exp\Big(\int_t^T (f_s - u_s) \cdot dW_s - \frac{1}{2}\int_t^T |f_s - u_s|^2 ds\Big) \,\big|\, \mathcal{F}_t \vee \mathcal{F}_T^{(X_0, \boldsymbol{W}^0, \boldsymbol{\mu})}\Big]\Big].
\end{aligned}$$

Therefore, by (1.43), we obtain:

$$\begin{aligned}
&\mathbb{E}^{\mathbb{P}^{\beta}}\Big[\mathbf{1}_C \mathbf{1}_D \mathbf{1}_E \exp\Big(\int_0^T f_s \cdot dW_s^{\beta} - \frac{1}{2}\int_0^T |f_s|^2 ds\Big)\Big] \\
&= \mathbb{E}^{\mathbb{P}}\Big[\mathbf{1}_C \mathbf{1}_D \mathbf{1}_E \exp\Big(\int_0^t (f_s - u_s) \cdot dW_s - \frac{1}{2}\int_0^t |f_s - u_s|^2 ds\Big)\Big] \\
&= \mathbb{E}^{\mathbb{P}}\Big[\mathbf{1}_C \mathbf{1}_E \exp\Big(\int_0^t (f_s - u_s) \cdot dW_s - \frac{1}{2}\int_0^t |f_s - u_s|^2 ds\Big) \mathbb{P}\big(D | \mathcal{F}_t^{(X_0, \boldsymbol{W}^0, \boldsymbol{\mu}, \boldsymbol{W})}\big)\Big].
\end{aligned}$$

By independence of $(X_0, \boldsymbol{W}^0, \boldsymbol{\mu})$ and $\boldsymbol{W}$, we must have:

$$\mathbb{P}[D | \mathcal{F}_t^{(X_0, \boldsymbol{W}^0, \boldsymbol{\mu}, \boldsymbol{W})}] = \mathbb{P}[D | \mathcal{F}_t^{(X_0, \boldsymbol{W}^0, \boldsymbol{\mu})}],$$

which is also almost surely equal to $\mathbb{P}(D | \mathcal{F}_{t+}^{\mathrm{nat}, (X_0, \boldsymbol{W}^0, \boldsymbol{\mu})})$. We deduce that:

$$\begin{aligned}
&\mathbb{E}^{\mathbb{P}^{\beta}}\Big[\mathbf{1}_C \mathbf{1}_D \mathbf{1}_E \exp\Big(\int_0^T f_s \cdot dW_s^{\beta} - \frac{1}{2}\int_0^T |f_s|^2 ds\Big)\Big] \\
&= \mathbb{E}^{\mathbb{P}^{\beta}}\Big[\mathbf{1}_C \mathbf{1}_E \exp\Big(\int_0^t f_s \cdot dW_s^{\beta} - \frac{1}{2}\int_0^t |f_s|^2 ds\Big) \mathbb{P}\big(D | \mathcal{F}_{t+}^{\mathrm{nat}, (X_0, \boldsymbol{W}^0, \boldsymbol{\mu})}\big)\Big] \\
&= \mathbb{E}^{\mathbb{P}^{\beta}}\Big[\mathbf{1}_C \mathbf{1}_E \exp\Big(\int_0^T f_s \cdot dW_s^{\beta} - \frac{1}{2}\int_0^T |f_s|^2 ds\Big) \mathbb{P}\big(D | \mathcal{F}_{t+}^{\mathrm{nat}, (X_0, \boldsymbol{W}^0, \boldsymbol{\mu})}\big)\Big].
\end{aligned}$$

Once again, we use the fact that the collection of random variables of the form $\exp(\int_0^T f_s \cdot dW_s^{\beta} - \frac{1}{2}\int_0^T |f_s|^2 ds)$, with $f \in L^2([0, T]; \mathbb{R}^d)$, is total in $L^2(\Omega, \sigma\{W_s^{\beta};\ s \leq T\}, \mathbb{P}^{\beta}; \mathbb{R})$. By an approximation argument, we deduce that, for any $D' \in \sigma\{W_s^{\beta};\ s \leq T\}$,

$$\mathbb{P}^{\beta}\big[C \cap D \cap D' \cap E\big] = \mathbb{E}^{\mathbb{P}^{\beta}}\Big[\mathbf{1}_C \mathbf{1}_E \mathbf{1}_{D'} \mathbb{P}\big[D | \mathcal{F}_{t+}^{\mathrm{nat}, (X_0, \boldsymbol{W}^0, \boldsymbol{\mu})}\big]\Big],$$

which clearly implies the required compatibility condition. This shows that, $(\Omega, \mathcal{F}, \mathbb{F}, \mathbb{P}^\beta)$ with input $(X_0, \boldsymbol{W}^0, \boldsymbol{\mu}, \boldsymbol{W}^\beta)$ is an admissible set-up.

Fifth Step. It remains to check that, under $\mathbb{P}^\beta$, $(\boldsymbol{X}^\beta, \boldsymbol{Y}^\beta, \boldsymbol{Z}^\beta, \boldsymbol{Z}^{0,\beta}, \boldsymbol{M}^\beta)$ is the solution of the FBSDE (1.39), when driven by $(X_0, \boldsymbol{W}^0, \boldsymbol{\mu}, \boldsymbol{W}^\beta)$. We first notice that, by construction,

$$\mathbb{E}^{\mathbb{P}}\bigg[\int_0^T (|Z_t^\beta|^2 + |Z_t^{0,\beta}|^2)dt + \sup_{0\le t\le T} |M_t^\beta|^2\bigg] < \infty.$$

Therefore, for any $p \in [1, 2)$,

$$\mathbb{E}^{\mathbb{P}^\beta}\bigg[\bigg(\int_0^T (|Z_t^\beta|^2 + |Z_t^{0,\beta}|^2)dt\bigg)^{p/2} + \sup_{0\le t\le T} |M_t^\beta|^p\bigg] < \infty.$$

Also, by stochastic integration by parts and by orthogonality of $\boldsymbol{W}$ and $\boldsymbol{M}^\beta$ (under $\mathbb{P}$), we may check that $(\mathcal{E}_t^\beta M_t^\beta)_{0\le t\le T}$, is an $\mathbb{F}$-martingale under $\mathbb{P}$, so that $\boldsymbol{M}^\beta$ is a martingale under $\mathbb{P}^\beta$. Returning to (1.40), taking the conditional expectation given $\mathcal{F}_t$ for any $t \in [0, T]$, and recalling that $(\boldsymbol{W}^0, \boldsymbol{W}^\beta)$ is a $2d$-dimensional Brownian motion under $\mathbb{P}^\beta$, it is quite standard to deduce that $\boldsymbol{Y}^\beta$ is a bounded process under $\mathbb{P}^\beta$ (and thus under $\mathbb{P}$ as well). In particular, for any $p \ge 1$,

$$\mathbb{E}^{\mathbb{P}^\beta}\bigg[\sup_{0\le t\le T}\bigg(\int_0^T Z_s^\beta \cdot dW_s^\beta + \int_0^T Z_s^{0,\beta} \cdot dW_s^0 + M_T^\beta\bigg)^{2p}\bigg] < \infty. \tag{1.44}$$

Now, by orthogonality of $\boldsymbol{W}$ and $\boldsymbol{M}^\beta$ (under $\mathbb{P}$), we may check that $(\mathcal{E}_t^\beta M_t^\beta W_t^\beta)_{0\le t\le T}$ are $\mathbb{F}$-martingales under $\mathbb{P}$, so that $[\boldsymbol{M}^\beta, \boldsymbol{W}^\beta]_\cdot \equiv 0$ under $\mathbb{P}^\beta$. Similarly, $[\boldsymbol{M}^\beta, \boldsymbol{W}^0]_\cdot \equiv 0$ under $\mathbb{P}^\beta$. Therefore, by (1.44) and by Burkholder-Davis-Gundy inequalities (for possibly discontinuous martingales), we deduce that, for any $p \ge 1$, $\mathbb{E}^{\mathbb{P}^\beta}[(\int_0^T |Z_t^\beta|^2 dt)^p]$, $\mathbb{E}^{\mathbb{P}^\beta}[(\int_0^T |Z_t^{0,\beta}|^2 dt)^p]$ and $\mathbb{E}^{\mathbb{P}^\beta}[\sup_{0\le t\le T} |M_t^\beta|^p]$ are finite.

It now remains to check that the forward and backward equations in (1.39) are satisfied pathwise by $(\boldsymbol{X}^\beta, \boldsymbol{Y}^\beta, \boldsymbol{Z}^\beta, \boldsymbol{Z}^{0,\beta}, \boldsymbol{M}^\beta)$ under $(X_0, \boldsymbol{W}^0, \boldsymbol{\mu}, \boldsymbol{W}^\beta)$, but this is a mere consequence of the definition of $\boldsymbol{W}^\beta$. □

End of the Proof of Theorem 1.57

In order to complete the proof of Theorem 1.57, we prove that the assumptions of Proposition 1.58 are satisfied.

Proof of Theorem 1.57. There are two important things that we need to prove: 1) first that the FBSDE (1.39) with any choice of cut-off functions is strongly uniquely solvable; 2) second, that the process $\boldsymbol{Z} = (Z_t)_{0\le t\le T}$ is bounded independently of the choice of the cut-off functions. Given these two ingredients, it is clear that (1.37) has a solution with $\boldsymbol{Z}$ bounded. In order to prove the strong unique solvability, we shall make use of Proposition 1.52, while we shall follow the proof of Lemma (Vol I)-4.11 to establish that $\boldsymbol{Z}$ is bounded. As we shall see, the key point for proving both claims is to show that we can bound the Lipschitz constant of the decoupling field independently of the cut-off functions in (1.39).

First Step. Obviously, the coefficients of (1.39) satisfy assumption **Lipschitz FBSDE in Random Environment**. In order to apply Proposition 1.52, the main difficulty is thus to check that assumption **Iteration of Lipschitz FBSDE** holds. Without any loss of generality, we can check it for $t = 0$ only. As a by-product of our proof, we shall have that the decoupling field is bounded, uniformly in all the variables and independently of the choice of the cut-off functions.

We thus consider an admissible set-up $(\Omega, \mathcal{F}, \mathbb{F}, \mathbb{P})$ with input $(\boldsymbol{W}^0, \boldsymbol{\mu}, \boldsymbol{W})$ and no initial information. On this set-up, we assume that there exists a solution $(\boldsymbol{X}, \boldsymbol{Y}, \boldsymbol{Z}, \boldsymbol{Z}^0, \boldsymbol{M})$ to (1.39) with some deterministic initial condition $x \in \mathbb{R}^d$ at $t = 0$. Given this solution, we let $(\mathcal{E}_t)_{0 \le t \le T}$ be the exponential local martingale associated with the stochastic integral:

$$\left(- \int_0^t \left[\psi(Z_s)(\sigma^{-1}b)\big(s, X_s, \mu_s, \phi(\hat{\alpha}_s)\big) \right] \cdot dW_s \right)_{0 \le t \le T},$$

i.e. the Doléans exponential of this stochastic integral, where $\hat{\alpha}_s$ is understood as $\hat{\alpha}_s = \hat{\alpha}(s, X_s, \mu_s, \sigma(s, X_s, \mu_s)^{-1\dagger} Z_s)$. Since the integrand is bounded, $(\mathcal{E}_t)_{0 \le t \le T}$ is a true martingale and we can define the probability measure $\mathbb{Q} = \mathcal{E}_T \cdot \mathbb{P}$. Under $\mathbb{Q}$, the process

$$\left(W_t^{\mathbb{Q}} = W_t + \int_0^t \psi(Z_s)(\sigma^{-1}b)\big(s, X_s, \mu_s, \phi(\hat{\alpha}_s)\big) ds \right)_{0 \le t \le T}$$

is a d-dimensional Brownian motion which, like $\boldsymbol{W}$ under $\mathbb{P}$, is orthogonal to $\boldsymbol{M}$ under $\mathbb{Q}$. Following the steps of the proof of Proposition 1.58, we learn that under $\mathbb{Q}$, $(\Omega, \mathcal{F}, \mathbb{F}, \mathbb{Q})$ equipped with $(\boldsymbol{W}^0, \boldsymbol{\mu}, \boldsymbol{W}^{\mathbb{Q}})$ is an admissible set-up, and that the pair $(\boldsymbol{W}^0, \boldsymbol{\mu})$ has the same law under $\mathbb{Q}$ and under $\mathbb{P}$. Moreover, on the probabilistic set-up $(\Omega, \mathcal{F}, \mathbb{F}, \mathbb{Q})$ equipped with $(\boldsymbol{W}^0, \boldsymbol{\mu}, \boldsymbol{W}^{\mathbb{Q}})$, $(X_t, Y_t, Z_t, Z_t^0, M_t)_{0 \le t \le T}$ is a solution of the forward-backward SDE:

$$\begin{cases} dX_t = \sigma(t, X_t, \mu_t) dW_t^{\mathbb{Q}} + \sigma^0(t, X_t, \mu_t) dW_t^0, \\ dY_t = -\psi(Z_t) f\big(t, X_t, \mu_t, \phi\big(\hat{\alpha}(t, X_t, \mu_t, \sigma(t, X_t, \mu_t)^{-1\dagger} Z_t)\big)\big) dt \\ \qquad -\psi(Z_t) Z_t \cdot (\sigma^{-1}b)\big(t, X_t, \mu_t, \phi\big(\hat{\alpha}(t, X_t, \mu_t, \sigma(t, X_t, \mu_t)^{-1\dagger} Z_t)\big)\big) dt \\ \qquad + Z_t \cdot dW_t^{\mathbb{Q}} + Z_t^0 \cdot dW_t^0 + dM_t, \end{cases} \tag{1.45}$$

over the interval $[0, T]$, with the same terminal condition as before. For any initial condition, the forward-backward SDE (1.45) with no initial information is strongly uniquely solvable. Notice that there is no need to assume that T is small enough since the forward equation is decoupled, and the backward equation can be treated as a standard BSDE. In particular, calling U_0 the decoupling field of (1.45) at time 0, we deduce that, $\mathbb{Q}$-almost surely, $Y_0 = U_0(x, \mathbb{Q} \circ (\boldsymbol{W}^0, \boldsymbol{\mu})^{-1}, (\boldsymbol{W}^0, \boldsymbol{\mu}))$. Since $\mathbb{Q}$ and $\mathbb{P}$ are equivalent and $\mathbb{Q} \circ (\boldsymbol{W}^0, \boldsymbol{\mu})^{-1} = \mathbb{P} \circ (\boldsymbol{W}^0, \boldsymbol{\mu})^{-1}$, we deduce that, $\mathbb{P}$-almost surely, $Y_0 = U_0(x, \mathbb{P} \circ (\boldsymbol{W}^0, \boldsymbol{\mu})^{-1}, (\boldsymbol{W}^0, \boldsymbol{\mu}))$. Therefore, in order to check assumption **Iteration of Lipschitz FBSDE**, it suffices to control the Lipschitz constant of U_0 in space.

Second Step. We now use the fact that the decoupling field is independent of the probabilistic set-up used to construct a solution to (1.45). We thus consider the same FBSDE as in (1.45), but on the original set-up with $\boldsymbol{W}^{\mathbb{Q}}$ replaced by the original $\boldsymbol{W}$:

$$\left\{\begin{aligned} dX_t &= \sigma(t, X_t, \mu_t)dW_t + \sigma^0(t, X_t, \mu_t)dW_t^0, \\ dY_t &= -\psi(Z_t)f\big(t, X_t, \mu_t, \phi\big(\hat{\alpha}(t, X_t, \mu_t, \sigma(t, X_t, \mu_t)^{-1\dagger}Z_t)\big)\big)dt \\ &\quad -\psi(Z_t)Z_t \cdot (\sigma^{-1}b)\big(t, X_t, \mu_t, \phi\big(\hat{\alpha}(t, X_t, \mu_t, \sigma(t, X_t, \mu_t)^{-1\dagger}Z_t)\big)\big)dt \\ &\quad +Z_t \cdot dW_t + Z_t^0 \cdot dW_t^0 + dM_t, \end{aligned}\right. \tag{1.46}$$

for $t \in [0, T]$, with x as deterministic initial condition. We then denote the solution by $(\boldsymbol{X}^x, \boldsymbol{Y}^x, \boldsymbol{Z}^x, \boldsymbol{Z}^{0,x}, \boldsymbol{M}^x)$.

The goal is then to prove that there exists a constant Γ, only depending upon L and T (and independent of the cut-off functions) such that, for all $x, x' \in \mathbb{R}^d$:

$$\mathbb{P}\big[|Y_0^{x'} - Y_0^x| \leq \Gamma|x' - x|\big] = 1,$$

which by Proposition 1.46 is enough to control the Lipschitz constant of the decoupling field. Fixing the values of x and x' and letting:

$$\begin{aligned} &\big(\delta X_t, \delta Y_t, \delta Z_t, \delta Z_t^0, \delta M_t\big) \\ &\quad = \big(X_t^{x'} - X_t^x, Y_t^{x'} - Y_t^x, Z_t^{x'} - Z_t^x, Z_t^{0,x'} - Z_t^{0,x}, M_t^{x'} - M_t^x\big), \quad t \in [0, T], \end{aligned}$$

we can write:

$$d\delta X_t = \big[\delta\sigma_t \delta X_t\big]dW_t + \big[\delta\sigma_t^0 \delta X_t\big]dW_t^0, \quad t \in [0, T], \tag{1.47}$$

where we used the same notation as in Theorem (Vol I)-4.45, namely $\delta\sigma_t\delta X_t$ and $\delta\sigma_t^0\delta X_t$ read as square matrices of dimension d, with the following entries:

$$\begin{aligned} \big(\delta\sigma_t \delta X_t\big)_{i,j} &= \sum_{\ell=1}^d \big(\delta\sigma_t\big)_{i,j,\ell}\big(\delta X_t\big)_\ell, \\ \big(\delta\sigma_t^0 \delta X_t\big)_{i,j} &= \sum_{\ell=1}^d \big(\delta\sigma_t^0\big)_{i,j,\ell}\big(\delta X_t\big)_\ell, \quad i, j \in \{1, \cdots, d\}^2, \end{aligned}$$

where $(\delta X_t)_\ell$ is the ℓth coordinate of δX_t and

$$\begin{aligned} \big(\delta\sigma_t\big)_{i,j,\ell} &= \frac{\sigma_{i,j}\big(t, X_t^{\ell-1;x\leftrightsquigarrow x'}, \mu_t\big) - \sigma_{i,j}\big(t, X_t^{\ell;x\leftrightsquigarrow x'}, \mu_t\big)}{(\delta X_t)_\ell}\mathbf{1}_{\{(\delta X_t)_\ell \neq 0\}}, \\ \big(\delta\sigma_t^0\big)_{i,j,\ell} &= \frac{\sigma_{i,j}^0\big(t, X_t^{\ell-1;x\leftrightsquigarrow x'}, \mu_t\big) - \sigma_{i,j}^0\big(t, X_t^{\ell;x\leftrightsquigarrow x'}, \mu_t\big)}{(\delta X_t)_\ell}\mathbf{1}_{\{(\delta X_t)_\ell \neq 0\}}, \end{aligned}$$

with:

$$X_t^{\ell;x\leftrightsquigarrow x'} = \big((X_t^x)_1, \cdots, (X_t^x)_\ell, (X_t^{x'})_{\ell+1}, \cdots, (X_t^{x'})_d\big).$$

From the Lipschitz property of σ and σ^0 in the space variable, we deduce that the processes $(\delta\sigma_t)_{0\le t\le T}$ and $(\delta\sigma_t^0)_{0\le t\le T}$ are bounded by a constant Γ only depending upon L in the assumption. As in the proof of Theorem (Vol I)-4.45, $(\delta\sigma_t\delta X_t)_{i,j}$ reads as the inner product of $((\delta\sigma_t)_{i,j,\ell})_{1\le\ell\le d}$ and $(\delta X_t)_\ell)_{1\le\ell\le d}$, but, because of the additional indices (i,j), we decided not to indicate the inner product explicitly in the notation while we shall let it appear for the terms coming below.

Indeed, in a similar fashion, the pair $(\delta Y_t, \delta Z_t, \delta Z_t^0)_{0\le t\le T}$ satisfies a backward equation of the form:

$$\begin{aligned}\delta Y_t &= \delta g_T \cdot \delta X_T + \int_t^T \big(\delta F_s^{(1)} \cdot \delta X_s + \delta F_s^{(2)} \cdot \delta Z_s\big) ds \\ &\quad - \int_t^T \big(\delta Z_s \cdot dW_s + \delta Z_s^0 \cdot dW_s^0\big) + \big(\delta M_T - \delta M_t\big),\end{aligned} \tag{1.48}$$

where δg_T is an $\mathbb{R}^d$-valued random variable bounded by Γ while $\delta \boldsymbol{F}^{(1)} = (\delta F_t^{(1)})_{0\le t\le T}$ and $\delta \boldsymbol{F}^{(2)} = (\delta F_t^{(2)})_{0\le t\le T}$ are bounded and progressively measurable $\mathbb{R}^d$-valued processes, whose bounds depend upon the details of the cut-off functions ϕ and ψ. As in the proof of Theorem (Vol I)-4.45, "$\cdot$" denotes the inner product acting on elements of $\mathbb{R}^d$. Notice also that, as a uniform bound on the growth of $\delta \boldsymbol{F}^{(1)}$ and $\delta \boldsymbol{F}^{(2)}$, we have:

$$\begin{aligned}|\delta F_t^{(1)}| &\le \Gamma(1 + |Z_t^x|^2 + |Z_t^{x'}|^2), \\ |\delta F_t^{(2)}| &\le \Gamma(1 + |Z_t^x| + |Z_t^{x'}|) \quad t \in [0,T],\end{aligned}$$

the constant Γ only depending on L in the assumption.

Using the fact $\delta \boldsymbol{F}^{(2)}$ is bounded, we introduce a probability $\mathbb{Q}$ (we use again the letter $\mathbb{Q}$ although this probability is different from the one introduced in the first step), equivalent to $\mathbb{P}$, under which $(W_t^{\mathbb{Q}} = W_t - \int_0^t \delta F_s^{(2)} ds)_{0\le t\le T}$ is a Brownian motion. Following the proof of Proposition 1.58, we know that $\delta \boldsymbol{M}$ and $\int_0^\cdot \delta Z_s^0 \cdot dW_s^0$ remain $\mathbb{F}$-martingales under $\mathbb{Q}$. Moreover, it is easily checked that $(\int_0^t \delta Z_s \cdot dW_s^{\mathbb{Q}})_{0\le t\le T}$ is also an $\mathbb{F}$-martingale under $\mathbb{Q}$. Therefore,

$$\mathbb{E}\big[\delta Y_0 | \mathcal{F}_0\big] = \mathbb{E}^{\mathbb{Q}}\big[\delta Y_0 | \mathcal{F}_0\big] = \mathbb{E}^{\mathbb{Q}}\bigg[\bigg(\delta g_T \cdot \delta X_T + \int_0^T \delta F_s^{(1)} \cdot \delta X_s \, ds\bigg) \,\Big|\, \mathcal{F}_0\bigg]. \tag{1.49}$$

In order to handle the right-hand side in (1.49), we need to investigate $d\mathbb{Q}/d\mathbb{P}$. This requires to go back to (1.46). The usual trick is to take the exponential of the solution and to expand $(\exp(\lambda Y_t))_{0\le t\le T}$ for a well-chosen λ. Taking advantage of the convexity of the exponential and of the fact that g, f, and b are bounded in x, we get, for any $x \in \mathbb{R}^d$,

$$\begin{aligned}&\exp(\lambda Y_t^x) \\ &\quad + \frac{\lambda^2}{2}\bigg(\int_t^T \exp(\lambda Y_s^x)\big(|Z_s^x|^2 + |Z_s^{0,x}|^2\big) ds + \int_t^T \exp(\lambda Y_{s-}^x) d\big[(\boldsymbol{M}^x)^c\big]_s\bigg) \\ &\le \exp(\lambda Y_T^x) + C\lambda \int_t^T \exp(\lambda Y_s^x)(1 + |Z_s^x|^2) ds \\ &\quad - \lambda \int_t^T \exp(\lambda Y_{s-}^x)(Z_s^x \cdot dW_s + Z_s^{0,x} \cdot dW_s^0 + dM_s^x),\end{aligned} \tag{1.50}$$

where $(\boldsymbol{M}^x)^c$ is the continuous part of $\boldsymbol{M}^x$.

Choosing $|\lambda|$ large enough and then taking the conditional expectation given $\mathcal{F}_t$, we deduce that $(Y_t^x)_{0\le t\le T}$ is bounded by Γ, where we allow the constant Γ to increase from line to line, while still remaining independent of x. In particular, any continuous version of the decoupling field $\mathbb{R}^d \ni x \mapsto U_0(x, \mathbb{P}\circ(\boldsymbol{W}^0,\boldsymbol{\mu})^{-1},\cdot)$ (such a continuous version exists by Proposition 1.46 but the value of the Lipschitz constant therein depends on the details of ϕ and ψ) must be bounded by Γ with probability 1 under $\mathbb{P}\circ(\boldsymbol{W}^0,\boldsymbol{\mu})^{-1}$.

Third Step. Actually, the bound (1.50) says more. It says that the stochastic integral $(\int_0^t Z_s^x \cdot dW_s)_{0\le t\le T}$ is of bounded mean oscillation (or BMO for short). See Definition (Vol I)-4.17. That is, for any $\mathbb{F}$-stopping time τ, $\mathbb{E}[\int_\tau^T |Z_s^x|^2 ds|\mathcal{F}_\tau] \le C'$, for some constant C' that only depends on the maturity time T and the constant C in the assumption. Luckily, the same holds by replacing Z_s^x by $\delta F_s^{(2)}$ from (1.48), as $|\delta F_s^{(2)}| \le \Gamma(1+|Z_s^x|+|Z_s^{x'}|)$. By Proposition (Vol I)-4.18, the BMO property implies that there exists an exponent $r>1$, only depending on L and T, such that the r-moment of the Doléans exponential martingale associated with $(\int_0^t \delta F_s^{(2)}\cdot dW_s)_{0\le t\le T}$ is bounded by Γ (for a new value of Γ only depending on L and T), namely:

$$\mathbb{E}\Big[\Big(\frac{d\mathbb{Q}}{d\mathbb{P}}\Big)^r\Big] = \mathbb{E}\Big[\exp\Big(r\int_0^T \delta F_s^{(2)}\cdot dW_s - \frac{r}{2}\int_0^T |\delta F_s^{(2)}|^2 ds\Big)\Big] \le \Gamma.$$

We claim that the same result holds conditional on $\mathcal{F}_0$, that is, $\mathbb{P}$-almost surely,

$$\mathbb{E}\Big[\Big(\frac{d\mathbb{Q}}{d\mathbb{P}}\Big)^r \,|\, \mathcal{F}_0\Big] = \mathbb{E}\Big[\exp\Big(r\int_0^T \delta F_s^{(2)}\cdot dW_s - \frac{r}{2}\int_0^T |\delta F_s^{(2)}|^2 ds\Big) \,|\, \mathcal{F}_0\Big] \le \Gamma.$$

Indeed, for any event $E\in\mathcal{F}_0$ with $\mathbb{P}(E)>0$, we may apply Proposition (Vol I)-4.18 with the conditional probability $\mathbb{P}[\cdot|E]$. Under this probability, $\boldsymbol{W}$ is an $\mathbb{F}$-Brownian motion and the stochastic integral $\int_0^\cdot Z_s\cdot dW_s$ coincides with that constructed under $\mathbb{P}$.

Now by (1.47), we deduce that for any $p\ge 1$, there exists a constant C_p independent of the choice of the cut-off functions ϕ and ψ, such that $\mathbb{E}[\sup_{0\le t\le T}|\delta X_s|^p|\mathcal{F}_0]^{1/p} \le C_p|x-x'|$. Therefore, applying Hölder's inequality, (1.49) and the bound for the r-conditional moment of $d\mathbb{Q}/d\mathbb{P}$ given $\mathcal{F}_0$, we obtain:

$$|\delta Y_0| \le \Gamma|x-x'|\Big\{1+\mathbb{E}\Big[\Big(\int_0^T (|Z_s^x|^2+|Z_s^{x'}|^2)ds\Big)^\varrho |\mathcal{F}_0\Big]^{1/\varrho}\Big\}, \tag{1.51}$$

for some $\varrho>1$. In order to estimate the right-hand side, we invoke Proposition (Vol I)-4.18 again, using the same trick as above in order to handle the conditioning. It guarantees that the ϱ-conditional moment of $\int_0^T(|Z_s^x|^2+|Z_s^{x'}|^2)ds$ given $\mathcal{F}_0$ is bounded by a constant that only depends upon ϱ and the BMO norms of the martingales $(\int_0^t Z_s^x\cdot dW_s)_{0\le t\le T}$ and $(\int_0^t Z_s^{x'}\cdot dW_s)_{0\le t\le T}$. We deduce that, with probability 1 under $\mathbb{P}$,

$$|\delta Y_0| \le \Gamma|x-x'|,$$

for a new value of the constant Γ, only depending upon L and T. This proves the required estimate for the Lipschitz constant of the decoupling field associated with the system (1.46). Using Proposition 1.46, we complete the proof of the strong unique solvability of (1.39).

Fourth Step. Consider now a solution $(\boldsymbol{X},\boldsymbol{Y},\boldsymbol{Z},\boldsymbol{Z}^0,\boldsymbol{M})$ of (1.39) on some admissible set-up $(\Omega,\mathcal{F},\mathbb{P})$ equipped with some $(X_0,\boldsymbol{W}^0,\boldsymbol{\mu},\boldsymbol{W})$, X_0 denoting the initial condition of (1.39).

It then remains to prove that the martingale integrand $\boldsymbol{Z}$ is bounded, independently of the choice of the cut-off functions. The proof is inspired by the proof of Lemma (Vol I)-4.11.

Arguing as in the first and second steps and in particular, making use of Girsanov's transform in the same way, it suffices to work with the solution of (1.45). By uniqueness in law, it even suffices to provide a bound for the martingale integrand in the solution of (1.46), the advantage being that (1.46) is constructed on the original set-up. Using the above notations, we denote the driver of the backward equation by:

$$\begin{aligned} F\big(t, X_t, \mu_t, Z_t\big) &= \psi(Z_t) f\big(t, X_t, \mu_t, \phi\big(\hat{\alpha}(t, X_t, \mu_t, \sigma(t, X_t, \mu_t)^{-1\dagger} Z_t)\big)\big) \\ &\quad - \psi(Z_t) Z_t \cdot (\sigma^{-1} b)\big(t, X_t, \mu_t, \phi\big(\hat{\alpha}(t, X_t, \mu_t, \sigma(t, X_t, \mu_t)^{-1\dagger} Z_t)\big)\big). \end{aligned}$$

From the assumptions we have on f and b, we may reproduce the argument used in (1.48) and write:

$$F\big(t, X_t, \mu_t, Z_t\big) = F\big(t, X_t, \mu_t, 0\big) + \delta F_t \cdot Z_t,$$

where $\delta \boldsymbol{F} = (\delta F_t)_{0 \le t \le T}$ is a bounded and progressively measurable $\mathbb{R}^d$-valued process, the bound depending on the cut-off functions ϕ and ψ. As a uniform control on the growth of $\delta \boldsymbol{F}$, we have:

$$|\delta F_t| \le \Gamma'\big(1 + |Z_t|\big), \quad t \in [0, T],$$

for a constant Γ' only depending on T and L in the assumption. Without any loss of generality, we can also assume that:

$$|F\big(t, X_t, \mu_t, 0| \le \Gamma', \quad t \in [0, T].$$

In particular, for a given $t \in [0, T]$, we can introduce a new probability measure $\mathbb{Q}'_t$, given by:

$$\frac{d\mathbb{Q}'_t}{d\mathbb{P}} = \exp\bigg(\int_t^T \delta F_s \cdot dW_s - \frac{1}{2}\int_t^T |\delta F_s|^2 ds\bigg).$$

Under $\mathbb{Q}'_t$, the process $(W_s^{\mathbb{Q}'_t} = W_s - \int_t^s \delta F_u du)_{t \le s \le T}$ is a Brownian motion. Then, by the same argument as above, based on (1.50) and the theory of BMO martingales, we can find an exponent $r' > 1$ such that:

$$\mathbb{E}\bigg[\Big(\frac{d\mathbb{Q}'_t}{d\mathbb{P}}\Big)^{r'} | \mathcal{F}_t\bigg] \le \Gamma', \tag{1.52}$$

for a possibly new value of Γ'. Under the probability $\mathbb{Q}'_t$, the backward equation in (1.46) takes the form:

$$dY_s = -F(s, X_s, \mu_s, 0)dt + Z_s \cdot dW_s^{\mathbb{Q}'_t} + Z_s^0 \cdot dW_s^0 + dM_s, \quad t \le s \le T, \tag{1.53}$$

on the t-initialized set-up $(\Omega, (\mathcal{F}_s)_{t\le s\le T}, \mathbb{Q}'_t)$ with $(\sigma\{X_0, W^0_s, \mu_s, W_s;\ s \le t\}, (W^0_s - W^0_t, \mu_s, W^{\mathbb{Q}'_t}_s - W^{\mathbb{Q}'_t}_t)_{t\le s\le T})$ as input. Admissibility of the set-up is proved in two steps. We check first that the set-up $(\Omega, (\mathcal{F}_s)_{t\le s\le T}, \mathbb{P})$ equipped with $(\sigma\{X_0, W^0_s, \mu_s, W_s;\ s \le t\}, (W^0_s - W^0_t, \mu_s, W_s - W_t)_{t\le s\le T})$ is admissible, which basically follows from the arguments used to justify the induction principle preceding the statement of Proposition 1.52. Then, proceeding as in the first two steps, we deduce that the set-up obtained by application of Girsanov's transform is also admissible. Notice that we use the σ-field $\sigma\{X_0, W^0_s, \mu_s, W_s;\ s \le t\}$ as initial information, while the initial condition X_t is only measurable with respect to the completion of $\sigma\{X_0, W^0_s, \mu_s, W_s;\ s \le t\}$. Of course, it suffices to restart from a version of X_t which is $\sigma\{X_0, W^0_s, \mu_s, W_s;\ s \le t\}$-measurable in order to guarantee the measurability property of the initial condition with respect to the initial information. Also, recall from the same arguments as in the first two steps that, under the probability $\mathbb{Q}'_t$, $\boldsymbol{M}^x$ remains a martingale, orthogonal to $\boldsymbol{W}^{\mathbb{Q}'_t}$ and $\boldsymbol{W}^0$.

Then, for $t+\Delta t \in [t,T]$, we can multiply both sides of (1.53) by $\int_t^{t+\Delta t} Z_s \cdot dW^{\mathbb{Q}'_t}_s$. Recall indeed that $\int_0^T |Z_s|^2 ds$ has finite moments of any order under $\mathbb{P}$ and then, thanks to (1.52), under $\mathbb{Q}'_t$. Since $\boldsymbol{M}^x$ and $\boldsymbol{W}^0$ are orthogonal to $\boldsymbol{W}^{\mathbb{Q}'_t}$ under $\mathbb{Q}'_t$ and since $(F(s, X_s, \mu_s, 0)_{t\le s\le T}$ is bounded by Γ', we deduce that:

$$\mathbb{E}^{\mathbb{Q}'_t}\Big[\int_t^{t+\Delta t} |Z^x_s|^2 ds \mid \mathcal{F}_t\Big] \le \Gamma'\Delta t + \mathbb{E}^{\mathbb{Q}'_t}\Big[Y^x_{t+\Delta t}\int_t^{t+\Delta t} Z^x_s \cdot dW^{\mathbb{Q}'_t}_s \mid \mathcal{F}_t\Big].$$

Now comes another crucial fact that we proved in (1.43): $\int_t^{t+\Delta t} Z_s \cdot dW^{\mathbb{Q}'_t}_s$ is orthogonal to $\mathcal{F}_t \vee \sigma\{W^0_s, \mu_s;\ t \le s \le T\}$ under $\mathbb{Q}'_t$. Introducing the decoupling field at time $t+\Delta t$ of (1.46), we get:

$$\begin{aligned}&\mathbb{E}^{\mathbb{Q}'_t}\Big[\int_t^{t+\Delta t} |Z_s|^2 ds \mid \mathcal{F}_t\Big]\\ &\le \Gamma'\Delta t + \mathbb{E}^{\mathbb{Q}'_t}\Big[\Big(Y_{t+\Delta t} - V_{t+\Delta t}\big(X_t, (X_0, \boldsymbol{W}^0, \boldsymbol{\mu})\big)\Big)\int_t^{t+\Delta t} Z_s \cdot dW^{\mathbb{Q}'_t}_s \mid \mathcal{F}_t\Big],\end{aligned}$$

where we let:

$$\begin{aligned}&V_{t+\Delta t}\big(x, (X_0, \boldsymbol{W}^0, \boldsymbol{\mu})\big)\\ &= U_{t+\Delta t}\Big(x, \mathcal{L}\big((W^0_s - W^0_t, \mu_s)_{t+\Delta t\le s\le T} | \mathcal{F}^{\mathrm{nat},(X_0,\boldsymbol{W}^0,\boldsymbol{\mu})}_{t+\Delta t}\big), (W^0_s - W^0_t, \mu_s)_{t+\Delta t\le s\le T}\Big).\end{aligned}$$

By Cauchy-Schwarz inequality and by the standard convexity inequality $ab \le 2a^2 + b^2/2$, we deduce that:

$$\mathbb{E}^{\mathbb{Q}'_t}\Big[\int_t^{t+\Delta t} |Z_s|^2 ds\, |\mathcal{F}_t\Big] \le \Gamma'\Delta t + \Gamma'\mathbb{E}^{\mathbb{Q}'_t}\Big[\big|Y_{t+\Delta t} - V_{t+\Delta t}\big(X_t, (X_0, \boldsymbol{W}^0, \boldsymbol{\mu})\big)\big|^2\, |\mathcal{F}_t\Big].$$

Then,

$$\begin{aligned}
&\mathbb{E}^{\mathbb{Q}'_t}\Big[\int_t^{t+\Delta t} |Z_s|^2 ds \,|\mathcal{F}_t\Big] \\
&\leq \Gamma'\Delta t + \Gamma'\mathbb{E}\Big[\big|Y_{t+\Delta t} - V_{t+\Delta t}\big(X_t,(X_0,\mathbf{W}^0,\boldsymbol{\mu})\big)\big|^{2r'/(r'-1)}\big|\,\mathcal{F}_t\Big]^{\frac{r'-1}{r'}} \\
&\leq \Gamma'\Delta t,
\end{aligned}$$

where we used the identity $Y_{t+\Delta t} = V_{t+\Delta t}(X_{t+\Delta t},(X_0,\mathbf{W}^0,\boldsymbol{\mu}))$ and the fact that $V_{t+\Delta t}(\cdot,(X_0,\mathbf{W}^0,\boldsymbol{\mu}))$ is Lipschitz continuous. As usual, the constant Γ' is allowed to increase from line to line. Multiplying both sides by:

$$\vartheta_t \frac{d\mathbb{Q}'_0}{d\mathbb{P}}|_{\mathcal{F}_t},$$

for a bounded $\mathcal{F}_t$-measurable random variable ϑ_t with values in $\mathbb{R}_+$, we get:

$$\mathbb{E}^{\mathbb{Q}'_0}\Big[\vartheta_t\int_t^{t+\Delta t} |Z_s|^2 ds\Big] \leq \Gamma'\mathbb{E}^{\mathbb{Q}'_0}\Big[\vartheta_t\int_t^{t+\Delta t} ds\Big].$$

Now, by a standard approximation argument, we deduce that, for any bounded and progressively measurable process $(\vartheta_s)_{0\leq s\leq T}$ with values in $\mathbb{R}_+$, we have:

$$\mathbb{E}^{\mathbb{Q}'_0}\Big[\int_0^T \vartheta_s|Z_s|^2 ds\Big] \leq \Gamma'\mathbb{E}^{\mathbb{Q}'_0}\Big[\int_0^T \vartheta_s ds\Big].$$

Choosing $\vartheta_s = \mathbf{1}_{\{|Z_s|^2>\Gamma'\}}$, we complete the proof. □

1.4.4 Revisiting the Pontryagin Stochastic Maximum Principle

The derivation of the above representation of the value function requires invertibility of the diffusion matrix σ. When σ is not invertible, if the coefficients are differentiable, we may use a stochastic version of the Pontryagin maximum principle instead of the first prong of the probabilistic approach described above.

Using the same set-up for the environment, together with the full Hamiltonian H as defined for example in (1.34), the approach based on the Pontryagin stochastic maximum principle leads to the FBSDE:

$$\begin{cases}
dX_t = b\big(t,X_t,\mu_t,\hat{\alpha}(t,X_t,\mu_t,Y_t)\big)dt \\
\qquad\quad +\sigma\big(t,X_t,\mu_t\big)dW_t + \sigma^0\big(t,X_t,\mu_t\big)dW_t^0, \\
dY_t = -\partial_x H\big(t,X_t,\mu_t,Y_t,Z_t,Z_t^0,\hat{\alpha}(t,X_t,\mu_t,Y_t)\big)dt \\
\qquad\quad +Z_t dW_t + Z_t^0 dW_t^0 + dM_t, \\
Y_T = \partial_x g(X_T,\mu_T),
\end{cases} \tag{1.54}$$

where $\boldsymbol{M} = (M_t)_{0\le t\le T}$ is a square-integrable, mean-zero continuous martingale, of zero cross-variation with $(\boldsymbol{W}^0, \boldsymbol{W})$. We refer to (1.34) for the definition of the Hamiltonian H. Since the control $\boldsymbol{\alpha} = (\alpha_t)_{0\le t\le T}$ does not appear in the volatilities σ and σ^0, we can use the reduced Hamiltonian $H^{(r)}$ defined as:

$$H^{(r)}(t,x,\mu,y,\alpha) = b(t,x,\mu,\alpha)\cdot y + f(t,x,\mu,\alpha),$$

for $t \in [0,T]$, $x, y \in \mathbb{R}^d$, $\mu \in \mathcal{X}$ and $\alpha \in A$, to determine which minimizer $\hat{\alpha}$ to use. Recall (1.32) for a previous discussion of this fact. So as before, we denote by $\hat{\alpha}(t,x,\mu,y)$ a minimizer of the Hamiltonian, namely:

$$\hat{\alpha}(t,x,\mu,y) \in \text{argmin}_{\alpha\in A} H^{(r)}(t,x,\mu,y,\alpha) = \text{argmin}_{\alpha\in A} H(t,x,\mu,y,z,z^0,\alpha),$$

and again, existence, uniqueness, and smoothness properties of a minimizer can be derived under suitable convexity conditions, taking full advantage of the fact that A is a closed convex subset of $\mathbb{R}^k$. See for instance Lemma 1.56 for a typical result.

Necessary Condition

Since the framework is nonstandard, we provide a proof of the necessary part of the Pontryagin stochastic maximum for optimal control problems in random environment, very much in the spirit of Subsection 1.4.1.

Assumption (Necessary SMP in Random Environment).

(A1) The functions b and f are differentiable with respect to (x,α), the mappings $\mathbb{R}^d \times A \ni (x,\alpha) \mapsto \partial_x(b,f)(t,x,\mu,\alpha)$ and $\mathbb{R}^d \times A \ni (x,\alpha) \mapsto \partial_\alpha(b,f)(t,x,\mu,\alpha)$ being continuous for each $(t,\mu) \in [0,T]\times\mathcal{X}$. Similarly, the functions σ, σ^0 and g are differentiable with respect to x, the mapping $\mathbb{R}^d \ni x \mapsto \partial_x(\sigma,\sigma^0)(t,x,\mu)$ being continuous for each $(t,\mu) \in [0,T]\times\mathcal{X}$, and $\mathbb{R}^d \ni (x,\mu) \mapsto \partial_x g(x,\mu)$ being continuous for each $\mu \in \mathcal{X}$.

(A2) The functions $[0,T] \ni t \mapsto (b,f)(t,0,0_{\mathcal{X}},0_A)$ and $[0,T] \ni t \mapsto (\sigma,\sigma^0)(t,0,0_{\mathcal{X}})$ are uniformly bounded, for some points $0_{\mathcal{X}} \in \mathcal{X}$ and $0_A \in A$. The derivatives $\partial_{(x,\alpha)}b$ and $\partial_x(\sigma,\sigma^0)$ are uniformly bounded. There exists a constant L such that, for any $R \ge 0$ and any (t,x,μ,α) with $|x| \le R$, $d(0_{\mathcal{X}},\mu) \le R$ and $|\alpha| \le R$, $|\partial_x f(t,x,\mu,\alpha)|$, $|\partial_x g(x,\mu)|$ and $|\partial_\alpha f(t,x,\mu,\alpha)|$ are bounded by $L(1+R)$.

Following the approach we took to generalize the standard Pontryagin stochastic maximum principle to the case of the optimal control of McKean-Vlasov diffusion processes in Theorem (Vol I)-6.14, we arrive at the following form of the necessary part of the principle. Clearly, it is tailored to the random environment framework considered in this part of the book.

Theorem 1.59 *Under assumption* **Necessary SMP in Random Environment**, *if we assume further that the Hamiltonian H is convex in $\alpha \in A$, then if the admissible control $\boldsymbol{\alpha} = (\alpha_t)_{0 \le t \le T} \in \mathbb{A}$ is optimal, for the associated controlled state $\boldsymbol{X} = (X_t)_{0 \le t \le T}$, and the corresponding solution $(\boldsymbol{Y}, \boldsymbol{Z}, \boldsymbol{Z}^0, \boldsymbol{M}) = (Y_t, Z_t, Z_t^0, M_t)_{0 \le t \le T}$ of the adjoint backward SDE:*

$$\begin{cases} dY_t = -\partial_x H\big(t, X_t, \mu_t, Y_t, Z_t, Z_t^0, \alpha_t\big)dt \\ \qquad\qquad + Z_t dW_t + Z_t^0 dW_t^0 + dM_t, \quad t \in [0, T], \\ Y_T = \partial_x g(X_T, \mu_T), \end{cases} \tag{1.55}$$

we have:

$$\forall \alpha \in A, \quad H^{(r)}(t, X_t, \mu_t, Y_t, \alpha_t) \le H^{(r)}(t, X_t, \mu_t, Y_t, \alpha), \tag{1.56}$$

$\mathrm{Leb}_1 \otimes \mathbb{P}$ *almost everywhere.*

Proof. The proof is similar to the proof of Theorem (Vol I)-6.14 in Subsection (Vol I)-6.3.1, so we only provide the main steps, and focus on the main differences. We first introduce some notation. For $\boldsymbol{\alpha} \in \mathbb{A}$ and $\boldsymbol{\beta} \in \mathbb{H}^{2,k}$ such that $\boldsymbol{\alpha} + \epsilon \boldsymbol{\beta} \in \mathbb{A}$ for $\epsilon > 0$ small enough, we let $\boldsymbol{\theta} = (\theta_t = (X_t, \mu_t, \alpha_t))_{0 \le t \le T}$ and we define the variation process $\boldsymbol{V} = (V_t)_{0 \le t \le T}$ as the solution of the stochastic differential equation:

$$dV_t = \big[\partial_x b(t, \theta_t) \cdot V_t + \partial_\alpha b(t, \theta_t) \cdot \beta_t\big]dt + \big[\partial_x \sigma(t, \theta_t) \cdot V_t\big]dW_t + \big[\partial_x \sigma^0(t, \theta_t) \cdot V_t\big]dW_t^0,$$

for $t \in [0, T]$, with $V_0 = 0$. Notice that we write $\sigma(t, \theta)$ and $\sigma^0(t, \theta)$ with $\theta = (x, \mu, \alpha)$, even though we assume that σ and σ^0 do not depend upon the control parameter α. If we let $\boldsymbol{\alpha}^\epsilon = (\alpha_t^\epsilon = \alpha_t + \epsilon \beta_t)_{0 \le t \le T}$ and $\boldsymbol{X}^\epsilon = \boldsymbol{X}^{\boldsymbol{\alpha}^\epsilon}$, then, repeating the computations of Lemma (Vol I)-6.10, we have:

$$\lim_{\epsilon \searrow 0} \mathbb{E}\bigg[\sup_{0 \le t \le T} \bigg|\frac{X_t^\epsilon - X_t}{\epsilon} - V_t\bigg|^2\bigg] = 0. \tag{1.57}$$

Recalling the definition of $J^\mu(\boldsymbol{\alpha})$ in (1.31), in analogy with Lemma (Vol I)-6.11, we also have that the function $\mathbb{A} \ni \boldsymbol{\alpha} \mapsto J^\mu(\boldsymbol{\alpha})$ is Gâteaux differentiable in the direction $\boldsymbol{\beta}$ and its derivative is given by:

$$\begin{aligned} &\frac{d}{d\epsilon} J^\mu(\boldsymbol{\alpha} + \epsilon \boldsymbol{\beta})\big|_{\epsilon=0} \\ &= \mathbb{E}\bigg[\int_0^T \big[\partial_x f(t, \theta_t) \cdot V_t + \partial_\alpha f(t, \theta_t) \cdot \beta_t\big]dt + \partial_x g(X_T, \mu_T) \cdot V_T\bigg]. \end{aligned} \tag{1.58}$$

Consider now the BSDE (1.55) by treating X_t as part of the randomness of the driver of this BSDE. The discussion of Example 1.20 says that it admits a unique solution $(\boldsymbol{Y}, \boldsymbol{Z}, \boldsymbol{Z}^0, \boldsymbol{M})$ on $(\Omega, \mathcal{F}, \mathbb{F}, \mathbb{P})$. Then we check, as in Lemma (Vol I)-6.12, that:

$$\mathbb{E}[Y_T \cdot V_T] = \mathbb{E} \int_0^T \big[Y_t \cdot \big(\partial_\alpha b(t, \theta_t) \cdot \beta_t\big) - \partial_x f(t, \theta_t) \cdot V_t \big] dt. \tag{1.59}$$

As in the proof of Lemma (Vol I)-6.12, the argument relies on an integration by parts. However, some care is needed because of the additional presence of $\boldsymbol{M}$. The proof now requires the use of stochastic integration by parts for discontinuous semi-martingales. In particular, one has to check that $\int_0^{\cdot} Y_{t-} \cdot dM_t$ is a true martingale. The standard argument is to prove that the supremum over $[0, T]$ is integrable by applying Burkholder-Davies-Gundy inequality together with the a priori estimate:

$$\mathbb{E}\bigg[\bigg(\int_0^T |Y_{t-}|^2 d\Big(\text{trace}\big([M]_t\big)\Big)\bigg)^{1/2}\bigg] \leq \mathbb{E}\big[\sup_{t\in[0,T]} |Y_t|^2\big]^{1/2} \mathbb{E}\big[\text{trace}\big([M]_T\big)\big]^{1/2} < \infty.$$

Putting together the duality relation (1.59) and (1.58), and using the terminal condition for Y_T, we deduce that the Gâteaux derivative of J^μ at $\boldsymbol{\alpha}$ in the direction $\boldsymbol{\beta}$ can be written as:

$$\frac{d}{d\epsilon} J^\mu(\boldsymbol{\alpha} + \epsilon \boldsymbol{\beta})\big|_{\epsilon=0} = \mathbb{E} \int_0^T \partial_\alpha H^{(r)}(t, X_t, \mu_t, Y_t, \alpha_t) \cdot \beta_t \, dt. \tag{1.60}$$

Finally, we conclude as we did in the proof of Theorem (Vol I)-6.14. □

Sufficiency

We now turn to a convenient form of the sufficient condition for the Pontryagin stochastic maximum principle, very much in the spirit of Theorem (Vol I)-3.17. In order to do so, we introduce the following set of assumptions:

Assumption (Sufficient SMP in Random Environment). Assume that there exist two constants $L \geq 0$ and $\lambda > 0$ such that:

(A1) The function b has the form $b(t, x, \mu, \alpha) = b_0(t, \mu) + b_1(t)x + b_2(t)\alpha$, where b_0, b_1 and b_2 are measurable mappings with values in $\mathbb{R}^d$, $\mathbb{R}^{d\times d}$ and $\mathbb{R}^{d\times k}$ respectively, and satisfy:

$$|b_0(t, \mu)| \leq L\big(1 + d(0_{\mathcal{X}}, \mu)\big), \quad |b_1(t)|, |b_2(t)| \leq L,$$

for some point $0_{\mathcal{X}} \in \mathcal{X}$. Similarly, there exist measurable functions $[0, T] \times \mathcal{X} \ni (t, \mu) \mapsto (\sigma_0, \sigma_0^0)(t, \mu)$ and $[0, T] \ni t \mapsto (\sigma_1, \sigma_1^0)(t)$, with values in $(\mathbb{R}^{d\times d})^2$ and $(\mathbb{R}^{d\times d\times d})^2$, such that:

$$\sigma(t, x, \mu) = \sigma_0(t, \mu) + \sigma_1(t)x, \quad \sigma^0(t, x, \mu) = \sigma_0^0(t, \mu) + \sigma_1^0(t)x,$$

where, for all $t \in [0, T]$ and $\mu \in \mathcal{X}$,

$$|(\sigma_0, \sigma_0^0)(t, \mu)| \leq L\big(1 + d(0_{\mathcal{X}}, \mu)\big), \quad |(\sigma_1, \sigma_1^0)(t)| \leq L.$$

(continued)

(A2) The function f is measurable and, for any $t \in [0, T]$ and $\mu \in \mathcal{X}$, the function $\mathbb{R}^d \times A \ni (x, \alpha) \mapsto f(t, x, \mu, \alpha)$ is continuously differentiable with L-Lipschitz continuous derivatives, and f is convex and uniformly λ-convex in α in the sense that:

$$f(t, x', \mu, \alpha') - f(t, x, \mu, \alpha) - (x' - x, \alpha' - \alpha) \cdot \partial_{(x,\alpha)} f(t, x, \mu, \alpha) \geq \lambda |\alpha' - \alpha|^2,$$

where $\partial_{(x,\alpha)} f$ stands for the gradient in the joint variables (x, α). Moreover,

$$|f(t, x, \mu, \alpha)| \leq L\big(1 + |x|^2 + [d(0_{\mathcal{X}}, \mu)]^2 + |\alpha|^2\big),$$
$$|(\partial_x f, \partial_\alpha f)(t, x, \mu, \alpha)| \leq L\big(1 + |x| + d(0_{\mathcal{X}}, \mu) + |\alpha|\big).$$

(A3) The function g is measurable and, for any $\mu \in \mathcal{X}$, the function $\mathbb{R}^d \ni x \mapsto g(x, \mu)$ is continuously differentiable and convex, and has a L-Lipschitz-continuous derivative. Moreover,

$$|g(x, \mu)| \leq L\big(1 + |x|^2 + [d(0_{\mathcal{X}}, \mu)]^2\big),$$
$$|\partial_x g(x, \mu)| \leq L\big(1 + |x| + d(0_{\mathcal{X}}, \mu)\big).$$

Notice that assumption **Sufficient SMP in Random Environment** subsumes assumption **Hamiltonian Minimization in Random Environment**. In particular, Lemma 1.56 addressing the regularity of $\hat{\alpha}$ applies.

Here is the corresponding version of Theorem (Vol I)-3.17.

Theorem 1.60 *Under assumption* **Sufficient SMP in Random Environment**, *consider an admissible set-up* $(\Omega, \mathcal{F}, \mathbb{F}, \mathbb{P})$ *for some input* $(X_0, \boldsymbol{W}^0, \boldsymbol{\mu}, \boldsymbol{W})$. *The forward-backward system:*

$$\begin{cases} dX_t = b\big(t, X_t, \mu_t, \hat{\alpha}(t, X_t, \mu_t, Y_t)\big)dt \\ \qquad\quad + \sigma\big(t, X_t, \mu_t\big)dW_t + \sigma^0\big(t, X_t, \mu_t\big)dW_t^0, \\ dY_t = -\partial_x H\big(t, X_t, \mu_t, Y_t, Z_t, Z_t^0, \hat{\alpha}(t, X_t, \mu_t, Y_t)\big)dt \\ \qquad\quad + Z_t dW_t + Z_t^0 dW_t^0 + dM_t, \end{cases} \tag{1.61}$$

with X_0 *as initial condition and* $Y_T = \partial_x g(X_T, \mu_T)$ *as terminal condition, where* $(M_t)_{0 \leq t \leq T}$ *is a square-integrable càd-làg martingale, with* $M_0 = 0$ *and zero cross-variation with* $(\boldsymbol{W}^0, \boldsymbol{W})$, *has a unique solution* $(\hat{X}_t, \hat{Y}_t, \hat{Z}_t, \hat{Z}_t^0, \hat{M}_t)_{0 \leq t \leq T}$ *such that:*

$$\mathbb{E}\Big[\sup_{0 \leq t \leq T} \big(|\hat{X}_t|^2 + |\hat{Y}_t|^2 + |\hat{M}_t|^2\big) + \int_0^T \big(|\hat{Z}_t|^2 + |\hat{Z}_t^0|^2\big)dt\Big] < +\infty, \tag{1.62}$$

and if we set $\hat{\boldsymbol{\alpha}} = (\hat{\alpha}_t = \hat{\alpha}(t, \hat{X}_t, \mu_t, \hat{Y}_t))_{0 \le t \le T}$, *then for any progressively measurable control* $\boldsymbol{\beta} = (\beta_t)_{0 \le t \le T}$ *satisfying* $\mathbb{E}\int_0^T |\beta_t|^2 dt < \infty$, *it holds:*

$$J^{\mu}(\hat{\boldsymbol{\alpha}}) + \lambda \mathbb{E} \int_0^T |\beta_t - \hat{\alpha}_t|^2 dt \le J^{\mu}(\boldsymbol{\beta}). \tag{1.63}$$

As by Proposition 1.52, the FBSDE (1.61) *admits a decoupling field U which is C-Lipschitz-continuous in x uniformly in the other variables, for a constant C which only depends on L and T, and in particular, which is independent of* $t \in [0, T]$. *As a result,* $\hat{\boldsymbol{Y}}$ *may be represented as a function of* $\hat{\boldsymbol{X}}$ *as in Proposition 1.50.*

Also, for any $t \in [0, T]$, $(x, \bar{\mathbb{P}}^0) \in \mathbb{R}^d \times \mathcal{P}_2(\bar{\Omega}^{0,t})$ *with* $\bar{\Omega}^{0,t} = \mathcal{C}([t, T]; \mathbb{R}^d) \times \mathcal{D}([t, T]; \mathcal{X})$, *such that* $(w_s^0)_{t \le s \le T}$ *is a d-dimensional Brownian motion starting from zero under* $\bar{\mathbb{P}}^0$ *with respect to the natural filtration generated by the canonical process* $(w_s^0, \nu_s)_{t \le s \le T}$, *it holds:*

$$|U_t(x, \bar{\mathbb{P}}^0, \cdot)| \le C\Big(1 + |x| + \mathbb{E}^{\bar{\mathbb{P}}^0}\big[\sup_{t \le s \le T} d(0_{\mathcal{X}}, \nu_s)^2 \,|\, \mathcal{G}_{t+}\big]^{1/2}\Big),$$

where the σ*-field* $\mathcal{G}_{t+}$ *is defined as* $\mathcal{G}_{t+} = \cap_{\varepsilon > 0} \sigma\{w_s^0, \nu_s;\ t \le s \le t + \varepsilon\}$.

Proof. Provided that (1.61) is indeed solvable, the proof of (1.63) is a straightforward replication of the proof of Theorem (Vol I)-3.17 with the additional appeal, as in the proof of Theorem 1.59, to the stochastic integration by parts formula for discontinuous martingales in order to handle the fact that $\boldsymbol{M}$ is discontinuous.

So we only need to prove the unique strong solvability of (1.61). By Lemma 1.56, we first observe that (1.61) has Lipschitz continuous coefficients. In order to complete the proof, it is thus sufficient to check the assumption of Proposition 1.52. Again the proof is a replication of the case when $\boldsymbol{\mu}$ is deterministic, see for instance Lemma (Vol I)-4.56. We reproduce it here for the sake of completeness. Given $x, y \in \mathbb{R}^d$, we consider on some t-initialized set-up $(\Omega, \mathcal{F}, (\mathcal{F}_s)_{t \le s \le T}, \mathbb{P})$ for an input $(W_s^0, \mu_s, W_s)_{t \le s \le T}$ and no initial information, two solutions of (1.61), if they do exist, $(\boldsymbol{X}^{t,x}, \boldsymbol{Y}^{t,x}, \boldsymbol{Z}^{t,x}, \boldsymbol{Z}^{0,t,x}, \boldsymbol{M}^{t,x})$ and $(\boldsymbol{X}^{t,y}, \boldsymbol{Y}^{t,y}, \boldsymbol{Z}^{t,y}, \boldsymbol{Z}^{0,t,y}, \boldsymbol{M}^{t,y})$ with x and y as initial conditions at time t. Repeating the proof of (1.63), but taking into account the fact that the initial conditions may be different and using conditional expectations given $\mathcal{F}_t$ instead of expectations, we get:

$$\begin{aligned}(y - x) \cdot Y_t^{t,x} &+ \mathbb{E}_t\Big[\int_t^T f(r, X_r^{t,x}, \mu_r, \hat{\alpha}_r^{t,x}) dr + g(X_T^{t,x}, \mu_T)\Big] + \lambda \mathbb{E}_t\Big[\int_t^T |\hat{\alpha}_r^{t,x} - \hat{\alpha}_r^{t,y}|^2 dr\Big] \\ &\le \mathbb{E}_t\Big[\int_t^T f(r, X_r^{t,y}, \mu_r, \hat{\alpha}_r^{t,y}) dr + g(X_T^{t,y}, \mu_T)\Big],\end{aligned}$$

where we used the short notations $(\hat{\alpha}_s^{t,\xi} = \hat{\alpha}(s, X_s^{t,\xi}, \mu_s, Y_s^{t,\xi}))_{t \le s \le T}$, for $\xi = x$ or y, and $\mathbb{E}_t[\cdot] = \mathbb{E}[\cdot|\mathcal{F}_t]$. Exchanging the roles of x and y, we get:

$$(x-y)\cdot Y_t^{t,y} + \mathbb{E}_t\Big[\int_t^T f(r, X_r^{t,y}, \mu_r, \hat{\alpha}_r^{t,y})dr + g(X_T^{t,y}, \mu_T)\Big] + \lambda \mathbb{E}_t\Big[\int_t^T |\hat{\alpha}_r^{t,y} - \hat{\alpha}_r^{t,y}|^2 dr\Big]$$
$$\le \mathbb{E}_t\Big[\int_t^T f(r, X_r^{t,x}, \mu_r, \hat{\alpha}_r^{t,x})dr + g(X_T^{t,x}, \mu_T)\Big].$$

Adding these two inequalities, we deduce that:

$$2\lambda \mathbb{E}_t\Big[\int_t^T |\hat{\alpha}_r^{t,y} - \hat{\alpha}_r^{t,x}|^2\Big] \le (y-x)\cdot (Y_t^{t,y} - Y_t^{t,x}). \tag{1.64}$$

Now, one can check that:

$$\mathbb{E}_t\big[\sup_{t\le s\le T} |X_s^{t,x} - X_s^{t,y}|^2\big] \le C\Big(|x-y|^2 + \mathbb{E}_t\Big[\int_t^T |\hat{\alpha}_r^{t,x} - \hat{\alpha}_r^{t,y}|^2 dr\Big]\Big),$$

and, then:

$$\mathbb{E}_t\big[\sup_{t\le s\le T} |Y_s^{t,x} - Y_s^{t,y}|^2\big] \le C\Big(|x-y|^2 + \mathbb{E}_t\Big[\int_t^T |\hat{\alpha}_r^{t,x} - \hat{\alpha}_r^{t,y}|^2 dr\Big]\Big), \tag{1.65}$$

for some constant C depending only L, λ, and T in the assumptions. By (1.64) and (1.65), we easily deduce that the assumptions of Proposition 1.52 are satisfied.

We now discuss the growth of the decoupling field U_t at time t. Going back to (1.63) and choosing $x = 0$ and $\boldsymbol{\beta} \equiv 0_A$ therein, for some point $0_A \in A$, we see that:

$$\mathbb{E}_t\Big[\int_t^T f\big(s, X_s^{t,0}, \mu_s, \hat{\alpha}_s^{t,0}\big)ds + g\big(X_T^{t,0}, \mu_T\big) + \lambda \int_t^T |\hat{\alpha}_s^{t,0} - 0_A|^2 ds\Big] \le J^{\mu}(0_A), \tag{1.66}$$

the dynamics associated with the null control being given by the solution of:

$$dX'_s = b\big(s, X'_s, \mu_s, 0_A\big)ds + \sigma\big(s, X'_s, \mu_s\big)dW_s + \sigma^0\big(s, X'_s, \mu_s\big)dW_s^0, \quad s \in [t,T]\ ;\ X'_t = 0.$$

It is clear that:

$$\mathbb{E}_t\big[\sup_{t\le s\le T} |X'_s|^2\big] \le C'\Big(1 + \mathbb{E}_t\big[\sup_{0\le s\le T} d(0_{\mathcal{X}}, \mu_s)^2\big]\Big),$$

for a constant C' independent of t, where $0_{\mathcal{X}}$ denotes any fixed point in $\mathcal{X}$. Therefore, by (1.66), we get:

$$\mathbb{E}_t\Big[\int_t^T f\big(s, X_s^{t,0}, \mu_s, \hat{\alpha}_s^{t,0}\big)ds + g\big(X_T^{t,0}, \mu_T\big) + \lambda \int_t^T |\hat{\alpha}_s^{t,0}|^2 ds\Big]$$
$$\le C'\Big(1 + \mathbb{E}_t\big[\sup_{0\le s\le T} d(0_{\mathcal{X}}, \mu_s)^2\big]\Big).$$

Using the fact that g and f are convex in (x, α), we get for any $\epsilon \in (0,1)$:

$$\mathbb{E}_t\Big[\int_t^T f(s,0,\mu_s,0_A)ds + g(0,\mu_T) + \lambda \int_t^T |\hat{\alpha}_s^{t,0}|^2 ds\Big]$$
$$\le C'\Big[1 + \epsilon^{-1}\mathbb{E}_t\big(\sup_{0\le s\le T} d(0_{\mathcal{X}},\mu_s)^2\big) + \epsilon\mathbb{E}_t\big(\sup_{t\le s\le T}|X_s^{t,0}|^2\big) + \epsilon\mathbb{E}_t\int_t^T |\hat{\alpha}_s^{t,0}|^2 ds\Big].$$

Allowing the constant C' to increase from line to line, it is then quite straightforward to deduce that:

$$\mathbb{E}_t\Big[\int_t^T |\hat{\alpha}_s^{t,0}|^2 ds\Big] \le C'\Big[1 + \mathbb{E}_t\big(\sup_{0\le s\le T} d(0_{\mathcal{X}},\mu_s)^2\big)\Big],$$

and then:

$$\mathbb{E}_t\big[\sup_{t\le s\le T}\big(|X_s^{t,0}|^2 + |Y_s^{t,0}|^2\big)\big] \le C'\big(1 + \mathbb{E}_t\big[\sup_{0\le s\le T} d(0_{\mathcal{X}},\mu_s)^2\big]\big), \tag{1.67}$$

from which we conclude that, $\mathbb{P}$-almost surely:

$$\Big|U_t\Big(0, \mathcal{L}\big((W_s^0,\mu_s)_{t\le s\le T}|\mu_t\big), (W_s^0,\mu_s)_{t\le s\le T}\Big)\Big| \le C'\Big[1 + \big(\mathbb{E}_t\big[\sup_{0\le s\le T} d(0_{\mathcal{X}},\mu_s)^2\big]\big)^{1/2}\Big].$$

Using the Lipschitz property of U_t, we deduce that, $\mathbb{P}$-almost surely, for all $x \in \mathbb{R}^d$:

$$\begin{aligned}&\Big|U_t\Big(x, \mathcal{L}\big((W_s^0,\mu_s)_{t\le s\le T}|\mu_t\big), (W_s^0,\mu_s)_{t\le s\le T}\Big)\Big| \\ &\qquad \le C'\Big[1 + |x| + \big(\mathbb{E}_t\big[\sup_{0\le s\le T} d(0_{\mathcal{X}},\mu_s)^2\big]\big)^{1/2}\Big].\end{aligned} \tag{1.68}$$

The argument may be reproduced on any admissible set-up. In particular, we may choose the canonical set-up introduced in the proof of Proposition 1.46 with any distribution $\bar{\mathbb{P}}^0$ for $(w_s^0, \nu_s)_{t\le s\le T}$, provided that it satisfies the conditions required therein. □

Remark 1.61 *Under the assumption of Proposition 1.55, Itô-Wentzell formula provides another expression for the optimal solution. Therefore, if uniqueness of the optimal solution holds, the two decoupling fields constructed in this section should satisfy:*

$$U_t\big(x, \mathcal{L}\big((W_s^0 - W_t^0, \mu_s)_{t\le s\le T} \mid \mathcal{F}_t^{\mathrm{nat},W^0}\big), \mathcal{F}_t^{\mathrm{nat},W^0}\big) = \partial_x U^{\mu}(t,x),$$

for $t \in [0,T]$ and $x \in \mathbb{R}^d$, where in the left-hand side, we used the same form for the decoupling field as in Remark 1.49.

1.5 Notes & Complements

The notion of FBSDEs in a random environment is fully justified by our desire to handle mean field games with a common noise, whose role will be played by $\boldsymbol{W}^0$ in the next chapters while $\boldsymbol{W}$ will still denote, as in Volume I, the idiosyncratic noise. The environment $\boldsymbol{\mu} = (\mu_t)_{0 \le t \le T}$ in the coefficients of the equation will account for the random state of the population when subject to systemic shocks.

Because of the random nature of $\boldsymbol{\mu}$, it is then necessary to specify the correlations between $(X_0, \boldsymbol{W}^0, \boldsymbol{\mu}, \boldsymbol{W})$ and the filtration $\mathbb{F}$ supporting the FBSDE. While the pair $(\boldsymbol{W}^0, \boldsymbol{W})$ is explicitly assumed to be an $\mathbb{F}$-Brownian motion, the measurability properties of the process $\boldsymbol{\mu}$ with respect to $\mathbb{F}$ may take several forms. The easiest case is when $\boldsymbol{\mu}$ is adapted to the filtration generated by $\boldsymbol{W}^0$, possibly augmented with some initial information; in such a case, the FBSDE is somewhat standard despite the random nature of the coefficients. However, this case is limited for the applications we have in mind: When solving mean field games with a common noise, we shall face cases where $\boldsymbol{\mu}$ involves an additional source of randomness beyond μ_0 and $\boldsymbol{W}^0$. This is the rationale for introducing the compatibility condition in Subsection 1.1.1. The property used to define immersion of filtrations was introduced by Brémaud and Yor under the name of (H)-hypothesis in [68]. The two equivalences merged into the characterization of compatibility given in Proposition 1.10 can be found in Lemma 2.17 of [216] by Jacod and Mémin and in Theorem 3 of [68] of Brémaud and Yor. Many different names and characterizations are associated with this property of a filtration enlargement. Jeanblanc and Le Cam call it *immersion* in [219], Jacod and Mémin use the notion of *very good extensions* in [216], and El Karoui, Nguyen, and Jeanblanc use instead the notion of *natural extensions* in [225]. We borrowed the terminology *compatible* from Kurtz [247], as in [100, 255]. As explained in the text, compatibility imposes a form of fairness in the observations: Given the realization of $(X_0, \boldsymbol{W}^0, \boldsymbol{\mu}, \boldsymbol{W})$ up until time t, the observation of $\mathcal{F}_t$ does not introduce any bias in the realization of $(X_0, \boldsymbol{W}^0, \boldsymbol{\mu}, \boldsymbol{W})$ after time t. The interest of the notion of compatibility is especially visible in the statement of Theorem 1.33: Provided that strong uniqueness holds true, the law of the solution to the FBSDE in environment $\boldsymbol{\mu}$ only depends on the joint distribution of the input. This is a crucial observation as it permits to choose, in a somewhat systematic way, the canonical space as underlying probability space.

We shall use the notion of compatibility throughout this volume and shed new light on it in the last chapter, when dealing with mean field games of timing, see Section 7.2.

To the best of our knowledge, the use of compatibility conditions for handling backward SDEs with random coefficients goes back to the papers by Buckdahn, Engelbert, and Răşcanu [78] and Buckdahn and Engelbert [76,77] on weak solutions to backward SDEs. Therein, the compatibility condition is expressed in terms of a martingale property, in the spirit of Lemma 1.9. In this regard, it is worth mentioning that this notion of compatibility, although named differently as we mentioned earlier, was first introduced by Jacod [215] and Jacod and Mémin [216] in order to address the connection between weak and strong solutions to a pretty general class of stochastic differential equations. The papers by Kurtz [246, 247], from which the term *compatible* is borrowed, also deal with weak and strong solutions to stochastic models.

In full generality, this notion of compatibility, although named differently, was first introduced by Jacod [215] and Jacod and Mémin [216] in order to address the connection between weak and strong solutions to a pretty general class of stochastic differential equations. The term *compatible* is borrowed from the papers by Kurtz [246, 247], which also deal with weak and strong solutions to stochastic models. In fact, it is fair to say that many different names and characterizations are associated with compatibility as a enlargement of filtration property: (H)-hypothesis [68], immersion [219], very good extensions [216], and natural extensions [225], while we borrow the term *compatible* from Kurtz.

The Kunita-Watanabe decomposition used in Subsection 1.1.3 goes back to the original paper by Kunita and Watanabe [244]. Within the framework of BSDEs, it was first used by El Karoui and Huang [224] to deal with filtrations that do not satisfy the martingale representation theorem, see also El Karoui, Peng, and Quenez [226]. For more recent examples of application, we refer to [29, 111, 240, 333].

Earlier version of the Yamada-Watanabe theorem for coupled FBSDEs with deterministic coefficients were proposed by Antonelli and Ma [26], Delarue [132], Kurtz [246], and Bahlali, Mezerdi, N'zi, and Ouknine [32]. To the best of our knowledge, the version within a random environment, as given in the text, see Theorem 1.33, is new.

It is worth noticing that the theory developed in this chapter for FBSDEs in a random environment addresses almost exclusively the notion of strong solutions, namely solutions that are adapted to the input. Theorem 1.33, our version of the Yamada-Watanabe theorem, is a case in point. In the next chapter, we shall face FBSDEs of McKean-Vlasov type whose solutions are not adapted to the input. As a result, we shall call them weak solutions. In order to select among all the weak solutions, those that are meaningful from the physical point of view, we shall require not only the input but also the output to be compatible with respect to the filtration $\mathbb{F}$ entering the definition of the underlying probabilistic set-up.

The concept of decoupling field for forward-backward stochastic differential equations with random coefficients is due to Ma, Wu, Zhang, and Zhang [272], while the induction scheme described under the header *General Mechanism* of Subsection 1.3.3 was introduced by Delarue in [132] and revisited next in [272].

The proof of the Souslin-Kuratowski theorem used in Proposition 1.32 may be found in Chapter 6 of Bogachev's monograph [64]. For the definition and the basic properties of the $J1$ Skorohod topology, we refer the reader to Billingsley's textbook [57].

Existence and properties of regular versions of conditional probabilities, as exposed in the statement of Theorem 1.1, can be found in many textbooks on probability and measure theory, for example [57, 64, 143, 301].

The use of random value functions and the verification result given in Proposition 1.55 are borrowed from Peng's paper [305]. Itô-Wentzell's formula, which is involved in the derivation of the verification argument based on the stochastic HJB equation, may be found in [105] or [243]. Existence and uniqueness of a classical solution to the stochastic HJB equation are discussed in Cardaliaguet et al. [86], see also Duboscq and Réveillac [142] for similar prospects. We refer to Ma, Yin, and Zhang [273] for the connection between backward SPDEs and forward-backward SDEs with random coefficients. As far as we know, the analysis in Subsection 1.4.3 is new. The formulation is tailor-made to mean field games with a common noise, as introduced in the next chapter. In analogy with our proof of Proposition 1.58, the author in [333] handles quadratic BSDEs when the filtration driving the backward equation may be larger than the filtration generated by a Wiener process. The use of the stochastic Pontryagin principle in Subsection 1.4.4 is more standard: the stochastic Pontryagin principle is known to be well fitted to stochastic optimal control problems with random coefficients.

MFGs with a Common Noise: Strong and Weak Solutions 2

Abstract
The purpose of this chapter is to introduce the notion of mean field game with a common noise. This terminology refers to the fact that in the finitely many player games from which the mean field game is derived, the states of the individual players are subject to correlated noise terms. In a typical model, each individual player feels an idiosyncratic noise as well as random shocks common to all the players. At the level of the mathematical analysis, the common noise introduces a randomization of most of the quantities and equations. In equilibrium, the statistical distribution of the population is no longer deterministic. One of the main feature of the chapter is the introduction and the analysis of the concepts of *weak* and *strong* solutions, very much in the spirit of the classical theory of stochastic differential equations.

2.1 Conditional Propagation of Chaos

Throughout the chapter, we consider game models in which players are subjected to two independent sources of noise: an idiosyncratic noise, independent from one individual to another, and a separate one, common to all the players, accounting for the common environment in which the individual states evolve. We extend the strategy implemented in the previous chapters using the asymptotic analysis of large N-player games to introduce mean field games. In order to do so, we first need to extend the McKean-Vlasov theory of propagation of chaos to accommodate the presence of the common noise.

R. Carmona, F. Delarue, *Probabilistic Theory of Mean Field Games with Applications II*, Probability Theory and Stochastic Modelling 84,
DOI 10.1007/978-3-319-56436-4_2

2.1.1 N-Player Games with a Common Noise

For the sake of simplicity, and because the theoretical results presented in this chapter are restricted to this case, we assume that the dynamics in $\mathbb{R}^d$, with $d \geq 1$, of the private state of player $i \in \{1, \cdots, N\}$ are given by stochastic differential equations (SDEs) of the form:

$$dX_t^i = b\big(t, X_t^i, \bar{\mu}_t^N, \alpha_t^i\big)dt + \sigma\big(t, X_t^i, \bar{\mu}_t^N, \alpha_t^i\big)dW_t^i + \sigma^0\big(t, X_t^i, \bar{\mu}_t^N, \alpha_t^i\big)dW_t^0, \tag{2.1}$$

where $\boldsymbol{W}^0, \boldsymbol{W}^1, \cdots, \boldsymbol{W}^N$ are $N+1$ independent d-dimensional Brownian motions defined on some complete filtered probability space $(\Omega, \mathcal{F}, \mathbb{F} = (\mathcal{F}_t)_{t \geq 0}, \mathbb{P})$. All the computations performed in this chapter apply as well when the dimension of the Brownian motions, say m, is different from d as long as the matrices σ and σ^0 are assumed to be $d \times m$-dimensional instead of $d \times d$. Since this generalization will only render the notation more cumbersome without bringing any new insight into the arguments, we restrict ourselves to Brownian motions with the same dimensions as the states. Also, we refer the reader to the Notes & Complements at the end of the chapter for a discussion of the general case of common noise given by a space-time white noise Gaussian measure. As in the previous chapters, the term $\bar{\mu}_t^N$ in (2.1) denotes the empirical distribution of the individual states at time t:

$$\bar{\mu}_t^N = \frac{1}{N} \sum_{i=1}^N \delta_{X_t^i}. \tag{2.2}$$

The processes $\boldsymbol{\alpha}^i = ((\alpha_t^i)_{t \geq 0})_{1 \leq i \leq N}$ are progressively measurable processes, with values in a Borel subset A of some Euclidean space $\mathbb{R}^k$, with $k \geq 1$, A being often taken to be a closed convex subset of $\mathbb{R}^k$. They stand for control processes. The coefficients b, σ and σ^0 are measurable functions defined on $[0, T] \times \mathbb{R}^d \times \mathcal{P}(\mathbb{R}^d) \times A$ with values in $\mathbb{R}^d$, $\mathbb{R}^{d \times d}$ and $\mathbb{R}^{d \times d}$ respectively. As in the previous chapters, the set $\mathcal{P}(\mathbb{R}^d)$ denotes the space of probability measures on $\mathbb{R}^d$ endowed with the topology of weak convergence.

Given this general set-up, the purpose of the chapter is to discuss asymptotic Nash equilibria when the number of players tends to infinity, each player attempting to minimize a cost functional of the same type as in the previous chapters, namely:

$$J^i(\boldsymbol{\alpha}^1, \cdots, \boldsymbol{\alpha}^N) = \mathbb{E}\bigg[\int_0^T f(t, X_t^i, \bar{\mu}_t^N, \alpha_t^i)dt + g(X_T^i, \bar{\mu}_T^N)\bigg].$$

Because of the presence of extra terms in the state dynamics, we need to reformulate the notion of mean field game in such a more general framework. We shall call these new game models, *mean field games with a common noise*.

Intuitively speaking, even in the limit $N \to \infty$, the equilibrium distribution of the population should still feel the influence of the common noise $\boldsymbol{W}^0$, and for that reason, it should not be deterministic. This forces us to revisit carefully the concepts introduced and analyzed in the previous chapters, and identify the right notion of *stochastic* solution to mean field games with a common noise. In the spirit of the time honored theory of classical stochastic differential equations, we introduce two concepts of solution: *strong* and *weak* solutions, differentiating them by whether or not the solution in question is adapted to the common noise.

Because of the presence of the common noise, we need to revisit several of the basic tools used so far in the analysis of mean field games, and understand what changes should be made when the input data of the models are parameterized by some form of external *random environment*. In line with the developments in the case of deterministic environment, we use the random distribution of the population in equilibrium as a code for the random environment. Because of this new twist in the set-up, we shall often appeal to the theory of FBSDEs and optimization in a random environment developed in the previous Chapter 1.

We first discuss conditional McKean-Vlasov SDEs. Subsequently, we introduce the notions of strong and weak solutions to MFGs with a common noise. We prove a suitable version of the classical Yamada-Watanabe Theorem in this framework. Recall that we already discussed a version of the Yamada-Watanabe Theorem for forward-backward SDEs in Chapter 1.

2.1.2 Set-Up for a Conditional McKean-Vlasov Theory

Before we can tackle the difficult problems raised by the equilibrium theory of large games, we need to understand the behavior as N tends to infinity of the solutions of symmetric systems of N stochastic equations coupled in a mean field way, and driven by idiosyncratic and common noises as in (2.1). In other words, we consider this asymptotic regime first ignoring the optimization component of the problem. For this purpose, we find it convenient to assume that all the models are defined on the same probability space supporting an infinite sequence of independent Wiener processes. So the model for a symmetric system of size N will be given by a system of N SDEs of the form:

$$dX_t^i = b\big(t, X_t^i, \bar{\mu}_t^N\big)dt + \sigma\big(t, X_t^i, \bar{\mu}_t^N\big)dW_t^i + \sigma^0\big(t, X_t^i, \bar{\mu}_t^N\big)dW_t^0 \qquad t \in [0,T],\ i \in \{1, \cdots, N\}, \tag{2.3}$$

for some $T > 0$, where $(\boldsymbol{W}^0, \boldsymbol{W}^1, \cdots, \boldsymbol{W}^N, \cdots)$ is a sequence of independent d-dimensional Wiener processes on some complete probability space $(\Omega, \mathcal{F}, \mathbb{P})$. In this section, the coefficients b, σ and σ^0 are measurable functions defined on $[0,T] \times \mathbb{R}^d \times \mathcal{P}(\mathbb{R}^d)$ with values in $\mathbb{R}^d$, $\mathbb{R}^{d \times d}$ and $\mathbb{R}^{d \times d}$ respectively, and as usual, $\bar{\mu}_t^N$ denotes for any $t \in [0, T]$, the empirical distribution (2.2) of the $(X_t^i)_{i=1,\cdots,N}$. Here, $\boldsymbol{W}^0$ is said to be the *common* source of noise, and as typical in the McKean-Vlasov theory, we shall often refer to the $(X_t^i)_{i=1,\cdots,N}$ as particles.

Exchangeable Sequences of Random Variables

When (2.3) is uniquely solvable and the initial conditions are exchangeable, the random variables $X_t^1, \cdots, X_t^N$ are exchangeable. Recall that a sequence $(X_n)_{n\geq 1}$ of random variables is said to be exchangeable if for every $n \geq 1$, the distribution of $(X_1, \cdots, X_n)$ is invariant under permutation of the indices. This simple remark has very useful consequences because of the fundamental result of De Finetti's theory of exchangeable sequences of random variables:

Theorem 2.1 *For any exchangeable sequence of random variables $(X_n)_{n\geq 1}$ of order 1, i.e., such that $\mathbb{E}[|X_1|] < \infty$, it holds, $\mathbb{P}$ almost surely,*

$$\lim_{n\to\infty} \frac{1}{n}\sum_{i=1}^{n} X_i = \mathbb{E}\big[X_1 | \mathcal{F}_\infty\big],$$

where $\mathcal{F}_\infty$ is the tail σ-field $\mathcal{F}_\infty = \bigcap_{n\geq 1} \sigma\{X_k, k \geq n\}$.

Intuitively, this form of law of large numbers says that, when exchangeability holds, the empirical measure of the random variables $(X_n)_{n\geq 1}$ behave asymptotically as if they were conditionally independent and identically distributed given the tail σ-field $\mathcal{F}_\infty$.

For that reason, and coming back to our current set-up, one may wonder about the convergence of the empirical measure $\bar{\mu}_t^N$ as N tends to ∞. When $\sigma^0 \equiv 0$, the standard theory of propagation of chaos says that, asymptotically, particles become independent and identically distributed, and $\bar{\mu}_t^N$ converges to their common asymptotic distribution. When σ^0 is not trivial, such a result cannot hold since, even in the limit $N \to \infty$, the particles must still keep track of the common noise $\boldsymbol{W}^0$, so they cannot become independent.

Although particles do not become asymptotically independent, in light of Theorem 2.1, it sounds reasonable to expect them to become asymptotically independent conditionally on the information generated by the common noise. So it is tempting to conjecture that, as N tends to ∞, the empirical distribution $\bar{\mu}_t^N$ converges towards the common conditional distribution of each particle given the common source of noise $\boldsymbol{W}^0$. We resolve this issue by a brute force computation in the next paragraph.

Nonlinear Stochastic Fokker-Planck Equation

In order to understand the asymptotic behavior of the empirical distributions $\bar{\mu}_t^N$, we consider their action on test functions. Fixing a smooth test function ϕ with compact support in $[0, T] \times \mathbb{R}^d$ and using a standard form of Itô's formula, we compute:

$$
\begin{aligned}
d\Big\langle \phi(t,\cdot), \frac{1}{N}\sum_{j=1}^{N}\delta_{X_t^j}\Big\rangle &= \frac{1}{N}\sum_{j=1}^{N} d\phi(t,X_t^j)\\
&= \frac{1}{N}\sum_{j=1}^{N}\Big(\partial_t\phi(t,X_t^j)dt + \partial_x\phi(t,X_t^j)\cdot dX_t^j + \frac{1}{2}\mathrm{trace}\big\{\partial^2_{xx}\phi(t,X_t^j)d\big[X^j,X^j\big]_t\big\}\Big)\\
&= \frac{1}{N}\sum_{j=1}^{N}\partial_t\phi(t,X_t^j)dt + \frac{1}{N}\sum_{j=1}^{N}\partial_x\phi(t,X_t^j)\cdot\big(\sigma\big(t,X_t^j,\bar\mu_t^N\big)dW_t^j\big)\\
&\quad+\frac{1}{N}\sum_{j=1}^{N}\partial_x\phi(t,X_t^j)\cdot b\big(t,X_t^j,\bar\mu_t^N\big)dt\\
&\quad+\frac{1}{N}\sum_{j=1}^{N}\partial_x\phi(t,X_t^j)\cdot\big(\sigma^0\big(t,X_t^j,\bar\mu_t^N\big)dW_t^0\big)\\
&\quad+\frac{1}{2N}\sum_{j=1}^{N}\mathrm{trace}\Big\{\Big([\sigma\sigma^\dagger](t,X_t^j,\bar\mu_t^N)+[\sigma^0\sigma^{0\dagger}](t,X_t^j,\bar\mu_t^N)\Big)\partial^2_{xx}\phi(t,X_t^j)\Big\}\,dt.
\end{aligned}
$$

Thinking about the limit as $N\to\infty$ and using the definition of the measures $\bar\mu_t^N$ we can rewrite the above equality as:

$$
\begin{aligned}
&\langle\phi(t,\cdot),\bar\mu_t^N\rangle - \langle\phi(0,\cdot),\bar\mu_0^N\rangle\\
&= \int_0^t\langle\partial_t\phi(s,\cdot),\bar\mu_s^N\rangle ds + \int_0^t\langle\partial_x\phi(s,\cdot)\cdot b\big(s,\cdot,\bar\mu_s^N\big),\bar\mu_s^N\rangle ds\\
&\quad+\frac{1}{2}\int_0^t\Big\langle\mathrm{trace}\Big\{\Big([\sigma\sigma^\dagger](s,\cdot,\bar\mu_s^N)+[\sigma^0\sigma^{0\dagger}](s,\cdot,\bar\mu_s^N)\Big)\partial^2_{xx}\phi(t,\cdot)\Big\},\bar\mu_s^N\Big\rangle ds\\
&\quad+\int_0^t\big\langle\partial_x\phi(s,\cdot)\cdot\sigma^0\big(s,\cdot,\bar\mu_s^N\big)dW_s^0,\bar\mu_s^N\big\rangle + O(N^{-1/2}),
\end{aligned}
$$

where the Landau notation $O(N^{-1/2})$ is to be understood, for example, in the L^2-norm sense if σ is bounded. So if, as $N\to\infty$, the limit

$$
\Big(\mu_t = \lim_{N\to\infty}\bar\mu_t^N\Big)_{0\le t\le T}
$$

exists in the weak functional sense, the above calculation shows that the limit $\boldsymbol{\mu} = (\mu_t)_{0\le t\le T}$ solves (at least formally) the Stochastic Partial Differential Equation (SPDE):

$$
\begin{aligned}
d\mu_t &= -\partial_x \cdot \big[b(t,\cdot,\mu_t)\mu_t\big]dt - \partial_x \cdot \Big(\big[\sigma^0(t,\cdot,\mu_t)dW_t^0\big]\mu_t\Big) \\
&\quad + \frac{1}{2}\mathrm{trace}\Big\{\partial^2_{xx}\Big[\Big(\big[\sigma\sigma^\dagger\big](t,\cdot,\mu_t) + \big[\sigma^0\sigma^{0\dagger}\big](t,\cdot,\mu_t)\Big)\mu_t\Big]\Big\}dt.
\end{aligned}
\tag{2.4}
$$

This SPDE reads as a nonlinear stochastic Fokker-Planck equation. Next, we identify the nonlinear forward SDE (most likely of the McKean-Vlasov type) leading to such a stochastic Fokker-Planck equation.

2.1.3 Formulation of the Limit Problem

While we started with a discussion of games with finitely many players, we now switch to the mathematical set-up of mean field games where the state of a single representative player is the object of interest. In order to disentangle the relative effects of the different sources of noise, we follow the strategy suggested in Subsection 1.2.3, see in particular (1.16). We introducc two complete probability spaces:

$$
(\Omega^0, \mathcal{F}^0, \mathbb{P}^0) \quad \text{and} \quad (\Omega^1, \mathcal{F}^1, \mathbb{P}^1),
$$

endowed with two right-continuous and complete filtrations $\mathbb{F}^0 = (\mathcal{F}_t^0)_{t\geq 0}$ and $\mathbb{F}^1 = (\mathcal{F}_t^1)_{t\geq 0}$. We shall assume that the common noise $\boldsymbol{W}^0$ is constructed on the space $(\Omega^0, \mathbb{F}^0, \mathbb{P}^0)$ and the idiosyncratic noises $(\boldsymbol{W}^n)_{n\geq 1}$ are constructed on the space $(\Omega^1, \mathbb{F}^1, \mathbb{P}^1)$. By convention, the index 0 always refers to the common noise and the index 1 to the idiosyncratic ones. We then define the product structure

$$
\Omega = \Omega^0 \times \Omega^1, \quad \mathcal{F}, \quad \mathbb{F} = (\mathcal{F}_t)_{t\geq 0}, \quad \mathbb{P}, \tag{2.5}
$$

where $(\mathcal{F}, \mathbb{P})$ is the completion of $(\mathcal{F}^0 \otimes \mathcal{F}^1, \mathbb{P}^0 \otimes \mathbb{P}^1)$ and $\mathbb{F}$ is the complete and right continuous augmentation of $(\mathcal{F}_t^0 \otimes \mathcal{F}_t^1)_{t\geq 0}$. Generic elements of Ω are denoted $\omega = (\omega^0, \omega^1)$ with $\omega^0 \in \Omega^0$ and $\omega^1 \in \Omega^1$. Given such a set-up, we shall construct the solution to (2.3) on the product space Ω.

Remark 2.2 *Like in our discussion following Definition 1.28 in Subsection 1.2.3, there is no need to impose a compatibility condition in the construction of the probabilistic set-up $(\Omega, \mathcal{F}, \mathbb{P})$. The reason is that there is no random environment entering the definition of this set-up. Indeed, for any $\mathcal{F}_0$-measurable $\mathbb{R}^d$-valued random variable X_0 accounting for some initial information, the process $(X_0, \boldsymbol{W}^0, \boldsymbol{W})$ is automatically compatible with respect to the filtration $\mathbb{F}$. This follows from the fact that $(\boldsymbol{W}^0, \boldsymbol{W})$ is a 2d-dimensional Brownian motion with respect to $\mathbb{F}$. Here, like most everywhere in the book, we use $\boldsymbol{W}$ for $\boldsymbol{W}^1$ for the sake of simplicity.*

Remark 2.3 *With a slight abuse of notation, we shall not distinguish a random variable X constructed on $(\Omega^0, \mathcal{F}^0, \mathbb{P}^0)$ (resp. $(\Omega^1, \mathcal{F}^1, \mathbb{P}^1)$) with its natural extension $\tilde{X} : (\omega^0, \omega^1) \mapsto X(\omega^0)$ (resp. $\tilde{X} : (\omega^0, \omega^1) \mapsto X(\omega^1)$) on $(\Omega, \mathcal{F}, \mathbb{P})$. Similarly, for a sub-$\sigma$-algebra $\mathcal{G}^0$ of $\mathcal{F}^0$ (resp. $\mathcal{G}^1$ of $\mathcal{F}^1$), we shall often just write $\mathcal{G}^0$ (resp. $\mathcal{G}^1$) for the sub-σ-algebra $\mathcal{G}^0 \otimes \{\emptyset, \Omega^1\}$ (resp. $\{\emptyset, \Omega^0\} \otimes \mathcal{G}^1$) of $\mathcal{F}$.*

Conditional Distributions Given $\mathcal{F}^0$

In order to formulate rigorously the McKean-Vlasov limit, we recall the following useful result, which is nothing but a refinement of Fubini's theorem. Given an $\mathbb{R}^d$-valued random variable X on Ω equipped with the σ-algebra $\mathcal{F}^0 \otimes \mathcal{F}^1$, we have that, for any $\omega^0 \in \Omega^0$, the section $X(\omega^0, \cdot) : \Omega^1 \ni \omega^1 \mapsto X(\omega^0, \omega^1)$ is a random variable on Ω^1 and we may consider the law of $X(\omega^0, \cdot)$, namely $\mathcal{L}(X(\omega^0, \cdot))$, as the realization of a mapping from Ω^0 into $\mathcal{P}(\mathbb{R}^d)$. Below, this mapping is shown to be a random variable from $(\Omega^0, \mathcal{F}^0, \mathbb{P}^0)$ into $\mathcal{P}(\mathbb{R}^d)$ endowed with its Borel σ-field.

However, this fact may not remain true when $\mathcal{F}^0 \otimes \mathcal{F}^1$ is replaced by its completion. A standard counter example is $X(\omega^0, \omega^1) = \mathbf{1}_{C^0}(\omega^0)\mathbf{1}_{C^1}(\omega^1)$ where $C^0 \in \mathcal{F}^0$ has zero $\mathbb{P}^0$-probability and C^1 is a nonmeasurable subset of Ω^1. Indeed, X is clearly measurable from Ω into $\mathbb{R}$ with respect to the completion of $\mathcal{F}^0 \otimes \mathcal{F}^1$. However, for any $\omega^0 \in C^0$, there is no way to compute the law of the section as it is not a random variable on Ω^1 equipped with $\mathcal{F}^1$.

This observation is rather annoying as we decided to work systematically with complete probability spaces and with complete filtrations. In order to overcome this difficulty, we notice that, in the example above, $X(\omega^0, \cdot)$ is not a random variable on $(\Omega^1, \mathcal{F}^1, \mathbb{P}^1)$ for ω^0 in an exceptional event only. It turns out that this fact may be generalized into a more general statement, which is quite classical in measure theory.

> *Given an $\mathbb{R}^d$-valued random variable X on Ω equipped with the completion $\mathcal{F}$ of $\mathcal{F}^0 \otimes \mathcal{F}^1$, for $\mathbb{P}^0$-a.e. $\omega^0 \in \Omega^0$, $X(\omega^0, \cdot)$ is a random variable on $(\Omega^1, \mathcal{F}^1, \mathbb{P}^1)$.*

In particular, we may define $\mathcal{L}(X^0(\omega^0, \cdot))$ for almost every $\omega^0 \in \Omega^0$. On the exceptional event where $\mathcal{L}(X^0(\omega^0, \cdot))$ cannot be computed, we may assign it arbitrary values in $\mathcal{P}(\mathbb{R}^d)$. We claim that the resulting mapping $\mathcal{L}^1(X^0) : \Omega^0 \ni \omega^0 \mapsto \mathcal{L}(X^0(\omega^0, \cdot))$ is a random variable from Ω^0 to $\mathcal{P}(\mathbb{R})$.

Lemma 2.4 *Given a random variable X from Ω, equipped as above with the completion $\mathcal{F}$ of $\mathcal{F}^0 \otimes \mathcal{F}^1$, into $\mathbb{R}^d$, the mapping $\mathcal{L}^1(X) : \Omega^0 \ni \omega^0 \mapsto \mathcal{L}(X(\omega^0, \cdot))$ is almost surely well defined under $\mathbb{P}^0$, and forms a random variable from $(\Omega^0, \mathcal{F}^0, \mathbb{P}^0)$ into $\mathcal{P}(\mathbb{R}^d)$ endowed with its Borel σ-field, as defined in the statement of Proposition (Vol I)-5.7. The random variable $\mathcal{L}^1(X)$ provides a conditional law of X given $\mathcal{F}^0$.*

Moreover, if Y is a version of X which is measurable from $(\Omega, \mathcal{F}^0 \otimes \mathcal{F}^1)$ into $\mathbb{R}^d$, then $\mathcal{L}^1(Y)$ coincides with $\mathcal{L}^1(X)$ with probability 1 under $\mathbb{P}^0$.

Note that, in practice, we shall never specify the definition of $\mathcal{L}^1(X)$ on the exceptional event where it cannot be defined. Also, notice that any other conditional law of X given $\mathcal{F}^0$ is $\mathbb{P}^0$ almost surely equal to $\mathcal{L}^1(X)$. This follows from the fact that $\mathcal{B}(\mathbb{R}^d)$ is generated by a countable π-system and fits the remark given after the statement of Theorem 1.1.

Observe finally that the statement may be easily extended to the case when X takes values in a Polish space $\mathcal{S}$.

Proof. By definition of the completion, there exists a random variable $Y : \Omega \to \mathbb{R}^d$, measurable with respect to $\mathcal{F}^0 \otimes \mathcal{F}^1$, such that $\mathbb{P}[X = Y] = 1$. We then check that $\Omega^0 \ni \omega^0 \mapsto \mathcal{L}(Y(\omega^0, \cdot))$ is a random variable from $(\Omega^0, \mathcal{F}^0)$ into $\mathcal{P}(\mathbb{R}^d)$. By Proposition (Vol I)-5.7, it suffices to prove that, for any $D \in \mathcal{B}(\mathbb{R}^d)$, the mapping $\Omega^0 \ni \omega^0 \mapsto [\mathcal{L}(Y(\omega^0, \cdot))](D)$ is measurable. Noticing that, for all $\omega^0 \in \Omega$,

$$\big[\mathcal{L}\big(Y(\omega^0, \cdot)\big)\big](D) = \mathbb{P}^1\big[Y(\omega^0, \cdot) \in D\big],$$

the result easily follows from the standard version of Fubini-Tonelli theorem. Observe also that, for another version $\tilde{Y}$ of Y (that is $\tilde{Y}$ is also measurable with respect to $\mathcal{F}^0 \otimes \mathcal{F}^1$ and $\mathbb{P}[Y = \tilde{Y}] = 1$), then $\mathcal{L}^1(Y)$ and $\mathcal{L}^1(\tilde{Y})$ are $\mathbb{P}^0$-almost surely equal.

We now denote by $\Omega^{0,\text{well defined}}$ the collection of $\omega^0 \in \Omega^0$ such that $X(\omega^0, \cdot)$ is a random variable from $(\Omega^1, \mathcal{F}^1, \mathbb{P}^1)$ into $\mathbb{R}^d$. We then let $\mathcal{L}^1(X)(\omega^0) = \mathcal{L}(X(\omega^0, \cdot))$ for $\omega^0 \in \Omega^{0,\text{well-defined}}$. And, for $\omega^0 \notin \Omega^{0,\text{well defined}}$, we assign a fixed arbitrary value in $\mathcal{P}(\mathbb{R}^d)$ to $\mathcal{L}^1(X)(\omega^0)$. Now, it is a known fact in measure theory that the complementary of $\Omega^{0,\text{well defined}}$ is a null set and that, for ω^0 in an event of probability 1 under $\mathbb{P}^0$, $\mathbb{P}^1[X(\omega^0, \cdot) = Y(\omega^0, \cdot)] = 1$. In particular, we can find an event in $\mathcal{F}^0$, of measure 1 under $\mathbb{P}^0$, on which $\mathcal{L}^1(X) = \mathcal{L}^1(Y)$. Since $\mathcal{F}^0$ is complete, we deduce that $\mathcal{L}^1(X)$ is a random variable.

In order to prove that $\mathcal{L}^1(X)$ provides a conditional law of X given $\mathcal{F}^0$, we consider $C \in \mathcal{F}^0$ and $D \in \mathcal{B}(\mathbb{R}^d)$. We have

$$\begin{aligned}
\mathbb{E}\big[\mathbf{1}_C \mathbf{1}_D(X)\big] &= \mathbb{E}\big[\mathbf{1}_C \mathbf{1}_D(Y)\big] \\
&= \mathbb{E}^0\big[\mathbf{1}_C \mathbb{P}^1\big(Y \in D\big)\big] = \mathbb{E}^0\big[\mathbf{1}_C \mathcal{L}^1(Y)(D)\big] = \mathbb{E}^0\big[\mathbf{1}_C \mathcal{L}^1(X)(D)\big],
\end{aligned}$$

which completes the proof. □

We apply the same argument in order to formulate the limit problem associated with (2.3). Precisely, using the same notation as in Lemma 2.4 for the conditional distribution, we associate with (2.3) the conditional McKean-Vlasov SDE (MKV SDE for short):

$$dX_t = b\big(t, X_t, \mathcal{L}^1(X_t)\big)dt + \sigma\big(t, X_t, \mathcal{L}^1(X_t)\big)dW_t + \sigma^0\big(t, X_t, \mathcal{L}^1(X_t)\big)dW_t^0, \tag{2.6}$$

for $t \in [0, T]$, this equation being set on the product space $\Omega = \Omega^0 \times \Omega^1$. Because of the augmentation of $\mathbb{F}$ in (2.5), we face the same kind of difficulty as in Lemma 2.4. In other words, nothing guarantees *a priori* that the flow $(\mathcal{L}^1(X_t))_{0 \le t \le T}$ is adapted to $\mathbb{F}^0$. Fortunately, we can appeal to the following result.

Lemma 2.5 *Given an* $\mathbb{R}^d$*-valued process* $(X_t)_{t\geq 0}$*, adapted to the filtration* $\mathbb{F}$*, consider for any* $t \geq 0$*, a version of* $\mathcal{L}^1(X_t)$ *as defined in Lemma 2.4. Then, the* $\mathcal{P}(\mathbb{R}^d)$*-valued process* $(\mathcal{L}^1(X_t))_{t\geq 0}$ *is adapted to* $\mathbb{F}^0$*. If, moreover,* $(X_t)_{t\geq 0}$ *has continuous paths and satisfies* $\mathbb{E}[\sup_{0\leq t\leq T}|X_t|^2] < \infty$ *for all* $T > 0$*, then we can find a version of each* $\mathcal{L}^1(X_t)$*,* $t \geq 0$*, such that the process* $(\mathcal{L}^1(X_t))_{t\geq 0}$ *has continuous paths in* $\mathcal{P}_2(\mathbb{R}^d)$ *and is* $\mathbb{F}^0$*-adapted.*

Proof. Given $t \geq 0$, we know that X_t is measurable with respect to the completion of $\mathcal{F}^0_{t+\varepsilon} \otimes \mathcal{F}^1_{t+\varepsilon}$ for any $\varepsilon > 0$. We then apply Lemma 2.4, but with $\mathcal{F}^0_{t+\varepsilon}$ instead of $\mathcal{F}^0$ and $\mathcal{F}^1_{t+\varepsilon}$ instead of $\mathcal{F}^1$. We deduce that (any version of) $\mathcal{L}^1(X_t)$ is measurable with respect to $\mathcal{F}^0_{t+\varepsilon}$. Letting ε tend to 0, we get that (any version of) $\mathcal{L}^1(X_t)$ is measurable with respect to $\mathcal{F}^0_t$.

In order to prove the second claim, it suffices to construct, for any $T > 0$, a version of each $\mathcal{L}^1(X_t)$, with $t \in [0, T]$, such that the process $(\mathcal{L}^1(X_t))_{0\leq t\leq T}$ has continuous paths from $[0, T]$ into $\mathcal{P}_2(\mathbb{R}^d)$. Since $\boldsymbol{X} = (X_t)_{0\leq t\leq T}$ is assumed to have continuous paths, we can find another $\mathbb{R}^d$-valued process, denoted by $\boldsymbol{Y} = (Y_t)_{0\leq t\leq T}$, with continuous paths, such that $\boldsymbol{Y}$ is a random variable from $(\Omega, \mathcal{F}^0 \otimes \mathcal{F}^1)$ with values in $\mathcal{C}([0, T]; \mathbb{R}^d)$ and $\mathbb{P}[\boldsymbol{X} = \boldsymbol{Y}] = 1$. Since $\mathbb{E}[\sup_{0\leq t\leq T}|Y_t|^2] < \infty$, $\mathbb{E}^1[\sup_{0\leq t\leq T}|Y_t|^2]$ is finite with probability 1 under $\mathbb{P}^0$. From Theorem (Vol I)-5.5 with $p = 2$, we easily deduce that, $\mathbb{P}^0$-almost surely, the mapping $[0, T] \ni t \mapsto \mathcal{L}^1(Y_t) \in \mathcal{P}_2(\mathbb{R}^d)$ is continuous. By Lemma 2.4, $\mathcal{L}^1(Y_t)$ is a version of $\mathcal{L}^1(X_t)$. Since $\mathcal{F}^0_t$ is complete, $\mathcal{L}^1(Y_t)$ is $\mathcal{F}^0_t$-measurable. □

As a byproduct of the proof, observe that we can find a common exceptional event $N \in \mathcal{F}^0$ such that, outside N, for all $t \in [0, T]$, $\mathcal{L}^1(X_t)$ is defined as $\mathcal{L}^1(Y_t)$ and is thus "well defined". Put differently, for ω^0 outside N, it makes sense to consider the entire flow $(\mathcal{L}^1(X_t))_{0\leq t\leq T}$.

Solving the Conditional McKean-Vlasov SDE

Based on our discussion in the previous paragraph, we can introduce the following definition.

Definition 2.6 *On the probabilistic set-up* $(\Omega, \mathcal{F}, \mathbb{F}, \mathbb{P})$ *defined in* (2.5)*, for any square integrable* $\mathcal{F}_0$*-measurable initial condition* X_0 *with values in* $\mathbb{R}^d$*, we call a solution to the conditional McKean-Vlasov SDE* (2.6) *on the interval* $[0, T]$ *an* $(\mathcal{F}_t)_{0\leq t\leq T}$*-adapted process* $\boldsymbol{X} = (X_t)_{0\leq t\leq T}$*, with continuous paths, such that* $\mathbb{E}[\sup_{0\leq t\leq T}|X_t|^2] < \infty$*,*

$$\mathbb{E}\int_0^T \Big(\big|b\big(t, X_t, \mathcal{L}^1(X_t)\big)\big| + \big|\sigma\big(t, X_t, \mathcal{L}^1(X_t)\big)\big|^2 + \big|\sigma^0\big(t, X_t, \mathcal{L}^1(X_t)\big)\big|^2 \Big) dt < \infty,$$

and the process $\boldsymbol{X}$*, together with a continuous* $\mathcal{P}_2(\mathbb{R}^d)$*-valued and* $\mathbb{F}^0$*-adapted version of* $(\mathcal{L}^1(X_t))_{0\leq t\leq T}$*, satisfy* (2.6) *with probability* 1 *under* $\mathbb{P}$*.*

We can now identify the SPDE (2.4) as the stochastic Fokker-Planck equation satisfied by the process $(\mathcal{L}^1(X_t))_{0\leq t\leq T}$ for any solution $\boldsymbol{X} = (X_t)_{0\leq t\leq T}$ of (2.6).

Proposition 2.7 *Let* $X = (X_t)_{0 \le t \le T}$ *be a solution of* (2.6) *on* $(\Omega, \mathcal{F}, \mathbb{F}, \mathbb{P})$ *and let* $\mu = (\mu_t = \mathcal{L}^1(X_t))_{0 \le t \le T}$ *be the flow of marginal conditional distributions of* X *given the common source of noise. Then, with* $\mathbb{P}^0$*-probability* 1, μ *satisfies* (2.4) *in the sense of distributions when acting on smooth functions from* $\mathbb{R}^d$ *to* $\mathbb{R}$ *that tend to* 0 *at infinity.*

Proof. Given a test function ϕ on $\mathbb{R}^d$ in a dense countable subset of the set $\mathcal{C}_0^\infty(\mathbb{R}^d; \mathbb{R})$ of smooth functions from $\mathbb{R}^d$ to $\mathbb{R}$ that tend to 0 at infinity, it suffices to expand $(\phi(X_t))_{0 \le t \le T}$ by means of Itô's formula and take expectation under $\mathbb{P}^1$ of both sides of the expansion to compute $\mathbb{E}^1[\phi(X_t)]$ for any $t \in [0, T]$. Observe that, for any $t \in [0, T]$, $\mathbb{E}^1[\phi(X_t)]$ is well defined up to an exceptional event in $\mathcal{F}^0$. Writing X_t as $e_t(X)$ where X is the path $(X_s)_{0 \le s \le T}$ regarded as a random variable with values in $\mathcal{C}([0, T]; \mathbb{R}^d)$ and e_t is the mapping $\mathcal{C}([0, T]; \mathbb{R}^d) \ni x \mapsto x_t$, we can easily assume that the exceptional event in $\mathcal{F}^0$ on which $\mathbb{E}^1[\phi(X_t)]$ is not well defined is in fact the same for all $t \in [0, T]$.

Now, we recall that, for all $t \in [0, T]$, $\mathbb{P}^0$-almost surely, $\mathbb{E}^1[\phi(X_t)]$ is equal to $\langle \phi, \mu_t \rangle$. Since both quantities are continuous in time, this permits to identify, with probability 1 under $\mathbb{P}^0$, the path $(\langle \phi, \mu_t \rangle)_{0 \le t \le T}$ for all ϕ in a dense countable subset of $\mathcal{C}_0^\infty(\mathbb{R}^d; \mathbb{R})$. □

Since we shall restrict the analysis to square integrable initial conditions, it makes sense to limit ourselves to the case when b, σ and σ^0 are defined on $[0, T] \times \mathbb{R}^d \times \mathcal{P}_2(\mathbb{R}^d)$ only, as opposed to the entire $[0, T] \times \mathbb{R}^d \times \mathcal{P}(\mathbb{R}^d)$. In complete analogy with assumption **MKV SDE** in Subsection (Vol I)-4.2.1, we shall assume:

Assumption (Conditional MKV SDE). The functions b, σ and σ^0 are bounded on bounded subsets of $[0, T] \times \mathbb{R}^d \times \mathcal{P}_2(\mathbb{R}^d)$, and are Lipschitz continuous in x and μ, uniformly in $t \in [0, T]$, $\mathbb{R}^d$ being equipped with the Euclidean norm and $\mathcal{P}_2(\mathbb{R}^d)$ with the 2-Wasserstein distance.

The definition of the 2-Wasserstein distance denoted by W_2 was introduced in Chapter (Vol I)-5. Along the lines of Theorem (Vol I)-4.21, we prove:

Proposition 2.8 *Let assumption* **Conditional MKV SDE** *be in force. Then, given a square integrable* $\mathcal{F}_0$*-measurable initial condition* X_0*, there exists a unique solution to the conditional McKean-Vlasov SDE* (2.6) *on the filtered probability space* $(\Omega, \mathcal{F}, \mathbb{F}, \mathbb{P})$.

Proof. The proof is just a variant of the proof of Theorem (Vol I)-4.21, but we give it for the sake of completeness. It consists in a new application of the contraction mapping theorem. We consider the space $\mathbb{S}^{2,d}$ of all $\mathbb{R}^d$-valued $\mathbb{F}$-progressively measurable processes satisfying:

$$\mathbb{E}\big[\sup_{0 \le t \le T} |X_t|^2\big] < \infty,$$

and we equip it with the norm:

$$\|X\|_{\mathbb{S}}^2 = \mathbb{E}\big[\sup_{0 \le t \le T} |X_t|^2\big] < \infty.$$

The space $\mathbb{S}^{2,d}$, equipped with $\|\boldsymbol{X}\|_{\mathbb{S}}$, is a Banach space. Furthermore, for all $\boldsymbol{X} \in \mathbb{S}^{2,d}$, we can find a $\mathcal{P}_2(\mathbb{R}^d)$-valued and $\mathbb{F}^0$-adapted version of $(\mathcal{L}^1(X_t))_{0\le t\le T}$ with continuous paths. In particular, we can define:

$$\begin{aligned} U_t = \xi &+ \int_0^t b\big(s, X_s, \mathcal{L}^1(X_s)\big)ds \\ &+ \int_0^t \sigma\big(s, X_s, \mathcal{L}^1(X_s)\big)dW_s + \int_0^t \sigma^0\big(s, X_s, \mathcal{L}^1(X_s)\big)dW_s^0, \quad 0 \le t \le T. \end{aligned}$$

It is easy to show that $\boldsymbol{U} \in \mathbb{S}^{2,d}$. If we fix $\boldsymbol{X}$ and $\boldsymbol{X}'$ in $\mathbb{S}^{2,d}$ and we denote by $\boldsymbol{U}$ and $\boldsymbol{U}'$ the processes defined via the above equality from $\boldsymbol{X}$ and $\boldsymbol{X}'$ respectively, we have for any $t \in [0, T]$:

$$\begin{aligned} &\mathbb{E}\bigg[\sup_{0\le s\le t}\bigg|\int_0^s \Big[b\big(r, X_r', \mathcal{L}^1(X_r')\big) - b\big(r, X_r, \mathcal{L}^1(X_r)\big)\Big]dr\bigg|^2\bigg] \\ &\quad \le C(T)\mathbb{E}\bigg[\int_0^t \Big(|X_s' - X_s|^2 + W_2\big(\mathcal{L}^1(X_s'), \mathcal{L}^1(X_s)\big)^2\Big)ds\bigg] \\ &\quad \le C(T)\int_0^t \mathbb{E}\big[|X_s' - X_s|^2\big]ds, \end{aligned}$$

where $C(T)$ is a constant that only depends upon T and the Lipschitz constant of b, whose value is allowed to vary from line to line. Above, we used the obvious bound:

$$\mathbb{E}\Big[W_2\big(\mathcal{L}^1(X_s'), \mathcal{L}^1(X_s)\big)^2\Big] \le \mathbb{E}\Big[\mathbb{E}^1\big[|X_s' - X_s|^2\big]\Big] = \mathbb{E}\big[|X_s' - X_s|^2\big],$$

$\mathbb{E}^1$ denoting the expectation with respect to $\omega^1 \in \Omega^1$. Burkholder-Davis-Gundy's inequality provides the same type of estimates for the stochastic integrals. This yields, allowing the constant $C(T)$ to depend on the Lipschitz constants of σ and σ^0:

$$\mathbb{E}\big[\sup_{0\le s\le t}|U_s' - U_s|^2\big] \le C(T)\int_0^t \mathbb{E}\big[|X_s' - X_s|^2\big]ds.$$

Calling Φ the mapping $\mathbb{S}^{2,d} \ni \boldsymbol{X} \mapsto \boldsymbol{U} \in \mathbb{S}^{2,d}$ and iterating the above inequality, we get, for any integer $k > 1$:

$$\begin{aligned} \mathbb{E}\big[\sup_{0\le t\le T}|\Phi^k(\boldsymbol{X}')_t - \Phi^k(\boldsymbol{X})_t|^2\big] &\le c(T)^k \int_0^T \frac{(T-s)^{k-1}}{(k-1)!}\mathbb{E}\big[|X_s' - X_s|^2\big]ds \\ &\le \frac{c(T)^k T^k}{k!}\|\boldsymbol{X}' - \boldsymbol{X}\|_{\mathbb{S}^2}, \end{aligned}$$

where Φ^k denotes the k-th composition of the mapping Φ with itself. This shows that for k large enough, Φ^k is a strict contraction. Hence Φ admits a unique fixed point. □

More About the Flow of Conditional Marginal Distributions $\boldsymbol{\mu}$

As expected, we can check that, whenever X_0 is just constructed on Ω^1, the flow of conditional marginal distributions $\boldsymbol{\mu} = (\mathcal{L}^1(X_t))_{0 \le t \le T}$ of a solution $\boldsymbol{X}$ coincides with the conditional marginal distributions of the solution given the smaller information generated by the common noise $\boldsymbol{W}^0$. More generally, we claim:

Proposition 2.9 *Let assumption* **Conditional MKV SDE** *be in force. For a given initial condition $X_0 \in L^2(\Omega, \mathcal{F}_0, \mathbb{P}; \mathbb{R}^d)$, let $\boldsymbol{X} = (X_t)_{0 \le t \le T}$ be the unique solution of the conditional McKean-Vlasov SDE* (2.6)*. Then, for any $t \in [0, T]$, with $\mathbb{P}^0$-probability* 1*, $\mathcal{L}^1(X_t)$ provides a version of the conditional distribution of X_t given the σ-field $\mathcal{F}_0^0 \vee \sigma\{\boldsymbol{W}^0\}$, regarded as a sub-$\sigma$-algebra of $\mathcal{F}$.*

Proof. By Fubini's theorem – see Lemma 2.4 for the way to handle the completion of the product σ-field – we clearly have, for any event $C_0^0 \in \mathcal{F}_0^0$, $\mathcal{F}_0^0$ being here regarded as a sub-σ-algebra of $\mathcal{F}^0$, any bounded and measurable function F on the space $\mathcal{C}([0, T]; \mathbb{R}^d)$, and any bounded and measurable function φ on $\mathbb{R}^d$,

$$\begin{aligned}\mathbb{E}\big[\mathbf{1}_{C_0^0 \times \Omega^1} F(\boldsymbol{W}^0)\varphi(X_t)\big] &= \mathbb{E}^0\big[\mathbf{1}_{C_0^0} F(\boldsymbol{W}^0)\mathbb{E}^1\big(\varphi(X_t)\big)\big] \\ &= \mathbb{E}^0\bigg[\mathbf{1}_{C_0^0} F(\boldsymbol{W}^0)\int_{\mathbb{R}^d}\varphi(x)\mu_t(dx)\bigg],\end{aligned}$$

where $\mu_t(dx)$ is understood as $\mu_t(\omega^0)(dx)$ with $\mu_t(\omega^0) = \mathcal{L}(X_t(\omega^0, \cdot))$. In order to complete the proof, it suffices to check that the mapping $\Omega^0 \ni \omega^0 \mapsto \mu_t(\omega^0) \in \mathcal{P}_2(\mathbb{R}^d)$ is measurable with respect to the completion of the σ-field generated by $\mathcal{F}_0^0$ and $\boldsymbol{W}^0$.

In order to check this last point, we may equip Ω^0 with the filtration $\mathbb{F}^{0,(\mathcal{F}_0^0, \boldsymbol{W}^0)}$ generated by $\mathcal{F}_0^0$ and $\boldsymbol{W}^0$, which is known to be right-continuous by Blumenthal's zero-one law. We then apply Proposition 2.8 on the resulting probabilistic set-up. Combined with Lemma 2.5, this permits to construct a solution $\boldsymbol{X}' = (X'_t)_{0 \le t \le T}$ to (2.6) such that $\Omega^0 \ni \omega^0 \mapsto \mathcal{L}^1(X'_t(\omega^0, \cdot))$ is measurable with respect to $\mathbb{F}^{0,(\mathcal{F}_0^0, \boldsymbol{W}^0)}$. The key point is then to notice that $\boldsymbol{X}'$ is also a solution to (2.6) on the original set-up equipped with $\mathbb{F}^0$ instead of $\mathbb{F}^{0,(\mathcal{F}_0^0, \boldsymbol{W}^0)}$. By the uniqueness part in Proposition 2.8, which holds on any canonical set-up, $\boldsymbol{X}'$ must coincide with $\boldsymbol{X}$. This completes the proof. □

Remark 2.10 *Here are several specific cases of interest in the sequel:*

1. *If X_0 is just defined on $(\Omega^1, \mathcal{F}_0^1, \mathbb{P}^1)$, we can assume that $\mathcal{F}_0^0$ is almost surely trivial. In particular, by repeating the proof of Proposition 2.9, we get that $\mathcal{L}^1\{X_t\}$ is a version of the conditional law of X_t given $\boldsymbol{W}^0$. This fits the aforementioned case where $\boldsymbol{W}^0$ plays the role of a common or systemic noise.*
2. *If X_0 is defined on $(\Omega^0, \mathcal{F}_0^0, \mathbb{P}^0)$, we can assume that $\mathcal{F}_0^0$ is the completion of $\sigma\{X_0\}$. Then, $\mathcal{L}^1(X_t)$ is a version of the conditional law of X_t given $(X_0, \boldsymbol{W}^0)$. In this setting, the full-fledged common noise is no more $\boldsymbol{W}^0$ but the entire $(X_0, \boldsymbol{W}^0)$* "initial condition-common noise."
3. *More generally, if X_0 is measurable with respect to $\sigma\{X_0^0, X_0^1\}$ with X_0^0 being constructed on $(\Omega^0, \mathcal{F}_0^0, \mathbb{P}^0)$ and taking values in a Polish space S^0 and X_0^1 being constructed on $(\Omega^1, \mathcal{F}_0^1, \mathbb{P}^1)$ and taking values in a Polish space S^1, we can work*

with $\mathcal{F}_0^0$ being given by the completion of $\sigma\{X_0^0\}$. In particular, $\mathcal{L}^1(X_t)$ is a version of the conditional law of X_t given $(X_0^0, \boldsymbol{W}^0)$ and $(X_0^0, \boldsymbol{W}^0)$ ends up playing the role of systemic noise.

4. *When the common noise is not present (i.e., $\sigma^0 \equiv 0$ or $\boldsymbol{W}^0 \equiv 0$), the conditional distribution $\mathcal{L}^1(X_t)$ reduces to the standard marginal distribution $\mathcal{L}(X_t)$ and the SPDE (2.4) reduces to a deterministic PDE. It is the Fokker-Planck equation associated with the nonlinear McKean-Vlasov diffusion process $\boldsymbol{X} = (X_t)_{0\le t\le T}$. We then recover the framework investigated in Chapter (Vol I)-4.*

Proposition 2.9 and Remark 2.10 suggest that another type of uniqueness should hold true, at the intersection between weak and strong according to the terminology introduced in Chapter 1. Indeed, with the same notation as in the third item in Remark 2.10, we may expect that the flow of conditional distributions $(\mathcal{L}^1(X_t))_{0\le t\le T}$ remains the same whenever $(X_0^1, \boldsymbol{W})$ is replaced by another pair $(X_0^{1\prime}, \boldsymbol{W}')$ with the same distribution, as long as $(X_0^0, \boldsymbol{W}^0)$ is kept untouched.

Proposition 2.11 *On top of assumption* **Conditional MKV SDE**, *assume further that X_0 is almost surely equal to $\psi(X_0^0, X_0^1)$, for a measurable mapping $\psi : \mathcal{S}^0 \times \mathcal{S}^1 \to \mathbb{R}^d$ where $\mathcal{S}^0$ and $\mathcal{S}^1$ are Polish spaces equipped with their Borel σ-fields, and for two random variables X_0^0 and X_0^1 constructed on $(\Omega^0, \mathcal{F}_0^0, \mathbb{P}^0)$ and $(\Omega^1, \mathcal{F}_0^1, \mathbb{P}^1)$ respectively and taking values in $\mathcal{S}^0$ and $\mathcal{S}^1$ respectively. On $(\Omega^1, \mathcal{F}^1, \mathbb{P}^1)$, consider another random variable $X_0^{1\prime}$ from $(\Omega^1, \mathcal{F}_0^1, \mathbb{P}^1)$ to $\mathcal{S}^1$ with the same law as X_0^1 and another $\mathbb{F}^1$-Brownian motion $\boldsymbol{W}' = (W_t')_{0\le t\le T}$ with values in $\mathbb{R}^d$, let $X_0 = \psi(X_0^0, X_0^1)$ and $X_0' = \psi(X_0^0, X_0^{1\prime})$, and denote by $\boldsymbol{X} = (X_t)_{0\le t\le T}$ and $\boldsymbol{X}' = (X_t')_{0\le t\le T}$ the solutions of the conditional McKean-Vlasov SDE (2.6), when driven by $(X_0, \boldsymbol{W}^0, \boldsymbol{W})$ and $(X_0', \boldsymbol{W}^0, \boldsymbol{W}')$ respectively. Then, with $\mathbb{P}^0$-probability 1, for any $t \in [0, T]$,*

$$\mathcal{L}^1(X_t) = \mathcal{L}^1(X_t').$$

Observe that we here defined the two Brownian motions $\boldsymbol{W}$ and $\boldsymbol{W}'$ on the same probability space $(\Omega^1, \mathcal{F}^1, \mathbb{F}^1, \mathbb{P}^1)$. Whenever $\boldsymbol{W}$ and $\boldsymbol{W}'$ are defined on distinct spaces, it suffices to use the product of the two underlying probability spaces to recover the framework used in the statement.

Proof. The proof consists in an application of the Yamada-Watanabe Theorem 1.33 proved in Chapter 1. Indeed, under the notation of the statement, we may call $\boldsymbol{\mu}' = (\mu_t')_{0\le t\le T}$ the flow of conditional marginal distributions of $\boldsymbol{X}'$ given $\sigma\{X_0^0, \boldsymbol{W}^0\}$, namely:

$$\mu_t' = \mathcal{L}^1(X_t'), \quad t \in [0, T].$$

Then, we can see the equation satisfied by $\boldsymbol{X}'$ as the forward equation of a uniquely solvable forward-backward stochastic differential equation with X_0' as initial condition and with coefficients driven by the environment $\boldsymbol{\mu}'$. By Proposition 2.9 and Remark 2.10, $\boldsymbol{\mu}'$ is adapted to $\sigma\{X_0^0, \boldsymbol{W}^0\}$ and, with the terminology used in Chapter 1, the set-up $(\Omega, \mathcal{F}, \mathbb{F}, \mathbb{P})$ used for the construction of $\boldsymbol{X}'$ is equipped with the input $(X_0, X_0^0, \boldsymbol{W}, \boldsymbol{W}^0)$. As explained in Remark 2.2, this set-up is admissible. Therefore, in order to apply Theorem 1.33, it suffices to choose $F \equiv G \equiv 0$ therein. Consequently, we can find a measurable map $\Phi : \mathbb{R}^d \times \mathcal{C}([0,T];\mathbb{R}^d) \times \mathcal{D}([0,T];\mathcal{P}_2(\mathbb{R}^d)) \times \mathcal{C}([0,T];\mathbb{R}^d) \to \mathcal{C}([0,T];\mathbb{R}^d)$ such that:

$$\mathbb{P}\Big[\boldsymbol{X}' = \Phi\big(\psi(X_0^0, X_0^{1\prime}), \boldsymbol{W}^0, \boldsymbol{\mu}', \boldsymbol{W}'\big)\Big] = 1.$$

Inspired by the proof of Theorem 2.8, we may consider the auxiliary SDE:

$$U_t = X_0 + \int_0^t b\big(s, U_s, \mu_s'\big)ds + \int_0^t \sigma\big(s, U_s, \mu_s'\big)dW_s + \int_0^t \sigma^0\big(s, U_s, \mu_s'\big)dW_s^0,$$

for $0 \le t \le T$. It makes sense since $\boldsymbol{\mu}'$ is $\mathbb{F}^0$-adapted. Then, Theorem 1.33 says that, for the same Φ as above,

$$\mathbb{P}\Big[\boldsymbol{U} = \Phi\big(\psi(X_0^0, X_0^1), \boldsymbol{W}^0, \boldsymbol{\mu}', \boldsymbol{W}\big)\Big] = 1.$$

Since $(X_0^{1\prime}, \boldsymbol{W}')$ has the same law as $(X_0^1, \boldsymbol{W})$ on $(\Omega^1, \mathcal{F}^1, \mathbb{P}^1)$, we deduce that, for almost every $\omega^0 \in \Omega^0$ under $\mathbb{P}^0$,

$$\forall t \in [0,T], \quad \mathcal{L}^1(U_t)(\omega^0) = \mathcal{L}^1(X_t')(\omega^0) = \mu_t'(\omega^0).$$

This proves that $\boldsymbol{U}$ solves the conditional McKean-Vlasov SDE (2.6), when driven by $(X_0, \boldsymbol{W}^0, \boldsymbol{W})$. By the uniqueness part in Proposition 2.8, we deduce that:

$$\mathbb{P}\big[\boldsymbol{U} = \boldsymbol{X}\big] = 1,$$

so that, for almost every $\omega^0 \in \Omega^0$ under $\mathbb{P}^0$,

$$\forall t \in [0,T], \quad \mu_t(\omega^0) = \mathcal{L}\big(X_t(\omega^0, \cdot)\big) = \mathcal{L}\big(U_t(\omega^0, \cdot)\big) = \mu_t'(\omega^0),$$

which completes the proof. □

2.1.4 Conditional Propagation of Chaos

Going back to the particle system (2.3), the aim is now to prove that it converges in a suitable sense to the solution of the conditional McKean-Vlasov SDE (2.6), our goal being to extend the classical result of propagation of chaos for systems of particles in mean field interaction to the conditional case.

As in the standard case, we shall prove the conditional propagation of chaos through a coupling argument with an auxiliary system of particles. The set-up used to construct the auxiliary system is the same as in Subsection 2.1.3. We assume that the space $(\Omega^0, \mathcal{F}^0, \mathbb{F}^0, \mathbb{P}^0)$ carries the common noise $\boldsymbol{W}^0 = (W_t^0)_{0 \le t \le T}$ and the space $(\Omega^1, \mathcal{F}^1, \mathbb{F}^1, \mathbb{P}^1)$ carries the independent noises $(\boldsymbol{W}^n)_{n \ge 1}$. For the sake of simplicity, we also assume that the common noise reduces to the sole $\boldsymbol{W}^0$, meaning that the initial condition in (2.3) is supported by $(\Omega^1, \mathcal{F}^1, \mathbb{F}^1, \mathbb{P}^1)$; however the result below can be extended to more general forms of common noise, at least those covered by Remark 2.10. We thus assume that $(\Omega^1, \mathcal{F}^1, \mathbb{F}^1, \mathbb{P}^1)$ carries a family of identically distributed and independent $\mathcal{F}_0^1$-measurable random variables $(X_0^n)_{n\geq 1}$ with values in $\mathbb{R}^d$ and such that $\mathbb{E}^1[|X_0^1|^2] < \infty$. For any $n \geq 1$, we then call $\underline{\boldsymbol{X}}^n = (\underline{X}_t^n)_{0\leq t\leq T}$ the solution to the McKean-Vlasov SDE (2.6), but with $\boldsymbol{W}^n$ instead of $\boldsymbol{W}$ as driving noise, and X_0^n as initial condition. In other words:

$$d\underline{X}_t^n = b\big(t, \underline{X}_t^n, \mathcal{L}^1(\underline{X}_t^n)\big)dt + \sigma\big(t, \underline{X}_t^n, \mathcal{L}^1(\underline{X}_t^n)\big)dW_t^n$$
$$+ \sigma^0\big(t, \underline{X}_t^n, \mathcal{L}^1(\underline{X}_t^n)\big)dW_t^0.$$

for $t \in [0, T]$ with $\underline{X}_0^n = X_0^n$ as initial condition. For each $n \geq 1$, the above SDE, with the prescribed initial condition, is uniquely solvable under assumption **Conditional MKV SDE**. Moreover, by Proposition 2.11, we have, for all $n \geq 1$,

$$\mathbb{P}^0\Big[\forall t \in [0, T], \quad \mathcal{L}^1(\underline{X}_t^n) = \mathcal{L}^1(\underline{X}_t^1)\Big] = 1.$$

We then claim:

Theorem 2.12 *Within the above framework and under assumption* **Conditional MKV SDE**, *the system of particles* (2.3) *with* $(X_0^1, \cdots, X_0^N)$ *as initial condition has a unique solution for every* $N \geq 1$. *It is denoted by* $((X_t^{N,i})_{0\leq t\leq T})_{i=1,\cdots,N}$. *Moreover,*

$$\lim_{N\to\infty} \Big(\max_{1\leq i\leq N} \mathbb{E}\big[\sup_{0\leq t\leq T} |X_t^{N,i} - \underline{X}_t^i|^2\big] + \sup_{0\leq t\leq T} \mathbb{E}\Big[W_2\Big(\frac{1}{N}\sum_{i=1}^N \delta_{X_t^{N,i}}, \mathcal{L}^1(\underline{X}_t^1)\Big)^2\Big]\Big) = 0.$$

When $\mathbb{E}^1[|X_0^1|^q] < \infty$ *for some* $q > 4$, *there exists a constant* C, *only depending on* T, $\mathbb{E}^1[|X_0^1|^q]$ *and the Lipschitz constants of* b, σ *and* σ^0, *such that:*

$$\max_{1\leq i\leq N} \mathbb{E}\big[\sup_{0\leq t\leq T} |X_t^{N,i} - \underline{X}_t^i|^2\big] + \sup_{0\leq t\leq T} \mathbb{E}\Big[W_2\Big(\frac{1}{N}\sum_{i=1}^N \delta_{X_t^{N,i}}, \mathcal{L}^1(\underline{X}_t^1)\Big)^2\Big] \leq C\epsilon_N,$$

where $(\epsilon_N)_{N\geq 1}$ *satisfies:*

$$\epsilon_N = \begin{cases} N^{-1/2}, & \text{if } d < 4, \\ N^{-1/2}\log N, & \text{if } d = 4, \\ N^{-2/d}, & \text{if } d > 4. \end{cases} \tag{2.7}$$

From now on, we use ϵ_N for the function of N and the dimension d given in the above formula (2.7).

Proof.

First Step. Under assumption **Conditional MKV SDE**, it is easily checked that, for any $N \geq 1$, any $i \in \{1, \cdots, N\}$, and any $t \in [0, T]$, the mapping:

$$(\mathbb{R}^d)^N \ni \boldsymbol{x} = (x_1, \cdots, x_N) \mapsto \big(b, \sigma, \sigma^0\big)\big(t, x_i, \bar{\mu}^N_{\boldsymbol{x}}\big)$$

satisfies the Lipschitz property:

$$\forall \boldsymbol{x} = (x_1, \cdots, x_N),\ \boldsymbol{x}' = (x'_1, \cdots, x'_N) \in \big(\mathbb{R}^d\big)^N,$$

$$\Big|\big(b, \sigma, \sigma^0\big)\big(t, x_i, \bar{\mu}^N_{\boldsymbol{x}}\big) - \big(b, \sigma, \sigma^0\big)\big(t, x'_i, \bar{\mu}^N_{\boldsymbol{x}'}\big)\Big|^2 \leq C\Big(|x_i - x'_i|^2 + \frac{1}{N}\sum_{j=1}^N |x_j - x'_j|^2\Big),$$

for a constant C which is uniform in $t \in [0, T]$ and $N \geq 1$, and where we used the same notation $\bar{\mu}^N_{\boldsymbol{x}} = N^{-1}\sum_{i=1}^N \delta_{x_i}$ as in Subsection (Vol I)-5.3.2 for the uniform distribution on the set $\{x_1, \cdots, x_N\}$. This shows that (2.3) with $(X^1_0, \cdots, X^N_0)$ as initial condition has a unique solution for every $N \geq 1$.

Second Step. It is quite standard to prove, for any $i \in \{1, \cdots, N\}$ and any $t \in [0, T]$,

$$\mathbb{E}\big[\sup_{0 \leq s \leq t} |X^{N,i}_s - \underline{X}^i_s|^2\big] \leq C\Big(\int_0^t \mathbb{E}\Big[|X^{N,i}_s - \underline{X}^i_s|^2 + W_2\big(\bar{\mu}^N_s, \mathcal{L}^1(\underline{X}^i_s)\big)^2\Big] ds\Big), \tag{2.8}$$

where $\bar{\mu}^N_s$ is the empirical measure of $(X^{N,1}_s, \cdots, X^{N,N}_s)$. Now,

$$\begin{aligned} W_2\big(\bar{\mu}^N_s, \mathcal{L}^1(\underline{X}^i_s)\big)^2 &= W_2\big(\bar{\mu}^N_s, \mathcal{L}^1(\underline{X}^1_s)\big)^2 \\ &\leq 2W_2\Big(\bar{\mu}^N_s, \frac{1}{N}\sum_{j=1}^N \delta_{\underline{X}^j_s}\Big)^2 + 2W_2\Big(\frac{1}{N}\sum_{j=1}^N \delta_{\underline{X}^j_s}, \mathcal{L}^1(\underline{X}^1_s)\Big)^2 \\ &\leq \frac{2}{N}\sum_{j=1}^N |X^{N,j}_s - \underline{X}^j_s|^2 + 2W_2\Big(\frac{1}{N}\sum_{j=1}^N \delta_{\underline{X}^j_s}, \mathcal{L}^1(\underline{X}^1_s)\Big)^2. \end{aligned} \tag{2.9}$$

Following the proof of Proposition 2.11, we can find a measurable map:

$$\Phi : \mathbb{R}^d \times \mathcal{C}([0, T]; \mathbb{R}^d) \times \mathcal{D}([0, T]; \mathcal{P}_2(\mathbb{R}^d)) \times \mathcal{C}([0, T]; \mathbb{R}^d) \to \mathcal{C}([0, T]; \mathbb{R}^d),$$

such that, for any $i \in \{1, \cdots, N\}$,

$$\mathbb{P}\Big[\underline{\boldsymbol{X}}^i = \Phi\big(X^i_0, \boldsymbol{W}^0, \mathcal{L}^1(\underline{\boldsymbol{X}}^1), \boldsymbol{W}^i\big)\Big] = 1. \tag{2.10}$$

In the same way, we can find a measurable map:

$$\Phi_N : \big(\mathbb{R}^d\big)^N \times \Big(\mathcal{C}([0, T]; \mathbb{R}^d)\Big)^{N+1} \to \Big(\mathcal{C}([0, T]; \mathbb{R}^d)\Big)^N,$$

such that:

$$\big(\boldsymbol{X}^{N,1}, \cdots, \boldsymbol{X}^{N,N}\big) = \Phi_N\Big(\big(X^1_0, \cdots, X^N_0\big), \big(\boldsymbol{W}^0, \boldsymbol{W}^1, \cdots, \boldsymbol{W}^N\big)\Big).$$

More generally, by taking advantage of the symmetry in the structure of the system of particles (2.3), we get, for any permutation ς of $\{1, \cdots, N\}$:

$$(\boldsymbol{X}^{N,\varsigma(1)}, \cdots, \boldsymbol{X}^{N,\varsigma(N)}) = \Phi_N\big((X_0^{\varsigma(1)}, \cdots, X_0^{\varsigma(N)}), \boldsymbol{W}^0, \boldsymbol{W}^{\varsigma(1)}, \cdots, \boldsymbol{W}^{\varsigma(N)}\big).$$

Together with (2.10), this implies that the processes $((\boldsymbol{X}^{N,i}, \underline{\boldsymbol{X}}^i))_{1 \le i \le N}$ are identically distributed. Therefore, (2.8) and (2.9) yield:

$$\mathbb{E}\big[\sup_{0 \le s \le t} |X_s^{N,1} - \underline{X}_s^1|^2\big] \le C\Big(\int_0^t \mathbb{E}\Big[|X_s^{N,1} - \underline{X}_s^1|^2 + W_2\Big(\frac{1}{N}\sum_{j=1}^N \delta_{\underline{X}_s^j}, \mathcal{L}^1(\underline{X}_s^1)\Big)^2\Big] ds\Big).$$

By Gronwall's lemma, we get:

$$\mathbb{E}\big[\sup_{0 \le s \le t} |X_s^{N,1} - \underline{X}_s^1|^2\big] \le C \int_0^t \mathbb{E}\Big[W_2\Big(\frac{1}{N}\sum_{j=1}^N \delta_{\underline{X}_s^j}, \mathcal{L}^1(\underline{X}_s^1)\Big)^2\Big] ds, \tag{2.11}$$

where the constant C is allowed to change from line to line.

Third Step. By (2.10), it is clear that the processes $(\underline{\boldsymbol{X}}^i)_{1 \le i \le N}$ are conditionally independent and identically distributed given $\boldsymbol{W}^0$ (or $\mathcal{F}^0$). In particular, by (5.19) in Chapter (Vol I)-5, we have, for any $s \in [0, T]$,

$$\mathbb{P}^0\Big[\lim_{N \to \infty} \mathbb{E}^1\Big[W_2\Big(\frac{1}{N}\sum_{j=1}^N \delta_{\underline{X}_s^j}, \mathcal{L}^1(\underline{X}_s^1)\Big)^2\Big] = 0\Big] = 1. \tag{2.12}$$

Now,

$$\begin{aligned}
&\mathbb{E}^1\Big[W_2\Big(\frac{1}{N}\sum_{j=1}^N \delta_{\underline{X}_s^j}, \mathcal{L}^1(\underline{X}_s^1)\Big)^2\Big] \\
&\le 2\mathbb{E}^1\Big[W_2\Big(\frac{1}{N}\sum_{j=1}^N \delta_{\underline{X}_s^j}, \delta_0\Big)^2\Big] + 2\mathbb{E}^1\Big[W_2\big(\delta_0, \mathcal{L}^1(\underline{X}_s^1)\big)^2\Big] \\
&\le \frac{2}{N}\mathbb{E}^1\Big[\sum_{j=1}^N |\underline{X}_s^j|^2\Big] + 2\mathbb{E}^1\big[|\underline{X}_s^1|^2\big] = 4\mathbb{E}^1\big[|\underline{X}_s^1|^2\big],
\end{aligned} \tag{2.13}$$

so that, from the bound $\mathbb{E}^0[\mathbb{E}^1[|\underline{X}_s^1|^2]] = \mathbb{E}[|\underline{X}_s^1|^2] < \infty$ and by Lebesgue dominated convergence theorem, we deduce from (2.12) that, for any $s \in [0, T]$,

$$\lim_{N \to \infty} \mathbb{E}\Big[W_2\Big(\frac{1}{N}\sum_{j=1}^N \delta_{\underline{X}_s^j}, \mathcal{L}^1(\underline{X}_s^1)\Big)^2\Big] = 0. \tag{2.14}$$

We claim that the convergence is uniform in $s \in [0, T]$. Indeed, by Cauchy-Schwarz inequality, we have, for any $s, t \in [0, T]$,

$$\left|\mathbb{E}\Big[W_2\Big(\frac{1}{N}\sum_{j=1}^N \delta_{\underline{X}_t^j}, \mathcal{L}^1(\underline{X}_t^1)\Big)^2\Big] - \mathbb{E}\Big[W_2\Big(\frac{1}{N}\sum_{j=1}^N \delta_{\underline{X}_s^j}, \mathcal{L}^1(\underline{X}_s^1)\Big)^2\Big]\right|$$

$$\leq \mathbb{E}\Big[\Big(W_2\Big(\frac{1}{N}\sum_{j=1}^N \delta_{\underline{X}_t^j}, \mathcal{L}^1(\underline{X}_t^1)\Big) - W_2\Big(\frac{1}{N}\sum_{j=1}^N \delta_{\underline{X}_s^j}, \mathcal{L}^1(\underline{X}_s^1)\Big)\Big)^2\Big]^{1/2}$$

$$\times \mathbb{E}\Big[\Big(W_2\Big(\frac{1}{N}\sum_{j=1}^N \delta_{\underline{X}_t^j}, \mathcal{L}^1(\underline{X}_t^1)\Big) + W_2\Big(\frac{1}{N}\sum_{j=1}^N \delta_{\underline{X}_s^j}, \mathcal{L}^1(\underline{X}_s^1)\Big)\Big)^2\Big]^{1/2}.$$

So, by (2.13),

$$\left|\mathbb{E}\Big[W_2\Big(\frac{1}{N}\sum_{j=1}^N \delta_{\underline{X}_t^j}, \mathcal{L}^1(\underline{X}_t^1)\Big)^2\Big] - \mathbb{E}\Big[W_2\Big(\frac{1}{N}\sum_{j=1}^N \delta_{\underline{X}_s^j}, \mathcal{L}^1(\underline{X}_s^1)\Big)^2\Big]\right|$$

$$\leq C\mathbb{E}\Big[\Big(W_2\Big(\frac{1}{N}\sum_{j=1}^N \delta_{\underline{X}_t^j}, \mathcal{L}^1(\underline{X}_t^1)\Big) - W_2\Big(\frac{1}{N}\sum_{j=1}^N \delta_{\underline{X}_s^j}, \mathcal{L}^1(\underline{X}_s^1)\Big)\Big)^2\Big]^{1/2}.$$

By the triangular inequality, we end up with:

$$\left|\mathbb{E}\Big[W_2\Big(\frac{1}{N}\sum_{j=1}^N \delta_{\underline{X}_t^j}, \mathcal{L}^1(\underline{X}_t^1)\Big)^2\Big] - \mathbb{E}\Big[W_2\Big(\frac{1}{N}\sum_{j=1}^N \delta_{\underline{X}_s^j}, \mathcal{L}^1(\underline{X}_s^1)\Big)^2\Big]\right|$$

$$\leq C\mathbb{E}\Big[W_2\Big(\frac{1}{N}\sum_{j=1}^N \delta_{\underline{X}_t^j}, \frac{1}{N}\sum_{j=1}^N \delta_{\underline{X}_s^j}\Big)^2\Big]^{1/2} + C\mathbb{E}\Big[W_2\Big(\mathcal{L}^1(\underline{X}_t^1), \mathcal{L}^1(\underline{X}_s^1)\Big)^2\Big]^{1/2}$$

$$\leq C\mathbb{E}\big[|\underline{X}_t^1 - \underline{X}_s^1|^2\big]^{1/2} \leq C|t-s|^{1/2},$$

where we used the Lipschitz property of the coefficients together with the bound

$$\mathbb{E}[\sup_{0\leq r\leq T} |\underline{X}_r^1|^2] \leq C,$$

to get the last inequality, and as before, we allowed the constant C to change from line to line. Therefore, by an equicontinuity argument, the convergence in (2.14) must be uniform in $s \in [0, T]$, in other words:

$$\lim_{N\to\infty} \sup_{0\leq s\leq T} \mathbb{E}\Big[W_2\Big(\frac{1}{N}\sum_{j=1}^N \delta_{\underline{X}_s^j}, \mathcal{L}^1(\underline{X}_s^1)\Big)^2\Big] = 0. \tag{2.15}$$

Going back to (2.11), we get, as a first a consequence,

$$\lim_{N\to\infty} \mathbb{E}\Big[\sup_{0\leq t\leq T} |X_t^{N,1} - \underline{X}_t^1|^2\Big] = 0. \tag{2.16}$$

As a second consequence of (2.15), we have

$$
\begin{aligned}
&\sup_{0\le t\le T}\mathbb{E}\Big[W_2\Big(\frac{1}{N}\sum_{j=1}^N \delta_{X_t^{N,j}}, \mathcal{L}^1(\underline{X}_t^1)\Big)^2\Big] \\
&\le 2\sup_{0\le t\le T}\mathbb{E}\Big[W_2\Big(\frac{1}{N}\sum_{j=1}^N \delta_{X_t^{N,j}}, \frac{1}{N}\sum_{j=1}^N \delta_{\underline{X}_t^j}\Big)^2\Big] + 2\sup_{0\le t\le T}\mathbb{E}\Big[W_2\Big(\frac{1}{N}\sum_{j=1}^N \delta_{\underline{X}_t^j}, \mathcal{L}^1(\underline{X}_t^1)\Big)^2\Big] \\
&\le 2\sup_{0\le t\le T}\mathbb{E}\Big[\frac{1}{N}\sum_{j=1}^N |X_t^{N,j}-\underline{X}_t^j|^2\Big] + 2\sup_{0\le t\le T}\mathbb{E}\Big[W_2\Big(\frac{1}{N}\sum_{j=1}^N \delta_{\underline{X}_t^j}, \mathcal{L}^1(\underline{X}_t^1)\Big)^2\Big],
\end{aligned}
\tag{2.17}
$$

so that, by (2.15) and (2.16), the left-hand side tends to 0.

Fourth Step. We now discuss the case where X_0^1 has a finite moment of order q, for some $q > 4$. As a preliminary remark, we observe that this implies

$$\mathbb{E}\big[\sup_{0\le t\le T}|\underline{X}_t^1|^q\big] < \infty,$$

which follows from the growth property of the coefficients under assumption **Conditional MKV SDE**. Then, by Theorem (Vol I)-5.8 and Remark (Vol I)-5.9, we can find a deterministic constant c, only depending upon d and q, such that, for any $t \in [0,T]$, $\mathbb{P}^0$-almost surely,

$$\mathbb{E}^1\Big[W_2\Big(\frac{1}{N}\sum_{j=1}^N \delta_{\underline{X}_t^j}, \mathcal{L}^1(\underline{X}_t^1)\Big)^2\Big] \le c\,\mathbb{E}^1\big[|\bar{X}_t^1|^q\big]^{2/q}\epsilon_N,$$

for the same sequence $(\varepsilon_N)_{N\ge 1}$ as in the statement. Taking expectation with respect to $\mathbb{P}^0$ and plugging the above bound into (2.11), we get the desired bound on $\mathbb{E}[\sup_{0\le t\le T}|X_t^{N,1}-\underline{X}_t^1|^2]$. Thanks to (2.17), we get a similar bound for:

$$\sup_{0\le t\le T}\mathbb{E}\Big[W_2\Big(\frac{1}{N}\sum_{i=1}^N \delta_{X_t^{N,i}}, \mathcal{L}^1(\underline{X}_t^1)\Big)^2\Big],$$

and this completes the proof. □

2.2 Strong Solutions to MFGs with Common Noise

2.2.1 Solution Strategy for Mean Field Games

We now return to the N-player game model formulated in the opening Subsection 2.1.1 and discuss the derivation of mean field games when players are subject to a common source of noise. Reproducing (2.1) for the sake of convenience, the states of the N players evolve and interact through the system of SDEs:

$$dX_t^i = b(t, X_t^i, \bar{\mu}_t^N, \alpha_t^i)dt + \sigma(t, X_t^i, \bar{\mu}_t^N, \alpha_t^i)dW_t^i + \sigma^0(t, X_t^i, \bar{\mu}_t^N, \alpha_t^i)dW_t^0,$$

for $t \in [0, T]$, and the players are assigned cost functionals of the form:

$$J^i(\boldsymbol{\alpha}^1, \cdots, \boldsymbol{\alpha}^N) = \mathbb{E}\left[\int_0^T f(t, X_t^i, \bar{\mu}_t^N, \alpha_t^i)dt + g(X_T^i, \bar{\mu}_T^N)\right],$$

where the control processes $\boldsymbol{\alpha}^1 = (\alpha_t^1)_{t\geq 0}, \cdots, \boldsymbol{\alpha}^N = (\alpha_t^N)_{t\geq 0}$ are taken in the set $\mathbb{A}$ of $\mathrm{Leb}_1 \otimes \mathbb{P}$-square-integrable $\mathbb{F}$-progressively measurable processes with values in a closed convex subset A of a Euclidean space $\mathbb{R}^k$ for some integer $k \geq 1$.

In the spirit of assumption **Optimization in a Random Environment** of Chapter 1, we shall assume throughout this subsection:

Assumption (Control).

(A1) The coefficients b, σ and σ^0 are Borel-measurable mappings from $[0, T] \times \mathbb{R}^d \times \mathcal{P}_2(\mathbb{R}^d) \times A$ to $\mathbb{R}^d$, $\mathbb{R}^{d\times d}$ and $\mathbb{R}^{d\times d}$ respectively. For any $t \in [0, T]$, the coefficients $b(t, \cdot, \cdot, \cdot)$, $\sigma(t, \cdot, \cdot, \cdot)$ and $\sigma^0(t, \cdot, \cdot, \cdot)$ are continuous on $\mathbb{R}^d \times \mathcal{P}_2(\mathbb{R}^d) \times A$. The coefficients $b(t, \cdot, \mu, \alpha)$, $\sigma(t, \cdot, \mu, \alpha)$ and $\sigma^0(t, \cdot, \mu, \alpha)$ are Lipschitz continuous in the x variable, uniformly in $(t, \mu, \alpha) \in [0, T] \times \mathcal{P}_2(\mathbb{R}^d) \times A$. Moreover, there exists a constant L such that:

$$|(b, \sigma, \sigma^0)(t, x, \mu, \alpha)| \leq L\big[1 + |x| + |\alpha| + M_2(\mu)\big],$$

where $M_2(\mu)^2$ stands for the second moment of μ.

(A2) The coefficients f and g are real-valued Borel-measurable mappings on $[0, T] \times \mathbb{R}^d \times \mathcal{P}_2(\mathbb{R}^d) \times A$ and $\mathbb{R}^d \times \mathcal{P}_2(\mathbb{R}^d)$ respectively. For any $t \in [0, T]$, $f(t, \cdot, \cdot, \cdot)$ and $g(\cdot, \cdot)$ are continuous on $\mathbb{R}^d \times \mathcal{P}_2(\mathbb{R}^d) \times A$ and $\mathbb{R}^d \times \mathcal{P}_2(\mathbb{R}^d)$ respectively. Moreover,

$$|f(t, x, \mu, \alpha)| + |g(x, \mu)| \leq L\big[1 + |x|^2 + |\alpha|^2 + \big(M_2(\mu)\big)^2\big].$$

As in Chapter (Vol I)-3, the question is to determine asymptotic equilibria when the size of the population N tends to ∞.

Imitating the strategy implemented in Chapters (Vol I)-3 and (Vol I)-4, the search for asymptotic Nash equilibria is to be performed by solving the optimization problem of one single player interacting with the limit of the flow of empirical measures $(\bar{\mu}_t^N)_{0\leq t\leq T}$. In view of Propositions 2.8 and 2.12, the limit of the flow of empirical measures $(\bar{\mu}_t^N)_{0\leq t\leq T}$ should match the flow of conditional marginal distributions of the optimal path given the common noise, where we recall from Remark 2.10 that the exact form of the common noise depends upon the dependence structure of the initial conditions. In order to fit the framework of Proposition 2.8, we assume throughout this short introduction that the initial conditions $X_0^1, \cdots, X_0^N$

of the players are independent and identically distributed according to some $\mu_0 \in \mathcal{P}_2(\mathbb{R}^d)$. In this case, the common noise is to be understood as the sole $\boldsymbol{W}^0$, the asymptotic empirical distribution of the initial states $X_0^1, \cdots, X_0^N$ matches μ_0, and is thus deterministic. Later on in the text, we shall discuss the case where μ_0 is allowed to be random.

When compared to the situation described in Chapters (Vol I)-3 and (Vol I)-4, the first striking difference is that the limit flow is now random as it keeps track of the common source of noise. For that reason, we shall often refer to the case investigated in Chapters (Vol I)-3 and (Vol I)-4 as *"deterministic"*, the word *deterministic* emphasizing the fact that the MFG equilibrium is characterized by a purely deterministic flow of measures. However, the reader needs to remain acutely aware of the possible misunderstanding due to this terminology. Indeed, despite the use of the word *deterministic*, the dynamics of the state of each single player in the *deterministic* case are stochastic!

In any case, following in the footsteps of Chapter (Vol I)-3, the search for an MFG equilibrium when subject to a common source of noise may be implemented in two major steps:

(i) Given an initial distribution $\mu_0 \in \mathcal{P}_2(\mathbb{R}^d)$, for any arbitrary continuous $\mathcal{P}_2(\mathbb{R}^d)$-valued adapted stochastic process $\boldsymbol{\mu} = (\mu_t)_{0 \le t \le T}$, solve the optimization problem:

$$\inf_{(\alpha_t)_{0 \le t \le T}} \mathbb{E}\left[\int_0^T f(t, X_t, \mu_t, \alpha_t) dt + g(X_T, \mu_T)\right], \tag{2.18}$$

subject to the dynamic constraint:

$$\begin{aligned} dX_t &= b\big(t, X_t, \mu_t, \alpha_t\big) dt + \sigma\big(t, X_t, \mu_t, \alpha_t\big) dW_t \\ &\quad + \sigma^0\big(t, X_t, \mu_t, \alpha_t\big) dW_t^0, \end{aligned} \tag{2.19}$$

over the time interval $t \in [0, T]$, with $X_0 \sim \mu_0$, and over controls which are adapted to both $\boldsymbol{W}$ and $\boldsymbol{W}^0$.

(ii) Determine the measure valued stochastic process $\boldsymbol{\mu} = (\mu_t)_{0 \le t \le T}$ so that the flow of conditional marginal distributions of one optimal path $(X_t)_{0 \le t \le T}$ given $\boldsymbol{W}^0$ is precisely $(\mu_t)_{0 \le t \le T}$ itself, i.e.,

$$\forall t \in [0, T], \quad \mu_t = \mathcal{L}\big(X_t \,|\, \sigma\{W_s^0, 0 \le s \le T\}\big). \tag{2.20}$$

Recall that we use freely the notation $X \sim \mu$ in lieu of $\mathcal{L}(X) = \mu$, where $\mathcal{L}(X)$ denotes the law of X, which we also denote by $\mathbb{P} \circ X^{-1}$ or by $\mathbb{P}_X$.
Obviously, in the absence of the common noise term $\boldsymbol{W}^0$, the measure valued adapted stochastic process $\boldsymbol{\mu}$ can be taken as a deterministic function $[0, T] \ni t \mapsto \mu_t \in \mathcal{P}_2(\mathbb{R}^d)$, and we are back to the situation investigated in Chapters (Vol I)-3 and (Vol I)-4.

Remark 2.13. *As explained in Remark (Vol I)-3.2, we shall restrict ourselves to the case where the optimal path* $(X_t)_{0 \le t \le T}$ *in step (ii) above is unique.*

2.2.2 Revisiting the Probabilistic Set-Up

In order to specify in a completely rigorous way the definition of an MFG equilibrium, we need to revisit the probabilistic set-up introduced earlier. The reason is twofold:

1. First, we learnt from Subsection 2.1 that, in order to handle conditional McKean-Vlasov constraints of the same type as (2.20), it might be convenient to disentangle the two sources of noise $\boldsymbol{W}$ and $\boldsymbol{W}^0$. Still, we said very little about the way to disentangle the systemic and idiosyncratic noises in the initial condition X_0 of (2.19). It is now necessary to elucidate the construction of X_0 whenever μ_0 in step (ii) of the search of an MFG equilibrium is allowed to be random, which is what happens when the conditional law of X_0 given the common noise is required to match a given random variable μ_0 with values in $\mathcal{P}_2(\mathbb{R}^d)$.
2. Second, we made clear in Chapter 1 the need to specify a Compatibility Condition when handling an optimal stochastic control problem in random environment. In Subsection 1.4.1, compatibility was formulated in terms of the filtration generated by the initial condition X_0, the two noises $\boldsymbol{W}^0$ and $\boldsymbol{W}$ and the environment $\boldsymbol{\mu}$. Now we must say what this compatibly condition becomes when defining the notion of equilibrium for mean field games with common noise.

In order to follow the approach used in Subsection 2.1, we shall work with two complete filtered probability spaces $(\Omega^0, \mathcal{F}^0, \mathbb{F}^0, \mathbb{P}^0)$ and $(\Omega^1, \mathcal{F}^1, \mathbb{F}^1, \mathbb{P}^1)$, the first one carrying $\boldsymbol{W}^0$ and the second carrying $\boldsymbol{W}$. We then equip the product space $\Omega = \Omega^0 \times \Omega^1$ with the completion $\mathcal{F}$ of the product σ-field under the product probability measure $\mathbb{P} = \mathbb{P}^0 \otimes \mathbb{P}^1$, the extension of $\mathbb{P}$ to $\mathcal{F}$ being still denoted by $\mathbb{P}$. The right-continuous and complete augmentation of the product filtration is denoted by $\mathbb{F}$.

Specification of the Initial Condition

The first question we address concerns the specification of the initial condition. As we just explained, the description we have given so far may not suffice for our purpose. Our aim is to identify the conditional distribution $\mathcal{L}^1(X_0)$ of the initial condition X_0 of (2.19) with the conditional law of X_0 given some information that may be observed at the macroscopic –or systemic– level, meaning an information that may be captured from the sole observation of the global population, and not of the private states of the players. In Remark 2.10, we provided several examples for this macroscopic initial information. Now we encapsulate all these examples into a single framework relevant to our analysis.

Throughout the analysis of MFG with common noise, we are given a random variable μ_0 on the probability space $(\Omega^0, \mathcal{F}_0^0, \mathbb{P}^0)$ with values in the space $(\mathcal{P}_2(\mathbb{R}^d), \mathcal{B}(\mathcal{P}_2(\mathbb{R}^d)))$. Along the line of step (ii) in the search of an MFG equilibrium, it is understood as the initial distribution of the population, with the difference that it is now allowed to be random. As a main requirement, we impose that the initial condition X_0 of (2.19) is an $\mathcal{F}_0$-measurable random variable from Ω into $\mathbb{R}^d$ satisfying $\mathcal{L}(X_0(\omega^0,\cdot)) = \mu_0(\omega^0)$ for $\mathbb{P}^0$-almost every $\omega^0 \in \Omega^0$, or equivalently,

$$\mathbb{P}^0\big[\mathcal{L}^1(X_0) = \mu_0\big] = 1. \tag{2.21}$$

The interpretation of (2.21) is twofold. On the one hand, it says that, given ω^0, the randomness used to sample a realization of the random variable $X_0(\omega^0,\cdot)$ is enclosed in $\mathcal{F}_0^1$ and is thus independent of μ_0, $\boldsymbol{W}^0$ and $\boldsymbol{W}$. According to the terminology introduced in Definition 1.6, this looks like a compatibility condition in the sense that $\mathcal{F}_T^{(\mu_0,\boldsymbol{W}^0,\boldsymbol{W})}$ can be shown to be independent of $\mathcal{F}_0^{(X_0,\mu_0)}$ given $\mathcal{F}_0^{\mu_0}$, the proof being given right below. On the other hand, (2.21) implies that, for any Borel subsets $B \subset \mathbb{R}^d$ and $C \subset \mathcal{P}_2(\mathbb{R}^d)$,

$$\mathbb{E}\big[\mathbf{1}_B(X_0)\mathbf{1}_C(\mu_0)\big] = \mathbb{E}^0\big[\mathbf{1}_C(\mu_0)\mu_0(B)\big],$$

proving that μ_0 is the conditional law of X_0 given $\sigma\{\mu_0\}$, namely, $\mathbb{P}$-almost surely,

$$\mathcal{L}\big(X_0 \,|\, \mu_0\big) = \mu_0. \tag{2.22}$$

Identity (2.22) is reminiscent of the fixed point condition (2.20) formulated in step (ii) of the search of an MFG equilibrium, except that μ_0 is now random.

The proof of the aforementioned compatibility condition is quite straightforward. It is based on the fact that, under $\mathbb{P}$, $\mathcal{F}_0$ is independent of $(\boldsymbol{W}^0,\boldsymbol{W})$. Then, for any Borel subsets $B_0 \subset \mathbb{R}^d$, $C_0 \subset \mathcal{P}_2(\mathbb{R}^d)$, $D \subset \mathcal{C}([0,T];\mathbb{R}^d) \times \mathcal{C}([0,T];\mathbb{R}^d)$,

$$\begin{aligned}\mathbb{E}\big[\mathbf{1}_{B_0}(X_0)\mathbf{1}_{C_0}(\mu_0)\mathbf{1}_D(\boldsymbol{W}^0,\boldsymbol{W})\big] &= \mathbb{E}\big[\mathbf{1}_{B_0}(X_0)\mathbf{1}_{C_0}(\mu_0)\big]\mathbb{E}\big[\mathbf{1}_D(\boldsymbol{W}^0,\boldsymbol{W})\big] \\ &= \mathbb{E}\big[\mu_0(B_0)\mathbf{1}_{C_0}(\mu_0)\big]\mathbb{E}\big[\mathbf{1}_D(\boldsymbol{W}^0,\boldsymbol{W})\big],\end{aligned}$$

which suffices to conclude that $\mathcal{F}_T^{(\mu_0,\boldsymbol{W}^0,\boldsymbol{W})}$ is independent of $\mathcal{F}_0^{(X_0,\mu_0)}$ given $\mathcal{F}_0^{\mu_0}$. Observe that, from Blumenthal's 0-1 law, $\mathcal{F}_0^{(X_0,\mu_0)}$ is also equal to $\mathcal{F}_0^{(X_0,\mu_0,\boldsymbol{W}^0,\boldsymbol{W})}$.

We now address the specific question of the construction of an initial condition X_0 satisfying (2.21) for a given random variable μ_0 from $(\Omega^0,\mathcal{F}_0^0,\mathbb{P}^0)$ into $\mathcal{P}_2(\mathbb{R}^d)$. Such a construction may be achieved by means of Lemma (Vol I)-5.29. Using the same function $\psi : [0,1] \times \mathcal{P}_2(\mathbb{R}^d) \to \mathbb{R}^d$ as in the statement of Lemma (Vol I)-5.29, and assuming that there exists a uniformly distributed random variable η from $(\Omega^1,\mathcal{F}_0^1,\mathbb{P}^1)$ into $[0,1)$, Lemma (Vol I)-5.29 ensures that X_0 defined by:

$$X_0(\omega^0,\omega^1) = \psi\big(\eta(\omega^1),\mu_0(\omega^0)\big), \quad (\omega^0,\omega^1) \in \Omega^0 \times \Omega^1, \tag{2.23}$$

is a random variable from the space $(\Omega,\mathcal{F}_0,\mathbb{P})$ into $\mathbb{R}^d$ such that, for any $\omega^0 \in \Omega^0$, $\mathcal{L}(X_0(\omega^0,\cdot))$ is precisely $\mu_0(\omega^0)$. It is worth mentioning that, in that case, X_0 is of the form specified in the third item of Remark 2.10.

Compatibility Condition in the Optimal Control Problem

So far, we addressed the question of compatibility between the initial condition X_0 and the filtration generated by μ_0, $\boldsymbol{W}^0$ and $\boldsymbol{W}$. However, we know from Subsection 1.4.1 that we also need to discuss the compatibility of the process $(X_0, \boldsymbol{W}^0, \boldsymbol{\mu}, \boldsymbol{W})$ with the filtration $\mathbb{F}$ with respect to which the control process in (2.18)–(2.19) is required to be progressively measurable.

With the same terminology as in Subsection 1.4.1, the unknown $\boldsymbol{\mu}$ in the search of an MFG equilibrium is referred to as *a random environment*. Whenever $\boldsymbol{\mu}$ satisfies the fixed point condition (2.20), compatibility is easily checked since $\boldsymbol{\mu}$ is adapted to the Brownian motion $\boldsymbol{W}^0$, see Remark 1.12. Anyhow, as we shall see next, there are cases for which condition (2.20) is too strong and need to be relaxed. Of course, so is the case if μ_0 is random, and we gave instances of this kind in Remark 2.10 within the simpler framework of *uncontrolled* McKean-Vlasov SDEs. However, the difficulty we shall face below is more substantial and, even in cases when μ_0 is deterministic, it will be needed to enlarge the filtration. In a nutshell, such a difficulty will occur in cases when $\boldsymbol{\mu}$ is not adapted to $\mathcal{F}_0^0 \vee \mathcal{F}^{\boldsymbol{W}^0}$. If so is the case, μ_t is no longer seen as the conditional law of X_t given $\boldsymbol{W}^0$ (nor $\mathcal{F}_0^0 \vee \mathcal{F}^{\boldsymbol{W}^0}$), but as the conditional law of X_t given a larger filtration.

When $\boldsymbol{\mu}$ is not adapted to $\boldsymbol{W}^0$, the question of compatibility really matters. In full analogy with our discussion in Subsection 1.4.1, it is mandatory to require the filtration $\mathbb{F}$ to be compatible with $(X_0, \boldsymbol{W}^0, \boldsymbol{\mu}, \boldsymbol{W})$. Unfortunately, this will not suffice for our purpose. In order to guarantee a weak form of stability of MFG equilibria, we shall demand more. Instead we shall enlarge the environment $\boldsymbol{\mu}$ into a process $\underline{\mathfrak{M}} = (\mathfrak{M}_t)_{0 \le t \le T}$, with values in a Polish space Ξ which is larger than $\mathcal{X} = \mathcal{P}_2(\mathbb{R}^d)$ in the sense that it is equipped with a family of *continuous projection mappings* $(\pi_t : \Xi \to \mathcal{X})_{0 \le t \le T}$ such that $\boldsymbol{\mu} = (\mu_t = \pi_t(\mathfrak{M}_t))_{0 \le t \le T}$. The process $\underline{\mathfrak{M}} = (\mathfrak{M}_t)_{0 \le t \le T}$ will be called a *lifting* of $\boldsymbol{\mu}$. Although $\underline{\mathfrak{M}}$ does not explicitly appear in the coefficients driving the optimal stochastic control problem (2.18)–(2.19), it will be part of the input as it will dictate the compatibility condition: the filtration $\mathbb{F}$ will be assumed to be compatible with $(X_0, \boldsymbol{W}^0, \underline{\mathfrak{M}}, \boldsymbol{W})$. In order to stress the fact that $\underline{\mathfrak{M}}$ is a lifting of $\boldsymbol{\mu}$, we will sometimes write that $\mathbb{F}$ is compatible with $(X_0, \boldsymbol{W}^0, (\underline{\mathfrak{M}}, \boldsymbol{\mu}), \boldsymbol{W})$.

A typical example for the triple $(\mathcal{X}, \Xi, (\pi_t)_{0 \le t \le T})$ is $\mathcal{X} = \mathcal{P}_2(\mathbb{R}^d)$, $\Xi = \mathcal{P}_2(\mathcal{C}([0,T];\mathbb{R}^d))$ and $\pi_t : \mathcal{P}_2(\mathcal{C}([0,T];\mathbb{R}^d)) \ni \mathfrak{m} \mapsto \nu_t = \mathfrak{m} \circ e_t^{-1}$, where e_t is the evaluation mapping at time t on $\mathcal{C}([0,T];\mathbb{R}^d)$. Then, for a random variable $\mathfrak{M}$ with values in Ξ, we may define $\mathfrak{M}_t$ as the image of $\mathfrak{M}$ by the mapping $\mathcal{C}([0,T];\mathbb{R}^d) \ni \boldsymbol{x} \mapsto \boldsymbol{x}_{\cdot \wedge t} \in \mathcal{C}([0,T];\mathbb{R}^d)$. Below, we use a slightly different version. Still with $\mathcal{X} = \mathcal{P}_2(\mathbb{R}^d)$, we take $\Xi = \mathcal{P}_2([\mathcal{C}([0,T];\mathbb{R}^d)]^2) = \mathcal{P}_2(\mathcal{C}([0,T];\mathbb{R}^{2d}))$. Writing a vector in $\mathbb{R}^{2d}$ in the form $(x,w) \in \mathbb{R}^d \times \mathbb{R}^d$, we take $\pi_t : \mathcal{P}_2(\mathcal{C}([0,T];\mathbb{R}^{2d})) \ni \mathfrak{m} \mapsto \nu_t = \mathfrak{m} \circ (e_t^x)^{-1}$, where e_t^x is the composition of e_t with the projection mapping $\mathbb{R}^d \times \mathbb{R}^d \ni (x,w) \mapsto x$ (e_t being now regarded as a mapping from $\mathcal{C}([0,T];\mathbb{R}^{2d})$ into $\mathbb{R}^{2d}$). And, for a random variable $\mathfrak{M}$ with values in Ξ, we define $\mathfrak{M}_t$ as the image of $\mathfrak{M}$ by the mapping $[\mathcal{C}([0,T];\mathbb{R}^d)]^2 \ni (\boldsymbol{x},\boldsymbol{w}) \mapsto (\boldsymbol{x},\boldsymbol{w})_{\cdot \wedge t} \in \mathcal{C}([0,T];\mathbb{R}^d) \times \mathcal{C}([0,T];\mathbb{R}^d)$. The notion of *lifting* is then easily understood. We view flows of probability measures on $\mathbb{R}^d$ as flows of marginal distributions induced by the first d coordinates of a continuous process with values in $\mathbb{R}^{2d}$. In such a case, $\mathfrak{M}_t$ is the law of the process stopped at time t and μ_t is the marginal law of the first d coordinates at time t. Of course, it is not *always* possible for a flow of probability measures to be the flow of marginal distributions of a continuous process, but, in all the cases we handle below, it is indeed the case.

Note that extending the environment is by no means a limitation for applying the results of Subsection 1.4. Indeed, it will suffice to replace $\mathcal{X}$ by Ξ and $\boldsymbol{\mu}$ by $\underline{\mathfrak{M}}$ to apply them.

Summary

In order to proceed with the search of an MFG equilibrium, we often use two probability spaces

$$(\Omega^0, \mathcal{F}^0, \mathbb{P}^0) \quad \text{and} \quad (\Omega^1, \mathcal{F}^1, \mathbb{P}^1),$$

equipped with two right-continuous and complete filtrations $\mathbb{F}^0 = (\mathcal{F}_t^0)_{0 \le t \le T}$ and $\mathbb{F}^1 = (\mathcal{F}_t^1)_{0 \le t \le T}$. The common noise $\boldsymbol{W}^0 = (W_t^0)_{0 \le t \le T}$ is constructed on the space $(\Omega^0, \mathbb{F}^0, \mathbb{P}^0)$, while the idiosyncratic noise $\boldsymbol{W} = (W_t)_{0 \le t \le T}$ is constructed on $(\Omega^1, \mathbb{F}^1, \mathbb{P}^1)$. By convention, the index 0 always refers to the common noise and the index 1 to the idiosyncratic one. We then define the product structure

$$\Omega = \Omega^0 \times \Omega^1, \quad \mathcal{F}, \quad \mathbb{F} = (\mathcal{F}_t)_{0 \le t \le T}, \quad \mathbb{P}, \tag{2.24}$$

where $(\mathcal{F}, \mathbb{P})$ is the completion of $(\mathcal{F}^0 \otimes \mathcal{F}^1, \mathbb{P}^0 \otimes \mathbb{P}^1)$ and $\mathbb{F}$ is the complete and right continuous augmentation of $(\mathcal{F}_t^0 \otimes \mathcal{F}_t^1)_{0 \le t \le T}$ Generic elements of Ω are denoted $\omega = (\omega^0, \omega^1)$ with $\omega^0 \in \Omega^0$ and $\omega^1 \in \Omega^1$.

Given such a set-up, we shall search for an MFG equilibrium on the product space Ω for an initial random distribution $\mu_0 : (\Omega^0, \mathcal{F}_0^0) \to (\mathcal{P}_2(\mathbb{R}^d), \mathcal{B}(\mathcal{P}_2(\mathbb{R}^d)))$ and an initial private state $X_0 \in L^2(\Omega, \mathcal{F}_0, \mathbb{P}; \mathbb{R}^d)$ satisfying the prescription $\mathcal{L}(X_0(\omega^0, \cdot)) = \mu_0(\omega^0)$ for $\mathbb{P}^0$-almost every $\omega^0 \in \Omega^0$.

2.2.3 FBSDE Formulation of First Step in the Search for a Solution

In this subsection, we look at the optimal control problem defined in step (i) of the search of an MFG equilibrium. In order to stay with the framework introduced in Chapter 1 to handle stochastic optimal control problems in a random environment, we do not require the probability space to be of the aforementioned product form (2.24).

For the purpose of the analysis of MFGs with a common noise, we formulate in a systematic way the optimal control problem in (2.18) and (2.19) by means of an FBSDE in a random environment. Depending upon the case, we may use Theorem 1.57 or Theorem 1.60 in order to identify the optimal path of the underlying optimal control problem as the forward component of the solution of an FBSDE.

Although the strategy is very simple, the fact that the system used to represent the optimal path may involve the coefficients in the cost functional, as in Theorem 1.57, or their derivatives, as in Theorem 1.60, may render the exposition slightly cumbersome. In order to capture the commonalities of the two approaches and highlight the underpinnings of the methodology, we shall make a suitable assumption in order to merge the two possible cases.

The statement of assumption **FBSDE** below is based upon the notion of lifting introduced in the previous Subsection 2.2.2. We now formalize this notion of lifting in a precise definition.

Definition 2.14. *On a complete probability space $(\Omega, \mathcal{F}, \mathbb{F}, \mathbb{P})$, possibly of a non-product form, equipped with a complete and right-continuous filtration $\mathbb{F}$ and a tuple $(X_0, \boldsymbol{W}^0, \boldsymbol{\mu}, \boldsymbol{W})$ satisfying:*

1. $X_0 \in L^2(\Omega, \mathcal{F}_0, \mathbb{P}; \mathbb{R}^d)$,
2. *$\boldsymbol{\mu}$ is a càd-làg $\mathbb{F}$-adapted process with values in $\mathcal{X} = \mathcal{P}_2(\mathbb{R}^d)$ satisfying $\mathbb{E}[\sup_{0 \le t \le T} M_2(\mu_t)^2] < \infty$,*
3. *$(\boldsymbol{W}^0, \boldsymbol{W})$ is a 2d-Brownian motion with respect to $\mathbb{F}$ under $\mathbb{P}$,*
4. *$(X_0, \boldsymbol{W}^0, \boldsymbol{\mu})$ is independent of $\boldsymbol{W}$ under $\mathbb{P}$,*

we call lifting of $\boldsymbol{\mu}$, a càd-làg $\mathbb{F}$-adapted process $\underline{\mathfrak{M}} = (\mathfrak{M}_t)_{0 \le t \le T}$ taking values in a larger Polish metric space Ξ containing $\mathcal{P}_2(\mathbb{R}^d)$, such that:

$$\forall t \in [0, T], \quad \mu_t = \pi_t(\mathfrak{M}_t),$$

where $(\pi_t)_{0 \le t \le T}$ is a family of continuous projection mappings from Ξ into $\mathcal{P}_2(\mathbb{R}^d)$.

We then say that $(\Omega, \mathcal{F}, \mathbb{F}, \mathbb{P})$ equipped with $(X_0, \boldsymbol{W}^0, (\underline{\mathfrak{M}}, \boldsymbol{\mu}), \boldsymbol{W})$ is an admissible lifting if $(\Omega, \mathcal{F}, \mathbb{F}, \mathbb{P})$ equipped with $(X_0, \boldsymbol{W}^0, \underline{\mathfrak{M}}, \boldsymbol{W})$ is admissible in the sense of Definition 1.13 in Chapter 1, or equivalently if $(X_0, \boldsymbol{W}^0, \underline{\mathfrak{M}}, \boldsymbol{W})$ is compatible with $\mathbb{F}$ in the sense of Definition 1.6. Using the same abuse of terminology as before, we shall often say that $(X_0, \boldsymbol{W}^0, (\underline{\mathfrak{M}}, \boldsymbol{\mu}), \boldsymbol{W})$ is compatible with the filtration $\mathbb{F}$.

*Before stating the form of assumption **FBSDE** which is relevant to the investigations of this chapter, we warn the reader that, as we did in Chapter 1, we shall restrict ourselves to volatility coefficients σ and σ^0 which do not depend upon the control variable α.*

Assumption (FBSDE). On top of assumption **Control**, there exist an integer $m \ge 1$ together with deterministic measurable functions B from $[0, T] \times \mathbb{R}^d \times \mathcal{P}_2(\mathbb{R}^d) \times \mathbb{R}^m \times \mathbb{R}^{m \times d}$ into $\mathbb{R}^d$, F from $[0, T] \times \mathbb{R}^d \times \mathcal{P}_2(\mathbb{R}^d) \times \mathbb{R}^m \times (\mathbb{R}^{m \times d})^2$ into $\mathbb{R}^m$ and G from $\mathbb{R}^d \times \mathcal{P}_2(\mathbb{R}^d)$ into $\mathbb{R}^m$, such that, for any probabilistic set-up $(\Omega, \mathcal{F}, \mathbb{F}, \mathbb{P})$ equipped with a compatible lifting $(X_0, \boldsymbol{W}^0, (\underline{\mathfrak{M}}, \boldsymbol{\mu}), \boldsymbol{W})$ as above, it holds that:

(continued)

(A1) The optimal control problem defined in (2.18) and (2.19), namely:

$$\min_{\boldsymbol{\alpha}} J^{\mu}(\boldsymbol{\alpha}), \quad J^{\mu}(\boldsymbol{\alpha}) = \mathbb{E}\left[\int_0^T f(s, X_s^{\boldsymbol{\alpha}}, \mu_s, \alpha_s)ds + g(X_T^{\boldsymbol{\alpha}}, \mu_T)\right],$$

where $\boldsymbol{\alpha} = (\alpha_t)_{0 \le t \le T}$ is an $\mathbb{F}$-progressively measurable square-integrable A-valued process and $\boldsymbol{X}^{\boldsymbol{\alpha}} = (X_t^{\boldsymbol{\alpha}})_{0 \le t \le T}$ solves:

$$dX_t^{\boldsymbol{\alpha}} = b\big(t, X_t^{\boldsymbol{\alpha}}, \mu_t, \alpha_t\big)dt + \sigma(t, X_t^{\boldsymbol{\alpha}}, \mu_t)dW_t + \sigma^0(t, X_t^{\boldsymbol{\alpha}}, \mu_t)dW_t^0, \quad t \in [0, T],$$

with $X_0^{\boldsymbol{\alpha}} = X_0$ as initial condition, has a unique solution, characterized as the forward component of the solution of a strongly uniquely solvable FBSDE:

$$\begin{cases} dX_t = B(t, X_t, \mu_t, Y_t, Z_t)dt \\ \qquad + \sigma(t, X_t, \mu_t)dW_t + \sigma^0(t, X_t, \mu_t)dW_t^0, \\ dY_t = -F(t, X_t, \mu_t, Y_t, Z_t, Z_t^0)dt \\ \qquad + Z_t dW_t + Z_t^0 dW_t^0 + dM_t, \quad t \in [0, T], \end{cases} \tag{2.25}$$

with X_0 as initial condition for $\boldsymbol{X} = (X_t)_{0 \le t \le T}$ and $Y_T = G(X_T, \mu_T)$ as terminal condition for $\boldsymbol{Y} = (Y_t)_{0 \le t \le T}$, where $\boldsymbol{M} = (M_t)_{0 \le t \le T}$ is a *càd-làg* martingale with respect to the filtration $\mathbb{F}$, of zero cross variation with $(\boldsymbol{W}^0, \boldsymbol{W})$ and with initial condition $M_0 = 0$.

Once again, we refer the reader to Subsection 1.4, and more generally to Chapter 1, for the FBSDE characterization of a stochastic control problem in a random environment. Obviously, the coefficients (B, F, G) need to be connected to the original coefficients $(b, \sigma, \sigma^0, f, g)$. For starters, we require that B is of a specific form.

(A2) There exists a deterministic measurable function $\check{\alpha}$, with values in A such that:

$$B(t, x, \mu, y, z) = b\big(t, x, \mu, \check{\alpha}(t, x, \mu, y, z)\big),$$

for all $(t, x, \mu, y, z) \in [0, T] \times \mathbb{R}^d \times \mathcal{P}_2(\mathbb{R}^d) \times \mathbb{R}^m \times \mathbb{R}^{m \times d}$.

This innocent looking assumption is not without consequences. In particular, it implies that the optimal state process $\boldsymbol{X}$ is the controlled process associated with the optimal control $\hat{\boldsymbol{\alpha}}$ given by $\hat{\alpha}_t = (\check{\alpha}(t, X_t, \mu_t, Y_t, Z_t))_{0 \le t \le T}$.

The typical examples we have in mind are:

1. the FBSDE is associated with the probabilistic representation of the value function of the control problem, see Theorem 1.57. Notice that in this case, F does not depend on the variable z^0;
2. the FBSDE derives from the stochastic Pontryagin principle, see Theorem 1.60.

Given that the volatilities σ and σ^0 do not depend upon the control parameter α, when it comes to minimizing the Hamiltonian with respect to the control parameter, in order to make our life easier, we may minimize the reduced Hamiltonian $H^{(r)}$ instead of the full Hamiltonian H. Denoting as usual by $\hat{\alpha}(t, x, \mu, y)$ the minimizer of the reduced Hamiltonian $A \ni \alpha \mapsto H^{(r)}(t, x, \mu, y, \alpha)$, in the two important cases above, the function $\check{\alpha}$ is given by:

$$\check{\alpha}(t, x, \mu, y, z) = \begin{cases} \hat{\alpha}\big(t, x, \mu, \sigma(t, x, \mu)^{-1\dagger} z\big) & \text{in the first case;} \\ \hat{\alpha}(t, x, \mu, y) & \text{in the second case.} \end{cases} \tag{2.26}$$

The form of the coefficients (B, F, G) is quite clear in both cases. Undoubtedly, B is completely determined by the coefficient b together with the function $\check{\alpha}$ because of **(A2)**. The coefficients F and G take different forms in each of the two cases and the processes $\boldsymbol{Y}$ and $\boldsymbol{Z}$ do not have the same meaning. In the first case, $\boldsymbol{Y}$ represents the optimal cost, while it is associated with the optimal control in the second case. In both cases, $\hat{\alpha}$ is the argument of the minimization of the reduced Hamiltonian.

Finally, we shall require the following two structural conditions:

(A3) There exists a constant $L \ge 0$ such that:

$$\begin{aligned} &|(B, \check{\alpha})(t, x, \mu, y, z)| \le L\big[1 + |x| + |y| + |z| + M_2(\mu)\big], \\ &|F(t, x, \mu, y, z, z^0)| + |G(x, \mu)| \\ &\qquad \le L\big[1 + |x| + |y| + |z|^2 + |z^0| + \big(M_2(\mu)\big)^2\big]. \end{aligned}$$

Recalling that the solutions to (2.25) are required to satisfy:

$$\mathbb{E}\Big[\sup_{0 \le t \le T}\big(|X_t|^2 + |Y_t|^2 + |M_t|^2\big) + \int_0^T \big(|Z_t|^2 + |Z_t^0|^2\big)dt\Big] < \infty,$$

we see that the adapted stochastic process $(\check{\alpha}(t, X_t, \mu_t, Y_t, Z_t))_{0 \le t \le T}$ is always square-integrable on $[0, T] \times \Omega$.

Note that there is no real loss of generality in requiring assumption **FBSDE** to hold on any probabilistic set-up. Indeed, as shown in Theorems 1.57 and 1.60, existence and uniqueness for FBSDEs of the type (2.25) may be guaranteed on any arbitrary probabilistic set-up under quite general conditions. We thus assume that the representation of the optimally controlled paths holds independently of the probabilistic set-up, which means in particular, that the coefficients (B, F, G) are independent of the set-up which is used, and thus of the initial condition. As a result, they remain the same if the initial condition X_0 is changed and/or the initial time 0 is replaced by any other time $t \in [0, T]$. Also they remain the same if the environment process $\boldsymbol{\mu}$ is changed as long as the prescriptions in Definition 2.14 are fulfilled.

2.2.4 Strong MFG Matching Problem: Solutions and Strong Solvability

In this subsection, we assume again that the probability space is of the product form (2.24).

Equilibrium on an Arbitrary Space

We now concentrate on step (ii) of the search for an MFG equilibrium, see (2.20), which we call the matching problem or the fixed point step.

Throughout the subsection, we work on a probabilistic set-up of the product form (2.24) equipped with an initial random distribution $\mu_0 : (\Omega^0, \mathcal{F}_0^0) \to (\mathcal{P}_2(\mathbb{R}^d), \mathcal{B}(\mathcal{P}_2(\mathbb{R}^d))$ and an initial private state $X_0 \in L^2(\Omega, \mathcal{F}_0, \mathbb{P}; \mathbb{R}^d)$ satisfying the prescription $\mathcal{L}(X_0(\omega^0, \cdot)) = \mu_0(\omega^0)$ for $\mathbb{P}^0$-almost every $\omega^0 \in \Omega^0$.

Whenever μ_0 is random, the fixed point step (2.20) must be revisited as the flow of marginal equilibrium measures $\boldsymbol{\mu} = (\mu_t)_{0 \le t \le T}$ cannot be adapted with respect to the filtration generated by the sole $\boldsymbol{W}^0$. A natural strategy is thus to require $\boldsymbol{\mu} = (\mu_t)_{0 \le t \le T}$ to be adapted with respect to the filtration generated by both μ_0 and $\boldsymbol{W}^0$. According to the standard terminology in the theory of stochastic differential equations, this sounds like a *strong* equilibrium, as the solution is required to be measurable with respect to the information generated by the input $(\mu_0, \boldsymbol{W}^0)$. However, pursuing the comparison with the theory of stochastic differential equations, we also guess that this may not suffice and that there might be more complicated cases for which $\boldsymbol{\mu}$ is not adapted to the filtration generated by $(\mu_0, \boldsymbol{W}^0)$. This prompts us to introduce a quite general definition of an equilibrium, allowing for solutions which are not necessarily adapted to the common source of noise.

In order to proceed, we must specify the form of the information generated by the equilibrium $\boldsymbol{\mu}$ itself. Recalling that we aim at lifting any equilibrium $\boldsymbol{\mu}$ with values in $\mathcal{C}([0, T]; \mathcal{P}_2(\mathbb{R}^d))$ into some $\mathfrak{M}$ with values in $\mathcal{P}_2([\mathcal{C}([0, T]; \mathbb{R}^d)]^2)$, we introduce first the following definition.

Definition 2.15. *Given a probability space $(\Omega, \mathcal{F}, \mathbb{P})$ equipped with a filtration $\mathbb{F}$, a random variable $\mathfrak{M}$ from $(\Omega, \mathcal{F})$ into $\mathcal{P}_2(\mathcal{C}([0, T]; \mathbb{R}^{2d}))$ for $T > 0$, is said to be $\mathbb{F}$-adapted if for any $t \in [0, T]$ and any $C \in \sigma\{x_s, w_s;\ s \le t\}$, where $(\boldsymbol{x}, \boldsymbol{w})$ is the* *canonical process on* $[\mathcal{C}([0,T];\mathbb{R}^d)]^2 \cong \mathcal{C}([0,T];\mathbb{R}^{2d})$, *the random variable* $\mathfrak{M}(C)$ *is* $\mathcal{F}_t$*-measurable. Equivalently, denoting by* $\mathcal{F}_t^{\mathrm{nat},\mathfrak{M}} = \sigma\{\mathfrak{M}(C);\ C \in \sigma\{x_s, w_s;\ s \leq t\}\}$, $\mathfrak{M}$ *is* $\mathbb{F}$*-adapted if and only if* $\mathcal{F}_t^{\mathrm{nat},\mathfrak{M}} \subset \mathcal{F}_t$ *for all* $t \in [0,T]$.

In the above statement, the space $\mathcal{C}([0,T];\mathbb{R}^{2d})$ is equipped with the topology of the uniform convergence on $[0,T]$, and the space $\mathcal{P}_2(\mathcal{C}([0,T];\mathbb{R}^{2d}))$ with the corresponding 2-Wasserstein distance. Moreover, $\mathfrak{M}(C)$ has to be understood as the measure of C under the random measure $\mathfrak{M}$. Sometimes, we shall write $\mathfrak{M}(\omega, C)$ in order to specify the underlying realization $\omega \in \Omega$. The reader may want to take a look at Proposition (Vol I)-5.7 for basic properties of the Borel σ-field on $\mathcal{P}_2(\mathcal{C}([0,T];\mathbb{R}^{2d}))$. Recall in particular that $\mathfrak{M}$ is a random variable with values in $\mathcal{P}_2(\mathcal{C}([0,T];\mathbb{R}^{2d}))$ if and only if, for any $C \in \mathcal{B}(\mathcal{C}([0,T];\mathbb{R}^{2d}))$, the mapping $\Omega \ni \omega \mapsto \mathfrak{M}(\omega, C)$ is a random variable. Our choice to denote the canonical process on $\mathcal{C}([0,T];\mathbb{R}^{2d})$ by $(\boldsymbol{x}, \boldsymbol{w})$ is consistent with the analysis provided in the sequel. Below, $\mathfrak{M}$ will stand for the conditional law (given some σ-field) of the pair process formed by the forward component $\boldsymbol{X}$ and the Brownian motion $\boldsymbol{W}$ in an FBSDE of the same form as (2.25).

The idea behind Definition 2.15 is to specify the measurability properties of a flow of marginal equilibrium measures $\boldsymbol{\mu}$ through a lifted random variable $\mathfrak{M}$ with values in $\mathcal{P}_2(\mathcal{C}([0,T];\mathbb{R}^{2d}))$ such that, for all $t \in [0,T]$, $\mu_t = \mathfrak{M} \circ (e_t^x)^{-1}$, where e_t^x denotes the evaluation map on $\mathcal{C}([0,T];\mathbb{R}^{2d})$, giving the first d coordinates at time t. The rationale for this choice will be made clear in Chapter 3 when we discuss the construction of MFG equilibria.

For the time being, we state a general definition for the solutions of MFG problems with a common noise (also called an *equilibrium*).

Definition 2.16. *Given a probabilistic set-up* $(\Omega, \mathcal{F}, \mathbb{F}, \mathbb{P})$ *as in* (2.24), *equipped with Brownian motions* $\mathbf{W}^0$ *and* $\mathbf{W}$, *an initial random distribution* μ_0 *and an initial private state* X_0, *satisfying in particular the constraint* $\mathcal{L}^1(X_0) = \mu_0$, *we say that an* $\mathcal{F}_T^0$*-measurable random variable* $\mathfrak{M}$ *with values in* $\mathcal{P}_2(\mathcal{C}([0,T];\mathbb{R}^{2d}))$ *induces a solution to the MFG problem (2.18)–(2.19)–(2.20), with* $\boldsymbol{\mu} = (\mu_t : \Omega^0 \ni \omega^0 \mapsto \mu_t(\omega^0) = \mathfrak{M}(\omega^0) \circ (e_t^x)^{-1} \in \mathcal{P}_2(\mathbb{R}^d))_{0 \leq t \leq T}$ *as flow of marginal equilibrium measures, if*

(i) The filtration $\mathbb{F}$ *is compatible with* $(X_0, \mathbf{W}^0, \mathfrak{M}, \mathbf{W})$ *in the sense of Definition 1.6, with* $\mathbb{F}^{\mathrm{nat},(X_0,\mathbf{W}^0,\mathfrak{M},\mathbf{W})}$ *therein being the filtration generated by* $(X_0, \mathbf{W}^0, \mathbf{W})$ *and* $\mathbb{F}^{\mathrm{nat},\mathfrak{M}} = (\mathcal{F}_t^{\mathrm{nat},\mathfrak{M}})_{0 \leq t \leq T}$.

(ii) There exists a $\mathrm{Leb}_1 \otimes \mathbb{P}$ *square-integrable and* $\mathbb{F}$*-progressively measurable process* $\boldsymbol{\alpha} = (\alpha_t)_{0 \leq t \leq T}$ *with values in A, such that the solution* $\boldsymbol{X} = (X_t)_{0 \leq t \leq T}$ *of the SDE:*

$$dX_t = b(t, X_t, \mu_t, \alpha_t)dt + \sigma(t, X_t, \mu_t)dW_t + \sigma^0(t, X_t, \mu_t)dW_t^0, \quad t \in [0,T],$$

with X_0 as initial condition, satisfies:

$$\mathbb{P}^0\Big[\omega^0 \in \Omega^0 : \mathfrak{M}(\omega^0) = \mathcal{L}\big(\boldsymbol{X}(\omega^0,\cdot), \boldsymbol{W}(\cdot)\big)\Big] = 1, \tag{2.27}$$

(iii) For any other square-integrable and $\mathbb{F}$-progressively measurable control process $\boldsymbol{\beta}$ with values in A, it holds that:

$$J^{\mu}(\boldsymbol{\alpha}) \leq J^{\mu}(\boldsymbol{\beta}), \tag{2.28}$$

where J^{μ} is defined as in **(A1)** *in assumption* **FBSDE**.

We then call the pair $(\boldsymbol{W}^0, \mathfrak{M})$ (or sometimes $\mathfrak{M}$ alone) an equilibrium. The flow $\boldsymbol{\mu}$ is called the associated flow of equilibrium marginal measures.

Definition 2.16 looks rather complicated at first. In comparison with the original formulation in Subsection 2.2.1, we changed the fixed point condition (2.20) on the conditional marginal measures into the fixed point condition (2.27) on the conditional distribution of the whole trajectory, which we denoted by $\mathcal{L}(\boldsymbol{X}(\omega^0,\cdot), \boldsymbol{W}(\cdot))$ and which we shall also denote by $\mathcal{L}^1(\boldsymbol{X}, \boldsymbol{W})(\omega^0)$. In this respect, we stress the fact that Lemma 2.4 may be easily extended to random variables taking values in a Polish space. In comparison with the formulation introduced in Subsection 2.2.1, we also specified the notion of compatibility used for handling the stochastic optimal control in the environment $\boldsymbol{\mu}$ induced by the lifting $\mathfrak{M}$.

Hopefully, the reader will understand in the next chapter that we performed these two changes in order to prove existence of equilibria. Somehow the Compatibility Condition and the fixed point step dictate the macroscopic information that an agent should observe on the population in order to implement the Nash strategy. Basically, we require the agent to observe $\boldsymbol{W}^0$ and the full-fledged conditional law of the whole path $(\boldsymbol{X}, \boldsymbol{W})$ given the realization ω^0. The useful information is thus much larger than that generated by $\boldsymbol{W}^0$ and the flow of marginal conditional laws $(\mathcal{L}^1(X_t))_{0\leq t\leq T}$ – which is nothing but $\boldsymbol{\mu}$ as we shall check below. As we shall see in the next chapter, this framework seems to be the right one for constructing solutions to MFG with a common noise. Except for some specific cases, we will not be able to construct solutions for which $\mathbb{F}$ is compatible with the sole $(X_0, \boldsymbol{W}^0, \boldsymbol{\mu}, \boldsymbol{W})$.

We insist on the fact that, despite this enlarged framework, the results of Subsection 1.4 still apply. In order to fit the framework used therein, it suffices to extend the environment. Instead of $\boldsymbol{\mu}$, we may look at the process $\underline{\mathfrak{M}} = (\mathfrak{M}_t = \mathfrak{M} \circ \mathcal{E}_t^{-1})_{0\leq t\leq T}$, where $\mathcal{E}_t : \mathcal{C}([0,T];\mathbb{R}^{2d}) \ni (\boldsymbol{x}, \boldsymbol{w}) \mapsto (x_{s\wedge t}, w_{s\wedge t})_{0\leq s\leq T} \in \mathcal{C}([0,T];\mathbb{R}^{2d})$. Equivalently, this amounts to choosing $\mathcal{X}$ in Subsection 1.4 as $\mathcal{P}_2(\mathcal{C}([0,T];\mathbb{R}^{2d}))$. It is plain to check that $\underline{\mathfrak{M}}$ is a continuous process with values in $\mathcal{C}([0,T];\mathbb{R}^{2d})$ and that its canonical filtration coincides with $\mathcal{F}^{\mathrm{nat},\mathfrak{M}}$. In particular, the compatibility requirement in Definition 2.16 may be rewritten as a Compatibility Condition between $\mathbb{F}$ and the filtration generated by $(X_0, \boldsymbol{W}^0, \underline{\mathfrak{M}}, \boldsymbol{W})$. Obviously, the sequence $(\mathfrak{M}_t)_{0\leq t\leq T}$ subsumes $(\mu_t)_{0\leq t\leq T}$ since μ_t is then understood as $\mu_t = \mathfrak{M}_t \circ (e_t^x)^{-1}$ with e_t^x being the evaluation map providing the first d coordinates at time t on $\mathcal{C}([0,T];\mathbb{R}^{2d})$.

The identification of the canonical filtration generated by $\underline{\mathfrak{M}}$ follows from the following observation. Notice first that the σ-field $\sigma\{\mathcal{E}_t\}$ generated by $\mathcal{E}_t$ on $\mathcal{C}([0,T];\mathbb{R}^{2d})$ is $\sigma\{x_s, w_s;\ s \leq t\}$ where $(\boldsymbol{x}, \boldsymbol{w})$ denotes the canonical process on $[\mathcal{C}([0,T];\mathbb{R}^d)]^2 \cong \mathcal{C}([0,T];\mathbb{R}^{2d})$. Owing to Proposition (Vol I)-5.7, the σ-field generated by the random variable $\mathfrak{M}_t$ has the form $\sigma\{\mathfrak{M}(C);\ C \in \sigma\{x_s, w_s;\ s \leq t\}\}$, which is exactly $\mathcal{F}_t^{\mathrm{nat},\mathfrak{M}}$.

Notice that we shall often use $\mathfrak{M}$ and $\underline{\mathfrak{M}}$ alternatively since they can be easily identified mathematically speaking. We now make the connection with the matching problem (ii) articulated in Subsection 2.2.1:

Proposition 2.17. *With the notation of Definition 2.16, for $\mathbb{P}^0$-almost every $\omega^0 \in \Omega^0$, it holds:*

$$\forall t \in [0,T], \quad \mu_t(\omega^0) = \mathcal{L}\big(X_t(\omega^0,\cdot)\big).$$

Moreover, for any $t \in [0,T]$, μ_t provides a conditional law of X_t given $\mathcal{F}^0$ and of X_t given $\mathcal{F}_t^0$, both σ-algebras being regarded as sub-σ-algebras of $\mathcal{F}$, see Remark 2.2. In particular, μ_t is the conditional law of X_t given $\mathcal{F}_t^{\mathrm{nat},(\boldsymbol{W}^0,\mathfrak{M})}$, the latter being also regarded as a sub-σ-algebra of $\mathcal{F}$.

Proof. The first point is rather straightforward. Recall that, for a given $t \in [0,T]$, e_t^x denotes the evaluation map providing the first d coordinates at time t on $\mathcal{C}([0,T];\mathbb{R}^{2d})$. It is easy to check that the mapping $\mathcal{P}_2(\mathcal{C}([0,T];\mathbb{R}^{2d})) \ni \mathfrak{m} \mapsto \mathfrak{m} \circ (e_t^x)^{-1} \in \mathcal{P}_2(\mathbb{R}^d)$ is measurable (it is even continuous). Therefore, both $\Omega^0 \ni \omega^0 \mapsto \mu_t(\omega^0)$ and $\Omega^0 \ni \omega^0 \mapsto \mathcal{L}(\boldsymbol{X}(\omega^0,\cdot),\boldsymbol{W}) \circ (e_t^x)^{-1}$ are random variables. Of course, they coincide $\mathbb{P}^0$-almost surely. Now, for $\mathbb{P}^0$-almost every $\omega^0 \in \Omega^0$ and for any $C \in \mathcal{B}(\mathbb{R}^d)$ in a countable generating π-system,

$$\begin{aligned}
\big[\mathcal{L}\big(\boldsymbol{X}(\omega^0,\cdot),\boldsymbol{W}\big) \circ (e_t^x)^{-1}\big](C) &= \mathbb{P}^1\Big[\omega^1 \in \Omega^1 : \big(\boldsymbol{X}(\omega^0,\omega^1),\boldsymbol{W}(\omega^1)\big) \in (e_t^x)^{-1}(C)\Big] \\
&= \mathbb{P}^1\big[X_t(\omega^0,\cdot) \in C\big] \\
&= \big[\mathcal{L}\big(X_t(\omega^0,\cdot)\big)\big](C).
\end{aligned}$$

We now turn to the second claim. The identification with a conditional law given $\mathcal{F}^0$ follows from Lemma 2.4. In order to complete the proof, it suffices to notice, from Lemma 2.5, that the mapping $\Omega^0 \ni \omega^0 \mapsto \mu_t(\omega^0)$ is $\mathcal{F}_t^0$-measurable and is $\mathcal{F}_t^{\mathrm{nat},(\boldsymbol{W}^0,\mathfrak{M})}$-measurable. □

When assumption **FBSDE** is in force (whatever the probabilistic set-up is), we use the following characterization of a solution.

Proposition 2.18. *When assumption* **FBSDE** *is in force, there exists a solution to the MFG problem (2.18)–(2.19)–(2.20) on the same probabilistic set-up $(\Omega,\mathcal{F},\mathbb{F},\mathbb{P})$ as in Definition 2.16, if and only if there exists an $\mathbb{F}$-adapted $\mathbb{R}^d$-valued continuous process $\boldsymbol{X}$, with X_0 as initial condition, such that $\mathbb{F}$ is compatible with $(X_0,\boldsymbol{W}^0,\mathcal{L}^1(\boldsymbol{X},\boldsymbol{W}),\boldsymbol{W})$ and $\boldsymbol{X}$ solves, together with some triple $(\boldsymbol{Y},\boldsymbol{Z},\boldsymbol{Z}^0,\boldsymbol{M})$, the following McKean-Vlasov FBSDE on $(\Omega,\mathcal{F},\mathbb{F},\mathbb{P})$:*

$$\begin{cases} dX_t = B\big(t, X_t, \mathcal{L}^1(X_t), Y_t, Z_t\big)dt + \sigma\big(t, X_t, \mathcal{L}^1(X_t)\big)dW_t \\ \qquad +\sigma^0\big(t, X_t, \mathcal{L}^1(X_t)\big)dW_t^0, \\ dY_t = -F\big(t, X_t, \mathcal{L}^1(X_t), Y_t, Z_t, Z_t^0\big)dt \\ \qquad +Z_t dW_t + Z_t^0 dW_t^0 + dM_t, \quad t \in [0, T], \\ Y_T = G\big(X_T, \mathcal{L}^1(X_T)\big), \end{cases} \tag{2.29}$$

where $\boldsymbol{M} = (M_t)_{0 \le t \le T}$ is a càd-làg martingale with respect to the filtration $\mathbb{F}$, with $M_0 = 0$ and of zero cross variation with $(\boldsymbol{W}^0, \boldsymbol{W})$. In such a case, the equilibrium is given by $\mathfrak{M} = \mathcal{L}^1(\boldsymbol{X}, \boldsymbol{W})$, and its marginals by $\boldsymbol{\mu} = (\mathcal{L}^1(X_t))_{0 \le t \le T}$.

Remark 2.19. *The conclusion of Proposition 2.18 is particularly convenient. Indeed, when assumption* **FBSDE** *is satisfied, equation (2.29), when regarded as an FBSDE in the random environment formed by $\mathfrak{M} = \mathcal{L}^1(\boldsymbol{X}, \boldsymbol{W})$ and $\boldsymbol{\mu} = (\mu_t = \mathfrak{M} \circ (e_t^x)^{-1})_{0 \le t \le T}$, is uniquely solvable in the strong sense. So by Theorem 1.33, its forward component $\boldsymbol{X}$ must be a functional of $(X_0, \boldsymbol{W}^0, \mathfrak{M}, \boldsymbol{W})$. This says that we may completely disregard the original filtration $\mathbb{F}$, and just focus on the information enclosed in $(X_0, \boldsymbol{W}^0, \mathfrak{M}, \boldsymbol{W})$.*

In particular, it is worth mentioning that the process $(X_0, \boldsymbol{W}^0, \mathfrak{M}, \boldsymbol{W})$ is automatically compatible with $(\boldsymbol{W}^0, \mathfrak{M})$ since for all $t \in [0, T]$, $\mathcal{F}_T^{(\boldsymbol{W}^0, \mathfrak{M})}$ and $\mathcal{F}_t^{(X_0, \boldsymbol{W}^0, \mathfrak{M}, \boldsymbol{W})}$ are conditionally independent given $\mathcal{F}_t^{(\boldsymbol{W}^0, \mathfrak{M})}$, the definition of a σ-field generated by $\mathfrak{M}$ being the same as in Definition 2.16. In other words, $(X_0, \boldsymbol{W})$ is necessarily observed in a fair way as the observation of $(X_0, \boldsymbol{W})$ does not introduce any bias in the observation of $(\boldsymbol{W}^0, \mathfrak{M})$. The information enclosed in $(\boldsymbol{W}^0, \mathfrak{M})$ is somehow the canonical information needed to describe an equilibrium. We shall come back to this point in the next subsection.

We provide the proof of the compatibility property.

Proof. By Lemma 1.7, it suffices to prove that, for all $t \in [0, T]$, $\mathcal{F}_T^{\mathrm{nat},(\boldsymbol{W}^0, \mathfrak{M})}$ and $\mathcal{F}_t^{\mathrm{nat},(X_0, \boldsymbol{W}^0, \mathfrak{M}, \boldsymbol{W})}$ are conditionally independent given $\mathcal{F}_t^{\mathrm{nat},(\boldsymbol{W}^0, \mathfrak{M})}$.

For a given $t \in [0, T]$, consider three Borel subsets $C_t, C_T, E_t \subset \mathcal{C}([0, T]; \mathbb{R}^d)$, two Borel subsets $D_t, D_T \subset \mathcal{P}_2(\mathcal{C}([0, T]; \mathbb{R}^{2d}))$ and a Borel subset $B \subset \mathbb{R}^d$. Recalling the notation $\mathfrak{M}_t = \mathfrak{M} \circ \mathcal{E}_t^{-1}$, where $\mathcal{E}_t : \mathcal{C}([0, T]; \mathbb{R}^{2d}) \ni (\boldsymbol{x}, \boldsymbol{w}) \mapsto (x_{s \wedge t}, w_{s \wedge t})_{0 \le s \le T} \in \mathcal{C}([0, T]; \mathbb{R}^{2d})$, and denoting by $\boldsymbol{W}^0_{\cdot \wedge t}$ and $\boldsymbol{W}_{\cdot \wedge t}$ the processes $\boldsymbol{W}^0$ and $\boldsymbol{W}$ stopped at t, we have:

$$\begin{aligned} &\mathbb{E}\big[\mathbf{1}_B(X_0)\mathbf{1}_{C_t}(\boldsymbol{W}^0_{\cdot \wedge t})\mathbf{1}_{D_t}(\mathfrak{M}_t)\mathbf{1}_{E_t}(\boldsymbol{W}_{\cdot \wedge t})\mathbf{1}_{C_T}(\boldsymbol{W}^0)\mathbf{1}_{D_T}(\mathfrak{M}_T)\big] \\ &= \mathbb{E}^0\big[\mathbf{1}_{C_t}(\boldsymbol{W}^0_{\cdot \wedge t})\mathbf{1}_{D_t}(\mathfrak{M}_t)\mathbf{1}_{C_T}(\boldsymbol{W}^0)\mathbf{1}_{D_T}(\mathfrak{M}_T)\mathbb{E}^1\big[\mathbf{1}_B(X_0)\mathbf{1}_{E_t}(\boldsymbol{W}_{\cdot \wedge t})\big]\big]. \end{aligned}$$

Observe that the set $C = \{(\boldsymbol{x}, \boldsymbol{w}) \in \mathcal{C}([0, T], \mathbb{R}^{2d}) : (x_0, \boldsymbol{w}_{\cdot \wedge t}) \in B \times E_t\}$ belongs to $\sigma\{x_s, w_s;\ s \le t\}$. Moreover, by definition, we have, for $\mathbb{P}^0$-almost every $\omega^0 \in \Omega^0$,

$$\mathbb{E}^1\big[\mathbf{1}_B(X_0)\mathbf{1}_{E_t}(\boldsymbol{W}_{\cdot \wedge t})\big] = \mathfrak{M}(C) = \mathfrak{M}\big(\mathcal{E}_t^{-1}(C)\big) = \mathfrak{M}_t(C),$$

since $\mathcal{E}_t^{-1}(C) = C$. Therefore,

$$\begin{aligned}&\mathbb{E}\big[\mathbf{1}_B(X_0)\mathbf{1}_{C_t}(\boldsymbol{W}^0_{\cdot\wedge t})\mathbf{1}_{D_t}(\mathfrak{M}_t)\mathbf{1}_{E_t}(\boldsymbol{W}_{\cdot\wedge t})\mathbf{1}_{C_T}(\boldsymbol{W}^0)\mathbf{1}_{D_T}(\mathfrak{M}_T)\big]\\&= \mathbb{E}^0\big[\mathbf{1}_{C_t}(\boldsymbol{W}^0_{\cdot\wedge t})\mathbf{1}_{D_t}(\mathfrak{M}_t)\mathbf{1}_{C_T}(\boldsymbol{W}^0)\mathbf{1}_{D_T}(\mathfrak{M}_T)\mathfrak{M}_t(C)\big].\end{aligned}$$

Since $\mathfrak{M}_t(C)$ is measurable with respect to $\mathcal{F}_t^{\mathrm{nat},(\boldsymbol{W}^0,\mathfrak{M})}$, compatibility follows. □

Remark 2.20. *In contrast with the setting of Chapter 1, the compatibility condition in Proposition 2.18 concerns the output, through the random variable $\mathcal{L}^1(\boldsymbol{X},\boldsymbol{W})$, and not the input.*

Intuitively, the compatibility condition is here a way to select solutions that are somehow meaningful from the physical point of view. We provide in Chapter 3, see Subsection 3.5.5, an example of a solution that does not satisfy this compatibility condition.

Strong Solvability and Strong Solutions

Inspired by Definition 1.17, we introduce the following definition.

Definition 2.21. *We say that the MFG problem (2.18)–(2.19)–(2.20) is strongly solvable if it has a solution on any probabilistic set-up in the sense of Definition 2.16.*

The idea behind the notion of *strong solvability* is that the filtration $\mathbb{F}^0$ can be arbitrarily chosen. A typical example is to choose $\mathbb{F}^0$ as the usual augmentation of the canonical filtration generated by the common noise $\boldsymbol{W}^0$ and the initial condition μ_0, in which case a mere variation of Lemma 2.4 shows that $\mathfrak{M}$ is the conditional law of $(\boldsymbol{X},\boldsymbol{W})$ given $(\mu_0,\boldsymbol{W}^0)$ and μ_t is the conditional law of X_t given $(\mu_0,\boldsymbol{W}^0)$. We thus recover the intuitive description of an MFG equilibrium provided in the preliminary Subsections 2.1.1 and 2.2.1 before we introduced the notion of lifting.

The following definition is here to stress the importance of this example.

Definition 2.22. *A solution to the MFG problem (2.18)–(2.19)–(2.20) on a set-up $(\Omega,\mathcal{F},\mathbb{F},\mathbb{P})$ as in Definition 2.16 is said to be strong if it satisfies Definition 2.16 with $\mathbb{F}^0$ being the complete and right-continuous filtration generated by μ_0 and $\boldsymbol{W}^0$.*

Whenever solvability holds in the strong sense, a convenient way to construct a strong solution is to work with the *canonical* set-up $\bar{\Omega} = \bar{\Omega}^{00} \times \bar{\Omega}^1$, made of

$$\begin{aligned}\bar{\Omega}^{00} &= \mathcal{P}_2(\mathbb{R}^d) \times \mathcal{C}([0,T];\mathbb{R}^d),\\ \bar{\Omega}^1 &= [0,1) \times \mathcal{C}([0,T];\mathbb{R}^d),\end{aligned} \tag{2.30}$$

$\bar{\Omega}^{00}$ being equipped with the completion of the product measure $\mathcal{V}^0 \otimes \mathcal{W}_d$, where $\mathcal{V}^0$ is some initial distribution on the space $\mathcal{P}_2(\mathbb{R}^d)$ and $\mathcal{W}_d$ is the Wiener measure on

$\mathcal{C}([0,T];\mathbb{R}^d)$, and $\bar{\Omega}^1$ being equipped with the completion of the product measure $\text{Leb}_1 \otimes \mathcal{W}_d$, where Leb_1 is the Lebesgue measure on $[0,1)$. The canonical random variable on $\bar{\Omega}^{00}$ is denoted by $(\nu^0, \boldsymbol{w}^0)$ and the canonical random variable on $\bar{\Omega}^1$ is denoted by $(\eta, \boldsymbol{w}^1)$.

Under these conditions, the initial distribution of the population is ν^0 and the initial state X_0 is defined as:

$$X_0 = \psi(\eta, \nu^0), \tag{2.31}$$

with ψ as in Lemma (Vol I)-5.29, see also (2.23).

2.3 Weak Solutions for MFGs with Common Noise

2.3.1 Weak MFG Matching Problem

In parallel with the notion of strong solvability, we may also want to consider a weaker notion of solvability covering cases for which MFG equilibria only exist on some (but not necessarily all) probabilistic set-ups.

Definition 2.23. *Given $\mathcal{V}^0 \in \mathcal{P}_2(\mathcal{P}_2(\mathbb{R}^d))$, we say that the MFG problem admits a weak solution (or a weak equilibrium) with $\mathcal{V}^0$ as initial condition if there exists a probabilistic set-up $(\Omega, \mathcal{F}, \mathbb{F}, \mathbb{P})$ satisfying the same requirements as in Definition 2.16 on which μ_0 has $\mathcal{V}^0$ as distribution and the MFG problem has an $\mathcal{F}_T^0$-measurable solution $\mathfrak{M} : \Omega^0 \to \mathcal{P}_2(\mathcal{C}([0,T];\mathbb{R}^{2d}))$.*

When assumption **FBSDE** *is in force, we say that the MFG problem admits a weak equilibrium if we can find a probabilistic set-up of the same type as above, for which μ_0 has $\mathcal{V}^0$ as distribution and there exists an $\mathbb{F}$-adapted $\mathbb{R}^d$-valued continuous process $\boldsymbol{X}$ such that $\mathbb{F}$ is compatible with $(X_0, \boldsymbol{W}^0, \mathcal{L}^1(\boldsymbol{X}, \boldsymbol{W}), \boldsymbol{W})$ and $\boldsymbol{X}$ solves, together with some tuple $(\boldsymbol{Y}, \boldsymbol{Z}, \boldsymbol{Z}^0, \boldsymbol{M})$, the McKean-Vlasov FBSDE* (2.29).

We say that the MFG problem is weakly solvable if we can find a weak equilibrium in the sense of Definition 2.23. In such a case, we often say that the pair $(\boldsymbol{W}^0, \mathfrak{M})$ forms a weak solution. Notice that a weak equilibrium may not be measurable with respect to the σ-field generated by μ_0 and $\boldsymbol{W}^0$. As a result, we cannot ensure that μ_t is the conditional law of X_t given $(\mu_0, \boldsymbol{W}^0)$ as for a strong solution but only that μ_t is the conditional law of X_t given $(\boldsymbol{W}^0, \mathfrak{M})$, see Proposition 2.17.

Canonical Space and Distribution of an Equilibrium

For the remainder of the chapter we assume that assumption **FBSDE** holds on any probabilistic set-up, and we work on the *extended* canonical space:

$$\begin{aligned} \bar{\Omega}^0 &= \mathcal{P}_2(\mathbb{R}^d) \times \mathcal{C}([0,T];\mathbb{R}^d) \times \mathcal{P}_2(\mathcal{C}([0,T];\mathbb{R}^{2d})), \\ \bar{\Omega}^1 &= [0,1) \times \mathcal{C}([0,T];\mathbb{R}^d). \end{aligned} \tag{2.32}$$

Compared with the canonical space (2.30) introduced for strong MFG equilibria, we added an additional coordinate to $\bar{\Omega}^0$. This extra coordinate is intended to carry the equilibrium measure $\mathfrak{M}$. By Yamada-Watanabe Theorem 1.33 for FBSDEs, the law of the solution to (2.25) is the same irrespective of the underlying probabilistic set-up provided that the compatibility condition described in Definition 2.16 holds. In particular, in order to investigate the law of an equilibrium $\mathfrak{M}$ on the space $\mathcal{P}_2(\mathcal{C}([0,T];\mathbb{R}^{2d}))$, there is no loss of generality in working on the canonical space directly, see Lemma 2.25 below.

As above, the space $\bar{\Omega}^1$ is equipped with the measure $\bar{\mathbb{P}}^1 = \text{Leb}_1 \otimes \mathcal{W}_d$ and is then completed. The canonical random variable on $\bar{\Omega}^1$ is denoted by $(\eta, \boldsymbol{w} = (w_t)_{0 \le t \le T})$. The canonical random variable on $\bar{\Omega}^0$ is denoted by $(\nu^0, \boldsymbol{w}^0 = (w_t^0)_{0 \le t \le T}, \mathfrak{m})$, the associated flow of marginal measures being denoted by $\boldsymbol{\nu} = (\nu_t = \mathfrak{m} \circ (e_t^x)^{-1})_{0 \le t \le T}$, where as usual e_t^x denotes the evaluation map providing the first d coordinates at time t on $\mathcal{C}([0,T];\mathbb{R}^{2d})$. The canonical filtration on $\bar{\Omega}^0$ is the filtration generated by $(\nu^0, w_t^0, \mathfrak{m}_t)_{0 \le t \le T}$ where $\mathfrak{m}_t = \mathfrak{m} \circ \mathcal{E}_t^{-1}$, with $\mathcal{E}_t : \mathcal{C}([0,T];\mathbb{R}^{2d}) \ni (\boldsymbol{x}, \boldsymbol{w}) \mapsto (x_{s \wedge t}, w_{s \wedge t})_{0 \le s \le T} \in \mathcal{C}([0,T];\mathbb{R}^{2d})$. Note that, in contrast with the canonical space structure introduced in Subsection 1.2.2, the input $\underline{\mathfrak{m}} = (\mathfrak{m}_t)_{0 \le t \le T}$ is a continuous function of time in the present set-up.

The initial distribution of the population is ν^0 and the initial state X_0 is defined as $X_0 = \psi(\eta, \nu^0)$, with ψ as in Lemma (Vol I)-5.29, see also (2.23).

We now introduce a concept which will help us sort out the roles of probability measures on the canonical spaces.

Definition 2.24. *Given an initial law $\mathcal{V}^0 \in \mathcal{P}_2(\mathcal{P}_2(\mathbb{R}^d))$, we say that a probability measure $\mathcal{M}$ on $\bar{\Omega}^0$ equipped with its Borel σ-field is a distribution of an equilibrium if:*

1. *On the completion $(\bar{\Omega}^0, \bar{\mathcal{F}}^0, \bar{\mathbb{P}}^0)$ of $(\bar{\Omega}^0, \mathcal{B}(\bar{\Omega}^0), \mathcal{M})$, equipped with the complete and right-continuous augmentation $\mathbb{F}^0$ of the canonical filtration, ν^0 has $\mathcal{V}^0$ as distribution and the process $(w_t^0)_{0 \le t \le T}$ is a d-dimensional $\mathbb{F}^0$-Wiener process,*
2. *On the product probabilistic set-up $(\bar{\Omega} = \bar{\Omega}^0 \times \bar{\Omega}^1, \bar{\mathcal{F}}, \bar{\mathbb{F}}, \bar{\mathbb{P}})$ constructed as in (2.24), with $\bar{\mathbb{P}}^1 = \text{Leb}_1 \otimes \mathcal{W}_d$, the canonical process $(\mathfrak{m}_t)_{0 \le t \le T}$ is an equilibrium in the sense of Definition 2.16.*

We are well aware of the fact that the reader may find the terminology *distribution of an equilibrium* used in the definition above to be confusing. Indeed, what we already defined as a weak equilibrium (and even an *equilibrium measure* from time to time) is in principle different from this notion of *distribution of an equilibrium*. We ask the reader to bear with us just for a little while longer. Our choice of terminology will be unambiguously vindicated in Lemma 2.25 below. In the meantime, we stress the fact that, according to Definition 2.23 and the interpretation following its statement, a weak solution to the equilibrium problem, (or a weak equilibrium) is a pair $(\boldsymbol{W}^0, \mathfrak{M})$, where $\boldsymbol{W}^0$ is a d-dimensional Brownian motion and $\mathfrak{M}$ is a random variable with values in $\mathcal{P}_2(\mathcal{C}([0,T];\mathbb{R}^{2d}))$, that is a random measure on $\mathcal{C}([0,T];\mathbb{R}^{2d})$. On the other hand, a *distribution of an equilibrium* as we just defined it, is what we should expect from our experience with the classical theory of weak and strong solutions of SDEs. It should be a probability measure on a specific canonical space accommodating the initial condition as well as the noise sources, and for which the canonical process becomes a solution.

Observe also that on the canonical set-up, the compatibility condition required in Definition 2.16 is automatically satisfied. Indeed, following the proof of Lemma 1.7, in order to prove this claim, it suffices to check that, for any $t \in [0, T]$, $\mathcal{F}_t^{\mathrm{nat},\eta}$ and $\mathcal{F}_T^{\mathrm{nat},(\boldsymbol{W}^0,\mathfrak{M},\boldsymbol{W})}$ are conditionally independent given $\mathcal{F}_t^{\mathrm{nat},(X_0,\boldsymbol{W}^0,\mathfrak{M},\boldsymbol{W})}$. Since $(\eta, X_0, \boldsymbol{W}^0, \mathfrak{M})$ is independent of $\boldsymbol{W}$, it is actually sufficient to prove that, for any $t \in [0, T]$, $\mathcal{F}_t^{\mathrm{nat},\eta}$ and $\mathcal{F}_T^{\mathrm{nat},(\boldsymbol{W}^0,\mathfrak{M})}$ are conditionally independent given $\mathcal{F}_t^{\mathrm{nat},(X_0,\boldsymbol{W}^0,\mathfrak{M})}$. Now, for any bounded Borel-measurable functions θ_t^0 and θ_T^0 from $\mathcal{C}([0,T];\mathbb{R}^d) \times \mathcal{P}_2(\mathcal{C}([0,T];\mathbb{R}^{2d}))$ into $\mathbb{R}$, θ_t^1 from $[0,1)$ into $\mathbb{R}$ and θ_0 from $\mathbb{R}^d$ into $\mathbb{R}$, we have:

$$
\begin{aligned}
&\mathbb{E}\Big[\theta_0(X_0)\theta_t^1(\eta)\theta_t^0(\boldsymbol{W}^0,\mathfrak{M})\theta_T^0(\boldsymbol{W}^0,\mathfrak{M})\Big] \\
&= \int_0^1 \theta_t^1(s)\mathbb{E}\Big[\theta_0\big(\psi(s,\nu^0)\big)\theta_t^0(\boldsymbol{W}^0,\mathfrak{M})\theta_T^0(\boldsymbol{W}^0,\mathfrak{M})\Big]ds \\
&= \int_0^1 \theta_t^1(s)\mathbb{E}\Big[\theta_0\big(\psi(s,\nu^0)\big)\theta_t^0(\boldsymbol{W}^0,\mathfrak{M})\mathbb{E}\big[\theta_T^0(\boldsymbol{W}^0,\mathfrak{M}) \mid \mathcal{F}_t^{\mathrm{nat},(\boldsymbol{W}^0,\mathfrak{M})}\big]\Big]ds \\
&= \mathbb{E}\Big[\theta_0(X_0)\theta_t^1(\eta)\theta_t^0(\boldsymbol{W}^0,\mathfrak{M})\mathbb{E}\big[\theta_T^0(\boldsymbol{W}^0,\mathfrak{M}) \mid \mathcal{F}_t^{\mathrm{nat},(\boldsymbol{W}^0,\mathfrak{M})}\big]\Big],
\end{aligned}
\tag{2.33}
$$

which proves what we wanted to check.

The following lemma says that every *weak equilibrium* induces a *distribution of an equilibrium*. Because of this result, we are free to use the terminologies *weak solution* and *distribution of an equilibrium* interchangeably.

Lemma 2.25. *Under assumption* **FBSDE**, *if $\mathfrak{M}$ is an MFG equilibrium constructed on some probabilistic set-up $(\Omega, \mathcal{F}, \mathbb{F}, \mathbb{P})$ equipped with a triplet $(X_0, \boldsymbol{W}^0, \boldsymbol{W})$ and satisfying the prescription of Definition 2.16, and if we define $\mathcal{M}$ as the distribution of $(\mathfrak{M} \circ (e_0^x)^{-1}, \boldsymbol{W}^0, \mathfrak{M})$ on $\bar{\Omega}^0$ equipped with its Borel σ-field, then $\mathcal{M}$ is a distribution of an equilibrium.*

Proof. If (2.25) is solved on an arbitrary admissible set-up with the constraint that

$$
\mathbb{P}^0\big[\omega^0 \in \Omega^0 : \mathfrak{M}(\omega^0) = \mathcal{L}(X(\omega^0,\cdot),\boldsymbol{W})\big] = 1,
$$

then we can solve the FBSDE (2.25) on the canonical set-up choosing the measure $\bar{\mathbb{P}}^0$ in Definition 2.24 as the completion of $\mathcal{L}(\mu_0, \boldsymbol{W}_0, \mathfrak{M})$, with $\mu_0 = \mathfrak{M} \circ e_0^{-1}$. We then define $\bar{\mathbb{P}}$ in Definition 2.24 accordingly.

By Theorem 1.33, we know that we can find a measurable mapping ϕ from $\mathbb{R}^d \times \mathcal{C}([0,T];\mathbb{R}^d) \times \mathcal{P}_2(\mathcal{C}([0,T];\mathbb{R}^{2d})) \times \mathcal{C}([0,T];\mathbb{R}^d)$ such that the forward component of the solution of the FBSDE is almost surely equal to $\phi(X_0, \boldsymbol{W}^0, \mathfrak{M}, \boldsymbol{W})$ on the original set-up and to $\phi(\psi(\eta,\nu^0), \boldsymbol{w}^0, \mathrm{m}, \boldsymbol{w})$ on the canonical set-up. In particular, letting $\boldsymbol{X} = \phi(X_0, \boldsymbol{W}^0, \mathfrak{M}, \boldsymbol{W})$

and $\boldsymbol{x} = \phi(\psi(\eta, \nu^0), \boldsymbol{w}^0, \mathfrak{m}, \boldsymbol{w})$, the law of $(X_0, \mu_0, \boldsymbol{W}^0, \mathfrak{M}, \boldsymbol{W}, \boldsymbol{X})$ under $\mathbb{P}$ is the same as the law of $(\psi(\eta, \nu^0), \nu^0, \boldsymbol{w}^0, \mathfrak{m}, \boldsymbol{w}, \boldsymbol{x})$ under $\bar{\mathbb{P}}$. We then conclude by Lemma 2.26 below by noticing that:

$$\bar{\mathbb{E}}\big[\mathbf{1}_C(\boldsymbol{X}, \boldsymbol{W})\mathbf{1}_{C^0}(\nu^0, \boldsymbol{w}^0, \mathfrak{m})\big] = \bar{\mathbb{E}}^0\big[\mathfrak{m}(C)\mathbf{1}_{C^0}(\nu^0, \boldsymbol{w}^0, \mathfrak{m})\big],$$

for all Borel subsets C^0 and C of $\bar{\Omega}^0$ and $\mathcal{C}([0,T];\mathbb{R}^{2d})$, which follows from the fact that $(X_0, \boldsymbol{W}^0, \mathfrak{M}, \boldsymbol{W})$ forms an equilibrium. Pay attention to the fact that, in the statement of Lemma 2.26 below, we use the capital letter $\boldsymbol{X}$ instead of $\boldsymbol{x}$ for the forward component of the solution of the FBSDE (2.25) as we usually reserve lower cases for canonical processes. □

The last step in the proof of Lemma 2.25 can be formalized and its consequence strengthened in a quite systematic manner.

Lemma 2.26. *Under assumption* **FBSDE**, *$\mathcal{M}$ is a distribution of an equilibrium if, on the same space $\bar{\Omega}$ as in Definition 2.24, the solution $(X_t, Y_t, Z_t, Z_t^0, M_t)_{0\le t\le T}$ to the FBSDE (2.25) with $\boldsymbol{W}^0 = \boldsymbol{w}^0$, $\boldsymbol{W} = \boldsymbol{w}$ and $\boldsymbol{\mu} = \boldsymbol{\nu} = (\nu_t = \mathfrak{m} \circ (e_t^x)^{-1})_{0\le t\le T}$, satisfies for any Borel subset $C \subset \mathcal{C}([0,T];\mathbb{R}^{2d})$ in a countable generating π-system of $\mathcal{B}(\mathcal{C}([0,T];\mathbb{R}^{2d}))$, and any Borel subset $C^0 \subset \bar{\Omega}^0$ in a countable generating π-system of $\mathcal{B}(\bar{\Omega}^0)$,*

$$\bar{\mathbb{E}}\big[\mathbf{1}_C(\boldsymbol{X}, \boldsymbol{w})\mathbf{1}_{C^0}(\nu^0, \boldsymbol{w}^0, \mathfrak{m})\big] = \bar{\mathbb{E}}^0\big[\mathfrak{m}(C)\mathbf{1}_{C^0}(\nu^0, \boldsymbol{w}^0, \mathfrak{m})\big]. \tag{2.34}$$

Being a sufficient condition, equation (2.34) is required for countable families of sets C and C^0 generating the σ-fields and closed under pairwise intersection. However, a simple monotone class argument can be used to prove that it would be equivalent to assume that (2.34) holds for all the sets C and C^0 in these σ-fields.

Proof. As $(\nu^0, \boldsymbol{w}^0, \boldsymbol{\nu})$ is constructed on $\bar{\Omega}^0$, we deduce from Fubini's theorem that:

$$\bar{\mathbb{E}}\big[\mathbf{1}_C(\boldsymbol{X}, \boldsymbol{W})\mathbf{1}_{C^0}(\nu^0, \boldsymbol{w}^0, \mathfrak{m})\big] = \bar{\mathbb{E}}^0\big[\bar{\mathbb{E}}^1\big[\mathbf{1}_C(\boldsymbol{X}, \boldsymbol{w})\big]\mathbf{1}_{C^0}(\nu^0, \boldsymbol{w}^0, \mathfrak{m})\big],$$

which proves by (2.34) that $\bar{\mathbb{P}}^0$-almost surely, for any C in a countable generating π-system of the Borel σ-field $\mathcal{B}(\mathcal{C}([0,T];\mathbb{R}^{2d}))$,

$$\bar{\mathbb{E}}^1\big[\mathbf{1}_C(\boldsymbol{X}, \boldsymbol{w})\big] = \mathfrak{m}(C),$$

which says exactly that, for $\bar{\mathbb{P}}^0$-almost every $\bar{\omega}^0 \in \bar{\Omega}^0$, $\mathcal{L}(\boldsymbol{X}(\bar{\omega}^0, \cdot), \boldsymbol{w}) = \mathfrak{m}(\bar{\omega}^0)$. □

2.3.2 Yamada-Watanabe Theorem for MFG Equilibria

From a mathematical standpoint, weak solutions are much more flexible than strong solutions. However, a weak solution may not be entirely satisfactory from the practical point of view since the equilibrium may incorporate an extra source of noise in addition to the initial condition and the common noise. It is thus very important to have sufficient conditions to guarantee *a posteriori* that a weak solution is actually strong in the sense that it is adapted to the initial condition and to the common source of noise. The discussion of Subsection 1.2.3 strongly suggests the desirability of a Yamada-Watanabe type result for MFG equilibria. In order to derive such a result, we first specify the notion of strong uniqueness.

Definition 2.27. *We say that strong uniqueness holds for the MFG equilibrium if, for any filtered probabilistic set-up $(\Omega, \mathcal{F}, \mathbb{F}, \mathbb{P})$ as in Definition 2.16, any solutions $\mathfrak{M}^a$ and $\mathfrak{M}^b$ on $(\Omega, \mathcal{F}, \mathbb{F}, \mathbb{P})$, with the same initial random distribution μ_0 having some $\mathcal{V}^0 \in \mathcal{P}_2(\mathcal{P}_2(\mathbb{R}^d))$ as law, satisfy:*

$$\mathbb{P}^0\big[\mathfrak{M}^a = \mathfrak{M}^b\big] = 1.$$

Following Definition 1.18, we shall sometimes specialize the definition by saying that strong uniqueness holds but only for a prescribed value of $\mathcal{L}(\mathcal{V}^0)$.

Not surprisingly, weak uniqueness is related to uniqueness in law.

Definition 2.28. *We say that weak uniqueness holds for the MFG equilibrium if any two weak solutions, possibly defined on different filtered probabilistic set-ups but driven by the same initial law $\mathcal{V}^0 \in \mathcal{P}_2(\mathcal{P}_2(\mathbb{R}^d))$, have the same distribution on $\bar{\Omega}^0$, the notion of distribution of a solution being defined as in the statement of Lemma 2.25.*

Following Definition 2.27, we shall sometimes restrict weak uniqueness to a given prescribed value of $\mathcal{L}(\mathcal{V}^0)$.

With these two notions of uniqueness in hand, we can state and prove the main result of this section, which may be summarized as *strong uniqueness implies weak uniqueness*.

Theorem 2.29. *Assume that strong uniqueness, as defined in Definition 2.27, holds true for a specific choice of $\mathcal{V}^0 \in \mathcal{P}_2(\mathcal{P}_2(\mathbb{R}^d))$. If there exists a weak solution with $\mathcal{V}^0$ as initial condition, then there exists a measurable mapping*

$$\Psi : \mathcal{P}_2(\mathbb{R}^d) \times \mathcal{C}([0,T];\mathbb{R}^d) \to \mathcal{P}_2\big([\mathcal{C}([0,T];\mathbb{R}^d)]^2\big)$$

such that, for any weak solution $(\mathbf{W}^0, \mathfrak{M})$ on a space $(\Omega, \mathcal{F}, \mathbb{F}, \mathbb{P})$ with initial random distribution μ_0 distributed as $\mathcal{V}^0$, it holds:

$$\mathbb{P}^0\big[\mathfrak{M} = \Psi(\mu_0, \mathbf{W}^0)\big] = 1.$$

In particular, this implies that any weak solution is a strong solution on the appropriate probability structure.

Moreover, on any probabilistic set-up $(\Omega, \mathcal{F}, \mathbb{F}, \mathbb{P})$ *as in Definition 2.16, the initial random distribution* μ_0 *having* $\mathcal{V}^0$ *as distribution, the random variable* $\mathfrak{M} = \Psi(\mu_0, \boldsymbol{W}^0)$ *is an equilibrium. So in that case, weak existence implies strong existence and thus existence of a solution on the canonical space* (2.30).

Like for Theorem 1.33, the proof of Theorem 2.29 is an adaptation of the argument used in the original Yamada-Watanabe theorem for stochastic differential equations. The starting point is the following lemma.

Lemma 2.30. *Given two weak solutions with the same initial condition* $\mathcal{V}^0 \in \mathcal{P}_2(\mathcal{P}_2(\mathbb{R}^d))$*, denoted by their distributions* $\mathcal{M}^a$ *and* $\mathcal{M}^b$ *on the space* $\bar{\Omega}^0$*, one can construct a probabilistic set-up* $(\Omega, \mathcal{F}, \mathbb{F}, \mathbb{P})$ *satisfying the prescription of Definition 2.16 and equipped with a tuple* $(\mu_0, \boldsymbol{W}^0, \boldsymbol{W})$ *and with two equilibria* $\mathfrak{M}^a$ *and* $\mathfrak{M}^b$ *such that* $(\mu_0, \boldsymbol{W}^0, \mathfrak{M}^a)$ *and* $(\mu_0, \boldsymbol{W}^0, \mathfrak{M}^b)$ *have distributions* $\mathcal{M}^a$ *and* $\mathcal{M}^b$ *respectively.*

Proof.

First Step. Following the proof of Theorem 1.33, we define $\Omega^0_{\text{input}} = \mathcal{P}_2(\mathbb{R}^d) \times \mathcal{C}([0,T];\mathbb{R}^d)$. We equip it with its Borel σ-field and the product probability measure $\mathcal{V}^0 \otimes \mathcal{W}_d$, which we denote $\mathbb{Q}^0_{\text{input}}$. The canonical process is denoted by $(\nu^0, \boldsymbol{w}^0)$ and the canonical filtration by $\mathbb{G}^{0,\text{nat},\text{input}} = (\mathcal{G}_t^{0,\text{nat},\text{input}})_{0 \le t \le T}$. The completed filtered space is denoted by $(\Omega^0_{\text{input}}, \mathcal{G}^0_{\text{input}}, \mathbb{G}^{0,\text{input}}, \mathbb{Q}^0_{\text{input}})$. We will use the notation ω^0_{input} for a generic element of Ω^0_{input}.

Then, we may regard $\bar{\Omega}^0$ as $\Omega^0_{\text{input}} \times \Omega^0_{\text{output}}$, with $\Omega^0_{\text{output}} = \mathcal{P}_2([\mathcal{C}([0,T];\mathbb{R}^d)]^2)$. The canonical random variable on Ω^0_{output} is denoted by $\mathfrak{m}$ and the canonical filtration it generates along the lines of Definition 2.15 is denoted by $\mathbb{G}^{0,\text{nat},\text{output}} = (\mathcal{G}_t^{0,\text{nat},\text{output}})_{0 \le t \le T}$. We use the notation ω^0_{output} for a generic element of Ω^0_{output}.

We equip the space $\bar{\Omega}^0$ with its Borel σ-field, and with the measure $\mathcal{M}^a$ (resp. $\mathcal{M}^b$). We denote by $(\nu^0, \boldsymbol{w}^0, \mathfrak{m})$ the canonical random variable on $\bar{\Omega}^0$ and we call q^a (resp. q^b) the conditional law of $\mathfrak{m}$ on the space $\Omega^0_{\text{output}} = \mathcal{P}_2([\mathcal{C}([0,T];\mathbb{R}^d)]^2)$ given $\sigma\{\nu^0, \boldsymbol{w}^0\}$. We then consider the extended space:

$$\begin{aligned}\hat{\Omega}^0 &= \mathcal{P}_2(\mathbb{R}^d) \times \mathcal{C}([0,T];\mathbb{R}^d) \times \mathcal{P}_2\big([\mathcal{C}([0,T];\mathbb{R}^d)]^2\big) \times \mathcal{P}_2\big([\mathcal{C}([0,T];\mathbb{R}^d)]^2\big)\\ &= \Omega^0_{\text{input}} \times \big[\Omega^0_{\text{output}}\big]^2.\end{aligned}$$

We endow $\hat{\Omega}^0$ with its Borel σ-field and we denote by $(\nu^0, \boldsymbol{w}^0, \mathfrak{m}^a, \mathfrak{m}^b)$ the canonical random variable. The canonical filtration, as in Definition 2.15, is denoted by $\hat{\mathbb{F}}^{0,\text{nat}}$. We then define the probability measure $\hat{\mathbb{P}}^0$ by:

$$\hat{\mathbb{P}}^0(C \times D^a \times D^b) = \int_C q^a(\omega^0_{\text{input}}, D^a) q^b(\omega^0_{\text{input}}, D^b) d\mathbb{Q}^0_{\text{input}}(\omega^0_{\text{input}}), \tag{2.35}$$

for $C \in \mathcal{B}(\Omega^0_{\text{input}})$ and $D^a, D^b \in \mathcal{B}(\mathcal{P}_2([\mathcal{C}([0,T];\mathbb{R}^d)]^2))$. We denote by $\hat{\mathcal{F}}^0$ the completion of the Borel σ-field on $\hat{\Omega}^0$ under the probability $\hat{\mathbb{P}}^0$ and by $\hat{\mathbb{F}}^0$ the right-continuous and completed augmentation of the canonical filtration. The extension of $\hat{\mathbb{P}}^0$ to the completed σ-field is still denoted by $\hat{\mathbb{P}}^0$.

For a continuous process $\boldsymbol{\theta} = (\theta_t)_{0 \leq t \leq T}$ on $\hat{\Omega}^0$ with values in a Polish space $\mathcal{S}$, we denote by $\mathbb{F}^{0,\mathrm{nat},\boldsymbol{\theta}} = (\mathcal{F}_t^{0,\mathrm{nat},\boldsymbol{\theta}})_{0 \leq t \leq T}$ the canonical filtration generated by $\boldsymbol{\theta}$ and we denote by $\hat{\mathbb{F}}^{0,\boldsymbol{\theta}}$ the completion under $\hat{\mathbb{P}}^0$.

Second Step. We claim that, for any $t \in [0,T]$ and any $D \in \mathcal{G}_t^{0,\mathrm{nat},\mathrm{output}}$, the random variables $q^a((\nu^0, \boldsymbol{w}^0), D)$ and $q^b((\nu^0, \boldsymbol{w}^0), D)$ are measurable with respect to the completion $\hat{\mathcal{F}}_t^{0,(\nu^0,\boldsymbol{w}^0)}$ of the σ-field $\mathcal{F}_{t+}^{0,\mathrm{nat},(\nu^0,\boldsymbol{w}^0)}$ under $\hat{\mathbb{P}}^0$, the latter being regarded as a sub-σ-field of $\hat{\mathcal{F}}^0$.

Noting that $\boldsymbol{w}^0$ is a Brownian motion under $\mathcal{M}^a$ with respect to the filtration generated by $(\nu^0, \boldsymbol{w}^0, \mathfrak{m}^a)$, we deduce that, for any $t \in [0,T]$, $C \in \mathcal{G}_T^{0,\mathrm{nat},\mathrm{input}}$ and $D \in \mathcal{G}_t^{0,\mathrm{nat},\mathrm{output}}$:

$$
\begin{aligned}
\mathcal{M}^a(C \times D) &= \mathbb{E}^{\mathcal{M}^a}\big[\mathcal{M}^a(C \times \Omega^0_{\mathrm{output}} \,|\, \mathcal{F}_{t+}^{0,\mathrm{nat},(\nu^0,\boldsymbol{w}^0)})\mathcal{M}^a(\Omega^0_{\mathrm{input}} \times D \,|\, \mathcal{F}_{t+}^{0,\mathrm{nat},(\nu^0,\boldsymbol{w}^0)})\big] \\
&= \mathbb{E}^{\mathcal{M}^a}\big[\mathbf{1}_{C \times \Omega^0_{\mathrm{output}}} \mathcal{M}^a(\Omega^0_{\mathrm{input}} \times D \,|\, \mathcal{F}_{t+}^{0,\mathrm{nat},(\nu^0,\boldsymbol{w}^0)})\big],
\end{aligned}
$$

where we here regarded $\mathbb{F}^{0,\mathrm{nat},(\nu^0,\boldsymbol{w}^0)}$ as a filtration on $\bar{\Omega}^0$ instead of $\hat{\Omega}^0$.

Regarding $\mathcal{M}^a(\Omega^0_{\mathrm{input}} \times D \,|\, \mathcal{F}_{t+}^{0,\mathrm{nat},(\nu^0,\boldsymbol{w}^0)})$ as a $\mathcal{G}_{t+}^{0,\mathrm{nat},\mathrm{input}}$-measurable random variable θ : $\Omega^0_{\mathrm{input}} \to \mathbb{R}$, we then have $\mathcal{M}^a(C \times D) = \mathbb{E}^{\mathbb{Q}^0_{\mathrm{input}}}[\mathbf{1}_C \theta]$. Now, we can also write:

$$
\mathcal{M}^a(C \times D) = \int_C q^a(\omega^0_{\mathrm{input}}, D) d\mathbb{Q}^0_{\mathrm{input}}(\omega^0_{\mathrm{input}}) = \mathbb{E}^{\mathbb{Q}^0_{\mathrm{input}}}\big[\mathbf{1}_C q^a(\cdot, D)\big],
$$

where $q^a(\cdot, D)$ is understood as a real valued $\mathcal{G}_T^{0,\mathrm{nat},\mathrm{input}}$-measurable random variable on $\Omega^0_{\mathrm{input}}$. We deduce that $\mathbb{Q}^0_{\mathrm{input}}$ almost surely, $q^a(\cdot, D) = \theta$. In particular, with $\hat{\mathbb{P}}^0$-probability 1, $q^a((\nu^0, \boldsymbol{w}^0), D) = \theta(\nu^0, \boldsymbol{w}^0)$. Since the random variable $\theta(\nu^0, \boldsymbol{w}^0)$ is $\mathcal{F}_{t+}^{0,\mathrm{nat},(\nu^0,\boldsymbol{w}^0)}$-measurable, $q^a((\nu^0, \boldsymbol{w}^0), D)$ is $\hat{\mathcal{F}}_t^{0,(\nu^0,\boldsymbol{w}^0)}$-measurable.

Third Step. We now check that $\boldsymbol{w}^0$ is a d-dimensional Brownian motion with respect to $\hat{\mathbb{F}}^0$. To do so, we notice once again that $\boldsymbol{w}^0$ is a d-dimensional Brownian motion with respect to $\hat{\mathbb{F}}^{0,(\nu^0,\boldsymbol{w}^0)}$ under $\hat{\mathbb{P}}^0$. Then, we consider $C \in \mathcal{G}_t^{0,\mathrm{nat},\mathrm{input}}$, $C' \in \sigma\{w^0_s - w^0_t;\ t \leq s \leq T\}$ and $D^a, D^b \in \mathcal{G}_t^{0,\mathrm{nat},\mathrm{output}}$. Then, identifying C' with a Borel subset of $\Omega^0_{\mathrm{input}}$, we have by (2.35):

$$
\begin{aligned}
\hat{\mathbb{P}}^0\big[(C \cap C') \times D^a \times D^b\big] &= \int_{C \cap C'} q^a\big(\omega^0_{\mathrm{input}}, D^a\big) q^b\big(\omega^0_{\mathrm{input}}, D^b\big) d\mathbb{Q}^0_{\mathrm{input}}(\omega^0_{\mathrm{input}}) \\
&= \mathbb{Q}^0_{\mathrm{input}}(C') \int_C q^a\big(\omega^0_{\mathrm{input}}, D^a\big) q^b\big(\omega^0_{\mathrm{input}}, D^b\big) d\mathbb{Q}^0_{\mathrm{input}}(\omega^0_{\mathrm{input}}),
\end{aligned}
$$

from which we easily deduce that $(w^0_s - w^0_t)_{t \leq s \leq T}$ is independent of $\hat{\mathcal{F}}^0_t$. Above we used the measurability properties of the kernels q^a and q^b established in the second step.

We now prove that $(\nu^0, \boldsymbol{w}^0, \mathfrak{m}^a)$ is compatible with the filtration $\hat{\mathbb{F}}^0$, the argument being the same for $(\nu^0, \boldsymbol{w}^0, \mathfrak{m}^b)$. Given $C \in \mathcal{G}_T^{0,\mathrm{nat},\mathrm{input}}$ and $D^a, D^b \in \mathcal{G}_T^{0,\mathrm{nat},\mathrm{output}}$, it holds that:

$$
\begin{aligned}
\hat{\mathbb{P}}^0\big(C \times D^a \times D^b\big) &= \int_C q^a(\omega^0_{\mathrm{input}}, D^a) q^b(\omega^0_{\mathrm{input}}, D^b) d\mathbb{Q}^0_{\mathrm{input}}(\omega^0_{\mathrm{input}}) \\
&= \int_{C \times D^a} q^b(\omega^0_{\mathrm{input}}, D^b) d\mathcal{M}^a(\omega^0_{\mathrm{input}}, \omega^0_{\mathrm{output}}).
\end{aligned}
$$

We now recall that $q^b(\cdot, D^b)$ is $\mathcal{G}_t^{0,\mathrm{input}}$-measurable when $D^b \in \mathcal{G}_t^{\mathrm{nat},\mathrm{output}}$ for some $t \in [0,T]$. We thus recover the statement of Proposition 1.10 with $\mathbb{F}^1$ the completion (under $\mathcal{M}^a$) of the right-continuous augmentation of the canonical filtration on $\bar{\Omega}^0$ and $\mathbb{F}^2 = \mathbb{G}^{\mathrm{nat},\mathrm{output}}$. We deduce that for any $t \in [0,T]$, $\hat{\mathcal{F}}_t^{0,\mathrm{nat}}$ and $\hat{\mathcal{F}}_T^{0,(\nu^0,w^0,\mathrm{m}^a)}$ are conditionally independent given $\hat{\mathcal{F}}_t^{0,(\nu^0,w^0,\mathrm{m}^a)}$. Proceeding as in the proof of Lemma 1.7, we deduce that for any $t \in [0,T]$, $\hat{\mathcal{F}}_t^0$ and $\hat{\mathcal{F}}_T^{0,(\nu^0,w^0,\mathrm{m}^a)}$ are conditionally independent given $\hat{\mathcal{F}}_t^{0,(\nu^0,w^0,\mathrm{m}^a)}$, which is to say that $(\nu^0, \boldsymbol{w}^0, \mathrm{m}^a)$ is compatible with the filtration $\hat{\mathbb{F}}^0$.

Fourth Step. We now introduce $(\bar{\Omega}^1, \bar{\mathcal{F}}^1, \bar{\mathbb{F}}^1, \bar{\mathbb{P}}^1)$ as in (2.32), $\bar{\mathcal{F}}^1$ denoting the completion of the Borel σ-field and $\bar{\mathbb{F}}^1$ the complete and right-continuous augmentation of the canonical filtration under $\bar{\mathbb{P}}^1$. The canonical random variable on $\bar{\Omega}^1$ is denoted by $(\eta, \boldsymbol{w})$. From $(\hat{\Omega}^0, \hat{\mathcal{F}}^0, \hat{\mathbb{F}}^0, \hat{\mathbb{P}}^0)$ and $(\bar{\Omega}^1, \bar{\mathcal{F}}^1, \bar{\mathbb{F}}^1, \bar{\mathbb{P}}^1)$, we can construct the product probabilistic set-up $(\hat{\Omega}, \hat{\mathcal{F}}, \hat{\mathbb{F}}, \hat{\mathbb{P}})$ by following the procedure outlined in (2.24). Recall in particular that $(\hat{\mathcal{F}}, \hat{\mathbb{P}})$ is obtained by completion of $(\hat{\mathcal{F}}^0 \otimes \bar{\mathcal{F}}^1, \hat{\mathbb{P}}^0 \otimes \bar{\mathbb{P}}^1)$. The initial condition X_0 is defined by $X_0 = \psi(\eta, \nu^0)$, with ψ as in Lemma (Vol I)-5.29, see also (2.23).

From the second step, we know that $\boldsymbol{w}^0$ is an $\hat{\mathbb{F}}^0$-Brownian motion. It is quite straightforward to deduce that $(\boldsymbol{w}^0, \boldsymbol{w})$ is an $\hat{\mathbb{F}}$-Brownian motion. We also know that $(\nu^0, \boldsymbol{w}^0, \mathrm{m}^a)$ is compatible with the filtration $\hat{\mathbb{F}}^0$. Reproducing (2.33), we deduce that $(X_0, \boldsymbol{w}^0, \mathrm{m}^a, \boldsymbol{w})$ is compatible with $\hat{\mathbb{F}}$. Similarly, $(X_0, \boldsymbol{w}^0, \mathrm{m}^b, \boldsymbol{w})$ is compatible with $\hat{\mathbb{F}}$.

Fifth Step. So far, we have proved that the probabilistic set-up $(\hat{\Omega}, \hat{\mathcal{F}}, \hat{\mathbb{F}}, \hat{\mathbb{P}})$ equipped with the tuple $(X_0, \boldsymbol{w}^0, \mathrm{m}^a, \boldsymbol{w})$ (resp. $(X_0, \boldsymbol{w}^0, \mathrm{m}^b, \boldsymbol{w})$) satisfies the first point in the Definition 2.16 of an MFG equilibrium.

On $\hat{\Omega}$, we can define $(\boldsymbol{X}^a, \boldsymbol{Y}^a, \boldsymbol{Z}^a, \boldsymbol{Z}^{0,a}, \boldsymbol{M}^a)$ (resp. $(\boldsymbol{X}^b, \boldsymbol{Y}^b, \boldsymbol{Z}^b, \boldsymbol{Z}^{0,b}, \boldsymbol{M}^b)$) as the solution of the forward-backward system (2.25) with $\mathfrak{M} = (\mathrm{m}^a \circ \mathcal{E}_t^{-1})_{0 \le t \le T}$ and $\boldsymbol{\mu} = (\mathrm{m}^a \circ (e_t^x)^{-1})_{0 \le t \le T}$ (resp. $\mathfrak{M} = (\mathrm{m}^b \circ \mathcal{E}_t^{-1})_{0 \le t \le T}$ and $\boldsymbol{\mu} = (\mathrm{m}^b \circ (e_t^x)^{-1})_{0 \le t \le T}$), where we recall that $\mathcal{E}_t : \mathcal{C}([0,T];\mathbb{R}^{2d}) \ni (\boldsymbol{x}, \boldsymbol{w}) \mapsto (x_{s\wedge t}, w_{s\wedge t})_{0 \le s \le T} \in \mathcal{C}([0,T];\mathbb{R}^{2d})$ and that e_t^x is the mapping evaluating the first d coordinates at time t on $\mathcal{C}([0,T];\mathbb{R}^{2d})$. With our construction of the initial condition, it holds that $X_0^a = X_0^b = \psi(\eta, \nu^0) = X_0$.

From Theorem 1.33, we know the existence of a (measurable) functional Φ^a (resp. Φ^b), defined on $\mathbb{R}^d \times \mathcal{C}([0,T];\mathbb{R}^d) \times \mathcal{P}_2([\mathcal{C}([0,T];\mathbb{R}^d)]^2) \times \mathcal{C}([0,T];\mathbb{R}^d)$ and with values in $\mathcal{C}([0,T];\mathbb{R}^d)$, such that, $\hat{\mathbb{P}}$-almost surely,

$$\boldsymbol{X}^a = \Phi^a\big(X_0, \boldsymbol{w}^0, \mathrm{m}^a, \boldsymbol{w}\big) = \Phi^a\big(\psi(\eta, \nu^0), \boldsymbol{w}^0, \mathrm{m}^a, \boldsymbol{w}\big),$$

$$\big(\text{respectively } \boldsymbol{X}^b = \Phi^b\big(X_0, \boldsymbol{w}^0, \mathrm{m}^b, \boldsymbol{w}\big) = \Phi^b\big(\psi(\eta, \nu^0), \boldsymbol{w}^0, \mathrm{m}^b, \boldsymbol{w}\big)\ \big).$$

Importantly, the mapping Φ^a (resp. Φ^b) is independent of the set-up used to solve the FBSDE (2.25) as long as the law of the input is fixed. Proceeding as in the proof of Lemma 2.25, this allows to show that $\hat{\mathbb{P}}^0$ almost surely, $\mathcal{L}(\boldsymbol{X}^a, \boldsymbol{w}) = \mathrm{m}^a$, which proves that $(\nu^0, \boldsymbol{w}^0, \mathrm{m}^a)$ is a solution with $\mathcal{M}^a$ as distribution. The same holds for $(\boldsymbol{w}^0, \mathrm{m}^b)$. □

We are now in a position to complete the proof of Theorem 2.29.

Proof of Theorem 2.29.

First Step. We use the same notation as in the proof of Lemma 2.30. Given two weak solutions with the same initial condition $\mathcal{V}^0 \in \mathcal{P}_2(\mathcal{P}_2(\mathbb{R}^d))$ and denoting by $\mathcal{M}^a$ and $\mathcal{M}^b$ their distributions, strong uniqueness implies:

$$\hat{\mathbb{P}}^0\big[\mathrm{m}^a = \mathrm{m}^b\big] = 1, \tag{2.36}$$

and thus $(\nu^0, \boldsymbol{w}^0, \mathrm{m}^a) = (\nu^0, \boldsymbol{w}^0, \mathrm{m}^b)$ with $\hat{\mathbb{P}}^0$-probability 1, from which we deduce that $\mathcal{M}^a = \mathcal{M}^b$.

Second Step. Assume further that, for a given initial condition $\mathcal{V}^0 \in \mathcal{P}_2(\mathcal{P}_2(\mathbb{R}^d))$, there exists a weak solution $(\boldsymbol{W}^0, \mathfrak{M})$ on some $(\Omega, \mathcal{F}, \mathbb{F}, \mathbb{P})$, with $\mathcal{V}^0$ as initial condition. Denoting by $\mathcal{M}$ the law of the weak solution, and choosing $\mathcal{M}^a = \mathcal{M}^b = \mathcal{M}$, so that we may denote by q the common value of q^a and q^b, we have, for any C, D^a and D^b as in (2.35),

$$\begin{aligned}
&\mathbb{P}^0\big[C \times D^a \times D^b\big] \\
&= \int_C \bigg[\int_{D^a \times D^b} \mathbf{1}_{\{\omega^{0,a}_{\mathrm{output}} = \omega^{0,b}_{\mathrm{output}}\}} q(\omega^0_{\mathrm{input}}, d\omega^{0,a}_{\mathrm{output}}) q(\omega^0_{\mathrm{input}}, d\omega^{0,b}_{\mathrm{output}})\bigg] d\mathbb{Q}^0_{\mathrm{input}}(\omega^0_{\mathrm{input}}),
\end{aligned}$$

from which we deduce that, for almost every $\omega^0_{\mathrm{input}}$ under $\mathbb{Q}^0_{\mathrm{input}}$,

$$\big[q(\omega^0_{\mathrm{input}}, \cdot) \otimes q(\omega^0_{\mathrm{input}}, \cdot)\big]\Big\{(\omega^{0,a}_{\mathrm{output}}, \omega^{0,b}_{\mathrm{output}}) \in \big[\Omega^0_{\mathrm{output}}\big]^2 : \omega^{0,a}_{\mathrm{output}} = \omega^{0,b}_{\mathrm{output}}\Big\} = 1,$$

The only way for the canonical variables $\mathrm{m}^a : [\Omega^0_{\mathrm{output}}]^2 \ni (\omega^{0,a}_{\mathrm{output}}, \omega^{0,b}_{\mathrm{output}}) \mapsto \omega^{0,a}_{\mathrm{output}} \in \Omega^0_{\mathrm{output}}$ and $\mathrm{m}^b : [\Omega^0_{\mathrm{output}}]^2 \ni (\omega^{0,a}_{\mathrm{output}}, \omega^{0,b}_{\mathrm{output}}) \mapsto \omega^{0,a}_{\mathrm{output}} \in \Omega^0_{\mathrm{output}}$ to be independent and almost surely equal is that they are almost surely constant, with $\Psi(\omega^0_{\mathrm{input}})$ as common value defined by:

$$\forall D \in \mathcal{B}\big([\mathcal{C}([0,T];\mathbb{R})]^2\big), \quad \big[\Psi(\omega^0_{\mathrm{input}})\big](D) = \int_{\Omega^0_{\mathrm{output}}} \omega^0_{\mathrm{output}}(D) q\big(\omega^0_{\mathrm{input}}, d\omega^0_{\mathrm{output}}\big).$$

This proves that:

$$\mathbb{P}^0\big[\mathfrak{M} = \Psi(\mu_0, \boldsymbol{W}^0)\big] = 1.$$

By weak uniqueness, the above is obviously true for any other weak solution.

Third Step. We still assume that there exists a weak solution with $\mathcal{V}^0$ as initial condition, but we do not denote it anymore by $(\boldsymbol{W}^0, \mathfrak{M})$. We then prove that, on any probabilistic set-up $(\Omega, \mathcal{F}, \mathbb{F}, \mathbb{P})$ as in Definition 2.16, the initial random distribution μ_0 having $\mathcal{V}^0$ as distribution, the random variable $\mathfrak{M} = \Psi(\mu_0, \boldsymbol{W}^0)$ is an equilibrium.

Recalling that $(\boldsymbol{x}, \boldsymbol{w}) = (x_t, w_t)_{0 \le t \le T}$ denotes the canonical process on $[\mathcal{C}([0,T];\mathbb{R}^d)]^2$, we observe, by taking $D \in \sigma\{x_s, w_s;\ s \le t\}$ for some $t \in [0,T]$, that the random variable $\Omega^0_{\mathrm{output}} \ni \omega^0_{\mathrm{output}} \mapsto \omega^0_{\mathrm{output}}(D)$ is $\mathcal{G}^{\mathrm{nat,output}}_t$-measurable. Then, we deduce from the second step of the proof of Lemma 2.30 that the mapping $\Omega^0_{\mathrm{input}} \ni \omega^0_{\mathrm{input}} \mapsto [\Psi(\omega^0_{\mathrm{input}})](D)$ is $\mathcal{G}^{0,\mathrm{input}}_t$-measurable. Therefore, the mapping $\Omega^0_{\mathrm{input}} \ni \omega^0_{\mathrm{input}} \mapsto [\Psi(\omega^0_{\mathrm{input}})] \circ \mathcal{E}_t^{-1} \in \mathcal{P}_2([\mathcal{C}([0,T];\mathbb{R}^d)]^2)$ is also $\mathcal{G}^{0,\mathrm{input}}_t$-measurable, where $\mathcal{E}_t$ is as in the fifth step of the proof of Lemma 2.30. We deduce that the process $\boldsymbol{\mathfrak{M}} = (\mathfrak{M}_t = \mathfrak{M} \circ \mathcal{E}_t^{-1})_{0 \le t \le T}$ is $\mathbb{F}^0$-adapted, and thus $\mathbb{F}^0$-progressively measurable since it is continuous.

The compatibility condition in Definition 2.16 is easily checked. Moreover, solving the FBSDE (2.25) with $(\mathfrak{M}_t)_{0\leq t\leq T}$ as input and X_0 as initial condition, we know, as in the fifth step of Lemma 2.30, that the forward process $\boldsymbol{X}$ may be written as

$$\boldsymbol{X} = \phi\big(X_0, \boldsymbol{W}^0, \mathfrak{M}, \boldsymbol{W}\big).$$

Invoking Lemma 2.26 as in the proof of Lemma 2.25 and as in the fifth step of Lemma 2.30, we deduce that $\mathcal{L}^1(\boldsymbol{X}, \boldsymbol{W}) = \mathfrak{M}$, which proves that $\mathfrak{M}$ induces an equilibrium. □

2.3.3 Infinite Dimensional Stochastic FBSDEs

We close the chapter with a parallel to the analysis provided in Subsection (Vol I)-3.1.5 in the absence of common noise. In the same way we argued that the system of PDEs (Vol I)-(3.12) was the cornerstone of the analytic approach to MFGs, the results of this chapter show that in the presence of a common noise, one should be able to extract equilibria from a system of stochastic PDEs, one of the Hamilton-Jacobi-Bellman type, the other of the Kolmogorov-Fokker-Planck type. In fact, we already introduced in Subsections 1.4.2 and 2.1.2 all the necessary ingredients to formulate the SPDE forward-backward system.

Using the minimizer $\hat{\alpha}$ of the reduced Hamiltonian introduced in (1.32) and (1.33), (1.36) says that the counterpart of the backward PDE of the system (Vol I)-(3.12) is given by the backward stochastic PDE:

$$\begin{aligned} d_t U(t,\cdot) = -\Big[\frac{1}{2}\mathrm{trace}\big[\big(\sigma\sigma^\dagger + \sigma^0(\sigma^0)^\dagger\big)(t,\cdot,\mu_t)\partial^2_{xx}U(t,\cdot)\big] \\ + H\big[t,\cdot,\mu_t,\partial_x U(t,\cdot),\hat{\alpha}\big(t,\cdot,\mu_t,\partial_x U(t,\cdot)\big)\big] \\ + \mathrm{trace}\big(\sigma^0(t,\cdot,\mu_t)\partial_x V(t,\cdot)\big)\Big]dt + V(t,\cdot)\cdot dW^0_t, \end{aligned} \tag{2.37}$$

with $U(T,x) = g(x,\mu_T)$ as terminal condition, and we deduce from (2.4) that the forward equation takes the form:

$$\begin{aligned} d\mu_t = -\partial_x \cdot \big[b\big(t,\cdot,\mu_t,\hat{\alpha}(t,\cdot,\mu_t,\partial_x U(t,\cdot))\big)\mu_t\big]dt - \partial_x\cdot\Big(\sigma^0\big(t,\cdot,\mu_t\big)dW^0_t\mu_t\Big) \\ + \frac{1}{2}\mathrm{trace}\Big\{\partial^2_{xx}\Big[\Big(\big[\sigma\sigma^\dagger\big](t,\cdot,\mu_t) + \big[\sigma^0\sigma^{0\dagger}\big](t,\cdot,\mu_t)\Big)\mu_t\Big]\Big\}dt. \end{aligned} \tag{2.38}$$

As we already explained in Subsection 1.4.2, the role of the random field V in the backward SPDE is to guarantee that the value function $U(t,\cdot)$ is adapted to the common noise up until time t.

In this regard, it is worth mentioning that here, the flow of conditional marginal measures $\boldsymbol{\mu} = (\mu_t)_{0\leq t\leq T}$ is implicitly required to be adapted to the filtration generated by $\boldsymbol{W}^0$ and the initial condition μ_0 whenever the latter is random.

According to the terminology for solutions of MFG problems introduced in this chapter, and in particular Definition 2.22, the pair (2.37)–(2.38) provides an SPDE formulation of the search for strong solutions.

2.4 Notes & Complements

Examples of mean field games with a common noise were considered by Carmona, Fouque, and Sun in [102], Gomes and Saude in [182] and Guéant, Lasry, and Lions in [189]. The reader may also find examples in Chapter 4.

For practical applications of the theory of MFG to social and biological phenomena such as herding, bird flocking, fish schooling, ..., it is more realistic to model the common source of random shocks from the environment by a zero-mean Gaussian white noise field $\boldsymbol{W}^0 = (W^0(\Lambda, B))_{\Lambda, B}$, parameterized by the Borel subsets Λ of a Polish space Ξ (most often, $\Xi = \mathbb{R}^\ell$ is the most natural choice), and the Borel subsets B of $[0,\infty)$, such that:

$$\mathbb{E}\big[W^0(\Lambda, B)W^0(\Lambda', B')\big] = \lambda(\Lambda \cap \Lambda')|B \cap B'|,$$

where we use the notation $|B|$ for the Lebesgue measure of a Borel subset of $[0,\infty)$. Here λ is a nonnegative measure on Ξ, called the spatial intensity of $\boldsymbol{W}^0$. In this case, the dynamics of the private states are given by SDEs of the form:

$$\begin{aligned} dX_t^i &= b\big(t, X_t^i, \bar{\mu}_t^N, \alpha_t^i\big)dt + \sigma\big(t, X_t^i, \bar{\mu}_t^N, \alpha_t^i\big)dW_t^i \\ &\quad + \int_{\Xi} \sigma^0\big(t, X_t^i, \bar{\mu}_t^N, \alpha_t^i, \xi\big) W^0(d\xi, dt). \end{aligned} \tag{2.39}$$

If we think of $W^0(d\xi, dt)$ as a random noise which is white in time (to provide the time derivative of a Brownian motion) and colored in space (the spectrum of the color being given by the Fourier transform of λ), then the motivating example one should keep in mind is a function σ^0 of the form $\sigma^0(t, x, \mu, \alpha, \xi) \sim \sigma^0(t, x, \mu, \alpha)\delta(x - \xi)$ with $\Xi = \mathbb{R}^d$, where δ is a mollified version of the delta function which we treat as the actual point mass at 0 for the purpose of this informal discussion. In this case, integration with respect to the spatial part of the random measure $\boldsymbol{W}^0$ gives:

$$\int_{\mathbb{R}^d} \sigma^0(t, X_t^i, \bar{\mu}_t^N, \alpha_t^i, \xi) W^0(d\xi, dt) = \sigma^0(t, X_t^i, \bar{\mu}_t^N, \alpha_t^i) W^0(X_t^i, dt), \tag{2.40}$$

which says that, at time t, the private state of player i is subject to several sources of random shocks: its own idiosyncratic noise W_t^i, but also, an independent white noise shock picked up at the very location/value of her/his own private state.

The case treated in the text corresponds to σ^0 being independent of ξ, as the random measure $\boldsymbol{W}^0$ may as well be independent of the spatial component so that we can assume that $W^0(d\xi, dt) = W^0(dt) = dW_t^0$, for an extra Wiener process $\boldsymbol{W}^0$ independent of the space location ξ and of the idiosyncratic noise terms $(\boldsymbol{W}^i)_{1\leq i\leq N}$. We did not use this general formulation in the text for two main reasons: a) the technicalities needed to handle the stochastic integrals with respect to the white noise measure $\boldsymbol{W}^0$ can distract from the understanding of the technology brought to bear to study games with common noise; b) to keep the presentation to a reasonable level of technicality, we chose to state and prove the theoretical results of this chapter in the case of common random shocks given by a standard Wiener process $\boldsymbol{W}^0$.

In our construction of a probabilistic set-up for carrying idiosyncratic and common noises, our choice for writing the space Ω as the product of Ω^0 and Ω^1 is mostly for convenience. As explained in the statement of Lemma 2.4, it permits to reduce the conditional law of a random variable X given $\mathcal{F}^0$ to the law $\mathcal{L}^1(X)$ of the section of X on Ω^1. However, we could also work with a single probability space Ω instead of the two Ω^0 and Ω^1, and then with conditional probabilities instead of $\mathcal{L}^1$. For subtleties about the application of Fubini's theorem on the completion of product spaces, we refer to any monograph on measure theory and integration, see for instance Rudin [322].

Our presentation of conditional McKean-Vlasov equations is inspired by Sznitman's seminal lecture notes [325], where he treats the non-conditional case. The analysis of the conditional case, including the derivation of the stochastic limiting Fokker-Planck equation with possibly more complex forms of the common noise, was carried out in Vaillancourt [335], Dawson and Vaillancourt [130], Kurtz and Xiong [249, 250] and Coghi and Flandoli [120]. The same problem is addressed in Kolokoltsov and Troeva [237], but with a different point of view, much closer to the one we adopted in Subsection (Vol I)-5.7.4. De Finetti's theory, including the statement of Theorem 2.1, may be found in Aldous' lectures notes [16] and in Kingman's article [230]. See also Billingsley's monograph for a shorter account [58].

The notions of weak and strong solutions for mean field games with a common noise are taken from the work [100] of Carmona, Delarue, and Lacker where they were based on the notion of relaxed controls. Here, we tried to pattern the notions of weak and strong solutions very much in the spirit of the standard notions of weak and strong solutions used in the analysis of stochastic differential equations. The rationale for lifting the environment from $\boldsymbol{\mu}$ to $\mathfrak{M}$ in the search of an equilibrium, as we did in Subsection 2.2.2, will be made clear in the next chapter. As explained in the text, the information enclosed in $(\boldsymbol{W}^0, \mathfrak{M})$ is somehow the canonical information needed to describe an equilibrium and this fact will be explicitly used to guarantee a weak form of stability of the resulting MFG equilibria. As for the notions of weak and strong equilibria, this idea is borrowed from [100].

The system formed by the forward equation (2.38) and the backward equation (2.37) reads as an infinite dimensional forward-backward stochastic differential equation, with $(\mu_t)_{0\leq t\leq T}$ as forward component and $(U(t,\cdot))_{0\leq t\leq T}$ as backward one. Standard FBSDE theory says that, provided that existence and uniqueness of a solution hold true, there should be a *decoupling field* permitting to express the backward variable in terms of the forward one, namely $U(t,x)$ should write $\mathcal{U}(t,x,\mu_t)$ for some function $\mathcal{U} : [0,T] \times \mathbb{R}^d \times \mathcal{P}_2(\mathbb{R}^d) \ni (t,x,\mu) \mapsto \mathcal{U}(t,x,\mu) \in \mathbb{R}$. Moreover, in the spirit of the finite dimensional theory, $\mathcal{U}$, if it exists, is expected to satisfy a partial differential equation on its domain of definition. In accordance with the terminology introduced in Subsection (Vol I)-5.7.2, this PDE will be called the *master equation*. It is the purpose of Chapters 4 and 5 to provide sufficient conditions under which $\mathcal{U}$ is properly defined and satisfies the *master equation* on $[0,T] \times \mathbb{R}^d \times \mathcal{P}_2(\mathbb{R}^d)$. Existence and uniqueness of a classical solution to (2.37)–(2.38) were addressed by Chassagneux, Crisan, and Delarue [114] and by Cardaliaguet, Delarue, Lasry, and Lions in [86]. We shall revisit both in Chapter 5.

Solving MFGs with a Common Noise

3

Abstract

The lion's share of this chapter is devoted to the construction of equilibria for mean field games with a common noise. We develop a general two-step strategy for the search of weak solutions. The first step is to apply Schauder's theorem in order to prove the existence of strong solutions to mean field games driven by a discretized version of the common noise. The second step is to make use of a general stability property of weak equilibria in order to pass to the limit along these discretized equilibria. We also present several criteria for strong uniqueness, in which cases weak equilibria are known to be strong.

3.1 Introduction

3.1.1 Road Map to Weak Solutions

We learnt from Chapter 2 that solutions to mean field games with a common noise could be characterized through forward-backward stochastic differential equations of the conditional McKean-Vlasov type. It is then reasonable to expect that an approach along the lines of the strategy developed in Chapters (Vol I)-3 and (Vol I)-4 for the solution of mean field games without a common noise could be feasible in the present situation. In particular, one could imagine an existence proof based on Schauder's fixed point theorem. However, as we shall see next, the presence of the common noise renders such an approach much more intricate. Indeed, equilibria are now *randomized* and they *live* in a much bigger space than in the deterministic case (i.e., when there is no common noise). This makes compactness arguments much more difficult to come by.

In order to circumvent this difficulty, we discretize the common source of noise so that the probability space on which it is defined is of finite cardinality. In this way, the search for equilibria can take place in a space which is merely a finite

R. Carmona, F. Delarue, *Probabilistic Theory of Mean Field Games with Applications II*, Probability Theory and Stochastic Modelling 84,
DOI 10.1007/978-3-319-56436-4_3

product of copies of the space used in the deterministic case, and we can use similar compactness criteria to solve these approximate equilibrium models. Once equilibria have been constructed for finite sources of common random shocks, we show that one can extract weakly converging subsequences as the mesh of the discretization gets finer and finer. Under suitable conditions, we shall prove that any weak limit solves the original mean field game problem in the weak sense according to the terminology introduced in Chapter 2.

Making use of the results from Subsection 1.4 on the connection between optimization in random environment and FBSDEs, we then exhibit three sets of conditions under which mean field games with a common noise have a weak solution. We also provide two explicit criteria under which uniqueness holds in the strong sense. In such a case, we know from the Yamada-Watanabe theorem for MFG equilibria proven in Chapter 2, that weak solutions to the MFG problem are also strong solutions. The first strong uniqueness result is based on the Lasry-Lions monotonicity condition already used in Chapter (Vol I)-3. The second one provides an interesting instance of failure of uniqueness in the absence of a common noise, uniqueness being *restored* by the presence of the common noise.

3.1.2 Statement of the Problem

The set-up is the same as in Definition 2.16. We are given:

1. an initial condition $\mathcal{V}_0 \in \mathcal{P}_2(\mathcal{P}_2(\mathbb{R}^d))$, a complete probability space $(\Omega^0, \mathcal{F}^0, \mathbb{P}^0)$, endowed with a complete and right-continuous filtration $\mathbb{F}^0 = (\mathcal{F}_t^0)_{0 \le t \le T}$, an $\mathcal{F}_0^0$-measurable initial random probability measure μ_0 on $\mathbb{R}^d$ with $\mathcal{V}_0$ as distribution, and a d-dimensional $\mathbb{F}^0$-Brownian motion $\boldsymbol{W}^0 = (W_t^0)_{0 \le t \le T}$,
2. a complete probability space $(\Omega^1, \mathcal{F}^1, \mathbb{P}^1)$ endowed with a complete and right-continuous filtration $\mathbb{F}^1 = (\mathcal{F}_t^1)_{0 \le t \le T}$ and a d-dimensional $\mathbb{F}^1$-Brownian motion $\boldsymbol{W} = (W_t)_{0 \le t \le T}$.

We then denote by $(\Omega, \mathcal{F}, \mathbb{P})$ the completion of the product space $(\Omega^0 \times \Omega^1, \mathcal{F}^0 \otimes \mathcal{F}^1, \mathbb{P}^0 \otimes \mathbb{P}^1)$ endowed with the filtration $\mathbb{F} = (\mathcal{F}_t)_{0 \le t \le T}$ obtained by augmenting the product filtration $\mathbb{F}^0 \otimes \mathbb{F}^1$ in a right-continuous way and by completing it.

As in the previous chapter, we shall make extensive use of the useful notation $\mathcal{L}^1(X)(\omega^0) = \mathcal{L}(X(\omega^0, \cdot))$ for $\omega^0 \in \Omega^0$ and a random variable X on Ω, see Subsection 2.1.3.

For a drift b from $[0, T] \times \mathbb{R}^d \times \mathcal{P}_2(\mathbb{R}^d) \times A$ to $\mathbb{R}^d$ where A is a closed convex subset of $\mathbb{R}^k$, two (uncontrolled) volatility coefficients σ and σ^0 from $[0, T] \times \mathbb{R}^d \times \mathcal{P}_2(\mathbb{R}^d)$ to $\mathbb{R}^{d \times d}$, and for cost functions f and g from $[0, T] \times \mathbb{R}^d \times \mathcal{P}_2(\mathbb{R}^d) \times A$ to $\mathbb{R}$

and from $\mathbb{R}^d \times \mathcal{P}_2(\mathbb{R}^d)$ to $\mathbb{R}$, the search for an MFG equilibrium along the lines of Definition 2.16 may be summarized as follows:

(i) Given an $\mathcal{F}_0^0$-measurable random variable $\mu_0 : \Omega^0 \to \mathcal{P}_2(\mathbb{R}^d)$, with $\mathcal{V}_0$ as distribution, an initial condition $X_0 : \Omega \to \mathbb{R}^d$ such that $\mathcal{L}^1(X_0) = \mu_0$, and an $\mathcal{F}_T^0$-measurable random variable $\mathfrak{M}$ with values in $\mathcal{P}_2(\mathcal{C}([0,T];\mathbb{R}^{2d}))$ such that $\mathbb{F}$ is compatible with $(X_0, \boldsymbol{W}^0, \mathfrak{M}, \boldsymbol{W})$ and $\boldsymbol{\mu}_0 = \mathfrak{M} \circ (e_0^x)^{-1}$, where e_t^x is the evaluation map providing the first d coordinates at time t on $\mathcal{C}([0,T];\mathbb{R}^{2d})$, solve the (standard) stochastic control problem (with random coefficients):

$$\inf_{(\alpha_t)_{0\le t\le T}} \mathbb{E}\left[\int_0^T f(t, X_t, \mu_t, \alpha_t)dt + g(X_T, \mu_T)\right], \tag{3.1}$$

subject to:

$$dX_t = b\big(t, X_t, \mu_t, \alpha_t\big)dt + \sigma\big(t, X_t, \mu_t\big)dW_t + \sigma^0\big(t, X_t, \mu_t\big)dW_t^0, \tag{3.2}$$

for $t \in [0,T]$, with X_0 as initial condition and with $\mu_t = \mathfrak{M} \circ (e_t^x)^{-1}$.

(ii) Determine the input $\mathfrak{M}$ and one solution $\boldsymbol{X} = (X_t)_{0\le t\le T}$ of the above optimal control problem so that:

$$\mathfrak{M} = \mathcal{L}^1\big(\boldsymbol{X}, \boldsymbol{W}\big), \qquad \mathbb{P}^0 - a.s. \ . \tag{3.3}$$

Recall that, whenever $\mathbb{F}^0$ is the complete and right-continuous augmentation of the filtration generated by μ_0 and $\boldsymbol{W}^0$, solutions are said to be strong. Also, recall that as explained in Remark 1.12, the compatibility condition is automatically satisfied in that case.

3.1.3 Overview of the Strategy

In Chapter 2, we introduced assumption **FBSDE**, under which the optimal paths of the stochastic optimal control problem (3.1) can be characterized as the forward component of the solution of an FBSDE of the form:

$$\begin{cases} dX_t = B(t, X_t, \mu_t, Y_t, Z_t)dt \\ \qquad\quad + \sigma(t, X_t, \mu_t)dW_t + \sigma^0(t, X_t, \mu_t)dW_t^0, \\ dY_t = -F(t, X_t, \mu_t, Y_t, Z_t, Z_t^0)dt \\ \qquad\quad + Z_t dW_t + Z_t^0 dW_t^0 + dM_t, \quad t \in [0,T], \\ Y_T = G(X_T, \mu_T), \end{cases} \tag{3.4}$$

where $(M_t)_{0\le t\le T}$ is an $\mathbb{F}$-martingale, with $M_0 = 0$, $[\boldsymbol{M}, \boldsymbol{W}]_\cdot \equiv 0$ and $[\boldsymbol{M}, \boldsymbol{W}^0]_\cdot \equiv 0$. As in (3.1), $\boldsymbol{\mu}$ reads as $\boldsymbol{\mu} = (\mu_t = \mathfrak{M} \circ (e_t^x)^{-1})_{0\le t\le T}$.

In analogy with the strategy developed in Chapters (Vol I)-3 and (Vol I)-4 for solving MFGs without common noise, a natural approach is to apply a fixed point argument of Schauder type in order to get the existence of a solution, and then to focus on the uniqueness part separately. In the present situation, the fixed point argument should concern the mapping

$$\Phi : \text{input} = \big(\mathfrak{M}(\omega^0)\big)_{\omega^0 \in \Omega^0} \mapsto \text{output} = \big(\mathcal{L}^1(\boldsymbol{X}, \boldsymbol{W})(\omega^0)\big)_{\omega^0 \in \Omega^0},$$

where $\boldsymbol{X}$ is the forward component in the triplet solving FBSDE (3.4) in the random super-environment $\mathfrak{M}$ and the random sub-environment $(\mu_t = \mathfrak{M} \circ (e_t^x)^{-1})_{0 \le t \le T}$. Accordingly, the fixed point may be characterized through the solution of a McKean-Vlasov FBSDE of the conditional type, see Proposition 2.18.

The limitation of such a strategy is quite clear. At first sight, it would seem quite tempting to solve the fixed point for each ω^0, but this would be completely meaningless, since the solution of the backward equation in essence relies on a martingale representation property and thus integrates ω^0 under $\mathbb{P}^0$. Alternatively, we could see the whole process as a single input, and then seek a fixed point in a larger space. However, with such a point of view, we could hardly hope being able to use a tractable compactness criterion.

Our approach tries to combine the benefits of both alternatives. It consists in discretizing the randomness encapsulated in ω^0 in order to reduce the size of the space on which the fixed point problem is to be solved. In a nutshell, we first prove that the discretized problem admits strong solutions, and passing to the limit along a converging subsequence, we establish existence of a weak solution.

3.1.4 Assumption and Main Statement

The purpose of this chapter is to prove the existence of weak equilibria under a suitable reinforcement of assumption **FBSDE** introduced in Chapter 2.

The required assumption may be split into three parts, each one concerning: 1) the regularity properties of the coefficients $(b, \sigma, \sigma^0, f, g)$; 2) the stochastic control problem, in the spirit of assumption **FBSDE**; 3) the regularity properties of the coefficients entering assumption **FBSDE**.

We start with:

Assumption (Coefficients MFG with a Common Noise). There exists a constant $L > 0$ such that:

(A1) The drift b has the form:

$$b(t, x, \mu, \alpha) = b_1(t, x, \mu) + b_2(t)\alpha,$$

where $[0, T] \ni t \mapsto b_2(t) \in \mathbb{R}^{d \times k}$ is measurable and bounded by L.

(continued)

(A2) The coefficients b_1, σ and σ^0 are Borel-measurable mappings from $[0,T] \times \mathbb{R}^d \times \mathcal{P}_2(\mathbb{R}^d)$ into $\mathbb{R}^d$, $\mathbb{R}^{d\times d}$ and $\mathbb{R}^{d\times d}$ respectively. For any $t \in [0,T]$, the coefficients $b_1(t,\cdot,\cdot)$, $\sigma(t,\cdot,\cdot)$ and $\sigma^0(t,\cdot,\cdot)$ are continuous on $\mathbb{R}^d \times \mathcal{P}_2(\mathbb{R}^d)$ and, for any $\mu \in \mathcal{P}_2(\mathbb{R}^d)$, the functions $b_1(t,\cdot,\mu)$, $\sigma(t,\cdot,\mu)$ and $\sigma^0(t,\cdot,\mu)$ are continuously differentiable (with respect to x). Moreover,

$$|(b_1,\sigma,\sigma^0)(t,x,\mu)| \le L\big[1 + |x| + M_2(\mu)\big],$$
$$|\partial_x(b_1,\sigma,\sigma^0)(t,x,\mu)| \le L.$$

(A3) The coefficients f and g are Borel-measurable mappings from $[0,T] \times \mathbb{R}^d \times \mathcal{P}_2(\mathbb{R}^d) \times A$ to $\mathbb{R}$ and from $\mathbb{R}^d \times \mathcal{P}_2(\mathbb{R}^d)$ to $\mathbb{R}$ respectively. For any $t \in [0,T]$, the coefficients $f(t,\cdot,\cdot,\cdot)$ and $g(\cdot,\cdot)$ are continuous on $\mathbb{R}^d \times \mathcal{P}_2(\mathbb{R}^d) \times A$ and $\mathbb{R}^d \times \mathcal{P}_2(\mathbb{R}^d)$ respectively. For any $t \in [0,T]$ and $\mu \in \mathcal{P}_2(\mathbb{R}^d)$, the function $f(t,\cdot,\mu,\cdot)$ is continuously differentiable (in (x,α)) and the function $g(\cdot,\mu)$ is continuously differentiable (in x). Moreover,

$$|f(t,x,\mu,\alpha)| + |g(x,\mu)| \le L\big[1 + |x|^2 + |\alpha|^2 + \big(M_2(\mu)\big)^2\big],$$
$$|\partial_x f(t,x,\mu,\alpha)| + |\partial_\alpha f(t,x,\mu,\alpha)| + |\partial_x g(x,\mu)|$$
$$\le L\big[1 + |x| + |\alpha| + M_2(\mu)\big].$$

Also, the function $\partial_\alpha f$ is L-Lipschitz-continuous in x.

(A4) f satisfies the following form of L^{-1}-convexity property:

$$f(t,x,\mu,\alpha') - f(t,x,\mu,\alpha) - (\alpha' - \alpha)\cdot \partial_\alpha f(t,x,\mu,\alpha) \ge L^{-1}|\alpha' - \alpha|^2.$$

We recall that $M_2(\mu)$ is defined as $(\int_{\mathbb{R}^d} |x|^2 d\mu(x))^{1/2}$.

Here is now the refinement of assumption **FBSDE** from Chapter 2. The reader may want to review part of Subsection 2.2.2 for the notion of *lifting*:

Assumption (FBSDE MFG with a Common Noise). On top of assumption **Coefficients MFG with a Common Noise**, there exist an integer $m \ge 1$ together with deterministic measurable coefficients $B : [0,T] \times \mathbb{R}^d \times \mathcal{P}_2(\mathbb{R}^d) \times \mathbb{R}^m \times \mathbb{R}^{m\times d} \to \mathbb{R}^d$, $F : [0,T] \times \mathbb{R}^d \times \mathcal{P}_2(\mathbb{R}^d) \times \mathbb{R}^m \times (\mathbb{R}^{m\times d})^2 \to \mathbb{R}^m$ and $G : \mathbb{R}^d \times \mathcal{P}_2(\mathbb{R}^d) \to \mathbb{R}^m$, such that, for any probabilistic set-up $(\Omega, \mathcal{F}, \mathbb{F}, \mathbb{P})$ equipped with a compatible lifting $(X_0, \boldsymbol{W}^0, (\mathfrak{M}, \boldsymbol{\mu}), \boldsymbol{W})$, one has:

(continued)

(A1) The optimal control problem defined in (3.1) and (3.2) with X_0 as initial condition, has a unique solution, characterized as the forward component of the unique strong solution of the FBSDE:

$$\begin{cases} dX_t = B(t, X_t, \mu_t, Y_t, Z_t)dt \\ \qquad\qquad + \sigma(t, X_t, \mu_t)dW_t + \sigma^0(t, X_t, \mu_t)dW_t^0, \\ dY_t = -F(t, X_t, \mu_t, Y_t, Z_t, Z_t^0)dt \\ \qquad\qquad + Z_t dW_t + Z_t^0 dW_t^0 + dM_t, \quad t \in [0, T], \end{cases} \tag{3.5}$$

with X_0 as initial condition for $\boldsymbol{X} = (X_t)_{0 \le t \le T}$ and $Y_T = G(X_T, \mu_T)$ as terminal condition for $\boldsymbol{Y} = (Y_t)_{0 \le t \le T}$, where $\boldsymbol{M} = (M_t)_{0 \le t \le T}$ is a *càd-làg* martingale with respect to the filtration $\mathbb{F}$, of zero cross variation with $(\boldsymbol{W}^0, \boldsymbol{W})$ and with initial condition $M_0 = 0$.

(A2) For any other sub-environment $\boldsymbol{\mu}'$ defined from $\mathfrak{M}$ as in the statement of the stability Theorem 1.53, the solution $(\boldsymbol{X}', \boldsymbol{Y}', \boldsymbol{Z}', \boldsymbol{Z}^{0\prime}, \boldsymbol{M}')$ to (3.5) with X_0 as initial condition and with $\boldsymbol{\mu}$ replaced by $\boldsymbol{\mu}'$ satisfies, together with $(\boldsymbol{X}, \boldsymbol{Y}, \boldsymbol{Z}, \boldsymbol{Z}^0, \boldsymbol{M})$, the stability estimate (1.19) stated in Theorem 1.53.

(A3) There exists a deterministic measurable function $\breve{\alpha}$ from $[0, T] \times \mathbb{R}^d \times \mathcal{P}_2(\mathbb{R}^d) \times \mathbb{R}^m \times \mathbb{R}^{m\times d}$ into A, such that the optimal control $\hat{\boldsymbol{\alpha}} = (\hat{\alpha}_t)_{0 \le t \le T}$ associated with the problem (3.1)–(3.2) and the initial condition X_0 has the form $(\hat{\alpha}_t = \breve{\alpha}(t, X_t, \mu_t, Y_t, Z_t))_{0 \le t \le T}$, where $(\boldsymbol{X}, \boldsymbol{Y}, \boldsymbol{Z}, \boldsymbol{Z}^0, \boldsymbol{M})$ is the solution of (3.5). Also, $(B(t, X_t, \mu_t, Y_t, Z_t))_{0 \le t \le T}$ is equal to $(b(t, X_t, \mu_t, \breve{\alpha}(t, X_t, \mu_t, Y_t, Z_t)))_{0 \le t \le T}$.

Of course, in comparison with assumption **FBSDE**, the novelty lies in (**A2**), in which we require stability of the solutions with respect to the sub-environment.

Regarding the coefficients B, F, and G, we require:

There exists a constant $L \ge 0$, such that

(A4) For any $t \in [0, T]$, the coefficients $B(t, \cdot, \cdot, \cdot, \cdot)$, $F(t, \cdot, \cdot, \cdot, \cdot, \cdot)$ are continuous on $\mathbb{R}^d \times \mathcal{P}_2(\mathbb{R}^d) \times \mathbb{R}^m \times \mathbb{R}^{m\times d}$ and $\mathbb{R}^d \times \mathcal{P}_2(\mathbb{R}^d) \times \mathbb{R}^m \times (\mathbb{R}^{m\times d})^2$. Similarly, G is continuous.

(A5) For any $x \in \mathbb{R}^d$, $y \in \mathbb{R}^m$, $z \in \mathbb{R}^{m\times d}$ and $\mu \in \mathcal{P}_2(\mathbb{R}^d)$,

$$|B(t, x, \mu, y, z)| \le L\big[1 + |x| + M_2(\mu) + |y| + |z|\big],$$
$$|\breve{\alpha}(t, x, \mu, y, z)| \le L\big[1 + |x| + M_2(\mu) + |y| + |z|\big],$$
$$|F(t, x, \mu, y, z, z^0)| + |G(x, \mu)| \le L\big[1 + |y| + |z| + |z^0|\big].$$

(continued)

(A6) For any $x, x' \in \mathbb{R}^d$, $y, y' \in \mathbb{R}^m$, $z, z', z^0, z^{0\prime} \in \mathbb{R}^{m\times d}$ and $\mu \in \mathcal{P}_2(\mathbb{R}^d)$,

$$\begin{aligned}
&\big|(B,F)(t,x',\mu,y',z',z^{0\prime}) - (B,F)(t,x,\mu,y,z,z^0)\big| \\
&\quad \le L\big(|x'-x| + |y'-y| + |z'-z| + |z^{0\prime} - z^0|\big), \\
&\big|G(x',\mu) - G(x,\mu)\big| \le L|x'-x|.
\end{aligned}$$

Of course, the above assumptions are tailor-made to the results obtained in Subsection 1.4 on stochastic control problems in a random environment. Typical instances in which they are satisfied are given in Section 3.4.

We can now state the main result of the chapter.

Theorem 3.1 *Under assumption* **FBSDE MFG with a Common Noise** *(which includes assumption* **Coefficients MFG with a Common Noise***), the mean field game* (3.1)–(3.2)–(3.3) *with* $\mathcal{V}^0 \in \mathcal{P}_2(\mathcal{P}_2(\mathbb{R}^d))$ *as initial condition has a weak solution in the sense of Definition 2.23.*

3.2 Stability of Weak Equilibria

The framework of this section is different and more general than in the previous section. In particular, we do not require assumption **Coefficients MFG with a Common Noise** *nor assumption* **FBSDE MFG with a Common Noise**.

3.2.1 Passing to the Limit in a Sequence of Weak Equilibria

Our strategy should be clear by now. Undoubtedly, it requires a suitable toolbox for passing to the limit in weak equilibria. This is what we call *stability of weak equilibria*. Basically, we must be able to check that weak limits of weak equilibria remain weak equilibria! Obviously, this raises crucial questions regarding the topology used for the convergence, and the criteria used in the characterization of the weak limits as weak equilibria. In short, the convergence must preserve the basic features which make it possible to identify a measure-valued process with the solution of an MFG.

Because of our choice to solve underlying stochastic optimal control problems by means of forward-backward stochastic differential equations, we understand the need to introduce a topology which accommodates quite well forward-backward systems. Unfortunately, it is well known that forward-backward SDEs are not well adapted to weak convergence. Generally speaking, it is rather difficult to identify a convenient topology for which the standard tightness criteria are checked by the integrand processes $(Z_t)_{0\leq t\leq T}$ and $(Z_t^0)_{0\leq t\leq T}$ appearing in the backward martingales. In order to overcome this difficulty, we shall make a systematic use of the necessary condition in the Pontryagin principle in order to represent the control process associated with a weak equilibrium in terms of the component $(Y_t)_{0\leq t\leq T}$ of an FBSDE instead of $(Z_t)_{0\leq t\leq T}$. This will provide us with a more robust structure when discussing tightness properties of the control processes.

As if the obstacles identified above were not enough of a hindrance, we shall face another serious difficulty. Generally speaking, it is highly problematic to pass to the limit in the compatibility conditions underpinning FBSDEs in a random environment. To overcome this extra hurdle, we shall exploit the full-fledged conditioning rule involving the enlarged environment $\mathfrak{M}$ in the definition of a weak solution in order to guarantee that compatibility is preserved in the limit.

3.2.2 Meyer-Zheng Topology

Most of the results mentioned in this subsection are stated without proof. We refer the reader to the Notes & Complements at the end of the chapter for references and a detailed bibliography.

As we said in the previous subsection, we shall prove tightness of the optimal controls associated with a sequence of weak equilibria by taking advantage of the necessary condition in the stochastic Pontryagin principle. In this way, any of the optimal controls under consideration will be given through the backward component $\boldsymbol{Y} = (Y_t)_{0\leq t\leq T}$ of the solution to an FBSDE. Such a process being a càd-làg semi-martingale, we need a topology on the space $\mathcal{D}([0,T];\mathbb{R}^m)$ (or larger) for $m \geq 1$, and tightness criteria which can easily be verified for sequences of semi-martingales.

As explained in the Notes & Complements below, this question has been already addressed in the literature. One frequent suggestion, when dealing with convergence of backward SDEs, is to work with the so-called *Meyer-Zheng topology*. Here, we provide its definition and its basic properties. Once again, we refer to the Notes & Complements at the end of the chapter for a complete bibliography on the subject.

Description of the Topology

We start with the following definition.

Definition 3.2 *Given $m \geq 1$, we call $\mathcal{M}([0,T];\mathbb{R}^m)$ the space of equivalence classes of Borel-measurable functions from $[0,T]$ to $\mathbb{R}^m$ (two functions being equivalent if they are equal for almost every $t \in [0,T]$ for the Lebesgue measure). We equip $\mathcal{M}([0,T];\mathbb{R}^m)$ with the distance:*

$$d_{\mathcal{M}}(\boldsymbol{x},\boldsymbol{x}') = \int_0^T \min\big(1, |x(t) - x'(t)|\big)dt,$$

for two elements $\boldsymbol{x} = (x(t))_{0\leq t\leq T}$ and $\boldsymbol{x}' = (x'(t))_{0\leq t\leq T}$ in $\mathcal{M}([0,T];\mathbb{R}^m)$.

Notice that convergence in the metric $d_{\mathcal{M}}$ is the convergence in dt-measure: a sequence of functions $(\boldsymbol{x}^n)_{n\in\mathbb{N}}$ converges to $\boldsymbol{x}$ for $d_{\mathcal{M}}$ if and only if:

$$\forall \varepsilon > 0, \quad \lim_{n\to\infty} \mathrm{Leb}_1\big(t \in [0,T] : |x_t^n - x_t| \geq \varepsilon\big) = 0,$$

where Leb_1 denotes the Lebesgue measure on $[0,T]$.

Obviously, any sequence of continuous functions $(\boldsymbol{x}^n)_{n\geq 0}$ from $[0,T]$ to $\mathbb{R}^m$ converging in $\mathcal{C}([0,T];\mathbb{R}^m)$ equipped with the topology of the uniform convergence is convergent in $\mathcal{M}([0,T];\mathbb{R}^m)$. Similarly, any sequence of càd-làg functions $(\boldsymbol{x}^n)_{n\geq 0}$ from $[0,T]$ to $\mathbb{R}^m$ converging in $\mathcal{D}([0,T];\mathbb{R}^m)$ equipped with the Skorohod $J1$ topology is convergent in $\mathcal{M}([0,T];\mathbb{R}^m)$.

We now list without proof a few properties of the space $\mathcal{M}([0,T];\mathbb{R}^m)$.

Lemma 3.3 *The space $\mathcal{M}([0,T];\mathbb{R}^m)$ equipped with the distance $d_{\mathcal{M}}$ is complete and separable. In particular, it is a Polish space.*

We call Meyer-Zheng topology, the topology of the weak convergence on the space $\mathcal{P}(\mathcal{M}([0,T];\mathbb{R}^m))$ of probability measures on $\mathcal{M}([0,T];\mathbb{R}^m)$. The second claim identifies the compact subsets of $(\mathcal{M}([0,T];\mathbb{R}^m), d_{\mathcal{M}})$.

Lemma 3.4 *A subset $\boldsymbol{A} \subset \mathcal{M}([0,T];\mathbb{R}^m)$ is relatively compact for the topology induced by $d_{\mathcal{M}}$ if and only if the following conditions hold true:*

$$\limsup_{a\to\infty} \sup_{x\in A} \mathrm{Leb}_1\big(t \in [0,T] : |x(t)| \geq a\big) = 0,$$

$$\limsup_{h\searrow 0} \sup_{x\in A} \int_0^T \min\big(1, |x((t+h)\wedge T) - x(t)|\big)dt = 0.$$

As an exercise, we let the reader check the following statement.

Lemma 3.5 *If a measurable function $\theta : [0,T]\times\mathbb{R}^m \to \mathbb{R}$ is such that for each $t \in [0,T]$, the function $\mathbb{R}^m \ni x \mapsto \theta(t,x)$ is continuous, then the function $(\mathcal{M}([0,T];\mathbb{R}^m), d_{\mathcal{M}}) \ni \boldsymbol{x} \mapsto (\theta(t,x_t))_{0\leq t\leq T} \in (\mathcal{M}([0,T];\mathbb{R}), d_{\mathcal{M}})$ is continuous.*

Also, if $(\boldsymbol{x}^n)_{n\geq 0}$ is a sequence of functions in $(\mathcal{M}([0,T];\mathbb{R}^m), d_{\mathcal{M}})$ converging to $\boldsymbol{x}$ for the distance $d_{\mathcal{M}}$, and if furthermore the uniform integrability property:

$$\limsup_{a\to\infty} \sup_{n\in\mathbb{N}} \int_0^T \big|\theta\big(t, x^n(t)\big)\big| \mathbf{1}_{\{|\theta(t,x^n(t))|\geq a\}} dt = 0,$$

holds, then we have:

$$\lim_{n\to\infty} \int_0^T \big|\theta\big(t, x(t)\big) - \theta\big(t, x^n(t)\big)\big| dt = 0.$$

Stochastic Processes and $\mathscr{M}([0,T];\mathbb{R}^m)$-Valued Random Variables

We use the above properties to prove a simple tightness criterion for $\mathbb{R}^m$-valued stochastic processes. First, we identify $\mathbb{R}^m$-valued stochastic processes with random variables taking values in $\mathscr{M}([0,T];\mathbb{R}^m)$. Here, we say that $\boldsymbol{X} = (X_t)_{0\le t\le T}$ is a stochastic process if it is a measurable mapping

$$\boldsymbol{X} : [0,T] \times \Omega \ni (t,\omega) \mapsto X_t(\omega),$$

the domain $[0,T]\times\Omega$ being equipped with the product σ-field $\mathcal{B}([0,T]) \otimes \mathcal{F}$ and the range $\mathbb{R}^m$ with its Borel σ-field.

For such a stochastic process $\boldsymbol{X}$ on a probability space $(\Omega,\mathcal{F},\mathbb{P})$, we know that, for any $\omega \in \Omega$, the mapping $[0,T] \ni t \mapsto X_t(\omega)$ is measurable. We denote by $\bar{\boldsymbol{X}}(\omega)$ its equivalence class for the almost everywhere equality under the Lebesgue measure. In order to prove that $\bar{\boldsymbol{X}}$ can be viewed as a random variable with values in $\mathscr{M}([0,T];\mathbb{R}^m)$, it suffices to check that for any $a > 0$ and any $\boldsymbol{x} \in \mathscr{M}([0,T];\mathbb{R}^m)$, the set:

$$\Big\{\omega \in \Omega : \int_0^T \min\big(1, |\bar{X}_t(\omega) - x(t)|\big)dt < a\Big\}$$

belongs to $\mathcal{F}$. Of course, this is quite clear since:

$$\int_0^T \min\big(1, |\bar{X}_t(\omega) - x(t)|\big)dt = \int_0^T \min\big(1, |X_t(\omega) - x(t)|\big)dt.$$

By Fubini's theorem, the right-hand side is an $\mathbb{R}$-valued random variable. Clearly, $\bar{\boldsymbol{X}}$ and $\bar{\boldsymbol{Y}}$ are almost surely equal under the probability $\mathbb{P}$ if and only if $\boldsymbol{X}$ and $\boldsymbol{Y}$ are almost everywhere equal under the probability $\mathrm{Leb}_1 \otimes \mathbb{P}$.

We now proceed with the converse. To any random variable $\bar{\boldsymbol{X}}$ with values in $\mathscr{M}([0,T];\mathbb{R}^m)$, we try to associate a canonical representative of the equivalence class induced by $\bar{\boldsymbol{X}}(\omega)$ for each $\omega \in \Omega$. The strategy is similar to that used in Lemma 1.27. Indeed, it is quite tempting to use, as we did before,

$$X_t(\omega) = \begin{cases} \lim_{n\to\infty} n \int_t^{(t+1/n)\wedge T} \bar{X}_s(\omega)ds & \text{whenever the limit exists,} \\ 0 \quad \text{otherwise.} \end{cases}$$

Unfortunately, this approach is not feasible since nothing is known regarding the integrability of $\boldsymbol{X}$ in time. In order to overcome this difficulty, for each integer $p \ge 1$, we denote by $\pi_p : \mathbb{R}^m \to \mathbb{R}^m$ the orthogonal projection onto the closed ball of center 0 and of radius p, and for any $t \in [0,T]$ and $n \ge 1$ we set:

$$X_t^p(\omega) = \begin{cases} \lim_{n\to\infty} n \int_t^{(t+1/n)\wedge T} \pi_p\big(\bar{X}_s(\omega)\big)ds & \text{whenever the limit exists,} \\ 0 \quad \text{otherwise.} \end{cases}$$

The mapping $\Omega \ni \omega \mapsto n\int_t^{(t+1/n)\wedge T} \pi_p(\bar{X}_s(\omega))ds$ is an $\mathbb{R}^m$-valued random variable. For each $p \geq 1$, since it is continuous in t for ω fixed, the mapping:

$$[0,T] \times \Omega \ni (t,\omega) \mapsto n\int_t^{(t+1/n)\wedge T} \pi_p(\bar{X}_s(\omega))ds \in \mathbb{R}^m$$

is measurable with respect to $\mathcal{B}([0,T]) \otimes \mathcal{F}$. In particular, $\boldsymbol{X}^p = (X_t^p)_{0\leq t\leq T}$ forms a (jointly measurable) stochastic process. Moreover, by Lebesgue's differentiation theorem, for any $\omega \in \Omega$, $\text{Leb}_1(t \in [0,T];\, X_t^p(\omega) \neq \pi_p(\bar{X}_t(\omega))) = 0$. Finally, we let:

$$X_t(\omega) = \begin{cases} \lim_{p\to\infty} X_t^p(\omega) & \text{whenever the limit exists,} \\ 0 & \text{otherwise.} \end{cases} \tag{3.6}$$

Clearly, for any $\omega \in \Omega$ and for almost every $t \in [0,T]$, the sequence $(X_t^p(\omega))_{p\geq 0}$ becomes constant for large enough indices. In particular, for any $\omega \in \Omega$, the limit in (3.6) exists for almost every $t \in [0,T]$, and $\bar{\boldsymbol{X}}(\omega)$ is the equivalence class of $\boldsymbol{X}(\omega)$.

As a result of the above equivalence, we can associate a probability measure on $\mathcal{M}([0,T];\mathbb{R}^m)$ to each stochastic process $\boldsymbol{X}$. This probability measure is nothing but the law of the random variable $\bar{\boldsymbol{X}}$, where $\bar{\boldsymbol{X}}(\omega)$ is the equivalence class of $[0,T] \ni t \mapsto X_t(\omega)$. One of the first result of the general theory of stochastic processes is that two processes $\boldsymbol{X}$ and $\boldsymbol{Y}$ induce the same distribution on $\mathcal{M}([0,T];\mathbb{R}^m)$ if and only if, for any integer $n \geq 1$, for almost every $(t_1,\cdots,t_n) \in [0,T]^n$, $(X_{t_1},\cdots,X_{t_n})$ and $(Y_{t_1},\cdots,Y_{t_n})$ have the same distribution on $(\mathbb{R}^m)^n$.

Weak Convergence

Now that we have a correspondence between (jointly measurable) stochastic processes and random variables with values in $\mathcal{M}([0,T];\mathbb{R}^m)$, we can address the weak convergence of stochastic processes for the Meyer-Zheng topology on the space $\mathcal{M}([0,T];\mathbb{R}^m)$.

As a starter, we prove the following analogue of Lemma 3.5.

Lemma 3.6 *If $\theta : [0,T] \times \mathbb{R}^m \to \mathbb{R}$ is a measurable function such that for each $t \in [0,T]$, the function $\mathbb{R}^m \ni x \mapsto \theta(t,x)$ is continuous, if $(\boldsymbol{X}^n = (X_t^n)_{0\leq t\leq T})_{n\in\mathbb{N}\cup\{\infty\}}$ is a sequence of stochastic processes, $\boldsymbol{X}^n$ being defined on a probability space $(\Omega^n, \mathcal{F}^n, \mathbb{P}^n)$, in such a way that $\mathbb{P}^\infty \circ (\boldsymbol{X}^\infty)^{-1}$ is the limit of the sequence $(\mathbb{P}^n \circ (\boldsymbol{X}^n)^{-1})_{n\geq 0}$ for the Meyer-Zheng topology, and if we assume further that:*

$$\forall \varepsilon > 0, \quad \limsup_{a\to\infty} \sup_{n\geq 0} \mathbb{P}^n\left[\int_0^T \left|\theta\big(t, X^n(t)\big)\right| \mathbf{1}_{\{|\theta(t,X^n(t))|\geq a\}} dt \geq \varepsilon\right] = 0, \tag{3.7}$$

then $\mathbb{P}^\infty[\int_0^T |\theta(t, X_t^\infty)|dt < \infty] = 1$, *and the sequence* $(\mathbb{P}^n \circ (\boldsymbol{\Theta}^n)^{-1})_{n\geq 0}$ *converges to* $\mathbb{P}^\infty \circ (\boldsymbol{\Theta}^\infty)^{-1}$ *on* $\mathcal{C}([0,T];\mathbb{R})$, *where for any* $n \in \mathbb{N} \cup \{\infty\}$, *the process* $\boldsymbol{\Theta}^n$ *is defined for any* $t \in [0,T]$ *by:*

$$\Theta_t^n = \int_0^t \theta(s, X_s^n)ds.$$

Proof.

First Step. Notice first that by assumption (3.7), $\boldsymbol{\Theta}^n$ is well defined with $\mathbb{P}^n$-probability 1 for each $n \geq 0$ since $\mathbb{P}^n[\int_0^T |\theta(t, X_t^n)|dt < \infty] = 1$. When θ is bounded, the proof is a direct consequence of Lemma 3.5 since, in that case, the function $\mathcal{M}([0,T];\mathbb{R}^m) \ni \boldsymbol{x} \mapsto ([0,T] \ni t \mapsto \int_0^t \theta(s, x_s)ds) \in \mathcal{C}([0,T];\mathbb{R})$ is continuous.

Second Step. When θ is not bounded, we approximate θ by $\phi_a \circ \theta$, where for $a > 0$, $\phi_a : \mathbb{R} \to \mathbb{R}$ coincides with the identity on $[-a, a]$ and with the function $a \times \text{sign}(\cdot)$ outside $[-a, a]$. By the uniform integrability assumption, we can choose, for any $\varepsilon > 0$, a large enough such that, for any $n \geq 0$,

$$\mathbb{P}^n\left[\int_0^T \left|(\theta - \phi_a \circ \theta)(t, X_t^n)\right|dt > \varepsilon\right] \leq \varepsilon. \tag{3.8}$$

We claim that this also holds true for $n = \infty$. Indeed we have, for any $a < a'$,

$$\mathbb{P}^n\left[\int_0^T \left|(\phi_{a'} \circ \theta - \phi_a \circ \theta)(t, X_t^n)\right|dt > \varepsilon\right] \leq \varepsilon.$$

By the first step, $(\mathbb{P}^n \circ (\int_0^T |(\phi_{a'} \circ \theta - \phi_a \circ \theta)(t, X_t^n)|dt)^{-1})_{n\geq 0}$ converges in the weak sense to $\mathbb{P}^\infty \circ (\int_0^T |(\phi_{a'} \circ \theta - \phi_a \circ \theta)(t, X_t^\infty)|dt)^{-1}$. Then, by the porte-manteau theorem, we get:

$$\mathbb{P}^\infty\left[\int_0^T \left|(\phi_{a'} \circ \theta - \phi_a \circ \theta)(t, X_t^\infty)\right|dt > \varepsilon\right] \leq \varepsilon. \tag{3.9}$$

In particular, for all $\varepsilon > 0$, there exists a large enough such that for $a' > a$:

$$\mathbb{P}^\infty\left[\int_0^T \left|(\phi_{a'} \circ \theta)(t, X_t^\infty)\right|dt > Ta + \varepsilon\right] \leq \varepsilon,$$

and then,

$$\mathbb{P}^\infty\left[\int_0^T \left|\theta(t, X_t^\infty)\right|dt > Ta + \varepsilon\right] \leq \varepsilon,$$

which shows that:

$$\mathbb{P}^\infty\left[\int_0^T \left|\theta(t, X_t^\infty)\right|dt < \infty\right] = 1.$$

Returning to (3.9) and letting a' tend to ∞, we deduce that (3.8) holds for $n = \infty$.

Third Step. It remains to take a bounded and uniformly continuous function ψ from $\mathcal{C}([0,T];\mathbb{R})$ into $\mathbb{R}$. Then, for all $a > 0$,

$$\begin{aligned}
&\Big|\mathbb{E}^n[\psi(\boldsymbol{\Theta}^n)] - \mathbb{E}^\infty[\psi(\boldsymbol{\Theta}^\infty)]\Big| \\
&\le \Big|\mathbb{E}^n\Big[\psi\Big(\int_0^\cdot (\phi_a\circ\theta)(t,X_t^n)dt\Big)\Big] - \mathbb{E}^\infty\Big[\psi\Big(\int_0^\cdot (\phi_a\circ\theta)(t,X_t^\infty)dt\Big)\Big]\Big| \\
&\quad + \mathbb{E}^n\Big[m\Big(\int_0^T \big|(\phi_a\circ\theta-\theta)(t,X_t^n)\big|dt\Big)\Big] + \mathbb{E}^\infty\Big[m\Big(\int_0^T \big|(\phi_a\circ\theta-\theta)(t,X_t^\infty)\big|dt\Big)\Big],
\end{aligned}$$

where m is the modulus of continuity of ψ. According to the second step, we can choose a large enough such that the last line above is as small as desired, uniformly in $n \ge 0$. Choosing n large enough to handle the first term in the right-hand side, we complete the proof. □

Now, we provide a simple tightness criterion for semi-martingales. Generally speaking, the description of relatively compact subsets of the set of probability measures on $\mathcal{M}([0,T];\mathbb{R}^m)$ is given by Prokhorov's theorem since this space is Polish. As a consequence of Lemma 3.4, the following characterization holds true:

Proposition 3.7 *Let $(\boldsymbol{X}^n = (X_t^n)_{0\le t\le T})_{n\ge 0}$ be a sequence of $\mathbb{R}^m$-valued processes, each $\boldsymbol{X}^n$ being defined on a probability space $(\Omega^n,\mathcal{F}^n,\mathbb{P}^n)$. Then the collection of distributions $(\mathbb{P}^n\circ(\boldsymbol{X}^n)^{-1})_{n\ge 0}$ on $\mathcal{M}([0,T];\mathbb{R}^m)$ is relatively compact in the space of probability measures on $\mathcal{M}([0,T];\mathbb{R}^m)$ if and only if*

$$\lim_{a\to\infty}\sup_{n\ge 0}\int_0^T \mathbb{P}^n\big(|X_t^n|\ge a\big)dt = 0,$$

$$\limsup_{h\searrow 0}\sup_{n\ge 0}\mathbb{E}^n\int_0^T \min\big(1,|X^n((t+h)\wedge T) - X^n(t)|\big)dt = 0.$$

Of course, whenever the distributions $(\mathbb{P}^n\circ(\boldsymbol{X}^n)^{-1})_{n\ge 0}$, each $\boldsymbol{X}^n$ being regarded as a random variable with values in $\mathcal{M}([0,T];\mathbb{R}^m)$, forms a tight family on $\mathcal{M}([0,T];\mathbb{R}^m)$, we may associate with each weak limit a random variable with values in $\mathcal{M}([0,T];\mathbb{R}^m)$, this random variable being constructed on some probability space $(\Omega,\mathcal{F},\mathbb{P})$, say the canonical one. Then, following the procedure described above, we may reconstruct a stochastic process $\boldsymbol{X}$. With a slight abuse of terminology, we will say that the sequence $(\mathbb{P}^n\circ(\boldsymbol{X}^n)^{-1})_{n\ge 0}$ is tight on $\mathcal{M}([0,T];\mathbb{R}^m)$ equipped with the distance $d_{\mathcal{M}}$, and that $\mathbb{P}\circ\boldsymbol{X}^{-1}$ is a weak limit of $(\mathbb{P}^n\circ(\boldsymbol{X}^n)^{-1})_{n\ge 0}$.

Another simple, but useful, observation is that a sequence $(\mathbb{P}^n\circ(\boldsymbol{X}^n)^{-1})_{n\ge 0}$ is tight on $\mathcal{M}([0,T];\mathbb{R}^m)$ if and only if, for any $i\in\{1,\cdots,m\}$, the collection of laws formed by the ith coordinate of the $(\boldsymbol{X}^n)_{n\ge 0}$ is tight on $\mathcal{M}([0,T];\mathbb{R})$.

As usual with stochastic processes, the difficult part in Proposition 3.7 is to control the increments appearing in the second condition. This is precisely where the theory of martingales comes in. There exists a quite simple criterion for tightness of semi-martingales, based on the following definition:

Definition 3.8 *Let $(X = (X_t)_{0 \le t \le T})$ be a real-valued stochastic process adapted to a complete and right-continuous filtration $\mathbb{F} = (\mathcal{F}_t)_{0 \le t \le T}$ on a complete probability space $(\Omega, \mathcal{F}, \mathbb{P})$. Assume that the paths of X are in $\mathcal{D}([0,T];\mathbb{R})$ and that, for any $t \in [0,T]$, X_t is integrable. Then, X is said to have a finite conditional variation on $[0,T]$ if the quantity:*

$$\mathcal{V}_T(X) = \sup_{N \ge 1} \sup_{0=t_0<\cdots<t_N=T} \mathbb{E}\Big[\sum_{i=0}^{N-1} \big|\mathbb{E}[X_{t_{i+1}} - X_{t_i} | \mathcal{F}_{t_i}]\big|\Big] \tag{3.10}$$

is finite, in which case we call $\mathcal{V}_T(X)$ the conditional variation of X (with respect to $\mathbb{F}$).

Of course, whenever X is a martingale, the conditional variation is null, which is the key observation for understanding the interest of the following tightness criterion.

Theorem 3.9 *Let $(X^n = (X^n_t)_{0 \le t \le T})_{n \ge 0}$ be a sequence of càd-làg real-valued adapted processes, each X^n being defined on a complete probability space $(\Omega^n, \mathcal{F}^n, \mathbb{P}^n)$ equipped with a right-continuous and complete filtration $\mathbb{F}^n$. Denoting by $\mathcal{V}^n_T(X^n)$ the conditional variation of X^n with respect to $\mathbb{F}^n$ under $\mathbb{P}^n$, and assuming that:*

$$\sup_{n \ge 0} \big[\mathbb{E}^n\big(|X^n_T|\big) + \mathcal{V}^n_T(X^n)\big] < \infty,$$

then the collection of distributions induced by the $(X^n)_{n \ge 0}$'s is tight on $\mathcal{M}([0,T];\mathbb{R})$ and any limit reads as the distribution induced by a càd-làg process on $\mathcal{M}([0,T];\mathbb{R})$.

3.2.3 Back to MFGs with Common Noise and Main Result

We now return to our original motivation: the stability of weak equilibria for MFGs with a common noise.

Sequence of Stochastic Optimal Control Problems

In order to make things clear, we start with an optimal stochastic control problem in a random environment of the type considered in Chapter 1, see for instance the presentation in Subsection 1.4.1. However, in contrast with the framework of Chapter 1, we shall not consider a single environment $\boldsymbol{\mu}$ but instead, a collection of environments $(\boldsymbol{\mu}^n)_{n\geq 0}$. Then, we shall address the question of the asymptotic behavior of the optimal paths associated with each $\boldsymbol{\mu}^n$, $n \geq 0$, as n tends to ∞.

For each $n \geq 0$, we thus consider a complete probability space $(\Omega^n, \mathcal{F}^n, \mathbb{P}^n)$ equipped with a right-continuous and complete filtration $\mathbb{F}^n$, and also a four-tuple $(X_0^n, \boldsymbol{W}^{0,n}, \boldsymbol{\mu}^n, \boldsymbol{W}^n)$, X_0^n being $\mathcal{F}_0^n$ measurable, the pair $(\boldsymbol{W}^{0,n}, \boldsymbol{W}^n)$ forming a $2d$-Brownian motion with respect to the filtration $\mathbb{F}^n$, the process $\boldsymbol{\mu}^n$ having paths in $\mathcal{D}([0,T];\mathcal{X})$ for some Polish space $(\mathcal{X}, d_{\mathcal{X}})$ and satisfying $\mathbb{E}^n[\sup_{0\leq t\leq T} d_{\mathcal{X}}(0_{\mathcal{X}}, \mu_t^n)^2] < \infty$, and the triple $(X_0^n, \boldsymbol{W}^{0,n}, \boldsymbol{\mu}^n)$ being independent of $\boldsymbol{W}^n$. Importantly, we also require that $\mathbb{F}^n$ is compatible with the complete and right-continuous filtration generated by $(X_0^n, \boldsymbol{W}^{0,n}, \boldsymbol{\mu}^n, \boldsymbol{W}^n)$, see Subsection 1.1.1.

We are also given a closed convex subset $A \subset \mathbb{R}^k$, for some $k \geq 1$, together with coefficients $(b^n, \sigma^n, \sigma^{0,n}, f^n, g^n)_{n\geq 0}$ satisfying:

Assumption (Sequence of Optimization Problems). For any integer $n \geq 0$, the coefficients $b^n : [0,T] \times \mathbb{R}^d \times \mathcal{X} \times A \to \mathbb{R}^d$, $\sigma^n : [0,T] \times \mathbb{R}^d \times \mathcal{X} \to \mathbb{R}^{d\times d}$, $\sigma^{0,n} : [0,T] \times \mathbb{R}^d \times \mathcal{X} \to \mathbb{R}^{d\times d}$, $f^n : [0,T] \times \mathbb{R}^d \times \mathcal{X} \times A \to \mathbb{R}$ and $g^n : \mathbb{R}^d \times \mathcal{X} \to \mathbb{R}$ are Borel-measurable. Moreover, there exists a constant $L \geq 0$ such that, for any integer $n \geq 0$,

(A1) b^n, σ^n and $\sigma^{0,n}$ are L-Lipschitz continuous in x, uniformly in $t \in [0,T]$, $\mu \in \mathcal{X}$ and $\alpha \in A$.

(A2) b^n, σ^n and $\sigma^{0,n}$ are at most of L-linear growth in (x, μ, α), uniformly in $t \in [0,T]$, *i.e.* for all $t \in [0,T]$, $x \in \mathbb{R}^d$, $\mu \in \mathcal{X}$ and $\alpha \in A$,

$$|b^n(t,x,\mu,\alpha)| \leq L\big(1 + |x| + d_{\mathcal{X}}(0_{\mathcal{X}}, \mu) + |\alpha|\big),$$

$$|(\sigma^n, \sigma^{0,n})(t,x,\mu)| \leq L\big(1 + |x| + d_{\mathcal{X}}(0_{\mathcal{X}}, \mu)\big),$$

where $0_{\mathcal{X}}$ is some arbitrary point in $\mathcal{X}$.

(A3) f^n and g^n are at most of L-quadratic growth in (x, μ, α), uniformly in $t \in [0,T]$, i.e. for all $t \in [0,T]$, $x \in \mathbb{R}^d$, $\mu \in \mathcal{X}$ and $\alpha \in A$,

$$|f^n(t,x,\mu,\alpha)| \leq L\big(1 + |x|^2 + d_{\mathcal{X}}(0_{\mathcal{X}}, \mu)^2 + |\alpha|^2\big),$$

$$|g^n(x,\mu)| \leq L\big(1 + |x|^2 + d_{\mathcal{X}}(0_{\mathcal{X}}, \mu)^2\big).$$

We shall also assume:

(A4) The sequence of coefficients $(b^n, \sigma^n, \sigma^{0,n}, f^n, g^n)_{n\geq 0}$ converge to some $(b, \sigma, \sigma^0, f, g)$, uniformly on any compact subset of the underlying

(continued)

domain. For any $t \in [0, T]$, the functions $b(t, \cdot, \cdot, \cdot)$ and $f(t, \cdot, \cdot, \cdot)$ are continuous on $\mathbb{R}^d \times \mathcal{X} \times A$ (with values in $\mathbb{R}^d$ and $\mathbb{R}$ respectively), the functions $\sigma(t, \cdot, \cdot)$ and $\sigma^0(t, \cdot, \cdot)$ are continuous on $\mathbb{R}^d \times \mathcal{X}$ with values in $\mathbb{R}^{d\times d}$, and the function g is continuous on $\mathbb{R}^d \times \mathcal{X}$ with values in $\mathbb{R}$.

It is worth noting that the bounds on the coefficients $(b^n, \sigma^n, \sigma^{0,n}, f^n, g^n)_{n\geq 0}$ survive the limit $n \to \infty$ and still hold for the coefficients $(b, \sigma, \sigma^0, f, g)$. Now, for each integer $n \geq 0$, we may consider, on the filtered probability space $(\Omega^n, \mathcal{F}^n, \mathbb{F}^n, \mathbb{P}^n)$, controlled processes of the form:

$$dX_t = b^n\big(t, X_t, \mu_t^n, \alpha_t\big)dt + \sigma^n(t, X_t, \mu_t^n)dW_t^n + \sigma^{0,n}(t, X_t, \mu_t^n)dW_t^{0,n}, \tag{3.11}$$

for $t \in [0, T]$, with the same X_0^n as above as initial condition, where $(\alpha_t)_{0\leq t\leq T}$ is an $\mathbb{F}^n$-progressively measurable process with values in A such that:

$$\mathbb{E}^n \int_0^T |\alpha_t|^2 dt < \infty,$$

$\mathbb{E}^n$ standing for the expectation with respect to $\mathbb{P}^n$. To $\boldsymbol{\alpha} = (\alpha_t)_{0\leq t\leq T}$, we associate the cost:

$$J^{n,\mu^n}(\boldsymbol{\alpha}) = \mathbb{E}^n\bigg[\int_0^T f^n(s, X_s, \mu_s^n, \alpha_s)ds + g^n(X_T, \mu_T^n)\bigg], \tag{3.12}$$

which depends upon n, not only through the environment $\boldsymbol{\mu}^n$, but also through the coefficients $(b^n, \sigma^n, \sigma^{0,n}, f^n, g^n)$. We then assume that, for each $n \in \mathbb{N}$, J^{n,μ^n} has a unique minimizer $\hat{\boldsymbol{\alpha}}^n = (\hat{\alpha}_t^n)_{0\leq t\leq T}$. We refer to Chapter 1 for conditions under which such an optimizer exists and is unique. The question is then to investigate the asymptotic behavior of $(\hat{\boldsymbol{\alpha}}^n)_{n\geq 0}$ whenever the sequence of triples $((X_0^n, \boldsymbol{W}^{0,n}, \boldsymbol{\mu}^n))_{n\geq 0}$ converges in the weak sense on $\mathbb{R}^d \times \mathcal{C}([0, T]; \mathbb{R}^d) \times \mathcal{D}([0, T]; \mathcal{X})$ as n tends to ∞.

In order to do so, we shall assume that the necessary condition in the stochastic Pontryagin principle holds, see Theorem 1.59. Following (Vol I)-(3.5) and (1.32), we introduce, for any $n \geq 0$, the reduced Hamiltonian:

$$H^{(r),n}\big(t, x, \mu, y, \alpha\big) = b^n(t, x, \mu, \alpha) \cdot y + f^n(t, x, \mu, \alpha), \tag{3.13}$$

for $t \in [0, T]$, $x \in \mathbb{R}^d$, $\mu \in \mathcal{X}$, $y \in \mathbb{R}^d$, and $\alpha \in A$. A minimizer, if and when it exists, then reads:

$$\hat{\alpha}^n\big(t, x, \mu, y\big) \in \text{argmin}_{\alpha\in A} H^{(r),n}\big(t, x, \mu, y, \alpha\big). \tag{3.14}$$

In order to implement the stochastic Pontryagin principle, we make the following assumption:

Assumption (Sequence of Necessary SMP). There exists a constant $L > 0$ such that:

(A1) For any integer $n \geq 0$, the drift b^n has the form:

$$b^n(t, x, \mu, \alpha) = b_1^n(t, x, \mu) + b_2^n(t)\alpha,$$

where the mapping $[0, T] \ni t \mapsto b_2^n(t) \in \mathbb{R}^{d\times k}$ is measurable and bounded by L.

(A2) For any integer $n \geq 0$, the functions b_1^n and f^n are differentiable with respect to x and (x, α) respectively, the mappings $\mathbb{R}^d \ni x \mapsto \partial_x b_1^n(t, x, \mu)$, $\mathbb{R}^d \times A \ni (x, \alpha) \mapsto \partial_x f^n(t, x, \mu, \alpha)$ and $\mathbb{R}^d \times A \ni (x, \alpha) \mapsto \partial_\alpha f^n(t, x, \mu, \alpha)$ being continuous for each $(t, \mu) \in [0, T] \times \mathcal{X}$. Similarly, the functions σ^n, $\sigma^{0,n}$ and g^n are differentiable with respect to x, the function $\mathbb{R}^d \ni x \mapsto \partial_x(\sigma^n, \sigma^{0,n})(t, x, \mu)$ being continuous for each $(t, \mu) \in [0, T] \times \mathcal{X}$, and the function $\mathbb{R}^d \ni x \mapsto \partial_x g^n(x, \mu)$ being continuous for each $\mu \in \mathcal{X}$.

(A3) For any integer $n \geq 0$, any $R \geq 0$, and any (t, x, μ, α) with $|x| \leq R$, $d_{\mathcal{X}}(0_{\mathcal{X}}, \mu) \leq R$ and $|\alpha| \leq R$, $|\partial_x f^n(t, x, \mu, \alpha)|$, $|\partial_\alpha f^n(t, x, \mu, \alpha)|$ and $|\partial_x g^n(x, \mu)|$ are bounded by $L(1 + R)$. Moreover, the function $\partial_\alpha f^n$ is L-Lipschitz-continuous in x.

(A4) For any integer $n \geq 0$, f^n satisfies the convexity property:

$$\begin{aligned} f^n(t, x, \mu, \alpha') - f^n(t, x, \mu, \alpha) - (\alpha' - \alpha) \cdot \partial_\alpha f^n(t, x, \mu, \alpha) \\ \geq L^{-1}|\alpha' - \alpha|^2, \end{aligned}$$

for all $(t, x, \mu, \alpha, \alpha') \in [0, T] \times \mathbb{R}^d \times \mathcal{X} \times A \times A$.

Following **(A4)** in assumption **Sequence of Optimization Problems**, we shall also assume:

(A5) For any $(t, x, \mu) \in [0, T] \times \mathbb{R}^d \times \mathcal{X}$, the function $A \ni \alpha \mapsto f(t, x, \mu, \alpha)$ is continuously differentiable and the coefficients $(\partial_\alpha f^n)_{n\geq 0}$ converge to $\partial_\alpha f$, uniformly on compact subsets.

Implicitly, **(A5)** needs **(A4)** in assumption **Sequence of Optimization Problems**, as otherwise f would not make sense. Below, we never assume that assumption

Sequence of Necessary SMP holds true without demanding assumption **Sequence of Optimization Problems** to hold true as well. Importantly, the bounds we have on the coefficients $(\partial_\alpha f^n)_{n\geq 0}$ pass to the limiting coefficient $\partial_\alpha f$. In particular, f is convex in α.

Of course, the rationale for **(A1)** and **(A4)** is to guarantee that the Hamiltonian $H^{(r),n}$ in (3.13) are strictly convex with respect to α, uniformly in n. By Lemmas (Vol I)-3.3 and 1.56 (see also Lemma (Vol I)-4.43), the minimizer $\hat{\alpha}^n$ of $H^{(r),n}$ in (3.14) is uniquely defined. Similarly, we may associate the Hamiltonian $H^{(r)}$ with the limiting coefficients (b, σ, σ^0, f) through the analogue of (3.13). We then call $\hat{\alpha}$ the unique minimizer of $H^{(r)}$.

We then claim:

Lemma 3.10 *Under assumptions* **Sequence of Necessary SMP** *and* **Sequence of Optimization Problems**, *the sequence of minimizers* $(\hat{\alpha}^n)_{n\geq 0}$ *converges to* $\hat{\alpha}$ *uniformly on compact subsets of* $[0,T]\times\mathbb{R}^d\times\mathcal{X}\times\mathbb{R}^d$. *Moreover, for any* $t\in[0,T]$, *the function* $\mathbb{R}^d\times\mathcal{X}\times\mathbb{R}^d \ni (x,\mu,y)\mapsto \hat{\alpha}(t,x,\mu,y)$ *is continuous and there exists a constant* C *such that, for any* $n\geq 0$, *any* $t\in[0,T]$, *any* $x,x',y,y'\in\mathbb{R}^d$ *and any* $\mu\in\mathcal{X}$,

$$\begin{aligned}
&\big|\hat{\alpha}^n(t,x,\mu,y)\big| \leq C\big(1+|x|+d_{\mathcal{X}}(0_{\mathcal{X}},\mu)+|y|\big),\\
&\big|\hat{\alpha}^n(t,x',\mu,y')-\hat{\alpha}^n(t,x,\mu,y)\big| \leq C\big(|x'-x|+|y'-y|\big).
\end{aligned} \tag{3.15}$$

By letting n tend to ∞, observe that (3.15) is also satisfied by the limiting minimizer $\hat{\alpha}$.

Proof.

First Step. Following the proof of Lemma (Vol I)-3.3, we have, for any integer $n\geq 0$ and any $(t,x,\mu,y)\in[0,T]\times\mathbb{R}^d\times\mathcal{X}\times\mathbb{R}^d$,

$$\begin{aligned}
&\big(\hat{\alpha}^n(t,x,\mu,y)-\hat{\alpha}(t,x,\mu,y)\big)\\
&\quad\cdot\Big(\partial_\alpha H^{(r),n}\big(t,x,\mu,y,\hat{\alpha}^n(t,x,\mu,y)\big)-\partial_\alpha H^{(r)}\big(t,x,\mu,y,\hat{\alpha}(t,x,\mu,y)\big)\Big)\leq 0.
\end{aligned}$$

Using the convexity of $H^{(r),n}$ and following once again the proof of Lemma (Vol I)-3.3, we deduce that:

$$L^{-1}\big|\hat{\alpha}^n(t,x,\mu,y)-\hat{\alpha}(t,x,\mu,y)\big| \leq \big|\big(\partial_\alpha H^{(r)}-\partial_\alpha H^{(r),n}\big)\big(t,x,\mu,y,\hat{\alpha}(t,x,\mu,y)\big)\big|.$$

By assumption, the right-hand side tends to 0, which proves the first claim in the statement.

Second Step. The claim regarding the growth and the (x,y)-Lipschitz continuity of the $(\hat{\alpha}^n)_{n\geq 0}$'s is a direct consequence of Lemma (Vol I)-3.3. The continuity of $\hat{\alpha}(t,\cdot)$, for each $t\in[0,T]$, may be proved by a standard compactness argument. □

Assumption (Control Bounds). With the same notation as above for the sequence of filtered probability spaces $((\Omega^n, \mathcal{F}^n, \mathbb{F}^n, \mathbb{P}^n))_{n\geq 0}$, each of them being equipped with a compatible input $(X_0^n, \boldsymbol{W}^{0,n})$, it holds that:

(A1) The sequences of probability measures $(\mathbb{P}^n \circ (\sup_{0\leq t\leq T} |X_t^n|)^{-1})_{n\geq 0}$ and $(\mathbb{P}^n \circ (\sup_{0\leq t\leq T} d(0_{\mathcal{X}}, \mu_t^n))^{-1})_{n\geq 0}$ are uniformly square-integrable. In particular,

$$\sup_{n\geq 0} \mathbb{E}\Big[\sup_{0\leq t\leq T} |X_t^n|^2 + \sup_{0\leq t\leq T} \big(d_{\mathcal{X}}(0_{\mathcal{X}}, \mu_t^n)^2\big)\Big] < \infty.$$

Moreover, the sequence of probability measures $(\mathbb{P}^n \circ (\boldsymbol{\mu}^n)^{-1})_{n\geq 0}$ is tight on $\mathcal{D}([0,T],\mathcal{X})$ when equipped with the $J1$ Skorohod topology.

(A2) For any $n \geq 0$, there exists a minimizer $\hat{\boldsymbol{\alpha}}^n = (\hat{\alpha}_t^n)_{0\leq t\leq T}$ for the cost functional $J^{n,\boldsymbol{\mu}^n}$ such that:

$$\sup_{n\geq 0} \mathbb{E}^n\Big[\int_0^T |\hat{\alpha}^n|^2 dt\Big] < \infty.$$

In particular, for any $n \geq 0$, for any $\mathbb{F}^n$-progressively measurable and square-integrable control $(\boldsymbol{\beta} = (\beta_t))_{0\leq t\leq T}$, the following holds true:

$$J^{n,\boldsymbol{\mu}^n}(\hat{\boldsymbol{\alpha}}^n) \leq J^{n,\boldsymbol{\mu}^n}(\boldsymbol{\beta}).$$

In (**A1**) above, we used the following standard definition: A sequence $(m^n)_{n\geq 0}$ of probabilities measures on $(\mathbb{R}, \mathcal{B}(\mathbb{R}))$ is said to be uniformly p-integrable, for a real $p > 0$, if

$$\lim_{a\to\infty} \sup_{n\geq 0} \int_{|x|\geq a} |x|^p dm^n(x) = 0.$$

Also, the minimizers $(\hat{\boldsymbol{\alpha}}^n)_{n\geq 0}$ are unique if assumption **FBSDE** in Chapter 2 or assumption **FBSDE MFG with a Common Noise** is in force.

Here is now the first real result of this section.

Proposition 3.11 *Under assumptions* **Sequence of Optimization Problems**, **Sequence of Necessary SMP**, *and* **Control Bounds**, *for each* $n \geq 0$, *denote by* $\boldsymbol{X}^n = (X_t^n)_{0\leq t\leq T}$ *the optimally controlled process driven by* $\hat{\boldsymbol{\alpha}}^n$, *namely the solution of* (3.11) *with* $\boldsymbol{\alpha} = \hat{\boldsymbol{\alpha}}^n$. *Then, the sequence* $(\mathbb{P}^n \circ (\boldsymbol{X}^n)^{-1})_{n\geq 0}$ *is tight on* $\mathcal{C}([0,T];\mathbb{R}^d)$. *In particular, the sequence* $(\mathbb{P}^n \circ (X_0^n, \boldsymbol{W}^{0,n}, \boldsymbol{\mu}^n, \boldsymbol{W}^n, \boldsymbol{X}^n)^{-1})_{n\geq 0}$ *is tight on the space* $\Omega_{input} \times \mathcal{C}([0,T];\mathbb{R}^d)$.

Moreover, if $(X_0^\infty, \boldsymbol{W}^{0,\infty}, \boldsymbol{\mu}^\infty, \boldsymbol{W}^\infty, \boldsymbol{X}^\infty)$ *is an* $\Omega_{input} \times \mathcal{C}([0,T];\mathbb{R}^d)$*-valued process on a complete probability space* $(\Omega^\infty, \mathcal{F}^\infty, \mathbb{P}^\infty)$ *such that the probability measure* $\mathbb{P}^\infty \circ (X_0^\infty, \boldsymbol{W}^{0,\infty}, \boldsymbol{\mu}^\infty, \boldsymbol{W}^\infty, \boldsymbol{X}^\infty)^{-1}$ *is a weak limit of the sequence* $(\mathbb{P}^n \circ (X_0^n, \boldsymbol{W}^{0,n}, \boldsymbol{\mu}^n, \boldsymbol{W}^n, \boldsymbol{X}^n)^{-1})_{n\geq 0}$*, we can associate with the complete and right-continuous filtration* $\mathbb{F}^\infty$ *generated by* $(X_0^\infty, \boldsymbol{W}^{0,\infty}, \boldsymbol{\mu}^\infty, \boldsymbol{W}^\infty, \boldsymbol{X}^\infty)$*, the stochastic control problem given by the cost functional:*

$$J^{\boldsymbol{\mu}^\infty}(\boldsymbol{\beta}) = \mathbb{E}^\infty\left[\int_0^T f(s, X_s, \mu_s^\infty, \beta_s)ds + g(X_T, \mu_T^\infty)\right], \tag{3.16}$$

under the dynamic constraint:

$$dX_t = b\big(t, X_t, \mu_t^\infty, \beta_t\big)dt + \sigma(t, X_t, \mu_t^\infty)dW_t^\infty + \sigma^0(t, X_t, \mu_t^\infty)dW_t^{0,\infty}, \tag{3.17}$$

for $t \in [0,T]$ *with* X_0^∞ *as initial condition, for* $\mathbb{F}^\infty$*-progressively measurable square-integrable control processes* $\boldsymbol{\beta} = (\beta_t)_{0\leq t\leq T}$ *with values in* A.

If the filtration $\mathbb{F}^\infty$ *is compatible with the process* $(X_0^\infty, \boldsymbol{W}^{0,\infty}, \boldsymbol{\mu}^\infty, \boldsymbol{W}^\infty)$*, then the process* $\boldsymbol{X}^\infty$ *is an optimal path for the control problem* (3.16)–(3.17) *when considered on the compatible set-up* $(\Omega^\infty, \mathcal{F}^\infty, \mathbb{F}^\infty, \mathbb{P}^\infty)$ *equipped with the input* $(X_0^\infty, \boldsymbol{W}^{0,\infty}, \boldsymbol{\mu}^\infty, \boldsymbol{W}^\infty)$.

Recall that:

$$\Omega_{\text{input}} = \mathbb{R}^d \times \mathcal{C}([0,T];\mathbb{R}^d) \times \mathcal{D}([0,T];\mathcal{X}) \times \mathcal{C}([0,T];\mathbb{R}^d), \tag{3.18}$$

$\mathcal{C}([0,T];\mathbb{R}^d)$ being equipped with the topology of the uniform convergence, and $\mathcal{D}([0,T];\mathcal{X})$ with the $J1$ Skorohod topology. The proof of Proposition 3.11 is deferred to Subsection 3.2.4 below.

Application to MFGs

One serious difficulty when we try to identify the limit process $\boldsymbol{X}^\infty$ with an optimal path for the cost functional $J^{\boldsymbol{\mu}^\infty}$ is to check the compatibility condition. Fortunately, this compatibility is given for free when dealing with weak solutions of MFG with a common noise, if we lift each environment $\boldsymbol{\mu}^n$ –seen as *càd-làg* processes with values in $\mathcal{P}_2(\mathbb{R}^d)$– into a random variable $\mathfrak{M}^n$ with values in $\mathcal{P}_2(\mathcal{C}([0,T];\mathbb{R}^{2d}))$, according to the lifting procedure described in Definition 2.16. Compatibility is then understood as compatibility between $\mathbb{F}^n$ and the lifted process $(X_0^n, \boldsymbol{W}^{0,n}, \mathfrak{M}^n, \boldsymbol{W}^n)$. **This key observation is the rationale for the lifting procedure underpinning the definition of an MFG equilibrium.**

Here is a first statement that reflects this idea.

Proposition 3.12 *Under the assumptions of Proposition 3.11, if* $\mathcal{X} = \mathcal{P}_2(\mathbb{R}^d)$*, and if for any* $n \in \mathbb{N}$*, there exists a random variable* $\mathfrak{M}^n$ *from* $(\Omega^n, \mathcal{F}^n)$ *into* $\mathcal{P}_2(\mathcal{C}([0,T];\mathbb{R}^{2d}))$ *such that:*

1. *the filtration* $\mathbb{F}^n$ *is compatible with* $(X_0^n, \boldsymbol{W}^{0,n}, \mathfrak{M}^n, \boldsymbol{W}^n)$ *in the sense of Definition 2.16,*
2. *for any* $t \in [0,T]$, μ_t^n *is* $\mathbb{P}^n$*-almost surely equal to* $\mathfrak{M}^n \circ (e_t^x)^{-1}$, *where* e_t^x *is the evaluation map providing the first d coordinates at time t on* $\mathcal{C}([0,T];\mathbb{R}^{2d})$,
3. *the conditional law of* $(\boldsymbol{X}^n, \boldsymbol{W}^n)$ *given* $(\boldsymbol{W}^{0,n}, \mathfrak{M}^n)$ *is* $\mathfrak{M}^n$,

then, the sequence $(\mathbb{P}^n \circ (\mathfrak{M}^n)^{-1})_{n\geq 0}$ *is tight on* $\mathcal{P}_2(\mathcal{C}([0,T];\mathbb{R}^{2d}))$.

Moreover, on any complete probability space $(\Omega^\infty, \mathcal{F}^\infty, \mathbb{P}^\infty)$, *equipped with a random process* $(X_0^\infty, \boldsymbol{W}^{0,\infty}, \mathfrak{M}^\infty, \boldsymbol{W}^\infty, \boldsymbol{X}^\infty)$ *such that the probability measure* $\mathbb{P}^\infty \circ (X_0^\infty, \boldsymbol{W}^{0,\infty}, \mathfrak{M}^\infty, \boldsymbol{W}^\infty, \boldsymbol{X}^\infty)^{-1}$ *is a weak limit of the sequence* $(\mathbb{P}^n \circ (X_0^n, \boldsymbol{W}^{0,n}, \mathfrak{M}^n, \boldsymbol{W}^n, \boldsymbol{X}^n)^{-1})_{n\geq 0}$, *the complete and right-continuous filtration* $\mathbb{F}^\infty$ *generated by* $(X_0^\infty, \boldsymbol{W}^{0,\infty}, \mathfrak{M}^\infty, \boldsymbol{W}^\infty, \boldsymbol{X}^\infty)$ *is compatible with the process* $(X_0^\infty, \boldsymbol{W}^{0,\infty}, \mathfrak{M}^\infty, \boldsymbol{W}^\infty)$. *In particular,* $\boldsymbol{X}^\infty$ *is an optimal path for the stochastic optimal control problem* (3.16)–(3.17), *when it is understood for the filtration* $\mathbb{F}^\infty$, *the super-environment* $\mathfrak{M}^\infty$, *and the sub-environment* $\boldsymbol{\mu}^\infty = (\mu_t^\infty = \mathfrak{M}^\infty \circ (e_t^x)^{-1})_{0\leq t\leq T}$.

Observe that, in the above statement, we completely disregard the weak convergence of the sequence $(\boldsymbol{\mu}^n)_{n\geq 0}$ and the measurability (or compatibility) properties of any weak limit $\boldsymbol{\mu}^\infty$. The reason is that the functional that maps $\mathfrak{M}^n$ (for $n \in \mathbb{N} \cup \{\infty\}$) onto $\boldsymbol{\mu}^n$ according to the principle stated in Definition 2.16 is continuous. Indeed, the mapping $\boldsymbol{e} : \mathcal{P}_2(\mathcal{C}([0,T];\mathbb{R}^{2d})) \ni \mathrm{m} \mapsto (\mathrm{m} \circ (e_t^x)^{-1})_{0\leq t\leq T} \in \mathcal{C}([0,T], \mathcal{P}_2(\mathbb{R}^d))$, which maps each $\mathfrak{M}^n$ onto $\boldsymbol{\mu}^n$ is 1-Lipschitz continuous, since, for any two m and m' in $\mathcal{P}_2(\mathcal{C}([0,T];\mathbb{R}^{2d}))$, it holds:

$$
\begin{aligned}
\sup_{0\leq t\leq T} W_2\big((\boldsymbol{e}(\mathrm{m}))_t, (\boldsymbol{e}(\mathrm{m}'))_t\big) &= \sup_{0\leq t\leq T} W_2\big(\mathrm{m} \circ (e_t^x)^{-1}, \mathrm{m}' \circ (e_t^x)^{-1}\big) \\
&\leq W_2\big(\mathrm{m}, \mathrm{m}'\big),
\end{aligned}
$$

where, in the first two arguments, W_2 is the 2-Wasserstein distance on $\mathcal{P}_2(\mathbb{R}^d)$ whereas it denotes the 2-Wasserstein on $\mathcal{P}_2(\mathcal{C}([0,T];\mathbb{R}^{2d}))$ on the second line.

A quick comparison with Definition 2.16 shows that the three items in the statement of Proposition 3.12 are characteristic features of an MFG equilibrium. This says that we should not be far from our original purpose, namely proving a stability property for weak solutions to MFG problems. However, in order to do so, we must fit the framework introduced in Chapter 2, especially that of Definition 2.16. For this reason, we now require each Ω^n to be of the product form $\Omega^{0,n} \times \Omega^{1,n}$, where $(\Omega^{0,n}, \mathcal{F}^{0,n}, \mathbb{F}^{0,n}, \mathbb{P}^{0,n})$ and $(\Omega^{1,n}, \mathcal{F}^{1,n}, \mathbb{F}^{1,n}, \mathbb{P}^{1,n})$ are two complete filtered probability spaces, the filtrations being complete and right-continuous. We then define $(\Omega^n, \mathcal{F}^n, \mathbb{P}^n)$ as the completion of the product space $(\Omega^{0,n} \times \Omega^{1,n}, \mathcal{F}^{0,n} \otimes \mathcal{F}^{1,n}, \mathbb{P}^{0,n} \otimes \mathbb{P}^{1,n})$ and $\mathbb{F}^n$ as the complete and right-continuous augmentation of the product of the two filtrations. The Brownian motions $\boldsymbol{W}^{0,n}$ and $\boldsymbol{W}^n$ are assumed to be constructed on $(\Omega^{0,n}, \mathcal{F}^{0,n}, \mathbb{F}^{0,n}, \mathbb{P}^{0,n})$ and $(\Omega^{1,n}, \mathcal{F}^{1,n}, \mathbb{F}^{1,n}, \mathbb{P}^{1,n})$ respectively.

In order to guarantee a similar decomposition in the limiting setting, we shall appeal to assumption **FBSDE** introduced in Chapter 2, as made clear by the main result of this section.

Theorem 3.13 *On top of assumptions* **Sequence of Optimization Problems** *and* **Sequence of Necessary SMP** *right above and* **FBSDE** *in Chapter 2, assume further that for any $n \geq 0$, there exists a random variable $\mathfrak{M}^n$ from $(\Omega^{n,0}, \mathcal{F}^{n,0})$ into $\mathcal{P}_2(\mathcal{C}([0,T];\mathbb{R}^{2d}))$ such that the filtration $\mathbb{F}^n$ is compatible with the process $(X_0^n, \boldsymbol{W}^{0,n}, \mathfrak{M}^n, \boldsymbol{W}^n)$.*

Letting $\boldsymbol{\mu}^n = (\mu_t^n = \mathfrak{M}^n \circ (e_t^x)^{-1})_{0\leq t\leq T}$, assume that for each $n \geq 0$, there exists an optimal control $\hat{\boldsymbol{\alpha}}^n = (\hat{\alpha}_t^n)_{0\leq t\leq T}$, with $\boldsymbol{X}^n = (X_t^n)_{0\leq t\leq T}$ as associated optimal path, to the optimal stochastic control problem (3.11)–(3.12)*, such that $\mathfrak{M}^n$ coincides with $\mathcal{L}^1(\boldsymbol{X}^n, \boldsymbol{W}^n)$ with $\mathbb{P}^{0,n}$-probability* 1.

If the resulting sequence $(\hat{\boldsymbol{\alpha}}^n)_{n\geq 0}$ satisfies (**A2**) *in assumption* **Control Bounds** *and the measures $(\mathbb{P}^n \circ (\sup_{0\leq t\leq T}|X_t^n|)^{-1})_{n\geq 0}$ are uniformly square-integrable, then the family of probability measures $(\mathbb{P}^{0,n} \circ (\mu_0^n, \boldsymbol{W}^{0,n}, \mathfrak{M}^n)^{-1})_{n\geq 0}$ is tight on $\mathcal{P}_2(\mathbb{R}^d) \times \mathcal{C}([0,T];\mathbb{R}^d) \times \mathcal{P}_2(\mathcal{C}([0,T];\mathbb{R}^{2d}))$ and any weak limit generates a weak solution in the sense of Definition 2.23 and a distribution of an equilibrium in the sense of Definition 2.24 (for the problem driven by the coefficients (b,σ,σ^0,f,g)).*

Obviously, for any $n \in \mathbb{N}$, $\mathfrak{M}^n$, as defined in the statement, is an MFG equilibrium.

3.2.4 Proof of the Weak Stability of Optimal Paths

We first prove Proposition 3.11.

Throughout the section, we let the assumptions of Proposition 3.11 be in force, namely assumptions **Sequence of Optimization Problems**, **Sequence of Necessary SMP** and **Control Bounds**. We shall also use the same notation as in the statement. The proof of Proposition 3.11 comprises several lemmas. Lemma 3.14 below establishes the first part of the statement in Proposition 3.11.

Lemma 3.14 *The family $(\mathbb{P}^n \circ (\boldsymbol{X}^n)^{-1})_{n\geq 0}$ is tight on $\mathcal{C}([0,T];\mathbb{R}^d)$ equipped with the topology of uniform convergence. Moreover, the sequence $(\mathbb{P}^n \circ (\hat{\boldsymbol{\alpha}}^n)^{-1})_{n\geq 0}$, with $\hat{\boldsymbol{\alpha}}^n = (\hat{\alpha}_t^n)_{t\in[0,T]}$ for each $n \geq 0$, is tight on $\mathcal{M}([0,T];\mathbb{R}^k)$ equipped with the Meyer-Zheng topology. Any weak limit of $(\mathbb{P}^n \circ (\hat{\boldsymbol{\alpha}}^n)^{-1})_{n\geq 0}$ may be regarded as the law of an A-valued process.*

Proof.

First Step. We start with the following observation. By Theorem 1.59, we know that, for every $n \geq 0$, the optimal control $(\hat{\alpha}_t^n)_{0\leq t\leq T}$ associated with the optimal control problem (3.11)–(3.12) can be expressed through the backward component of the following forward-backward system:

$$\begin{cases} dX_t^n = b^n\big(t, X_t^n, \mu_t^n, \hat{\alpha}_t^n\big)dt \\ \qquad +\sigma^n(t, X_t^n, \mu_t^n)dW_t^n + \sigma^{0,n}(t, X_t^n, \mu_t^n)dW_t^{0,n}, \\ dY_t^n = -\partial_x H^n\big(t, X_t^n, \mu_t^n, Y_t^n, Z_t^n, Z_t^{0,n}, \hat{\alpha}_t^n\big)dt \\ \qquad +Z_t^n dW_t^n + Z_t^{0,n} dW_t^{0,n} + dM_t^n, \quad t \in [0, T], \end{cases} \tag{3.19}$$

with the terminal condition $Y_T^n = \partial_x g^n(X_T^n, \mu_T^n)$, where H^n is the full-fledged Hamiltonian:

$$\begin{aligned} H^n\big(t, x, \mu, y, z, z^0, \alpha\big) &= b^n(t, x, \mu, \alpha) \cdot y + f^n(t, x, \mu, \alpha) \\ &\quad + \text{trace}\big[\sigma^n(t, x, \mu)z^\dagger + \sigma^{0,n}(t, x, \mu)(z^0)^\dagger\big], \end{aligned}$$

for $t \in [0, T]$, $x \in \mathbb{R}^d$, $\mu \in \mathcal{X}$, $\alpha \in A$, $y \in \mathbb{R}^d$, and $z, z^0 \in \mathbb{R}^{d\times d}$.

The above equation (3.19) is set on the probabilistic set-up $(\Omega^n, \mathbb{F}^n, \mathbb{P}^n)$, the filtration $\mathbb{F}^n$ being compatible with $(X_0^n, \boldsymbol{W}^{0,n}, \boldsymbol{\mu}^n, \boldsymbol{W}^n)$. Recall that $\boldsymbol{M}^n = (M_t^n)_{0\le t\le T}$ denotes a square-integrable martingale with respect to $\mathbb{F}^n$ with 0 as initial condition and of zero bracket with $(\boldsymbol{W}^{0,n}, \boldsymbol{W}^n)$. We refer to Chapter 1 for a complete account.

Taking the square and then the expectation in the backward equation, using (**A2**) in **Sequence of Optimization Problems**, (**A1**) and (**A3**) in **Sequence of Necessary SMP** and (**A1**) and (**A2**) in **Control Bounds**, we deduce that:

$$\sup_{n\ge 0}\left(\mathbb{E}^n\big[\sup_{0\le t\le T}|Y_t^n|^2\big] + \mathbb{E}^n\int_0^T \big(|Z_t^n|^2 + |Z_t^{0,n}|^2\big)dt\right) < \infty, \tag{3.20}$$

see for instance the proof of Example 1.20.

Second Step. We now prove that the family $(\mathbb{P}^n \circ (\boldsymbol{X}^n)^{-1})_{n\ge 0}$ is tight in $\mathcal{C}([0, T]; \mathbb{R}^d)$, using Aldous' criterion. For an integer $n \ge 0$, a stopping time τ with respect to filtration $\mathbb{F}^n$ and a real $\delta > 0$, we have:

$$\begin{aligned} \mathbb{E}^n\big[|X^n_{(\tau+\delta)\wedge T} - X^n_\tau|\big] &\le \mathbb{E}^n\bigg[\int_\tau^{(\tau+\delta)\wedge T} \big|b(s, X_s^n, \mu_s^n, \hat{\alpha}_s^n)\big|ds\bigg] \\ &\quad + \mathbb{E}^n\bigg[\bigg|\int_\tau^{(\tau+\delta)\wedge T} \sigma(s, X_s^n, \mu_s^n)dW_s^n\bigg|\bigg] \\ &\quad + \mathbb{E}^n\bigg[\bigg|\int_\tau^{(\tau+\delta)\wedge T} \sigma^0(s, X_s^n, \mu_s^n)dW_s^{0,n}\bigg|\bigg]. \end{aligned}$$

Thanks to the bound on the growth of the coefficients, we can find a constant C, independent of n, δ and τ, such that:

$$\begin{aligned} \mathbb{E}^n\big[|X^n_{(\tau+\delta)\wedge T} - X^n_\tau|\big] &\le C\delta^{1/2}\bigg[\mathbb{E}^n\int_\tau^{(\tau+\delta)\wedge T} \big(1 + |X_s^n| + d_{\mathcal{X}}(0_{\mathcal{X}}, \mu_s^n) + |\hat{\alpha}_s^n|\big)^2 ds\bigg]^{1/2} \\ &\quad + C\mathbb{E}^n\bigg[\bigg(\int_\tau^{(\tau+\delta)\wedge T} \big(1 + |X_s^n| + d_{\mathcal{X}}(0_{\mathcal{X}}, \mu_s^n)\big)^2 ds\bigg)^{1/2}\bigg]. \end{aligned}$$

Plugging the bounds we have on the processes $\boldsymbol{X}^n$, $\boldsymbol{\mu}^n$ and $\hat{\boldsymbol{\alpha}}^n$, we deduce (with a possibly new value of the constant C):

$$\mathbb{E}^n\big[|X^n_{(\tau+\delta)\wedge T} - X^n_\tau|\big] \leq C\delta^{1/2},$$

which shows that Aldous' criterion is satisfied.

Therefore, the sequence of measures $(\mathbb{P}^n \circ (\boldsymbol{X}^n)^{-1})_{n\geq 0}$ is tight on $\mathcal{D}([0,T];\mathbb{R}^d)$ equipped with the $J1$ Skorohod topology, and thus in $\mathcal{C}([0,T];\mathbb{R}^d)$ since the processes $(\boldsymbol{X}^n)_{n\geq 0}$ have continuous sample paths.

Third Step. We now prove that the sequence $(\mathbb{P}^n \circ (\boldsymbol{Y}^n)^{-1})_{n\geq 0}$ is tight on $\mathscr{M}([0,T];\mathbb{R}^d)$ equipped with the Meyer-Zheng topology.

Going back to the backward equation (3.19), observe that, for any sequence $0 = t_0 < t_1 < \cdots < t_N = T$, the conditional variation of $\boldsymbol{Y}^n$ along the grid $t_0, t_1, \cdots, t_N$ can be bounded as follows:

$$\begin{aligned}
\sum_{k=0}^{N-1} \mathbb{E}^n\Big[\big|\mathbb{E}^n\big(Y^n_{t_{k+1}} - Y^n_{t_k}|\mathcal{F}_{t_k}\big)\big|\Big] &\leq \sum_{k=0}^{N-1} \mathbb{E}^n\bigg[\int_{t_k}^{t_{k+1}} \big|\partial_x H^n\big(s, X^n_s, \mu^n_s, Y^n_s, Z^n_s, Z^{0,n}_s, \hat{\alpha}^n_s\big)\big| ds\bigg] \\
&= \mathbb{E}^n \int_0^T \big|\partial_x H^n\big(s, X^n_s, \mu^n_s, Y^n_s, Z^n_s, Z^{0,n}_s, \hat{\alpha}^n_s\big)\big| ds \\
&\leq C\mathbb{E}^n \int_0^T \big(1 + |X^n_s| + |Y^n_s| + |Z^n_s| + |Z^{0,n}_s| + |\hat{\alpha}^n_s|\big) ds,
\end{aligned}$$

the constant C being independent of n thanks to assumption **Sequence of Necessary SMP**. Together with (3.20) and assumption **Control Bounds**, we get:

$$\sup_{n\geq 0} \sup_{N\geq 1} \sup_{0=t_0<\cdots<t_N=T} \sum_{k=0}^{N-1} \mathbb{E}^n\big[|\mathbb{E}^n\big(Y^n_{t_{k+1}} - Y^n_{t_k}|\mathcal{F}_{t_k}\big)|\big] < \infty.$$

By Theorem 3.9 (tightness criterion for the Meyez-Zheng topology), we deduce that the family of probability measures $(\mathbb{P}^n \circ (\boldsymbol{Y}^n)^{-1})_{n\geq 0}$, with $\boldsymbol{Y}^n = (Y^n_t)_{0\leq t\leq T}$, is tight on $\mathcal{M}([0,T];\mathbb{R}^d)$ equipped with the Meyer-Zheng topology. Any weak limit is supported by $\mathcal{D}([0,T];\mathbb{R}^d)$.

Fourth Step. In order to complete the proof, we recall that, for each $n \geq 0$ and each $t \in [0,T]$, $\hat{\alpha}^n_t = \hat{\alpha}^n(t, X^n_t, \mu^n_t, Y^n_t)$. We know that the sequence of functions $(\hat{\alpha}^n)_{n\geq 0}$ converges to $\hat{\alpha}$, uniformly on compact subsets of $[0,T] \times \mathbb{R}^d \times \mathcal{X} \times \mathbb{R}^m$ to $\hat{\alpha}$, see Lemma 3.10.

Then, we may expand $\hat{\alpha}^n_t$ as

$$\hat{\alpha}^n_t = \hat{\alpha}^n(t, X^n_t, \mu^n_t, Y^n_t) = \big[\hat{\alpha}^n - \hat{\alpha}\big](t, X^n_t, \mu^n_t, Y^n_t) + \hat{\alpha}(t, X^n_t, \mu^n_t, Y^n_t).$$

For a given compact set $K \subset \mathbb{R}^d \times \mathcal{X} \times \mathbb{R}^m$, we can write:

$$\begin{aligned}
&\int_0^T \min\Big(1, \big|\big[\hat{\alpha}^n - \hat{\alpha}\big](t, X^n_t, \mu^n_t, Y^n_t)\big|\Big) dt \\
&\quad \leq T \sup_{(t,x,\mu,\alpha)\in[0,T]\times K} \big|\big(\hat{\alpha}^n - \hat{\alpha}\big)(t, x, \mu, \alpha)\big| + \int_0^T \mathbf{1}_{\{(X^n_t, \mu^n_t, Y^n_t)\notin K\}} dt.
\end{aligned} \tag{3.21}$$

For K being fixed, we know that the first term in the right-hand side tends to 0 as n tends to ∞. Regarding the second term in the right-hand side, we have:

$$\mathbb{E}^n \int_0^T \mathbf{1}_{\{(X_t^n,\mu_t^n,Y_t^n)\notin K\}} dt = \int_0^T \mathbb{P}^n\big[(X_t^n,\mu_t^n,Y_t^n)\notin K\big]dt.$$

Choosing K of the form $K = K_x \times K_\mu \times K_y$, with K_x being a compact subset of $\mathbb{R}^d$, K_μ a compact subset of $\mathcal{X}$, and K_y a compact subset of $\mathbb{R}^m$, we have the bound:

$$\mathbb{E}^n \int_0^T \mathbf{1}_{\{(X_t^n,\mu_t^n,Y_t^n)\notin K\}} dt \leq \int_0^T \big(\mathbb{P}\big[X_t^n \notin K_x\big] + \mathbb{P}\big[\mu_t^n \notin K_\mu\big] + \mathbb{P}\big[Y_t^n \notin K_y\big]\big)dt.$$

Owing to Proposition 3.7, we know that, for any $\varepsilon > 0$, we can choose K_y such that:

$$\int_0^T \mathbb{P}\big[Y_t^n \notin K_y\big]dt \leq \varepsilon.$$

Similarly, from the properties of the uniform topology on $\mathcal{C}([0,T];\mathbb{R}^d)$ and of the $J1$ Skorohod topology on $\mathcal{D}([0,T];\mathbb{R}^m)$, we can choose, for any $\varepsilon > 0$, K_x and K_μ such that:

$$\int_0^T \big(\mathbb{P}\big[X_t^n \notin K_x\big] + \mathbb{P}\big[\mu_t^n \notin K_\mu\big]\big)dt \leq \varepsilon.$$

Returning to (3.21), we deduce that:

$$\lim_{n\to\infty} \mathbb{E}^n \int_0^T \min\Big(1, \big|[\hat{\alpha}^n - \hat{\alpha}](t, X_t^n, \mu_t^n, Y_t^n)\big|\Big)dt = 0,$$

that is:

$$\forall \varepsilon > 0, \quad \lim_{n\to\infty} \mathbb{P}^n\Big[d_{\mathcal{M}}\big(\hat{\boldsymbol{\alpha}}^n, (\hat{\alpha}(t, X_t^n, \mu_t^n, Y_t^n))_{0\leq t\leq T}\big) \geq \varepsilon\Big] = 0.$$

In order to complete the proof, it remains to check that the sequence of probability measures $(\mathbb{P}^n \circ (\hat{\alpha}(t, X_t^n, \mu_t^n, Y_t^n))_{0\leq t\leq T}^{-1})_{n\geq 0}$ converges to $\mathbb{P}^\infty \circ (\hat{\alpha}(t, X_t^\infty, \mu_t^\infty, Y_t^\infty))_{0\leq t\leq T}^{-1}$ provided that $(\mathbb{P}^n \circ (X_t^n, \mu_t^n, Y_t^n)_{0\leq t\leq T}^{-1})_{n\geq 0}$ converges to $\mathbb{P}^\infty \circ (X_t^\infty, \mu_t^\infty, Y_t^\infty)_{0\leq t\leq T}^{-1}$, the latter being obviously true up to a subsequence. This follows from the continuous mapping theorem, using the fact that $\hat{\alpha}(t,\cdot)$ is continuous, for each $t \in [0,T]$, see Lemma 3.10. □

We now turn to the second part of the statement of Proposition 3.11.

Lemma 3.15 *With the same notation and assumptions as in the statement of Lemma 3.14, consider a complete probability space* $(\Omega^\infty, \mathcal{F}^\infty, \mathbb{P}^\infty)$ *equipped with a process* $(X_0^\infty, \boldsymbol{W}^{0,\infty}, \boldsymbol{\mu}^\infty, \boldsymbol{W}^\infty, \boldsymbol{X}^\infty, \hat{\boldsymbol{\alpha}}^\infty)$ *whose law under* $\mathbb{P}^\infty$ *is a weak limit of the sequence* $(\mathbb{P}^n \circ (X_0^n, \boldsymbol{W}^{0,n}, \boldsymbol{\mu}^n, \boldsymbol{W}^n, \boldsymbol{X}^n, \hat{\boldsymbol{\alpha}}^n)^{-1})_{n\geq 0}$ *on the space* $\Omega_{input} \times \mathcal{C}([0,T];\mathbb{R}^d) \times \mathcal{M}([0,T];\mathbb{R}^k)$, *with* Ω_{input} *as in (3.18). Denote by* $\mathbb{F}^\infty$ *the complete and right-continuous filtration generated by* $(X_0^\infty, \boldsymbol{W}^{0,\infty}, \boldsymbol{\mu}^\infty, \boldsymbol{W}^\infty, \boldsymbol{X}^\infty)$ *on* $(\Omega^\infty, \mathcal{F}^\infty, \mathbb{P}^\infty)$.

Then, up to a modification, $\hat{\boldsymbol{\alpha}}^\infty$ is an $\mathbb{F}^\infty$-progressively measurable square-integrable process with values in A and the pair $(\boldsymbol{X}^\infty, \hat{\boldsymbol{\alpha}}^\infty)$ solves the SDE (3.17). Moreover, for any $\mathbb{F}^\infty$-progressively measurable and square-integrable process $\boldsymbol{\beta} = (\beta_t)_{0\le t\le T}$ with values in A, it holds that:

$$J^{\mu^\infty}\big(\hat{\boldsymbol{\alpha}}^\infty\big) \le J^{\mu^\infty}\big(\boldsymbol{\beta}\big),$$

where J^{μ^∞} denotes the same cost functional as in (3.16).

Therefore, if the filtration $\mathbb{F}^\infty$ is compatible with $(X_0^\infty, \boldsymbol{W}^{0,\infty}, \boldsymbol{\mu}^\infty, \boldsymbol{W}^\infty)$, $\hat{\boldsymbol{\alpha}}^\infty$ is a minimizer of the optimal stochastic control problem (3.16)–(3.17) when considered on the compatible set-up $(\Omega^\infty, \mathcal{F}^\infty, \mathbb{F}^\infty, \mathbb{P}^\infty)$ equipped with the input $(X_0^\infty, \boldsymbol{W}^{0,\infty}, \boldsymbol{\mu}^\infty, \boldsymbol{W}^\infty)$.

Proof.

First Step. We first observe that, by assumption, the sequences $(\mathbb{P}^n \circ (X_0^n)^{-1})_{n\ge 0}$ and $(\mathbb{P}^n \circ (\boldsymbol{\mu}^n)^{-1})_{n\ge 0}$ are tight on the spaces $\mathbb{R}^d$ and $\mathcal{D}([0,T];\mathcal{X})$ respectively. Moreover, by Lemma 3.14, the sequences $(\mathbb{P}^n \circ (\boldsymbol{X}^n)^{-1})_{n\ge 0}$ and $(\mathbb{P}^n \circ (\hat{\boldsymbol{\alpha}}^n)^{-1})_{n\ge 0}$ are tight on $\mathcal{C}([0,T];\mathbb{R}^d)$ and $\mathscr{M}([0,T];\mathbb{R}^k)$ respectively. It is quite standard to deduce that that the sequence $(\mathbb{P}^n \circ (\boldsymbol{\Theta}^n)^{-1})_{n\ge 0}$, with

$$\boldsymbol{\Theta}^n = (X_0^n, \boldsymbol{W}^{0,n}, \boldsymbol{\mu}^n, \boldsymbol{W}^n, \boldsymbol{X}^n, \hat{\boldsymbol{\alpha}}^n),$$

is tight on the space indicated in the statement.

Therefore, we are allowed to consider a weak limit. On some complete probability space $(\Omega^\infty, \mathcal{F}^\infty, \mathbb{P}^\infty)$, we call $\boldsymbol{\Theta}^\infty = (X_0^\infty, \boldsymbol{W}^{0,\infty}, \boldsymbol{\mu}^\infty, \boldsymbol{W}^\infty, \boldsymbol{X}^\infty, \hat{\boldsymbol{\alpha}}^\infty)$ a process distributed according to the weak limit, the process $\hat{\boldsymbol{\alpha}}^\infty$ being reconstructed as an A-valued process by the same argument as in Lemma 3.14, namely $\hat{\boldsymbol{\alpha}}^\infty = (\hat{\alpha}(t, X_t^\infty, \mu_t^\infty, Y_t^\infty))_{0\le t\le T}$ for another càd-làg process $\boldsymbol{Y}^\infty = (Y_t^\infty)_{0\le t\le T}$ with values in $\mathbb{R}^d$. We then denote by $\mathbb{G}^\infty$ the complete and right-continuous augmentation of the filtration generated by $\boldsymbol{\Theta}^\infty$. Pay attention that, at this stage of the proof, we do not know whether $\mathbb{G}^\infty$ is strictly larger than $\mathbb{F}^\infty$ or not, with $\mathbb{F}^\infty$ defined as in the statement.

The first step is to check that the limit process satisfies (3.17). In this regard, it is quite obvious that the initial condition of $\boldsymbol{X}^\infty$ is X_0^∞; it is also pretty standard to check that $(\boldsymbol{W}^{0,\infty}, \boldsymbol{W}^\infty)$ is a $2d$-dimensional Brownian motion with respect to the filtration $\mathbb{G}^\infty$. Thanks to **(A1)** and **(A2)** in assumption **Control Bounds**, we also have that:

$$\mathbb{E}^\infty\Big[\sup_{0\le t\le T} |X_t^\infty|^2 + \int_0^T |\hat{\alpha}_s^\infty|^2 ds\Big] < \infty,$$

where, to handle the second term, we made use of Lemma 3.6 with θ therein being a bounded approximation of the function $|\hat{\alpha}(t,\cdot,\cdot,\cdot)|^2$. In order to identify the limit process with a solution of the SDE (3.17), we introduce, for any $n \in \mathbb{N}\cup\{\infty\}$, the auxiliary processes:

$$\boldsymbol{B}^n = \Big(B_t^n = \int_0^t b^n(s, X_s^n, \mu_s^n, \hat{\alpha}_s^n) ds\Big)_{0\le t\le T},$$

$$\boldsymbol{\Sigma}^n = \Big(\Sigma_t^n = \int_0^t \sigma^n(s, X_s^n, \mu_s^n) dW_s^n\Big)_{0\le t\le T},$$

$$\boldsymbol{\Sigma}^{0,n} = \Big(\Sigma_t^{0,n} = \int_0^t \sigma^{0,n}(s, X_s^n, \mu_s^n) dW_s^{0,n}\Big)_{0\le t\le T}.$$

We claim that $(\mathbb{P}^n \circ (\boldsymbol{B}^n)^{-1})_{n\geq 0}$ weakly converges to $\mathbb{P}^\infty \circ (\boldsymbol{B}^\infty)^{-1}$ on $\mathcal{C}([0,T];\mathbb{R}^d)$. Indeed, by the same argument as in the fourth step of Lemma 3.14, the sequences $(\mathbb{P}^n \circ (b^n(t,X_t^n,\mu_t^n,\hat{\alpha}_t^n))_{0\leq t\leq T}^{-1})_{n\geq 0}$ and $\mathbb{P}^\infty \circ (b(t,X_t^n,\mu_t^n,\hat{\alpha}_t^n))_{0\leq t\leq T}^{-1}$ have the same weak limits on $\mathscr{M}([0,T];\mathbb{R}^d)$. Now, by Lemma 3.6 –using the growth condition (**A2**) and the continuity condition (**A4**) in assumption **Sequence of Optimization Problems** together with (**A1**) and (**A2**) in assumption **Control Bounds** in order to check the uniform integrability condition–, we deduce that $(\mathbb{P}^n \circ (\boldsymbol{B}^n)^{-1})_{n\geq 0}$ converges weakly to $\mathbb{P}^\infty \circ (\boldsymbol{B}^\infty)^{-1}$ on $\mathcal{C}([0,T];\mathbb{R}^d)$.

We now address the same question for the sequences $(\boldsymbol{\Sigma}^n)_{n\geq 0}$ and $(\boldsymbol{\Sigma}^{0,n})_{n\geq 0}$. For simplicity, we only provide the analysis for the sequence $(\boldsymbol{\Sigma}^n)_{n\geq 0}$. By the same argument as above, we know that $(\mathbb{P}^n \circ (\sigma^n(t,X_t^n,\mu_t^n))_{0\leq t\leq T}^{-1})_{n\geq 0}$ converges weakly to $\mathbb{P}^\infty \circ (\sigma(t,X_t^\infty,\mu_t^\infty)^{-1})_{0\leq t\leq T}$ on $\mathscr{M}([0,T];\mathbb{R}^{d\times d})$. Replace for a while σ^n by

$$\sigma^{n,h}(t,x,\mu) = \frac{1}{h}\int_{(t-h)_+}^{t} \sigma^n(s,x,\mu)ds, \quad t\in[0,T],\ x\in\mathbb{R}^d,\ \mu\in\mathcal{X},$$

for some $h > 0$, and define $\sigma^{\infty,h}$ in the same way, with σ^n in the right-hand side being replaced by σ. Then, for each integer $n \geq 0$, the function $\sigma^{n,h}$ is jointly continuous in all the parameters, and similarly for the function $\sigma^{\infty,h}$. Once again, we have that $(\mathbb{P}^n \circ (\sigma^{\infty,h}(t,X_t^n,\mu_t^n))_{0\leq t\leq T}^{-1})_{n\geq 0}$ converges weakly to $\mathbb{P}^\infty \circ (\sigma^h(t,X_t^\infty,\mu_t^\infty))_{0\leq t\leq T}^{-1}$, but on $\mathcal{D}([0,T];\mathbb{R}^{d\times d})$ equipped with $J1$. Thanks to the growth condition (**A2**) in assumption **Sequence of Optimization Problems** and to the bounds (**A1**) in assumption **Control Bounds** and (3.19), we know from a result by Kurtz and Protter, see the Notes & Complements in the appendix, that $(\mathbb{P}^n \circ (\boldsymbol{\Sigma}^{n,h})^{-1})_{n\geq 0}$ weakly converges to $\mathbb{P}^\infty \circ (\boldsymbol{\Sigma}^{\infty,h})^{-1}$ on $\mathcal{C}([0,T];\mathbb{R}^d)$, with a quite obvious definition of $\boldsymbol{\Sigma}^{n,h}$ and $\boldsymbol{\Sigma}^{\infty,h}$. In order to prove the same result for $h=0$, we consider the differences $\boldsymbol{\Sigma}^{n,h} - \boldsymbol{\Sigma}^n$ and $\boldsymbol{\Sigma}^{\infty,h} - \boldsymbol{\Sigma}^\infty$. In order to do so, we observe that, for any $(x,\mu)\in\mathbb{R}^d\times\mathcal{X}$,

$$\lim_{h\to 0}\int_0^T |\sigma^{\infty,h}(t,x,\mu) - \sigma(t,x,\mu)|dt = 0,$$

which is a consequence of Lebesgue's differentiation theorem. Now, for any compact subset $K\subset\mathbb{R}^d\times\mathcal{X}$, we call:

$$w_K(t,\delta) = \sup\left\{|\sigma(t,x',\mu') - \sigma(t,x,\mu)|;\quad |x-x'|\leq\delta,\ d_{\mathcal{X}}(\mu,\mu')\leq\delta\right\}$$

the modulus of continuity of $\sigma(t,\cdot)$ on K. Then, for any $\delta > 0$ and any δ-net $(x_i,\mu_i)_{1\leq i\leq N}$ of K,

$$\begin{aligned}
&\int_0^T \sup_{(x,\mu)\in K} |\sigma^{\infty,h}(t,x,\mu) - \sigma(t,x,\mu)|dt \\
&\quad\leq \sum_{i=1}^N \int_0^T |\sigma^{\infty,h}(t,x_i,\mu_i) - \sigma(t,x_i,\mu_i)|dt + \int_0^T w_K(t,\delta)dt.
\end{aligned}$$

From the growth condition (**A2**) in assumption **Sequence of Optimization Problems** and by Lebesgue's dominated convergence theorem, the second term in the right-hand side tends to 0 as δ tends to 0. Therefore, the left-hand side tends to 0 as h tends to 0. Now, since $(\sigma^n)_{n\geq 0}$ converges to σ, uniformly on any compact subset of $[0,T]\times\mathbb{R}^d\times\mathcal{X}$, we also have:

$$\lim_{h \searrow 0} \sup_{n \geq 0} \int_0^T \sup_{(x,\mu) \in K} |\sigma^{n,h}(t,x,\mu) - \sigma^n(t,x,\mu)| dt = 0.$$

Then, by a standard uniform integrability argument based on (**A1**) in assumption **Control Bounds**, we easily get that:

$$\lim_{h \searrow 0} \sup_{n \geq 0} \mathbb{E}^n \Big[\sup_{0 \leq t \leq T} |\Sigma_t^{n,h} - \Sigma_t^n| \Big] = 0, \quad \lim_{h \searrow 0} \mathbb{E}^\infty \Big[\sup_{0 \leq t \leq T} |\Sigma_t^{\infty,h} - \Sigma_t^\infty| \Big] = 0.$$

This suffices to prove that $(\mathbb{P}^n \circ (\boldsymbol{\Sigma}^n)^{-1})_{n \geq 0}$ converges to $\mathbb{P}^\infty \circ (\boldsymbol{\Sigma}^\infty)^{-1}$. Importantly, convergence holds in the joint sense, namely the whole $(\mathbb{P}^n \circ (\boldsymbol{\Theta}^n, \boldsymbol{B}^n, \boldsymbol{\Sigma}^n, \boldsymbol{\Sigma}^{0,n})^{-1})_{n \geq 0}$ converges to $\mathbb{P}^\infty \circ (\boldsymbol{\Theta}^\infty, \boldsymbol{B}^\infty, \boldsymbol{\Sigma}^\infty, \boldsymbol{\Sigma}^{0,\infty})^{-1}$ in the weak sense.

Second Step. Passing to the limit in (3.11), we deduce that, with $\mathbb{P}^\infty$-probability 1, for all $t \in [0,T]$,

$$\begin{aligned} X_t^\infty &= X_0^\infty + B_t^\infty + \Sigma_t^\infty + \Sigma_t^{0,\infty} \\ &= X_0^\infty + \int_0^t b(s, X_s^\infty, \mu_s^\infty, \hat{\alpha}_s^\infty) ds + \int_0^t \sigma(s, X_s^\infty, \mu_s^\infty) dW_s^\infty \\ &\quad + \int_0^t \sigma^0(s, X_s^\infty, \mu_s^\infty) dW_s^{0,\infty}. \end{aligned} \tag{3.22}$$

We deduce that the process $(\int_0^t b(s, X_s^\infty, \mu_s^\infty, \hat{\alpha}_s^\infty) ds)_{0 \leq t \leq T}$ is $\mathbb{F}^\infty$-adapted. Recalling the form of b specified in (**A1**) in assumption **Sequence of Necessary SMP**, we deduce that $(\int_0^t b_2(s) \hat{\alpha}_s^\infty ds)_{0 \leq t \leq T}$ is $\mathbb{F}^\infty$-adapted

Now, we have, for almost every $t \in [0,T]$, with $\mathbb{P}^\infty$-probability 1,

$$b_2(t) \hat{\alpha}_t^\infty = \lim_{p \to \infty} p \int_{(t-1/p)+}^t b_2(s) \hat{\alpha}_s^\infty ds,$$

Since $\mathbb{F}^\infty$ is complete, we obtain that, for t outside a subset of $[0,T]$ of zero Lebesgue measure, $b_2(t) \hat{\alpha}_t^\infty$ is $\mathcal{F}_t^\infty$-measurable.

Calling $(\hat{\alpha}_t^{o,\infty})_{0 \leq t \leq T}$ the optional projection of $(\hat{\alpha}_t^\infty)_{0 \leq t \leq T}$ given the filtration $\mathbb{F}^\infty$, which takes values in A by convexity of A, we deduce that, for almost every $t \in [0,T]$, with probability 1 under $\mathbb{P}^\infty$,

$$b_2(t) \hat{\alpha}_t^\infty = \mathbb{E}^\infty \big[b_2(t) \hat{\alpha}_t^\infty | \mathcal{F}_t^\infty \big] = b_2(t) \hat{\alpha}_t^{o,\infty}.$$

Plugging into (3.22), we get:

$$\begin{aligned} X_t^\infty = X_0^\infty + \int_0^t b(s, X_s^\infty, \mu_s^\infty, \hat{\alpha}_s^{o,\infty}) ds \\ + \int_0^t \sigma(s, X_s^\infty, \mu_s^\infty) dW_s + \int_0^t \sigma^0(s, X_s^\infty, \mu_s^\infty) dW_s^0. \end{aligned} \tag{3.23}$$

Third Step. We now pass to the limit in the inequality (**A3**) of assumption **Control Bounds**.

We proceed as follows. Since the process $\boldsymbol{\mu}^\infty$ is càd-làg, we can find a countable subset $\mathcal{T} \subset [0,T]$, such that, with probability 1 under $\mathbb{P}^\infty$, $\boldsymbol{\mu}^\infty$ is continuous at any point $t \in [0,T] \setminus \mathcal{T}$. In particular, for any integer $N \geq 1$, we can find an increasing sequence

$0 = t_0^N < t_1^N < \cdots < t_N^N = T$, of step-size less than $2T/N$, such that $t_i^N \in [0, T] \setminus \mathcal{T}$. Given such a sequence, we also consider a family $(\Phi(t_i^N, \cdot))_{i=0,\cdots,N-1}$ of bounded and continuous functions from $\Omega_{\text{input}} \times \mathcal{C}([0, T]; \mathbb{R}^d)$ into A, with Ω_{input} as in (3.18).

For any $n \in \mathbb{N} \cup \{\infty\}$, we define the process:

$$\beta_t^{n,N} = \sum_{i=0}^{N-1} \mathbf{1}_{[t_i^N, t_{i+1}^N)}(t) \Phi\big(t_i^N, X_0^n, \boldsymbol{W}^{0,n}_{\cdot \wedge t_i^N}, \boldsymbol{\mu}^n_{\cdot \wedge t_i^N}, \boldsymbol{W}^n_{\cdot \wedge t_i^N}, \boldsymbol{X}^n_{\cdot \wedge t_i^N}\big), \quad t \in [0, T], \tag{3.24}$$

where the notation $\cdot \wedge t_i^N$ in $(\boldsymbol{W}^{0,n}_{\cdot \wedge t_i^N}, \boldsymbol{\mu}^n_{\cdot \wedge t_i^N}, \boldsymbol{W}^n_{\cdot \wedge t_i^N}, \boldsymbol{X}^n_{\cdot \wedge t_i^N})$ indicates that the processes are stopped at t_i^N. By **(A2)** in **Control Bounds**, we know that, for any $n \geq 0$,

$$J^{n,\mu^n}(\hat{\boldsymbol{\alpha}}^n) \leq J^{n,\mu^n}(\boldsymbol{\beta}^{n,N}),$$

with $\boldsymbol{\beta}^{n,N} = (\beta_t^{n,N})_{0 \leq t \leq T}$.

We now observe that, for each $i \in \{0, \cdots, N\}$, $(\boldsymbol{\mu}^n_{\cdot \wedge t_i^N})_{n \geq 0}$ converges weakly to $\boldsymbol{\mu}^\infty_{\cdot \wedge t_i^N}$. This follows from the fact that, for any $t \in [0, T]$, the mapping $\mathcal{D}([0, T]; \mathcal{X}) \ni \boldsymbol{\nu} \mapsto \boldsymbol{\nu}_{\cdot \wedge t}$ is continuous at any path $\boldsymbol{\nu}$ that admits t as point of continuity. As above, weak convergence holds jointly with that of the other processes involved in the analysis. We deduce that the sequence $(\boldsymbol{\beta}^{n,N})_{n \geq 0}$ converges weakly to $\boldsymbol{\beta}^{\infty,N}$ in $\mathcal{D}([0, T]; A)$, the convergence holding jointly with that of the other processes.

Recalling **(A1)** in assumption **Control Bounds** and (3.20), we notice from the growth property of the minimizers $(\hat{\alpha}^n)_{n \geq 0}$ in Lemma 3.10 that $\sup_{n \geq 0} \mathbb{E}^n[\sup_{0 \leq t \leq T} |\hat{\alpha}_t^n|^2] < \infty$. Therefore, by Lemma 3.6 and the growth condition **(A3)** in assumption **Sequence of Optimization Problems**, we deduce that:

$$\left(g(X_T^n, \mu_T^n) + \int_0^T f(s, X_s^n, \mu_s^n, \hat{\alpha}_s^n) ds \right)_{n \geq 0} \tag{3.25}$$

converges in law toward:

$$g(X_T^\infty, \mu_T^\infty) + \int_0^T f(s, X_s^\infty, \mu_s^\infty, \hat{\alpha}_s^\infty) ds. \tag{3.26}$$

Now, in order to handle the fact that we do not have any square uniform integrability property on the processes $(\boldsymbol{\alpha}^n)_{n \geq 0}$, we split the second term in (3.25) into:

$$\begin{aligned} \int_0^T f(s, X_s^n, \mu_s^n, \hat{\alpha}_s^n) ds &= \int_0^T \Big(f(s, X_s^n, \mu_s^n, 0) + \partial_\alpha f(s, X_s^n, \mu_s^n, 0) \cdot \hat{\alpha}_s^n \Big) ds \\ &\quad + \int_0^T \Big(f(s, X_s^n, \mu_s^n, \hat{\alpha}_s^n) - f(s, X_s^n, \mu_s^n, 0) - \partial_\alpha f(s, X_s^n, \mu_s^n, 0) \cdot \hat{\alpha}_s^n \Big) ds. \end{aligned}$$

Using the various bounds on the coefficients together with **(A1)** in assumption **Control Bounds**, we deduce from a standard uniform integrability argument that:

$$\begin{aligned} &\lim_{n \to \infty} \mathbb{E}^n \int_0^T \Big(f(s, X_s^n, \mu_s^n, 0) + \partial_\alpha f(s, X_s^n, \mu_s^n, 0) \cdot \hat{\alpha}_s^n \Big) ds \\ &= \mathbb{E}^\infty \int_0^T \Big(f(s, X_s^\infty, \mu_s^\infty, 0) + \partial_\alpha f(s, X_s^\infty, \mu_s^\infty, 0) \cdot \hat{\alpha}_s^\infty \Big) ds. \end{aligned}$$

By Fatou's lemma for weak convergence, we also have:

$$\liminf_{n\to\infty} \mathbb{E}^n\Big[\int_0^T \Big(f(s,X_s^n,\mu_s^n,\hat{\alpha}_s^n) - f(s,X_s^n,\mu_s^n,0) - \partial_\alpha f(X_s^n,\mu_s^n,0)\cdot\hat{\alpha}_s^n\Big)ds\Big]$$
$$\geq \mathbb{E}^\infty\Big[\int_0^T \Big(f(s,X_s^\infty,\mu_s^\infty,\hat{\alpha}_s^\infty) - f(s,X_s^\infty,\mu_s^\infty,0) - \partial_\alpha f(s,X_s^\infty,\mu_s^\infty,0)\cdot\hat{\alpha}_s^\infty\Big)ds\Big],$$

since, by convexity of f with respect to α, the integrands in the left and right-hand sides are nonnegative.

Returning to (3.25) and (3.26) and observing that the remaining terms may be easily managed by a uniform integrability argument, we deduce that:

$$J^{\mu^\infty}(\hat{\boldsymbol{\alpha}}^\infty) \leq \liminf_{n\to\infty} J^{n,\mu^n}(\boldsymbol{\beta}^{n,N}). \tag{3.27}$$

In order to pass to the limit in the right-hand side, we need to introduce, for each $n \in \mathbb{N}\cup\{\infty\}$, the solution $\boldsymbol{X}^{n,N}$ to the SDE (3.11) (including the case $n = \infty$) when set on $(\Omega^n,\mathcal{F}^n,\mathbb{P}^n)$ with $\boldsymbol{\alpha} = \boldsymbol{\beta}^{n,N}$. Arguing as above, we can prove that the sequence of tuples $(X_0^n,\boldsymbol{W}^{0,n},\boldsymbol{\mu}^n,\boldsymbol{W}^n,\boldsymbol{X}^{n,N},\boldsymbol{\beta}^{n,N})_{n\geq 0}$ converges in distribution toward $(X_0^\infty,\boldsymbol{W}^{0,\infty},\boldsymbol{\mu}^\infty,\boldsymbol{W}^\infty,\boldsymbol{X}^{\infty,N},\boldsymbol{\beta}^{\infty,N})$. Implicitly, we chose to construct the weak limit of $(X_0^n,\boldsymbol{W}^{0,n},\boldsymbol{\mu}^n,\boldsymbol{W}^n,\boldsymbol{X}^{n,N},\boldsymbol{\beta}^{n,N})_{n\geq 0}$ on the same probability space $(\Omega^\infty,\mathcal{F}^\infty,\mathbb{P}^\infty)$ as above. This is possible because the SDE (3.17) is uniquely solvable in the strong sense when the input $(X_0^\infty,\boldsymbol{W}^{0,\infty},\boldsymbol{\mu}^\infty,\boldsymbol{W}^\infty,\boldsymbol{\beta}^{\infty,N})$ is given. In particular, whatever the complete filtered probability space $(\Omega,\mathcal{F},\mathbb{G},\mathbb{P})$ carrying a copy $(X_0,\boldsymbol{W}^0,\boldsymbol{\mu},\boldsymbol{W},\boldsymbol{\beta}^N)$ of the input $(X_0^\infty,\boldsymbol{W}^{0,\infty},\boldsymbol{\mu}^\infty,\boldsymbol{W}^\infty,\boldsymbol{\beta}^{\infty,N})$ –with the constraint that $(\boldsymbol{W}^0,\boldsymbol{W})$ is a $2d$-Brownian motion with respect to $\mathbb{G}$ and $(X_0,\boldsymbol{W}^0,\boldsymbol{\mu},\boldsymbol{W},\boldsymbol{\beta}^N)$ is $\mathbb{G}$-progressively measurable–, the solution $\boldsymbol{X}^N$ to (3.17) is necessarily progressively measurable with respect to the complete and right-continuous augmentation $\mathbb{F}$ of the filtration generated by $(X_0,\boldsymbol{W}^0,\boldsymbol{\mu},\boldsymbol{W},\boldsymbol{\beta}^N)$. To prove it, it suffices to solve the equation on the space $(\Omega,\mathcal{F},\mathbb{F},\mathbb{P})$; by a standard uniqueness argument based on Gronwall's lemma, the solution coincides with $\boldsymbol{X}^N$. In comparison with Theorem 1.33, which shows the analogue for FBSDEs, there is no need to discuss any compatibility condition: What really matters here is that $(\boldsymbol{W}^0,\boldsymbol{W})$ is a $\mathbb{G}$-Brownian motion. Once $\boldsymbol{X}^N$ is known to be $\mathbb{F}$-progressively measurable, we can invoke Theorem 1.33 restricted to the special case when the backward equation is trivial, with $(\Omega,\mathcal{F},\mathbb{F},\mathbb{P})$ as set-up and $(\boldsymbol{\mu},\boldsymbol{\beta}^N)$ as environment. Now, the compatibility condition required in the statement of Theorem 1.33 holds since $\mathbb{F}$ is generated by $(X_0,\boldsymbol{W}^0,\boldsymbol{\mu},\boldsymbol{W},\boldsymbol{\beta}^N)$. This says that $\boldsymbol{X}^N$ is a measurable function of the input and that its law is independent of the probability space on which it is constructed. In particular, we can reconstruct it for free on the space $(\Omega^\infty,\mathcal{F}^\infty,\mathbb{F}^\infty,\mathbb{P}^\infty)$.

As above, we deduce that:

$$\Big(\int_0^T f(s,X_s^{n,N},\mu_s^n,\beta_s^{n,N})ds + g(X_T^{n,N},\mu_T^n)\Big)_{n\geq 0}$$

converges in law toward:

$$\int_0^T f(s,X_s^{\infty,N},\mu_s^\infty,\beta_s^{\infty,N})ds + g(X_T^{\infty,N},\mu_T^\infty).$$

In order to complete the proof, it suffices to prove a uniform integrability argument. Recalling that the processes $(\boldsymbol{\beta}^{n,N})_{n\geq 0}$ are uniformly bounded in n and taking advantage of assumption **Sequence of Optimization Problems**, there exists a constant C, independent of n but depending on N, such that:

$$\begin{aligned}
&|g(X_T^{n,N},\mu_T^n)| + \int_0^T |f(t,X_t^{n,N},\mu_t^n,\beta_t^{n,N})|dt \\
&\quad \leq C\Big(1 + \sup_{0\leq t\leq T} |X_t^{n,N}|^2 + \sup_{0\leq t\leq T} \big(d_{\mathcal{X}}(0_{\mathcal{X}},\mu_t^n)\big)^2\Big).
\end{aligned} \tag{3.28}$$

Now, by a standard stability argument for SDEs, we can find a constant C, independent of n but depending on N, such that:

$$\sup_{0\leq t\leq T} |X_t^{n,N}|^2 \leq C\Big(|X_0^n|^2 + \sup_{0\leq t\leq T} \big(d_{\mathcal{X}}(0_{\mathcal{X}},\mu_t^n)\big)^2\Big).$$

Together with (**A1**) in assumption **Control Bounds**, this implies that the sequence $(\mathbb{P}^n \circ (\sup_{0\leq t\leq T} |X_t^{n,N}|)^{-1})_{n\geq 0}$ is uniformly square-integrable. Therefore, considering the law under $\mathbb{P}^n$ of the left-hand side in (3.28), we get a sequence of uniformly integrable measures (uniformly with respect to $n \geq 0$). We deduce that the sequence $(J^{n,\mu^n}(\boldsymbol{\beta}^{n,N}))_{n\geq 0}$ converges to $J^{\mu^\infty}(\boldsymbol{\beta}^{\infty,N})$, so that (3.27) yields:

$$J^{\mu^\infty}(\hat{\boldsymbol{\alpha}}^\infty) \leq J^{\mu^\infty}(\boldsymbol{\beta}^{\infty,N}). \tag{3.29}$$

Now, we use the fact that any $\mathbb{F}^\infty$-progressively measurable and square integrable control $\boldsymbol{\beta} = (\beta_t)_{0\leq t\leq T}$ with values in A may be approximated in $L^2([0,T]\times\Omega^\infty;\mathrm{Leb}_1\otimes\mathbb{P}^\infty)$ by a sequence of controls $(\boldsymbol{\beta}^{\infty,N})_{N\geq 1}$ for a suitable choice, for each $N \geq 1$, of the functions $(\Phi(t_i^N,\cdot))_{1\leq i\leq N}$ in (3.24). Under the growth and regularity conditions in assumption **Sequence of Optimization Problems**, it is then standard to check that the sequence $(J^{\mu^\infty}(\boldsymbol{\beta}^{\infty,N}))_{N\geq 1}$ converges to $J^{\mu^\infty}(\boldsymbol{\beta})$, which shows that (3.29) holds true with $J^{\mu^\infty}(\boldsymbol{\beta}^{\infty,N})$ replaced by $J^{\mu^\infty}(\boldsymbol{\beta})$.

The strategy for constructing the functions $(\Phi(t_i^N,\cdot))_{1\leq i\leq N}$ is well known. Denoting by π_R the orthogonal projection from $\mathbb{R}^k$ onto the intersection $A\cap B(0,R)$ where $B(0,R)$ is the closed ball of center 0 and of radius $R>0$, we first approximate $\boldsymbol{\beta}$ in $L^2([0,T]\times\Omega^\infty;\mathrm{Leb}_1\otimes\mathbb{P}^\infty)$ by $(\pi_R(\beta_t))_{0\leq t\leq T}$. Notice that, since A is closed and convex, so is $A\cap B(0,R)$; therefore, the projection is well defined. This says that $\boldsymbol{\beta}$ may be assumed to be bounded. Then, considering an approximating process of the form:

$$\left(\frac{1}{h}\int_{t-h}^t \beta_s ds\right)_{0\leq t\leq T},$$

for some $h>0$, we may assume that $\boldsymbol{\beta}$ has Lipschitz-continuous paths. Above, $\boldsymbol{\beta}$ is extended to negative times by assigning to them a fixed arbitrary value in A. The important point is that the above approximation is A-valued since A is closed and convex. Then, for any mesh $(t_i^N)_{i=0,\cdots,N}$ as in (3.24), we approximate $\boldsymbol{\beta}$ by the process:

$$\left(\sum_{i=1}^N \beta_{t_{i-1}^N}\mathbf{1}_{[t_{i-1}^N,t_i^N)}(t)\right)_{0\leq t\leq T}.$$

By time continuity of the process $\boldsymbol{\beta}$, each β_{t_i} is measurable with respect to the completion of the σ-field $\sigma\{X_0^\infty, W_s^{0,\infty}, \mu_s^\infty, W_s^\infty, X_s^\infty; s \leq t_i^N\}$. Thus, we can find a version which is measurable with respect to $\sigma\{X_0^\infty, W_s^{0,\infty}, \mu_s^\infty, W_s^\infty, X_s^\infty; s \leq t_i^N\}$, from which we deduce that, for each $i \in \{0, \cdots, N-1\}$, there exists a bounded and measurable function $\Phi(t_i^N, \cdot)$ from $\Omega_{\text{input}} \times \mathcal{C}([0,T];\mathbb{R}^d)$ into A such that:

$$\beta_{t_i^N} = \Phi(t_i^N, X_0^\infty, \boldsymbol{W}^{0,\infty}_{\cdot\wedge t_i^N}, \boldsymbol{\mu}^\infty_{\cdot\wedge t_i^N}, \boldsymbol{W}^\infty_{\cdot\wedge t_i^N}, \boldsymbol{X}^\infty_{\cdot\wedge t_i^N}).$$

Using another approximation argument, this time in the Hilbert space:

$$L^2\Big(\Omega_{\text{input}} \times \mathcal{C}([0,T];\mathbb{R}^d), \mathbb{P}^\infty \circ (X_0^\infty, \boldsymbol{W}^{0,\infty}_{\cdot\wedge t_i^N}, \boldsymbol{\mu}^\infty_{\cdot\wedge t_i^N}, \boldsymbol{W}^\infty_{\cdot\wedge t_i^N}, \boldsymbol{X}^\infty_{\cdot\wedge t_i^N})^{-1}\Big),$$

we can assume that each $\Phi(t_{i-1}^N, \cdot)$ is continuous. To do so, we may invoke a generalization of Lusin's theorem to Euclidean-space-valued functions defined on a Polish space, see the Notes & Complements below, and then compose by the projection on the convex set A.

Fourth Step. As a consequence of (3.29), we deduce, with the same notation as in (3.23) for the optional projection $(\hat{\alpha}_t^{o,\infty})_{0\leq t\leq T}$ of $(\hat{\alpha}_t^\infty)_{0\leq t\leq T}$:

$$J^{\mu^\infty}(\hat{\boldsymbol{\alpha}}^\infty) \leq J^{\mu^\infty}(\hat{\boldsymbol{\alpha}}^{o,\infty}). \tag{3.30}$$

Recalling the SDE (3.23) and taking advantage of the convexity of f with respect to α, we observe that:

$$\begin{aligned}
J^{\mu^\infty}(\hat{\boldsymbol{\alpha}}^\infty) &= \mathbb{E}^\infty\bigg[\int_0^T f(s, X_s^\infty, \mu_s^\infty, \hat{\alpha}_s^\infty)ds + g(X_T^\infty, \mu_T^\infty)\bigg] \\
&\geq \mathbb{E}^\infty\bigg[\int_0^T f(s, X_s^\infty, \mu_s^\infty, \hat{\alpha}_s^{o,\infty})ds + g(X_T^\infty, \mu_T^\infty)\bigg] \\
&\quad + \mathbb{E}^\infty\bigg[\int_0^T (\hat{\alpha}_s^\infty - \hat{\alpha}_s^{o,\infty}) \cdot \partial_\alpha f(s, X_s^\infty, \mu_s^\infty, \hat{\alpha}_s^{o,\infty})ds\bigg] \\
&\quad + L^{-1}\mathbb{E}^\infty\bigg[\int_0^T |\hat{\alpha}_s^\infty - \hat{\alpha}_s^{o,\infty}|^2 ds\bigg].
\end{aligned}$$

By definition of the optional projection, we have $\mathbb{E}^\infty[(\hat{\alpha}_s^\infty - \hat{\alpha}_s^{o,\infty}) \cdot \partial_\alpha f(s, X_s^\infty, \mu_s^\infty, \hat{\alpha}_s^{o,\infty})] = 0$ for all $s \in [0,T]$. By (3.30), this shows that:

$$\mathbb{E}^\infty \int_0^T |\hat{\alpha}_s^\infty - \hat{\alpha}_s^{o,\infty}|^2 ds = 0,$$

and thus $\hat{\boldsymbol{\alpha}}^\infty$ has an $\mathbb{F}^\infty$-progressively measurable modification.

Last Step. Whenever compatibility holds true, the third claim in the statement is an obvious consequence of the second claim. □

3.2.5 Proof of the Solvability Theorem

We now turn to the proof of Theorem 3.13. We start with the proof of Proposition 3.12.

Proof of Proposition 3.12

The first step of the proof is to show that the random variables $(\mathfrak{M}^n)_{n\geq 0}$ are tight. This will follow from the following general lemma:

Lemma 3.16 *Assume that for each integer $n \geq 0$, $\boldsymbol{X}^n = (X_t^n)_{0\leq t\leq T}$ is a process on a complete probability space $(\Omega^n, \mathcal{F}^n, \mathbb{P}^n)$ with paths in $\mathcal{C}([0,T];\mathbb{R}^d)$, such that the family of processes $(\mathbb{P}^n \circ (\boldsymbol{X}^n)^{-1})_{n\geq 0}$ is tight and $(\|\boldsymbol{X}^n\|_\infty = \sup_{0\leq t\leq T}|X_t^n|)_{n\geq 0}$ is uniformly square-integrable, in the sense that:*

$$\lim_{a\to\infty}\sup_{n\geq 0}\mathbb{E}^n\Big(\|\boldsymbol{X}^n\|_\infty^2\mathbf{1}_{\{\|X^n\|_\infty\geq a\}}\Big) = 0.$$

Assume also that, for any $n \geq 0$, there exist a sub-σ-field $\mathcal{G}^n \subset \mathcal{F}^n$ together with a random variable $\mathfrak{M}^n$ from $(\Omega^n, \mathcal{G}^n, \mathbb{P}^n)$ into $\mathcal{P}_2(\mathcal{C}([0,T];\mathbb{R}^{2d}))$ which provides a version of the conditional law of $(\boldsymbol{X}^n, \boldsymbol{W}^n)$ given $\mathcal{G}^n$. Then, the random variables $(\mathfrak{M}^n)_{n\geq 0}$ are tight in $\mathcal{P}_2(\mathcal{C}([0,T];\mathbb{R}^{2d}))$.

Proof. We first prove that the random variables $(\mathfrak{M}^n)_{n\geq 0}$ are tight on the space $\mathcal{P}(\mathcal{C}([0,T];\mathbb{R}^{2d}))$, equipped with the topology of weak convergence.

Since the family $(\boldsymbol{X}^n, \boldsymbol{W}^n)_{n\geq 0}$ is tight on $\mathcal{C}([0,T];\mathbb{R}^{2d})$, we can find, for any $\varepsilon > 0$, a sequence of compact subsets $(\mathcal{K}_p \subset \mathcal{C}([0,T];\mathbb{R}^{2d}))_{p\in\mathbb{N}}$ such that:

$$\forall p \in \mathbb{N}, \quad \sup_{n\geq 0}\mathbb{P}^n\big((\boldsymbol{X}^n, \boldsymbol{W}^n) \notin \mathcal{K}_p\big) \leq \frac{\varepsilon}{4^p}.$$

Therefore,

$$\forall p \in \mathbb{N}, \quad \sup_{n\geq 0}\mathbb{E}^n\big(\mathfrak{M}^n(\mathcal{K}_p^\complement)\big) \leq \frac{\varepsilon}{4^p}$$

and then,

$$\forall p \in \mathbb{N}, \quad \sup_{n\geq 0}\mathbb{P}^n\Big(\mathfrak{M}^n(\mathcal{K}_p^\complement) \geq \frac{1}{2^p}\Big) \leq \frac{\varepsilon}{2^p}.$$

Summing over $p \in \mathbb{N}$, we deduce that:

$$\sup_{n\geq 0}\mathbb{P}^n\Big(\bigcup_{p\in\mathbb{N}}\{\mathfrak{M}^n(\mathcal{K}_p^\complement) \geq \frac{1}{2^p}\}\Big) \leq 2\varepsilon.$$

Clearly the collection of probability measures $\mathfrak{m} \in \mathcal{P}(\mathcal{C}([0,T];\mathbb{R}^{2d})$ such that $\mathfrak{m}(\mathcal{K}_p^\complement) \leq 2^{-p}$ for all $p \geq 0$ is relatively compact for the topology of weak convergence. This proves that the sequence $(\mathfrak{M}^n)_{n\geq 0}$ of random measures is tight on $\mathcal{P}(\mathcal{C}([0,T];\mathbb{R}^{2d}))$ equipped with the weak topology.

In order to prove that tightness holds on $\mathcal{P}_2(\mathcal{C}([0,T];\mathbb{R}^{2d}))$ equipped with the Wasserstein distance, we make use of the second assumption in the statement. We know that the sequence of probability measures $(\mathbb{P}^n \circ (\sup_{0\le t\le T}|(X_t^n, W_t^n)|)^{-1})_{n\geq 0}$ is uniformly square-integrable. Therefore, proceeding as above, we can find, for any $\varepsilon > 0$, a sequence of positive reals $(a_p)_{p\in\mathbb{N}}$ such that:

$$\forall p \in \mathbb{N}, \quad \sup_{n\geq 0} \mathbb{E}^n\Big(\sup_{0\le t\le T} |(X_t^n, W_t^n)|^2 \mathbf{1}_{\{\sup_{0\le t\le T}|(X_t^n,W_t^n)|\geq a_p\}}\Big) \le \frac{\varepsilon}{4^p},$$

from which we get:

$$\forall p \in \mathbb{N}, \quad \sup_{n\geq 0} \mathbb{E}^n\bigg(\int_{\|(x,w)\|_\infty \geq a_p} \|(x,w)\|_\infty^2 d\mathfrak{M}^n(x,w)\bigg) \le \frac{\varepsilon}{4^p}.$$

Therefore,

$$\forall p \in \mathbb{N}, \quad \sup_{n\geq 0} \mathbb{P}^n\bigg(\int_{\|(x,w)\|_\infty \geq a_p} \|(x,w)\|_\infty^2 d\mathfrak{M}^n(x,w) \geq 2^{-p}\bigg) \le \frac{\varepsilon}{2^p}.$$

Finally,

$$\sup_{n\geq 0} \mathbb{P}^n\bigg(\bigcup_{p\in\mathbb{N}} \bigg\{\int_{\|(x,w)\|_\infty \geq a_p} \|(x,w)\|_\infty^2 d\mathfrak{M}^n(x,w) \geq 2^{-p}\bigg\}\bigg) \le 2\varepsilon.$$

The result follows from Corollary (Vol I)-5.6. □

We now turn to:

Proof of Proposition 3.12. The tightness property of the random variables $(\mathfrak{M}^n)_{n\geq 0}$ is a direct consequence of Lemma 3.16.

It thus remains to prove the compatibility property. To this end, we notice that, for all real-valued bounded and continuous functions φ on $\mathcal{C}([0,T];\mathbb{R}^{2d})$ and ψ on $\mathcal{C}([0,T];\mathbb{R}^d) \times \mathcal{P}_2(\mathcal{C}([0,T];\mathbb{R}^{2d}))$,

$$\mathbb{E}^n\big[\varphi(\boldsymbol{X}^n, \boldsymbol{W}^n)\psi(\boldsymbol{W}^{0,n}, \mathfrak{M}^n)\big] = \mathbb{E}^n\bigg[\bigg(\int_{\mathcal{C}([0,T];\mathbb{R}^{2d})} \varphi(x,w)d\mathfrak{M}^n(x,w)\bigg)\psi(\boldsymbol{W}^{0,n}, \mathfrak{M}^n)\bigg].$$

Since the mapping $\mathcal{P}_2(\mathcal{C}([0,T];\mathbb{R}^{2d})) \ni \mathrm{m} \mapsto \int_{\mathcal{C}([0,T];\mathbb{R}^{2d})} \varphi(x,w)d\mathrm{m}(x,w)$ is continuous, we may pass to the limit as n tends to ∞. We denote by $(X_0^\infty, \boldsymbol{W}^{0,\infty}, \mathfrak{M}^\infty, \boldsymbol{W}^\infty, \boldsymbol{X}^\infty)$ a process whose law under $\mathbb{P}^\infty$ is a weak limit of the sequence $(\mathbb{P}^n \circ (X_0^n, \boldsymbol{W}^{0,n}, \mathfrak{M}^n, \boldsymbol{W}^n, \boldsymbol{X}^n)^{-1})_{n\geq 0}$; then, we get:

$$\begin{aligned}
&\mathbb{E}^\infty\big[\varphi(\boldsymbol{X}^\infty, \boldsymbol{W}^\infty)\psi(\boldsymbol{W}^{0,\infty}, \mathfrak{M}^\infty)\big] \\
&\quad = \mathbb{E}^\infty\bigg[\bigg(\int_{\mathcal{C}([0,T];\mathbb{R}^{2d})} \varphi(x,w)d\mathfrak{M}^\infty(x,w)\bigg)\psi(\boldsymbol{W}^{0,\infty}, \mathfrak{M}^\infty)\bigg],
\end{aligned}$$

so that, in the limiting setting as well, $\mathfrak{M}^\infty$ is a version of the conditional law of $(\boldsymbol{X}^\infty, \boldsymbol{W}^\infty)$ given $(\boldsymbol{W}^{0,\infty}, \mathfrak{M}^\infty)$.

Now, compatibility follows from the same argument as in the proof of Remark 2.19. We do it again in order to make the argument self-contained. Following the proof of Lemma 1.7, it suffices to prove that, for all $t \in [0,T]$, $\mathcal{F}_T^{\mathrm{nat},(\boldsymbol{W}^{0,\infty},\mathfrak{M}^\infty)}$ and $\mathcal{F}_t^{\mathrm{nat},(\boldsymbol{W}^{0,\infty},\mathfrak{M}^\infty,\boldsymbol{X}^\infty,\boldsymbol{W}^\infty)}$, are conditionally independent given $\mathcal{F}_t^{\mathrm{nat},(W^{0,\infty},\mathfrak{M}^\infty)}$. Observe indeed that X_0^∞, which usually appears in compatibility conditions, is here encoded by $\boldsymbol{X}^\infty$.

For a given $t \in [0,T]$, consider four Borel subsets $C_t, C_T, B_t, E_t \subset \mathcal{C}([0,T];\mathbb{R}^d)$ and two Borel subsets $D_t, D_T \subset \mathcal{P}_2(\mathcal{C}([0,T];\mathbb{R}^{2d}))$. Letting $\mathfrak{M}_t^\infty = \mathfrak{M}^\infty \circ \mathcal{E}_t^{-1}$, where $\mathcal{E}_t : \mathcal{C}([0,T];\mathbb{R}^{2d}) \ni (\boldsymbol{x},\boldsymbol{w}) \mapsto (x_{s\wedge t}, w_{s\wedge t})_{0\le s\le T} \in \mathcal{C}([0,T];\mathbb{R}^d)$, we have:

$$
\begin{aligned}
&\mathbb{E}\big[\mathbf{1}_{C_t}(\boldsymbol{W}^{0,\infty}_{\cdot\wedge t})\mathbf{1}_{D_t}(\mathfrak{M}_t^\infty)\mathbf{1}_{B_t}(\boldsymbol{X}^\infty_{\cdot\wedge t})\mathbf{1}_{E_t}(\boldsymbol{W}^\infty_{\cdot\wedge t})\mathbf{1}_{C_T}(\boldsymbol{W}^{0,\infty})\mathbf{1}_{D_T}(\mathfrak{M}_T^\infty)\big] \\
&= \mathbb{E}\Big[\mathbf{1}_{C_t}(\boldsymbol{W}^{0,\infty}_{\cdot\wedge t})\mathbf{1}_{D_t}(\mathfrak{M}_t^\infty)\mathbf{1}_{C_T}(\boldsymbol{W}^{0,\infty})\mathbf{1}_{D_T}(\mathfrak{M}_T^\infty)\mathfrak{M}^\infty\big[\mathcal{E}_t^{-1}(B_t\times E_t)\big]\Big] \\
&= \mathbb{E}\Big[\mathbf{1}_{C_t}(\boldsymbol{W}^{0,\infty}_{\cdot\wedge t})\mathbf{1}_{D_t}(\mathfrak{M}_t^\infty)\mathbf{1}_{C_T}(\boldsymbol{W}^{0,\infty})\mathbf{1}_{D_T}(\mathfrak{M}_T^\infty)\mathfrak{M}_t^\infty(B_t\times E_t)\Big].
\end{aligned}
$$

Since $\mathfrak{M}_t^\infty(B_t \times E_t)$ is measurable with respect to $\mathcal{F}_t^{\mathrm{nat},(W^{0,\infty},\mathfrak{M}^\infty)}$, compatibility easily follows.

The conclusion follows from Proposition 3.11, applied with $(\mathfrak{M}^n)_{n\ge 0}$ as sequence of environments and using the fact that the mapping $\mathcal{P}_2(\mathcal{C}([0,T];\mathbb{R}^{2d})) \ni \mathfrak{m} \mapsto (\mathfrak{m}\circ(e_t^x)^{-1})_{0\le t\le T} \in \mathcal{C}([0,T];\mathcal{P}_2(\mathbb{R}^d))$ is continuous. □

Conclusion

We now have all the ingredients to complete the proof of Theorem 3.13:

Proof of Theorem 3.13. The first point is to check that, under the assumptions of Theorem 3.13, the assumptions of Proposition 3.12 are satisfied. The main point is to observe from Lemma 3.16 that the sequence $(\mathbb{P}^{0,n}\circ(\mathfrak{M}^n)^{-1})_{n\ge 0}$ is tight on $\mathcal{P}_2(\mathcal{C}([0,T];\mathbb{R}^{2d}))$. As a consequence, the sequence $(\mathbb{P}^{0,n}\circ(\boldsymbol{\mu}^n)^{-1})_{n\ge 0}$ is tight on $\mathcal{C}([0,T];\mathcal{P}_2(\mathbb{R}^d))$, where $\boldsymbol{\mu}^n = (\mu_t^n = \mathfrak{M}^n\circ(e_t^x)^{-1})_{0\le t\le T}$. This follows from the fact the mapping $\mathcal{P}_2(\mathcal{C}([0,T];\mathbb{R}^{2d})) \ni \mathfrak{m} \mapsto (\mathfrak{m}\circ(e_t^x)^{-1})_{0\le t\le T} \in \mathcal{C}([0,T];\mathcal{P}_2(\mathbb{R}^d))$ is continuous. Moreover, since the measures $(\mathbb{P}^n \circ (\sup_{0\le t\le T}|X_t^n|)^{-1})_{n\ge 0}$ are assumed to be uniformly square-integrable, it is straightforward to check that the sequence $(\mathbb{P}^{0,n}\circ(\sup_{0\le t\le T}\|\mu_t^n\|_2)^{-1})_{n\ge 0}$ is also uniformly square-integrable. This shows that assumption **Control Bounds** is satisfied.

We now apply Proposition 3.12. There is a complete probability space $(\Omega^\infty,\mathcal{F}^\infty,\mathbb{P}^\infty)$, equipped with a random process $(X_0^\infty, \boldsymbol{W}^{0,\infty}, \mathfrak{M}^\infty, \boldsymbol{W}^\infty, \boldsymbol{X}^\infty)$ and with the complete and right-continuous filtration $\mathbb{F}^\infty$ generated by $(X_0^\infty, \boldsymbol{W}^{0,\infty}, \mathfrak{M}^\infty, \boldsymbol{W}^\infty, \boldsymbol{X}^\infty)$, for which $\mathbb{F}^\infty$ is compatible with $(X_0^\infty, \boldsymbol{W}^{0,\infty}, \mathfrak{M}^\infty, \boldsymbol{W}^\infty)$ and $\boldsymbol{X}^\infty$ is an optimal path for the stochastic optimal control problem (3.16)–(3.17), when it is understood with respect to the filtration $\mathbb{F}^\infty$, to the super-environment $\mathfrak{M}^\infty$ and to the sub-environment $\boldsymbol{\mu}^\infty = (\mu_t^\infty = \mathfrak{M}^\infty \circ (e_t^x)^{-1})_{0\le t\le T}$. Moreover, as shown by the proof of Proposition 3.12, $\mathfrak{M}^\infty$ provides a version of the conditional law of $(\boldsymbol{X}^\infty, \boldsymbol{W}^\infty)$ given $(\boldsymbol{W}^{0,\infty}, \mathfrak{M}^\infty)$.

The end of the proof is a mere adaptation of the argument used to establish Lemma 2.25. We duplicate it for the sake of completeness. To do so, we appeal to assumption **FBSDE**. We deduce that the process $\boldsymbol{X}^\infty$ is the unique solution of a uniquely solvable FBSDE, called $(\star)$, set on the probability space $(\Omega^\infty,\mathcal{F}^\infty,\mathbb{P}^\infty)$ equipped with the tuple $(X_0^\infty, \boldsymbol{W}^{0,\infty}, \mathfrak{M}^\infty, \boldsymbol{W}^\infty)$. We transfer this solution to the *extended* canonical space introduced in Chapter 2, see (2.32):

$$
\begin{aligned}
\bar{\Omega}^0 &= \mathcal{P}_2(\mathbb{R}^d)\times\mathcal{C}([0,T];\mathbb{R}^d)\times\mathcal{P}_2(\mathcal{C}([0,T];\mathbb{R}^{2d})), \\
\bar{\Omega}^1 &= [0,1)\times\mathcal{C}([0,T];\mathbb{R}^d).
\end{aligned}
$$

We equip $\bar{\Omega}^0$ with the law of the process $(\mu_0^\infty, \boldsymbol{W}^{0,\infty}, \mathfrak{M}^\infty)$ under $\mathbb{P}^\infty$ and we call $(\bar{\Omega}^0, \bar{\mathcal{F}}^0, \bar{\mathbb{P}}^0)$ the completed space. The complete and right-continuous augmentation of the canonical filtration is denoted by $\bar{\mathbb{F}}^0$ and the canonical random variable is denoted by $(\nu^0, \boldsymbol{w}, \mathfrak{m})$. We then denote by $\boldsymbol{\nu} = (\nu_t = \mathfrak{m} \circ (e_t^x)^{-1})_{0 \le t \le T}$ the sub-environment deriving from $\mathfrak{m}$. As in (2.32), we equip $\bar{\Omega}^1$ with the product of the uniform and the Wiener laws and we call $(\bar{\Omega}^1, \bar{\mathcal{F}}^1, \bar{\mathbb{P}}^1)$ the completed space. The complete and right-continuous augmentation of the canonical filtration is denoted by $\bar{\mathbb{F}}^1$ and the canonical random variable is denoted by $(\eta, \boldsymbol{w})$. Following Definition 2.24, the completion of the product of the spaces $(\bar{\Omega}^0, \bar{\mathcal{F}}^0, \bar{\mathbb{F}}^0, \bar{\mathbb{P}}^0)$ and $(\bar{\Omega}^1, \bar{\mathcal{F}}^1, \bar{\mathbb{F}}^1, \bar{\mathbb{P}}^1)$ is denoted by $(\bar{\Omega}, \bar{\mathcal{F}}, \bar{\mathbb{F}}, \bar{\mathbb{P}})$. On $(\bar{\Omega}, \bar{\mathcal{F}}, \bar{\mathbb{F}}, \bar{\mathbb{P}})$, we may solve the FBSDE $(\star)$ with $X_0 = \psi(\eta, \nu^0)$ as initial conditions, with ψ as in (2.23). The forward component of the solution is denoted by $\boldsymbol{X} = (X_t)_{0 \le t \le T}$. By assumption **FBSDE**, $\boldsymbol{X} = (X_t)_{0 \le t \le T}$ is the unique optimal path of the analogue of the stochastic control problem (3.16)–(3.17), but set on $(\bar{\Omega}, \bar{\mathcal{F}}, \bar{\mathbb{F}}, \bar{\mathbb{P}})$ equipped with $(X_0, \boldsymbol{w}^0, (\mathfrak{m}, \boldsymbol{\nu}), \boldsymbol{w})$.

Importantly, by Theorem 1.33, the two distributions $\bar{\mathbb{P}} \circ (\nu^0, \boldsymbol{w}^0, \mathfrak{m}, \boldsymbol{w}, \boldsymbol{X})^{-1}$ and $\mathbb{P}^\infty \circ (\mu_0^\infty, \boldsymbol{W}^{0,\infty}, \mathfrak{M}^\infty, \boldsymbol{W}^\infty, \boldsymbol{X}^\infty)^{-1}$ are equal. Since $\mu_0^\infty = \mathfrak{M}^\infty \circ (e_0^x)^{-1}$ and $\mathfrak{M}^\infty$ is the conditional law of $(\boldsymbol{X}^\infty, \boldsymbol{W}^\infty)$ given $(\boldsymbol{W}^{0,\infty}, \mathfrak{M}^\infty)$ under $\mathbb{P}^\infty$, we also have that $\mathfrak{m}$ is the conditional law of $(\boldsymbol{X}, \boldsymbol{w})$ given $(\boldsymbol{w}^0, \mathfrak{m})$ under $\bar{\mathbb{P}}$. Equivalently, for $\bar{\mathbb{P}}^0$-almost every $\omega^0 \in \bar{\Omega}^0$, $\mathcal{L}(\boldsymbol{X}(\omega^0, \cdot), \boldsymbol{w}) = \mathfrak{m}(\omega^0)$, which completes the proof. □

3.3 Solving MFGs by Constructing Approximate Equilibria

We now turn to the proof of Theorem 3.1. The strategy consists in two main steps:

(i) We first solve an approximate problem in which the conditioning is discrete;
(ii) Next, we extract a converging subsequence by means of a tightness argument, using the tools introduced in the previous section.

Throughout the section, $\mathcal{V}^0$ denotes an initial distribution in $\mathcal{P}_2(\mathcal{P}_2(\mathbb{R}^d))$; also, we assume that assumptions **Coefficients MFG with a Common Noise** and **FBSDE MFG with a Common Noise** are in force.

3.3.1 Approximate Problem

We let $\bar{\Omega}^{00}$ and $\bar{\Omega}^1$ be the canonical spaces $\mathcal{P}_2(\mathbb{R}^d) \times \mathcal{C}([0,T];\mathbb{R}^d)$ and $[0,1) \times \mathcal{C}([0,T];\mathbb{R}^d)$. We denote by $(\nu^0, \boldsymbol{w}^0 = (w_t^0)_{0 \le t \le T})$ the canonical process on $\bar{\Omega}^{00}$ and by $(\eta, \boldsymbol{w} = (w_t)_{0 \le t \le T})$ the canonical process on $\bar{\Omega}^1$. The completion of $\bar{\Omega}^{00}$ equipped with the Borel σ-field and with the product measure $\mathcal{V}^0 \otimes \mathcal{W}_d$ is denoted by $(\bar{\Omega}^0, \bar{\mathcal{F}}^0, \bar{\mathbb{P}}^0)$. We then call $\bar{\mathbb{F}}^0$ the complete and right-continuous augmentation of the canonical filtration. Similarly, the completion of $\bar{\Omega}^1$ equipped with the Borel σ-field and with the product measure $\text{Leb}_1 \otimes \mathcal{W}_d$ is denoted by $(\bar{\Omega}^1, \bar{\mathcal{F}}^1, \bar{\mathbb{P}}^1)$. The complete and right-continuous augmentation of the canonical filtration is denoted by $\bar{\mathbb{F}}^1$. The completion of the product space is denoted by $(\bar{\Omega} = \bar{\Omega}^{00} \times \bar{\Omega}^1, \bar{\mathcal{F}}, \bar{\mathbb{F}}, \bar{\mathbb{P}})$ along the same lines as in (2.5).

Discretization of the Common Noise

In order to construct the discrete conditioning, we choose two integers $\ell, n \geq 1$, ℓ referring to the step size of the space grid and n to the step size of the time grid. Denoting by $\lfloor x \rfloor$ the floor part of x, we then let $\Pi_\Lambda^{(1)}$ be the projection from $\mathbb{R}$ into itself defined by:

$$\Pi_\Lambda^{(1)} : \mathbb{R} \ni x \mapsto \begin{cases} \Lambda^{-1}\lfloor \Lambda x \rfloor & \text{if } |x| \leq \Lambda, \\ \Lambda \text{sign}(x) & \text{if } |x| > \Lambda, \end{cases}$$

with $\Lambda = 2^\ell$, and

$$\Pi_\Lambda^{(d)} : \mathbb{R}^d \ni x = (x_1, \cdots, x_d) \mapsto \Pi_\Lambda^{(d)}(x) = \big(\Pi_\Lambda^{(1)}(x_1), \cdots, \Pi_\Lambda^{(1)}(x_d)\big).$$

For any integer $j \geq 1$, we also consider the projection $\Pi_{\Lambda,j}^{(d)}$ from $(\mathbb{R}^d)^j$ into itself defined iteratively by:

$$\Pi_{\Lambda,1}^{(d)} \equiv \Pi_\Lambda^{(d)},$$
$$\Pi_{\Lambda,j+1}^{(d)}(x^1, \cdots, x^{j+1}) = (y^1, \cdots, y^j, y^{j+1}), \quad \text{for } (x^1, \cdots, x^{j+1}) \in (\mathbb{R}^d)^{j+1},$$

where

$$(y^1, \cdots, y^j) = \Pi_{\Lambda,j}^{(d)}(x^1, \cdots, x^j) \in (\mathbb{R}^d)^j, \quad y^{j+1} = \Pi_\Lambda^{(d)}\big(y^j + x^{j+1} - x^j\big) \in \mathbb{R}^d.$$

The following lemma will be very useful in the sequel.

Lemma 3.17 *With the above notation, for $(x^1, \cdots, x^j) \in (\mathbb{R}^d)^j$ such that, for any $i \in \{1, \cdots, j\}$, $|x^i|_\infty \leq \Lambda - 1$, with $|x^i|_\infty = \sup_{k=1,\cdots,d} |(x^i)_k|$, let:*

$$(y^1, \cdots, y^j) = \Pi_{\Lambda,j}^{(d)}(x^1, \cdots, x^j).$$

Then, provided that $j \leq \Lambda$, it holds for any $i \in \{1, \cdots, j\}$:

$$|x^i - y^i|_\infty \leq \frac{i}{\Lambda}.$$

Proof. When $i = 1$, the result is obvious since $|x^1|_\infty \leq \Lambda - 1$. When $i \in \{2, \cdots, j\}$, for $j \leq \Lambda$, and $|y^{i-1} - x^{i-1}|_\infty \leq (i-1)/\Lambda$, we have:

$$|y^{i-1} + x^i - x^{i-1}|_\infty \leq \Lambda - 1 + \frac{i-1}{\Lambda} < \Lambda,$$

so that:

$$\big|(y^{i-1} + x^i - x^{i-1}) - \Pi_\Lambda^{(d)}(y^{i-1} + x^i - x^{i-1})\big|_\infty \leq \frac{1}{\Lambda}.$$

Therefore,

$$
\begin{aligned}
|x^i - y^i|_\infty &\leq \left|x^i - (y^{i-1} + x^i - x^{i-1})\right|_\infty + \left|(y^{i-1} + x^i - x^{i-1}) - \Pi_\Lambda^{(d)}(y^{i-1} + x^i - x^{i-1})\right|_\infty \\
&\leq \left|x^i - (y^{i-1} + x^i - x^{i-1})\right|_\infty + \frac{1}{\Lambda} = |x^{i-1} - y^{i-1}|_\infty + \frac{1}{\Lambda} \leq \frac{i}{\Lambda},
\end{aligned}
$$

and the result follows by induction. □

Given an integer n, we let $N = 2^n$, we consider the dyadic time mesh:

$$
t_i = \frac{iT}{N}, \quad i \in \{0, 1, \cdots, N\}, \tag{3.31}
$$

and we define the discrete random variables:

$$
(V_1, \cdots, V_{N-1}) = \Pi_{\Lambda,N-1}^{(d)}\left(w_{t_1}^0, \cdots, w_{t_{N-1}}^0\right). \tag{3.32}
$$

By independence of the increments of the Brownian motion, it is pretty clear that:

Lemma 3.18 *Given $i = 1, \cdots, N-1$, the random vector $(V_1, \cdots, V_i)$ has the whole $\mathbb{J}^i$ as support, with $\mathbb{J} = \{-\Lambda, -\Lambda + 1/\Lambda, -\Lambda + 2/\Lambda, \cdots, \Lambda - 1/\Lambda, \Lambda\}^d$.*

The random variables $V_1, \cdots, V_{N-1}$ must be understood as a discretization of the common noise $\boldsymbol{w}^0$.

Discretization of the Initial Distribution

We now proceed with the discretization of the initial distribution ν^0. With the same notation as above, we may consider $\nu^0 \circ (\Pi_\Lambda^{(d)})^{-1}$, which is a probability measure on $\mathbb{J}$.

Here, $\nu^0 \circ (\Pi_\Lambda^{(d)})^{-1}$ reads as a vector of weights indexed by the elements of $\mathbb{J}$, namely $([\nu^0 \circ (\Pi_\Lambda^{(d)})^{-1}](x))_{x \in \mathbb{J}}$. We then define new weights, with values in $\{0, 1/\Lambda^{2d+4}, \cdots, 1 - 1/\Lambda^{2d+4}, 1\}$:

$$
\begin{aligned}
\mu_0 &= \sum_{x \in \mathbb{J}} \Pi_{\Lambda^{2d+4}}^{(1)}\left(\left[\nu^0 \circ (\Pi_\Lambda^{(d)})^{-1}\right](x)\right)\delta_x \\
&\quad + \left(1 - \sum_{x \in \mathbb{J}} \Pi_{\Lambda^{2d+4}}^{(1)}\left(\left[\nu^0 \circ (\Pi_\Lambda^{(d)})^{-1}\right](x)\right)\right)\delta_0.
\end{aligned} \tag{3.33}
$$

Importantly, $\rho^0 = (\mu_0(x))_{x \in \mathbb{J}}$ forms a random variable with values in the set $\{0, 1/\Lambda^{2d+4}, \cdots, 1\}^{\mathbb{J}}$. Its law is denoted by $\mathcal{L}(\rho^0)$ and we denote by $\mathrm{Supp}(\mathcal{L}(\rho^0)) \subset \{0, 1/\Lambda^{2d+4}, \cdots, 1\}^{\mathbb{J}}$ its support. The law of μ_0, seen as a random variable with values in $\mathcal{P}_2(\mathbb{R}^d)$, is denoted by $\mathcal{L}(\mu_0)$ and its support is $\mathbb{S}^0 = \mathrm{Supp}(\mathcal{L}(\mu_0)) = \{\sum_{x \in \mathbb{J}} \varrho(x)\delta_x;\ \varrho \in \mathrm{Supp}(\mathcal{L}(\rho^0))\}$.

By independence of ν^0 and $\boldsymbol{w}^0$, the random variable $(\mu^0, V_1, \cdots, V_{N-1})$ has the whole finite product space $\mathbb{S}^0 \times \mathbb{J}^{N-1}$ as its topological support.

Discretized Input

Below, we call a *discretized input* a family $\boldsymbol{\vartheta} = (\vartheta^0, \cdots, \vartheta^{N-1})$ where, for any $i = 0, \cdots, N-1$,

$$\vartheta^i : \mathbb{S}^0 \times \mathbb{J}^i \to \mathcal{C}\big([t_i, t_{i+1}]; \mathcal{P}_2(\mathbb{R}^d)\big), \tag{3.34}$$

$\mathcal{P}_2(\mathbb{R}^d)$ being endowed with the 2-Wasserstein distance with the constraint that, when $i = 0$, $\vartheta^0(\varsigma) = \varsigma$ for all $\varsigma \in \mathbb{S}^0$. We then let, for all $(\varsigma, v_1, \cdots, v_{N-1}) \in \mathbb{S}^0 \times \mathbb{J}^{N-1}$ and $t \in [0, T]$,

$$\begin{cases} \vartheta_t(\varsigma, v_1, \cdots, v_{N-1}) = \big(\vartheta^i(\varsigma, v_1, \cdots, v_i)\big)_t, \\ \qquad\qquad\qquad\qquad t \in [t_i, t_{i+1}),\ i \in \{0, \cdots, N-1\}, \\ \vartheta_T(\varsigma, v_1, \cdots, v_{N-1}) = \big(\vartheta^{N-1}(\varsigma, v_1, \cdots, v_{N-1})\big)_T, \end{cases} \tag{3.35}$$

which permits to define the $\mathcal{P}_2(\mathbb{R}^d)$-valued càd-làg process:

$$\mu_t = \vartheta_t(\mu_0, V_1, \cdots, V_{N-1}), \quad t \in [0, T]. \tag{3.36}$$

Given such an input $\boldsymbol{\vartheta}$, we consider the associated forward-backward SDE (3.4) with $\boldsymbol{\mu} = (\mu_t)_{0 \le t \le T}$ given by (3.36). By assumption **FBSDE MFG with a Common Noise**, we know that, for a given discretized input $\boldsymbol{\vartheta}$, the FBSDE (3.4) set on the space $(\bar{\Omega}, \bar{\mathcal{F}}, \bar{\mathbb{F}}, \bar{\mathbb{P}})$ equipped with $(X_0 = \psi(\eta, \mu^0), \boldsymbol{w}^0, \boldsymbol{\mu}, \boldsymbol{w})$ has a unique solution, with ψ as in (2.23). Notice that there is no need to check any Compatibility Condition since the environment $\boldsymbol{\mu}$ is adapted with respect to the filtration generated by $(\nu^0, \boldsymbol{w}^0)$. For the same reason, pay attention that $\boldsymbol{M} \equiv 0$.

We then call *discretized output* the family $\Phi(\boldsymbol{\vartheta}) = (\varphi^0(\boldsymbol{\vartheta}), \cdots, \varphi^{N-1}(\boldsymbol{\vartheta}))$ of measures:

$$\begin{cases} \Phi_t(\varsigma, v_1, \cdots, v_{N-1}) = \big(\varphi^i(\boldsymbol{\vartheta})(\varsigma, v_1, \cdots, v_i)\big)_t, \\ \qquad\qquad\qquad\qquad t \in [t_i, t_{i+1}),\ i \in \{0, \cdots, N-1\}, \\ \Phi_T(\varsigma, v_1, \cdots, v_{N-1}) = \big(\varphi^{N-1}(\boldsymbol{\vartheta})(\varsigma, v_1, \cdots, v_{N-1})\big)_T, \end{cases}$$

with

$$\varphi^i(\boldsymbol{\vartheta})(\varsigma, v_1, \cdots, v_i) : [t_i, t_{i+1}] \ni t \mapsto \mathcal{L}\big(X_t \,|\, \mu_0 = \varsigma, V_1 = v_1, \cdots, V_i = v_i\big), \tag{3.37}$$

for $\varsigma \in \mathbb{S}^0$ and $v_1, \cdots, v_i \in \mathbb{J}$, which is well defined as a function from $\mathbb{S}^0 \times \mathbb{J}^i$ into $\mathcal{C}([t_i, t_{i+1}]; \mathcal{P}_2(\mathbb{R}^d))$ since the law of $(\mu_0, V_1, \cdots, V_i)$ has the whole $\mathbb{S}^0 \times \mathbb{J}^i$ as support.

Pay attention that $\Phi(\boldsymbol{\vartheta})$ and $(\Phi_t(\boldsymbol{\vartheta}))_{0\le t\le T}$ denote two slightly different objects, but we can easily associate $(\Phi_t(\boldsymbol{\vartheta}))_{0\le t\le T}$ with $\Phi(\boldsymbol{\vartheta})$ and conversely.

The purpose of the next section is to prove:

Theorem 3.19 *Within the above framework, including two integers $\ell, n \ge 1$, let $\Lambda = 2^\ell$ and $N = 2^n$. Then, the mapping Φ defined by* (3.37) *has a fixed point $\boldsymbol{\vartheta} = (\vartheta^0, \cdots, \vartheta^{N-1})$, $(\vartheta^0, \cdots, \vartheta^{N-1})$ forming a discretized input as in* (3.34). *It satisfies:*

$$\sup_{t\in[0,T]} \sup_{\varsigma\in\mathbb{S}^0} \sup_{v_1,\cdots,v_{N-1}\in\mathbb{J}} \left[\int_{\mathbb{R}^d} |x|^4 \big[\Phi_t(\boldsymbol{\vartheta})(\varsigma, v_1, \cdots, v_{N-1})\big](dx)\right] \le C,$$

and

$$\sup_{s,t\in[0,T]} \sup_{\varsigma\in\mathbb{S}^0} \sup_{v_1,\cdots,v_{N-1}\in\mathbb{J}} \left[W_2\Big(\Phi_t(\boldsymbol{\vartheta})(\varsigma, v_1, \cdots, v_{N-1}), \Phi_s(\boldsymbol{\vartheta})(\varsigma, v_1, \cdots, v_{N-1})\Big)\right]$$
$$\le C(t-s)^{1/2},$$

for a constant $C \ge 0$, possibly depending on ℓ and n.

3.3.2 Solvability of the Approximate Fixed Point Problem

This subsection is devoted to the proof of Theorem 3.19, the values of ℓ, Λ and n, N being fixed, with $\Lambda = 2^\ell$ and $N = 2^n$ throughout.

Preliminary Lemmas

The proof is divided into several lemmas. Given a discretized input $\boldsymbol{\vartheta}$, we first prove estimates which will be needed later on. We denote by $(\boldsymbol{X}, \boldsymbol{Y}, \boldsymbol{Z}, \boldsymbol{Z}^0, \boldsymbol{M} \equiv 0)$ the solution to the FBSDE (3.4) driven by $\boldsymbol{\mu}$ given by (3.36), and by the initial condition $X_0 = \psi(\eta, \mu_0)$, with μ_0 as in (3.33).

Lemma 3.20 *There exists a constant C, depending on ℓ and n but not on $\boldsymbol{\vartheta}$, such that, for any $i \in \{0, \cdots, N-1\}$, for any $s, t \in [t_i, t_{i+1}]$ and for any $\varsigma \in \mathbb{S}^0$ and $v_1, \cdots, v_i \in \mathbb{J}^i$,*

$$W_2\Big(\mathcal{L}(X_t \,|\, \mu_0 = \varsigma, V_1 = v_1, \cdots, V_i = v_i),\ \mathcal{L}(X_s \,|\, \mu_0 = \varsigma, V_1 = v_1, \cdots, V_i = v_i)\Big)$$
$$\le C(t-s)^{1/2}. \tag{3.38}$$

In particular, for each $i \in \{0, \cdots, N-1\}$, $\varphi^i(\boldsymbol{\vartheta})$ appears as a mapping from $\mathbb{S}^0 \times \mathbb{J}^i$ into $\mathcal{C}([t_i, t_{i+1}]; \mathcal{P}_2(\mathbb{R}^d))$, $\mathcal{P}_2(\mathbb{R}^d)$ being endowed with the 2-Wasserstein distance W_2.

Proof. For a given $(\varsigma, v_1, \cdots, v_{N-1}) \in \mathbb{S}^0 \times \mathbb{J}^{N-1}$, we know that $\mathbb{P}[\mu_0 = \varsigma, V_1 = v_1, \cdots, V_{N-1} = v_{N-1}] > 0$. Therefore, for any $i \in \{0, \cdots, N-1\}$, for any $s, t \in [t_i, t_{i+1}]$,

$$W_2\Big(\mathcal{L}(X_t \,|\, \mu_0 = \varsigma, V_1 = v_1, \cdots, V_i = v_i), \mathcal{L}(X_s \,|\, \mu_0 = \varsigma, V_1 = v_1, \cdots, V_i = v_i)\Big)$$

$$\leq \bar{\mathbb{E}}\big[|X_t - X_s|^2 \,|\, \mu_0 = \varsigma, V_1 = v_1, \cdots, V_i = v_i\big]^{1/2}$$

$$\leq \left(\frac{\bar{\mathbb{E}}\big[|X_t - X_s|^2\big]}{\bar{\mathbb{P}}\big[\mu_0 = \varsigma, V_1 = v_1, \cdots, V_i = v_i\big]}\right)^{1/2}.$$

Since the coefficients F and G satisfy the growth condition from (**A5**) in assumption **FBSDE MFG with a Common Noise**, we can find a constant C, independent of the input $\boldsymbol{\vartheta}$, such that:

$$\bar{\mathbb{E}}\left[\sup_{t \in [0,T]} |Y_t|^4 + \left(\int_0^T (|Z_t|^2 + |Z_t^0|^2)dt\right)^2\right] \leq C,$$

which is a quite standard inequality in BSDE theory since $\boldsymbol{M} \equiv 0$ here.

Using the fact that the coefficients B, σ, and σ^0 satisfy the growth condition from (**A5**) in assumption **FBSDE MFG with a Common Noise** and applying Gronwall's lemma, we deduce that there exists a constant C, independent of the input $\boldsymbol{\vartheta}$, such that:

$$\bar{\mathbb{E}}\Big[\sup_{t \in [0,T]} |X_t|^4\Big] \leq C, \tag{3.39}$$

where we used the fact that any realization of μ_0 has support included in $[-\Lambda, \Lambda]^d$.

Finally, using once again (**A5**) in assumption **FBSDE MFG with a Common Noise**, we can find a constant C such that, for any $s, t \in [0, T]$,

$$\bar{\mathbb{E}}\big[|X_t - X_s|^2\big] \leq C|t - s|.$$

Since

$$\inf_{\varsigma \in \mathbb{S}^0} \inf_{v_1, \cdots, v_{N-1} \in \mathbb{J}} \Big[\bar{\mathbb{P}}\big[\mu_0 = \varsigma, V_1 = v_1, \cdots, V_{N-1} = v_{N-1}\big]\Big] > 0,$$

we can easily complete the proof. □

With the notation introduced in (3.37), Lemma 3.20 says that:

$$\forall \varsigma \in \mathbb{S}^0,\ \forall v_1, \cdots, v_{N-1} \in \mathbb{J},\ \forall i \in \{0, \cdots, N-1\},\ \forall s, t \in [t_i, t_{i+1}],$$
$$\Big|\big[\varphi^i(\boldsymbol{\vartheta})(\varsigma, v_1, \cdots, v_i)\big]_t - \big[\varphi^i(\boldsymbol{\vartheta})(\varsigma, v_1, \cdots, v_i)\big]_s\Big| \leq C|t - s|^{1/2},$$

so that, for each $i \in \{0, \cdots, N-1\}$, $\varphi^i(\boldsymbol{\vartheta})(\varsigma, v_1, \cdots, v_i) \in \mathcal{C}([t_i, t_{i+1}]; \mathcal{P}_2(\mathbb{R}^d))$ or, equivalently, $\varphi^i(\boldsymbol{\vartheta}) \in [\mathcal{C}([t_i, t_{i+1}]; \mathcal{P}_2(\mathbb{R}^d))]^{\mathbb{S}^0 \times \mathbb{J}^i}$. Put differently, we can view $\Phi = (\varphi^0, \cdots, \varphi^{N-1})$ as a mapping from the set $[\mathcal{C}([t_i, t_{i+1}]; \mathcal{P}_2(\mathbb{R}^d))]^{\mathbb{S}^0 \times \mathbb{J}^i}$ into itself:

$$
\begin{aligned}
\Phi = \big(\varphi^0, \cdots, \varphi^{N-1}\big) :& \\
&\prod_{i=0}^{N-1} \big[\mathcal{C}([t_i, t_{i+1}]; \mathcal{P}_2(\mathbb{R}^d))\big]^{\mathbb{S}^0 \times \mathbb{J}^i} \rightarrow \prod_{i=0}^{N-1} \big[\mathcal{C}([t_i, t_{i+1}]; \mathcal{P}_2(\mathbb{R}^d))\big]^{\mathbb{S}^0 \times \mathbb{J}^i} \\
&(\vartheta^0, \cdots, \vartheta^{N-1}) = \boldsymbol{\vartheta} \mapsto \Phi(\boldsymbol{\vartheta}) = \big(\varphi^0(\boldsymbol{\vartheta}), \cdots, \varphi^{N-1}(\boldsymbol{\vartheta})\big).
\end{aligned}
$$

As a by-product, we get the following reformulation of Lemma 3.20:

Lemma 3.21 *There exists a constant C, depending on ℓ and n, such that, for any discretized input $\boldsymbol{\vartheta} = (\vartheta^0, \cdots, \vartheta^{N-1})$, the output $\Phi(\boldsymbol{\vartheta})$ satisfies:*

$$
\sup_{t \in [0,T]} \sup_{\varsigma \in \mathbb{S}^0} \sup_{v_1, \cdots, v_{N-1} \in \mathbb{J}} \left[\int_{\mathbb{R}^d} |x|^4 \big[\Phi_t(\boldsymbol{\vartheta})(\varsigma, v_1, \cdots, v_{N-1})\big](dx) \right] \leq C, \tag{3.40}
$$

and

$$
\begin{aligned}
\sup_{s,t \in [0,T]} \sup_{\varsigma \in \mathbb{S}^0} \sup_{v_1, \cdots, v_{N-1} \in \mathbb{J}} &\Big[W_2\Big(\Phi_t(\boldsymbol{\vartheta})(\varsigma, v_1, \cdots, v_{N-1}), \Phi_s(\boldsymbol{\vartheta})(\varsigma, v_1, \cdots, v_{N-1})\Big)\Big] \\
&\leq C(t-s)^{1/2}.
\end{aligned} \tag{3.41}
$$

Proof. The proof of (3.40) uses the same ingredients as the proof of Lemma 3.20. First, we observe that the support of $\mu_0, V_1, \cdots, V_{N-1}$ is of finite cardinality. Second, we notice from (3.39) that, for any input $\boldsymbol{\vartheta} \in \prod_{i=0}^{N-1} [\mathcal{C}([t_i, t_{i+1}]; \mathcal{P}_2(\mathbb{R}^d))]^{\mathbb{S}^0 \times \mathbb{J}^i}$, the forward component X of the system (3.4) satisfies:

$$
\sup_{0 \leq t \leq T} \bar{\mathbb{E}}\big[|X_t|^4\big] \leq C,
$$

for some constant C independent of the input $\boldsymbol{\vartheta}$. Inequality (3.40) easily follows. Inequality (3.41) then follows from (3.38). □

Final Step

Here is now the final step in the proof of Theorem 3.19.

Lemma 3.22 *With the same notation as in* (3.34) *and* (3.35), *consider the set $\mathcal{E}$ of discretized inputs $\boldsymbol{\vartheta}$ such that, for any $i \in \{0, \cdots, N-1\}$, any $t \in [t_i, t_{i+1}]$, any $\varsigma \in \mathbb{S}^0$ and any $v_1, \cdots, v_i \in \mathbb{J}^i$,*

$$
\int_{\mathbb{R}^d} |x|^4 \big[\vartheta^i(\varsigma, v_1, \cdots, v_i)\big]_t(dx) \leq C, \tag{3.42}
$$

where the constant C is the same as in Lemma 3.21. Then, the restriction of the mapping Φ to $\mathcal{E}$ has a fixed point.

Proof. Following the proof of Theorem (Vol I)-4.39, existence of a fixed point is proved by means of Schauder's fixed point theorem. To this end, we first notice that $\mathcal{E}$ is a convex subspace of the product space:

$$\prod_{i=0}^{N-1} \big[\mathcal{C}\big([t_i, t_{i+1}]; \mathcal{M}_f^1(\mathbb{R}^d)\big)\big]^{\mathbb{S}^0 \times \mathbb{J}^i},$$

where $\mathcal{M}_f^1(\mathbb{R}^d)$ denotes the set of finite signed measures μ over $\mathbb{R}^d$ such that $\mathbb{R}^d \ni x \mapsto |x|$ is integrable under $|\mu|$ equipped with the Kantorovich-Rubinstein norm $\|\mu\|_{\mathrm{KR}\star}$defined as

$$\|\mu\|_{\mathrm{KR}\star} = |\mu(\mathbb{R}^d)| + \sup\left\{ \int_{\mathbb{R}^d} h(x) d\mu(x); \quad h \in \mathrm{Lip}_1(\mathbb{R}^d),\ h(0) = 0 \right\}.$$

In order to apply Schauder's fixed point theorem, we prove that Φ is continuous and that its range has a compact closure.

Continuity follows from (**A2**) in assumption **FBSDE MFG with a Common Noise**, see also (1.19) and Theorem 1.53. Given two inputs $\boldsymbol{\vartheta}$ and $\boldsymbol{\vartheta}'$ in $\mathcal{E}$, we call $(\boldsymbol{X}, \boldsymbol{Y}, \boldsymbol{Z}, \boldsymbol{Z}^0, \boldsymbol{M})$ and $(\boldsymbol{X}', \boldsymbol{Y}', \boldsymbol{Z}', \boldsymbol{Z}^{0\prime}, \boldsymbol{M}')$ the corresponding solutions to the FBSDE (3.4) with $\boldsymbol{M} \equiv \boldsymbol{M}' \equiv 0$, when driven by $\boldsymbol{\mu}$ and $\boldsymbol{\mu}'$ associated with $\boldsymbol{\vartheta}$ and $\boldsymbol{\vartheta}'$ through (3.36) and by the same initial condition $X_0 = \psi(\eta, \mu_0)$. We then have:

$$\begin{aligned}
\bar{\mathbb{E}}\Big[\sup_{0 \le t \le T} |X_t - X_t'|^2\Big] \le C\bar{\mathbb{E}}\Big[&\big|G(X_T, \mu_T) - G(X_T, \mu_T')\big|^2 \\
&+ \int_0^T \big|(B, F, \sigma, \sigma^0)\big(t, X_t, \mu_t, Y_t, Z_t, Z_t^0\big) \\
&\quad - (B, F, \sigma, \sigma^0)\big(t, X_t, \mu_t', Y_t, Z_t, Z_t^0\big)\big|^2 dt\Big],
\end{aligned}$$

for a constant C independent of $\boldsymbol{\vartheta}$ and $\boldsymbol{\vartheta}'$. Observe that:

$$\begin{aligned}
\bar{\mathbb{E}}\Big[\big|G(X_T, \mu_T) - G(X_T, \mu_T')\big|^2\Big] \le \sum_{\varsigma \in \mathbb{S}^0} \sum_{v_1, \cdots, v_{N-1} \in \mathbb{J}} \bar{\mathbb{E}}\Big[&\big|G(X_T, [\vartheta^{N-1}(\varsigma, v_1, \cdots, v_{N-1})]_T) \\
&- G\big(X_T, [(\vartheta')^{N-1}(\varsigma, v_1, \cdots, v_{N-1})]_T\big)\big|^2\Big].
\end{aligned}$$

Similarly,

$$\begin{aligned}
&\bar{\mathbb{E}}\Big[\int_0^T \big|(B, F, \sigma, \sigma^0)\big(t, X_t, \mu_t, Y_t, Z_t, Z_t^0\big) - (B, F, \sigma, \sigma^0)\big(t, X_t, \mu_t', Y_t, Z_t, Z_t^0\big)\big|^2 dt\Big] \\
&\le \sum_{i=0}^{N-1} \sum_{\varsigma \in \mathbb{S}^0} \sum_{v_1, \cdots, v_{N-1} \in \mathbb{J}} \bar{\mathbb{E}}\Big[\int_{t_i}^{t_{i+1}} \big|(B, F, \sigma, \sigma^0)\big(t, X_t, [\vartheta^i(\varsigma, v_1, \cdots, v_i)]_t, Y_t, Z_t, Z_t^0\big) \\
&\qquad - (B, F, \sigma, \sigma^0)\big(t, X_t, [(\vartheta')^i(\varsigma, v_1, \cdots, v_i)]_t, Y_t, Z_t, Z_t^0\big)\big|^2 dt\Big].
\end{aligned}$$

Thanks to the growth and continuity properties in assumptions **FBSDE MFG with a Common Noise** and **Coefficients MFG with a Common Noise**, we deduce that:

$$\lim_{\vartheta' \to \vartheta} \bar{\mathbb{E}}\Big[\sup_{0 \le t \le T} |X_t - X'_t|^2 \Big] = 0.$$

Following the same argument as in the proof of Lemma 3.21, we can find a constant C, independent of ϑ and ϑ', such that, for any $i \in \{0, \cdots, N-1\}$, any $\varsigma \in \mathbb{S}^0$, any $v_1, \cdots, v_i \in \mathbb{J}$ and any $t \in [t_i, t_{i+1})$,

$$\begin{aligned} &W_2\big(\mathcal{L}(X_t \,|\, \mu_0 = \varsigma, V_1 = v_1, \cdots, V_i = v_i), \mathcal{L}(X'_t \,|\, \mu_0 = \varsigma, V_1 = v_1, \cdots, V_i = v_i)\big) \\ &\quad \le C\bar{\mathbb{E}}\big[|X_t - X'_t|^2\big]^{1/2}, \end{aligned}$$

from which we deduce that:

$$\lim_{\vartheta' \to \vartheta} \sup_{i=0,\cdots,N-1} \sup_{v_1,\cdots,v_i \in \mathbb{J}} \sup_{s \in [t_i, t_{i+1}]} W_2\Big(\big[\varphi^i(\vartheta)(\varsigma, v_1, \cdots, v_i)\big]_s, \big[\varphi^i(\vartheta')(\varsigma, v_1, \cdots, v_i)\big]_s \Big) = 0,$$

which shows that the restriction of Φ to $\mathcal{E}$ is continuous.

We now establish the relative compactness of the range of Φ. To this end, we are given an input $\vartheta \in \mathcal{E}$. We then observe that, for any $i \in \{0, \cdots, N-1\}$, $t \in [t_i, t_{i+1})$ and $A \in \mathcal{B}(\mathbb{R}^d)$,

$$\bar{\mathbb{P}}\big[X_t \in A\big] = \bar{\mathbb{E}}\big[\big(\Phi_t(\vartheta)(\mu_0, V_1, \cdots, V_{N-1})\big)(A)\big]. \tag{3.43}$$

Using again the fact that the support of the random vector $(\mu_0, V_1, \cdots, V_{N-1})$ has finite cardinality, we deduce that there exists a constant, still denoted by C, possibly depending upon ℓ and n, such that, for any $\varsigma \in \mathbb{S}^0$, any $v_1, \cdots, v_{N-1} \in \mathbb{J}$ and any $t \in [0, T]$,

$$\big(\Phi_t(\vartheta)(\varsigma, v_1, \cdots, v_{N-1})\big)(A) \le C\bar{\mathbb{P}}\big(X_t \in A\big).$$

By (3.39), we know that, for any $\varepsilon > 0$, we can find a compact subset $K \subset \mathbb{R}^d$ (K being independent of the input $\boldsymbol{v}$), such that:

$$\sup_{0 \le t \le T} \bar{\mathbb{P}}\big(X_t \notin K\big) < \varepsilon.$$

Therefore, for any $\varsigma \in \mathbb{S}^0$, any $v_1, \cdots, v_{N-1} \in \mathbb{J}$ and any $t \in [0, T]$,

$$\big(\Phi_t(\vartheta)(\varsigma, v_1, \cdots, v_{N-1})\big)\big(K^{\complement}\big) \le C\varepsilon.$$

It follows that there exists a subset $\mathcal{K} \subset \mathcal{P}(\mathbb{R}^d)$, independent of ϑ, compact for the topology of weak convergence, such that, for any $\varsigma \in \mathbb{S}^0$, any $v_1, \cdots, v_{N-1} \in \mathbb{J}$, and any $t \in [0, T]$,

$$\Phi_t(\vartheta)(\varsigma, v_1, \cdots, v_{N-1}) \in \mathcal{K}. \tag{3.44}$$

Using (3.40), we conclude that $\Phi_t(\vartheta)$ always (*i.e.* for any $t \in [0, T]$ and any input ϑ) lives in a compact subset, still denoted by $\mathcal{K}$, of $\mathcal{P}_2(\mathbb{R}^d)$ endowed with the Wasserstein distance W_2. We emphasize the fact that $\mathcal{K}$ is independent of ϑ.

Finally, (3.41) allows us to use Arzelà-Ascoli theorem and infer that, for any $i \in \{0, \cdots, N-1\}$, there exists a compact subset $\mathcal{C}([t_i, t_{i+1}]; \mathcal{P}_2(\mathbb{R}^d))$ that contains the mapping

$$[t_i, t_{i+1}] \ni t \mapsto \varphi^i(\boldsymbol{\vartheta})(\varsigma, v_1, \cdots, v_i) \in \mathcal{C}\big([t_i, t_{i+1}]; \mathcal{P}_2(\mathbb{R}^d)\big),$$

for all input $\boldsymbol{\vartheta} \in \mathcal{E}$, any $\varsigma \in \mathbb{S}^0$ and any $v_1, \cdots, v_{N-1} \in \mathbb{J}^{N-1}$. Since $\varsigma, v_1, \cdots, v_{N-1}$ are restricted to a finite set, we conclude that the range of Φ has a compact closure in $\mathcal{E}$. □

3.3.3 Tightness of the Approximating Solutions

The goal is now to let the parameters ℓ and n in the statement of Theorem 3.19 vary while still using the notations $\Lambda = 2^\ell$ and $N = 2^n$. Accordingly, in the definition (3.37), we now specify the dependence of the various terms upon the parameters ℓ and n defined in Subsection 3.3.1. So we write $\Phi^{\ell,n}$ for Φ, μ_0^ℓ for μ_0, and $(V_i^{\ell,n})_{1 \le i \le N-1}$ for $(V_i)_{1 \le i \le N-1}$. We call $\boldsymbol{\vartheta}^{\ell,n}$ a fixed point of $\Phi^{\ell,n}$ and we let:

$$\boldsymbol{\mu}^{\ell,n} = \Big(\mu_t^{\ell,n} = \vartheta_t^{\ell,n}\big(\mu_0^\ell, (V_1^{\ell,n}, \cdots, V_{N-1}^{\ell,n})\big)\Big)_{0 \le t \le T}. \tag{3.45}$$

We then denote by $(\boldsymbol{X}^{\ell,n}, \boldsymbol{Y}^{\ell,n}, \boldsymbol{Z}^{\ell,n}, \boldsymbol{Z}^{0,\ell,n}, \boldsymbol{M}^{\ell,n} \equiv 0)$ the solution of the FBSDE (3.4) driven by $\boldsymbol{\mu}^{\ell,n}$ and with the initial condition $X_0^\ell = \psi(\eta, \mu_0^\ell)$, the function ψ being defined as in (2.23), see also Lemma (Vol I)-5.29. So $(\boldsymbol{X}^{\ell,n}, \boldsymbol{Y}^{\ell,n}, \boldsymbol{Z}^{\ell,n}, \boldsymbol{Z}^{0,\ell,n})$ solves the forward-backward SDE of the conditional McKean-Vlasov type:

$$\begin{cases} dX_t^{\ell,n} = B\big(t, X_t^{\ell,n}, \mu_t^{\ell,n}, Y_t^{\ell,n}, Z_t^{\ell,n}\big)dt \\ \qquad\qquad + \sigma\big(t, X_t^{\ell,n}, \mu_t^{\ell,n}\big)dw_t + \sigma^0\big(t, X_t^{\ell,n}, \mu_t^{\ell,n}\big)dw_t^0, \\ dY_t^{\ell,n} = -F\big(t, X_t^{\ell,n}, \mu_t^{\ell,n}, Y_t^{\ell,n}, Z_t^{\ell,n}, Z_t^{0,\ell,n}\big)dt \\ \qquad\qquad + Z_t^{\ell,n}dw_t + Z_t^{0,\ell,n}dw_t^0, \quad t \in [0, T], \end{cases} \tag{3.46}$$

with $Y_T^{\ell,n} = G(X_T^{\ell,n}, \mu_T^{\ell,n})$ as terminal condition. Recall that in the present situation, there is no additional martingale term $\boldsymbol{M}^{\ell,n} = (M_t^{\ell,n})_{0 \le t \le T}$ since the environment is adapted to the noise $\boldsymbol{w}^0$.

Saying that $\boldsymbol{\mu}^{\ell,n}$ satisfies (3.45) is the same as saying that $\boldsymbol{\mu}^{\ell,n} = (\mu_t^{\ell,n})_{0 \le t \le T}$ satisfies the *discrete* McKean-Vlasov constraint:

$$\forall i \in \{0, \cdots, N-1\}, \ \forall t \in [t_i, t_{i+1}), \quad \mu_t^{\ell,n} = \mathcal{L}\big(X_t^{\ell,n} \,|\, \mu_0^\ell, V_1^{\ell,n}, \cdots, V_i^{\ell,n}\big),$$

the equality remaining true when $i = N-1$ and $t = t_N = T$.

The purpose of this paragraph is to investigate the tightness properties of the families

$$\Big(\bar{\mathbb{P}} \circ \big(\boldsymbol{\mu}^{\ell,n} = (\mu_t^{\ell,n})_{0 \le t \le T}\big)^{-1}\Big)_{\ell,n \ge 1} \quad \text{and} \quad \Big(\bar{\mathbb{P}} \circ \big(\boldsymbol{X}^{\ell,n} = (X_t^{\ell,n})_{0 \le t \le T}\big)^{-1}\Big)_{\ell,n \ge 1}.$$

Tightness of $(X^{\ell,n})_{\ell,n\geq 1}$

We start with an important *a priori* estimate.

Lemma 3.23 *There exists a constant C such that, for any $\ell, n \geq 1$,*

$$\bar{\mathbb{E}}\big[\sup_{0\leq t\leq T} |X_t^{\ell,n}|^2\big] \leq C.$$

Moreover, the family $(\bar{\mathbb{P}} \circ (\boldsymbol{X}^{\ell,n})^{-1})_{\ell,n\geq 1}$ is tight on the space $\mathcal{C}([0,T];\mathbb{R}^d)$ and the family $(\sup_{0\leq t\leq T} |X_t^{\ell,n}|)_{\ell,n\geq 1}$ is uniformly square-integrable under $\bar{\mathbb{P}}$.

Proof.

First Step. We shall use the following inequality repeatedly. From the definition (3.33), we have, for any $a \geq 0$,

$$\begin{aligned}
&\int_{\mathbb{R}^d} |x|_\infty^2 \mathbf{1}_{\{|x|_\infty \geq a\}} d\mu_0^\ell(x) \\
&\leq \int_{\mathbb{R}^d} |x|_\infty^2 \mathbf{1}_{\{|x|_\infty \geq a\}} d\big[\nu^0 \circ \big(\Pi_\Lambda^{(d)}\big)^{-1}\big](x) \qquad (3.47)\\
&= \int_{\mathbb{R}^d} \big|\Pi_\Lambda^{(d)}(x)\big|_\infty^2 \mathbf{1}_{\{|\Pi_\Lambda^{(d)}(x)|_\infty \geq a\}} d\nu^0(x) \leq \int_{\mathbb{R}^d} \big(1+|x|\big)^2 \mathbf{1}_{\{|x|\geq a-1\}} d\nu^0(x).
\end{aligned}$$

To pass from the first to the second line, we used the fact that the weight of any nonzero x in $\mathbb{J}$ is less under μ_0^ℓ than under $\nu^0 \circ (\Pi_\Lambda^{(d)})^{-1}$, see (3.33).

Therefore,

$$\bar{\mathbb{E}} \int_{\mathbb{R}^d} |x|_\infty^2 \mathbf{1}_{\{|x|_\infty \geq a\}} d\mu_0^\ell(x) \leq \bar{\mathbb{E}} \int_{\mathbb{R}^d} \big(1+|x|\big)^2 \mathbf{1}_{\{|x|\geq a-1\}} d\nu^0(x). \qquad (3.48)$$

Now, observe that:

$$\bar{\mathbb{E}} \int_{\mathbb{R}^d} \big(1+|x|\big)^2 d\nu^0(x) \leq 2\Big(1 + \bar{\mathbb{E}}\big(M_2(\nu^0)^2\big)\Big) < \infty,$$

the last inequality following from the fact that $\mathcal{V}^0 \in \mathcal{P}_2(\mathcal{P}_2(\mathbb{R}^d))$. We deduce that:

$$\sup_{\ell\geq 1} \bar{\mathbb{E}} \int_{\mathbb{R}^d} |x|^2 d\mu_0^\ell(x) < \infty. \qquad (3.49)$$

Moreover, by Lebesgue's dominated convergence theorem,

$$\lim_{a\to\infty} \bar{\mathbb{E}} \int_{\mathbb{R}^d} \big(1+|x|\big)^2 \mathbf{1}_{\{|x|\geq a-1\}} d\nu^0(x) = 0,$$

and thus

$$\lim_{a\to\infty} \sup_{\ell\geq 1} \bar{\mathbb{E}} \int_{\mathbb{R}^d} |x|^2 \mathbf{1}_{\{|x|\geq a\}} d\mu_0^\ell(x) = 0. \qquad (3.50)$$

Second Step. From (3.49), we notice that:

$$\sup_{\ell \geq 1} \bar{\mathbb{E}}\big[|X_0^\ell|^2\big] < \infty.$$

Now, thanks to (**A5**) in assumption **FBSDE MFG with a Common Noise**, we can prove that there exists a constant C, independent of ℓ and n, such that:

$$\bar{\mathbb{E}}\Big[\sup_{t\in[0,T]} |Y_t^{\ell,n}|^4 + \Big(\int_0^T \big(|Z_t^{\ell,n}|^2 + |Z_t^{0,\ell,n}|^2\big)dt\Big)^2\Big] \leq C, \tag{3.51}$$

where $(\boldsymbol{X}^{\ell,n}, \boldsymbol{Y}^{\ell,n}, \boldsymbol{Z}^{\ell,n}, \boldsymbol{Z}^{0,\ell,n})$ is the solution of FBSDE (3.46). The above inequality is similar to that used in the proof of Lemma 3.20, but here we emphasize that the constant C therein is independent of the parameters n and ℓ.

Plugging (3.51) and (3.49) into the forward equation satisfied by $\boldsymbol{X}^{\ell,n}$ in the FBSDE (3.4) and using (**A5**) in assumption **FBSDE MFG with a Common Noise** and (**A2**) in assumption **Coefficients MFG with a Common Noise**, we deduce that:

$$\sup_{\ell,n\geq 1} \bar{\mathbb{E}}\big[\sup_{0\leq t\leq T} |X_t^{\ell,n}|^2\big] < \infty.$$

Tightness of the family $(\bar{\mathbb{P}} \circ (\boldsymbol{X}^{\ell,n})^{-1})_{\ell,n\geq 1}$ is then proved by means of Aldous' criterion as in the proof of Lemma 3.14.

Third Step. We now prove that the family $(\sup_{0\leq t\leq T} |X_t^{\ell,n}|)_{\ell,n\geq 1}$ is uniformly square-integrable.

To do so, we observe first that the family $(|X_0^\ell|)_{\ell\geq 1}$ is uniformly square-integrable, namely,

$$\lim_{a\to\infty} \sup_{\ell\geq 1} \bar{\mathbb{E}}\big[|X_0^\ell|^2 \mathbf{1}_{\{|X_0^\ell|\geq a\}}\big] = 0, \tag{3.52}$$

which follows from (3.50) together with the fact that μ_0^ℓ is the conditional law of X_0^ℓ given μ_0^ℓ.

In order to formulate our second observation, we define $\bar{\mathbb{F}}^{0,\ell,n}$ as the filtration:

$$\begin{aligned}
\bar{\mathcal{F}}_t^{0,\ell,n} &= \sigma\big\{\mu_0^\ell, V_1^{\ell,n}, \cdots, V_i^{\ell,n}\big\}, \quad t \in [t_i, t_{i+1}), \quad i \in \{1, \cdots, N\},\\
\bar{\mathcal{F}}_T^{0,\ell,n} &= \sigma\big\{\mu_0^\ell, V_1^{\ell,n}, \cdots, V_{N-1}^{\ell,n}\big\}.
\end{aligned}$$

Notice that $\bar{\mathcal{F}}_0^{0,\ell,n}$ may be just denoted by $\bar{\mathcal{F}}_0^{0,\ell}$. Then, we claim that for any $t \in [0,T]$:

$$\begin{aligned}
\bar{\mathbb{E}}\Big[M_2(\mu_t^{\ell,n})^4 \,|\, \bar{\mathcal{F}}_0^{0,\ell}\Big] &= \bar{\mathbb{E}}\Big[\bar{\mathbb{E}}\big[|X_t^{\ell,n}|^2 \,|\, \bar{\mathcal{F}}_t^{0,\ell,n}\big]^2 \,|\, \bar{\mathcal{F}}_0^{0,\ell}\Big]\\
&= \bar{\mathbb{E}}\Big[\bar{\mathbb{E}}\Big[\bar{\mathbb{E}}\big[|X_t^{\ell,n}|^2 \,|\, \bar{\mathcal{F}}_t^0\big] \,|\, \bar{\mathcal{F}}_t^{0,\ell,n}\Big]^2 \,|\, \bar{\mathcal{F}}_0^{0,\ell}\Big]\\
&\leq \bar{\mathbb{E}}\Big[\bar{\mathbb{E}}\Big[\bar{\mathbb{E}}\big[|X_t^{\ell,n}|^2 \,|\, \bar{\mathcal{F}}_t^0\big]^2 \,|\, \bar{\mathcal{F}}_t^{0,\ell,n}\Big] \,|\, \bar{\mathcal{F}}_0^{0,\ell}\Big]\\
&= \bar{\mathbb{E}}\Big[\bar{\mathbb{E}}\big[|X_t^{\ell,n}|^2 \,|\, \bar{\mathcal{F}}_t^0\big]^2 \,|\, \bar{\mathcal{F}}_0^{0,\ell}\Big].
\end{aligned} \tag{3.53}$$

Also, recall from Proposition 2.17 that $\bar{\mathbb{E}}[|X_t^{\ell,n}|^2 \,|\, \bar{\mathcal{F}}_t^0]$ merely writes $\bar{\mathbb{E}}^1[|X_t^{\ell,n}|^2]$, up to a $\bar{\mathbb{P}}^0$ exceptional event.

Now, applying Itô's formula and taking advantage of the growth conditions on the coefficients, we can find a constant C, independent of ℓ and n, such that for all $t \in [0, T]$:

$$
\begin{aligned}
\bar{\mathbb{E}}^1\big[|X_t^{\ell,n}|^2\big] \leq C\Big[& 1 + \bar{\mathbb{E}}^1\big[|X_0^{\ell}|^2\big] + \int_0^t \Big(\bar{\mathbb{E}}^1\big[|X_s^{\ell,n}|^2\big] + M_2(\mu_s^{\ell,n})^2\Big) ds \\
& + \int_0^t \Big(\bar{\mathbb{E}}^1\big[|Y_s^{\ell,n}|^2 + |Z_s^{\ell,n}|^2\big]\Big) ds\Big] \\
& + 2\bar{\mathbb{E}}^1\Big(\int_0^t X_s^{\ell,n} \cdot \sigma^0(s, X_s^{\ell,n}, \mu_s^{\ell,n}) dW_s^0\Big).
\end{aligned}
\tag{3.54}
$$

By stochastic Fubini's theorem, the last term is equal to:

$$
\bar{\mathbb{E}}^1\Big(\int_0^t X_s^{\ell,n} \cdot \sigma^0(s, X_s^{\ell,n}, \mu_s^{\ell,n}) dW_s^0\Big) = \int_0^t \bar{\mathbb{E}}^1\Big[\big(\sigma^0(s, X_s^{\ell,n}, \mu_s^{\ell,n})\big)^\dagger X_s^{\ell,n}\Big] \cdot dW_s^0.
$$

Our goal now is to take the square and then the conditional expectation given $\bar{\mathcal{F}}_0^{0,\ell}$ under $\bar{\mathbb{P}}^0$ in (3.54). As a preliminary step, notice that:

$$
\begin{aligned}
& \bar{\mathbb{E}}^0\Big[\Big(\int_0^t \bar{\mathbb{E}}^1\Big[\big(\sigma^0(s, X_s^{\ell,n}, \mu_s^{\ell,n})\big)^\dagger X_s^{\ell,n}\Big] \cdot dW_s^0\Big)^2 \,|\, \bar{\mathcal{F}}_0^{0,\ell}\Big] \\
& \leq C\bar{\mathbb{E}}^0\Big[\int_0^t \Big|\bar{\mathbb{E}}^1\Big[\big(\sigma^0(s, X_s^{\ell,n}, \mu_s^{\ell,n})\big)^\dagger X_s^{\ell,n}\Big]\Big|^2 ds \,|\, \bar{\mathcal{F}}_0^{0,\ell}\Big] \\
& \leq C\bar{\mathbb{E}}^0\Big[\int_0^t \Big[1 + \bar{\mathbb{E}}^1\big[|X_s^{\ell,n}|^2\big]^2 + M_2(\mu_s^{\ell,n})^4\Big] ds \,|\, \bar{\mathcal{F}}_0^{0,\ell}\Big],
\end{aligned}
$$

the constant C being allowed to vary from line to line provided that it remains independent of ℓ and n.

Returning to (3.54), we get:

$$
\begin{aligned}
\bar{\mathbb{E}}^0\Big[\bar{\mathbb{E}}^1\big[|X_t^{\ell,n}|^2\big]^2 \,|\, \bar{\mathcal{F}}_0^{0,\ell}\Big] \leq C\Big[& 1 + \bar{\mathbb{E}}^0\Big[\bar{\mathbb{E}}^1\big[|X_0^{\ell}|^2\big]^2 \,|\, \bar{\mathcal{F}}_0^{0,\ell}\Big] \\
& + \int_0^t \Big(\bar{\mathbb{E}}^0\Big[\bar{\mathbb{E}}^1\big[|X_s^{\ell,n}|^2\big]^2 + M_2(\mu_s^{\ell,n})^4 \,|\, \bar{\mathcal{F}}_0^{0,\ell}\Big]\Big) ds \\
& + \bar{\mathbb{E}}^0\Big[\Big(\int_0^T \Big(\bar{\mathbb{E}}^1\big[|Y_s^{\ell,n}|^2 + |Z_s^{\ell,n}|^2\big]\Big) ds\Big)^2 \,|\, \bar{\mathcal{F}}_0^{0,\ell}\Big]\Big].
\end{aligned}
$$

Now, by (3.53), we deduce that for any $t \in [0, T]$:

$$
\begin{aligned}
\bar{\mathbb{E}}^0\Big[\bar{\mathbb{E}}^1\big[|X_t^{\ell,n}|^2\big]^2 \,|\, \bar{\mathcal{F}}_0^{0,\ell}\Big] \leq C\Big[& 1 + \bar{\mathbb{E}}^0\Big[\bar{\mathbb{E}}^1\big[|X_0^{\ell}|^2\big]^2 \,|\, \bar{\mathcal{F}}_0^{0,\ell}\Big] \\
& + \int_0^t \bar{\mathbb{E}}^0\Big[\bar{\mathbb{E}}^1\big[|X_s^{\ell,n}|^2\big]^2 \,|\, \bar{\mathcal{F}}_0^{0,\ell}\Big] ds \\
& + \bar{\mathbb{E}}^0\Big[\Big(\int_0^T \Big(\bar{\mathbb{E}}^1\big[|Y_s^{\ell,n}|^2 + |Z_s^{\ell,n}|^2\big]\Big) ds\Big)^2 \,|\, \bar{\mathcal{F}}_0^{0,\ell}\Big]\Big].
\end{aligned}
$$

By a standard application of Gronwall's lemma, we obtain:

$$\sup_{0\le t\le T} \bar{\mathbb{E}}^0\Big[\bar{\mathbb{E}}^1\big[|X_t^{\ell,n}|^2\big]^2 \,|\, \bar{\mathcal{F}}_0^{0,\ell}\Big] \le C\Big[1 + \bar{\mathbb{E}}^0\Big[\bar{\mathbb{E}}^1\big[|X_0^{\ell}|^2\big]^2 \,|\, \bar{\mathcal{F}}_0^{0,\ell}\Big] + \bar{\mathbb{E}}^0\Big[\Big(\int_0^T \Big(\bar{\mathbb{E}}^1\big[|Y_s^{\ell,n}|^2 + |Z_s^{\ell,n}|^2\big]\Big)ds\Big)^2 \,|\, \bar{\mathcal{F}}_0^{0,\ell}\Big]\Big].$$

Observe that:

$$\bar{\mathbb{E}}^0\Big[\bar{\mathbb{E}}^1\big[|X_0^{\ell}|^2\big]^2 \,|\, \bar{\mathcal{F}}_0^{0,\ell}\Big] = \Big(\int_{\mathbb{R}^d} |x|^2 d\mu_0^{\ell}(x)\Big)^2 = M_2(\mu_0^{\ell})^4.$$

Therefore, by invoking (3.53) again, we conclude that for any $t \in [0, T]$:

$$\sup_{0\le t\le T} \bar{\mathbb{E}}^0\Big[M_2(\mu_t^{n,\ell})^4 \,|\, \bar{\mathcal{F}}_0^{0,\ell}\Big] \le C\Big[1 + M_2(\mu_0^{\ell})^4 + \bar{\mathbb{E}}^0\Big[\Big(\int_0^T \Big(\bar{\mathbb{E}}^1\big[|Y_s^{\ell,n}|^2 + |Z_s^{\ell,n}|^2\big]\Big)ds\Big)^2 \,|\, \bar{\mathcal{F}}_0^{0,\ell}\Big]\Big].$$

Fourth Step. Observe that, in the above inequality, the term in the left-hand side is also equal to:

$$\sup_{0\le t\le T} \bar{\mathbb{E}}^0\Big[M_2(\mu_t^{n,\ell})^4 \,|\, \bar{\mathcal{F}}_0^{0,\ell}\Big] = \sup_{0\le t\le T} \bar{\mathbb{E}}\Big[M_2(\mu_t^{n,\ell})^4 \,|\, \bar{\mathcal{F}}_0^{\ell}\Big],$$

where $\bar{\mathcal{F}}_0^{\ell} = \sigma\{\mu_0^{\ell}, \eta\}$ is a sub-σ-field of $\bar{\mathcal{F}}_0$.

Using the conclusion of the third step and reapplying Gronwall's lemma, we obtain:

$$\bar{\mathbb{E}}\Big[\sup_{0\le t\le T} |X_t^{\ell,n}|^4 \,|\, \bar{\mathcal{F}}_0^{\ell}\Big]^{1/2} \le C\Big(1 + |X_0^{\ell}|^2 + M_2(\mu_0^{\ell})^2 + \bar{\mathbb{E}}^0\Big[\Big(\int_0^T \Big(\bar{\mathbb{E}}^1\big[|Y_s^{\ell,n}|^2 + |Z_s^{\ell,n}|^2\big]\Big)ds\Big)^2 \,|\, \bar{\mathcal{F}}_0^{0,\ell}\Big]^{1/2} + \bar{\mathbb{E}}\Big[\Big(\int_0^T (|Y_t^{\ell,n}|^2 + |Z_t^{\ell,n}|^2)dt\Big)^2 \,|\, \bar{\mathcal{F}}_0^{\ell}\Big]^{1/2}\Big).$$

By (3.47), the sequence $(M_2(\mu_0^{\ell})^2)_{\ell\ge 1}$ is bounded by $C(1 + M_2(\nu^0)^2)$. Also, from (3.51), we know that:

$$\sup_{\ell,n\ge 1} \bar{\mathbb{E}}\Big[\bar{\mathbb{E}}^0\Big[\Big(\int_0^T \Big(\bar{\mathbb{E}}^1\big[|Y_s^{\ell,n}|^2+|Z_s^{\ell,n}|^2\big]\Big)ds\Big)^2 \,|\, \bar{\mathcal{F}}_0^{0,\ell}\Big] + \bar{\mathbb{E}}\Big[\Big(\int_0^T (|Y_t^{\ell,n}|^2+|Z_t^{\ell,n}|^2)dt\Big)^2 \,|\, \bar{\mathcal{F}}_0^{\ell}\Big]\Big] < \infty.$$

Thus, recalling (3.52), the family $(\bar{\mathbb{E}}[\sup_{0\le t\le T} |X_t^{\ell,n}|^4 \,|\, \bar{\mathcal{F}}_0^{\ell}]^{1/2})_{\ell,n\ge 1}$ is uniformly integrable. We deduce that the family $(\sup_{0\le t\le T} |X_t^{\ell,n}|^2)_{\ell,n\ge 1}$ is uniformly integrable, see for instance (Vol I)-(4.45). □

Tightness of $(\mu^{\ell,n})_{\ell,n\geq 1}$

We now look at the flows of equilibrium measures $((\mu_t^{\ell,n})_{0\leq t\leq T})_{\ell,n\geq 1}$. They are only right continuous in time. For that reason, tightness must be investigated in the space $\mathcal{D}([0,T];\mathcal{P}_2(\mathbb{R}^d))$. In order to avoid some of the unpleasant idiosyncrasies of the topology of that space, we work with continuous interpolations of $(\mu_t^{\ell,n})_{0\leq t\leq T}$, prove that they are tight on $\mathcal{C}([0,T];\mathcal{P}_2(\mathbb{R}^d))$, and then show that $(\mu_t^{\ell,n})_{0\leq t\leq T}$ stays close enough to its interpolation as ℓ and n tend to infinity.

We first define a *lifting* of each $(\mu_t^{\ell,n})_{0\leq t\leq T}$. Imitating the role of $\mathfrak{M}$ in Definition 2.16 of an equilibrium, we let:

$$\mathfrak{M}^{\ell,n} = \mathcal{L}\big((X_t^{\ell,n}, w_t)_{0\leq t\leq T} \,|\, \mu_0^{\ell}, V_1^{\ell,n}, \cdots, V_{N-1}^{\ell,n}\big). \tag{3.55}$$

For any $t \in [0,T]$, we define:

$$\bar{\mu}_t^{\ell,n} = \mathfrak{M}^{\ell,n} \circ (e_t^x)^{-1} = \mathcal{L}\big(X_t^{\ell,n} \,|\, \mu_0^{\ell}, V_1^{\ell,n}, \cdots, V_{N-1}^{\ell,n}\big),$$

and we prove the following tightness result:

Lemma 3.24 *The family $(\bar{\mathbb{P}} \circ (\mathfrak{M}^{\ell,n})^{-1})_{\ell,n\geq 1}$ is tight on $\mathcal{P}_2(\mathcal{C}([0,T];\mathbb{R}^{2d}))$. In particular, the family $(\bar{\mathbb{P}} \circ (\bar{\boldsymbol{\mu}}^{\ell,n})^{-1})_{\ell,n\geq 1}$ is tight on $\mathcal{C}([0,T];\mathcal{P}_2(\mathbb{R}^d))$.*

Proof. The first claim is a straightforward adaptation of Lemma 3.16. The second claim follows from the fact that the mapping $\mathcal{P}_2(\mathcal{C}([0,T];\mathbb{R}^{2d})) \ni \mathrm{m} \mapsto (\mathrm{m} \circ (e_t^x)^{-1})_{0\leq t\leq T} \in \mathcal{C}([0,T];\mathcal{P}_2(\mathbb{R}^d))$ is continuous. □

The distance between the interpolation $\bar{\boldsymbol{\mu}}^{\ell,n}$ and $\boldsymbol{\mu}^{\ell,n}$ is controlled by the following lemma:

Lemma 3.25 *The following limit holds true:*

$$\lim_{n\to\infty} \sup_{\ell\geq 1} \bar{\mathbb{E}}\Big[\sup_{0\leq t\leq T} W_2(\mu_t^{\ell,n}, \bar{\mu}_t^{\ell,n})^2\Big]^{1/2} = 0.$$

Proof.

First Step. We first notice that, for any $i \in \{0,\cdots,N-1\}$, $X_{t_i}^{\ell,n}$ is independent of the σ-algebra generated by $(w_s^0 - w_{t_i}^0, w_s - w_{t_i})$, $s \in [t_i,T]$. Therefore, we have:

$$\mu_{t_i}^{\ell,n} = \mathcal{L}\Big(X_{t_i}^{\ell,n} \,|\, \mu_0^{\ell}, V_1^{\ell,n}, \cdots, V_i^{\ell,n},\, w^0_{(i+1)T/N} - w^0_{iT/N}, w^0_{(i+2)T/N} - w^0_{(i+1)T/N}, w^0_T - w^0_{(N-1)T/N}\Big).$$

By construction, see (3.32), $V_{i+1}^{\ell,n}, \cdots, V_{N-1}^{\ell,n}$ belong to the σ-field:

$$\sigma\Big\{V_1^{\ell,n}, \cdots, V_i^{\ell,n}, w^0_{(i+1)T/N} - w^0_{iT/N}, w^0_{(i+2)T/N} - w^0_{(i+1)T/N}, w^0_T - w^0_{(N-1)T/N}\Big\},$$

from which we deduce that:

$$\mu_{t_i}^{\ell,n} = \mathcal{L}\big(X_{t_i}^{\ell,n} \mid \mu_0^\ell, V_1^{\ell,n}, \cdots, V_{N-1}^{\ell,n}\big) = \bar{\mu}_{t_i}^{\ell,n}.$$

Second Step. Therefore, for any $i \in \{0, \cdots, N-1\}$,

$$\begin{aligned}
\sup_{t\in[t_i,t_{i+1}]} W_2\big(\bar{\mu}_t^{\ell,n}, \mu_t^{\ell,n}\big)^2 &\le 2 \sup_{t\in[t_i,t_{i+1}]} W_2\big(\bar{\mu}_t^{\ell,n}, \bar{\mu}_{t_i}^{\ell,n}\big)^2 + 2 \sup_{t\in[t_i,t_{i+1}]} W_2\big(\mu_t^{\ell,n}, \mu_{t_i}^{\ell,n}\big)^2 \\
&\le 2\bar{\mathbb{E}}\Big[\sup_{t\in[t_i,t_{i+1}]} |X_t^{\ell,n} - X_{t_i}^{\ell,n}|^2 \mid \mu_0^\ell, V_1^{\ell,n}, \cdots, V_{N-1}^{\ell,n}\Big] \\
&\quad + 2\bar{\mathbb{E}}\Big[\sup_{t\in[t_i,t_{i+1}]} |X_t^{\ell,n} - X_{t_i}^{\ell,n}|^2 \mid \mu_0^\ell, V_1^{\ell,n}, \cdots, V_i^{\ell,n}\Big],
\end{aligned}$$

and then:

$$\begin{aligned}
\sup_{t\in[0,T]} W_2\big(\bar{\mu}_t^{\ell,n}, \mu_t^{\ell,n}\big)^2 &\le 2\bar{\mathbb{E}}\Big[\sup_{|t-s|\le 2^{-n}} |X_t^{\ell,n} - X_s^{\ell,n}|^2 \mid \mu_0^\ell, V_1^{\ell,n}, \cdots, V_{N-1}^{\ell,n}\Big] \\
&\quad + 2 \sup_{i=0,\cdots,N-1} \bar{\mathbb{E}}\Big[\sup_{|t-s|\le 2^{-n}} |X_t^{\ell,n} - X_s^{\ell,n}|^2 \mid \mu_0^\ell, V_1^{\ell,n}, \cdots, V_i^{\ell,n}\Big].
\end{aligned}$$

Taking expectations and applying Doob's inequality, we get:

$$\bar{\mathbb{E}}\Big[\sup_{t\in[0,T]} W_2\big(\bar{\mu}_t^{\ell,n}, \mu_t^{\ell,n}\big)^2\Big] \le C\bar{\mathbb{E}}\Big[\sup_{|t-s|\le 2^{-n}} |X_t^{\ell,n} - X_s^{\ell,n}|^2\Big].$$

By Lemma 3.23 and by a standard uniform integrability argument, we get that the right-hand side tends to 0 as n tends to ∞, uniformly in $\ell \ge 1$. □

3.3.4 Extraction of a Subsequence

By Lemmas 3.23 and 3.24, the sequence $(\bar{\mathbb{P}} \circ (\boldsymbol{w}^0, \mathfrak{M}^{\ell,n}, \boldsymbol{X}^{\ell,n}, \boldsymbol{w})^{-1})_{\ell,n\ge 1}$ is tight on the space

$$\mathcal{C}([0,T];\mathbb{R}^d) \times \mathcal{P}_2\big(\mathcal{C}([0,T];\mathbb{R}^{2d})\big) \times \mathcal{C}([0,T];\mathbb{R}^d) \times \mathcal{C}([0,T];\mathbb{R}^d).$$

We now aim at making use of Theorem 3.13.

In order to do so, we must check that the family of optimal controls $(\hat{\boldsymbol{\alpha}}^{\ell,n})_{\ell,n\ge 1}$ associated with the optimal paths $(\boldsymbol{X}^{\ell,n})_{\ell,n\ge 1}$ satisfy (**A2**) in assumption **Control Bounds**:

Lemma 3.26 *There exists a constant C such that, for any $\ell, n \ge 1$,*

$$\bar{\mathbb{E}}\int_0^T |\hat{\alpha}_t^{\ell,n}|^2 dt \le C.$$

Proof. The proof is based on the fact that, for each $\ell, n \geq 1$, $\hat{\boldsymbol{\alpha}}^{\ell,n}$ is equal to

$$\hat{\boldsymbol{\alpha}}^{\ell,n} = \big(\breve{\alpha}(t, X_t^{\ell,n}, \mu_t^{\ell,n}, Y_t^{\ell,n}, Z_t^{\ell,n})\big)_{0 \leq t \leq T}.$$

We complete it by combining **(A5)** in assumption **FBSDE MFG with a Common Noise** with the bounds for $(\boldsymbol{X}^{\ell,n}, \boldsymbol{Y}^{\ell,n}, \boldsymbol{Z}^{\ell,n})_{\ell,n\geq 1}$ in the statement of Lemma 3.23 and in its proof, see for instance (3.51). □

We also need to guarantee that, asymptotically, the initial measure μ_0^ℓ gets to ν^0:

Lemma 3.27 *As $\ell \to \infty$, $W_2(\mu_0^\ell, \nu^0)$ converges in probability to 0.*

Proof. We recall the definition (3.33) of μ_0^ℓ, for $\ell \geq 1$:

$$\mu_0^\ell = \sum_{x \in \mathbb{J}^\ell} \Pi^{(1)}_{\Lambda^{2d+4}}\Big(\big[\nu^0 \circ (\Pi_\Lambda^{(d)})^{-1}\big](x)\Big)\delta_x + \bigg(1 - \sum_{x \in \mathbb{J}^\ell} \Pi^{(1)}_{\Lambda^{2d+4}}\Big(\big[\nu^0 \circ (\Pi_\Lambda^{(d)})^{-1}\big](x)\Big)\bigg)\delta_0,$$

where we put the superscript ℓ in $\mathbb{J}^{(\ell)}$ in order to emphasize the dependence upon ℓ.

Let, for any $\ell \geq 1$,

$$\nu^{0,\ell} = \sum_{x \in \mathbb{J}^{(\ell)}} \big[\nu^0 \circ (\Pi_\Lambda^{(d)})^{-1}\big](x)\delta_x.$$

Then,

$$W_2(\mu_0^\ell, \nu^{0,\ell})^2 \leq \sum_{x \in \mathbb{J}^{(\ell)}} |x|^2 \Big[\big[\nu^0 \circ (\Pi_\Lambda^{(d)})^{-1}\big](x) - \Pi^{(1)}_{\Lambda^{2d+4}}\Big(\big[\nu^0 \circ (\Pi_\Lambda^{(d)})^{-1}\big](x)\Big)\Big].$$

Observe that the cardinality of $\mathbb{J}^{(\ell)}$ is less than $C\Lambda^{2d}$, for a constant C independent of ℓ. We easily deduce that $W_2(\mu_0^\ell, \nu^{0,\ell})^2 \leq \Lambda^{-2}$.

Now, it is easily checked that for a new value of C:

$$W_2(\nu^0, \nu^{0,\ell})^2 \leq \int_{\mathbb{R}^d} \big|\Pi_\Lambda^{(d)}(x) - x\big|^2 d\nu^0(x) \leq C\Lambda^{-2} + C\int_{\mathbb{R}^d} \mathbf{1}_{\{|x| \geq \Lambda\}}(1 + |x|^2) d\nu^0(x),$$

which tends to 0 as ℓ tends to ∞. □

Finally, the last point to check is that, asymptotically, $\mathfrak{M}^{\ell,n}$ gets closer and closer to the conditional law of $(\boldsymbol{X}^{\ell,n}, \boldsymbol{w})$ given $(\nu^0, \boldsymbol{w}^0)$. Observe indeed that here, differently from the framework used in the statement of Theorem 3.13, $\mathfrak{M}^{\ell,n}$ is not the conditional law of $(\boldsymbol{X}^{\ell,n}, \boldsymbol{w})$ given $(\boldsymbol{w}^0, \mathfrak{M}^{\ell,n})$, but the conditional law of $(\boldsymbol{X}^{\ell,n}, \boldsymbol{w})$ given a discretization of $(\nu^0, \boldsymbol{w}^0)$.

Lemma 3.28 *Consider a complete probability space $(\Omega^\infty, \mathcal{F}^\infty, \mathbb{P}^\infty)$ equipped with a process $(\boldsymbol{W}^{0,\infty}, \mathfrak{M}^\infty, \boldsymbol{W}^\infty, \boldsymbol{X}^\infty)$ whose law under $\mathbb{P}^\infty$ is a weak limit of*

$(\bar{\mathbb{P}}\circ(\boldsymbol{w}^0, \mathfrak{M}^{\ell,n}, \boldsymbol{w}, \boldsymbol{X}^{\ell,n})^{-1})_{\ell,n\geq 1}$ *on* $\mathcal{C}([0,T];\mathbb{R}^d)\times\mathcal{P}_2(\mathcal{C}([0,T];\mathbb{R}^{2d}))\times\mathcal{C}([0,T];\mathbb{R}^d)\times\mathcal{C}([0,T];\mathbb{R}^d)$ *as* ℓ *and* n *tend to* ∞ *with the prescription* $\ell = 2n$*. Then,* $\mathfrak{M}^\infty$ *is the conditional law of* $(\boldsymbol{X}^\infty, \boldsymbol{W}^\infty)$ *given* $(\boldsymbol{W}^{0,\infty}, \mathfrak{M}^\infty)$*.*

Proof. For two bounded uniformly continuous functions h^0 and h^1, h^0 from $\mathcal{C}([0,T];\mathbb{R}^d)\times\mathcal{P}_2(\mathcal{C}([0,T];\mathbb{R}^{2d}))$ into $\mathbb{R}$ and h^1 from $\mathcal{C}([0,T];\mathbb{R}^d)\times\mathcal{C}([0,T];\mathbb{R}^d)$ into $\mathbb{R}$, we have by (3.55), for any $\ell, n \geq 1$:

$$\bar{\mathbb{E}}\Big[h^0(\bar{\boldsymbol{V}}^{\ell,n}, \mathfrak{M}^{\ell,n})h^1(\boldsymbol{X}^{\ell,n}, \boldsymbol{w})\Big] = \bar{\mathbb{E}}\Big[h^0(\bar{\boldsymbol{V}}^{\ell,n}, \mathfrak{M}^{\ell,n})\int_{\mathcal{C}([0,T];\mathbb{R}^{2d})} h^1(\boldsymbol{v}, \mathrm{m})d\mathfrak{M}^{\ell,n}(\boldsymbol{v}, \mathrm{m})\Big], \tag{3.56}$$

where we let $\bar{V}^{\ell,n}_t = V^{\ell,n}_{i-1}$ if $t \in [(i-1)T/N, iT/N) = [(i-1)T/2^n, iT/2^n)$ and $\bar{V}^{\ell,n}_T = V^{\ell,n}_{N-1}$.

Here comes the prescription $\ell = 2n$. By Lemma 3.17, we know that, on the event

$$\Big\{\sup_{0\leq t\leq T} |w^0_t| \leq \Lambda - 1 = 2^\ell - 1 = 4^n - 1\Big\},$$

it holds, for any $i \in \{1, \cdots, N-1\}$,

$$\big|V^{\ell,n}_i - w^0_{iT/N}\big|_\infty \leq \frac{i}{\Lambda} \leq \frac{2^n}{4^n} = \frac{1}{2^n},$$

where we used the fact that $N = 2^n < 4^n = 2^\ell = \Lambda$. Thus,

$$\lim_{n\to\infty} \bar{\mathbb{P}}\Big[\sup_{0\leq t\leq T} \big|\bar{V}^{\ell,n}_t - w^0_t\big|_\infty \leq \frac{1}{2^n} + \sup_{0\leq s,t\leq T, |t-s|\leq 1/2^n} \big|w^0_s - w^0_t\big|\Big] = 1.$$

This shows that, in (3.56),

$$\lim_{n\to\infty}\Big|\bar{\mathbb{E}}\Big[h^0(\bar{\boldsymbol{V}}^{\ell,n}, \mathfrak{M}^{\ell,n})h^1(\boldsymbol{X}^{\ell,n}, \boldsymbol{w})\Big] - \bar{\mathbb{E}}\Big[h^0(\boldsymbol{w}^0, \mathfrak{M}^{\ell,n})h^1(\boldsymbol{X}^{\ell,n}, \boldsymbol{w})\Big]\Big| = 0,$$

$$\begin{aligned}\lim_{n\to\infty}\Big|&\bar{\mathbb{E}}\Big[h^0(\bar{\boldsymbol{V}}^{\ell,n}, \mathfrak{M}^{\ell,n})\int_{\mathcal{C}([0,T];\mathbb{R}^{2d})} h^1(\boldsymbol{v}, \mathrm{m})d\mathfrak{M}^{\ell,n}(\boldsymbol{v}, \mathrm{m})\Big] \\ &- \bar{\mathbb{E}}\Big[h^0(\boldsymbol{w}^0, \mathfrak{M}^{\ell,n})\int_{\mathcal{C}([0,T];\mathbb{R}^{2d})} h^1(\boldsymbol{v}, \mathrm{m})d\mathfrak{M}^{\ell,n}(\boldsymbol{v}, \mathrm{m})\Big]\Big| = 0.\end{aligned}$$

Therefore, passing to the limit in both sides of (3.56), we obtain:

$$\begin{aligned}&\mathbb{E}^\infty\Big[h^0(\boldsymbol{W}^{0,\infty}, \mathfrak{M}^\infty)h^1(\boldsymbol{X}^\infty, \boldsymbol{W}^\infty)\Big] \\ &= \mathbb{E}^\infty\Big[h^0(\boldsymbol{W}^{0,\infty}, \mathfrak{M}^\infty)\int_{\mathcal{C}([0,T];\mathbb{R}^{2d})} h^1(\boldsymbol{v}, \mathrm{m})d\mathfrak{M}^\infty(\boldsymbol{v}, \mathrm{m})\Big].\end{aligned}$$

This suffices to complete the proof. □

Conclusion

We now have all the ingredients to complete the proof of Theorem 3.1. Up to the fact that $\mathfrak{M}^{\ell,n}$ only fits the conditional law of $(\boldsymbol{X}^{\ell,n}, \boldsymbol{w})$ given $(\boldsymbol{w}^0, \mathfrak{M}^{\ell,n})$ in the limit $\ell, n \to \infty$, we are in a similar framework to that of the statement of Theorem 3.13, and this suffices to conclude. In this respect, Lemma 3.25 is used to pass to the limit in the approximating coefficients and Lemma 3.27 guarantees that the limiting initial condition has the right distribution.

3.4 Explicit Solvability Results

Our goal is now to identify conditions under which Theorem 3.1 applies.

The examples we have in mind are those specified right below the statement of assumption **FBSDE** in Subsection 2.2.3:

1. the FBSDE (3.5) is constructed by means of Theorem 1.57 and is intended to describe the dynamics of the value function of the underlying stochastic control problem;
2. the FBSDE (3.5) is constructed by means of the stochastic Pontryagin maximum principle, as explained in the statement of Theorem 1.60.

3.4.1 Using the Representation of the Value Function

A first strategy is to characterize the value function of the optimal stochastic control problem (3.1)–(3.2) as the value function of an FBSDE in the spirit of Theorem 1.57 and Proposition 1.58.

Assumption (MFG with a Common Noise HJB). There exist two constants $L \geq 0$ and $\lambda > 0$ such that:

(A1) The drift b has the form:

$$b(t, x, \mu, \alpha) = b_1(t, x, \mu) + b_2(t)\alpha,$$

where the mapping $[0, T] \ni t \mapsto b_2(t) \in \mathbb{R}^{d \times k}$ is measurable and bounded by L.

(A2) The coefficients b_1, σ, and σ^0 are Borel-measurable mappings from $[0, T] \times \mathbb{R}^d \times \mathcal{P}_2(\mathbb{R}^d)$ into $\mathbb{R}^d$, $\mathbb{R}^{d \times d}$, and $\mathbb{R}^{d \times d}$ respectively. For any $t \in [0, T]$, the functions $b_1(t, \cdot, \cdot)$, $\sigma(t, \cdot, \cdot)$, and $\sigma^0(t, \cdot, \cdot)$ are continuous on $\mathbb{R}^d \times \mathcal{P}_2(\mathbb{R}^d)$ and, for any $\mu \in \mathcal{P}_2(\mathbb{R}^d)$, the functions $b_1(t, \cdot, \mu)$, $\sigma(t, \cdot, \mu)$ and $\sigma^0(t, \cdot, \mu)$ are continuously differentiable with respect to x. Moreover,

$$|(b_1, \sigma, \sigma^{-1}, \sigma^0)(t, x, \mu)| \leq L,$$

$$|\partial_x(b_1, \sigma, \sigma^0)(t, x, \mu)| \leq L.$$

In particular, σ is invertible.

(continued)

(A3) The coefficients f and g are Borel-measurable mappings from $[0,T] \times \mathbb{R}^d \times \mathcal{P}_2(\mathbb{R}^d) \times A$ to $\mathbb{R}$ and from $\mathbb{R}^d \times \mathcal{P}_2(\mathbb{R}^d)$ to $\mathbb{R}$ respectively. For any $t \in [0,T]$, the functions $f(t,\cdot,\cdot,\cdot)$ and $g(\cdot,\cdot)$ are continuous on $\mathbb{R}^d \times \mathcal{P}_2(\mathbb{R}^d) \times A$ and $\mathbb{R}^d \times \mathcal{P}_2(\mathbb{R}^d)$ respectively. For any $t \in [0,T]$ and $\mu \in \mathcal{P}_2(\mathbb{R}^d)$, the function $f(t,\cdot,\mu,\cdot)$ is continuously differentiable (in (x,α)) and the function $g(\cdot,\mu)$ is continuously differentiable in x. Moreover,

$$|f(t,x,\mu,\alpha)| \le L\big(1+|\alpha|^2\big), \quad |g(x,\mu)| \le L,$$

$$|\partial_x f(t,x,\mu,\alpha)| + |\partial_x g(x,\mu)| \le L, \quad |\partial_\alpha f(t,x,\mu,\alpha)| \le L\big(1+|\alpha|\big),$$

and the function $\partial_\alpha f$ is L-Lipschitz-continuous in x.

(A4) f satisfies the uniform λ-convexity property:

$$f(t,x,\mu,\alpha') - f(t,x,\mu,\alpha) - (\alpha'-\alpha)\cdot\partial_\alpha f(t,x,\mu,\alpha) \ge \lambda|\alpha'-\alpha|^2.$$

In full analogy with Lemmas (Vol I)-3.3 and 3.10, there exists a minimizer $\hat{\alpha}$ of the reduced Hamiltonian $H^{(r)}$, whose definition is similar to (3.13). For any $t \in [0,T]$, the function $\mathbb{R}^d \times \mathcal{P}_2(\mathbb{R}^d) \times \mathbb{R}^d \ni (x,\mu,y) \mapsto \hat{\alpha}(t,x,\mu,y)$ is continuous and there exists a constant C such that, for any $t \in [0,T]$, any $x,x',y,y' \in \mathbb{R}^d$ and any $\mu \in \mathcal{P}_2(\mathbb{R}^d)$,

$$\begin{aligned}
&\big|\hat{\alpha}(t,x,\mu,y)\big| \le C\big(1+|y|\big),\\
&\big|\hat{\alpha}(t,x',\mu,y') - \hat{\alpha}(t,x,\mu,y)\big| \le C\big(|x'-x| + |y'-y|\big).
\end{aligned} \tag{3.57}$$

Here, the boundedness of $\hat{\alpha}$ in (x,μ) follows from that of $\partial_\alpha f$.

We know that, under assumption **MFG with a Common Noise HJB**, the assumptions of Proposition 1.58 and Theorem 1.57 are satisfied. For any super-environment $\mathfrak{M}$ and any sub-environment $\boldsymbol{\mu} = (\mu_t)_{0\le t\le T}$, the solution of the stochastic optimal control problem (3.1)–(3.2) may be characterized as the unique solution of the forward-backward system (3.4) with the following truncated coefficients:

$$\begin{aligned}
&B(t,x,\mu,y,z) = \psi(z) b\big(t,x,\mu,\phi\big(\hat{\alpha}\big(t,x,\mu,\sigma(t,x,\mu)^{-1\dagger}z\big)\big)\big),\\
&F(t,x,\mu,y,z,z^0) = \psi(z) f\big(t,x,\mu,\phi\big(\hat{\alpha}\big(t,x,\mu,\sigma(t,x,\mu)^{-1\dagger}z\big)\big)\big),\\
&G(x,\mu) = g(x,\mu).
\end{aligned} \tag{3.58}$$

which are thus independent of y. Above, the functions ϕ and ψ are the same cut-off functions as in the statement of Proposition 1.58. They satisfy $\phi(\alpha) = \alpha$ for $|\alpha| \le C(1+R)$ and $\psi(z) = 1$ for $|z| \le R$ and $\psi(z) = 0$ for $|z| \ge 2R$, with

$|\psi(z) - \psi(z')| \leq (4|z - z'|)/\min(|z|, |z'|)$, for the same R as in the statement of Theorem 1.57. By Theorem 1.57, ϕ, ψ, and R may be chosen independently of the input $\boldsymbol{\mu} = (\mu_t)_{0 \leq t \leq T}$.

By Theorem 3.1, we get:

Theorem 3.29 *Under assumption* **MFG with a Common Noise HJB**, *the mean field game* (3.1)–(3.2)–(3.3) *with* $\mathcal{V}^0 \in \mathcal{P}_2(\mathcal{P}_2(\mathbb{R}^d))$ *as initial condition has a weak solution in the sense of Definition 2.23.*

Of course, the solution to the mean field game may be characterized through an FBSDE of the conditional McKean-Vlasov type, whose form derives from the FBSDE (3.5) along the procedure described in the statement of Proposition 2.18.

3.4.2 Using the Stochastic Pontryagin Principle

We now discuss the case when the solutions of the optimal stochastic control problem (3.1)–(3.2) are characterized by means of the stochastic maximum principle in Theorem 1.60.

Assumption (MFG with a Common Noise SMP). There exist two constants $L \geq 0$ and $\lambda > 0$ such that:

(A1) The drift b is an affine function of (x, α) in the sense that it is of the form:

$$b(t, x, \mu, \alpha) = b_0(t, \mu) + b_1(t)x + b_2(t)\alpha, \tag{3.59}$$

where b_0, b_1 and b_2 are $\mathbb{R}^d$, $\mathbb{R}^{d\times d}$ and $\mathbb{R}^{d\times k}$ valued respectively, and are measurable. and satisfy:

$$|b_0(t, \mu)| \leq L\big(1 + M_2(\mu)\big), \quad |(b_1, b_2)(t)| \leq L.$$

Similarly, σ and σ^0 have the form:

$$\begin{aligned} \sigma(t, x, \mu) &= \sigma_0(t, \mu) + \sigma_1(t)x, \\ \sigma^0(t, x, \mu) &= \sigma_0^0(t, \mu) + \sigma_1^0(t)x, \end{aligned} \tag{3.60}$$

where $[0, T] \times \mathcal{P}_2(\mathbb{R}^d) \ni (t, \mu) \mapsto (\sigma_0, \sigma_0^0)(t, \mu)$ and $[0, T] \ni t \mapsto (\sigma_1, \sigma_1^0)(t)$ are measurable mappings with values in $(\mathbb{R}^{d\times d})^2$ and $(\mathbb{R}^{d\times d\times d})^2$ and satisfy:

$$|(\sigma_0, \sigma_0^0)(t, \mu)| \leq L\big(1 + M_2(\mu)\big), \quad |(\sigma_1, \sigma_1^0)(t)| \leq L.$$

(A2) For any $t \in [0, T]$, the function $\mathcal{P}_2(\mathbb{R}^d) \ni \mu \mapsto (b_0, \sigma_0, \sigma_0^0)(t, \mu)$ is continuous.

(continued)

(A3) The coefficients f and g are measurable. For any $t \in [0,T]$, the functions $f(t,\cdot,\cdot,\cdot)$ and $g(\cdot,\cdot)$ are continuous on $\mathbb{R}^d \times \mathcal{P}_2(\mathbb{R}^d) \times A$ and $\mathbb{R}^d \times \mathcal{P}_2(\mathbb{R}^d)$ respectively. For any $t \in [0,T]$ and $\mu \in \mathcal{P}_2(\mathbb{R}^d)$, the functions $\mathbb{R}^d \times A \ni (x,\alpha) \mapsto f(t,x,\mu,\alpha)$ and $\mathbb{R}^d \ni x \mapsto g(x,\mu)$ are once continuously differentiable. Moreover,

$$|\partial_x f(t,x,\mu,\alpha)|,\ |\partial_x g(x,\mu)| \le L, \quad |\partial_\alpha f(t,x,\mu,\alpha)| \le L\big(1+|\alpha|\big).$$

For any $t \in [0,T]$, the functions $\partial_x f(t,\cdot,\cdot,\cdot)$, $\partial_\alpha f(t,\cdot,\cdot,\cdot)$ and $\partial_x g$ are continuous in (x,μ,α) and in (x,μ) respectively, and are L-Lipschitz continuous in (x,α) and in x.

(A4) The two cost functions f and g are at most of quadratic growth:

$$|f(t,x,\mu,\alpha)| + |g(x,\mu)| \le L\Big(1+|x|^2+\big(M_2(\mu)\big)^2+|\alpha|^2\Big).$$

(A5) The terminal cost function g is convex in the variable x and the running cost function f is convex and uniformly λ-convex in α in the sense that:

$$f(t,x',\mu,\alpha')-f(t,x,\mu,\alpha)-(x'-x,\alpha'-\alpha)\cdot\partial_{(x,\alpha)}f(t,x,\mu,\alpha) \ge \lambda|\alpha'-\alpha|^2,$$

where $\partial_{(x,\alpha)}f$ denotes the gradient of f in the variables (x,α).

Observe from (**A3**) that the growth condition (**A4**) could be strengthened for free. However, we prefer (**A4**) in its current form as we shall make use of it right below.

By Lemmas (Vol I)-3.3 and 3.10, there exists a minimizer $\hat{\alpha}$ to the reduced Hamiltonian $H^{(r)}$ given by:

$$H^{(r)}(t,x,\mu,y,\alpha) = b(t,x,\mu,\alpha)\cdot y + f(t,x,\mu,\alpha).$$

The minimizer $\hat{\alpha}$ satisfies (3.57).

Moreover, by Theorem 1.60, we know that, for any super-environment $\mathfrak{M}$ and any sub-environment $\boldsymbol{\mu} = (\mu_t)_{0\le t\le T}$, the solution of the stochastic optimal control problem (3.1)–(3.2) may be characterized as the unique solution of the forward-backward system (3.4) with the following coefficients:

$$\begin{aligned}
&B(t,x,\mu,y) = b\big(t,x,\mu,\hat{\alpha}(t,x,\mu,y)\big),\\
&F(t,x,\mu,y,z,z^0) = \partial_x H\big(t,x,\mu,y,z,z^0,\alpha\big)\Big|_{\alpha=\hat{\alpha}(t,x,\mu,y)},\\
&G(x,\mu) = \partial_x g(x,\mu),
\end{aligned} \tag{3.61}$$

where H is the full Hamiltonian defined as:

$$H(t,x,\mu,y,z,\alpha) = b(t,x,\mu,\alpha)\cdot y + f(t,x,\mu,\alpha) + \mathrm{trace}\big[\sigma(t,x,\mu)z^\dagger\big] + \mathrm{trace}\big[\sigma^0(t,x,\mu)(z^0)^\dagger\big].$$

By Theorem 3.1, we get:

Theorem 3.30 *Under assumption* **MFG with a Common Noise SMP**, *the mean field game* (3.1)–(3.2)–(3.3) *with* $\mathcal{V}^0 \in \mathcal{P}_2(\mathcal{P}_2(\mathbb{R}^d))$ *as initial condition has a weak solution in the sense of Definition 2.23.*

As in the previous subsection, the solution to the mean field game may be characterized through an FBSDE of the conditional McKean-Vlasov type, whose form derives from the FBSDE (3.5) along the procedure described in the statement of Proposition 2.18.

3.4.3 Allowing for Quadratic Cost Functionals

Theorem 3.30 may be compared with Proposition (Vol I)-4.57, which applies without common noise when $\partial_x f$ and $\partial_x g$ are bounded. Our goal is now to relax the boundedness assumption in full analogy with Theorem (Vol I)-4.53 when viewed as a relaxation of Proposition (Vol I)-4.57. Allowing $\partial_x f$ and $\partial_x g$ to be unbounded is particularly important if we want to handle quadratic cost functionals.

Not surprisingly, we shall use the same type of assumptions as those introduced in the statement of Theorem (Vol I)-4.53, namely:

Assumption (MFG with a Common Noise SMP Relaxed). We assume that $A = \mathbb{R}^k$ and that there exist two constants $L \geq 0$ and $\lambda > 0$ such that:

(A1) Assumption **MFG with a Common Noise SMP** is in force except for the requirement

$$\big|\partial_x f(t,x,\mu,\alpha)\big| \leq L, \quad \big|\partial_x g(x,\mu)\big| \leq L,$$

in **(A4)**, even though all the other requirements are maintained.

(A2) For all $t \in [0,T]$, $x,x' \in \mathbb{R}^d$, $\alpha,\alpha' \in \mathbb{R}^k$ and $\mu,\mu' \in \mathcal{P}_2(\mathbb{R}^d)$, it holds:

$$\begin{aligned}\big|f(t,x',\mu',\alpha') - f(t,x,\mu,\alpha)\big| + \big|g(x',\mu') - g(x,\mu)\big| \\ \leq L\big[1 + |(x',\alpha')| + |(x,\alpha)| + M_2(\mu) + M_2(\mu')\big] \\ \times \big[|(x',\alpha') - (x,\alpha)| + W_2(\mu',\mu)\big].\end{aligned}$$

(continued)

(A3) For all $(t, x) \in [0, T] \times \mathbb{R}^d$,

$$x \cdot \partial_x f(t, 0, \delta_x, 0) \geq -L(1 + |x|), \quad x \cdot \partial_x g(0, \delta_x) \geq -L(1 + |x|).$$

(A4) In (3.59) and (3.60), the functions b_0, σ_0 and σ_0^0 are bounded by L and the function σ_1 is identically zero.

Once again, by Lemmas (Vol I)-3.3 and 3.10, there exists a minimizer $\hat{\alpha}$ to the reduced Hamiltonian $H^{(r)}$. The minimizer $\hat{\alpha}$ satisfies the Lipschitz and growth conditions in (3.57).

The desired extension of Theorem 3.30 is the following.

Theorem 3.31 *Under assumption* **MFG with a Common Noise SMP Relaxed**, *the mean field game* (3.1)–(3.2)–(3.3) *with* $\mathcal{V}^0 \in \mathcal{P}_2(\mathcal{P}_2(\mathbb{R}^d))$ *as initial condition has a weak solution in the sense of Definition 2.23.*

The strategy is the same as for the proof of Theorem (Vol I)-4.53. It relies on a suitable approximation procedure. The first step is an adaptation of Lemma (Vol I)-4.58.

Lemma 3.32 *Assume that we can find two sequences of functions* $(f^n)_{n\geq 1}$ *and* $(g^n)_{n\geq 1}$ *satisfying:*

(i) there exist two constants L' *and* $\lambda' > 0$ *such that, for any* $n \geq 1$, *the coefficients* $(b, \sigma, \sigma^0, f^n, g^n)$ *satisfy assumption* **MFG with a Common Noise SMP Relaxed** *with respect to* L' *and* λ'*;*

(ii) $(f^n, \partial_x f^n, \partial_\alpha f^n)$ *(resp.* $(g^n, \partial_x g^n)$*) converges toward* $(f, \partial_x f, \partial_\alpha f)$ *(resp.* $(g, \partial_x g)$*) uniformly on bounded subsets of* $[0, T] \times \mathbb{R}^d \times \mathcal{P}_2(\mathbb{R}^d) \times \mathbb{R}^k$ *(resp.* $\mathbb{R}^d \times \mathcal{P}_2(\mathbb{R}^d)$*);*

(iii) for any $n \geq 1$, *the mean field game* (3.1)–(3.2)–(3.3), *driven by* $(b, \sigma, \sigma^0, f^n, g^n)$ *instead of* $(b, \sigma, \sigma^0, f, g)$ *and with* $\mathcal{V}^0 \in \mathcal{P}_2(\mathcal{P}_2(\mathbb{R}^d))$ *as initial condition, has a weak solution.*

Under these conditions, the mean field game (3.1)–(3.2)–(3.3), *with coefficients* $(b, \sigma, \sigma^0, f, g)$ *and initial condition* $\mathcal{V}^0 \in \mathcal{P}_2(\mathcal{P}_2(\mathbb{R}^d))$, *has a weak solution.*

Taking for granted the result of Lemma 3.32, the key point is to let, for any integer $p \geq 1$ and any $t \in [0, T]$, $x \in \mathbb{R}^d$, $\mu \in \mathcal{P}_2(\mathbb{R}^d)$, and $\alpha \in \mathbb{R}^k$,

$$f^p(t, x, \mu, \alpha) = f(t, x, \mu, \alpha) + \frac{1}{p}|x|^2; \quad g^p(x, \mu) = g(x, \mu) + \frac{1}{p}|x|^2,$$

so that the functions f^p and g^p are strictly convex in the joint variable (x, α). Then, by Lemma (Vol I)-4.59, we can find, for any $p \geq 1$, two sequences of functions $(f^{p,n})_{n\geq 1}$ and $(g^{p,n})_{n\geq 1}$ such that the sequence $(b, \sigma, \sigma^0, f^{p,n}, g^{p,n})_{n\geq 1}$

satisfies assumptions (i) and (ii) in the statement of Lemma 3.32 and, for each $n \geq 1$, the tuple $(b, \sigma, \sigma^0, f^{p,n}, g^{p,n})$ satisfies assumption **MFG with a Common Noise SMP**. Therefore, by combining Theorem 3.30 and Lemma 3.32, we deduce that the mean field game (3.1)–(3.2)–(3.3) driven by $(b, \sigma, \sigma^0, f^p, g^p)$ instead of $(b, \sigma, \sigma^0, f, g)$ and with $\mathcal{V}^0 \in \mathcal{P}_2(\mathcal{P}_2(\mathbb{R}^d))$ as initial condition, has a weak solution. Reapplying Lemma 3.32, we deduce that the mean field game (3.1)–(3.2)–(3.3), driven by $(b, \sigma, \sigma^0, f, g)$ and with $\mathcal{V}^0 \in \mathcal{P}_2(\mathcal{P}_2(\mathbb{R}^d))$ as initial condition, is weakly solvable, from which Theorem 3.31 easily follows.

3.4.4 Proof of the Approximation Lemma

In order to complete the proof of Theorem 3.31, it remains to prove Lemma 3.32. The proof is split into several steps. Throughout the analysis, we work, for every $n \geq 1$, with a probability space $(\Omega^n, \mathcal{F}^n, \mathbb{F}^n, \mathbb{P}^n)$ of the product form as in Definition 2.16, where Ω^n is equal to $\Omega^{0,n} \times \Omega^{1,n}$ and $(\Omega^{0,n}, \mathcal{F}^{0,n}, \mathbb{F}^{0,n}, \mathbb{P}^{0,n})$ and $(\Omega^{1,n}, \mathcal{F}^{1,n}, \mathbb{F}^{1,n}, \mathbb{P}^{1,n})$ are two complete filtered probability spaces, the filtrations being complete and right-continuous. We then ask $(\Omega^n, \mathcal{F}^n, \mathbb{P}^n)$ to be the completion of the product space $(\Omega^{0,n} \times \Omega^{1,n}, \mathcal{F}^{0,n} \otimes \mathcal{F}^{1,n}, \mathbb{P}^{0,n} \otimes \mathbb{P}^{1,n})$ and $\mathbb{F}^n$ to be the complete and right-continuous augmentation of the product of the two filtrations. We then equip $(\Omega^{0,n}, \mathcal{F}^{0,n}, \mathbb{F}^{0,n}, \mathbb{P}^{0,n})$ with an $\mathbb{F}^{0,n}$-Brownian motion $\boldsymbol{W}^{0,n}$ with values in $\mathbb{R}^d$ and with an $\mathcal{F}_0^{0,n}$-measurable random variable μ_0^n with values in $\mathcal{P}_2(\mathbb{R}^d)$ and with distribution $\mathcal{V}^0 \in \mathcal{P}_2(\mathcal{P}_2(\mathbb{R}^d))$. Similarly, we equip $(\Omega^{1,n}, \mathcal{F}^{1,n}, \mathbb{F}^{1,n}, \mathbb{P}^{1,n})$ with an $\mathbb{F}^{1,n}$-Brownian motion $\boldsymbol{W}^n$ with values in $\mathbb{R}^d$. We also equip $(\Omega^n, \mathcal{F}^n, \mathbb{F}^n, \mathbb{P}^n)$ with a random variable X_0^n with values in $\mathbb{R}^d$ such that $\mu_0^n = \mathcal{L}^{1,n}(X_0^n)$, where $\mathcal{L}^{1,n}$ denotes the marginal law on $(\Omega^{1,n}, \mathcal{F}^{1,n}, \mathbb{F}^{1,n}, \mathbb{P}^{1,n})$.

We assume that, for any $n \geq 1$, there exists a random variable $\mathfrak{M}^n$, constructed on the space $(\Omega^{0,n}, \mathcal{F}^{0,n}, \mathbb{F}^{0,n}, \mathbb{P}^{0,n})$ and with values in $\mathcal{P}_2(\mathcal{C}([0,T]; \mathbb{R}^{2d}))$, that induces, together with $(X_0^n, \boldsymbol{W}^{0,n}, \boldsymbol{W}^n)$, a solution to the mean field game (3.1)–(3.2)–(3.3) with $(b, \sigma, \sigma^0, f^n, g^n)$ as coefficients and $\mathcal{V}^0 \in \mathcal{P}_2(\mathcal{P}_2(\mathbb{R}^d))$ as initial condition.

Observe that, from Lemma 2.25, all the solutions may be constructed on the same canonical space. In particular, we can assume that Ω^n, $\Omega^{0,n}$ and $\Omega^{1,n}$ are independent of n. We can even assume that $\mathcal{F}^{1,n}$, $\mathbb{F}^{1,n}$ and $\mathbb{P}^{1,n}$ are also independent of n. Accordingly, we can construct X_0^n, $\boldsymbol{W}^{0,n}$ and $\boldsymbol{W}^n$ independently of n. As a result, we shall drop the index n in all the aforementioned quantities Ω^n, $\Omega^{0,n}$, $\Omega^{1,n}$, $\mathcal{F}^{1,n}$, $\mathbb{F}^{1,n}$, $\mathbb{P}^{1,n}$, X_0^n, $\boldsymbol{W}^{0,n}$ and $\boldsymbol{W}^n$, but not in the others. As we work on the canonical space, we could do the same for $\mathfrak{M}^n$, but, for the sake of clarity, we feel better to keep the index n in the notation. In order to avoid any confusion between $\mathbb{P}^n$ and $\mathbb{P}^{1,n}$ when dropping out the index n, for $n = 1$, we shall write $\mathbb{P}^{1,1}$ for $\mathbb{P}^{1,n}$, and similarly for $\mathcal{F}^{1,n}$ and $\mathbb{F}^{1,n}$.

For each $n \geq 1$, the MFG equilibrium $\mathfrak{M}^n$ may be represented through a tuple $(\boldsymbol{X}^n, \boldsymbol{Y}^n, \boldsymbol{Z}^n, \boldsymbol{Z}^{0,n}, \boldsymbol{M}^n)$ solving an FBSDE of the form (3.5) on $(\Omega, \mathcal{F}^n, \mathbb{F}^n, \mathbb{P}^n)$ equipped with $(X_0, \boldsymbol{W}^0, \mathfrak{M}^n, \boldsymbol{W})$ and with $\boldsymbol{\mu}^n = (\mathfrak{M}^n \circ (e_t^x)^{-1})_{0 \leq t \leq T}$ as sub-environment. For each $n \geq 1$, the coefficients of the FBSDE read:

$$B(t,x,\mu,y) = b\big(t,x,\mu,\hat{\alpha}^n(t,x,\mu,y)\big),$$

$$F(t,x,\mu,y,z,z^0) = \partial_x H^n\big(t,x,\mu,y,z,z^0,\alpha\big)\Big|_{\alpha=\hat{\alpha}^n(t,x,\mu,y)}, \tag{3.62}$$

$$G(x,\mu) = \partial_x g^n(x,\mu),$$

where:

$$H^n(t,x,\mu,y,z,z^0,\alpha) = b(t,x,\mu,\alpha)\cdot y + f^n(t,x,\mu,\alpha) + \text{trace}\big[\sigma(t,x,\mu)z^\dagger\big] + \text{trace}\big[\sigma^0(t,x,\mu)(z^0)^\dagger\big],$$

and

$$\hat{\alpha}^n(t,x,\mu,y) = \text{argmin}_{\alpha\in\mathbb{R}^k} H^{(r),n}(t,x,\mu,y,\alpha),$$

with

$$H^{(r),n}(t,x,\mu,y,\alpha) = b(t,x,\mu,\alpha)\cdot y + f^n(t,x,\mu,\alpha).$$

Compactness of the Approximating MFG Equilibria

As an application, we have:

Lemma 3.33 *Under the assumption of Lemma 3.32, the sequence of probability measures* $(\mathbb{P}^n \circ (\boldsymbol{X}^n)^{-1})_{n\geq 1}$ *is tight on* $\mathcal{C}([0,T];\mathbb{R}^d)$ *and the sequence* $(\mathbb{P}^n \circ (\sup_{0\leq t\leq T}|X_t^n|^2)^{-1})_{n\geq 1}$ *is uniformly square-integrable. Moreover,*

$$\sup_{n\geq 1}\mathbb{E}^n\bigg[\sup_{0\leq t\leq T}|Y_t^n|^2 + \int_0^T \big(|Z_t^n|^2 + |Z_t^{0,n}|^2\big)dt\bigg] < \infty. \tag{3.63}$$

Proof. The proof is an adaptation of the proof of Lemma (Vol I)-4.58. For that reason, we shall only give a sketch. For any $n \geq 1$ and $t \in [0,T]$, we let $\hat{\alpha}_t^n = \hat{\alpha}^n(t, X_t^n, \mu_t^n, Y_t^n)$. The crucial point is to prove that:

$$\sup_{n\geq 1}\mathbb{E}^n\bigg[\int_0^T |\hat{\alpha}_s^n|^2 ds\bigg] < \infty. \tag{3.64}$$

We then apply Theorem 1.60 with two different choices for $\boldsymbol{\alpha}$. Using the letter $\boldsymbol{\beta}^n$ to denote the effective choice of $\boldsymbol{\alpha}$ and calling $\boldsymbol{U}^n$ the corresponding controlled paths, we shall consider the two cases:

$$(i)\ \beta_s^n = \mathbb{E}^{1,1}(\hat{\alpha}_s^n) \quad \text{for } t\leq s\leq T; \quad (ii)\ \beta_s^n = 0 \quad \text{for } t \leq s\leq T, \tag{3.65}$$

for some $t \in [0,T]$.

We compare the cost to each of these controls with the optimal cost in order to derive useful information on the optimal control $(\hat{\alpha}^n_s)_{0\le s\le T}$. The comparison relies on the stochastic Pontryagin principle proved in Theorem 1.60 and, more precisely, on the fact that a similar version to (1.63) holds true, but with the initial time 0 being replaced by some $t \in [0,T]$ and with the expectation being replaced by the conditional expectation given $\mathcal{F}^n_t$. We already accounted for this extension in the proof of Theorem 1.60. Throughout the proof, we shall write $\mathbb{E}^n_t$ for the conditional expectation $\mathbb{E}^n[\cdot\,|\,\mathcal{F}^n_t]$ and $\mathbb{E}^{0,n}_t$ for the conditional expectation $\mathbb{E}^n[\cdot\,|\,\mathcal{F}^{0,n}_t]$, which is also equal to $\mathbb{E}^{1,1}\mathbb{E}^n_t$ and to $\mathbb{E}^n_t\mathbb{E}^{1,1}$.

First Step. We first consider (*i*) in (3.65), for some $t \in [0,T]$. In this case,

$$
\begin{aligned}
U^n_s &= X^n_t + \int_t^s \big[b_0(r,\mu^n_r) + b_1(r)U^n_r + b_2(r)\mathbb{E}^{1,1}(\hat{\alpha}^n_r)\big]dr \\
&\quad + \int_t^s \sigma_0(r,\mu^n_r)dW_r + \int_t^s \big[\sigma^0_0(r,\mu^n_r) + \sigma^0_1(r)U^n_r\big]dW^0_r, \quad s \in [t,T].
\end{aligned}
\tag{3.66}
$$

By taking expectation under $\mathbb{P}^{1,1}$ on both sides of (3.66), we have $\mathbb{E}^{1,1}(U^n_s) = \mathbb{E}^{1,1}(X^n_s)$, for $s \in [t,T]$, since $(\mathbb{E}^{1,1}(U^n_s))_{t\le s\le T}$ and $(\mathbb{E}^{1,1}(X^n_s))_{t\le s\le T}$ satisfy the same SDE. Moreover,

$$
\begin{aligned}
\big[U^n_s - \mathbb{E}^{1,1}(U^n_s)\big] &= \big[X^n_t - \mathbb{E}^{1,1}(X^n_t)\big] + \int_t^s b_1(r)\big[U^n_r - \mathbb{E}^{1,1}(U^n_r)\big]ds \\
&\quad + \int_t^s \sigma_0(r,\mu^n_r)dW_r \\
&\quad + \int_t^s \big[\sigma^0_1(r)\big(U^n_r - \mathbb{E}^{1,1}(U^n_r)\big)\big]dW^0_r, \quad s \in [t,T],
\end{aligned}
$$

from which it easily follows that there exists a constant C such that, for all $n \ge 1$,

$$
\sup_{t\le s\le T} \mathbb{E}^{0,n}_t\big[|U^n_s - \mathbb{E}^{1,1}(U^n_s)|^2\big] \le C\Big(1 + \mathbb{E}^{0,n}_t\big[|X^n_t - \mathbb{E}^{1,1}(X^n_t)|^2\big]\Big).
\tag{3.67}
$$

By the conditional version of Theorem 1.60 at time t, with $g^n(\cdot,\mu^n_T)$ as terminal cost and $(f^n(s,\cdot,\mu^n_s,\cdot))_{t\le s\le T}$ as running cost, we get, taking the expectation under $\mathbb{E}^{1,1}$ and then following (Vol I)-(4.87),

$$
\begin{aligned}
&\mathbb{E}^{0,n}_t\big[g^n(\mathbb{E}^{1,1}(U^n_T),\mu^n_T)\big] \\
&\quad + \mathbb{E}^{0,n}_t \int_t^T \Big[\lambda'|\hat{\alpha}^n_s - \mathbb{E}^{1,1}(\hat{\alpha}^n_s)|^2 + f^n\big(s,\mathbb{E}^{1,1}(U^n_s),\mu^n_s,\mathbb{E}^{1,1}(\hat{\alpha}^n_s)\big)\Big]ds \\
&\le \mathbb{E}^{0,n}_t\Big[g^n(U^n_T,\mu^n_T) + \int_t^T f^n\big(s,U^n_s,\mu^n_s,\mathbb{E}^{1,1}(\hat{\alpha}^n_s)\big)ds\Big].
\end{aligned}
\tag{3.68}
$$

Using (**A2**) in assumption **MFG with Common Noise SMP Relaxed**, we deduce that there exists a constant C (independent of n) such that:

$$
\begin{aligned}
&\mathbb{E}_t^{0,n} \int_t^T |\hat{\alpha}_s^n - \mathbb{E}^{1,1}(\hat{\alpha}_s^n)|^2 ds \\
&\quad \le C\Big(1 + \mathbb{E}_t^{0,n}\big[|U_T^n|^2\big]^{1/2} + \mathbb{E}_t^{0,n}\big[|X_T^n|^2\big]^{1/2}\Big)\mathbb{E}_t^{0,n}\big[|U_T^n - \mathbb{E}^{1,1}(U_T^n)|^2\big]^{1/2} \\
&\qquad + C\int_t^T \Big[\Big(1 + \mathbb{E}_t^{0,n}\big[|U_s^n|^2\big]^{1/2} + \mathbb{E}_t^{0,n}\big[|X_s^n|^2\big]^{1/2} + \mathbb{E}_t^{0,n}\big[|\hat{\alpha}_s^n|^2\big]^{1/2}\Big) \\
&\qquad\qquad \times \mathbb{E}_t^{0,n}\big[|U_s^n - \mathbb{E}^{1,1}(U_s^n)|^2\big]^{1/2}\Big] ds,
\end{aligned}
$$

where we used the fact that:

$$
\mathbb{E}_t^{0,n}[M_2(\mu_s^n)^2] = \mathbb{E}_t^{0,n}\big[\mathbb{E}^{1,1}[|X_s^n|^2]\big] = \mathbb{E}^n\big[\mathbb{E}^n[|X_s^n|^2 \mid \mathcal{F}_T^{0,n}] \mid \mathcal{F}_t^{0,n}\big] = \mathbb{E}_t^{0,n}[|X_s^n|^2].
$$

From (3.67) together with the identity $\mathbb{E}^{1,1}(U_s^n) = \mathbb{E}^{1,1}(X_s^n)$, for all $s \in [t, T]$, we obtain:

$$
\begin{aligned}
\mathbb{E}_t^{0,n} \int_t^T |\hat{\alpha}_s^n - \mathbb{E}^{1,1}(\hat{\alpha}_s^n)|^2 ds &\le C\Big[1 + \sup_{t \le s \le T} \mathbb{E}_t^{0,n}[|X_s^n|^2]^{1/2} + \Big(\mathbb{E}_t^{0,n}\int_t^T |\hat{\alpha}_s^n|^2 ds\Big)^{1/2}\Big] \\
&\qquad \times \Big(1 + \mathbb{E}_t^{0,n}\big[|X_t^n - \mathbb{E}^{1,1}(X_t^n)|^2\big]^{1/2}\Big).
\end{aligned}
\tag{3.69}
$$

Next, we observe from the growth conditions on the coefficients that:

$$
\sup_{t \le s \le T} \mathbb{E}_t^{0,n}[|X_s^n|^2] \le C\Big[1 + \mathbb{E}_t^{0,n}[|X_t^n|^2] + \mathbb{E}_t^{0,n}\int_t^T |\hat{\alpha}_s^n|^2 ds\Big]. \tag{3.70}
$$

Also, duplicating the proof of (3.67), we have:

$$
\begin{aligned}
&\sup_{t \le s \le T} \mathbb{E}_t^{0,n}\big[|X_s^n - \mathbb{E}^{1,1}(X_s^n)|^2\big] \\
&\quad \le C\Big[1 + \mathbb{E}_t^{0,n}\big[|X_t^n - \mathbb{E}^{1,1}(X_t^n)|^2\big] + \mathbb{E}_t^{0,n}\int_t^T |\hat{\alpha}_s^n - \mathbb{E}^{1,1}(\hat{\alpha}_s^n)|^2 ds\Big].
\end{aligned}
$$

Therefore,

$$
\begin{aligned}
\sup_{t \le s \le T} \mathbb{E}_t^{0,n}\big[|X_s^n - \mathbb{E}^{1,1}(X_s^n)|^2\big] &\le C\Big[1 + \mathbb{E}_t^{0,n}\big[|X_t^n|^2\big]^{1/2} + \Big(\mathbb{E}^n\int_t^T |\hat{\alpha}_s^n|^2 ds\Big)^{1/2}\Big] \\
&\qquad \times \Big(1 + \mathbb{E}_t^{0,n}\big[|X_t^n - \mathbb{E}^{1,1}(X_t^n)|^2\big]^{1/2}\Big),
\end{aligned}
\tag{3.71}
$$

which is similar to (Vol I)-(4.89).

Second Step. We now compare X^n to the process controlled by the null control. So we consider case (*ii*) in (3.65), and now,

$$U_s^n = X_t^n + \int_t^s \big[b_0(r,\mu_r^n) + b_1(r)U_r^n\big]dr$$

$$+ \int_t^s \sigma_0(r,\mu_r^n)dW_r + \int_t^s \big(\sigma_0^0(r,\mu_r^n) + \sigma_1^0(r)U_r^n\big)dW_r^0,$$

with $s \in [t,T]$. Thanks to the growth conditions on the coefficients together with (3.70), we have $\sup_{n\geq 1}\mathbb{E}_t^{0,n}[\sup_{t\leq s\leq T}|U_s^n|^2] \leq C(1+\mathbb{E}_t^{0,n}[|X_t^n|^2])$. Using Theorem 1.60 as before in the derivation of (3.68), we get:

$$\mathbb{E}_t^{0,n}\big[g^n(\mathbb{E}^{1,1}(X_T^n),\mu_T^n)\big] + \mathbb{E}_t^{0,n}\int_t^T \big[\lambda'|\hat{\alpha}_s^n|^2 + f^n\big(s,\mathbb{E}^{1,1}(X_s^n),\mu_s^n,\mathbb{E}^{1,1}(\hat{\alpha}_s^n)\big)\big]ds$$

$$\leq \mathbb{E}_t^{0,n}\bigg[g^n\big(U_T^n,\mu_T^n\big) + \int_t^T f^n\big(s,U_s^n,\mu_s^n,0\big)ds\bigg].$$

By convexity of f^n with respect to α and by the growth conditions on $\partial_\alpha f$ in assumption **MFG with a Common Noise SMP**, we have:

$$\mathbb{E}_t^{0,n}\big[g^n(\mathbb{E}^{1,1}(X_T^n),\mu_T^n)\big] + \mathbb{E}_t^{0,n}\int_t^T \big[\lambda'|\hat{\alpha}_s^n|^2 + f^n\big(s,\mathbb{E}^{1,1}(X_s^n),\mu_s^n,0\big)\big]ds$$

$$\leq \mathbb{E}_t^{0,n}\bigg[g^n\big(U_T^n,\mu_T^n\big) + \int_t^T f^n\big(s,U_s^n,\mu_s^n,0\big)ds\bigg] + C\mathbb{E}_t^{0,n}\int_t^T |\hat{\alpha}_s^n|ds,$$

for some constant C, independent of n. Using **(A2)** in assumption **MFG with a Common Noise SMP Relaxed**, we obtain:

$$\mathbb{E}_t^{0,n}\big[g^n(\mathbb{E}^{1,1}(X_T^n),\delta_{\mathbb{E}^{1,1}(X_T^n)})\big] + \mathbb{E}_t^{0,n}\int_t^T \big[\lambda'|\hat{\alpha}_s^n|^2 + f^n\big(s,\mathbb{E}^{1,1}(X_s^n),\delta_{\mathbb{E}^{1,1}(X_s^n)},0\big)\big]ds$$

$$\leq \mathbb{E}_t^{0,n}\bigg[g^n\big(0,\delta_{\mathbb{E}^{1,1}(X_T^n)}\big) + \int_t^T f^n\big(s,0,\delta_{\mathbb{E}^{1,1}(X_s^n)},0\big)ds\bigg] + C\mathbb{E}_t^{0,n}\int_t^T |\hat{\alpha}_s^n|ds$$

$$+ C\Big(1 + \sup_{t\leq s\leq T}\big[\mathbb{E}_t^{0,n}[|X_s^n|^2]^{1/2}\big]\Big)$$

$$\times\Big(1 + \mathbb{E}_t^{0,n}\big[|X_t^n|^2\big]^{1/2} + \sup_{t\leq s\leq T}\mathbb{E}_t^{0,n}\big[|X_s^n - \mathbb{E}^{1,1}(X_s^n)|^2\big]^{1/2}\Big),$$

where we used the fact that $\mathbb{E}^{1,1}[X_t^n] = \int_{\mathbb{R}^d} x d\mu_t^n(x)$ and that $W_2(\mu_s^n,\delta_{\mathbb{E}^{1,1}(X_s^n)})^2 \leq \mathbb{E}^{1,1}[|X_s^n - \mathbb{E}^{1,1}(X_s^n)|^2]$. Using (3.70), we have:

$$\mathbb{E}_t^{0,n}\big[g^n\big(\mathbb{E}^{1,1}(X_T^n),\delta_{\mathbb{E}^{1,1}(X_T^n)}\big)\big]+\mathbb{E}_t^{0,n}\int_t^T\Big[\frac{\lambda'}{2}|\hat{\alpha}_s^n|^2+f^n\big(s,\mathbb{E}^{1,1}(X_s^n),\delta_{\mathbb{E}^{1,1}(X_s^n)},0\big)\Big]ds$$
$$\leq \mathbb{E}_t^{0,n}\bigg[g^n\big(0,\delta_{\mathbb{E}^{1,1}(X_T^n)}\big)+\int_t^T f^n\big(s,0,\delta_{\mathbb{E}^{1,1}(X_s^n)},0\big)ds\bigg]$$
$$+C\Big(1+\mathbb{E}_t^{0,n}\big[|X_t^n|^2\big]+\sup_{t\leq s\leq T}\mathbb{E}_t^{0,n}\big[|X_s^n-\mathbb{E}^{1,1}(X_s^n)|^2\big]\Big).$$

By (3.71) and Young's inequality, we obtain:

$$\mathbb{E}_t^{0,n}\big[g^n\big(\mathbb{E}^{1,1}(X_T^n),\delta_{\mathbb{E}^{1,1}(X_T^n)}\big)\big]+\mathbb{E}_t^{0,n}\int_t^T\big[\frac{\lambda'}{4}|\hat{\alpha}_s^n|^2+f^n\big(s,\mathbb{E}^{1,1}(X_s^n),\delta_{\mathbb{E}^{1,1}(X_s^n)},0\big)\big]ds$$
$$\leq \mathbb{E}_t^{0,n}\bigg[g^n\big(0,\delta_{\mathbb{E}^{1,1}(X_T^n)}\big)+\int_0^T f^n\big(s,0,\delta_{\mathbb{E}^{1,1}(X_s^n)},0\big)ds\bigg]+C\Big(1+\mathbb{E}_t^{0,n}\big[|X_t^n|^2\big]\Big).$$

By convexity of g^n and f^n in x, we obtain:

$$\mathbb{E}_t^{0,n}\big[\mathbb{E}^{1,1}(X_T^n)\cdot\partial_x g^n\big(0,\delta_{\mathbb{E}^{1,1}(X_T^n)}\big)\big]+\mathbb{E}_t^{0,n}\int_t^T\big[\frac{\lambda'}{4}|\hat{\alpha}_s^n|^2+\mathbb{E}^{1,1}(X_s^n)\cdot\partial_x f^n\big(s,0,\delta_{\mathbb{E}^{1,1}(X_s^n)},0\big)\big]ds$$
$$\leq C\Big(1+\mathbb{E}_t^{0,n}\big[|X_t^n|^2\big]\Big).$$

Using (**A3**) in assumption **MFG with Common Noise SMP Relaxed** together with (3.70) and following (4.91) in the proof of Lemma (Vol I)-4.58, we deduce that:

$$\mathbb{E}_t^{0,n}\big[\sup_{t\leq s\leq T}|X_s^n|^2\big]+\mathbb{E}_t^{0,n}\bigg[\int_t^T|\hat{\alpha}_s^n|^2ds\bigg]\leq C\Big(1+\mathbb{E}_t^{0,n}\big[|X_t^n|^2\big]\Big). \tag{3.72}$$

As a consequence, it also holds that:

$$\mathbb{E}_t^{0,n}\big[\sup_{t\leq s\leq T}M_2(\mu_s^n)^2\big]\leq C\Big(1+\mathbb{E}_t^{0,n}\big[|X_t^n|^2\big]\Big). \tag{3.73}$$

The estimates (3.72) and (3.73) are specially relevant. Recall indeed from Lemma 3.10 that $|\hat{\alpha}_t^n|\leq C(1+|X_t^n|+M_2(\mu_t^n)+|Y_t^n|)$, for $t\in[0,T]$. Expressing Y_t^n in terms of the decoupling field and using Theorem 1.60 in order to control the growth of the decoupling field, see in particular (1.68), we know that, for any $t\in[0,T]$,

$$|Y_t^n|\leq C\Big(1+|X_t^n|+\mathbb{E}_t^{0,n}\big[\sup_{0\leq s\leq T}M_2(\mu_s^n)^2\big]^{1/2}\Big),$$

for a constant C independent of n. Therefore,

$$|\hat{\alpha}_t^n|\leq C\Big(1+|X_t^n|+\sup_{0\leq s\leq t}M_2(\mu_s^n)+\mathbb{E}_t^{0,n}\big[\sup_{t\leq s\leq T}M_2(\mu_s^n)^2\big]^{1/2}\Big).$$

In particular, by (3.73), it holds, $\mathrm{Leb}_1 \otimes \mathbb{P}^n$ almost everywhere,

$$\begin{aligned}|\hat{\alpha}_t^n| &\le C\Big(1 + |X_t^n| + \sup_{0\le s\le t} M_2(\mu_s^n) + \mathbb{E}_t^{0,n}\big[\sup_{t\le s\le T} M_2(\mu_s^n)^2\big]^{1/2}\Big)\\ &\le C\Big(1 + |X_t^n| + \sup_{0\le s\le t} M_2(\mu_s^n) + \mathbb{E}_t^{0,n}\big[|X_t^n|^2\big]^{1/2}\Big),\end{aligned}$$

where we used the fact that $\sup_{t\le s\le T} M_2(\mu_s^n)$ is $\mathcal{F}_T^{0,n}$-measurable in order to identify the conditional expectation $\mathbb{E}_t^n[\sup_{t\le s\le T} M_2(\mu_s^n)^2]$ with $\mathbb{E}_t^{0,n}[\sup_{t\le s\le T} M_2(\mu_s^n)^2]$.

Now, we recall from the identity $\mathbb{E}_t^{0,n} = \mathbb{E}^{1,1}\mathbb{E}_t^n$ that $\mathbb{E}_t^{0,n}[|X_t^n|^2]^{1/2}$ coincides with $\mathbb{E}^{1,1}[|X_t^n|^2]^{1/2}$ since X_t^n is $\mathcal{F}_t^n$-measurable. Moreover, $M_2(\mu_s^n)$ is equal to $\mathbb{E}^{1,1}[|X_s^n|^2]^{1/2}$. We deduce that:

$$|\hat{\alpha}_t^n| \le C\Big(1 + |X_t^n| + \sup_{0\le s\le t} \mathbb{E}^{1,1}\big[|X_s^n|^2\big]^{1/2}\Big). \tag{3.74}$$

Third Step. Returning to the SDE satisfied by the forward process $\boldsymbol{X}^n$, for any $n \ge 1$, and plugging (3.74) therein, it is quite straightforward to deduce, by means of Itô's formula, that:

$$\begin{aligned}\mathbb{E}^{1,1}\big[|X_t^n|^2\big] &\le \mathbb{E}^{1,1}\big[|X_0|^2\big] + C\int_0^t \Big(1 + \sup_{0\le r\le s} \mathbb{E}^{1,1}\big[|X_r^n|^2\big]\Big)ds\\ &\quad + 2\int_0^t \mathbb{E}^{1,1}\big[\big(\sigma_0^0(r,\mu_r^n) + \sigma_1^0(r)X_r^n\big)^{\dagger} X_r^n\big]\cdot dW_r^0.\end{aligned} \tag{3.75}$$

for a constant C independent of n.

Taking the square of the supremum between 0 and t and then the conditional expectation given $\mathcal{F}_0^{0,n}$, we deduce, by Gronwall's lemma, that:

$$\mathbb{E}_0^{0,n}\Big(\sup_{0\le s\le T} \mathbb{E}^{1,1}\big[|X_s^n|^2\big]^2\Big)^{1/2} \le C\Big(1 + \mathbb{E}^{1,1}\big[|X_0|^2\big]\Big),$$

the constant C being allowed to increase from line to line. Injecting the above bound into (3.74) and duplicating the argument used to prove (3.75), we get:

$$\mathbb{E}_0^n\big[\sup_{0\le s\le t} |X_s^n|^4\big]^{1/2} \le C\Big(1 + |X_0|^2 + \mathbb{E}^{1,1}\big[|X_0|^2\big]\Big),$$

which, together with the fact that X_0 is square integrable, suffices to prove the uniform square-integrability of the family $(\mathbb{P}^n \circ (\sup_{0\le s\le T} |X_s^n|^2)^{-1})_{n\ge 1}$ by means of the same argument as in (Vol I)-(4.46). Namely, for any event D and any $\varepsilon > 0$,

$$\begin{aligned}\mathbb{E}^n\big[\sup_{0\le t\le T} |X_t^n|^2 \mathbf{1}_D\big] &\le \mathbb{E}^n\Big[\mathbb{E}_0^n\big[\sup_{0\le t\le T} |X_t^n|^4\big]^{1/2}\big[\mathbb{P}^n(D|\mathcal{F}_0^n)\big]^{1/2}\Big]\\ &\le C\mathbb{E}^n\Big[\Big(1 + |X_0|^2 + \mathbb{E}^{1,1}\big[|X_0|^2\big]\Big)\big[\mathbb{P}^n(D|\mathcal{F}_0^n)\big]^{1/2}\Big]\\ &\le C\Big(\varepsilon\mathbb{E}^n\big[1 + |X_0|^2\big] + \frac{1}{\varepsilon}\mathbb{E}^n\Big[\Big(1 + |X_0|^2 + \mathbb{E}^{1,1}\big[|X_0|^2\big]\Big)\mathbb{P}^n(D|\mathcal{F}_0^n)\Big]\Big)\\ &= C\Big(\varepsilon\mathbb{E}^n\big[1 + |X_0|^2\big] + \frac{1}{\varepsilon}\mathbb{E}^n\big[(1 + |X_0|^2 + \mathbb{E}^{1,1}[|X_0|^2])\mathbf{1}_D\big]\Big).\end{aligned}$$

Since the law of X_0 under $\mathbb{P}^n$ is independent of n, we deduce that:

$$\lim_{\delta \searrow 0} \sup_{n\geq 1} \sup_{D\in\mathcal{F}:\mathbb{P}^n(D)\leq\delta} \mathbb{E}^n\Big[\sup_{0\leq t\leq T} |X_t^n|^2 \mathbf{1}_D\Big] = 0.$$

This shows that:

$$\lim_{a\to\infty} \sup_{n\geq 1} \mathbb{E}^n\Big[\sup_{0\leq t\leq T} |X_t^n|^2 \mathbf{1}_{\{\sup_{0\leq t\leq T} |X_t^n|^2 \geq a\}}\Big] = 0,$$

where we used the obvious fact that:

$$\sup_{n\geq 1} \mathbb{E}^n\Big[\sup_{0\leq t\leq T} |X_t^n|^2\Big] \tag{3.76}$$

is finite. Now, tightness of the sequence $(\mathbb{P}^n \circ (\boldsymbol{X}^n)^{-1})_{n\geq 0}$ follows from Aldous' criterion, as in the proof of Lemma 3.14.

Finally, (3.64) is a consequence of (3.74) and (3.63) may be proved by combining the bound for (3.76) with standard estimates for BSDEs, see for instance Example 1.20. □

End of the Proof of Lemma 3.32

Now, Lemma 3.32 follows from Theorem 3.13. Observe that assumption **FBSDE** is guaranteed by Theorem 1.60.

3.5 Uniqueness of Strong Solutions

We now discuss two cases of strong uniqueness. The first one relies on the Lasry and Lions monotonicity condition introduced in Section (Vol I)-3.4 in the absence of a common noise. The second one is of a different nature. It provides an interesting example of a model for which the common noise restores strong uniqueness even though uniqueness does not hold in its absence.

3.5.1 Lasry-Lions Monotonicity Condition

Proposition 3.34 *On top of assumption* **Coefficients MFG with a Common Noise**, *assume that b, σ and σ^0 are independent of μ and that f has the form:*

$$f(t,x,\mu,\alpha) = f_0(t,x,\mu) + f_1(t,x,\alpha), \quad t\in[0,T],\ x\in\mathbb{R}^d,\ \alpha\in A,\ \mu\in\mathcal{P}_2(\mathbb{R}^d),$$

$f_0(t,\cdot,\cdot)$ and g satisfying the Lasry-Lions monotonicity condition of Definition (Vol I)-3.28 for any $t\in[0,T]$, which we recall below for the sake of completeness.

Assume further that, for any probabilistic set-up $(\Omega,\mathcal{F},\mathbb{F},\mathbb{P})$ of the same product form as in Definition 2.16, equipped with two Brownian motions $\boldsymbol{W}^0$ and $\boldsymbol{W}$, an initial random distribution μ_0 and an initial private state X_0 satisfying the *constraint* $\mathcal{L}^1(X_0) = \mu_0$, *for any* $\mathcal{F}_T^0$*-measurable random variable* $\mathfrak{M}$ *with values in* $\mathcal{P}_2(\mathcal{C}([0,T];\mathbb{R}^{2d}))$ *such that* $\mathbb{F}$ *is compatible with* $(X_0, \mathbf{W}^0, \mathfrak{M}, \mathbf{W})$, *the optimization problem* (3.1)–(3.2), *with* $\boldsymbol{\mu} = (\mu_t = \mathfrak{M} \circ (e_t^x)^{-1})_{0 \le t \le T}$, *has a unique minimizer* $\hat{\boldsymbol{\alpha}}$.

Then, strong uniqueness holds for the mean field game (3.1)–(3.2)–(3.3), *in the sense of Definition 2.27. Moreover, the weak solutions given by Theorems 3.29, 3.30, and 3.31 are actually strong solutions, in the sense of Definition 2.22.*

Recall that a real valued function h on $\mathbb{R}^d \times \mathcal{P}_2(\mathbb{R}^d)$ is said to be monotone (in the sense of Lasry and Lions) if, for all $\mu \in \mathcal{P}_2(\mathbb{R}^d)$, the mapping $\mathbb{R}^d \ni x \mapsto h(x,\mu)$ is at most of quadratic growth, and for all $\mu, \mu' \in \mathcal{P}_2(\mathbb{R}^d)$, we have:

$$\int_{\mathbb{R}^d} [h(x,\mu) - h(x,\mu')] \, d(\mu - \mu')(x) \ \geq 0. \tag{3.77}$$

Proof. We first introduce some necessary new notation. Given a general set-up as in the statement of the proposition, we consider two solutions $\mathfrak{M}$ and $\mathfrak{M}'$ to the mean field game. We then call $\boldsymbol{\mu} = (\mu_t)_{0\le t\le T}$ and $\boldsymbol{\mu}' = (\mu'_t)_{0\le t\le T}$ the corresponding sub-environments. We also denote by $\hat{\boldsymbol{\alpha}}^{\mu} = (\hat{\alpha}_t^{\mu})_{0\le t\le T}$ and $\hat{\boldsymbol{\alpha}}^{\mu'} = (\hat{\alpha}_t^{\mu'})_{0\le t\le T}$ the respective optimal controls in the environments $(\mathfrak{M}, \boldsymbol{\mu})$ and $(\mathfrak{M}', \boldsymbol{\mu}')$, and by $\boldsymbol{X}^{\mu} = (X_t^{\mu})_{0\le t\le T}$ and $\boldsymbol{X}^{\mu'} = (X_t^{\mu'})_{0\le t\le T}$ the corresponding optimal paths. We also denote by $J^{\mu}(\hat{\boldsymbol{\alpha}}^{\mu})$ and $J^{\mu'}(\hat{\boldsymbol{\alpha}}^{\mu'})$ the associated costs.

We assume that the random variables $\mathfrak{M}$ and $\mathfrak{M}'$ differ on an event of positive probability. Then, the processes $\hat{\boldsymbol{\alpha}}^{\mu}$ and $\hat{\boldsymbol{\alpha}}^{\mu'}$ must differ on a measurable subset of $[0,T] \otimes \mathcal{F}$ of positive $\mathrm{Leb}_1 \otimes \mathbb{P}$-measure, as otherwise $\boldsymbol{X}^{\mu}$ and then $\boldsymbol{X}^{\mu'}$ would coincide up to a $\mathbb{P}$-null set which would imply that $\mathfrak{M}$ and $\mathfrak{M}'$ also coincide.

Therefore, by strict optimality of $\hat{\boldsymbol{\alpha}}^{\mu}$ in the environment $\boldsymbol{\mu}$, we have:

$$\mathbb{E}\biggl[\int_0^T f(t, X_t^{\mu}, \mu_t, \hat{\alpha}_t^{\mu})dt + g(X_T^{\mu}, \mu_T)\biggr] < \mathbb{E}\biggl[\int_0^T f(t, X_t^{\mu'}, \mu_t, \hat{\alpha}_t^{\mu'})dt + g(X_T^{\mu'}, \mu_T)\biggr],$$

where we used the fact that, in the environment $\boldsymbol{\mu}$, the process driven by $\hat{\boldsymbol{\alpha}}^{\mu'}$ is exactly $\boldsymbol{X}^{\mu'}$ since the drift b and the diffusion coefficients σ and σ^0 do not depend on the measure argument. Similarly, we have:

$$\mathbb{E}\biggl[\int_0^T f(t, X_t^{\mu'}, \mu'_t, \hat{\alpha}_t^{\mu'})dt + g(X_T^{\mu'}, \mu'_T)\biggr] < \mathbb{E}\biggl[\int_0^T f(t, X_t^{\mu}, \mu'_t, \hat{\alpha}_t^{\mu})dt + g(X_T^{\mu}, \mu'_T)\biggr].$$

Adding the two inequalities, we get:

$$\begin{aligned}
&\mathbb{E}\biggl[\int_0^T f(t, X_t^{\mu}, \mu_t, \hat{\alpha}_t^{\mu})dt + g(X_T^{\mu}, \mu_T)\biggr] - \mathbb{E}\biggl[\int_0^T f(t, X_t^{\mu}, \mu'_t, \hat{\alpha}_t^{\mu})dt + g(X_T^{\mu}, \mu'_T)\biggr] \\
&< \mathbb{E}\biggl[\int_0^T f(t, X_t^{\mu'}, \mu_t, \hat{\alpha}_t^{\mu'})dt + g(X_T^{\mu'}, \mu_T)\biggr] - \mathbb{E}\biggl[\int_0^T f(t, X_t^{\mu'}, \mu'_t, \hat{\alpha}_t^{\mu'})dt + g(X_T^{\mu'}, \mu'_T)\biggr],
\end{aligned}$$

which reads:

$$\mathbb{E}^0\Big[\int_0^T \Big(\int_{\mathbb{R}^d} (f_0(t,x,\mu_t) - f_0(t,x,\mu_t'))d\mu_t(x)\Big)dt + \int_{\mathbb{R}^d} (g(x,\mu_T) - g(x,\mu_T'))d\mu_T(x)\Big]$$

$$< \mathbb{E}^0\Big[\int_0^T \Big(\int_{\mathbb{R}^d} (f_0(t,x,\mu_t) - f_0(t,x,\mu_t'))d\mu_t'(x)\Big)dt + \int_{\mathbb{R}^d} (g(x,\mu_T) - g(x,\mu_T'))d\mu_T'(x)\Big],$$

which contradicts the monotonicity assumption.

The second part of the statement is a mere consequence of Theorem 2.29. □

3.5.2 Common Noise and Restoration of Uniqueness

An interesting question, both from the theoretical and practical points of view, is to investigate whether the common noise can contribute to strong uniqueness. It is indeed a well-known fact that, for finite dimensional stochastic differential equations, the presence of noise can restore strong uniqueness for systems driven by singular coefficients. The reader is referred to the Notes & Complements at the end of the chapter for references.

In the current context, the question seems really challenging, as the noise is finite-dimensional while the unknown, namely the equilibrium measure, evolves in an infinite dimensional space. One should expect that a very strong hypoellipticity property would be needed for the noise to propagate throughout the whole system and force uniqueness of the equilibrium.

Quite surprisingly, we provide here a simple model for which such a phenomenon does occur. The example is designed in such a way that the optimal feedback can be explicitly computed from the sole knowledge of the mean of the equilibrium. In this way, the required hypoellipticity property in the space of measures just consists in a standard ellipticity property in dimension one.

Using the notation of the general set-up introduced in Subsection 3.1.2, we assume that the dynamics of $\boldsymbol{X}$ are one dimensional and take the form:

$$dX_t = \big(b(t,\mu_t) + X_t + \alpha_t\big)dt + dW_t + \sigma^0 dW_t^0, \quad t \in [0,T], \tag{3.78}$$

with some (deterministic) initial condition $X_0 = x_0 \in \mathbb{R}$, where $\boldsymbol{\mu} = (\mu_t)_{0\le t\le T}$ denotes a flow of square-integrable random probability measures on $\mathbb{R}$, and $\boldsymbol{\alpha} = (\alpha_t)_{0\le t\le T}$ a one-dimensional control with values in $A = \mathbb{R}$. Here $b : [0,T] \times \mathcal{P}_2(\mathbb{R}) \to \mathbb{R}$ is measurable, bounded, and W_2-Lipschitz continuous in the measure argument, uniformly in time $t \in [0,T]$. Finally, σ^0 is a nonnegative real constant. We use a cost functional of the form:

$$J^{\mu}(\boldsymbol{\alpha}) = \mathbb{E}\Big[\int_0^T \frac{1}{2}\big[\big(X_t + f(t,\mu_t)\big)^2 + \alpha_t^2\big]dt + \frac{1}{2}\big(X_T + g(\mu_T)\big)^2\Big], \tag{3.79}$$

where $f : [0,T] \times \mathcal{P}_2(\mathbb{R}) \to \mathbb{R}$ and $g : \mathcal{P}_2(\mathbb{R}) \to \mathbb{R}$ are bounded, measurable, and W_2-Lipschitz continuous in the measure argument, uniformly in time $t \in [0,T]$.

The reduced Hamiltonian from Subsection 3.4.1 takes the simple form:

$$H^{(r)}(t,x,\mu,\alpha,y) = \big(b(t,\mu) + x + \alpha\big)y + \frac{1}{2}\big(x + f(t,\mu)\big)^2 + \frac{1}{2}\alpha^2,$$

for $t \in [0,T]$, $x, y, \alpha \in \mathbb{R}$, and $\mu \in \mathcal{P}_2(\mathbb{R})$. Given the values of x, y and μ, the minimizer of H is easily computed and simply reads $\hat{\alpha}(t,x,\mu,y) = -y$.

It is easily checked that the assumptions of the stochastic Pontryagin principle in Theorem 1.60 hold true. For any given flow of random measures $\boldsymbol{\mu} = (\mu_t)_{0\le t\le T}$ deriving from some super-environment $\mathfrak{M}$ such that $(\boldsymbol{W}^0, \mathfrak{M}, \boldsymbol{W})$ is compatible with $\mathbb{F}$, the forward-backward system derived from the stochastic Pontryagin principle reads:

$$\begin{cases} dX_t = \big(b(t,\mu_t) + X_t - Y_t\big)dt + dW_t + \sigma^0 dW_t^0, \\ dY_t = -\big(X_t + Y_t + f(t,\mu_t)\big)dt + Z_t dW_t + Z_t^0 dW_t^0 + dM_t, \end{cases} \tag{3.80}$$

where $(M_t)_{0\le t\le T}$ is a square-integrable càd-làg martingale, of zero cross-variation with $(\boldsymbol{W}^0, \boldsymbol{W})$ and with $M_0 = 0$, the terminal condition being:

$$Y_T = X_T + g(\mu_T). \tag{3.81}$$

It is easily checked that assumption **MFG with a Common Noise SMP Relaxed** holds. For this reason, the remainder of the section is devoted to the analysis of uniqueness.

Theorem 3.35 *On top of the above assumption, assume that $\sigma^0 > 0$. Then, the conditional McKean-Vlasov problem consisting of the forward-backward system* (3.80), *with terminal condition* (3.81) *and constraint:*

$$\forall t \in [0,T], \quad \mu_t = \mathfrak{M} \circ (e_t^x)^{-1}, \quad \textit{with } \mathfrak{M} = \mathcal{L}^1\big(\boldsymbol{X}, \boldsymbol{W}\big), \tag{3.82}$$

admits a unique strong solution. In particular the mean field game (3.1)–(3.2)–(3.3) *with the above coefficients admits a unique strong equilibrium in the sense of Definition 2.22.*

The next proposition shows that, in the above statement, strong uniqueness is due to the presence of the common noise $\boldsymbol{W}^0$:

Proposition 3.36 *Consider the same framework as in the statement of Theorem 3.35, but assume that $\sigma^0 = 0$. Then, we may find x_0, b, f, and g, for which the mean field game* (3.1)–(3.2)–(3.3) *admits an infinite number of deterministic equilibria.*

The proofs of Theorem 3.35 and Proposition 3.36 rely on the following characterization of the equilibria.

Theorem 3.37 *Consider the framework of Theorem 3.35 with $\sigma^0 \geq 0$. Then, we can find coefficients:*

$$\bar{b}, \bar{f} : [0,T] \times \mathbb{R} \to \mathbb{R}, \quad \bar{g} : \mathbb{R} \to \mathbb{R},$$

which are bounded on the whole space, Lipschitz continuous in the space variable uniformly in time, and which only depend upon b, f, and g respectively (in particular they are independent of the initial condition x_0 and the viscosity parameter σ^0), such that a 5-tuple $(\boldsymbol{X}, \boldsymbol{Y}, \boldsymbol{Z}, \boldsymbol{Z}^0, \boldsymbol{M}) = (X_t, Y_t, Z_t, Z_t^0, M_t)_{0 \leq t \leq T}$ is a solution on the set-up $(\Omega, \mathcal{F}, \mathbb{F}, \mathbb{P})$ equipped with $(\mathbf{W}^0, \mathfrak{M}, \mathbf{W})$, for some super-environment $\mathfrak{M}$, of the conditional McKean-Vlasov problem consisting of (3.80)-(3.81)-(3.82) *if and only if the process $\boldsymbol{Y} = (Y_t)_{0 \leq t \leq T}$ has the form:*

$$Y_t = \eta_t X_t + \chi_t, \quad t \in [0,T],$$

where $\boldsymbol{\eta} = (\eta_t)_{0 \leq t \leq T}$ solves the well-posed Riccati equation:

$$\dot{\eta}_t = \eta_t^2 - 2\eta_t - 1, \quad \eta_T = 1, \tag{3.83}$$

the process $\boldsymbol{X} = (X_t)_{0 \leq t \leq T}$ solves the forward equation:

$$dX_t = \big(\bar{b}(t, \bar{\mu}_t) + X_t - Y_t\big)dt + dW_t + \sigma^0 dW_t^0, \quad t \in [0,T]\,; \quad X_0 = x_0, \tag{3.84}$$

and the pair $(\bar{\boldsymbol{\mu}}, \boldsymbol{\chi}) = (\bar{\mu}_t, \chi_t)_{0 \leq t \leq T}$ is $\mathbb{F}^0$-progressively measurable and solves on the space $(\Omega^0, \mathcal{F}^0, \mathbb{F}^0, \mathbb{P}^0)$ the forward-backward stochastic differential equation:

$$\begin{cases} d\bar{\mu}_t = \Big(\bar{b}\big(t, \bar{\mu}_t\big) + (1 - \eta_t)\bar{\mu}_t - \chi_t\Big)dt + \sigma^0 dW_t^0, \\ d\chi_t = -\big((1 - \eta_t)\chi_t + \bar{f}(t, \bar{\mu}_t) + \bar{b}(t, \bar{\mu}_t)\eta_t\big)dt \\ \qquad\quad + dm_t^0, \quad t \in [0,T], \\ \bar{\mu}_0 = x_0, \quad \chi_T = \bar{g}(\bar{\mu}_T), \end{cases} \tag{3.85}$$

where $\boldsymbol{m}^0 = (m_t^0)_{0 \leq t \leq T}$ is a square-integrable càd-làg martingale on the space $(\Omega^0, \mathcal{F}^0, \mathbb{F}^0, \mathbb{P}^0)$ with $m_0^0 = 0$ as initial condition, and $(\mathbf{W}^0, \bar{\boldsymbol{\mu}})$ is compatible with $\mathbb{F}^0$. Moreover, it must hold:

$$\forall t \in [0,T], \quad \mathbb{P}^0\big[\bar{\mu}_t = \mathbb{E}^1[X_t]\big] = 1. \tag{3.86}$$

Theorem 3.37 says that equilibria to the MFG problem associated with the dynamics (3.78) and with the cost functional (3.79) may be characterized through the auxiliary forward-backward system (3.85) satisfied by their means. Put it differently, an equilibrium may be characterized by its mean only. Indeed, as we shall see in the proof of Theorem 3.37, equilibria must be Gaussian given the realization of $\boldsymbol{W}^0$, their conditional variance being explicitly determined through the linear-quadratic structure of the dynamics (3.78) and the cost functional (3.85), independently of the choice of b, f, g, σ, and x_0.

The fact that equilibria may be entirely described through their conditional means trivialize the analysis of the hypoellipticity property of the process $(\bar{\mu}_t)_{0 \le t \le T}$.

Taking Theorem 3.37 for granted momentarily, the proofs of Theorem 3.35 and Proposition 3.36 will follow from standard results for classical forward-backward SDEs:

1. When $\sigma^0 > 0$, the forward-backward system enters the so-called *nondegenerate regime* in which the noise forces uniqueness of a solution.
2. When $\sigma^0 = 0$, the auxiliary forward-backward system is deterministic (or *inviscid*) and may develop discontinuities or, using the same language as in the theory of hyperbolic equations, *shocks*. Because of that, we may observe, for a relevant choice of the coefficients, several solutions.

Remark 3.38 *The compatibility condition required in the statement of Theorem 3.37 is a way to select a weak solution to the FBSDE* (3.85) *which makes sense from the physical point of view. We accounted for a similar fact in Remark 2.20. In particular, we shall prove that, for the selected solution, the martingale $\boldsymbol{m}^0 = (m_t^0)_{0 \le t \le T}$ can be represented as a stochastic integral with respect to $\boldsymbol{W}^0$. We provide a counter-example in Subsection 3.5.5 below.*

3.5.3 Auxiliary Results for the Restoration of Uniqueness

The proofs of Theorem 3.35 and Proposition 3.36 rely on the following characterization of the equilibria.

Lemma 3.39 *Under the assumptions and notation of Theorem 3.35, but allowing $\sigma^0 \ge 0$, for any $\mathcal{F}_T^0$-measurable super-environment $\mathfrak{M}$ and the associated $\mathbb{F}^0$-progressively measurable sub-environment $\boldsymbol{\mu} = (\mu_t = \mathfrak{M} \circ (e_t^x)^{-1})_{0 \le t \le T}$ with values in $\mathcal{P}_2(\mathbb{R})$, a 5-tuple $(\boldsymbol{X}, \boldsymbol{Y}, \boldsymbol{Z}, \boldsymbol{Z}^0, \boldsymbol{M}) = (X_t, Y_t, Z_t, Z_t^0, M_t)_{0 \le t \le T}$ is a solution of the forward-backward system* (3.80) *with terminal condition* (3.81) *if and only if the process $\boldsymbol{Y}$ is of the form:*

$$Y_t = \eta_t X_t + \chi_t, \quad t \in [0, T],$$

where $\boldsymbol{\eta} = (\eta_t)_{0 \le t \le T}$ is the unique solution of the Riccati equation (3.83), *and $\boldsymbol{\chi} = (\chi_t)_{0 \le t \le T}$ solves the backward stochastic differential equation:*

$$d\chi_t = -\big((1 - \eta_t)\chi_t + f(t, \mu_t) + b(t, \mu_t)\eta_t\big)dt + dm_t^0, \quad \chi_T = g(\mu_T), \tag{3.87}$$

where $\boldsymbol{m}^0 = (m_t^0)_{0 \le t \le T}$ is a square-integrable càd-làg martingale on the space $(\Omega^0, \mathcal{F}^0, \mathbb{F}^0, \mathbb{P}^0)$, with 0 as initial condition.

Proof. Referring to Subsection (Vol I)-3.5.1, see also (Vol I)-(2.48)-(2.49), we first notice that the Riccati equation is uniquely solvable on the whole $[0, T]$. Alternatively, unique solvability can be checked directly by noticing that any local solution must remain in the interval $[1-\sqrt{2}, 1+\sqrt{2}]$, whose end points are the roots of the characteristic quadratic equation. Now, consider a pair $(\boldsymbol{X}, \boldsymbol{Y})$ such that:

$$dX_t = \big(b(t, \mu_t) + X_t - Y_t\big)dt + dW_t + \sigma^0 dW_t^0,$$

with

$$Y_t = \eta_t X_t + \chi_t, \quad t \in [0, T],$$

for some process $(\chi_t)_{0 \le t \le T}$. Then, $\boldsymbol{Y} = (Y_t)_{0 \le t \le T}$ is a semi-martingale if and only if $\boldsymbol{\chi} = (\chi_t)_{0 \le t \le T}$ is also a semi-martingale. Moreover,

$$\begin{aligned}
d\chi_t &= dY_t - \dot{\eta}_t X_t dt - \eta_t dX_t \\
&= dY_t - \dot{\eta}_t X_t dt - \eta_t\big(b(t, \mu_t) + X_t - \eta_t X_t - \chi_t\big)dt - \eta_t\big(dW_t + \sigma^0 dW_t^0\big) \\
&= dY_t - \big(\eta_t^2 - 2\eta_t - 1\big)X_t dt - \eta_t\big(b(t, \mu_t) + X_t - \eta_t X_t - \chi_t\big)dt \\
&\quad - \eta_t\big(dW_t + \sigma^0 dW_t^0\big) \\
&= dY_t + \big[(\eta_t + 1)X_t - \eta_t\big(b(t, \mu_t) - \chi_t\big)\big]dt - \eta_t\big(dW_t + \sigma^0 dW_t^0\big) \\
&= dY_t + \big[Y_t + X_t - \eta_t b(t, \mu_t) + (\eta_t - 1)\chi_t\big]dt - \eta_t\big(dW_t + \sigma^0 dW_t^0\big).
\end{aligned}$$

In particular, $(M_t' = Y_t + \int_0^t [X_s + Y_s + f(s, \mu_s)]ds)_{0 \le t \le T}$ is a local martingale if and only if $(m_t^0 = \chi_t + \int_0^t [f(s, \mu_s) + \eta_s b(s, \eta_s) + (1 - \eta_s)\chi_s]ds)_{0 \le t \le T}$ is also a local martingale.

Moreover, if $\mathbb{E}[\sup_{0 \le t \le T} |Y_t|^2]$ is finite, then $\mathbb{E}[\sup_{0 \le t \le T} |X_t|^2]$ is also finite because of the equation satisfied by $\boldsymbol{X}$, so that $\mathbb{E}[\sup_{0 \le t \le T} |\chi_t|^2]$ is finite as well by definition of $\boldsymbol{\chi}$. Conversely, if $\mathbb{E}[\sup_{0 \le t \le T} |\chi_t|^2]$ is finite, so is $\mathbb{E}[\sup_{0 \le t \le T} |X_t|^2]$ as we can see by plugging the definition of $\boldsymbol{Y}$ in terms of $\boldsymbol{\chi}$ in the SDE satisfied by $\boldsymbol{X}$. We then deduce that $\mathbb{E}[\sup_{0 \le t \le T} |Y_t|^2]$ is finite. In particular, $(M_t')_{0 \le t \le T}$ is a square-integrable martingale if and only if $(m_t^0)_{0 \le t \le T}$ is a square-integrable martingale. This completes the proof. □

We can now turn to the characterization of the equilibria.

Proof of Theorem 3.37. By Lemma 3.39, we know that solutions of the conditional McKean-Vlasov problem (3.80)-(3.81)-(3.82) are characterized by the forward-backward equation:

$$\begin{cases}
dX_t = \big(b(t, \mu_t) + (1 - \eta_t)X_t - \chi_t\big)dt + dW_t + \sigma^0 dW_t^0, \\
d\chi_t = -\big((1 - \eta_t)\chi_t + f(t, \mu_t) + b(t, \mu_t)\eta_t\big)dt + dm_t^0,
\end{cases} \tag{3.88}$$

with the terminal condition $\chi_T = g(\mu_T)$ and the McKean-Vlasov constraint:

$$\forall t \in [0, T], \quad \mu_t = \mathfrak{M} \circ (e_t^x)^{-1}, \quad \mathfrak{M} = \mathcal{L}^1(\boldsymbol{X}, \boldsymbol{W}).$$

In order to proceed with the analysis of the system (3.88), we make the following two observations. The first one is that the forward component of the system admits an explicit factorization (in terms of $\boldsymbol{\chi} = (\chi_t)_{0\le t\le T}$):

$$X_t = \exp\left(\int_0^t (1-\eta_s)ds\right)\left[x_0 + \int_0^t \exp\left(-\int_0^s (1-\eta_r)dr\right)\left(b(s,\mu_s)-\chi_s\right)ds\right.$$
$$+\int_0^t \exp\left(-\int_0^s (1-\eta_r)dr\right)dW_s \tag{3.89}$$
$$\left.+\,\sigma^0\int_0^t \exp\left(-\int_0^s (1-\eta_r)dr\right)dW_s^0\right],$$

for $t \in [0,T]$. The second observation is that, whenever the flow of random measures $\boldsymbol{\mu} = (\mu_t)_{0\le t\le T}$ is frozen, the backward equation in (3.88) is uniquely solvable and the unique solution may be constructed on the space $(\Omega^0, \mathcal{F}^0, \mathbb{P}^0)$ equipped with $(\boldsymbol{W}^0, \mathfrak{M})$. Indeed, we know from Lemma 1.15 that $\mathbb{F}^0$ is compatible with $(\boldsymbol{W}^0, \mathfrak{M})$, which shows that it makes sense to solve the equation on $(\Omega^0, \mathcal{F}^0, \mathbb{F}^0, \mathbb{P}^0)$ equipped with $(\boldsymbol{W}^0, \mathfrak{M})$. The solution takes a very simple form. Letting:

$$\varpi_t = \exp\left(-\int_0^t (1-\eta_s)ds\right), \quad t \in [0,T],$$

we have:

$$\chi_t = \varpi_t \mathbb{E}^0\left[\int_t^T \varpi_s^{-1}\left(f(s,\mu_s) + b(s,\mu_s)\eta_s\right)ds + \varpi_T^{-1} g(\mu_T)\,|\,\mathcal{F}_t^0\right].$$

The compatibility condition is then especially useful to compute the above conditional expectation. In order to transfer the resulting solution $(\boldsymbol{\chi}, \boldsymbol{m}^0)$ to $(\Omega, \mathcal{F}, \mathbb{F}, \mathbb{P})$ equipped with $(\boldsymbol{W}^0, \mathfrak{M}, \boldsymbol{W})$, it suffices to notice that $\boldsymbol{m}^0$ is automatically an $\mathbb{F}$-martingale, which is a special case of Proposition 1.10. Put it differently, the solution $(\boldsymbol{\chi}, \boldsymbol{m}^0) = (\chi_t, m_t^0)_{0\le t\le T}$ to the backward equation (3.87), when solved on $(\Omega, \mathcal{F}, \mathbb{F}, \mathbb{P})$ equipped with $(\boldsymbol{W}^0, \mathfrak{M}, \boldsymbol{W})$, is a progressively measurable process on the space $(\Omega^0, \mathcal{F}^0, \mathbb{F}^0, \mathbb{P}^0)$. As a consequence, (3.89) says that the process $(X_t)_{0\le t\le T}$ is, conditional on $\mathcal{F}^0$, an Ornstein-Uhlenbeck process with conditional mean and variance functions:

$$\bar{\mu}_t = \exp\left(\int_0^t (1-\eta_s)ds\right)\left[x_0 + \int_0^t \exp\left(-\int_0^s (1-\eta_r)dr\right)\left(b(s,\mu_s)-\chi_s\right)ds\right.$$
$$\left.+\,\sigma^0\int_0^t \exp\left(-\int_0^s (1-\eta_r)dr\right)dW_s^0\right],$$
$$\bar{\sigma}_t^2 = \exp\left(2\int_0^t (1-\eta_s)ds\right)\int_0^t \exp\left(-2\int_0^s (1-\eta_r)dr\right)ds,$$

for $t \in [0,T]$. Notice that the stochastic integral in the definition of the mean can be defined pathwise by means of an integration by parts.

Therefore, under the McKean-Vlasov constraint (3.82), μ_t is, for any $t \in [0,T]$, a normal distribution with $\bar{\mu}_t$ as random mean and $\bar{\sigma}_t^2$ as deterministic variance. In particular, the variance at time t is entirely determined, independently of the initial condition x_0. We thus let, for any real $\bar{\mu} \in \mathbb{R}$,

$$(\bar{b},\bar{f})(t,\bar{\mu}) = (b,f)\big(t,\mathcal{N}(\bar{\mu},\bar{\sigma}_t^2)\big), \quad t \in [0,T]\,; \quad \bar{g}(\bar{\mu}) = g\big(\mathcal{N}(\bar{\mu},\bar{\sigma}_T^2)\big).$$

From these definitions, it is easily checked that $\bar{b}$, $\bar{f}$, and $\bar{g}$ are bounded and are Lipschitz continuous in $\bar{\mu}$ uniformly in time. Moreover, it is plain to check that the backward equation in (3.85) is satisfied. Taking the mean under $\mathbb{P}^1$ in the constraint (3.82), it is clear that (3.86) holds true. Similarly, by taking the mean under $\mathbb{P}^1$ in the forward equation in (3.88), we deduce that the forward equation in (3.85) is satisfied.

Conversely, if the auxiliary SDE (3.84) and the forward-backward system (3.85) are satisfied with $Y_t = \eta_t X_t + \chi_t$, then $\boldsymbol{X}$ must have an expansion similar to (3.89). In particular, it has Gaussian marginal distributions with variance $\bar{\sigma}_t^2$ at time $t \in [0,T]$. Moreover, computing the conditional mean in (3.84) using in addition the fact that Y_t expands in terms of (X_t, χ_t) and χ_t is $\mathcal{F}^0$-measurable, we check that $(\mathbb{E}^1[X_t])_{0 \le t \le T}$ and $(\bar{\mu}_t)_{0 \le t \le T}$ solve the same linear (random coefficients) SDE, with the same initial condition, and are thus equal. This shows that the additional constraint (3.86) holds and that the McKean-Vlasov constraint (3.82) holds with $\mathfrak{M} = \mathcal{L}^1(\boldsymbol{X}, \boldsymbol{W})$ and

$$\mu_t = \mathcal{N}(\bar{\mu}_t, \bar{\sigma}_t^2), \quad t \in [0,T].$$

This permits to rewrite the coefficients $\bar{b}, \bar{f}$, and $\bar{g}$ of the variable $\bar{\mu}_t$ in (3.84) and (3.85) as functions b, f, and g of μ_t. In particular, we deduce from equation (3.84) that the forward equation (3.88) is satisfied, which completes the proof. Then, it only remains to observe that the filtration generated by $(\boldsymbol{W}^0, \mathfrak{M})$, recall Definition 2.15, coincides with the one generated by $(\boldsymbol{W}^0, \bar{\boldsymbol{\mu}})$; thus, it is compatible with $\mathbb{F}^0$ if and only if the other is compatible. The backward equation in (3.88) easily follows. □

3.5.4 Proof of the Restoration of Uniqueness

Case with a Common Noise

We start with:

Proof of Theorem 3.35. By Theorem 3.37, it suffices to prove that the auxiliary forward-backward system (3.85) has a unique solution. Recall that it takes the form:

$$\begin{cases} d\bar{\mu}_t = \Big(\bar{b}\big(t,\bar{\mu}_t\big) + (1-\eta_t)\bar{\mu}_t - \chi_t\Big)dt + \sigma^0 dW_t^0, \\ d\chi_t = -\Big(\big(1-\eta_t\big)\chi_t + \bar{f}(t,\bar{\mu}_t) + \bar{b}(t,\bar{\mu}_t)\eta_t\Big)dt + dm_t^0, \quad t \in [0,T], \\ \chi_T = \bar{g}(\bar{\mu}_T), \end{cases} \tag{3.90}$$

We would like to apply Theorem (Vol I)-4.12, but unfortunately, the coefficients of the forward equation are not bounded. In order to overcome this difficulty, we use the following change of unknown:

$$\hat{\mu}_t = \varpi_t \bar{\mu}_t, \quad \varpi_t = \exp\left(- \int_0^t (1-\eta_s) ds \right), \quad t \in [0,T],$$

and set:

$$\hat{b}(t,x) = \varpi_t \bar{b}(t, \varpi_t^{-1} x), \quad \hat{f}(t,x) = \bar{f}(t, \varpi_t^{-1} x), \quad \hat{g}(x) = \bar{g}(\varpi_T^{-1} x),$$

for $t \in [0,T]$, and $x \in \mathbb{R}$. Since the weights $(\varpi_t)_{0 \le t \le T}$ are bounded from below and from above by constants that depend on T only, the coefficients:

$$\hat{b}, \hat{f} : [0,T] \times \mathbb{R} \to \mathbb{R}, \quad \hat{g} : \mathbb{R} \to \mathbb{R}$$

are bounded on the whole domain and Lipschitz continuous in the space variable. Therefore, the pair $(\bar{\mu}_t, \chi_t)_{0 \le t \le T}$ solves the system (3.90) if and only if the pair $(\hat{\mu}_t, \chi_t)_{0 \le t \le T}$ solves the system:

$$\begin{cases} d\hat{\mu}_t = \left(\hat{b}(t, \hat{\mu}_t) - \varpi_t \chi_t \right) dt + \varpi_t^0 \sigma dW_t^0, \\ d\chi_t = -\left((1-\eta_t)\chi_t + \hat{f}(t, \hat{\mu}_t) + \varpi_t^{-1} \hat{b}(t, \hat{\mu}_t) \eta_t \right) dt + dm_t^0, \quad t \in [0,T], \\ \chi_T = \hat{g}(\hat{\mu}_T). \end{cases}$$

By Theorem (Vol I)-4.12, there exists a solution such that $(m_t^0)_{0 \le t \le T}$ writes as a stochastic integral with respect to $\boldsymbol{W}^0$, and it is the only one to have this property. Moreover, there exists a decoupling field which is Lipschitz continuous in space, uniformly in time. Duplicating the proof of Proposition 1.52, this says that uniqueness holds among the most general class of solutions for which $(m_t^0)_{0 \le t \le T}$ is a general martingale. Most importantly, the solution is progressively measurable with respect to the complete and right-continuous augmentation of the filtration generated by $\boldsymbol{W}^0$. In particular, $(\boldsymbol{W}^0, \bar{\boldsymbol{\mu}})$ is compatible with $\mathbb{F}^0$. □

The Case Without Common Noise

Proof of Proposition 3.36. The starting point is the same as in the proof of Theorem 3.35. It suffices to exhibit coefficients b, f, and g such that the system (3.90) (with $\sigma^0 = 0$) has several solutions.

The key point is to use the same kind of change of unknown as in the proof of Theorem 3.35:

$$\hat{\mu}_t = \varpi_t \bar{\mu}_t, \quad \hat{\chi}_t = \varpi_t^{-1} \chi_t, \quad \varpi_t = \exp\left(- \int_0^t (1-\eta_s) ds \right), \quad t \in [0,T].$$

Then, $(\bar{\mu}_t, \bar{\chi}_t)_{0 \le t \le T}$ is a (deterministic) solution to (3.90) if and only if $(\hat{\mu}_t, \hat{\chi}_t)_{0 \le t \le T}$ is a deterministic solution to:

$$\begin{cases} d\hat{\mu}_t = \left(\hat{b}(t, \hat{\mu}_t) - \varpi_t^2 \hat{\chi}_t \right) dt \\ d\hat{\chi}_t = -\varpi_t^{-1} \left(\hat{f}(t, \hat{\mu}_t) + \varpi_t^{-1} \hat{b}(t, \hat{\mu}_t) \eta_t \right) dt, \quad t \in [0,T], \\ \hat{\chi}_T = \varpi_T^{-1} \hat{g}(\hat{\mu}_T), \end{cases} \tag{3.91}$$

with the same definition of the coefficients $\hat{b}, \hat{f}$ and $\hat{g}$ as in the proof of Theorem 3.35. Notice that we removed the martingale part as we are just looking for deterministic solutions of the ordinary forward-backward system above. Assume now that $b = 0$, so that $\bar{b}$ and $\hat{b}$ are also zero, and choose f and g in such a way that:

$$\hat{f}(t,x) = -\varpi_t^3 x, \quad \hat{g}(x) = \varpi_T x, \quad x \in [-R, R],$$

for some large enough real $R > 0$. Then, the system reduces to:

$$d\hat{\mu}_t = -\varpi_t^2 \hat{\chi}_t dt, \quad d\hat{\chi}_t = \varpi_t^2 \hat{\mu}_t dt, \quad t \in [0,T]\,; \quad \hat{\chi}_T = \hat{\mu}_T, \tag{3.92}$$

provided that the solution $(\hat{\mu}_t, \hat{\chi}_t)_{0 \le t \le T}$ remains in $[-R,R]^2$.

With the solution φ of the ordinary differential equation $\dot{\varphi}(t) = \varpi_{\varphi(t)}^{-2}$, for $t \in [0,T]$ and with $\varphi(0) = 0$, the existence and uniqueness of which are established right below, we finally get that $(\hat{\mu}_t, \hat{\chi}_t)_{0 \le t \le T}$ solves (3.92) if and only if the time changed curves $(\hat{\mu}_{\varphi(t)}, \hat{\chi}_{\varphi(t)})_{0 \le t \le T}$ satisfy:

$$d\big[\hat{\mu}_{\varphi(t)}\big] = -\hat{\chi}_{\varphi(t)} dt, \quad d\big[\hat{\chi}_{\varphi(t)}\big] = \hat{\mu}_{\varphi(t)} dt, \quad t \in [0,T]\,; \quad \hat{\chi}_{\varphi(T)} = \hat{\mu}_{\varphi(T)}.$$

Choosing $T = \varphi^{-1}(\pi/4)$, which is shown below to be possible, we recover exactly the example discussed in Subsection (Vol I)-4.3.4. We deduce that, whenever $x_0 = 0$, there exist infinitely many solutions which remain in $[-R,R]^2$.

We now address the existence and uniqueness of a solution φ to the ordinary differential equation $\dot{\varphi}(t) = \varpi_{\varphi(t)}^{-2}$, for $t \in [0,T]$, with the initial condition $\varphi(0) = 0$. Of course, we observe that, for a given extension of $(\eta_t)_{0 \le t \le T}$ to times $t < 0$ and $t > T$, the equation is locally uniquely solvable. The local solution satisfies $\varphi(t) \ge 0$ for t in the interval of definition of the solution, which shows in particular that the extension of the function $(\eta_t)_{0 \le t \le T}$ to negative times plays no role. Now, we recall that $(\eta_t)_{0 \le t \le T}$ satisfies the Riccati equation $\dot{\eta}_t = \eta_t^2 - 2\eta_t - 1 = (\eta_t - 1)^2 - 2$. Recalling that $\eta_t \in [1 - \sqrt{2}, 1 + \sqrt{2}]$ for all $t \in [0,T]$, we deduce that $(\eta_t)_{0 \le t \le T}$ is nonincreasing on $[0,T]$. Since $\eta_T = 1$, this shows that $(\eta_t)_{0 \le t \le T}$ remains in $[1, 1 + \sqrt{2}]$. As a by-product, for all $t \in [0,T]$, $\varpi_t^{-2} \le 1$. In particular, on any interval $[0,S]$ on which φ is defined and satisfies $\varphi(S) \le T$, it holds that $\dot{\varphi} \le 1$, from which we deduce that, for any $t \in [0,S]$, $\varphi(t) \le t$. As a by-product, φ can be uniquely defined on the entire $[0,T]$ and it satisfies $\varphi([0,T]) \subset [0,T]$, proving that the extension of $(\eta_t)_{0 \le t \le T}$ outside $[0,T]$ does not matter.

It then remains to prove that we can choose T such that $\varphi(T) = \pi/4$. Notice that it is not completely trivial because φ depends on T itself through the solution of the Riccati equation. To emphasize this fact, we write φ^T for φ and η^T for η. Using the fact that the derivatives of φ^T and η^T on $[0,T]$ are in $[0,1]$ and $[-2,0]$ respectively, we can use a compactness argument to prove that the mapping $T \mapsto \varphi^T(T)$ is continuous. In particular, it suffices to show that there exists T such that $\varphi^T(T) \ge \pi/4$ to prove that there exists T such that $\varphi^T(T) = \pi/4$. We argue by contradiction. If $\varphi^T(T) < \pi/4$ for all $T > 0$, then, for all $T > 0$ and all $t \in [0,T]$,

$$\dot{\varphi}^T(t) \ge \exp\bigg(-2 \int_0^{\pi/4} (1 - \eta_s^T) ds \bigg) \ge \exp\big(-\frac{\sqrt{2}\pi}{2} \big),$$

and then $\varphi^T(t) \ge \exp(-\frac{\sqrt{2}\pi}{2})T$. The contradiction easily follows. □

3.5.5 Further Developments on Weak Solutions

To close this part of the book devoted to mean field games with a common noise, we revisit the previous example through the lens of the notion of weak equilibrium introduced in Chapter 2.

Weak Solutions That Are Not Strong

We now adapt the previous analysis in order to provide an example for which the mean field game is uniquely solvable in the weak sense but not in the strong sense. In lieu of (3.78), consider instead:

$$dX_t = \big(b(t,\mu_t) + X_t + \alpha_t\big)dt + dW_t + \text{sign}(\bar{\mu}_t)dW_t^0, \quad t \in [0,T], \tag{3.93}$$

where, for a flow $\boldsymbol{\mu} = (\mu_t)_{0\le t\le T}$ of square integrable probability measures on $\mathbb{R}$, $\bar{\boldsymbol{\mu}}$ stands for the mean function $(\bar{\mu}_t = \int_{\mathbb{R}} x d\mu_t(x))_{0\le t\le T}$. Here and only here, we define the sign function as $\text{sign}(x) = 1$ if $x \ge 0$, $\text{sign}(x) = -1$ if $x < 0$, so that, in contrast with our previous definition of the sign function, $\text{sign}(0) \ne 0$.

For the same cost functional as in (3.79), we can easily adapt the statement of Theorem 3.37 in order to characterize the solutions of the corresponding mean field game through the forward-backward system:

$$\begin{cases} d\bar{\mu}_t = \Big(\bar{b}(t,\bar{\mu}_t) + (1-\eta_t)\bar{\mu}_t - \chi_t\Big)dt + \text{sign}(\bar{\mu}_t)dW_t^0 \\ d\chi_t = -((1-\eta_t)\chi_t + \bar{f}(t,\bar{\mu}_t) + \bar{b}(t,\bar{\mu}_t)\eta_t)dt \\ \qquad\quad + dm_t^0, \quad t \in [0,T], \\ \bar{\mu}_0 = x_0, \quad \chi_T = \bar{g}(\bar{\mu}_T), \end{cases} \tag{3.94}$$

which is obviously similar to (3.85). Here the coefficients $\bar{b}, \bar{f}$, and $\bar{g}$ are the same as in Theorem 3.37 with $\sigma^0 = 1$. The proof is easily adapted to the present situation. It suffices to observe that the process $(\int_0^t \text{sign}(\bar{\mu}_s)dW_s^0)_{0\le t\le T}$ is a Brownian motion, which permits to recover the setting used in the proof of Theorem 3.37.

Under the same assumption as in the statement of Theorem 3.37, it is quite simple to provide weak solutions to (3.94). It suffices to solve (3.94) with $(\int_0^t \text{sign}(\bar{\mu}_s)dW_s^0)_{0\le t\le T}$ replaced by a prescribed $\mathbb{F}^0$-Brownian motion $\boldsymbol{W}^{0\prime}$, and then let $(W_t^0 = \int_0^t \text{sign}(\bar{\mu}_s)dW_s^{0\prime})_{0\le t\le T}$. Solvability of the system driven by $\boldsymbol{W}^{0\prime}$ is tackled by the same argument as in the proof of Theorem 3.35. The resulting solution is adapted with respect to the filtration generated by $\boldsymbol{W}^{0\prime}$. In particular, the filtration generated by $(\boldsymbol{W}^0, \boldsymbol{\mu})$ coincides with that generated by $\boldsymbol{W}^{0\prime}$ and is thus compatible with $\mathbb{F}^0$.

Actually, weak solutions must have the same law. Put differently, uniqueness must hold in law. Indeed, by the same argument as above, any solution to (3.94) may be regarded as a solution to an equation of the same form as (3.90), but driven by $(W_t^{0\prime} = \int_0^t \text{sign}(\bar{\mu}_s)dW_s^0)_{0\le t\le T}$. Arguing as in the proof of Theorem 3.35, solutions are written as a common function of $\boldsymbol{W}^{0\prime}$. Therefore, they all have the same law.

Most importantly, there exists a deterministic decoupling field $u : [0,T]\times\mathbb{R} \to \mathbb{R}$ such that, for all $t \in [0,T]$, $\chi_t = u(t,\bar{\mu}_t)$. Indeed, by the same change of Brownian motion as above, (3.94) reduces to (3.90). We showed in the proof of Theorem 3.35 that the solution to (3.90) was given by Theorem (Vol I)-4.12, which provides the decoupling field.

Now, we claim that there is no strong solution when $\bar{b}$, $\bar{f}$, and $\bar{g}$ are odd and $\bar{\mu}_0 = 0$. Indeed in this case, for any solution $(\bar{\boldsymbol{\mu}}, \boldsymbol{\chi}, \boldsymbol{m}^0)$, $(-\bar{\boldsymbol{\mu}}, -\boldsymbol{\chi}, -\boldsymbol{m}^0)$ is also a solution to (3.94). In this regard, observe that this holds true despite our definition of the sign function, which is not odd. Indeed, using the fact that u is bounded and invoking Girsanov theorem, it is easily checked that, for any $t \in (0,T]$, $\mathbb{P}^0[\bar{\mu}_t = 0] = 0$. Therefore, we must have $-u(t,x) = u(t,-x)$ for any $t \in [0,T]$ and $x \in \mathbb{R}$. Therefore, for any solution to (3.94), $\bar{\boldsymbol{\mu}}$ solves:

$$d\bar{\mu}_t = \bar{B}(t,\bar{\mu}_t)dt + \mathrm{sign}(\bar{\mu}_t)dW_t^0, \quad t \in [0,T] \; ; \; \bar{\mu}_0 = 0,$$

where $\bar{B}$ is odd in space. By Tanaka's formula,

$$d|\bar{\mu}_t| = \mathrm{sign}(\bar{\mu}_t)\bar{B}(t,\bar{\mu}_t)dt + dW_t^0 + dL_t^{\bar{\boldsymbol{\mu}}}, \quad t \in [0,T],$$

where $(L_t^{\bar{\boldsymbol{\mu}}})_{0\le t\le T}$ is the local time of $\bar{\boldsymbol{\mu}}$ at 0. We then argue as in the analysis of the standard Tanaka SDE. Since $\bar{B}$ is odd, we have $\mathrm{sign}(\bar{\mu}_t)\bar{B}(t,\bar{\mu}_t) = \bar{B}(t,|\bar{\mu}_t|)$. Therefore, $\boldsymbol{W}^0$ must be measurable with respect to the filtration generated by $|\bar{\boldsymbol{\mu}}|$, which is strictly smaller than the filtration generated by $\bar{\boldsymbol{\mu}}$. This shows that $\bar{\boldsymbol{\mu}}$ cannot be adapted to $\boldsymbol{W}^0$, proving that we cannot have a strong solution.

More About the Compatibility Condition

We close this section with an example for which the compatibility condition between $(\bar{\boldsymbol{\mu}}, \boldsymbol{\chi}) = (\bar{\mu}_t, \chi_t)_{0\le t\le T}$ and $\mathbb{F}^0$ in the statement in Theorem 3.37 is not satisfied. This example highlights the *practical* meaning of the Compatibility Condition.

With the same notation as in the proof of Proposition 3.36, consider the forward-backward system:

$$\begin{cases} d\hat{\mu}_t = -\varpi_t^2\hat{\chi}_t dt, \\ d\hat{\chi}_t = -\varpi_t^{-2}\mathrm{sign}(\hat{\mu}_t)dt, \quad t \in [0,1], \\ \hat{\mu}_0 = 0, \quad \hat{\chi}_1 = -4\varpi_1^{-2}\hat{\mu}_1, \end{cases} \tag{3.95}$$

where the sign function is as before except that we let $\mathrm{sign}(0) = 0$. For a symmetric random variable ε defined on some $(\Omega^0, \mathcal{F}^0, \mathbb{P}^0)$ and taking values in $\{-1,1\}$, it is easy to check that the functions $\hat{\boldsymbol{\chi}}$ and $\hat{\boldsymbol{\mu}}$ defined by:

$$\begin{cases} \hat{\chi}_t = -\varpi_t^{-2}(t-\frac{1}{2})_+\varepsilon, \\ \hat{\mu}_t = \frac{1}{2}\left(t-\frac{1}{2}\right)_+^2\varepsilon, \quad t \in [0,1], \end{cases}$$

solve (3.95). The solution is obviously measurable with respect to the completion of the σ-field generated by ε. Without any loss of generality, we may assume that $\mathcal{F}^0$ coincides with this σ-field. We then let $\mathbb{F}^0 = (\mathcal{F}_t^0 = \mathcal{F}^0)_{0\le t\le 1}$.

We then recover a system of the form (3.91) with $\hat{b} \equiv 0$. Following the proof of Theorem 3.35, we can associate with $(\hat{\chi}_t, \hat{\mu}_t)_{0\le t\le 1}$ a solution to (3.90) with $\bar{b} \equiv 0$. Following the proof of Lemma 3.39, we may construct a pair $(\boldsymbol{X}, \boldsymbol{Y})$ solving a system of the form:

$$\begin{cases} dX_t = \big(X_t - Y_t\big)dt, \\ dY_t = -\big(X_t + Y_t + f(t, \mu_t)\big)dt, \quad t \in [0, 1], \\ X_0 = 0, \quad Y_1 = X_1 + g(\mu_1). \end{cases}$$

The process $\boldsymbol{Y}$ reads $Y_t = \eta_t X_t + \varpi_t \hat{\chi}_t$, for all $t \in [0, 1]$, with $\chi_t = 0$ for $t \in [0, 1/2]$. In particular, $\boldsymbol{X}$ may follow two paths, according to the value of ε. Clearly, for $t \in [0, 1/2]$, $X_t = 0$ and then $\mathfrak{M}_t = 0$, where the process $\boldsymbol{\mathfrak{M}} = (\mathfrak{M}_t)_{0\le t\le 1}$ has the same definition as in Subsection 2.2.4. In particular, $\mathcal{F}_0^{\mathfrak{M}}$ reduces to the null sets. Moreover, $\mathcal{F}_1^{\mathfrak{M}}$ is equal to the completion of the σ-field generated by ε, namely $\mathcal{F}^0 = \mathcal{F}_0^0$. In particular, conditional on $\mathcal{F}_0^{\mathfrak{M}}$, $\mathcal{F}_1^{\mathfrak{M}}$ and $\mathcal{F}_0^0$ are not independent. Therefore, the compatibility condition fails, which had to be expected since $\mathcal{F}_0^0$ anticipates on the future of the environment: $\hat{\mu}_1$ is $\mathcal{F}_0^0$-measurable but is not observable before $t = 1/2$.

3.6 Notes & Complements

To the best of our knowledge, there are very few published results on the solvability of mean field games with a common noise. We believe that the recent paper [100] by Carmona, Delarue, and Lacker is the only work which addresses the problem in a general framework. Under the additional condition that the coefficients satisfy monotonicity conditions, in the sense of Lasry-Lions or in the sense of Definition (Vol I)-3.31, existence and uniqueness have directly been investigated in the papers by Ahuja [11] and by Cardaliaguet, Delarue, Lasry, and Lions [86]. In the latter, the authors used a continuation method very much in the spirit of that implemented in Chapter (Vol I)-6. A more specific existence result has been established by Lacker and Webster [257] under the assumption that the coefficients satisfy suitable properties of invariance by translation.

The strategy used in this chapter for proving Theorem 3.1 shares some common features with that developed in [100]. Generally speaking, both approaches rely on the same discretization procedure of the conditioning upon the realizations of the common noise. In both cases, this first step is essential since it permits to apply the same fixed point argument as when there is no common noise. However, these two approaches differ from one another in the method used to pass to the limit along the relaxation of the discretization in the conditioning. Whatever the strategy, the crucial point is to establish the compactness of the sequence of equilibrium optimal controls associated with each of the discretized conditioning. In the text, we rely on the Pontryagin stochastic maximum principle to prove tightness of this sequence with respect to the so-called Meyer-Zheng topology. An alternative strategy is proposed in [100] by relaxing the notion of controls, in analogy with what we did in Section (Vol I)-6.6 to handle optimal control problems of McKean-Vlasov diffusion processes. Generally speaking, the notion of relaxed controls is very useful as it provides simple compactness criteria. However, in order to maintain consistency across the different chapters, we chose not to invoke this weaker notion, and to perform the analysis of the weak solvability of MFGs with a common noise by means of forward-backward stochastic differential equations only. In the end, the structural conditions used in [100] to prove existence of a weak solution with a control in the *strong (or classical) sense* may be compared to those used in the text. On one hand, in order to use the maximum principle, we demand more on the regularity properties of the coefficients. On the other hand, the use of the non-degeneracy property in the assumption of Theorem 3.29 allows us to avoid any convexity assumption for the cost functionals in the direction x, while convexity in the direction x is required in [100]. Of course, weak solutions with relaxed controls, that is to say *very weak solutions*, exist under much weaker conditions than those used in this chapter.

In the proof of Theorem 3.1, a major technical difficulty for passing to the limit along the discretization procedure is to check that in the limit, the probabilistic set-up satisfies the required compatibility condition. This is especially clear from the statement of Proposition 3.11. As explained in the text, compatibility is necessary to identify the limit process with an equilibrium. Here, it follows for free from the procedure introduced in Definition 2.16 for lifting the environment into a super-environment. The idea for this trick is borrowed from [100].

The Meyer-Zheng topology used for proving tightness of the adjoint processes in the Pontryagin stochastic maximum principle goes back to the earlier paper [280] by Meyer and Zheng. Part of our presentation was inspired by the paper by Kurtz [245]. Meyer-Zheng's topology has been widely used in the analysis of backward SDEs. See for example [78,275,296]. Throughout the chapter, the theorem used for passing to the limit in the various stochastic integrals is taken from Kurtz and Protter [248]. The generalization of Lusin's theorem invoked in the proof of Lemma 3.15 may be found in Bogachev [64], see Theorem 7.1.13 therein.

As for the last part of the chapter about the restoration of uniqueness by the common noise forcing, we make the following observations. First, we recall that this question has been widely discussed within the classical theory of stochastic differential equations since the early works of Zvonkin [346] and Veretennikov [336]. We refer to Flandoli's monograph [156] for an overview of the subject, including a discussion on the infinite dimensional case. In the case of mean field games, the specific example discussed in the paper is inspired by the note [159] by Foguen Tchuendom. Proposition 3.36 raises interesting questions about the possible selection of *physical* equilibria by letting the viscosity σ tend to 0. Our guess is that, for the counter-example studied in the proof of Proposition 3.36, it should be possible to carry out an analysis similar to that performed by Bafico and Baldi in [30] for standard stochastic differential equations with a vanishing viscosity.

Part II

The Master Equation, Convergence, and Approximation Problems

The Master Field and the Master Equation

4

Abstract

We introduce the concept of master field within the framework of mean field games with a common noise. We present it as the decoupling field of an infinite dimensional forward-backward system of stochastic partial differential equations characterizing the equilibria. The forward equation is a stochastic Fokker-Planck equation and the backward equation a stochastic Hamilton-Jacobi-Bellman equation. We show that whenever existence and uniqueness of equilibria hold for any initial condition, the master field is a viscosity solution of Lions' master equation.

4.1 Introduction and Construction of the Master Field

4.1.1 General Objective

As explained in Chapters (Vol I)-3 and 2, mean field game equilibria may be described by means of a system of partial differential equations (PDEs) where time runs in opposite directions. The forward PDE is a Fokker-Planck equation describing the dynamics of the statistical distribution of the population in equilibrium. The backward PDE is a Hamilton-Jacobi-Bellman equation describing the evolution of the optimal expected costs in equilibrium. Both equations become stochastic when the state dynamics depend upon an additional source of randomness common to all the players. This extra source of random shocks was referred to as a *common noise* in Chapters 2 and 3.

Such a system of (possibly stochastic) PDEs is reminiscent of the standard theory of finite dimensional forward-backward stochastic differential equations we presented in Chapter (Vol I)-4. A remarkable feature of these systems is the existence of a so-called *decoupling field* which gives the solution of the backward

R. Carmona, F. Delarue, *Probabilistic Theory of Mean Field Games with Applications II*, Probability Theory and Stochastic Modelling 84,
DOI 10.1007/978-3-319-56436-4_4

equation at any given time, as a function of the solution at the same time, of the forward equation. Our first goal is to extend this connection to the forward-backward systems used to characterize mean field games. The corresponding decoupling field will be called *master field*.

In the existing literature, the decoupling field is constructed mostly when the corresponding forward-backward system is uniquely solvable. As shown in Chapter (Vol I)-4, when the coefficients of the system are deterministic, the decoupling field is deterministic as well, so that the value of the backward component at a given time reads as a function of the sole value of the forward component at the same time. This remarkable fact reflects the Markovian nature of the forward component of the solution. The objective here is to adapt the approach to mean field games, and to construct the master field when an equilibrium exists and is unique. Uniqueness, which will be understood in a weak sense according to the terminology introduced in Chapter 2, could be viewed as a proxy for the Markov property in the finite dimensional case with deterministic coefficients. Explicit examples were already given in Chapter 3.

Another key feature of decoupling fields in finite dimension is the fact that they solve a nonlinear PDE. Generally speaking, solvability holds in the viscosity sense but, under suitable regularity conditions on the coefficients, the decoupling field may be smooth, and thus solve the aforementioned PDE in the classical sense as well. In the case of mean field games, the master field is expected to solve an equation called the *master equation*, a PDE on the product of the state space and the space of probability measures on the state space. Within the framework of mean field games without common noise, we already encountered instances of the master equation in Chapter (Vol I)-5 as examples of applications of the chain rule for smooth functions on the Wasserstein space $\mathcal{P}_2(\mathbb{R}^d)$. In this chapter, we prove that the master field is a viscosity solution of the master equation, the key argument consisting in a suitable version of the dynamic programming principle. Smoothness properties of the master field will be investigated in Chapter 5.

The formulation of the master equation relies heavily on the differential calculus on the Wasserstein space introduced in Chapter (Vol I)-5. The chain rule for flows of probability measures established in Section (Vol I)-5.6 plays a crucial role in its derivation. To be more specific, an extended version that holds for flows of random (or conditional) measures will be needed if we want to handle models with a common source of noise.

The concept of master equation and the use of the terminology *master* go back to the seminal lectures of P. L. Lions at the *Collège de France*. The word *master* emphasizes the fact that all the information needed to describe the equilibria of the game is contained in a single equation, namely the master equation. This equation plays the same role as the Chapman-Kolmorogov equation, also called "master equation" in physics, for the evolution of a Markov semi-group. As we shall see, this connection between the master equation for a mean field game and the Chapman-Kolmorogov equation for a Markov semi-group can be made precise: The generator of the master equation coincides with the generator of the semi-group associated with the McKean-Vlasov SDE satisfied by the state of a representative player when the population is in equilibrium; we already introduced the latter generator in Subsection (Vol I)-5.7.4.

The outline of the chapter is as follows. We first provide a systematic construction of the master field under the assumption that the mean field game admits a unique weak equilibrium in the sense of Chapter 2. This master field is then shown to satisfy the dynamic programming principle. Next, by means of an extended version of the chain rule proved in Chapter (Vol I)-5, we identify the form of the master equation and show that the master field is a viscosity solution. Finally, we revisit some of the examples discussed in the first chapter in light of these new analytic tools.

4.1.2 General Set-Up

The construction of the master field given in this chapter applies to the general framework of mean field games with a common noise introduced in Chapter 2, a review of which is given below for the reader's convenience, even if it is short on specifics at times.

The set-up is the same as in Definition 2.16. We are given:

1. an initial condition $\mathcal{V}_0 \in \mathcal{P}_2(\mathcal{P}_2(\mathbb{R}^d))$, a complete probability space $(\Omega^0, \mathcal{F}^0, \mathbb{P}^0)$, endowed with a complete and right-continuous filtration $\mathbb{F}^0 = (\mathcal{F}_t^0)_{0 \le t \le T}$ and a d-dimensional $\mathbb{F}^0$-Brownian motion $\boldsymbol{W}^0 = (W_t^0)_{0 \le t \le T}$,
2. a complete probability space $(\Omega^1, \mathcal{F}^1, \mathbb{P}^1)$ endowed with a complete and right-continuous filtration $\mathbb{F}^1 = (\mathcal{F}_t^1)_{0 \le t \le T}$ and a d-dimensional $\mathbb{F}^1$-Brownian motion $\boldsymbol{W} = (W_t)_{0 \le t \le T}$.

$(\Omega, \mathcal{F}, \mathbb{P})$ will be the completion of the product space $(\Omega^0 \times \Omega^1, \mathcal{F}^0 \otimes \mathcal{F}^1, \mathbb{P}^0 \otimes \mathbb{P}^1)$ endowed with the filtration $\mathbb{F} = (\mathcal{F}_t)_{0 \le t \le T}$ obtained by augmenting the product filtration $\mathbb{F}^0 \otimes \mathbb{F}^1$ in a right-continuous way and completing it.

We recall the useful notation $\mathcal{L}^1(X)(\omega^0) = \mathcal{L}(X(\omega^0, \cdot))$ when $\omega^0 \in \Omega^0$ and X is a random variable on Ω which was introduced in Subsection 2.1.3.

For a drift b from $[0, T] \times \mathbb{R}^d \times \mathcal{P}_2(\mathbb{R}^d) \times A$ with values in $\mathbb{R}^d$, where A is a closed convex subset of $\mathbb{R}^k$, two (uncontrolled) volatility functions σ and σ^0 from $[0, T] \times \mathbb{R}^d \times \mathcal{P}_2(\mathbb{R}^d)$ to $\mathbb{R}^{d \times d}$ and two running and terminal scalar cost functions f and g defined on $[0, T] \times \mathbb{R}^d \times \mathcal{P}_2(\mathbb{R}^d) \times A$ and $\mathbb{R}^d \times \mathcal{P}_2(\mathbb{R}^d)$ respectively, the search for an MFG equilibrium along the lines of Definition 2.16 consists in the following two-step procedure:

(i) Given an $\mathcal{F}_0$-measurable random variable $\mu_0 : \Omega^0 \to \mathcal{P}_2(\mathbb{R}^d)$, with $\mathcal{V}_0$ as distribution, an initial condition $X_0 : \Omega \to \mathbb{R}^d$ such that $\mathcal{L}^1(X_0) = \mu_0$, and an $\mathcal{F}_T^0$-measurable random variable $\mathfrak{M}$ with values in $\mathcal{P}_2(\mathcal{C}([0,T];\mathbb{R}^{2d}))$ such that $\mathbb{F}$ is compatible with $(X_0, \boldsymbol{W}^0, \mathfrak{M}, \boldsymbol{W})$ and $\boldsymbol{\mu}_0 = \mathfrak{M} \circ (e_0^x)^{-1}$, where e_t^x is the mapping evaluating the first d coordinates at time t on $\mathcal{C}([0,T];\mathbb{R}^{2d})$, solve the (standard) stochastic control problem (with random coefficients):

$$\inf_{(\alpha_s)_{0\leq s\leq T}} \mathbb{E}\left[\int_0^T f(s, X_s, \mu_s, \alpha_s)ds + g(X_T, \mu_T)\right] \tag{4.1}$$

subject to:

$$dX_s = b\big(s, X_s, \mu_s, \alpha_s\big)ds + \sigma\big(s, X_s, \mu_s\big)dW_s + \sigma^0\big(s, X_s, \mu_s\big)dW_s^0, \tag{4.2}$$

with X_0 as initial condition and with $\mu_s = \mathfrak{M} \circ (e_s^x)^{-1}$, for $0 \leq s \leq T$.

(ii) Determine the input $\mathfrak{M}$ so that, for one optimal path $(X_s)_{0\leq \leq T}$, it holds that:

$$\mathfrak{M} = \mathcal{L}^1\big(\boldsymbol{X}, \boldsymbol{W}\big). \tag{4.3}$$

In order to guarantee the well posedness of the cost functional (4.1) and the unique solvability of (4.2), we shall assume the following throughout the chapter.

Assumption (Control). There exists a constant $L \geq 0$ such that:

(A1) For any $t \in [0,T]$, the coefficients $b(t,\cdot,\cdot,\cdot)$ and $(\sigma,\sigma^0)(t,\cdot,\cdot)$ are continuous on $\mathbb{R}^d \times \mathcal{P}_2(\mathbb{R}^d) \times A$ and $\mathbb{R}^d \times \mathcal{P}_2(\mathbb{R}^d)$ respectively. The coefficients $b(t,\cdot,\mu,\alpha)$, $\sigma(t,\cdot,\mu)$ and $\sigma^0(t,\cdot,\mu)$ are L-Lipschitz continuous in the x variable, uniformly in $(t,\mu,\alpha) \in [0,T] \times \mathcal{P}_2(\mathbb{R}^d) \times A$. Moreover,

$$|b(t,x,\mu,\alpha)| + |(\sigma,\sigma^0)(t,x,\mu)| \leq L\big[1 + |x| + |\alpha| + M_2(\mu)\big],$$

where, as usual, $M_2(\mu)^2$ denotes the second moment of μ.

(A2) f and g are Borel-measurable scalar functions on $[0,T]\times\mathbb{R}^d\times\mathcal{P}_2(\mathbb{R}^d)\times A$ and $\mathbb{R}^d \times \mathcal{P}_2(\mathbb{R}^d)$ respectively. For any $t \in [0,T]$, the functions $f(t,\cdot,\cdot,\cdot)$ and $g(\cdot,\cdot)$ are continuous on $\mathbb{R}^d \times \mathcal{P}_2(\mathbb{R}^d) \times A$ and $\mathbb{R}^d \times \mathcal{P}_2(\mathbb{R}^d)$ respectively. Moreover,

$$|f(t,x,\mu,\alpha)| + |g(x,\mu)| \leq \Gamma\big[1 + |x|^2 + |\alpha|^2 + M_2(\mu)^2\big].$$

Arbitrary Initial Time. Throughout the chapter, we shall consider equilibria that may be initialized at a time $t \in [0,T]$ different from 0. Of course, this requires a suitable adaptation of the definition of the optimization problem (4.1)-(4.2), and in particular of the cost functional, with t in lieu of 0 in the integral of the running cost. This also requires a redefinition of the notion of equilibrium: we let the reader adapt Definition 2.16 accordingly. In the new definition, we shall force the Brownian motions $\boldsymbol{W}^0$ and $\boldsymbol{W}$, which are then defined on $[t, T]$, to start from 0 at time t. Moreover, the super-environment $\mathfrak{M}$ is required to take values in $\mathcal{P}_2(\mathcal{C}([t, T]; \mathbb{R}^{2d}))$, and its canonical filtration is defined along the lines of Definition 2.15, the canonical process therein being defined on $[t, T]$ in lieu of $[0, T]$. The compatibility condition is now understood as compatibility between $\mathbb{F}$ and $(X_t, (W_s^0)_{t \le s \le T}, \mathfrak{M}, (W_s)_{t \le s \le T})$, $\mathfrak{M}$ being regarded as a random variable with values in $\mathcal{P}_2(\mathcal{C}([t, T]; \mathbb{R}^{2d}))$. Following Definition 1.39, we shall sometimes say that the set-up $(\Omega, \mathcal{F}, \mathbb{F}, \mathbb{P})$ is t-initialized, the initial information being generated by the sole initial private state X_t.

It is plain to check that all the results of Chapter 2 are easily adapted to these t-initialized set-ups.

A Primer on the Master Field

According to the discussion in Subsection 2.3.3 of Chapter 2, the search for an optimum in (4.1)–(4.2) may be connected to the solvability of an infinite dimensional FBSDE of the form (2.37)–(2.38), with a stochastic Fokker-Planck equation as forward equation and a stochastic HJB equation as backward equation. This is especially meaningful if we seek strong solutions, namely equilibria $\boldsymbol{\mu}$ that are adapted to the noise $\boldsymbol{W}^0$, in which case there is no need to require compatibility between $\mathbb{F}$ and $(X_0, \boldsymbol{W}^0, \mathfrak{M}, \boldsymbol{W})$. If existence and uniqueness hold, it sounds reasonable to expect a generalization of the concept of decoupling field, already discussed in Subsection (Vol I)-4.1 in the finite dimensional setting. Roughly speaking, the existence of a decoupling field says that the backward component must read as a deterministic function of the forward component, which in our case, and with the same kind of notations as in (2.37)–(2.38), reads:

$$\text{for a.e. } \omega^0 \in \Omega^0, \quad \text{for all } t \in [0, T], \quad U^{\boldsymbol{\mu}}(t, \cdot, \omega^0) = \mathcal{U}\big(t, \mu_t(\omega^0)\big)(\cdot),$$

where $U^{\boldsymbol{\mu}}(t, \cdot, \omega^0)$ is the random value function of the optimization problem in the random environment $\boldsymbol{\mu}$. We emphasize that the reason for appealing to an analogue of the notion of decoupling field introduced in Subsection (Vol I)-4.1 rather than its generalization discussed in Subsection 1.3 is the fact that here, the coefficients are deterministic (although they are infinite dimensional).

The present discussion being set at a rather informal level, there is no real need to specify the space of functions to which $U^{\boldsymbol{\mu}}(t, \cdot, \omega^0)$ is expected to belong. Moreover, $\boldsymbol{\mu} = (\mu_t)_{0 \le t \le T}$ is the solution of the forward Fokker-Planck equation describing the evolution of the population in equilibrium. The definition of the function $\mathcal{U}$ does not depend upon $\boldsymbol{\mu}$. It is a function from $[0, T] \times \mathcal{P}_2(\mathbb{R}^d)$ into the space of functions of the same type as $U^{\boldsymbol{\mu}}(t, \cdot, \omega^0)$, since $\mathcal{P}_2(\mathbb{R}^d)$ appears as the state space of the forward component. Specifying the values $U^{\boldsymbol{\mu}}(t, \cdot, \omega^0)$ at points x of the physical state space $\mathbb{R}^d$ (which makes sense for example if $U^{\boldsymbol{\mu}}(t, \cdot, \omega^0)$ is a continuous function), this relationship also writes:

for a.e. $\omega^0 \in \Omega^0$, for all $(t,x) \in [0,T] \times \mathbb{R}^d$, $U^{\mu}(t,x,\omega^0) = \mathcal{U}\big(t,\mu_t(\omega^0)\big)(x)$, or, for a.e. $\omega^0 \in \Omega^0$, for all $(t,x) \in [0,T] \times \mathbb{R}^d$,

$$U^{\mu}(t,x,\omega^0) = \mathcal{U}\big(t,x,\mu_t(\omega^0)\big). \tag{4.4}$$

Basically, $\mathcal{U}$ is the object we are tracking in this chapter. It is touted as the master field of the game.

4.1.3 Construction of the Master Field

We now provide a rigorous construction of the master field which does not require the analysis of the infinite dimensional forward-backward system (2.37)–(2.38) describing the equilibria of the game. Our strategy is to make systematic use of the forward-backward stochastic differential equations that describe the optimal paths of the underlying optimization problem.

Throughout this section, we use the same assumptions as in Subsection 2.2.3, but with an arbitrary initial time $t \in [0,T]$ in lieu of 0.

Assumption (FBSDE). On top of assumption **Control**, there exist an integer $m \geq 1$ together with deterministic measurable functions $B : [0,T] \times \mathbb{R}^d \times \mathcal{P}_2(\mathbb{R}^d) \times \mathbb{R}^m \times \mathbb{R}^{m\times d} \to \mathbb{R}^d$, $F : [0,T] \times \mathbb{R}^d \times \mathcal{P}_2(\mathbb{R}^d) \times \mathbb{R}^m \times \mathbb{R}^{m\times d} \to \mathbb{R}^m$, $G : \mathbb{R}^d \times \mathcal{P}_2(\mathbb{R}^d) \to \mathbb{R}^m$ and $\breve{\alpha} : [0,T] \times \mathbb{R}^d \times \mathcal{P}_2(\mathbb{R}^d) \times \mathbb{R}^m \times \mathbb{R}^{m\times d} \to A$, such that:

(A1) For any $t \in [0,T]$, and any t-initialized probabilistic set-up $(\Omega, \mathcal{F}, \mathbb{F}, \mathbb{P})$ equipped with a compatible lifting $(X_t, \boldsymbol{W}^0, (\mathfrak{M}, \mu), \boldsymbol{W})$ as in Subsection 4.1.2, the conditional law $\mathcal{L}^1(X_t)$ of the initial condition X_t possibly differing from μ_t, the optimal control problem defined in (4.1)–(4.2), namely:

$$\min_{\alpha} J^{\mu}(\alpha), \quad J^{\mu}(\alpha) = \mathbb{E}\bigg[\int_t^T f(s, X_s^{\alpha}, \mu_s, \alpha_s)ds + g(X_T^{\alpha}, \mu_T)\bigg],$$

where $\boldsymbol{\alpha} = (\alpha_s)_{t\leq s\leq T}$ is an $\mathbb{F}$-progressively measurable square-integrable A-valued control process and $\boldsymbol{X}^{\alpha} = (X_s^{\alpha})_{t\leq s\leq T}$ solves:

$$dX_s^{\alpha} = b\big(s, X_s^{\alpha}, \mu_s, \alpha_s\big)ds + \sigma(s, X_s^{\alpha}, \mu_s)dW_s + \sigma^0(s, X_s^{\alpha}, \mu_s)dW_s^0,$$

for $s \in [t,T]$ and with $X_t^{\alpha} = X_t$ as initial condition, has a unique solution, characterized as the forward component of the unique solution to the strongly uniquely solvable FBSDE:

(continued)

$$\begin{cases} dX_s = B(s, X_s, \mu_s, Y_s, Z_s)ds \\ \qquad\quad +\sigma(s, X_s, \mu_s)dW_s + \sigma^0(s, X_s, \mu_s)dW_s^0, \\ dY_s = -F(s, X_s, \mu_s, Y_s, Z_s, Z_s^0)ds \\ \qquad\quad +Z_s dW_s + Z_s^0 dW_s^0 + dM_s, \quad s \in [t, T], \end{cases} \tag{4.5}$$

with X_t as initial condition for $\boldsymbol{X} = (X_s)_{t \le s \le T}$ and $Y_T = G(X_T, \mu_T)$ as terminal condition for $\boldsymbol{Y} = (Y_s)_{t \le s \le T}$, where $\boldsymbol{M} = (M_s)_{t \le s \le T}$ is a *càd-làg* martingale with respect to the filtration $\mathbb{F}$, of zero cross variation with $(\boldsymbol{W}^0, \boldsymbol{W})$ and with initial condition $M_t = 0$.

(A2) For all $(t, x, \mu, y, z) \in [0, T] \times \mathbb{R}^d \times \mathcal{P}_2(\mathbb{R}^d) \times \mathbb{R}^m \times \mathbb{R}^{m \times d}$,

$$B(t, x, \mu, y, z) = b\big(t, x, \mu, \breve{\alpha}(t, x, \mu, y, z)\big),$$

(A3) There exists a constant $L \ge 0$ such that:

$$\begin{aligned} &|(\sigma, \sigma^0)(t, x, \mu)| \le L\big[1 + |x| + M_2(\mu)\big], \\ &|(B, \breve{\alpha})(t, x, \mu, y, z)| \le L\big[1 + |x| + |y| + |z| + M_2(\mu)\big], \\ &|F(t, x, \mu, y, z, z^0)| + |G(x, \mu)| \\ &\qquad\qquad \le L\big[1 + |x| + |y| + |z|^2 + |z^0|^2 + \big(M_2(\mu)\big)^2\big], \end{aligned}$$

for all $(t, x, \mu, y, z) \in [0, T] \times \mathbb{R}^d \times \mathcal{P}_2(\mathbb{R}^d) \times \mathbb{R}^m \times \mathbb{R}^{m \times d}$.

Recall that the typical examples we have in mind for the FBSDE (4.5) are:

1. the FBSDE is associated with the value function of the control problem, as in Theorem 1.57, in which case F does not depend on z^0;
2. the FBSDE derives from the stochastic Pontryagin principle, as in Theorem 1.60.

As we already emphasized in formula (2.26) of Subsection 2.2.3, if we denote by $\hat{\alpha}(t, x, \mu, y)$ the minimizer of the reduced Hamiltonian $H^{(r)}(t, x, \mu, y, \cdot)$, the function $\breve{\alpha}$ is given by $\breve{\alpha}(t, x, \mu, y, z) = \hat{\alpha}(t, x, \mu, \sigma(t, x, \mu)^{-1\dagger} z)$ in the first case and $\breve{\alpha}(t, x, \mu, y, z) = \hat{\alpha}(t, x, \mu, y)$ in the second.

The construction of the master field is now quite simple, although its description still requires some more notations. The intuitive idea is the following: assume that, for any fixed initial distribution $\mu \in \mathcal{P}_2(\mathbb{R}^d)$, there exists a unique weak equilibrium to the mean field game, the population having μ as initial condition at time t, in the sense that we can find a general probabilistic set-up on which we can construct a weak solution as in Chapter 2, and that any other solution has the same distribution.

Then $\mathcal{U}(t,x,\mu)$ must be the optimal cost of the optimization problem (4.1), when μ therein denotes the equilibrium of the game and $\boldsymbol{X}$ is forced to start from x at time t. Now since we have a characterization of such an optimal cost in terms of the initial value of the backward component of an FBSDE, we can represent it in a systematic way.

The best way to standardize the procedure is to transfer the weak solutions to the canonical space, as already discussed in Chapter 2, see Definition 2.24 and Lemma 2.25. The principle here is the same, except that now, the initial time may be any time $t \in [0,T]$. Assume that, for some $(t,\mu) \in [0,T] \times \mathcal{P}_2(\mathbb{R}^d)$, there exists a unique weak equilibrium to the MFG problem when the population starts from the distribution μ at time t. Following the statement of Lemma 2.25, we call $\mathcal{M}^{t,\mu}$ the (unique) distribution of the triplet $(\mu_t, \boldsymbol{W}^0, \mathfrak{M}) \cong (\mu_t, W_s^0, \mathfrak{M}_s)_{t\le s\le T}$ on $\mathcal{P}_2(\mathbb{R}^d) \times \mathcal{C}([t,T];\mathbb{R}^d) \times \mathcal{P}_2(\mathcal{C}([t,T];\mathbb{R}^{2d}))$, for any weak solution $\mathfrak{M}$ initialized with $\mathfrak{M} \circ (e_t^x)^{-1} = \mu$ and constructed on some t-initialized probabilistic set-up $(\Omega,\mathcal{F},\mathbb{F},\mathbb{P})$. Here, $(\Omega,\mathcal{F},\mathbb{F},\mathbb{P})$ is required to satisfy the same prescription as in Definition 2.16 and in particular, to be equipped with some d-dimensional Brownian motion $\boldsymbol{W}^0$, which is regarded as the common noise. Of course, the first marginal of $\mathcal{M}^{t,\mu}$ on $\mathcal{P}_2(\mathbb{R}^d)$ is δ_μ, the Dirac mass at μ. Once again, uniqueness follows from the uniqueness in law of the equilibrium: any weak solution, with the prescribed initial condition μ, generates the same distribution $\mathcal{M}^{t,\mu}$. Now, we can consider the same canonical space $\bar{\Omega}$ as in Chapter 2, see (2.32) and Definition 2.24, namely (pay attention that the spaces depend on t, which is fixed in the present situation):

$$\begin{aligned}
&\bar{\Omega}^t = \bar{\Omega}^{0,t} \times \bar{\Omega}^{1,t},\\
&\bar{\Omega}^{0,t} = \mathcal{P}_2(\mathbb{R}^d) \times \mathcal{C}([t,T];\mathbb{R}^d) \times \mathcal{P}_2\big(\mathcal{C}([t,T];\mathbb{R}^{2d})\big),\\
&\bar{\Omega}^{1,t} = [0,1) \times \mathcal{C}([t,T];\mathbb{R}^d).
\end{aligned} \tag{4.6}$$

As in Definition 2.24, we can denote by $(\bar{\Omega}^{0,t}, \mathcal{F}^{0,t,\mu}, \mathbb{P}^{0,t,\mu})$ the completion of $\bar{\Omega}^{0,t}$ equipped with its Borel σ-field and with the probability measure $\mathcal{M}^{t,\mu}$. Similarly, we can denote by $(\bar{\Omega}^{1,t}, \mathcal{F}^{1,t}, \mathbb{P}^{1,t})$ the completion of $\bar{\Omega}^{1,t}$ equipped with its Borel σ-field and with the probability measure $\mathrm{Leb}_1 \otimes \mathcal{W}_d^t$, where Leb_1 is the Lebesgue measure on $[0,1)$ and $\mathcal{W}_d^t$ is the d-dimensional Wiener measure on $\mathcal{C}([t,T];\mathbb{R}^d)$. We then call $(\bar{\Omega}^t, \mathcal{F}^{t,\mu}, \mathbb{P}^{t,\mu})$ the completion of the product space $(\bar{\Omega}^{0,t} \times \bar{\Omega}^{1,t}, \mathcal{F}^{0,t,\mu} \otimes \mathcal{F}^{1,t}, \mathbb{P}^{0,t,\mu} \otimes \mathbb{P}^{1,t})$. When t is equal to 0, we shall systematically remove the symbol t in the superscripts $(0,t,\mu)$ and $(1,t)$ (so that $\bar{\Omega}^{0,t}$ is just denoted by $\bar{\Omega}^0$ and so on...)

The canonical random variable on $\bar{\Omega}^{1,t}$ is denoted by $(\eta, \boldsymbol{w} = (w_s)_{t\le s\le T})$. The complete and right-continuous augmentation of the filtration generated by $(\eta,\boldsymbol{w})$ is denoted by $\mathbb{F}^{1,t} = (\mathcal{F}_s^{1,t})_{t\le s\le T}$. The canonical random variable on $\bar{\Omega}^{0,t}$ is denoted by $(\nu^0, \boldsymbol{w}^0 = (w_s^0)_{t\le s\le T}, \mathfrak{m})$, the associated flow of marginal measures being denoted by $\boldsymbol{\nu} = (\nu_s = \mathfrak{m} \circ (e_s^x)^{-1})_{t\le s\le T}$. We call $\mathbb{F}^{0,t,\mu} = (\mathcal{F}_s^{0,t,\mu})_{t\le s\le T}$ the filtration generated by $(\nu^0, w_s^0, \mathfrak{m}_s)_{t\le s\le T}$ where $\mathfrak{m}_s = \mathfrak{m} \circ (\mathcal{E}_s^t)^{-1}$, with $\mathcal{E}_s^t : \mathcal{C}([t,T];\mathbb{R}^{2d}) \ni (\boldsymbol{x},\boldsymbol{w}) \mapsto (x_{r\wedge s}, w_{r\wedge s})_{t\le s\le T} \in \mathcal{C}([t,T];\mathbb{R}^{2d})$. On $\bar{\Omega}^t$, we let $\mathbb{F}^{t,\mu}$ be the complete and right-continuous augmentation of the product filtration.

Finally, the initial state x_t of the population is then defined as $x_t = \psi(\eta, \nu^0)$, with ψ as in (2.23) and Lemma (Vol I)-5.29.

At this stage of the discussion, the presence of $\mathcal{P}_2(\mathbb{R}^d)$ in $\bar{\Omega}^{0,t}$ in the definition of $\bar{\Omega}^{0,t}$ may seem like an overkill as μ is a fixed deterministic probability measure. Indeed, we could use $\bar{\Omega}^{00,t} = \mathcal{C}([t,T];\mathbb{R}^d) \times \mathcal{P}_2(\mathcal{C}([t,T];\mathbb{R}^{2d}))$ as canonical space instead of $\bar{\Omega}^{0,t}$. The rationale for working with $\bar{\Omega}^{0,t}$ will be made clear below, when we work with equilibria starting from initial random states with values in $\mathcal{P}_2(\mathbb{R}^d)$.

On such a canonical set-up, we can solve the forward-backward system:

$$\begin{cases} dx_s = B(s, x_s, \nu_s, y_s, z_s)ds + \sigma(s, x_s, \nu_s)dw_s + \sigma^0(s, x_s, \nu_s)dw_s^0, \\ dy_s = -F(s, x_s, \nu_s, y_s, z_s, z_s^0)ds + z_s dw_s + z_s^0 dw_s^0 + dm_s, \end{cases}$$

with $x_t = \psi(\eta, \nu^0)$ as initial condition at time t and with $y_T = G(x_T, \nu_T)$ as terminal condition, where $\boldsymbol{m}$ has zero cross variation with $(\boldsymbol{w}^0, \boldsymbol{w})$ and $m_t = 0$ as initial value.

More generally, for any random variable $\xi \in L^2(\bar{\Omega}^t, \sigma\{\nu^0, \eta\}, \mathbb{P}^{t,\mu}; \mathbb{R}^d)$, we can solve, on the set-up $(\bar{\Omega}^t, \mathcal{F}^{t,\mu}, \mathbb{F}^{t,\mu}, \mathbb{P}^{t,\mu})$ equipped with $(\xi, \boldsymbol{w}^0, \mathfrak{m}, \boldsymbol{w})$, the FBSDE:

$$\begin{cases} dx_s^{t,\xi} = B\big(s, x_s^{t,\xi}, \nu_s, y_s^{t,\xi}, z_s^{t,\xi}\big)ds \\ \qquad +\sigma\big(s, x_s^{t,\xi}, \nu_s\big)dw_s + \sigma^0\big(s, x_s^{t,\xi}, \nu_s\big)dw_s^0, \\ dy_s^{t,\xi} = -F\big(s, x_s^{t,\xi}, \nu_s, y_s^{t,\xi}, z_s^{t,\xi}, z_s^{0,t,\xi}\big)ds \\ \qquad +z_s^{t,\xi}dw_s + z_s^{0,t,\xi}dw_s^0 + dm_s^{t,\xi}, \end{cases} \tag{4.7}$$

with $x_t^{t,\xi} = \xi$ as initial condition at time t and with $y_T^{t,\xi} = G(x_T^{t,\xi}, \nu_T)$ as terminal condition, where $\boldsymbol{m}^{t,\xi}$ has zero cross variation with $(\boldsymbol{w}^0, \boldsymbol{w})$ and $m_t^{t,\xi} = 0$ as initial value. Compatibility between $\mathbb{F}^{t,\mu}$ and $(\xi, \boldsymbol{w}^0, \mathfrak{m}, \boldsymbol{w})$ may be checked along the line of identity (2.33) following Definition 2.24 in Chapter 2.

From Theorem 1.33, the law of $(\boldsymbol{x}^{t,\xi}, \boldsymbol{y}^{t,\xi}, \boldsymbol{z}^{t,\xi}, \boldsymbol{z}^{0,t,\xi}, \boldsymbol{m}^{t,\xi})$ is uniquely determined by the law of the input $(\xi, \boldsymbol{w}^0, \mathfrak{m}, \boldsymbol{w})$. Therefore, the law of the pair $(\xi, y_t^{t,\xi})$ is uniquely determined as well. The guess is that, when $\xi = x_t$, $y_t^{t,\xi}$ should be a function of the sole x_t, and that their relationship should be of the form $y_t = \mathcal{U}(t, x_t, \nu^0) = \mathcal{U}(t, x_t, \mu)$ for the same $\mathcal{U}$ as in (4.4). In order to access the function $\mathcal{U}$, one has to use conditioning on the value of X_t or, equivalently, to solve (4.7) but under the prescription that $\xi = x$ for some deterministic x.

This suggests that we take a special look at the case $\xi = x$ in (4.7), leading to what can be regarded as the definition of the master field.

Definition 4.1 *Let us assume that weak existence and uniqueness hold for any initial condition $(t, \mathcal{V}) \in [0,T] \times \mathcal{P}_2(\mathcal{P}_2(\mathbb{R}^d))$ in the sense that all the weak solutions of the mean field game with $(t, \mathcal{V})$ as initial condition have the same distribution. Let us also assume that assumption* **FBSDE** *holds. For any $(t, \mu) \in [0,T] \times \mathcal{P}_2(\mathbb{R}^d)$ we denote by $\mathcal{M}^{t,\mu}$ the law of the equilibrium on the space $\mathcal{P}_2(\mathbb{R}^d) \times \mathcal{C}([t,T];\mathbb{R}^d) \times \mathcal{P}_2(\mathcal{C}([t,T];\mathbb{R}^{2d}))$ with $\mathcal{V} = \delta_\mu$ as initial condition, and we define the function $\mathcal{U}$ on $[0,T] \times \mathbb{R}^d \times \mathcal{P}_2(\mathbb{R}^d)$ by:*

$$
\begin{aligned}
&\mathcal{U}(t,x,\mu) \\
&= \int \Big[\int_t^T f\big(s, x_s^{t,x}, \nu_s, \check{\alpha}(s, x_s^{t,x}, \nu_s, y_s^{t,x}, z_s^{t,x})\big) ds + g(x_T^{t,x}, \nu_T) \Big] d\Big[\overline{\mathcal{M}^{t,\mu} \otimes \mathcal{W}_d^t} \Big],
\end{aligned}
$$

where the integral is over the set $\mathcal{P}_2(\mathbb{R}^d) \times \mathcal{C}([t,T];\mathbb{R}^d) \times \mathcal{P}_2(\mathcal{C}([t,T];\mathbb{R}^{2d})) \times \mathcal{C}([t,T];\mathbb{R}^d)$, $\boldsymbol{\nu}$ *is understood as* $\boldsymbol{\nu}=(\nu_s=\mathfrak{m} \circ (e_s^x)^{-1})_{t \le s \le T}$, *and where* $(x_s^{t,x}, y_s^{t,x}, z_s^{t,x}, z_s^{0,t,x})_{t \le s \le T}$ *is the solution of* (4.7) *with* $\xi \equiv x$.

Above, we denoted by $\overline{\mathcal{M}^{t,\mu} \otimes \mathcal{W}_d^t}$ the completion of the measure $\mathcal{M}^{t,\mu} \otimes \mathcal{W}_d^t$, namely the extension of $\mathcal{M}^{t,\mu} \otimes \mathcal{W}_d^t$ to the completion of the Borel σ-field on $\mathcal{P}_2(\mathbb{R}^d) \times \mathcal{C}([t,T];\mathbb{R}^d) \times \mathcal{P}_2(\mathcal{C}([t,T];\mathbb{R}^d)) \times \mathcal{C}([t,T];\mathbb{R}^d)$.

The interpretation of $\mathcal{U}(t,x,\mu)$ is quite clear: it is the optimal cost of the optimization problem (4.1)–(4.2) in the random environment $(\mathfrak{m}_s)_{t \le s \le T}$, with x as initial condition at time t and with $(w_s^0, \mathfrak{m}_s, w_s)_{t \le s \le T} = (w_s^0 - w_t^0, \mathfrak{m}_s, w_s - w_t)_{t \le s \le T}$ as t-initialized set-up. Importantly, this interpretation is independent of the probabilistic set-up used to construct the MFG equilibrium and to solve the FBSDE (4.7). Indeed, by assumption, the law of the equilibrium is independent of the underlying probabilistic set-up and, by assumption **FBSDE**, the joint law of the MFG equilibrium and the solution of the FBSDE is also independent of the underlying probabilistic set-up, see Theorem 1.33.

We refer the reader to Theorems 3.29, 3.30, and 3.31 in Chapter 3 for examples of existence of weak solutions. We refer to Proposition 3.34 and Theorem 3.35 in the same chapter for examples of strong and thus weak uniqueness, and to Subsection 3.5.5 for an example for which weak uniqueness holds while strong does not. Some of these examples will be revisited in paragraph 4.4.4 below.

4.1.4 Dynamic Programming Principle

In standard forward-backward stochastic differential equations, the decoupling field makes the connection between the backward component of the solution and the forward component at any time. Equation (4.4) is the expected counterpart in the current infinite dimensional framework. A natural question is to determine whether the mapping $\mathcal{U}$ constructed in Definition 4.1 indeed satisfies such a principle (provided uniqueness of the equilibria holds and assumption **FBSDE** is in force).

Another way to formulate (4.4) is to demand that, along the equilibrium $\boldsymbol{\mu} = (\mu_t)_{0 \le t \le T}$ and the optimal path $\boldsymbol{X} = (X_t)_{0 \le t \le T}$, $\mathcal{U}(t, X_t, \mu_t)$ be, at any time $t \in [0,T]$, the optimal *remaining* cost from t to T. This would express, under the uniqueness property (and the technical assumption **FBSDE**), the Markovian nature of solutions: *at any time, the optimal remaining cost would be a deterministic function of the private (random) state of the player and of the collective (random) state of the population.*

We prove below that this is indeed the case. As a byproduct, we shall derive a version of the dynamic programming principle for mean field games.

Of course, the intuitive idea behind the Markovian nature alluded to above is that we can restart the analysis from the initial condition (X_t, μ_t), despite the fact that μ_t is a random measure. Because of that, we shall often consider weak solutions with random initial conditions as we did in Definition 2.24. Given an initial time $t \in [0, T]$ and a distribution $\mathcal{V} \in \mathcal{P}_2(\mathcal{P}_2(\mathbb{R}^d))$, a weak solution is a probability measure $\mathcal{M}^{t,\mathcal{V}}$ on the space $\bar{\Omega}^{0,t}$. When $\mathcal{V} = \delta_\mu$ for some $\mu \in \mathcal{P}_2(\mathbb{R}^d)$, it holds that $\mathcal{M}^{t,\mathcal{V}} = \mathcal{M}^{t,\mu}$. Of course, we shall often write $\mathcal{M}^{t,\mu}$ for $\mathcal{M}^{t,\delta_\mu}$.

For such $\mathcal{V}$ and $\mathcal{M}^{t,\mathcal{V}}$, we call $(\bar{\Omega}^{0,t}, \mathcal{F}^{0,t,\mathcal{V}}, \mathbb{P}^{0,t,\mathcal{V}})$ the completion of $\bar{\Omega}^{0,t}$ equipped with its Borel σ-field and $\mathcal{M}^{t,\mathcal{V}}$. As above, we can equip it with a filtration $\mathbb{F}^{0,t,\mathcal{V}}$. We then define $(\bar{\Omega}^t, \mathcal{F}^{t,\mathcal{V}}, \mathbb{F}^{t,\mathcal{V}}, \mathbb{P}^{t,\mathcal{V}})$ accordingly.

We start with a quite technical (but necessary) lemma for ensuring that $\mathcal{U}$ is indeed a measurable function.

Proposition 4.2 *Assume that weak existence and uniqueness hold for any initial condition $(t, \mathcal{V}) \in [0, T] \times \mathcal{P}_2(\mathcal{P}_2(\mathbb{R}^d))$ and that assumption* **FBSDE** *is in force. Then, for any $t \in [0, T]$, the function $\mathbb{R}^d \times \mathcal{P}_2(\mathbb{R}^d) \ni (x, \mu) \mapsto \mathcal{U}(t, x, \mu)$ is measurable.*

Proof. The strategy relies on the Souslin-Kuratowski theorem, see Proposition 1.32. Throughout the proof, we fix the value of $t \in [0, T]$.

First Step. The idea is to introduce the set $\mathcal{S}$ of probability measures $\mathcal{M} \in \mathcal{P}_2(\bar{\Omega}^{0,t})$ that generate an equilibrium after time t with a Dirac initial condition δ_μ, for some $\mu \in \mathcal{P}_2(\mathbb{R}^d)$.

In order to make clear the definition of $\mathcal{S}$, we need some preliminary notations. As usual, we denote by $(\nu^0, \boldsymbol{w}^0, \mathfrak{m})$ the canonical random variable on $\bar{\Omega}^{0,t}$. Then, we let $\boldsymbol{\nu} = (\nu_s = \mathfrak{m} \circ (e_s^x)^{-1})_{t \le s \le T}$. The canonical filtration generated by $(\nu^0, \boldsymbol{w}^0, \mathfrak{m})$ along the lines of Definition 2.15 is denoted by $\mathbb{F}^{0,t,\text{nat}} = (\mathcal{F}_s^{0,t,\text{nat}})_{t \le s \le T}$. For $\mathcal{M} \in \mathcal{P}(\bar{\Omega}^{0,t})$, we call $(\bar{\Omega}^{0,t}, \mathcal{F}^{0,t,\mathcal{M}}, \mathbb{P}^{0,t,\mathcal{M}})$ the completion of $(\bar{\Omega}^{0,t}, \mathcal{B}(\bar{\Omega}^{0,t}), \mathcal{M})$ and we denote by $\mathbb{F}^{0,t,\mathcal{M}}$ the completion and right-continuous augmentation of $\mathbb{F}^{0,t,\text{nat}}$ under $\mathcal{M}$.

We then equip the product space $\bar{\Omega}^{0,t} \times \bar{\Omega}^{1,t}$ with its Borel σ-field and the product measure $\mathcal{M} \otimes (\text{Leb}_1 \otimes \mathcal{W}_d^t)$. The canonical random variable on $\bar{\Omega}^{0,t} \times \bar{\Omega}^{1,t}$ is denoted by $(\boldsymbol{w}^0, \mathfrak{m}, \eta, \boldsymbol{w})$, the canonical filtration being denoted by $\mathbb{F}^{t,\text{nat}} = (\mathcal{F}_s^{t,\text{nat}})_{t \le s \le T}$. The completed space is denoted by $(\bar{\Omega}^{0,t} \times \bar{\Omega}^{1,t}, \mathcal{F}^{t,\mathcal{M}}, \mathbb{P}^{t,\mathcal{M}})$ and the complete and right-continuous augmentation of the canonical filtration is denoted by $\mathbb{F}^{t,\mathcal{M}}$.

We now consider $\mathcal{M} \in \mathcal{P}_2(\bar{\Omega}^{0,t})$ such that under $\mathcal{M}$:

1. the process $\boldsymbol{w}^0 = (w_s^0)_{t \le s \le T}$ is a d-dimensional Brownian motion with respect to $(\mathcal{F}_s^{0,t,\text{nat}})_{t \le s \le T}$ starting from $w_t^0 = 0$;
2. the random variable ν^0 is almost surely constant, which means that, with probability 1 under $\mathcal{M}$, for any $B \in \mathcal{B}(\mathbb{R}^d)$,

$$\nu^0(B) = \int_{\bar{\Omega}^{0,t}} \nu^0(B) \mathcal{M}(d\nu^0, d\boldsymbol{w}^0, d\mathfrak{m}) \, ; \tag{4.8}$$

3. the random variable $\mathfrak{m}_t = \mathfrak{m} \circ \mathcal{E}_t^{-1}$ satisfies, with probability 1, $\mathfrak{m} \circ (e_t^x)^{-1} = \nu^0$.

Using the same argument as in (2.33), we deduce that the filtration $(\mathbb{F}_s^{t,\mathcal{M}})_{t\le s\le T}$ is compatible with the process $(x_t, (w_s^0, \mathfrak{m}_s, w_s)_{t\le s\le T})$, where x_t is defined as $x_t = \psi(\eta, \nu^0)$. In particular, we may regard $(\bar{\Omega}^{0,t}\times\bar{\Omega}^{1,t}, \mathcal{F}^{t,\mathcal{M}}, (\mathcal{F}_s^{t,\mathcal{M}})_{t\le s\le T}, \mathbb{P}^{t,\mathcal{M}})$ equipped with $(x_t, (w_s^0, \mathfrak{m}_s, w_s)_{t\le s\le T})$ as a t-initialized set-up. On this t-initialized set-up, we can consider the forward-backward system:

$$\begin{cases} dx_s = B\big(s, x_s, \nu_s, y_s, z_s\big)ds + \sigma\big(s, x_s, \nu_s\big)dw_s + \sigma^0\big(s, x_s, \nu_s\big)dw_s^0, \\ dy_s = -F\big(s, x_s, \nu_s, y_s, z_s, z_s^0\big)ds + z_s dw_s + z_s^0 dw_s^0 + dm_s, \quad t \le s \le T, \end{cases} \tag{4.9}$$

with $x_t = \psi(\eta, \nu^0)$ as initial condition and with $[\boldsymbol{m}, \boldsymbol{w}]_\cdot \equiv 0$ and $[\boldsymbol{m}, \boldsymbol{w}^0]_\cdot \equiv 0$ on $[t, T]$ and $m_t = 0$.

We then say that $\mathcal{M} \in \mathcal{S}$ if, in addition to 1, 2 and 3 above, we also have:

4. $\mathbb{P}^{t,\mathcal{M}}[\mathfrak{m} = \mathcal{L}^1(\boldsymbol{x}, \boldsymbol{w})] = 1$.

Following Lemma 2.26, condition 4 above may be reformulated as follows:

4'. For any Borel subset $C \subset \mathcal{C}([t, T]; \mathbb{R}^{2d})$ in a countable generating π-system of $\mathcal{B}(\mathcal{C}([t, T]; \mathbb{R}^{2d}))$ and any Borel subset $C^0 \subset \bar{\Omega}^{0,t}$ in a countable generating π-system of $\mathcal{B}(\bar{\Omega}^{0,t})$,

$$\begin{aligned} \mathbb{P}^{t,\mathcal{M}} \circ (\nu^0, \boldsymbol{w}^0, \mathfrak{m}, \boldsymbol{x}, \boldsymbol{w})^{-1}(C^0 \times C) &= \mathbb{E}^{t,\mathcal{M}}\big[\mathbf{1}_C(\boldsymbol{x}, \boldsymbol{w})\mathbf{1}_{C^0}(\nu^0, \boldsymbol{w}^0, \mathfrak{m})\big] \\ &= \int_{\bar{\Omega}^{0,t}} \mathfrak{m}(C)\mathbf{1}_{C^0}(\nu^0, \boldsymbol{w}^0, \mathfrak{m})d\mathcal{M}, \end{aligned}$$

where $\boldsymbol{x} = (x_s)_{t\le s\le T}$.

Clearly, the set $\mathcal{S}$ is the set of distributions of all the equilibria starting from a deterministic initial condition at time $t \in [0, T]$.

Second Step. We now notice that the set of probability measures $\mathcal{M}$ such that point 1 above holds is a closed subset of $\mathcal{P}_2(\bar{\Omega}^{0,t})$. Similarly, recalling that the mapping $\mathcal{P}_2(\mathbb{R}^d) \ni \nu^0 \mapsto \nu^0(B)$ is measurable for any Borel subset $B \in \mathcal{B}(\mathbb{R}^d)$, we deduce that, for any $B \in \mathcal{B}(\mathbb{R}^d)$, the condition (4.8) defines a Borel subset of $\mathcal{P}_2(\bar{\Omega}^{0,t})$. Choosing B in a generating countable π-system of $\mathcal{B}(\mathbb{R}^d)$, we deduce that the set of probability measures $\mathcal{M}$ such that point 2 holds is also a Borel subset of $\mathcal{P}_2(\bar{\Omega}^{0,t})$. Regarding point 3, we notice that the set of pairs $(\nu^0, \mathfrak{m}) \in \mathcal{P}_2(\mathbb{R}^d) \times \mathcal{P}_2(\mathcal{C}([0, T]; \mathbb{R}^{2d}))$, such that $\nu^0 = \mathfrak{m} \circ (e_t^x)^{-1}$ is closed. Therefore, the set of probability measures $\mathcal{M}$ such that point 3 holds is a closed subset of $\mathcal{P}_2(\bar{\Omega}^{0,t})$.

It remains to do the same with point 4'. By uniqueness of the solution to the FBSDE (4.9), we know from a mere adaptation of Proposition 1.31 that the law of $(\nu^0, \boldsymbol{w}^0, \mathfrak{m}, \boldsymbol{x}_s, \boldsymbol{w})$ on $\bar{\Omega}^{0,t} \times [\mathcal{C}([t, T]; \mathbb{R}^d)]^2$ depends on the law $\mathcal{M}$ in a measurable way. Therefore, the set of probability measures $\mathcal{M}$ such that point 4' holds true is a Borel subset of $\mathcal{P}_2(\bar{\Omega}^{0,t})$. Hence, $\mathcal{S}$ is a Borel subset of $\mathcal{P}_2(\bar{\Omega}^{0,t})$.

Now, we consider the mapping:

$$\mathcal{H} : \mathcal{S} \ni \mathcal{M} \mapsto \int_{\bar{\Omega}^{0,t}} \nu^0 d\mathcal{M},$$

where, for any $C \in \mathcal{B}(\mathbb{R}^d)$, $[\int_{\bar{\Omega}^{0,t}} \nu^0 d\mathcal{M}](C) = \int_{\bar{\Omega}^{0,t}} \nu^0(C) d\mathcal{M}$. Weak existence and uniqueness say that, for any $\mu \in \mathcal{P}_2(\mathbb{R}^d)$, there exists a unique $\mathcal{M}$ such that $\mathcal{H}(\mathcal{M}) = \mu$. Put differently, $\mathcal{H}$ is one-to-one and onto. By Proposition 1.32, the mapping $\mathcal{P}_2(\mathbb{R}^d) \ni \mu \mapsto \mathcal{M}^{t,\mu}$ is measurable, where $\mathcal{M}^{t,\mu}$ denotes the weak solution, or equivalently the law of the equilibrium, starting from μ at time t.

Third Step. Now, the claim follows from a new application of Proposition 1.31, which guarantees that the law of $(\boldsymbol{x}^{t,x}, \boldsymbol{y}^{t,x}, \int_t^{\cdot} (z_s^{t,x}, z_s^{0,t,x}) ds, \mathfrak{m})$, regarded as an element of $\mathcal{P}_2(\mathcal{C}([t,T];\mathbb{R}^d) \times \mathcal{D}([t,T];\mathbb{R}^m) \times \mathcal{C}([t,T];\mathbb{R}^{2(m\times d)}) \times \mathcal{P}_2(\mathcal{C}([t,T];\mathbb{R}^{2d})))$, is the image, by a measurable function, of $(x, \mathcal{M}^{t,\mu})$, regarded as an element of $\mathbb{R}^d \times \mathcal{P}_2(\bar{\Omega}^{0,t})$. By the second step, it is also the image, by a measurable function, of (x,μ), regarded as an element of $\mathbb{R}^d \times \mathcal{P}_2(\mathbb{R}^d)$. The proof is easily completed. □

Here is now the first main result of this section.

Proposition 4.3 *Assume that weak existence and uniqueness hold for any initial condition $(t, \mathcal{V}) \in [0,T] \times \mathcal{P}_2(\mathcal{P}_2(\mathbb{R}^d))$ and that assumption* **FBSDE** *is in force. Then, for every MFG solution $(X_0, \boldsymbol{W}^0, \mathfrak{M}, \boldsymbol{W})$ on a set-up $(\Omega, \mathcal{F}, \mathbb{F} = (\mathcal{F}_t)_{0\le t\le T}, \mathbb{P})$ as in Definition 2.16, with $\boldsymbol{X} = (X_t)_{0\le t\le T}$ as optimal path, it holds, for any control $\boldsymbol{\beta} = (\beta_t)_{0\le t\le T}$ constructed on the same set-up and progressively measurable with respect to the complete and right-continuous augmentation of the filtration generated by $(X_0, \boldsymbol{W}^0, \mathfrak{M}, \boldsymbol{W})$, for any $t \in [0,T]$ and $\mathbb{P}$-almost surely:*

$$\mathcal{U}(t, X_t, \mu_t) \le \mathbb{E}\bigg[\int_t^T f(s, X_s^{t,\boldsymbol{\beta}}, \mu_s, \beta_s) ds + g(X_T^{t,\boldsymbol{\beta}}, \mu_T) \,\Big|\, \mathcal{F}_t^{\mathrm{nat},(X_0,\boldsymbol{W}^0,\mathfrak{M},\boldsymbol{W})}\bigg],$$

where $\boldsymbol{X}^{t,\boldsymbol{\beta}} = (X_s^{t,\boldsymbol{\beta}})_{t\le s\le T}$ is the solution of the controlled SDE (4.2) driven by $(\beta_s)_{t\le s\le T}$ and initialized with $X_t^{\boldsymbol{\beta}} = X_t$ at time t.

Furthermore, equality holds when $\boldsymbol{\beta} = (\check{\alpha}(s, X_s, \mu_s, Y_s, Z_s))_{t\le s\le T}$, and the result remains true if the weak solution starts from some initial time $t_0 \ne 0$.

As usual, the control process $\boldsymbol{\beta}$ is required to be A-valued, $\mathbb{F}$-measurable and square-integrable, see Chapter 1. We also refer to **(A1)** in assumption **FBSDE** for the form of $\boldsymbol{X}^{\boldsymbol{\beta}}$.

Proof. Following Definition 2.24 (see also the proof of Lemma 2.30), any weak solution with some $\mathcal{V}$ as initial condition generates a solution $\mathcal{M}$ on the canonical space $\bar{\Omega}^0 = \mathcal{P}_2(\mathbb{R}^d) \times \mathcal{C}([0,T];\mathbb{R}^d) \times \mathcal{P}_2(\mathcal{C}([0,T];\mathbb{R}^{2d}))$. On the enlarged probability space $\bar{\Omega} = \bar{\Omega}^0 \times \bar{\Omega}^1$, with $(\nu^0, \boldsymbol{w}^0, \mathfrak{m}, \eta, \boldsymbol{w})$ as canonical process, equipped with the completion $\bar{\mathbb{P}}$ of the probability measure $\mathcal{M} \otimes (\mathrm{Leb}_1 \otimes \mathcal{W}_d)$ and with the complete and right-continuous augmentation $\bar{\mathbb{F}}$ of the canonical filtration according to the procedure described in Subsection 4.1.3, we can consider the forward-backward system

$$dx_t = B(t, x_t, \nu_t, y_t, z_t)dt + \sigma(t, x_t, \nu_t)dw_t + \sigma^0(t, x_t, \nu_t)dw_t^0,$$

$$dy_t = -F(t, x_t, \nu_t, y_t, z_t, z_t^0)dt + z_t dw_t + z_t^0 dw_t^0 + dm_t,$$

with $x_0 = \psi(\eta, \nu^0)$ as initial condition and $y_T = G(x_T, \nu_T)$ as terminal boundary condition, where $\boldsymbol{m} = (m_t)_{0\le t\le T}$ is a càd-làg martingale with $[\boldsymbol{m}, \boldsymbol{w}]_\cdot \equiv 0$, $[\boldsymbol{m}, \boldsymbol{w}^0]_\cdot \equiv 0$ and $m_0 = 0$.

First Step. We now fix some $t \in [0, T]$. Then, we consider the following regular conditional probability measures on $(\bar{\Omega}, \mathcal{B}(\bar{\Omega}))$:

(a) We call $(\mathbb{P}^{t,\mathcal{B}}_{\omega})_{\omega \in \bar{\Omega}}$ the regular conditional probability measure of $\mathcal{M} \otimes (\text{Leb}_1 \otimes \mathcal{W}_d)$ given the σ-field $\sigma\{\nu^0, w^0_s, \mathfrak{m}_s, \eta, w_s;\ s \le t\}$. For any realization $\omega \in \bar{\Omega}$, we call $(\bar{\Omega}, \mathcal{F}^t_{\omega}, \mathbb{P}^t_{\omega})$ the completion of $(\bar{\Omega}, \mathcal{B}(\bar{\Omega}), \mathbb{P}^{t,\mathcal{B}}_{\omega})$ and then $\mathbb{F}^t_{\omega}$ the complete and right-continuous augmentation of the canonical filtration;

(b) We let $(\mathbb{Q}^{t,\mathcal{B}}_{\omega})_{\omega \in \bar{\Omega}}$ be the regular conditional probability measure of $\mathcal{M} \otimes (\text{Leb}_1 \otimes \mathcal{W}_d)$ given the σ-field $\sigma\{\nu^0, w^0_s, \mathfrak{m}_s;\ s \le t\}$. For any realization $\omega \in \bar{\Omega}$, we call $(\bar{\Omega}, \mathcal{G}^t_{\omega}, \mathbb{Q}^t_{\omega})$ the completion of $(\bar{\Omega}, \mathcal{B}(\bar{\Omega}), \mathbb{Q}^{t,\mathcal{B}}_{\omega})$ and then $\mathbb{G}^t_{\omega}$ the complete and right-continuous augmentation of the canonical filtration.

By a mere adaptation of the proof of Lemma 1.43, we can prove that, for $\bar{\mathbb{P}}$-almost every $\omega \in \bar{\Omega}$, on the space $\bar{\Omega}$ equipped with either $(\mathcal{F}^t_{\omega}, \mathbb{P}^t_{\omega})$ or $(\mathcal{G}^t_{\omega}, \mathbb{Q}^t_{\omega})$, it holds that:

(c) The processes $(\nu^0, \boldsymbol{w}^0, \mathfrak{m})$ and $(\eta, \boldsymbol{w})$ are independent;

(d) The process $(w^0_s - w^0_t, w_s - w_t)_{t \le s \le T}$ forms a $2d$-dimensional Brownian motion with respect to the filtration generated by $(\nu^0, \boldsymbol{w}^0, \mathfrak{m}, \eta, \boldsymbol{w})$, and thus with respect to $\mathbb{F}^t_{\omega}$ or, respectively, $\mathbb{G}^t_{\omega}$;

(e) The process $(x_t, w^0_s - w^0_t, \mathfrak{m}_s, w_s - w_t)_{t \le s \le T}$ is compatible with $(\mathcal{F}^t_{s,\omega})_{t \le s \le T}$ or, respectively, $(\mathcal{G}^t_{s,\omega})_{t \le s \le T}$;

(f) The process $(x_s, y_s, z_s, z^0_s, m_s - m_t)_{t \le s \le T}$ satisfies the FBSDE:

$$\begin{aligned} dx_s &= B(s, x_s, \nu_s, y_s, z_s)ds + \sigma(s, x_s, \nu_s)dw_s + \sigma^0(s, x_s, \nu_s)dw^0_s, \\ dy_s &= -F(s, x_s, \nu_s, y_s, z_s, z^0_s)ds + z_s dw_s + z^0_s dw^0_s + d\big(m_s - m_t\big), \end{aligned} \tag{4.10}$$

for $s \in [t, T]$, with the terminal condition $y_T = G(x_T, \nu_T)$, the process $(m_s - m_t)_{t \le s \le T}$ being a square integrable martingale of zero cross variation with $(w^0_s - w^0_t, w_s - w_t)_{t \le s \le T}$.

Importantly, observe that the law of the initial condition in (4.10) is different under $\mathbb{P}^t_{\omega}$ and $\mathbb{Q}^t_{\omega}$. Indeed, under $\mathbb{P}^t_{\omega}$, it is concentrated on $x_t(\omega)$, namely $\mathbb{P}^t_{\omega}[\omega' \in \bar{\Omega};\ x_t(\omega') = x_t(\omega)] = 1$, while under $\mathbb{Q}^t_{\omega}$, x_t has the distribution $\nu_t(\omega)$. The first claim is quite obvious. We refer to the proof of Lemma 1.43 for a detailed account on the way to handle the fact that x_t is measurable with respect to the completion of the Borel σ-field. The second assertion follows from the more general fact that, for $\bar{\mathbb{P}}$-almost every $\omega \in \bar{\Omega}$, for any $C^0 \in \mathcal{B}(\bar{\Omega}^0)$ and $C \in \mathcal{B}(\mathcal{C}([0, T]; \mathbb{R}^{2d}))$,

$$\mathbb{Q}^t_{\omega}\Big[\big\{(\nu^0, \boldsymbol{w}^0, \mathfrak{m}) \in C^0\big\} \cap \big\{(\boldsymbol{x}, \boldsymbol{w}) \in C\big\}\Big] = \mathbb{E}^{\mathbb{Q}^{\omega}_t}\Big[\mathbf{1}_{C^0}\big(\nu^0, \boldsymbol{w}^0, \mathfrak{m}\big)\mathfrak{m}(C)\Big]. \tag{4.11}$$

In order to prove (4.11), it suffices to notice that, for any $C^{0,t} \in \sigma\{\nu^0, w^0_s, \mathfrak{m}_s;\ s \le t\}$,

$$\begin{aligned} \int_{C^{0,t}} \mathbb{E}^{\mathbb{Q}^t_{\omega}}\big[\mathbf{1}_{C^0}(\nu^0, \boldsymbol{w}^0, \mathfrak{m})\mathbf{1}_C(\boldsymbol{x}, \boldsymbol{w})\big]d\bar{\mathbb{P}}(\omega) &= \mathbb{E}^{\bar{\mathbb{P}}}\Big[\mathbf{1}_{C^{0,t}}\mathbf{1}_{C^0}\big(\nu^0, \boldsymbol{w}^0, \mathfrak{m}\big)\mathbf{1}_C(\boldsymbol{x}, \boldsymbol{w})\Big] \\ &= \mathbb{E}^{\bar{\mathbb{P}}}\Big[\mathbf{1}_{C^{0,t}}\mathbf{1}_{C^0}\big(\nu^0, \boldsymbol{w}^0, \mathfrak{m}\big)\mathfrak{m}(C)\Big] \\ &= \int_{C^{0,t}} \mathbb{E}^{\mathbb{Q}^t_{\omega}}\big[\mathbf{1}_{C^0}\big(\nu^0, \boldsymbol{w}^0, \mathfrak{m}\big)\mathfrak{m}(C)\big]d\bar{\mathbb{P}}(\omega). \end{aligned}$$

Choosing C^0 and C in countable generating π-systems, we easily complete the proof of (4.11).

Second Step. We now return to (e) and (f). They say that, for $\bar{\mathbb{P}}$-almost every $\omega \in \bar{\Omega}$, under $\mathbb{Q}^t_\omega$, $(x_s)_{t\le s\le T}$ is an optimal path of the stochastic optimal control problem (4.1)–(4.2) set on the t-initialized set-up $(\bar{\Omega}, \mathcal{F}^t_{T,\omega}, \mathbb{F}^t_\omega, \mathbb{Q}^t_\omega)$ equipped with $(x_t, (w^0_s - w^0_t, \mathfrak{m}_s, w_s - w_t)_{t\le s\le T})$, provided that, obviously, the running cost is then integrated from t to T in (4.1).

Combined with (4.11), we shall deduce that, under $\mathbb{Q}^t_\omega$, $(\nu_t, w^0_s - w^0_t, \mathfrak{m}_s)_{t\le s\le T}$ is a solution to the MFG problem set on $[t,T]$ instead of $[0,T]$ and with $\nu_t(\omega)$ as initial distribution. Regarding the latter point, notice indeed that, with probability 1 under $\bar{\mathbb{P}}$, $\mathbb{Q}^t_\omega[\omega' \in \bar{\Omega} : \nu_t(\omega') = \nu_t(\omega)] = 1$. In order to fit the exact definition of an equilibrium, we have to check that $\mathbb{Q}^t_\omega$ has a product form. To do so, we call $(\mathbb{P}^{0,t,\mathcal{B}}_{\omega^0})_{\omega^0\in\bar{\Omega}^0}$ the regular conditional probability measure on $(\bar{\Omega}^0, \mathcal{B}(\bar{\Omega}^0))$ of $\mathcal{M}$ given $\sigma\{\nu^0, w^0_s, \mathfrak{m}_s;\, s \le t\}$. It is easily checked that, for $\bar{\mathbb{P}}$-almost every $\omega \in \bar{\Omega}$, $\mathbb{Q}^{t,\mathcal{B}}_\omega = \mathbb{P}^{0,t,\mathcal{B}}_{\omega^0} \otimes (\mathrm{Leb}_1 \otimes \mathcal{W}_d)$. In particular, calling $(\bar{\Omega}^0, \mathcal{F}^{0,t}_{\omega^0}, \mathbb{P}^{0,t}_{\omega^0})$ the completion of $(\bar{\Omega}^0, \mathcal{B}(\bar{\Omega}^0), \mathbb{P}^{0,t,\mathcal{B}}_{\omega^0})$, for any realization $\omega^0 \in \bar{\Omega}^0$, we recover the setting used in the Definition 2.16 of an MFG solution.

Third Step. Consider now a square-integrable and $(\bar{\mathcal{F}}_s)_{t\le s\le T}$-progressively measurable control process $\bar{\boldsymbol{\beta}} = (\bar{\beta}_s)_{t\le s\le T}$ and then denote by $\boldsymbol{x}^{\bar{\boldsymbol{\beta}}} = (x^{\bar{\boldsymbol{\beta}}}_s)_{t\le s\le T}$ the solution, on $\bar{\Omega}$ equipped with $(\bar{\mathbb{F}}, \bar{\mathbb{P}})$, of the SDE:

$$dx^{\bar{\boldsymbol{\beta}}}_s = b\big(s, x^{\bar{\boldsymbol{\beta}}}_s, \nu_s, \bar{\beta}_s\big)ds + \sigma\big(s, x^{\bar{\boldsymbol{\beta}}}_s, \nu_s\big)dw_s + \sigma^0\big(s, x^{\bar{\boldsymbol{\beta}}}_s, \nu_s\big)dw^0_s, \quad s \in [t,T],$$

with $x^{\bar{\boldsymbol{\beta}}}_t = x_t$ as initial condition.

For a while, we assume $\bar{\boldsymbol{\beta}}$ to be progressively measurable with respect to the filtration generated by $(x_0, \boldsymbol{w}^0, \mathfrak{m}, \boldsymbol{w})$. Following the proof of Lemma 1.40, it is clear that, for $\bar{\mathbb{P}}$-almost every $\omega \in \bar{\Omega}$, $(x^{\bar{\boldsymbol{\beta}}}_s, \bar{\beta}_s)_{t\le s\le T}$ solves, under the probability measure $\mathbb{P}^t_\omega$, the SDE:

$$dx^{\bar{\boldsymbol{\beta}}}_s = b\big(s, x^{\bar{\boldsymbol{\beta}}}_s, \nu_s, \bar{\beta}_s\big)ds + \sigma\big(s, x^{\bar{\boldsymbol{\beta}}}_s, \nu_s\big)dw_s + \sigma^0\big(s, x^{\bar{\boldsymbol{\beta}}}_s, \nu_s\big)dw^0_s, \quad s \in [t,T],$$

with initial condition $x_t(\omega)$ in the sense that $\mathbb{P}^t_\omega[\omega' \in \bar{\Omega};\, x^{\bar{\boldsymbol{\beta}}}_t(\omega') = x_t(\omega)] = 1$. In particular, $(x^{\bar{\boldsymbol{\beta}}}_s)_{t\le s\le T}$ can be regarded as a controlled path starting from $x_t(\omega)$ on the t-initialized set-up $(\bar{\Omega}, \mathcal{F}^t_{\omega,T}, (\mathcal{F}^t_{\omega,s})_{t\le s\le T}, \mathbb{P}^t_\omega)$ equipped with $(w^0_s - w^0_t, \mathfrak{m}_s, w_s - w_t)_{t\le s\le T}$.

We know that the optimal path for the stochastic optimal control problem (4.1)–(4.2) starting from $x_t(\omega)$ on the t-initialized set-up $(\bar{\Omega}, \mathcal{F}^t_\omega, (\mathcal{F}^t_{\omega,s})_{t\le s\le T}, \mathbb{P}^t_\omega)$ equipped with $(w^0_s - w^0_t, \mathfrak{m}_s, w_s - w_t)_{t\le s\le T}$ is given by the solution of the FBSDE (4.10), so:

$$\mathbb{E}^{\mathbb{P}^t_\omega}\bigg[\int_t^T f\big(s, x_s, \nu_s, \hat{\alpha}_s\big)ds + g(x_T, \nu_T)\bigg] \le \mathbb{E}^{\mathbb{P}^t_\omega}\bigg[\int_t^T f\big(s, x^{\bar{\boldsymbol{\beta}}}_s, \nu_s, \bar{\beta}_s\big)ds + g(x^{\bar{\boldsymbol{\beta}}}_T, \nu_T)\bigg], \tag{4.12}$$

where $\hat{\alpha}_s = \breve{\alpha}(s, x_s, \nu_s, y_s, z_s)$.

Fourth Step. By independence of $(\nu^0, \boldsymbol{w}^0, \mathfrak{m})$ and $(\eta, \boldsymbol{w})$ under $\bar{\mathbb{P}}$, we observe that, for almost every $\omega \in \bar{\Omega}$ under $\bar{\mathbb{P}}$, the marginal laws of $(\nu^0, \boldsymbol{w}^0, \mathfrak{m})$ under $\mathbb{P}^t_\omega$ and $\mathbb{Q}^t_\omega$ are the same. In particular, under $\mathbb{P}^t_\omega$, $(\nu_t, w^0_s - w^0_t, \mathfrak{m}_s)_{t\le s\le T}$ has the law $\mathcal{M}^{t,\nu_t(\omega)}$ of the MFG equilibrium with $\nu_t(\omega)$ as initial condition, where we used the same notation as in Definition 4.1. Therefore, under $\mathbb{P}^t_\omega$, the input $(\nu_t, w^0_s - w^0_t, \mathfrak{m}_s, w_s - w_t)_{t\le s\le T}$ of the FBSDE (4.10) is distributed according to $\mathcal{M}^{t,\nu_t(\omega)} \otimes \mathcal{W}^t_d$. We deduce that the left-hand side in (4.12) coincides with $\mathcal{U}(t, x_t(\omega), \nu_t(\omega))$.

Moreover, we know that the right-hand side may be identified with the conditional expectation:

$$\mathbb{E}^{\bar{\mathbb{P}}}\left[\int_t^T f(s, x_s^{\bar{\beta}}, \nu_s, \bar{\beta}_s)ds + g(x_T^{\bar{\beta}}, \nu_T) \,\Big|\, \sigma\{\nu^0, w_s^0, \mathfrak{m}_s, \eta, w_s;\ s \le t\}\right],$$

evaluated at ω. Indeed, for a random variable χ from $(\bar{\Omega}, \bar{\mathcal{F}}_T)$ into $\mathbb{R}$, we can find a random variable $\tilde{\chi}$ from $(\bar{\Omega}, \mathcal{B}(\bar{\Omega}))$ into $\mathbb{R}$ such that $\bar{\mathbb{P}}[\chi = \tilde{\chi}] = 1$. From the proof of Lemma 1.43, we learn that, for $\bar{\mathbb{P}}$-almost every $\omega \in \bar{\Omega}$, $\tilde{\chi}$ is $\mathcal{F}_{\omega,T}^t$-measurable and $\mathbb{P}_\omega^t[\chi = \tilde{\chi}] = 1$. Therefore, for $\bar{\mathbb{P}}$-almost every $\omega \in \bar{\Omega}$,

$$\begin{aligned}
\mathbb{E}^{\bar{\mathbb{P}}}\big[\chi \,|\, \sigma\{\nu^0, w_s^0, \mathfrak{m}_s, \eta, w_s;\ s \le t\}\big](\omega) &= \mathbb{E}^{\bar{\mathbb{P}}}\big[\tilde{\chi} \,|\, \sigma\{\nu^0, w_s^0, \mathfrak{m}_s, \eta, w_s;\ s \le t\}\big](\omega) \\
&= \mathbb{E}^{\mathbb{P}_\omega^t}\big[\tilde{\chi}\big] = \mathbb{E}^{\mathbb{P}_\omega^t}\big[\chi\big].
\end{aligned}$$

Now, using the fact that $\mathcal{U}(t,\cdot,\cdot)$ is measurable, by conditioning (4.12) with respect to the σ-field $\sigma\{x_0, w_s^0, \mathfrak{m}_s, w_s;\ s \le t\} \subset \sigma\{x_0, w_s^0, \mathfrak{m}_s, \eta, w_s;\ s \le t\}$, we get:

$$\begin{aligned}
&\mathbb{E}^{\bar{\mathbb{P}}}\Big[\mathcal{U}\big(t, x_t, \nu_t\big) \,|\, \sigma\big\{x_0, w_s^0, \mathfrak{m}_s, w_s;\ s \le t\big\}\Big] \\
&\quad \le \mathbb{E}^{\bar{\mathbb{P}}}\left[\int_t^T f(s, x_s^{\bar{\beta}}, \nu_s, \bar{\beta}_s)ds + g(x_T^{\bar{\beta}}, \nu_T) \,\Big|\, \sigma\big\{x_0, w_s^0, \mathfrak{m}_s, w_s;\ s \le t\big\}\right].
\end{aligned}$$

Last Step. By progressive measurability with respect to the filtration generated by the process $(x_0, \boldsymbol{w}^0, \mathfrak{m}, \boldsymbol{w})$, there exists a progressively measurable mapping $\mathfrak{b}$ from the product space $\mathbb{R}^d \times \mathcal{C}([0,T];\mathbb{R}^d) \times \mathcal{P}_2(\mathcal{C}([0,T];\mathbb{R}^{2d})) \times \mathcal{C}([0,T];\mathbb{R}^d)$ equipped with the canonical filtration into A such that $\bar{\boldsymbol{\beta}} = \mathfrak{b}(x_0, \boldsymbol{w}^0, \mathfrak{m}, \boldsymbol{w})$. It is thus quite straightforward to transfer $\bar{\boldsymbol{\beta}}$ into some $\boldsymbol{\beta} = (\beta_s)_{t \le s \le T}$ onto the original space $(\Omega, \mathcal{F}, \mathbb{F}, \mathbb{P})$. Then, we may denote $\boldsymbol{X}^{\boldsymbol{\beta}} = (X_s^{\boldsymbol{\beta}})_{t \le s \le T}$ the solution to the controlled SDE (4.2) on the original space $(\Omega, \mathcal{F}, \mathbb{F}, \mathbb{P})$, with $X_t^{\boldsymbol{\beta}} = X_t$ as initial condition at time t. Since the solution may be constructed by a Picard iteration, it is quite straightforward to prove that $(X_0, \boldsymbol{W}^0, \mathfrak{M}, \boldsymbol{W}, \int_t^\cdot \beta_s ds, \boldsymbol{X}^{\boldsymbol{\beta}})$ and $(x_0, \boldsymbol{w}^0, \mathfrak{m}, \boldsymbol{w}, \int_t^\cdot \bar{\beta}_s ds, \boldsymbol{x}^{\bar{\boldsymbol{\beta}}})$ have the same law. We deduce that:

$$\begin{aligned}
&\mathbb{E}\Big[\mathcal{U}(t, X_t, \mu_t) \,\big|\, \mathcal{F}_t^{\mathrm{nat},(X_0, \boldsymbol{W}^0, \mathfrak{M}, \boldsymbol{W})}\Big] \\
&\quad \le \mathbb{E}\left[\int_t^T f(s, X_s^{\boldsymbol{\beta}}, \mu_s, \beta_s)ds + g(X_T^{\boldsymbol{\beta}}, \mu_T) \,\Big|\, \mathcal{F}_t^{\mathrm{nat},(X_0, \boldsymbol{W}^0, \mathfrak{W}, \boldsymbol{W})}\right].
\end{aligned}$$

Since X_t is measurable with respect to the completion of $\mathcal{F}_t^{\mathrm{nat},(X_0, \boldsymbol{W}^0, \mathfrak{M}, \boldsymbol{W})}$ under $\mathbb{P}$, this completes the proof of the upper bound when $(\beta_s)_{t \le s \le T}$ is progressively measurable with respect to the filtration generated by $(x_0, \boldsymbol{w}^0, \mathfrak{m}, \boldsymbol{w})$.

When $\boldsymbol{\beta} = (\beta_s)_{t \le s \le T}$ is general, we claim that we can find a process $\tilde{\boldsymbol{\beta}} = (\tilde{\beta}_s)_{t \le s \le T}$, progressively measurable with respect to the filtration generated by $(X_0, \boldsymbol{W}^0, \mathfrak{M}, \boldsymbol{W})$, such that $\tilde{\boldsymbol{\beta}}$ and $\boldsymbol{\beta}$ are almost everywhere equal under $\mathrm{Leb}_1 \otimes \bar{\mathbb{P}}$, which suffices to complete the proof of the upper bound.

From the interpretation of the left-hand side in (4.12), we deduce that the equality holds when $\boldsymbol{\beta} = (\breve{\alpha}(s, X_s, \mu_s, Y_s, Z_s))_{t \le s \le T}$. □

Corollary 4.4 *Assume that weak existence and uniqueness hold for any initial condition* $(t, \mathcal{V}) \in [0, T] \times \mathcal{P}_2(\mathcal{P}_2(\mathbb{R}^d))$ *and that assumption* **FBSDE** *is in force. Then, for any probability space* $(\Omega, \mathcal{F}, \mathbb{P})$ *equipped with two random variables* μ *and* ξ*, with values in* $\mathcal{P}_2(\mathbb{R}^d)$ *and* $\mathbb{R}^d$ *respectively,* μ *having distribution* $\mathcal{V}$ *and* ξ *being square-integrable, it holds:*

$$\mathbb{E}\big[|\mathcal{U}(t, \xi, \mu)|\big] < \infty.$$

Proof. As usual, we call $\mathcal{M}$ the law of the equilibrium initialized with the law $\mathcal{V}$ at time t. We also call $(q(\nu, \cdot))_{\nu \in \mathcal{P}_2(\mathbb{R}^d)}$ the conditional law of ξ given the realization of μ.

Without any loss of generality, we assume that $t = 0$. We then use the same notation as in the proof of Proposition 4.3. On $\bar{\Omega}$, equipped with the completion $\bar{\mathbb{P}}$ of the product measure $\mathcal{M} \otimes (\mathrm{Leb}_1 \otimes \mathcal{W}_d)$ on $\bar{\Omega}^0 \times \bar{\Omega}^1$, we consider the random variable $x_0 = \psi(\eta, q(\nu^0, \cdot))$. By definition of ψ in (2.23), $\mathcal{L}^1(x_0)$ is distributed according to $q(\nu^0, \cdot)$. Therefore, the pair (ν^0, x_0) is distributed according to $\mathbb{P} \circ (\mu, \xi)^{-1}$.

Now, we consider the solution $(\boldsymbol{x}, \boldsymbol{y}, \boldsymbol{z}, \boldsymbol{z}^0, \boldsymbol{m})$ of the FBSDE (4.10) with x_0 as initial condition. By repeating the arguments in the proof of Proposition 4.3, we see that, for $\bar{\mathbb{P}}$-almost every $\omega \in \bar{\Omega}$, under $\mathbb{P}^t_\omega$, with $t = 0$,

$$\mathcal{U}\big(0, x_0(\omega), \nu^0(\omega)\big) = \mathbb{E}^{\mathbb{P}^0_\omega}\bigg[\int_0^T f\big(s, x_s, \nu_s, \hat{\alpha}_s\big)ds + g(x_T, \nu_T)\bigg],$$

so that:

$$\big|\mathcal{U}\big(0, x_0(\omega), \nu^0(\omega)\big)\big| \le \mathbb{E}^{\mathbb{P}^0_\omega}\bigg[\int_0^T \big|f\big(s, x_s, \nu_s, \hat{\alpha}_s\big)\big|ds + \big|g(x_T, \nu_T)\big|\bigg].$$

Owing to the growth properties of f and g and to the integrability conditions of the processes $\boldsymbol{x}$, $\boldsymbol{\nu}$ and $\hat{\boldsymbol{\alpha}}$, the right-hand side is integrable, which completes the proof. □

The following is the main result of this section.

Theorem 4.5 *Assume that weak existence and uniqueness hold for any initial condition* $(t, \mathcal{V}) \in [0, T] \times \mathcal{P}_2(\mathcal{P}_2(\mathbb{R}^d))$ *and that assumption* **FBSDE** *is in force. Then, for every solution* $(X_0, \boldsymbol{W}^0, \mathfrak{M}, \boldsymbol{W})$ *on a set-up* $(\Omega, \mathcal{F}, \mathbb{F} = (\mathcal{F}_t)_{0 \le t \le T}, \mathbb{P})$ *as in Definition 2.16, with* $\boldsymbol{X} = (X_t)_{0 \le t \le T}$ *as optimal path, it holds, for any control* $\boldsymbol{\beta} = (\beta_t)_{0 \le t \le T}$ *constructed on the same set-up and progressively measurable with respect to the complete and right-continuous augmentation of the filtration generated by* $(X_0, \boldsymbol{W}^0, \mathfrak{M}, \boldsymbol{W})$*, for any* $0 \le t \le t + h \le T$ *and* $\mathbb{P}$*-almost surely:*

$$\begin{aligned}&\mathcal{U}\big(t, X_t, \mu_t\big)\\&= \mathbb{E}\bigg[\int_t^{t+h} f(s, X_s^{\boldsymbol{\beta}}, \mu_s, \beta_s)ds + \mathcal{U}(t+h, X_{t+h}^{\boldsymbol{\beta}}, \mu_{t+h})\,\Big|\, \mathcal{F}_t^{\mathrm{nat},(X_0, \boldsymbol{W}^0, \mathfrak{M}, \boldsymbol{W})}\bigg],\end{aligned}$$

where $\boldsymbol{X}^{t,\boldsymbol{\beta}} = (X_s^{t,\boldsymbol{\beta}})_{t \le s \le T}$ *is the solution of the controlled SDE (4.2) driven by* $(\beta_s)_{t \le s \le T}$ *and initialized with* $X_t^{\boldsymbol{\beta}} = X_t$ *at time* t.

Moreover, equality holds when $\boldsymbol{\beta} = (\check{\alpha}(s, X_s, \mu_s, Y_s, Z_s))_{t \le s \le T}$, *and the result remains true if the weak solution starts from some initial time* $t_0 \neq 0$.

Proof. The proof is based on the arguments used to show Proposition 4.3. Naturally, we use the same notation.

First Step. We first work on the canonical space $\bar{\Omega}$. Recalling the interpretation of the left-hand side in (4.12), we have, for $\bar{\mathbb{P}}$-almost every $\omega \in \bar{\Omega}$:

$$\begin{aligned}\mathcal{U}\big(t, x_t(\omega), \nu_t(\omega)\big) &= \mathbb{E}^{\mathbb{P}^t_\omega}\bigg[\int_t^T f\big(s, x_s, \nu_s, \hat{\alpha}_s\big)ds + g(x_T, \nu_T)\bigg] \\ &= \mathbb{E}^{\mathbb{P}^t_\omega}\bigg[\int_t^{t+h} f\big(s, x_s, \nu_s, \hat{\alpha}_s\big)ds + \int_{t+h}^T f\big(s, x_s, \nu_s, \hat{\alpha}_s\big)ds + g(x_T, \nu_T)\bigg].\end{aligned}$$

Now, by taking conditional expectation given $\sigma\{\nu^0, w^0_s, \mathfrak{m}_s, \eta, w_s;\ s \le t+h\}$ inside the expectation in the right-hand side, we deduce from Proposition 4.3 that:

$$\mathcal{U}\big(t, x_t(\omega), \nu_t(\omega)\big) \ge \mathbb{E}^{\mathbb{P}^t_\omega}\bigg[\int_t^{t+h} f\big(s, x_s, \nu_s, \hat{\alpha}_s\big)ds + \mathcal{U}(t+h, x_{t+h}, \nu_{t+h})\bigg],$$

which shows the lower bound in the statement, as the above can be transferred onto the original space by the same argument as in the proof of Proposition 4.3.

Second Step. To prove the converse inequality, we shall work under the conditional probability $\mathbb{P}^{t+h}_\omega$. Given some control $\boldsymbol{\beta} = (\beta_s)_{t \le s \le T}$, progressively measurable with respect to the canonical filtration generated by $(x_0, \boldsymbol{w}^0, \mathfrak{m}, \boldsymbol{w})$, we call $(x^{\boldsymbol{\beta}}_s)_{t \le s \le T}$ the solution of:

$$dx^{\boldsymbol{\beta}}_s = b\big(s, x^{\boldsymbol{\beta}}_s, \nu_s, \beta_s\big)ds + \sigma\big(s, x^{\boldsymbol{\beta}}_s, \nu_s\big)dw_s + \sigma^0\big(s, x^{\boldsymbol{\beta}}_s, \nu_s\big)dw^0_s, \quad t \le s \le T,$$

with $x^{\boldsymbol{\beta}}_t = x_t$ as initial condition. On $(\bar{\Omega}, \mathcal{F}, (\mathcal{F}_s)_{t+h \le s \le T}, \bar{\mathbb{P}})$ equipped with the $t+h$ initialized set-up $(\mathcal{F}^{\mathrm{nat},(\nu^0,\boldsymbol{w}^0,\mathfrak{m},\eta,\boldsymbol{w})}_{t+h}, (w^0_s - w^0_{t+h}, \mathfrak{m}_s, w_s - w_t)_{t+h \le s \le T}), \mathcal{F}^{\mathrm{nat},(\nu^0,\boldsymbol{w}^0,\mathfrak{m},\eta,\boldsymbol{w})}_{t+h}$ here playing the role of the initial information, see Chapter 1, we then solve the same FBSDE as in (4.10), but with a different initial condition at time $t+h$ instead of t, namely

$$\begin{aligned}dx^{t+h}_s &= B\big(s, x^{t+h}_s, \nu_s, y^{t+h}_s, z^{t+h}_s\big)ds + \sigma\big(s, x^{t+h}_s, \nu_s\big)dw_s + \sigma^0\big(s, x^{t+h}_s, \nu_s\big)dw^0_s, \\ dy^{t+h}_s &= -F\big(s, x^{t+h}_s, \nu_s, y^{t+h}_s, z^{t+h}_s, z^{0,t+h}_s\big)ds + z^{t+h}_s dw_s + z^{0,t+h}_s dw^0_s + dm^{t+h}_s,\end{aligned}$$

for $s \in [t+h, T]$, with $x^{t+h}_{t+h} = x^{\boldsymbol{\beta}}_{t+h}$ as initial condition and $y^{t+h}_T = G(x^{t+h}_T, \nu_T)$ as terminal condition, where $[\bar{\boldsymbol{m}}^{t+h}, \boldsymbol{w}]_\cdot \equiv 0$, $[\bar{\boldsymbol{m}}^{t+h}, \boldsymbol{w}^0]_\cdot \equiv 0$ and $m^{t+h}_{t+h} = 0$. Then, as in the first step of the proof of Proposition 4.3, we know that, for $\bar{\mathbb{P}}$-almost every $\omega \in \bar{\Omega}$, the above FBSDE still holds true on the $(t+h)$-initialized set-up $(\bar{\Omega}, \mathcal{F}^{t+h}_\omega, \mathbb{F}^{t+h}_\omega, \mathbb{P}^{t+h}_\omega)$ equipped with $(w^0_s - w^0_{t+h}, \mathfrak{m}_s, w_s - w_t)_{t+h \le s \le T}$. The initial condition is given by $\mathbb{P}^{t+h}_\omega[\omega' \in \bar{\Omega} : x^{t+h}_{t+h}(\omega') = x^{\boldsymbol{\beta}}_{t+h}(\omega)] = 1$. Following once again the interpretation of the left-hand side in (4.12), we know that:

$$\mathcal{U}\big(t+h, x^{\boldsymbol{\beta}}_{t+h}(\omega), \nu_{t+h}(\omega)\big) = \mathbb{E}^{\mathbb{P}^{t+h}_\omega}\bigg[\int_{t+h}^T f\big(s, x^{t+h}_s, \nu_s, \hat{\alpha}^{t+h}_s\big)ds + g\big(x^{t+h}_T, \nu_T\big)\bigg],$$

with $\hat{\alpha}_s^{t+h} = \breve{\alpha}(s, x_s^{t+h}, \nu_s, y_s^{t+h}, z_s^{t+h})$. Now, we let:

$$(\bar{x}_s, \bar{\alpha}_s) = \begin{cases} (x_s^{\beta}, \beta_s), & s \in [t, t+h), \\ (x_s^{t+h}, \hat{\alpha}_s^{t+h}), & s \in [t+h, T]. \end{cases}$$

The process $(\bar{x}_s)_{t \le s \le T}$ is a controlled process, controlled by $(\bar{\alpha}_s)_{t \le s \le T}$. Applying Proposition 4.3 and making use of the above identity, we deduce that:

$$\begin{aligned} \mathcal{U}\big(t, x_t(\omega), \nu_t(\omega)\big) &\le \mathbb{E}^{\mathbb{P}^t_\omega}\bigg[\int_{t+h}^{T} f(s, \bar{x}_s, \nu_s, \bar{\alpha}_s) ds + \int_t^{t+h} f(s, \bar{x}_s, \nu_s, \bar{\alpha}_s) ds + g(\bar{x}_T, \nu_T)\bigg] \\ &= \mathbb{E}^{\mathbb{P}^t_\omega}\bigg[\int_t^{t+h} f(s, x_s^{\beta}, \nu_s, \beta_s) ds + \mathcal{U}(t+h, x_{t+h}^{\beta}, \nu_{t+h})\bigg], \end{aligned}$$

from which we complete the proof by transferring the inequality onto the original space by the same argument as above. □

Remark 4.6 *We shall not discuss the case when $\boldsymbol{\beta}$ is just progressively measurable with respect to the larger filtration $\mathbb{F}$, as the current version of Theorem 4.5 will suffice for our purpose. This would require to extend the canonical space with an additional factor carrying the control process $\bar{\boldsymbol{\beta}}$.*

4.2 Master Field and Optimal Feedback

Our purpose here is to shed new light on the connection between the master field $\mathcal{U}$ and the optimal control $(\breve{\alpha}(s, x_s^{t,x}, \nu_s, y_s^{t,x}, z_s^{t,x}))_{t \le s \le T}$ appearing in the Definition 4.1 of $\mathcal{U}$. Recall that since σ and σ^0 are assumed to be independent of the control parameter α, in the two most important cases of interest, the function $\breve{\alpha}$ is given explicitly by formula (2.26) in terms of the minimizer $\hat{\alpha}$ of the reduced Hamiltonian:

$$H^{(r)}(t, x, \mu, y, \alpha) = b(t, x, \mu, \alpha) \cdot y + f(t, x, \mu, \alpha),$$

for $x, y \in \mathbb{R}^d$, $\mu \in \mathcal{P}_2(\mathbb{R}^d)$, $\alpha \in A$, namely:

$$\hat{\alpha}(t, x, \mu, y) = \mathrm{argmin}_{\alpha \in A} H^{(r)}(t, x, \mu, y, \alpha). \tag{4.13}$$

From classical results of the standard theory of stochastic optimal control, we expect that the optimal control of the optimization problem (4.1) in the environment $\boldsymbol{\mu} = (\mu_t)_{0 \le t \le T}$ is given by a feedback function which, in the present situation, should be of the form:

$$[0, T] \times \mathbb{R}^d \ni (t, x) \mapsto \hat{\alpha}\big(t, x, \mu_t, \partial_x U^{\boldsymbol{\mu}}(t, x)\big), \tag{4.14}$$

where U^{μ} stands for the value function in the environment μ as in the relationship (4.4), see for instance Proposition 1.55. As a result, once we choose to work on a specific set-up, the value of the optimal control $(\hat{\alpha}_t)_{0\le t\le T}$ should be given by:

$$\hat{\alpha}_t = \breve{\alpha}\big(t, X_t, \mu_t, Y_t, Z_t\big) = \hat{\alpha}\big(t, X_t, \mu_t, \partial_x\mathcal{U}(t, X_t, \mu_t)\big), \tag{4.15}$$

since we expect from the relationship (4.4) that:

$$\partial_x U^{\mu}(t,x) = \partial_x\mathcal{U}(t,x,\mu). \tag{4.16}$$

The objective of this section is to make the relationships (4.15) and (4.16) rigorous.

4.2.1 Master Field and Pontryagin Stochastic Maximum Principle

In order to establish (4.16), we shall make use of the necessary part of the Pontryagin stochastic maximum principle, as given for example in Theorem 1.59. But first, we recast assumption **Necessary SMP in Random Environment** of Chapter 1 for the present context.

Assumption (Necessary SMP Master).

(A1) The functions b and f are differentiable with respect to (x,α), the mappings $\mathbb{R}^d \times A \ni (x,\alpha) \mapsto \partial_x(b,f)(t,x,\mu,\alpha)$ and $\mathbb{R}^d \times A \ni (x,\alpha) \mapsto \partial_\alpha(b,f)(t,x,\mu,\alpha)$ being continuous for each $(t,\mu) \in [0,T] \times \mathcal{P}_2(\mathbb{R}^d)$. Similarly, the functions σ, σ^0 and g are differentiable with respect to x, the mapping $\mathbb{R}^d \ni x \mapsto \partial_x(\sigma,\sigma^0)(t,x,\mu)$ being continuous for each $(t,\mu) \in [0,T] \times \mathcal{P}_2(\mathbb{R}^d)$, and the mapping $\mathbb{R}^d \ni (x,\mu) \mapsto \partial_x g(x,\mu)$ being continuous for each $\mu \in \mathcal{P}_2(\mathbb{R}^d)$.

(A2) The function $[0,T] \ni t \mapsto (b,\sigma,\sigma^0,f)(t,0,\delta_0,0_A)$ is uniformly bounded, for some point $0_A \in A$. The derivatives $\partial_x(b,\sigma,\sigma^0)$ and $\partial_\alpha b$ are uniformly bounded. There exists a constant L such that, for any $R \ge 0$ and any (t,x,μ,α) with $|x| \le R$, $M_2(\mu) \le R$ and $|\alpha| \le R$, $|\partial_x f(t,x,\mu,\alpha)|$, $|\partial_x g(x,\mu)|$ and $|\partial_\alpha f(t,x,\mu,\alpha)|$ are bounded by $L(1+R)$.

In addition, we shall also require the following condition on the Hamiltonian.

(A3) For any $(t,x,\mu,y) \in [0,T] \times \mathbb{R}^d \times \mathcal{P}_2(\mathbb{R}^d) \times \mathbb{R}^d$, the function $A \ni \alpha \mapsto H^{(r)}(t,x,\mu,y,\alpha)$ is convex in the variable α, and admits a unique minimizer $\hat{\alpha}(t,x,\mu,y)$. It satisfies, for any $\beta \in A$,

$$\partial_\alpha H^{(r)}\big(t,x,\mu,y,\hat{\alpha}(t,x,\mu,y)\big)\cdot\big(\beta - \hat{\alpha}(t,x,\mu,y)\big) \ge 0. \tag{4.17}$$

Again, since σ and σ^0 are independent of α, condition (**A3**) would be exactly the same if we were to write it in terms of the full Hamiltonian instead of the reduced Hamiltonian. We reformulate Theorem 1.59 in the following way.

Proposition 4.7 *Let the assumptions and notations of Definition 4.1 be in force, and let us assume further that assumption* **Necessary SMP Master** *holds. Then, for any $t \in [0, T)$ and $x \in \mathbb{R}^d$, we must have:*

$$\check{\alpha}(s, x_s^{t,x}, \nu_s, y_s^{t,x}, z_s^{t,x}) = \hat{\alpha}(s, x_s^{t,x}, \nu_s, \upsilon_s^{t,x}), \quad ds \otimes d\mathbb{P}^{t,\mu} \text{ a.e.},$$

where, on the t-initialized set-up $(\bar{\Omega}^t, \mathcal{F}^{t,\mu}, \mathbb{F}^{t,\mu}, \mathbb{P}^{t,\mu})$ equipped with $(\boldsymbol{w}^0, \mathfrak{m}, \boldsymbol{w})$, the process $\boldsymbol{\upsilon}^{t,x} = (\upsilon_s^{t,x})_{t \le s \le T}$ solves, the backward equation:

$$\begin{aligned} d\upsilon_s^{t,x} &= -\partial_x H\big(s, x_s^{t,x}, \nu_s, \upsilon_s^{t,x}, \zeta_s^{t,x}, \zeta_s^{0,t,x}, \check{\alpha}(s, x_s^{t,x}, \nu_s, y_s^{t,x}, z_s^{t,x})\big) ds \\ &\quad + \zeta_s^{t,x} dw_s + \zeta_s^{0,t,x} dw_s^0 + dn_s^{t,x}, \end{aligned} \tag{4.18}$$

with $\upsilon_T^{t,x} = \partial_x g(x_T^{t,x}, \nu_T)$, where $(n_s^{t,x})_{t \le s \le T}$ is a martingale with respect to $\mathbb{F}^{t,\mu}$ with $[\boldsymbol{n}^{t,x}, \boldsymbol{w}^0]_\cdot \equiv 0$, $[\boldsymbol{n}^{t,x}, \boldsymbol{w}]_\cdot \equiv 0$ and $n_t^{t,x} = 0$. In particular,

$$\mathbb{E}^{t,\mu}\Big[\sup_{t \le s \le T} |\upsilon_s^{t,x}|^2\Big] < \infty.$$

It is important for practical purposes that the mapping $\mathbb{R}^d \ni x \mapsto \mathbb{E}^{t,\mu}[\upsilon^{t,x}]$ is continuous. The following lemma shows that this is indeed the case provided that the solution $(\boldsymbol{x}^{t,x}, \boldsymbol{y}^{t,x}, \boldsymbol{z}^{t,x}, \boldsymbol{z}^{0,t,x}, \boldsymbol{m}^{t,x})$ to (4.7) with $\xi = x$ is continuous (in a suitable sense) with respect to x.

Lemma 4.8 *Under the same assumptions and notations as in Proposition 4.7, there exists a constant $C \ge 0$ such that, for any pair $(t, \mu) \in [0, T] \times \mathbb{R}^d$ and any $x, x' \in \mathbb{R}^d$,*

$$\begin{aligned} &\big|\mathbb{E}^{t,\mu}[\upsilon_t^{t,x}] - \mathbb{E}^{t,\mu}[\upsilon_t^{t,x'}]\big|^2 \\ &\le C\bigg[\mathbb{E}^{t,\mu}\int_t^T \big|\partial_x H\big(s, x_s^{t,x}, \nu_s, \upsilon_s^{t,x}, \zeta_s^{t,x}, \zeta_s^{0,t,x}, \check{\alpha}(s, x_s^{t,x}, \nu_s, y_s^{t,x}, z_s^{t,x})\big) \\ &\qquad\qquad - \partial_x H\big(s, x_s^{t,x'}, \nu_s, \upsilon_s^{t,x}, \zeta_s^{t,x}, \zeta_s^{0,t,x}, \check{\alpha}(s, x_s^{t,x'}, \nu_s, y_s^{t,x'}, z_s^{t,x'})\big)\big|^2 ds \\ &\qquad + \mathbb{E}^{t,\mu}\Big[\big|\partial_x g(x_T^{t,x}, \nu_T) - \partial_x g(x_T^{t,x'}, \nu_T)\big|^2\Big]\bigg]. \end{aligned}$$

Proof. The proof just follows from standard stability estimates for BSDEs similar to those used in Example 1.20. They are based on the fact that, by boundedness of $\partial_x b$, $\partial_x \sigma$ and $\partial_x \sigma^0$, the auxiliary BSDE (4.18) has a Lipschitz continuous driver in υ, ζ, and ζ^0. $\square$

Remark 4.9 *From* (4.15), *we may expect $\upsilon_t^{t,x}$ to be almost surely constant under $\mathbb{P}^{t,\mu}$. We shall return to this point in Corollary 4.11 below.*

4.2.2 Space Derivative of the Master Field

Since our goal is to prove (4.15), we need to identify conditions under which the master field is differentiable in the space direction. In order to do so, we add the following assumption.

Assumption (Decoupling Master). On top of Assumption **FBSDE**, we also assume that:

(A1) For any $t \in [0,T]$, any t-initialized probabilistic set-up $(\Omega, \mathcal{F}, \mathbb{F}, \mathbb{P})$ equipped with a compatible lifting $(\boldsymbol{W}^0, (\mathfrak{M}, \mu), \boldsymbol{W})$ as in Subsection 4.1.2, and any $x, x' \in \mathbb{R}^d$, the solutions $(\boldsymbol{X}, \boldsymbol{Y}, \boldsymbol{Z}, \boldsymbol{Z}^0, \boldsymbol{M})$ and $(\boldsymbol{X}', \boldsymbol{Y}', \boldsymbol{Z}', \boldsymbol{Z}^{0\prime}, \boldsymbol{M}')$ to (4.5) with x and x' as initial conditions, satisfy the stability estimate (1.19) stated in Theorem 1.53, for a constant Γ independent of x and x'.

(A2) For any $(t, \mu) \in [0,T] \times \mathcal{P}_2(\mathbb{R}^d)$, the function $\mathbb{R}^d \times \mathbb{R}^m \times \mathbb{R}^{m \times d} \ni (x, y, z) \mapsto \check{\alpha}(t, x, \mu, y, z)$ is continuous.

We are now able to identify the space derivative of the master field.

Theorem 4.10 *Let the assumptions and notation of Definition 4.1 be in force, and let us assume further that assumptions* **Necessary SMP Master** *and* **Decoupling Master** *hold.*

Then, for any $(t, \mu) \in [0,T] \times \mathcal{P}_2(\mathbb{R}^d)$, the mapping $\mathbb{R}^d \ni x \mapsto \mathcal{U}(t, x, \mu)$ is continuously differentiable and

$$\forall x \in \mathbb{R}^d, \quad \partial_x \mathcal{U}(t, x, \mu) = \mathbb{E}^{t,\mu}[\upsilon_t^{t,x}],$$

where the right-hand side is as in the statement of Proposition 4.7.

Proof. Throughout the proof, we use the same notations as in Definition 4.1. The strategy is then a variant of that used for proving Theorem 1.59.

First Step. The time index t and the measure μ being fixed, we consider, on the canonical set-up $\bar{\Omega}^t$, a square-integrable $\mathbb{F}^t$-progressively measurable process $\boldsymbol{\beta} = (\beta_s)_{t \le s \le T}$ with values in A. For any $x \in \mathbb{R}^d$, we then call $\boldsymbol{x}^{t,x,\boldsymbol{\beta}} = (x_s^{t,x,\boldsymbol{\beta}})_{t \le s \le T}$ the solution of the SDE:

$$dx_s^{t,x,\boldsymbol{\beta}} = b(s, x_s^{t,x,\boldsymbol{\beta}}, \nu_s, \beta_s)ds + \sigma(s, x_s^{t,x,\boldsymbol{\beta}}, \nu_s)dw_s + \sigma^0(s, x_s^{t,x,\boldsymbol{\beta}}, \nu_s)dw_s^0, \quad s \in [t,T],$$

with the initial condition $x_t^{t,x,\boldsymbol{\beta}} = x$. We also write $\hat{\boldsymbol{\alpha}}^{t,x} = (\check{\alpha}(s, x_s^{t,x}, \nu_s, y_s^{t,x}, z_s^{t,x}))_{t \le s \le T}$ for the optimal feedback under the initial condition (t, x) and in the environment $\mathfrak{m}$. With these notations, we recall that $\boldsymbol{x}^{t,x} = \boldsymbol{x}^{t,x,\hat{\boldsymbol{\alpha}}^{t,x}}$.

Then, following the proof of Theorem 1.59, see in particular (1.57), we have for any unit vector $e \in \mathbb{R}^d$:

$$\lim_{\epsilon \to 0} \mathbb{E}^{t,\mu}\Big[\sup_{t \le s \le T} \Big| \frac{1}{\epsilon}\big(x_s^{t,x+\epsilon e,\hat{\boldsymbol{\alpha}}^{t,x}} - x_s^{t,x}\big) - \partial_e x_s^{t,x} \Big|^2 \Big] = 0,$$

where

$$\begin{aligned} d\big[\partial_e x_s^{t,x}\big] &= \partial_x b\big(s, x_s^{t,x}, \nu_s, \hat{\alpha}_s^{t,x}\big)\partial_e x_s^{t,x} ds \\ &\quad + \partial_x \sigma\big(s, x_s^{t,x}, \nu_s\big)\partial_e x_s^{t,x} dw_s + \partial_x \sigma^0\big(s, x_s^{t,x}, \nu_s\big)\partial_e x_s^{t,x} dw_s^0, \end{aligned} \tag{4.19}$$

for $s \in [t, T]$, and with $\partial_e x_t^{t,x} = e$ as initial condition.

Second Step. By Proposition 4.3, using the fact that $\hat{\boldsymbol{\alpha}}^{t,x}$ is progressively measurable with respect to the complete and right-continuous augmentation of the filtration generated by $(\boldsymbol{w}^0, \mathfrak{m}, \boldsymbol{w})$, we have, for any real $h > 0$,

$$\begin{aligned} \mathcal{U}(t, x + he, \mu) - \mathcal{U}(t, x, \mu) \le \mathbb{E}^{t,\mu}\Big[\int_t^T & \big[f\big(s, x_s^{t,x+he,\hat{\boldsymbol{\alpha}}^{t,x}}, \nu_s, \hat{\alpha}_s^{t,x}\big) - f\big(s, x_s^{t,x}, \nu_s, \hat{\alpha}_s^{t,x}\big)\big] ds \\ & + g\big(x_T^{t,x+he,\hat{\boldsymbol{\alpha}}^{t,x}}, \nu_T\big) - g\big(x_T^{t,x}, \nu_T\big) \Big]. \end{aligned}$$

Dividing by h and letting $h \searrow 0$, we deduce that:

$$\begin{aligned} &\limsup_{h \searrow 0} \frac{1}{h}\big[\mathcal{U}(t, x + he, \mu) - \mathcal{U}(t, x, \mu)\big] \\ &\quad \le \mathbb{E}^{t,\mu}\Big[\int_t^T \partial_x f\big(s, x_s^{t,x}, \nu_s, \hat{\alpha}_s^{t,x}\big) \cdot \partial_e x_s^{t,x} ds + \partial_e g(x_T^{t,x}, \nu_T) \cdot \partial_e x_T^{t,x} \Big]. \end{aligned}$$

Now, by a standard application of Itô's formula, which is reminiscent of the proof of Theorem 1.59, see (1.58) and (1.59), we have:

$$\begin{aligned} &\mathbb{E}^{t,\mu}\Big[\int_t^T \partial_x f\big(s, x_s^{t,x}, \nu_s, \hat{\alpha}_s^{t,x}\big) \cdot \partial_e x_s^{t,x} ds + \partial_x g(x_T^{t,x}, \nu_T) \cdot \partial_e x_T^{t,x} \Big] \\ &= \mathbb{E}^{t,\mu}\Big[\int_t^T \partial_x f\big(s, x_s^{t,x}, \nu_s, \hat{\alpha}_s^{t,x}\big) \cdot \partial_e x_s^{t,x} ds + \upsilon_T^{t,x} \cdot \partial_e x_T^{t,x} \Big] = \mathbb{E}^{t,\mu}\big[\upsilon_t^{t,x}\big] \cdot e, \end{aligned}$$

from which we deduce that:

$$\limsup_{h \searrow 0} \frac{1}{h}\big[\mathcal{U}(t, x + he, \mu) - \mathcal{U}(t, x, \mu)\big] \le \mathbb{E}^{t,\mu}[\upsilon_t^{t,x}] \cdot e. \tag{4.20}$$

Using the same type of argument with $h < 0$, we get in a similar way that:

$$\liminf_{h \nearrow 0} \frac{1}{h}\big[\mathcal{U}(t, x + he, \mu) - \mathcal{U}(t, x, \mu)\big] \ge \mathbb{E}^{t,\mu}[\upsilon_t^{t,x}] \cdot e. \tag{4.21}$$

Third Step. Assuming that $\mathbb{R}^d \ni x \mapsto \mathcal{U}(t,x,\mu)$ and $\mathbb{R}^d \ni x \mapsto \mathbb{E}^{t,\mu}[v_t^{t,x}]$ are continuous, we use (4.20) and (4.21) to prove that $\mathcal{U}(t,\cdot,\mu)$ is continuously differentiable.

For the same unit vector e as above, consider the curve:

$$\gamma : \mathbb{R} \ni s \mapsto \mathcal{U}(t, x+se, \mu) - \mathcal{U}(t,x,\mu) - \int_0^s \Big(\mathbb{E}^{t,\mu}\big[v_t^{t,x+re}\big]\cdot e\Big)dr.$$

By assumption, γ is continuous. Moreover, by (4.20) and (4.21), it satisfies:

$$\forall s \in \mathbb{R}, \quad \limsup_{h \searrow 0} \frac{\gamma(s+h)-\gamma(s)}{h} \le 0, \quad \liminf_{h \nearrow 0} \frac{\gamma(s+h)-\gamma(s)}{h} \ge 0. \tag{4.22}$$

Considering a mollification kernel $\rho : \mathbb{R} \to \mathbb{R}$ with compact support, we know that the mollified curve:

$$\gamma * \rho : \mathbb{R} \ni s \mapsto \int_{\mathbb{R}} \gamma(s-r)\rho(r)dr,$$

is continuously differentiable. Clearly, it must satisfy (4.22). In particular, $\gamma * \rho$ must be nonincreasing and nondecreasing. It is thus constant, which proves that γ is also constant and thus:

$$\lim_{h\to 0}\frac{1}{h}\big[\mathcal{U}(t,x+he,\mu) - \mathcal{U}(t,x,\mu)\big] = \mathbb{E}^{t,\mu}[v_t^{t,x}]\cdot e,$$

which shows, by continuity of the right-hand side, that $\mathcal{U}$ is continuously differentiable in x.

Last Step. It remains to check that, for (t,μ) fixed, the mappings $\mathbb{R}^d \ni x \mapsto \mathcal{U}(t,x,\mu)$ and $\mathbb{R}^d \ni x \mapsto \mathbb{E}^{t,\mu}[v_t^x]$ are continuous.

We make use of the stability estimate in Theorem 1.53. It says that, for any $x, x' \in \mathbb{R}^d$:

$$\mathbb{E}^{t,\mu}\bigg[\sup_{t\le s\le T}\big[|x_s^{t,x} - x_s^{t,x'}|^2 + |y_s^{t,x} - y_s^{t,x'}|^2\big] + \int_t^T |z_s^{t,x} - z_s^{t,x'}|^2 ds\bigg] \le C|x-x'|^2.$$

Plugging these bounds into the Definition 4.1 of the master field and making use of the regularity properties of f, g and $\check{\alpha}$, we deduce that $\mathbb{R}^d \ni x \mapsto \mathcal{U}(t,x,\mu)$ is continuous. Similarly, from Lemma 4.8, we deduce that the mapping $\mathbb{R}^d \ni x \mapsto \mathbb{E}^{t,\mu}[v_t^{t,x}]$ is continuous. □

The following corollary gives a representation of the solution of the backward SDE (4.18) in Proposition 4.7 in terms of the master field evaluated along the solution of the forward state equation.

Corollary 4.11 *Under the assumption of Theorem 4.10, for any $(t,\mu) \in [0,T] \times \mathcal{P}_2(\mathbb{R}^d)$, and any $x \in \mathbb{R}^d$, there exists an at most countable subset $\mathcal{Q} \subset [t,T]$ such that, for any $s \in [t,T] \setminus \mathcal{Q}$, with probability 1 under $\mathbb{P}^{t,\mu}$,*

$$v_s^{t,x} = \partial_x \mathcal{U}(s, x_s^{t,x}, \nu_s).$$

According to Proposition 1.46, $\partial_x \mathcal{U}(s,\cdot,\nu_s)$ plays, at time $s \in [t,T]\setminus \mathcal{Q}$, the role of the decoupling field of the backward equation (4.18) when coupled with the forward equation (4.7), with $\xi = x$.

Proof. Instead of using the general framework of decoupling fields, as defined in Proposition 1.46, we provide a direct proof. Working on the canonical set-up and fixing $s \in [t,T]$, we have by Proposition 4.3:

$$\begin{aligned}
&\mathcal{U}(s, x_s^{t,x}, \nu_s) \\
&= \mathbb{E}^{t,\mu}\biggl[\int_s^T f\bigl(r, x_r^{t,x}, \nu_r, \hat{\alpha}_r^{t,x}\bigr)dr + g(x_T^{t,x}, \nu_T) \,\big|\, \sigma\bigl\{w_r^0, \mathfrak{m}_r, w_r;\ t \le r \le s\bigr\}\biggr],
\end{aligned} \tag{4.23}$$

and for any $h \in \mathbb{R}$ and unitary vector $e \in \mathbb{R}^d$,

$$\begin{aligned}
\mathcal{U}(s, x_s^{t,x+he}, \nu_s) \le \mathbb{E}^{t,\mu}\biggl[&\int_s^T f\bigl(r, x_r^{t,x+he,\hat{\boldsymbol{\alpha}}^{t,x}}, \nu_r, \hat{\alpha}_r^{t,x}\bigr)dr \\
&+ g(x_T^{t,x+he,\hat{\boldsymbol{\alpha}}^{t,x}}, \nu_T) \,\big|\, \sigma\bigl\{w_r^0, \mathfrak{m}_r, w_r;\ t \le r \le s\bigr\}\biggr],
\end{aligned} \tag{4.24}$$

where we used the same notation as in the proof of Theorem 4.10. Therefore, subtracting (4.23) to (4.24), dividing by $h > 0$ and letting h tend to 0, we must have:

$$\begin{aligned}
\partial_x \mathcal{U}(s, x_s^{t,x}, \nu_s) \cdot \partial_e x_s^{t,x} \le \mathbb{E}^{t,\mu}\biggl[&\int_s^T \partial_x f\bigl(r, x_r^{t,x}, \nu_r, \hat{\alpha}_r^{t,x}\bigr) \cdot \partial_e x_r^{t,x} dr \\
&+ \partial_e g(x_T^{t,x}, \nu_T) \cdot \partial_e x_T^{t,x} \,\big|\, \sigma\bigl(w_r^0, \mathfrak{m}_r, w_r; t \le r \le s\bigr)\biggr].
\end{aligned}$$

Following the proof of Theorem 4.10, we can prove by integration by parts that the right-hand side is equal to $\upsilon_s^{t,x} \cdot \partial_e x_s^{t,x}$. Therefore,

$$\begin{aligned}
\partial_x \mathcal{U}(s, x_s^{t,x}, \nu_s) \cdot \partial_e x_s^{t,x} &\le \mathbb{E}^{t,\mu}\Bigl[\upsilon_s^{t,x} \cdot \partial_e x_s^{t,x} \,\big|\, \sigma\bigl(w_r^0, \mathfrak{m}_r, w_r; t \le r \le s\bigr)\Bigr] \\
&= \mathbb{E}^{t,\mu}\Bigl[\upsilon_s^{t,x} \,\big|\, \sigma\bigl(w_r^0, \mathfrak{m}_r, w_r; t \le r \le s\bigr)\Bigr] \cdot \partial_e x_s^{t,x},
\end{aligned}$$

the last line following from the fact that $\partial_e x_s^{t,x}$ is measurable with respect to the completion of the σ-field $\sigma\{w_r^0, \mathfrak{m}_r, w_r;\ t \le r \le s\}$, as it is the limit in L^2 of random variables that are measurable with respect to the completion of $\sigma\{w_r^0, \mathfrak{m}_r, w_r;\ t \le r \le s\}$. Changing $h > 0$ into $h < 0$ and repeating the argument, we get that the inequality is actually an equality:

$$\partial_x \mathcal{U}(s, x_s^{t,x}, \nu_s) \cdot \partial_e x_s^{t,x} = \mathbb{E}^{t,\mu}\Bigl[\upsilon_s^{t,x} \,\big|\, \sigma\bigl\{w_r^0, \mathfrak{m}_r, w_r;\ t \le r \le s\bigr\}\Bigr] \cdot \partial_e x_s^{t,x}.$$

The above holds true, for any $s \in [t,T]$ and any unitary vector $e \in \mathbb{R}^d$, with probability 1 under $\mathbb{P}_\mu^t$.

It is a standard procedure to check that $\partial_e x_s^{t,x}$ reads $\nabla x_s^{t,x} e$ where $\nabla x_s^{t,x}$ solves the linearized equation (4.19), but in $\mathbb{R}^{d\times d}$ instead of $\mathbb{R}^d$, with the identity matrix as initial condition. From the theory of linear SDEs (see the Notes & Complements at the end of the chapter for references), $\nabla x_s^{t,x}$ is invertible with $\mathbb{P}^{t,\mu}$-probability 1, from which we get, for all $s \in [t, T]$, with probability 1 under $\mathbb{P}^{t,\mu}$,

$$\partial_x \mathcal{U}(s, x_s^{t,x}, \nu_s) = \mathbb{E}^{t,\mu}\Big[\upsilon_s^{t,x} \,|\, \sigma\{w_r^0, \mathfrak{m}_r, w_r;\ t \le r \le s\}\Big].$$

Recall now that $\boldsymbol{\upsilon} = (\upsilon_s)_{t \le s \le T}$ is càd-làg. Therefore, we can find an at most countable subset $\mathcal{Q} \subset [t, T]$, such that, for any $s \in [t, T] \setminus \mathcal{Q}$, $\mathbb{P}^{t,\mu}[\lim_{r \nearrow s} \upsilon_r^{t,x} = \upsilon_s^{t,x}] = 1$. Now, for any $s \in [t, T]$ and $r \in [t, s)$, $\upsilon_r^{t,x}$ is measurable with respect to the completion of $\sigma\{w_u^0, \mathfrak{m}_u, w_u;\ t \le u \le s\}$. Therefore, for $s \in [t, T] \setminus \mathcal{Q}$, $\upsilon_s^{t,x}$ is measurable with respect to the completion of $\sigma\{w_r^0, \mathfrak{m}_r, w_r;\ t \le r \le s\}$. We deduce that, for such an s, with $\mathbb{P}^{t,\mu}$-probability 1:

$$\partial_x \mathcal{U}(s, x_s^{t,x}, \nu_s) = \upsilon_s^{t,x},$$

which completes the proof. □

We refer the reader to Subsection 4.4.4 below for a set of self-contained assumptions under which Theorem 4.10 holds.

4.3 Itô's Formula Along a Flow of Conditional Measures

Throughout the next two sections, our goal is to prove that the master field solves, in a suitable sense, a partial differential equation on the space of probability measures, referred to as the *master equation* in the sequel. In the spirit of the examples discussed in Section (Vol I)-5.7, the derivation of such a PDE relies on a suitable use of the chain rule for functions defined on the Wasserstein space $\mathcal{P}_2(\mathbb{R}^d)$. For that reason, we now extend the discussion initiated in Chapter (Vol I)-5 on the differentiation of a smooth functional along a flow of measures, the main objective being to generalize the Itô formula proved in Section (Vol I)-5.6 to the case when the measures may be random.

4.3.1 Conditional Measures of an Itô Process Subject to Two Noises

The typical framework we have in mind is the following: $\boldsymbol{X} = (X_t)_{t \ge 0}$ is an Itô process of the form

$$dX_t = B_t dt + \Sigma_t dW_t + \Sigma_t^0 dW_t^0, \quad t \ge 0, \tag{4.25}$$

for two d-dimensional independent Brownian motions $\boldsymbol{W} = (W_t)_{t \ge 0}$ and $\boldsymbol{W}^0 = (W_t^0)_{t \ge 0}$ defined on a general set-up of the same form as in Subsection 4.1.2, namely:

1. we are given complete probability spaces $(\Omega^0, \mathcal{F}^0, \mathbb{P}^0)$ and $(\Omega^1, \mathcal{F}^1, \mathbb{P}^1)$, the space $(\Omega^1, \mathcal{F}^1, \mathbb{P}^1)$ being the completion of a probability space with a countably generated σ-field, endowed with complete and right-continuous filtrations $\mathbb{F}^0 = (\mathcal{F}_t^0)_{t \ge 0}$ and $\mathbb{F}^1 = (\mathcal{F}_t^1)_{t \ge 0}$ and d-dimensional $\mathbb{F}^0$ and $\mathbb{F}^1$ Brownian motions $\boldsymbol{W}^0 = (W_t^0)_{t \ge 0}$, and $\boldsymbol{W} = (W_t)_{t \ge 0}$ respectively.

2. we denote by $(\Omega, \mathcal{F}, \mathbb{P})$ the completion of the product space $(\Omega^0 \times \Omega^1, \mathcal{F}^0 \otimes \mathcal{F}^1, \mathbb{P}^0 \otimes \mathbb{P}^1)$ endowed with the filtration $\mathbb{F} = (\mathcal{F}_t)_{t \geq 0}$ obtained by augmenting the product filtration $\mathbb{F}^0 \otimes \mathbb{F}^1$ in a right-continuous way and by completing it.

As usual, we denote by $\mathbb{E}^0$ the expectation on the first space, by $\mathbb{E}^1$ the expectation on the second space and by $\mathbb{E}$ the expectation on the product space.

In that framework, $(B_t)_{t \geq 0}$, $(\Sigma_t)_{t \geq 0}$ and $(\Sigma_t^0)_{t \geq 0}$ are progressively measurable processes with respect to $\mathbb{F}$, with values in $\mathbb{R}^d$, $\mathbb{R}^{d \times d}$ and $\mathbb{R}^{d \times d}$ respectively, such that for any finite horizon $T > 0$,

$$\mathbb{E}\left[\int_0^T \left(|B_t|^2 + |\Sigma_t|^4 + |\Sigma_t^0|^4\right) dt\right] < \infty. \tag{4.26}$$

Because of Lemmas 2.4 and 2.5, we know that $\mathbb{P}^0$-almost surely, for any $T > 0$, the quantity $\mathbb{E}^1[\sup_{0 \leq t \leq T} |X_t|^2]$ is finite, proving that, for $\mathbb{P}^0$ almost every $\omega^0 \in \Omega^0$, for any $t \in [0, +\infty)$, the random variable $X_t(\omega^0, \cdot)$ is in $L^2(\Omega^1, \mathcal{F}^1, \mathbb{P}^1)$. We can thus define:

$$\mu_t(\omega^0) = \mathcal{L}^1(X_t)(\omega^0) = \mathcal{L}\big(X_t(\omega^0, \cdot)\big).$$

In the spirit of Section (Vol I)-5.6, we investigate the possible time differentiability of quantities of the form $(U(\mu_t))_{t \geq 0}$, for a given smooth function U on the space $\mathcal{P}_2(\mathbb{R}^d)$.

Progressively Measurable Versions of Conditional Expectations

Throughout the analysis, we shall associate with real valued $\mathbb{F}$-progressively measurable processes $(\theta_t)_{t \geq 0}$, conditional expectations of the form $(\mathbb{E}^1[\theta_t])_{t \geq 0}$ whenever sufficient integrability conditions are in force. For this reason, we first collect several conditions under which such a process is well defined and may be regarded as an $\mathbb{F}^0$-progressively measurable process.

Joint Measurability. We start with the following observation, which may be proved by a mere adaptation of Lemmas 2.4 and 2.5. Let $\boldsymbol{\theta} : [0, +\infty) \times \Omega \to \mathbb{R}$ be a $\mathcal{B}([0, +\infty)) \otimes \mathcal{F}$-measurable mapping. Since $\mathcal{B}([0, +\infty)) \otimes \mathcal{F}$ is included in the completion of $\mathcal{B}([0, +\infty)) \otimes (\mathcal{F}^0 \otimes \mathcal{F}^1)$ under $\mathrm{Leb}_1 \otimes \mathbb{P}^0 \otimes \mathbb{P}^1$, there exists a mapping $\tilde{\boldsymbol{\theta}} : [0, +\infty) \times \Omega \to \mathbb{R}$, $\mathcal{B}([0, +\infty)) \otimes (\mathcal{F}^0 \otimes \mathcal{F}^1)$-measurable, which is almost everywhere equal to $\boldsymbol{\theta}$ under the completion of $\mathrm{Leb}_1 \otimes \mathbb{P}^0 \otimes \mathbb{P}^1$.

Now, if $\boldsymbol{\theta}$ takes values in $\mathbb{R}_+$, we may consider $(\mathbb{E}^1[\theta_t])_{t \geq 0}$, each $\mathbb{E}^1[\theta_t]$ being regarded as an $\mathcal{F}^0$-measurable random variable with values in $[0, +\infty]$ uniquely defined up to a null subset. Then, up to a null set under $\mathrm{Leb}_1 \otimes \mathbb{P}^0$, $\mathbb{E}^1[\theta_t] = \mathbb{E}^1[\tilde{\theta}_t]$; moreover, the mapping $[0, +\infty) \times \Omega^0 \ni (t, \omega^0) \mapsto \mathbb{E}^1[\theta_t](\omega^0)$ is measurable with respect to the completion of $\mathcal{B}([0, +\infty)) \otimes \mathcal{F}^0$ under $\mathrm{Leb}_1 \otimes \mathbb{P}^0$. Arguing as Subsection 2.1.3, we deduce that, for $\mathbb{P}^0$-almost every $\omega^0 \in \Omega^0$, the mapping $[0, +\infty) \ni t \mapsto \mathbb{E}^1[\theta_t]$ is measurable (with respect to the completion of the Borel σ-field) and for all $t \geq 0$,

$$\int_0^t \mathbb{E}^1[\theta_s]ds = \int_0^t \mathbb{E}^1[\tilde{\theta}_s]ds.$$

In the more general case when $\boldsymbol{\theta}$ takes values in $\mathbb{R}$, we assume in addition that, for $\mathbb{P}^0$-almost every $\omega^0 \in \Omega^0$, for all $t \geq 0$,

$$\int_0^t \mathbb{E}^1[|\theta_s|]ds < \infty,$$

which makes sense since, for $\mathbb{P}^0$-almost every $\omega^0 \in \Omega^0$, the mapping $[0, +\infty) \ni t \mapsto \mathbb{E}^1[|\theta_t|]$ is measurable. Then, up to a null set under $\text{Leb}_1 \otimes \mathbb{P}^0$, the mappings $[0, +\infty) \times \Omega^0 \ni (t, \omega^0) \mapsto \mathbb{E}^1[(\theta_t)_+](\omega^0)$ and $[0, +\infty) \times \Omega^0 \ni (t, \omega^0) \mapsto \mathbb{E}^1[(\theta_t)_-](\omega^0)$ are finite, where as usual we denote by x_+ and x_- the positive and negative parts of any real number x. The map $[0, +\infty) \times \Omega^0 \ni (t, \omega^0) \mapsto \mathbb{E}^1[\theta_t](\omega^0)\mathbf{1}_{\{\mathbb{E}^1[|\theta_t|](\omega^0)<\infty\}}$ is measurable with respect to the completion of the σ-field $\mathcal{B}([0, +\infty)) \otimes \mathcal{F}^0$ under $\text{Leb}_1 \otimes \mathbb{P}^0$. As above, up to a null set under $\text{Leb}_1 \otimes \mathbb{P}^0$, $\mathbb{E}^1[\theta_t] = \mathbb{E}^1[\tilde{\theta}_t]$. Moreover, for all $t \geq 0$, we also have:

$$\int_0^t \mathbb{E}^1[\theta_s]ds = \int_0^t \mathbb{E}^1[\tilde{\theta}_s]ds.$$

This says that we can redefine $(\mathbb{E}^1[\theta_t])_{t\geq 0}$, up to a $\text{Leb}_1 \otimes \mathbb{P}^0$ null set so that it becomes $\mathcal{B}([0, +\infty)) \otimes \mathbb{F}^0$-measurable.

When $\boldsymbol{\theta} = (\theta_t)_{t\geq 0}$ is progressively measurable, we know that for any $t \geq 0$, $\mathbb{E}^1[\theta_t]$ is $\mathcal{F}_t^0$-measurable. If we call $(\theta_t^0)_{t\geq 0}$ a $\mathcal{B}([0, +\infty)) \otimes \mathbb{F}^0$-measurable version of $(\mathbb{E}^1[\theta_t])_{t\geq 0}$, then for all $t \geq 0$ except in a Borel subset $N \subset [0, +\infty)$ of zero Lebesgue measure, $\mathbb{P}^0[\theta_t^0 = \mathbb{E}^1[\theta_t]] = 1$, which proves that θ_t^0 is $\mathcal{F}_t^0$-measurable for all $t \notin N$. Now, the mapping $[0, +\infty) \times \Omega^0 \ni (t, \omega^0) \mapsto \theta_t^0 \mathbf{1}_N(t)$ is jointly measurable and forms an adapted process with respect to $\mathbb{F}^0$. By standard results in the general theory of stochastic processes, we can find an $\mathbb{F}^0$-measurable modification. This says that we can a redefine $(\mathbb{E}^1[\theta_t])_{t\geq 0}$, up to a $\text{Leb}_1 \otimes \mathbb{P}^0$ - null set, so that it becomes $\mathbb{F}^0$-progressively measurable.

Measurability with Values in L^2. We now assume that $\text{Leb}_1 \otimes \mathbb{P}^0[(s, \omega^0) \in [0, +\infty) \times \Omega^0 : \mathbb{E}^1[\theta_s^2](\omega^0) = \infty] = 0$. Then, we know that for any random variable $Z \in L^2(\Omega^1, \mathcal{F}^1, \mathbb{P}^1; \mathbb{R})$, the mapping $[0, +\infty) \times \Omega^0 \ni (t, \omega^0) \mapsto \mathbb{E}^1[|\theta_t - Z|^2](\omega^0)\mathbf{1}_{\{\mathbb{E}^1[|\theta_t|^2](\omega^0)<\infty\}}$ is measurable with respect to the completion of $\mathcal{B}([0, +\infty)) \otimes \mathcal{F}^0$ under $\text{Leb}_1 \otimes \mathbb{P}^0$ and coincides with $(t, \omega^0) \mapsto \mathbb{E}^1[|\tilde{\theta}_t - Z|^2](\omega^0)\mathbf{1}_{\{\mathbb{E}^1[|\tilde{\theta}_t|^2](\omega^0)<\infty\}}$, for the same $(\tilde{\theta}_t)_{t\geq 0}$ as above, up to a $\text{Leb}_1 \otimes \mathbb{P}^0$ - null set in $\mathcal{B}([0, +\infty)) \otimes \mathcal{F}^0$. Since $(\Omega^1, \mathcal{F}^1, \mathbb{P}^1)$ is the completion of a countably generated probability space, we know that $L^2(\Omega^1, \mathcal{F}^1, \mathbb{P}^1; \mathbb{R})$ is separable. Hence, the mapping $[0, +\infty) \times \Omega^0 \ni (t, \omega^0) \mapsto \theta_t(\omega^0, \cdot)\mathbf{1}_{\{\mathbb{E}^1[|\theta_t|^2](\omega^0)<\infty\}} \in L^2(\Omega^1, \mathcal{F}^1, \mathbb{P}^1; \mathbb{R})$ is measurable with respect to the completion of $\mathcal{B}([0, +\infty)) \otimes \mathcal{F}^0$ under $\text{Leb}_1 \otimes \mathbb{P}^0$, and it coincides with the mapping $[0, +\infty) \times \Omega^0 \ni (t, \omega^0) \mapsto \tilde{\theta}_t(\omega^0, \cdot)\mathbf{1}_{\{\mathbb{E}^1[|\tilde{\theta}_t|^2](\omega^0)<\infty\}} \in L^2(\Omega^1, \mathcal{F}^1, \mathbb{P}^1; \mathbb{R})$ up to a null set in $\mathcal{B}([0, +\infty)) \otimes \mathcal{F}^0$ under $\text{Leb}_1 \otimes \mathbb{P}^0$.

4.3.2 Refresher on $\mathcal{C}^2$-Regularity

In Chapter (Vol I)-5, we introduced two notions of $\mathcal{C}^2$ regularity for a function of measures. We called the strongest one *full $\mathcal{C}^2$ regularity* and the weakest one *partial $\mathcal{C}^2$ regularity*. We then proved that partial $\mathcal{C}^2$ regularity was sufficient to expand a function along a deterministic flow of measures. In the present situation, saying that the flow $(\mu_t)_{t\geq 0}$ is deterministic means that the dynamics of $(X_t)_{t\geq 0}$ only feel the noise $(W_t)_{t\geq 0}$. Note also that, $(\mu_t)_{t\geq 0}$ being deterministic, the second-order partial derivatives in the direction μ played no role in the expansion.

Full $\mathcal{C}^2$-Regularity

Things are different when $(\mu_t)_{t\geq 0}$ is random. As in the standard chain rule for Itô processes, the expansion is expected to rely on the second-order derivatives in the direction of the measure. This suggests that we may need to assume full $\mathcal{C}^2$ regularity, whose definition we recall from Chapter (Vol I)-5.

Assumption (Full $\mathcal{C}^2$ Regularity). We shall say that a real valued function u on $\mathcal{P}_2(\mathbb{R}^d)$ is fully $\mathcal{C}^2$ regular if:

(A1) The function u is $\mathcal{C}^1$ in the sense of L-differentiation, and its first derivative has a jointly continuous version $\mathcal{P}_2(\mathbb{R}^d) \times \mathbb{R}^d \ni (\mu, v) \mapsto \partial_\mu u(\mu)(v) \in \mathbb{R}^d$.

(A2) For each fixed $\mu \in \mathcal{P}_2(\mathbb{R}^d)$, the version of $\mathbb{R}^d \ni v \mapsto \partial_\mu u(\mu)(v) \in \mathbb{R}^d$ used in **(A1)** is differentiable on $\mathbb{R}^d$ in the classical sense and its derivative is given by a jointly continuous function $\mathcal{P}_2(\mathbb{R}^d) \times \mathbb{R}^d \ni (\mu, v) \mapsto \partial_v \partial_\mu u(\mu)(v) \in \mathbb{R}^{d\times d}$.

(A3) For each fixed $v \in \mathbb{R}^d$, the version of $\mathcal{P}_2(\mathbb{R}^d) \ni \mu \mapsto \partial_\mu u(\mu)(v) \in \mathbb{R}^d$ used in **(A1)** is L-differentiable component by component, with a derivative given by a function $(\mu, v, v') \mapsto \partial^2_\mu u(\mu)(v)(v') \in \mathbb{R}^{d\times d}$ such that for any $\mu \in \mathcal{P}_2(\mathbb{R}^d)$ and any $X \in L^2(\Omega, \mathcal{F}, \mathbb{P}; \mathbb{R}^d)$ with $\mathcal{L}(X) = \mu$ over an atomless Polish probability space $(\Omega, \mathcal{F}, \mathbb{P})$, $\partial^2_\mu u(\mu)(x)(X)$ gives the Fréchet derivative at X of $L^2(\Omega, \mathcal{F}, \mathbb{P}; \mathbb{R}^d) \ni X' \mapsto \partial_\mu u(\mathcal{L}(X'))(v)$, for every $v \in \mathbb{R}^d$. Denoting $\partial^2_\mu u(\mu)(v)(v')$ by $\partial^2_\mu u(\mu)(v, v')$, the map $\mathcal{P}_2(\mathbb{R}^d) \times \mathbb{R}^d \times \mathbb{R}^d \ni (\mu, v, v') \mapsto \partial^2_\mu u(\mu)(v, v')$ is also assumed to be continuous for the product topology.

It is worth recalling Remark (Vol I)-5.82.

Remark 4.12 *The following observations may be useful.*

1. *Under* **(A1)**, *there exists one and only one version of* $\partial_\mu u(\mu)(\cdot) \in L^2(\mathbb{R}^d, \mu; \mathbb{R}^d)$ *for each* $\mu \in \mathcal{P}_2(\mathbb{R}^d)$ *such that the mapping* $\mathcal{P}_2(\mathbb{R}^d) \times \mathbb{R}^d \ni (\mu, v) \mapsto \partial_\mu u(\mu)(v) \in \mathbb{R}^d$ *is jointly continuous.*

2. *Under* **(A2)**, *there exists one and only one version of* $\partial_\mu u(\mu)(\cdot)$ *for each* $\mu \in \mathcal{P}_2(\mathbb{R}^d)$ *such that* $\mathbb{R}^d \ni v \mapsto \partial_\mu u(\mu)(v)$ *is differentiable for each* $\mu \in \mathcal{P}_2(\mathbb{R}^d)$ *and the mapping* $\mathcal{P}_2(\mathbb{R}^d) \times \mathbb{R}^d \ni (\mu, v) \mapsto \partial_v \partial_\mu u(\mu)(v)$ *is jointly continuous. In particular, the values of the derivatives* $\partial_v \partial_\mu u(\mu)(v)$, *for* $\mu \in \mathcal{P}_2(\mathbb{R}^d)$ *and* $v \in \mathbb{R}^d$, *are uniquely determined.*
3. *Under* **(A3)**, *there exists one and only one continuous version of* $\partial_\mu u(\mu)(\cdot)$ *for each* $\mu \in \mathcal{P}_2(\mathbb{R}^d)$ *such that for each fixed* $v \in \mathbb{R}^d$, *the mapping* $\mathcal{P}_2(\mathbb{R}^d) \ni \mu \mapsto \partial_\mu u(\mu)(v)$ *is L-continuously differentiable and the derivative* $\mathcal{P}_2(\mathbb{R}^d) \times \mathbb{R}^d \times \mathbb{R}^d \ni (\mu, v, v') \mapsto \partial^2_\mu u(\mu)(v, v')$ *is jointly continuous. Also, the values of* $\partial^2_\mu u$ *are uniquely determined.*

For the sake of definiteness, we also remind the reader of the following important result involving this notion of regularity. Recall Proposition (Vol I)-5.91.

Proposition 4.13 *Let us assume that the function u is fully* C^2 *regular. Then, for each integer* $N \geq 1$, *its empirical projection* u^N *defined as the function* $u^N : (\mathbb{R}^d)^N \ni (x_1, \cdots, x_N) \mapsto u(N^{-1} \sum_{i=1}^N \delta_{x_i})$ *is* C^2 *on* $(\mathbb{R}^d)^N$ *and, for all* $x^1, \cdots, x^N \in \mathbb{R}^d$,

$$\begin{aligned}
&\partial^2_{x^i x^j} u^N(x^1, \cdots, x^N) \\
&\quad = \frac{1}{N} \partial_v \partial_\mu u\left(\frac{1}{N} \sum_{\ell=1}^N \delta_{x^\ell}\right)(x^i)\delta_{i,j} + \frac{1}{N^2} \partial^2_\mu u\left(\frac{1}{N} \sum_{\ell=1}^N \delta_{x^\ell}\right)(x^i, x^j),
\end{aligned}$$

the equality being an equality between elements of $\mathbb{R}^{d \times d}$.

Simple C^2-Regularity

Inspired by the notion of partial C^2 regularity used in Chapter (Vol I)-5, we shall slightly weaken the assumption **Full** C^2 **Regularity**, by requiring continuity of the second-order derivatives only at points (v, μ) and (v, v', μ) such that v and v' belong to the support of μ. This is what we shall call *simple* C^2 *regularity*:

Assumption (Simple C^2 **Regularity).** We say that a real valued function u on $\mathcal{P}_2(\mathbb{R}^d)$ is simply C^2 regular if the following three conditions are satisfied.

(A1) The function u is C^1 in the sense of L-differentiation, and its first derivative has a version $\mathcal{P}_2(\mathbb{R}^d) \times \mathbb{R}^d \ni (\mu, v) \mapsto \partial_\mu u(\mu)(v) \in \mathbb{R}^d$ which is locally bounded and is continuous at any (μ, v) such that $v \in \mathrm{Supp}(\mu)$.

(A2) The version of $\mathbb{R}^d \ni v \mapsto \partial_\mu u(\mu)(v) \in \mathbb{R}^d$ used in **(A1)** for each $\mu \in \mathcal{P}_2(\mathbb{R}^d)$ is differentiable on $\mathbb{R}^d$ in the classical sense and its derivative forms a global map $\mathcal{P}_2(\mathbb{R}^d) \times \mathbb{R}^d \ni (\mu, v) \mapsto \partial_v \partial_\mu u(\mu)(v) \in \mathbb{R}^d$ which is locally bounded and is jointly continuous at any point (μ, v) such that $v \in \mathrm{Supp}(\mu)$.

(continued)

(A3) The version of $\mathbb{R}^d \ni v \mapsto \partial_\mu u(\mu)(v) \in \mathbb{R}^d$ used in **(A1)** for each $\mu \in \mathcal{P}_2(\mathbb{R}^d)$ is such that for any $v \in \mathbb{R}^d$, the mapping $\mathcal{P}_2(\mathbb{R}^d) \ni \mu \mapsto \partial_\mu u(\mu)(v) \in \mathbb{R}^d$ is L-differentiable component by component, with a derivative given by a function $(\mu, v') \mapsto \partial^2_\mu u(\mu)(v)(v') \in \mathbb{R}^{d\times d}$. Denoting $\partial^2_\mu u(\mu)(v)(v')$ by $\partial^2_\mu u(\mu)(v, v')$, we can find a version of each $\partial^2_\mu u(\mu)(v, \cdot)$ such that the global map $\mathcal{P}_2(\mathbb{R}^d) \times \mathbb{R}^d \times \mathbb{R}^d \ni (\mu, v, v') \mapsto \partial^2_\mu u(\mu)(v, v')$ is locally bounded and is jointly continuous at any point (μ, v, v') such that $v, v' \in \text{Supp}(\mu)$.

We stress the fact that under assumption **Simple $\mathcal{C}^2$ Regularity**, for any $\mu \in \mathcal{P}_2(\mathbb{R}^d)$, the functions $\partial_\mu u(\mu)(\cdot)$ and $\partial_v \partial_\mu u(\mu)(\cdot)$ are uniquely defined on the support $\text{Supp}(\mu)$ of μ; also the function $\partial^2_\mu u(\mu)(\cdot, \cdot)$ is uniquely defined on $[\text{Supp}(\mu)]^2$. Observe also that **(A3)** is demanding. In contrast with what we have done so far, it requires to consider $\partial_\mu u(\mu)(v)$ at pairs $(\mu, v) \in \mathcal{P}_2(\mathbb{R}^d) \times \mathbb{R}^d$ for which v may not be in the support of μ.

4.3.3 Chain Rule Under $\mathcal{C}^2$-Regularity

We consider an $\mathbb{R}^d$-valued Itô process of the same form as (4.25). In order to extend the chain rule proven in Theorem (Vol I)-5.92 to the current framework, we need a copy $(\tilde{\Omega}^1, \tilde{\mathcal{F}}^1, \tilde{\mathbb{P}}^1)$ of the probability space $(\Omega^1, \mathcal{F}^1, \mathbb{P}^1)$, the expectation under $\tilde{\mathbb{P}}^1$ being denoted by $\tilde{\mathbb{E}}^1$.

Given such a copy, we consider, for any random variable X defined on $\Omega = \Omega^0 \times \Omega^1$, the random variable $\tilde{X}$, defined as a copy of X on the space $\tilde{\Omega} = \Omega^0 \times \tilde{\Omega}^1$. In particular, we shall consider the copies $(\tilde{X}_t)_{t\geq 0}$, $(\tilde{B}_t)_{t\geq 0}$, $(\tilde{\Sigma}_t)_{t\geq 0}$ and $(\tilde{\Sigma}^0_t)_{t\geq 0}$ of the processes $(X_t)_{t\geq 0}$, $(B_t)_{t\geq 0}$, $(\Sigma_t)_{t\geq 0}$ and $(\Sigma^0_t)_{t\geq 0}$.

Theorem 4.14 *Assume that u is simply $\mathcal{C}^2$ and that, for any compact subset $\mathcal{K} \subset \mathcal{P}_2(\mathbb{R}^d)$,*

$$\sup_{\mu\in\mathcal{K}} \bigg[\int_{\mathbb{R}^d} |\partial_\mu u(\mu)(v)|^2 d\mu(v) + \int_{\mathbb{R}^d} \Big|\partial_v \partial_\mu u(\mu)(v)\Big|^2 d\mu(v) + \int_{\mathbb{R}^d} \int_{\mathbb{R}^d} \Big|\partial^2_\mu u(\mu)(v, v')\Big|^2 d\mu(v) d\mu(v') \bigg] < +\infty. \tag{4.27}$$

Then, letting

$$\mu_t(\omega^0) = \mathcal{L}^1(X_t)(\omega^0), \quad \omega^0 \in \Omega^0, \quad t \geq 0,$$

with $\boldsymbol{X}$ satisfying (4.25), *it holds that, $\mathbb{P}^0$-almost surely, for all $t \geq 0$,*

$$
\begin{aligned}
u(\mu_t) = u(\mu_0) &+ \int_0^t \mathbb{E}^1\big[\partial_\mu u(\mu_s)(X_s) \cdot B_s\big] ds \\
&+ \int_0^t \mathbb{E}^1\big[(\Sigma_s^0)^\dagger \partial_\mu u(\mu_s)(X_s)\big] \cdot dW_s^0 \\
&+ \frac{1}{2} \int_0^t \mathbb{E}^1\Big[\mathrm{trace}\big\{\partial_v \partial_\mu u(\mu_s)(X_s) \Sigma_s \Sigma_s^\dagger\big\}\Big] ds \\
&+ \frac{1}{2} \int_0^t \mathbb{E}^1\Big[\mathrm{trace}\big\{\partial_v \partial_\mu u(\mu_s)(X_s) \Sigma_s^0 (\Sigma_s^0)^\dagger\big\}\Big] ds \\
&+ \frac{1}{2} \int_0^t \mathbb{E}^1 \tilde{\mathbb{E}}^1\Big[\mathrm{trace}\big\{\partial_\mu^2 u(\mu_s)(X_s, \tilde{X}_s) \Sigma_s^0 (\tilde{\Sigma}_s^0)^\dagger\big\}\Big] ds.
\end{aligned}
\tag{4.28}
$$

Notice that except for the last one, all the expectations under $\mathbb{E}^1$ could be rewritten as expectations under $\tilde{\mathbb{E}}^1$, using the copies $(\tilde{X}_t)_{t\geq 0}$, $(\tilde{B}_t)_{t\geq 0}$, $(\tilde{\Sigma}_t)_{t\geq 0}$ and $(\tilde{\Sigma}_t^0)_{t\geq 0}$ of the processes $(X_t)_{t\geq 0}$, $(B_t)_{t\geq 0}$, $(\Sigma_t)_{t\geq 0}$ and $(\Sigma_t^0)_{t\geq 0}$. In the last term, the expectation is taken under both $\mathbb{E}^1$ and $\tilde{\mathbb{E}}^1$: Notice also that, conditional on $\mathbb{F}^0 = (\mathcal{F}_t^0)_{t\geq 0}$, the processes $(X_t, \Sigma_t^0)_{t\geq 0}$ and $(\tilde{X}_t, \tilde{\Sigma}_t^0)_{t\geq 0}$ are independent on $\Omega^0 \times \Omega^1 \times \tilde{\Omega}^1$.

Importantly, we observe that all the integrands that appear in the right-hand side of (4.27) have versions, up to a null subset in $\mathcal{B}([0, +\infty)) \otimes \mathcal{F}^0$, that are $\mathbb{F}^0$-progressively measurable. The proof is as follows. First we observe that, for each $\mu \in \mathcal{P}_2(\mathbb{R}^d)$, the mappings $\mathbb{R}^d \ni v \mapsto \partial_\mu u(\mu)(v)\mathbf{1}_{\mathrm{Supp}(\mu)}(v)$, $\mathbb{R}^d \ni v \mapsto \partial_v \partial_\mu u(\mu)(v)\mathbf{1}_{\mathrm{Supp}(\mu)}(v)$ and $\mathbb{R}^d \times \mathbb{R}^d \ni (v, v') \mapsto \partial_v \partial_\mu u(\mu)(v, v')\mathbf{1}_{\mathrm{Supp}(\mu)^2}(v, v')$ are measurable as the restrictions to the support of μ are continuous. Then, we know that for any compactly supported smooth functions ϱ_d from $\mathbb{R}^d$ into itself and $\varrho_{d\times d}$ from $\mathbb{R}^{d\times d}$ into itself, the mappings:

$$
\begin{aligned}
&L^2(\Omega^1, \mathcal{F}^1, \mathbb{P}^1; \mathbb{R}^d) \ni X \mapsto \varrho_d\big(\partial_\mu u(\mathcal{L}(X))(X)\big) \in L^2(\Omega^1, \mathcal{F}^1, \mathbb{P}^1; \mathbb{R}^d) \\
&L^2(\Omega^1, \mathcal{F}^1, \mathbb{P}^1; \mathbb{R}^d) \ni X \mapsto \varrho_{d\times d}\big(\partial_v \partial_\mu u(\mathcal{L}(X))(X)\big) \in L^2(\Omega^1, \mathcal{F}^1, \mathbb{P}^1; \mathbb{R}^{d\times d}), \\
&L^2(\Omega^1, \mathcal{F}^1, \mathbb{P}^1; \mathbb{R}^d) \ni X \\
&\qquad \mapsto \varrho_{d\times d}\big(\partial_\mu^2 u(\mathcal{L}(X))(X, \tilde{X})\big) \in L^2(\Omega^1 \times \tilde{\Omega}^1, \mathcal{F}^1 \otimes \tilde{\mathcal{F}}^1, \mathbb{P}^1 \otimes \tilde{\mathbb{P}}^1; \mathbb{R}^{d\times d}),
\end{aligned}
$$

are continuous. For the latter, we use the fact that the mapping $L^2(\Omega^1, \mathcal{F}^1, \mathbb{P}^1; \mathbb{R}^d) \ni X \mapsto \tilde{X} \in L^2(\tilde{\Omega}^1, \tilde{\mathcal{F}}^1, \tilde{\mathbb{P}}^1; \mathbb{R}^d)$ is obviously continuous as $(\tilde{\Omega}^1, \tilde{\mathcal{F}}^1, \tilde{\mathbb{P}}^1)$ just consists in a copy of $(\tilde{\Omega}^1, \tilde{\mathcal{F}}^1, \tilde{\mathbb{P}}^1)$. By the bound $\mathbb{E}[\sup_{0\leq t\leq T} |X_t|^2] < \infty$, which holds true for $T > 0$, we know that $\mathbb{P}^0[\forall T > 0,\ \mathbb{E}^1[\sup_{0\leq t\leq T} |X_t|^2] < \infty] = 1$. We deduce that the process:

$$
[0, +\infty) \times \Omega^0 \ni (t, \omega^0) \mapsto X_t(\omega^0, \cdot) \in L^2(\Omega^1, \mathcal{F}^1, \mathbb{P}^1; \mathbb{R}^d)
$$

has a version with continuous trajectories. Following Lemma 2.5 and the discussion in Subsection 4.3.1, the latter is also $\mathbb{F}^0$-adapted. Therefore, the mapping:

$$[0,+\infty) \times \Omega^0 \ni (t,\omega^0) \mapsto \varrho_d\big(\partial_\mu u(\mathcal{L}^1(X_t))(X_t)\big)(\omega^0,\cdot) \in L^2(\Omega^1,\mathcal{F}^1,\mathbb{P}^1;\mathbb{R}^d)$$

is continuous and $\mathbb{F}^0$-adapted. By the discussion in Subsection 4.3.1, we know that the mapping:

$$[0,+\infty) \times \Omega^0 \ni (t,\omega^0) \mapsto B_t(\omega^0,\cdot) \in L^2(\Omega^1,\mathcal{F}^1,\mathbb{P}^1;\mathbb{R}^d)$$

admits an $\mathbb{F}^0$-progressively measurable version. By continuity of the inner product in $L^2(\Omega^1,\mathcal{F}^1,\mathbb{P}^1;\mathbb{R}^d)$, we deduce that:

$$[0,+\infty) \times \Omega^0 \ni (t,\omega^0) \mapsto \mathbb{E}^1\big[\varrho_d\big(\partial_\mu u\big(\mathcal{L}^1(X_t)\big)(X_t)\big)\cdot B_t\big](\omega^0)$$

admits an $\mathbb{F}^0$-progressively measurable version. By letting ϱ_d tend to the identity function, we complete the analysis of the second term in the first line of (4.28). The same arguments hold for the other terms.

Also, from the equality $\mathbb{P}^0[\forall T>0,\ \mathbb{E}^1[\sup_{0\le t\le T}|X_t|^2] < \infty] = 1$, we know that, with $\mathbb{P}^0$-probability 1, for all $T>0$, there exists a compact subset of $\mathcal{P}_2(\mathbb{R}^d)$ containing the family $(\mu_t)_{0\le t\le T}$, see Corollary (Vol I)-5.6. Therefore,

$$\mathbb{P}^0\bigg[\sup_{0\le t\le T}\int_{\mathbb{R}^d}|\partial_v u(\mu_t)(v)|^2 d\mu_t(v) < \infty\bigg] = 1,$$

that is

$$\mathbb{P}^0\bigg[\sup_{0\le t\le T}\mathbb{E}^1\big[|\partial_v u(\mu_t)(X_t)|^2\big] < \infty\bigg] = 1.$$

This shows that $(\int_0^t \mathbb{E}^1[\partial_v u(\mu_s)(X_s)\cdot B_s]ds)_{t\ge 0}$ defines an $\mathbb{F}^0$-adapted continuous process. A similar argument can be used for the other terms in (4.28). Importantly, observe that the integrand of the stochastic integral therein satisfies:

$$\mathbb{P}^0\bigg[\forall t\ge 0,\ \int_0^t \Big|\mathbb{E}^1\big[(\Sigma_s^0)^\dagger\partial_\mu u(\mu_s)(X_s)\big]\Big|^2 ds < \infty\bigg] = 1,$$

which shows that the stochastic integral is well defined as a local martingale, though under the standing assumptions, it may not be square-integrable.

As for Theorem (Vol I)-5.92, the proof of Theorem 4.14 relies on a mollification argument, whose statement and proof are given first. The reader may want to compare this result with the similar Lemma (Vol I)-5.95.

Lemma 4.15 *Assume that the chain rule holds for any function u that is fully $\mathcal{C}^2$ with first and second order derivatives that are bounded and uniformly continuous with respect to the space and measure arguments. Then, it also holds for any function u that is simply $\mathcal{C}^2$.*

Proof. Throughout the proof, we use the same copy $(\tilde{\Omega}^1, \tilde{\mathcal{F}}^1, \tilde{\mathbb{P}}^1)$ of the space $(\Omega^1, \mathcal{F}^1, \mathbb{P}^1)$ as in Theorem 4.14. For a random variable X defined on $(\Omega^1, \mathcal{F}^1, \mathbb{P}^1)$ (respectively on $(\Omega, \mathcal{F}, \mathbb{P})$), we denote by $\tilde{X}$ the copy of X on $(\tilde{\Omega}^1, \tilde{\mathcal{F}}^1, \tilde{\mathbb{P}}^1)$ (respectively on $(\tilde{\Omega}, \tilde{\mathcal{F}}, \tilde{\mathbb{P}})$). Functions of the measure argument will be systematically lifted onto $L^2(\tilde{\Omega}^1, \tilde{\mathcal{F}}^1, \tilde{\mathbb{P}}^1; \mathbb{R}^d)$.

First Step. We assume that there exists a sequence of functions $(u_n)_{n\geq 1}$ from $\mathcal{P}_2(\mathbb{R}^d)$ into $\mathbb{R}^d$ such that, for each $n \geq 1$, the chain rule can be applied to u_n. Moreover, we assume that for any μ and any $v, v' \in \mathrm{Supp}(\mu)$, the following limits hold as n tends to $+\infty$:

$$\begin{aligned}
&u_n(\mu) \to u(\mu), \quad \partial_\mu u_n(\mu)(v) \to \partial_\mu u(\mu)(v),\\
&\partial_v \partial_\mu u_n(\mu)(v) \to \partial_v \partial_\mu u(\mu)(v),\\
&\partial^2_\mu u_n(\mu)(v, v') \to \partial^2_\mu u(\mu)(v, v').
\end{aligned} \tag{4.29}$$

We also assume that, for any compact subset $\mathcal{K} \subset \mathcal{P}_2(\mathbb{R}^d)$,

$$\begin{aligned}
\sup_{\mu\in\mathcal{K}} \sup_{n\geq 1} \biggl[\int_{\mathbb{R}^d} \bigl|\partial_\mu u_n(\mu)(v)\bigr|^2 d\mu(v) + \int_{\mathbb{R}^d} \bigl|\partial_v \partial_\mu u_n(\mu)(v)\bigr|^2 d\mu(v)&\\
+ \int_{\mathbb{R}^d}\int_{\mathbb{R}^d} \bigl|\partial^2_\mu u_n(\mu)(v, v')\bigr|^2 d\mu(v) d\mu(v') \biggr] < \infty.&
\end{aligned} \tag{4.30}$$

Now, we recall that for (t, ω^0) outside a null Borel subset of $[0, +\infty) \times \Omega^0$, we have:

$$\mathbb{E}^1\Bigl[|X_t(\omega^0, \cdot)|^2 + |B_t(\omega^0, \cdot)|^2 + \bigl|\Sigma^0_t(\omega^0, \cdot)\bigr|^2 + \bigl|(\Sigma_t \Sigma_t^\dagger)(\omega^0, \cdot)\bigr|^2 + \bigl|(\Sigma^0_t(\Sigma^0_t)^\dagger)(\omega^0, \cdot)\bigr|^2\Bigr] < \infty.$$

Therefore, by a standard uniform integrability argument, for (t, ω^0) outside a null Borel subset of $[0, +\infty) \times \Omega^0$, we have:

$$\begin{aligned}
&\lim_{n\to\infty} \mathbb{E}^1\Bigl[\partial_\mu u_n(\mathcal{L}(X_t))(X_t) \cdot B_t\Bigr] = \mathbb{E}^1\Bigl[\partial_\mu u(\mathcal{L}(X_t))(X_t) \cdot B_t\Bigr],\\
&\lim_{n\to\infty} \mathbb{E}^1\Bigl[(\Sigma^0_t)^\dagger \partial_\mu u_n(\mathcal{L}(X_t))(X_t)\Bigr] = \mathbb{E}^1\Bigl[(\Sigma^0_t)^\dagger \partial_\mu u(\mathcal{L}(X_t))(X_t)\Bigr],\\
&\lim_{n\to\infty} \mathbb{E}^1\Bigl[\mathrm{trace}\Bigl\{\partial_v \partial_\mu u_n(\mathcal{L}(X_t))(X_t) \Sigma_t (\Sigma_t)^\dagger\Bigr\}\Bigr] = \mathbb{E}^1\Bigl[\mathrm{trace}\Bigl\{\partial_v \partial_\mu u(\mathcal{L}(X_t))(X_t) \Sigma_t (\Sigma_t)^\dagger\Bigr\}\Bigr],\\
&\lim_{n\to\infty} \mathbb{E}^1\Bigl[\mathrm{trace}\Bigl\{\partial_v \partial_\mu u_n(\mathcal{L}(X_t))(X_t) \Sigma^0_t (\Sigma^0_t)^\dagger\Bigr\}\Bigr] = \mathbb{E}^1\Bigl[\mathrm{trace}\Bigl\{\partial_v \partial_\mu u(\mathcal{L}(X_t))(X_t) \Sigma^0_t (\Sigma^0_t)^\dagger\Bigr\}\Bigr].
\end{aligned}$$

Similarly,

$$\begin{aligned}
&\lim_{n\to\infty} \mathbb{E}^1\tilde{\mathbb{E}}^1\Bigl[\mathrm{trace}\Bigl\{\partial^2_\mu u_n(\mathcal{L}(X_t))(X_t, \tilde{X}_t) \Sigma^0_t (\tilde{\Sigma}^0_t)^\dagger\Bigr\}\Bigr]\\
&\qquad = \mathbb{E}^1\tilde{\mathbb{E}}^1\Bigl[\mathrm{trace}\Bigl\{\partial^2_\mu u(\mathcal{L}(X_t))(X_t, \tilde{X}_t) \Sigma^0_t (\tilde{\Sigma}^0_t)^\dagger\Bigr\}\Bigr].
\end{aligned}$$

In order to pass to the limit in the chain rule itself, we must pass to the limit in the integrals. To do so, we recall that, for any $T > 0$,

$$\mathbb{E}\Big[\sup_{0\le t\le T}|X_t|^2\Big] < \infty,$$

from which we get that, for $\mathbb{P}^0$-almost every $\omega^0 \in \Omega^0$,

$$\mathbb{E}^1\big[\sup_{0\le t\le T}|X_t|^2\big](\omega^0) < \infty.$$

Therefore, by Corollary (Vol I)-5.6, we deduce that, with $\mathbb{P}^0$-probability 1, there exists a compact subset of $\mathcal{P}_2(\mathbb{R}^d)$ containing the family $(\mu_t)_{0\le t\le T}$. In particular, on an event of $\mathbb{P}^0$-probability 1, we have by (4.30):

$$\begin{aligned}\sup_{n\ge 1}\sup_{0\le t\le T}\Big[&\mathbb{E}^1\Big[\partial_\mu u_n\big(\mathcal{L}(X_t)\big)(X_t)\cdot B_t\Big] + \mathbb{E}^1\Big[(\Sigma_t^0)^\dagger\partial_\mu u_n\big(\mathcal{L}(X_t)\big)(X_t)\Big]\\
&+\mathbb{E}^1\Big[\mathrm{trace}\Big\{\partial_v\partial_\mu u_n\big(\mathcal{L}(X_t)\big)(X_t)\Sigma_t(\Sigma_t)^\dagger\Big\}\Big]\\
&+\mathbb{E}^1\Big[\mathrm{trace}\Big\{\partial_v\partial_\mu u_n\big(\mathcal{L}(X_t)\big)(X_t)\Sigma_t^0(\Sigma_t^0)^\dagger\Big\}\Big]\\
&+\mathbb{E}^1\tilde{\mathbb{E}}^1\Big[\mathrm{trace}\Big\{\partial_\mu^2 u_n\big(\mathcal{L}(X_t)\big)(X_t,\tilde X_t)\Sigma_t^0(\tilde\Sigma_t^0)^\dagger\Big\}\Big]\Big] < \infty.\end{aligned}$$

This suffices to apply Lebesgue's dominated convergence theorem in order to pass to the limit in the four Lebesgue integrals in (4.28). So as n tends to $+\infty$,

$$\sup_{0\le t\le T}\left|\int_0^t \mathbb{E}^1\big[\partial_\mu u_n(\mu_s)(X_s)\cdot B_s\big]ds - \int_0^t \mathbb{E}^1\big[\partial_\mu u(\mu_s)(X_s)\cdot B_s\big]ds\right| \to 0$$

in probability under $\mathbb{P}^0$, with similar results for the other terms.

In fact, the same result also holds true for the stochastic integral. Indeed, considering the random variable $\Omega^0 \ni \omega^0 \mapsto \mathcal{L}^1(\sup_{0\le t\le T}|X_t|)(\omega^0) \in \mathcal{P}_2([0,+\infty))$ for a given $T > 0$, we can find, for any $\varepsilon > 0$, a compact subset $\mathcal{K}_\varepsilon$ of $\mathcal{P}_2([0,+\infty))$ such that $\mathbb{P}^0[\mathcal{L}^1(\sup_{0\le t\le T}|X_t|) \in \mathcal{K}_\varepsilon] \ge 1-\varepsilon$. As relative compactness in $\mathcal{P}_2([0,+\infty))$ is mostly described in terms of tails of distributions, we can assume that, for any $[0,+\infty)$-valued random variables ζ and ζ' such that $\zeta \le \zeta'$, $\mathcal{L}(\zeta') \in \mathcal{K}_\varepsilon \Rightarrow \mathcal{L}(\zeta) \in \mathcal{K}_\varepsilon$. In particular, letting:

$$\tau_\epsilon = \inf\Big\{t \ge 0 : \mathcal{L}^1\big(\sup_{0\le s\le t}|X_s|\big) \notin \mathcal{K}_\epsilon\Big\} \wedge T,$$

we clearly have that $\mathbb{P}^0[\tau_\varepsilon < T] \le \varepsilon$. Also, since we may regard $(\mathcal{L}^1(\sup_{0\le s\le t}|X_s|))_{t\ge 0}$ as an $\mathbb{F}^0$-adapted continuous process with values in $\mathcal{P}_2([0,+\infty))$, τ_ε is a stopping time. Then, by repeating the same argument as above, using in addition Lebesgue's dominated convergence theorem, we have, for any $\epsilon > 0$:

$$\lim_{n\to\infty}\mathbb{E}^0\Big[\int_0^{\tau_\epsilon}\Big|\mathbb{E}^1\Big[(\Sigma_s^0)^\dagger\partial_\mu u_n\big(\mathcal{L}(X_s)\big)(X_s)\Big] - \mathbb{E}^1\Big[(\Sigma_s^0)^\dagger\partial_\mu u\big(\mathcal{L}(X_s)\big)(X_s)\Big]\Big|^2 ds\Big]=0,$$

from which we easily deduce that, as n tends to $+\infty$,

$$\sup_{0\le t\le T}\left|\int_0^t \mathbb{E}^1\Big[(\Sigma_s^0)^\dagger \partial_\mu u_n\big(\mathcal{L}(X_s)\big)(X_s)\Big]dW_s^0 - \int_0^t \mathbb{E}^1\Big[(\Sigma_s^0)^\dagger \partial_\mu u\big(\mathcal{L}(X_s)\big)(X_s)\Big]dW_s^0\right|,$$

converges to 0 in probability under $\mathbb{P}^0$. This suffices to derive (4.28).

Second Step. It remains to construct the sequence $(u_n)_{n\ge 1}$. We use the same mollification procedure as in the proofs of Lemma (Vol I)-5.95 and Theorem (Vol I)-5.99. We proceed in several steps. First, for a given smooth function $\rho : \mathbb{R}^d \to \mathbb{R}^d$ with compact support, we let:

$$\forall \mu \in \mathcal{P}_2(\mathbb{R}^d), \quad (u \star \rho)(\mu) = u\big(\mu \circ \rho^{-1}\big),$$

where $\mu \circ \rho^{-1}$ denotes the push forward image of μ by ρ. It is pretty clear that the lifted version of $u \star \rho$ is $\tilde{u} \circ \rho$, where $\tilde{u}$ is the lift of u and ρ is canonically lifted as $\tilde{\rho} : L^2(\tilde{\Omega}^1, \tilde{\mathcal{F}}^1, \tilde{\mathbb{P}}^1; \mathbb{R}^d) \ni \tilde{X} \mapsto \rho(\tilde{X})$. Following if needed (Vol I)-(5.87), we have (with streamlined notations):

$$\begin{aligned}
&\partial_\mu\big[u \star \rho\big](\mu)(v) = \partial_\mu u\big(\mu \circ \rho^{-1}\big)\big(\rho(v)\big)\partial\rho(v),\\
&\partial_\mu^2\big[u \star \rho\big](\mu)(v,w) = \partial_\mu^2 u\big(\mu \circ \rho^{-1}\big)\big(\rho(v),\rho(w)\big)\partial\rho(v)\otimes\partial\rho(w).\\
&\partial_v\partial_\mu\big[u \star \rho\big](\mu)(v)\\
&\quad = \partial_\mu u\big(\mu \circ \rho^{-1}\big)\big(\rho(v)\big)\partial^2\rho(v) + \partial_v\big[\partial_\mu u\big(\mu \circ \rho^{-1}\big)\big]\big(\rho(v)\big)\partial\rho(v)\otimes\partial\rho(v).
\end{aligned} \tag{4.31}$$

Above, we used versions of $\partial_\mu u$ for which the derivatives appearing on each line do exist.

Then, following the proofs of Lemma (Vol I)-5.95 and Theorem (Vol I)-5.99, we deduce that $u \star \rho$ and its first and second order derivatives are bounded and are continuous at points (μ, v) such that v is in the support of μ. Moreover, if we choose a sequence of compactly supported smooth functions $(\rho_n)_{n\ge 1}$ such that $|\rho_n(v)| \le C|v|$, $|\partial\rho_n(v)| \le C$ and $|\partial^2\rho_n(v)| \le C$, for any $n \ge 1$ and $v \in \mathbb{R}^d$ and for a constant $C \ge 0$, and $\rho_n(v) = v$ for any $n \ge 1$ and v with $|v| \le n$, then the sequence $(u_n = u \star \rho_n)_{n\ge 1}$ satisfies (4.29) and (4.30). The net result is that we can assume u to be bounded without any loss of generality.

Now, arguing as in the proof of Theorem (Vol I)-5.99 and denoting by φ the density of the standard Gaussian distribution of dimension d, the function $\mathcal{P}_2(\mathbb{R}^d) \ni \mu \mapsto u(\mu * \varphi)$ satisfies:

$$\partial_\mu^2\big[u(\mu * \varphi)\big](v,w) = \int_{\mathbb{R}^d}\int_{\mathbb{R}^d} \partial_\mu^2 u(\mu * \varphi)(v - v', w - w')\varphi(v')\varphi(w')dv'dw'.$$

By the same argument as in the proof of Theorem (Vol I)-5.99, the map $(\mu, v, w) \mapsto \partial_\mu^2[u(\mu * \varphi)](v,w)$ is continuous. The same holds for $\mathcal{P}_2(\mathbb{R}^d) \times \mathbb{R}^d \times \mathbb{R}^d \ni (\mu, v, w) \mapsto \partial_\mu^2[u(\mu * \varphi_\epsilon)](v,w)$, where φ_ϵ is the density of $N(0, \epsilon I_d)$. Moreover, for any $\mu \in \mathcal{P}_2(\mathbb{R})^d$ and $v, w \in \mathrm{Supp}(\mu)$, $\partial_\mu^2[u(\mu * \varphi_\epsilon)](v,w) \to \partial_\mu^2 u(\mu)(v,w)$ as ε tends to 0, a similar result holding for the other derivatives as well. See the proof of Theorem (Vol I)-5.99. This shows (4.29) with $(u_\epsilon = u(\mu * \varphi_\epsilon))_{\epsilon>0}$. For the same family, (4.30) is easily checked since, for any compact subset $\mathcal{K} \subset \mathcal{P}_2(\mathbb{R}^d)$, the set $\{\mu * \varphi_\epsilon;\ \mu \in \mathcal{K},\ \epsilon \in (0,1)\}$ is included in another compact subset.

In particular, we can assume that u and its first and second order derivatives are bounded and continuous. Repeating the first approximation argument, we can restrict the arguments in u and in its derivatives to compact subsets; as a byproduct, we can assume that u and its first and second order derivatives are bounded and uniformly continuous, see Lemma (Vol I)-(5.94) if necessary. □

We now turn to the proof of Theorem 4.14, which goes along the same lines as the proof of Theorem (Vol I)-5.92.

Proof of Theorem 4.14. By Lemma 4.15, we can assume that u and its first and second order derivatives are bounded and uniformly continuous. Also, recalling that u can be replaced by $u\star\varphi$, for some compactly supported smooth function φ, we can replace $(X_t)_{t\geq 0}$ by $(\varphi(X_t))_{t\geq 0}$. This says that we can assume that $(X_t)_{t\geq 0}$ is a bounded Itô process. In fact, repeating the proof of Lemma 4.15, we can even assume that $(B_t)_{t\geq 0}$, $(\Sigma_t)_{t\geq 0}$ and $(\Sigma_t^0)_{t\geq 0}$ are all bounded. Indeed, it suffices to prove the chain rule when $(X_t)_{t\geq 0}$ is driven by truncated processes and then pass to the limit along the truncation.

First Step. For each integer $N \geq 1$, we construct N copies $(\boldsymbol{X}^\ell = (X_t^\ell)_{t\geq 0})_{\ell=1,\cdots,N}$ of $(X_t)_{t\geq 0}$ driven by independent idiosyncratic noises $(\boldsymbol{W}^\ell = (W_t^\ell)_{t\geq 0})_{\ell=1,\cdots,N}$ in lieu of $\boldsymbol{W} = (W_t)_{t\geq 0}$, and the same common noise $\boldsymbol{W}^0 = (W_t^0)_{t\geq 0}$. This requires to define copies of the initial conditions and the coefficients of the Itô process $\boldsymbol{X}$. Recalling that the filtered probability space $(\Omega, \mathcal{F}, \mathbb{F}, \mathbb{P})$ on which the process $\boldsymbol{X}$ is constructed is the completion of $(\Omega^0\times\Omega^1, \mathcal{F}^0\otimes \mathcal{F}^1, \mathbb{P}^0\otimes\mathbb{P}^1)$ equipped with the complete and right-continuous augmentation of $\mathbb{F}^0\otimes\mathbb{F}^1$, we introduce a family $((\Omega^{1,\ell}, \mathcal{F}^{1,\ell}, \mathbb{F}^{1,\ell}, \mathbb{P}^{1,\ell}))_{\ell=1,\cdots,N}$ of copies of the space $(\Omega^1, \mathcal{F}^1, \mathbb{F}^1, \mathbb{P}^1)$ and then let for any $\ell \in \{1,\cdots,N\}$, $(\bar\Omega^\ell, \bar{\mathcal{F}}^\ell, \bar{\mathbb{F}}^\ell, \bar{\mathbb{P}}^\ell)$ be the completion of $(\Omega^0\times\Omega^{1,\ell}, \mathcal{F}^0\otimes \mathcal{F}^{1,\ell}, \mathbb{P}^0\otimes\mathbb{P}^{1,\ell})$ equipped with the complete and right-continuous augmentation of $\mathbb{F}^0\otimes \mathbb{F}^{1,\ell}$, where we put a "bar" on the different symbols so that there is no confusion between $(\bar\Omega^1, \bar{\mathcal{F}}^1, \bar{\mathbb{F}}^1, \bar{\mathbb{P}}^1)$ and $(\Omega^1, \mathcal{F}^1, \mathbb{F}^1, \mathbb{P}^1)$. On each $(\bar\Omega^\ell, \bar{\mathcal{F}}^\ell, \bar{\mathbb{F}}^\ell, \bar{\mathbb{P}}^\ell)$ with $\ell \in \{1,\cdots,N\}$, it is now quite simple to copy the initial condition and the coefficients of the Itô process $\boldsymbol{X}$. As X_0 reads as a measurable mapping $X_0 : \Omega^0\times\Omega^1 \ni (\omega^0,\omega^1) \mapsto X_0(\omega^0,\omega^1) \in \mathbb{R}^d$, the copy X_0^ℓ, for any $\ell \in \{1,\cdots,N\}$, reads $X_0^\ell : \Omega^0\times\Omega^{1,\ell} \ni (\omega^0,\omega^{1,\ell}) \mapsto X_0^\ell(\omega^0,\omega^{1,\ell}) = X_0(\omega^0,\omega^{1,\ell}) \in \mathbb{R}^d$, and similarly for $(B_t)_{t\geq 0}$, $(\Sigma_t)_{t\geq 0}$ and $(\Sigma_t^0)_{t\geq 0}$, the copies of which are respectively denoted by $\boldsymbol{B}^\ell = (B_t^\ell)_{t\geq 0}$, $\boldsymbol{\Sigma}^\ell = (\Sigma_t^\ell)_{t\geq 0}$ and $\boldsymbol{\Sigma}^{0,\ell} = (\Sigma_t^{0,\ell})_{t\geq 0}$. In the end, for any $\ell \in \{1,\cdots,N\}$, the copy $\boldsymbol{X}^\ell = (X_t^\ell)_{t\geq 0}$ of $\boldsymbol{X}$ has dynamics:

$$dX_t^\ell = B_t^\ell dt + \Sigma_t^\ell dW_t^\ell + \Sigma_t^{0,\ell} dW_t^0, \quad t \geq 0.$$

We now collect all the N copies on a single probability space $(\bar\Omega^{1\ldots N}, \bar{\mathcal{F}}^{1\ldots N}, \bar{\mathbb{P}}^{1\ldots N})$, obtained as the completion of $(\Omega^0\times\Omega^{1,1\ldots N}, \mathcal{F}^0\otimes\mathcal{F}^{1,1\ldots N}, \mathbb{P}^0\otimes\mathbb{P}^{1,1\ldots N})$, with:

$$\Omega^{1,1\ldots N} = \prod_{\ell=1}^N \Omega^{1,\ell}, \quad \mathcal{F}^{1,1\ldots N} = \bigotimes_{\ell=1}^N \mathcal{F}^{1,\ell}, \quad \mathbb{F}^{1,1\ldots N} = \bigotimes_{\ell=1}^N \mathbb{F}^{1,\ell}, \quad \mathbb{P}^{1,1\ldots N} = \bigotimes_{\ell=1}^N \mathbb{P}^{1,\ell},$$

and equipped with the complete and right-continuous augmentation $\bar{\mathbb{F}}^{1\ldots N}$ of $\mathbb{F}^0\otimes\mathbb{F}^{1,1\ldots N}$.

All the various random variables constructed on the different spaces Ω^0, $\Omega^{1,1}$, $\bar\Omega^1,\cdots,\Omega^{1,N}$, $\bar\Omega^N$, can then be canonically transferred to $(\bar\Omega^{1\ldots N}, \bar{\mathcal{F}}^{1\ldots N}, \bar{\mathbb{F}}^{1\ldots N}, \bar{\mathbb{P}}^{1\ldots N})$. Indeed, for any $\ell \in \{1,\cdots,N\}$ and any event E in $\bar{\mathcal{F}}^\ell$, the set $\{(\omega^0,\omega^1,\cdots,\omega^N) \in \bar\Omega^{1\ldots N} : (\omega^0,\omega^\ell) \in E\}$ belongs to $\bar{\mathcal{F}}^{1\ldots N}$, which is obviously true if $E \in \mathcal{F}^0\otimes\mathcal{F}^{1,\ell}$ and which is also true if E is a null subset of $(\bar{\Omega}^\ell, \bar{\mathcal{F}}^\ell, \bar{\mathbb{P}}^\ell)$. This allows to define, on the same space, the flow of marginal empirical measures:

$$\bar{\mu}_t^N = \frac{1}{N}\sum_{\ell=1}^N \delta_{X_t^\ell}, \quad t \geq 0.$$

Notice that, for any $\omega^0 \in \Omega^0$, the processes $(\boldsymbol{X}^\ell(\omega^0,\cdot) = (X_t^\ell(\omega^0,\cdot))_{t\geq 0})_{\ell=1,\cdots,N}$ are independent and identically distributed.

We now have all the required ingredients to apply Itô's formula to $u^N(X_t^1,\cdots,X_t^N)$, where, as in Chapter (Vol I)-5, u^N is the empirical projection defined as $u^N(x_1,\cdots,x_N) = u(N^{-1}\sum_{i=1}^N \delta_{x_i})$. Applying the classical Itô formula to u^N and expressing the result in terms of L-derivatives using Proposition 4.13, we deduce that $\bar{\mathbb{P}}^{1\ldots N}$-almost surely on the space $(\bar{\Omega}^{1\ldots N}, \bar{\mathcal{F}}^{1\ldots N}, \bar{\mathbb{P}}^{1\ldots N})$, for any $t \geq 0$:

$$\begin{aligned}
u^N(X_t^1,\cdots,X_t^N) &= u^N(X_0^1,\cdots,X_0^N) \; + \frac{1}{N}\sum_{\ell=1}^N \int_0^t \partial_\mu u(\bar{\mu}_s^N)(X_s^\ell)\cdot B_s^\ell ds \\
&+ \frac{1}{N}\sum_{\ell=1}^N \int_0^t \partial_\mu u(\bar{\mu}_s^N)(X_s^\ell)\cdot (\Sigma_s^\ell dW_s^\ell) + \frac{1}{N}\sum_{\ell=1}^N \int_0^t \partial_\mu u(\bar{\mu}_s^N)(X_s^\ell)\cdot (\Sigma_s^{0,\ell} dW_s^0) \\
&+ \frac{1}{2N}\biggl[\sum_{\ell=1}^N \int_0^t \mathrm{trace}\{\partial_v\partial_\mu u(\bar{\mu}_s^N)(X_s^\ell)A_s^\ell\}ds + \sum_{\ell=1}^N \int_0^t \mathrm{trace}\{\partial_v\partial_\mu u(\bar{\mu}_s^N)(X_s^\ell)A_s^{0,\ell}\}ds\biggr] \\
&+ \frac{1}{2N^2}\biggl[\sum_{\ell=1}^N \int_0^t \mathrm{trace}\{\partial_\mu^2 u(\bar{\mu}_s^N)(X_s^\ell, X_s^\ell)A_s^\ell\}ds + \sum_{\ell,\ell'=1}^N \int_0^t \mathrm{trace}\{\partial_\mu^2 u(\bar{\mu}_s^N)(X_s^\ell, X_s^{\ell'})A_s^{0,\ell,\ell'}\}ds\biggr],
\end{aligned}$$

with $A_s^\ell = \Sigma_s^\ell(\Sigma_s^\ell)^\dagger$, $A_s^{0,\ell,\ell'} = \Sigma_s^{0,\ell}(\Sigma_s^{0,\ell'})^\dagger$ and $A_s^{0,\ell} = A_s^{0,\ell,\ell}$.

We now proceed as in the proof of Theorem (Vol I)-5.92, paying attention to the fact that, for $\mathrm{Leb}_1 \otimes \mathbb{P}^0$ almost every $(t,\omega^0) \in [0,+\infty) \times \Omega^0$, all the vectors:

$$\Bigl(\bigl(\bar{\mu}_t^N(\omega^0,\cdot), X_t^\ell(\omega^0,\cdot), B_t^\ell(\omega^0,\cdot), \Sigma_t^\ell(\omega^0,\cdot), \Sigma_t^{0,\ell}(\omega^0,\cdot), A_t^\ell(\omega^0,\cdot), A_t^{0,\ell}(\omega^0,\cdot)\bigr)\Bigr)_{\ell=1,\cdots,N}$$

have the same law on $(\bar{\Omega}^{1,1\ldots N}, \bar{\mathcal{F}}^{1,1\ldots N}, \bar{\mathbb{P}}^{1,1\ldots N})$, the same being true for the vectors:

$$\Bigl(\bigl(\bar{\mu}_t^N(\omega^0,\cdot), X_t^\ell(\omega^0,\cdot), X_t^{\ell'}(\omega^0,\cdot), A_t^{0,\ell,\ell'}(\omega^0,\cdot)\bigr)\Bigr)_{\ell,\ell'=1,\cdots,N,\ell\neq\ell'}.$$

Therefore, taking expectations with respect to $\bar{\mathbb{P}}^{1,1\ldots N}$ in the above expansion, we get, for any $t \geq 0$ and almost every $\omega^0 \in \Omega^0$ under $\mathbb{P}^0$,

$$\begin{aligned}
\bar{\mathbb{E}}^{1,1\ldots N}\bigl[u(\bar{\mu}_t^N(\omega^0,\cdot))\bigr] &= \bar{\mathbb{E}}^{1,1\ldots N}\bigl[u(\bar{\mu}_0^N(\omega^0,\cdot))\bigr] \\
&+ \int_0^t \bar{\mathbb{E}}^{1,1\ldots N}\bigl[\partial_\mu u(\bar{\mu}_s^N(\omega^0,\cdot))(X_s^1(\omega^0,\cdot))\cdot B_s^1(\omega^0,\cdot)\bigr]ds
\end{aligned}$$

$$
\begin{aligned}
&+ \int_0^t \bar{\mathbb{E}}^{1,1\ldots N}\big[(\Sigma_s^{0,1}(\omega^0,\cdot))^\dagger \partial_\mu u(\bar{\mu}_s^N(\omega^0,\cdot))(X_s^1(\omega^0,\cdot))\big] \cdot dW_s^0 \qquad (4.32)\\
&+ \frac{1}{2}\int_0^t \bar{\mathbb{E}}^{1,1\ldots N}\Big[\mathrm{trace}\big\{\partial_v \partial_\mu u(\bar{\mu}_s^N(\omega^0,\cdot))(X_s^1(\omega^0,\cdot))A_s^1(\omega^0,\cdot)\big\}\Big]ds\\
&+ \frac{1}{2}\int_0^t \bar{\mathbb{E}}^{1,1\ldots N}\Big[\mathrm{trace}\big\{\partial_v \partial_\mu u(\bar{\mu}_s^N(\omega^0,\cdot))(X_s^1(\omega^0,\cdot))A_s^{0,1}(\omega^0,\cdot)\big\}\Big]ds\\
&+ \frac{1}{2}\int_0^t \bar{\mathbb{E}}^{1,1\ldots N}\Big[\mathrm{trace}\big\{\partial_\mu^2 u(\bar{\mu}_s^N(\omega^0,\cdot))(X_s^1(\omega^0,\cdot),X_s^2(\omega^0,\cdot))A_s^{0,1,2}(\omega^0,\cdot)\big\}\Big]ds\\
&+ O(1/N),
\end{aligned}
$$

where $O(\cdot)$ stands for the Landau notation, the underlying constant being uniform in $\omega^0 \in \Omega^0$ and in t in a compact subset of $[0,+\infty)$.

Second Step. We now investigate the convergence of the flow of measures $(\bar{\mu}_t^N(\omega^0,\cdot))_{t\geq 0}$ to $(\mu_t^1(\omega^0))_{t\geq 0}$. We proceed as follows. We use the fact that, under the expectation $\mathbb{E}$, for any $T>0$,

$$
\forall s,s' \in [0,T], \quad \mathbb{E}\big[|X_s - X_{s'}|^p\big] \leq C_p|s'-s|^{p/2},
$$

for a constant C_p depending only on p and T. By Kolmogorov's continuity theorem, we can find for any $\alpha \in (0,1/2)$, a random variable ξ defined on $(\Omega,\mathcal{F},\mathbb{P})$, with finite moments of any order, such that for every $\omega = (\omega^0,\omega^1) \in \Omega = \Omega^0 \times \Omega^1$,

$$
\forall s,s' \in [0,T], \quad \big|X_{s'}(\omega^0,\omega^1) - X_s(\omega^0,\omega^1)\big| \leq \xi(\omega^0,\omega^1)|s'-s|^\alpha.
$$

Denoting by ξ^ℓ the copy of ξ on $\Omega^0 \times \Omega^{1,\ell}$, constructed for each $\ell = 1,\cdots,N$ with the procedure described earlier, we have for every $(\omega^0,\omega^{1,\ell}) \in \bar{\Omega}^\ell = \Omega^0 \times \Omega^{1,\ell}$,

$$
\forall s,s' \in [0,T], \quad \big|X_{s'}^\ell(\omega^0,\omega^{1,\ell}) - X_s^\ell(\omega^0,\omega^{1,\ell})\big| \leq \xi^\ell(\omega^0,\omega^{1,\ell})|s'-s|^\alpha.
$$

In the sequel, the variables $(\xi^\ell)_{\ell=1,\cdots,N}$ are extended in an obvious way to the whole $\bar{\Omega}^{1\ldots N} = \Omega^0 \times \bar{\Omega}^{1,1\ldots N}$. For $\mathbb{P}^0$-almost every $\omega^0 \in \Omega^0$, the variables $(\xi^\ell(\omega^0,\cdot))_{\ell=1,\cdots,N}$ are independent and identically distributed under $\bar{\mathbb{P}}^{1,1\ldots N}$.

The real $T>0$ being fixed, we consider the subdivision $(s_q = qT/Q)_{q=0,\cdots,Q}$ of $[0,T]$, for an integer $Q \geq 1$. For a given $s \in [0,T)$, we set $\bar{s} = qT/Q$ whenever $qT/Q \leq s < (q+1)T/Q$. When $s=T$, we let $\bar{s}=T$. Then, for $\mathbb{P}^0$-almost every $\omega^0 \in \Omega^0$,

$$
\begin{aligned}
&\bar{\mathbb{E}}^{1,1\ldots N}\Big[\sup_{0\leq s\leq T}\Big(W_2\big(\bar{\mu}_s^N(\omega^0,\cdot),\mu_s^1(\omega^0)\big)\Big)^2\Big]\\
&\leq C\bigg(\bar{\mathbb{E}}^{1,1\ldots N}\Big[\sup_{0\leq s\leq T}\Big(W_2\big(\bar{\mu}_s^N(\omega^0,\cdot),\bar{\mu}_{\bar{s}}^N(\omega^0,\cdot)\big)\Big)^2\Big]\\
&\quad + \sup_{0\leq s\leq T}\Big(W_2\big(\mu_s^1(\omega^0),\mu_{\bar{s}}^1(\omega^0)\big)\Big)^2 \qquad (4.33)\\
&\quad + \bar{\mathbb{E}}^{1,1\ldots N}\Big[\sup_{q=0,\cdots,Q}\Big(W_2\big(\bar{\mu}_{s_q}^N(\omega^0,\cdot),\mu_{s_q}^1(\omega^0)\big)\Big)^2\Big]\bigg),
\end{aligned}
$$

for a universal constant C, and where we let, for every $s \in [0, T]$,

$$\mu_s^1(\omega^0) = \mathcal{L}\big(X_s^1(\omega^0, \cdot)\big).$$

We notice that, in (4.33),

$$\begin{aligned}
&\bar{\mathbb{E}}^{1,1\ldots N}\Big[\sup_{0\le s\le T} W_2\Big(\bar{\mu}_s^N(\omega^0,\cdot), \bar{\mu}_{\bar{s}}^N(\omega^0,\cdot)\Big)^2\Big] \\
&\le \frac{1}{N}\bar{\mathbb{E}}^{1,1\ldots N}\Big[\sup_{0\le s\le T}\sum_{\ell=1}^N \big|X_s^\ell(\omega^0,\cdot) - X_{\bar{s}}^\ell(\omega^0,\cdot)\big|^2\Big] \\
&\le \frac{1}{N}\Big(\frac{T}{Q}\Big)^{2\alpha}\bar{\mathbb{E}}^{1,1\ldots N}\Big[\sum_{\ell=1}^N \big(\xi^\ell(\omega^0,\cdot)\big)^2\Big] = \Big(\frac{T}{Q}\Big)^{2\alpha}\mathbb{E}^1\big[\big(\xi(\omega^0,\cdot)\big)^2\big],
\end{aligned}$$

so that, for $\mathbb{P}^0$-almost every $\omega^0 \in \Omega^0$,

$$\limsup_{N\to\infty}\bar{\mathbb{E}}^{1,1\ldots N}\Big[\sup_{0\le s\le T} W_2\Big(\bar{\mu}_s^N(\omega^0,\cdot), \bar{\mu}_{\bar{s}}^N(\omega^0,\cdot)\Big)^2\Big] \le \Big(\frac{T}{Q}\Big)^{2\alpha}\mathbb{E}^1\big[\big(\xi(\omega^0,\cdot)\big)^2\big].$$

Similarly,

$$\begin{aligned}
\sup_{0\le s\le T} W_2\Big(\mu_s^1(\omega^0), \mu_{\bar{s}}^1(\omega^0)\Big)^2 &\le \sup_{0\le s\le T}\mathbb{E}^1\big[|X_s(\omega^0,\cdot) - X_{\bar{s}}(\omega^0,\cdot)|^2\big] \\
&\le \Big(\frac{T}{Q}\Big)^{2\alpha}\mathbb{E}^1\big[\big(\xi(\omega^0,\cdot)\big)^2\big],
\end{aligned}$$

so that, for $\mathbb{P}^0$-almost every $\omega^0 \in \Omega^0$,

$$\limsup_{N\to\infty}\sup_{0\le s\le T} W_2\Big(\mu_s^1(\omega^0), \mu_{\bar{s}}^1(\omega^0)\Big)^2 \le \Big(\frac{T}{Q}\Big)^{2\alpha}\mathbb{E}^1\big[\big(\xi(\omega^0,\cdot)\big)^2\big].$$

Taking the lim sup over N in (4.33) and recalling that the 2-Wasserstein distance between a distribution and the empirical distribution of any of its sample tends to 0 in L^2 norm as the size of the sample tends to ∞, see (Vol I)-(5.19) if needed, we get:

$$\limsup_{N\to\infty}\bar{\mathbb{E}}^{1,1\ldots N}\Big[\sup_{0\le s\le T} W_2\Big(\bar{\mu}_s^N(\omega^0,\cdot), \mu_s^1(\omega^0)\Big)^2\Big] \le C\Big(\frac{T}{Q}\Big)^{2\alpha}\mathbb{E}^1\big[\big(\xi(\omega^0,\cdot)\big)^2\big].$$

Recalling that, with $\mathbb{P}^0$-probability 1, $\mathbb{E}^1[(\xi(\omega^0,\cdot))^2] < \infty$ and letting $Q \to \infty$, we deduce that, for $\mathbb{P}^0$- almost every $\omega^0 \in \Omega^0$,

$$\limsup_{N\to\infty}\bar{\mathbb{E}}^{1,1\ldots N}\Big[\sup_{0\le s\le T} W_2\Big(\bar{\mu}_s^N(\omega^0,\cdot), \mu_s^1(\omega^0)\Big)^2\Big] = 0.$$

Third Step. Using the boundedness and the continuity of the coefficients it is now plain to pass to the limit in (4.32). The value of $t \ge 0$ being fixed, all the terms in (4.32) converge in $\mathbb{P}_0$ probability. In the limit, we get the identity:

$$
\begin{aligned}
\bar{\mathbb{E}}^1\big[u(\mu_t^1(\omega^0))\big] &= \bar{\mathbb{E}}^1\big[u(\mu_0^1(\omega^0))\big] + \int_0^t \bar{\mathbb{E}}^1\big[\partial_\mu u(\mu_s^1(\omega^0))(X_s^1(\omega^0,\cdot))B_s^1(\omega^0,\cdot)\big]ds \\
&+ \int_0^t \bar{\mathbb{E}}^1\big[(\Sigma_s^{0,1}(\omega^0,\cdot))^\dagger \partial_\mu u(\mu_s^1(\omega^0))(X_s^1(\omega^0,\cdot))\big]\cdot dW_s^0 \\
&+ \frac{1}{2}\int_0^t \bar{\mathbb{E}}^1\Big[\mathrm{trace}\big\{\partial_v \partial_\mu u(\mu_s^1(\omega^0))(X_s^1(\omega^0,\cdot))A_s^1(\omega^0,\cdot)\big\}\Big]ds \qquad (4.34)\\
&+ \frac{1}{2}\int_0^t \bar{\mathbb{E}}^1\Big[\mathrm{trace}\big\{\partial_v \partial_\mu u(\mu_s^1(\omega^0))(X_s^1(\omega^0,\cdot))A_s^{0,1}(\omega^0,\cdot)\big\}\Big]ds \\
&+ \frac{1}{2}\int_0^t \bar{\mathbb{E}}^1\Big[\mathrm{trace}\big\{\partial_\mu^2 u(\mu_s^1(\omega^0))(X_s^1(\omega^0,\cdot),X_s^2(\omega^0,\cdot))A_s^{0,1,2}(\omega^0,\cdot)\big\}\Big]ds.
\end{aligned}
$$

By pathwise continuity (in t) of the various terms in the above identity, (4.34) holds $\mathbb{P}^0$-almost surely, for every $t \in [0,T]$. To conclude the proof notice that the process $(\mu_t, X_t, B_t, \Sigma_t, \Sigma^0)_{0\le t\le T}$ has the same distribution as $(\mu_t^1, X_t^1, B_t^1, \Sigma_t, \Sigma_t^{0,1})_{0\le t\le T}$ and that the stochastic process $(\mu_t^1, X_t^1, X_t^2, \Sigma_t^{0,1}, \Sigma_t^{0,2})_{0\le t\le T}$ has the same distribution as $(\mu_t, X_t, \tilde{X}_t, \Sigma_t^0, \tilde{\Sigma}_t^0)_{0\le t\le T}$. □

4.3.4 Chain Rule in Both the State and the Measure Variables

For the purpose of the next section, we prove an extension of the chain rule to functions u which depend on both (t,x) and the measure argument μ. Recall that t stands for time, and x for the state variable in the physical space on which the probability measure μ is defined.

For a given $T > 0$, we say that a function

$$
u : [0,T] \times \mathbb{R}^d \times \mathcal{P}_2(\mathbb{R}^d) \ni (t,x,\mu) \mapsto u(t,x,\mu) \in \mathbb{R}
$$

is *simply* $\mathcal{C}^{1,2,2}$ if the conditions (**A1**)–(**A4**) below are satisfied.

Assumption (Joint Chain Rule Common Noise). For a given $T > 0$, the function u is a continuous function from $[0,T]\times\mathbb{R}^d\times\mathcal{P}_2(\mathbb{R}^d)$ to $\mathbb{R}$ that satisfies:

(**A1**) u is differentiable with respect to t and the partial derivative $\partial_t u : [0,T]\times \mathbb{R}^d \times \mathcal{P}_2(\mathbb{R}^d) \to \mathbb{R}$ is continuous.

(**A2**) u is twice differentiable with respect to x and the partial derivatives $\partial_x u : [0,T] \times \mathbb{R}^d \times \mathcal{P}_2(\mathbb{R}^d) \to \mathbb{R}^d$ and $\partial_{xx}^2 u : [0,T] \times \mathbb{R}^d \times \mathcal{P}_2(\mathbb{R}^d) \to \mathbb{R}^{d\times d}$ are continuous.

(**A3**) For any $(t,x) \in [0,T] \times \mathbb{R}^d$, the mapping $\mathcal{P}_2(\mathbb{R}^d) \ni \mu \mapsto u(t,x,\mu)$ is *simply* $\mathcal{C}^2$; moreover, the versions of $\partial_\mu u(t,x,\mu)(\cdot)$ and of its derivatives used for each $\mu \in \mathcal{P}_2(\mathbb{R}^d)$ in the *simple* $\mathcal{C}^2$ property is such that the global maps $[0,T]\times\mathbb{R}^d\times\mathcal{P}_2(\mathbb{R}^d)\times\mathbb{R}^d \ni (t,x,\mu,v) \mapsto \partial_\mu u(t,x,\mu)(v) \in$

(continued)

$\mathbb{R}^d$, $[0,T] \times \mathbb{R}^d \times \mathcal{P}_2(\mathbb{R}^d) \times \mathbb{R}^d \ni (t,x,\mu,v) \mapsto \partial_v \partial_\mu u(t,x,\mu)(v) \in \mathbb{R}^{d\times d}$, and $[0,T] \times \mathbb{R}^d \times \mathcal{P}_2(\mathbb{R}^d) \times \mathbb{R}^d \times \mathbb{R}^d \ni (t,x,\mu,v,v') \mapsto \partial^2_\mu u(t,x,\mu)(v,v') \in \mathbb{R}^{d\times d}$ are locally bounded and continuous at any points (t,x,μ,v) such that $v \in \mathrm{Supp}(\mu)$ and (t,x,μ,v,v') such that $v,v' \in \mathrm{Supp}(\mu)$.

(A4) For the same version of the derivative of u in μ as in **(A3)**, the global map $[0,T] \times \mathbb{R}^d \times \mathcal{P}_2(\mathbb{R}^d) \times \mathbb{R}^d \ni (t,x,\mu,v) \mapsto \partial_\mu u(t,x,\mu)(v) \in \mathbb{R}^d$ is differentiable in x, the partial derivative $\partial_x \partial_\mu u : [0,T] \times \mathbb{R}^d \times \mathcal{P}_2(\mathbb{R}^d) \times \mathbb{R}^d \ni (t,x,\mu,v) \mapsto \partial_x \partial_\mu u(t,x,\mu)(v) \in \mathbb{R}^{d\times d}$ being locally bounded and continuous at any point (t,x,μ,v) such that $v \in \mathrm{Supp}(\mu)$.

Remark 4.16 *When $\partial_x \partial_\mu u$ is bounded, the mapping:*

$$[0,T] \times \mathbb{R}^d \times L^2(\Omega^1, \mathcal{F}^1, \mathbb{P}^1; \mathbb{R}^d) \ni (t,x,X) \\ \mapsto \partial_x \partial_\mu u(t,x,\mathcal{L}^1(X))(X) \in L^2(\Omega^1, \mathcal{F}^1, \mathbb{P}^1; \mathbb{R}^d)$$

is continuous. In particular, it is easy to check that Schwarz' theorem applies in such a framework. To be more specific, it permits to exchange the derivative in $\mathbb{R}^d$ and the Fréchet derivative in $L^2(\Omega^1, \mathcal{F}^1, \mathbb{P}^1; \mathbb{R}^d)$. Hence, for any $(t,x) \in [0,T] \times \mathbb{R}^d$, the function $\mathcal{P}_2(\mathbb{R}^d) \ni \mu \mapsto \partial_x u(t,x,\mu)$ is L-differentiable with respect to μ and

$$\partial_\mu \partial_x u(t,x,\mu)(\cdot) = \Big[\partial_x \partial_\mu u(t,x,\mu)(\cdot)\Big]^\dagger.$$

The following theorem is the generalization of the chain rule proved in Proposition (Vol I)-5.102 for deterministic flows of marginal distributions. In order to state the appropriate chain rule, we consider two $\mathbb{R}^d$-valued Itô processes $\boldsymbol{X}^0 = (X^0_t)_{0\le t\le T}$ and $\boldsymbol{X} = (X_t)_{0\le t\le T}$. The process $\boldsymbol{X} = (X_t)_{0\le t\le T}$ satisfies the same assumption as in Subsection 4.3.1. In particular, its dynamics are given by (4.25), but on $[0,T]$ in lieu of the whole $[0,+\infty)$, with coefficients satisfying (4.26). The touted Itô's expansion will be computed along the flow of its conditional marginal laws. In this sense, it plays the role played by the Itô process $\boldsymbol{\xi} = (\xi_t)_{0\le t\le T}$ in Proposition (Vol I)-5.102.

Similarly, the Itô process $\boldsymbol{X}^0 = (X^0_t)_{0\le t\le T}$ is assumed to satisfy:

$$dX^0_t = b_t dt + \sigma_t dW_t + \sigma^0_t dW^0_t, \quad t \in [0,T], \tag{4.35}$$

with $(b_t)_{0\le t\le T}$, $(\sigma_t)_{0\le t\le T}$ and $(\sigma^0_t)_{0\le t\le T}$ satisfying the same assumptions as the coefficients $(B_t)_{0\le t\le T}$, $(\Sigma_t)_{0\le t\le T}$ and $(\Sigma^0_t)_{0\le t\le T}$ in (4.26). For the purpose of the chain rule we are about to prove, this process plays the role of the Itô process $\boldsymbol{X} = (X_t)_{0\le t\le T}$ in Proposition (Vol I)-5.102. The main difference is that because of the presence of the *common noise* $\boldsymbol{W}^0$ in the dynamics of both Itô processes, we consider the flow of conditional marginal distributions given the common noise, and as a result, the new chain rule will involve second order L-derivatives which were not present in the original chain rule of Proposition (Vol I)-5.102.

Theorem 4.17 *Let us assume that the Itô processes* $\boldsymbol{X}^0$ *and* $\boldsymbol{X}$ *are as above, and that* u *is a simply* $\mathcal{C}^{1,2,2}$ *function from* $[0,T] \times \mathbb{R}^d \times \mathcal{P}_2(\mathbb{R}^d)$ *to* $\mathbb{R}$ *such that, for any compact subset* $\mathcal{K} \subset \mathbb{R}^d \times \mathcal{P}_2(\mathbb{R}^d)$,

$$
\begin{aligned}
\sup_{(t,x,\mu)\in[0,T]\times\mathcal{K}} \Bigg[& \int_{\mathbb{R}^d} \big|\partial_\mu u(t,x,\mu)(v)\big|^2 d\mu(v) + \int_{\mathbb{R}^d} \Big|\partial_v\big[\partial_\mu u(t,x,\mu)\big](v)\Big|^2 d\mu(v) \\
& + \int_{\mathbb{R}^d}\int_{\mathbb{R}^d} \Big|\partial_x \partial_\mu u(t,x,\mu)(v)\Big|^2 d\mu(v) \qquad (4.36) \\
& + \int_{\mathbb{R}^d}\int_{\mathbb{R}^d} \Big|\partial_\mu^2 u(t,x,\mu)(v,v')\Big|^2 d\mu(v) d\mu(v') \Bigg] < \infty,
\end{aligned}
$$

Then, letting:

$$
\mu_t(\omega^0) = \mathcal{L}\big(X_t(\omega^0,\cdot)\big), \quad \omega^0 \in \Omega^0, \quad t \in [0,T],
$$

it holds that, $\mathbb{P}$*-almost surely, for all* $t \in [0,T]$,

$$
\begin{aligned}
u(t,X_t^0,\mu_t) &= u(0,X_0^0,\mu_0) + \int_0^t \partial_t u(s,X_s^0,\mu_s) ds + \int_0^t \partial_x u(s,X_s^0,\mu_s)\cdot b_s ds \\
&+ \int_0^t \tilde{\mathbb{E}}^1\big[\partial_\mu u(s,X_s^0,\mu_s)(\tilde{X}_s)\cdot \tilde{B}_s\big] ds + \int_0^t \partial_x u(s,X_s^0,\mu_s)\cdot \big(\sigma_s dW_s\big) \\
&+ \int_0^t \partial_x u(s,X_s^0,\mu_s)\cdot \big(\sigma_s^0 dW_s^0\big) + \int_0^t \tilde{\mathbb{E}}^1\big[(\tilde{\Sigma}_s^0)^\dagger \partial_\mu u\big(s,X_s^0,\mu_s\big)(\tilde{X}_s)\big]\cdot dW_s^0 \\
&+ \frac{1}{2}\int_0^t \mathrm{trace}\big[\partial_{xx}^2 u(s,X_s^0,\mu_s)\sigma_s \sigma_s^\dagger\big] ds + \frac{1}{2}\int_0^t \mathrm{trace}\big[\partial_{xx}^2 u(s,X_s^0,\mu_s)\sigma_s^0 (\sigma_s^0)^\dagger\big] ds \\
&+ \frac{1}{2}\int_0^t \tilde{\mathbb{E}}^1\Big[\mathrm{trace}\big\{\partial_v \partial_\mu u(s,X_s^0,\mu_s)(\tilde{X}_s)\tilde{\Sigma}_s \tilde{\Sigma}_s^\dagger\big\}\Big] ds \qquad (4.37) \\
&+ \frac{1}{2}\int_0^t \tilde{\mathbb{E}}^1\Big[\mathrm{trace}\big\{\partial_v \partial_\mu u(s,X_s^0,\mu_s)(\tilde{X}_s)\tilde{\Sigma}_s^0 \big(\tilde{\Sigma}_s^0\big)^\dagger\big\}\Big] ds \\
&+ \frac{1}{2}\int_0^t \tilde{\mathbb{E}}^1\tilde{\tilde{\mathbb{E}}}^1\Big[\mathrm{trace}\big\{\partial_\mu^2 u(s,X_s^0,\mu_s)(\tilde{X}_s,\tilde{\tilde{X}}_s)\tilde{\Sigma}_s^0 \big(\tilde{\tilde{\Sigma}}_s^0\big)^\dagger\big\}\Big] ds \\
&+ \int_0^t \tilde{\mathbb{E}}^1\Big[\mathrm{trace}\big\{\partial_x \partial_\mu u(s,X_s^0,\mu_s)(\tilde{X}_s)\sigma_s^0 \big(\tilde{\Sigma}_s^0\big)^\dagger\big\}\Big] ds,
\end{aligned}
$$

where, as in the statement of Theorem 4.14, the processes $(\tilde{X}_t)_{t\geq 0}$, $(\tilde{B}_t)_{t\geq 0}$, $(\tilde{\Sigma}_t)_{t\geq 0}$ *and* $(\tilde{\Sigma}^0_t)_{t\geq 0}$ *are copies of the processes* $(X_t)_{t\geq 0}$, $(B_t)_{t\geq 0}$, $(\Sigma_t)_{t\geq 0}$ *and* $(\Sigma^0_t)_{t\geq 0}$ *on the space* $\tilde{\Omega} = \Omega^0 \times \tilde{\Omega}^1$. *In the penultimate term,* $\tilde{\tilde{\Omega}}^1$ *is a new copy of* Ω^1 *and* $(\tilde{\tilde{X}}_t)_{t\geq 0}$ *and* $(\tilde{\tilde{\Sigma}}^0_t)_{t\geq 0}$ *are copies of the processes* $(X_t)_{t\geq 0}$ *and* $(\Sigma_t)_{t\geq 0}$ *on the space* $\tilde{\tilde{\Omega}} = \Omega^0 \times \tilde{\tilde{\Omega}}^1$.

By adapting the discussion right after the statement of Theorem 4.14, see also Remark (Vol I)-5.103, we notice that all the integrands appearing in the chain rule have versions, up a to null subset in $\mathcal{B}([0,T]) \otimes \mathcal{F}$, that are $\mathbb{F}$-progressively measurable. Moreover, with $\mathbb{P}$-probability 1, all the Lebesgue integrals in time are well defined, and the stochastic integrals form local martingales.

The various terms in (4.37) may be interpreted as follows. The third and fourth terms in the right-hand side read as a "drift" term; the fifth, sixth and seventh terms form a local martingale; the terms on the fourth line read as the brackets deriving from $\boldsymbol{X}^0$; the fifth, sixth and seventh lines correspond to the brackets deriving from $\boldsymbol{X}$; the last line is the bracket between $\boldsymbol{X}^0$ and $\boldsymbol{X}$.

Proof. A reasonable strategy would be to repeat the arguments used in the proof of Theorem 4.14. However, for pedagogical reasons, we choose to give a slightly different proof.

First Step. First we remark that it is enough to prove the chain rule when u does not depend upon time as the extension to the time inhomogeneous case is straightforward. Indeed, we may incorporate time as an additional component of the Itô process $\boldsymbol{X}^0$. When u is twice differentiable in (t, x, μ) in the sense that it satisfies assumption **Joint Chain Rule Common Noise**, but with t incorporated as an additional space variable, then the time-dependent version of the chain rule follows from a mere application of the time-independent version. When u is not twice differentiable with respect to the time variable, we may mollify u it in that direction and use an approximation argument similar to the one used in the first step of the proof of Lemma 4.15.

We now restrict ourselves to the time-independent case, in which case we can use another approximation argument, replacing $\mathbb{R}^d \times \mathcal{P}_2(\mathbb{R}^d) \ni (x,\mu) \mapsto u(x,\mu)$ by $\mathbb{R}^d \times \mathcal{P}_2(\mathbb{R}^d) \ni (x,\mu) \mapsto u(\rho(x), (\mu * \varphi_\epsilon) \circ \rho^{-1})$, for ρ denoting a smooth function from $\mathbb{R}^d$ into $\mathbb{R}^d$ with compact support and for φ_ϵ denoting the density of the Gaussian distribution $N(0, \epsilon I_d)$. Arguing again as in the first step of the proof of Lemma 4.15, it suffices to prove the chain rule for $\mathbb{R}^d \times \mathcal{P}_2(\mathbb{R}^d) \ni (x,\mu) \mapsto u(\rho(x), (\mu * \varphi_\epsilon) \circ \rho^{-1})$ and then to pass to the limit along a sequence of functions $(\rho_n)_{n\geq 1}$ converging in a suitable sense to the identity function on $\mathbb{R}^d$, and $\epsilon \searrow 0$.

The rationale for considering $\mathbb{R}^d \times \mathcal{P}_2(\mathbb{R}^d) \ni (x,\mu) \mapsto u(\rho(x), (\mu * \varphi_\epsilon) \circ \rho^{-1})$ in lieu of $\mathbb{R}^d \times \mathcal{P}_2(\mathbb{R}^d) \ni (x,\mu) \mapsto u$ is clear. Owing to assumption **Joint Chain Rule with Common Noise** and repeating the computations performed in the proof of Theorem 4.14, we indeed observe that $\mathbb{R}^d \times \mathcal{P}_2(\mathbb{R}^d) \ni (x,\mu) \mapsto u(\rho(x), (\mu * \varphi_\epsilon) \circ \rho^{-1})$ satisfies the same assumption, but, in addition, all the derivatives of order one and two are bounded and continuous on the whole space.

Second Step. From now on, we thus assume that $u : \mathbb{R}^d \times \mathcal{P}_2(\mathbb{R}^d) \to \mathbb{R}^d$ is time-independent and that all the derivatives of order 1 and 2 appearing in assumption **Joint Chain Rule with Common Noise** are bounded and continuous on the whole space. In particular, this says that the functions:

$$
\begin{aligned}
&\mathbb{R}^d \times L^2(\Omega^1, \mathcal{F}^1, \mathbb{P}^1; \mathbb{R}^d) \ni (x, X) \\
&\qquad \mapsto \partial_\mu u\big(x, \mathcal{L}^1(X)\big)(X) \in L^2(\Omega^1, \mathcal{F}^1, \mathbb{P}^1; \mathbb{R}^d), \\
&\mathbb{R}^d \times L^2(\Omega^1, \mathcal{F}^1, \mathbb{P}^1; \mathbb{R}^d) \ni (x, X) \\
&\qquad \mapsto \partial_x \partial_\mu u\big(x, \mathcal{L}^1(X)\big)(X) \in L^2(\Omega^1, \mathcal{F}^1, \mathbb{P}^1; \mathbb{R}^{d\times d}), \\
&\mathbb{R}^d \times L^2(\Omega^1, \mathcal{F}^1, \mathbb{P}^1; \mathbb{R}^d) \ni (x, X) \\
&\qquad \mapsto \partial_v \partial_\mu u\big(x, \mathcal{L}^1(X)\big)(X) \in L^2(\Omega^1, \mathcal{F}^1, \mathbb{P}^1; \mathbb{R}^{d\times d}), \\
&\mathbb{R}^d \times L^2(\Omega^1, \mathcal{F}^1, \mathbb{P}^1; \mathbb{R}^d) \ni (x, X) \\
&\qquad \mapsto \partial_\mu^2 u\big(x, \mathcal{L}^1(X)\big)(X, \tilde{X}) \in L^2(\Omega^1 \times \tilde{\Omega}^1, \mathcal{F}^1 \otimes \mathcal{F}^1, \mathbb{P}^1 \otimes \tilde{\mathbb{P}}^1; \mathbb{R}^{d\times d}), \\
&\mathbb{R}^d \times \mathbb{R}^d \times L^2(\Omega^1, \mathcal{F}^1, \mathbb{P}^1; \mathbb{R}^d) \ni (x, v, X) \\
&\qquad \mapsto \partial_\mu^2 u\big(x, \mathcal{L}^1(X)\big)(v, X) \in L^2(\Omega^1, \mathcal{F}^1, \mathbb{P}^1; \mathbb{R}^{d\times d}),
\end{aligned}
\tag{4.38}
$$

are continuous. The next step is to replace $(X_t^0)_{t\geq 0}$ by $(\delta_{X_t^0})_{t\geq 0}$ so that it can be viewed as a measure-valued process. In this way, the entire noise $(\boldsymbol{W}, \boldsymbol{W}^0) = (W_t, W_t^0)_{t\geq 0}$ can be treated as the *common noise*, so that $\delta_{X_t^0}$ is indeed the conditional law of X_t^0 given the common noise.

The reader might object to the rationale of such a point of view. After all, the common noise in the dynamics of the Itô process associated with $(\mu_t)_{t\geq 0}$ is $\boldsymbol{W}^0$ only, and not $(\boldsymbol{W}, \boldsymbol{W}^0)$. Actually, in the dynamics of the Itô process $(X_t)_{t\geq 0}$, $\boldsymbol{W}$ may be replaced by an independent copy, say $\tilde{\boldsymbol{W}}$, which may be constructed by considering the completion of the enlarged probability space $(\Omega \times \tilde{\Omega}^1 = \Omega^0 \times \Omega^1 \times \tilde{\Omega}^1, \mathcal{F} \times \tilde{\mathcal{F}}^1, \mathbb{P} \otimes \tilde{\mathbb{P}}^1)$, with $(\tilde{\Omega}^1, \tilde{\mathcal{F}}^1, \tilde{\mathbb{P}}^1)$ denoting, as usual, a copy of $(\Omega^1, \mathcal{F}^1, \mathbb{P}^1)$. This does not affect the dynamics of $(\mu_t)_{t\geq 0}$ and thus guarantees that there is no conflict in choosing $(\boldsymbol{W}, \boldsymbol{W}^0)$ as the common noise in the dynamics of $\boldsymbol{X}^0$.

With such an approach, the function u, originally defined on $\mathbb{R}^d \times \mathcal{P}_2(\mathbb{R}^d)$, is lifted onto the space $\mathcal{P}_2(\mathbb{R}^{2d})$. The lift reads:

$$
\mathcal{V}(\nu) = u\Big(\int_{\mathbb{R}^d} x d\nu_1(x), \nu_2 \Big), \qquad \nu \in \mathcal{P}_2(\mathbb{R}^{2d}),
$$

where ν_1 denotes the image of ν by the mapping $\mathbb{R}^d \times \mathbb{R}^d \ni (x_1, x_2) \mapsto x_1$ and ν_2 the image of ν by the mapping $\mathbb{R}^d \times \mathbb{R}^d \ni (x_1, x_2) \mapsto x_2$.

It is then an easy exercise to compute the derivatives of $\mathcal{V}$ in terms of those of u. Denoting by $\tilde{\mathcal{V}}$ the lift of $\mathcal{V}$ on $L^2(\Omega^1, \mathcal{F}^1, \mathbb{P}^1; \mathbb{R}^{2d})$, we have, for any random variables $X, Y \in L^2(\Omega^1, \mathcal{F}^1, \mathbb{P}^1; \mathbb{R}^{2d})$ and any $\varepsilon > 0$,

$$
\begin{aligned}
\tilde{\mathcal{V}}(X + \varepsilon Y) &= \mathcal{V}\big[\mathcal{L}^1(X + \varepsilon Y)\big] \\
&= u\Big(\mathbb{E}^1(X_1) + \varepsilon \mathbb{E}^1(Y_1), \mathcal{L}^1\big(X_2 + \varepsilon Y_2\big)\Big),
\end{aligned}
$$

where X_1 and X_2 are the d-dimensional coordinates of X, seen as a random vector with values in $\mathbb{R}^d \times \mathbb{R}^d$, and similarly for Y_1 and Y_2.

Then, differentiating with respect to ε, we easily get that $\tilde{\mathcal{V}}$ is differentiable in the direction Y at X, with:

$$\frac{d}{d\varepsilon}_{|\varepsilon=0} \tilde{\mathcal{V}}(X+\varepsilon Y) = \mathbb{E}^1\Big[\partial_x u\big(\mathbb{E}^1(X_1), \mathcal{L}^1(X_2)\big)Y_1 + \partial_\mu u\big(\mathbb{E}^1(X_1), \mathcal{L}(X_2)\big)(X_2)Y_2\Big].$$

Since the two mappings:

$$\begin{aligned}
&\big[L^2(\Omega^1,\mathcal{F}^1,\mathbb{P}^1;\mathbb{R}^d)\big]^2 \ni (X_1,X_2) \mapsto \partial_x u\big(\mathbb{E}^1(X_1), \mathcal{L}^1(X_2)\big) \in L^2(\Omega^1,\mathcal{F}^1,\mathbb{P}^1;\mathbb{R}^d),\\
&\big[L^2(\Omega^1,\mathcal{F}^1,\mathbb{P}^1;\mathbb{R}^d)\big]^2 \ni (X_1,X_2) \mapsto \partial_\mu u\big(\mathbb{E}^1(X_1), \mathcal{L}(X_2)\big)(X_2) \in L^2(\Omega^1,\mathcal{F}^1,\mathbb{P}^1;\mathbb{R}^d),
\end{aligned}$$

are continuous, we deduce that $\tilde{\mathcal{V}}$ is Fréchet differentiable and that, for any $(v_1,v_2) \in \mathbb{R}^d \times \mathbb{R}^d$, $\partial_\mu \mathcal{V}(\nu)(v_1,v_2)$ is the $2d$-dimensional vector with d-dimensional coordinates given by:

$$\partial_\nu \mathcal{V}(\nu)(v_1,v_2) = \bigg(\partial_x u\Big(\int_{\mathbb{R}^d} x d\nu_1(x), \nu_2\Big), \partial_\mu u\Big(\int_{\mathbb{R}^d} x d\nu_1(x), \nu_2\Big)(v_2)\bigg),$$

the first coordinate reading as $\partial_{\nu_1}\mathcal{V}(\nu)(v_1,v_2)$ and the second one as $\partial_{\nu_2}\mathcal{V}(\nu)(v_1,v_2)$. Therefore, differentiating with respect to ν again, we have that, for any $(v_1,v_2) \in \mathbb{R}^d \times \mathbb{R}^d$ and any $(v_1',v_2') \in \mathbb{R}^d \times \mathbb{R}^d$, $\partial_\nu^2 \mathcal{V}(\nu)((v_1,v_2),(v_1',v_2'))$ reads as a vector in $(\mathbb{R}^d \times \mathbb{R}^d)^2$, with entries:

$$\begin{aligned}
\partial^2_{\nu_1}\mathcal{V}(\nu)\big((v_1,v_2),(v_1',v_2')\big) &= \partial^2_{xx} u\Big(\int_{\mathbb{R}^d} x d\nu_1(x), \nu_2\Big),\\
\partial_{\nu_1}\partial_{\nu_2}\mathcal{V}(\nu)\big((v_1,v_2),(v_1',v_2')\big) &= \partial_x\partial_\mu u\Big(\int_{\mathbb{R}^d} x d\nu_1(x), \nu_2\Big)(v_2),\\
\partial_{\nu_2}\partial_{\nu_1}\mathcal{V}(\nu)\big((v_1,v_2),(v_1',v_2')\big) &= \Big[\partial_x\partial_\mu u\Big(\int_{\mathbb{R}^d} x d\nu_1(x), \nu_2\Big)(v_2')\Big]^\dagger,\\
\partial^2_{\nu_2}\mathcal{V}(\nu)\big((v_1,v_2),(v_1',v_2')\big) &= \partial^2_\mu u\Big(\int_{\mathbb{R}^d} x d\nu_1(x), \nu_2\Big)(v_2,v_2').
\end{aligned} \tag{4.39}$$

In order to fully justify (4.39), we use the same argument as above. Namely, we first prove that directional derivatives exist. Then, we use the continuity of the various mappings in (4.38) to deduce that differentiability holds true in the Fréchet sense. Observe that we used Remark 4.16 in the third line.

Finally, differentiating $\mathbb{R}^d \times \mathbb{R}^d \ni (v_1,v_2) \mapsto \partial_\nu \mathcal{V}(\nu)(v_1,v_2)$ with respect to v_1, v_2, we identify the second order derivative $\partial_v\partial_\nu \mathcal{V}(\nu)(v_1,v_2)$ with an element of $(\mathbb{R}^d \times \mathbb{R}^d)^2$ with the entries:

$$\begin{aligned}
&\partial_{v_1}\partial_{\nu_1}\mathcal{V}(\nu)(v_1,v_2) = \partial_{v_2}\partial_{\nu_1}\mathcal{V}(\nu)(v_1,v_2) = \partial_{v_1}\partial_{\nu_2}\mathcal{V}(\nu)(v_1,v_2) = 0,\\
&\partial_{v_2}\partial_{\nu_2}\mathcal{V}(\nu)(v_1,v_2) = \partial_v\partial_\mu u\Big(\int_{\mathbb{R}^d} x d\nu_1(x), \nu_2\Big)(v_1,v_2).
\end{aligned}$$

Clearly, $\mathcal{V}$ is fully $\mathcal{C}^2$.

Third Step. The final form of the chain rule is then obtained by applying Theorem 4.14 to $\mathcal{V}$, the idiosyncratic noise in the dynamics of $\boldsymbol{X}$ being replaced by a copy $\tilde{\boldsymbol{W}}$ of $\boldsymbol{W}$ defined on $(\tilde{\Omega}^1,\tilde{\mathcal{F}}^1,\tilde{\mathbb{P}}^1)$. The various symbols $\mathbb{E}^1$ in the statement of Theorem 4.14 are to be replaced by the expectation symbol $\tilde{\mathbb{E}}^1$, acting on the copies $\tilde{\boldsymbol{X}}$, $\tilde{\boldsymbol{B}}$, $\tilde{\boldsymbol{\Sigma}}$ and $\tilde{\boldsymbol{\Sigma}}^0$ (see the definition of the statement) of $\boldsymbol{X}$, $\boldsymbol{B}$, $\boldsymbol{\Sigma}$ and $\boldsymbol{\Sigma}^0$. Also, the symbol $\tilde{\mathbb{E}}^1$ in (4.28) should be replaced by $\tilde{\tilde{\mathbb{E}}}^1$, accounting for the expectation on a second copy $(\tilde{\tilde{\Omega}}^1, \tilde{\tilde{\mathcal{F}}}^1, \tilde{\tilde{\mathbb{P}}}^1)$; similarly, we should replace $\tilde{\boldsymbol{X}}$ and $\tilde{\boldsymbol{\Sigma}}^0$ by copies $\tilde{\tilde{\boldsymbol{X}}}$ and $\tilde{\tilde{\boldsymbol{\Sigma}}}^0$. Since the dynamics of $\boldsymbol{X}^0$ do not incorporate the idiosyncratic noise $\tilde{\boldsymbol{W}}$, notice finally that $\boldsymbol{X}^0$ is never integrated with respect to $\tilde{\mathbb{E}}^1$.

We now explain how the various terms in (4.37) follow from the derivatives we just computed. First, we notice that in the function $\mathcal{V}$, the measure ν_1 is taken as the Dirac measure at X_t^0 for t varying in $[0, T]$. Then, the first-order terms in (4.37) are well understood as they correspond to a standard first-order expansion. The structure of the second-order terms is slightly more subtle. The second-order derivative in ν_1 yields the standard second-order derivative in the direction x and the second-order derivatives in (υ_2, ν_2) give the same second-order structure as in Theorem 4.14. Finally, the second-order cross-derivatives in ν_1 and ν_2 yield $\partial_x \partial_\mu$, each with a coefficient $1/2$. □

4.4 The Master Equation

Our goal is now to prove that the master field $\mathcal{U}$, as defined in Definition 4.1 of Section 4.1, satisfies a PDE on the enlarged state space comprising the physical states in $\mathbb{R}^d$ and probability measures on $\mathbb{R}^d$ to account for their statistical distributions.

4.4.1 General Principle for Deriving the Equation

The fact that $\mathcal{U}$ solves a PDE is reminiscent of the theory of finite dimensional forward-backward stochastic differential equations. The construction of the master field performed in Section 4.1 was designed in such a way that the value function describing the dynamics of the cost functional in equilibrium reads as a function of the solution of the forward Fokker-Planck equation describing the dynamics of the population in equilibrium. As the value function solves some backward Hamilton-Jacobi-Bellman equation, see (2.37)–(2.38), the master field reads as a generalization of the notion of decoupling field introduced in the theory of finite dimensional forward-backward stochastic differential equations, which is known to provide a sort of nonlinear Feynman-Kac formula.

If it should not come as a surprise that $\mathcal{U}$ might solve a PDE, the real questions are in fact to determine which PDE, and in which sense it is satisfied. In order to guess the form of the PDE, it suffices to combine the dynamic programming principle proven in Theorem 4.5 with the chain rule established in Theorem 4.17. The dynamic programming principle indeed says that, along a weak equilibrium of the mean field game, the process:

$$\left(\mathcal{U}(t, X_t, \mu_t) - \int_0^t f\big(s, X_s, \mu_s, \check{\alpha}(s, X_s, \mu_s, Y_s, Z_s)\big)ds\right)_{0 \le t \le T} \tag{4.40}$$

should be a martingale whenever $(X_t, Y_t, Z_t, Z_t^0, M_t)_{0\le t\le T}$ is the solution of the forward-backward system (4.5) characterizing the optimal path under the flow of measures $(\mu_t)_{0\le t\le T}$. Expanding the process $(\mathcal{U}(t, X_t, \mu_t))_{0\le t\le T}$ by means of Itô's formula, one expects to uncover the PDE by identifying to 0 the bounded variation term in the Itô expansion of the above martingale.

Obviously, the main shortcoming of this approach is the requirement that $\mathcal{U}$ be differentiable with respect to the space and measure arguments in order for the application of Itô's formula to be licit. Therefore, we should first prove that $\mathcal{U}$ is indeed differentiable. We shall not attempt to do that at this stage because it involves technical issues which we postpone to the next Chapter 5. Without any differentiability property of $\mathcal{U}$, the most we can hope for is to prove that the PDE holds in the viscosity sense, for a suitable notion of viscosity solution which needs to be identified.

The main advantage of the concept of viscosity solution is that existence is almost for free once the dynamic programming principle and Itô's formula are available. On the other hand, uniqueness is a more challenging issue that we shall not address in this book.

Still, before we introduce the notion of viscosity solution and prove the existence of a such a solution, a few remarks concerning the equation are in order. It is tempting to believe that the PDE ought to be stochastic because of the presence of the common noise in the dynamics of the players. However, our experience with the theory of finite-dimensional forward-backward stochastic differential equations tells us that this should not be the case. Indeed, in the case of finite-dimensional forward-backward stochastic differential equations with deterministic coefficients, the decoupling field is deterministic and satisfies, at least formally, a deterministic PDE. The reasons remain the same in the current framework. The coefficients of the infinite dimensional forward-backward system comprising the Fokker-Planck and the Hamilton-Jacobi-Bellman equations are deterministic functions of the unknowns, namely the equilibrium measure (which is random) and the value function (which is random as well). For that reason, the PDE satisfied by the master field is expected to be deterministic. The presence of the common noise $\boldsymbol{W}^0$ in the dynamics of the players is merely the reason for the existence of a nontrivial *second-order term* in the PDE.

To close this informal discussion, we stress that the PDE satisfied by the *master field* will be called the *master equation*, according to the terminology introduced by Lions in his lectures at the *Collège de France*.

4.4.2 Formal Derivation of the Master Equation

The astute reader is presumably already aware of the fact that the formal computation based on the martingale property of (4.40) works perfectly well provided that the optimal control $(\breve{\alpha}(t, X_t, \mu_t, Y_t, Z_t))_{0\le t\le T}$ can be expressed in terms of some feedback function, depending on $\mathcal{U}$, evaluated along the optimal path $(X_t)_{0\le t\le T}$.

The identification performed in Proposition 4.7 and Corollary 4.11 indicates that the following should be true:

$$\check{\alpha}\big(t, X_t, \mu_t, Y_t, Z_t\big) = \hat{\alpha}\big(t, X_t, \mu_t, \partial_x \mathcal{U}(t, X_t, \mu_t)\big),$$

where, as above,

$$\hat{\alpha}(t, x, \mu, y) = \text{argmin}_{\alpha \in A} H^{(r)}(t, x, \mu, y, \alpha),$$

the reduced Hamiltonian $H^{(r)}$ taking the shortened form:

$$H^{(r)}(t, x, \mu, y, \alpha) = b(t, x, \mu, \alpha) \cdot y + f(t, x, \mu, \alpha),$$

for $x, y \in \mathbb{R}^d$, $\mu \in \mathcal{P}_2(\mathbb{R}^d)$, $\alpha \in A$, since both σ and σ^0 are independent of the control.

So when $\mathcal{U}$ is smooth, we have all the necessary ingredients to derive the form of the master equation. By means of the chain rule in Theorem 4.17, it must hold that:

$$\begin{aligned}
&\partial_t \mathcal{U}(t, x, \mu) + b\big(t, x, \mu, \hat{\alpha}(t, x, \mu, \partial_x \mathcal{U}(t, x, \mu))\big) \cdot \partial_x \mathcal{U}(t, x, \mu) \\
&\quad + \int_{\mathbb{R}^d} b\big(t, v, \mu, \hat{\alpha}(t, v, \mu, \partial_x \mathcal{U}(t, v, \mu))\big) \cdot \partial_\mu \mathcal{U}(t, x, \mu)(v) d\mu(v) \\
&\quad + \frac{1}{2} \text{trace}\Big[\big(\sigma\sigma^\dagger + \sigma^0(\sigma^0)^\dagger\big)(t, x, \mu) \partial^2_{xx} \mathcal{U}(t, x, \mu)\Big] \\
&\quad + \frac{1}{2} \int_{\mathbb{R}^d} \text{trace}\Big[\big(\sigma\sigma^\dagger + \sigma^0(\sigma^0)^\dagger\big)(t, v, \mu) \partial_v \partial_\mu \mathcal{U}(t, x, \mu)(v)\Big] d\mu(v) \\
&\quad + \frac{1}{2} \int_{\mathbb{R}^{2d}} \text{trace}\Big[\sigma^0(t, v, \mu)(\sigma^0)^\dagger(t, v', \mu) \partial^2_\mu \mathcal{U}(s, x, \mu)(v, v')\Big] d\mu(v) d\mu(v') \\
&\quad + \int_{\mathbb{R}^d} \text{trace}\Big[\sigma^0(t, x, \mu)(\sigma^0)^\dagger(t, v, \mu) \partial_x \partial_\mu \mathcal{U}(t, x, \mu)(v)\Big] d\mu(v) \\
&\quad + f\big(t, x, \mu, \hat{\alpha}(t, x, \mu, \partial_x \mathcal{U}(t, x, \mu))\big) = 0,
\end{aligned} \tag{4.41}$$

for $t \in [0, T]$, $x \in \mathbb{R}^d$ and $\mu \in \mathcal{P}_2(\mathbb{R}^d)$, and with the terminal condition $\mathcal{U}(T, x, \mu) = g(x, \mu)$.

This equation can be rearranged in order to compare it with the Hamilton-Jacobi-Bellman equation associated with the underlying optimization problem. Denoting by H^* the *minimized* Hamiltonian:

$$\begin{aligned}
H^*(t, x, \mu, p) &= b\big(t, x, \mu, \hat{\alpha}(t, x, \mu, p)\big) \cdot p + f\big(t, x, \mu, \hat{\alpha}(t, x, \mu, p)\big), \\
&= \inf_{\alpha \in \mathbb{R}^d} \big[b(t, x, \mu, \alpha) \cdot p + f(t, x, \mu, \alpha)\big] \\
&= \inf_{\alpha \in \mathbb{R}^d} H^{(r)}(t, x, \mu, p, \alpha),
\end{aligned}$$

for $t \in [0, T]$, $x \in \mathbb{R}^d$, $\mu \in \mathcal{P}_2(\mathbb{R}^d)$, and $p \in \mathbb{R}^d$, and where $H^{(r)}$ is the reduced Hamiltonian, the equation may be rewritten in the form:

$$\begin{aligned}
&\partial_t \mathcal{U}(t,x,\mu) + H^*\big(t,x,\mu,\partial_x \mathcal{U}(t,x,\mu)\big)\\
&\quad + \frac{1}{2}\text{trace}\Big[\big(\sigma\sigma^\dagger + \sigma^0(\sigma^0)^\dagger\big)(t,x,\mu)\partial^2_{xx}\mathcal{U}(t,x,\mu)\Big]\\
&\quad + \int_{\mathbb{R}^d} b\big(t,v,\mu,\hat{\alpha}(t,v,\mu,\partial_x\mathcal{U}(t,v,\mu))\big)\cdot\partial_\mu\mathcal{U}(t,x,\mu)(v)d\mu(v)\\
&\quad + \frac{1}{2}\int_{\mathbb{R}^d}\text{trace}\Big[\big(\sigma\sigma^\dagger + \sigma^0(\sigma^0)^\dagger\big)(t,v,\mu)\partial_v\partial_\mu\mathcal{U}(t,x,\mu)(v)\Big]d\mu(v)\\
&\quad + \frac{1}{2}\int_{\mathbb{R}^{2d}}\text{trace}\Big[\sigma^0(t,v,\mu)(\sigma^0)^\dagger(t,v',\mu)\partial^2_\mu\mathcal{U}(s,x,\mu)(v,v')\Big]d\mu(v)d\mu(v')\\
&\quad + \int_{\mathbb{R}^d}\text{trace}\Big[\sigma^0(t,x,\mu)\big(\sigma^0\big)^\dagger(t,v,\mu)\partial_x\partial_\mu\mathcal{U}(t,x,\mu)(v)\Big]d\mu(v) = 0.
\end{aligned} \tag{4.42}$$

Notice that the terms in the first two lines of the equation are exactly the same as in the standard HJB equation associated with the optimization part of the MFG problem. The remaining terms, in the fourth remaining lines, are connected with the dynamics of the population themselves.

The master equation is expected to encapsulate all the information needed to describe the solution of the associated MFG. We shall prove later on that it is indeed the case.

First-Order Master Equation

The master equation is simpler in the absence of the common noise. Indeed, if $\sigma^0 = 0$, the equation reduces to:

$$\begin{aligned}
&\partial_t \mathcal{U}(t,x,\mu) + b\big(t,x,\mu,\hat{\alpha}(t,x,\mu,\partial_x\mathcal{U}(t,x,\mu))\big)\cdot\partial_x\mathcal{U}(t,x,\mu)\\
&\quad + \int_{\mathbb{R}^d} b\big(t,v,\mu,\hat{\alpha}(t,v,\mu,\partial_x\mathcal{U}(t,v,\mu))\big)\cdot\partial_\mu\mathcal{U}(t,x,\mu)(v)d\mu(v)\\
&\quad + \frac{1}{2}\text{trace}\Big[\big(\sigma\sigma^\dagger\big)(t,x,\mu)\partial^2_{xx}\mathcal{U}(t,x,\mu)\Big]\\
&\quad + \frac{1}{2}\int_{\mathbb{R}^d}\text{trace}\Big[\big(\sigma\sigma^\dagger\big)(t,v,\mu)\partial_v\partial_\mu\mathcal{U}(t,x,\mu)(v)\Big]d\mu(v)\\
&\quad + f\big(t,x,\mu,\hat{\alpha}(t,x,\mu,\partial_x\mathcal{U}(t,x,\mu))\big) = 0,
\end{aligned} \tag{4.43}$$

for $t \in [0, T]$, $x \in \mathbb{R}^d$ and $\mu \in \mathcal{P}_2(\mathbb{R}^d)$, with $\mathcal{U}(T, x, \mu) = g(x, \mu)$ as terminal condition. A remarkable feature of this equation is the absence of second-order partial derivatives ∂^2_μ. For that reason, we shall say that this equation is of the *first order*, although the terminology could be misleading since it remains of the second-order in x as long as $\sigma \neq 0$.

Importantly, observe that in this case, we recover the equation postulated in the verification principle established in Proposition (Vol I)-5.106, whose statement may now be reformulated as follows: any classical solution of the master equation yields an equilibrium to the corresponding MFG problem.

As before, equation (4.43) can be rewritten in terms of the *minimized Hamiltonian* H^* in order to compare it with a standard Hamilton-Jacobi-Bellman equation:

$$\begin{aligned}
&\partial_t \mathcal{U}(t,x,\mu) + H^*\big(t,x,\mu,\partial_x\mathcal{U}(t,x,\mu)\big) + \frac{1}{2}\text{trace}\Big[\big(\sigma\sigma^\dagger\big)(t,x,\mu)\partial^2_{xx}\mathcal{U}(t,x,\mu)\Big] \\
&+ \int_{\mathbb{R}^d} b\big(t,v,\mu,\hat{\alpha}(t,v,\mu,\partial_x\mathcal{U}(t,v,\mu))\big)\cdot\partial_\mu\mathcal{U}(t,x,\mu)(v)d\mu(v) \\
&+ \frac{1}{2}\int_{\mathbb{R}^d}\text{trace}\Big[\big(\sigma\sigma^\dagger\big)(t,v,\mu)\partial_v\partial_\mu\mathcal{U}(t,x,\mu)(v)\Big]d\mu(v) = 0,
\end{aligned}$$

for $t \in [0,T]$, $x \in \mathbb{R}^d$ and $\mu \in \mathcal{P}_2(\mathbb{R}^d)$, with the terminal condition $\mathcal{U}(T,x,\mu) = g(x,\mu)$.

4.4.3 The Master Field as a Viscosity Solution

We now prove that the master field defined in Definition 4.1 is a viscosity solution of the master equation (4.41). In preparation for the definition of a viscosity solution of (4.41), we first define the class of test functions used for that purpose.

Definition 4.18 *A function $\phi : [0,T]\times\mathbb{R}^d\times\mathcal{P}_2(\mathbb{R}^d)\to\mathbb{R}$ which is simply $\mathcal{C}^{1,2,2}$ in the sense of the definition given in assumption* **Joint Chain Rule Common Noise** *in Subsection 4.3.4, is said to be a test function if the quantities:*

$$\begin{aligned}
&\int_{\mathbb{R}^d}\Big[\big|\partial_\mu\phi(t,x,\mu)(v)\big|^2 + \big|\partial_x\partial_\mu\phi(t,x,\mu)(v)\big|^2\Big]d\mu(v) \\
&+ \int_{\mathbb{R}^d}\int_{\mathbb{R}^d}\big|\partial^2_\mu\phi(t,x,\mu)(v,v')\big|^2 d\mu(v)d\mu(v')
\end{aligned} \tag{4.44}$$

and

$$\sup_{v\in\mathbb{R}^d}\big|\partial_v\partial_\mu\phi(t,x,\mu)(v)\big| \tag{4.45}$$

are finite, uniformly in (t,x,μ) in compact subsets of $[0,T]\times\mathbb{R}^d\times\mathcal{P}_2(\mathbb{R}^d)$.

Observe that any function ϕ as in the statement of Definition 4.18 satisfies:

$$\begin{aligned}
|\partial_\mu\phi(t,x,\mu)(v)| \le &\int_{\mathbb{R}^d}|\partial_\mu\phi(t,x,\mu)(v')|d\mu(v') \\
&+ \sup_{v'\in\mathbb{R}^d}\big|\partial_v\partial_\mu\phi(t,x,\mu)(v')\big|\int_{\mathbb{R}^d}|v-v'|d\mu(v')
\end{aligned}$$

$$\leq \Big(\int_{\mathbb{R}^d} |\partial_\mu \phi(t,x,\mu)(v')|^2 d\mu(v') \Big)^{1/2} \\ + \sup_{v' \in \mathbb{R}^d} \big| \partial_v \partial_\mu \phi(t,x,\mu)(v') \big(|v| + M_2(\mu) \big), \tag{4.46}$$

so that the mapping $\mathbb{R}^d \ni v \mapsto \partial_\mu \phi(t,x,\mu)(v)$ is at most of linear growth (in v), uniformly in (t,x,μ) in compact subsets.

We are now ready to define what we mean by a viscosity solution of (4.41).

Definition 4.19 *Let assumption* **FBSDE** *be in force, the coefficients $b, f, g, \sigma, \sigma^0$, and $\breve{\alpha}$ that appear therein being jointly continuous in all their variables. Let us assume further that the optimizer $\hat{\alpha}$ in* (4.13) *is uniquely defined, jointly continuous in all its variables, and is at most of linear growth in (x,y), uniformly in (t,μ) in compact subsets.*

We then say that a real valued continuous function u on $[0,T] \times \mathbb{R}^d \times \mathcal{P}_2(\mathbb{R}^d)$, continuously differentiable in the direction x (that is $\partial_x u : [0,T] \times \mathbb{R}^d \times \mathcal{P}_2(\mathbb{R}^d) \ni (t,x,\mu) \mapsto \partial_x u(t,x,\mu)$ is continuous) is a viscosity solution of the master equation (4.41) *if:*

1. *the function $[0,T] \times \mathcal{P}_2(\mathbb{R}^d) \ni (t,\mu) \mapsto \int_{\mathbb{R}^d} |\partial_x u(t,v,\mu)|^2 d\mu(v)$ is bounded on any compact subset,*
2. *for any $(t,x,\mu) \in [0,T) \times \mathbb{R}^d \times \mathcal{P}_2(\mathbb{R}^d)$ and any test function $\phi : [0,T] \times \mathbb{R}^d \times \mathcal{P}_2(\mathbb{R}^d) \to \mathbb{R}$ such that $u(t,x,\mu) = \phi(t,x,\mu)$ and $u(s,y,\nu) \geq \phi(s,y,\nu)$ (resp. $u(s,y,\nu) \leq \phi(s,y,\nu)$) for all (s,y,ν) in a neighborhood of (t,x,μ), it holds that*

$$\begin{aligned}
&\partial_t \phi(t,x,\mu) + b\big(t,x,\mu,\hat{\alpha}(t,x,\mu,\partial_x u(t,x,\mu))\big) \cdot \partial_x \phi(t,x,\mu) \\
&+ \int_{\mathbb{R}^d} b\big(t,v,\mu,\hat{\alpha}(t,v,\mu,\partial_x u(t,v,\mu))\big) \cdot \partial_\mu \phi(t,x,\mu)(v) d\mu(v) \\
&+ \frac{1}{2} \mathrm{trace}\Big[\big(\sigma\sigma^\dagger + \sigma^0 (\sigma^0)^\dagger \big)(t,x,\mu) \partial^2_{xx} \phi(t,x,\mu) \Big] \\
&+ \frac{1}{2} \int_{\mathbb{R}^d} \mathrm{trace}\Big[\big(\sigma\sigma^\dagger + \sigma^0 (\sigma^0)^\dagger \big)(t,v,\mu) \partial_v \partial_\mu \phi(t,x,\mu)(v) \Big] d\mu(v) \qquad (4.47) \\
&+ \frac{1}{2} \int_{\mathbb{R}^{2d}} \mathrm{trace}\Big[\sigma^0(t,v,\mu) (\sigma^0)^\dagger (t,v',\mu) \partial^2_\mu \phi(t,x,\mu)(v,v') \Big] d\mu(v) d\mu(v') \\
&+ \int_{\mathbb{R}^d} \mathrm{trace}\Big[\sigma^0(t,x,\mu) \big(\sigma^0\big)^\dagger (t,v,\mu) \partial_x \partial_\mu \phi(t,x,\mu)(v) \Big] d\mu(v) \\
&+ f\big(t,x,\mu,\hat{\alpha}(t,x,\mu,\partial_x u(t,x,\mu))\big) \leq 0 \quad (\text{resp.} \geq 0),
\end{aligned}$$

3. *$u(T,x,\mu) = g(x,\mu)$, for any $(x,\mu) \in \mathbb{R}^d \times \mathcal{P}_2(\mathbb{R}^d)$.*

The reader must pay attention to the fact that, contrary to the usual definition of viscosity solutions of parabolic equations, inequality (4.47) keeps track of the derivative of u in the direction x. In fact, using the fact that (t, x, μ) is a minimum (respectively maximum) point of $u - \phi$ and that u is continuously differentiable in the space variable, it holds that

$$\partial_x u(t, x, \mu) = \partial_x \phi(t, x, \mu),$$

which shows that, in the first and last lines of the equation, the gradient of u in x can be replaced by the gradient of ϕ in x.

However, we emphasize that it is not possible to replace the gradient in x of u by the gradient of ϕ in x in the second line of the equation since the term that appears there is *nonlocal*. Indeed, it relies on the values of $\partial_x u(t, \cdot, \mu)$ on the whole support of μ. The simple comparison of u and ϕ in the neighborhood of (t, x, μ) is not sufficient to compare $\partial_x u(t, v, \mu)$ with $\partial_x \phi(t, v, \mu)$ for v away from x. This is the reason why, in the definition, u is assumed to be continuously differentiable in x. In that framework, condition 1 in the definition guarantees that the second line in (4.47) is well defined.

Moreover, while condition (4.44) is reminiscent of (4.36) in the statement of Theorem 4.17, condition (4.45) looks more stringent. Actually, it is mandatory in order to give a sense to the integral that appears in the fourth line of (4.47), at least when σ and σ^0 are of linear growth in x, which is the case under assumptions **MFG with a Common Noise SMP** and **MFG with a Common Noise SMP Relaxed** in Section 3.4. The fact that the bound involves a supremum over $v \in \mathbb{R}^d$ may seem rather restrictive. In fact, thanks to Proposition (Vol I)-5.36, we have a systematic tool in order to identify a Lipschitz continuous version (in the direction v) of the derivative $\partial_\mu \phi$, in which case $\partial_v \partial_\mu \phi$ is indeed bounded. When σ and σ^0 are bounded, such a constraint may be relaxed.

Note finally that, in comparison with the standard definition of viscosity solutions of parabolic equations, the signs in (4.47) are reversed since the equation (4.41) (or equivalently (4.47)) is set backward in time.

Implementing the dynamic programming principle proven in Proposition 4.5 together with the chain rule established in Theorem 4.17, we get:

Proposition 4.20 *Let assumption* **FBSDE** *be in force, the coefficients $b, f, g, \sigma, \sigma^0$, and $\check{\alpha}$ that appear therein being jointly continuous in all the parameters, and let us assume further that the optimizer $\hat{\alpha}$ in* (4.13) *is uniquely defined, jointly continuous in all the parameters and at most of linear growth in (x, y) uniformly in (t, μ) in compact subsets.*

Suppose also that weak existence and uniqueness to the mean field game hold for any initial condition $(t, \mathcal{V}) \in [0, T] \times \mathcal{P}_2(\mathcal{P}_2(\mathbb{R}^d))$ and that the master field $\mathcal{U}$ is jointly continuous in the three parameters (t, x, μ) and is differentiable in x, the partial derivative $\partial_x \mathcal{U}$ being jointly continuous in the three parameters (t, x, μ).

Finally, assume that, with the same notations as in Definition 4.1, with $\mathbb{P}^{t,\mu}$-probability 1, *and for almost every $s \in [t, T]$:*

$$\check{\alpha}(s, x_s^{t,x}, \nu_s, y_s^{t,x}, z_s^{t,x}) = \hat{\alpha}\big(s, x_s^{t,x}, \nu_s, \partial_x \mathcal{U}(s, x_s^{t,x}, \nu_s)\big). \tag{4.48}$$

Then, $\mathcal{U}$ is a viscosity solution of the master equation (4.41).

We refer the reader to Proposition 4.7 and Corollary 4.11 for sufficient conditions under which the identification (4.48) of the optimal feedback holds. Examples will be given at the end of the section.

Proof. The proof is quite similar to that for standard stochastic optimal control problems, but some attention must be paid to the fact that the measure argument is infinite dimensional. We first introduce a new notation for the purpose of the proof. Given $(t, x, \mu) \in [0, T] \times \mathbb{R}^d \times \mathcal{P}_2(\mathbb{R}^d)$ and a test function ϕ as in Definition 4.18, we denote by $\Psi[\phi](t, x, \mu)$ the left-hand side of (4.47).

First Step. We check that the mapping $[0, T] \times \mathbb{R}^d \times \mathcal{P}_2(\mathbb{R}^d) \ni (t, x, \mu) \mapsto \Psi[\phi](t, x, \mu)$ is continuous. A quick glance at the left-hand side in (4.47) shows that the only difficulty is to check the continuity of the *integral terms* which appear on the second, fourth, fifth, and sixth lines. To this end, we notice that the terms appearing on the second, fourth, and sixth lines are of the form $\mathbb{E}[\zeta(t, x, \mathcal{L}(\xi), \xi) \cdot \partial\rho(t, x, \mathcal{L}(\xi))(\xi)]$, where ξ is a random variable defined on some auxiliary probability space, which may be $(\Omega, \mathcal{F}, \mathbb{P})$ itself, with μ as distribution, $\zeta(t, x, \mu, v)$ being a coefficient that may be $b(t, v, \mu, \hat{\alpha}(t, v, \mu, \partial_x \mathcal{U}(t, v, \mu)))$, $(\sigma\sigma^\dagger)(t, v, \mu)$, $(\sigma^0(\sigma^0)^\dagger)(t, v, \mu)$ or $\sigma^0(t, x, \mu)(\sigma^0(t, v, \mu))^\dagger$ and, similarly, $\partial\rho(t, x, \mu)(v)$ possibly matching $\partial_\mu \phi(t, x, \mu)(v)$, $\partial_v \partial_\mu \phi(t, x, \mu)(v)$ or $\partial_x \partial_\mu \phi(t, x, \mu)(v)$.

We now consider a sequence $(t_n, x_n, \xi_n)_{n \geq 1}$ that converges to (t, x, ξ), $(\xi_n)_{n \geq 1}$ converging on the L^2 space defined over the auxiliary probability space. In order to establish the continuity of the terms on the second, fourth, and sixth lines, it suffices to show that, with the same convention as above:

$$\begin{aligned} &\lim_{n \to \infty} \mathbb{E}\big[\zeta\big(t_n, x_n, \mathcal{L}(\xi_n), \xi_n\big) \cdot \partial\rho\big(t_n, x_n, \mathcal{L}(\xi_n)\big)(\xi_n)\big] \\ &\quad = \mathbb{E}\big[\zeta\big(t, x, \mathcal{L}(\xi), \xi\big) \cdot \partial\rho\big(t, x, \mathcal{L}(\xi)\big)(\xi)\big]. \end{aligned} \tag{4.49}$$

When $\zeta(t, x, \mu, v)$ has the form $b(t, v, \mu, \hat{\alpha}(t, v, \mu, \partial_x \mathcal{U}(t, v, \mu)))$, we know that as n tends to ∞,

$$\begin{aligned} &b\big(t_n, \xi_n, \mathcal{L}(\xi_n), \hat{\alpha}\big(t_n, \xi_n, \mathcal{L}(\xi_n), \partial_x \mathcal{U}(t_n, \xi_n, \mathcal{L}(\xi_n))\big)\big) \\ &\quad \to b\big(t, \xi, \mathcal{L}(\xi), \hat{\alpha}\big(t, \xi, \mathcal{L}(\xi), \partial_x \mathcal{U}(t, \xi, \mathcal{L}(\xi))\big)\big), \end{aligned}$$

in probability. Moreover, because of point 1 in Definition 4.19 and the linear growth property of b and $\hat{\alpha}$ in the variable y, all the terms above are uniformly bounded in L^2. Now, by assumption on ρ, see in particular (4.46), the sequence $(\partial\rho(t_n, x_n, \mathcal{L}(\xi_n))(\xi_n))_{n \geq 1}$ converges in probability to $\partial\rho(t, x, \mathcal{L}(\xi))(\xi)$ and is uniformly integrable in L^2. By a standard uniform integrability argument, we deduce that, for ψ and $\partial\rho$ as on the second line of (4.47), (4.49) holds true, so that the term on the second line of (4.47) is continuous. A similar argument holds for the term on the sixth line, since σ^0 is at most of linear growth in the space variable, which guarantees that the sequence $((\sigma^0)^\dagger(t_n,\xi_n,\mathcal{L}(\xi_n)))_{n\geq 1}$ is uniformly square-integrable. For the terms on the fourth line, we may also proceed in the same way, using in addition (4.47) and noticing once again that the sequences $((\sigma\sigma^\dagger)(t_n,\xi_n,\mathcal{L}(\xi_n)))_{n\geq 1}$ and $(((\sigma^0)(\sigma^0)^\dagger)(t_n,\xi_n,\mathcal{L}(\xi_n)))_{n\geq 1}$ are uniformly integrable since σ and σ^0 are most of linear growth.

Finally, the term on the fifth line may be represented in the form:

$$\mathbb{E}\big[\sigma^0(t,\xi,\mathcal{L}(\xi))(\sigma^0(t,\xi',\mathcal{L}(\xi')))^\dagger\partial^2_\mu\phi(t,x,\mathcal{L}(\xi))(\xi,\xi')\big],$$

where ξ and ξ' are independent random variables that both have distribution μ. This means that, in order to investigate the analogue of (4.49), we must consider two sequence $(\xi_n)_{n\geq 1}$ and $(\xi'_n)_{n\geq 1}$ converging in L^2 to ξ and ξ' respectively, with the prescription that, for each $n\geq 1$, ξ_n and ξ'_n are independent and identically distributed. By independence, the sequence $(\sigma^0(t_n,\xi_n,\mathcal{L}(\xi_n))(\sigma^0(t,\xi'_n,\mathcal{L}(\xi'_n)))^\dagger)_{n\geq 1}$ is uniformly integrable in L^2, which is enough to apply the same argument as above.

Second Step. Given $(t,x,\mu)\in[0,T]\times\mathbb{R}^d\times\mathcal{P}_2(\mathbb{R}^d)$, we consider the same canonical set-up $\bar{\Omega}^t=\bar{\Omega}^{0,t}\times\bar{\Omega}^{1,t}$ as in the Definition 4.1 of the master field. The canonical process on $\bar{\Omega}^{0,t}$ is denoted by $(\nu^0,\boldsymbol{w}^0,\boldsymbol{\nu})$ and the canonical process on $\bar{\Omega}^{1,t}$ by $(\eta,\boldsymbol{w})$. Denoting by $\mathcal{M}^{t,\mu}$ the law of the equilibrium starting from μ at time t, we let $\mathbb{P}^{0,t,\mu}$ be the completion of $\delta_\mu\otimes\mathcal{M}^{t,\mu}$ and then $\mathbb{P}^{1,t}$ be the completion of $\text{Leb}_1\otimes\mathcal{W}^t_d$. As in subsection 4.1.3, the completion of the product probability measure $\mathbb{P}^{0,t,\mu}\otimes\mathbb{P}^{1,t}$ on $\bar{\Omega}^t$ is denoted by $\mathbb{P}^{t,\mu}$. We then denote by $(\boldsymbol{x}^{t,x},\boldsymbol{y}^{t,x},\boldsymbol{z}^{t,x},\boldsymbol{m}^{t,x})$ the solution of the forward-backward system (4.7) with $\xi=x$.

The dynamic programming principle Proposition 4.3 implies that, for any $0\leq t\leq s<s+h\leq T$,

$$\mathbb{E}^{t,\mu}\bigg[\int_s^{s+h}f(r,x_r^{t,x},\nu_r,\breve{\alpha}(r,x_r^{t,x},\nu_r,y_r^{t,x},z_r^{t,x}))dr+\mathcal{U}(s+h,x_{s+h}^{t,x},\nu_s)\big|\mathcal{F}_s^{\text{nat},t,(\psi(\eta,\nu^0),\boldsymbol{w}^0,\text{m},\boldsymbol{w})}\bigg]$$
$$=\mathcal{U}(s,x_s^{t,x},\nu_s),$$

where $\mathcal{F}_s^{\text{nat},t,(\psi(\eta,\nu^0),\boldsymbol{w}^0,\text{m},\boldsymbol{w})}=\mathcal{F}_s^{\text{nat},(\psi(\eta,\nu^0),w_r^0,\text{m}_r,w_r)_{t\leq r\leq T}}$.

Applying the above identity over the interval $(s+\varepsilon,s+h)$ in lieu of $(s,s+h)$, for $0<\varepsilon<h$, letting ε tend to 0 and using the continuity of $\mathcal{U}$, we deduce that:

$$\mathbb{E}^{t,\mu}\bigg[\int_s^{s+h}f(r,x_r^{t,x},\nu_r,\breve{\alpha}(r,x_r^{t,x},\nu_r,y_r^{t,x},z_r^{t,x}))dr+\mathcal{U}(s+h,x_{s+h}^{t,x},\nu_{s+h})\big|\mathcal{F}_s^{t,(\psi(\eta,\nu^0),\boldsymbol{w}^0,\text{m},\boldsymbol{w})}\bigg]$$
$$=\mathcal{U}(s,x_s^{t,x},\nu_s),$$

where $\mathcal{F}_s^{t,(\psi(\eta,\nu^0),\boldsymbol{w}^0,\text{m},\boldsymbol{w})}$ is the completion of $\cap_{\varepsilon>0}\mathcal{F}_{s+\varepsilon}^{\text{nat},t,(\psi(\eta,\nu^0),\boldsymbol{w}^0,\text{m},\boldsymbol{w})}$.

Now, since the process $(\boldsymbol{x}^{t,x},\boldsymbol{y}^{t,x},\boldsymbol{z}^{t,x})$ is progressively measurable with respect to the filtration $\mathbb{F}^{t,(\psi(\eta,\nu^0),\boldsymbol{w}^0,\text{m},\boldsymbol{w})}=(\mathcal{F}_s^{t,(\psi(\eta,\nu^0),\boldsymbol{w}^0,\text{m},\boldsymbol{w})})_{t\leq s\leq T}$, the process:

$$\bigg(\int_t^s f(r,x_r^{t,x},\nu_r,\breve{\alpha}(r,x_r^{t,x},\nu_r,y_r^{t,x},z_r^{t,x}))dr+\mathcal{U}(s+h,x_{s+h}^{t,x},\nu_s)\bigg)_{t\leq s\leq T}$$

is a martingale with respect to $\mathbb{F}^{t,(\psi(\eta,\nu^0),\boldsymbol{w}^0,\text{m},\boldsymbol{w})}$ under $\mathbb{P}^{t,\mu}$.

Therefore, for any $\mathbb{F}^{t,(\psi(\eta,\nu^0),\boldsymbol{w}^0,\mathrm{m},\boldsymbol{w})}$-stopping time τ with values in $[t,T]$,

$$\mathcal{U}(t,x,\mu) = \mathbb{E}^{t,\mu}\left[\int_t^\tau f(r, x_r^{t,x}, \nu_r, \breve{\alpha}(r, x_r^{t,x}, \nu_r, y_r^{t,x}, z_r^{t,x}))dr + \mathcal{U}(\tau, x_\tau^{t,x}, \nu_\tau)\right].$$

By assumption, we can identify $\breve{\alpha}(s, x_s^{t,x}, \nu_s, y_s^{t,x}, z_s^{t,x})$ with $\hat{\alpha}(s, x_s^{t,x}, \nu_s, \partial_x\mathcal{U}(s, x_s^{t,x}, \nu_s))$, so we get:

$$\mathcal{U}(t,x,\mu) = \mathbb{E}^{t,\mu}\left[\int_t^\tau f\Big(s, x_s^{t,x}, \nu_s, \hat{\alpha}\big(s, x_s^{t,x}, \nu_s, \partial_x\mathcal{U}(s, x_s^{t,x}, \nu_s)\big)\Big)ds + \mathcal{U}(\tau, x_\tau^{t,x}, \nu_\tau)\right].$$

Take now a test function ϕ such that $\mathcal{U}(t,x,\mu) = \phi(t,x,\mu)$ and assume that, under $\mathbb{P}^{t,\mu}$, the process $(x_s^{t,x}, \nu_s)_{t\le s\le\tau}$ lives in a compact set and satisfies $\mathcal{U}(\tau, x_\tau^{t,x}, \nu_\tau) \ge \phi(\tau, x_\tau^{t,x}, \nu_\tau)$. We then intend to apply Theorem 4.17 and to take expectation in the subsequent expansion, with $u \equiv \phi$, $(X_s^0)_{t\le s\le T} = (x_{s\wedge\tau}^{t,x})_{t\le s\le T}$ and $(X_s)_{t\le s\le T} = (x_s^{t,\xi})_{t\le s\le T}$, with $\xi = \psi(\eta, \nu^0)$. Observe from the growth properties of the coefficients under assumption **FBSDE** that:

$$\mathbb{E}^{t,\mu}\left[\int_t^\tau \big(|B(s, x_s^{t,x}, \nu_s, y_s^{t,x}, z_s^{t,x})|^2 + |\sigma(s, x_s^{t,x}, \nu_s)|^4 + |\sigma^0(s, x_s^{t,x}, \nu_s)|^4\big)ds\right] < \infty,$$

which shows that $(X_s^0 = x_{s\wedge\tau}^{t,x})_{t\le s\le T}$ here satisfies (4.26). Unfortunately, $(X_s = x_s^{t,\xi})_{t\le s\le T}$ does not, which prevents us from applying Theorem 4.17 in a straightforward manner. Still, we claim that $(\phi(s, x_s^{t,x}, \nu_s))_{t\le s\le\tau}$ may be expanded as in (4.37). The detailed argument is as follows, the key point therein being to use the fact that (4.44) and (4.45) are finite. To make it clear, we first apply our generalized form of Itô's formula to $(\phi(s, x_{s\wedge\tau}^{t,x}, (\nu_s * \varphi_\epsilon)\circ\rho^{-1}))_{t\le s\le T}$, where φ_ϵ is the d-dimensional Gaussian kernel of variance ϵI_d and ρ is a smooth function with compact support, see the proof of Lemma 4.15. Since $(x_{s\wedge\tau}^{t,x})_{t\le s\le T}$ lives in a compact set, all the derivatives that are involved in the expansion (4.37) are bounded, and the expansion can be fully justified by using the same approximation arguments as in the proofs of Lemma 4.15 and Theorem 4.17. Then, we take expectation in the expansion and let ϵ tend to 0 first and then ρ tend to the identity uniformly on compact subsets. This can be done by combining the arguments used in the first step, the boundedness of the two quantities (4.44) and (4.45), and the expressions (4.31) for the derivatives of the approximating coefficients, assuming in addition that $|\partial^2\rho(v)| \le C(1+|v|)^{-1}$. As a result, we get:

$$0 \ge \mathbb{E}^{t,\mu}\left[\int_t^\tau \Psi[\phi](s, x_s^{t,x}, \nu_s)ds\right].$$

Third Step. We now assume that $\mathcal{U}$ and ϕ satisfy $\mathcal{U}(s,y,\nu) \ge \phi(s,y,\nu)$ for all (s,y,ν) such that:

$$|s-t| + |y-x| + W_2(\nu,\mu) \le \delta. \tag{4.50}$$

The goal is to prove that $\Psi[\phi](t,x,\mu) \le 0$. We proceed by contradiction assuming that $\Psi[\phi](t,x,\mu) > 0$. By continuity of $\Psi[\phi]$, we can change δ in such a way that:

$$\Psi[\phi](s,y,\nu) > 0, \tag{4.51}$$

for all $(s, y, \nu) \in [0, T] \times \mathbb{R}^d \times \mathcal{P}_2(\mathbb{R}^d)$ such that $|s - t| + |y - x| + W_2(\mu, \nu) < \delta$. Since the pair $(\boldsymbol{x}^{t,x}, \boldsymbol{\nu})$ is continuous in time under $\mathbb{P}^{t,\mu}$, we may find, for a given $\epsilon \in (0, 1/2)$, a real $h \in (0, \delta)$ and a compact subset $\mathcal{K} \subset \mathcal{P}_2(\mathbb{R}^d)$ such that:

$$\mathbb{P}^{t,\mu}\Big[\sup_{t \le s \le t+h} \big(|x_s^{t,x} - x| + W_2(\nu_s, \mu)\big) > \delta\Big] \le \epsilon, \quad \mathbb{P}^{t,\mu}\Big[\forall s \in [t, T],\ \nu_s \notin \mathcal{K}\Big] \le \epsilon. \tag{4.52}$$

Choose τ as the first exit time:

$$\tau = \inf\big\{s \ge t : \quad |x_s^{t,x} - x| + W_2(\nu_s, \mu) \ge \delta,\ \nu_s \notin \mathcal{K}\big\} \wedge (t + h).$$

Since the filtration $\mathbb{F}^{0,t}$ satisfies the usual conditions, the first exit time $\inf\{s \ge t : \nu_s \notin \mathcal{K}\}$ is an $\mathbb{F}^{0,t}$ stopping time. By continuity of the process $(\boldsymbol{x}^{t,x}, \boldsymbol{\nu})$, the first exit time $\inf\{s \ge t : |x_s^{t,x} - x| + W_2(\nu_s, \mu) \ge \delta\}$ is an $\mathbb{F}^t$ stopping time. We deduce that τ is an $\mathbb{F}^t$ stopping time.

Since all the conditions of the second step are now satisfied, we conclude that:

$$0 \ge \mathbb{E}^{t,\mu}\Big[\int_t^\tau \Psi[\phi](s, x_s^{t,x}, \nu_s) ds\Big].$$

By (4.51), this says that $\mathbb{P}^{t,\mu}\big[\tau_K = t\big] = 1$, which is a contradiction with (4.52), from which it follows that

$$\mathbb{P}^{t,\mu}\big[\tau_K < t + h\big] < 2\epsilon < 1.$$

This completes the proof. □

4.4.4 Revisiting the Existence and Uniqueness Results of Chapter 3

We now return to the existence and uniqueness results proven in Chapter 3 and show that in all these cases, the master field is well defined and is a continuous viscosity solution of the master equation in the sense provided in the previous subsection.

Using Assumptions MFG with a Common Noise HJB and Lasry-Lions Monotonicity

Our first result is the following.

Theorem 4.21 *Let assumption* **MFG with a Common Noise HJB** *from Subsection 3.4.1 hold. Let us assume moreover that b, σ and σ^0 are independent of μ, that f has the form*

$$f(t, x, \mu, \alpha) = f_0(t, x, \mu) + f_1(t, x, \alpha), \quad t \in [0, T],\ x \in \mathbb{R}^d,\ \alpha \in A,\ \mu \in \mathcal{P}_2(\mathbb{R}^d),$$

$f_0(t, \cdot, \cdot)$ and g satisfying the Lasry-Lions monotonicity property (3.77) *and being L-Lipschitz in the measure argument for the same L as in assumption* **MFG with a Common Noise HJB**, *and that the coefficients b, f, g, σ, and σ^0 are jointly continuous in all their variables and that their derivatives in x and α, which exist*

under assumption **MFG with a Common Noise HJB**, *are jointly continuous. Then, the master field* $\mathcal{U}$ *is well defined and is a continuous viscosity solution of the master equation* (4.41).

Proof. The proof is rather long, so it is important to keep the gory details in check. We are trying to apply Proposition 4.20, and most of our efforts will be devoted to checking the continuity of the master field and its derivative with respect to the state variable x.

First Step. We first recall from the proof of Theorem 3.29 in Subsection 3.4.1 and from Subsection 2.2.3 that, under assumption **MFG with a Common Noise HJB**, assumption **FBSDE** holds, the coefficients (B, F, G) driving the system (4.5) being given by (3.58). In particular, $\breve{\alpha}(t, x, \mu, y, z)$ is equal to $\hat{\alpha}(t, x, \mu, \sigma(t, x)^{-1\dagger}z)$ where $\hat{\alpha}(t, x, \mu, y)$ is the optimizer of the reduced Hamiltonian, see for instance (4.14). Since b, σ and σ^0 are independent of μ and f has a separated structure, $\hat{\alpha}$ is independent of μ. We shall denote it by $\hat{\alpha}(t, x, y)$. As already explained in the proof of Theorem 3.29, convexity of the Hamiltonian ensures that the minimizer $\hat{\alpha}$ is continuous in all its variables, including time since all the coefficients are time continuous. Moreover, by (3.57), $\hat{\alpha}$ is at most of linear growth, which proves that the preliminary assumptions in both Definition 4.19 and Proposition 4.20 are satisfied.

Now, by Theorem 3.29, we know that, for any initial condition $(t, \mu) \in [0, T] \times \mathcal{P}_2(\mathbb{R}^d)$, the MFG problem admits a weak equilibrium. By Proposition 3.34, the equilibrium is strongly unique so that by Lemma 2.30, it is also unique in law. So the master field is well defined by Definition 4.1.

We shall prove point 1 in Definition 4.19 below. In fact, we prove that $\mathcal{U}$ is Lipschitz continuous in the space variable, uniformly in the time and measure arguments.

For the time being, we check that the identity (4.48) in Proposition 4.20 is satisfied. To this end, it suffices to check that assumption **Necessary SMP Master** and **Decoupling Master** hold. Assumption **Necessary SMP Master** is easily checked under assumption **MFG with a Common Noise HJB**, using in addition the fact that the derivatives of the coefficients in x and α are here assumed to be jointly continuous. In assumption **Decoupling Master**, **(A2)** is already known. The most demanding assumption is **(A1)**, but this follows from the combination of Theorem 1.57 and Theorem 1.53. Therefore, Theorem 4.10 and Corollary 4.11 apply, and we can argue that assumption (4.48) is satisfied.

Second Step. It now remains to discuss the continuity of the master field and of its derivative in space. Given an initial condition $(t, x, \mu) \in [0, T] \times \mathbb{R}^d \times \mathcal{P}_2(\mathbb{R}^d)$, we use the same representation of $\mathcal{U}(t, x, \mu)$ as in Definition 4.1. Namely, denoting by $\mathcal{M}^{t,\mu}$ the law of the equilibrium on the space $\mathcal{P}_2(\mathbb{R}^d) \times \mathcal{C}([t, T]; \mathbb{R}^d) \times \mathcal{P}_2(\mathcal{C}([t, T]; \mathbb{R}^{2d}))$, we have:

$$\mathcal{U}(t, x, \mu) = \int \biggl[\int_t^T f\Bigl(s, x_s^x, \nu_s, \hat{\alpha}\bigl(s, x_s^x, \sigma^{-1\dagger}(s, x_s^x) z_s^x\bigr)\Bigr) ds + g(x_T^x, \nu_T) \biggr] d\Bigl[\overline{\mathcal{M}^{t,\mu} \otimes \mathcal{W}_d^t}\Bigr],$$

the integral being over the space $\mathcal{P}_2(\mathbb{R}^d) \times \mathcal{C}([t, T]; \mathbb{R}^d) \times \mathcal{P}_2(\mathcal{C}([t, T]; \mathbb{R}^{2d})) \times \mathcal{C}([t, T]; \mathbb{R}^d)$, and the process $(x_s^x, y_s^x, z_s^x, z_s^{0,x}, m_s^x)_{t \le s \le T}$ satisfying the forward-backward system:

$$\begin{cases} dx_s^x = B\bigl(s, x_s^x, y_s^x, z_s^x\bigr) ds + \sigma(s, x_s^x) dw_s + \sigma^0(s, x_s^x) dw_s^0, \\ dy_s^x = -F\bigl(s, x_s^x, \nu_s, y_s^x, z_s^x, z_s^{0,x}\bigr) ds + z_s^x dw_s + z_s^{0,x} dw_s^0 + dm_s^x, \end{cases} \tag{4.53}$$

for $s \in [t, T]$, on the same probabilistic set-up as in Definition 4.1, with x as initial condition for x_t and $G(x_T^x, \nu_T)$ as terminal condition for y_T^x, and where $\boldsymbol{m}^x = (m_s^x)_{t \le s \le T}$ is a càd-làg martingale of zero cross-variation with $(\boldsymbol{w}^0, \boldsymbol{w})$ and with $m_t^x = 0$ as initial value. Remember in particular that, on such a probabilistic set-up, the random variable $(\nu^0, \boldsymbol{w}^0, \mathfrak{m}^0)$ is forced to be distributed according to the law of the equilibrium starting from μ at time t. Pay attention that we removed the measure argument in the coefficients of the forward equation.

Recalling the definition of $\bar{\Omega}^t$, $\bar{\Omega}^{0,t}$ and $\bar{\Omega}^{1,t}$, see (4.6), we let $\mathbb{P}^{0,t,\mu}$ be the completion of $\delta_\mu \otimes \mathcal{M}^{t,\mu}$ on $\bar{\Omega}^{0,t}$ and then $\mathbb{P}^{1,t}$ be the completion of $\mathrm{Leb}_1 \otimes \mathcal{W}_d^t$ on $\bar{\Omega}^{1,t}$. As in subsection 4.1.3, the completion of the product probability $\mathbb{P}^{0,t,\mu} \otimes \mathbb{P}^{1,t}$ measure on $\bar{\Omega}^t$ is denoted by $\mathbb{P}^{t,\mu}$.

In the framework of assumption **MFG with a Common Noise HJB**, we know from (3.58) that the coefficients driving the auxiliary FBSDE (4.53) (or equivalently (4.5)) are exactly those that appear in the definition of $\mathcal{U}$ (after removal of the cut-off functions φ and ψ in (3.58)). In particular, taking the expectation in the backward part of (4.5), we deduce that $\mathcal{U}(t, x, \mu)$ coincides with $\mathbb{E}^{t,\mu}[y_t^x]$. Now, $\mathbb{E}^{t,\mu}[y_t^x]$ may be represented as the expectation of the value at time (t, x) of the decoupling field of the FBSDE (4.53) on the t-initialized set-up $(\bar{\Omega}^t, \mathcal{F}^{t,\mu}, \mathbb{F}^{t,\mu}, \mathbb{P}^{t,\mu})$ equipped with $(\boldsymbol{w}^0, \boldsymbol{m}, \boldsymbol{w})$. From Proposition 1.57, we know that the decoupling field is Lipschitz continuous in the direction x, uniformly in time and in the environment. In particular, $\mathcal{U}$ is Lipschitz continuous in the direction x, uniformly in time t and in the measure argument μ.

Third Step. In order to investigate the smoothness in the direction of μ, we make use of the Lasry-Lions monotonicity condition. The first step is to notice that, under the standing assumption, the reduced Hamiltonian $H^{(r)}$ defined in assumption **MFG with a Common Noise MFG** is strictly convex in the direction of the control α. The key point is then to insert such a convexity bound into (1.41).

In order to proceed, we start with an helpful observation: Strong uniqueness of the MFG holds true, see Proposition 3.34. As a by-product, we know from Proposition 2.29 that, for a given $\mu \in \mathcal{P}_2(\mathbb{R}^d)$, under $\mathcal{M}^{t,\mu}$, the equilibrium $\mathfrak{m}$ is almost surely equal to a function of $(\nu^0, \boldsymbol{w}^0)$. Since ν^0 is almost surely equal to μ, this says that we can simply use $\mathcal{C}([t, T]; \mathbb{R}^d)$ as canonical space for the construction of the equilibrium. Instead of $\bar{\Omega}^{0,t}$, we shall use $\hat{\Omega}^{0,t} = \mathcal{C}([t, T]; \mathbb{R}^d)$ equipped with the completion $(\mathcal{F}^{0,t}, \mathbb{P}^{0,t})$ of $(\mathcal{B}(\hat{\Omega}^{0,t}), \mathcal{W}_d^t)$ and with the complete (and automatically right-continuous) augmentation $\mathbb{F}^{0,t}$ of the canonical filtration. We then construct $(\hat{\Omega}^t, \mathcal{F}^t, \mathbb{P}^t)$ as the completion of $(\hat{\Omega}^{0,t} \times \bar{\Omega}^{1,t}, \mathcal{F}^{0,t} \otimes \mathcal{F}^{1,t}, \mathbb{P}^{0,t} \otimes \mathbb{P}^{1,t})$. As usual, we equip it with the complete and right-continuous augmentation $\mathbb{F}^t = (\mathcal{F}_s^t)_{t \le s \le T}$ of the product filtration. On this simpler setting, we let $\mathfrak{m}^\mu = \Psi(\mu, \boldsymbol{w}^0)$, with Ψ as in Proposition 2.29 but on $[t, T]$ instead of $[0, T]$. Notice that we removed the parameter μ from the various notations used for the σ-fields and the probability measures. Moreover, whenever there is no possible confusion, we shall also forget the superscript μ in $\mathfrak{m}$ and we shall just write $\mathfrak{m}$ for $\mathfrak{m}^\mu$. Given the definition of $\mathfrak{m}$, we may define $\boldsymbol{\nu}$ as above and then solve (4.53) on $(\hat{\Omega}^t, \mathcal{F}^t, \mathbb{F}^t, \mathbb{P}^t)$. Importantly, there is no need to check any compatibility condition since the environment $\mathfrak{m}$ is, up to null sets, adapted to the noise $\boldsymbol{w}^0$. Below, we shall use the same notation $(\boldsymbol{x}^x, \boldsymbol{y}^x, \boldsymbol{z}^x, \boldsymbol{z}^{0,x}, \boldsymbol{m}^x)$ for the solution to (4.53). This is licit since the solution constructed on $(\hat{\Omega}^t, \mathcal{F}^t, \mathbb{F}^t, \mathbb{P}^t)$ may be canonically embedded into $(\bar{\Omega}^t, \mathcal{F}^{t,\mu}, \mathbb{F}^{t,\mu}, \mathbb{P}^{t,\mu})$. Observe that $\boldsymbol{m}^x$ is null since $\mathfrak{m}$ is adapted to the noise $\boldsymbol{w}^0$. We now consider a square integrable $\mathcal{F}_t^t$-measurable random variable ξ with values in $\mathbb{R}^d$ and a square-integrable $\mathbb{F}^t$-progressively measurable control process $\boldsymbol{\beta} = (\beta_s)_{t \le s \le T}$ with values in A. Then, we consider the solution $\boldsymbol{x}^{\xi,\boldsymbol{\beta}} = (x_s^{\xi,\boldsymbol{\beta}})_{t \le s \le T}$ of the SDE:

$$dx_s^{\xi,\boldsymbol{\beta}} = b\big(s, x_s^{\xi,\boldsymbol{\beta}}, \beta_s\big)ds + \sigma\big(s, x_s^{\xi,\boldsymbol{\beta}}\big)dw_s + \sigma^0\big(s, x_s^{\xi,\boldsymbol{\beta}}\big)dw_s^0, \quad x_t^{\xi,\boldsymbol{\beta}} = \xi. \tag{4.54}$$

Recall that, under the standing assumption **Lasry-Lions Monotonicity**, the coefficient b is independent of the measure argument and the coefficients σ and σ^0 are independent of both the control parameter and the measure argument. With $\boldsymbol{x}^{\xi,\boldsymbol{\beta}}$ as above and with some initial $\mu \in \mathcal{P}_2(\mathbb{R}^d)$, we associate the cost functional (for simplicity, we do not specify the time t in the left-hand side right below):

$$J^{\xi,\mu}(\boldsymbol{\beta}) = \mathbb{E}^t\bigg[\int_t^T f(s, x_s^{\xi,\boldsymbol{\beta}}, \nu_s, \beta_s)ds + g(x_T^{\xi,\boldsymbol{\beta}}, \nu_T)\bigg]. \tag{4.55}$$

Below, we shall focus on the special, though important, case $\xi = \psi(\eta, \mu)$, with ψ as in (2.23), the set-up $(\hat{\Omega}^t, \mathcal{F}^t, \mathbb{F}^t, \mathbb{P}^t)$ being equipped with $(\eta, \boldsymbol{w}^0, \boldsymbol{w})$. We then just denote $\boldsymbol{x}^{\xi,\boldsymbol{\beta}}$ by $\boldsymbol{x}^{\boldsymbol{\beta}}$. For the same value of ξ, we denote by $\hat{\boldsymbol{\alpha}} = (\hat{\alpha}_s)_{t\le s\le T}$ the process $(\hat{\alpha}(s, x_s, \sigma^{-1\dagger}(s, x_s)z_s))_{t\le s\le T}$, where instead of x, we put ξ as initial condition in the five-tuple $(\boldsymbol{x}^\xi, \boldsymbol{y}^\xi, \boldsymbol{z}^\xi, \boldsymbol{z}^{0,\xi}, \boldsymbol{m}^\xi \equiv 0)$ solving (4.53), the equation being defined on $(\hat{\Omega}^t, \mathcal{F}^t, \mathbb{F}^t, \mathbb{P}^t)$, and where we wrote $\boldsymbol{x}$ for $\boldsymbol{x}^\xi$. In particular, with our notation, we have $\boldsymbol{x} = \boldsymbol{x}^{\hat{\boldsymbol{\alpha}}}$. From assumption **MFG with a Common Noise HJB**, $\check{\alpha}(s, x_s, \nu_s, y_s, z_s)$ is here equal to $\hat{\alpha}(s, x_s, \sigma(s, x_s)^{-1\dagger}z_s)$. Then, using the bound (1.41), we deduce that, whenever $\boldsymbol{\beta}$ is bounded,

$$J^{\xi,\mu}(\boldsymbol{\beta}) - J^{\xi,\mu}(\hat{\boldsymbol{\alpha}}) \ge \lambda\mathbb{E}^t\int_t^T |\hat{\alpha}_s - \beta_s|^2 ds, \tag{4.56}$$

for the same constant λ as in assumption **MFG with a Common Noise HJB**. From Proposition 1.57 together with (3.57), we know that $\hat{\boldsymbol{\alpha}}$ is bounded by a constant C, which is independent of t, ξ and μ. Therefore, from assumption **MFG with a Common Noise HJB**, we must have, recalling that the coefficients (b, σ, σ^0) are independent of μ and allowing the value of the constant C to increase from line to line,

$$\mathbb{E}^t\big[\sup_{t\le s\le T}|x_s|^2\big] \le C\big(1 + \mathbb{E}^t\big[|\xi|^2\big]\big). \tag{4.57}$$

By our special choice of ξ, we know that $(x_s)_{t\le s\le T}$ describes the optimal path of a typical player when the population is in equilibrium, that is, for any $s \in [t, T]$, for $\mathbb{P}^{0,t}$ almost every $\omega^0 \in \hat{\Omega}^{0,t}$,

$$\nu_s(\omega^0) = \mathcal{L}\big(x_s(\omega^0, \cdot)\big).$$

Therefore, by (4.57), we deduce that:

$$\sup_{t\le s\le T}\mathbb{E}^t\big[\big(M_2(\nu_s)\big)^2\big] \le C\big[1 + \big(M_2(\mu)\big)^2\big]. \tag{4.58}$$

Fourth Step. The goal is now to consider another initial condition $\mu' \in \mathcal{P}_2(\mathbb{R}^d)$ for the equilibrium. We let $\mathfrak{m}' = \mathfrak{m}^{\mu'}$ and we define $\boldsymbol{\nu}'$ accordingly.

We now define the initial private states associated with the two equilibria. We proceed differently from above. We call $\hat{\psi} : [0,1) \times \mathcal{P}_2(\mathbb{R}^{2d}) \to \mathbb{R}^{2d}$ the analogue of the mapping ψ given by (2.23), see also Lemma (Vol I)-5.29, but in dimension $2d$ instead of d. Then, for some coupling π between μ and μ', we let $(x_t, x'_t) = \hat{\psi}(\eta, \pi)$; each coordinate x_t and x'_t being of dimension d, x_t being distributed according to μ and x'_t being distributed according to μ'. On $(\hat{\Omega}^t, \mathcal{F}^t, \mathbb{F}^t, \mathbb{P}^t)$ we can solve (4.53) with x_t and x'_t as respective initial conditions.

We call $(\boldsymbol{x}, \boldsymbol{y}, \boldsymbol{z}, \boldsymbol{z}^0, \boldsymbol{m} \equiv 0)$ and $(\boldsymbol{x}', \boldsymbol{y}', \boldsymbol{z}', \boldsymbol{z}^{0\prime}, \boldsymbol{m}' \equiv 0)$ the respective solutions, the superscript *prime* in the second notation indicating that the environment is $\mathfrak{m}'$ instead of $\mathfrak{m}$. Then, as in the second step, we let $\hat{\boldsymbol{\alpha}} = (\hat{\alpha}_s)_{t \le s \le T} = (\hat{\alpha}(s, x_s, \sigma(s, x_s)^{-1\dagger} z_s))_{t \le s \le T}$ and $\hat{\boldsymbol{\alpha}}' = (\hat{\alpha}'_s)_{t \le s \le T} = (\hat{\alpha}(s, x'_s, \sigma(s, x'_s)^{-1\dagger} z'_s))_{t \le s \le T}$.

Now, given a square-integrable $\mathbb{F}^t$-progressively measurable control process $\boldsymbol{\beta}$ with values in A, we denote by $\boldsymbol{x}^{\boldsymbol{\beta}}$ and $\boldsymbol{x}^{\boldsymbol{\beta},\prime}$ the solutions (on the extended space) of the SDE (4.54) with x_t and x'_t as respective initial conditions. In particular,

$$\boldsymbol{x} = \boldsymbol{x}^{\hat{\boldsymbol{\alpha}}} \quad \text{and} \quad \boldsymbol{x}' = \boldsymbol{x}^{\hat{\boldsymbol{\alpha}}',\prime}.$$

Similar to (4.55), we now let:

$$\begin{aligned} J(\boldsymbol{\beta}) &= \mathbb{E}^t\biggl[\int_t^T f(s, x_s^{\boldsymbol{\beta}}, \nu_s, \beta_s)ds + g(x_T^{\boldsymbol{\beta}}, \nu_T)\biggr], \\ J'(\boldsymbol{\beta}) &= \mathbb{E}^t\biggl[\int_t^T f(s, x_s^{\boldsymbol{\beta},\prime}, \nu'_s, \beta_s)ds + g(x_T^{\boldsymbol{\beta},\prime}, \nu'_T)\biggr]. \end{aligned} \tag{4.59}$$

Choosing $\boldsymbol{\beta} = \boldsymbol{\alpha}'$ in (4.56), which is licit since the latter is known to be bounded, see Proposition 1.57, we get:

$$J(\hat{\boldsymbol{\alpha}}') - J(\hat{\boldsymbol{\alpha}}) \ge \lambda \mathbb{E}^t \int_t^T |\hat{\alpha}'_s - \hat{\alpha}_s|^2 ds,$$

that is:

$$J'(\hat{\boldsymbol{\alpha}}') - J(\hat{\boldsymbol{\alpha}}) \ge \lambda \mathbb{E}^t \int_t^T |\hat{\alpha}'_s - \hat{\alpha}_s|^2 ds + J'(\hat{\boldsymbol{\alpha}}') - J(\hat{\boldsymbol{\alpha}}'). \tag{4.60}$$

Now, recalling the decomposition $f(t, x, \mu, \alpha) = f_0(t, x, \mu) + f_1(t, x, \alpha)$, we write:

$$\begin{aligned} &J'(\hat{\boldsymbol{\alpha}}') - J(\hat{\boldsymbol{\alpha}}') \\ &= \mathbb{E}^t\biggl[\int_t^T \bigl[f_0\bigl(s, x'_s, \nu'_s\bigr) - f_0\bigl(s, x'_s, \nu_s\bigr)\bigr]ds + g\bigl(x'_T, \nu'_T\bigr) - g\bigl(x'_T, \nu_T\bigr)\biggr] \\ &\quad + \mathbb{E}^t\biggl[\int_t^T \bigl[f\bigl(s, x'_s, \nu_s, \hat{\alpha}'_s\bigr) - f\bigl(s, x_s^{\hat{\boldsymbol{\alpha}}'}, \nu_s, \hat{\alpha}'_s\bigr)\bigr]ds + g\bigl(x'_T, \nu_T\bigr) - g\bigl(x_T^{\hat{\boldsymbol{\alpha}}'}, \nu_T\bigr)\biggr]. \end{aligned} \tag{4.61}$$

By definition of an equilibrium, we also have, for any $s \in [t, T]$ and for $\mathbb{P}^{0,t}$-almost every $\omega^0 \in \hat{\Omega}^{0,t}$:

$$\nu_s(\omega^0) = \mathcal{L}\bigl(x_s(\omega^0, \cdot)\bigr), \quad \nu'_s(\omega^0) = \mathcal{L}\bigl(x'_s(\omega^0, \cdot)\bigr). \tag{4.62}$$

Therefore, the first line in the right-hand side of (4.61) reads:

$$\mathbb{E}'\Big[\int_t^T [f_0(s,x'_s,\nu'_s) - f_0(s,x'_s,\nu_s)]ds + g(x'_T,\nu'_T) - g(x'_T,\nu_T)\Big] \tag{4.63}$$
$$= \mathbb{E}^{0,t}\Big[\int_t^T \int_{\mathbb{R}^d} [f_0(s,x,\nu'_s) - f_0(s,x,\nu_s)]d\nu'_s(x) + \int_{\mathbb{R}^d} (g(x,\nu'_T) - g(x,\nu_T))d\nu'_T(x)\Big].$$

Now, using the Lipschitz property of the coefficients f and g in the space argument, see assumption **MFG with a Common Noise HJB**, the second term in the right-hand side of (4.61) satisfies the bound:

$$\Big|\mathbb{E}'\Big[\int_t^T [f(s,x'_s,\nu_s,\hat{\alpha}'_s) - f(s,x_s^{\hat{\boldsymbol{\alpha}}'},\nu_s,\hat{\alpha}'_s)]ds + g(x'_T,\nu_T) - g(x_T^{\hat{\boldsymbol{\alpha}}'},\nu_T)\Big]\Big|$$
$$\leq C \sup_{t\leq s\leq T} \mathbb{E}'\big[|x'_s - x_s^{\hat{\boldsymbol{\alpha}}'}|^2\big]^{1/2}.$$

Recalling that $\boldsymbol{x}' = \boldsymbol{x}^{\hat{\boldsymbol{\alpha}}',\prime}$, we easily deduce from the Lipschitz property of the coefficients b, σ and σ^0 in the space direction that:

$$\Big|\mathbb{E}'\Big[\int_t^T [f(s,x'_s,\nu_s,\hat{\alpha}'_s) - f(s,x_s^{\hat{\boldsymbol{\alpha}}'},\nu_s,\hat{\alpha}'_s)]ds + g(x'_T,\nu_T) - g(x_T^{\hat{\boldsymbol{\alpha}}'},\nu_T)\Big]\Big|$$
$$\leq C\mathbb{E}'[|x'_t - x_t|^2]^{1/2}. \tag{4.64}$$

Therefore, from (4.60), (4.61), (4.63), and (4.64), we finally deduce that:

$$\lambda\mathbb{E}'\int_t^T |\hat{\alpha}'_s - \hat{\alpha}_s|^2 ds + \mathbb{E}'\Big[\int_t^T \int_{\mathbb{R}^d} [f_0(s,x,\nu'_s) - f_0(s,x,\nu_s)]d\nu'_s(x)$$
$$+ \int_{\mathbb{R}^d} (g(x,\nu'_T) - g(x,\nu_T))d\nu'_T(x)\Big] \leq C\mathbb{E}'[|x'_t - x_t|^2]^{1/2} + J'(\hat{\boldsymbol{\alpha}}') - J(\hat{\boldsymbol{\alpha}}).$$

Exchanging the roles of (x_t,μ) and (x'_t,μ'), summing the two resulting inequalities, and using the Lasry-Lions monotonicity condition from Subsection 3.5.1, we get:

$$\lambda\mathbb{E}'\int_t^T |\hat{\alpha}'_s - \hat{\alpha}_s|^2 ds \leq C\mathbb{E}'\big[|x'_t - x_t|^2\big]^{1/2}. \tag{4.65}$$

In particular, we must also have:

$$\mathbb{E}'\Big[\sup_{t\leq s\leq T} |x_s - x'_s|^2\Big] \leq C\Big(\mathbb{E}'\big[|x_t - x'_t|^2\big]^{1/2} + \mathbb{E}'\big[|x_t - x'_t|^2\big]\Big).$$

Therefore, by (4.62),

$$\mathbb{E}^{0,t}\Big[\sup_{t\leq s\leq T} \big(W_2(\nu_s,\nu'_s)\big)^2\Big] \leq C\Big(\mathbb{E}'\big[|x_t - x'_t|^2\big]^{1/2} + \mathbb{E}'\big[|x_t - x'_t|^2\big]\Big). \tag{4.66}$$

Fifth Step. We consider the FBSDE (4.53) with initial condition x at time t, and with $\mathfrak{m}$ or $\mathfrak{m}'$ as environments. By Theorem 1.57, both FBSDEs have a unique solution with a bounded martingale integrand. We call $(\boldsymbol{x}^x, \boldsymbol{y}^x, \boldsymbol{z}^x, \boldsymbol{z}^{0,x}, \boldsymbol{m}^x)$ and $(\boldsymbol{x}^{x,\prime}, \boldsymbol{y}^{x,\prime}, \boldsymbol{z}^{x,\prime}, \boldsymbol{z}^{0,x,\prime}, \boldsymbol{m}^{x,\prime})$ the respective solutions. Importantly, we regard both solutions as solutions on $(\hat{\Omega}^t, \mathcal{F}^t, \mathbb{F}^t, \mathbb{P}^t)$ equipped with $(\mathfrak{m}, \mathfrak{m}')$ as single super-environment. Since the compatibility constraint is trivially satisfied, we deduce from Theorem 1.57 the following stability estimate:

$$\begin{aligned}
&\mathbb{E}^t\Big[\sup_{t\le s\le T}|x_s^x - x_s^{x,\prime}|^2 + \int_t^T |z_s^x - z_s^{x,\prime}|^2 ds\Big] \\
&\quad \le C\Big(\mathbb{E}^t\big[|x_t - x_t'|^2\big]^{1/2} + \mathbb{E}^t\big[|x_t - x_t'|^2\big]\Big).
\end{aligned} \tag{4.67}$$

Recall that:

$$\begin{aligned}
\mathcal{U}(t,x,\mu) &= \mathbb{E}^t\Big[\int_t^T f\big(s, x_s^x, \nu_s, \hat{\alpha}(s, x_s^x, \sigma(s,x_s^x)^{-1\dagger} z_s^x)\big)ds + g(x_T^x, \nu_T)\Big],\\
\mathcal{U}(t,x,\mu') &= \mathbb{E}^t\Big[\int_t^T f\big(s, x_s^{x,\prime}, \nu_s', \hat{\alpha}(s, x_s^{x,\prime}, \sigma(s,x_s^{x,\prime})^{-1\dagger} z_s^{x,\prime})\big)ds + g(x_T^{x,\prime}, \nu_T')\Big],
\end{aligned}$$

and that because of Proposition 1.57, the processes z^x and $z^{x,\prime}$ are bounded by a constant C, independent of x, μ, and μ'. Therefore, from (3.57), the optimal control processes $(\hat{\alpha}(s, x_s^x, \sigma(s,x_s^x)^{-1\dagger} z_s^x))_{t\le s\le T}$ and $(\hat{\alpha}(s, x_s^{x,\prime}, \sigma(s,x_s^{x,\prime})^{-1\dagger} z_s^{x,\prime}))_{t\le s\le T}$ are bounded independently of x, μ' and t. Moreover, the standing assumption implies that g is Lipschitz continuous in the space and measure arguments and f is locally Lipschitz continuous in the space, measure, and control arguments, the Lipschitz constant being at most of linear growth in the control argument, uniformly in time. From this, (4.66) and (4.67) we deduce that:

$$|\mathcal{U}(t,x,\mu) - \mathcal{U}(t,x,\mu')| \le C\Big(\mathbb{E}^t\big[|x_t - x_t'|^2\big]^{1/2} + \mathbb{E}^t\big[|x_t - x_t'|^2\big]\Big)^{1/2}.$$

If we denote by π the joint distribution of (x_t, x_t') as before, π is any coupling between μ and μ', we get:

$$|\mathcal{U}(t,x,\mu) - \mathcal{U}(t,x,\mu')| \le C\Big[\Big(\int_{\mathbb{R}^{2d}} |\zeta - \zeta'|^2 d\pi(\zeta,\zeta')\Big)^{1/2} + \int_{\mathbb{R}^{2d}} |\zeta - \zeta'|^2 d\pi(\zeta,\zeta')\Big]^{1/2}.$$

Since the constant C is independent of π, we can take the infimum over π and conclude that:

$$|\mathcal{U}(t,x,\mu) - \mathcal{U}(t,x,\mu')| \le C\Big(\big(W_2(\mu,\mu')\big)^{1/2} + W_2(\mu,\mu')\Big). \tag{4.68}$$

Now, the time regularity of $\mathcal{U}$ easily follows. Indeed, by Theorem 4.5,

$$\begin{aligned}
&\mathcal{U}(t,x,\mu) - \mathcal{U}(t+h,x,\mu) \\
&= \mathbb{E}^t\Big[\mathcal{U}(t+h, x_{t+h}^x, \nu_{t+h}) - \mathcal{U}(t+h,x,\mu) + \int_t^{t+h} f\big(s, x_s^x, \nu_s, \hat{\alpha}(s, x_s^x, \sigma(s,x_s^x)^{-1\dagger} z_s^x)\big)ds\Big].
\end{aligned}$$

Since the control process $(\hat{\alpha}(s, x_s^x, \sigma(s, x_s^x)^{-1\dagger} z_s^x))_{t \le s \le T}$ is bounded independently of x, μ, and t, we deduce from the Lipschitz property of $\mathcal{U}$ in the space argument and from the 1/2-Hölder property (4.68) in the measure argument, that:

$$\begin{aligned} &\big|\mathcal{U}(t, x, \mu) - \mathcal{U}(t+h, x, \mu)\big| \\ &\qquad \le C\Big(h + \mathbb{E}^t\big[|x_{t+h}^x - x|\big] + \mathbb{E}^t\big[\big(W_2(\nu_{t+h}, \mu)\big)^{1/2} + W_2(\nu_{t+h}, \mu)\big]\Big), \end{aligned}$$

from which we easily get:

$$\big|\mathcal{U}(t, x, \mu) - \mathcal{U}(t+h, x, \mu)\big| \le Ch^{1/4}.$$

Last Step. We now discuss the regularity of $\partial_x \mathcal{U}$. To do so, since assumptions **Necessary SMP Master** and **Decoupling Master** hold in the present situation, we use the representation of $\partial_x \mathcal{U}$ provided by Theorem 4.10. So with the same notations as above, we consider on the space $(\hat{\Omega}^t, \mathcal{F}^t, \mathbb{F}^t, \mathbb{P}^t)$, the two backward SDEs with Lipschitz coefficients:

$$\begin{aligned} dv_s^x &= -\partial_x H\big(s, x_s^x, \nu_s, v_s^x, \zeta_s^x, \zeta_s^{0,x}, \hat{\alpha}_s^x\big)ds + \zeta_s^x dw_s + \zeta_s^{0,x} dw_s^0, \\ dv_s^{x',\prime} &= -\partial_x H\big(s, x_s^{\prime,x'}, \nu_s', v_s^{x',\prime}, \zeta_s^{x',\prime}, \zeta_s^{0,x',\prime}, \hat{\alpha}_s^{x',\prime}\big)ds + \zeta_s^{x',\prime} dw_s + \zeta_s^{0,x',\prime} dw_s^0, \end{aligned} \tag{4.69}$$

for $t \le s \le T$, with $v_T^x = \partial_x g(x_T^x, \nu_T)$, $v_T^{x',\prime} = \partial_x g(x_T^{x',\prime}, \nu_T')$. Observe as above that there is no additional orthogonal martingale in the dynamics since the filtration is Brownian. Above, we used the notations:

$$\begin{aligned} \hat{\alpha}_s^x &= \hat{\alpha}\big(s, x_s^x, \sigma(s, x_s^x)^{-1\dagger} z_s^x\big), \\ \hat{\alpha}_s^{\prime,x'} &= \hat{\alpha}\big(s, x_s^{x',\prime}, \sigma(s, x_s^{x',\prime})^{-1\dagger} z_s^{x',\prime}\big), \quad t \le s \le T. \end{aligned}$$

Using the fact that all the processes appearing in the coefficients are adapted with respect to the completion of the filtration generated by the noises $\boldsymbol{w}^0$ and $\boldsymbol{w}$, the solutions of both equations may be easily regarded as the solutions of similar equations but set onto $(\bar{\Omega}^t, \mathcal{F}^{t,\mu}, \mathbb{F}^{t,\mu}, \mathbb{P}^{t,\mu})$ and $(\bar{\Omega}^t, \mathcal{F}^{t,\mu'}, \mathbb{F}^{t,\mu'}, \mathbb{P}^{t,\mu'})$. In this regard, both solutions fit the setting used in the statement of Theorem 4.10. We deduce that:

$$\partial_x \mathcal{U}(t, x, \mu) = \mathbb{E}^t\big[v_t^x\big], \quad \partial_x \mathcal{U}(t, x', \mu') = \mathbb{E}^t\big[v_t^{x',\prime}\big].$$

Actually, Corollary 4.11 applies as well, from which we derive the identification $v_s^x = \partial_x \mathcal{U}(s, x_s^x, \nu_s)$ and $v_s^{x',\prime} = \partial_x \mathcal{U}(s, x_s^{x',\prime}, \nu_s')$. We get that:

$$(v_s^x)_{t \le s \le T} \quad \text{and} \quad (v_s^{x',\prime})_{t \le s \le T}$$

are bounded by a constant C, which is independent of t, x, x', μ and μ'. We stress the fact that both bounds hold with $\mathbb{P}^t$-probability 1 for all $s \in [t, T]$. This follows from the fact that both $(v_s^x)_{t \le s \le T}$ and $(v_s^{x',\prime})_{t \le s \le T}$ are continuous since we work in a Brownian filtration. Also, up to a modification of C, we have:

$$\mathbb{E}^t\bigg[\bigg(\int_t^T \big(|\zeta_s^x|^2 + |\zeta_s^{0,x}|^2\big)ds\bigg)^2\bigg] \le C. \tag{4.70}$$

Moreover, because of assumption (**A1**) of **MFG with a Common Noise HJB**, the partial derivative $\partial_x H(t,x,\mu,y,z,z^0,\alpha)$ is affine in y, z, and z^0, the matrix coefficient multiplying y being equal to $\partial_x b_1(t,x)$, and the coefficient multiplying z and z^0 being equal to $\partial_x\sigma(t,x)$ and $\partial_x\sigma^0(t,x)$ respectively. All of them are uniformly bounded in (t,x) by (**A2**) of the same assumption. Therefore, by standard BSDE stability property, we get that:

$$\begin{aligned}
&\mathbb{E}^t\big[\sup_{t\le s\le T}|v_s^x - v_s^{x',\prime}|^2\big] \\
&\le C\Big[\mathbb{E}^t\big[|\partial_x g(x_T^x,\nu_T) - \partial_x g(x_T^{x',\prime},\nu_T')|^2\big] \\
&\quad + \mathbb{E}^t\int_t^T \Big(|\partial_x f(s,x_s^x,\nu_s,\alpha_s^x) - \partial_x f(s,x_s^{x',\prime},\nu_s',\alpha_s^{x',\prime})|^2 \\
&\quad + |\partial_x b(s,x_s^x) - \partial_x b(s,x_s^{x',\prime})|^2 \\
&\quad + |\partial_x(\sigma,\sigma^0)(s,x_s^x) - \partial_x(\sigma,\sigma^0)(s,x_s^{x',\prime})|^2\big(|\zeta_s^x|^2 + |\zeta_s^{0,x}|^2\big)\Big)ds\Big].
\end{aligned} \tag{4.71}$$

Importantly, we observe from (4.70) and from the fact that $\partial_x\sigma$ and $\partial_x\sigma^0$ are bounded that the last term in the right-hand side can be bounded by:

$$\begin{aligned}
&\mathbb{E}^t\Big[\int_t^T \Big(|\partial_x(\sigma,\sigma^0)(s,x_s^x) - \partial_x(\sigma,\sigma^0)(s,x_s^{x',\prime})|^2\big(|\zeta_s^x|^2 + |\zeta_s^{0,x}|^2\big)\Big)ds\Big] \\
&\quad \le C\mathbb{E}^t\Big[\sup_{t\le s\le T}\big|\partial_x(\sigma,\sigma^0)(s,x_s^x) - \partial_x(\sigma,\sigma^0)(s,x_s^{x',\prime})\big|^4\Big]^{1/2}.
\end{aligned} \tag{4.72}$$

We now choose the coupling π between μ and μ' in the introduction of the fourth step as an optimal coupling for the 2-Wasserstein distance, so that $\mathbb{E}^t[|x_t - x_t'|^2] = W_2(\mu,\mu')^2$. By combining Theorem 1.53, which holds true since both environments here derive from the common source of noise $\boldsymbol{w}^0$, (4.65), (4.66) and (4.67), we get that:

$$\begin{aligned}
&\mathbb{E}^t\Big[\sup_{t\le s\le T}|x_s^x - x_s^{x',\prime}|^2 + \int_t^T |\hat\alpha_s^x - \hat\alpha_s^{x',\prime}|^2 ds\Big] \\
&\qquad \le C\Big(|x-x'|^2 + W_2(\mu,\mu') + \big(W_2(\mu,\mu')\big)^{1/2}\Big).
\end{aligned} \tag{4.73}$$

Recall moreover (4.66):

$$\mathbb{E}^{0,t}\Big[\sup_{t\le s\le T}\big(W_2(\nu_s,\nu_s')\big)^2\Big] \le C\Big(W_2(\mu,\mu') + \big(W_2(\mu,\mu')\big)^{1/2}\Big). \tag{4.74}$$

From (4.71), (4.72), and (4.74) and by boundedness and continuity of the coefficients in (4.71), we easily deduce that, for any $t \in [0,T]$,

$$\lim_{(x',\mu')\to(x,\mu)} \partial_x\mathcal{U}(t,x',\mu') = \partial_x\mathcal{U}(t,x,\mu).$$

We shall prove that the convergence is uniform in t when (x,μ) and (x',μ') are restricted to a compact subset $\mathcal{K} \subset \mathbb{R}^d \times \mathcal{P}_2(\mathbb{R}^d)$, with $\mathcal{K} = \mathcal{K}_x \times \mathcal{K}_\mu$, $\mathcal{K}_x$ and $\mathcal{K}_\mu$ being compact subsets of $\mathbb{R}^d$ and $\mathcal{P}_2(\mathbb{R}^d)$ respectively. The strategy relies on a tightness argument. Since b is bounded in (t,x) and is at most of linear growth in α and since $z^{t,x,\mu}$ is bounded, independently of t, x, and μ, it is indeed clear that the family of distributions $(\mathbb{P}^t \circ (x_s^{t,x,\mu})^{-1})_{t\in[0,T],s\in[t,T],(x,\mu)\in\mathcal{K}}$ is tight on $\mathbb{R}^d$. Here, the superscript (t,x,μ) in $(\boldsymbol{x}^{t,x,\mu}, \boldsymbol{y}^{t,x,\mu}, \boldsymbol{z}^{t,x,\mu}, \boldsymbol{z}^{0,t,x,\mu}, \boldsymbol{m}^{t,x,\mu})$ indicates that the FBSDE (4.53) is initialized with x at time t and that, in the coefficients of the FBSDE, the environment $\mathfrak{m}$ is initialized with μ at time t, namely $\mathfrak{m} = \mathfrak{m}^{t,\mu}$, where, in the latter notation, we also indicate the dependence upon t. Moreover, by a similar argument, for any $t \in [0,T]$, any $\mu \in \mathcal{K}_\mu$ and any $D \in \mathcal{F}^t$, it holds, for all $a > 0$,

$$\mathbb{E}^t\Big[\sup_{t\le s\le T} |x_s^{t,\mu}|^2 \mathbf{1}_D\Big] \le C\Big(\mathbb{E}^t\big[|x_t^{t,\mu}|^2\mathbf{1}_D\big] + \big(\mathbb{P}^t(D)\big)^{1/2}\Big)$$

$$\le C\Big(a^2\mathbb{P}^t(D) + \int_{\{|x|\ge a\}} |x|^2 d\mu(x) + \big(\mathbb{P}^t(D)\big)^{1/2}\Big),$$

where the second term in the right-hand side of the first line follows from Cauchy-Schwarz' inequality. Here, $(\boldsymbol{x}^{t,\mu}, \boldsymbol{y}^{t,\mu}, \boldsymbol{z}^{t,\mu}, \boldsymbol{m}^{t,\mu})$ denotes the solution to (4.53) with $x_t = \psi(\eta,\mu)$ as initial condition at time t and with $\mathfrak{m} = \mathfrak{m}^{t,\mu}$ as environment, see (2.23).

This proves that:

$$\lim_{\delta\to 0} \sup_{0\le t\le T} \sup_{\mu\in\mathcal{K}_\mu} \sup_{D\in\mathcal{F}^t:\mathbb{P}^t(D)\le\delta} \mathbb{E}^t\Big[\sup_{t\le s\le T} |x_s^{t,\mu}|^2\mathbf{1}_D\Big] = 0.$$

Recalling that $\nu_s^{t,\mu} = \mathcal{L}^1(X_s^{t,\mu})$ and following Lemma 3.16, we deduce that the family of distributions:

$$\big(\mathbb{P}^t \circ (\nu_s^{t,\mu})^{-1}\big)_{t\in[0,T],s\in[t,T],\mu\in\mathcal{K}_\mu}$$

is tight on $\mathcal{P}_2(\mathbb{R}^d)$. Therefore, for any $\varepsilon > 0$, there exists a compact subset $\mathcal{K}^\varepsilon \subset \mathbb{R}^d \times \mathcal{P}_2(\mathbb{R}^d)$, with $\mathcal{K}^\varepsilon = \mathcal{K}_x^\varepsilon \times \mathcal{K}_\mu^\varepsilon$, $\mathcal{K}_x^\varepsilon$ and $\mathcal{K}_\mu^\varepsilon$ being compact subsets of $\mathbb{R}^d$ and $\mathcal{P}_2(\mathbb{R}^d)$ respectively, such that

$$\forall (x,\mu)\in\mathcal{K}, \quad \begin{array}{l} \sup_{0\le t\le T}\sup_{t\le s\le T} \mathbb{P}^t\big((x_s^{t,x,\mu}, \nu_s^{t,\mu}) \notin \mathcal{K}^\varepsilon\big) \le \varepsilon, \\ \sup_{0\le t\le T} \mathbb{P}^t\big(\exists s\in[t,T] : x_s^{t,x,\mu} \notin \mathcal{K}_x^\varepsilon\big) \le \varepsilon. \end{array} \tag{4.75}$$

Returning to (4.71), (4.72), and (4.73) and introducing the modulii of continuity of the bounded and continuous functions $\partial_x g$, $\partial_x f$, $\partial_x b_1$, $\partial_x \sigma$ and $\partial_x \sigma^0$ on the compact sets $\mathcal{K}^\varepsilon$, $[0,T]\times\mathcal{K}^\varepsilon\times\{\alpha\in A : |\alpha|\le C\}$, $[0,T]\times\mathcal{K}^\varepsilon$ and $[0,T]\times\mathcal{K}_x^\varepsilon$ respectively, with C a common bound to all the processes $\hat{\boldsymbol{\alpha}}^{t,x,\mu} = (\hat{\alpha}(s, x_s^{t,x,\mu}, \sigma(s, x_s^{t,x,\mu})^{-1\dagger} z_s^{t,x,\mu}))_{t\le s\le T}$, we can find a bounded (measurable) function $w_\varepsilon : \mathbb{R}_+ \to \mathbb{R}_+$, with:

$$\lim_{\delta\searrow 0} w_\varepsilon(\delta) = 0,$$

such that, for any $\delta > 0$ and any $(x,\mu), (x',\mu') \in \mathcal{K}$,

$$\mathbb{E}^t\Big[\sup_{t\le s\le T} |\upsilon_s^{t,x,\mu} - \upsilon_s^{t,x',\mu'}|^2\Big]$$
$$\le C\Big[\varepsilon + w_\varepsilon(\delta) + \mathbb{P}^t\Big(\sup_{t\le s\le T} |x_s^{t,x,\mu} - x_s^{t,x',\mu'}| + W_2(\upsilon_T^{t,\mu}, \upsilon_T^{t,\mu'}) > \delta\Big)$$
$$+ \int_t^T \mathbb{P}^t\big(|x_s^{t,x,\mu} - x_s^{t,x',\mu'}| + W_2(\upsilon_s^{t,\mu}, \upsilon_s^{t,\mu'}) + |\alpha_s^{t,x,\mu} - \alpha_s^{t,x',\mu'}| > \delta\big)ds\Big]^{1/2},$$

where we used the same notation convention as above for $\boldsymbol{\upsilon}^{t,x,\mu}$ and $\hat{\boldsymbol{\alpha}}^{t,x,\mu}$ and where we used the fact that, in (4.71), the tuple $(\boldsymbol{x}^{x',\prime}, \boldsymbol{\nu}', \boldsymbol{\upsilon}^{x',\prime}, \boldsymbol{\zeta}^{x',\prime}, \boldsymbol{\zeta}^{0,x',\prime}, \hat{\boldsymbol{\alpha}}^{x',\prime})$ matches $(\boldsymbol{x}^{t,x',\mu'}, \boldsymbol{\nu}^{t,\mu'}, \boldsymbol{\upsilon}^{t,x',\mu'}, \boldsymbol{\zeta}^{t,x',\mu'} \boldsymbol{\zeta}^{t,x',\mu'} \hat{\boldsymbol{\alpha}}^{t,x',\mu'})$. From (4.73) and (4.74), this shows that:

$$\lim_{(x',\mu')\to(x,\mu)} \sup_{0\le s\le T} \sup_{(x,\mu),(x',\mu')\in\mathcal{K}} |\partial_x \mathcal{U}(s,x,\mu) - \partial_x \mathcal{U}(s,x',\mu')| = 0. \tag{4.76}$$

Following the analysis of the time regularity of $\mathcal{U}$, we finally get:

$$\partial_x \mathcal{U}(t,x,\mu) - \partial_x \mathcal{U}(t+h,x,\mu)$$
$$= \mathbb{E}^t\Big[\partial_x \mathcal{U}\big(t+h, x_{t+h}^{t,x,\mu}, \nu_{t+h}^{t,\mu}\big) - \partial_x \mathcal{U}(t+h,x,\mu)$$
$$+ \int_t^{t+h} \partial_x H\Big(s, x_s^{t,x,\mu}, \nu_s^{t,\mu}, \upsilon_s^{t,x,\mu}, \zeta_s^{t,x,\mu}, \zeta_s^{0,t,x,\mu}, \hat{\alpha}\big(s, x_s^{t,x,\mu}, \sigma(s, x_s^{t,x,\mu})^{-1\dagger} z_s^{t,x,\mu}\big)\Big)ds\Big].$$

In order to handle the first term in the right-hand side, we use (4.76), together with the tightness property (4.75) and the fact that $\partial_x \mathcal{U}$ is bounded. Recalling that the integral on the last line is bounded by $C(h + h^{1/2}\mathbb{E}[\int_t^T (|\zeta_s^{t,x,\mu}|^2 + |\zeta_s^{0,t,x,\mu}|^2)ds]^{1/2})$, we thus deduce that, for any $\varepsilon, \delta > 0$ and any $(x,\mu) \in \mathcal{K}$,

$$\big|\partial_x \mathcal{U}(t,x,\mu) - \partial_x \mathcal{U}(t+h,x,\mu)\big|$$
$$\le C\Big(\varepsilon + w_\varepsilon(\delta) + \mathbb{P}^t\big(|x_{t+h}^{t,x,\mu} - x| + W_2(\nu_{t+h}^{t,\mu}, \mu) > \delta\big) + h^{1/2}\Big)$$
$$\le C\Big(\varepsilon + w_\varepsilon(\delta) + \delta^{-1}h^{1/2}\Big),$$

where $w_\varepsilon(\delta) \to 0$ as $\delta \to 0$. Time continuity of $\partial_x \mathcal{U}$ easily follows. □

Remark 4.22 *Under the assumptions of Theorem 4.21, we know, as shown in the third step of the proof, that, for any $\mathcal{V} \in \mathcal{P}_2(\mathcal{P}_2(\mathbb{R}^d))$, the equilibrium $\mathfrak{M}$ initialized with the distribution $\mathcal{V}$ at (say) time 0 and constructed on some probability space $(\Omega, \mathcal{F}, \mathbb{F}, \mathbb{P})$ of the product form, equipped with $(X_0, \mathbf{W}^0, \mu_0, \mathbf{W})$, with $\mu_0 \sim \mathcal{V}$ and $\mathcal{L}^1(X_0) = \mu_0$, is adapted with respect to the filtration generated by the initial distribution μ_0 of the population and by the common noise $\mathbf{W}^0$. In particular, Blumenthal's zero-one law says that, in the FBSDE* (4.5) *(or equivalently* (4.7)*, with $t = 0$ and $\xi = X_0$), with $\mu_s = \mathfrak{M} \circ (e_s^x)^{-1}$, the completion of the filtration generated* *by* $(X_0, \boldsymbol{W}^0, \mathfrak{M}, \boldsymbol{W})$ *is right-continuous. Therefore, in the conditional expectation appearing in the statement of Proposition 4.3, we may replace* $\mathcal{F}_t^{\mathrm{nat},(X_0,\boldsymbol{W}^0,\mathfrak{M},\boldsymbol{W})}$ *by* $\mathcal{F}_t^{(X_0,\boldsymbol{W}^0,\mu_0,\boldsymbol{W})}$. *Then, the conditional expectation identifies with* Y_t *when* $\boldsymbol{\beta} = \hat{\boldsymbol{\alpha}}$, *which shows that* $Y_t = \mathcal{U}(t, X_t, \mu_t)$. *Hence, at any time* $t \in [0, T]$, $\mathcal{U}(t, \cdot, \mu_t)$ *reads as the decoupling field, at time t, of the FBSDE* (4.5).

Using Assumption MFG with a Common Noise SMP Relaxed and Lasry-Lions Monotonicity Condition

In the same way as above, we claim:

Theorem 4.23 *Let assumption* **MFG with a Common Noise SMP Relaxed** *in Subsection 3.4.3 together with the Lasry-Lions monotonicity conditions in Subsection 3.5.1 be in force, namely b,* σ, *and* σ^0 *are independent of* μ, *f has the form:*

$$f(t, x, \mu, \alpha) = f_0(t, x, \mu) + f_1(t, x, \alpha), \quad t \in [0, T], \; x \in \mathbb{R}^d, \; \alpha \in A, \; \mu \in \mathcal{P}_2(\mathbb{R}^d),$$

$f_0(t, \cdot, \cdot)$ *and g satisfying the monotonicity property* (3.77). *Assume moreover that the coefficients b, f, g,* σ, *and* σ^0 *are jointly continuous in all the parameters and that their derivatives in x and* α, *which exist under assumption* **MFG with a Common Noise SMP Relaxed**, *are jointly continuous in all its variables. Then, the master field* $\mathcal{U}$ *is well defined and is a continuous viscosity solution of the master equation* (4.41).

Proof. The proof goes along the same lines as in Theorem 4.21. So we just point out the main differences. Throughout the whole discussion, the time index t is kept fixed. Moreover, we shall assume that the probability measures μ and μ' and the random variables x_t and x_t' that appear in the proof of Theorem 4.21 have a bounded second-order moment, less than some arbitrarily fixed constant c. Similarly, we can assume that the initial positions x and x' that we shall consider have norms not greater than c. With such a c at hand, all the constants C that appear below may depend on c.

For starters, we notice that the coefficients driving the forward-backward system (4.53) are no longer given by (3.58) but by (3.61). In particular, $\mathcal{U}(t, x, \mu)$ does not coincide anymore with y_t^x in (4.53). From Theorem 4.10, y_t^x, which is also equal to v_t^x, matches $\partial_x \mathcal{U}(t, x, \mu)$. Notice that we can argue that it is possible to remove $\mathbb{E}^{t,\mu}$ in Theorem 4.10. Indeed, for the same reasons as in the proof of Theorem 4.21, the filtration used for solving the FBSDE is the completion of the filtration generated by the two noises $\boldsymbol{w}^0$ and $\boldsymbol{w}$, and as a consequence, the initial value of the backward component is almost surely deterministic. Observe also that assumptions **Necessary SMP Master** and **Decoupling Master** are satisfied, which makes licit the application of Theorem 4.10 and Corollary 4.11. Importantly, (**A1**) is ensured by Theorem 1.60.

For the time being, we focus on the proof of (4.66). By (1.63), the lower bound (4.56) remains true with $\hat{\boldsymbol{\alpha}}^\xi = (\hat{\alpha}(s, x_s, y_s))_{t \le s \le T}$, where $(\boldsymbol{x}, \boldsymbol{y}) = (\boldsymbol{x}^{\xi,\hat{\boldsymbol{\alpha}}}, \boldsymbol{y}^{\xi,\hat{\boldsymbol{\alpha}}})$ with $\xi = \psi(\eta, v^0)$. Similarly, following the third step in the proof of Lemma 3.33, it is pretty clear that (4.57) and (4.58) remain valid as well. However, it is not true anymore that $\hat{\boldsymbol{\alpha}}$ is bounded. Indeed, the proof of (3.74) merely implies that, for any $s \in [t, T]$, $|\hat{\alpha}_s| \le C(1 + |x_s| +$

$\sup_{t \le r \le s} \mathbb{E}^{1,t}[|x_r|^2]^{1/2})$. Still, together with the local Lipschitz property of the coefficients f and g –see assumption **MFG with a Common Noise SMP Relaxed**–, this is sufficient to repeat the proof of (4.64) and (4.66).

Now, we can investigate $|\mathcal{U}(t,x,\mu) - \mathcal{U}(t,x',\mu)|$, for $|x|,|x'| \le c$ and $M_2(\mu) \le c$. Using the same notation as in the fifth step of the proof of Theorem 4.21 and applying Theorem 1.53, we obtain:

$$\mathbb{E}^t\Big[\sup_{t \le s \le T}\big(|x_s^x - x_s^{x'}|^2 + |y_s^x - y_s^{x'}|^2\big)\Big] \le C|x-x'|^2.$$

The key point is to use this bound with the definitions of $\mathcal{U}(t,x,\mu)$ and $\mathcal{U}(t,x',\mu)$ from Definition 4.1, which take the form:

$$\mathcal{U}(t,x,\mu) = \mathbb{E}^t\bigg[\int_t^T f\big(s, x_s^x, \nu_s, \hat{\alpha}(s,x_s^x,y_s^x)\big)ds + g(x_T^x, \nu_T)\bigg],$$

$$\mathcal{U}(t,x',\mu) = \mathbb{E}^t\bigg[\int_t^T f\big(s, x_s^{x'}, \nu_s, \hat{\alpha}(s,x_s^{x'},y_s^{x'})\big)ds + g(x_T^{x'}, \nu_T)\bigg].$$

Following (3.63) in the statement of Lemma 3.33 (see (3.72) for the proof) and using the fact that $\hat{\alpha}$ is at most of linear growth, we claim that:

$$\mathbb{E}^t\Big[\sup_{t \le s \le T}\big(|\hat{\alpha}(s,x_s^x,y_s^x)|^2 + |\hat{\alpha}(s,x_s^{x'},y_s^{x'})|^2\big)ds\Big] \le C.$$

Therefore, using the local Lipschitz continuity properties of f and g together with the Lipschitz continuity of $\hat{\alpha}$ and the bound (4.58), we easily deduce that $\mathcal{U}$ is Lipschitz continuous in the direction x on any bounded subset of $[0,T] \times \mathbb{R}^d \times \mathcal{P}_2(\mathbb{R}^d)$.

In order to investigate continuity in the directions t and μ, we proceed as in the fifth step of the proof of Theorem 4.21, with the proviso that the constant C that appears in all the estimates may depend on c.

It then remains to investigate the regularity of $\partial_x \mathcal{U}$. Again, we may proceed as in the last step of the proof of Theorem 4.21, noticing that the setting here is easier to handle since $\partial_x b$ is constant in x. We then get an analogue of (4.71), but without the terms involving $\partial_x b$, $\partial_x \sigma$ and $\partial_x \sigma^0$ since the latter are constant functions. Then, using the fact that, under assumption **MFG with a Common Noise SMP Relaxed** (see **(A4)** therein), $\partial_x g$ and $\partial_x f$ are Lipschitz continuous in (x,μ) and (x,μ,α), we complete the proof. □

Remark 4.24 *As pointed out in the proof of Theorem 4.23, we know from Theorem 4.10 that y_t^x in (4.53) coincides with $\partial_x \mathcal{U}(t,x,\mu)$.*

4.5 Revisiting the Examples of Chapter 1 of the First Volume

4.5.1 Revisiting the Coupled OUs Game

As starters, we revisit the benchmark example of the Ornstein-Uhlenbeck state processes coupled through their empirical mean introduced in Chapter (Vol I)-1 and solved in several forms in Section (Vol I)-2.5. Taking advantage of the fact that we constructed exact Nash equilibria in Chapter (Vol I)-2, we can approach the master equation and its solution by passing to the limit in the equations satisfied by the value functions of the finite games, such a strategy being somewhat reminiscent of Lemma (Vol I)-1.2.

Throughout this subsection, we use the same notations as in Section (Vol I)-2.5. We invite the reader to check that section for the definitions and meanings of the objects and quantities we use below.

The Limit $N \to \infty$ of the N-Player Game

Our starting point here is the set of value functions constructed by the PDE method based on the solution of a system of coupled HJB equations. We emphasize the dependence upon the number N of players and we now write:

- η_t^N and χ_t^N for the solutions η_t and χ_t, at time t, of the system (Vol I)-2.82,
- $V^{i,N}(t,x) = (\eta_t^N/2)(\bar{x} - x^i)^2 + \chi_t^N$ for the value function of player i in the N player game, when evaluated at the point (t,x).

Clearly,

$$\lim_{N\to\infty} \eta_t^N = \eta_t^\infty, \qquad \text{and} \qquad \lim_{N\to\infty} \chi_t^N = \chi_t^\infty,$$

where the functions η_t^∞ and χ_t^∞ solve the system:

$$\begin{cases} \dot{\eta}_t^\infty = 2(a+q)\eta_t^\infty + (\eta_t^\infty)^2 - (\epsilon - q^2), \\ \dot{\chi}_t^\infty = -\dfrac{1}{2}\sigma^2(1-\rho^2)\eta_t^\infty, \end{cases} \tag{4.77}$$

which is solved in the same way as when N is finite. Thanks to the remark on Riccati equations in Subsection (Vol I)-2.5.1, we find:

$$\eta_t^\infty = \frac{-(\epsilon - q^2)\big(e^{(\delta^+ - \delta^-)(T-t)} - 1\big) - c\big(\delta^+ e^{(\delta^+ - \delta^-)(T-t)} - \delta^-\big)}{\big(\delta^- e^{(\delta^+ - \delta^-)(T-t)} - \delta^+\big) - c\,\big(e^{(\delta^+ - \delta^-)(T-t)} - 1\big)}, \tag{4.78}$$

and

$$\chi_t^\infty = \frac{1}{2}\sigma^2(1-\rho^2)\int_t^T \eta_s^\infty \, ds, \tag{4.79}$$

where:

$$\delta^\pm = -(a+q) \pm \sqrt{R}, \qquad \text{with} \qquad R = (a+q)^2 + \epsilon - q^2 > 0. \tag{4.80}$$

Next we consider the equilibrium behavior of the players' value functions $(V^{i,N})_{i=1,\cdots,N}$. For the purpose of the present discussion we notice that the value functions $(V^{i,N})_{i=1,\cdots,N}$ of all the players in the N player game can be written as:

$$V^{i,N}\big(t,(x^1,\cdots,x^N)\big) = V^N\Big(t,x^i,\frac{1}{N}\sum_{j=1}^{N}\delta_{x^j}\Big),$$

for $t \in [0,T]$ and $(x^1,\cdots,x^N) \in \mathbb{R}^N$, where the single function V^N is defined as:

$$V^N(t,x,\mu) = \frac{\eta_t^N}{2}\Big(x - \int_{\mathbb{R}} x d\mu(x)\Big)^2 + \chi_t^N, \quad (t,x,\mu) \in [0,T]\times\mathbb{R}\times\mathcal{P}_1(\mathbb{R}).$$

Since the dependence upon the measure is only through the mean of the measure, we shall often use the function:

$$v^N(t,x,\bar{\mu}) = \frac{\eta_t^N}{2}(x-\bar{\mu})^2 + \chi_t^N, \quad (t,x,\bar{\mu}) \in [0,T]\times\mathbb{R}\times\mathbb{R}.$$

Notice that, at least for $(t,x,\bar{\mu})$ fixed, we have:

$$\lim_{N\to\infty} v^N(t,x,\bar{\mu}) = v^\infty(t,x,\bar{\mu})$$

where:

$$v^\infty(t,x,\bar{\mu}) = \frac{\eta_t^\infty}{2}(x-\bar{\mu})^2 + \chi_t^\infty, \qquad (t,x,\bar{\mu}) \in [0,T]\times\mathbb{R}\times\mathbb{R}.$$

Similarly, all the optimal strategies in (Vol I)-(2.86) may be expressed through a single feedback function $\hat{\alpha}^N(t,x,\bar{\mu}) = [q + (1-1/N)\eta_t^N](\bar{\mu}-x)$ since $\hat{\alpha}_t^i = \hat{\alpha}^N(t,X_t^i,1/N\sum_{i=1}^N X_t^i)$. Clearly,

$$\lim_{N\to\infty} \hat{\alpha}^N(t,x,\mu) = \hat{\alpha}^\infty(t,x,\bar{\mu}),$$

where $\hat{\alpha}^\infty(t,x,\bar{\mu}) = [q+\eta_t^\infty](\bar{\mu}-x)$.

Search for an Asymptotic Equilibrium

We now consider a filtered probability space $(\Omega,\mathcal{F},\mathbb{P})$ obtained as above as the completion of $(\Omega^0\times\Omega^1,\mathcal{F}^0\otimes\mathcal{F}^1,\mathbb{F}^0\otimes\mathbb{F}^1,\mathbb{P}^0\otimes\mathbb{P}^1)$, where $(\Omega^0,\mathcal{F}^0,\mathbb{P}^0)$ and $(\Omega^1,\mathcal{F}^1,\mathbb{P}^1)$ are two complete probability spaces equipped respectively with two complete and right-continuous filtrations $\mathbb{F}^0$ and $\mathbb{F}^1$ and with two 1-dimensional Brownian motions $\boldsymbol{W}^0$ and $\boldsymbol{W}$. Also, $(\Omega^1,\mathcal{F}^1,\mathbb{P}^1)$ is equipped with an $\mathcal{F}_0^1$-measurable random variable X_0 satisfying $\mathbb{E}^1[|X_0|^2] < \infty$.

Now that we have a candidate for the asymptotic optimal feedback function, we consider the solution of the McKean-Vlasov equation:

$$
\begin{aligned}
d\hat{X}_t &= \Big[a(\bar{\mu}_t - \hat{X}_t) + \alpha^\infty(t, \hat{X}_t, \bar{\mu}_t)\Big]dt + \sigma\left(\rho dW_t^0 + \sqrt{1-\rho^2}dW_t\right) \\
&= \left(a + q + \eta_t^\infty\right)(\bar{\mu}_t - \hat{X}_t)dt + \sigma\left(\rho dW_t^0 + \sqrt{1-\rho^2}dW_t\right),
\end{aligned}
\tag{4.81}
$$

subject to the initial condition $\hat{X}_0 = X_0$, for an $\mathcal{F}_0^1$-measurable square-integrable random variable X_0, and to the condition $\bar{\mu}_t = \mathbb{E}^1[\hat{X}_t]$ (so that $\bar{\mu}_0 = \mathbb{E}^1[X_0]$). Implementing the form of $\hat{\alpha}^\infty(t, x, \bar{\mu})$, we get:

$$
d\bar{\mu}_t = \sigma\rho dW_t^0, \quad t \in [0, T], \tag{4.82}
$$

which suggests the investigation of conditional equilibria with $(\bar{\mu}_t)_{0\le t\le T}$ as flow of conditional means. In order to proceed, we observe that $\hat{\boldsymbol{X}}$ can be written as a progressively measurable function of X_0, $\boldsymbol{W}^0$, $\boldsymbol{W}$. In particular, $\mathfrak{M} = \mathcal{L}^1(\hat{\boldsymbol{X}}, \boldsymbol{W})$ is measurable with respect to $\sigma\{\boldsymbol{W}^0\}$ and the 4-tuple $(X_0, \boldsymbol{W}^0, \mathfrak{M}, \boldsymbol{W})$ is obviously compatible with the filtration $\mathbb{F}$. This shows that the set-up induced by our candidate for solving the mean field game is admissible.

On such a set-up, we consider an optimization problem in the random environment $(\bar{\mu}_t)_{0\le t\le T}$, along the lines of Section 1.4. Namely, we minimize the cost functional:

$$
J^{\bar{\mu}}(\boldsymbol{\alpha}) = \mathbb{E}\left[\frac{c}{2}\left(\bar{\mu}_T - X_T^{\boldsymbol{\alpha}}\right)^2 + \int_0^T \left[\frac{\epsilon}{2}\left(\bar{\mu}_t - X_t^{\boldsymbol{\alpha}}\right)^2 - q\alpha_t(\bar{\mu}_t - X_t^{\boldsymbol{\alpha}}) + \frac{1}{2}\left(\alpha_t\right)^2\right]dt\right],
$$

over square-integrable progressively measurable scalar-valued stochastic processes $\boldsymbol{\alpha} = (\alpha_t)_{0\le t\le T}$, the dynamics of the process $(X_t^{\boldsymbol{\alpha}})_{0\le t\le T}$ being subject to $X_0^{\boldsymbol{\alpha}} = X_0$ and to:

$$
dX_t^{\boldsymbol{\alpha}} = a(\bar{\mu}_t - X_t^{\boldsymbol{\alpha}})dt + \alpha_t dt + \sigma\left(\rho dW_t^0 + \sqrt{1-\rho^2}dW_t\right), \quad t \in [0, T].
$$

The reduced Hamiltonian reads:

$$
H^{(r)}(x, \bar{\mu}, y, \alpha) = \left[a(\bar{\mu} - x) + \alpha\right]y + \frac{\alpha^2}{2} - q\alpha(\bar{\mu} - x) + \frac{\epsilon}{2}(\bar{\mu} - x)^2,
$$

for $(x, \bar{\mu}, y) \in \mathbb{R}^3$, the minimizer being given by $\hat{\alpha}(t, x, \bar{\mu}, y) = q(\bar{\mu} - x) - y$. We notice that:

$$
\hat{\alpha}^\infty(t, x, \bar{\mu}) = \hat{\alpha}\big(t, x, \bar{\mu}, \eta_t^\infty(x - \bar{\mu})\big).
$$

This suggests the use of the Pontryagin stochastic maximum principle with $(Y_t = \eta_t^\infty(\hat{X}_t - \bar{\mu}_t))_{0\le t\le T}$ in order to check that $(\hat{\alpha}^\infty(t, \hat{X}_t, \bar{\mu}_t))_{0\le t\le T}$ is the optimal control process and $(\hat{X}_t)_{0\le t\le T}$ is the optimal state associated with the minimization of the functional $J^{\bar{\mu}}$ in the environment $(\bar{\mu}_t)_{0\le t\le T}$. Thanks to the condition $\epsilon \ge q^2$,

H is convex in (x, α), so that Theorem 1.60 indeed applies. From the Riccati equation (4.77) satisfied by $(\eta_t^\infty)_{0 \le t \le T}$, we get:

$$
\begin{aligned}
d\big[\eta_t^\infty(\hat{X}_t - \bar{\mu}_t)\big] &= \big[\dot{\eta}_t^\infty - (a+q)\eta_t^\infty - (\eta_t^\infty)^2\big](\hat{X}_t - \bar{\mu}_t)dt + \sigma\eta_t^\infty\sqrt{1-\rho^2}dW_t \\
&= \big[(a+q)\eta_t^\infty - (\epsilon - q^2)\big](\hat{X}_t - \bar{\mu}_t)dt + \sigma\eta_t^\infty\sqrt{1-\rho^2}dW_t.
\end{aligned}
$$

Now,

$$
\partial_x H^{(r)}(x, \bar{\mu}, y, \alpha) = q\alpha - ay - \epsilon(\bar{\mu} - x),
$$

so that:

$$
\begin{aligned}
\partial_x H^{(r)}\big(x, \bar{\mu}, \eta_t^\infty(x - \bar{\mu}), \hat{\alpha}^\infty(t, x, \bar{\mu})\big) &= (q^2 - \epsilon)(\bar{\mu} - x) - (a+q)\eta_t^\infty(x - \bar{\mu}) \\
&= -\big[(a+q)\eta_t^\infty - (\epsilon - q^2)\big](x - \bar{\mu}).
\end{aligned}
$$

Observe that it is perfectly licit to formulate the Pontryagin principle by means of the sole reduced Hamiltonian since the volatility coefficients are constant. This proves that:

$$
d\big[\eta_t^\infty(\hat{X}_t - \bar{\mu}_t)\big] = -\partial_x H\big(\hat{X}_t, \bar{\mu}_t, \eta_t^\infty(\hat{X}_t - \bar{\mu}_t), \hat{\alpha}^\infty(t, \hat{X}_t, \bar{\mu}_t)\big)dt + \sigma\eta_t^\infty\sqrt{1-\rho^2}dW_t,
$$

for $t \in [0, T]$. This guarantees that, in the environment $(\bar{\mu}_t)_{0 \le t \le T}$ (or in the super-environment $\mathfrak{M}$), the solution $(\hat{X}_t)_{0 \le t \le T}$ to (4.81) is the optimal path. By (4.82), this proves that the flow of conditional marginal measures of $(\hat{X}_t)_{0 \le t \le T}$ forms an equilibrium. The equilibrium is strong since it is adapted to the filtration generated by $\boldsymbol{W}^0$.

Search for the Master Field

From the previous analysis, one clearly sees that, in the environment $(\bar{\mu}_t)_{0 \le t \le T}$, the random field $[0, T] \times \mathbb{R} \ni (t, x) \mapsto v^\infty(t, x, \bar{\mu}_t)$ is the (random) value function of the optimization problem with random coefficients. Therefore, a natural candidate for the master field is the mapping:

$$
\mathcal{U}^\infty : [0, T] \times \mathbb{R} \times \mathcal{P}_2(\mathbb{R}) \ni (t, x, \mu) \mapsto v^\infty\Big(t, x, \int_{\mathbb{R}} x d\mu(x)\Big) = \frac{\eta_t^\infty}{2}(x - \bar{\mu})^2 + \chi_t^\infty,
$$

where we use the generic notation $\bar{\mu}$ for the mean of μ. Notice that $\bar{\mu}$ is now a deterministic number while it was a random number in the previous subsection. Notice also that we defined $\mathcal{U}$ on $[0, T] \times \mathbb{R} \times \mathcal{P}_2(\mathbb{R})$ although the measure argument could be taken in the larger space $\mathcal{P}_1(\mathbb{R})$. Our choice is motivated by the desire to fit the framework of Definition 4.1.

We now check that $\mathcal{U}^\infty$ solves the master equation. From (4.42), the latter reads:

$$\begin{aligned}
&\partial_t \mathcal{U}(t,x,\mu) + H^*\big(x,\mu,\partial_x \mathcal{U}(t,x,\mu)\big) \\
&\quad + \frac{\sigma^2}{2}\partial^2_{xx}\mathcal{U}(t,x,\mu) + \frac{\rho^2\sigma^2}{2}\int_{\mathbb{R}^2} \partial^2_\mu \mathcal{U}(s,x,\mu)(v,v')d\mu(v)d\mu(v') \\
&\quad + \frac{\sigma^2}{2}\int_{\mathbb{R}} \partial_v\partial_\mu \mathcal{U}(t,x,\mu)(v)d\mu(v) + \sigma^2\rho^2 \int_{\mathbb{R}} \partial_x\partial_\mu \mathcal{U}(t,x,\mu)(v)d\mu(v) \\
&\quad + \int_{\mathbb{R}} (a+q+\eta_t^\infty)(\bar{\mu}-v)\partial_\mu \mathcal{U}(t,x,\mu)(v)d\mu(v) = 0,
\end{aligned}$$

where the minimized Hamiltonian is given by:

$$H^*(x,\bar{\mu},y) = (a+q)(\bar{\mu}-x)y - \frac{1}{2}y^2 + \frac{1}{2}(\epsilon - q^2)(\bar{\mu}-x)^2.$$

In order to proceed, we compute all the terms appearing in the master equation one by one. We first notice that:

$$\begin{aligned}
&\partial_t \mathcal{U}^\infty(t,x,\mu) + H^*\big(x,\mu,\partial_x \mathcal{U}^\infty(t,x,\mu)\big) \\
&= \frac{\dot{\eta}_t^\infty}{2}(x-\bar{\mu})^2 + \dot{\chi}_t^\infty - (a+q)(x-\bar{\mu})^2\eta_t^\infty - \frac{1}{2}\big(\eta_t^\infty\big)^2(x-\bar{\mu})^2 + \frac{1}{2}(\epsilon-q^2)(x-\bar{\mu})^2 \\
&= \dot{\chi}_t^\infty,
\end{aligned} \tag{4.83}$$

where we used the Riccati equation (4.77) to pass from the second to the third line. Now, using the differential calculus developed in Chapter (Vol I)-5, we get:

$$\begin{aligned}
&\partial^2_{xx}\mathcal{U}^\infty(t,x,\mu) = \eta_t^\infty, \quad \partial_\mu \mathcal{U}^\infty(t,x,\mu)(v) = \eta_t^\infty(\bar{\mu}-x), \\
&\partial_x\partial_\mu \mathcal{U}^\infty(t,x,\mu)(v) = -\eta_t^\infty, \quad \partial_v\partial_\mu \mathcal{U}^\infty(t,x,\mu)(v) = 0, \\
&\partial^2_\mu \mathcal{U}^\infty(t,x,\mu)(v,v') = \eta_t^\infty,
\end{aligned}$$

for $(t,x,\mu,v,v') \in [0,T]\times\mathbb{R}\times\mathcal{P}_2(\mathbb{R})\times\mathbb{R}\times\mathbb{R}$. Therefore,

$$\begin{aligned}
&\frac{\sigma^2}{2}\partial^2_{xx}\mathcal{U}^\infty(t,x,\mu) + \frac{\rho^2\sigma^2}{2}\int_{\mathbb{R}^2} \partial^2_\mu \mathcal{U}^\infty(s,x,\mu)(v,v')d\mu(v)d\mu(v') \\
&\quad + \frac{\sigma^2}{2}\int_{\mathbb{R}} \partial_v\partial_\mu \mathcal{U}^\infty(t,x,\mu)(v)d\mu(v) + \sigma^2\rho^2\int_{\mathbb{R}} \partial_x\partial_\mu \mathcal{U}^\infty(t,x,\mu)(v)d\mu(v) \\
&= \frac{\sigma^2}{2}(1-\rho^2)\eta_t^\infty = -\dot{\chi}_t^\infty,
\end{aligned} \tag{4.84}$$

the last line following from the second equation in (4.77). Finally, thanks to the form of $\partial_\mu \mathcal{U}^\infty$,

$$\int_{\mathbb{R}} (a + q + \eta_t^\infty)(\bar{\mu} - v)\partial_\mu \mathcal{U}^\infty(t, x, \mu)(v) d\mu(v) = 0. \tag{4.85}$$

The sum of the terms in (4.83), (4.84), and (4.85) is zero, which indeed shows that $\mathcal{U}^\infty$ solves the master equation.

4.5.2 Revisiting the First Macro-Economic Growth Model

We apply the strategy and the tools developed in this chapter to the first growth model discussed in Subsection (Vol I)-1.4.1.

Throughout this subsection, we use the notations of Subsection (Vol I)-1.4.1. We refer the reader to that subsection for the definitions and meanings of the objects we use below. Actually, we shall just use the fact that a, b, c, and E below are positive constants and that $p > 1$. Also, we recall that we are dealing with a maximization instead of minimization problem; in this respect, the theoretical results established before remain true provided that the Hamiltonian is defined accordingly.

Form of the Master Equation

The reduced Hamiltonian of the system reads:

$$H^{(r)}(x, \mu, y, \alpha) = \alpha y + c\frac{x^a}{[(d\mu/dx)(x)]^b} - \frac{E}{p}\frac{\alpha^p}{[\mu([x, \infty))]^b},$$

for $x, \alpha \geq 0$, $y \in \mathbb{R}$, and $\mu \in \mathcal{P}_2(\mathbb{R})$, which makes sense when μ has a density which does not vanish at x. As in Chapter (Vol I)-1, we use the convention that the first term is set to 0 when the density is not defined or is itself 0 and that the second term is set to 0 when μ does not charge the interval $[x, \infty)$.

The value $\hat{\alpha}(x, \mu, y)$ of α maximizing $H^{(r)}$ is given, at least for $y \geq 0$, by:

$$\hat{\alpha}(x, \mu, y) = \left(\frac{y}{E}\big[\mu([x, \infty))\big]^b\right)^{1/(p-1)} \tag{4.86}$$

so that:

$$\begin{aligned}
H^{(r)}\big(x, \mu, y, \hat{\alpha}(x, \mu, y)\big) &= \left(\frac{y}{E}\big[\mu([x, \infty))\big]^b\right)^{1/(p-1)} y + c\frac{x^a}{[(d\mu/dx)(x)]^b} \\
&\quad - \frac{E}{p}\frac{\Big((y/E)[\mu([x, \infty))]^b\Big)^{p/(p-1)}}{[\mu([x, \infty))]^b} \\
&= \frac{p-1}{p}E^{-1/(p-1)} y^{p/(p-1)} \big[\mu([x, \infty))\big]^{b/(p-1)} + c\frac{x^a}{[(d\mu/dx)(x)]^b}.
\end{aligned}$$

Therefore, from (4.42), we deduce that the master equation takes the form (at least at points (t, x, μ) for which μ has a density that does not vanish at x and for which $\partial_x \mathcal{U}(t, x, \mu) \geq 0$):

$$
\begin{aligned}
&\partial_t \mathcal{U}(t,x,\mu) + \frac{p-1}{pE^{1/(p-1)}} \big(\partial_x \mathcal{U}(t,x,\mu)\big)^{p/(p-1)} \big[\mu([x,\infty))\big]^{b/(p-1)} \\
&+ c \frac{x^a}{[(d\mu/dx)(x)]^b} + \frac{\sigma^2}{2} x^2 \partial^2_{xx} \mathcal{U}(t,x,\mu) \\
&+ \int_{\mathbb{R}} \left(\frac{\partial_x \mathcal{U}(t,v,\mu)}{E} \big[\mu([x,\infty))\big]^b \right)^{1/(p-1)} \partial_\mu \mathcal{U}(t,x,\mu)(v) d\mu(v) \\
&+ \frac{\sigma^2}{2} \int_{\mathbb{R}} v^2 \partial_v \partial_\mu \mathcal{U}(t,x,\mu)(v) d\mu(v) + \frac{\sigma^2}{2} \int_{\mathbb{R}^2} v v' \partial^2_\mu \mathcal{U}(s,x,\mu)(v,v') d\mu(v) d\mu(v') \\
&+ \sigma^2 \int_{\mathbb{R}} x v \partial_x \partial_\mu \mathcal{U}(t,x,\mu)(v) d\mu(v) = 0,
\end{aligned}
$$

with the terminal condition $\mathcal{U}(T, \cdot, \cdot) \equiv 0$.

We now specialize the equation when μ is a Pareto distribution of parameter $q > 0$, in which case we write $\mu^{(q)}$ instead of μ. Using the explicit formula (Vol I)-(1.26) for the density of $\mu^{(q)}$ and the fact that:

$$
\mu^{(q)}([x,\infty)) = 1 \wedge \frac{q^k}{x^k},
$$

we get:

$$
\begin{aligned}
f\big(x, \mu^{(q)}, \alpha\big) &= c \frac{x^a}{(kq^k/x^{k+1})^b} \mathbf{1}_{\{x \geq q\}} - \frac{E}{p} \frac{\alpha^p}{1 \wedge (q^{kb}/x^{kb})} \\
&= \frac{c}{k^b q^{kb}} x^{a+b(k+1)} \mathbf{1}_{\{x \geq q\}} - \frac{E}{pq^{kb}} \alpha^p \big(x^{kb} \vee q^{kb}\big),
\end{aligned}
$$

and

$$
\hat{\alpha}(x,\mu,y) = \left[\frac{y}{E} \Big(\frac{q^{kb}}{x^{kb}} \wedge 1 \Big) \right]^{1/(p-1)}, \tag{4.87}
$$

so that:

$$
\begin{aligned}
&H^{(r)}\big(x, \mu^{(q)}, y, \hat{\alpha}(x,\mu,y)\big) \\
&= \frac{p-1}{p} E^{-1/(p-1)} y^{p/(p-1)} \Big(\frac{q^{kb/(p-1)}}{x^{kb/(p-1)}} \wedge 1 \Big) + c \frac{x^{a+(k+1)b}}{k^b q^{kb}} \mathbf{1}_{\{x \geq q\}}.
\end{aligned}
$$

Therefore, the master equation becomes:

$$
\begin{aligned}
&\partial_t \mathcal{U}\big(t,x,\mu^{(q)}\big) \\
&\quad + \frac{p-1}{pE^{1/(p-1)}}\big(\partial_x \mathcal{U}(t,x,\mu^{(q)})\big)^{p/(p-1)}\Big(\frac{q^{kb/(p-1)}}{x^{kb/(p-1)}} \wedge 1\Big) + c\frac{x^{a+(k+1)b}}{k^b q^{kb}}\mathbf{1}_{\{x\geq q\}} \\
&\quad + \frac{\sigma^2}{2}x^2\partial^2_{xx}\mathcal{U}\big(t,x,\mu^{(q)}\big) \\
&\quad + \int_{\mathbb{R}}\bigg(\frac{\partial_x \mathcal{U}(t,v,\mu)}{E}\Big(\frac{q^{kb}}{v^{kb}} \wedge 1\Big)\bigg)^{1/(p-1)} \partial_\mu \mathcal{U}\big(t,x,\mu^{(q)}\big)(v)d\mu^{(q)}(v) \\
&\quad + \frac{\sigma^2}{2}\int_{\mathbb{R}} v^2 \partial_v \partial_\mu \mathcal{U}\big(t,x,\mu^{(q)}\big)(v)d\mu^{(q)}(v) \\
&\quad + \sigma^2\int_{\mathbb{R}} xv\partial_x\partial_\mu \mathcal{U}\big(t,x,\mu^{(q)}\big)(v)d\mu^{(q)}(v) \\
&\quad + \frac{\sigma^2}{2}\int_{\mathbb{R}^2} vv'\partial^2_\mu \mathcal{U}\big(s,x,\mu^{(q)}\big)(v,v')d\mu^{(q)}(v)d\mu^{(q)}(v') = 0.
\end{aligned}
\tag{4.88}
$$

Master Equation Along Pareto Distributions

Assuming that the initial distribution of the values of the state is given by the Pareto distribution $\mu^{(1)}$, we restrict the search for equilibria to Pareto distributions. This means that the description of the equilibrium flow of measures, denoted by $(\mu_t)_{0\leq t\leq T}$ throughout the analysis, reduces to the description of the flow of corresponding Pareto parameters, denoted by $(q_t)_{0\leq t\leq T}$.

By (4.86), the optimal feedback function must read:

$$
\begin{aligned}
(t,x) \mapsto &\bigg[\frac{\partial_x \mathcal{U}(t,x,\mu_t)}{E}\Big(\frac{q_t^{kb}}{x^{kb}} \wedge 1\Big)\bigg]^{1/(p-1)} \\
&= \bigg[\frac{\partial_x \mathcal{U}(t,x,\mu^{(q_t)})}{E}\Big(\frac{q_t^{kb}}{x^{kb}} \wedge 1\Big)\bigg]^{1/(p-1)},
\end{aligned}
$$

where $\mu^{(q_t)}$ denotes the Pareto distribution of parameter q_t. In order to guarantee that the equilibrium flow of measures is of Pareto type, it must satisfy the condition:

$$
\gamma_t x = \bigg(\frac{\partial_x \mathcal{U}(t,x,\mu^{(q_t)})}{E}\frac{q_t^{kb}}{x^{kb}}\bigg)^{1/(p-1)}, \quad x \geq q_t. \tag{4.89}
$$

for some mapping $[0,T] \ni t \mapsto \gamma_t \in (0,+\infty)$. There is no need to check the condition for $x < q_t$ as the path driven by the Pareto distribution is always greater than or equal to $(q_t)_{t\geq 0}$.

Since we focus on equilibria of Pareto type, we compute the master field $\mathcal{U}$ at Pareto distributions only. Given a parameter $q > 0$, we make the formal change of unknown:

$$\mathcal{V}(t,x,q) = \mathcal{U}\big(t,x,\mu^{(q)}\big),$$

where, as above, $\mu^{(q)}$ stands for the Pareto distribution of parameter q. We then compute, at least formally, the derivatives of $\mathcal{V}$ in terms of those of $\mathcal{U}$, so that we can reformulate the master equation for $\mathcal{U}$ as a PDE for $\mathcal{V}$. We clearly have (we shall specify later to which domain the triple (t,x,q) should belong):

$$\begin{aligned}
\partial_t \mathcal{V}(t,x,q) &= \partial_t \mathcal{U}\big(t,x,\mu^{(q)}\big),\\
\partial_x \mathcal{V}(t,x,q) &= \partial_x \mathcal{U}\big(t,x,\mu^{(q)}\big),\\
\partial^2_{xx} \mathcal{V}(t,x,q) &= \partial^2_{xx} \mathcal{U}\big(t,x,\mu^{(q)}\big).
\end{aligned}$$

In order to compute the derivative of $\mathcal{V}$ with respect to q, we must make the connection with the derivatives of $\mathcal{U}$ in the direction of μ. The key point is then to differentiate the mapping $q \mapsto \mathcal{U}(t,x,\mu^{(q)})$. Recalling that, for a random variable X with Pareto distribution of parameter 1, $qX \sim \mu^{(q)}$, the trick is to notice that:

$$\mathcal{V}(t,x,q) = \mathcal{U}\big(t,x,\mathcal{L}(qX)\big),$$

so that:

$$\partial_q \mathcal{V}(t,x,q) = \mathbb{E}\big[\partial_\mu \mathcal{U}\big(t,x,\mu^{(q)}\big)(qX)X\big] = \frac{1}{q}\int_{\mathbb{R}} \partial_\mu \mathcal{U}\big(t,x,\mu^{(q)}\big)(v) v d\mu^{(q)}(v).$$

In particular,

$$\partial^2_{xq} \mathcal{V}(t,x,q) = \frac{1}{q}\int_{\mathbb{R}} \partial_x \partial_\mu \mathcal{U}\big(t,x,\mu^{(q)}\big)(v) v d\mu^{(q)}(v).$$

Similarly,

$$\begin{aligned}
\partial^2_q \mathcal{V}(t,x,q) &= \frac{1}{q^2}\int_{\mathbb{R}} \partial_v \partial_\mu \mathcal{U}\big(t,x,\mu^{(q)}\big)(v) v^2 d\mu^{(q)}(v)\\
&\quad + \frac{1}{q^2}\int_{\mathbb{R}}\int_{\mathbb{R}} \partial^2_\mu \mathcal{U}\big(t,x,\mu^{(q)}\big)(v) v v' d\mu^{(q)}(v) d\mu^{(q)}(v').
\end{aligned}$$

Moreover, the relationship (4.89) takes the form:

$$\gamma_t x = \left(\frac{\partial_x \mathcal{V}(t,x,q_t)}{E}\frac{q_t^{kb}}{x^{kb}}\right)^{1/(p-1)}, \qquad x \geq q_t.$$

Rewriting (4.88), we get:

$$\begin{aligned}
&\partial_t \mathcal{V}(t,x,q) + \frac{p-1}{pE^{1/(p-1)}}\big(\partial_x \mathcal{V}(t,x,q)\big)^{p/(p-1)}\Big(\frac{q^{kb/(p-1)}}{x^{kb/(p-1)}} \wedge 1\Big) \\
&\quad + c\frac{x^{a+(k+1)b}}{k^b q^{kb}}\mathbf{1}_{\{x\ge q\}} + \gamma_t q \partial_q \mathcal{V}(t,x,q) + \frac{1}{2}\sigma^2\big[x^2\partial_x^2 \mathcal{V}(t,x,q) \\
&\quad + q^2\partial_q^2\mathcal{V}(t,x,q) + 2xq\partial^2_{xq}\mathcal{V}(t,x,q)\big] = 0.
\end{aligned} \tag{4.90}$$

Now we look for a continuously differentiable path $[0,T] \ni t \mapsto B_t \in [0,+\infty)$, with $B_T = 0$, such that:

$$\mathcal{V}(t,x,q) = B_t \frac{x^{p+bk}}{q^{bk}}, \tag{4.91}$$

solves the parameterized master equation (4.90) on the set $\{x \ge q\}$. Under the additional condition that $a+b=p$, B must be the solution of the equation:

$$\dot{B}_t + \frac{p-1}{pE^{1/(p-1)}}\big(B_t(p+bk)\big)^{p/(p-1)} + \frac{c}{k^b} - \gamma_t B_t bk + \frac{\sigma^2}{2}B_t p(p-1) = 0.$$

Condition (4.89) reads:

$$\gamma_t = \Big(\frac{B_t(p+bk)}{E}\Big)^{1/(p-1)}, \tag{4.92}$$

so that the above equation for B becomes:

$$\dot{B}_t + \frac{(p+bk)^{1/(p-1)}}{E^{1/(p-1)}}\Big(p-1-\frac{bk}{p}\Big)B_t^{p/(p-1)} + \frac{\sigma^2}{2}p(p-1)B_t + \frac{c}{k^b} = 0, \tag{4.93}$$

for $t \in [0,T]$, with the terminal condition $B_T = 0$. This equation is locally uniquely solvable. The key point is that the local solution is nonnegative. If B vanishes at some time t, necessarily $\dot{B}_t < 0$, so that $B_{t-\delta} > 0$ for δ small enough. Moreover, if $p(p-1) < bk$, the local solution cannot exceed the smallest positive root B of the equation:

$$\frac{(p+bk)^{1/(p-1)}}{E^{1/(p-1)}}\Big(p-1-\frac{bk}{p}\Big)B^{p/(p-1)} + \frac{\sigma^2}{2}p(p-1)B + \frac{c}{k^b} = 0,$$

so that the local solution to (4.93) can be extended to the entire $[0,T]$, proving that the ODE admits a unique solution.

Equilibrium Given by Pareto Distributions

The above analysis suggests that, whenever X_0 has a Pareto distribution of parameter q_0, the flow of marginal conditional distributions (given $(W_t^0)_{0\le t\le T}$) of the process:

$$dX_t = \gamma_t X_t dt + \sigma X_t dW_t^0, \quad t \in [0, T],$$

generates an MFG equilibrium with $(\gamma_t)_{0\le t\le T}$ as in (4.92), for $(B_t)_{0\le t\le T}$ solving the ODE (4.93). Denoting by μ_t the conditional distribution of X_t (given the common noise), it holds that $\mu_t = \mu^{(q_t)}$, where:

$$q_t = \exp\Big(\int_0^t \gamma_s ds - \frac{\sigma^2}{2}t + \sigma W_t^0\Big), \quad t \in [0, T].$$

In order to check that $(X_t)_{0\le t\le T}$ maximizes the reward functional associated with the running reward f given by:

$$f(x, \mu, \alpha) = c\frac{x^a}{[(d\mu/dx)(x)]^b} - \frac{E}{p}\frac{\alpha^p}{[\mu([x, \infty))]^b},$$

we first check that:

$$\Big(\mathcal{V}(t, X_t, q_t) + \int_0^t f(X_s, \mu_s, \gamma_s X_s) ds\Big)_{0\le t\le T}, \quad \text{for} \quad s \in [0, T],$$

is a martingale. The proof follows from a straightforward application of Itô's formula combined with the PDE (4.90). The fact that (4.90) is satisfied for $x \ge q$ only is not a problem since with probability 1, $X_t > q_t$ for any $t \in [0, T]$. Notice that the equality $X_t = q_t$ holds for scenarios for which $X_0 = q_0$, which are of zero probability.

Given this martingale property, it still remains to check that for any Lipschitz continuous feedback function $\alpha : [0, T] \times \mathbb{R}_+ \to \mathbb{R}_+$, the process:

$$\Big(\mathcal{V}(t, X_t^{\boldsymbol{\alpha}}, q_t) + \int_0^t f(X_s^{\boldsymbol{\alpha}}, \mu_s, \alpha_s) ds\Big)_{0\le t\le T} \tag{4.94}$$

is a super-martingale where $(X_t^{\boldsymbol{\alpha}})_{0\le t\le T}$ solves:

$$dX_t^{\boldsymbol{\alpha}} = \alpha(t, X_t^{\boldsymbol{\alpha}})dt + \sigma X_t^{\boldsymbol{\alpha}} dW_t^0, \quad t \in [0, T],$$

with $X_0^{\boldsymbol{\alpha}} = X_0$ and $\alpha_t = \alpha(t, X_t^{\boldsymbol{\alpha}})$ for $t \in [0, T]$. We shall prove this assertion when T is small enough. The proof goes along the lines of Proposition 1.55 and relies once again on Itô's formula. The only difficulty is that $X_t^{\boldsymbol{\alpha}}$ might be smaller than q_t for some $t \in [0, T]$. In other words, we are facing the fact that $\mathcal{V}$ satisfies the PDE (4.90) on the set $[0, T] \times \{x \ge q\}$ only. In order to circumvent this obstacle, a possible strategy is to replace $\mathcal{V}$ by:

$$\mathcal{V}(t,x,q) = B_t x^p \Big(\frac{x^{bk}}{q^{bk}} \vee 1 \Big),$$

for the same constant B_t as above. Obviously, the PDE (4.90) is not satisfied when $x < q$, but $\mathcal{V}$ defines, at least in small time, a super-solution on the set $[0,T] \times \{0 \leq x < q\}$, as (4.90) holds with $=0$ replaced by ≤ 0 when T is small enough. Generally speaking, this follows from the simple fact that B_t tends to 0 when t tends to T. Computations are rather straightforward. It suffices to observe that the left-hand side in (4.90) becomes, for $x < q$,

$$\Big(\dot{B}_t + \frac{p-1}{pE^{1/(p-1)}} \big(\frac{p}{p-1} B_t \big)^{p/(p-1)} + \frac{1}{2}\sigma^2 p(p-1) B_t \Big) x^p.$$

Inserting the ODE (4.93) satisfied by $(B_t)_{0 \leq t \leq T}$, this may be rewritten under the form:

$$\Big(h(B_t) - \frac{c}{k^b} \Big) x^p,$$

where $h : \mathbb{R} \to \mathbb{R}$ is a continuous function matching 0 in 0.

Heuristically, this should suffice for our purpose, but the justification requires some modicum of care as the function $\mathcal{V}$, when extended as above to the set $[0,T] \times \{0 \leq x < q\}$, is not $\mathcal{C}^{1,2}$ (which is the standard condition needed in order to apply Itô's expansion), the first-order derivatives in (x,q) being discontinuous on the diagonal $\{x = q\}$. The argument for justifying the Itô expansion is a bit technical so that we only sketch it in broad strokes. We write $\mathcal{V}(t, X_t^{\boldsymbol{\alpha}}, q_t) = B_t (X_t^{\boldsymbol{\alpha}})^p [\varphi(X_t^{\boldsymbol{\alpha}}/q_t)]^{bk}$, with $\varphi(r) = \max(1,r)$. The key point is that $(X_t^{\boldsymbol{\alpha}}/q_t)_{0 \leq t \leq T}$ is always a bounded variation process, so that the expansion of $(\varphi(X_t^{\boldsymbol{\alpha}}/q_t))_{0 \leq t \leq T}$ only requires to control φ' and not φ''. Then, we can regularize φ by a sequence $(\varphi_n)_{n \geq 1}$ such that $(\varphi_n)'(r) = 0$, for $r \leq 1 - 1/n$, $(\varphi_n)'(r) = 1$, for $r \geq 1$ and $(\varphi_n)'(r) \in [0,1]$ for $r \in [1 - 1/n, 1]$. The fact that $(\varphi_n)'(r)$ is uniformly bounded in n permits to expand $(B_t (X_t^{\boldsymbol{\alpha}})^p [\varphi_n(X_t^{\boldsymbol{\alpha}}/q_t)]^{bk})_{0 \leq t \leq T}$ and then to pass to the limit. The super-martingale property shows that

$$\int_{\mathbb{R}^d} \mathcal{V}(0,x,q_0) d\mu^{(q_0)}(x) \geq \sup_{(\alpha_t)_{0 \leq t \leq T}} \mathbb{E}\Big[\int_0^T f(X_t^{\boldsymbol{\alpha}}, q_t, \alpha_t) dt \Big], \tag{4.95}$$

which, together with the martingale property along $(X_t)_{0 \leq t \leq T}$, shows that equality holds and that the Pareto distributions $(\mu_t = \mu^{(q_t)})_{0 \leq t \leq T}$ form an MFG equilibrium.

4.6 Notes & Complements

As already stated in Chapter 1, the concept of decoupling field for finite dimensional forward-backward SDEs with random coefficients is due to Ma, Wu, Zhang, and Zhang [272]. For a flow of random measures $(\mu_t)_{0 \leq t \leq T}$ given by the conditional distributions of the state of a population in equilibrium given the realization of the common noise, the random field $[0,T]\times\mathbb{R}^d \ni (t,x) \mapsto \mathcal{U}(t,x,\mu_t)$ coincides with the value function of the underlying optimal control problem in the random environment $(\mu_t)_{0\le t\le T}$, see Remark 4.22. In particular, this random field can be identified with the decoupling field of the finite-dimensional forward-backward SDE representing the value function. This makes the connection between the notion of master field introduced in this chapter, and the notion of decoupling field introduced in [272] and investigated in Chapter 1.

In any case, the master field must be regarded as an infinite-dimensional generalization of the notion of decoupling field used in the standard theory of finite-dimensional forward-backward SDEs with deterministic, instead of random, coefficients. Actually, the terminology *decoupling field* did not appear in the earlier works on the subject. See for instance the works [297, 298] by Pardoux and Peng together with the references in the more recent monograph by Pardoux and Răşcanu [299]. Instead, forward-backward SDEs with deterministic coefficients were regarded as *nonlinear Feynman-Kac formulas*. In this respect, our introduction of the master field can be interpreted as an infinite-dimensional *nonlinear Feynman-Kac formula*.

The definition of the master field provided in this chapter has been suggested by Carmona and Delarue in [97], but the analysis of the master field provided in the first section of the chapter is entirely new. In particular, the dynamic programming principle given in Proposition 4.2 is new. Here, our formulation is explicitly based upon the representation of the underlying optimal control problem by means of a forward-backward system. Obviously, it would be interesting to address directly the validity of the dynamic programming principle.

Although our definition of the master field together with its subsequent analysis is somewhat new, the general concept goes back to Lasry and Lions. In his lectures [265] at *Collège de France*, Lions introduced the master equation presented in Section 4.4. The connection between the master field and the master equation is pretty clear: as highlighted in this chapter, the solution of the master equation is the master field. Here, we took a typically probabilistic road for connecting both: we started from the value function of the game, which we called the master field, and then we derived the equation satisfied in the viscosity sense by the value function. In this regard, the dynamic programming principle played a crucial role. In the end, the forward-backward system of the McKean-Vlasov type (4.7) provides a Lagrangian description of the mean field game while the master equation (4.41) reads as an Eulerian description. Equivalently, the forward and backward components of the system (4.7) form the *stochastic characteristics* of the master equation. Lasry and Lions' point of view for introducing the master equation is slightly different: when there is no common noise, they call system of characteristics the mean field game system (Vol I)-(3.12) in Chapter (Vol I)-3 formed by the Fokker-Planck equation and the Hamilton-Jacobi-Bellman equation, in which case the characteristics are deterministic.

The derivation of the master equation from the dynamic programming principle is quite standard in the theory of optimal control problems. We refer the reader to the monographs by Fleming and Soner [157] and by Yong and Zhou [343] for a general overview of the use of viscosity solutions in the theory of stochastic optimal control. The notion of viscosity solutions for equations on spaces of probability measures is connected with the wider literature on viscosity solutions for HJB equations in infinite dimension, as discussed in the series of papers by Crandall and Lions [121–124]. We also refer to the forthcoming monograph by Fabbri, Gozzi, and Swiech on stochastic optimal control in infinite dimensions, see [150]. Within the framework of mean field games, we refer to the notes by Cardaliaguet [83] for an instance of viscosity solutions of the master equation. Therein, Cardaliaguet discusses the particular case of games with deterministic state dynamics. Accordingly, the solutions of the master equation are understood in the viscosity sense, very much in the spirit of Definition 4.19.

As noticed in the articles [50,97], the notion of master equation may be extended to other types of stochastic control problems, including the control of McKean-Vlasov diffusion processes presented in Chapter (Vol I)-6. Recently, Pham and Wei [311,312] investigated the corresponding form of the dynamic programming principle together with the existence and uniqueness of viscosity solutions to the corresponding master equation, when reformulated as a PDE in a Hilbert space. In both papers, part of the analysis relies on the same chain rule as the one we use here, see Theorem 4.17, which is inspired by the chain rule originally established in the papers by Buckdahn, Li, Peng, and Rainer [79] and by Chassagneux, Crisan and Delarue [114].

We refer to the next Chapter 5 for further results and further references on the master equation and, more on the existence and uniqueness of classical solutions.

Results on the invertibility of the gradient of the flow formed by the solution of an SDE, as used in the proof of Corollary 4.11, may be found in Protter [315].

Classical Solutions to the Master Equation 5

Abstract

This chapter is concerned with existence and uniqueness of classical solutions to the master equation. The importance of classical solutions will be demonstrated in the next chapter where they play a crucial role in proving the convergence of games with finitely many players to mean field games. We propose constructions based on the differentiability properties of the flow generated by the solutions of the forward-backward system of the McKean-Vlasov type representing the equilibrium of the mean field game on an L^2-space. Existence of a classical solution is first established for small time. It is then extended to arbitrary finite time horizons under the additional Lasry-Lions monotonicity condition.

5.1 Master Field of a McKean-Vlasov FBSDE

5.1.1 General Prospect

The starting point of our analysis is a mild generalization of the notion of master field introduced and studied in the previous Chapter 4.

So far, the master field $\mathcal{U} : [0,T] \times \mathbb{R}^d \times \mathcal{P}_2(\mathbb{R}^d) \to \mathbb{R}$ has been regarded as the value function of the underlying mean field game. For an initial condition (t, x, μ), x standing for the private state of the representative player at time t and μ for the conditional distribution of the population at time t given the realization of the systemic noise, $\mathcal{U}(t, x, \mu)$ is defined as the equilibrium expected future cost to the representative player at time t.

Below, we shall exploit another, though indissolubly connected, interpretation of $\mathcal{U}(t, x, \mu)$. As we already accounted for in Remark 4.22, whenever equilibria are represented by means of the forward-backward system associated with the Hamilton-Jacobi-Bellman formulation of the optimization problem – as opposed

R. Carmona, F. Delarue, *Probabilistic Theory of Mean Field Games with Applications II*, Probability Theory and Stochastic Modelling 84,
DOI 10.1007/978-3-319-56436-4_5

to the formulation based on the Pontryagin maximum principle – $\mathcal{U}(t, x, \mu)$ may also be written as the initial value, at time t, of the backward component $(Y_s)_{t \le s \le T}$ of the conditional McKean-Vlasov FBSDE characterizing the equilibrium. Under such an identification, the master field appears as a *decoupling field* that permits to express the realization of the backward component of the underlying McKean-Vlasov FBSDE in terms of the realization of the forward component and of its distribution. The fact that the master field acts both on the private state of the representative player and on the conditional distribution of the population is reminiscent of the fact that, here, the state space for the forward component of the solution of the conditional McKean-Vlasov FBSDE is the whole $\mathbb{R}^d \times \mathcal{P}_2(\mathbb{R}^d)$. In this way, writing $\mathcal{U}(t, \cdot, \cdot)$ as a function defined on the enlarged state space $\mathbb{R}^d \times \mathcal{P}_2(\mathbb{R}^d)$ sounds as a natural extension of the situation encountered with *standard* FBSDEs in Chapter (Vol I)-4 for which the *decoupling field* at time t writes as a function defined on the state space $\mathbb{R}^d$.

We already commented on these observations in Chapter 4. Here, we go one step further and extend the notion of master field to any FBSDE of the conditional McKean-Vlasov type provided that it is uniquely solvable in the strong sense. We cast the first objectives of this chapter in the following terms. We address the smoothness of the master field of a general FBSDE of the conditional McKean-Vlasov type by investigating the smoothness of the flow formed by the solution of the FBSDE in the space $L^2(\Omega, \mathcal{F}, \mathbb{P}; \mathbb{R}^d)$, where $(\Omega, \mathcal{F}, \mathbb{P})$ denotes the probability space on which the forward-backward system is defined. When the McKean-Vlasov FBSDE derives from a mean field game, these general results will establish the required differentiability properties of the master field underpinning the mean field game. In particular, whenever enough smoothness holds, this will imply that the master field is not only a viscosity solution of the master equation as proved in the previous chapter, but also a classical solution.

One critical feature of this approach is the fact that the flow property is investigated in $L^2(\Omega, \mathcal{F}, \mathbb{P}; \mathbb{R}^d)$. This is reminiscent of the construction of the L-differential calculus defined in Chapter (Vol I)-5 by lifting functions of probability measures into functions of random variables. Similarly, we discuss the smoothness properties of the master field at the level of random variables. This is especially convenient for us since the forward component of a McKean-Vlasov FBSDE takes values in $L^2(\Omega, \mathcal{F}, \mathbb{P}; \mathbb{R}^d)$. We refer the reader to the Notes & Complements at the end of the chapter for references from which this approach is borrowed.

The analysis is split into two main steps. Imitating the induction argument used in Chapter (Vol I)-4 to solve a general FBSDE of the McKean-Vlasov type, we first consider the case of a small enough time horizon T. In the special case of mean field games, we then succeed in applying the short time analysis iteratively on a time horizon of arbitrary length provided that the Lasry-Lions monotonicity condition is in force.

5.1.2 Definition of the Master Field

Throughout the chapter, we use the same probabilistic set-up as in Definition 2.16.

We are given:

1. a complete probability space $(\Omega^0, \mathcal{F}^0, \mathbb{P}^0)$, endowed with a complete and right-continuous filtration $\mathbb{F}^0 = (\mathcal{F}_t^0)_{0 \le t \le T}$ and a d-dimensional $\mathbb{F}^0$-Brownian motion $\boldsymbol{W}^0 = (W_t^0)_{0 \le t \le T}$,
2. the completion $(\Omega^1, \mathcal{F}^1, \mathbb{P}^1)$ of a countably generated space, endowed with a complete and right-continuous filtration $\mathbb{F}^1 = (\mathcal{F}_t^1)_{0 \le t \le T}$ and a d-dimensional $\mathbb{F}^1$-Brownian motion $\boldsymbol{W} = (W_t)_{0 \le t \le T}$.

We then denote by $(\Omega, \mathcal{F}, \mathbb{P})$ the completion of the product space $(\Omega^0 \otimes \Omega^1, \mathcal{F}^0 \otimes \mathcal{F}^1, \mathbb{P}^0 \otimes \mathbb{P}^1)$ endowed with the filtration $\mathbb{F} = (\mathcal{F}_t)_{0 \le t \le T}$ obtained by augmenting the product filtration $\mathbb{F}^0 \otimes \mathbb{F}^1$ in a right-continuous way and by completing it. We are also given an $\mathcal{F}_0$-measurable $\mathbb{R}^d$-valued random variable ξ.

We recall the useful notation $\mathcal{L}^1(X)(\omega^0) = \mathcal{L}(X(\omega^0, \cdot))$ for $\omega^0 \in \Omega^0$ and a random variable X on Ω, see Subsection 2.1.3.

Importantly, we shall assume that $(\Omega^1, \mathcal{F}_0^1, \mathbb{P}^1)$ is rich enough so that, for any distribution $\nu \in \mathcal{P}_2(\mathbb{R}^q)$, with $q \ge 1$, we can construct an $\mathcal{F}_0^1$-measurable-random variable with ν as distribution. We refer to (2.30)–(2.31) in Chapter 2 for a possible construction of such a space.

Forward-Backward System of the Conditional McKean-Vlasov Type

We are now given a maturity time $T > 0$ together with coefficients:

$$
\begin{aligned}
&B : [0, T] \times \mathbb{R}^d \times \mathcal{P}_2(\mathbb{R}^d) \times \mathbb{R}^m \times \mathbb{R}^{m \times d} \times \mathbb{R}^{m \times d} \to \mathbb{R}^d, \\
&F : [0, T] \times \mathbb{R}^d \times \mathcal{P}_2(\mathbb{R}^d) \times \mathbb{R}^m \times \mathbb{R}^{m \times d} \times \mathbb{R}^{m \times d} \to \mathbb{R}^m, \\
&G : \mathbb{R}^d \times \mathcal{P}_2(\mathbb{R}^d) \to \mathbb{R}^m, \\
&\sigma, \sigma^0 : [0, T] \times \mathbb{R}^d \times \mathcal{P}_2(\mathbb{R}^d) \to \mathbb{R}^{d \times d},
\end{aligned}
$$

for two integers $d, m \ge 1$. For some initial condition $\xi \in L^2(\Omega, \mathcal{F}_0, \mathbb{P}; \mathbb{R}^d)$, we then consider the following system of equations:

$$\begin{cases} dX_t = B\big(t, X_t, \mathcal{L}^1(X_t), Y_t, Z_t, Z_t^0\big)dt \\ \qquad +\sigma\big(t, X_t, \mathcal{L}^1(X_t)\big)dW_t + \sigma^0\big(t, X_t, \mathcal{L}^1(X_t)\big)dW_t^0, \\ dY_t = -F\big(t, X_t, \mathcal{L}^1(X_t), Y_t, Z_t, Z_t^0\big)dt \\ \qquad +Z_t dW_t + Z_t^0 dW_t^0, \quad t \in [0, T], \\ X_0 = \xi, \quad Y_T = G\big(X_T, \mathcal{L}^1(X_T)\big). \end{cases} \tag{5.1}$$

The unknowns $\boldsymbol{X} = (X_t)_{0 \le t \le T}$, $\boldsymbol{Y} = (Y_t)_{0 \le t \le T}$, $\boldsymbol{Z} = (Z_t)_{0 \le t \le T}$ and $\boldsymbol{Z}^0 = (Z_t^0)_{0 \le t \le T}$ have dimensions d, m, $m \times d$, and $m \times d$ respectively.

Definition 5.1 *In the above probabilistic set-up, we call a solution to* (5.1) *any four-tuple* $(\boldsymbol{X}, \boldsymbol{Y}, \boldsymbol{Z}, \boldsymbol{Z}^0)$ *of* $\mathbb{F}$*-progressively measurable processes, with values in* $\mathbb{R}^d$, $\mathbb{R}^m$, $\mathbb{R}^{m\times d}$ *and* $\mathbb{R}^{m\times d}$ *respectively,* $\boldsymbol{X}$ *and* $\boldsymbol{Y}$ *having continuous paths, such that:*

$$\mathbb{E}\bigg[\sup_{0 \le t \le T}\Big(|X_t|^2 + |Y_t|^2\Big) + \int_0^T \big(|Z_t|^2 + |Z_t^0|^2\big)dt\bigg] < \infty,$$

and such that (5.1) *holds with* $\mathbb{P}$*-probability* 1.

In the framework of Definition 5.1, the construction of $(\mathcal{L}^1(X_t))_{0 \le t \le T}$ in (5.1) is made clear by Lemma 2.5. It is worth mentioning that system (5.1) matches system (2.29) in Chapter 2, except for the fact that in the backward equation, the martingale term is now written as the sum of two stochastic integrals with respect to $\boldsymbol{W}^0$ and $\boldsymbol{W}$. The rationale for requiring the martingale part to be of this specific form is that we shall only work with strong solutions, namely with solutions that are progressively measurable with respect to the completion of the filtration generated by the initial condition ξ and $(\boldsymbol{W}^0, \boldsymbol{W})$, in which case martingales can be represented as stochastic integrals. As a by-product, the process $\boldsymbol{Y}$ is necessarily continuous in time; in contrast, observe that $\boldsymbol{Y}$ may be discontinuous in the framework discussed in Definition 1.17. In this regard, we check below that, if $\boldsymbol{X}$ is progressively measurable with respect to $\mathbb{F}^{(\xi, \boldsymbol{W}^0, \boldsymbol{W})}$, then $(\mathcal{L}^1(X_t))_{0 \le t \le T}$ is progressively measurable with respect to the filtration $\mathbb{F}^{0,(\mathcal{L}^1(\xi), \boldsymbol{W}^0)}$ on $(\Omega^0, \mathcal{F}^0, \mathbb{P}^0)$. In particular, all the processes appearing in (5.1) are $\mathbb{F}^{(\xi, \boldsymbol{W}^0, \boldsymbol{W})}$-progressively measurable whenever $(\boldsymbol{X}, \boldsymbol{Y}, \boldsymbol{Z}, \boldsymbol{Z}^0)$ is $\mathbb{F}^{(\xi, \boldsymbol{W}^0, \boldsymbol{W})}$-progressively measurable, which makes licit the application of the martingale representation theorem.

As already stated, our goal is to show that, under appropriate assumptions, there exists not only a unique solution to (5.1), but also a continuous *decoupling field* $\mathcal{U} : [0, T] \times \mathbb{R}^d \times \mathcal{P}_2(\mathbb{R}^d) \to \mathbb{R}^m$ such that $Y_s = \mathcal{U}(s, X_s, \mathcal{L}^1(X_s))$, $0 \le s \le T$. Pushing further the analysis initiated in Chapter 4, we shall provide explicit conditions under which $\mathcal{U}$ is a *classical* solution (instead of a viscosity solution) to some *master equation* on $[0, T] \times \mathbb{R}^d \times \mathcal{P}_2(\mathbb{R}^d)$.

Master Field of a Conditional McKean-Vlasov FBSDE

Somehow, the strategy for constructing the master field is similar to the one we used in Chapter (Vol I)-4 to define the decoupling field of a standard FBSDE. In any case, we must let the initial condition of the FBSDE (5.1) vary. This prompts us to consider the following version of (5.1), appropriately initialized at (t,ξ), for some $t \in [0,T]$ and $\xi \in L^2(\Omega, \mathcal{F}_t, \mathbb{P}; \mathbb{R}^d)$:

$$\begin{cases} dX_s^{t,\xi} = B\big(s, X_s^{t,\xi}, \mathcal{L}^1(X_s^{t,\xi}), Y_s^{t,\xi}, Z_s^{t,\xi}, Z_s^{0;t,\xi}\big)ds \\ \qquad\quad + \sigma\big(s, X_s^{t,\xi}, \mathcal{L}^1(X_s^{t,\xi})\big)dW_s + \sigma^0\big(s, X_s^{t,\xi}, \mathcal{L}^1(X_s^{t,\xi})\big)dW_s^0, \\ dY_s^{t,\xi} = -F\big(s, X_s^{t,\xi}, \mathcal{L}^1(X_s^{t,\xi}), Y_s^{t,\xi}, Z_s^{t,\xi}, Z_s^{0;t,\xi}\big)ds \\ \qquad\quad + Z_s^{t,\xi} dW_s + Z_s^{0;t,\xi} dW_s^0, \quad s \in [t,T], \\ X_t^{t,\xi} = \xi, \quad Y_T^{t,\xi} = G\big(X_T^{t,\xi}, \mathcal{L}^1(X_T^{t,\xi})\big). \end{cases} \tag{5.2}$$

As a typical example for ξ, we shall consider $\xi \in L^2(\Omega^1, \mathcal{F}_t^1, \mathbb{P}^1; \mathbb{R}^d)$ and $\mathcal{L}^1(\xi) = \mu$, for some given $\mu \in \mathcal{P}_2(\mathbb{R}^d)$. This choice is possible thanks to the preliminary assumption we made in the first lines of the subsection: the space $(\Omega^1, \mathcal{F}_0^1, \mathbb{P}^1)$ is assumed to be rich enough so that, for any $\mu \in \mathcal{P}_2(\mathbb{R}^d)$, there exists a random variable $\xi \in L^2(\Omega^1, \mathcal{F}_0^1, \mathbb{P}^1; \mathbb{R}^d)$ with $\mathcal{L}^1(\xi) = \mu$.

Of course, what really matters here is μ and not ξ. According to the terminology introduced in Chapter (Vol I)-5, ξ is just a lifting used to represent μ. For a given $\mu \in \mathcal{P}_2(\mathbb{R}^d)$, we may find several ξ's such that $\mathcal{L}^1(\xi) = \mu$. However, it is important to notice that under the prescription $\mathcal{L}^1(\xi) = \mu$, the construction of the decoupling field is somehow oblivious to the particular choice of ξ.

The latter observation is based upon the following weak uniqueness property, which is reminiscent of Proposition 2.11 for conditional McKean-Vlasov SDEs. Assume indeed that (5.2) has a unique $(\mathcal{F}_s)_{t \le s \le T}$-progressively measurable solution $(\boldsymbol{X}^{t,\xi}, \boldsymbol{Y}^{t,\xi}, \boldsymbol{Z}^{t,\xi}, \boldsymbol{Z}^{0;t,\xi})$, with a solution which is progressively measurable with respect to the completion of the filtration generated by $(\xi, W_s^0 - W_t^0, W_s - W_t)_{t \le s \le T}$. Then, for any $s \in [t,T]$, there exists a function $\Phi_s : \mathbb{R}^d \times \mathcal{C}([t,s]; \mathbb{R}^d) \times \mathcal{C}([t,s]; \mathbb{R}^d) \to \mathbb{R}^d$ such that:

$$\mathbb{P}\Big[X_s^{t,\xi} = \Phi_s\Big(\xi, (W_r^0 - W_t^0)_{t \le r \le s}, (W_r - W_t)_{t \le r \le s}\Big)\Big] = 1.$$

Therefore, with probability 1 under $\mathbb{P}^0$,

$$\mathcal{L}^1(X_s^{t,\xi}) = (\mu \otimes \mathcal{W}_d^t) \circ \Big[\Phi_s\Big(\cdot, (W_r^0 - W_t^0)_{t \le r \le s}, \cdot\Big)\Big]^{-1}. \tag{5.3}$$

Of course, observe that Φ_s may depend on the law of ξ. Following the proof of Lemma 2.5, $(\mathcal{L}^1(X_s^{t,\xi}))_{t \le s \le T}$ is $\mathcal{F}_s^{0,(W_r^0 - W_t^0)_{t \le r \le T}}$-measurable. By continuity of the trajectories, it is progressively measurable with respect to the completion of the filtration generated by $(W_s^0 - W_t^0)_{t \le s \le T}$. In particular, we may regard it as an environment: obviously $(\Omega, \mathcal{F}, (\mathcal{F}_s)_{t \le s \le T}, \mathbb{P})$ is compatible with $(\xi, W^0_s - W^0_t, W_s - W_t)_{t \le s \le T}$ and thus with the process $(\xi, W^0_s - W^0_t, \mathcal{L}^1(X^{t,\xi}_s), W_s - W_t)_{t \le s \le T}$. Therefore, if for any $\mathcal{P}_2(\mathbb{R}^d)$-valued environment $\boldsymbol{\mu} = (\mu_s)_{t \le s \le T}$ and any $\mathcal{F}_t$-measurable initial condition X_t such that $(X_t, (W^0_s - W^0_t, \mu_s, W_s - W_t)_{t \le s \le T})$ is compatible with $(\mathcal{F}_s)_{t \le s \le T}$, the auxiliary system:

$$\begin{cases} dX^{\boldsymbol{\mu}}_s = B\big(s, X^{\boldsymbol{\mu}}_s, \mu_s, Y^{\boldsymbol{\mu}}_s, Z^{\boldsymbol{\mu}}_s, Z^{0;\boldsymbol{\mu}}_s\big)ds \\ \qquad + \sigma\big(s, X^{\boldsymbol{\mu}}_s, \mu_s\big)dW_s + \sigma^0\big(s, X^{\boldsymbol{\mu}}_s, \mu_s\big)dW^0_s, \\ dY^{\boldsymbol{\mu}}_s = -F\big(s, X^{\boldsymbol{\mu}}_s, \mu_s, Y^{\boldsymbol{\mu}}_s, Z^{\boldsymbol{\mu}}_s, Z^{0;\boldsymbol{\mu}}_s\big)ds \\ \qquad + Z^{\boldsymbol{\mu}}_s dW_s + Z^0_s dW^{0,\boldsymbol{\mu}}_s + dM^{\boldsymbol{\mu}}_s, \quad s \in [t, T], \\ X^{\boldsymbol{\mu}}_t = X_t, \quad Y^{\boldsymbol{\mu}}_T = G\big(X^{\boldsymbol{\mu}}_T, \mu_T\big), \end{cases} \tag{5.4}$$

has the strong uniqueness property as defined in Definition 1.18, then, by Theorem 1.33, we can find a measurable function $\Phi : \mathbb{R}^d \times \mathcal{C}([t,T];\mathbb{R}^d) \times \mathcal{D}([t,T];\mathcal{P}_2(\mathbb{R}^d)) \times \mathcal{C}([t,T];\mathbb{R}^d) \to \mathcal{C}([t,T];\mathbb{R}^d)$ such that:

$$\mathbb{P}\Big[\boldsymbol{X}^{t,\xi} = \Phi\Big(\xi, \big(W^0_r - W^0_t\big)_{t \le r \le T}, \big(\mathcal{L}^1(X^{t,\xi}_r)\big)_{t \le r \le T}, \big(W_r - W_t\big)_{t \le r \le T}\Big)\Big] = 1.$$

Now, if ξ' is another lifting of μ constructed on $(\Omega^1, \mathcal{F}^1_t, \mathbb{P}^1)$ such that $\mathcal{L}^1(\xi') = \mu$, then we may let:

$$\boldsymbol{X}' = \Phi\Big(\xi', \big(W^0_r - W^0_t\big)_{t \le r \le T}, \big(\mathcal{L}^1(X^{t,\xi}_r)\big)_{t \le r \le T}, \big(W_r - W_t\big)_{t \le r \le T}\Big),$$

where we put ξ' instead of ξ in the first component. Since the argument in the function in the right-hand side has the same law, under $\mathbb{P}$, as the process $(\xi, (W^0_r - W^0_t)_{t \le r \le T}, (\mathcal{L}^1(X^{t,\xi}_r))_{t \le r \le T}, (W_r - W_t)_{t \le r \le T})$, we deduce from a new application of Theorem 1.33 that $\boldsymbol{X}'$ is the forward component of the solution of the system (5.2) when regarded as an auxiliary system of the same type as (5.4) with $\boldsymbol{\mu} = (\mathcal{L}^1(X^{t,\xi}_r))_{t \le r \le T}$ as environment but with $X_t = \xi'$ in lieu of $X_t = \xi$ as initial condition. Obviously, by construction of $\boldsymbol{X}'$, with probability 1 under $\mathbb{P}^0$,

$$\forall r \in [t,T], \quad \mathcal{L}^1(X'_r) = \mathcal{L}^1(X^{t,\xi}_r),$$

from which we get that $\boldsymbol{X}'$ is the forward component of a solution of (5.2) with ξ' in lieu of ξ. If, as above, we assume that uniqueness holds true for (5.2), then $\boldsymbol{X}'$ coincides with $\boldsymbol{X}^{t,\xi'}$. This shows that, with probability 1 under $\mathbb{P}^0$,

$$\mathbb{P}^0\Big[\forall s \in [t,T],\ \mathcal{L}^1(X^{t,\xi}_s) = \mathcal{L}^1(X^{t,\xi'}_s)\Big] = 1, \tag{5.5}$$

which is the required ingredient to define the master field.

Thanks to (5.5), it now makes sense, for a given $\mu \in \mathcal{P}_2(\mathbb{R}^d)$, to consider $(\mathcal{L}^1(X_r^{t,\xi}))_{t\le r\le T}$ without specifying the choice of the lifted random variable ξ that has μ as conditional distribution given $\boldsymbol{W}^0$. We then introduce, for any $x \in \mathbb{R}^d$,

$$\begin{cases} dX_s^{t,x,\mu} = B\big(s, X_s^{t,x,\mu}, \mathcal{L}^1(X_s^{t,\xi}), Y_s^{t,x,\mu}, Z_s^{t,x,\mu}, Z_s^{0;t,x,\mu}\big)ds \\ \qquad\qquad +\sigma\big(s, X_s^{t,x,\mu}, \mathcal{L}^1(X_s^{t,\xi})\big)dW_s \\ \qquad\qquad +\sigma^0\big(s, X_s^{t,x,\mu}, \mathcal{L}^1(X_s^{t,\xi})\big)dW_s^0, \\ dY_s^{t,x,\mu} = -F\big(s, X_s^{t,x,\mu}, \mathcal{L}^1(X_s^{t,\xi}), Y_s^{t,x,\mu}, Z_s^{t,x,\mu}, Z_s^{0;t,x,\mu}\big)ds \\ \qquad\qquad +Z_s^{t,x,\mu}dW_s + Z_s^{0;t,x,\mu}dW_s^0, \quad s \in [t,T], \\ X_t^{t,x,\mu} = x, \quad Y_T^{t,x,\mu} = G\big(X_T^{t,x,\mu}, \mathcal{L}^1(X_T^{t,\xi})\big). \end{cases} \tag{5.6}$$

Under the assumption that (5.2) is uniquely solvable and that its solution is progressively measurable with respect to the completion of the filtration generated by $(\xi, (W_s^0 - W_t^0)_{t\le s\le T}, (W_s - W_t)_{t\le s\le T})$, existence and uniqueness of a solution to (5.6) hold true provided that the auxiliary system (5.4) is strongly uniquely solvable along the lines of Definition 1.19. Indeed, since the environment $\boldsymbol{\mu} = (\mathcal{L}^1(X_s^{t,\xi}))_{t\le s\le T}$ is $\mathbb{F}^{(W_s^0 - W_t^0)_{t\le s\le T}}$-progressively measurable, the auxiliary system can be solved with respect to the smaller filtration $\mathbb{F}^{(W_s^0 - W_t^0, W_s - W_t)_{t\le s\le T}}$ instead of $(\mathcal{F}_s)_{t\le s\le T}$. The resulting solution then coincides with the solution obtained by working with the larger filtration. As a consequence, the martingale $\boldsymbol{M}^{\mu}$ in (5.4) is 0 since it has 0 bracket with $(W_s^0 - W_t^0, W_s - W_t)_{t\le s\le T}$. We recover (5.6).

This prompts us to specify the definition of the master field of the FBSDE (5.2) of conditional McKean-Vlasov type as follows.

Definition 5.2 *Assume that for any* $(t,x,\mu) \in [0,T] \times \mathbb{R}^d \times \mathcal{P}_2(\mathbb{R}^d)$ *and any random variable* $\xi \in L^2(\Omega^1, \mathcal{F}_t^1, \mathbb{P}^1; \mathbb{R}^d)$ *with distribution* μ, (5.2) *has a unique progressively measurable solution* $(X_s^{t,\xi}, Y_s^{t,\xi}, Z_s^{t,\xi}, Z_s^{0;t,\xi})_{s\in[t,T]}$, *which is progressively measurable with respect to the completion of the filtration generated by* $(\xi, W_s^0 - W_t^0, W_s - W_t)_{t\le s\le T}$. *Furthermore, assume that the auxiliary system* (5.4) *is strongly uniquely solvable in the sense of Definition 1.19.*

Then, under these conditions, (5.6) *has a unique solution, which we shall denote by* $(X_s^{t,x,\mu}, Y_s^{t,x,\mu}, Z_s^{t,x,\mu}, Z^{0,t,x,\mu})_{s\in[t,T]}$. *It is independent of the choice of the lifting* ξ *of* μ. *Under these conditions, we call master field of* (5.2) *(or of* (5.6)*) the function:*

$$\mathcal{U} : [0,T] \times \mathbb{R}^d \times \mathcal{P}_2(\mathbb{R}^d) \ni (t,x,\mu) \mapsto Y_t^{t,x,\mu} \in \mathbb{R}^m. \tag{5.7}$$

Observe that $Y_t^{t,x,\mu}$ in (5.7) is almost surely deterministic since (5.6) can be solved by equipping the probabilistic set-up with the filtration generated by $(W_s^0 - W_t^0)_{t\le s\le T}$ and $(W_s - W_t)_{t\le s\le T}$.

We already discussed conditions under which existence and uniqueness hold:

1. For example, when the coefficients B, F, G, and σ are Lipschitz continuous in all the variables except possibly time, and there is no common noise, then Theorem (Vol I)-4.24 ensures that (5.2) and (5.6) are uniquely solvable in small time. This short time solvability result may be easily adapted to the case when there is a common noise. We shall come back to this point below.
2. Also, when the coefficients B, F, G, σ, and σ^0 derive from a mean field game, either through the method based on the representation of the value function or through the stochastic Pontryagin principle, Theorems 3.29, 3.30, and 3.31 together with Proposition 3.34 ensure, though under different sets of conditions, that (5.2) is uniquely solvable. In all cases, (5.6) is uniquely solvable as well, since the unique solvability property is part of assumption **FBSDE** in Subsection 2.2.3, which is known to hold true under the assumptions of Theorems 3.29, 3.30, and 3.31.

Remark 5.3 *Regarding our preliminary discussion in Subsection 5.1 on the interpretation of the master field of a mean field game as the decoupling field of an FBSDE of the McKean-Vlasov type, it is worth emphasizing one more time that when the McKean Vlasov FBSDE used to characterize the solution of a mean field game derives from the stochastic Pontryagin principle, the decoupling field of the FBSDE does not coincide with the master field of the game but instead, with its gradient in space. We already made this point in Subsection 4.2.2, and we will come back to it in Section 5.4.*

5.1.3 Short Time Analysis

Our analysis of the master field, as defined in Definition 5.2, is performed first for a short enough time horizon T. It is only in a second part, when we concentrate on conditional McKean-Vlasov FBSDEs derived from mean field games, that we switch to the case of an arbitrary T. In doing so, we shall benefit from the existence and uniqueness results proved in Chapter 4.

We start with the same kind of assumption as in Subsection (Vol I)-4.2.3.

Assumption (Conditional MKV FBSDE in Small Time). There exist two constants $\Gamma, L \geq 0$ such that:

(A1) The mappings B, F, G, σ, and σ^0 are measurable from $[0,T] \times \mathbb{R}^d \times \mathcal{P}_2(\mathbb{R}^d) \times \mathbb{R}^m \times \mathbb{R}^{m\times d} \times \mathbb{R}^{m\times d}$ to $\mathbb{R}^d$, from $[0,T] \times \mathbb{R}^d \times \mathcal{P}_2(\mathbb{R}^d) \times \mathbb{R}^m \times \mathbb{R}^{m\times d} \times \mathbb{R}^{m\times d}$ to $\mathbb{R}^m$, from $\mathbb{R}^d \times \mathcal{P}_2(\mathbb{R}^d)$ to $\mathbb{R}^m$, from $[0,T] \times \mathbb{R}^d \times \mathcal{P}_2(\mathbb{R}^d)$ to $\mathbb{R}^{d\times d}$ and from $[0,T] \times \mathbb{R}^d \times \mathcal{P}_2(\mathbb{R}^d)$ to $\mathbb{R}^{d\times d}$ respectively.

(continued)

(A2) For all $t \in [0,T]$, the coefficients $B(t,0,\delta_0,0,0,0)$, $F(t,0,\delta_0,0,0,0)$, $\sigma(t,0,\delta_0)$ and $\sigma^0(t,0,\delta_0)$ are bounded by Γ. Similarly, $G(0,\delta_0)$ is bounded by Γ.

(A3) $\forall t \in [0,T]$, $\forall x,x' \in \mathbb{R}^d$, $\forall y,y' \in \mathbb{R}^m$, $\forall z,z',z^0,z^{0\prime} \in \mathbb{R}^{m\times d}$, $\forall \mu,\mu' \in \mathcal{P}_2(\mathbb{R}^d)$,

$$
\begin{aligned}
&|(B,F)(t,x,\mu,y,z,z^0) - (B,F)(t,x',\mu',y',z',z^{0\prime})| \\
&\quad \le L\big[|x-x'| + |y-y'| + |z-z'| + |z^0 - z^{0\prime}| + W_2(\mu,\mu')\big],
\end{aligned}
$$

$$
\begin{aligned}
&|(\sigma,\sigma^0)(t,x,\mu) - (\sigma,\sigma^0)(t,x',\mu')| \le L\big[|x-x'| + W_2(\mu,\mu')\big], \\
&|G(x,\mu) - G(x',\mu')| \le L\big[|x-x'| + W_2(\mu,\mu')\big].
\end{aligned}
$$

In full similarity with Theorems (Vol I)-4.24 and 1.45, we prove the following result:

Theorem 5.4 *Under assumption* **Conditional MKV FBSDE in Small Time**, *there exist two constants $c > 0$ and $C \ge 0$, only depending upon the parameter L in the assumption, such that for $T \le c$, for any initial condition (t,ξ), with $t \in [0,T]$ and $\xi \in L^2(\Omega,\mathcal{F}_t,\mathbb{P};\mathbb{R}^d)$, the FBSDE* (5.2) *has a unique solution $(\boldsymbol{X}^{t,\xi},\boldsymbol{Y}^{t,\xi},\boldsymbol{Z}^{t,\xi},\boldsymbol{Z}^{0;t,\xi})$. This solution is progressively measurable with respect to the completion of the filtration generated by $(\xi,\mathcal{L}^1(\xi),W_s^0 - W_t^0, W_s - W_t)_{t\le s\le T}$.*

Moreover, for any $\xi' \in L^2(\Omega,\mathcal{F}_t,\mathbb{P};\mathbb{R}^d)$ and any tuple $(B',F',G',\sigma',\sigma^{0\prime})$ satisfying assumption **Conditional MKV FBSDE in Small Time**, *the solution $(\boldsymbol{X}',\boldsymbol{Y}',\boldsymbol{Z}',\boldsymbol{Z}^{0;\prime})$ to* (5.2) *with $(B',F',G',\sigma',\sigma^{0\prime})$ in lieu of (B,F,G,σ,σ^0) as coefficients and (t,ξ') in lieu of (t,ξ) as initial condition, satisfies:*

$$
\begin{aligned}
&\mathbb{E}\bigg[\sup_{t\le s\le T}\big(|X_s - X_s'|^2 + |Y_s - Y_s'|^2\big) + \int_t^T \big(|Z_s - Z_s'|^2 + |Z_s^0 - Z_s^{0\prime}|^2\big)ds \,\Big|\, \mathcal{F}_t\bigg] \\
&\quad \le C\bigg\{\mathbb{E}\bigg[|\xi-\xi'|^2 + \big|(G-G')\big(X_T,\mathcal{L}^1(X_T)\big)\big|^2 \\
&\qquad\qquad + \int_t^T \big|\big(B-B',F-F',\sigma-\sigma',\sigma^0-\sigma^{0\prime}\big)(s,\theta_s)\big|^2 ds \,\Big|\, \mathcal{F}_t\bigg] \\
&\qquad + \mathbb{E}\bigg[|\xi-\xi'|^2 + \big|(G-G')\big(X_T,\mathcal{L}^1(X_T)\big)\big|^2 \\
&\qquad\qquad + \int_t^T \big|\big(B-B',F-F',\sigma-\sigma',\sigma^0-\sigma^{0\prime}\big)(s,\theta_s)\big|^2 ds \,\Big|\, \mathcal{F}_t^0\bigg]\bigg\},
\end{aligned}
\tag{5.8}
$$

where we used the notations $(\boldsymbol{X},\boldsymbol{Y},\boldsymbol{Z},\boldsymbol{Z}^0)$ for $(\boldsymbol{X}^{t,\xi},\boldsymbol{Y}^{t,\xi},\boldsymbol{Z}^{t,\xi},\boldsymbol{Z}^{0;t,\xi})$, and $\boldsymbol{\theta}$ for $(\theta_s = (X_s,\mathcal{L}^1(X_s),Y_s,Z_s,Z_s^0))_{t\le s\le T}$.

Obviously, there is a slight abuse of notation in (5.8) since $\sigma - \sigma'$ and $\sigma^0 - \sigma^{0\prime}$ are independent of the variables y and z.

Proof. The first part of the proof is a mere adaptation of the arguments used to prove Theorem (Vol I)-4.24. Given a proxy for the solution of the forward equation, we solve the backward equation and, plugging this solution of the backward equation into the coefficients of the forward equation, we get a new proxy for the solution of the forward equation. The goal is to prove that this procedure creates a contraction when T is small enough.

For a given proxy $(\boldsymbol{Y}, \boldsymbol{Z}, \boldsymbol{Z}^0) = (Y_s, Z_s, Z_s^0)_{t \le s \le T}$, with $(\boldsymbol{Y}, \boldsymbol{Z}, \boldsymbol{Z}^0)$ being progressively measurable with respect to $\mathbb{F}^{(\xi, \mathcal{L}^1(\xi), W_s^0 - W_t^0, W_s - W_t)_{t \le s \le T}}$, the forward equation takes the same form as in (5.2), namely:

$$
\begin{aligned}
dX_s &= B\big(s, X_s, \mathcal{L}^1(X_s), Y_s, Z_s, Z_s^0\big)ds \\
&\quad + \sigma\big(s, X_s, \mathcal{L}^1(X_s)\big)dW_s + \sigma^0\big(s, X_s, \mathcal{L}^1(X_s)\big)dW_s^0, \quad s \in [t, T],
\end{aligned}
\tag{5.9}
$$

with $X_t = \xi$ as initial condition at time t.

In order to solve the forward equation (5.9), one must invoke an obvious generalization of Proposition 2.8 that allows for random coefficients b, σ, and σ^0, using the same notation as in the statement of Proposition 2.8. Additionally, we must pay attention to the fact that $(\mathcal{L}^1(X_s))_{t \le s \le T}$ is progressively measurable with respect to the completion of the filtration generated by $(\mathcal{L}^1(\xi), W_s^0 - W_t^0)_{t \le s \le T}$, which guarantees that the coefficients in the backward equation that has to be solved next are indeed progressively measurable with respect to the completion of the filtration generated by $(\xi, \mathcal{L}^1(\xi), W_s^0 - W_t^0, W_s - W_t)_{t \le s \le T}$. Adaptedness follows from the fact that, for a given proxy $(\boldsymbol{Y}, \boldsymbol{Z}, \boldsymbol{Z}^0)$, the solution to the equation (5.9) is constructed by means of a Picard iteration. At each step of the iteration and at any time $s \in [t, T]$, we can find a measurable mapping Φ_s from $\mathbb{R}^d \times \mathcal{P}_2(\mathbb{R}^d) \times \mathcal{C}([t, s]; \mathbb{R}^d) \times \mathcal{C}([t, s]; \mathbb{R}^d)$ such that the next outcome at time s in the iteration reads $X_s = \Phi_s(\xi, \mathcal{L}^1(\xi), (W_r^0 - W_t^0)_{t \le r \le s}, (W_r - W_t)_{t \le r \le T})$ with probability 1, from which we get that X_s is $\mathcal{F}_s^{(\xi, \mathcal{L}^1(\xi), (W_r^0 - W_t^0)_{t \le r \le s}, (W_r - W_t)_{t \le r \le s})}$-measurable. Moreover, with probability 1 under $\mathbb{P}^0$,

$$
\mathcal{L}^1(X_s) = \big(\mathcal{L}^1(\xi) \otimes \mathcal{W}_d^t\big) \circ \Big[\Phi_s\big(\cdot, \mathcal{L}^1(\xi), (W_r^0 - W_t^0)_{t \le r \le s}, \cdot\big)\Big]^{-1}.
$$

It is standard to deduce that $\mathcal{L}^1(X_s)$ is $\mathcal{F}_s^{0, (\mathcal{L}^1(\xi), (W_r^0 - W_t^0)_{t \le r \le s})}$-measurable, see for instance Lemma 2.4 or Proposition (Vol I)-5.7.

The proof of the stability property (5.8) goes along the lines of the proof of Theorem 1.45. It must be divided into two main steps. The first one is to regard $(\mathcal{L}^1(X_s))_{t \le s \le T}$ and $(\mathcal{L}^1(X_s'))_{t \le s \le T}$ as environments, and then to apply Theorem 1.45. This provides the following estimate:

$$
\begin{aligned}
&\mathbb{E}\bigg[\sup_{t \le s \le T}\big(|X_s - X_s'|^2 + |Y_s - Y_s'|^2\big) + \int_t^T \big(|Z_s - Z_s'|^2 + |Z_s^0 - Z_s^{0\prime}|^2\big)ds \,\Big|\, \mathcal{F}_t\bigg] \\
&\le C\mathbb{E}\bigg[|\xi - \xi'|^2 + \big|G(X_T, \mathcal{L}^1(X_T)) - G'(X_T, \mathcal{L}^1(X_T))\big|^2 \\
&\qquad + \int_t^T \big|(B, F, \sigma, \sigma^0)(s, \theta_s) - (B', F', \sigma', \sigma^{0\prime})(s, X_s, \mathcal{L}^1(X_s'), Y_s, Z_s)\big|^2 ds \,\Big|\, \mathcal{F}_t\bigg].
\end{aligned}
$$

Using the Lipschitz property of the coefficients, we get:

$$
\begin{aligned}
&\mathbb{E}\Big[\sup_{t\le s\le T}\big(|X_s-X'_s|^2+|Y_s-Y'_s|^2\big)+\int_t^T\big(|Z_s-Z'_s|^2+|Z^0_s-Z^{0\prime}_s|^2\big)ds\,\big|\,\mathcal{F}_t\Big]\\
&\quad\le C\mathbb{E}\Big[\mathbb{E}^1\big[\sup_{t\le s\le T}|X_s-X'_s|^2\big]+|\xi-\xi'|^2+\big|(G-G')\big(X_T,\mathcal{L}^1(X_T)\big)\big|^2 \qquad (5.10)\\
&\qquad+\int_t^T\big|(B-B',F-F',\sigma-\sigma',\sigma^0-\sigma^{0\prime})(s,\theta_s)\big|^2ds\,\Big|\,\mathcal{F}_t\Big].
\end{aligned}
$$

Plugging the above bound in the forward equation satisfied by $\boldsymbol{X}-\boldsymbol{X}'$, we easily get the same type of bound, but for $\boldsymbol{X}-\boldsymbol{X}'$, with an additional T in front of $\mathbb{E}^1[\sup_{t\le s\le T}|X_s-X'_s|^2]$, namely:

$$
\begin{aligned}
\mathbb{E}\Big[\sup_{t\le s\le T}|X_s-X'_s|^2\,|\,\mathcal{F}_t\Big]&\le CT\mathbb{E}\Big[\mathbb{E}^1\big[\sup_{t\le s\le T}|X_s-X'_s|^2\big]\,|\,\mathcal{F}_t\Big]\\
&\quad+C\mathbb{E}\Big[|\xi-\xi'|^2+\big|(G-G')\big(X_T,\mathcal{L}^1(X_T)\big)\big|^2 \qquad (5.11)\\
&\quad+\int_t^T\big|(B-B',F-F',\sigma-\sigma',\sigma^0-\sigma^{0\prime})(s,\theta_s)\big|^2ds\,\Big|\,\mathcal{F}_t\Big].
\end{aligned}
$$

Now, we observe that, for a real-valued integrable random variable ζ on $(\Omega^0,\mathcal{F}^0,\mathbb{P}^0)$, $\mathbb{E}[\zeta|\mathcal{F}_t]=\mathbb{E}^0[\zeta|\mathcal{F}^0_t]$, where, in the left-hand side, ζ is regarded as extended to $(\Omega,\mathcal{F},\mathbb{P})$ in the usual way. Moreover, for a real-valued integrable random variable ζ on $(\Omega,\mathcal{F},\mathbb{P})$, $\mathbb{E}^1[\mathbb{E}[\zeta|\mathcal{F}_t]]=\mathbb{E}^0[\mathbb{E}^1(\zeta)|\mathcal{F}^0_t]=\mathbb{E}[\mathbb{E}^1(\zeta)|\mathcal{F}_t]=\mathbb{E}[\zeta|\mathcal{F}^0_t]$, see Lemma 5.16 below.

Therefore, taking the expectation under $\mathbb{P}^1$ in (5.11), we get, for T small enough,

$$
\begin{aligned}
&\mathbb{E}\Big[\mathbb{E}^1\big[\sup_{t\le s\le T}|X_s-X'_s|^2\big]\,|\,\mathcal{F}_t\Big]\\
&\quad\le C\mathbb{E}\Big[|\xi-\xi'|^2+\big|(G-G')\big(X_T,\mathcal{L}^1(X_T)\big)\big|^2\\
&\qquad+\int_t^T\big|(B-B',F-F',\sigma-\sigma',\sigma^0-\sigma^{0\prime})(s,\theta_s)\big|^2ds\,\Big|\,\mathcal{F}^0_t\Big].
\end{aligned}
$$

Plugging the above bound into (5.10), we complete the proof. □

Remark 5.5 *We stress the fact that, in the statement of Theorem 5.4, the random variable $\mathcal{L}^1(\xi)$ may not be measurable with respect to $\sigma\{\xi\}$. Consider for instance $\xi=\mathbf{1}_{C^0\times C^1}$, with $C^0\in\mathcal{F}^0_t$, $C^1\in\mathcal{F}^1_t$ and $\mathbb{P}^0(C^0)=\mathbb{P}^1(C^1)=1/2$. Then, $\sigma\{\xi\}=\{\emptyset,\Omega,C^0\times C^1,(C^0\times C^1)^{\complement}\}$, while $\mathcal{L}^1(\xi)(\omega^0)$ is the Bernoulli distribution of parameter $1/2$ if $\omega^0\in C^0$ and is the Bernoulli distribution of parameter 0 if $\omega^0\notin C^0$. In particular, $\sigma\{\mathcal{L}^1(\xi)\}=\{\emptyset,\Omega^0,C^0,(C^0)^{\complement}\}$.*

Remark 5.6 *By adapting the argument used in Example 1.20, we can show that uniqueness in Theorem 5.4 holds true for a larger class of equations. Namely, any solution of the system:*

$$\begin{cases} dX_s^{t,\xi} = B\big(s, X_s^{t,\xi}, \mathcal{L}^1(X_s^{t,\xi}), Y_s^{t,\xi}, Z_s^{t,\xi}, Z_s^{0;t,\xi}\big)ds \\ \qquad + \sigma\big(s, X_s^{t,\xi}, \mathcal{L}^1(X_s^{t,\xi})\big)dW_s + \sigma^0\big(s, X_s^{t,\xi}, \mathcal{L}^1(X_s^{t,\xi})\big)dW_s^0, \\ dY_s^{t,\xi} = -F\big(s, X_s^{t,\xi}, \mathcal{L}^1(X_s^{t,\xi}), Y_s^{t,\xi}, Z_s^{t,\xi}, Z_s^{0;t,\xi}\big)ds \\ \qquad + Z_s^{t,\xi} dW_s + Z_s^{0;t,\xi} dW_s^0 + dM_s^{t,\xi}, \quad s \in [t,T], \\ X_t^{t,\xi} = \xi, \quad Y_T^{t,\xi} = G\big(X_T^{t,\xi}, \mathcal{L}^1(X_T^{t,\xi})\big), \end{cases}$$

where $(M_s^{t,\xi})_{t \le s \le T}$ *in the second equation is an m-dimensional càd-làg martingale starting from 0 at time t and of zero bracket with* $(W_s - W_t)_{t \le s \le T}$ *and* $(W_s^0 - W_t^0)_{t \le s \le T}$, *coincides in short time with the solution of (5.2), as given by Theorem 5.4. In particular,* $(M_s^{t,\xi})_{t \le s \le T}$ *must be zero.*

Thanks to Theorems 1.45 and 5.4, we have all the necessary ingredients needed in the definition of the master field as stated in Definition 5.2. By a mere variation of the proof of Lemma (Vol I)-4.5, we have:

Proposition 5.7 *Under assumption* **Conditional MKV FBSDE in Small Time** *and with the notation of Theorem 5.4, there exists a constant* C' *such that, for* $T \le c$, *for all* (t, x, μ) *and* (t', x', μ') *in* $[0, T] \times \mathbb{R}^d \times \mathcal{P}_2(\mathbb{R}^d)$,

$$\begin{aligned} &|\mathcal{U}(t, x, \mu)| \le C'\big(1 + |x| + M_2(\mu)\big), \\ &|\mathcal{U}(t, x, \mu) - \mathcal{U}(t', x', \mu')| \\ &\qquad \le C'\big(|x - x'| + W_2(\mu, \mu')\big) + \big(1 + |x| + M_2(\mu)\big)|t - t'|^{1/2}. \end{aligned}$$

As a consequence of Proposition 5.7, we get the fact that $\mathcal{U}$ is indeed a decoupling field for the FBSDE 5.1.

Proposition 5.8 *Under assumption* **Conditional MKV FBSDE in Small Time** *and with the same notation as in the statement of Theorem 5.4, consider a square-integrable* $\mathcal{F}_t$*-measurable initial condition* ξ *at time t. Then, for* $T \le c$, *with C as in the statement of Theorem 5.4, it holds that:*

$$Y_t^{t,\xi} = \mathcal{U}\big(t, \xi, \mathcal{L}^1(\xi)\big).$$

In particular, with probability 1 under $\mathbb{P}$, *for all* $s \in [t, T]$,

$$Y_s^{t,\xi} = \mathcal{U}\big(s, X_s^{t,\xi}, \mathcal{L}^1(X_s^{t,\xi})\big).$$

Proof. The proof is a variant of the conditioning arguments used in Chapter 1, see for instance Lemma 1.40, but it bypasses any use of regular conditional probabilities. Instead, we take full advantage of the smoothness of $\mathcal{U}$.

First Step. Without any loss of generality, we assume that ξ is $\mathcal{F}_t^0 \otimes \mathcal{F}_t^1$-measurable. Under the standing assumption, we know that $L^2(\Omega^1, \mathcal{F}^1, \mathbb{P}^1; \mathbb{R}^d)$ is separable. Regarding ξ as a random variable from $(\Omega^0, \mathcal{F}_t^0, \mathbb{P}^0)$ with values in $L^2(\Omega^1, \mathcal{F}^1, \mathbb{P}^1; \mathbb{R}^d)$, which is licit since for any $Z \in L^2(\Omega^1, \mathcal{F}_t^1, \mathbb{P}^1; \mathbb{R}^d)$, $\Omega^0 \ni \omega^0 \mapsto \mathbb{E}^1[|Z - \xi|^2](\omega^0)$ is a random variable, we deduce that for any $N \geq 1$, there exists a compact subset $\mathcal{K}_N \subset L^2(\Omega^1, \mathcal{F}_t^1, \mathbb{P}^1; \mathbb{R}^d)$ such that $\mathbb{P}^0[\xi \notin \mathcal{K}_N] \leq 1/N$.

Now, for each $N \geq 1$, we can find a covering of $\mathcal{K}_N$ with n_N pairwise disjoint Borel subsets $B_1^N, \cdots, B_N^N$ of diameter less than $1/N$. For each $i \in \{1, \cdots, n_N\}$, we choose one point X_i^N in B_i^N and we define:

$$\xi^N(\omega^0, \cdot) = \sum_{i=1}^{n_N} X_i^N \mathbf{1}_{B_i^N}(\xi(\omega^0, \cdot)),$$

so that:

$$\begin{aligned}\|\xi^N(\omega^0, \cdot) - \xi(\omega^0, \cdot)\|_{L^2(\Omega^1, \mathcal{F}^1, \mathbb{P}^1; \mathbb{R}^d)} &\leq \frac{1}{N} \mathbf{1}_{\mathcal{K}_N}(\xi(\omega^0, \cdot)) \\ &\quad + \|\xi(\omega^0, \cdot)\|_{L^2(\Omega^1, \mathcal{F}_t^1, \mathbb{P}^1; \mathbb{R}^d)} \mathbf{1}_{\mathcal{K}_N^{\complement}}(\xi(\omega^0, \cdot)),\end{aligned}$$

and thus,

$$\lim_{N \to \infty} \mathbb{E}^0\Big[\|\xi^N - \xi\|^2_{L^2(\Omega^1, \mathcal{F}_t^1, \mathbb{P}^1; \mathbb{R}^d)}\Big] = 0.$$

Second Step. We now approximate each X_i^N by a simple random variable $\bar{X}_i^N$ at most at distance $1/N$ from X_i^N in $L^2(\Omega^1, \mathcal{F}_t^1, \mathbb{P}^1; \mathbb{R}^d)$. Writing:

$$\bar{X}_i^N(\omega^1) = \sum_{j=1}^{m_i^N} \bar{x}_{i,j}^N \mathbf{1}_{A_{i,j}^N}(\omega^1), \quad \omega^1 \in \Omega^1,$$

where $\bar{x}_{i,1}^N, \cdots, \bar{x}_{i,m_i^N}^N \in \mathbb{R}^d$ and $A_{i,1}^N, \cdots, A_{i,m_i^N}^N$ are N pairwise disjoint events in $\mathcal{F}_t^1$ covering Ω^1, we let:

$$\bar{\xi}^N(\omega^0, \omega^1) = \sum_{i=1}^{n_N} \sum_{j=1}^{m_i^N} \bar{x}_{i,j}^N \mathbf{1}_{A_{i,j}^N}(\omega^1) \mathbf{1}_{B_i^N}(\xi(\omega^0, \cdot)), \quad \omega^0 \in \Omega^0, \ \omega^1 \in \Omega^1,$$

or, equivalently,

$$\bar{\xi}^N(\omega^0, \cdot) = \sum_{i=1}^{n_N} \bar{X}_i^N \mathbf{1}_{B_i^N}(\xi(\omega^0, \cdot)), \quad \omega^0 \in \Omega^0.$$

Obviously,

$$\lim_{N \to \infty} \mathbb{E}^0\Big[\|\bar{\xi}^N - \xi\|^2_{L^2(\Omega^1, \mathcal{F}_t^1, \mathbb{P}^1; \mathbb{R}^d)}\Big] = 0.$$

Third Step. Now, we observe that, for $\mathbb{P}$-almost every $\omega = (\omega^0, \omega^1) \in \Omega^0 \times \Omega^1$,

$$\boldsymbol{Y}^{t, \bar{\xi}^N}(\omega^0, \omega^1) = \sum_{i=1}^{n_N} \boldsymbol{Y}^{t, \bar{X}_i^N}(\omega^0, \omega^1) \mathbf{1}_{B_i^N}(\xi(\omega^0, \cdot)) + Y_t^{t,0} \mathbf{1}_{B_0^N}(\xi(\omega^0, \cdot)),$$

where $B_0^N = \Omega^1 \setminus (\cup_{i=1}^N B_i^N)$, and,

$$\boldsymbol{Y}^{t,\bar{X}_i^N}(\omega^0,\omega^1) = \sum_{j=1}^{m_i^N} \boldsymbol{Y}^{t,\bar{x}_{i,j}^N,\mathcal{L}^1(\bar{X}_i^N)}(\omega^0,\omega^1)\mathbf{1}_{A_{i,j}^N}(\omega^1),$$

from which we deduce that:

$$Y_t^{t,\bar{\xi}^N}(\omega^0,\omega^1) = \sum_{i=1}^{n_N}\sum_{j=1}^{m_i^N} \mathcal{U}(t,\bar{x}_{i,j}^N,\mathcal{L}^1(\bar{X}_i^N))\mathbf{1}_{B_i^N}(\xi(\omega^0,\cdot))\mathbf{1}_{A_{i,j}^N}(\omega^1) + \mathcal{U}(t,0,\delta_0)\mathbf{1}_{B_0^N}(\xi(\omega^0,\cdot)),$$

and then, for $\mathbb{P}$-almost every $\omega = (\omega^0,\omega^1) \in \Omega^0 \times \Omega^1$,

$$Y_t^{t,\bar{\xi}^N}(\omega^0,\omega^1) = \mathcal{U}\Big(t,\bar{\xi}^N(\omega^0,\omega^1),\mathcal{L}^1\big(\bar{\xi}^N(\omega^0,\cdot)\big)\Big).$$

By the stability Theorem 5.4, $(Y_t^{t,\bar{\xi}^N})_{N\geq 1}$ converges to $Y_t^{t,\xi}$ in $L^2(\Omega,\mathcal{F},\mathbb{P};\mathbb{R}^m)$. Thanks to the regularity of $\mathcal{U}$, the right-hand side converges in probability to $\mathcal{U}(t,\xi,\mathcal{L}^1(\xi))$. This completes the first part of the proof. The second part, concerning the representation of $Y_s^{t,\xi}$, easily follows. □

5.1.4 Solution of a Master PDE

Our goal is to prove that, under suitable regularity properties of the coefficients, the master field $\mathcal{U}$ satisfies the assumptions of the Itô formula proved in Chapter 4 for functions of an Itô process and the marginal laws of a possibly different Itô process. In other words, we are looking for conditions under which $\mathcal{U}$ is $\mathcal{C}^{1,2,2}$ over $[0,T] \times \mathbb{R}^d \times \mathcal{P}_2(\mathbb{R}^d)$ (with values in $\mathbb{R}^m$) in the sense defined in assumption **Joint Chain Rule Common Noise** in Subsection 4.3.4.

Assume for a while that this is indeed the case and then apply Itô's formula in Theorem 4.17 with:

$$\boldsymbol{X}^0 = \big(X_s^{t,x,\mu}\big)_{t\leq s\leq T}, \quad \mu = \big(\mathcal{L}^1(X_s^{t,\xi})_{t\leq s\leq T}\big),$$

for $t \in [0,T]$, $x \in \mathbb{R}^d$, $\mu \in \mathcal{P}_2(\mathbb{R}^d)$ and $\xi \in L^2(\Omega^1,\mathcal{F}_t^1,\mathbb{P}^1;\mathbb{R}^d)$ with $\mathcal{L}^1(\xi) = \mu$. Then, (4.37) yields:

$$\begin{aligned}
&\mathcal{U}(t+h,X_{t+h}^{t,x,\mu},\mathcal{L}^1(X_{t+h}^{t,\xi})) = \mathcal{U}(t,x,\mu)\\
&\quad + \int_t^{t+h} \partial_x\mathcal{U}\big(s,X_s^{t,x,\mu},\mathcal{L}^1(X_s^{t,\xi})\big)\cdot B\big(s,X_s^{t,x,\mu},\mathcal{L}^1(X_s^{t,\xi}),Y_s^{t,x,\mu},Z_s^{t,x,\mu},Z_s^{0;t,x,\mu}\big)ds\\
&\quad + \int_t^{t+h} \tilde{\mathbb{E}}^1\Big[\partial_\mu\mathcal{U}\big(s,X_s^{t,x,\mu},\mathcal{L}^1(X_s^{t,\xi})\big)(\tilde{X}_s^{t,\xi})\cdot B\big(s,\tilde{X}_s^{t,\xi},\mathcal{L}^1(X_s^{t,\xi}),\tilde{Y}_s^{t,\xi},\tilde{Z}_s^{t,\xi},\tilde{Z}_s^{0;t,\xi}\big)\Big]ds
\end{aligned}$$

$$
\begin{aligned}
&+\int_t^{t+h} \partial_x \mathcal{U}\big(s, X_s^{t,x,\mu}, \mathcal{L}^1(X_s^{t,\xi})\big)\cdot\big[\sigma\big(s, X_s^{t,x,\mu}, \mathcal{L}^1(X_s^{t,\xi})\big)dW_s\big] \\
&+\int_t^{t+h} \partial_x \mathcal{U}\big(s, X_s^{t,x,\mu}, \mathcal{L}^1(X_s^{t,\xi})\big)\cdot\big[\sigma^0\big(s, X_s^{t,x,\mu}, \mathcal{L}^1(X_s^{t,\xi})\big)dW_s^0\big] \\
&+\int_t^{t+h} \tilde{\mathbb{E}}^1\big[(\sigma^0)^\dagger\big(s, \tilde{X}_s^{t,\xi}, \mathcal{L}^1(X_s^{t,\xi})\big)\partial_\mu \mathcal{U}\big(s, X_s^{t,x,\mu}, \mathcal{L}^1(X_s^{t,\xi})\big)(\tilde{X}_s^{t,\xi})\big]\cdot dW_s^0 \\
&+\frac{1}{2}\int_t^{t+h} \mathrm{trace}\big[\partial_{xx}^2 \mathcal{U}\big(s, X_s^{t,x,\mu}, \mathcal{L}^1(X_s^{t,\xi})\big) \\
&\qquad\qquad \times\big(\sigma\sigma^\dagger+\sigma^0(\sigma^0)^\dagger\big)\big(s, X_s^{t,x,\mu}, \mathcal{L}^1(X_s^{t,\xi})\big)\big]ds \\
&+\frac{1}{2}\int_t^{t+h} \tilde{\mathbb{E}}^1\big[\mathrm{trace}\big(\partial_v\partial_\mu \mathcal{U}\big(s, X_s^{t,x,\mu}, \mathcal{L}^1(X_s^{t,\xi})\big)(\tilde{X}_s^{t,\xi}) \\
&\qquad\qquad \times\big(\sigma\sigma^\dagger+\sigma^0(\sigma^0)^\dagger\big)\big(s, \tilde{X}_s^{t,\xi}, \mathcal{L}^1(X_s^{t,\xi})\big)\big)\big]ds \\
&+\frac{1}{2}\int_t^{t+h} \tilde{\mathbb{E}}^1\tilde{\tilde{\mathbb{E}}}^1\Big[\mathrm{trace}\big\{\partial_\mu^2 \mathcal{U}\big(s, X_s^{t,x,\mu}, \mathcal{L}^1(X_s^{t,\xi})\big)(\tilde{X}_s^{t,\xi}, \tilde{\tilde{X}}_s^{t,\xi}) \\
&\qquad\qquad \times\big(\sigma^0\big(s, \tilde{X}_s^{t,\xi}, \mathcal{L}^1(X_s^{t,\xi})\big)(\sigma^0)^\dagger\big(s, \tilde{\tilde{X}}_s^{t,\xi}, \mathcal{L}^1(X_s^{t,\xi})\big)\big)\big\}\Big]ds \\
&+\int_t^{t+h} \tilde{\mathbb{E}}^1\Big[\mathrm{trace}\big\{\partial_x\partial_\mu \mathcal{U}\big(s, X_s^{t,x,\mu}, \mathcal{L}^1(X_s^{t,\xi})\big)(\tilde{X}_s^{t,\xi}) \\
&\qquad\qquad \times\sigma^0\big(s, X_s^{t,x,\mu}, \mathcal{L}^1(X_s^{t,\xi})\big)\big(\sigma^0\big(s, X_s^{t,\xi}, \mathcal{L}^1(X_s^{t,\xi})\big)\big)^\dagger\big\}\Big]ds.
\end{aligned}
\tag{5.12}
$$

Identifying the left-hand side with $Y_{t+h}^{t,x,\mu}$ and recalling the FBSDE (5.6) satisfied by the process $(\boldsymbol{X}^{t,x,\mu}, \boldsymbol{Y}^{t,x,\mu}, \boldsymbol{Z}^{t,x,\mu}, \boldsymbol{Z}^{0;t,x,\mu})$, we deduce that, $\mathrm{Leb}_1 \otimes \mathbb{P}$ almost-everywhere,

$$
\begin{aligned}
Z_s^{t,x,\mu} &= \sigma^\dagger\big(s, X_s^{t,x,\mu}, \mathcal{L}^1(X_s^{t,\xi})\big)\partial_x \mathcal{U}\big(s, X_s^{t,x,\mu}, \mathcal{L}^1(X_s^{t,\xi})\big), \\
Z_s^{0;t,x,\mu} &= \big(\sigma^0\big)^\dagger\big(s, X_s^{t,x,\mu}, \mathcal{L}^1(X_s^{t,\xi})\big)\partial_x \mathcal{U}\big(s, X_s^{t,x,\mu}, \mathcal{L}^1(X_s^{t,\xi})\big) \\
&\quad + \tilde{\mathbb{E}}^1\big[(\sigma^0)^\dagger\big(s, \tilde{X}_s^{t,\xi}, \mathcal{L}^1(X_s^{t,\xi})\big)\partial_\mu \mathcal{U}\big(s, X_s^{t,x,\mu}, \mathcal{L}^1(X_s^{t,\xi})\big)(\tilde{X}_s^{t,\xi})\big],
\end{aligned}
$$

and replacing x by ξ, we get:

$$
\begin{aligned}
Z_s^{t,\xi} &= \sigma^\dagger\big(s, X_s^{t,\xi}, \mathcal{L}^1(X_s^{t,\xi})\big)\partial_x \mathcal{U}\big(s, X_s^{t,\xi}, \mathcal{L}^1(X_s^{t,\xi})\big), \\
Z_s^{0;t,\xi} &= \big(\sigma^0\big)^\dagger\big(s, X_s^{t,\xi}, \mathcal{L}^1(X_s^{t,\xi})\big)\partial_x \mathcal{U}\big(s, X_s^{t,\xi}, \mathcal{L}^1(X_s^{t,\xi})\big) \\
&\quad + \tilde{\mathbb{E}}^1\big[(\sigma^0)^\dagger\big(s, \tilde{X}_s^{t,\xi}, \mathcal{L}^1(X_s^{t,\xi})\big)\partial_\mu \mathcal{U}\big(s, X_s^{t,\xi}, \mathcal{L}^1(X_s^{t,\xi})\big)(\tilde{X}_s^{t,\xi})\big].
\end{aligned}
$$

Moreover, $\mathcal{U}$ satisfies the PDE:

$$
\begin{aligned}
&\partial_t \mathcal{U}(t,x,\mu) \\
&\quad + B\big(t,x,\mu,\mathcal{U}(t,x,\mu),\partial_x^\sigma \mathcal{U}(t,x,\mu),\partial_{(x,\mu)}^{\sigma^0}\mathcal{U}(t,x,\mu)\big)\cdot \partial_x \mathcal{U}(t,x,\mu) \\
&\quad + \int_{\mathbb{R}^d} B\big(t,v,\mu,\mathcal{U}(t,v,\mu),\partial_x^\sigma \mathcal{U}(t,v,\mu),\partial_{(x,\mu)}^{\sigma^0}\mathcal{U}(t,v,\mu)\big)\cdot \partial_\mu \mathcal{U}(t,x,\mu)(v) d\mu(v) \\
&\quad + \frac{1}{2}\mathrm{trace}\Big[\big(\sigma\sigma^\dagger + \sigma^0(\sigma^0)^\dagger\big)(t,x,\mu)\partial_{xx}^2 \mathcal{U}(t,x,\mu)\Big] \\
&\quad + \frac{1}{2}\int_{\mathbb{R}^d} \mathrm{trace}\Big[\big(\sigma\sigma^\dagger + \sigma^0(\sigma^0)^\dagger\big)(t,v,\mu)\partial_v\partial_\mu \mathcal{U}(t,x,\mu)(v)\Big] d\mu(v) \qquad (5.13) \\
&\quad + \frac{1}{2}\int_{\mathbb{R}^{2d}} \mathrm{trace}\Big[\sigma^0(t,v,\mu)(\sigma^0)^\dagger(t,v',\mu)\partial_\mu^2 \mathcal{U}(t,x,\mu)(v,v')\Big] d\mu(v) d\mu(v') \\
&\quad + \int_{\mathbb{R}^d} \mathrm{trace}\Big[\sigma^0(t,x,\mu)\big(\sigma^0\big)^\dagger(t,v,\mu)\partial_x\partial_\mu \mathcal{U}(t,x,\mu)(v)\Big] d\mu(v) \\
&\quad + F\big(t,x,\mu,\mathcal{U}(t,x,\mu),\partial_x^\sigma \mathcal{U}(t,x,\mu),\partial_{(x,\mu)}^{\sigma^0}\mathcal{U}(t,x,\mu)\big) = 0,
\end{aligned}
$$

with the terminal condition $\mathcal{U}(T,x,\mu) = G(x,\mu)$, where we have let:

$$
\begin{aligned}
&\partial_x^\sigma \mathcal{U}(t,x,\mu) = \sigma^\dagger(t,x,\mu)\partial_x \mathcal{U}(t,x,\mu), \\
&\partial_x^{\sigma^0} \mathcal{U}(t,x,\mu) = (\sigma^0)^\dagger(t,x,\mu)\partial_x \mathcal{U}(t,x,\mu), \\
&\partial_\mu^{\sigma^0} \mathcal{U}(t,x,\mu) = \int_{\mathbb{R}^d} (\sigma^0)^\dagger(t,v,\mu)\partial_\mu \mathcal{U}(t,x,\mu)(v) d\mu(v), \\
&\partial_{(x,\mu)}^{\sigma^0} \mathcal{U}(t,x,\mu) = \partial_x^{\sigma^0} \mathcal{U}(t,x,\mu) + \partial_\mu^{\sigma^0} \mathcal{U}(t,x,\mu).
\end{aligned}
$$

Our first objective in this chapter is to prove that, for small time, and for sufficiently smooth coefficients B, F, G, σ, and σ^0, $\mathcal{U}$ satisfies the required regularity conditions to apply the chain rule. This will prove that, at least for small time, $\mathcal{U}$ is a *classical solution* of (5.13), see Theorem 5.10 below. We shall also prove that it is the unique solution to satisfy suitable growth conditions, see Theorem 5.11. Next, our second objective is to extend the result to time intervals of arbitrary lengths when the coefficients B, F, and G come from a mean field game along the lines of one of the examples discussed in Subsection 4.4.4.

5.1.5 Statements of the Main Results

In order to simplify the analysis, most of the results below are just stated in the case when σ and σ^0 are constant and B and F are independent of z and z^0.

Here is the statement of the first main result of this chapter: whenever T is small enough and the coefficients B, F, G, σ, and σ^0 are smooth enough, the master field is a classical solution of the master equation.

As emphasized in the above *warning*, we shall restrict the proof to the case of coefficients B and F independent of z and z^0, and constant volatilities σ and σ^0. In the Notes & Complements at the end of the chapter, we provide references to papers in which the result is shown to hold under more general conditions.

Although seemingly restrictive, this assumption on the structure of the coefficients will suffice to establish the existence of a smooth solution to the master equation deriving from mean field games, at least whenever the coefficients of the game are of a specific form. See Section 5.4.

Smoothness Conditions on the Coefficients

The smoothness of $\mathcal{U}$ will be established in Sections 5.2 and 5.3 when T is small enough and the following assumptions are in force:

Assumption (Smooth Coefficients Order 2). The functions σ and σ^0 are constant and the coefficients B and F are independent of the variables z and z^0. Moreover, there exist two constants $\Gamma, L \geq 0$ such that, for $h : [0,T] \times \mathbb{R}^q \times \mathcal{P}_2(\mathbb{R}^d) \ni (t,w,\mu) \mapsto h(t,w,\mu) \in \mathbb{R}^l$ being B, F, or G, with $q = d+m$ in the first two cases and $q = d$ in the last case, $l = d$ in the first case, $l = m$ in the second and third cases, and h being also independent of t when equal to G, it holds that:

(A1) $h : [0,T] \times \mathbb{R}^q \times \mathcal{P}_2(\mathbb{R}^d) \to \mathbb{R}^l$ is continuous, twice differentiable with respect to w, and the partial derivatives $\partial_w h : [0,T] \times \mathbb{R}^q \times \mathcal{P}_2(\mathbb{R}^d) \to \mathbb{R}^{l\times q}$ and $\partial^2_w h : [0,T] \times \mathbb{R}^q \times \mathcal{P}_2(\mathbb{R}^d) \to \mathbb{R}^{l\times q\times q}$ are continuous and bounded by L and Γ respectively; moreover, $|h(t,0,\delta_0)| \leq \Gamma$.

(A2) For any $(t,w) \in [0,T] \times \mathbb{R}^q$, the mapping $\mathcal{P}_2(\mathbb{R}^d) \ni \mu \mapsto h(t,w,\mu)$ is fully $\mathcal{C}^2$; moreover, the function $[0,T] \times \mathbb{R}^q \times \mathcal{P}_2(\mathbb{R}^d) \times \mathbb{R}^d \ni (t,w,\mu,v) \mapsto \partial_\mu h(t,w,\mu)(v) \in \mathbb{R}^{l\times d}$ is continuous and bounded by L, and the functions $[0,T] \times \mathbb{R}^q \times \mathcal{P}_2(\mathbb{R}^d) \times \mathbb{R}^d \ni (t,w,\mu,v) \mapsto \partial_v \partial_\mu h(t,w,\mu)(v) \in \mathbb{R}^{l\times d\times d}$ and $[0,T] \times \mathbb{R}^q \times \mathcal{P}_2(\mathbb{R}^d) \times \mathbb{R}^d \times \mathbb{R}^d \ni (t,w,\mu,v,v') \mapsto \partial^2_\mu h(t,w,\mu)(v,v') \in \mathbb{R}^{l\times d\times d}$ are continuous and bounded by Γ.

(A3) The version of the derivative of h with respect to μ used in the $\mathcal{C}^2$ property is such that the global map $[0,T] \times \mathbb{R}^q \times \mathcal{P}_2(\mathbb{R}^d) \times \mathbb{R}^d \ni (t,w,\mu,v) \mapsto \partial_\mu h(t,w,\mu)(v) \in \mathbb{R}^{l\times d}$ is differentiable in w, the partial derivative $\partial_w \partial_\mu h : [0,T] \times \mathcal{P}_2(\mathbb{R}^d) \times \mathbb{R}^d \ni (t,w,\mu,v) \mapsto \partial_w \partial_\mu h(t,w,\mu)(v) \in \mathbb{R}^{l\times d\times q}$ being continuous and bounded by Γ.

(continued)

(A4) For any $t \in [0,T]$, for any $w, w' \in \mathbb{R}^q$, $v, v', \tilde{v}, \tilde{v}' \in \mathbb{R}^d$ and $\mu, \mu' \in \mathcal{P}_2(\mathbb{R}^d)$,

$$
\begin{aligned}
&|\partial_w^2 h(t,w,\mu) - \partial_w^2 h(t,w',\mu')| \leq \Gamma\big(|w-w'| + W_1(\mu,\mu')\big),\\
&|(\partial_v \partial_\mu h, \partial_w \partial_\mu h)(t,w,\mu)(v) - (\partial_v \partial_\mu h, \partial_w \partial_\mu h)(t,w',\mu')(v')|\\
&\qquad\qquad \leq \Gamma\big(|w-w'| + |v-v'| + W_1(\mu,\mu')\big),\\
&|\partial_\mu^2 h(t,w,\mu)(v,\tilde{v}) - \partial_\mu^2 h(t,w',\mu')(v',\tilde{v}')|\\
&\qquad\qquad \leq \Gamma\big(|w-w'| + |v-v'| + |\tilde{v}-\tilde{v}'| + W_1(\mu,\mu')\big),
\end{aligned}
$$

where, in the last two lines, we used the versions of the derivatives provided by **(A2)** and **(A3)**.

(A5) The two constants σ and σ^0 are bounded by Γ.

Notice actually that, in **(A2)** and **(A3)**, there is one and only one globally continuous version of each of the derivatives $\partial_\mu h$, $\partial_v \partial_\mu h$ and $\partial_\mu^2 h$, $\partial_w \partial_\mu h$, see for instance Remarks (Vol I)-5.82 and 4.12.

Observe that assumption **Smooth Coefficients Order 2** subsumes assumption **Conditional MKV FBSDE in Small Time**. Indeed, since $\partial_w h$ and $\partial_\mu h$ are assumed to be bounded by L, we have for all $t \in [0,T]$, $x, x' \in \mathbb{R}^d$, $y, y' \in \mathbb{R}^m$ and $\mu, \mu' \in \mathcal{P}_2(\mathbb{R}^d)$,

$$
\begin{aligned}
&|(B,F)(t,x,\mu,y) - (B,F)(t,x',\mu',y')|\\
&\qquad\qquad \leq L\big(|x-x'| + |y-y'| + W_1(\mu,\mu')\big), \qquad (5.14)\\
&|G(x,\mu) - G(x',\mu')| \leq L\big(|x-x'| + W_1(\mu,\mu')\big).
\end{aligned}
$$

The fact the above bounds hold with respect to the 1-Wasserstein distance W_1 in lieu of the 2-Wasserstein distance W_2 as in assumption **Conditional MKV FBSDE in Small Time** will play a key role in the analysis. It says that the coefficients B and F can be extended to the whole $[0,T] \times \mathbb{R}^d \times \mathcal{P}_1(\mathbb{R}^d) \times \mathbb{R}^m \supset [0,T] \times \mathbb{R}^d \times \mathcal{P}_2(\mathbb{R}^d) \times \mathbb{R}^m$, and similarly for G. However, we shall not make use of any of these extensions.

Also, for the same reasons as in Remark 4.16, Schwarz' theorem can be applied to h as in assumption **Smooth Coefficients Order 2**: for any $(t,w) \in [0,T] \times \mathbb{R}^q$, the mapping $\mathcal{P}_2(\mathbb{R}^d) \ni \mu \mapsto \partial_w h(t,w,\mu)$ is L-differentiable with respect to μ and $\partial_\mu \partial_w h(t,w,\mu)(\cdot) = [\partial_w \partial_\mu h(t,w,\mu)(\cdot)]^\dagger$. In particular, we can find a version of $\partial_\mu \partial_w h(t,w,\mu)(\cdot)$ which is bounded by Γ and which satisfies **(A4)**. As a consequence, all the second derivatives of h are bounded and we easily deduce that $\partial_w h$ and $\partial_\mu h$ satisfy the same Lipschitz properties as $\partial_w^2 h$ and $\partial_v \partial_\mu h$ in **(A4)**. In particular, h satisfies assumption **Joint Chain Rule Common Noise** in Subsection 4.3.4.

Obviously, assumption **Smooth Coefficients Order 2** is rather restrictive, though the mollification procedure used in the proof of Lemma 4.15, see (4.31) and the lines after, provides a systematic way to construct coefficients satisfying this assumption, see also Lemma (Vol I)-5.94.

Finally, observe that the distinction between L and Γ in assumption **Smooth Coefficients Order 2** is important: L will dictate the length of the interval on which we shall prove the existence of a classical solution to the master equation, while Γ will not play any role in this regard.

Statements

Inspired by assumption **Smooth Coefficients Order 2**, we let:

Definition 5.9 *For an integer $m \geq 1$, we denote by $\mathfrak{S}_m$ the space of functions $V : [0,T] \times \mathbb{R}^d \times \mathcal{P}_2(\mathbb{R}^d) \ni (t,x,\mu) \mapsto V(t,x,\mu) \in \mathbb{R}^m$ for which we can find a constant $C \geq 0$ such that:*

(i) V satisfies the same properties as h in assumption **Smooth Coefficients Order 2**, *but with L and Γ replaced by C and with $q = d$ and $l = m$, namely:*

(i.a) V is continuous and is twice differentiable with respect to x, and $\partial_x V$ and $\partial_x^2 V$ are continuous and bounded by C; moreover, $|V(t,0,\delta_0)| \leq C$.

(i.b) For any $(t,x) \in [0,T]\times\mathbb{R}^d$, the map $\mathcal{P}_2(\mathbb{R}^d) \ni \mu \mapsto V(t,x,\mu)$ is fully $\mathcal{C}^2$; the function $[0,T]\times\mathbb{R}^d\times\mathcal{P}_2(\mathbb{R}^d)\times\mathbb{R}^d \ni (t,x,\mu,v) \mapsto \partial_\mu V(t,w,\mu)(v) \in \mathbb{R}^{d\times d}$ is continuous and bounded by C, and the functions $[0,T] \times \mathbb{R}^d \times \mathcal{P}_2(\mathbb{R}^d) \times \mathbb{R}^d \ni (t,x,\mu,v) \mapsto \partial_v \partial_\mu V(t,x,\mu)(v) \in \mathbb{R}^{m\times d\times d}$ and $[0,T] \times \mathbb{R}^d\times\mathcal{P}_2(\mathbb{R}^d)\times\mathbb{R}^d\times\mathbb{R}^d \ni (t,w,\mu,v,v') \mapsto \partial_\mu^2 V(t,x,\mu)(v,v') \in \mathbb{R}^{m\times d\times d}$ are continuous and bounded by C.

(i.c) The map $[0,T] \times \mathbb{R}^d \times \mathcal{P}_2(\mathbb{R}^d) \times \mathbb{R}^d \ni (t,x,\mu,v) \mapsto \partial_\mu V(t,x,\mu)(v) \in \mathbb{R}^{m\times d}$ is differentiable in x, the partial derivative $\partial_x\partial_\mu V : [0,T]\times\mathcal{P}_2(\mathbb{R}^d)\times\mathbb{R}^d \ni (t,x,\mu,v) \mapsto \partial_x\partial_\mu V(t,w,\mu)(v) \in \mathbb{R}^{m\times d\times d}$ being continuous and bounded by C.

(i.d) For any $t \in [0,T]$, for any $x,x' \in \mathbb{R}^d$, $v,v',\tilde{v},\tilde{v}' \in \mathbb{R}^d$ and $\mu,\mu' \in \mathcal{P}_2(\mathbb{R}^d)$,

$$|\partial_x^2 V(t,x,\mu) - \partial_x^2 V(t,x',\mu')| \leq C\big(|x-x'| + W_1(\mu,\mu')\big),$$

$$\begin{aligned}&|(\partial_v\partial_\mu V, \partial_x\partial_\mu V)(t,x,\mu)(v) - (\partial_v\partial_\mu V, \partial_x\partial_\mu V)(t,x',\mu')(v')| \\ &\qquad\qquad \leq C\big(|x-x'| + |v-v'| + W_1(\mu,\mu')\big),\end{aligned}$$

$$\begin{aligned}&|\partial_\mu^2 V(t,x,\mu)(v,\tilde{v}) - \partial_\mu^2 V(t,x',\mu')(v',\tilde{v}')| \\ &\qquad\qquad \leq C\big(|x-x'| + |v-v'| + |\tilde{v}-\tilde{v}'| + W_1(\mu,\mu')\big).\end{aligned}$$

(ii) For any $x \in \mathbb{R}^d$ and $\mu \in \mathcal{P}_2(\mathbb{R}^d)$, the function $[0,T] \ni t \mapsto V(t,x,\mu)$ is continuously differentiable and $\partial_t V$ is at most of linear growth in (x,μ), uniformly in t, and is jointly continuous in all the variables.

The set $\mathfrak{S}_m$ is the space we use below for investigating existence and uniqueness of a solution to (5.13). In short time, our main result now takes the following form:

Theorem 5.10 *Under assumption* **Smooth Coefficients Order 2**, *there exists a constant $c = c(L)$, c not depending upon Γ, such that, for $T \leq c$, the function $\mathcal{U}$ defined in* (5.7) *is in $\mathfrak{S}_m$ and satisfies the corresponding form of the PDE* (5.13) *when the coefficients are independent of the variables z and z^0, and σ and σ^0 are constant.*

Furthermore, uniqueness also holds in the class $\mathfrak{S}_m$.

Theorem 5.11 *Under assumption* **Conditional MKV FBSDE in Small Time**, *and provided that σ and σ^0 are bounded, there exists at most one solution to the PDE* (5.13) *in the class $\mathfrak{S}_m$, whatever the length T is.*

Notice that Theorem 5.11 still holds under the weaker assumption **Conditional MKV FBSDE in Small Time**. In particular, σ and σ^0 may not be constant and B and F may depend on (z, z^0).

Extension to arbitrary time intervals will be discussed in Section 5.4. The principle for extending the result from small to long horizons has been already discussed in Chapter (Vol I)-4. Basically, it is still the same: the goal is to prove that, throughout the induction used to extend the result, the master field remains in a space of admissible boundary conditions for which the length of the interval of solvability can be bounded from below. Generally speaking, this requires, first to isolate a class of functions on $\mathbb{R}^d \times \mathcal{P}_2(\mathbb{R}^d)$ in which the master field remains at any time, and second, to control the Lipschitz constant of the master field, uniformly along the induction. In our case, the Lipschitz constant means the Lipschitz constant in both the space variable and the measure argument. In Section 5.4, we give two examples, taken from Subsection 4.4.4, for which the Lipschitz constant of the master field can indeed be controlled. In both cases, the forward-backward system derives from a mean field game.

5.2 First-Order Smoothness for Small Time Horizons

The purpose of this section is to prove that, in small time, the mapping $\mathcal{U}$ given in Definition 5.2 satisfies the smoothness property required to apply the chain rule. We prove this by showing that the stochastic flows defined in (5.2) and (5.6) are differentiable with respect to $\xi \in L^2(\Omega^1, \mathcal{F}_t^1, \mathbb{P}^1; \mathbb{R}^d)$, $x \in \mathbb{R}^d$ and $\mu \in \mathcal{P}_2(\mathbb{R}^d)$.

Since we now assume that σ and σ^0 are constant and B and F are independent of Meme Remarque concernant z^0, we rewrite these equations for the sake of definiteness:

$$\begin{cases} dX_s^{t,\xi} = B\big(s, X_s^{t,\xi}, \mathcal{L}^1(X_s^{t,\xi}), Y_s^{t,\xi}\big)ds + \sigma dW_s + \sigma^0 dW_s^0, \\ dY_s^{t,\xi} = -F\big(s, X_s^{t,x,\mu}, \mathcal{L}^1(X_s^{t,\xi}), Y_s^{t,\xi}\big)ds + Z_s^{t,\xi} dW_s \\ \qquad\qquad + Z_s^{0;t,\xi} dW_s^0, \quad s \in [t,T], \\ X_t^{t,\xi} = \xi, \quad Y_T^{t,\xi} = G\big(X_T^{t,\xi}, \mathcal{L}^1(X_T^{t,\xi})\big), \end{cases}$$

and

$$\begin{cases} dX_s^{t,x,\mu} = B\big(s, X_s^{t,x,\mu}, \mathcal{L}^1(X_s^{t,\xi}), Y_s^{t,x,\mu}\big)ds + \sigma dW_s + \sigma^0 dW_s^0, \\ dY_s^{t,x,\mu} = -F\big(s, X_s^{t,x,\mu}, \mathcal{L}^1(X_s^{t,\xi}), Y_s^{t,x,\mu}\big)ds \\ \qquad\qquad + Z_s^{t,x,\mu} dW_s + Z_s^{0;t,x,\mu} dW_s^0, \quad s \in [t,T], \\ X_t^{t,x,\mu} = x, \quad Y_T^{t,x,\mu} = G\big(X_T^{t,x,\mu}, \mathcal{L}^1(X_T^{t,\xi})\big), \end{cases}$$

where $\mu = \mathcal{L}^1(\xi)$.

Taking advantage of the fact that the coefficients are independent of z and z^0, we shall drop the processes $\boldsymbol{Z}^{t,\xi}$, $\boldsymbol{Z}^{0;t,\xi}$, $\boldsymbol{Z}^{t,x,\mu}$ and $\boldsymbol{Z}^{0;t,x,\mu}$ in the above equations. Instead, using the convenient notations $\mathbb{E}_s = \mathbb{E}[\cdot \mid \mathcal{F}_s]$ and $\mathbb{E}_s^0 = \mathbb{E}[\cdot \mid \mathcal{F}_s^0]$, we shall write:

$$\begin{cases} X_s^{t,\xi} = \xi + \displaystyle\int_t^s B\big(r, X_r^{t,\xi}, \mathcal{L}^1(X_r^{t,\xi}), Y_r^{t,\xi}\big)dr + \sigma W_s + \sigma^0 W_s^0, \\ Y_s^{t,\xi} = \mathbb{E}_s\bigg[\displaystyle\int_s^T F\big(r, X_r^{t,\xi}, \mathcal{L}^1(X_r^{t,\xi}), Y_r^{t,\xi}\big)dr + G\big(X_T^{t,\xi}, \mathcal{L}^1(X_T^{t,\xi})\big)\bigg], \end{cases} \tag{5.15}$$

for $s \in [t,T]$, and

$$\begin{cases} X_s^{t,x,\mu} = x + \displaystyle\int_t^s B\big(r, X_r^{t,x,\mu}, \mathcal{L}^1(X_r^{t,\xi}), Y_r^{t,x,\mu}\big)dr + \sigma W_s + \sigma^0 W_s^0, \\ Y_s^{t,x,\mu} = \mathbb{E}_s\bigg[\displaystyle\int_s^T F\big(r, X_r^{t,x,\mu}, \mathcal{L}^1(X_r^{t,\xi}), Y_r^{t,x,\mu}\big)dr + G\big(X_T^{t,x,\mu}, \mathcal{L}^1(X_T^{t,\xi})\big)\bigg], \end{cases} \tag{5.16}$$

for $s \in [t,T]$, where $\mu = \mathcal{L}^1(\xi)$. Obviously, the mapping $\mathcal{P}_2(\mathbb{R}^d) \ni \mu \mapsto (\boldsymbol{X}^{t,x,\mu}, \boldsymbol{Y}^{t,x,\mu}) \in \mathbb{S}^2([t,T];\mathbb{R}^d) \times \mathbb{S}^2([t,T];\mathbb{R}^m)$ may be canonically lifted into $L^2(\Omega^1, \mathcal{F}_t^1, \mathbb{P}^1; \mathbb{R}^d) \ni \xi \mapsto (\boldsymbol{X}^{t,x,\xi}, \boldsymbol{Y}^{t,x,\xi})$, the image of which just reads as a new notation for the solution of the FBSDE (5.16). This new notation will be useful as it permits to keep track of the dependence upon the original random variable ξ which has μ as distribution.

Throughout the proof, we use the notation $\|\chi\|_p$ for the L^p norm, $p \in [1,\infty]$, of a random variable $\chi \in L^p(\Omega^1, \mathcal{F}^1, \mathbb{P}^1; \mathbb{R}^d)$. Also, we assume $T \leq 1$.

5.2.1 Warm-Up: Linearization

We start with a weaker version of assumption **Smooth Coefficients Order 2**, in which coefficients are just required to be once continuously differentiable.

François: when we say that h could be B of F, shouldn't we mention 'up to re-ordering of the variables x, μ and y and regrouping (x, y) into a single variable w'. If we do that, we may want to do it earlier as well.

Assumption (Smooth Coefficients Order 1). There exist two constants $\Gamma, L \geq 0$ such that, for $h : [0, T] \times \mathbb{R}^q \times \mathcal{P}_2(\mathbb{R}^d) \ni (t, w, \mu) \mapsto h(t, w, \mu) \in \mathbb{R}^l$ being B, F, or G with $q = d + m$ in the first two cases and $q = d$ in the last cases, $l = d$ in the first case, $l = m$ in the second and third cases, and h being independent of t when equal to G, it holds that:

(A1) h is once differentiable with respect to w and the partial derivative $\partial_w h : [0, T] \times \mathbb{R}^q \times \mathcal{P}_2(\mathbb{R}^d) \to \mathbb{R}^{l \times q}$ is continuous and bounded by L.

(A2) For any $(t, w) \in [0, T] \times \mathbb{R}^q$, the mapping $\mathcal{P}_2(\mathbb{R}^d) \ni \mu \mapsto h(t, w, \mu)$ is L-differentiable; for all $(t, w, \mu) \in [0, T] \times \mathbb{R}^q \times \mathcal{P}_2(\mathbb{R}^d)$, the function $\mathbb{R}^d \ni v \mapsto \partial_\mu h(t, w, \mu)(v)$ has a version such that the mapping $[0, T] \times \mathbb{R}^q \times \mathcal{P}_2(\mathbb{R}^d) \times \mathbb{R}^d \ni (t, w, \mu, v) \mapsto \partial_\mu h(t, w, \mu)(v) \in \mathbb{R}^{l \times d}$ is continuous and bounded by L.

(A3) For any $t \in [0, T]$, the function $\mathbb{R}^q \times \mathcal{P}_2(\mathbb{R}^d) \ni (w, \mu) \mapsto \partial_w h(t, w, \mu)$ is Γ-Lipschitz, namely, for all $w, w' \in \mathbb{R}^q$ and all $\mu, \mu' \in \mathcal{P}_2(\mathbb{R}^d)$,

$$|\partial_w h(t, w, \mu) - \partial_w h(t, w', \mu')| \leq \Gamma\big(|w - w'| + W_1(\mu, \mu')\big).$$

Similarly, for all $v, v' \in \mathbb{R}^d$,

$$|\partial_\mu h(t, w, \mu)(v) - \partial_\mu h(t, w', \mu')(v')|] \leq \Gamma\big(|w - w'| + |v - v'| + W_1(\mu, \mu')\big).$$

(A4) $|h(t, 0, \delta_0)|$ is bounded by Γ.

From an argument already used in Subsection 5.1.5, we see that assumption **Smooth Coefficients Order 1** implies (5.14).

Revisiting the Stability Estimates for the Original FBSDE System

As a consequence of Theorem 5.4, we have the following bounds:

Lemma 5.12 *There exist two constants $c = c(L) > 0$ and $C \geq 0$ such that, for $T \leq c$, for any $t \in [0, T]$, $x \in \mathbb{R}^d$, $\xi \in L^2(\Omega^1, \mathcal{F}_t^1, \mathbb{P}^1; \mathbb{R}^d)$ and $\mu \in \mathcal{P}_2(\mathbb{R}^d)$,*

$$\mathbb{E}_t\Big[\sup_{t\le s\le T}\big(|X_s^{t,\xi}|^2+|Y_s^{t,\xi}|^2\big)\Big]^{1/2}\le C\big(1+|\xi|+\|\xi\|_2\big),$$
$$\mathbb{E}\Big[\sup_{t\le s\le T}\big(|X_s^{t,x,\mu}|^2+|Y_s^{t,x,\mu}|^2\big)\Big]^{1/2}\le C\Big(1+|x|+\|\xi\|_2\Big), \tag{5.17}$$

and, for any $x'\in\mathbb{R}^d$, $\xi'\in L^2(\Omega^1,\mathcal{F}_t^1,\mathbb{P}^1;\mathbb{R}^d)$ *and* $\mu'\in\mathcal{P}_2(\mathbb{R}^d)$,

$$\begin{aligned}\mathbb{E}_t\Big[\sup_{t\le s\le T}\big(|X_s^{t,\xi}-X_s^{t,\xi'}|^2+|Y_s^{t,\xi}-Y_s^{t,\xi'}|^2\big)\Big]^{1/2}&\\ \le C\big(|\xi-\xi'|+\|\xi-\xi'\|_2\big),&\\ \mathbb{E}\Big[\sup_{t\le s\le T}\big(|X_s^{t,x,\mu}-X_s^{t,x',\mu'}|^2+|Y_s^{t,x,\mu}-Y_s^{t,x,\mu'}|^2\big)\Big]^{1/2}&\\ \le C\big(|x-x'|+W_2(\mu,\mu')\big).&\end{aligned} \tag{5.18}$$

Observe in fact that the application of Theorem 5.4 is not entirely licit since $(\boldsymbol{X}^{t,x,\mu},\boldsymbol{Y}^{t,x,\mu},\boldsymbol{Z}^{t,x,\mu},\boldsymbol{Z}^{0;t,x,\mu})$ is not the solution of a McKean-Vlasov FBSDE, the flow $(\mathcal{L}^1(X_s^{t,\xi}))_{t\le s\le T}$ in the coefficients reading as a parameter. However, the proof of Theorem 5.4 may be easily adapted. This will be clarified in the proof of Proposition 5.13 below.

As made clear by the statement of Lemma 5.12, it will be useful to use the notation $\mathbb{S}^2([t,T];\mathbb{R}^q)$ for the space of continuous and $(\mathcal{F}_s)_{t\le s\le T}$-adapted process with values in $\mathbb{R}^q$.

In the statement above, the notation $c=c(L)$ emphasizes the fact that c only depends on the Lipschitz constant L of assumption **Smooth Coefficients Order 1**. In contrast, the constant C may depend upon the other parameter Γ appearing in (**A3**) and (**A4**), but there is no need to keep track of this dependence for our current purposes.

We now make a crucial observation: the two bounds in (5.18) are not optimal. Thanks to the fact that the coefficients σ and σ^0 are constant and to the fact that the derivatives of the coefficients in the direction μ are bounded in L^∞, the two bounds (5.18) in expectations can be turned into pointwise bounds.

Proposition 5.13 *There exist two constants* $c=c(L)>0$ *and* $C\ge 0$ *such that, for* $T\le c$, *for any* $t\in[0,T]$, $x,x'\in\mathbb{R}^d$, $\xi,\xi'\in L^2(\Omega^1,\mathcal{F}_t^1,\mathbb{P}^1;\mathbb{R}^d)$ *and* $\mu,\mu'\in\mathcal{P}_2(\mathbb{R}^d)$, *with* $\mathbb{P}$*-probability* 1 *it holds:*

$$\sup_{t\le s\le T}\big(|X_s^{t,\xi}-X_s^{t,\xi'}|+|Y_s^{t,\xi}-Y_s^{t,\xi'}|\big)\le C\big(|\xi-\xi'|+\|\xi-\xi'\|_1\big),$$
$$\sup_{t\le s\le T}\big(|X_s^{t,x,\mu}-X_s^{t,x',\mu'}|+|Y_s^{t,x,\mu}-Y_s^{t,x',\mu'}|\big)\le C\big(|x-x'|+W_1(\mu,\mu')\big).$$

Proof. The proof relies on two main observations. The first one is to notice that the coefficients are L-Lipschitz continuous in the direction μ with respect to the 1-Wasserstein distance W_1, which is of course stronger than the regularity assumption used in Theorem 5.4, where coefficients are just assumed to be Lipschitz continuous with respect to W_2. The second observation is that σ and σ^0 are constant.

First Step. Based on these observations, we compute the difference between the forward components $(X_s^{t,\xi})_{t\le s\le T}$ and $(X_s^{t,\xi'})_{t\le s\le T}$. By Gronwall's lemma we get, for $T \le 1$, for a constant C only depending on L and for $t \le s \le T$,

$$\begin{aligned}\sup_{s\le u\le T} &|X_u^{t,\xi} - X_u^{t,\xi'}| \\ &\le |X_s^{t,\xi} - X_s^{t,\xi'}| + C\int_s^T \Big(|Y_u^{t,\xi} - Y_u^{t,\xi'}| + \mathbb{E}^1\big[|X_u^{t,\xi} - X_u^{t,\xi'}|\big]\Big)du.\end{aligned} \tag{5.19}$$

Taking the expectation under $\mathbb{P}^1$, we obtain, for $T \le c$, with $c > 0$ only depending on L,

$$\begin{aligned}\sup_{s\le u\le T} |X_u^{t,\xi} - X_u^{t,\xi'}| &\le |X_s^{t,\xi} - X_s^{t,\xi'}| + \mathbb{E}^1\big[|X_s^{t,\xi} - X_s^{t,\xi'}|\big] \\ &\quad + C\int_s^T \Big(|Y_u^{t,\xi} - Y_u^{t,\xi'}| + \mathbb{E}^1\big[|Y_u^{t,\xi} - Y_u^{t,\xi'}|\big]\Big)du,\end{aligned} \tag{5.20}$$

where we allow the constant C to increase from line to line.

Second Step. For $t \le r \le s \le T$, we now write:

$$\begin{aligned}\mathbb{E}_r\big[|Y_s^{t,\xi} - Y_s^{t,\xi'}|\big] &\le \mathbb{E}_r\Big[\big|G(X_T^{t,\xi}, \mathcal{L}^1(X_T^{t,\xi})) - G(X_T^{t,\xi'}, \mathcal{L}^1(X_T^{t,\xi'}))\big|\Big] \\ &\quad + \mathbb{E}_r\Big[\int_s^T \big|F(u, X_u^{t,\xi}, \mathcal{L}^1(X_u^{t,\xi}), Y_u^{t,\xi}) - F(u, X_u^{t,\xi'}, \mathcal{L}^1(X_u^{t,\xi'}), Y_u^{t,\xi'})\big|du\Big],\end{aligned}$$

from which we get:

$$\begin{aligned}\mathbb{E}_r\big[|Y_s^{t,\xi} - Y_s^{t,\xi'}|\big] &\le C\Big(\sup_{r\le u\le T} \mathbb{E}_r\Big[|X_u^{t,\xi} - X_u^{t,\xi'}| + \mathbb{E}^1\big[|X_u^{t,\xi} - X_u^{t,\xi'}|\big]\Big]\Big) \\ &\quad + C\int_s^T \mathbb{E}_r\big[|Y_u^{t,\xi} - Y_u^{t,\xi'}|\big]du.\end{aligned}$$

By (5.20), we get:

$$\begin{aligned}\mathbb{E}_r\big[|Y_s^{t,\xi} - Y_s^{t,\xi'}|\big] &\le C\Big(|X_r^{t,\xi} - X_r^{t,\xi'}| + \mathbb{E}^1\big[|X_r^{t,\xi} - X_r^{t,\xi'}|\big]\Big) \\ &\quad + C\int_s^T \mathbb{E}_r\Big[|Y_u^{t,\xi} - Y_u^{t,\xi'}| + \mathbb{E}^1\big[|Y_u^{t,\xi} - Y_u^{t,\xi'}|\big]\Big]du.\end{aligned}$$

Taking the expectation under $\mathbb{P}^1$ and appealing to Gronwall's lemma once again, we get:

$$\sup_{r\le s\le T} \mathbb{E}_r\big[|Y_s^{t,\xi} - Y_s^{t,\xi'}|\big] \le C\Big(|X_r^{t,\xi} - X_r^{t,\xi'}| + \mathbb{E}^1\big[|X_r^{t,\xi} - X_r^{t,\xi'}|\big]\Big),$$

where we used the fact that $\mathbb{E}^1[\mathbb{E}[\cdot\,|\,\mathcal{F}_r]] = \mathbb{E}[\mathbb{E}^1[\cdot]|\mathcal{F}_r]$, see Lemma 5.16 at the end of the subsection for a proof. In particular, choosing $r = s$ in the above left-hand side, we get, for T small enough:

$$|Y_s^{t,\xi} - Y_s^{t,\xi'}| \leq C\Big(|X_s^{t,\xi} - X_s^{t,\xi'}| + \mathbb{E}^1\big[|X_s^{t,\xi} - X_s^{t,\xi'}|\big]\Big).$$

Plugging this estimate into (5.20), we get:

$$\sup_{s\leq u\leq T} |X_u^{t,\xi} - X_u^{t,\xi'}| \leq C\Big(|X_s^{t,\xi} - X_s^{t,\xi'}| + \mathbb{E}^1\big[|X_s^{t,\xi} - X_s^{t,\xi'}|\big]\Big). \tag{5.21}$$

Choosing $s = t$, we easily complete the proof of the first inequality. The second inequality easily follows from the same strategy. □

The strategy for investigating the derivatives of the solutions to (5.15) and (5.16) is standard. We identify the derivatives with the solutions of linearized systems, obtained by formal differentiation of the original equation. For that reason, the analysis of the differentiability relies on some preliminary stability estimates for linear FBSDEs of the conditional McKean-Vlasov type.

Linearized System

Generally speaking, we are dealing with a linear FBSDE of the form:

$$\begin{aligned}
\mathcal{X}_s &= \eta + \int_t^s \delta B\big(r, \theta_r, \tilde{X}_r'\big)\big(\vartheta_r, \tilde{\mathcal{X}}_r'\big)dr, \\
\mathcal{Y}_s &= \mathbb{E}_s\bigg[\int_s^T \delta F\big(r, \theta_r, \tilde{X}_r'\big)\big(\vartheta_r, \tilde{\mathcal{X}}_r'\big)dr + \delta G\big(X_T, \tilde{X}_T'\big)\big(\mathcal{X}_T, \tilde{\mathcal{X}}_T'\big)\bigg],
\end{aligned} \tag{5.22}$$

for $s \in [t, T]$, where η is an initial condition in $L^2(\Omega^1, \mathcal{F}_t^1, \mathbb{P}^1; \mathbb{R}^d)$, $\boldsymbol{\theta} = (\boldsymbol{X}, \boldsymbol{Y})$ and $\boldsymbol{\theta}' = (\boldsymbol{X}', \boldsymbol{Y}')$ are solutions of (5.15) or (5.16), $\boldsymbol{\vartheta} = (\boldsymbol{\mathcal{X}}, \boldsymbol{\mathcal{Y}})$ denotes the unknowns in the above equation and $\boldsymbol{\mathcal{X}}'$ is an auxiliary process, which may be $\boldsymbol{\mathcal{X}}$ itself (in which case it is unknown). As usual, the symbol $\sim$ denotes an independent copy on $(\tilde{\Omega}^1, \tilde{\mathcal{F}}^1, \tilde{\mathbb{P}}^1)$. The processes $\boldsymbol{X}$, $\boldsymbol{X}'$, $\boldsymbol{\mathcal{X}}$ and $\boldsymbol{\mathcal{X}}'$ have the same dimension, the same being true for the processes $\boldsymbol{Y}$, $\boldsymbol{Y}'$ and $\boldsymbol{\mathcal{Y}}$. The coefficients read as measurable mappings:

$$\begin{aligned}
\delta B : [0,T] \times \Big[\big(\mathbb{R}^d \times \mathbb{R}^m\big) \times L^2(\tilde{\Omega}^1, \tilde{\mathcal{F}}^1, \tilde{\mathbb{P}}^1; \mathbb{R}^d)\Big]^2 &\to \mathbb{R}^d \\
\big(r, ((x,y), \tilde{X}), ((u,v), \tilde{U})\big) &\mapsto \delta B\big(r, (x,y), \tilde{X}\big)\big((u,v), \tilde{U}\big), \\
\delta F : [0,T] \times \Big[\big(\mathbb{R}^d \times \mathbb{R}^m\big) \times L^2(\tilde{\Omega}^1, \tilde{\mathcal{F}}^1, \tilde{\mathbb{P}}^1; \mathbb{R}^d)\Big]^2 &\to \mathbb{R}^m \\
\big(r, ((x,y), \tilde{X}), ((u,v), \tilde{U})\big) &\mapsto \delta F\big(r, (x,y), \tilde{X}\big)\big((u,v), \tilde{U}\big), \\
\delta G : \Big[\mathbb{R}^d \times L^2(\tilde{\Omega}^1, \tilde{\mathcal{F}}^1, \tilde{\mathbb{P}}^1; \mathbb{R}^d)\Big]^2 &\to \mathbb{R}^m \\
\big(((x,y), \tilde{X}), ((u,v), \tilde{U})\big) &\mapsto \delta G\big((x,y), \tilde{X}\big)\big((u,v), \tilde{U}\big).
\end{aligned}$$

Here is an example for δB, δF, and δG.

Example 5.14. As a typical example for the coefficients δB, δF, and δG, we may think of the derivatives, with respect to some parameter λ, of the original coefficients B, F, and G when computed along some couple $\boldsymbol{\theta}^\lambda = (\boldsymbol{X}^\lambda, \boldsymbol{Y}^\lambda)$ solving (5.1). As for the parameterization by λ, we may think of the parameterization with respect to the initial condition which is applied to the entire system.

The exact form of the coefficients δB, δF, and δG can then be derived by replacing B, F, and G by a generic continuously differentiable Lipschitz function $h : [0,T] \times (\mathbb{R}^d \times \mathbb{R}^m) \times \mathcal{P}_2(\mathbb{R}^d) \to \mathbb{R}^l$ as in assumption **Smooth Coefficients Order 1**, and by applying the following procedure. Given such a generic h, we consider a process of the form:

$$\Big(h\big(r, \theta_r^\lambda, \mathcal{L}^1(X_r^\lambda)\big)\Big)_{r\in[t,T]},$$

where $\mathbb{R} \ni \lambda \mapsto (\theta_r^\lambda)_{r\in[t,T]} \in \mathbb{S}^2([t,T];\mathbb{R}^d) \times \mathbb{S}^2([t,T];\mathbb{R}^m)$ is differentiable with respect to λ, with derivatives taken in the aforementioned space:

$$\theta^\lambda_{r\,|\lambda=0} = \theta_r, \qquad \frac{d}{d\lambda}_{|\lambda=0} \theta_r^\lambda = \vartheta_r, \qquad r \in [t,T],$$

the process $(\vartheta_r)_{r\in[t,T]}$ taking its values in $\mathbb{R}^d \times \mathbb{R}^m$ and, for the moment, having nothing to do with the solution of (5.22). Then, it is easy to check that the mapping $\mathbb{R} \ni \lambda \mapsto (h(r, \theta_r^\lambda, \mathcal{L}^1(X_r^\lambda)))_{r\in[t,T]} \in \mathbb{S}^2([t,T];\mathbb{R}^q)$ is differentiable and that the derivative reads as follows:

$$\begin{aligned} &\partial h(r, \theta_r, \tilde{X}_r)(\vartheta_r, \tilde{\mathcal{X}}_r) \\ &\quad = \partial_w h\big(r, \theta_r, \mathcal{L}^1(X_r)\big)\vartheta_r + \tilde{\mathbb{E}}^1\big[\partial_\mu h\big(r, \theta_r, \mathcal{L}^1(X_r)\big)(\tilde{X}_r)\tilde{\mathcal{X}}_r\big], \end{aligned} \tag{5.23}$$

where $\mathcal{X}_r$ denotes the first d-dimensional component of $\mathcal{V}_r$. Of course, if h only acts on $(X_r, \mathcal{L}^1(X_r))_{r\in[t,T]}$ instead of $(\theta_r, \mathcal{L}^1(X_r))_{r\in[t,T]}$, then differentiability is understood accordingly.

In Example 5.14, the coefficients δB, δF, and δG are obtained by replacing h by B, F, and G and by computing ∂B, ∂F and ∂G accordingly. We then have $(\delta B, \delta F, \delta G) = (\partial B, \partial F, \partial G)$.

Leaving Example 5.14 and going back to the general case, we apply the same procedure: in order to specify the form of δB, δF, and δG together with the assumptions they satisfy, it suffices to make explicit the generic form of a function δH that may be δB, δF, or δG and to detail the assumptions it satisfies. Given a square-integrable process $\boldsymbol{V} = (V_r)_{r\in[t,T]}$, possibly matching $(X_r)_{r\in[t,T]}$ or $(\theta_r)_{r\in[t,T]}$, together with another square-integrable process $(\mathcal{V}_r)_{r\in[t,T]}$, possibly matching $(\mathcal{X}_r)_{r\in[t,T]}$ or $(\vartheta_r)_{r\in[t,T]}$, we assume that $\delta H(r, V_r, \tilde{X}_r')$ acts on $(\mathcal{V}_r, \tilde{\mathcal{X}}_r')$ in the following way:

$$\delta H\big(r, V_r, \tilde{X}_r'\big)\big(\mathcal{V}_r, \tilde{\mathcal{X}}_r'\big) = \delta H_\ell\big(r, V_r, \tilde{X}_r'\big)\big(\mathcal{V}_r, \tilde{\mathcal{X}}_r'\big) + \delta H_a(r), \tag{5.24}$$

where $\delta H_a(r)$ is square-integrable and $\delta H_\ell(r, V_r, \tilde{X}_r')$ acts linearly on $(\mathcal{V}_r, \tilde{\mathcal{X}}_r')$ in the following sense:

$$\delta H_\ell(r, V_r, \tilde{X}_r')(\mathcal{V}_r, \tilde{\mathcal{X}}_r') = \delta h_\ell(r, V_r, \tilde{X}_r')\mathcal{V}_r + \tilde{\mathbb{E}}^1\big[\delta \tilde{H}_\ell(r, V_r, \tilde{X}_r')\tilde{\mathcal{X}}_r'\big]. \tag{5.25}$$

Here $\delta h_\ell(r, \cdot)$ and $\delta \tilde{H}_\ell(r, \cdot)$ are maps from $\mathbb{R}^q \times L^2(\tilde{\Omega}^1, \tilde{\mathcal{F}}^1, \tilde{\mathbb{P}}^1; \mathbb{R}^d)$ into $\mathbb{R}^{l \times q}$ and from $\mathbb{R}^q \times L^2(\tilde{\Omega}^1, \tilde{\mathcal{F}}^1, \tilde{\mathbb{P}}^1; \mathbb{R}^d)$ into $L^2(\tilde{\Omega}^1, \tilde{\mathcal{F}}^1, \tilde{\mathbb{P}}^1; \mathbb{R}^{l \times d})$ respectively, for suitable q and l. Moreover, there exist two constants $\Gamma, L \geq 0$ such that, for $r \in [0, T]$, $w, w' \in \mathbb{R}^q$ and $\tilde{X}, \tilde{X}' \in L^2(\tilde{\Omega}^1, \tilde{\mathcal{F}}^1, \tilde{\mathbb{P}}^1; \mathbb{R}^d)$,

$$\big|\delta h_\ell(r, w, \tilde{X})\big| \leq L, \quad |\delta \tilde{H}_\ell(r, w, \tilde{X})| \leq L, \tag{5.26}$$

the second bound holding with $\tilde{\mathbb{P}}^1$-probability 1, and,

$$\begin{aligned}
&\big|\delta h_\ell(r, w, \tilde{X}) - \delta h_\ell(r, w', \tilde{X}')\big| \leq \Gamma\Big[|w - w'| + \tilde{\mathbb{E}}^1\big[|\tilde{X} - \tilde{X}'|\big]\Big], \\
&|\delta \tilde{H}_\ell(r, w, \tilde{X}) - \delta \tilde{H}_\ell(r, w', \tilde{X}')| \\
&\quad \leq \Gamma\Big[|w - w'| + |\tilde{X} - \tilde{X}'| + \tilde{\mathbb{E}}^1\big[|\tilde{X} - \tilde{X}'|\big]\Big],
\end{aligned} \tag{5.27}$$

the second bound holding with $\tilde{\mathbb{P}}^1$-probability 1. Conditions (5.26) and (5.27) must be compared with assumption **Smooth Coefficients Order 1**, the constant L in (5.26) playing the role of L in assumption **Smooth Coefficients Order 1**. It is worth mentioning that the constant L has a major role in the sequel as it dictates the size of the time interval on which all the estimates derived in this section hold.

The comparison between (5.26)–(5.27) and assumption **Smooth Coefficients Order 1** may be made more explicit within the framework of Example 5.14.

Example 5.15 (Continuation of Example 5.14). Conditions (5.26) and (5.27) read as follows when $\delta h_\ell(r, V_r, \tilde{X}_r') = \partial_w h(r, \theta_r, \mathcal{L}^1(X_r))$ and $\delta \tilde{H}_\ell(r, V_r, \tilde{X}_r') = \partial_\mu h(r, \theta_r, \mathcal{L}^1(X_r))(\tilde{X}_r')$ in the decomposition (5.23):

1. Equation (5.26) expresses the fact that the derivatives of h in the directions w and μ are bounded in L^∞. Importantly and as already suggested, the constant L corresponds to L in assumption **Smooth Coefficients Order 1**.
2. Equation (5.27) says that the derivatives in the directions w and μ are Lipschitz continuous.

Conditioning on the Product Space

We now provide several technical lemmas that we shall use repeatedly throughout the section.

Lemma 5.16 *Let ζ be an $\mathbb{R}$-valued integrable random variable defined on $(\Omega, \mathcal{F}, \mathbb{P})$. Then, $\mathbb{P}$ - almost surely, it holds that:*

$$\mathbb{E}^1\big[\mathbb{E}[\zeta|\mathcal{F}_t]\big] = \mathbb{E}^0\big[\mathbb{E}^1(\zeta)|\mathcal{F}_t^0\big] = \mathbb{E}\big[\mathbb{E}^1(\zeta)|\mathcal{F}_t\big] = \mathbb{E}[\zeta|\mathcal{F}_t^0].$$

Proof. Without any loss of generality, we can assume that ζ is $\mathcal{F}^0 \otimes \mathcal{F}^1$-measurable.

Consider a bounded $\mathcal{F}_t^0 \otimes \mathcal{F}_t^1$-measurable random variable χ with values in $\mathbb{R}$. Integrating first with respect to the variable ω^1, we have:

$$\mathbb{E}\big[\chi \mathbb{E}^1(\zeta)\big] = \mathbb{E}\big[\mathbb{E}^1(\chi)\zeta\big].$$

Notice that $\mathbb{E}^1(\chi)$, seen as a random variable from $\Omega^0 \times \Omega^1$ to $\mathbb{R}$, is $\mathcal{F}_t$-measurable. Therefore,

$$\mathbb{E}\big[\chi \mathbb{E}^1(\zeta)\big] = \mathbb{E}\big[\mathbb{E}^1(\chi)\mathbb{E}[\zeta|\mathcal{F}_t]\big].$$

Again, integrating first with respect to ω^1 in the right-hand side, we get:

$$\mathbb{E}\big[\chi \mathbb{E}^1(\zeta)\big] = \mathbb{E}\big[\chi \mathbb{E}^1\big(\mathbb{E}[\zeta|\mathcal{F}_t]\big)\big].$$

By the same argument as above, $\mathbb{E}^1(\mathbb{E}[\zeta|\mathcal{F}_t])$ is $\mathcal{F}_t$-measurable. This shows that:

$$\mathbb{E}\big[\mathbb{E}^1(\zeta)|\mathcal{F}_t\big] = \mathbb{E}^1\big[\mathbb{E}[\zeta|\mathcal{F}_t]\big].$$

Now, $\mathbb{E}^1(\zeta)$ may be regarded as a random variable on $(\Omega^0, \mathcal{F}^0, \mathbb{P}^0)$. Therefore, for any $B^0 \in \mathcal{F}_t^0$ and $B^1 \in \mathcal{F}_t^1$,

$$\begin{aligned}
\mathbb{E}\big[\mathbf{1}_{B^0 \times B^1}\mathbb{E}^1(\zeta)\big] &= \mathbb{E}^0\big[\mathbf{1}_{B^0}\mathbb{E}^1(\zeta)\big]\mathbb{P}^1[B^1] \\
&= \mathbb{E}^0\big[\mathbf{1}_{B^0}\mathbb{E}^0[\mathbb{E}^1(\zeta)|\mathcal{F}_t^0]\big]\mathbb{P}^1[B^1].
\end{aligned}$$

Regarding $\mathbb{E}^0[\mathbb{E}^1(\zeta)|\mathcal{F}_t^0]$ as a random variable on $(\Omega, \mathcal{F}, \mathbb{P})$, we deduce that $\mathbb{E}^0[\mathbb{E}^1(\zeta)|\mathcal{F}_t^0] = \mathbb{E}[\mathbb{E}^1(\zeta)|\mathcal{F}_t]$. Finally, choosing $B^1 = \Omega^1$ in the above equality, we get $\mathbb{E}^0[\mathbb{E}^1(\zeta)|\mathcal{F}_t^0] = \mathbb{E}[\zeta|\mathcal{F}_t^0]$, hence completing the proof. □

5.2.2 Estimate of the Solution to the Linearized Sysem

Part of our analysis relies on stability estimates for systems of a more general form than (5.22). Indeed we need to analyze systems of the form:

$$\begin{aligned}
\mathcal{X}_s &= \eta + \int_t^s \delta B\big(r, \bar{\theta}_r, \tilde{X}_r''\big)\big(\bar{\vartheta}_r, \tilde{\mathcal{X}}_r''\big)dr, \\
\mathcal{Y}_s &= \mathbb{E}_s\bigg[\int_s^T \delta F\big(r, \bar{\theta}_r, \tilde{X}_r''\big)\big(\bar{\vartheta}_r, \tilde{\mathcal{X}}_r''\big)dr + \delta G\big(X_T, \tilde{X}_T'\big)\big(\mathcal{X}_T, \tilde{\mathcal{X}}_T'\big)\bigg],
\end{aligned} \tag{5.28}$$

the difference between (5.28) and (5.22) being that the coefficients (except for the terminal condition) may depend on processes $\bar{\boldsymbol{\vartheta}}$ and $\boldsymbol{\mathcal{X}}''$ other than $\bar{\boldsymbol{\theta}}$, $\boldsymbol{X}''$, though of the same dimensions. Observe that the unknown $\mathcal{X}$ appears only once, in δG.

Notation. For $\gamma \in [T, 1]$, a pair $\boldsymbol{\vartheta} = (\vartheta_s = (\mathcal{X}_s, \mathcal{Y}_s))_{s\in[t,T]}$ with values in $\mathbb{S}^2([t,T];\mathbb{R}^d) \times \mathbb{S}^2([t,T];\mathbb{R}^m)$ and a pair of random variables (X, χ) with values in a Euclidean space, we let:

$$\mathcal{E}_s(\boldsymbol{\vartheta}) = |\mathcal{X}_s| + \gamma|\mathcal{Y}_s|. \tag{5.29}$$

Note that $\mathcal{E}_s(\boldsymbol{\vartheta})$ depends on γ, even though we omit this dependence in the notation. We also let:

$$\begin{aligned}
\mathcal{R}_a(s) &= \mathbb{E}_s\Big[\gamma|\delta G_a(T)| + \int_t^T \big|\big(\delta B_a(r), \delta F_a(r)\big)\big|dr\Big] \\
&= \mathbb{E}_s\Big[\gamma|\delta G_a(T)| + \int_s^T \big|\big(\delta B_a(r), \delta F_a(r)\big)\big|dr\Big] \\
&\quad + \int_t^s \big|\big(\delta B_a(r), \delta F_a(r)\big)\big|dr.
\end{aligned} \tag{5.30}$$

We claim that:

Lemma 5.17 *There exists a constant C, only depending on L, such that for any solution $\boldsymbol{\vartheta} = (\boldsymbol{\mathcal{X}}, \boldsymbol{\mathcal{Y}})$ to a system of the same type as (5.28), it holds, with probability 1,*

$$\mathcal{E}_s(\boldsymbol{\vartheta}) \le C\Big(|\eta| + \mathcal{R}_a(s) + \mathbb{E}_s\Big[\gamma\mathbb{E}^1\big[|\mathcal{X}'_T|\big] + \int_t^T \big(|\bar{\vartheta}_r| + \mathbb{E}^1\big[|\mathcal{X}''_r|\big]\big)dr\Big]\Big). \tag{5.31}$$

Proof.

First Step. We start with the trivial case when the coefficients δB_ℓ and δF_ℓ and $\delta\tilde{G}_\ell$ are null. See (5.24) and (5.25) for the notations replacing H therein by B, F or G. Then, (5.28) reads as a system driven by the linear part δg_ℓ which appears in the decomposition of δG in the form (5.25), plus a remainder involving δB_a, δF_a and δG_a. Then, we have, for all $s \in [t, T]$,

$$\sup_{s\le r\le T} |\mathcal{X}_r| \le |\mathcal{X}_s| + \int_s^T |\delta B_a(r)|dr,$$

while,

$$|\mathcal{Y}_s| \le L\mathbb{E}_s\big[|\mathcal{X}_T|\big] + \mathbb{E}_s\big[|\delta G_a(T)|\big] + \int_s^T \mathbb{E}_s\big[|\delta F_a(r)|\big]dr,$$

so that:

$$\begin{aligned}
|\mathcal{Y}_s| &\le L|\mathcal{X}_s| + \mathbb{E}_s\big[|\delta G_a(T)|\,|\big] + \int_s^T \mathbb{E}_s\big[|\delta B_a(r)| + |\delta F_a(r)|\big]dr \\
&\le L|\eta| + C\Big(\int_t^s |\delta B_a(r)|dr + \mathbb{E}_s\big[|\delta G_a(T)|\big] + \int_s^T \mathbb{E}_s\big[|\delta B_a(r)| + |\delta F_a(r)|\big]dr\Big),
\end{aligned}$$

for a constant C only depending on L. Therefore,

$$\mathcal{E}_s(\boldsymbol{\vartheta}) \le C\big(|\eta| + \mathcal{R}_a(s)\big).$$

Second Step. When δB_ℓ and δF_ℓ are nonzero, we view them as parts of δB_a and δF_a when evaluated along the values of $(\bar{\boldsymbol{\theta}}, \boldsymbol{X}'', \bar{\boldsymbol{\vartheta}}, \boldsymbol{\mathcal{X}}'')$. Similarly, we can see $\delta\tilde{G}_\ell$ as a part of δG_a when evaluated along the values of $(X_T, \tilde{X}'_T, \mathcal{X}'_T)$. We are thus led back to the previous case, but with a generalized version of the remainder term $\mathcal{R}_a$. The analysis of the new remainder may be split into three pieces: a first term involving δb_ℓ and δf_ℓ; a second term involving $\delta\tilde{B}_\ell$, $\delta\tilde{F}_\ell$ and $\delta\tilde{G}_\ell$; a last term involving δB_a, δF_a and δG_a, corresponding to the original $\mathcal{R}_a$. As a final bound, we get:

$$\begin{aligned}
\mathcal{E}_s(\vartheta) \le C\bigg(&|\eta| + \gamma\mathbb{E}_s\Big[\big|\tilde{\mathbb{E}}^1\big[\delta\tilde{G}_\ell(X_T, X'_T)\tilde{\mathcal{X}}'_T\big]\big| + \big|\delta G_a(T)\big|\Big] \\
&+ \mathbb{E}_s\bigg[\int_t^T \Big(\big|(\delta b_\ell, \delta f_\ell)(s, \bar{\theta}_r, \tilde{X}''_r)\bar{\vartheta}_r\big| + \big|\tilde{\mathbb{E}}^1\big[(\delta\tilde{B}_\ell, \delta\tilde{F}_\ell)(\bar{\theta}_s, \tilde{X}''_s)\tilde{\mathcal{X}}''_s\big]\big| \\
&\qquad + \big|(\delta B_a, \delta F_a)(r)\big|\Big)dr\bigg]\bigg).
\end{aligned}$$

By boundedness of $(\delta b_\ell, \delta f_\ell)$ and of $(\delta\tilde{B}_\ell, \delta\tilde{F}_\ell, \delta\tilde{G}_\ell)$, we get:

$$\mathcal{E}_s(\vartheta) \le C\bigg(|\eta| + \mathcal{R}_a(s) + \mathbb{E}_s\Big[\gamma\tilde{\mathbb{E}}^1\big[|\tilde{\mathcal{X}}'_T|\big] + \int_t^T \mathbb{E}_s\Big[|\bar{\vartheta}_r| + \tilde{\mathbb{E}}^1\big[|\tilde{\mathcal{X}}''_r|\big]dr\Big]\bigg), \tag{5.32}$$

which completes the proof. □

In particular, we have the following useful result for systems of the form (5.22) obtained by requiring $\boldsymbol{\vartheta} = \bar{\boldsymbol{\vartheta}}$ together with $\boldsymbol{\mathcal{X}}' = \boldsymbol{\mathcal{X}}''$ in (5.31), and setting γ small enough.

Corollary 5.18 *There exist two constants $c, C > 0$, only depending on L, such that, for $T \le \gamma^2 \le c$ and for any solution ϑ to a system of the type* (5.22), *it holds with $\mathbb{P}$-probability* 1*:*

$$\begin{aligned}
\mathcal{E}_s(\boldsymbol{\vartheta}) \le C\bigg(&|\eta| + \mathcal{R}_a(s) + \mathbb{E}_s\Big[\gamma\mathbb{E}^1\big[|\mathcal{X}'_T|\big] + \int_t^T \mathbb{E}^1\big[|\mathcal{X}'_u|\big]du\Big]\bigg) \\
&+ \frac{C}{\gamma}\int_t^T \bigg(|\eta| + \mathcal{R}_a(r) + \mathbb{E}_r\Big[\gamma\mathbb{E}^1\big[|\mathcal{X}'_T|\big] + \int_t^T \mathbb{E}^1\big[|\mathcal{X}'_u|\big]du\Big]\bigg)dr,
\end{aligned}$$

$$\begin{aligned}
\sup_{t\le s\le T} \mathcal{E}_s(\boldsymbol{\vartheta}) \le C\bigg(&|\eta| + \sup_{t\le s\le T} \mathcal{R}_a(s) \\
&+ \sup_{t\le s\le T} \mathbb{E}_s\Big[\gamma\mathbb{E}^1\big[|\mathcal{X}'_T|\big] + \int_t^T \mathbb{E}^1\big[|\mathcal{X}'_u|\big]du\Big]\bigg).
\end{aligned} \tag{5.33}$$

When $\boldsymbol{\vartheta} = \boldsymbol{\vartheta}'$, *it holds with* $\mathbb{P}$*-probability* 1,

$$\mathcal{E}_s(\boldsymbol{\vartheta}) \leq C\Big[|\eta| + \|\eta\|_1 + \mathcal{R}_a(s) + \mathbb{E}^1\big[\mathcal{R}_a(s)\big]\Big] + \frac{C}{\gamma}\int_t^T \Big[|\eta| + \|\eta\|_1 + \mathcal{R}_a(r) + \mathbb{E}^1\big[\mathcal{R}_a(r)\big]\Big]dr, \tag{5.34}$$

$$\sup_{t\leq s\leq T} \mathcal{E}_s(\boldsymbol{\vartheta}) \leq C\Big[|\eta| + \|\eta\|_1 + \sup_{t\leq s\leq T}\Big(\mathcal{R}_a(s) + \mathbb{E}^1\big[\mathcal{R}_a(s)\big]\Big)\Big].$$

Proof.

First Step. We start with the proof of (5.33). We rewrite (5.31) with the prescription $\boldsymbol{\vartheta} = \bar{\boldsymbol{\vartheta}}$ and $\boldsymbol{\mathcal{X}}' = \boldsymbol{\mathcal{X}}''$:

$$\mathcal{E}_s(\boldsymbol{\vartheta}) \leq C\bigg(|\eta| + \mathcal{R}_a(s) + \mathbb{E}_s\bigg[\gamma\mathbb{E}^1\big[|\mathcal{X}'_T|\big] + \int_t^T \big(|\vartheta_u| + \mathbb{E}^1\big[|\mathcal{X}'_u|\big]\big)du\bigg]\bigg).$$

For $t \leq r \leq s \leq T$, taking conditional expectation given $\mathcal{F}_r$, we get:

$$\begin{aligned}\mathbb{E}_r\big[\mathcal{E}_s(\boldsymbol{\vartheta})\big] &\leq C\bigg(|\eta| + \mathbb{E}_r\big[\mathcal{R}_a(s)\big] + \mathbb{E}_r\bigg[\gamma\mathbb{E}^1\big[|\mathcal{X}'_T|\big] + \int_t^T \big(|\vartheta_u| + \mathbb{E}^1\big[|\mathcal{X}'_u|\big]\big)du\bigg]\bigg)\\ &\leq C\bigg(|\eta| + \mathcal{R}_a(r) + \mathbb{E}_r\bigg[\gamma\mathbb{E}^1\big[|\mathcal{X}'_T|\big] + \int_t^T \big(|\vartheta_u| + \mathbb{E}^1\big[|\mathcal{X}'_u|\big]\big)du\bigg]\bigg),\end{aligned} \tag{5.35}$$

where we used the equality $\mathbb{E}_r[\mathcal{R}_a(s)] = \mathcal{R}_a(r)$. Observe now from (5.29) that:

$$T\sup_{r\leq u\leq T}\mathbb{E}_r\big[|\vartheta_u|\big] \leq \gamma^2 \sup_{r\leq u\leq T}\mathbb{E}_r\big[|\vartheta_u|\big] \leq \gamma \sup_{r\leq u\leq T}\mathbb{E}_r\big[\mathcal{E}_u(\boldsymbol{\vartheta})\big],$$

from which we get:

$$\begin{aligned}\sup_{r\leq s\leq T}\mathbb{E}_r\big[\mathcal{E}_s(\boldsymbol{\vartheta})\big] &\leq C\bigg(|\eta| + \mathcal{R}_a(r) + \mathbb{E}_r\bigg[\gamma\mathbb{E}^1\big[|\mathcal{X}'_T|\big] + \int_t^r |\vartheta_u|du + \int_t^T \mathbb{E}^1\big[|\mathcal{X}'_u|\big]du\bigg]\bigg)\\ &\quad + C\gamma\sup_{r\leq u\leq T}\mathbb{E}_r\big[|\mathcal{E}_u(\boldsymbol{\vartheta})|\big].\end{aligned}$$

Choosing γ small enough in (5.31), we obtain:

$$\sup_{r\leq s\leq T}\mathbb{E}_r\big[\mathcal{E}_s(\boldsymbol{\vartheta})\big] \leq C\bigg(|\eta| + \mathcal{R}_a(r) + \mathbb{E}_r\bigg[\gamma\mathbb{E}^1\big[|\mathcal{X}'_T|\big] + \int_t^r |\vartheta_u|du + \int_t^T \mathbb{E}^1\big[|\mathcal{X}'_u|\big]du\bigg]\bigg).$$

Choosing $r = s$ and using the same trick as above for controlling $\boldsymbol{\vartheta}$, we finally have:

$$\begin{aligned}\mathcal{E}_s(\boldsymbol{\vartheta}) &\leq C\Big(|\eta| + \mathcal{R}_a(s) + \mathbb{E}_s\Big[\gamma\mathbb{E}^1[|\mathcal{X}'_T|] + \int_t^T \mathbb{E}^1[|\mathcal{X}'_u|]du\Big]\Big) + C\int_t^s |\vartheta_u|du \\ &\leq C\Big(|\eta| + \mathcal{R}_a(s) + \mathbb{E}_s\Big[\gamma\mathbb{E}^1[|\mathcal{X}'_T|] + \int_t^T \mathbb{E}^1[|\mathcal{X}'_u|]du\Big]\Big) + \frac{C}{\gamma}\int_t^s \mathcal{E}_u(\boldsymbol{\vartheta})du.\end{aligned}$$

By Gronwall's lemma, we deduce:

$$\begin{aligned}\mathcal{E}_s(\boldsymbol{\vartheta}) &\leq C\Big(|\eta| + \mathcal{R}_a(s) + \mathbb{E}_s\Big[\gamma\mathbb{E}^1[|\mathcal{X}'_T|] + \int_t^T \mathbb{E}^1[|\mathcal{X}'_u|]du\Big]\Big) \\ &\quad + \frac{C}{\gamma}\int_t^T \Big(|\eta| + \mathcal{R}_a(r) + \mathbb{E}_r\Big[\gamma\mathbb{E}^1[|\mathcal{X}'_T|] + \int_t^T \mathbb{E}^1[|\mathcal{X}'_u|]du\Big]\Big)dr,\end{aligned}$$

where we used the fact that $T \leq \gamma^2$. Then, taking the supremum over s, we get:

$$\sup_{t\leq s\leq T} \mathcal{E}_s(\boldsymbol{\vartheta}) \leq C\Big(|\eta| + \sup_{t\leq s\leq T} \mathcal{R}_a(s) + \sup_{t\leq s\leq T} \mathbb{E}_s\Big[\gamma\mathbb{E}^1[|\mathcal{X}'_T|] + \int_t^T \mathbb{E}^1[|\mathcal{X}'_u|]du\Big]\Big).$$

Second Step. Whenever $\boldsymbol{\mathcal{X}} = \boldsymbol{\mathcal{X}}'$, (5.35) reads:

$$\mathbb{E}_r\big[\mathcal{E}_s(\boldsymbol{\vartheta})\big] \leq C\Big(|\eta| + \mathcal{R}_a(r) + \mathbb{E}_r\Big[\gamma\mathbb{E}^1[|\mathcal{X}_T|] + \int_t^T (|\vartheta_u| + \mathbb{E}^1[|\mathcal{X}_u|])du\Big]\Big).$$

Taking expectation under $\mathbb{P}^1$ and invoking Lemma 5.16, we get:

$$\mathbb{E}_r\Big[\mathbb{E}^1\big[\mathcal{E}_s(\boldsymbol{\vartheta})\big]\Big] \leq C\Big(\|\eta\|_1 + \mathbb{E}^1\big[\mathcal{R}_a(r)\big] + \mathbb{E}_r\Big[\gamma\mathbb{E}^1[|\mathcal{X}_T|] + \int_t^T \mathbb{E}^1[|\vartheta_u|]du\Big]\Big).$$

Adding the two inequalities, we obtain:

$$\begin{aligned}\mathbb{E}_r\big[\mathcal{E}_s(\boldsymbol{\vartheta})\big] + \mathbb{E}_r\Big[\mathbb{E}^1\big[\mathcal{E}_s(\boldsymbol{\vartheta})\big]\Big] &\leq C\Big(|\eta| + \|\eta\|_1 + \mathcal{R}_a(r) + \mathbb{E}^1[\mathcal{R}_a(r)] \\ &\quad + \mathbb{E}_r\Big[\gamma\mathbb{E}^1[|\mathcal{X}_T|] + \int_t^T (|\vartheta_u| + \mathbb{E}^1[|\vartheta_u|])du\Big]\Big).\end{aligned}$$

Choosing $s = T$ and then γ small enough, we get a bound for $\mathbb{E}^1[|\mathcal{X}_T|]$, which basically says that we can remove $\mathbb{E}^1[|\mathcal{X}_T|]$ in the above right-hand side. Repeating the same computations as those developed in the first step to handle (5.35), we deduce:

$$\begin{aligned}\mathcal{E}_s(\boldsymbol{\vartheta}) &\leq C\Big(|\eta| + \|\eta\|_1 + \mathcal{R}_a(s) + \mathbb{E}^1\big[\mathcal{R}_a(s)\big]\Big) \\ &\quad + \frac{C}{\gamma}\int_t^T \Big(|\eta| + \mathcal{R}_a(r) + \|\eta\|_1 + \mathbb{E}^1\big[\mathcal{R}_a(r)\big]\Big)dr,\end{aligned}$$

together with:

$$\sup_{t\leq s\leq T} \Big(\mathcal{E}_s(\boldsymbol{\vartheta}) + \mathbb{E}^1\big[\mathcal{E}_s(\boldsymbol{\vartheta})\big]\Big) \leq C\Big[|\eta| + \|\eta\|_1 + \sup_{t\leq s\leq T} \Big(\mathcal{R}_a(s) + \mathbb{E}^1\big[\mathcal{R}_a(s)\big]\Big)\Big],$$

which completes the proof. □

5.2.3 Stability Estimates

The next step is to compare two solutions of (5.28), say $\boldsymbol{\vartheta}^1$ and $\boldsymbol{\vartheta}^2$, driven by two different tuples of coefficients $(\delta B^1, \delta F^1, \delta G^1)$ and $(\delta B^2, \delta F^2, \delta G^2)$ satisfying:

$$\delta b_\ell^1 \equiv \delta b_\ell^2,\ \delta f_\ell^1 \equiv \delta f_\ell^2,\ \delta g_\ell^1 \equiv \delta g_\ell^2,$$
$$\delta \tilde{B}_\ell^1 \equiv \delta \tilde{B}_\ell^2,\ \delta \tilde{F}_\ell^1 \equiv \delta \tilde{F}_\ell^2,\ \delta \tilde{G}_\ell^1 \equiv \delta \tilde{G}_\ell^2,$$

and by two different sets of inputs:

$$(\bar{\boldsymbol{\theta}}^1, \boldsymbol{X}'^{,1}, \boldsymbol{X}''^{,1}, \bar{\boldsymbol{\vartheta}}^1, \boldsymbol{\mathcal{X}}'^{,1}, \boldsymbol{\mathcal{X}}''^{,1}) \text{ and } (\bar{\boldsymbol{\theta}}^2, \boldsymbol{X}'^{,2}, \boldsymbol{X}''^{,2}, \bar{\boldsymbol{\vartheta}}^2, \boldsymbol{\mathcal{X}}'^{,2}, \boldsymbol{\mathcal{X}}''^{,2}),$$

and with the same starting point η.

We denote by $(\Delta\mathcal{R}_a(s))_{t\le s\le T}$ the process:

$$\begin{aligned}
&\Delta\mathcal{R}_a(s)\\
&= \mathbb{E}_s\Big[\big|(\delta G_a^1 - \delta G_a^2)(T)\big| + \int_t^T \big|(\delta B_a^1 - \delta B_a^2, \delta F_a^1 - \delta F_a^2)(r)\big| dr\Big].
\end{aligned} \tag{5.36}$$

Recall (5.30) for the definition of $\mathcal{R}_a(s)$. Also, we denote by $\mathcal{R}_a^1$ and $\mathcal{R}_a^2$ the remainders associated with the tuples $(\delta B^1, \delta F^1, \delta G^1)$ and $(\delta B^2, \delta F^2, \delta G^2)$ through (5.30). Then, we have:

Lemma 5.19 *There exist three constants c, C, and K, with c and C only depending on L, and K only depending on T, L, and Γ, such that, for $T \le c$,*

$$\begin{aligned}
&\mathcal{E}_s(\boldsymbol{\vartheta}^1 - \boldsymbol{\vartheta}^2) \le C\Delta\mathcal{R}_a(s)\\
&\quad + C\mathbb{E}_s\Big[\gamma\mathbb{E}^1\big[|\mathcal{X}_T'^{,1} - \mathcal{X}_T'^{,2}|\big] + \int_t^T \Big(|\bar{\vartheta}_u^1 - \bar{\vartheta}_u^2| + \mathbb{E}^1\big[|\mathcal{X}_u''^{,1} - \mathcal{X}_u''^{,2}|\big]\Big) du\Big]\\
&\quad + K\big(\mathbb{E}_s + \mathbb{E}_s^0\big)\Big[\Big\{1 \wedge \Big(|X_T^1 - X_T^2| + |X_T'^{,1} - X_T'^{,2}| + \mathbb{E}^1\big[|X_T'^{,1} - X_T'^{,2}|\big]\Big)\Big\}\\
&\qquad\qquad \times \big(|\mathcal{X}_T^2| + |\mathcal{X}_T'^{,2}|\big)\\
&\qquad + \int_t^T \Big[\Big\{1 \wedge \Big(|\bar{\theta}_u^1 - \bar{\theta}_u^2| + |X_u''^{,1} - X_u''^{,2}| + \mathbb{E}^1\big[|X_u''^{,1} - X_u''^{,2}|\big]\Big)\Big\}\\
&\qquad\qquad \times \big(|\bar{\vartheta}_u^2| + |\mathcal{X}_u''^{,2}|\big)\Big] du\Big],
\end{aligned} \tag{5.37}$$

where the symbol $\mathbb{E}_s + \mathbb{E}_s^0$ accounts for the fact that we consider the sum of the two conditional expectations under $\mathbb{P}$ with respect to $\mathcal{F}_s$ and $\mathcal{F}_s^0$.

Notice from Lemma 5.16 that, in (5.37), $\mathbb{E}_s^0[\cdot]$ also reads as $\mathbb{E}_s[\mathbb{E}^1[\cdot]]$.

Remark 5.20 *Specialized to the particular case* $\boldsymbol{\theta}^1 = \bar{\boldsymbol{\theta}}^1 = \boldsymbol{\theta}^2 = \bar{\boldsymbol{\theta}}^2$, $\boldsymbol{X}^{\prime,1} = \boldsymbol{X}^{\prime\prime,1} = \boldsymbol{X}^{\prime,2} = \boldsymbol{X}^{\prime\prime,2}$, $\boldsymbol{\vartheta}^1 = \bar{\boldsymbol{\vartheta}}^1$, $\boldsymbol{\vartheta}^2 = \bar{\boldsymbol{\vartheta}}^2$, $\boldsymbol{\mathcal{X}}^1 = \boldsymbol{\mathcal{X}}^{\prime,1} = \boldsymbol{\mathcal{X}}^{\prime\prime,1}$, $\boldsymbol{\mathcal{X}}^2 = \boldsymbol{\mathcal{X}}^{\prime,2} = \boldsymbol{\mathcal{X}}^{\prime\prime,2}$ *and* $\Delta\mathcal{R}_a = 0$, *Lemma 5.19 reads as a short time uniqueness result for* (5.22) *when* $\boldsymbol{\mathcal{X}}' = \boldsymbol{\mathcal{X}}$.

Proof. We use linearity in computing the difference of the two systems of the form (5.28) satisfied by $\boldsymbol{\vartheta}^1$ and $\boldsymbol{\vartheta}^2$. The resulting system is linear in $\Delta\boldsymbol{\vartheta} = \boldsymbol{\vartheta}^1 - \boldsymbol{\vartheta}^2$, $\Delta\bar{\boldsymbol{\vartheta}} = \bar{\boldsymbol{\vartheta}}^1 - \bar{\boldsymbol{\vartheta}}^2$, $\Delta\boldsymbol{\mathcal{X}}' = \boldsymbol{\mathcal{X}}^{\prime,1} - \boldsymbol{\mathcal{X}}^{\prime,2}$, and $\Delta\boldsymbol{\mathcal{X}}'' = \boldsymbol{\mathcal{X}}^{\prime\prime,1} - \boldsymbol{\mathcal{X}}^{\prime\prime,2}$, but contains some remainders. We denote these remainders by $\Delta\delta B_a$, $\Delta\delta F_a$, and $\Delta\delta G_a$. Using the notations introduced in (5.24) and (5.25), they may be expanded as:

$$
\begin{aligned}
\Delta\delta H_a(s) &= \big(\delta h^1_\ell(s, \bar{\theta}^1_s, \tilde{X}^{\prime\prime,1}_s) - \delta h^1_\ell(s, \bar{\theta}^2_s, \tilde{X}^{\prime\prime,2}_s)\big)\bar{\vartheta}^2_s \\
&\quad + \tilde{\mathbb{E}}^1\big[\big(\delta\tilde{H}^1_\ell(s, \bar{\theta}^1_s, \tilde{X}^{\prime\prime,1}_s) - \delta\tilde{H}^1_\ell(s, \bar{\theta}^2_s, \tilde{X}^{\prime\prime,2}_s)\big)\tilde{\mathcal{X}}^{\prime\prime,2}_s\big] \\
&\quad + \delta H^1_a(s) - \delta H^2_a(s), \\
\delta\Delta G_a(T) &= \big(\delta g^1_\ell(X^1_T, \tilde{X}^{1,\prime}_T) - \delta g^1_\ell(X^2_T, \tilde{X}^{\prime,2}_T)\big)\mathcal{X}^2_T \\
&\quad + \tilde{\mathbb{E}}^1\big[\big(\delta\tilde{G}^1_\ell(X^1_T, \tilde{X}^{\prime,1}_T) - \delta\tilde{G}^1_\ell(X^2_T, \tilde{X}^{\prime,2}_T)\big)\tilde{\mathcal{X}}^{\prime,2}_T\big] \\
&\quad + \delta G^1_a(T) - \delta G^2_a(T),
\end{aligned}
\tag{5.38}
$$

where $\Delta\delta H_a$ may stand for $\Delta\delta B_a$ or $\Delta\delta G_a$ and δH^1 may stand for δB^1 or δF^1, with a corresponding meaning for δh^1_ℓ, $\delta\tilde{H}^1_\ell$ and δH^1_a: δh^1_ℓ may be δb^1_ℓ or δf^1_ℓ; $\delta\tilde{H}^1_\ell$ may be $\delta\tilde{B}^1_\ell$ or $\delta\tilde{F}^1_\ell$; δH^1_a may be δB^1_a or δF^1_a; and δH^2_a may be δB^2_a or δF^2_a. With these notations in hand, the terms $\Delta\delta H_a(s)$ and $\Delta\delta G_a(T)$ come from (recall (5.24)):

$$
\begin{aligned}
&\delta H^1\big(r, \bar{\theta}^1_r, \tilde{X}^{\prime\prime,1}_r\big)\big(\bar{\vartheta}^1_r, \tilde{\mathcal{X}}^{\prime\prime,1}_r\big) - \delta H^2\big(r, \bar{\theta}^2_r, \tilde{X}^{\prime\prime,2}_r\big)\big(\bar{\vartheta}^2_r, \tilde{\mathcal{X}}^{\prime\prime,2}_r\big) \\
&\quad = \delta H^1_\ell\big(r, \bar{\theta}^1_r, \tilde{X}^{\prime\prime,1}_r\big)\big(\Delta\bar{\vartheta}_r, \Delta\tilde{\mathcal{X}}''_r\big) + \Delta\delta H_a(r), \\
&\delta G^1\big(X^1_T, \tilde{X}^{\prime,1}_T\big)\big(\mathcal{X}^1_T, \tilde{\mathcal{X}}^{\prime,1}_T\big) - \delta G^2\big(X^2_T, \tilde{X}^{\prime,2}_T\big)\big(\mathcal{X}^2_T, \tilde{\mathcal{X}}^{\prime,2}_T\big) \\
&\quad = \delta G^1_\ell\big(X^1_T, \tilde{X}^{\prime,1}_T\big)\big(\Delta\mathcal{X}_T, \Delta\tilde{\mathcal{X}}'_T\big) + \Delta\delta G_a(T).
\end{aligned}
\tag{5.39}
$$

Our goal is to apply Lemma 5.17. In the statement of Lemma 5.17, we see from (5.39) that $\boldsymbol{\vartheta}$ must be understood as $\Delta\boldsymbol{\vartheta}$, $\bar{\boldsymbol{\vartheta}}$ as $\Delta\bar{\boldsymbol{\vartheta}}$, and similarly for the processes labeled with "prime" and "double prime". Moreover, the remainder $(\delta B_a, \delta F_a, \delta G_a)$ in the statement must be understood as $(\Delta\delta B_a, \Delta\delta F_a, \Delta\delta G_a)$.

We estimate the remainder terms in (5.31), recalling (5.30) for the meaning of the remainder in the stability estimate. By (5.38), the remainder can be split into three pieces according to δh^1_ℓ, $\delta\tilde{H}^1_\ell$ and $\delta H^1_a - \delta H^2_a$.

First Step. We first provide an upper bound for the terms involving $(\delta b^1_\ell, \delta f^1_\ell)$ and δg^1_ℓ. We make use of assumption (5.27) and of the conditional Cauchy-Schwarz' inequality. Getting rid of the constant γ in front of $|\delta G_a(T)|$ in (5.30), we let:

$$
\begin{aligned}
\Delta r_\ell(s) = \mathbb{E}_s\bigg[&\big|\big(\delta g^1_\ell(X^1_T, \tilde{X}^{\prime,1}_T) - \delta g^1_\ell(X^2_T, \tilde{X}^{\prime,2}_T)\big)\mathcal{X}^2_T\big| \\
&+ \int_t^T \big|\big((\delta b^1_\ell, \delta f^1_\ell)(u, \bar{\theta}^1_u, \tilde{X}^{\prime\prime,1}_u) - (\delta b^1_\ell, \delta f^1_\ell)(u, \bar{\theta}^2_u, \tilde{X}^{\prime\prime,2}_u)\big)\bar{\vartheta}^2_u\big| du\bigg].
\end{aligned}
$$

Recalling the bound (5.26) together with the Lipschitz property (5.27), we know that for a generic function δh_ℓ^1, which may be δb_ℓ^1 or δf_ℓ^1,

$$\begin{aligned}&\big|(\delta h_\ell^1(u,\bar\theta_u^1,\tilde X_u^{\prime\prime,1})-\delta h_\ell^1(u,\bar\theta_u^2,\tilde X_u^{\prime\prime,2}))\bar\vartheta_u^2\big|\\&\quad\le\Big[L\wedge\Big\{\Gamma\Big(|\bar\theta_u^1-\bar\theta_u^2|+\mathbb{E}^1\big[|X_u^{\prime\prime,1}-X_u^{\prime\prime,2}|\big]\Big)\Big\}\Big]|\bar\vartheta_u^2|.\end{aligned}\tag{5.40}$$

Therefore, we get, for a constant K, allowed to vary from line to line,

$$\begin{aligned}&\mathbb{E}_s\Big[\int_t^T\big|((\delta b_\ell^1,\delta f_\ell^1)(u,\bar\theta_u^1,\tilde X_u^{\prime\prime,1})-(\delta b_\ell^1,\delta f_\ell^1)(u,\bar\theta_u^2,\tilde X_u^{\prime\prime,2}))\bar\vartheta_u^2\big|du\Big]\\&\quad\le K\mathbb{E}_s\Big[\int_t^T\Big\{1\wedge\Big(|\bar\theta_u^1-\bar\theta_u^2|+\mathbb{E}^1\big[|X_u^{\prime\prime,1}-X_u^{\prime\prime,2}|\big]\Big)\Big\}|\bar\vartheta_u^2|du\Big].\end{aligned}$$

Finally, the term involving δg_ℓ^1 can also be handled in a similar way, provided the "bar" process is replaced by the "non-bar" process and the "double prime" process by the "prime" process. We thus get:

$$\begin{aligned}\Delta r_\ell(s)\le K\mathbb{E}_s\Big[&\Big\{1\wedge\Big(|X_T^1-X_T^2|+\mathbb{E}^1\big[|X_T^{\prime,1}-X_T^{\prime,2}|\big]\Big)\Big\}|\mathcal{X}_T^2|\\&+\int_t^T\Big[\Big\{1\wedge\Big(|\bar\theta_u^1-\bar\theta_u^2|+\mathbb{E}^1\big[|X_u^{\prime\prime,1}-X_u^{\prime\prime,2}|\big]\Big)\Big\}|\bar\vartheta_u^2|\Big]du\Big].\end{aligned}$$

Second Step. We now provide an upper bound for the terms involving $\delta\tilde B_\ell^1$, $\delta\tilde F_\ell^1$ or $\delta\tilde G_\ell^1$. We can make use of the bound (5.26) or the Lipschitz property (5.27). For a generic function $\delta\tilde H_\ell^1$, which may be $\delta\tilde B_\ell^1$ or $\delta\tilde F_\ell^1$, we get:

$$\begin{aligned}&\big|\tilde{\mathbb{E}}^1\big[(\delta\tilde H_\ell^1(u,\bar\theta_u^1,\tilde X_u^{\prime\prime,1})-\delta\tilde H_\ell^1(u,\bar\theta_u^2,\tilde X_u^{\prime\prime,2}))\tilde{\mathcal{X}}_u^{\prime\prime,2}\big]\big|\\&\quad\le\Gamma\mathbb{E}^1\Big[\Big\{1\wedge\Big(|\bar\theta_u^1-\bar\theta_u^2|+|X_u^{\prime\prime,1}-X_u^{\prime\prime,2}|+\mathbb{E}^1\big[|X_u^{\prime\prime,1}-X_u^{\prime\prime,2}|\big]\Big)\Big\}|\mathcal{X}_u^{\prime\prime,2}|\Big].\end{aligned}\tag{5.41}$$

Obviously, we have a similar bound for the term involving $\tilde G_\ell$. Letting:

$$\begin{aligned}\Delta\tilde R_\ell(s)=\mathbb{E}_s\Big[&\big|\tilde{\mathbb{E}}^1\big[(\delta\tilde G_\ell^1(X_T^1,\tilde X_T^{\prime,1})-\delta\tilde G_\ell^1(X_T^2,\tilde X_T^{\prime,2}))\tilde{\mathcal{X}}_T^{\prime,2}\big]\big|\\&+\int_t^T\big|\tilde{\mathbb{E}}^1\big[((\delta\tilde B_\ell^1,\delta\tilde F_\ell^1)(u,\bar\theta_u^1,\tilde X_u^{\prime\prime,1})-(\delta\tilde B_\ell^1,\delta\tilde F_\ell^1)(u,\bar\theta_u^2,\tilde X_u^{\prime\prime,2}))\tilde{\mathcal{X}}_u^{\prime\prime,2}\big]\big|du\Big],\end{aligned}$$

we get:

$$\begin{aligned}\Delta\tilde R_\ell(s)\le K\mathbb{E}_s^0\Big[&\Big\{1\wedge\Big(|X_T^1-X_T^2|+|X_T^{\prime,1}-X_T^{\prime,2}|+\mathbb{E}^1\big[|X_T^{\prime,1}-X_T^{\prime,2}|\big]\Big)\Big\}|\mathcal{X}_T^{\prime,2}|\\&+\int_t^T\Big[\Big\{1\wedge\Big(|\bar\theta_u^1-\bar\theta_u^2|+|X_u^{\prime\prime,1}-X_u^{\prime\prime,2}|+\mathbb{E}^1\big[|X_u^{\prime\prime,1}-X_u^{\prime\prime,2}|\big]\Big)\Big\}|\mathcal{X}_u^{\prime\prime,2}|\Big]du\Big].\end{aligned}$$

Conclusion. In order to complete the proof of the first part, notice that the terms labeled by a directly give the remainder $\Delta\mathcal{R}_a(s)$ in (5.37). □

Remark 5.21 *As the reader may guess, the terms of the form* $\boldsymbol{\vartheta}^2$, $\boldsymbol{\mathcal{X}}'^{,2}$ *and* $\boldsymbol{\mathcal{X}}''^{,2}$ *in* (5.37) *will be handled by means of Corollary 5.18. However, we note that when compared with* $|\mathcal{X}_s| + |\mathcal{Y}_s|$, *the term* $\mathcal{E}_s(\boldsymbol{\vartheta})$ *used in Corollary 5.18 incorporates an additional pre-factor* γ, *see* (5.29)*. Roughly speaking, both quantities are 'equivalent' provided* γ *is not too small. In the sequel, we often choose* γ *exactly equal to* c, *so that* $|\mathcal{X}_s| + |\mathcal{Y}_s|$ *and* $\mathcal{E}_s(\boldsymbol{\vartheta})$ *are comparable.*

Corollary 5.22 *Consider a family of progressively measurable random paths:*

$$\big((\boldsymbol{\theta}^{\xi}, \boldsymbol{X}'^{,\xi}) : [t,T] \ni s \mapsto (\theta_s^{\xi}, X_s'^{,\xi})\big)_{\xi},$$

parameterized by $\xi \in L^2(\Omega^1, \mathcal{F}_t^1, \mathbb{P}^1; \mathbb{R}^d)$, *and assume that there exists a constant* κ *such that, for all* ξ_1 *and* ξ_2*:*

$$\sup_{s\in[t,T]} \Big[|\theta_s^{\xi_1} - \theta_s^{\xi_2}| + |X_s'^{,\xi_1} - X_s'^{,\xi_2}| \Big] \le \kappa\big(|\xi_1 - \xi_2| + \|\xi_1 - \xi_2\|_1\big). \tag{5.42}$$

Then, there exist two constants c *and* K', *with* c *only depending on* L, *and* K' *only depending on* L, Γ, *and* κ, *such that for* $T \le \gamma^2 \le c$, *choosing in the statement of Lemma 5.19:*

$$(\bar{\boldsymbol{\theta}}^i, \bar{\boldsymbol{\vartheta}}^i) = (\boldsymbol{\theta}^i, \boldsymbol{\vartheta}^i), \quad \boldsymbol{X}''^{,i} = \boldsymbol{X}'^{,i}, \ \boldsymbol{\mathcal{X}}''^{,i} = \boldsymbol{\mathcal{X}}'^{,i},$$

with $(\boldsymbol{\theta}^i, \boldsymbol{X}'^{,i}) = (\boldsymbol{\theta}^{\xi_i}, \boldsymbol{X}'^{,\xi_i})$ *for* $i = 1, 2$, *with probability* 1 *and for all* $s \in [t,T]$, *it holds:*

$$\begin{aligned}
&\sup_{t\le s\le T} \mathcal{E}_s(\boldsymbol{\vartheta}^1 - \boldsymbol{\vartheta}^2) \\
&\quad \le K' \bigg[\sup_{t\le s\le T} \Big(\Delta\mathcal{R}_a(s) + \mathbb{E}_s\big[\mathcal{T} + \mathbb{E}^1[\mathcal{T}]\big]\Big) \\
&\qquad + \sup_{t\le s\le T} \mathbb{E}_s\Big[\gamma \mathbb{E}^1\big[|\mathcal{X}_T'^{,1} - \mathcal{X}_T'^{,2}|\big] + \int_t^T \mathbb{E}^1\big[|\mathcal{X}_u'^{,1} - \mathcal{X}_u'^{,2}|\big] du \Big]\bigg],
\end{aligned} \tag{5.43}$$

with:

$$\begin{aligned}
\mathcal{T} &= \Big\{ 1 \wedge \Big(|\xi_1 - \xi_2| + \|\xi_1 - \xi_2\|_1 \Big)\Big\} \big(|\mathcal{X}_T^2| + |\mathcal{X}_T'^{,2}| \big) \\
&\quad + \int_t^T \Big[\Big\{ 1 \wedge \Big(|\xi_1 - \xi_2| + \|\xi_1 - \xi_2\|_1 \Big)\Big\} \big(|\vartheta_u^2| + |\mathcal{X}_u'^{,2}| \big) \Big] du.
\end{aligned}$$

When $\boldsymbol{\vartheta}^i = (\mathcal{X}^i, \mathcal{Y}^i)$, with $\mathcal{X}^i = \mathcal{X}^{i,\prime}$, we have, modifying the values of c and K' if necessary,

$$\sup_{t \le s \le T} \mathcal{E}_s(\boldsymbol{\vartheta}^1 - \boldsymbol{\vartheta}^2) \tag{5.44}$$
$$\le K'\Big(\sup_{t \le s \le T} \Delta\mathcal{R}_a(s) + \mathbb{E}^1\Big[\sup_{t \le s \le T} \Delta\mathcal{R}_a(s)\Big] + \sup_{t \le s \le T} \mathbb{E}_s\big[\mathcal{T} + \mathbb{E}^1[\mathcal{T}]\big]\Big),$$

where:

$$\begin{aligned}\mathcal{T} + \mathbb{E}^1\big[\mathcal{T}\big] \le C\Big\{ &|\eta| + \|\eta\|_1 + \sup_{t \le r \le T} \mathcal{R}_a^2(r) + \mathbb{E}^1\big[\sup_{t \le r \le T} \mathcal{R}_a^2(r)\big]\Big\} \\ &\wedge \Big\{\mathbb{E}^1\Big[|\xi_1 - \xi_2|\Big(|\eta| + \sup_{t \le r \le T} \mathcal{R}_a^2(r)\Big)\Big] \\ &+ \Big(|\xi_1 - \xi_2| + \|\xi_1 - \xi_2\|_1\Big) \\ &\quad \times \Big(|\eta| + \|\eta\|_1 + \sup_{t \le r \le T} \mathcal{R}_a^2(r) + \mathbb{E}^1\Big[\sup_{t \le r \le T} \mathcal{R}_a^2(r)\Big]\Big)\Big\}.\end{aligned}$$

Proof.

First Step. The strategy is to make use of Lemma 5.19 and to insert (5.42) in (5.37). Since we consider the case $(\bar{\boldsymbol{\theta}}^i, \bar{\boldsymbol{\vartheta}}^i) = (\boldsymbol{\theta}^i, \boldsymbol{\vartheta}^i)$ and $(\boldsymbol{X}^{\prime\prime,i}, \boldsymbol{\mathcal{X}}^{\prime\prime,i}) = (\boldsymbol{X}^{\prime,i}, \boldsymbol{\mathcal{X}}^{\prime,i})$, (5.37) yields:

$$\begin{aligned}\mathcal{E}_s(\boldsymbol{\vartheta}^1 - \boldsymbol{\vartheta}^2) &\le C\Delta\mathcal{R}_a(s) \\ &+ C\mathbb{E}_s\Big[\gamma\mathbb{E}^1[|\mathcal{X}_T^{\prime,1} - \mathcal{X}_T^{\prime,2}|] + \int_t^T \Big(|\vartheta_u^1 - \vartheta_u^2| + \mathbb{E}^1[|\mathcal{X}_u^{\prime,1} - \mathcal{X}_u^{\prime,2}|]\Big)du\Big] \\ &+ K'(\mathbb{E}_s + \mathbb{E}_s^0)\Big[\Big\{1 \wedge \Big(|\xi_1 - \xi_2| + \|\xi_1 - \xi_2\|_1\Big)\Big\}(|\mathcal{X}_T^2| + |\mathcal{X}_T^{\prime,2}|) \\ &\quad + \int_t^T \Big[\Big\{1 \wedge \Big(|\xi_1 - \xi_2| + \|\xi_1 - \xi_2\|_1\Big)\Big\}(|\vartheta_u^2| + |\mathcal{X}_u^{\prime,2}|)\Big]du\Big],\end{aligned} \tag{5.45}$$

where K' may depend on K and κ. Letting:

$$\begin{aligned}\mathcal{T} = &\Big\{1 \wedge \Big(|\xi_1 - \xi_2| + \|\xi_1 - \xi_2\|_1\Big)\Big\}(|\mathcal{X}_T^2| + |\mathcal{X}_T^{\prime,2}|) \\ &+ \int_t^T \Big[\Big\{1 \wedge \Big(|\xi_1 - \xi_2| + \|\xi_1 - \xi_2\|_1\Big)\Big\}(|\vartheta_u^2| + |\mathcal{X}_u^{\prime,2}|)\Big]du,\end{aligned}$$

we can rewrite (5.45) as follows:

$$\begin{aligned}\mathcal{E}_s(\boldsymbol{\vartheta}^1 - \boldsymbol{\vartheta}^2) &\le C\Delta\mathcal{R}_a(s) + K'\mathbb{E}_s\big[\mathcal{T} + \mathbb{E}^1[\mathcal{T}]\big] \\ &+ C\mathbb{E}_s\Big[\gamma\mathbb{E}^1[|\mathcal{X}_T^{\prime,1} - \mathcal{X}_T^{\prime,2}|] + \int_t^T \Big(|\vartheta_u^1 - \vartheta_u^2| + \mathbb{E}^1[|\mathcal{X}_u^{\prime,1} - \mathcal{X}_u^{\prime,2}|]\Big)du\Big].\end{aligned} \tag{5.46}$$

We thus recover the same setting as in the first step of the proof of Corollary 5.18. Since $T \leq \gamma^2$, we deduce that there exist two constants $c' > 0$ and $K' \geq 0$, with c' only depending on L and with K' only depending on L, Γ, and κ, such that:

$$\sup_{t \leq s \leq T} \mathcal{E}_s(\boldsymbol{\vartheta}^1 - \boldsymbol{\vartheta}^2) \leq K' \Big[\sup_{t \leq s \leq T} \big(\Delta \mathcal{R}_a(s) + \mathbb{E}_s[\mathcal{T} + \mathbb{E}^1[\mathcal{T}]] \big)$$
$$+ \sup_{t \leq s \leq T} \mathbb{E}_s \Big[\gamma \mathbb{E}^1[|\mathcal{X}'^{,1}_T - \mathcal{X}'^{,2}_T|] + \int_t^T \mathbb{E}^1[|\mathcal{X}'^{,1}_u - \mathcal{X}'^{,2}_u|] du \Big].$$

Second Step. We now prove (5.44) when $\boldsymbol{\mathcal{X}}^i = \boldsymbol{\mathcal{X}}'^{,i}$, for $i = 1, 2$. Returning to (5.46), we then recover the same setting as in the second step of the proof of Corollary 5.18. We get:

$$\sup_{t \leq s \leq T} \mathcal{E}_s(\boldsymbol{\vartheta}^1 - \boldsymbol{\vartheta}^2) \leq K' \Big(\sup_{t \leq s \leq T} \Delta \mathcal{R}_a(s) + \mathbb{E}^1 \Big[\sup_{t \leq s \leq T} \Delta \mathcal{R}_a(s) \Big] + \sup_{t \leq s \leq T} \mathbb{E}_s[\mathcal{T} + \mathbb{E}^1[\mathcal{T}]] \Big).$$

We now make use of (5.34) in the statement of Corollary 5.18, choosing $\gamma^2 = c$ therein. Following Remark 5.21, this provides a bound not only for $\mathcal{E}_s(\boldsymbol{\vartheta}^2)$ but also for $|\vartheta_s^2|$. We deduce that:

$$\mathcal{T} \leq C \Big\{ 1 \wedge \Big(|\xi_1 - \xi_2| + \|\xi_1 - \xi_2\|_1 \Big) \Big\} \Big[|\eta| + \|\eta\|_1 + \sup_{t \leq s \leq T} \Big(\mathcal{R}_a^2(s) + \mathbb{E}^1[\mathcal{R}_a^2(s)] \Big) \Big].$$

Then, using the fact that, for two random variables τ and τ', $\mathbb{E}^1[\tau \wedge \tau'] \leq \mathbb{E}^1[\tau] \wedge \mathbb{E}^1[\tau']$, we have:

$$\mathbb{E}^1[\mathcal{T}] \leq C \Big\{ \|\eta\|_1 + \mathbb{E}^1 [\sup_{t \leq s \leq T} \mathcal{R}_a^2(s)] \Big\}$$
$$\wedge \Big\{ \mathbb{E}^1 \Big[|\xi_1 - \xi_2| \Big(|\eta| + \sup_{t \leq s \leq T} \mathcal{R}_a^2(s) \Big) \Big] + \|\xi_1 - \xi_2\|_1 \Big(\|\eta\|_1 + \mathbb{E}^1 \Big[\sup_{t \leq s \leq T} \mathcal{R}_a^2(s) \Big] \Big) \Big\}.$$

Finally,

$$\mathcal{T} + \mathbb{E}^1[\mathcal{T}] \leq C \Big\{ |\eta| + \|\eta\|_1 + \sup_{t \leq s \leq T} \mathcal{R}_a^2(s) + \mathbb{E}^1 [\sup_{t \leq s \leq T} \mathcal{R}_a^2(s)] \Big\}$$
$$\wedge \Big\{ \mathbb{E}^1 \Big[|\xi_1 - \xi_2| \Big(|\eta| + \sup_{t \leq s \leq T} \mathcal{R}_a^2(s) \Big) \Big]$$
$$+ \Big(|\xi_1 - \xi_2| + \|\xi_1 - \xi_2\|_1 \Big)$$
$$\times \Big(|\eta| + \|\eta\|_1 + \sup_{t \leq s \leq T} \mathcal{R}_a^2(s) + \mathbb{E}^1 \Big[\sup_{t \leq s \leq T} \mathcal{R}_a^2(s) \Big] \Big) \Big\}.$$

The result easily follows. □

5.2.4 Analysis of the First-Order Derivatives

First-Order Derivatives of the McKean-Vlasov System

As we already explained in Examples 5.14 and 5.15, the form of the system (5.22) has been chosen to allow the investigation of the derivative of the system of the original FBSDE in the direction of the measure. Thus, we shall make use of the results from Subsection 5.2.3.

To make things clear, we also recall the identification of h_ℓ, $\tilde{H}_\ell$ and H_a in (5.23):

$$\begin{aligned} \delta h_\ell(t,w,\tilde{\mathcal{X}}) &= \partial_w h\big(t,w,\mathcal{L}^1(\tilde{\mathcal{X}})\big), \\ \delta \tilde{H}_\ell(t,w,\tilde{\mathcal{X}}) &= \partial_\mu h\big(t,w,\mathcal{L}^1(\tilde{\mathcal{X}})\big)(\tilde{\mathcal{X}}), \ H_a = 0. \end{aligned} \tag{5.47}$$

The next results state the first order differentiability of the McKean-Vlasov system (5.15).

Lemma 5.23 *Given a continuously differentiable path of initial conditions $\mathbb{R} \ni \lambda \mapsto \xi^\lambda \in L^2(\Omega^1, \mathcal{F}_t^1, \mathbb{P}^1; \mathbb{R}^d)$, t standing for the initial time in $[0,T]$, we can find a constant $c = c(L) > 0$ such that, for $T \leq c$, the path $\mathbb{R} \ni \lambda \mapsto \boldsymbol{\theta}^\lambda = (\boldsymbol{X}^\lambda, \boldsymbol{Y}^\lambda) = \boldsymbol{\theta}^{t,\xi^\lambda} \in \mathbb{S}^2([t,T];\mathbb{R}^d) \times \mathbb{S}^2([t,T];\mathbb{R}^m)$ is continuously differentiable.*

Proof.

First Step. Under assumption **Smooth Coefficients Order 1**, existence and uniqueness of a solution to (5.1) may be proved for a small time horizon T by a contraction argument, see Theorem 5.4. We can approximate $(\boldsymbol{X}^\lambda, \boldsymbol{Y}^\lambda)$ as the limit of a Picard sequence $\boldsymbol{\theta}^{n,\lambda} = (\boldsymbol{X}^{n,\lambda}, \boldsymbol{Y}^{n,\lambda})$, defined by:

$$X_s^{n+1,\lambda} = \xi^\lambda + \int_t^s B\big(r, \theta_r^{n,\lambda}, \mathcal{L}^1(X_r^{n,\lambda})\big)dr + \sigma\big(W_s - W_t\big) + \sigma^0\big(W_s^0 - W_t^0\big),$$

$$Y_s^{n+1,\lambda} = \mathbb{E}_s\bigg[G\big(X_T^{n+1,\lambda}, \mathcal{L}^1(X_T^{n+1,\lambda})\big) + \int_s^T F\big(r, \theta_r^{n,\lambda}, \mathcal{L}^1(X_r^{n,\lambda})\big)dr\bigg],$$

with the initialization $\boldsymbol{\theta}^{0,\lambda} \equiv 0$. We can prove by induction that, for any $n \geq 0$, the mapping $\mathbb{R} \ni \lambda \mapsto \boldsymbol{\theta}^{n,\lambda} = (\boldsymbol{X}^{n,\lambda}, \boldsymbol{Y}^{n,\lambda}) \in \mathbb{S}^2([t,T];\mathbb{R}^d) \times \mathbb{S}^2([t,T];\mathbb{R}^m)$ is continuously differentiable. We give just a sketch of the proof. For the forward component, this follows from the fact that given a continuously differentiable path $\mathbb{R} \ni \lambda \mapsto h^\lambda \in \mathbb{S}^2([t,T];\mathbb{R})$, the path $\mathbb{R} \ni \lambda \mapsto (\int_t^s h_r^\lambda dr)_{s\in[t,T]}$ with values in $\mathbb{S}^2([t,T];\mathbb{R})$ is continuously differentiable. To handle the backward component, it suffices to prove first that the path $\mathbb{R} \ni \lambda \mapsto (\mathbb{E}_s[h_T^\lambda])_{s\in[t,T]}$, with values in $\mathbb{S}^2([t,T];\mathbb{R})$, is continuously differentiable, which is straightforward by means of Doob's inequality. This is enough to handle the terminal condition and also the driver since we can split the integral from s to T into an integral from t to s, to which we can apply the result used for the forward component, and an integral from t to T, which can be seen as a new h_T. In this way, we can prove that $\mathbb{R} \ni \lambda \mapsto \boldsymbol{Y}^{n+1,\lambda}$ is continuously differentiable from $\mathbb{R}$ to $\mathbb{S}^2([t,T];\mathbb{R}^m)$.

The derivatives, denoted by $(\boldsymbol{\mathcal{X}}^{n,\lambda}, \boldsymbol{\mathcal{Y}}^{n,\lambda})$, satisfy the system:

$$
\begin{aligned}
\mathcal{X}_s^{n+1,\lambda} &= \chi^\lambda + \int_t^s \partial B\big(r, \theta_r^{n,\lambda}, \tilde{X}_r^{n,\lambda}\big)\big(\vartheta_r^{n,\lambda}, \tilde{\mathcal{X}}_r^{n,\lambda}\big)dr, \\
\mathcal{Y}_s^{n+1,\lambda} &= \mathbb{E}_s\bigg[\partial G\big(X_T^{n+1,\lambda}, \tilde{X}_T^{n+1,\lambda}\big)\big(\mathcal{X}_T^{n+1,\lambda}, \tilde{\mathcal{X}}_T^{n+1,\lambda}\big) \\
&\quad + \int_s^T \partial F\big(r, \theta_r^{n,\lambda}, \tilde{X}_r^{n,\lambda}\big)\big(\vartheta_r^{n,\lambda}, \tilde{\mathcal{X}}_r^{n,\lambda}\big)dr\bigg],
\end{aligned}
\tag{5.48}
$$

where we have used the notations $\chi^\lambda = [d/d\lambda]\xi^\lambda$, $\boldsymbol{\vartheta}^{n,\lambda} = (\boldsymbol{\mathcal{X}}^{n,\lambda}, \boldsymbol{\mathcal{Y}}^{n,\lambda})$ and where ∂B, ∂F, and ∂G are defined according to (5.23). We thus obtain a system of the form (5.28) with $\boldsymbol{\theta} = \boldsymbol{\theta}^{n+1,\lambda}$, $\boldsymbol{X} = \boldsymbol{X}' = \boldsymbol{X}^{n+1,\lambda}$, $\bar{\boldsymbol{\theta}} = \boldsymbol{\theta}^{n,\lambda}$, $\boldsymbol{X}'' = \boldsymbol{X}^{n,\lambda}$, $\boldsymbol{\vartheta} = \boldsymbol{\vartheta}^{n+1,\lambda}$, $\boldsymbol{\mathcal{X}} = \boldsymbol{\mathcal{X}}' = \boldsymbol{\mathcal{X}}^{n+1,\lambda}$, $\bar{\boldsymbol{\vartheta}} = \boldsymbol{\vartheta}^{n,\lambda}$, $\boldsymbol{\mathcal{X}}'' = \boldsymbol{\mathcal{X}}^{n,\lambda}$, and χ^λ playing the role of η.

Second Step. We now apply Lemma 5.17, noticing that the remainder $\mathcal{R}_a$ therein is zero, see (5.47). We get:

$$
\mathcal{E}_s(\boldsymbol{\vartheta}^{n+1,\lambda}) \leq C\bigg(|\chi^\lambda| + \mathbb{E}_s\bigg[\gamma\mathbb{E}^1[|\mathcal{X}_T^{n+1,\lambda}|] + \int_t^T [|\boldsymbol{\vartheta}_r^{n,\lambda}| + \mathbb{E}^1[|\mathcal{X}_r^{n,\lambda}|]]dr\bigg]\bigg).
$$

Applying the same strategy as in the proof of Corollary 5.18, we obtain, for $t \leq r \leq s \leq T$,

$$
\mathbb{E}_r[\mathcal{E}_s(\boldsymbol{\vartheta}^{n+1,\lambda})] \leq C\bigg(|\chi^\lambda| + \mathbb{E}_r\bigg[\gamma\mathbb{E}^1[|\mathcal{X}_T^{n+1,\lambda}|] + \int_t^T \Big[|\boldsymbol{\vartheta}_u^{n,\lambda}| + \mathbb{E}^1[|\mathcal{X}_u^{n,\lambda}|]\Big]du\bigg]\bigg).
\tag{5.49}
$$

Taking the expectation under $\mathbb{P}^1$, we deduce that:

$$
\mathbb{E}_r\Big[\mathbb{E}^1[|\mathcal{X}_s^{n+1,\lambda}|]\Big] \leq C\bigg(\|\chi^\lambda\|_1 + \mathbb{E}_r\bigg[\gamma\mathbb{E}^1[|\mathcal{X}_T^{n+1,\lambda}|] + \int_t^T \mathbb{E}^1[|\boldsymbol{\vartheta}_u^{n,\lambda}|]du\bigg]\bigg).
$$

Choosing $s = T$ in the left-hand side and then γ small enough and allowing the constant C to increase from line to line, this yields:

$$
\mathbb{E}_r\Big[\mathbb{E}^1[|\mathcal{X}_T^{n+1,\lambda}|]\Big] \leq C\bigg(\|\chi^\lambda\|_1 + \mathbb{E}_r\bigg[\int_t^T \mathbb{E}^1[|\boldsymbol{\vartheta}_u^{n,\lambda}|]du\bigg]\bigg).
$$

By plugging into (5.49), we get:

$$
\mathbb{E}_r[\mathcal{E}_s(\boldsymbol{\vartheta}^{n+1,\lambda})] \leq C\bigg(|\chi^\lambda| + \|\chi^\lambda\|_1 + \mathbb{E}_r\bigg[\int_t^T \Big(|\boldsymbol{\vartheta}_u^{n,\lambda}| + \mathbb{E}^1[|\boldsymbol{\vartheta}_u^{n,\lambda}|]\Big)du\bigg]\bigg),
$$

and then, by taking the expectation under $\mathbb{P}^1$ once again,

$$
\begin{aligned}
&\mathbb{E}_r\Big[\mathcal{E}_s(\boldsymbol{\vartheta}^{n+1,\lambda}) + \mathbb{E}^1[\mathcal{E}_s(\boldsymbol{\vartheta}^{n+1,\lambda})]\Big] \\
&\quad \leq C\bigg(|\chi^\lambda| + \|\chi^\lambda\|_1 + \mathbb{E}_r\bigg[\int_t^T \Big(|\boldsymbol{\vartheta}_u^{n,\lambda}| + \mathbb{E}^1[|\boldsymbol{\vartheta}_u^{n,\lambda}|]\Big)du\bigg]\bigg).
\end{aligned}
$$

In particular, for $T \leq \gamma^2$,

$$\begin{aligned}&\mathbb{E}_r\Big[\mathcal{E}_s(\boldsymbol{\vartheta}^{n+1,\lambda}) + \mathbb{E}^1\big[\mathcal{E}_s(\boldsymbol{\vartheta}^{n+1,\lambda})\big]\Big] \\ &\quad\leq C\Big[|\chi^\lambda| + \|\chi^\lambda\|_1 + \gamma \sup_{t\leq u\leq T}\Big(\mathbb{E}_r\Big[\mathcal{E}_u(\boldsymbol{\vartheta}^{n,\lambda}) + \mathbb{E}^1\big[\mathcal{E}_u(\boldsymbol{\vartheta}^{n,\lambda})\big]\Big]\Big)\Big],\end{aligned}$$

and, passing the supremum inside the conditional expectation and choosing $r = s$ and entering the supremum inside the conditional expectation in the right-hand side,

$$\begin{aligned}&\mathcal{E}_s(\boldsymbol{\vartheta}^{n+1,\lambda}) + \mathbb{E}^1\big[\mathcal{E}_s(\boldsymbol{\vartheta}^{n+1,\lambda})\big] \\ &\quad\leq C\Big[|\chi^\lambda| + \|\chi^\lambda\|_1 + \gamma\mathbb{E}_s\Big[\sup_{t\leq u\leq T}\Big(\mathcal{E}_u(\boldsymbol{\vartheta}^{n,\lambda}) + \mathbb{E}^1\big[\mathcal{E}_u(\boldsymbol{\vartheta}^{n,\lambda})\big]\Big)\Big]\Big].\end{aligned}$$

By induction, we deduce that, for $C\gamma \leq 1/2$, with probability 1 under $\mathbb{P}$,

$$\begin{aligned}&\sup_{t\leq s\leq T}\Big[\mathcal{E}_s(\boldsymbol{\vartheta}^{n,\lambda}) + \mathbb{E}^1\big[\mathcal{E}_s(\boldsymbol{\vartheta}^{n,\lambda})\big]\Big] \\ &\quad\leq \sum_{i=0}^{n-1} C(C\gamma)^i\Big(|\chi^\lambda| + \|\chi^\lambda\|_1\Big) \leq 2C\Big(|\chi^\lambda| + \|\chi^\lambda\|_1\Big),\end{aligned}$$

where we used the fact that $\boldsymbol{\vartheta}^{0,\lambda} = 0$. Choosing γ exactly equal to $1/(2C)$, we get, for a new constant C',

$$\sup_{t\leq s\leq T}\Big[|\vartheta_s^{n,\lambda}| + \mathbb{E}^1\big[|\vartheta_s^{n,\lambda}|\big]\Big] \leq 2C'\Big(|\chi^\lambda| + \|\chi^\lambda\|_1\Big).$$

Third Step. We now make use of (5.37) in Lemma 5.19 in order to compare $\boldsymbol{\vartheta}^{n,\lambda}$ and $\boldsymbol{\vartheta}^{n+j,\lambda}$, for $j \geq 1$. Clearly, the remainder $\Delta\mathcal{R}_a$ in (5.36) is zero since the $\mathcal{R}_a$ terms are here equal to zero, recall (5.47). By the above argument, $|\vartheta_s^{n,\lambda}|$ and $\mathbb{E}^1[|\vartheta_s^{n,\lambda}|]$ are less than $C'(|\chi^\lambda| + \mathbb{E}^1[|\chi^\lambda|])$. Therefore, (5.37) with $\boldsymbol{\vartheta}^1 = \boldsymbol{\vartheta}^{n+j,\lambda}$, $\boldsymbol{\vartheta}^2 = \boldsymbol{\vartheta}^{n,\lambda}$, $\bar{\boldsymbol{\vartheta}}^1 = \boldsymbol{\vartheta}^{n+j-1,\lambda}$, $\bar{\boldsymbol{\vartheta}}^2 = \boldsymbol{\vartheta}^{n-1,\lambda}$, $\boldsymbol{\mathcal{X}}^1 = \boldsymbol{\mathcal{X}}'^{,1} = \boldsymbol{\mathcal{X}}^{n+j,\lambda}$, $\boldsymbol{\mathcal{X}}^2 = \boldsymbol{\mathcal{X}}'^{,2} = \boldsymbol{\mathcal{X}}^{n,\lambda}$, $\boldsymbol{\mathcal{X}}''^{,1} = \boldsymbol{\mathcal{X}}^{n+j-1,\lambda}$, $\boldsymbol{\mathcal{X}}''^{,2} = \boldsymbol{\mathcal{X}}^{n-j,\lambda}$, $\boldsymbol{\theta}^1 = \boldsymbol{\theta}^{n+j,\lambda}$, $\boldsymbol{\theta}^2 = \boldsymbol{\theta}^{n,\lambda}$, $\bar{\boldsymbol{\theta}}^1 = \boldsymbol{\theta}^{n+j-1,\lambda}$, $\bar{\boldsymbol{\theta}}^2 = \boldsymbol{\theta}^{n-1,\lambda}$, $\boldsymbol{X}^1 = \boldsymbol{X}'^{,1} = \boldsymbol{X}^{n+j,\lambda}$, $\boldsymbol{X}^2 = \boldsymbol{X}'^{,2} = \boldsymbol{X}^{n,\lambda}$, $\boldsymbol{X}''^{,1} = \boldsymbol{X}^{n+j-1,\lambda}$, $\boldsymbol{X}''^{,2} = \boldsymbol{X}^{n-1,\lambda}$, yields:

$$\begin{aligned}&\mathcal{E}_s(\boldsymbol{\vartheta}^{n+j,\lambda} - \boldsymbol{\vartheta}^{n,\lambda}) \\ &\leq C\mathbb{E}_s\Big[\gamma\mathbb{E}^1\big[|\mathcal{X}_T^{n+j,\lambda} - \mathcal{X}_T^{n,\lambda}|\big] + \int_t^T\Big(|\vartheta_u^{n+j-1,\lambda} - \vartheta_u^{n-1,\lambda}| + \mathbb{E}^1\big[|\mathcal{X}_u^{n+j-1,\lambda} - \mathcal{X}_u^{n-1,\lambda}|\big]\Big)du\Big] \\ &\quad + K'(\mathbb{E}_s + \mathbb{E}_s^0)\Big[\Big\{1 \wedge \Big(|X_T^{n+j,\lambda} - X_T^{n,\lambda}| + \mathbb{E}^1\big[|X_T^{n+j,\lambda} - X_T^{n,\lambda}|\big]\Big)\Big\} \times \Big(|\chi^\lambda| + \|\chi^\lambda\|_1\Big) \\ &\quad + \int_t^T\Big[\Big\{1 \wedge \Big(|\theta_u^{n+j-1,\lambda} - \theta_u^{n-1,\lambda}| + \mathbb{E}^1\big[|X_u^{n+j-1,\lambda} - X_u^{n-1,\lambda}|\big]\Big)\Big\}\Big(|\chi^\lambda| + \|\chi^\lambda\|_1\Big)\Big]du\Big],\end{aligned} \tag{5.50}$$

where K' is a constant only depending on L and Γ.

In order to complete the proof, it remains to estimate the difference $\boldsymbol{\theta}^{n+j,\lambda} - \boldsymbol{\theta}^{n,\lambda}$. We start with the case $j = 1$. To do so, we follow the proof of Proposition 5.13. In full analogy with (5.19), we claim that, for all $s \in [t, T]$, it holds that:

$$\sup_{t \le r \le s} |X_r^{n+1,\lambda} - X_r^{n,\lambda}| \le C \int_t^s \Big(|\theta_r^{n,\lambda} - \theta_r^{n-1,\lambda}| + \mathbb{E}^1\big[|X_r^{n,\lambda} - X_r^{n-1,\lambda}|\big] \Big) dr, \tag{5.51}$$

while, reproducing the second step of the proof Proposition 5.13, we obtain, for all $t \le r \le s \le T$,

$$\begin{aligned}\mathbb{E}_r\big[|Y_s^{n+1,\lambda} - Y_s^{n,\lambda}|\big] &\le C\mathbb{E}_r\Big[|X_T^{n+1,\lambda} - X_T^{n,\lambda}| + \mathbb{E}^1\big[|X_T^{n+1,\lambda} - X_T^{n,\lambda}|\big]\Big] \\ &\quad + C\int_s^T \mathbb{E}_r\Big[|\theta_u^{n,\lambda} - \theta_u^{n-1,\lambda}| + \mathbb{E}^1\big[|X_u^{n,\lambda} - X_u^{n-1,\lambda}|\big]\Big] du,\end{aligned}$$

where C only depends on L. Plugging (5.51) into the above inequality, we deduce that:

$$\mathbb{E}_r\big[|Y_s^{n+1,\lambda} - Y_s^{n,\lambda}|\big] \le C \int_t^T \mathbb{E}_r\Big[|\theta_u^{n,\lambda} - \theta_u^{n-1,\lambda}| + \mathbb{E}^1\big[|\theta_u^{n,\lambda} - \theta_u^{n-1,\lambda}|\big]\Big] du.$$

By taking the conditional expectation in (5.51) and allowing the constant C to increase from line to line, we get:

$$\begin{aligned}\mathbb{E}_r\big[|\theta_s^{n+1,\lambda} - \theta_s^{n,\lambda}|\big] &\le C \int_r^T \mathbb{E}_r\Big[|\theta_u^{n,\lambda} - \theta_u^{n-1,\lambda}| + \mathbb{E}^1\big[|\theta_u^{n,\lambda} - \theta_u^{n-1,\lambda}|\big]\Big] du \\ &\quad + C\int_t^r \Big[|\theta_u^{n,\lambda} - \theta_u^{n-1,\lambda}| + \mathbb{E}^1\big[|\theta_u^{n,\lambda} - \theta_u^{n-1,\lambda}|\big]\Big] du.\end{aligned}$$

By taking the expectation under $\mathbb{P}^1$, we finally have:

$$\begin{aligned}&\mathbb{E}_r\Big[|\theta_s^{n+1,\lambda} - \theta_s^{n,\lambda}| + \mathbb{E}^1\big[|\theta_s^{n+1,\lambda} - \theta_s^{n,\lambda}|\big]\Big] \\ &\le C \int_r^T \mathbb{E}_r\Big[|\theta_u^{n,\lambda} - \theta_u^{n-1,\lambda}| + \mathbb{E}^1\big[|\theta_u^{n,\lambda} - \theta_u^{n-1,\lambda}|\big]\Big] du \\ &\quad + C\int_t^r \Big[|\theta_u^{n,\lambda} - \theta_u^{n-1,\lambda}| + \mathbb{E}^1\big[|\theta_u^{n,\lambda} - \theta_u^{n-1,\lambda}|\big]\Big] du.\end{aligned}$$

Therefore, taking the supremum over $s \in [r, T]$ and then over $r \in [t, T]$, we obtain:

$$\begin{aligned}&\sup_{t \le r \le s \le T} \mathbb{E}_r\Big[|\theta_s^{n+1,\lambda} - \theta_s^{n,\lambda}| + \mathbb{E}^1\big[|\theta_s^{n+1,\lambda} - \theta_s^{n,\lambda}|\big]\Big] \\ &\le CT \sup_{t \le r \le u \le T} \mathbb{E}_r\Big[|\theta_u^{n,\lambda} - \theta_u^{n-1,\lambda}| + \mathbb{E}^1\big[|\theta_u^{n+1,\lambda} - \theta_u^{n,\lambda}|\big]\Big] \\ &\quad + CT \sup_{t \le u \le T} \Big[|\theta_u^{n,\lambda} - \theta_u^{n-1,\lambda}| + \mathbb{E}^1\big[|\theta_u^{n,\lambda} - \theta_u^{n-1,\lambda}|\big]\Big],\end{aligned} \tag{5.52}$$

where, to define the supremum on the first and second lines, we used the fact that, for a real-valued process $(\zeta_s)_{t \le s \le T}$ with continuous paths such that $(\sup_{t \le s \le T} |\zeta_s|)_{t \le s \le T}$ is integrable, we can construct a version of the conditional expectations $(\mathbb{E}_r[\zeta_s])_{t \le r \le s \le T}$ such that, with probability 1, for all $t \le r \le s \le T$,

$$\lim_{(s',r')\to(s,r),s'>s,r'>r} \mathbb{E}_{r'}\big[\zeta_{s'}\big] = \mathbb{E}_r\big[\zeta_s\big].$$

To do so, it suffices to construct first $(\mathbb{E}_r[\zeta_s])_{t\le r\le T}$ for s in a dense countable subset Q of $[t,T]$, each process $(\mathbb{E}_r[\zeta_s])_{t\le r\le T}$ being right-continuous. Then, denoting by w the pathwise modulus of continuity of the paths of ζ, we observe that, for any $r \in [t,T]$ and $s, s' \in Q$,

$$\big|\mathbb{E}_r[\zeta_s] - \mathbb{E}_r[\zeta_{s'}]\big| \le \sup_{t\le r\le T} \mathbb{E}_r\big[w(|s'-s|)\big].$$

In order to complete the construction, we notice that the family $(\sup_{t\le r\le T}\mathbb{E}_r[w(\varepsilon)])_{\varepsilon>0}$ converges to 0 almost surely since it is nondecreasing in ε and converges to 0 in probability as ε tends to 0. Therefore, on a common event of probability 1, all the mappings $Q \ni s \mapsto \mathbb{E}_r[\zeta_s]$, for $r \in [t,T]$, extends into a continuous mapping and the resulting collection of mappings $([t,T] \ni s \mapsto \mathbb{E}_r[\zeta_s])_{r\in[t,T]}$ is uniformly continuous.

Also, since $\mathbb{E}_s[\zeta_s] = \zeta_s$ for all $s \in Q$, the version we just constructed must satisfy, by an obvious continuity argument, $\mathbb{E}_s[\zeta_s] = \zeta_s$ for all $s \in [t,T]$, on a common event of probability 1. In particular, $\sup_{t\le r\le s\le T}\mathbb{E}_r[\zeta_s] \ge \sup_{t\le s\le T}\zeta_s$.

Therefore, going back to (5.52) and choosing $CT \le 1/4$ therein, we deduce that:

$$\sup_{t\le s\le T}\Big[|\theta_s^{n+1,\lambda} - \theta_s^{n,\lambda}| + \mathbb{E}^1\big[|\theta_s^{n+1,\lambda} - \theta_s^{n,\lambda}|\big]\Big] \le \frac{1}{2^n}\psi^\lambda,$$

for $n \ge 0$, with:

$$\psi^\lambda = \sup_{t\le r\le u\le T} \mathbb{E}_r\Big[|\theta_u^{1,\lambda}| + \mathbb{E}^1\big[|\theta_u^{1,\lambda}|\big]\Big].$$

Importantly, observe that $\mathbb{E}[(\psi^\lambda)^2] < \infty$, the bound being uniform with respect to λ in compact subsets of $\mathbb{R}$. Also, by the triangle inequality, $\boldsymbol{\theta}^{n+j} - \boldsymbol{\theta}^n$ satisfies the same bound as $\boldsymbol{\theta}^{n+1} - \boldsymbol{\theta}^n$, for any $j \ge 1$, up to an additional multiplicative factor 2 in the right-hand side.

Returning to (5.50) and plugging the above bound, we deduce that, for a new value of the constant K':

$$\begin{aligned}
&\mathcal{E}_s\big(\boldsymbol{\vartheta}^{n+j,\lambda} - \boldsymbol{\vartheta}^{n,\lambda}\big) \\
&\le C\mathbb{E}_s\Big[\gamma\mathbb{E}^1\big[|\mathcal{X}_T^{n+j,\lambda} - \mathcal{X}_T^{n,\lambda}|\big] + \int_t^T \Big(|\vartheta_u^{n+j-1,\lambda} - \vartheta_u^{n-1,\lambda}| + \mathbb{E}^1\big[|\mathcal{X}_u^{n+j-1,\lambda} - \mathcal{X}_u^{n-1,\lambda}|\big]\Big)du\Big] \\
&\quad + K'\mathbb{E}_s\Big[\big(\psi^{n,\lambda} + \mathbb{E}^1\big[\psi^{n,\lambda}\big]\big)\big(|\chi^\lambda| + \|\chi^\lambda\|_1 + \mathbb{E}^1\big[\psi^{n,\lambda}|\chi^\lambda|\big]\big)\Big],
\end{aligned} \tag{5.53}$$

where we have let $\psi^{n,\lambda} = 1 \wedge (2^{-n}\psi^\lambda)$.

Fourth Step. Take the supremum with respect to s in (5.53) and then the expectation of the square under $\mathbb{P}$. By Doob's inequality and from the fact that $\gamma \le 1$, we deduce:

$$\begin{aligned}
&\mathbb{E}\Big[\sup_{t\le s\le T}\mathcal{E}_s\big(\boldsymbol{\vartheta}^{n+j,\lambda} - \boldsymbol{\vartheta}^{n,\lambda}\big)^2\Big] \\
&\le C\gamma^2\mathbb{E}\Big[\sup_{t\le s\le T}\mathcal{E}_s\big(\boldsymbol{\vartheta}^{n+j,\lambda} - \boldsymbol{\vartheta}^{n,\lambda}\big)^2\Big] + C\frac{T^2}{\gamma^2}\mathbb{E}\Big[\sup_{t\le s\le T}\mathcal{E}_s\big(\boldsymbol{\vartheta}^{n+j-1,\lambda} - \boldsymbol{\vartheta}_s^{n-1,\lambda}\big)^2\Big] \\
&\quad + K'\mathbb{E}\Big[\big(\psi^{n,\lambda} + \mathbb{E}^1\big[\psi^{n,\lambda}\big]\big)^2\Big(|\chi^\lambda| + \|\chi^\lambda\|_1 + \mathbb{E}^1\big[\psi^{n,\lambda}|\chi^\lambda|\big]\Big)^2\Big],
\end{aligned}$$

for possibly new values of C and K'. Choosing $C\gamma^2 \leq 1/3$ and $T \leq \gamma^2$, we finally have:

$$
\begin{aligned}
&\mathbb{E}\Big[\sup_{t\le s\le T} \mathcal{E}_s\big(\boldsymbol{\vartheta}^{n+j,\lambda} - \boldsymbol{\vartheta}^{n,\lambda}\big)^2\Big] \\
&\leq \frac{1}{2}\mathbb{E}\Big[\sup_{t\le s\le T} \mathcal{E}_s\big(\boldsymbol{\vartheta}^{n+j-1,\lambda} - \boldsymbol{\vartheta}_s^{n-1,\lambda}\big)^2\Big] \\
&\quad + K'\mathbb{E}\Big[\big(\psi^{n,\lambda} + \mathbb{E}^1[\psi^{n,\lambda}]\big)^2\Big(|\chi^\lambda| + \|\chi^\lambda\|_1 + \mathbb{E}^1\big[\psi^{n,\lambda}|\chi^\lambda|\big]\Big)^2\Big],
\end{aligned}
\tag{5.54}
$$

for a new value of K'.

Observe now that the sequence $(\mathbb{E}[(\psi^{n,\lambda})^2])_{n\geq 0}$ converges to 0, uniformly with respect to λ in compact subsets of $\mathbb{R}$. Observe also that the sequence $(\psi^{n,\lambda})_{n\geq 0}$ is bounded by 1. Moreover, by continuity of the map $\mathbb{R} \ni \lambda \mapsto \chi^\lambda \in L^2(\Omega^1, \mathcal{F}_t^1, \mathbb{P}^1; \mathbb{R}^d)$, the random variables $(\chi^\lambda)_{\lambda\in[a,b]}$ are uniformly square-integrable, for any $a < b$. Therefore, the left-hand side in (5.54) converges to 0 as n tends to ∞, the convergence being uniform with respect to λ in compact subsets and in $j \geq 1$. By a Cauchy argument, the proof is completed. □

We emphasize that the derivative process $[d/d\lambda]_{|\lambda=0}\theta^\lambda$ given by Lemma 5.23 satisfies (5.22) with $\eta = \chi$, with $\boldsymbol{X} = \boldsymbol{X}' = \boldsymbol{X}^0$ and $\boldsymbol{\mathcal{X}} = \boldsymbol{\mathcal{X}}' = [d/d\lambda]_{|\lambda=0}\boldsymbol{X}^\lambda$ and with the coefficients given in (5.47). In particular, for T small enough, the uniqueness of the solution to (5.22) (see Remark 5.20) ensures that the derivative process at $\lambda = 0$ depends only on the family $(\xi^\lambda)_{\lambda\in\mathbb{R}}$ through ξ^0 and $[d/d\lambda]_{|\lambda=0}\xi^\lambda$. Thus, when $\xi^0 = \xi$ and $[d/d\lambda]_{|\lambda=0}\xi^\lambda = \chi$, we may denote by $(\partial_\chi \boldsymbol{X}^{t,\xi}, \partial_\chi \boldsymbol{Y}^{t,\xi})$ the tangent process at ξ in the direction χ. It satisfies:

$$
\left\{
\begin{aligned}
\partial_\chi X_s^{t,\xi} &= \chi + \int_t^s \Big(\partial_x B\big(r, X_r^{t,\xi}, \mathcal{L}^1(X_r^{t,\xi}), Y_r^{t,\xi}\big)\partial_\chi X_r^{t,\xi} \\
&\qquad + \partial_y B\big(r, X_r^{t,\xi}, \mathcal{L}^1(X_r^{t,\xi}), Y_r^{t,\xi}\big)\partial_\chi Y_r^{t,\xi} \\
&\qquad + \tilde{\mathbb{E}}^1\big[\partial_\mu B\big(r, X_r^{t,\xi}, \mathcal{L}^1(X_r^{t,\xi}), Y_r^{t,\xi}\big)(\tilde{X}_r^{t,\xi})\partial_\chi \tilde{X}_r^{t,\xi}\big]\Big)dr, \\
\partial_\chi Y_s^{t,\xi} &= \mathbb{E}_s\Big[\partial_x G\big(X_T^{t,\xi}, \mathcal{L}^1(X_T^{t,\xi})\big)\partial_\chi X_T^{t,\xi} \\
&\qquad + \tilde{\mathbb{E}}^1\big[\partial_\mu G\big(X_T^{t,\xi}, \mathcal{L}^1(X_T^{t,\xi})\big)(\tilde{X}_T^{t,\xi})\partial_\chi \tilde{X}_T^{t,\xi}\big] \\
&\qquad + \int_s^T \Big(\partial_x F\big(r, X_r^{t,\xi}, \mathcal{L}^1(X_r^{t,\xi}), Y_r^{t,\xi}\big)\partial_\chi X_r^{t,\xi} \\
&\qquad + \partial_y F\big(r, X_r^{t,\xi}, \mathcal{L}^1(X_r^{t,\xi}), Y_r^{t,\xi}\big)\partial_\chi Y_r^{t,\xi} \\
&\qquad + \tilde{\mathbb{E}}^1\big[\partial_\mu F\big(r, X_r^{t,\xi}, \mathcal{L}^1(X_r^{t,\xi}), Y_r^{t,\xi}\big)(\tilde{X}_r^{t,\xi})\partial_\chi \tilde{X}_r^{t,\xi}\big]\Big)dr\Big],
\end{aligned}
\right.
\tag{5.55}
$$

or equivalently, using the notation (5.23),

$$
\left\{
\begin{aligned}
\partial_\chi X_s^{t,\xi} &= \chi + \int_t^s \partial B\big(r, \theta_r^{t,\xi}, \tilde{X}_r^{t,\xi}\big)\big(\partial_\chi \theta_r^{t,\xi}, \partial_\chi \tilde{X}_r^{t,\xi}\big)dr, \\
\partial_\chi Y_s^{t,\xi} &= \mathbb{E}_s\Big[\partial G\big(X_T^{t,\xi}, \tilde{X}_T^{t,\xi}\big)\big(\partial_\chi X_T^{t,\xi}, \partial_\chi \tilde{X}_T^{t,\xi}\big) \\
&\qquad + \int_s^T \Big(\partial F\big(r, \theta_r^{t,\xi}, \tilde{X}_r^{t,\xi}\big)\big(\partial_\chi \theta_r^{t,\xi}, \partial_\chi \tilde{X}_r^{t,\xi}\big)\Big)dr\Big],
\end{aligned}
\right.
\tag{5.56}
$$

where we let $\partial_\chi \boldsymbol{\theta}^{t,\xi} = (\partial_\chi \boldsymbol{X}^{t,\xi}, \partial_\chi \boldsymbol{Y}^{t,\xi})$. By linearity of the system, $\partial_\chi \boldsymbol{\theta}^{t,\xi}$ is linear in χ. By a direct application of Corollary 5.18 – recall $\delta H_a = 0$ in the current case – we have:

Lemma 5.24 *There exist two constants $c > 0$ and C, only depending on L, such that, for $T \leq c$, it holds, for any $\xi, \chi \in L^2(\Omega^1, \mathcal{F}^1, \mathbb{P}^1; \mathbb{R}^d)$, with probability 1 under $\mathbb{P}$,*

$$\sup_{t \leq s \leq T} |\partial_\chi \theta_s^{t,\xi}| \leq C\big(|\chi| + \|\chi\|_1\big), \tag{5.57}$$

and

$$\sup_{t \leq s \leq T} \mathbb{E}^1\big[|\partial_\chi \theta_s^{t,\xi}|\big] \leq C\|\chi\|_1. \tag{5.58}$$

By taking the expectation of the square in the inequality (5.57), we deduce that the map $L^2(\Omega^1, \mathcal{F}_t^1, \mathbb{P}^1; \mathbb{R}^d) \ni \chi \mapsto \partial_\chi \boldsymbol{\theta}^{t,\xi} \in \mathbb{S}^2([t,T]; \mathbb{R}^d) \times \mathbb{S}^2([t,T]; \mathbb{R}^m)$ is continuous, which proves that $L^2(\Omega^1, \mathcal{F}_t^1, \mathbb{P}^1; \mathbb{R}^d) \ni \xi \mapsto \boldsymbol{\theta}^{t,\xi} \in \mathbb{S}^2([t,T]; \mathbb{R}^d) \times \mathbb{S}^2([t,T]; \mathbb{R}^m)$ is Gâteaux differentiable. The next lemma shows that the Gâteaux derivative is continuous:

Lemma 5.25 *There exist two constants $c > 0$ and C, with c only depending on L and C only depending on L and Γ, such that, for $T \leq c$, it holds, for any $\xi_1, \xi_2, \chi \in L^2(\Omega^1, \mathcal{F}_t^1, \mathbb{P}^1; \mathbb{R}^d)$, with probability 1 under $\mathbb{P}$,*

$$\begin{aligned}
&\sup_{t \leq s \leq T} |\partial_\chi \theta_s^{t,\xi_1} - \partial_\chi \theta_s^{t,\xi_2}| \\
&\quad \leq C\Big(\mathbb{E}^1\big[|\xi_1 - \xi_2|\,|\chi|\big] + \big(|\xi_1 - \xi_2| + \|\xi_1 - \xi_2\|_1\big)\big(|\chi| + \|\chi\|_1\big)\Big), \\
&\mathbb{E}^1\big[\sup_{t \leq s \leq T} |\partial_\chi \theta_s^{t,\xi_1} - \partial_\chi \theta_s^{t,\xi_2}|\big] \leq C\Big(\mathbb{E}^1\big[|\xi_1 - \xi_2|\,|\chi|\big] + \|\xi_1 - \xi_2\|_1 \|\chi\|_1\Big).
\end{aligned}$$

Proof. The first inequality is a consequence of Proposition 5.13 and of (5.44) in Corollary 5.22, with $\mathcal{R}_a^2 = \Delta \mathcal{R}_a = 0$. The second inequality easily follows by taking expectation under $\mathbb{P}^1$. □

First-Order Derivatives of the Non-McKean-Vlasov System with Respect to μ

We reproduce the same analysis as above, but with the process $\theta^{t,x,\xi}$ instead of $\theta^{t,\xi}$ by taking advantage of the fact that the dependence of the coefficients of the system (5.6) upon the law of $\theta^{t,\xi}$ is already known to be smooth. This permits to discuss the differentiability of $\theta^{t,x,\xi}$ in a straightforward manner. Importantly, recall that $\theta^{t,x,\xi}$ coincides with $\theta^{t,x,\mu}$ if $\xi \sim \mu$. Throughout the analysis, we shall use both notations, choosing one or the other according to the context.

We mimic the strategy of the previous subsection. Considering a continuously differentiable mapping $\mathbb{R} \ni \lambda \mapsto \xi^\lambda \in L^2(\Omega^1, \mathcal{F}_t^1, \mathbb{P}^1; \mathbb{R}^d)$, we prove that the mapping $\mathbb{R} \ni \lambda \mapsto (\boldsymbol{X}^{t,x,\xi^\lambda}, \boldsymbol{Y}^{t,x,\xi^\lambda}) \in \mathbb{S}^2([t,T];\mathbb{R}^d) \times \mathbb{S}^2([t,T];\mathbb{R}^m)$ is continuously differentiable. The crucial insight is that, for any $\lambda \in \mathbb{R}$, the coefficients of the FBSDE (5.6) satisfied by θ^{t,x,ξ^λ} depend in a smooth way upon the solution θ^{t,ξ^λ} of the FBSDE (5.2). Since we have already established the continuous differentiability of the mapping $\lambda \mapsto \theta^{t,\xi^\lambda}$, it suffices now to prove that the solution of a standard FBSDE depending on a parameter λ in a continuously differentiable way is also continuously differentiable with respect to λ. The proof consists in a mere adaptation of the proof of Lemma 5.23. We shall omit it. When $\xi^0 = \xi$ and $[d/d\lambda]_{\lambda=0}\xi^\lambda = \chi$, we shall denote the directional derivative at ξ in the χ direction by:

$$\left(\partial_\chi X_s^{t,x,\xi}, \partial_\chi Y_s^{t,x,\xi}\right)_{s\in[t,T]},$$

seen as an element of $\mathbb{S}^2([t,T];\mathbb{R}^d) \times \mathbb{S}^2([t,T];\mathbb{R}^m)$. By the same argument as above, the pair process $(\partial_\chi \boldsymbol{X}^{t,x,\xi}, \partial_\chi \boldsymbol{Y}^{t,x,\xi})$ satisfies a "differentiated" system, of the type (5.22), namely:

$$\left\{\begin{aligned}
\partial_\chi X_s^{t,x,\xi} &= \int_t^s \Big(\partial_x B\big(r, X_r^{t,x,\xi}, \mathcal{L}^1(X_r^{t,\xi}), Y_r^{t,x,\xi}\big)\partial_\chi X_r^{t,x,\xi} \\
&\quad + \partial_y B\big(r, X_r^{t,x,\xi}, \mathcal{L}^1(X_r^{t,\xi}), Y_r^{t,x,\xi}\big)\partial_\chi Y_r^{t,x,\xi} \\
&\quad + \tilde{\mathbb{E}}^1\big[\partial_\mu B\big(r, X_r^{t,x,\xi}, \mathcal{L}^1(X_r^{t,\xi}), Y_r^{t,x,\xi}\big)(\tilde{X}_r^{t,\xi})\partial_\chi \tilde{X}_r^{t,\xi}\big]\Big) dr, \\
\partial_\chi Y_s^{t,x,\xi} &= \mathbb{E}_s\bigg[\partial_x G\big(X_T^{t,x,\xi}, \mathcal{L}^1(X_T^{t,\xi})\big)\partial_\chi X_T^{t,x,\xi} \\
&\quad + \tilde{\mathbb{E}}^1\big[\partial_\mu G\big(X_T^{t,x,\xi}, \mathcal{L}^1(X_T^{t,\xi})\big)(\tilde{X}_T^{t,\xi})\partial_\chi \tilde{X}_T^{t,\xi}\big] \\
&\quad + \int_s^T \Big(\partial_x F\big(r, X_r^{t,x,\xi}, \mathcal{L}^1(X_r^{t,\xi}), Y_r^{t,x,\xi}\big)\partial_\chi X_r^{t,x,\xi} \\
&\quad + \partial_y F\big(r, X_r^{t,x,\xi}, \mathcal{L}^1(X_r^{t,\xi}), Y_r^{t,x,\xi}\big)\partial_\chi Y_r^{t,x,\xi} \\
&\quad + \tilde{\mathbb{E}}^1\big[\partial_\mu F\big(r, X_r^{t,x,\xi}, \mathcal{L}^1(X_r^{t,\xi}), Y_r^{t,x,\xi}\big)(\tilde{X}_r^{t,\xi})\partial_\chi \tilde{X}_r^{t,\xi}\big]\Big) dr\bigg],
\end{aligned}\right. \tag{5.59}$$

for $s \in [t,T]$. Equivalently, using the notation (5.23),

$$\left\{\begin{aligned}
\partial_\chi X_s^{t,x,\xi} &= \int_t^s \partial B\big(r, \theta_r^{t,x,\xi}, \tilde{X}_r^{t,\xi}\big)\big(\partial_\chi \theta_r^{t,x,\xi}, \partial_\chi \tilde{X}_r^{t,\xi}\big) dr, \\
\partial_\chi Y_s^{t,\xi} &= \mathbb{E}_s\bigg[\partial G\big(X_T^{t,x,\xi}, \tilde{X}_T^{t,\xi}\big)\big(\partial_\chi X_T^{t,x,\xi}, \partial_\chi \tilde{X}_T^{t,\xi}\big) \\
&\qquad + \int_s^T \partial F\big(r, \theta_r^{t,x,\xi}, \tilde{X}_r^{t,\xi}\big)\big(\partial_\chi \theta_r^{t,x,\xi}, \partial_\chi \tilde{X}_r^{t,\xi}\big) dr\bigg],
\end{aligned}\right. \tag{5.60}$$

where, as usual, we have let $\partial_\chi \boldsymbol{\theta}^{t,x,\xi} = (\partial_\chi \boldsymbol{X}^{t,x,\xi}, \partial_\chi \boldsymbol{Y}^{t,x,\xi})$. Importantly, the law of the process $\partial_\chi \boldsymbol{\theta}^{t,x,\xi}$ depends on ξ through the joint law $\mathcal{L}^1(\xi,\chi)$; here, the shorten notation $\partial_\chi \boldsymbol{\theta}^{t,x,\mu}$, with $\mu = \mathcal{L}^1(\xi)$, would be meaningless.

As uniqueness holds in small time, the solution only depends on the family $(\xi^\lambda)_{\lambda\in\mathbb{R}}$ through the values of ξ and χ. In comparison with (5.22), we have $\eta = 0$, $\boldsymbol{\theta} = \boldsymbol{\theta}^{t,x,\mu}$, $\boldsymbol{X}' = \boldsymbol{X}^{t,\xi}$, $\boldsymbol{\vartheta} = \partial_\chi \boldsymbol{\theta}^{t,x,\xi}$ and $\boldsymbol{\mathcal{X}}' = \partial_\chi \boldsymbol{X}^{t,\xi}$, the tangent process $\partial_\chi \boldsymbol{X}^{t,\xi}$ being given by Lemma 5.23 and (5.55). The coefficients are of the general shape (5.24) and (5.25). When h stands for one of the functions B, F, or G and $\boldsymbol{V}$ for $\boldsymbol{\theta}^{t,x,\xi}$ or $\boldsymbol{X}^{t,x,\xi}$ and $\boldsymbol{X}'$ for $\boldsymbol{X}^{t,\xi}$, according to the cases, it holds, as in (5.47),

$$\begin{aligned}
&\delta h_\ell\big(s, V_s, \tilde{X}'_s\big) = \partial_w h\big(s, V_s, \mathcal{L}^1(\tilde{X}'_s)\big),\\
&\delta \tilde{H}_\ell\big(s, V_s, \tilde{X}'_s\big) = \partial_\mu h\big(s, V_s, \mathcal{L}^1(\tilde{X}'_s)\big)(\tilde{X}'_s),\\
&\delta H_a = 0.
\end{aligned}\tag{5.61}$$

Combining Proposition 5.13 with Corollaries 5.18 and 5.22, we deduce, in analogy with Lemma 5.25:

Lemma 5.26 *There exist two constants c and C, with c only depending on L and C only depending on Γ, such that, for $T \le c$, it holds, for all $\xi_1, \xi_2, \chi \in L^2(\Omega^1, \mathcal{F}^1_t, \mathbb{P}^1; \mathbb{R}^d)$, with probability 1 under $\mathbb{P}$,*

$$\begin{aligned}
&\sup_{t\le s\le T} |\partial_\chi \theta_s^{t,x,\xi}| \le C\|\chi\|_1,\\
&\sup_{t\le s\le T} |\partial_\chi \theta_s^{t,x,\xi_1} - \partial_\chi \theta_s^{t,x,\xi_2}| \le C\Big(\mathbb{E}^1\big[|\xi_1-\xi_2|\,|\chi|\big] + \|\xi_1-\xi_2\|_1\|\chi\|_1\Big).
\end{aligned}\tag{5.62}$$

Proof. The first bound in (5.62) is a direct consequence of (5.33) in Corollary 5.18, with $\eta = 0$, $\mathcal{R}_a = 0$ and $\boldsymbol{\mathcal{X}}' = \partial_\chi \boldsymbol{X}^{t,\xi}$, combined with (5.58) in Lemma 5.24.

The second bound follows from (5.43) in Corollary 5.22, with $\eta = 0$, $\mathcal{R}_a^2 = \Delta\mathcal{R}_a = 0$, $\boldsymbol{\theta}^{\xi_i} = \boldsymbol{\theta}^{t,x,\xi_i}$, $\boldsymbol{\vartheta}^{\xi_i} = \partial_\chi \boldsymbol{\theta}^{t,x,\xi_i}$, $\boldsymbol{X}'^{,\xi_i} = \boldsymbol{X}^{t,\xi_i}$ and $\boldsymbol{\mathcal{X}}'^{,i} = \partial_\chi \boldsymbol{X}^{t,\xi_i}$, for $i = 1, 2$, combined with Lemma 5.25. □

We deduce:

Lemma 5.27 *For $T \le c$, with $c > 0$ only depending on L, for any $t \in [0,T]$ and $x \in \mathbb{R}^d$, the function:*

$$L^2(\Omega^1, \mathcal{F}^1_t, \mathbb{P}^1; \mathbb{R}^d) \ni \xi \mapsto Y_t^{t,x,\xi}$$

is continuously Fréchet differentiable. In particular, the function $\mathcal{P}_2(\mathbb{R}^d) \ni \mu \mapsto \mathcal{U}(t,x,\mu) = Y_t^{t,x,\mu}$ is L-differentiable. Moreover, for all $x \in \mathbb{R}^d$, for all $\xi_1, \xi_2, \chi \in L^2(\Omega^1, \mathcal{F}^1, \mathbb{P}^1; \mathbb{R}^d)$, we have, with $\mu_1 = \mathcal{L}^1(\xi_1)$ and $\mu_2 = \mathcal{L}^1(\xi_2)$,

$$\begin{aligned}&\|\partial_\mu \mathcal{U}(t,x,\mu_1)(\xi_1)\|_\infty \le C,\\&\mathbb{E}^1\big[\big(\partial_\mu \mathcal{U}(t,x,\mu_1)(\xi_1) - \partial_\mu \mathcal{U}(t,x,\mu_2)(\xi_2)\big)\chi\big]\\&\quad \le C\Big(\|\xi_1-\xi_2\|_1\|\chi\|_1 + \mathbb{E}^1\big[|\xi_1-\xi_2|\,|\chi|\big]\Big).\end{aligned} \tag{5.63}$$

Proof. We know that, for every $(t,x) \in [0,T] \times \mathbb{R}^d$, the mapping:

$$L^2(\Omega^1, \mathcal{F}_t^1, \mathbb{P}^1; \mathbb{R}^d) \ni \xi \mapsto Y_t^{t,x,\xi}$$

is Gâteaux differentiable, the derivative in the direction χ at point ξ reading $\partial_\chi Y_t^{t,x,\xi}$. As already argued right below Lemma 5.24, the mapping $L^2(\Omega^1, \mathcal{F}_t^1, \mathbb{P}^1; \mathbb{R}^d) \ni \chi \mapsto \partial_\chi Y_t^{t,x,\xi} \in \mathbb{R}^m$ is linear and continuous. Hence, we can find a random variable, denoted $DY_t^{t,x,\xi} \in L^2(\Omega^1, \mathcal{F}_t^1, \mathbb{P}^1; \mathbb{R}^{m\times d})$ such that:

$$\partial_\chi Y_t^{t,x,\xi} = \mathbb{E}^1\big[DY_t^{t,x,\xi}\chi\big].$$

By the second line in (5.62), the mapping:

$$L^2(\Omega^1, \mathcal{F}_t^1, \mathbb{P}^1; \mathbb{R}^d) \ni \xi \mapsto DY_t^{t,x,\xi} \in L^2(\Omega^1, \mathcal{F}_t^1, \mathbb{P}^1; \mathbb{R}^{m\times d})$$

is continuous, from which we deduce that $L^2(\Omega^1, \mathcal{F}_t^1, \mathbb{P}^1; \mathbb{R}^d) \ni \xi \mapsto Y_t^{t,x,\xi}$ is continuously Fréchet differentiable.

Since $L^2(\Omega^1, \mathcal{F}_t^1, \mathbb{P}^1; \mathbb{R}^d) \ni \xi \mapsto Y_t^{t,x,\xi}$ is the lift of $\mathcal{P}_2(\mathbb{R}^d) \ni \mu \mapsto Y_t^{t,x,\mu} = \mathcal{U}(t,x,\mu)$, this shows that $\mathcal{P}_2(\mathbb{R}^d) \ni \mu \mapsto \mathcal{U}(t,x,\mu) \in \mathbb{R}^m$ is L-differentiable. Moreover, $DY_t^{t,x,\xi} = \partial_\mu \mathcal{U}(t,x,\mathcal{L}^1(\xi))(\xi)$. Now, the first line in (5.62) shows that

$$\Big|\mathbb{E}^1\big[\partial_\mu \mathcal{U}(t,x,\mathcal{L}^1(\xi))(\xi)\chi\big]\Big| \le C_1\|\chi\|_1,$$

which proves the first line (5.63). In the same way, the second claim in (5.63) follows from the second line in (5.62). □

We now discuss the Lipschitz property in x of $\partial_\mu \mathcal{U}(t,x,\mu)$:

Lemma 5.28 *For $T \le c$, with $c > 0$ only depending on L, we can find a constant C such that, for $\xi \in L^2(\Omega^1, \mathcal{F}_t^1, \mathbb{P}^1; \mathbb{R}^d)$ with μ as distribution,*

$$\forall x_1, x_2 \in \mathbb{R}^d, \quad \big\|\partial_\mu \mathcal{U}(t,x_1,\mu)(\xi) - \partial_\mu \mathcal{U}(t,x_2,\mu)(\xi)\big\|_\infty \le C|x_1-x_2|.$$

Proof. Thanks to the relationship $\partial_\chi Y_t^{t,x,\xi} = \mathbb{E}^1[\partial_\mu U(t,x,\mathcal{L}^1(\xi))(\xi)\chi]$, it suffices to discuss the Lipschitz property in x of the tangent process $(\partial_\chi X_s^{t,x,\xi}, \partial_\chi Y_s^{t,x,\xi})_{s\in[t,T]}$, seen as an element of $\mathbb{S}^2([t,T];\mathbb{R}^d) \times \mathbb{S}^2([t,T];\mathbb{R}^m)$, ξ and χ denoting elements of $L^2(\Omega^1, \mathcal{F}_t^1, \mathbb{P}^1; \mathbb{R}^d)$.

Basically, the strategy is the same as in the proofs of Lemmas 5.25 and 5.26. It is based on a tailored-made version of Corollary 5.22, obtained by applying Lemma 5.12 and Lemma 5.19 with $\boldsymbol{\theta}^1 = \bar{\boldsymbol{\theta}}^1 = \boldsymbol{\theta}^{t,x_1,\xi}$, $\boldsymbol{\theta}^2 = \bar{\boldsymbol{\theta}}^2 = \boldsymbol{\theta}^{t,x_2,\xi}$, $\boldsymbol{X}^{\prime,1} = \boldsymbol{X}^{\prime,2} = \boldsymbol{X}^{\prime\prime,1} = \boldsymbol{X}^{\prime\prime,2} = \boldsymbol{X}^{t,\xi}$ and $\boldsymbol{\mathcal{X}}^{\prime,1} = \boldsymbol{\mathcal{X}}^{\prime,2} = \boldsymbol{\mathcal{X}}^{\prime\prime,1} = \boldsymbol{\mathcal{X}}^{\prime\prime,2} = \partial_\chi \boldsymbol{\theta}^{t,\xi}$. Informally, it consists in choosing $\eta = 0$ and in replacing $|\xi_1 - \xi_2|$ by $|x_1 - x_2|$ and $\|\xi_1 - \xi_2\|_1$ by 0 in the statement of Corollary 5.22.

We end up with $|\partial_\chi Y_t^{t,x_1,\xi} - \partial_\chi Y_t^{t,x_2,\xi}| \le C|x_1-x_2|\,\|\chi\|_1$. □

Derivatives with Respect to the Space Argument

We now discuss the derivatives of $\mathcal{U}$ with respect to the variable x. Returning to (5.16), it is pretty clear that the process $(\boldsymbol{X}^{t,x,\mu}, \boldsymbol{Y}^{t,x,\mu})$ can be regarded as the solution of a standard FBSDE parameterized by the law μ of some random variable $\xi \in L^2(\Omega^1, \mathcal{F}_t^1, \mathbb{P}^1; \mathbb{R}^d)$. Therefore, the smoothness with respect to the parameter x can be tackled in a standard fashion, without worrying about the McKean-Vlasov structure. Equivalently, the result can be obtained by applying the results of Subsection 5.2.3, with the following version of $\delta H(r, \cdot)$ in (5.24):

$$
\begin{aligned}
&\delta H_\ell(r, V_r)(\mathcal{V}_r) = \partial_x h\big(r, V_r, \mathcal{L}^1(X_r^{t,\xi})\big)\mathcal{V}_r,\\
&\delta h_\ell(r, V_r) = \partial_x h\big(r, V_r, \mathcal{L}^1(X_r^{t,\xi})\big),\\
&\delta \tilde{H}_\ell = 0,\\
&\delta H_a(r) = 0.
\end{aligned}
\tag{5.64}
$$

Proceeding as before, we claim that the map $\mathbb{R}^d \ni x \mapsto (\boldsymbol{X}^{t,x,\mu}, \boldsymbol{Y}^{t,x,\mu}) \in \mathbb{S}^2([t,T]; \mathbb{R}^d) \times \mathbb{S}^2([t,T]; \mathbb{R}^m)$ is continuously differentiable. The tangent process $(\partial_x \boldsymbol{X}^{t,x,\mu}, \partial_x \boldsymbol{Y}^{t,x,\mu})$ is regarded as a process with paths in $\mathbb{S}^2([t,T]; \mathbb{R}^{d\times d}) \times \mathbb{S}^2([t,T]; \mathbb{R}^{m\times d})$. On the model of (5.59), it satisfies the FBSDE:

$$
\left\{
\begin{aligned}
\partial_x X_s^{t,x,\mu} &= I_d + \int_t^s \Big(\partial_x B\big(s, X_r^{t,x,\mu}, \mathcal{L}^1(X_r^{t,\xi}), Y_r^{t,x,\mu}\big)\partial_x X_r^{t,x,\mu}\\
&\qquad + \partial_y B\big(s, X_r^{t,x,\mu}, \mathcal{L}^1(X_r^{t,\xi}), Y_r^{t,x,\mu}\big)\partial_x Y_r^{t,x,\mu}\Big)dr,\\
\partial_x Y_s^{t,x,\mu} &= \mathbb{E}_s\bigg[\partial_x G\big(X_T^{t,x,\mu}, \mathcal{L}^1(X_T^{t,\xi})\big)\partial_x X_T^{t,x,\mu}\\
&\qquad + \int_s^T \Big(\partial_x F\big(r, X_r^{t,x,\mu}, \mathcal{L}^1(X_r^{t,\xi}), Y_r^{t,x,\mu}\big)\partial_x X_r^{t,x,\mu}\\
&\qquad + \partial_y F\big(r, X_r^{t,x,\mu}, \mathcal{L}^1(X_r^{t,\xi}), Y_r^{t,x,\mu}\big)\partial_x Y_r^{t,x,\mu}\Big)dr\bigg],
\end{aligned}
\right.
\tag{5.65}
$$

for $s \in [t,T]$.

As a consequence, we easily get from Corollary 5.18, for $T \leq c$, with c only depending on L, $\sup_{t\leq s\leq T} |\partial_x \boldsymbol{\theta}_s^{t,x,\mu}| \leq C$. Recalling the identity $\mathcal{U}(t,x,\mu) = \theta_t^{t,x,\mu}$, we deduce that $\mathbb{R}^d \ni x \mapsto \mathcal{U}(t,x,\mu)$ is continuously differentiable and that $\|\partial_x \mathcal{U}\|_\infty \leq C$, which is reminiscent of the bounds obtained in Chapter (Vol I)-4 for the Lipschitz constant of the decoupling field of a standard FBSDE. In the same way, we can adapt Lemmas 5.25 or 5.26 in order to investigate the difference $\partial_x \boldsymbol{\theta}^{t,x_1,\mu} - \partial_x \boldsymbol{\theta}^{t,x_2,\mu}$ for two elements $x_1, x_2 \in \mathbb{R}^d$. It can be checked that, for any $t \in [0,T]$, any $\mu \in \mathcal{P}_2(\mathbb{R}^d)$, the mapping $\mathbb{R}^d \ni x \mapsto \partial_x \mathcal{U}(t,x,\mu)$ is C-Lipschitz continuous. Intuitively, such a property is much simpler to prove than the continuity of $\partial_\mu \mathcal{U}$ because of the very simple structure of $H(r,\cdot)$ in (5.64), the function $\partial_x h$ being Lipschitz-continuous with respect to the first argument.

To get the smoothness of $\partial_x \mathcal{U}$ in the direction μ, we may investigate the difference $\partial_x \boldsymbol{\theta}^{t,x,\mu_1} - \partial_x \boldsymbol{\theta}^{t,x,\mu_2}$, with $\mu_i = \mathcal{L}^1(\xi_i)$, for $\xi_i \in L^2(\Omega^1, \mathcal{F}_t^1, \mathbb{P}^1; \mathbb{R}^d)$ and $i = 1, 2$. Reapplying Corollary 5.22, exactly in the same way as in the proof of Lemma 5.26, we deduce that, for all $x \in \mathbb{R}^d$, $\xi_1, \xi_2 \in L^2(\Omega^1, \mathcal{F}_t^1, \mathbb{P}^1; \mathbb{R}^d)$,

$$\big|\partial_x \mathcal{U}\big(t, x, \mathcal{L}^1(\xi_1)\big) - \partial_x \mathcal{U}\big(t, x, \mathcal{L}^1(\xi_2)\big)\big| \leq C\mathbb{E}^1\big[|\xi_1 - \xi_2|\big], \tag{5.66}$$

that is, for all $\mu_1, \mu_2 \in \mathcal{P}_2(\mathbb{R}^d)$,

$$\big|\partial_x \mathcal{U}(t, x, \mu_1) - \partial_x \mathcal{U}(t, x, \mu_2)\big| \leq C W_1(\mu_1, \mu_2).$$

Final Statement

The following provides a complete statement about the first-order differentiability:

Theorem 5.29 *For $T \leq c$, with c only depending on L, and $t \in [0, T]$, the function $\mathbb{R}^d \times \mathcal{P}_2(\mathbb{R}^d) \ni (x, \mu) \mapsto \mathcal{U}(t, x, \mu)$ is continuously differentiable and there exists a constant $C \geq 0$, such that for all $t \in [0, T]$, $x \in \mathbb{R}^d$ and $\mu \in \mathcal{P}_2(\mathbb{R}^d)$, $|\partial_x \mathcal{U}(t, x, \mu)|$ is bounded by C and, for all $x_1, x_2 \in \mathbb{R}^d$ and $\mu_1, \mu_2 \in \mathcal{P}_2(\mathbb{R}^d)$,*

$$\begin{aligned} &|\mathcal{U}(t, x_1, \mu_1) - \mathcal{U}(t, x_2, \mu_2)| + |\partial_x \mathcal{U}(t, x_1, \mu_1) - \partial_x \mathcal{U}(t, x_2, \mu_2)| \\ &\quad \leq C\big(|x_1 - x_2| + W_1(\mu_1, \mu_2)\big), \end{aligned} \tag{5.67}$$

Moreover, the function $[0, T] \times \mathbb{R}^d \times \mathcal{P}_2(\mathbb{R}^d) \ni (t, x, \mu) \mapsto \partial_x \mathcal{U}(t, x, \mu) \in \mathbb{R}^{m \times d}$ is continuous. Also, it holds, for all $t \in [0, T]$, $x \in \mathbb{R}^d$, $\xi \in L^2(\Omega^1, \mathcal{F}_t^1, \mathbb{P}^1; \mathbb{R}^d)$,

$$\big\|\partial_\mu \mathcal{U}\big(t, x, \mathcal{L}^1(\xi)\big)(\xi)\big\|_\infty \leq C, \tag{5.68}$$

and, for all $\xi_1, \xi_2, \chi \in L^2(\Omega^1, \mathcal{F}_t^1, \mathbb{P}^1; \mathbb{R}^d)$, with $\xi_1 \sim \mu_1$ and $\xi_2 \sim \mu_2$,

$$\begin{aligned} &\mathbb{E}^1\big[\big(\partial_\mu \mathcal{U}(t, x_1, \mu_1)(\xi_1) - \partial_\mu \mathcal{U}(t, x_2, \mu_2)(\xi_2)\big)\chi\big] \\ &\quad \leq C\Big[\|\chi\|_1\big(|x_1 - x_2| + \|\xi_1 - \xi_2\|_1\big) + \mathbb{E}^1\big[|\xi_1 - \xi_2|\,|\chi|\big]\Big]. \end{aligned} \tag{5.69}$$

In particular, for each $(t, x, \mu) \in [0, T] \times \mathbb{R}^d \times \mathcal{P}_2(\mathbb{R}^d)$, we can find a version (which is necessarily unique) in $L^2(\mathbb{R}^d, \mu; \mathbb{R}^{m \times d})$ of the function $\mathbb{R}^d \ni v \mapsto \partial_\mu \mathcal{U}(t, x, \mu)(v) \in \mathbb{R}^{m \times d}$ such that the map $(t, x, \mu, v) \mapsto \partial_\mu \mathcal{U}(t, x, \mu)(v)$ is continuous and bounded by C and satisfies for all $t \in [0, T]$, $x_1, x_2 \in \mathbb{R}^d$, $v_1, v_2 \in \mathbb{R}^d$ and $\mu_1, \mu_2 \in \mathcal{P}_2(\mathbb{R}^d)$:

$$\begin{aligned} &\big|\partial_\mu \mathcal{U}(t, x_1, \mu_1)(v_1) - \partial_\mu \mathcal{U}(t, x_2, \mu_2)(v_2)\big| \\ &\quad \leq C\big(|x_1 - x_2| + |v_1 - v_2| + W_1(\mu_1, \mu_2)\big). \end{aligned} \tag{5.70}$$

We refer the reader to Subsection (Vol I)-5.3.4 for an account on the notion of joint differentiability for functions defined on the product space $\mathbb{R}^d \times \mathcal{P}_2(\mathbb{R}^d)$.

According to the terminology introduced in Chapter (Vol I)-5, $\mathcal{U}(t,\cdot,\cdot)$ is said to be fully $\mathcal{C}^1$. Uniqueness of the jointly continuous version is discussed in Remark (Vol I)-5.82, see also Remark 4.12.

In order to prove (5.70), we shall make use of Lemma (Vol I)-5.41, the statement of which we recall under the new label (5.30).

Lemma 5.30 *Let $(u(x,\mu)(\cdot))_{x\in\mathbb{R}^n,\mu\in\mathcal{P}_2(\mathbb{R}^d)}$ be a collection of real-valued functions satisfying, for all $x \in \mathbb{R}^n$ and $\mu \in \mathcal{P}_2(\mathbb{R}^d)$, $u(x,\mu)(\cdot) \in L^\infty(\mathbb{R}^d,\mu;\mathbb{R})$, and for which there exists a constant C such that, for all $x, x' \in \mathbb{R}^n$, and $\xi, \xi', \chi \in L^2(\Omega,\mathcal{F},\mathbb{P};\mathbb{R}^d)$,*

$$\begin{aligned}&\mathbb{E}\Big[\big(u(x,\mathcal{L}(\xi))(\xi) - u(x',\mathcal{L}(\xi'))(\xi')\big)\chi\Big]\\&\quad\le C\Big[\|\chi\|_1\big(|x-x'| + \|\xi-\xi'\|_1\big) + \mathbb{E}\big[|\xi-\xi'|\,|\chi|\big]\Big],\end{aligned}$$

where $(\Omega,\mathcal{F},\mathbb{P})$ is an atomless probability space. Then, for each $(x,\mu) \in \mathbb{R}^n \times \mathcal{P}_2(\mathbb{R}^d)$, we can find a version of $u(x,\mu)(\cdot) \in L^\infty(\mathbb{R}^d,\mu;\mathbb{R})$ such that, for the same constant C as above, for all $x, x' \in \mathbb{R}^n$, $\mu, \mu' \in \mathcal{P}_2(\mathbb{R}^d)$ and $v, v' \in \mathbb{R}^d$,

$$\big|u(x,\mu)(v) - u(x',\mu')(v')\big| \le C\big(|x-x'| + W_1(\mu,\mu') + |v-v'|\big).$$

We now turn to the proof of Theorem 5.29.

Proof of Theorem 5.29.

First Step. The Lipschitz property of $\mathcal{U}(t,\cdot,\cdot)$ is a direct consequence of Proposition 5.13. The joint continuous differentiability is a consequence of the partial continuous differentiability and of the joint continuity properties of the derivatives, see Lemmas 5.27 and 5.28 together with (5.66) and the discussion right above (5.66). Obviously, (5.67) and (5.69) follow in the same way. Moreover, (5.68) is a straightforward consequence of (5.63).

Thus, for any $t \in [0,T]$, the existence of a version of $\partial_\mu\mathcal{U}(t,\cdot,\cdot)(\cdot)$ satisfying (5.70) follows from the auxiliary Lemma 5.30. Hence, the maps $\partial_x\mathcal{U}(t,\cdot,\cdot)$ and $\partial_\mu\mathcal{U}(t,\cdot,\cdot)(\cdot)$ are uniformly continuous, uniformly in the time parameter $t \in [0,T]$.

Second Step. We now discuss the time continuity of the derivatives. In order to make the analysis consistent, we shall require the generic initial condition ξ to be $\mathcal{F}_0^1$-measurable, or equivalently, to belong to $L^2(\Omega^1,\mathcal{F}_0^1,\mathbb{P}^1;\mathbb{R}^d)$. This is by no means a limitation since we assumed $(\Omega^1,\mathcal{F}_0^1,\mathbb{P}^1)$ to be rich enough so that, for any distribution $\nu \in \mathcal{P}_2(\mathbb{R}^q)$, with $q \ge 1$, there exists an $\mathcal{F}_0^1$-measurable-random variable with ν as distribution.

We start with the time continuity of $\partial_\mu\mathcal{U}(t,\cdot,\cdot)(\cdot)$. We claim that it suffices to prove that, for any $(x,\xi) \in \mathbb{R}^d \times L^2(\Omega^1,\mathcal{F}_0^1,\mathbb{P}^1;\mathbb{R}^d)$, the map $[0,T] \ni t \mapsto \partial_\mu\mathcal{U}(t,x,\mathcal{L}^1(\xi))(\xi) \in L^2(\Omega^1,\mathcal{F}_0^1,\mathbb{P}^1;\mathbb{R}^{m\times d})$ is continuous. Assume indeed that continuity holds true in this sense and use the fact that, by the uniform bound (5.68) and the uniform continuity property (5.70), the family of functions $(\partial_\mu\mathcal{U}(s,x,\mu)(\cdot))_{s\in[0,T]}$ is, for any $(x,\mu) \in \mathbb{R}^d \times \mathcal{P}_2(\mathbb{R}^d)$, relatively compact for the topology of uniform convergence on compact subsets of $\mathbb{R}^d$: The time continuity property in $L^2(\Omega^1,\mathcal{F}_0^1,\mathbb{P}^1;\mathbb{R}^{m\times d})$ says that any limit of $\partial_\mu\mathcal{U}(s,x,\mu)(\cdot)$ as s tends to t must coincide with a version of $\partial_\mu\mathcal{U}(t,x,\mu)(\cdot)$ in $L^2(\mathbb{R}^d,\mu;\mathbb{R}^{m\times d})$. In particular, when μ has full support, any limit as s tends to t, must coincide with $\partial_\mu\mathcal{U}(t,x,\mu)(\cdot)$, since, in that case, there is only one continuous version in $L^2(\mathbb{R}^d,\mu;\mathbb{R}^{m\times d})$. This proves that, for all $v \in \mathbb{R}^d$, $\partial_\mu\mathcal{U}(s,x,\mu)(v)$ tends to $\partial_\mu\mathcal{U}(t,x,\mu)(v)$ when μ has full support. When μ is no more of full support, we may approximate it, in $\mathcal{P}_2(\mathbb{R}^d)$, by a sequence of measures $(\mu_n)_{n\geq 0}$ with full support. Owing once again to (5.70), we can exchange the two limits taken over s tends to t and n tends to ∞, namely

$$\begin{aligned}
&\lim_{s\to t} \partial_\mu \mathcal{U}(s,x,\mu)(v) \\
&= \lim_{s\to t}\lim_{n\to\infty} \partial_\mu \mathcal{U}(s,x,\mu_n)(v) \\
&= \lim_{n\to\infty}\lim_{s\to t} \partial_\mu \mathcal{U}(s,x,\mu_n)(v) = \lim_{n\to\infty} \partial_\mu \mathcal{U}(t,x,\mu_n)(v) = \partial_\mu \mathcal{U}(t,x,\mu)(v),
\end{aligned}$$

which proves that, in any case, for all $v \in \mathbb{R}^d$, $\partial_\mu \mathcal{U}(s,x,\mu)(v)$ tends to $\partial_\mu \mathcal{U}(t,x,\mu)(v)$

Third Step. Following our plan, we now prove that, for all $(x,\xi) \in \mathbb{R}^d \times L^2(\Omega^1,\mathcal{F}_0^1,\mathbb{P}^1;\mathbb{R}^d)$, the map $[0,T] \ni t \mapsto \partial_\mu \mathcal{U}(t,x,\mu)(\xi) \in L^2(\Omega^1,\mathcal{F}_0^1,\mathbb{P}^1;\mathbb{R}^{m\times d})$ is continuous.

Considering $\xi,\chi \in L^2(\Omega^1,\mathcal{F}_0^1,\mathbb{P}^1;\mathbb{R}^d)$ together with $0 \leq t \leq s \leq T$, and letting $\mu = \mathcal{L}^1(\xi)$, it suffices to bound the time increment

$$\mathbb{E}^1\Big[\Big(\partial_\mu \mathcal{U}(t,x,\mu)(\xi) - \partial_\mu \mathcal{U}(s,x,\mu)(\xi)\Big)\chi\Big]$$

by $C(t,s)\|\chi\|_2$, the constant $C(t,s)$ being independent of χ and converging to 0 as $s-t$ tends to 0. We have:

$$\begin{aligned}
&\mathbb{E}^1\Big[\Big(\partial_\mu \mathcal{U}(t,x,\mu)(\xi) - \partial_\mu \mathcal{U}(s,x,\mu)(\xi)\Big)\chi\Big] \\
&= \tilde{\mathbb{E}}^1\Big[\Big(\partial_\mu \mathcal{U}(t,x,\mu)(\tilde{\xi}) - \partial_\mu \mathcal{U}(s,x,\mu)(\tilde{\xi})\Big)\tilde{\chi}\Big] \\
&= \mathbb{E}\tilde{\mathbb{E}}^1\Big[\Big(\partial_\mu \mathcal{U}\big(s,X_s^{t,x,\xi},\mathcal{L}^1(X_s^{t,\xi})\big)(\tilde{X}_s^{t,\xi}) - \partial_\mu \mathcal{U}(s,x,\mu)(\tilde{\xi})\Big)\tilde{\chi}\Big] \\
&\quad + \mathbb{E}\tilde{\mathbb{E}}^1\Big[\Big(\partial_\mu \mathcal{U}(t,x,\mu)(\tilde{\xi}) - \partial_\mu \mathcal{U}\big(s,X_s^{t,x,\xi},\mathcal{L}^1(X_s^{t,\xi})\big)(\tilde{X}_s^{t,\xi})\Big)\tilde{\chi}\Big].
\end{aligned} \tag{5.71}$$

By the smoothness property of $\partial_\mu \mathcal{U}(s,\cdot,\cdot)(\cdot)$, the first term in the right-hand side is bounded by $C(\mathbb{E}[|X_s^{t,x,\xi}-x|^2]^{1/2} + \mathbb{E}[|X_s^{t,\xi}-\xi|^2]^{1/2})\|\chi\|_2$, the constant C being allowed to increase from line to line. The coefficients of (5.2) and (5.6) being at most of linear growth, we deduce from (5.17) that $\mathbb{E}[|X_s^{t,\xi}-\xi|^2]^{1/2}$ and $\mathbb{E}[|X_s^{t,x,\xi}-x|^2]^{1/2}$ are less than $C(1+\|\xi\|_2)(s-t)^{1/2}$ and $C(1+|x|+\|\xi\|_2)(s-t)^{1/2}$ respectively. Therefore, the first term in the last line of (5.71) is bounded by:

$$C\big(1+|x|+\|\xi\|_2\big)(s-t)^{1/2}. \tag{5.72}$$

We now handle the second term in the last line of (5.71). Differentiating with respect to ξ in the direction χ the two relationships $\mathcal{U}(t,x,\mu) = Y_t^{t,x,\xi}$ and $\mathcal{U}(s,X_s^{t,x,\xi},\mathcal{L}^1(X_s^{t,\xi})) = Y_s^{t,x,\xi}$, we obtain:

$$\begin{aligned}
&\tilde{\mathbb{E}}^1\big[\partial_\mu \mathcal{U}(t,x,\mu)(\tilde{\xi})\tilde{\chi}\big] = \partial_\chi Y_t^{t,x,\xi}, \\
&\tilde{\mathbb{E}}^1\big[\partial_\mu \mathcal{U}\big(s,X_s^{t,x,\xi},\mathcal{L}^1(X_s^{t,\xi})\big)(\tilde{X}_s^{t,\xi})\partial_\chi \tilde{X}_s^{t,\xi}\big] \\
&\quad = \partial_\chi Y_s^{t,x,\xi} - \partial_x \mathcal{U}\big(s,X_s^{t,x,\xi},\mathcal{L}^1(X_s^{t,\xi})\big)\partial_\chi X_s^{t,x,\xi},
\end{aligned}$$

and then:

$$\begin{aligned}
&\mathbb{E}\tilde{\mathbb{E}}^1\Big[\big(\partial_\mu\mathcal{U}(t,x,\mu)(\tilde{\xi}) - \partial_\mu\mathcal{U}\big(s,X_s^{t,x,\xi},\mathcal{L}^1(X_s^{t,\xi})\big)(\tilde{X}_s^{t,\xi})\big)\tilde{\chi}\Big]\\
&= \mathbb{E}\big[\partial_\chi Y_t^{t,x,\xi} - \partial_\chi Y_s^{t,x,\xi}\big] + \mathbb{E}\Big[\partial_x\mathcal{U}\big(s,X_s^{t,x,\xi},\mathcal{L}^1(X_s^{t,\xi})\big)\partial_\chi X_s^{t,x,\xi}\Big]\\
&\quad + \mathbb{E}\tilde{\mathbb{E}}^1\Big[\partial_\mu\mathcal{U}\big(s,X_s^{t,x,\xi},\mathcal{L}^1(X_s^{t,\xi})\big)(\tilde{X}_s^{t,\xi})\big(\partial_\chi\tilde{X}_s^{t,\xi} - \tilde{\chi}\big)\Big].
\end{aligned}$$

The first term in the right-hand side writes $\mathbb{E}\int_t^s \partial F(r,\theta_r^{t,x,\xi},\tilde{X}_r^{t,\xi})(\partial_\chi\theta_r^{t,x,\xi},\partial_\chi\tilde{X}_r^{t,\xi})dr$, with the same notations as in (5.23). By assumption **Smooth Coefficients Order 1** and by Lemmas 5.24 and 5.26, it is bounded by $C(s-t)\|\chi\|_1$. Since $\partial_x\mathcal{U}$ is bounded, the second term is less than $C\mathbb{E}[|\partial_\chi X_s^{t,x,\xi}|] = C\mathbb{E}[|\partial_\chi X_s^{t,x,\xi} - \partial_\chi X_t^{t,x,\xi}|]$, where we used the fact that $\partial_\chi X_t^{t,x,\xi} = 0$. Owing to assumption **Smooth Coefficients Order 1** and Lemmas 5.24 and 5.26 again and taking advantage of the form of the linearized system (5.59), it is less than $C(s-t)\|\chi\|_1$. For the third term, we first recall the first bound in (5.63). We get that it is less than $C\mathbb{E}[|\partial_\chi X_s^{t,\xi} - \chi|]$. Then, combining the form of the linearized system (5.55) with assumption **Smooth Coefficients Order 1** and Lemma 5.24, we deduce that it is bounded by $C(s-t)\|\chi\|_1$. Continuity of $[0,T]\ni t\mapsto \partial_\mu\mathcal{U}(t,x,\mu)(\xi)\in L^2(\Omega^1,\mathcal{F}_0^1,\mathbb{P}^1;\mathbb{R}^{m\times d})$ easily follows.

Conclusion. Continuity of $[0,T]\ni t\mapsto \partial_x\mathcal{U}(t,x,\mu)\in\mathbb{R}^{m\times d}$, for $(x,\mu)\in\mathbb{R}^d\times\mathcal{P}_2(\mathbb{R}^d)$, is proved in the same way. □

5.3 Solutions to the Master Equation in Small Time

The goal of this section is to establish the short time solvability of the master equation, as announced in the statement of Theorem 5.10.

First, we prove that the decoupling field is twice differentiable in the space and measure arguments. The master equation is then derived according to the sketch provided in Subsection 5.1.4. Throughout the section, assumption **Smooth Coefficients Order 2** is in force and $T\le 1$.

5.3.1 Differentiability of the Linearized System

As we shall see, the main difficulty in the proof is to prove second order differentiability in the direction of the measure argument. In order to do so, we shall go back to the linearized systems (5.55) and (5.59) and prove that for $t\in[0,T]$ and $\chi\in L^\infty(\Omega^1,\mathcal{F}_t^1,\mathbb{P}^1;\mathbb{R}^d)$, $(\partial_\chi\boldsymbol{X}^{t,\xi},\partial_\chi\boldsymbol{Y}^{t,\xi})$ and $(\partial_\chi\boldsymbol{X}^{t,x,\xi},\partial_\chi\boldsymbol{Y}^{t,x,\xi})$ are differentiable with respect to ξ and x when regarded as random variables with values in $\mathbb{S}^2([t,T];\mathbb{R}^d)\times\mathbb{S}^2([t,T];\mathbb{R}^m)$.

General Strategy

The strategy is to regard the full-fledged tuples:

$$\big(\boldsymbol{X}^{t,\xi},\boldsymbol{Y}^{t,\xi},\partial_\chi\boldsymbol{X}^{t,\xi},\partial_\chi\boldsymbol{Y}^{t,\xi}\big)$$

and

$$\left(X^{t,x,\xi}, Y^{t,x,\xi}, \partial_\chi X^{t,x,\xi}, \partial_\chi Y^{t,x,\xi}\right)$$

as the solutions of a forward-backward system satisfying assumption **Smooth Coefficients Order 1** with two different initials conditions: $(X_t^{t,\xi}, \partial_\chi X_t^{t,\xi}) = (\xi, \chi)$ for the first and $(X_t^{t,x,\xi}, \partial_\chi X_t^{t,x,\xi}) = (x, 0)$ for the second. In this context, we may consider $(\boldsymbol{X}^{t,\xi}, \partial_\chi \boldsymbol{X}^{t,\xi})$ as a forward component of dimension $2d$ and $(\boldsymbol{Y}^{t,\xi}, \partial_\chi \boldsymbol{Y}^{t,\xi})$ as a backward component of dimension $2m$, and similarly for $(\boldsymbol{X}^{t,x,\xi}, \partial_\chi \boldsymbol{X}^{t,x,\xi})$ and $(\boldsymbol{Y}^{t,x,\xi}, \partial_\chi \boldsymbol{Y}^{t,x,\xi})$. If we follow this approach, in contrast with the original system, the two noise processes $\boldsymbol{W}$ and $\boldsymbol{W}^0$ do not have the same dimension as the forward component any longer. Obviously, this is not a limitation: we can complete artificially the forward equation satisfied by $\partial_\chi X$ with new Brownian motions of dimension d, by setting the corresponding volatilities to 0.

Once the problem is recast in this way, the whole drift $\boldsymbol{B}$ driving the pair (5.15)–(5.55) reads as a function with values in $(\mathbb{R}^d)^2$, with entries in $\mathbb{R}^d$ given by:

$$\boldsymbol{B}\big(t, (x, \partial x), \nu, (y, \partial y)\big) = \begin{cases} B\big(t, x, \mu, y\big), \\ DB\big(t, (x, \partial x), \nu, (y, \partial y)\big), \end{cases} \tag{5.73}$$

with

$$DB\big(t, (x, \partial x), \nu, (y, \partial y)\big) = \partial_x B(t, x, \mu, y)\partial x + \partial_y B(t, x, \mu, y)\partial y$$

$$+ \int_{\mathbb{R}^d \times \mathbb{R}^d} \partial_\mu B(t, x, \mu, y)(v)\partial v d\nu(v, \partial v),$$

where $(x, \partial x) \in \mathbb{R}^d \times \mathbb{R}^d$ is understood as the forward variable and $(y, \partial y) \in \mathbb{R}^m \times \mathbb{R}^m$ as the backward variable. Also, μ and $\partial\mu$ denote the first and second d-dimensional marginal measures of $\nu \in \mathcal{P}_2(\mathbb{R}^d \times \mathbb{R}^d)$. Similarly, we may define $\boldsymbol{F}$ and $\boldsymbol{G}$ as the driver and terminal condition of the system comprising (5.15) and (5.55).

Owing to assumption **Smooth Coefficients Order 2**, $\boldsymbol{B}$, $\boldsymbol{F}$ and $\boldsymbol{G}$ have at most linear growth in $(x, \partial x)$, $(y, \partial y)$ and ν. Moreover, they are differentiable with respect to x and y, and also in the direction of ν. For instance, the derivative of $\boldsymbol{B}$ with respect to ν reads as a function with values in $(\mathbb{R}^d \times \mathbb{R}^d)^2$, with $\mathbb{R}^d \times \mathbb{R}^d$ entries given by:

$$\partial_\mu \boldsymbol{B}\big(t, (x, \partial x), \nu, (y, \partial y)\big)(v, \partial v) = \begin{cases} \partial_\mu B\big(t, x, \mu, y\big)(v), \\ \partial_\mu DB\big(t, (x, \partial x), \nu, (y, \partial y)\big)(v, \partial v), \end{cases}$$

$$\partial_{\partial\mu} \boldsymbol{B}\big(t, (x, \partial x), \nu, (y, \partial y)\big)(v, \partial v) = \begin{cases} 0, \\ \partial_{\partial\mu} DB\big(t, (x, \partial x), \nu, (y, \partial y)\big)(v, \partial v), \end{cases}$$

where

$$\partial_\mu DB\big(t,(x,\partial x),\nu,(y,\partial y)\big)(v,\partial v)$$
$$= \partial_\mu\partial_x B(t,x,\mu,y)(v)\partial x + \partial_\mu\partial_y B(t,x,\mu,y)(v)\partial y$$
$$+ \int_{\mathbb{R}^d\times\mathbb{R}^d} \partial^2_\mu B(t,x,\mu,y)(v',v)\partial v' d\nu(v',\partial v') + \partial_v\partial_\mu B(t,x,\mu,y)(v)\partial v,$$
$$\partial_{\partial\mu} DB\big(t,(x,\partial x),\nu,(y,\partial y)\big)(v,\partial v) = \partial_\mu B(t,x,\mu,y)(v).$$

Differentiability in the direction ν may be argued in the following way. Differentiability of $\partial_x B$ and $\partial_y B$ in the direction μ follows from a suitable version of Schwarz' theorem, as explained in Remark 4.16. Differentiability of $(v,\partial v) \mapsto \int_{\mathbb{R}^d} \partial_\mu B(t,x,\mu,y)(v)\partial v d\nu(v,\partial v)$ with respect to ν follows from Examples 1 and 3 in Subsection (Vol I)-5.2.2. Continuity of the derivatives follows from assumption **Smooth Coefficients Order 2**.

Truncation of the Coefficients

However, it must be stressed that $\boldsymbol{B}$, $\boldsymbol{F}$, and $\boldsymbol{G}$ do not satisfy assumption **Smooth Coefficients Order 1**. Indeed, because of the linear part in $(\partial x, \partial y, \partial v)$ in the definition of the coefficients, the derivatives are not bounded. Fortunately, this apparent difficulty can easily be circumvented by noticing that the domain on which $\boldsymbol{B}$, $\boldsymbol{F}$ and $\boldsymbol{G}$ are computed in (5.15)–(5.55) is actually bounded. Indeed, we know from Lemmas 5.24 and 5.26 that, for $T \le c$, with $c > 0$ only depending on L, and for $\|\chi\|_\infty \le 1$,

$$\begin{aligned} &\Big\| \sup_{t\le s\le T} \Big(|\partial_\chi X_s^{t,\xi}| + |\partial_\chi Y_s^{t,\xi}| \Big) \Big\|_\infty \le C, \\ &\Big\| \sup_{t\le s\le T} \Big(|\partial_\chi X_s^{t,x,\xi}| + |\partial_\chi Y_s^{t,x,\xi}| \Big) \Big\|_\infty \le C, \end{aligned} \tag{5.74}$$

for all $(t,x,\xi) \in [0,T]\times\mathbb{R}^d\times L^2(\Omega^1,\mathcal{F}_t^1,\mathbb{P}^1;\mathbb{R}^d)$ and C only depending on L. As a consequence, we can write the drift driving the pair (5.15)–(5.55) as:

$$\boldsymbol{B}\Big(t,\big(x,\phi_d(\partial x)\big),\nu\circ(I_d,\phi_d)^{-1},\big(y,\phi_m(\partial y)\big)\Big), \tag{5.75}$$

where ϕ_d and ϕ_m are smooth cut-off functions on $\mathbb{R}^d$ and $\mathbb{R}^m$ equal to the identity on the ball of center 0 and radius C of the space on which they are defined, and bounded by $2C$ on the whole space. It is clear that such a $\boldsymbol{B}$ is at most of linear growth and has bounded derivatives with respect to x, y and ν. Proceeding similarly with $\boldsymbol{F}$ and $\boldsymbol{G}$, we can assume, without any loss of generality, that the coefficients $(\boldsymbol{B},\boldsymbol{F},\boldsymbol{G})$ satisfy assumption **Smooth Coefficients Order 1**.

By the first-order analysis performed in Subsection 5.2.4, see for instance Lemmas 5.24 and 5.58, we deduce that, for some $\chi \in L^\infty(\Omega^1,\mathcal{F}_t^1,\mathbb{P}^1;\mathbb{R}^d)$ with $\|\chi\|_\infty \le 1$, the mapping $L^2(\Omega^1,\mathcal{F}_t^1,\mathbb{P}^1;\mathbb{R}^d) \ni \xi \mapsto (\partial_\chi X_s^{t,\xi}, \partial_\chi Y_s^{t,\xi}) \in \mathbb{S}^2([0,T];\mathbb{R}^d)\times\mathbb{S}([0,T];\mathbb{R}^m)$ is Gâteaux differentiable. For each random variable ζ in the space $L^2(\Omega^1,\mathcal{F}_t^1,\mathbb{P}^1;\mathbb{R}^d)$, we denote by:

$$\partial^2_{\zeta,\chi}\boldsymbol{\theta}^{t,\xi} = \big(\partial^2_{\zeta,\chi}\boldsymbol{X}^{t,\xi},\partial^2_{\zeta,\chi}\boldsymbol{Y}^{t,\xi}\big) = \big(\partial^2_{\zeta,\chi}X_s^{t,\xi},\partial^2_{\zeta,\chi}Y_s^{t,\xi}\big)_{t\le s\le T}$$

the tangent process in the direction ζ. By differentiating (5.56), we see that this tangent process satisfies the forward-backward system:

$$\begin{cases} \partial^2_{\zeta,\chi} X_s^{t,\xi} = \displaystyle\int_t^s \Big[\partial B\big(r, \theta_r^{t,\xi}, \tilde{X}_r^{t,\xi}\big)\big(\partial^2_{\zeta,\chi}\theta_r^{t,\xi}, \partial^2_{\zeta,\chi}\tilde{X}_r^{t,\xi}\big) \\ \qquad + \partial^2 B\big(r, \theta_r^{t,\xi}, \tilde{X}_r^{t,\xi}\big)\big(\partial_\chi \theta_r^{t,\xi}, \partial_\zeta \theta_r^{t,\xi}, \partial_\chi \tilde{X}_r^{t,\xi}, \partial_\zeta \tilde{X}_r^{t,\xi}\big)\Big] dr, \\ \partial^2_{\zeta,\chi} Y_s^{t,\xi} = \mathbb{E}_s\Big[\partial G\big(X_T^{t,\xi}, \tilde{X}_T^{t,\xi}\big)\big(\partial^2_{\zeta,\chi} X_T^{t,\xi}, \partial^2_{\zeta,\chi}\tilde{X}_T^{t,\xi}\big) \\ \qquad + \partial^2 G\big(X_T^{t,\xi}, \tilde{X}_T^{t,\xi}\big)\big(\partial_\chi X_T^{t,\xi}, \partial_\zeta X_T^{t,\xi}, \partial_\chi \tilde{X}_T^{t,\xi}, \partial_\zeta \tilde{X}_T^{t,\xi}\big) \\ \qquad + \displaystyle\int_s^T \Big[\partial F\big(r, \theta_r^{t,\xi}, \tilde{X}_r^{t,\xi}\big)\big(\partial^2_{\zeta,\chi}\theta_r^{t,\xi}, \partial^2_{\zeta,\chi}\tilde{X}_r^{t,\xi}\big) \\ \qquad + \partial^2 F\big(s, \theta_r^{t,\xi}, \tilde{X}_r^{t,\xi}\big)\big(\partial_\chi \theta_r^{t,\xi}, \partial_\zeta \theta_r^{t,\xi}, \partial_\chi \tilde{X}_r^{t,\xi}, \partial_\zeta \tilde{X}_r^{t,\xi}\big)\Big] dr\Big], \end{cases} \tag{5.76}$$

where we use the following convention:

$$\begin{aligned} &\partial^2 h(r, w, X)(\vartheta, \vartheta', \mathcal{X}, \mathcal{X}') \\ &= \partial_w^2 h\big(r, w, \mathcal{L}(X)\big)\vartheta \otimes \vartheta' \\ &\quad + \tilde{\mathbb{E}}^1\big[\partial_w \partial_\mu h\big(r, w, \mathcal{L}(X)\big)(\tilde{X})\big(\tilde{\mathcal{X}}' \otimes \vartheta + \tilde{\mathcal{X}} \otimes \vartheta'\big)\big] \\ &\quad + \tilde{\mathbb{E}}^1\big[\partial_v \partial_\mu h\big(r, w, \mathcal{L}(X)\big)(\tilde{X})\tilde{\mathcal{X}} \otimes \mathcal{X}'\big] \\ &\quad + \tilde{\mathbb{E}}^1\tilde{\tilde{\mathbb{E}}}^1\big[\partial_\mu^2 h\big(r, w, \mathcal{L}(X)\big)(\tilde{X}, \tilde{\tilde{X}})\tilde{\mathcal{X}} \otimes \tilde{\tilde{\mathcal{X}}}'\big], \end{aligned} \tag{5.77}$$

where $(\tilde{\tilde{\Omega}}^1, \tilde{\tilde{\mathcal{F}}}^1, \tilde{\tilde{\mathbb{F}}}^1, \tilde{\tilde{\mathbb{P}}}^1)$ is a new copy of $(\Omega^1, \mathcal{F}^1, \mathbb{F}^1, \mathbb{P}^1)$, and $\otimes$ is the tensor product acting under the convention we introduced in Chapter (Vol I)-5, see (Vol I)-(5.80): the first element of the tensor product acts on the derivative which is performed first. Namely, for any $y, z \in \mathbb{R}^d$:

$$\partial_v \partial_\mu h(w, \mu)(v) y \otimes z = \Big(\sum_{i,j=1}^d \partial_{v_j}[\partial_\mu h^\ell(w, \mu)]_i(v) z_j y_i\Big)_{\ell=1,\cdots,l},$$

$$\partial_\mu^2 h(w, \mu)(v, v') y \otimes z = \Big(\sum_{i,j=1}^d \big(\partial_\mu[\partial_\mu h^\ell(w, \mu)]_i(v)\big)_j(v') z_j y_i \in \mathbb{R}\Big)_{\ell=1,\cdots,l},$$

where, as usual, $h : \mathbb{R}^q \times \mathcal{P}_2(\mathbb{R}^d) \to \mathbb{R}^l$. Similarly, for $y \in \mathbb{R}^d$ and $w, z \in \mathbb{R}^q$,

$$\partial_w \partial_\mu h(w, \mu)(v) y \otimes z = \Big(\sum_{i,j=1}^d \partial_{w_j}[\partial_\mu h^\ell(w, \mu)]_i(v) z_j y_i\Big)_{\ell=1,\cdots,l}. \tag{5.78}$$

Notice that, in the second line of the right-hand side of (5.77), we used Schwarz' theorem to identify the entries of $\partial_w\partial_\mu h$ and $[\partial_\mu\partial_w h]^\dagger$. As a result,

$$
\begin{aligned}
&\partial^2 B\big(r,\theta_r^{t,\xi},\tilde{X}_r^{t,\xi}\big)\big(\partial_\chi\theta_r^{t,\xi},\partial_\zeta\theta_r^{t,\xi},\partial_\chi\tilde{X}_r^{t,\xi},\partial_\zeta\tilde{X}_r^{t,\xi}\big)\\
&= \partial_w^2 B\big(r,\theta_r^{t,\xi},\tilde{X}_r^{t,\xi}\big)\partial_\chi\theta_r^{t,\xi}\otimes\partial_\zeta\theta_r^{t,\xi}\\
&\quad+\tilde{\mathbb{E}}^1\Big[\partial_w\partial_\mu B\big(r,X_r^{t,\xi},\mathcal{L}(X_r^{t,\xi}),Y_r^{t,\xi}\big)(\tilde{X}_r^{t,\xi})\\
&\qquad\qquad\times\big(\partial_\zeta\tilde{X}_r^{t,\xi}\otimes\partial_\chi\theta_r^{t,\xi}+\partial_\chi\tilde{X}_r^{t,\xi}\otimes\partial_\zeta\theta_r^{t,\xi}\big)\Big]\\
&\quad+\tilde{\mathbb{E}}^1\Big[\partial_v\partial_\mu B\big(r,X_r^{t,\xi},\mathcal{L}^1(X_r^{t,\xi}),Y_r^{t,\xi}\big)(\tilde{X}_r^{t,\xi})\partial_\chi\tilde{X}_r^{t,\xi}\otimes\partial_\zeta\tilde{X}_s^{t,\xi}\Big]\\
&\quad+\tilde{\mathbb{E}}^1\tilde{\tilde{\mathbb{E}}}^1\Big[\partial_\mu^2 B(r,X_r^{t,\xi},\mathcal{L}^1(X_r^{t,\xi}),Y_r^{t,\xi})(\tilde{X}_r^{t,\xi},\tilde{\tilde{X}}_r^{t,\xi})\partial_\chi\tilde{X}_r^{t,\xi}\otimes\partial_\zeta\tilde{\tilde{X}}_r^{t,\xi}\Big],
\end{aligned}
$$

and similarly for F and G.

Freezing the Initial Condition

Following the principle used to associate (5.16) with (5.15), we now freeze the initial condition of the process $(\boldsymbol{X}^{t,\xi},\partial_\chi\boldsymbol{X}^{t,\xi})$. When the frozen initial condition is $(x,0)\in\mathbb{R}^d\times\mathbb{R}^d$, the process associated with $((\boldsymbol{X}^{t,\xi},\partial_\chi\boldsymbol{X}^{t,\xi}),(\boldsymbol{Y}^{t,\xi},\partial_\chi\boldsymbol{Y}^{t,\xi}))$ is $((\boldsymbol{X}^{t,x,\mu}=\boldsymbol{X}^{t,x,\xi},\partial_\chi\boldsymbol{X}^{t,x,\xi}),(\boldsymbol{Y}^{t,x,\mu}=\boldsymbol{Y}^{t,x,\xi},\partial_\chi\boldsymbol{Y}^{t,x,\xi}))$, where μ denotes the law of ξ. Here the pair (ξ,χ) reads as the initial condition of the forward component $(\boldsymbol{X}^{t,\xi},\partial_\chi\boldsymbol{X}^{t,\xi})$. In particular, following the statement of Definition 5.2, the two-pair process $((\boldsymbol{X}^{t,x,\mu},\partial_\chi\boldsymbol{X}^{t,x,\xi}),(\boldsymbol{Y}^{t,x,\mu},\partial_\chi\boldsymbol{Y}^{t,x,\xi}))$ only depends on (ξ,χ) through $\mathcal{L}^1(\xi,\chi)$.

Then, proceeding as above, we prove a similar result with $\partial_\chi\boldsymbol{\theta}^{t,\xi}$ replaced by $\partial_\chi\boldsymbol{\theta}^{t,x,\xi}$, showing that the map $L^2(\Omega^1,\mathcal{F}_t^1,\mathbb{P}^1;\mathbb{R}^d)\ni\xi\mapsto(\partial_\chi X_s^{t,x,\xi},\partial_\chi Y_s^{t,x,\xi})\in\mathbb{S}^2([0,T];\mathbb{R}^d)\times\mathbb{S}([0,T];\mathbb{R}^m)$ is Gâteaux differentiable. Moreover, for any random variable $\zeta\in L^2(\Omega^1,\mathcal{F}_t^1,\mathbb{P}^1;\mathbb{R}^d)$, the tangent process:

$$
\partial^2_{\zeta,\chi}\boldsymbol{\theta}^{t,x,\xi}=\big(\partial^2_{\zeta,\chi}\boldsymbol{X}^{t,x,\xi},\partial^2_{\zeta,\chi}\boldsymbol{Y}^{t,x,\xi}\big)=\big(\partial^2_{\zeta,\chi}X_s^{t,x,\xi},\partial^2_{\zeta,\chi}Y_s^{t,x,\xi}\big)_{t\le s\le T},
$$

satisfies the forward-backward system:

$$
\left\{
\begin{aligned}
\partial^2_{\zeta,\chi}X_s^{t,x,\xi}&=\int_t^s\Big[\partial B\big(r,\theta_r^{t,x,\xi},\tilde{X}_r^{t,\xi}\big)\big(\partial^2_{\zeta,\chi}\theta_r^{t,x,\xi},\partial^2_{\zeta,\chi}\tilde{X}_r^{t,\xi}\big)\\
&\quad+\partial^2 B\big(r,\theta_r^{t,x,\xi},\tilde{X}_r^{t,\xi}\big)\big(\partial_\chi\theta_r^{t,x,\xi},\partial_\zeta\theta_r^{t,x,\xi},\partial_\chi\tilde{X}_r^{t,\xi},\partial_\zeta\tilde{X}_r^{t,\xi}\big)\Big]dr,\\
\partial^2_{\zeta,\chi}Y_s^{t,x,\xi}&=\mathbb{E}_s\Big[\partial G\big(X_T^{t,x,\xi},\tilde{X}_T^{t,\xi}\big)\big(\partial^2_{\zeta,\chi}X_T^{t,x,\xi},\partial^2_{\zeta,\chi}\tilde{X}_T^{t,\xi}\big)\\
&\quad+\partial^2 G\big(X_T^{t,x,\xi},\tilde{X}_T^{t,\xi}\big)\big(\partial_\chi X_T^{t,x,\xi},\partial_\zeta X_T^{t,x,\xi},\partial_\chi\tilde{X}_T^{t,\xi},\partial_\zeta\tilde{X}_T^{t,\xi}\big)\\
&\quad+\int_s^T\Big[\partial F\big(r,\theta_r^{t,x,\xi},\tilde{X}_r^{t,\xi}\big)\big(\partial^2_{\zeta,\chi}\theta_r^{t,x,\xi},\partial^2_{\zeta,\chi}\tilde{X}_r^{t,\xi}\big)\\
&\quad+\partial^2 F\big(r,\theta_r^{t,x,\xi},\tilde{X}_r^{t,\xi}\big)\big(\partial_\chi\theta_r^{t,x,\xi},\partial_\zeta\theta_r^{t,x,\xi},\partial_\chi\tilde{X}_r^{t,\xi},\partial_\zeta\tilde{X}_r^{t,\xi}\big)\Big]dr\Big],
\end{aligned}
\right.
\tag{5.79}
$$

Statements

Applying Theorem 5.29, we get the following result:

Proposition 5.31 *Under assumption* **Smooth Coefficients Order 2**, *there exists a constant* $c > 0$, *only depending on L, such that for* $T \leq c$, *for any* $t \in [0, T]$ *and* $\chi \in L^\infty(\Omega^1, \mathcal{F}_t^1, \mathbb{P}^1; \mathbb{R}^d)$ *with* $\|\chi\|_\infty \leq 1$, *the map* $\mathbb{R}^d \times L^2(\Omega^1, \mathcal{F}_t^1, \mathbb{P}^1; \mathbb{R}^d) \ni (x, \xi) \mapsto \partial_\chi Y_t^{t,x,\xi} = \mathbb{E}^1[\partial_\mu \mathcal{U}(t, x, \mu)(\tilde{\xi})\chi] \in \mathbb{R}^d$ *is continuously differentiable.*

Below, we denote by $\partial_x \partial_\chi Y_t^{t,x,\xi}$ the partial derivative with respect to x, and by $D_\xi \partial_\chi Y_t^{t,x,\xi}$ the partial derivative with respect to ξ. The quantity $\partial_x \partial_\chi Y_t^{t,x,\xi}$ takes values in $\mathbb{R}^{m\times d}$ while $D_\xi \partial_\chi Y_t^{t,x,\xi}$ is an element of $L^2(\Omega^1, \mathcal{F}_t^1, \mathbb{P}^1; \mathbb{R}^{m\times d})$ which satisfies $\partial^2_{\zeta,\chi} Y_t^{t,x,\xi} = \mathbb{E}^1[D_\xi \partial_\chi Y_t^{t,x,\xi} \zeta]$.

In fact, $D_\xi \partial_\chi Y_t^{t,x,\xi}$ may be represented as follows. We learned from the truncation procedure (5.75) that $\partial_\chi Y^{t,x,\xi}$ coincides with $\boldsymbol{U}(t, (x, 0), \nu)$, where $\nu = \mathcal{L}^1(\xi, \chi)$, for some mapping $\boldsymbol{U} : [0, T] \times (\mathbb{R}^d \times \mathbb{R}^d) \times \mathcal{P}_2(\mathbb{R}^d \times \mathbb{R}^d) \to \mathbb{R}^m$ reading as the decoupling field of the forward-backward system of the McKean-Vlasov type driven by $(\boldsymbol{B}, \boldsymbol{F}, \boldsymbol{G})$. As already explained, this system satisfies the assumption of Theorem 5.29 so that, for T small enough, $\boldsymbol{U}$ is L-differentiable with respect to the measure argument ν. Recalling that $\boldsymbol{U}$ takes values in $\mathbb{R}^m$, we notice that $\partial_\nu \boldsymbol{U}$ takes values in $\mathbb{R}^{m\times(2d)}$. Denoting by $\partial_\mu \boldsymbol{U}$ the first block of dimension $m \times d$ of $\partial_\nu \boldsymbol{U}$, we have the identification:

$$D_\xi \partial_\chi Y_t^{t,x,\xi} = \partial_\mu \boldsymbol{U}\big(t, (x, 0), \mathcal{L}(\xi, \chi)\big)(\xi, \chi). \tag{5.80}$$

Below, we shall just write $\boldsymbol{U}(t, x, \nu)$ for $\boldsymbol{U}(t, (x, 0), \nu)$.

From Lemma 5.27 and Theorem 5.29, we deduce the following result:

Proposition 5.32 *Under assumption* **Smooth Coefficients Order 2**, *there exist a constant* $c > 0$ *only depending on L, and a constant C only depending on L and* Γ, *such that for* $T \leq c$, *for any* $t \in [0, T]$ *and any* $\chi \in L^\infty(\Omega^1, \mathcal{F}_t^1, \mathbb{P}^1; \mathbb{R}^d)$ *with* $\|\chi\|_\infty \leq 1$, *it holds, for all* $x_1, x_2 \in \mathbb{R}^d$ *and* $\xi_1, \xi_2 \in L^2(\Omega^1, \mathcal{F}_t^1, \mathbb{P}^1; \mathbb{R}^d)$,

$$\big|\partial_x \partial_\chi Y_t^{t,x_1,\xi_1} - \partial_x \partial_\chi Y_t^{t,x_2,\xi_2}\big| \leq C\big(|x_1 - x_2| + \|\xi_1 - \xi_2\|_1\big),$$

and, for all $\zeta \in L^2(\Omega^1, \mathcal{F}_t^1, \mathbb{P}^1; \mathbb{R}^d)$,

$$\begin{aligned}
&\mathbb{E}^1\Big[\big(D_\xi \partial_\chi Y_t^{t,x_1,\xi_1} - D_\xi \partial_\chi Y_t^{t,x_2,\xi_2}\big)\zeta\Big] \\
&\quad \leq C\Big[\|\zeta\|_1\big(|x_2 - x_2| + \|\xi_1 - \xi_2\|_1\big) + \mathbb{E}^1\big[|\xi_1 - \xi_2|\,|\zeta|\big]\Big].
\end{aligned}$$

Moreover, for the same C as above, for any $t \in [0, T]$ and $\xi \in L^2(\Omega^1, \mathcal{F}_t^1, \mathbb{P}^1; \mathbb{R}^d)$,

$$\left\| D_\xi \partial_\chi Y_t^{t,x,\xi} \right\|_\infty \leq C.$$

5.3.2 Second-Order Differentiability of the Decoupling Field

Propositions 5.31 and 5.32 are one step forward into the proof of the second order differentiability of $\mathcal{U}$, but they do not suffice. The goal of this section is to fill part of the gap and to prove that $\partial_\mu U$ is differentiable with respect to μ and v.

Recall that, throughout the subsection, assumption **Smooth Coefficients Order 2** is in force.

Differentiability with Respect to v

We start with:

Proposition 5.33 *Under assumption* **Smooth Coefficients Order 2**, *there exist a constant c only depending on L, and a constant C only depending on L and Γ, such that for $T \leq c$ and any $(t, x, \mu) \in [0, T] \times \mathbb{R}^d \times \mathcal{P}_2(\mathbb{R}^d)$, the function $\mathbb{R}^d \ni v \mapsto \partial_\mu \mathcal{U}(t, x, \mu)(v) \in \mathbb{R}^d$, whose existence is guaranteed by Theorem 5.29, is continuously differentiable, the function $[0, T] \times \mathbb{R}^d \times \mathcal{P}_2(\mathbb{R}^d) \times \mathbb{R}^d \ni (t, x, \mu, v) \mapsto \partial_v \partial_\mu \mathcal{U}(t, x, \mu)(v) \in \mathbb{R}^{d \times d}$ being continuous in all the arguments and satisfying, for all $t \in [0, T]$, $x_1, x_2 \in \mathbb{R}^d$, $\mu_1, \mu_2 \in \mathcal{P}_2(\mathbb{R}^d)$ and $v_1, v_2 \in \mathbb{R}^d$,*

$$\begin{aligned} &\left| \partial_v \partial_\mu \mathcal{U}(t, x_1, \mu_1)(v_1) - \partial_v \partial_\mu \mathcal{U}(t, x_2, \mu_2)(v_2) \right| \\ &\qquad \leq C\big(|x_1 - x_2| + W_1(\mu_1, \mu_2) + |v_1 - v_2| \big). \end{aligned}$$

The proof is based on Theorem (Vol I)-5.104, the statement of which we recall right below under the new label 5.34. In order to do so, we recall first the following assumption:

Assumption (Sufficiency for Partial $\mathcal{C}^2$). The function $u : \mathcal{P}_2(\mathbb{R}^d) \to \mathbb{R}$ is L-continuously differentiable and, on $(\Omega^1, \mathcal{F}^1, \mathbb{P}^1)$, its lifted version $\tilde{u} : L^2(\Omega^1, \mathcal{F}^1, \mathbb{P}^1; \mathbb{R}^d) \ni X \mapsto u(\mathcal{L}^1(X)) \in \mathbb{R}$ satisfies:

(A1) For any $\chi \in L^2(\Omega^1, \mathcal{F}^1, \mathbb{P}^1; \mathbb{R}^d)$ and any continuously differentiable map $\mathbb{R} \ni \lambda \mapsto X^\lambda \in L^2(\Omega^1, \mathcal{F}^1, \mathbb{P}^1; \mathbb{R}^d)$ with the property that all the $(X^\lambda)_{\lambda \in \mathbb{R}}$ have the same distribution and that $|[d/d\lambda]X^\lambda| \leq 1$, the mapping:

$$\mathbb{R} \ni \lambda \mapsto D\tilde{u}(X^\lambda) \cdot \chi = \mathbb{E}\big[\partial_\mu u(\mathcal{L}^1(X^\lambda))(X^\lambda) \cdot \chi\big] \in \mathbb{R}$$

(continued)

is continuously differentiable, the derivative at $\lambda = 0$ only depending upon the family $(X^\lambda)_{\lambda\in\mathbb{R}}$ through the values of X^0 and $[d/d\lambda]_{|\lambda=0}X^\lambda$, and being denoted by:

$$\partial^2_{\zeta,\chi}\tilde{u}(X) = \frac{d}{d\lambda}_{|\lambda=0}\big[D\tilde{u}(X^\lambda)\cdot\chi\big],$$

$$\text{whenever} \quad X = X^0 \text{ and } \zeta = \frac{d}{d\lambda}_{|\lambda=0}X^\lambda.$$

(A2) There exists a constant C such that, for any X, X', χ and ζ in $L^2(\Omega^1,\mathcal{F}^1,\mathbb{P}^1;\mathbb{R}^d)$, with $X \sim X'$ and $|\zeta| \le 1$ (with probability 1), it holds:

$$\begin{aligned}
&(i) \quad |D\tilde{u}(X)\cdot\chi| + |\partial^2_{\zeta,\chi}u(X)| \le C\|\chi\|_2,\\
&(ii) \quad |D\tilde{u}(X)\cdot\chi - D\tilde{u}(X')\cdot\chi| + |\partial^2_{\zeta,\chi}\tilde{u}(X) - \partial^2_{\zeta,\chi}\tilde{u}(X')|\\
&\qquad \le C\|X - X'\|_2\|\chi\|_2.
\end{aligned}$$

Theorem 5.34 *Under assumption* **Sufficiency for Partial** $\mathcal{C}^2$, *u is partially $\mathcal{C}^2$, which means that:*

1. *The mapping $\mathcal{P}_2(\mathbb{R}^d) \times \mathbb{R}^d \ni (\mu, v) \mapsto \partial_\mu u(\mu)(v)$ is locally bounded (i.e., bounded on any compact subset) and is continuous at any (μ, v) such that $v \in \mathrm{Supp}(\mu)$.*
2. *For any $\mu \in \mathcal{P}_2(\mathbb{R}^d)$, the mapping $\mathbb{R}^d \ni v \mapsto \partial_\mu u(\mu)(v) \in \mathbb{R}^d$ is continuously differentiable and its derivative is locally bounded and is jointly continuous with respect to (μ, v) at any point (μ, v) such that $v \in \mathrm{Supp}(\mu)$, the derivative being denoted by $\mathbb{R}^d \ni v \mapsto \partial_v\partial_\mu u(\mu)(v) \in \mathbb{R}^{d\times d}$.*

Proof of Proposition 5.33.

First Step. We already know from the statement of Theorem 5.29 that, for T small enough, for any $(t,x,\mu) \in [0,T]\times\mathbb{R}^d\times\mathcal{P}_2(\mathbb{R}^d)$, there exists a version of the function $\mathbb{R}^d \ni v \mapsto \partial_\mu\mathcal{U}(t,x,\mu)(v) \in \mathbb{R}^d$ such that the function $[0,T]\times\mathbb{R}^d\times\mathcal{P}_2(\mathbb{R}^d)\times\mathbb{R}^d \ni (t,x,\mu,v) \mapsto \partial_\mu\mathcal{U}(t,x,\mu)(v) \in \mathbb{R}^d$ is continuous and is C-Lipschitz continuous with respect to the variables x, μ and v, the Lipschitz property in the direction μ holding in the sense of the W_1-Wasserstein distance.

Second Step. We now use Theorem 5.34. We start with the following observation. For a fixed (t,x) and for a continuously differentiable mapping $\mathbb{R} \ni s \mapsto \xi^s \in L^2(\Omega^1,\mathcal{F}^1_t,\mathbb{P}^1;\mathbb{R}^d)$ with the property that for all $s \in \mathbb{R}$, $|[d/ds]\xi^s| \le 1$ $\mathbb{P}^1$-almost surely, and for any other random variable $\chi \in L^2(\Omega^1,\mathcal{F}^1_t,\mathbb{P}^1;\mathbb{R}^d)$, we consider the map:

$$\mathbb{R}^2 \ni (r,s) \mapsto \varrho(r,s) = \mathcal{U}\big(t,x,\mathcal{L}^1(r\chi + \xi^s)\big).$$

Then, by Theorem 5.29, ϱ is jointly differentiable with respect to (r,s) with:

$$\partial_r\varrho(r,s) = \mathbb{E}^1\big[\partial_\mu\mathcal{U}\big(t,x,\mathcal{L}^1(r\chi+\xi^s)\big)\chi\big],$$
$$\partial_s\varrho(r,s) = \mathbb{E}^1\Big[\partial_\mu\mathcal{U}\big(t,x,\mathcal{L}^1(r\chi+\xi^s)\big)\frac{d}{ds}\xi^s\Big], \quad r,s\in\mathbb{R}.$$

Letting $\zeta^s = [d/ds]\xi^s$, we can write:

$$\partial_s\varrho(r,s) = \partial_{\zeta^s}Y_t^{t,x,r\chi+\xi^s}.$$

Since $\|\zeta^s\|_\infty \le 1$, we deduce from Proposition 5.31 that $\partial_s\varrho$ is differentiable with respect to r, with:

$$\partial_r\big[\partial_s\varrho(r,s)\big] = \mathbb{E}^1\big[D_\xi\partial_{\zeta^s}Y_t^{t,x,r\chi+\xi^s}\chi\big], \quad r,s\in\mathbb{R}.$$

Recalling the identity (5.80), we notice that $\partial_r[\partial_s\varrho]$ is jointly continuous. By Schwarz's theorem, we deduce that the mapping $\mathbb{R}\ni s\mapsto\partial_r\varrho(0,s)$ is continuously differentiable with:

$$\partial_s\big[\partial_r\varrho\big](0,s) = \mathbb{E}^1\big[D_\xi\partial_{\zeta^s}Y_t^{t,x,\xi^s}\chi\big].$$

In particular for $s=0$, the derivative takes the form:

$$\partial_s\big[\partial_r\varrho\big](0,0) = \mathbb{E}^1\big[D_\xi\partial_\zeta Y_t^{t,x,\xi}\chi\big],$$

where we let $\zeta=\zeta^0$ and $\xi=\xi^0$.

Following assumption **Sufficiency for Partial** $\mathcal{C}^2$, we now let, for a fixed $(t,x)\in[0,T]\times\mathbb{R}^d$:

$$u(\mu) = \mathcal{U}(t,x,\mu), \quad \mu\in\mathcal{P}_2(\mathbb{R}^d).$$

The above discussion shows that for any $\chi\in L^2(\Omega^1,\mathcal{F}^1,\mathbb{P}^1;\mathbb{R}^d)$ and any continuously differentiable map $\mathbb{R}\ni s\mapsto\xi^s\in L^2(\Omega^1,\mathcal{F}^1,\mathbb{P};\mathbb{R}^d)$ with the property that all the $(\xi^s)_{s\in\mathbb{R}}$ have the same distribution and that $\|[d/ds]\xi^s\|_\infty\le 1$, the mapping:

$$\mathbb{R}\ni s\mapsto\mathbb{E}^1\big[\partial_\mu u\big(\mathcal{L}^1(\xi^s)\big)(\xi^s)\chi\big]$$

is continuously differentiable with:

$$\partial^2_{\zeta,\chi}\tilde{u}(\xi) = \mathbb{E}^1\big[D_\xi\partial_\zeta Y_t^{t,x,\xi}\chi\big], \tag{5.81}$$

as derivative at $s=0$, where $\tilde{u}$ denotes the lifting of u and $\partial^2_{\zeta,\chi}\tilde{u}$ is as in assumption **Sufficiency for Partial** $\mathcal{C}^2$.

By Proposition 5.32, we deduce that (**A1**) and (**A2**) in assumption **Sufficiency for Partial** $\mathcal{C}^2$ hold. We deduce that for any $(t,x)\in[0,T]\times\mathbb{R}^d$ and $\mu\in\mathcal{P}_2(\mathbb{R}^d)$, there exists a version of $\mathbb{R}^d\ni v\mapsto\partial_\mu\mathcal{U}(t,x,\mu)(v)$ which is continuously differentiable.

Third Step. In order to complete the proof, we recall from (Vol I)-(5.115) that for $Z'\sim N(0,1)$, Z' being independent of (ξ,χ), it holds:

$$\partial^2_{\text{sign}(Z')e,\text{sign}(Z')\chi}\tilde{u}(\xi) = \mathbb{E}^1\Big[\partial_v\partial_\mu\mathcal{U}\big(t,x,\mathcal{L}^1(\xi)\big)(\xi)\chi\otimes e\Big]. \tag{5.82}$$

Thanks to (5.81) and Proposition 5.32, we deduce that for all $t \in [0, T]$, $x_1, x_2 \in \mathbb{R}^d$ and $\xi_1, \xi_2 \in L^2(\Omega^1, \mathcal{F}_t^1, \mathbb{P}^1; \mathbb{R}^d)$,

$$\mathbb{E}^1\Big[\Big(\partial_v \partial_\mu \mathcal{U}(t, x_1, \mathcal{L}^1(\xi_1)(\xi_1) - \partial_v \partial_\mu \mathcal{U}(t, x_2, \mathcal{L}^1(\xi_2)(\xi_2)\Big)\chi \otimes e\Big]$$
$$\leq C\Big[\|\chi\|_1\big(|x_1 - x_2| + \|\xi_1 - \xi_2\|_1\big) + \mathbb{E}^1\big[|\xi_1 - \xi_2|\,|\chi|\big]\Big].$$

Moreover, for all $\xi \in L^2(\Omega^1, \mathcal{F}_t^1, \mathbb{P}^1; \mathbb{R}^d)$,

$$\big\|\partial_v \partial_\mu \mathcal{U}\big(t, x, \mathcal{L}^1(\xi)\big)(\xi)\big\|_\infty \leq C.$$

By the auxiliary Lemma 5.30, we deduce that for each $(t, x, \mu) \in [0, T] \times \mathbb{R}^d \times \mathcal{P}_2(\mathbb{R}^d)$, there exists a version of $\partial_v \partial_\mu \mathcal{U}(t, x, \mu)(\cdot)$ in $L^2(\mathbb{R}^d, \mu; \mathbb{R}^{m \times (d \times d)})$ such that the map $\partial_v \partial_\mu \mathcal{U}(t, \cdot, \cdot)(\cdot)$ is continuous and satisfies the Lipschitz property in the statement. In order to complete the proof, it only remains to check that this globally continuous version of $\partial_v \partial_\mu \mathcal{U}(t, \cdot, \cdot)(\cdot)$ coincides with the partial derivative of $\partial_\mu \mathcal{U}(t, \cdot, \cdot)(\cdot)$ in the direction v on the whole space. When μ has full support, $\partial_v \partial_\mu \mathcal{U}(t, x, \mu)(\cdot)$ has a unique continuous version in $L^2(\mathbb{R}^d, \mu; \mathbb{R}^{m \times (d \times d)})$ and this version must coincide with the one provided by Lemma 5.30. Obviously, it is the partial derivative of $\partial_\mu \mathcal{U}(t, x, \mu)(\cdot)$ with respect to v. Since $\partial_\mu \mathcal{U}(t, \cdot, \cdot)(\cdot)$ is globally continuous, we deduce by a standard approximation argument that the same holds when the support of μ is strictly included in $\mathbb{R}^d$. This shows that the globally continuous version of $\partial_v \partial_\mu \mathcal{U}(t, \cdot, \cdot)(\cdot)$ provided by Lemma 5.30 is the partial derivative of $\partial_\mu \mathcal{U}(t, \cdot, \cdot)(\cdot)$ in the direction v.

Fourth Step. It remains to prove that the map $\partial_v \partial_\mu \mathcal{U}$ is globally continuous on $[0, T] \times \mathbb{R}^d \times \mathcal{P}_2(\mathbb{R}^d) \times \mathbb{R}^d$. To do so, we recall from the identification (5.80) that for any $\chi \in L^\infty$ with $\|\chi\|_\infty \leq 1$, the derivative with respect to ξ of the mapping:

$$(t, x, \xi) \mapsto \mathbb{E}^1\big[\partial_\mu \mathcal{U}\big(t, x, \mathcal{L}^1(\xi)\big)\chi\big]$$

satisfies the conclusion of Theorem 5.29 and is continuous in time. Choosing the direction of differentiation ζ of the same form as in (5.82), namely $\zeta = \mathrm{sign}(Z')e$, and χ of the form $\mathrm{sign}(Z')\chi$, where $Z' \sim N(0, 1)$ is independent of (ξ, χ) and $\|\chi\|_\infty \leq 1$, we deduce from the identities (5.80), (5.81), and (5.82), that for any $(x, \xi) \in \mathbb{R}^d \times L^2(\Omega^1, \mathcal{F}_0^1, \mathbb{P}^1; \mathbb{R}^d)$ and $\chi \in L^\infty(\Omega^1, \mathcal{F}_0^1, \mathbb{P}^1; \mathbb{R}^d)$ with $\|\chi\|_\infty \leq 1$, the mapping:

$$[0, T] \ni t \mapsto \mathbb{E}^1\big[\partial_v \partial_\mu \mathcal{U}\big(t, x, \mathcal{L}^1(\xi)\big)(\xi)\chi \otimes e\big]$$

is continuous. Proceeding as in the second step of the proof of Theorem 5.29, time continuity of the map $[0, T] \times \mathbb{R}^d \times \mathcal{P}_2(\mathbb{R}^d) \times \mathbb{R}^d \ni (t, x, \mu, v) \mapsto \partial_v \partial_\mu \mathcal{U}(t, x, \mu)(v)$ follows. □

Differentiability with Respect to μ in Bounded Directions

We now turn to the derivative in the direction μ. The analysis is divided into several lemmas. Part of the argument relies on the analysis of the quantity:

$$\big(\partial^2_{(1-\varepsilon)\zeta, \varepsilon\chi} \theta_s^{t,x,\xi}\big)_{t \leq s \leq T},$$

where $\xi, \zeta \in L^2(\Omega^1, \mathcal{F}_t^1, \mathbb{P}^1; \mathbb{R}^d)$, $\chi \in L^\infty(\Omega^1, \mathcal{F}_t^1, \mathbb{P}^1; \mathbb{R}^d)$ with $\|\chi\|_\infty \leq 1$, and ε is a symmetric Bernoulli random variable independent of (ξ, ζ, χ).

The rationale for considering such specific directions in the second-order derivatives may be explained as follows. If $\mathcal{U}$ is already known to be twice-differentiable in the measure argument, then it should hold:

$$
\begin{aligned}
\partial^2_{\zeta,\chi}\theta_t^{t,x,\xi} &= \frac{d}{ds}_{|s=0} \mathbb{E}^1\big[\partial_\mu \mathcal{U}\big(t,x,\mathcal{L}^1(\xi+s\zeta)\big)(\xi+s\zeta)\chi\big] \\
&= \mathbb{E}^1\big[\partial_v \partial_\mu \mathcal{U}\big(t,x,\mathcal{L}^1(\xi)\big)(\xi)\chi \otimes \zeta\big] \\
&\quad + \mathbb{E}^1\tilde{\mathbb{E}}^1\big[\partial^2_\mu \mathcal{U}\big(t,x,\mathcal{L}^1(\xi)\big)(\xi,\tilde{\xi})\chi \otimes \tilde{\zeta}\big].
\end{aligned}
\tag{5.83}
$$

Whenever (ζ,χ) is replaced by $((1-\varepsilon)\zeta,\varepsilon\chi)$, the above identity becomes:

$$
\begin{aligned}
\partial^2_{(1-\varepsilon)\zeta,\varepsilon\chi}\theta_t^{t,x,\xi} &= \mathbb{E}^1\big[\partial_v \partial_\mu \mathcal{U}\big(t,x,\mathcal{L}^1(\xi)\big)(\xi)(\varepsilon\chi) \otimes \big((1-\varepsilon)\zeta\big)\big] \\
&\quad + \mathbb{E}^1\tilde{\mathbb{E}}^1\big[\partial^2_\mu \mathcal{U}\big(t,x,\mathcal{L}^1(\xi)\big)(\xi,\tilde{\xi})(\varepsilon\chi) \otimes \big((1-\tilde{\varepsilon})\tilde{\zeta}\big)\big].
\end{aligned}
$$

Since the product $\varepsilon(1-\varepsilon)$ is always equal to 0 in the first term in the above right-hand side, and since ε is assumed to be independent of (ξ,χ,ζ), the above identity reduces to:

$$
\partial^2_{(1-\varepsilon)\zeta,\varepsilon\chi}\theta_t^{t,x,\xi} = \frac{1}{4}\mathbb{E}^1\tilde{\mathbb{E}}^1\big[\partial^2_\mu \mathcal{U}\big(t,x,\mathcal{L}^1(\xi)\big)(\xi,\tilde{\xi})\chi \otimes \tilde{\zeta}\big],
$$

which connects $\partial^2_{(1-\varepsilon)\zeta,\varepsilon\chi}\theta_t^{t,x,\chi}$ with the sole derivative $\partial^2_\mu \mathcal{U}$ instead of the full-fledge one as in (5.83).

In this perspective, here is the first lemma:

Lemma 5.35 *There exist a constant c, only depending on L, and a constant C, only depending on L and Γ, such that, for* $T \le c$, *for all* $(t,x) \in [0,T]\times\mathbb{R}^d$, *and* ξ,χ,ζ *and* ε *as above (in particular,* $\zeta \in L^2(\Omega^1,\mathcal{F}^1_t,\mathbb{P}^1;\mathbb{R}^d)$*), with the prescription that* $\chi \in L^\infty(\Omega^1,\mathcal{F}^1_t,\mathbb{P}^1;\mathbb{R}^d)$ *with* $\|\chi\|_\infty \le 1$, *it holds, with probability 1 under* $\mathbb{P}^1$:

$$
\sup_{t\le s\le T} |\partial^2_{(1-\varepsilon)\zeta,\varepsilon\chi}\theta_s^{t,x,\xi}| \le C\|\chi\|_1\|\zeta\|_1.
$$

Moreover, for all $x_1,x_2 \in \mathbb{R}^d$ *and* $\xi_1,\xi_2 \in L^2(\Omega^1,\mathcal{F}^1_t,\mathbb{P}^1;\mathbb{R}^d)$ *such that* ε *is independent of* (ξ_1,ξ_2,χ,ζ), *with probability 1 under* $\mathbb{P}^1$,

$$
\begin{aligned}
&\sup_{t\le s\le T} \big|\partial^2_{(1-\varepsilon)\zeta,\varepsilon\chi}\theta_s^{t,x_1,\xi_1} - \partial^2_{(1-\varepsilon)\zeta,\varepsilon\chi}\theta_s^{t,x_2,\xi_2}\big| \\
&\le C\Big(\|\zeta\|_1\|\chi\|_1\big(|x_1-x_2| + \|\xi_1-\xi_2\|_1\big) \\
&\qquad + \|\zeta\|_1\mathbb{E}^1\big[|\xi_1-\xi_2|\,|\chi|\big] + \|\chi\|_1\mathbb{E}^1\big[|\xi_1-\xi_2|\,|\zeta|\big]\Big).
\end{aligned}
$$

Proof of Lemma 5.35.

First Step. We start with the analysis of the same quantities as in the statement but with $\boldsymbol{\theta}^{t,\xi}$ in lieu of $\boldsymbol{\theta}^{t,x,\xi}$. Here is the first observation. By Lemma 5.24, we have:

$$\sup_{t\leq s\leq T} |\partial_{\varepsilon\chi}\theta_s^{t,\xi}| \leq C\big(\varepsilon|\chi| + \|\chi\|_1\big). \tag{5.84}$$

Therefore, appealing once again to Lemma 5.24, we get, with probability 1 under $\mathbb{P}$,

$$\sup_{t\leq s\leq T} |\partial_{\varepsilon\chi}\theta_s^{t,\xi} \otimes \partial_{(1-\varepsilon)\zeta}\theta_s^{t,\xi}| \leq C\big(\|\chi\|_1|\zeta| + |\chi|\|\zeta\|_1 + \|\chi\|_1\|\zeta\|_1\big), \tag{5.85}$$

where we used the fact that $\varepsilon(1-\varepsilon) = 0$.

Recalling the form of $\partial^2 B$ from (5.77) and taking advantage of the boundedness of the second order derivatives as stated in assumption **Smooth Coefficients Order 2**, we deduce that:

$$\begin{aligned}&\Big|\partial^2 B\big(s, \theta_s^{t,\xi}, \tilde{X}_s^{t,\xi}\big)\big(\partial_{\varepsilon\chi}\theta_s^{t,\xi}, \partial_{(1-\varepsilon)\zeta}\theta_s^{t,\xi}, \partial_{\varepsilon\chi}\tilde{X}_s^{t,\xi}, \partial_{(1-\varepsilon)\zeta}\tilde{X}_s^{t,\xi}\big)\Big| \\ &\quad \leq C\big(\|\chi\|_1|\zeta| + |\chi|\|\zeta\|_1 + \|\chi\|_1\|\zeta\|_1\big),\end{aligned}$$

with a similar bound with $\partial^2 F$ and $\partial^2 G$ in lieu of $\partial^2 B$.

Returning to (5.76) and regarding the system satisfied by $\partial^2_{\varepsilon\chi,(1-\varepsilon)\zeta}\boldsymbol{\theta}^{t,\xi}$ as a linear system of the type (5.28) and then appealing to the last inequality in Corollary 5.18, we deduce that, with probability 1 under $\mathbb{P}$,

$$\sup_{t\leq s\leq T} |\partial^2_{(1-\varepsilon)\zeta,\varepsilon\chi}\theta_s^{t,\xi}| \leq C\big(\|\chi\|_1|\zeta| + |\chi|\|\zeta\|_1 + \|\chi\|_1\|\zeta\|_1\big).$$

We now consider the difference $(\partial^2_{(1-\varepsilon)\zeta,\varepsilon\chi}\theta_s^{t,\xi_1} - \partial^2_{(1-\varepsilon)\zeta,\varepsilon\chi}\theta_s^{t,\xi_2})_{t\leq s\leq T}$. Lemma 5.25 yields:

$$\begin{aligned}&\sup_{t\leq s\leq T} |\partial_{\varepsilon\chi}\theta_s^{t,\xi_1} - \partial_{\varepsilon\chi}\theta_s^{t,\xi_2}| \\ &\quad \leq C\Big[\mathbb{E}^1\big[|\xi_1-\xi_2|\,|\chi|\big] + \big(|\xi_1-\xi_2| + \|\xi_1-\xi_2\|_1\big)\big(\varepsilon|\chi| + \|\chi\|_1\big)\Big].\end{aligned} \tag{5.86}$$

Therefore, by (5.84) with $(1-\varepsilon)\zeta$ in lieu of $\varepsilon\chi$, we deduce that:

$$\begin{aligned}&\sup_{t\leq s\leq T} \big(|\partial_{\varepsilon\chi}\theta_s^{t,\xi_1} - \partial_{\varepsilon\chi}\theta_s^{t,\xi_2}|\,|\partial_{(1-\varepsilon)\zeta}\theta_s^{t,\xi_1}|\big) \\ &\quad \leq C\Big[\mathbb{E}^1\big[|\xi_1-\xi_2|\,|\chi|\big] + \big(|\xi_1-\xi_2| + \|\xi_1-\xi_2\|_1\big)\big(\varepsilon|\chi| + \|\chi\|_1\big)\Big] \\ &\qquad\qquad \times \big[(1-\varepsilon)|\zeta| + \|\zeta\|_1\big],\end{aligned}$$

from which we get:

$$\begin{aligned}&\mathbb{E}^1\Big[\sup_{t\leq s\leq T} \big(|\partial_{\varepsilon\chi}\theta_s^{t,\xi_1} - \partial_{\varepsilon\chi}\theta_s^{t,\xi_2}|\,|\partial_{(1-\varepsilon)\zeta}\theta_s^{t,\xi_1}|\big)\Big] \\ &\quad \leq C\Big(\|\xi_1-\xi_2\|_1\|\zeta\|_1\|\chi\|_1 + \|\zeta\|_1\mathbb{E}^1\big[|\xi_1-\xi_2|\,|\chi|\big] + \|\chi\|_1\mathbb{E}^1\big[|\xi_1-\xi_2|\,|\zeta|\big]\Big),\end{aligned} \tag{5.87}$$

and similarly for $(|\partial_{(1-\varepsilon)\zeta}\theta_s^{t,\xi_1} - \partial_{(1-\varepsilon)\zeta}\theta_s^{t,\xi_2}||\partial_{\varepsilon\chi}\theta_s^{t,\xi_1}|)_{t\leq s\leq T}$.

We now make use of Corollary 5.22. In comparison, with Lemma 5.25, we have to take into account the fact the remainders $\mathcal{R}_a^1$ and $\mathcal{R}_a^2$ are not zero. Fortunately, they can be estimated by means of (5.85), (5.86), and (5.87). We get:

$$\mathbb{E}^1\big[\sup_{t\le s\le T} |\partial^2_{(1-\varepsilon)\zeta,\varepsilon\chi}\theta_s^{t,\xi_1} - \partial^2_{(1-\varepsilon)\zeta,\varepsilon\chi}\theta_s^{t,\xi_2}|\big]$$
$$\le C\Big(\|\zeta\|_1\|\chi\|_1\|\xi_1-\xi_2\|_1 + \|\zeta\|_1\mathbb{E}^1\big[|\xi_1-\xi_2|\,|\chi|\big] + \|\chi\|_1\mathbb{E}^1\big[|\xi_1-\xi_2|\,|\zeta|\big]\Big).$$

Second Step. We now complete the proof. In order to do so, we regard (5.79) as a system of the type (5.28). Recalling the first bound for $\partial_\chi\boldsymbol{\theta}^{t,x,\xi}$ and $\partial_\zeta\boldsymbol{\theta}^{t,x,\xi}$ in the statement of Lemma 5.24 and applying the last inequality in Corollary 5.18, the remainders being given by the McKean-Vlasov terms and by the first-order terms, we deduce from the first step that:

$$\sup_{t\le s\le T} |\partial^2_{(1-\varepsilon)\zeta,\varepsilon\chi}\theta_s^{t,x,\xi}| \le C\|\chi\|_1\|\zeta\|_1.$$

In order to estimate the difference $\partial^2_{(1-\varepsilon)\zeta,\varepsilon\chi}\boldsymbol{\theta}^{t,x_1,\xi_1} - \partial^2_{(1-\varepsilon)\zeta,\varepsilon\chi}\boldsymbol{\theta}^{t,x_2,\xi_2}$, we proceed as in the first step, making use of Corollary 5.22. Observe from Lemmas 5.26 and 5.28 that, on the model of (5.87),

$$\sup_{t\le s\le T}\big(|\partial_{\varepsilon\chi}\theta_s^{t,x_1,\xi_1} - \partial_{\varepsilon\chi}\theta_s^{t,x_2,\xi_2}|\,|\partial_{(1-\varepsilon)\zeta}\theta_s^{t,x_1,\xi_1}|\big)$$
$$\le C\Big(\big(|x_1-x_2| + \|\xi_1-\xi_2\|_1\big)\|\zeta\|_1\|\chi\|_1 + \|\zeta\|_1\mathbb{E}^1\big[|\xi_1-\xi_2|\,|\chi|\big]\Big),$$

with a similar bound for $\sup_{t\le s\le T}(|\partial_{(1-\varepsilon)\zeta}\theta_s^{t,x_1,\xi_1} - \partial_{(1-\varepsilon)\zeta}\theta_s^{t,x_2,\xi_2}|\,|\partial_{\varepsilon\chi}\theta_s^{t,x_1,\xi_1}|)$. We conclude as in the first step. □

We now identify the second-order derivative in the direction μ.

Lemma 5.36 *For $T \le c$, with c only depending on L, for $(t,x) \in [0,T]\times\mathbb{R}^d$, for $\xi \in L^2(\Omega^1,\mathcal{F}_t^1,\mathbb{P}^1;\mathbb{R}^d)$ and $\chi \in L^\infty(\Omega^1,\mathcal{F}_t^1,\mathbb{P}^1;\mathbb{R}^d)$ with $\|\chi\|_\infty \le 1$, it holds:*

$$\lim_{s\to 0}\Big\|\frac{1}{s}\Big(\partial_\mu\mathcal{U}\big(t,x,\mathcal{L}^1(\xi+s\chi)\big)(\xi) - \partial_\mu\mathcal{U}\big(t,x,\mathcal{L}(\xi)\big)(\xi)\Big) - \psi\big(t,x,\mathcal{L}^1(\xi,\chi)\big)(\xi)\Big\|_\infty = 0,$$

where $\psi : [0,T]\times\mathbb{R}^d\times\mathcal{P}_2(\mathbb{R}^{2d})\times\mathbb{R}^d \to \mathbb{R}^d$ satisfies, for all $t \in [0,T]$, $x \in \mathbb{R}^d$, $\xi \in L^2(\Omega^1,\mathcal{F}_t^1,\mathbb{P}^1;\mathbb{R}^d)$ and $\chi \in L^\infty(\Omega^1,\mathcal{F}_t^1,\mathbb{P}^1;\mathbb{R}^d)$ with $\|\chi\|_\infty \le 1$, $\psi(t,x,\mathcal{L}^1(\xi,\chi))(\cdot) \in L^\infty(\mathbb{R}^d,\mu;\mathbb{R}^d)$ and,

$$\big\|\psi\big(t,x,\mathcal{L}^1(\xi,\chi)\big)(\xi)\big\|_\infty \le C,$$

for a constant C only depending on L and Γ.

Proof.

First Step. We recall that, when $\|\chi\|_\infty \le 1$, $\partial_\chi Y^{t,x,\xi} = \mathbb{E}^1[\partial_\mu\mathcal{U}(t,x,\mathcal{L}^1(\xi))(\xi)\chi]$ coincides with $\boldsymbol{U}(t,x,\nu)$, where $\nu = \mathcal{L}^1(\xi,\chi)$, $\boldsymbol{U}(t,x,\cdot)$ being L-differentiable on $\mathcal{P}_2(\mathbb{R}^{2d})$ and satisfying the conclusion of Theorem 5.29 in short time, namely:

$$\|\partial_\nu\boldsymbol{U}(t,x,\nu)(\xi,\chi)\|_\infty \le C, \tag{5.88}$$

for $(t,x) \in [0,T] \times \mathbb{R}^d$, $\xi, \chi \in L^2(\Omega^1, \mathcal{F}_t^1, \mathbb{P}^1; \mathbb{R}^d)$ and $\nu = \mathcal{L}^1(\xi, \chi)$, and for all $\zeta \in L^1(\Omega^1, \mathcal{F}_t^1, \mathbb{P}^1; \mathbb{R}^d)$,

$$\begin{aligned}
&\Big|\mathbb{E}^1\big[\big(\partial_\mu \boldsymbol{U}(t,x,\nu)(\xi,\chi) - \partial_\mu \boldsymbol{U}(t,x',\nu')(\xi',\chi')\big)\zeta\big]\Big| \\
&\quad \le C\Big[\big(|x-x'| + \|\xi-\xi'\|_1 + \|\chi-\chi'\|_1\big)\|\zeta\|_1 + \mathbb{E}^1\big[\big(|\xi-\xi'| + |\chi-\chi'|\big)|\zeta|\big]\Big],
\end{aligned} \tag{5.89}$$

for all $x' \in \mathbb{R}^d$, $\xi', \chi' \in L^2(\Omega^1, \mathcal{F}_t^1, \mathbb{P}^1; \mathbb{R}^d)$ and $\nu' = \mathcal{L}^1(\xi', \chi')$.

By Lemma 5.30 we can find, for each $t \in [0,T]$, a C-bounded and C-Lipschitz continuous version of the mapping $\mathbb{R}^d \times \mathcal{P}_2(\mathbb{R}^{2d}) \times \mathbb{R}^{2d} \ni (x, \nu, (v,w)) \mapsto \partial_\mu \boldsymbol{U}(t,x,\nu)(v,w)$. We now recall that for $\xi, \chi, \zeta \in L^2(\Omega^1, \mathcal{F}_t^1, \mathbb{P}^1; \mathbb{R}^d)$,

$$\lim_{s \to 0} s^{-1}\Big[\boldsymbol{U}\big(t,x,\mathcal{L}^1(\xi + s\zeta, \chi)\big) - \boldsymbol{U}\big(t,x,\mathcal{L}^1(\xi,\chi)\big)\Big] = \mathbb{E}^1\big[\partial_\mu \boldsymbol{U}\big(t,x,\mathcal{L}^1(\xi,\chi)\big)(\xi,\chi)\zeta\big].$$

Second Step. Following the analysis of the derivatives of $\partial_\mu \mathcal{U}$ with respect to v, we shall apply Schwarz' theorem once again. For $\chi \in L^\infty(\Omega^1, \mathcal{F}_t^1, \mathbb{P}^1; \mathbb{R}^d)$, with $\|\chi\|_\infty \le 1$, and for $\zeta \in L^2(\Omega^1, \mathcal{F}_t^1, \mathbb{P}^1; \mathbb{R}^d)$, we let:

$$\varrho(r,s) = \mathcal{U}\big(t,x,\mathcal{L}^1(\xi + r\chi + s\zeta)\big), \quad r,s \in \mathbb{R}.$$

Then,

$$\begin{aligned}
\partial_r \varrho(r,s) &= \mathbb{E}^1\big[\partial_\mu \mathcal{U}\big(t,x,\mathcal{L}^1(\xi + r\chi + s\zeta)\big)(\xi + r\chi + s\zeta)\chi\big], \\
\partial_s \varrho(r,s) &= \mathbb{E}^1\big[\partial_\mu \mathcal{U}\big(t,x,\mathcal{L}^1(\xi + r\chi + s\zeta)\big)(\xi + r\chi + s\zeta)\zeta\big].
\end{aligned}$$

Then, we know that $\partial_r \varrho$ is differentiable with respect to s, with:

$$\partial_s \partial_r \varrho(r,s) = \mathbb{E}\Big[\partial_\mu \boldsymbol{U}\Big(t,x,\mathcal{L}^1\big(\xi + r\chi + s\zeta, \chi\big)\Big)\big(\xi + r\chi + s\zeta, \chi\big)\zeta\Big], \quad r,s \in \mathbb{R}.$$

By the first step, $\partial_s \partial_r \varrho$ is continuous. We deduce that $\partial_s \varrho(0,\cdot)$ is differentiable, with:

$$\begin{aligned}
&\mathbb{E}^1\big[\partial_\mu \boldsymbol{U}(t,x,\nu)(\xi,\chi)\zeta\big] \\
&= \lim_{r \to 0} r^{-1}\mathbb{E}^1\Big[\Big(\partial_\mu \mathcal{U}\big(t,x,\mathcal{L}^1(\xi + r\chi)\big)(\xi + r\chi) - \partial_\mu \mathcal{U}\big(t,x,\mathcal{L}^1(\xi)\big)(\xi)\Big)\zeta\Big] \\
&= \partial_r\big[\partial_s \varrho(r,0)\big]_{|r=0} \\
&= \partial_s\big[\partial_r \varrho(0,s)\big]_{|s=0} \\
&= \lim_{s \to 0} s^{-1}\mathbb{E}^1\Big[\Big(\partial_\mu \mathcal{U}\big(t,x,\mathcal{L}^1(\xi + s\zeta)\big)(\xi + s\zeta) - \partial_\mu \mathcal{U}\big(t,x,\mathcal{L}^1(\xi)\big)(\xi)\Big)\chi\Big].
\end{aligned} \tag{5.90}$$

By the first step again, we deduce that:

$$\begin{aligned}
&\mathbb{E}^1\Big[\Big(\partial_\mu \mathcal{U}\big(t,x,\mathcal{L}^1(\xi + s\chi)\big)(\xi + s\chi) - \partial_\mu \mathcal{U}\big(t,x,\mathcal{L}^1(\xi)\big)(\xi)\Big)\zeta\Big] \\
&= \int_0^s \mathbb{E}^1\big[\partial_\mu \boldsymbol{U}\big(t,x,\mathcal{L}^1(\xi + r\chi, \chi)\big)(\xi + r\chi, \chi)\zeta\big]dr \\
&= s\mathbb{E}^1\big[\partial_\mu \boldsymbol{U}(t,x,\nu)(\xi,\chi)\zeta\big] + o(s)\|\zeta\|_1,
\end{aligned} \tag{5.91}$$

with $\lim_{s\to 0}|o(s)/s| = 0$, uniformly in $\zeta \in L^2(\Omega^1, \mathcal{F}_t^1, \mathbb{P}^1; \mathbb{R}^d)$. Therefore,

$$\mathbb{E}\Big[\Big\{\frac{1}{s}\Big(\partial_\mu \mathcal{U}\big(t,x,\mathcal{L}^1(\xi+s\chi)\big)(\xi+s\chi) - \partial_\mu \mathcal{U}\big(t,x,\mathcal{L}^1(\xi)\big)(\xi)\Big) - \partial_\mu \boldsymbol{U}(t,x,\nu)(\xi,\chi)\Big\}\zeta\Big]$$
$$\leq \Big|\frac{o(s)}{s}\Big| \|\zeta\|_1.$$

And then, for $\xi \in L^2(\Omega^1, \mathcal{F}_t^1, \mathbb{P}^1; \mathbb{R}^d)$ and $\chi \in L^\infty(\Omega^1, \mathcal{F}_t^1, \mathbb{P}^1; \mathbb{R}^d)$ with $\|\chi\|_\infty \leq 1$, with $\mathbb{P}^1$-probability 1,

$$\lim_{s\to 0}\Big\|\frac{1}{s}\Big(\partial_\mu \mathcal{U}\big(t,x,\mathcal{L}^1(\xi+s\chi)\big)(\xi+s\chi) - \partial_\mu \mathcal{U}\big(t,x,\mathcal{L}^1(\xi)\big)(\xi)\Big) - \partial_\mu \boldsymbol{U}(t,x,\nu)(\xi,\chi)\Big\|_\infty = 0.$$

Third Step. For (t,x,ξ,χ) as above and for $s \neq 0$, we have:

$$\begin{aligned}
&\frac{1}{s}\Big(\partial_\mu \mathcal{U}\big(t,x,\mathcal{L}^1(\xi+s\chi)\big)(\xi) - \partial_\mu \mathcal{U}\big(t,x,\mathcal{L}^1(\xi)\big)(\xi)\Big)\\
&= \frac{1}{s}\Big(\partial_\mu \mathcal{U}\big(t,x,\mathcal{L}^1(\xi+s\chi)\big)(\xi+s\chi) - \partial_\mu \mathcal{U}\big(t,x,\mathcal{L}^1(\xi)\big)(\xi)\Big)\\
&\quad - \frac{1}{s}\Big(\partial_\mu \mathcal{U}\big(t,x,\mathcal{L}^1(\xi+s\chi)\big)(\xi+s\chi) - \partial_\mu \mathcal{U}\big(t,x,\mathcal{L}^1(\xi+s\chi)\big)(\xi)\Big).
\end{aligned}$$

By continuous differentiability of $\partial_\mu \mathcal{U}$ with respect to v, see Proposition 5.33, this can be rewritten as:

$$\begin{aligned}
&\frac{1}{s}\Big(\partial_\mu \mathcal{U}\big(t,x,\mathcal{L}^1(\xi+s\chi)\big)(\xi+s\chi) - \partial_\mu \mathcal{U}\big(t,x,\mathcal{L}^1(\xi)\big)(\xi)\Big)\\
&\quad - \frac{1}{s}\int_0^s \partial_v\partial_\mu \mathcal{U}\big(t,x,\mathcal{L}^1(\xi+s\chi)\big)(\xi+r\chi)\chi dr.
\end{aligned}$$

By the second step and by continuity of $\partial_v\partial_\mu \mathcal{U}$ with respect to its last two arguments, see again Proposition 5.33, we deduce that:

$$\begin{aligned}
\lim_{s\to 0}\Big\|&\frac{1}{s}\Big(\partial_\mu \mathcal{U}\big(t,x,\mathcal{L}^1(\xi+s\chi)\big)(\xi) - \partial_\mu \mathcal{U}\big(t,x,\mathcal{L}^1(\xi)\big)(\xi)\Big)\\
&- \big(\partial_\mu \boldsymbol{U}(t,x,\nu)(\xi,\chi) - \partial_v\partial_\mu \mathcal{U}(t,x,\mu)(\xi)\chi\big)\Big\|_\infty = 0,
\end{aligned}$$

which proves that the functions

$$\Big(\mathbb{R}^d \ni v \mapsto \frac{1}{s}\Big(\partial_\mu \mathcal{U}\big(t,x,\mathcal{L}^1(\xi+s\chi)\big)(v) - \partial_\mu \mathcal{U}\big(t,x,\mathcal{L}^1(\xi)\big)(v)\Big)\Big)_{s>0}$$

converge in $L^\infty(\mathbb{R}^d, \mu; \mathbb{R}^d)$ as s tends to 0, where $\mu = \mathcal{L}^1(\xi)$. Hence, there exists a function $\psi(t,x,\mathcal{L}^1(\xi,\chi))(\cdot) \in L^\infty(\mathbb{R}^d, \mu; \mathbb{R}^d)$ such that:

$$\lim_{s\to 0}\Big\|\frac{1}{s}\Big(\partial_\mu \mathcal{U}\big(t,x,\mathcal{L}^1(\xi+s\chi)\big)(\xi) - \partial_\mu \mathcal{U}\big(t,x,\mathcal{L}^1(\xi)\big)(\xi)\Big) - \psi\big(t,x,\mathcal{L}^1(\xi,\chi)\big)(\xi)\Big\|_\infty = 0,$$

which completes the proof. The second part of the statement follows from the identification $\psi(t,x,\mathcal{L}^1(\xi,\chi))(\xi) = \partial_\mu \boldsymbol{U}(t,x,\nu)(\xi,\chi) - \partial_v\partial_\mu \mathcal{U}(t,x,\mu)(\xi)\chi$. □

Importantly, ψ satisfies a useful symmetry property.

Lemma 5.37 *For $T \leq c$, with c only depending on L, for $(t,x) \in [0,T] \times \mathbb{R}^d$, for $\xi \in L^2(\Omega^1, \mathcal{F}_t^1, \mathbb{P}^1; \mathbb{R}^d)$ and $\chi, \zeta \in L^\infty(\Omega^1, \mathcal{F}_t^1, \mathbb{P}^1; \mathbb{R}^d)$, with $\|\chi\|_\infty, \|\zeta\|_\infty \leq 1$, it holds:*

$$\mathbb{E}^1\big[\psi\big(t,x,\mathcal{L}^1(\xi,\chi)\big)(\xi)\zeta\big] = \mathbb{E}^1\big[\psi\big(t,x,\mathcal{L}^1(\xi,\zeta)\big)(\xi)\chi\big].$$

Proof. We observe from (5.90) that:

$$\mathbb{E}^1\big[\partial_\mu \boldsymbol{U}\big(t,x,\mathcal{L}^1(\xi,\chi)\big)(\xi,\chi)\zeta\big] = \mathbb{E}^1\big[\partial_\mu \boldsymbol{U}\big(t,x,\mathcal{L}^1(\xi,\zeta)\big)(\xi,\zeta)\chi\big],$$

implying that:

$$\begin{aligned}
&\mathbb{E}^1\big[\partial_v \partial_\mu \mathcal{U}\big(t,x,\mathcal{L}^1(\xi)\big)(\xi)\chi \otimes \zeta + \psi\big(t,x,\mathcal{L}^1(\xi,\chi)\big)(\xi)\zeta\big] \\
&\quad = \mathbb{E}^1\big[\partial_v \partial_\mu \mathcal{U}\big(t,x,\mathcal{L}^1(\xi)\big)(\xi)\zeta \otimes \chi + \psi\big(t,x,\mathcal{L}^1(\xi,\zeta)\big)(\xi)\chi\big],
\end{aligned}$$

where we used the identification:

$$\psi\big(t,x,\mathcal{L}(\xi,\chi)\big)(\xi) = \partial_\mu \boldsymbol{U}(t,x,\nu)(\xi,\chi) - \partial_v \partial_\mu \mathcal{U}(t,x,\mu)(\xi)\chi$$

observed at the end of the proof of Lemma 5.36. The result follows by using the symmetry of $\partial_v \partial_\mu \mathcal{U}$, see Remark (Vol I)-5.98. □

From the above, we deduce the following identification:

Lemma 5.38 *For $T \leq c$, with c only depending on L, for all $(t,x) \in [0,T] \times \mathbb{R}^d$, $\xi \in L^2(\Omega^1, \mathcal{F}_t^1, \mathbb{P}^1; \mathbb{R}^d)$, $\chi, \zeta \in L^\infty(\Omega^1, \mathcal{F}_t^1, \mathbb{P}^1; \mathbb{R}^d)$ with $\|\chi\|_\infty, \|\zeta\|_\infty \leq 1$,*

$$\partial^2_{(1-\varepsilon)\zeta,\varepsilon\chi} \theta_t^{t,x,\xi} = \frac{1}{4}\mathbb{E}^1\big[\psi\big(t,x,\mathcal{L}^1(\xi,\chi)\big)(\xi)\zeta\big],$$

for any random variable ε constructed on $(\Omega^1, \mathcal{F}_t^1, \mathbb{P}^1)$ with a Bernoulli distribution of parameter $1/2$ and independent of (ξ,χ,ζ). Moreover,

$$\big|\mathbb{E}^1\big[\psi\big(t,x,\mathcal{L}^1(\xi,\chi)\big)(\xi)\zeta\big]\big| \leq C\|\chi\|_1\|\zeta\|_1,$$

and, for all $x_1, x_2 \in \mathbb{R}^d$ and $\xi_1, \xi_2 \in L^2(\Omega^1, \mathcal{F}_t^1, \mathbb{P}^1; \mathbb{R}^d)$,

$$\begin{aligned}
&\big|\mathbb{E}^1\big[\psi\big(t,x_1,\mathcal{L}^1(\xi_1,\chi)\big)(\xi_1)\zeta - \psi\big(t,x_2,\mathcal{L}^1(\xi_2,\chi)\big)(\xi_2)\zeta\big]\big| \\
&\quad \leq C\Big(\big(|x_1 - x_2| + \|\xi_1 - \xi_2\|_1\big)\|\zeta\|_1\|\chi\|_1 \\
&\qquad\qquad + \|\zeta\|_1\mathbb{E}^1\big[|\xi_1 - \xi_2|\,|\chi|\big] + \|\chi\|_1\mathbb{E}^1\big[|\xi_1 - \xi_2|\,|\zeta|\big]\Big).
\end{aligned}$$

Importantly, observe from the first bound that $\|\psi(t,x,\mathcal{L}^1(\xi,\chi))(\xi)\|_\infty \leq C\|\chi\|_1$.

Proof. From Proposition 5.31 and (5.80), we know that:

$$\partial^2_{(1-\varepsilon)\zeta,\varepsilon\chi} Y_t^{t,x,\xi} = \frac{d}{ds}_{|s=0} \mathbb{E}^1\big[\partial_\mu \mathcal{U}\big(t,x,\mathcal{L}^1(\xi+s(1-\varepsilon)\zeta)\big)(\xi+s(1-\varepsilon)\zeta)\varepsilon\chi\big],$$
$$= \mathbb{E}^1\big[\partial_\mu \boldsymbol{U}\big(t,x,\mathcal{L}^1(\xi,\varepsilon\chi)\big)(\xi,\varepsilon\chi)(1-\varepsilon)\zeta\big].$$

By Lemmas 5.36 and 5.37, we obtain:

$$\partial^2_{(1-\varepsilon)\zeta,\varepsilon\chi} Y_t^{t,x,\xi} = \mathbb{E}^1\big[\partial_v\partial_\mu U\big(t,x,\mathcal{L}^1(\xi)\big)(\xi)(\varepsilon\chi)\otimes\big((1-\varepsilon)\zeta\big)\big]$$
$$+ \mathbb{E}^1\big[\psi\big(t,x,\mathcal{L}^1(\xi,(1-\varepsilon)\zeta)\big)(\xi)\varepsilon\chi\big],$$

where we used the identification:

$$\psi\big(t,x,\mathcal{L}^1(\xi,\chi)\big)(\xi) = \partial_\mu \boldsymbol{U}(t,x,\nu)(\xi,\chi) - \partial_v\partial_\mu \mathcal{U}(t,x,\mu)(\xi)\chi$$

established at the end of the proof of Lemma 5.36.

Now, the first term in the right-hand side is 0. Regarding the second term, independence of ε and (ξ,χ) yields:

$$\partial^2_{(1-\varepsilon)\zeta,\varepsilon\chi} Y_t^{t,x,\xi} = \frac{1}{2}\mathbb{E}^1\big[\psi\big(t,x,\mathcal{L}^1(\xi,(1-\varepsilon)\zeta)\big)(\xi)\chi\big]$$
$$= \frac{1}{2}\mathbb{E}^1\big[\psi\big(t,x,\mathcal{L}^1(\xi,\chi)\big)(\xi)\big((1-\varepsilon)\zeta\big)\big],$$

where we used once again Lemma 5.37. Using independence one more time, the first claim easily follows. The two last claims are direct consequences of Lemma 5.35. □

Differentiability with Respect to μ along Unbounded Directions

As a consequence of Lemma 5.37, we deduce that for all χ,ζ in $L^\infty(\Omega^1,\mathcal{F}_t^1,\mathbb{P}^1;\mathbb{R}^d)$ with $\|\chi\|_\infty,\|\zeta\|_\infty \le 1$,

$$\mathbb{E}^1\big[\psi\big(t,x,\mathcal{L}^1(\xi,-\chi)\big)(\xi)\zeta\big] = -\mathbb{E}^1\big[\psi\big(t,x,\mathcal{L}^1(\xi,\zeta)\big)(\xi)\chi\big]$$
$$= -\mathbb{E}^1\big[\psi\big(t,x,\mathcal{L}^1(\xi,\chi)\big)(\xi)\zeta\big],$$

from which we deduce that:

$$\psi\big(t,x,\mathcal{L}^1(\xi,-\chi)\big)(\xi) = -\psi\big(t,x,\mathcal{L}^1(\xi,\chi)\big)(\xi). \tag{5.92}$$

As another consequence of Lemma 5.37, we get that for all $\lambda\in[0,1]$ and all χ,χ',ζ in $L^\infty(\Omega^1,\mathcal{F}_t^1,\mathbb{P}^1;\mathbb{R}^d)$ with $\|\chi\|_\infty,\|\chi'\|_\infty,\|\zeta\|_\infty \le 1$,

$$\mathbb{E}^1\big[\psi\big(t,x,\mathcal{L}^1(\xi,\lambda\chi+(1-\lambda)\chi')\big)(\xi)\zeta\big]$$
$$= \mathbb{E}^1\big[\psi\big(t,x,\mathcal{L}^1(\xi,\zeta)\big)(\xi)\big(\lambda\chi+(1-\lambda)\chi'\big)\big]$$
$$= \lambda\mathbb{E}^1\big[\psi\big(t,x,\mathcal{L}^1(\xi,\zeta)\big)(\xi)\chi\big] + (1-\lambda)\mathbb{E}^1\big[\psi\big(t,x,\mathcal{L}^1(\xi,\zeta)\big)(\xi)\chi'\big]$$
$$= \lambda\mathbb{E}^1\big[\psi\big(t,x,\mathcal{L}^1(\xi,\chi)\big)(\xi)\zeta\big] + (1-\lambda)\mathbb{E}^1\big[\psi\big(t,x,\mathcal{L}^1(\xi,\chi')\big)(\xi)\zeta\big],$$

that is, for all $\lambda \in [0,1]$ and $\chi, \chi' \in L^\infty(\Omega^1, \mathcal{F}_t^1, \mathbb{P}^1; \mathbb{R}^d)$, with $\|\chi\|_\infty, \|\chi'\|_\infty \leq 1$,

$$\begin{aligned}
&\psi\big(t, x, \mathcal{L}^1(\xi, \lambda\chi + (1-\lambda)\chi')\big)(\xi) \\
&= \lambda\psi\big(t, x, \mathcal{L}^1(\xi, \chi)\big)(\xi) + (1-\lambda)\psi\big(t, x, \mathcal{L}^1(\xi, \chi')\big)(\xi).
\end{aligned} \tag{5.93}$$

We then extend the definition of ψ as follows. For $\chi \in L^\infty(\Omega^1, \mathcal{F}_t^1, \mathbb{P}^1; \mathbb{R}^d)$ and $v \in \mathbb{R}^d$, we let:

$$\bar{\psi}\big(t, x, \mathcal{L}^1(\xi, \chi)\big)(v) = \begin{cases} \|\chi\|_\infty \psi\Big(t, x, \mathcal{L}^1\Big(\xi, \dfrac{\chi}{\|\chi\|_\infty}\Big)\Big)(v) & \text{if } \chi \neq 0, \\ 0 & \text{if } \chi = 0. \end{cases}$$

Observing from the last claim in Lemma 5.37 that $\psi(t, x, \mathcal{L}^1(\xi, \chi))(\xi) = 0$ whenever $\chi = 0$, we deduce from (5.93) that $\bar{\psi}(t, x, \mathcal{L}^1(\xi, \chi))(\xi)$ and $\psi(t, x, \mathcal{L}^1(\xi, \chi))(\xi)$ coincide whenever $\|\chi\|_\infty \leq 1$. Moreover, by definition, $\bar{\psi}(t, x, \mathcal{L}^1(\xi, \cdot))(\xi)$ satisfies:

$$\begin{aligned}
&\forall \lambda \geq 0,\ \chi \in L^\infty(\Omega^1, \mathcal{F}_t^1, \mathbb{P}^1; \mathbb{R}^d), \\
&\qquad \bar{\psi}\big(t, x, \mathcal{L}^1(\xi, \lambda\chi)\big)(\xi) = \lambda\bar{\psi}\big(t, x, \mathcal{L}^1(\xi, \chi)\big)(\xi).
\end{aligned}$$

By (5.92), this is still true when $\lambda < 0$. Moreover, choosing $\lambda = \|\chi\|_\infty + \|\chi'\|_\infty$ for $\chi, \chi' \in L^\infty(\Omega^1, \mathcal{F}^1, \mathbb{P}^1; \mathbb{R}^d)$, we get:

$$\bar{\psi}\big(t, x, \mathcal{L}^1(\xi, \chi + \chi')\big)(\xi) = \big(\|\chi\|_\infty + \|\chi'\|_\infty\big)\bar{\psi}\Big(t, x, \mathcal{L}^1\big(\xi, \frac{\chi + \chi'}{\|\chi\|_\infty + \|\chi'\|_\infty}\big)\Big)(\xi).$$

Using (5.93) with $\lambda = \|\chi\|_\infty / (\|\chi\|_\infty + \|\chi'\|_\infty)$, we deduce:

$$\begin{aligned}
\bar{\psi}\big(t, x, \mathcal{L}^1(\xi, \chi + \chi')\big)(\xi) &= \|\chi\|_\infty \bar{\psi}\Big(t, x, \mathcal{L}^1\big(\xi, \frac{\chi}{\|\chi\|_\infty}\big)\Big)(\xi) \\
&\quad + \|\chi'\|_\infty \bar{\psi}\Big(t, x, \mathcal{L}^1\big(\xi, \frac{\chi'}{\|\chi\|_\infty}\big)\Big)(\xi) \\
&= \bar{\psi}\big(t, x, \mathcal{L}^1(\xi, \chi)\big)(\xi) + \bar{\psi}\big(t, x, \mathcal{L}^1(\xi, \chi')\big)(\xi),
\end{aligned}$$

while, for all $\chi, \zeta \in L^\infty(\Omega^1, \mathcal{F}_t^1, \mathbb{P}^1; \mathbb{R}^d)$,

$$\begin{aligned}
\mathbb{E}^1\big[\bar{\psi}\big(t, x, \mathcal{L}^1(\xi, \chi)\big)(\xi)\zeta\big] &= \|\chi\|_\infty \mathbb{E}^1\big[\psi\big(t, x, \mathcal{L}^1(\xi, \frac{\chi}{\|\chi\|_\infty})\big)(\xi)\zeta\big] \\
&= \|\chi\|_\infty \|\zeta\|_\infty \mathbb{E}^1\big[\psi\big(t, x, \mathcal{L}^1(\xi, \frac{\chi}{\|\chi\|_\infty})\big)(\xi)\frac{\zeta}{\|\zeta\|_\infty}\big] \\
&= \mathbb{E}^1\big[\bar{\psi}\big(t, x, \mathcal{L}^1(\xi, \zeta)\big)(\xi)\chi\big],
\end{aligned} \tag{5.94}$$

where we used Lemma 5.37 to pass from the first to the second line. In particular, choosing ζ such that $\|\zeta\|_1 \leq 1$, we easily deduce from the first inequality in Lemma 5.38 that:

$$\left\| \bar{\psi}\big(t, x, \mathcal{L}^1(\xi, \chi)\big)(\xi) \right\|_\infty \leq C \|\chi\|_1. \tag{5.95}$$

So for any given $\xi \in L^2(\Omega^1, \mathcal{F}_t^1, \mathbb{P}^1; \mathbb{R}^d)$, the linear mapping:

$$L^\infty(\Omega^1, \mathcal{F}_t^1, \mathbb{P}^1; \mathbb{R}^d) \ni \chi \mapsto \bar{\psi}\big(t, x, \mathcal{L}^1(\xi, \chi)\big)(\xi) \in L^\infty(\Omega^1, \mathcal{F}_t^1, \mathbb{P}^1; \mathbb{R}^{m \times d})$$

extends by continuity to the whole $L^1(\Omega^1, \mathcal{F}_t^1, \mathbb{P}^1; \mathbb{R}^d)$.

By (5.94), (5.95), and the second inequality in Lemma 5.38, we also have, for all $x_1, x_2 \in \mathbb{R}^d$, $\xi_1, \xi_2 \in L^2(\Omega^1, \mathcal{F}_t^1, \mathbb{P}^1; \mathbb{R}^d)$ and $\chi_1, \chi_2, \zeta \in L^1(\Omega^1, \mathcal{F}_t^1, \mathbb{P}^1; \mathbb{R}^d)$,

$$\begin{aligned}
&\left| \mathbb{E}^1\big[\bar{\psi}\big(t, x_1, \mathcal{L}^1(\xi_1, \chi_1)\big)(\xi_1)\zeta - \bar{\psi}\big(t, x_2, \mathcal{L}^1(\xi_2, \chi_2)\big)(\xi_2)\zeta\big] \right| \\
&\quad \leq C\Big(\big(|x_1 - x_2| + \|\xi_1 - \xi_2\|_1\big) \|\zeta\|_1 \|\chi_1\|_1 + \|\zeta\|_1 \|\chi_1 - \chi_2\|_1 \\
&\qquad + \|\zeta\|_1 \mathbb{E}^1\big[|\xi_1 - \xi_2|\,|\chi_1|\big] + \|\chi_1\|_1 \mathbb{E}^1\big[|\xi_1 - \xi_2|\,|\zeta|\big]\Big).
\end{aligned} \tag{5.96}$$

By Lemma 5.40 below, which reads as a variation of Lemma 5.30, we can find, for all $(t, x) \in [0, T] \times \mathbb{R}^d$, $\xi \in L^2(\Omega^1, \mathcal{F}_t^1, \mathbb{P}^1; \mathbb{R}^d)$ and $\chi \in L^1(\Omega^1, \mathcal{F}_t^1, \mathbb{P}^1; \mathbb{R}^d)$, a version of each of $\bar{\psi}(t, x, \mathcal{L}^1(\xi, \chi))(\cdot)$ such that, for all $\xi_1, \xi_2 \in L^2(\Omega^1, \mathcal{F}_t^1, \mathbb{P}^1; \mathbb{R}^d)$, $\chi_1, \chi_2 \in L^2(\Omega^1, \mathcal{F}_t^1, \mathbb{P}^1; \mathbb{R}^d)$ and $x_1, x_2, v_1, v_2 \in \mathbb{R}^d$,

$$\begin{aligned}
&\left| \bar{\psi}\big(t, x_1, \mathcal{L}^1(\xi_1, \chi_1)\big)(v_1) - \bar{\psi}\big(t, x_2, \mathcal{L}^1(\xi_2, \chi_2)\big)(v_2) \right| \\
&\quad \leq C\Big(\|\chi_1\|_1 \big(|x_1 - x_2| + |v_1 - v_2| + \|\xi_1 - \xi_2\|_1\big) + \|\chi_1 - \chi_2\|_1 \\
&\qquad + \mathbb{E}^1\big[|\xi_1 - \xi_2|\,|\chi_1|\big]\Big).
\end{aligned} \tag{5.97}$$

The function $\bar{\psi}$ is the right object to identify the second-order derivative of $\mathcal{U}$ in the direction μ.

Proposition 5.39 *Under assumption* **Smooth Coefficients Order 2**, *there exist a constant c only depending on L, and a constant C only depending on L and Γ, such that for $T \leq c$ and any $(t, x, v) \in [0, T] \times \mathbb{R}^d \times \mathbb{R}^d$, the function $\mathcal{P}_2(\mathbb{R}^d) \ni \mu \mapsto \partial_\mu \mathcal{U}(t, x, \mu)(v) \in \mathbb{R}^d$ provided by Theorem 5.29 is L-differentiable. Moreover, for any $(t, x, v) \in [0, T] \times \mathbb{R}^d \times \mathbb{R}^d$ and $\mu \in \mathcal{P}_2(\mathbb{R}^d)$, we can find a version of $\mathbb{R}^d \ni v' \mapsto \partial_\mu^2 \mathcal{U}(t, x, \mu)(v, v') = \partial_\mu[\partial_\mu \mathcal{U}(t, x, \mu)(v)](v')$ such that the global function $[0, T] \times \mathbb{R}^d \times \mathcal{P}_2(\mathbb{R}^d) \times \mathbb{R}^d \times \mathbb{R}^d \ni (t, x, \mu, v, v') \mapsto \partial_\mu^2 \mathcal{U}(t, x, \mu)(v, v') \in \mathbb{R}^{d \times d}$ is jointly continuous in all the arguments and satisfy, for all $t \in [0, T]$, $x_1, x_2 \in \mathbb{R}^d$, $\mu_1, \mu_2 \in \mathcal{P}_2(\mathbb{R}^d)$ and $v_1, v_2, v_1', v_2' \in \mathbb{R}^d$,*

$$
\begin{aligned}
&\big|\partial_\mu^2 \mathcal{U}(t,x_1,\mu_1)(v_1,v_1') - \partial_\mu^2 \mathcal{U}(t,x_2,\mu_2)(v_2,v_2')\big| \\
&\qquad \le C\big(|x_1-x_2| + W_1(\mu_1,\mu_2) + |v_1-v_2| + |v_1'-v_2'|\big).
\end{aligned}
$$

Proof.

First Step. Recalling Lemma 5.36 and using in addition the linearity of $\bar{\psi}$ with respect to χ, we get for all $(t,x) \in [0,T] \times \mathbb{R}^d$, $\xi,\zeta \in L^2(\Omega^1,\mathcal{F}_t^1,\mathbb{P}^1;\mathbb{R}^d)$ and $\chi \in L^\infty(\Omega^1,\mathcal{F}_t^1,\mathbb{P}^1;\mathbb{R}^d)$ with $\|\chi\|_\infty$ possibly larger than 1,

$$
\begin{aligned}
&\lim_{s\to 0} \frac{1}{s}\mathbb{E}^1\Big[\Big(\partial_\mu \mathcal{U}\big(t,x,\mathcal{L}^1(\xi+s\chi)\big)(\xi+s\chi) - \partial_\mu \mathcal{U}\big(t,x,\mathcal{L}^1(\xi)\big)(\xi)\Big)\zeta\Big] \\
&\qquad = \mathbb{E}^1\big[\partial_v\partial_\mu \mathcal{U}(t,x,\mathcal{L}^1(\xi))(\xi)\zeta \otimes \chi + \bar{\psi}(t,x,\mathcal{L}^1(\xi,\chi))(\xi)\zeta\big],
\end{aligned}
$$

where again, we used the identity:

$$
\psi\big(t,x,\mathcal{L}(\xi,\chi)\big)(\xi) = \partial_\mu U(t,x,\nu)(\xi,\chi) - \partial_v\partial_\mu \mathcal{U}(t,x,\mu)(\xi)\chi,
$$

established at the end of the proof of Lemma 5.36. Then, for all $s \neq 0$,

$$
\begin{aligned}
&\mathbb{E}^1\Big[\frac{1}{s}\Big(\partial_\mu \mathcal{U}\big(t,x,\mathcal{L}^1(\xi+s\chi)\big)(\xi+s\chi) - \partial_\mu \mathcal{U}\big(t,x,\mathcal{L}^1(\xi)\big)(\xi)\Big)\zeta\Big] \\
&\qquad = \mathbb{E}^1\Big[\frac{1}{s}\Big(\int_0^s \Big[\bar{\psi}\big(t,x,\mathcal{L}^1(\xi+r\chi,\chi)\big)(\xi+r\chi) \\
&\qquad\qquad + \partial_v\partial_\mu \mathcal{U}\big(t,x,\mathcal{L}^1(\xi+r\chi)\big)(\xi+r\chi)\chi\Big]dr\Big)\zeta\Big],
\end{aligned}
$$

so that:

$$
\begin{aligned}
&\mathbb{E}^1\Big[\Big(\frac{1}{s}\Big(\partial_\mu \mathcal{U}\big(t,x,\mathcal{L}^1(\xi+s\chi)\big)(\xi) - \partial_\mu \mathcal{U}\big(t,x,\mathcal{L}^1(\xi)\big)(\xi)\Big) - \bar{\psi}\big(t,x,\mathcal{L}^1(\xi,\chi)\big)(\xi)\Big)\zeta\Big] \\
&\quad = \mathbb{E}^1\Big[\Big(\frac{1}{s}\int_0^s \Big[\Big(\bar{\psi}\big(t,x,\mathcal{L}^1(\xi+r\chi,\chi)\big)(\xi+r\chi) - \bar{\psi}\big(t,x,\mathcal{L}^1(\xi,\chi)\big)(\xi)\Big) \\
&\qquad + \Big(\partial_v\partial_\mu \mathcal{U}\big(t,x,\mathcal{L}^1(\xi+r\chi)\big)(\xi+r\chi) \\
&\qquad\qquad - \partial_v\partial_\mu \mathcal{U}\big(t,x,\mathcal{L}^1(\xi+s\chi)\big)(\xi+r\chi)\Big)\chi\Big]dr\Big)\zeta\Big].
\end{aligned}
$$

Thanks to Proposition 5.33 and (5.96), we deduce that:

$$
\begin{aligned}
&\Big|\mathbb{E}^1\Big[\Big(\frac{1}{s}\Big(\partial_\mu \mathcal{U}\big(t,x,\mathcal{L}^1(\xi+s\chi)\big)(\xi) - \partial_\mu \mathcal{U}\big(t,x,\mathcal{L}^1(\xi)\big)(\xi)\Big) - \bar{\psi}\big(t,x,\mathcal{L}^1(\xi,\chi)\big)(\xi)\Big)\zeta\Big]\Big| \\
&\qquad \le C\|\chi\|_\infty^2 \|\zeta\|_1,
\end{aligned}
$$

that is, for all $s \neq 0$, with $\mathbb{P}$-probability 1,

$$
\begin{aligned}
&\Big|\frac{1}{s}\Big(\partial_\mu \mathcal{U}\big(t,x,\mathcal{L}^1(\xi+s\chi)\big)(\xi) - \partial_\mu \mathcal{U}\big(t,x,\mathcal{L}^1(\xi)\big)(\xi)\Big) - \bar{\psi}\big(t,x,\mathcal{L}^1(\xi,\chi)\big)(\xi)\Big| \\
&\qquad \le Cs\|\chi\|_\infty^2.
\end{aligned}
$$

Therefore, when $\mathcal{L}^1(\xi)$ has full support, it holds for all $v \in \mathbb{R}^d$:

$$\lim_{s\to 0}\frac{1}{s}\Big(\partial_\mu\mathcal{U}\big(t,x,\mathcal{L}^1(\xi+s\chi)\big)(v)-\partial_\mu\mathcal{U}\big(t,x,\mathcal{L}^1(\xi)\big)(v)\Big)=\bar{\psi}\big(t,x,\mathcal{L}^1(\xi,\chi)\big)(v),$$

where we used the fact that the above functionals are continuous with respect to v. In particular, for a Gaussian random vector Z with an invertible covariance, Z being independent of (ξ,χ), we have, for all $v \in \mathbb{R}^d$,

$$\begin{aligned}&\lim_{s\to 0}\frac{1}{s}\Big(\partial_\mu\mathcal{U}\big(t,x,\mathcal{L}^1(\xi+Z+s\chi)\big)(v)-\partial_\mu\mathcal{U}\big(t,x,\mathcal{L}^1(\xi+Z)\big)(v)\Big)\\&=\bar{\psi}\big(t,x,\mathcal{L}^1(\xi+Z,\chi)\big)(v),\end{aligned}$$

or, equivalently,

$$\frac{d}{ds}\Big[\partial_\mu\mathcal{U}\big(t,x,\mathcal{L}^1(\xi+Z+s\chi)\big)(v)\Big]=\bar{\psi}\big(t,x,\mathcal{L}^1(\xi+Z+s\chi,\chi)\big)(v),$$

from which we deduce that:

$$\begin{aligned}&\partial_\mu\mathcal{U}\big(t,x,\mathcal{L}^1(\xi+Z+s\chi)\big)(v)-\partial_\mu\mathcal{U}\big(t,x,\mathcal{L}^1(\xi+Z)\big)(v)\\&=\int_0^s\bar{\psi}\big(t,x,\mathcal{L}^1(\xi+Z+r\chi,\chi)\big)(v)dr.\end{aligned}\tag{5.98}$$

Second Step. Recall from (5.97) that for $|r|\leq 1$ and for any random variables $\chi_1,\chi_2 \in L^2(\Omega^1,\mathcal{F}_t^1,\mathbb{P}^1;\mathbb{R}^d)$,

$$\begin{aligned}&\big|\bar{\psi}\big(t,x,\mathcal{L}^1(\xi+Z+r\chi_1,\chi_1)\big)(v)-\bar{\psi}\big(t,x,\mathcal{L}^1(\xi+Z+r\chi_2,\chi_2)\big)(v)\big|\\&\leq C\Big(\|\chi_1-\chi_2\|_1\|\chi_1\|_1+\mathbb{E}^1\big[|\chi_1-\chi_2|\,|\chi_1|\big]+\|\chi_1-\chi_2\|_1\Big),\end{aligned}$$

which shows, by a density argument, that (5.98) extends to any $\chi \in L^2(\Omega^1,\mathcal{F}_t^1,\mathbb{P}^1;\mathbb{R}^d)$ in lieu of $\chi \in L^\infty(\Omega^1,\mathcal{F}_t^1,\mathbb{P}^1;\mathbb{R}^d)$.

Observe that the identity (5.98) is generic in the following sense. When $\nu \in \mathcal{P}_2(\mathbb{R}^{2d})$ is given first, there is no difficulty for constructing a 3-tuple (ξ,χ,Z), with $(\xi,\chi)\sim\nu$ and with (ξ,χ) independent of Z. As a result, for all $\xi,\chi \in L^2(\Omega^1,\mathcal{F}_t^1,\mathbb{P}^1;\mathbb{R}^d)$, $v \in \mathbb{R}^d$ and $n \geq 1$,

$$\begin{aligned}&\frac{d}{ds}_{|s=0}\Big[\partial_\mu\mathcal{U}\big(t,x,\mathcal{L}^1(\xi+s\chi)*N_d(0,\tfrac{1}{n}I_d)\big)(v)\Big]\\&=\bar{\psi}\big(t,x,\mathcal{L}^1(\xi,\chi)*(N_d(0,\tfrac{1}{n}I_d),\delta_0)\big)(v),\end{aligned}$$

where $(N_d(0,\frac{1}{n}I_d),\delta_0)$ is a short notation for the law of $(Z,0)$ when $Z\sim N_d(0,\frac{1}{n}I_d)$. By linearity of $\bar{\psi}$, the right-hand side is easily shown to be linear with respect to χ. Moreover, by (5.95), it is continuous with respect to χ, seen as an element of $L^2(\Omega^1,\mathcal{F}_t^1,\mathbb{P}^1;\mathbb{R}^d)$, from which we deduce that the map $L^2(\Omega^1,\mathcal{F}_t^1,\mathbb{P}^1;\mathbb{R}^d)\ni\xi\mapsto\partial_\mu\mathcal{U}(t,x,\mathcal{L}^1(\xi)*N_d(0,n^{-1}I_d))(v)$ is Gâteaux differentiable.

To prove Fréchet differentiability, we use (5.97). For $x_1,x_2 \in \mathbb{R}^d$ and $\xi_1,\xi_2 \in L^2(\Omega^1,\mathcal{F}_t^1,\mathbb{P}^1;\mathbb{R}^d)$ and for $\chi \in L^2(\Omega^1,\mathcal{F}_t^1,\mathbb{P}^1;\mathbb{R}^d)$, we deduce from (5.97) that:

$$
\begin{aligned}
&\big|\bar{\psi}\big(t,x_1,\mathcal{L}^1(\xi_1,\chi)*(N_d(0,\tfrac{1}{n}I_d),\delta_0)\big)(v)-\bar{\psi}\big(t,x_2,\mathcal{L}^1(\xi_2,\chi)*(N_d(0,\tfrac{1}{n}I_d),\delta_0)\big)(v)\big|\\
&\quad\le C\Big(\big(|x_1-x_2|+\|\xi_1-\xi_2\|_1\big)\|\chi\|_1+\mathbb{E}^1\big[|\xi_1-\xi_2|\,|\chi|\big]\Big)\\
&\quad\le C\Big(|x_1-x_2|+\|\xi_1-\xi_2\|_2\Big)\|\chi\|_2,
\end{aligned}
$$

which suffices to prove that the Gâteaux derivative, regarded as an element of the space $L^2(\Omega^1,\mathcal{F}_t^1,\mathbb{P}^1;\mathbb{R}^d)$, is continuous with respect to the state variable ξ. Hence, for any $(t,x,v)\in[0,T]\times\mathbb{R}^d\times\mathbb{R}^d$, the map $L^2(\Omega^1,\mathcal{F}_t^1,\mathbb{P}^1;\mathbb{R}^d)\ni\xi\mapsto\partial_\mu\mathcal{U}(t,x,\mathcal{L}^1(\xi)*N_d(0,\frac{1}{n}I_d))(v)$ is Fréchet differentiable.

As a consequence, we may identify $\bar{\psi}(t,x,\mathcal{L}^1(\xi,\chi)*(N_d(0,\frac{1}{n}I_d),\delta_0))(v)$ with the L-derivative of $\partial_\mu\mathcal{U}(t,x,\mathcal{L}^1(\cdot)*N_d(0,\frac{1}{n}I_d))(v)$ computed at the point ξ in the direction χ. In particular, there exists a map $[0,T]\times\mathbb{R}^d\times\mathcal{P}_2(\mathbb{R}^d)\times\mathbb{R}^d\times\mathbb{R}^d\ni(t,x,\mu,v,v')\mapsto\mathcal{V}^n(t,x,\mu)(v,v')\in\mathbb{R}^{m\times d}$ such that:

$$
\bar{\psi}\big(t,x,\mathcal{L}^1(\xi,\chi)*(N_d(0,\tfrac{1}{n}I_d),\delta_0)\big)(v)=\mathbb{E}^1\big[\mathcal{V}^n\big(t,x,\mathcal{L}^1(\xi)\big)(v,\xi)\chi\big].
$$

By (5.95) and (5.97), and from the fact that $\mathcal{L}^1(\xi)*N_d(0,\frac{1}{n}I_d)$ has full support, we obtain:

$$
\big\|\mathcal{V}^n\big(t,x,\mathcal{L}^1(\xi)\big)(v,\xi)\big\|_\infty\le C,
$$

and, for all $x_1,x_2,v_1,v_2\in\mathbb{R}^d$ and $\xi_1,\xi_2\in L^2(\Omega^1,\mathcal{F}_t^1,\mathbb{P}^1;\mathbb{R}^d)$,

$$
\begin{aligned}
&\big|\mathbb{E}^1\big[\big(\mathcal{V}^n\big(t,x_1,\mathcal{L}^1(\xi_1)\big)(v_1,\xi_1)-\mathcal{V}^n\big(t,x_2,\mathcal{L}^1(\xi_2)\big)(v_2,\xi_2)\big)\chi\big]\big|\\
&\quad\le C\Big(\big(|x_1-x_2|+|v_1-v_2|+\|\xi_1-\xi_2\|_1\big)\|\chi\|_1+\mathbb{E}^1\big[|\xi_1-\xi_2|\,|\chi|\big]\Big).
\end{aligned}
$$

By the auxiliary Lemma 5.30, we can find, for each $(t,x,\mu,v)\in[0,T]\times\mathbb{R}^d\times\mathcal{P}_2(\mathbb{R}^d)\times\mathbb{R}^d$, a version of $\mathbb{R}^d\ni v'\mapsto\mathcal{V}^n(t,x,\mu)(v,v')\in L^\infty(\mathbb{R}^d,\mu;\mathbb{R}^{m\times(d\times d)})$ which satisfies, for all $v'\in\mathbb{R}^d$,

$$
\big|\mathcal{V}^n\big(t,x,\mathcal{L}^1(\xi)\big)(v,v')\big|\le C,
$$

and, for all $x_1,x_2,v_1,v_2,v_1',v_2'\in\mathbb{R}^d$ and $\xi_1,\xi_2\in L^2(\Omega^1,\mathcal{F}_t^1,\mathbb{P}^1;\mathbb{R}^d)$,

$$
\begin{aligned}
&\big|\mathcal{V}^n\big(t,x_1,\mathcal{L}^1(\xi_1)\big)(v_1,v_1')-\mathcal{V}^n\big(t,x_2,\mathcal{L}^1(\xi_2)\big)(v_2,v_2')\big|\\
&\quad\le C\big(|x_1-x_2|+|v_1-v_2|+|v_1'-v_2'|+\|\xi_1-\xi_2\|_1\big).
\end{aligned}
$$

In particular, the map $\mathcal{V}^n(t,\cdot,\cdot)(\cdot,\cdot)$ extends to $\mathbb{R}^d\times\mathcal{P}_1(\mathbb{R}^d)\times\mathbb{R}^d\times\mathbb{R}^d$. Since any bounded subset of $\mathcal{P}_2(\mathbb{R}^d)$ is a compact subset of $\mathcal{P}_1(\mathbb{R}^d)$, we deduce from the Arzelà-Ascoli theorem that we can extract a subsequence of the family $(\mathcal{V}^n(t,\cdot,\cdot)(\cdot,\cdot))_{n\ge1}$ which converges uniformly on any bounded subset of $\mathbb{R}^d\times\mathcal{P}_2(\mathbb{R}^d)\times\mathbb{R}^d\times\mathbb{R}^d$. We call $\mathcal{V}(t,\cdot,\cdot)(\cdot,\cdot)$ the limit. It satisfies the above uniform Lipschitz property. Moreover, passing to the limit in (5.98), we get:

$$
\partial_\mu\mathcal{U}\big(t,x,\mathcal{L}^1(\xi+s\chi)\big)(v)-\partial_\mu\mathcal{U}\big(t,x,\mathcal{L}^1(\xi)\big)(v)=\int_0^s\mathbb{E}^1\big[\mathcal{V}\big(t,x,\mathcal{L}^1(\xi+r\chi)\big)(v,\xi)\chi\big]dr,
$$

from which we easily deduce that for any $(t,x,v)\in[0,T]\times\mathbb{R}^d\times\mathbb{R}^d$, the mapping $\mathcal{P}_2(\mathbb{R}^d)\ni\mu\mapsto\partial_\mu\mathcal{U}(t,x,\mu)(v)$ is Fréchet differentiable. Obviously $\mathbb{R}^d\ni v'\mapsto\partial_\mu^2\mathcal{U}(t,x,\mu)(v,v')$ identifies with $\mathbb{R}^d\ni v'\mapsto\mathcal{V}(t,x,\mu)(v,v')$, which completes the proof.

Third Step. In order to complete the proof, it remains to prove the regularity of $\partial_\mu^2 \mathcal{U}$ in the time variable. We proceed as in the third step of the proof of Proposition 5.33. Indeed, we recall from Lemma 5.38 and from the above identification of $\bar{\psi}$ that, for a Bernoulli random variable ε with parameter $1/2$ independent of (ξ, χ, ζ), where $\xi \in L^2(\Omega^1, \mathcal{F}_0^1, \mathbb{P}^1; \mathbb{R}^d)$ and $\chi, \zeta \in L^\infty(\Omega^1, \mathcal{F}_0^1, \mathbb{P}^1; \mathbb{R}^d)$ with $\|\chi\|_\infty, \|\zeta\|_\infty \le 1$,

$$\partial^2_{(1-\varepsilon)\zeta,\varepsilon\chi} Y_t^{t,x,\xi} = \frac{1}{4}\mathbb{E}^1\tilde{\mathbb{E}}^1\big[\partial_\mu^2 \mathcal{U}\big(t, x, \mathcal{L}^1(\xi)\big)(\xi, \tilde{\xi})\chi \otimes \tilde{\zeta}\big]. \tag{5.99}$$

As explained in the fourth step of the proof of Proposition 5.33, the left-hand side is a continuous function of time. Therefore, so is the right-hand side. We conclude as in the second step of the proof of Theorem 5.29 or, equivalently, as in the proof of Proposition 5.33. □

Auxiliary Regularity Lemma

The following auxiliary lemma is a variant of Lemma 5.30.

Lemma 5.40 *Consider a collection $(u(x,\nu)(\cdot))_{x\in\mathbb{R}^n,\nu\in\mathcal{P}_2(\mathbb{R}^{2d})}$ of real-valued functions satisfying, for all $x \in \mathbb{R}^n$ and $\nu \in \mathcal{P}_2(\mathbb{R}^{2d})$, $u(x,\nu)(\cdot) \in L^\infty(\mathbb{R}^d, \mu; \mathbb{R})$, where μ denotes the first marginal of ν on $\mathbb{R}^d$, and for which there exists a constant C such that for all $x, x' \in \mathbb{R}^n$, $\xi, \xi', \chi, \chi' \in L^2(\Omega, \mathcal{F}, \mathbb{P}; \mathbb{R}^d)$ and $\zeta \in L^2(\Omega, \mathcal{F}, \mathbb{P}; \mathbb{R})$,*

$$\begin{aligned}
&\big|\mathbb{E}\big[\big(u(x, \mathcal{L}(\xi,\chi))(\xi) - u(x', \mathcal{L}(\xi',\chi'))(\xi')\big)\zeta\big]\big| \\
&\quad \le C\Big[\|\zeta\|_1\Big(\|\chi\|_1\big(|x-x'| + \|\xi-\xi'\|_1\big) + \mathbb{E}[|\xi-\xi'|\,|\chi|] + \|\chi-\chi'\|_1\Big) \\
&\qquad + \|\chi\|_1\mathbb{E}\big[|\xi-\xi'|\,|\zeta|\big]\Big],
\end{aligned}$$

for an atomless probability space $(\Omega, \mathcal{F}, \mathbb{P})$.

Then, for each $(x,\nu) \in \mathbb{R}^n \times \mathcal{P}_2(\mathbb{R}^{2d})$, we can find a version of $u(x,\nu)(\cdot) \in L^\infty(\mathbb{R}^d, \mu; \mathbb{R})$ such that, for the same constant C as above, for all $x, x' \in \mathbb{R}^n$, $\xi, \xi', \chi, \chi' \in L^2(\Omega, \mathcal{F}, \mathbb{P}; \mathbb{R}^d)$ and $v, v' \in \mathbb{R}^d$,

$$\begin{aligned}
&\big|u\big(x, \mathcal{L}(\xi,\chi)\big)(v) - u\big(x', \mathcal{L}(\xi',\chi)\big)(v')\big| \\
&\quad \le C\Big[\|\chi\|_1\big(|x-x'| + \|\xi-\xi'\|_1 + |v-v'|\big) + \|\chi-\chi'\|_1 + \mathbb{E}[|\xi-\xi'|\,|\chi|]\Big].
\end{aligned}$$

Proof. As a preliminary remark, we observe that the map $\mathbb{R}^d \times L^2(\Omega, \mathcal{F}, \mathbb{P}; \mathbb{R}^d) \times L^2(\Omega, \mathcal{F}, \mathbb{P}; \mathbb{R}^d) \ni (x, \xi, \chi) \mapsto u(x, \mathcal{L}(\xi,\chi))(\xi) \in L^2(\Omega, \mathcal{F}, \mathbb{P}; \mathbb{R})$ is continuous.

First Step. If $x, x' \in \mathbb{R}^n$ and $\xi, \xi', \chi, \chi' \in L^2(\Omega, \mathcal{F}, \mathbb{P}; \mathbb{R}^d)$, observe from the regularity assumption that with $\mathbb{P}$-probability 1,

$$\begin{aligned}
&\big|u\big(x, \mathcal{L}(\xi,\chi)\big)(\xi) - u\big(x', \mathcal{L}(\xi',\chi')\big)(\xi')\big| \\
&\quad \le C\Big[\|\chi\|_1\big(|x-x'| + \|\xi-\xi'\|_1 + |\xi-\xi'|\big) + \|\chi-\chi'\|_1 + \mathbb{E}\big[|\xi-\xi'|\,|\chi|\big]\Big].
\end{aligned}$$

In particular, if $p \ge 1$ is an integer and Z a Gaussian random variable $Z \sim N_d(0, I_d)$ independent of (ξ, ξ', χ, χ'), it holds that:

$$\left|u\big(x,\mathcal{L}(\xi+\tfrac{1}{p}Z,\chi)\big)(\xi+\tfrac{1}{p}Z)-u\big(x',\mathcal{L}(\xi'+\tfrac{1}{p}Z,\chi')\big)(\xi'+\tfrac{1}{p}Z)\right|$$
$$\leq C\Big[\|\chi\|_1\big(|x-x'|+\|\xi-\xi'\|_1+|\xi-\xi'|\big)+\|\chi-\chi'\|_1+\mathbb{E}\big[|\xi-\xi'||\chi|\big]\Big].$$

Integrating with respect to Z only, we get, with $\mathbb{P}$-probability 1,

$$\bigg|\int_{\mathbb{R}^d}u\big(x,\mathcal{L}(\xi+\tfrac{1}{p}Z,\chi)\big)(\xi+\tfrac{1}{p}z)\varphi_d(z)dz$$
$$-\int_{\mathbb{R}^d}u\big(x',\mathcal{L}(\xi'+\tfrac{1}{p}Z,\chi)\big)(\xi'+\tfrac{1}{p}z)\varphi_d(z)dz\bigg|$$
$$\leq C\Big[\|\chi\|_1\big(|x-x'|+|\xi-\xi'|+\|\xi-\xi'\|_1\big)+\|\chi-\chi'\|_1+\mathbb{E}\big[|\xi-\xi'||\chi|\big]\Big].$$

Observe that the integrals in the above left-hand side are well defined since the two functions $u(x,\mathcal{L}(\xi+\frac{1}{p}Z,\chi))(\cdot)$ and $u(x,\mathcal{L}(\xi'+\frac{1}{p}Z,\chi))(\cdot)$ belong to $L^\infty(\mathbb{R}^d,\mathrm{Leb}_d;\mathbb{R})$, which follows from the fact that $\mathcal{L}(\xi+\frac{1}{p}Z)$ and $\mathcal{L}(\xi+\frac{1}{p}Z)$ have positive densities. Setting:

$$u_p(x,\nu)(v)=\int_{\mathbb{R}^d}u\big(x,\nu*(N_d(0,\tfrac{1}{p^2}I_d),\delta_0)\big)(v+\tfrac{1}{n}z)\varphi_d(z)dz,$$

for $x\in\mathbb{R}^n$, $\nu\in\mathcal{P}_2(\mathbb{R}^{2d})$ and $v\in\mathbb{R}^d$, where $(N_d(0,\frac{1}{p^2}I_d),\delta_0)$ is a short notation for $\mathcal{L}(\frac{1}{p}Z,0)$, we conclude that $u_p(x,\nu)(\cdot)$ is continuous and satisfies with probability 1:

$$\left|u_p\big(x,\mathcal{L}(\xi,\chi)\big)(\xi)-u_p\big(x',\mathcal{L}(\xi',\chi')\big)(\xi')\right|$$
$$\leq C\Big[\|\chi\|_1\big(|x-x'|+|\xi-\xi'|+\|\xi-\xi'\|_1\big)+\|\chi-\chi'\|_1+\mathbb{E}\big[|\xi-\xi'||\chi|\big]\Big]. \quad (5.100)$$

with $\mathbb{P}$-probability 1.

Second Step. We now denote by ν the law of (ξ,χ), and by ν' the law of (ξ',χ'). Then, we call $\Pi(\nu,\nu')$ the set of probability measures π on $\mathcal{P}_2(\mathbb{R}^{4d})$ such that the image of π by the mapping $\mathbb{R}^{4d}\ni(v,v',w,w')\mapsto(v,w)$ is ν and the image of π by the mapping $\mathbb{R}^{4d}\ni(v,v',w,w')\mapsto(v',w')$ is ν'. We learned from Chapter (Vol I)-5 that $\Pi(\nu,\nu')$ is compact for the Wasserstein topology on $\mathcal{P}_2(\mathbb{R}^{4d})$. Since the function:

$$\mathcal{P}_2(\mathbb{R}^{4d})\ni\pi\mapsto\int_{\mathbb{R}^{4d}}|w|d\pi(v,v',w,w')\int_{\mathbb{R}^{4d}}|v-v'|d\pi(v,v',w,w')$$
$$+\int_{\mathbb{R}^{4d}}|w-w'|d\pi(v,v',w,w')+\int_{\mathbb{R}^{4d}}|v-v'||w|d\pi(v,v',w,w')$$

is continuous, we deduce that it has a minimum on $\Pi(\nu,\nu')$. Below, we call $\pi^\star$ a minimizer and (ξ,ξ',χ,χ') a 4-tuple with distribution $\pi^\star$.

Third Step. We now consider $\nu,\nu'\in\mathcal{P}_2(\mathbb{R}^d)$ with strictly positive smooth densities that decay at most exponentially fast at infinity and whose derivatives also decay exponentially at infinity.

Following the proof of Proposition (Vol I)-5.36, we can find four continuous mappings $\phi:(0,1)^d\to\mathbb{R}^d$, $\phi':(0,1)^d\to\mathbb{R}^d$, $\psi:(0,1)^d\times(0,1)^d\to\mathbb{R}^d$ and $\psi':(0,1)^d\times(0,1)^d\to\mathbb{R}^d$ such that, for any pair (ϖ,η) of independent and identically distributed random variables with uniform distribution on $(0,1)^d$, it holds that $(\psi(\eta,\varpi),\phi(\varpi))\sim\nu$

and $(\psi'(\eta',\varpi),\phi'(\varpi)) \sim \nu'$. Importantly, for any $w \in (0,1)^d$, the mappings $\psi(\cdot,w) : (0,1)^d \ni y \mapsto \psi(y,w) \in \mathbb{R}^d$ and $\psi'(\cdot,w) : (0,1)^d \ni y \mapsto \psi'(y,w) \in \mathbb{R}^d$ are one-to-one from $(0,1)^d$ onto $\mathbb{R}^d$. Moreover, the distribution $\mathrm{Leb}_1 \circ \psi(\cdot,w)^{-1}$ is the conditional law of $\xi\psi$ given $\chi = \phi(w)$. It has a positive density. Hence, for given values of v, v' in $\mathbb{R}^d$ and $w \in (0,1)^d$, we can find $y_0, y_0' \in (0,1)^d$ such that:

$$v = \psi(y_0, w), \quad v' = \psi(y_0', w).$$

Fix now ϖ and η two independent and identically distributed random variables with uniform distribution on $(0,1)^d$. For $\delta > 0$ such that $B(y_0,\delta) \subset (0,1)^d$ and $B(y_0',\delta) \subset (0,1)^d$, we let:

$$\eta' = \begin{cases} \eta & \text{if } \eta \notin B(y_0,\delta) \cup B(y_0',\delta), \\ \eta + y_0' - y_0 & \text{if } \eta \in B(y_0,\delta), \\ \eta + y_0 - y_0' & \text{if } \eta \in B(y_0',\delta), \end{cases}$$

where $B(y_0,\delta)$ denotes the d-dimensional open ball of center y_0 and radius δ. Clearly, η' is also uniformly distributed.

We let $(\bar{\xi},\bar{\chi}) = (\psi(\eta,\varpi),\phi(\varpi))$ and $(\bar{\xi}',\bar{\chi}') = (\psi'(\eta',\varpi),\phi'(\varpi))$. Obviously, $(\bar{\xi},\bar{\chi}) \sim \nu$ and $(\bar{\xi}',\bar{\chi}') \sim \nu'$.

Fourth Step. We now consider a Bernoulli random variable ε independent of (ξ,ξ',χ,χ') and (η,ϖ), and we let:

$$\begin{aligned}(\xi^\varepsilon,\chi^\varepsilon) &= \varepsilon(\xi,\chi) + (1-\varepsilon)(\bar{\xi},\bar{\chi}), \\ (\xi^{\varepsilon,\prime},\chi^{\varepsilon,\prime}) &= \varepsilon(\xi',\chi') + (1-\varepsilon)(\bar{\xi}',\bar{\chi}').\end{aligned}$$

Clearly, $(\xi^\varepsilon,\chi^\varepsilon)$ and $(\xi^{\varepsilon,\prime},\chi^{\varepsilon,\prime})$ have ν and ν' as distributions. Taking advantage of the conclusion of the first step, we deduce that, for all $x, x' \in \mathbb{R}^d$, with $\mathbb{P}$-probability 1,

$$\begin{aligned}&\big|u_p(x,\nu)(\xi^\varepsilon) - u_p(x',\nu')(\xi^{\varepsilon,\prime})\big| \\ &\leq C\Big(\|\chi\|_1\big(|x-x'|+\|\xi^\varepsilon-\xi^{\varepsilon,\prime}\|_1+|\xi^\varepsilon-\xi^{\varepsilon,\prime}|\big)+\|\chi^\varepsilon-\chi^{\varepsilon,\prime}\|_1+\mathbb{E}\big[|\xi^\varepsilon-\xi^{\varepsilon,\prime}|\,|\chi^\varepsilon|\big]\Big).\end{aligned}$$

Therefore, almost surely on the event $\{\varepsilon = 0\} \cap \{\eta \in B(y_0,\delta)\}$,

$$\begin{aligned}&\big|u_p(x,\nu)\big(\psi(\eta,\varpi)\big) - u_p(x',\nu')\big(\psi'(\eta+y_0'-y_0,\varpi)\big)\big| \\ &\quad\leq C\Big(\|\chi\|_1\big(|x-x'| + \|\xi^\varepsilon - \xi^{\varepsilon,\prime}\|_1 + |\psi(\eta,\varpi) - \psi'(\eta+y_0'-y_0,\varpi)|\big) \\ &\qquad + \|\chi^\varepsilon - \chi^{\varepsilon,\prime}\|_1 + \mathbb{E}\big[|\xi^\varepsilon - \xi^{\varepsilon,\prime}|\,|\chi^\varepsilon|\big]\Big).\end{aligned}$$

Therefore, we can find a sequence $(y^m, w^m)_{m\geq 1}$ converging to (y_0, w) such that:

$$\begin{aligned}&\big|u_p(x,\nu)\big(\psi(y^m,w^m)\big) - u_p(x',\nu')\big(\psi'(y^m+y_0'-y_0,w^m)\big)\big| \\ &\quad\leq C\Big(\|\chi\|_1\big(|x-x'| + \|\xi^\varepsilon - \xi^{\varepsilon,\prime}\|_1 + |\psi(y^n,w^m) - \psi'(y^n+y_0'-y_0,w^m)|\big) \\ &\qquad + \|\chi^\varepsilon - \chi^{\varepsilon,\prime}\|_1 + \mathbb{E}\big[|\xi^\varepsilon - \xi^{\varepsilon,\prime}|\,|\chi^\varepsilon|\big]\Big).\end{aligned}$$

By continuity of $u_p(x,\nu)(\cdot)$ and $u_p(x',\nu')(\cdot)$ and of ψ, and ψ', we get by taking the limit $m \to \infty$:

$$\begin{aligned}&\big|u_p(x,\nu)(v) - u_p(x',\nu')(v')\big| \\ &\leq C\Big(\|\chi\|_1\big(|x-x'|+\|\xi^\varepsilon - \xi^{\varepsilon,\prime}\|_1+|v-v'|\big)+\|\chi^\varepsilon-\chi^{\varepsilon,\prime}\|_1+\mathbb{E}[|\xi^\varepsilon-\xi^{\varepsilon,\prime}|\,|\chi^\varepsilon|]\Big),\end{aligned}$$

where we used the fact that $\psi(y_0,w) = v$ and $\psi'(y_0',w) = v'$. Letting the parameter of the Bernoulli random variable ε tend to 1, we deduce that:

$$\begin{aligned}&\big|u_p(x,\nu)(v) - u_p(x',\nu')(v')\big| \\ &\leq C\Big(\|\chi\|_1\big(|x-x'| + \|\xi - \xi'\|_1 + |v-v'|\big) + \|\chi-\chi'\|_1 + \mathbb{E}[|\xi-\xi'|\,|\chi|]\Big).\end{aligned} \tag{5.101}$$

Thanks to our construction of $\pi^\star$ in the second step, this may be rewritten as:

$$\begin{aligned}&\big|u_p(x,\nu)(v) - u_p(x',\nu')(v')\big| \\ &\leq C\Big(\int_{\mathbb{R}^{2d}} |\bar w| d\nu(\bar v,\bar w)\big(|x-x'| + |v-v'|\big) + \mathcal{W}(\nu,\nu')\Big),\end{aligned} \tag{5.102}$$

where

$$\begin{aligned}\mathcal{W}(\nu,\nu') = \inf_{\pi\in\Pi(\nu,\nu')}\Big[&\int_{\mathbb{R}^{4d}} |\bar w| d\pi(\bar v,\bar v',\bar w,\bar w')\int_{\mathbb{R}^{4d}} |\bar v - \bar v'| d\pi(\bar v,\bar v',\bar w,\bar w') \\ &+ \int_{\mathbb{R}^{4d}} |\bar w - \bar w'| d\pi(\bar v,\bar v',\bar w,\bar w') + \int_{\mathbb{R}^{4d}} |\bar v-\bar v'||\bar w| d\pi(\bar v,\bar v',\bar w,\bar w')\Big].\end{aligned}$$

In particular, for any $\xi,\xi',\chi,\chi' \in L^2(\Omega,\mathcal{F},\mathbb{P};\mathbb{R}^d)$, whatever the joint distribution of the 4-tuple (ξ,ξ',χ,χ'), (5.101) holds.

Inequalities (5.101) and (5.102) hold for probability measures ν,ν' with strictly positive smooth densities that together with their derivatives, decay exponentially fast at infinity. Since the set of such smooth probability measures is dense in $\mathcal{P}_2(\mathbb{R}^{2d})$, we deduce that the restriction of u_p to smooth probability measures extends by continuity to the whole $\mathbb{R}^n \times \mathcal{P}_2(\mathbb{R}^{2d}) \times \mathbb{R}^d$. We denote by $\bar u_p$ its continuous extension. It satisfies (5.101) and (5.102). By (5.100), for any $(x,\xi,\chi) \in \mathbb{R}^n \times L^2(\Omega,\mathcal{F},\mathbb{P};\mathbb{R}^d) \times L^2(\Omega,\mathcal{F},\mathbb{P};\mathbb{R}^d)$, it holds that $\mathbb{P}[u_p(x,\mathcal{L}(\xi,\chi))(\xi) = \bar u_p(x,\mathcal{L}(\xi,\chi))(\xi)] = 1$. Since $u_p(x,\mathcal{L}(\xi,\chi))(\cdot)$ and $\bar u_p(x,\mathcal{L}(\xi,\chi))(\cdot)$ are continuous, we conclude that for any $\nu \in \mathcal{P}_2(\mathbb{R}^{2d})$, $\bar u_p(x,\nu)(v)$ coincides with $u_p(x,\nu)(v)$ when v belongs to the support of μ, where again, μ is the first marginal law of ν on $\mathbb{R}^d$. Therefore, $\bar u_p(x,\nu)(\cdot)$ provides a version of $u_p(x,\nu)(\cdot)$ in $L^\infty(\mathbb{R}^d,\mu;\mathbb{R})$.

Fifth Step. In order to complete the proof, we observe from (5.101) with $(\xi,\chi) \equiv (0,0)$ that $\bar u_p$ is at most of linear growth:

$$|\bar u_p(x,\nu)(v)| \leq |\bar u_p(0,\delta_{(0,0)})(0)| + CM_1(\nu).$$

Since $u_p(0,\delta_{(0,0)})(0) = \mathbb{E}[u(0,\mathcal{L}(\frac{1}{p}Z,0))(\frac{1}{p}Z)]$ and since the map $\mathbb{R}^n \times L^2(\Omega,\mathcal{F},\mathbb{P};\mathbb{R}^d) \times L^2(\Omega,\mathcal{F},\mathbb{P};\mathbb{R}^d) \ni (x,\xi,\chi) \mapsto u(x,\mathcal{L}(\xi,\chi))(\xi) \in L^2(\Omega,\mathcal{F},\mathbb{P};\mathbb{R}^d)$ is continuous, the sequence $(\bar u_p(0,\delta_{(0,0)})(0))_{p\geq 1}$ is bounded, which shows that the functions $(\bar u_p)_{p\geq 1}$ are uniformly bounded.

Recalling that any bounded subset of $\mathcal{P}_4(\mathbb{R}^{2d})$ is a compact subset of $\mathcal{P}_2(\mathbb{R}^{2d})$, Arzelà-Ascoli's theorem implies that there exists a subsequence, still denoted by $(\bar{u}_p)_{p\geq 1}$, that converges uniformly on any bounded subset of $\mathbb{R}^n \times \mathcal{P}_4(\mathbb{R}^{2d}) \times \mathbb{R}^d$.

We then identify, for each $\nu \in \mathcal{P}_4(\mathbb{R}^{2d})$, the limit $\bar{u}(x,\nu)(\cdot)$ of $\bar{u}_p(x,\nu)(\cdot)$ with a version of $u(x,\nu)(\cdot)$ in $L^\infty(\mathbb{R}^d,\mu;\mathbb{R})$. This follows from the fact that for any bounded and measurable function $g : \mathbb{R}^d \to \mathbb{R}$,

$$\mathbb{E}\big[\bar{u}_p\big(x,\mathcal{L}(\xi,\chi)\big)(\xi)g(\xi)\big] = \mathbb{E}\big[u\big(x,\mathcal{L}(\xi+\tfrac{1}{p}Z,\chi)\big)(\xi+\tfrac{1}{p}Z)g(\xi)\big],$$

which implies, from the first inequality in the statement of Lemma 5.40, that:

$$\lim_{p\to\infty}\mathbb{E}\big[\bar{u}_p\big(x,\mathcal{L}(\xi,\chi)\big)(\xi)g(\xi)\big] = \mathbb{E}\big[u\big(x,\mathcal{L}(\xi,\chi)\big)(\xi)g(\xi)\big],$$

so that, passing to the limit in the left-hand side, we get:

$$\mathbb{E}[\bar{u}(x,\mathcal{L}(\xi,\chi))(\xi)g(\xi)] = \mathbb{E}[u(x,\mathcal{L}(\xi,\chi))(\xi)g(\xi)].$$

This provides, for each $(x,\nu) \in \mathbb{R}^n \times \mathcal{P}_4(\mathbb{R}^{2d})$, a version $\bar{u}(x,\nu)(\cdot)$ of $u(x,\nu)(\cdot)$ that satisfies the conclusion of Lemma 5.40. The map $\bar{u} : \mathbb{R}^n \times \mathcal{P}_4(\mathbb{R}^{2d}) \times \mathbb{R}^d \to \mathbb{R}$ constructed in this way is uniformly continuous (the second factor being equipped with W_2) on $\mathbb{R}^n \times \mathcal{K} \times \mathbb{R}^d$, for any subset $\mathcal{K} \subset \mathcal{P}_4(\mathbb{R}^{2d})$ which is bounded in $\mathcal{P}_2(\mathbb{R}^{2d})$. Therefore, it extends to the entire $\mathbb{R}^n \times \mathcal{P}_2(\mathbb{R}^{2d}) \times \mathbb{R}^d$. By the same argument as the one used for identifying the limit of the sequence $(\bar{u}_p)_{p\geq 1}$, we prove that for all $\nu \in \mathcal{P}_2(\mathbb{R}^{2d})$, $\bar{u}(x,\nu)(\cdot)$ is a version of $u(x,\nu)(\cdot)$ in $L^\infty(\mathbb{R}^d,\mu;\mathbb{R})$, which completes the proof. □

5.3.3 Derivation of the Master Equation

We now proceed with the derivation of the master equation. The proof is quite standard: Once the master field has been proved to be smooth enough, we can expand it along the forward component of the forward-backward system; by identifying the absolutely continuous part in the expansion with that in the backward component of the forward-backward system, we obtain that the master field indeed satisfies the master equation.

Regularity in the Other Directions

Actually, the first step is to complete the analysis of the smoothness of the master field. Indeed, it remains to discuss the existence of the other second order derivatives, namely $\partial_x^2\mathcal{U}$ and $\partial_x\partial_\mu\mathcal{U}$.

We claim:

Theorem 5.41 *For $T \leq c$, with c only depending on L, and $t \in [0,T]$, the function $\mathbb{R}^d \times \mathcal{P}_2(\mathbb{R}^d) \ni (x,\mu) \mapsto \partial_x\mathcal{U}(t,x,\mu)$ is continuously differentiable and there exists a constant $C \geq 0$, such that, for all $t \in [0,T]$, $x \in \mathbb{R}^d$ and $\mu \in \mathcal{P}_2(\mathbb{R}^d)$, $|\partial_x^2\mathcal{U}(t,x,\mu)|$ is bounded by C and, for all $x_1,x_2 \in \mathbb{R}^d$ and $\mu_1,\mu_2 \in \mathcal{P}_2(\mathbb{R}^d)$,*

$$|\partial_x^2\mathcal{U}(t,x_1,\mu_1) - \partial_x^2\mathcal{U}(t,x_2,\mu_2)| \leq C\big(|x_1-x_2| + W_1(\mu_1,\mu_2)\big),$$

Moreover, the function $[0,T] \times \mathbb{R}^d \times \mathcal{P}_2(\mathbb{R}^d) \ni (t,x,\mu) \mapsto \partial_x\mathcal{U}(t,x,\mu) \in \mathbb{R}^{m\times d}$ is continuous.

Also, for each $(t,x,\mu) \in [0,T] \times \mathbb{R}^d \times \mathcal{P}_2(\mathbb{R}^d)$, we can find a version of the map $\mathbb{R}^d \ni v \mapsto \partial_\mu \partial_x \mathcal{U}(t,x,\mu)(v) \in \mathbb{R}^{m\times(d\times d)}$ in $L^2(\mathbb{R}^d,\mu;\mathbb{R}^{m\times(d\times d)})$ such that the map $(t,x,\mu,v) \mapsto \partial_\mu \partial_x \mathcal{U}(t,x,\mu)(v)$ is continuous and bounded by C and satisfies, for all $t \in [0,T]$, $x_1,x_2 \in \mathbb{R}^d$, $v_1,v_2 \in \mathbb{R}^d$ and $\mu_1,\mu_2 \in \mathcal{P}_2(\mathbb{R}^d)$,

$$\begin{aligned}&\big|\partial_\mu \partial_x \mathcal{U}(t,x_1,\mu_1)(v_1) - \partial_\mu \partial_x \mathcal{U}(t,x_2,\mu_2)(v_2)\big| \\ &\quad \le C\big(|x_1-x_2| + |v_1-v_2| + W_1(\mu_1,\mu_2)\big).\end{aligned}$$

Moreover, the map $[0,T] \times \mathbb{R}^d \times \mathcal{P}_2(\mathbb{R}^d) \times \mathbb{R}^d \ni (t,x,\mu,v) \mapsto \partial_\mu \mathcal{U}(t,x,\mu)(v)$, as given by Theorem 5.29, is differentiable with respect to x and, for any $\ell \in \{1,\cdots,m\}$, $\partial_x \partial_\mu \mathcal{U}^\ell(t,x,\mu)(v)$ coincides with $[\partial_\mu \partial_x \mathcal{U}^\ell(t,x,\mu)(v)]^\dagger$.

Proof. The strategy is the same as above: We regard the pair $(\boldsymbol{\theta}^{t,x,\mu}, \partial_x \boldsymbol{\theta}^{t,x,\mu})$ as the solution of an enlarged forward-backward system. Similar to (5.73), the drift of the enlarged system should read:

$$\boldsymbol{B}\big(t,(x,\partial x),\mu,(y,\partial y)\big) = \begin{cases} B\big(t,x,\mu,y\big), \\ DB\big(t,(x,\partial x),\mu,(y,\partial y)\big), \end{cases}$$

where, although it refers to a different object, we use the same notation $\boldsymbol{B}$ as in (5.73) for denoting the enlarged drift and where DB is now given by:

$$DB\big(t,(x,\partial x),\mu,(y,\partial y)\big) = \partial_x B(t,x,\mu,y)\partial x + \partial_y B(t,x,\mu,y)\partial y.$$

Here, $(x,\partial x) \in \mathbb{R}^d \times \mathbb{R}^{d\times d}$ is understood as the forward variable and $(y,\partial y) \in \mathbb{R}^m \times \mathbb{R}^{m\times d}$ as the backward variable, and $\mu \in \mathcal{P}_2(\mathbb{R}^d)$ is understood as the marginal law of the first component of the forward variable on $\mathbb{R}^d$ in the corresponding McKean-Vlasov system.

By the same argument as in the proof of Lemma 5.26, we have the analogue of (5.74), namely:

$$\Big\| \sup_{t\le s\le T} \Big(|\partial_x X_s^{t,x,\mu}| + |\partial_x Y_s^{t,x,\mu}| \Big) \Big\|_\infty \le C,$$

for a constant C only depending on L (and in particular independent of Γ), provided that $T \le c$, where $c > 0$ is also a constant that only depends on L. In particular, we may regard the system satisfied by $(\boldsymbol{\theta}^{t,x,\mu}, \partial_x \boldsymbol{\theta}^{t,x,\mu})$ as a system driven by Lipschitz coefficients satisfying assumption **Smooth Coefficients Order 1**. Hence, the system satisfied by $(\boldsymbol{\theta}^{t,x,\mu}, \partial_x \boldsymbol{\theta}^{t,x,\mu})$ fulfills the assumption of Theorem 5.29. Therefore, for $T \le c$, for a new value of c, for any $t \in [0,T]$, the function $\mathbb{R}^d \times \mathcal{P}_2(\mathbb{R}^d) \ni (x,\mu) \mapsto \partial_x \mathcal{U}(t,x,\mu)$ is continuously differentiable and, for each $(x,\mu) \in \mathbb{R}^d \times \mathcal{P}_2(\mathbb{R}^d)$, we can find a version of $\mathcal{P}_2(\mathbb{R}^d) \ni \mu \mapsto \partial_\mu \partial_x \mathcal{U}(t,x,\mu)(v) \in L^2(\mathbb{R}^d,\mu;\mathbb{R}^{m\times d\times d})$, such that, for all $x_1,x_2,v_1,v_2 \in \mathbb{R}^d$ and $\mu_1,\mu_2 \in \mathcal{P}_2(\mathbb{R}^d)$,

$$\begin{aligned}&\big|\partial_x^2 \mathcal{U}(t,x_1,\mu_1) - \partial_x^2 \mathcal{U}(t,x_2,\mu_2)\big| \le C\big(|x_1-x_2| + W_1(\mu_1,\mu_2)\big), \\ &|\partial_\mu \partial_x \mathcal{U}(t,x_1,\mu_1)(v_1) - \partial_\mu \partial_x \mathcal{U}(t,x_2,\mu_2)(v_2)| \\ &\quad \le C\big(|x_1-x_2| + |v_1-v_2| + W_1(\mu_1,\mu_2)\big),\end{aligned} \tag{5.103}$$

for a constant C only depending on L and Γ. Moreover, for the same constant C, for any $t \in [0,T]$, $x,v \in \mathbb{R}^d$ and $\mu \in \mathcal{P}_2(\mathbb{R}^d)$,

$$\big|\partial_\mu \partial_x \mathcal{U}(t,x,\mu)(v)\big| \le C.$$

Finally, the functions $[0,T] \times \mathbb{R}^d \times \mathcal{P}_2(\mathbb{R}^d) \ni (t,x,\mu) \mapsto \partial_x^2 \mathcal{U}(t,x,\mu) \in \mathbb{R}^{m\times(d\times d)}$ and $[0,T] \times \mathbb{R}^d \times \mathcal{P}_2(\mathbb{R}^d) \times \mathbb{R}^d \ni (t,x,\mu,v) \mapsto \partial_\mu \partial_x \mathcal{U}(t,x,\mu)(v) \in \mathbb{R}^{m\times d\times d}$ are continuous.

In order to prove the existence of the cross-derivative $\partial_x \partial_\mu \mathcal{U}$, we may invoke Schwarz' theorem. Indeed, for $(t,x,\xi) \in [0,T] \times \mathbb{R}^d \times L^2(\Omega^1, \mathcal{F}_t^1, \mathbb{P}^1; \mathbb{R}^d)$ and $(y,\chi) \in \mathbb{R}^d \times L^2(\Omega^1, \mathcal{F}_t^1, \mathbb{P}^1; \mathbb{R}^d)$ with $|y| \le 1$, we may let:

$$\varrho : \mathbb{R}^2 \ni (r,s) \mapsto \mathcal{U}\big(t, x+ry, \mathcal{L}^1(\xi + s\chi)\big).$$

Clearly, ϱ is continuously differentiable, with:

$$\partial_r \varrho(r,s) = \partial_x \mathcal{U}\big(t,x+ry,\mathcal{L}^1(\xi+s\chi)\big)y,$$
$$\partial_s \varrho(r,s) = \mathbb{E}^1\big[\partial_\mu \mathcal{U}\big(t,x+ry,\mathcal{L}^1(\xi+s\chi)\big)(\xi+s\chi)\chi\big],$$

for $r,s \in \mathbb{R}^2$. The above analysis shows that $\partial_r \varrho$ is continuously differentiable. Therefore, by Schwarz' theorem, for any $s \in \mathbb{R}$, the mapping $\mathbb{R} \ni r \mapsto \partial_s \varrho(r,s)$ is differentiable with respect to r and

$$\partial_r \big[\partial_s \varrho(\cdot,s)\big]_{|\cdot=r} = \mathbb{E}^1\big[\partial_\mu \partial_x \mathcal{U}\big(t,x+ry,\mathcal{L}^1(\xi+s\chi)\big)(\xi+s\chi) y \otimes \chi\big].$$

In particular, when $s = 0$,

$$\begin{aligned}
&\mathbb{E}^1\big[\partial_\mu \mathcal{U}\big(t,x+ry,\mathcal{L}^1(\xi)\big)(\xi)\chi\big] - \mathbb{E}^1\big[\partial_\mu \mathcal{U}\big(t,x,\mathcal{L}^1(\xi)\big)(\xi)\chi\big] \\
&\quad = \int_0^r \mathbb{E}^1\big[\partial_\mu \partial_x \mathcal{U}\big(t,x+uy,\mathcal{L}^1(\xi)\big)(\xi) y \otimes \chi\big] du,
\end{aligned}$$

and then, for $\ell \in \{1,\cdots,m\}$,

$$\begin{aligned}
&\mathbb{E}^1\Big[\Big(\partial_\mu \mathcal{U}^\ell\big(t,x+ry,\mathcal{L}^1(\xi)\big)(\xi) - \partial_\mu \mathcal{U}^\ell\big(t,x,\mathcal{L}^1(\xi)\big)(\xi) - r\partial_\mu \partial_x \mathcal{U}^\ell\big(t,x,\mathcal{L}^1(\xi)\big)(\xi)y\Big)\chi\Big] \\
&\quad = \int_0^r \mathbb{E}^1\Big[\Big(\partial_\mu \partial_x \mathcal{U}^\ell\big(t,x+uy,\mathcal{L}^1(\xi)\big)(\xi) - \partial_\mu \partial_x \mathcal{U}^\ell\big(t,x,\mathcal{L}^1(\xi)\big)(\xi)\Big) y \otimes \chi\Big] du.
\end{aligned}$$

Thanks to (5.103), the right-hand side is less than $Cr^2\|\chi\|_1$. Therefore,

$$\begin{aligned}
&\Big\| \partial_\mu \mathcal{U}^\ell\big(t,x+ry,\mathcal{L}^1(\xi)\big)(\xi) - \partial_\mu \mathcal{U}^\ell\big(t,x,\mathcal{L}^1(\xi)\big)(\xi) - r\big[\partial_\mu \partial_x \mathcal{U}^\ell\big(t,x,\mathcal{L}^1(\xi)\big)(\xi)\big]^\dagger y \Big\|_\infty \\
&\quad \le Cr^2,
\end{aligned}$$

where we used the convention $\partial_\mu \partial_x \mathcal{U}^\ell(t,x,\mu)(v) = ([\partial_\mu \partial_{x_i} \mathcal{U}^\ell(t,x,\mu)(v)]_j)_{1\le i,j\le d}$.

When $\mu = \mathcal{L}^1(\xi)$ has full support, we get, for all $v \in \mathbb{R}^d$,

$$\Big| \partial_\mu \mathcal{U}^\ell(t,x+ry,\mu)(v) - \partial_\mu \mathcal{U}^\ell(t,x,\mu)(v) - r\big[\partial_\mu \partial_x \mathcal{U}^\ell(t,x,\mu)(v)\big]^\dagger y \Big| \le Cr^2.$$

Since the left-hand side is continuous with respect to (t,x,μ,v), we deduce that the above holds true for any $\mu \in \mathcal{P}_2(\mathbb{R}^d)$, whatever the support. This shows that, for any $t \in [0,T]$, $v \in \mathbb{R}^d$ and $\mu \in \mathcal{P}_2(\mathbb{R}^d)$, the mapping $\mathbb{R}^d \ni x \mapsto \partial_\mu \mathcal{U}(t,x,\mu)(v)$ is differentiable. Then, $\partial_x \partial_\mu \mathcal{U}^\ell$ and $[\partial_\mu \partial_x \mathcal{U}^\ell]^\dagger$ coincide. In particular, $\partial_x \partial_\mu \mathcal{U}$ satisfies (5.103). □

Derivation of the Master Equation: Proof of Theorem 5.10

We now prove that $\mathcal{U}$ solves the master equation, which will complete the proof of Theorem 5.10.

Proof. We consider $t \in [0,T]$ and $\epsilon > 0$ such that $t+\epsilon \in [0,T]$. Then, for all $(x,\mu) \in \mathbb{R}^d \times \mathcal{P}_2(\mathbb{R}^d)$,

$$\begin{aligned}\mathcal{U}(t+\epsilon,x,\mu) - \mathcal{U}(t,x,\mu) &= \mathcal{U}(t+\epsilon,x,\mu) - \mathcal{U}\big(t+\epsilon, X_{t+\epsilon}^{t,x,\mu}, \mathcal{L}^1(X_{t+\epsilon}^{t,\xi})\big) \\ &\quad + \mathcal{U}\big(t+\epsilon, X_{t+\epsilon}^{t,x,\mu}, \mathcal{L}^1(X_{t+\epsilon}^{t,\xi})\big) - \mathcal{U}(t,x,\mu),\end{aligned} \tag{5.104}$$

where $\xi \sim \mu$, with $\xi \in L^2(\Omega^1, \mathcal{F}_t^1, \mathbb{P}^1; \mathbb{R}^d)$.

Observe that the second line in (5.104) coincides with $Y_{t+\epsilon}^{t,x,\mu} - Y_t^{t,x,\mu}$. In particular,

$$\begin{aligned}&\mathbb{E}\big[\mathcal{U}\big(t+\epsilon, X_{t+\epsilon}^{t,x,\mu}, \mathcal{L}^1(X_{t+\epsilon}^{t,\xi})\big) - \mathcal{U}(t,x,\mu)\big] \\ &\quad = \mathbb{E}\big[Y_{t+\epsilon}^{t,x,\mu} - Y_t^{t,x,\mu}\big] = -\mathbb{E}\int_t^{t+\epsilon} F\big(s, X_s^{t,x,\mu}, \mathcal{L}^1(X_s^{t,\xi}), Y_s^{t,x,\mu}\big) ds.\end{aligned}$$

Taking advantage of the regularity property of F and of the time regularity of the processes $\boldsymbol{X}^{t,x,\mu} = (X_s^{t,x,\mu})_{t \le s \le T}$, $\boldsymbol{X}^{t,\xi} = (X_s^{t,\xi})_{t \le s \le T}$ and $\boldsymbol{Y}^{t,x,\mu} = (Y_s^{t,x,\mu} = \mathcal{U}(s, X_s^{t,x,\mu}, X_s^{t,\xi}))_{t \le s \le T}$, we easily deduce that:

$$\mathbb{E}\big[\mathcal{U}\big(t+\epsilon, X_{t+\epsilon}^{t,x,\mu}, \mathcal{L}^1(X_{t+\epsilon}^{t,\xi})\big) - \mathcal{U}(t,x,\mu)\big] = -\epsilon F\big(t,x,\mu,\mathcal{U}(t,x,\mu)\big) + o(\epsilon). \tag{5.105}$$

In order to handle the first line in (5.104), we shall make use of the chain rule proved in Theorem 4.17, but in the case when the function is time-homogeneous. Letting $\boldsymbol{X}^x = (X_s^x)_{t \le s \le T} = (X_s^{t,x,\mu})_{t \le s \le T}$, $\boldsymbol{X}^\xi = (X_s^\xi)_{t \le s \le T} = (X_s^{t,\xi})_{t \le s \le T}$, $\boldsymbol{\mu} = (\mu_s)_{t \le s \le T} = (\mathcal{L}^1(X_s^{t,\xi}))_{t \le s \le T}$, $\boldsymbol{B}^x = (B_s^x)_{t \le s \le T} = (B(s, X_s^{t,x,\mu}, \mathcal{L}^1(X_s^{t,\xi}), Y_s^{t,x,\mu}))_{t \le s \le T}$ and $\boldsymbol{B}^\xi = (B_s^\xi)_{t \le s \le T} = (B(s, X_s^{t,\xi}, \mathcal{L}^1(X_s^{t,\xi}), Y_s^{t,\xi}))_{t \le s \le T}$, we have:

$$\begin{aligned}&\mathbb{E}\big[\mathcal{U}(t+\epsilon, X_{t+\epsilon}^x, \mu_{t+\epsilon})\big] = \mathcal{U}(t,x,\mu) \\ &\quad + \mathbb{E}\int_t^{t+\epsilon} \partial_x \mathcal{U}(t+\epsilon, X_s^x, \mu_s) \cdot B_s^x ds + \mathbb{E}\int_t^{t+\epsilon} \tilde{\mathbb{E}}^1\Big[\partial_\mu \mathcal{U}(t+\epsilon, X_s^x, \mu_s)(\tilde{X}_s^\xi) \cdot \tilde{B}_s^\xi\Big] ds \\ &\quad + \frac{1}{2}\mathbb{E}\int_t^{t+\epsilon} \mathrm{trace}\Big[\partial_{xx}^2 \mathcal{U}(t+\epsilon, X_s^x, \mu_s)\Big(\sigma\sigma^\dagger + \sigma^0(\sigma^0)^\dagger\Big)\Big] ds \\ &\quad + \frac{1}{2}\mathbb{E}\int_t^{t+\epsilon} \tilde{\mathbb{E}}^1\Big[\mathrm{trace}\Big\{\partial_v \partial_\mu \mathcal{U}(t+\epsilon, X_s^x, \mu_s)(\tilde{X}_s^\xi)\Big(\sigma\sigma^\dagger + \sigma^0(\sigma^0)^\dagger\Big)\Big\}\Big] ds \\ &\qquad\qquad + \frac{1}{2}\mathbb{E}\int_t^{t+\epsilon} \tilde{\mathbb{E}}^1\tilde{\tilde{\mathbb{E}}}^1\Big[\mathrm{trace}\Big\{\partial_\mu^2 \mathcal{U}(t+\epsilon, X_s^x, \mu_s)(\tilde{X}_s^\xi, \tilde{\tilde{X}}_s^\xi)\sigma^0(\sigma^0)^\dagger\Big\}\Big] ds \\ &\quad + \mathbb{E}\int_t^{t+\epsilon} \tilde{\mathbb{E}}^1\Big[\mathrm{trace}\Big\{\partial_x \partial_\mu \mathcal{U}(t+\epsilon, X_s^x, \mu_s)(\tilde{X}_s^\xi)\sigma^0(\sigma^0)^\dagger\Big\}\Big] ds.\end{aligned}$$

Owing to the regularity of the derivatives of $\mathcal{U}$ with respect to the space variable, to the measure argument, and to the time parameter, we easily deduce that:

$$\begin{aligned}&\mathbb{E}\big[\mathcal{U}(t+\epsilon, X_{t+\epsilon}^x, \mu_{t+\epsilon})\big] = \mathcal{U}(t,x,\mu) \\ &\quad + \epsilon\Big(\partial_x \mathcal{U}(t,x,\mu) \cdot B\big(t,x,\mu,\mathcal{U}(t,x,\mu)\big) + \mathbb{E}^1\big[\partial_\mu \mathcal{U}(t,x,\mu)(\xi) \cdot B\big(t,\xi,\mu,\mathcal{U}(t,\xi,\mu)\big)\big] \\ &\quad + \frac{1}{2}\mathrm{trace}\big[\partial_{xx}^2 \mathcal{U}(t,x,\mu)\big(\sigma\sigma^\dagger + \sigma^0(\sigma^0)^\dagger\big)\end{aligned}$$

$$+\frac{1}{2}\mathbb{E}^1\Big[\mathrm{trace}\big\{\partial_v\partial_\mu\mathcal{U}(t,x,\mu)(\xi)\big(\sigma\sigma^\dagger+\sigma^0(\sigma^0)^\dagger\big)\big\}\Big] \qquad (5.106)$$
$$+\frac{1}{2}\mathbb{E}^1\tilde{\mathbb{E}}^1\Big[\mathrm{trace}\big\{\partial^2_\mu\mathcal{U}(t,x,\mu)(\xi,\tilde{\xi})\sigma^0(\sigma^0)^\dagger\big\}\Big]$$
$$+\mathbb{E}^1\Big[\mathrm{trace}\big\{\partial_x\partial_\mu\mathcal{U}(t,x,\mu)(\xi)\sigma^0(\sigma^0)^\dagger\big\}\Big]\Big)+\epsilon o(\epsilon).$$

Combining (5.104), (5.105), and (5.106), we deduce that $\mathcal{U}$ is right-differentiable in time. The right-derivative satisfies:

$$\begin{aligned}
&\partial_t\mathcal{U}(t,x,\mu)\\
&=-\Big(\partial_x\mathcal{U}(t,x,\mu)\cdot B\big(t,x,\mu,\mathcal{U}(t,x,\mu)\big)+\mathbb{E}^1\big[\partial_\mu\mathcal{U}(t,x,\mu)(\xi)\cdot B\big(t,\xi,\mu,\mathcal{U}(t,\xi,\mu)\big)\big]\\
&\quad+\frac{1}{2}\mathrm{trace}\big[\partial^2_{xx}\mathcal{U}(t,x,\mu)\big(\sigma\sigma^\dagger+\sigma^0(\sigma^0)^\dagger\big)\\
&\quad+\frac{1}{2}\mathbb{E}^1\Big[\mathrm{trace}\big\{\partial_v\partial_\mu\mathcal{U}(t,x,\mu)(\xi)\big(\sigma\sigma^\dagger+\sigma^0(\sigma^0)^\dagger\big)\big\}\Big]\\
&\qquad+\frac{1}{2}\mathbb{E}^1\tilde{\mathbb{E}}^1\Big[\mathrm{trace}\big\{\partial^2_\mu\mathcal{U}(t,x,\mu)(\xi,\tilde{\xi})\sigma^0(\sigma^0)^\dagger\big\}\Big]\\
&\quad+\mathbb{E}^1\Big[\mathrm{trace}\big\{\partial_x\partial_\mu\mathcal{U}(t,x,\mu)(\xi)\sigma^0(\sigma^0)^\dagger\big\}\Big]+F\big(t,x,\mu,\mathcal{U}(t,x,\mu)\big)\Big).
\end{aligned}$$

Since the right-hand side is continuous in all the variables, $\mathcal{U}$ is differentiable in time and $\partial_t\mathcal{U}$ is continuous in all the arguments. □

Uniqueness: Proof of Theorem 5.11

In this last paragraph, we do not assume anymore that the coefficients are independent of (z,z^0) and that σ and σ^0 are constant. Instead, we assume that the coefficients satisfy assumption **Conditional MKV FBSDE in Small Time** *and that σ and σ^0 are bounded.*

We prove that uniqueness holds true within the class $\mathfrak{S}_m$ defined in Definition 5.9.

The proof is based upon the fact that, under the existence of a classical solution in the class $\mathfrak{S}_m$, the McKean-Vlasov forward-backward system (5.1) has a unique solution.

Proposition 5.42 *Under assumption* **Conditional MKV FBSDE in Small Time,** *assume that there exists a solution $\mathcal{U}$ in the class $\mathfrak{S}_m$ to the master equation* (5.13). *Assume also that σ and σ^0 are bounded. Then, whatever the size of T, for any $t\in[0,T]$ and $\xi\in L^2(\Omega^1,\mathcal{F}^1_t,\mathbb{P}^1;\mathbb{R}^d)$, there exists a solution $(\boldsymbol{X},\boldsymbol{Y},\boldsymbol{Z},\boldsymbol{Z}^0)$ to* (5.2), *satisfying $X_t=\xi$ as initial condition at time t and, for all $s\in[t,T]$,*

$$\begin{aligned}
Y_s&=\mathcal{U}\big(s,X_s,\mathcal{L}^1(X_s)\big),\\
Z_s&=\sigma^\dagger\big(s,X_s,\mathcal{L}^1(X_s)\big)\partial_x\mathcal{U}\big(s,X_s,\mathcal{L}^1(X_s)\big),
\end{aligned}$$

$$Z_s^0 = (\sigma^0)^\dagger\big(s, X_s, \mathcal{L}^1(X_s)\big)\partial_x\mathcal{U}\big(s, X_s, \mathcal{L}^1(X_s)\big)$$
$$+ \tilde{\mathbb{E}}^1\big[(\sigma^0)^\dagger\big(s, \tilde{X}_s, \mathcal{L}^1(X_s)\big)\partial_\mu\mathcal{U}\big(s, X_s, \mathcal{L}^1(X_s)\big)(\tilde{X}_s)\big].$$

The resulting solution to (5.2) *is the unique one with the initial condition* $X_t = \xi$.

Proof. We recall the following useful notation:

$$\partial_x^\sigma\mathcal{U}(t, x, \mu) = \sigma^\dagger(t, x, \mu)\partial_x\mathcal{U}(t, x, \mu),$$
$$\partial_x^{\sigma^0}\mathcal{U}(t, x, \mu) = (\sigma^0)^\dagger(t, x, \mu)\partial_x\mathcal{U}(t, x, \mu),$$
$$\partial_\mu^{\sigma^0}\mathcal{U}(t, x, \mu) = \int_{\mathbb{R}^d}(\sigma^0)^\dagger(t, v, \mu)\partial_\mu\mathcal{U}(t, x, \mu)(v)d\mu(v),$$
$$\partial_{(x,\mu)}^{\sigma^0}\mathcal{U}(t, x, \mu) = \partial_x^{\sigma^0}\mathcal{U}(t, x, \mu) + \partial_\mu^{\sigma^0}\mathcal{U}(t, x, \mu),$$

for $t \in [0, T]$, $x, v \in \mathbb{R}^d$ and $\mu \in \mathcal{P}_2(\mathbb{R}^d)$.

First Step. We first prove the existence of a solution to (5.2). To do so, we consider the McKean-Vlasov SDE:

$$dX_s = B\Big(s, X_s, \mathcal{U}\big(s, X_s, \mathcal{L}^1(X_s)\big), \partial_x^\sigma\mathcal{U}\big(s, X_s, \mathcal{L}^1(X_s)\big), \partial_{(x,\mu)}^{\sigma^0}\mathcal{U}\big(s, X_s, \mathcal{L}^1(X_s)\big)\Big)ds$$
$$+ \sigma\big(s, X_s, \mathcal{L}^1(X_s)\big)dW_s + \sigma^0\big(s, X_s, \mathcal{L}^1(X_s)\big)dW_s^0.$$

Since $\mathcal{U}$ is assumed to belong to $\mathfrak{S}_m$, the coefficients of the equation are Lipschitz continuous. Therefore, for a given initial condition $\xi \in L^2(\Omega^1, \mathcal{F}_t^1, \mathbb{P}^1; \mathbb{R}^d)$ at time t, the equation has a unique solution $(X_s)_{t\le s\le T}$ satisfying $X_t = \xi$. It satisfies $\mathbb{E}[\sup_{t\le s\le T}|X_s|^2] < \infty$.

Defining $(Y_s)_{t\le s\le T}$, $(Z_s)_{t\le s\le T}$ and $(Z_s^0)_{t\le s\le T}$ as in the statement and observing that $Z_s = \partial_x^\sigma\mathcal{U}(s, X_s, \mathcal{L}^1(X_s))$ and $Z_s^0 = \partial_{(x,\mu)}^{\sigma^0}\mathcal{U}(s, X_s, \mathcal{L}^1(X_s))$ for all $s \in [t, T]$, we then deduce from Itô's formula in Theorem 4.17 that the 4-tuple $(X_s, Y_s, Z_s, Z_s^0)_{t\le s\le T}$ satisfies (5.2). Observe that, as required, the martingale integrands $(Z_s)_{t\le s\le T}$ and $(Z_s^0)_{t\le s\le T}$ are square-integrable. This follows from the fact that $\partial_x\mathcal{U}$ and $\partial_\mu\mathcal{U}$ are bounded and that σ and σ^0 are also bounded. This completes the proof.

Second Step. We now prove uniqueness. We call $(\boldsymbol{X}', \boldsymbol{Y}', \boldsymbol{Z}', \boldsymbol{Z}^{0,\prime}) = (X_s', Y_s', Z_s', Z_s^{0,\prime})_{t\le s\le T}$ another solution to (5.2) with $X_t' = \xi$ as initial condition.

The idea consists in decoupling the forward-backward system satisfied by the 4-tuple $(\boldsymbol{X}', \boldsymbol{Y}', \boldsymbol{Z}', \boldsymbol{Z}^{0,\prime})$ by letting $\bar{\boldsymbol{Y}}' = (\bar{Y}_s' = \mathcal{U}(s, X_s', \mathcal{L}^1(X_s')))_{t\le s\le T}$. Then, by applying Itô's formula once again and using the form of the PDE satisfied by $\mathcal{U}$, see (5.13), we deduce that:

$$d\bar{Y}_s' = \partial_x\mathcal{U}\big(s, X_s', \mathcal{L}^1(X_s')\big)\cdot\Big[B\big(s, X_s', \mathcal{L}^1(X_s'), Y_s', Z_s'\big) - B\big(s, X_s', \mathcal{L}^1(X_s'), \bar{Y}_s', \bar{Z}_s'\big)\Big]ds$$
$$+ \tilde{\mathbb{E}}^1\Big[\partial_\mu\mathcal{U}\big(s, X_s', \mathcal{L}^1(X_s')(\tilde{X}_s')$$
$$\cdot\Big(B\big(s, \tilde{X}_s', \mathcal{L}^1(X_s'), \tilde{Y}_s', \tilde{Z}_s'\big) - B\big(s, \tilde{X}_s', \mathcal{L}^1(X_s'), \tilde{\bar{Y}}_s', \tilde{\bar{Z}}_s'\big)\Big)\Big]ds$$
$$- F\big(s, X_s', \mathcal{L}^1(X_s'), \bar{Y}_s', \bar{Z}_s'\big)ds + \bar{Z}_s'dW_s + \bar{Z}_s^{0,\prime}dW_s^0,$$

with the terminal condition $\bar{Y}'_T = G(X'_T, \mathcal{L}(X'_T))$, where we let:

$$\begin{aligned}
\bar{Y}'_s &= \mathcal{U}\big(s, X'_s, \mathcal{L}^1(X'_s)\big), \\
\bar{Z}'_s &= \sigma^\dagger\big(s, X'_s, \mathcal{L}^1(X'_s)\big)\partial_x \mathcal{U}\big(s, X'_s, \mathcal{L}^1(X'_s)\big) = \partial_x^\sigma \mathcal{U}\big(s, X'_s, \mathcal{L}^1(X'_s)\big), \\
\bar{Z}^{0,\prime}_s &= \big(\sigma^0\big)^\dagger\big(s, X'_s, \mathcal{L}^1(X'_s)\big)\partial_x \mathcal{U}\big(s, X'_s, \mathcal{L}^1(X'_s)\big) \\
&\quad + \tilde{\mathbb{E}}^1\big[\big(\sigma^0\big)^\dagger\big(s, \tilde{X}'_s, \mathcal{L}^1(X'_s)\big)\partial_\mu \mathcal{U}\big(s, X'_s, \mathcal{L}^1(X'_s)\big)(\tilde{X}'_s)\big] \\
&= \partial^{\sigma^0}_{(x,\mu)} \mathcal{U}\big(s, X'_s, \mathcal{L}^1(X'_s)\big).
\end{aligned}$$

Now, we form the difference $(Y'_s - \bar{Y}'_s)_{t \le s \le T}$. Using the fact that $\partial_x \mathcal{U}$ and $\partial_\mu \mathcal{U}$ are bounded and that $\bar{Y}'_T = Y'_T$, and using standard stability arguments from the theory of backward SDEs, we easily get that:

$$\mathbb{E}\bigg[\sup_{t \le s \le T} |Y'_s - \bar{Y}'_s|^2 + \int_t^T |Z'_s - \bar{Z}'_s|^2 ds + \int_t^T |Z^{0,\prime}_s - \bar{Z}^{0,\prime}_s|^2\bigg] = 0,$$

which completes the proof of uniqueness. □

Now, the proof of Theorem 5.11 is rather straightforward. Any two solutions $\mathcal{U}$ and $\mathcal{U}'$ to the master equations induce the same solution to (5.2), when they are initialized with the same initial condition (t, ξ) with $t \in [0, T]$ and $\xi \in L^2(\Omega^1, \mathcal{F}^1_t, \mathbb{P}^1; \mathbb{R}^d)$. Therefore, $\mathcal{U}(t, \xi, \mathcal{L}^1(\xi)) = \mathcal{U}'(t, \xi, \mathcal{L}^1(\xi))$, which proves that, when μ has full support, $\mathcal{U}(t, x, \mu) = \mathcal{U}'(t, x, \mu)$ for all $x \in \mathbb{R}^d$. Then, by continuity of $\mathcal{U}$ and $\mathcal{U}'$, this is true for all $(t, x, \mu) \in [0, T] \times \mathbb{R}^d \times \mathcal{P}_2(\mathbb{R}^d)$.

5.4 Application to Mean Field Games

We now explain how to implement the previous results to the analysis of the master equation for mean field games. We split the discussion in two steps: The first one is devoted to the analysis of the master equation for mean field games in small time; in the second one, we provide explicit conditions under which the master equation has a classical solution over time intervals of arbitrary lengths.

Throughout the section, we use the same notation as in the general description of mean field games with common noise in the introduction of Chapter 4. Namely, we are given a complete probability space $(\Omega^0, \mathcal{F}^0, \mathbb{P}^0)$, endowed with a complete and right-continuous filtration $\mathbb{F}^0 = (\mathcal{F}^0_t)_{0 \le t \le T}$ and a d-dimensional $\mathbb{F}^0$-Brownian motion $\boldsymbol{W}^0 = (W^0_t)_{0 \le t \le T}$, and a complete probability space $(\Omega^1, \mathcal{F}^1, \mathbb{P}^1)$ endowed with a complete and right-continuous filtration $\mathbb{F}^1 = (\mathcal{F}^1_t)_{0 \le t \le T}$ and a d-dimensional $\mathbb{F}^1$-Brownian motion $\boldsymbol{W} = (W_t)_{0 \le t \le T}$. As usual, we denote by $(\Omega, \mathcal{F}, \mathbb{P})$ the completion of the product space $(\Omega, \mathcal{F}^0 \otimes \mathcal{F}^1, \mathbb{P}^0 \otimes \mathbb{P}^1)$ and we endow it with the filtration $\mathbb{F} = (\mathcal{F}_t)_{0 \le t \le T}$ obtained by augmenting the product filtration $\mathbb{F}^0 \otimes \mathbb{F}^1$ in a right-continuous way and by completing it.

For a drift b from $[0,T] \times \mathbb{R}^d \times \mathcal{P}_2(\mathbb{R}^d) \times A$ to $\mathbb{R}^d$, where A is a closed convex subset of $\mathbb{R}^k$, for two (uncontrolled) diffusion coefficients σ and σ^0 from $[0,T] \times \mathbb{R}^d \times \mathcal{P}_2(\mathbb{R}^d)$ to $\mathbb{R}^{d\times d}$ and for cost functions f and g from $[0,T] \times \mathbb{R}^d \times \mathcal{P}_2(\mathbb{R}^d) \times A$ to $\mathbb{R}$ and from $\mathbb{R}^d \times \mathcal{P}_2(\mathbb{R}^d)$ to $\mathbb{R}$, the search for an MFG equilibrium goes along the lines of Definition 2.16:

(i) Given an $\mathcal{F}_0^0$-measurable random variable $\mu_0 : \Omega^0 \to \mathcal{P}_2(\mathbb{R}^d)$, with $\mathcal{V}_0$ as distribution, an initial condition $X_0 : \Omega \to \mathbb{R}^d$ such that $\mathcal{L}^1(X_0) = \mu_0$, and an $\mathcal{F}_T^0$-measurable random variable $\mathfrak{M}$ with values in $\mathcal{P}_2(\mathcal{C}([0,T];\mathbb{R}^{2d}))$ such that $\mathbb{F}$ is compatible with $(X_0, \boldsymbol{W}^0, \mathfrak{M}, \boldsymbol{W})$ and $\boldsymbol{\mu}_0 = \mathfrak{M} \circ (e_0^x)^{-1}$, where e_t^x is the mapping evaluating the d first coordinates at time t on $\mathcal{C}([0,T];\mathbb{R}^{2d})$, solve the (standard) stochastic control problem (with random coefficients):

$$\inf_{(\alpha_s)_{0\le s\le T}} \mathbb{E}\left[\int_0^T f(s, X_s, \mu_s, \alpha_s)ds + g(X_T, \mu_T)\right], \tag{5.107}$$

subject to

$$dX_s = b\big(s, X_s, \mu_s, \alpha_s\big)ds + \sigma\big(s, X_s, \mu_s\big)dW_s + \sigma^0\big(s, X_s, \mu_s\big)dW_s^0, \tag{5.108}$$

for $s \in [0,T]$, with X_0 as initial condition and with $\mu_s = \mathfrak{M} \circ (e_s^x)^{-1}$.

(ii) Determine the input $\mathfrak{M}$ so that, for one optimal path $(X_s)_{0\le s\le T}$, it holds that

$$\mathfrak{M} = \mathcal{L}^1\big(\boldsymbol{X}, \boldsymbol{W}\big). \tag{5.109}$$

In order to guarantee the well posedness of the cost functional (5.107) and the unique solvability of (5.108), we recall the useful condition:

Assumption (Coefficients Growth). There exist two constants $\Gamma, L \ge 0$ such that:

(A1) For any $t \in [0,T]$, the coefficients $b(t,\cdot,\cdot,\cdot)$ and $(\sigma,\sigma^0)(t,\cdot,\cdot)$ are respectively continuous on $\mathbb{R}^d \times \mathcal{P}_2(\mathbb{R}^d) \times A$ and on $\mathbb{R}^d \times \mathcal{P}_2(\mathbb{R}^d)$. The coefficients $b(t,\cdot,\mu,\alpha)$, $\sigma(t,\cdot,\mu)$ and $\sigma^0(t,\cdot,\mu)$ are L-Lipschitz continuous in the x variable, uniformly in $(t,\mu,\alpha) \in [0,T]\times\mathcal{P}_2(\mathbb{R}^d)\times A$. Moreover,

$$|b(t,x,\mu,\alpha)| + |(\sigma,\sigma^0)(t,x,\mu)| \le \Gamma\big[1 + |x| + |\alpha| + M_2(\mu)\big],$$

(A2) The coefficients f and g are Borel-measurable mappings from $[0,T] \times \mathbb{R}^d \times \mathcal{P}_2(\mathbb{R}^d) \times A$ to $\mathbb{R}$ and from $\mathbb{R}^d \times \mathcal{P}_2(\mathbb{R}^d)$ to $\mathbb{R}$ respectively. For

(continued)

any $t \in [0, T]$, the coefficients $f(t, \cdot, \cdot, \cdot)$ and $g(\cdot, \cdot)$ are continuous on $\mathbb{R}^d \times \mathcal{P}_2(\mathbb{R}^d) \times A$ and $\mathbb{R}^d \times \mathcal{P}_2(\mathbb{R}^d)$ respectively. Moreover,

$$|f(t, x, \mu, \alpha)| + |g(x, \mu)| \leq \Gamma\big[1 + |x|^2 + |\alpha|^2 + M_2(\mu)^2\big].$$

5.4.1 Mean Field Games in Small Time

The purpose of this section is to prove that the master equation has a classical solution in small time, provided that the optimization problem (5.107) has a unique optimal path and that the coefficients b, σ, σ^0, f, and g driving the optimization problem are smooth enough.

In order to do so, we shall make use of the results of the previous section on the master equation associated with a general FBSDE of the McKean-Vlasov type. Actually, this strategy makes sense if we are able to characterize the equilibria of the mean field game through a McKean-Vlasov forward-backward system driven by coefficients that satisfy assumption **Smooth Coefficients Order 2** in Subsection 5.1.5. A quick glance at the results of Chapters 1 and 2 suggests to use the representation based upon the stochastic Pontryagin principle.

In order to fit the framework used in the previous sections, we shall assume:

Assumption (MFG Master Pontryagin).

(A1) The coefficients σ and σ^0 are constant; moreover, the coefficients b, f, and g are continuous with respect to all the variables, the space $\mathcal{P}_2(\mathbb{R}^d)$ being equipped with the 2-Wasserstein distance W_2.

In order to apply results from Chapter 2, we shall assume that assumptions **FBSDE** and **Decoupling Master** in Subsections 4.1.3 and 4.2.2 are in force, the latter being especially useful to make the connection between the derivative of the master field and the Pontryagin adjoint system:

(A2) Assumptions **FBSDE** and **Decoupling Master** are in force.

In particular, for any $t \in [0, T]$, any $X_t \in L^2(\Omega, \mathcal{F}_t, \mathbb{P}; \mathbb{R}^d)$ and any super-environment $(\mathfrak{M}_s)_{t \leq s \leq T}$ such that $(X_t, (W_s^0 - W_t^0, \mathfrak{M}_s, W_s - W_t)_{t \leq s \leq T})$ is compatible with $(\mathcal{F}_s)_{t \leq s \leq T}$, the optimization problem (5.107)–(5.108) with X_t as initial condition at time t has an optimal path. As assumed in assumption **FBSDE**, it may be represented by means of a forward-backward system.

In order to implement the stochastic Pontryagin principle, we also assume:

There exist three constants $L, \Gamma \geq 0$ and $\lambda > 0$ such that:

(A3) The drift b is an affine function of α in the sense that it is of the form

$$b(t,x,\mu,\alpha) = b_1(t,x,\mu) + b_2(t)\alpha,$$

where the function $[0,T] \ni t \mapsto b_2(t) \in \mathbb{R}^{d\times k}$ is continuous and bounded by L, and the function $[0,T] \ni (t,x,\mu) \mapsto b_1(t,x,\mu) \in \mathbb{R}^d$ is continuous, is L-Lipschitz continuous with respect to (x,μ) and is differentiable with respect to x, the derivative $\mathbb{R}^d \ni x \mapsto \partial_x b_1(t,x,\mu)$ being continuous for each $(t,\mu) \in [0,T] \times \mathcal{P}_2(\mathbb{R}^d)$.

(A4) For any $t \in [0,T]$ and $\mu \in \mathcal{P}_2(\mathbb{R}^d)$, the function $\mathbb{R}^d \times A \ni (x,\alpha) \mapsto f(t,x,\mu,\alpha)$ is once continuously differentiable and the function $[0,T] \times \mathbb{R}^d \times \mathcal{P}_2(\mathbb{R}^d) \times A \ni (t,x,\mu,\alpha) \mapsto (\partial_x f, \partial_\alpha f)(t,x,\alpha,\mu)$ is bounded by Γ at $(t,0,\delta_0,0_A)$, for any $t \in [0,T]$ and for some fixed $0_A \in A$, and is L-Lipschitz continuous in (x,μ,α). Moreover, f satisfies the convexity assumption

$$f(t,x,\mu,\alpha') - f(t,x,\mu,\alpha) - (\alpha' - \alpha)\cdot \partial_\alpha f(t,x,\mu,\alpha) \geq \lambda|\alpha' - \alpha|^2.$$

(A5) The function g is differentiable with respect to x and the derivative $\mathbb{R}^d \times \mathcal{P}_2(\mathbb{R}^d) \ni (x,\mu) \mapsto \partial_x g(x,\mu)$ is L-Lipschitz continuous.

Obviously, assumption **MFG Master Pontryagin** subsumes assumption **Coefficients Growth** provided that $f(t,0,\mu,0_A)$ and $g(0,\mu)$ are at most of quadratic growth in μ.

Moreover, following Lemma 1.56, see also Lemma (Vol I)-6.18 for the way to handle the regularity in the direction μ, we know that there exists a unique minimizer $\hat{\alpha}(t,x,\mu,y)$ of $A \ni \alpha \mapsto H(t,x,\mu,y) = b(t,x,\mu,\alpha)\cdot y + f(t,x,\mu,\alpha)$ and that the function $[0,T] \times \mathbb{R}^d \times \mathcal{P}_2(\mathbb{R}^d) \times A \ni (t,x,\mu,y) \mapsto \hat{\alpha}(t,x,\mu,y)$ is measurable, locally bounded and Lipschitz continuous with respect to (x,μ,y), uniformly in $t \in [0,T]$, the Lipschitz constant depending only upon L and λ. Moreover, there exists a constant $C > 0$, only depending on L, Γ and λ, such that

$$\forall t \in [0,T],\ x,y \in \mathbb{R}^d,\ \mu \in \mathcal{P}_2(\mathbb{R}^d), \quad \big|\hat{\alpha}(t,x,\mu,y)\big| \leq C\big(1 + |x| + M_2(\mu) + |y|\big).$$

Remark 5.43 *Since σ and σ^0 are constant, we prefer to denote the reduced Hamiltonian $H^{(r)}$ by H.*

Implementation of the Pontryagin Maximum Principle

By the necessary part of the Pontryagin maximum principle, see Theorem 1.59, we know that, for a super-environment $\mathfrak{M}$ satisfying the usual condition of compatibility, the solution to the optimal control problem (5.107)–(5.108), with X_t as initial condition, satisfies the forward-backward system:

$$\begin{cases} dX_s = b\big(s, X_s, \mu_s, \hat{\alpha}(s, X_s, \mu_s, Y_s)\big)ds + \sigma dW_s + \sigma^0 dW_s^0, \\ dY_s = -\partial_x H\big(s, X_s, \mu_s, Y_s, \hat{\alpha}(s, X_s, \mu_s, Y_s)\big)ds \\ \qquad\qquad + Z_s dW_s + Z_s^0 dW_s^0 + dM_s, \quad s \in [t, T], \\ Y_T = \partial_x g\big(X_T, \mu_T\big). \end{cases} \tag{5.110}$$

Above $\boldsymbol{M} = (M_s)_{t \le s \le T}$ is a square-integrable càd-làg martingale starting from 0, of zero bracket with $(W_s^0 - W_t^0, W_s - W_t)_{t \le s \le T}$. Pay attention that the solution of the optimal control problem (5.107)–(5.108) is known to exist and to be unique thanks to assumption **FBSDE**.

Whenever (5.110) is uniquely solvable for any super-environment $\mathfrak{M}$, it provides a characterization of the solution of the optimal control problem (5.107)–(5.108). In this framework, MFG equilibria may be characterized as the solutions of the McKean-Vlasov forward-backward system:

$$\begin{cases} dX_s = b\big(s, X_s, \mathcal{L}^1(X_s), \hat{\alpha}(s, X_s, \mathcal{L}^1(X_s), Y_s)\big)dt + \sigma dW_s + \sigma^0 dW_s^0, \\ dY_s = -\partial_x H\big(s, X_s, \mathcal{L}^1(X_s), Y_s, \hat{\alpha}(s, X_s, \mathcal{L}^1(X_s), Y_s)\big)ds \\ \qquad\qquad + Z_s dW_s + Z_s^0 dW_s^0 + dM_s, \quad s \in [t, T], \\ Y_T = \partial_x g\big(X_T, \mathcal{L}^1(X_T)\big), \end{cases} \tag{5.111}$$

with X_t as initial condition, in which case the equilibrium is given by $\mathfrak{M} = \mathcal{L}^1(\boldsymbol{X}, \boldsymbol{W})$. This representation reads as the analogue of Proposition 2.18.

We plan to apply the results of the previous section to (5.111), see in particular Theorem 5.10. However, it must be observed that, due to the structure of $\partial_x H$, (5.111) may not fit the Cauchy-Lipschitz assumption **Conditional MKV FBSDE in Small Time** in Subsection 5.1.3. This is a serious drawback. However, it can be easily circumvented in the typical cases under which **(A2)** in assumption **MFG Master Pontryagin** has been proved to hold in Chapter 1: see Theorems 1.57 and 1.60. In the first case, $\partial_x b$, $\partial_x f$ and $\partial_x g$ are assumed to be bounded; so, the process $\boldsymbol{Y} = (Y_s)_{t \le s \le T}$ must be bounded as well, by a constant only depending on the bounds of $\partial_x b$, $\partial_x f$ and $\partial_x g$. In the second case, $\partial_x b$ is a matrix depending only on t; so $\partial_x H$ is Lipschitz continuous in (x, μ, y, α) uniformly in time. Therefore, in both cases, there exists a smooth function $\phi : \mathbb{R}^d \to \mathbb{R}^d$, with bounded derivatives of all orders, such that (5.110) holds true with $\partial_x H(s, X_s, \mu_s, \phi(Y_s), \hat{\alpha}(s, X_s, \mu_s, Y_s))$ in lieu of $\partial_x H(s, X_s, \mu_s, Y_s, \hat{\alpha}(s, X_s, \mu_s, Y_s))$ and the function:

$$[0,T] \times \mathbb{R}^d \times \mathcal{P}_2(\mathbb{R}^d) \times \mathbb{R}^d \times A \ni (t, x, \mu, y, \alpha) \mapsto \partial_x H(t, x, \mu, \phi(y), \alpha)$$

is Lipschitz continuous in (x, μ, y, α) uniformly in $t \in [0, T]$. The Lipschitz constant only depends on the Lipschitz constant of ϕ and on L, T and Γ in assumption **MFG Master Pontryagin**; it is non-decreasing in T.

Therefore, we shall complement assumption **MFG Master Pontryagin** with:

(A6) There exists a smooth function ϕ, with bounded derivatives of any order, such that:

– the function ϕ is compactly supported unless $\partial_x b$ is a constant function; in particular, the function $[0,T] \times \mathbb{R}^d \times \mathcal{P}_2(\mathbb{R}^d) \times \mathbb{R}^d \times A \ni (t, x, \mu, y, \alpha) \mapsto \partial_x H(t, x, \mu, \phi(y), \alpha)$ is Lipschitz continuous in (x, μ, y, α) uniformly in $t \in [0, T]$,

– for any $t \in [0, T]$, any $X_t \in L^2(\Omega, \mathcal{F}_t, \mathbb{P}; \mathbb{R}^d)$ and any super-environment $(\mathfrak{M}_s)_{t \leq s \leq T}$ such that $(X_t, (W_s^0 - W_t^0, \mathfrak{M}_s, W_s - W_t)_{t \leq s \leq T})$ is compatible with $(\mathcal{F}_s)_{t \leq s \leq T}$, the optimal path $(X_s)_{t \leq s \leq T}$ together with the adjoint process $(Y_s)_{t \leq s \leq T}$ satisfy $\mathbb{P}[\forall s \in [t, T],\ \phi(Y_s) = Y_s] = 1$.

Once the function ϕ has been given, we know from Theorem 1.45 that, for T small enough, the system:

$$\begin{cases} dX_s = b\big(s, X_s, \mu_s, \hat{\alpha}(s, X_s, \mu_s, Y_s)\big)ds + \sigma dW_s + \sigma^0 dW_s^0, \\ dY_s = -\partial_x H\big(s, X_s, \mu_s, \phi(Y_s), \hat{\alpha}(s, X_s, \mu_s, Y_s)\big)ds \\ \qquad\qquad + Z_s dW_s + Z_s^0 dW^0 + dM_s, \quad s \in [t, T], \\ Y_T = \partial_x g(X_T, \mu_T), \end{cases} \tag{5.112}$$

has a unique solution for any initial condition $X_t \in L^2(\Omega, \mathcal{F}_t, \mathbb{P}; \mathbb{R}^d)$ and any super-environment $\mathfrak{M}$. As explained above, the difference between (5.110) and (5.112) lies in the additional ϕ in $\partial_x H$, but, with our choice for ϕ, the unique solution of (5.112) is also the unique solution of (5.110). This shows that optimal paths of the optimal control problem (5.107)–(5.108) are characterized as the unique solutions of (5.112).

Now, we know from Theorem 5.4 and Remark 5.6 that, in small time, the McKean-Vlasov forward-backward system:

$$\begin{cases} dX_s = b\big(s, X_s, \mathcal{L}^1(X_s), \hat{\alpha}(s, X_s, \mathcal{L}^1(X_s), Y_s)\big)ds + \sigma dW_s + \sigma^0 dW_s^0, \\ dY_s = -\partial_x H\big(s, X_s, \mathcal{L}^1(X_s), \phi(Y_s), \hat{\alpha}(s, X_s, \mathcal{L}^1(X_s), Y_s)\big)ds \\ \qquad\qquad + Z_s dW_s + Z_s^0 dW_s^0 + dM_s, \quad s \in [t, T], \\ Y_T = \partial_x g\big(X_T, \mathcal{L}^1(X_T)\big), \end{cases} \tag{5.113}$$

with X_t as initial condition, has a unique solution and it is adapted to the completion of the filtration generated by $(\xi, \mathcal{L}^1(\xi), (W_s^0 - W_t^0)_{t \le s \le T}, (W_s - W_t)_{t \le s \le T})$. Thanks to the choice of the function ϕ, this unique solution is the unique solution of (5.111), which shows that, for any initial condition, the mean field game has a unique strong solution. Invoking Theorem 5.4 once again, $\mathcal{L}^1((X_s, W_s - W_s^0)_{t \le s \le T})$ forms a strong equilibrium as it is adapted with respect to the filtration generated by the initial state and by the common noise $\boldsymbol{W}^0$. In particular, following Definition 4.1 in Chapter 4, we may associate a master field $\mathcal{U}$ with the mean field game.

Observe that the length c of the interval on which existence and uniqueness hold true is dictated by the Lipschitz constant of the coefficients in (5.112) and (5.113), namely the Lipschitz constant of $(\partial_x H(s, \cdot, \cdot, \phi(\cdot), \hat{\alpha}(s, \cdot, \cdot, \cdot)))_{s \in [0,T]}$ and the Lipschitz constant of $\partial_x g$.

Representation of the Master Field

The master field may be represented as follows, provided that $T \le c$, for the same constant c as above. For any $t \in [0, T]$ and $\xi \in L^2(\Omega^1, \mathcal{F}_t^1, \mathbb{P}^1; \mathbb{R}^d)$, the system:

$$\begin{cases} dX_s^{t,\xi} = b\big(s, X_s^{t,\xi}, \mathcal{L}^1(X_s^{t,\xi}), \hat{\alpha}(s, X_s^{t,\xi}, \mathcal{L}^1(X_s^{t,\xi}), Y_s^{t,\xi})\big)ds \\ \qquad\qquad + \sigma dW_s + \sigma^0 dW_s^0, \\ dY_s^{t,\xi} = -\partial_x H\big(s, X_s^{t,\xi}, \mathcal{L}^1(X_s^{t,\xi}), Y_s^{t,\xi}, \hat{\alpha}(s, X_s^{t,\xi}, \mathcal{L}^1(X_s^{t,\xi}), Y_s^{t,\xi})\big)ds \\ \qquad\qquad + Z_s^{t,\xi} dW_s + Z_s^{0;t,\xi} dW_s^0, \quad s \in [t, T], \\ X_t^{t,\xi} = \xi, \quad Y_T^{t,\xi} = \partial_x g\big(X_T^{t,\xi}, \mathcal{L}^1(X_T^{t,\xi})\big), \end{cases} \tag{5.114}$$

has a unique solution.

Following (5.6), we then consider, for all $(t, x, \mu) \in [0, T] \times \mathbb{R}^d \times \mathcal{P}_2(\mathbb{R}^d)$, the uniquely solvable system:

$$\begin{cases} dX_s^{t,x,\mu} = b\big(s, X_s^{t,x,\mu}, \mathcal{L}^1(X_s^{t,\xi}), \hat{\alpha}(s, X_s^{t,x,\mu}, \mathcal{L}^1(X_s^{t,\xi}), Y_s^{t,x,\mu})\big)ds \\ \qquad + \sigma dW_s + \sigma^0 dW_s^0, \\ dY_s^{t,x,\mu} = -\partial_x H\big(s, X_s^{t,x,\mu}, \mathcal{L}^1(X_s^{t,\xi}), Y_s^{t,x,\mu}, \hat{\alpha}(t, X_s^{t,x,\mu}, \mathcal{L}^1(X_s^{t,\xi}), Y_s^{t,x,\mu})\big)ds \\ \qquad + Z_s^{t,x,\mu} dW_s + Z_t^{0;t,x,\mu} dW_s^0, \quad s \in [t, T], \\ X_t^{t,x,\mu} = x, \quad Y_T^{t,x,\mu} = \partial_x g\big(X_T^{t,x,\mu}, \mathcal{L}^1(X_T^{t,\xi})\big), \end{cases} \tag{5.115}$$

with $\xi \sim \mu$.

Here come two crucial observations. By Theorem 4.10, the master field $\mathcal{V}$ of the pair (5.114)–(5.115) is nothing but $\partial_x \mathcal{U}$, $\mathcal{U}$ denoting the master field of the mean field game. In the statement of Theorem 4.10, the master field is constructed on the canonical space, but, by strong uniqueness, a similar representation holds true on any space. Regarding the master field, Definition 4.1 becomes:

$$\mathcal{U}(t, x, \mu) = \mathbb{E}\bigg[\int_t^T f\big(s, X_s^{t,x,\mu}, \mathcal{L}^1(X_s^{t,\xi}), \hat{\alpha}(s, X_s^{t,x,\mu}, \mathcal{L}^1(X_s^{t,\xi}), Y_s^{t,x,\mu})\big)ds \\ + g\big(X_T^{t,x,\mu}, \mathcal{L}^1(X_T^{t,\xi})\big)\bigg], \tag{5.116}$$

for any $(t, x, \mu) \in [0, T] \times \mathbb{R}^d \times \mathcal{P}_2(\mathbb{R}^d)$, where $\xi \sim \mu$.

Notice that, if the coefficients of (5.111) are smooth enough, then Theorem 5.10 applies. To make this statement clear, we let:

Assumption (MFG Smooth Coefficients). The set of controls A is the entire $\mathbb{R}^k$. Moreover, there exist constants L and Γ such that:

(A1) The function $b_1 : [0, T] \times \mathbb{R}^d \times \mathcal{P}_2(\mathbb{R}^d) \ni (t, x, \mu) \mapsto b_1(t, x, \mu) \in \mathbb{R}^{d \times d}$ satisfies the same assumption as h in assumption **Smooth Coefficients Order 2** in Subsection 5.1.5 with respect to L and Γ, with $w = x \in \mathbb{R}^d$ and $l = d^2$. The function $\partial_x b_1 : [0, T] \times \mathbb{R}^d \times \mathcal{P}_2(\mathbb{R}^d) \to \mathbb{R}^{(d \times d) \times d}$ also satisfies the same assumption as h in assumption **Smooth Coefficients Order 2** with respect to L and Γ, but with $w = x \in \mathbb{R}^d$ and $l = d^3$.

(A2) For $h : [0, T] \times \mathbb{R}^q \times \mathcal{P}_2(\mathbb{R}^d) \ni (t, w, \mu) \mapsto h(t, w, \mu) \in \mathbb{R}$ being f or g, with $q = d + k$ and $w = (x, \alpha)$ when $h = f$ and with $q = d$ and $w = x$ when $h = g$, h being independent of t in the last case, the function h is differentiable with respect to w and $\partial_w h$ satisfies the

(continued)

same assumption as h in assumption **Smooth Coefficients Order 2** with respect to L and Γ, with $q = d + k$ and $l = d + k$ when $h = f$ and with $q = l = d$ when $h = g$. Moreover, for any $(t, w) \in [0, T] \times \mathbb{R}^q$, the mapping $\mathcal{P}_2(\mathbb{R}^d) \ni \mu \mapsto h(t, w, \mu)$ is fully $\mathcal{C}^2$, and the functions

$$[0, T] \times \mathbb{R}^q \times \mathcal{P}_2(\mathbb{R}^d) \times \mathbb{R}^d \ni (t, w, \mu, v)$$
$$\mapsto \frac{1}{1+|w|+M_1(\mu)}\big(\partial_\mu h(t, w, \mu)(v), \partial_v \partial_\mu h(t, w, \mu)(v)\big) \in \mathbb{R}^d \times \mathbb{R}^{d\times d},$$
$$[0, T] \times \mathbb{R}^q \times \mathcal{P}_2(\mathbb{R}^d) \times \mathbb{R}^d \times \mathbb{R}^d \ni (t, w, \mu, v, v')$$
$$\mapsto \frac{1}{1 + |w| + M_1(\mu)}\partial^2_\mu h(t, w, \mu)(v, v') \in \mathbb{R}^{d\times d},$$

are bounded by Γ and jointly continuous in (t, w, μ, v, v') and are Γ-Lipschitz continuous with respect to (w, μ, v) and to (w, μ, v, v').

Importantly, observe from (**A2**) and Schwarz' theorem that $\partial_\mu f$ is differentiable with respect to x and α and that its derivatives with respect to x and α coincide respectively with the transposes of the derivatives, with respect to μ, of $\partial_x f$ and $\partial_\alpha f$, see Theorem 5.41. Similarly, $\partial_\mu g$ is differentiable with respect to x and $\partial_x \partial_\mu g$ coincides with $[\partial_\mu \partial_x g]^\dagger$. Pay also attention that, under assumption **MFG Smooth Coefficients**, the functions f and g may not satisfy assumption **Smooth Coefficients Order 2**. Indeed, it is pretty clear that $\partial_\mu f$ and $\partial_\mu g$ are just assumed to be locally bounded and locally Lipschitz continuous. Similarly, $\partial_x f$, $\partial_\alpha f$ or $\partial_x g$ may not be bounded as they are just assumed to be of linear growth. Lastly, observe that assumptions **MFG Master Pontryagin** and **MFG Smooth Coefficients** subsume assumption **Coefficients Growth**.

The rationale for assumption **MFG Smooth Coefficients** is as follows. The first step of the analysis is to investigate the smoothness of the flow of the solutions of the systems (5.114)–(5.115) with respect to ξ, x, and μ by means of Theorem 5.10. The second step is to return to formula (5.116) in order to deduce the expected properties of $\mathcal{U}$.

We now claim:

Lemma 5.44 *Under assumptions* **MFG Master Pontryagin** *and* **MFG Smooth Coefficients**, *the function* $\hat{\alpha} : [0, T] \times \mathbb{R}^d \times \mathcal{P}_2(\mathbb{R}^d) \times \mathbb{R}^d \ni (t, x, \mu, y) \mapsto \hat{\alpha}(t, x, \mu, y)$ *satisfies the same assumption as* h *in assumption* **Smooth Coefficients Order 2**, *with* $w = (x, y) \in \mathbb{R}^d \times \mathbb{R}^d$ *with respect to constants* L' *and* Γ' *in lieu of* L *and* Γ *such that* L' *only depends on* λ *and* L, *and* Γ' *only depends on* λ, L, *and* Γ.

Proof. As already observed above Remark 5.43, the function $\hat{\alpha}$ is Lipschitz-continuous with respect to (x, μ, y), uniformly with respect to t.

Now, the proof of differentiability follows from a straightforward application of the implicit function theorem. Indeed, since $A = \mathbb{R}^k$, we know that, for any $(t, x, \mu, y) \in [0, T] \times \mathbb{R}^d \times \mathcal{P}_2(\mathbb{R}^d) \times \mathbb{R}^d$, $\hat{\alpha}(t, x, \mu, y)$ is the unique root of the equation:

$$\big(b_2(t)\big)^{\dagger} y + \partial_\alpha f\big(t, x, \mu, \hat{\alpha}(t, x, \mu, y)\big) = 0.$$

In particular, for $\xi, \chi \in L^2(\Omega^1, \mathcal{F}^1, \mathbb{P}^1; \mathbb{R}^d)$, we have, for any $s \in \mathbb{R}$,

$$\big(b_2(t)\big)^{\dagger} y + \partial_\alpha f\big(t, x, \mathcal{L}^1(\xi + s\chi), \hat{\alpha}(t, x, \mathcal{L}^1(\xi + s\chi), y)\big) = 0.$$

By strict convexity, the matrix $\partial^2_\alpha f(t, x, \mu, \hat{\alpha}(t, x, \mu, y))$ is invertible, uniformly in (t, x, μ, y). By the implicit function theorem, we deduce that the function $\mathbb{R} \ni s \mapsto \hat{\alpha}(t, x, \mathcal{L}^1(\xi + s\chi), y)$ is differentiable, with:

$$\begin{aligned}
&\mathbb{E}^1\big[\partial_\mu \partial_\alpha f\big(t, x, \mathcal{L}^1(\xi), \hat{\alpha}(t, \mathcal{L}^1(\xi), y)\big)(\xi)\chi\big] \\
&\quad + \partial^2_\alpha f\big(t, x, \mathcal{L}^1(\xi), \hat{\alpha}(t, x, \mathcal{L}^1(\xi), y)\big) \frac{d}{ds}_{|s=0}\big[\hat{\alpha}\big(t, x, \mathcal{L}^1(\xi + s\chi), y\big)\big] = 0,
\end{aligned}$$

where we recall the convention $\partial_\mu \partial_\alpha f = ([\partial_\mu \partial_{\alpha_i} f]_j)_{1 \le i,j \le d}$. We easily deduce that the function $L^2(\Omega^1, \mathcal{F}^1, \mathbb{P}^1; \mathbb{R}^d) \ni \xi \mapsto \hat{\alpha}(t, x, \mathcal{L}^1(\xi), y)$ is Gâteaux differentiable. By continuity of the Gâteaux derivative, differentiability holds in the Fréchet sense. In particular, the function $\mathcal{P}_2(\mathbb{R}^d) \ni \mu \mapsto \hat{\alpha}(t, x, \mu, y)$ is L-differentiable and, for all $\mu \in \mathcal{P}_2(\mathbb{R}^d)$, a version of $\partial_\mu \hat{\alpha}(t, x, \mu, y)(\cdot)$ is given by:

$$\partial_\mu \hat{\alpha}(t, x, \mu, y)(v) = \big[\partial^2_\alpha f\big(t, x, \mu, \hat{\alpha}(t, x, \mu, y)\big)\big]^{-1} \partial_\mu \partial_\alpha f\big(t, x, \mu, \hat{\alpha}(t, x, \mu, y)\big)(v), \quad v \in \mathbb{R}^d,$$

from which we deduce that $\partial_\mu \hat{\alpha}$ is bounded and is Lipchitz continuous with respect to (x, μ, y, v). Thanks to **(A2)** in assumption **MFG Smooth Coefficients**, $\partial_\alpha f$ satisfies assumption **Smooth Coefficients Order 2** and the above equality permits to differentiate once more with respect to μ and v and to prove that the derivatives are also bounded and Lipschitz continuous with respect to all the variables except time.

The differentiability of $\partial_\mu \hat{\alpha}$ with respect to x, y easily follows once we have established that $\hat{\alpha}$ is differentiable with respect to x and y. The proof of the differentiability in the directions x and y can be achieved by the same argument as above. □

Therefore, under assumptions **MFG Master Pontryagin** and **MFG Smooth Coefficients**, the function $[0, T] \times \mathbb{R}^d \times \mathcal{P}_2(\mathbb{R}^d) \times \mathbb{R}^d \times \mathbb{R}^k \ni (t, x, \mu, y, \alpha) \mapsto \partial_x H(t, x, \mu, \phi(y), \alpha) \in \mathbb{R}^d$ satisfies the same assumption as h in assumption **Smooth Coefficients Order 2** in Subsection 5.1.5, with $w = (x, y, \alpha) \in \mathbb{R}^d \times \mathbb{R}^d \times \mathbb{R}^k$ and $l = d$ and with respect to constants L' and Γ' respectively depending on L and ϕ, and L, Γ and ϕ. By composition, the function $[0, T] \times \mathbb{R}^d \times \mathcal{P}_2(\mathbb{R}^d) \times \mathbb{R}^d \ni (t, x, \mu, y) \mapsto \partial_x H(t, x, \mu, \phi(y), \hat{\alpha}(t, x, \mu, y))$ satisfies the same assumption as h in assumption **Smooth Coefficients Order 2** with $w = (x, y) \in \mathbb{R}^d \times \mathbb{R}^d$ and $l = d$ and with respect to constants L', Γ' respectively depending on L, λ and ϕ, and L, Γ, λ and ϕ. Hence, Theorem 5.10 says that, for $T \le c$, with $c > 0$ only depending on L, λ and ϕ, the decoupling field $\mathcal{V} = \partial_x \mathcal{U}$ of the system (5.111) belongs to the class

$\mathfrak{S}_d$ defined in Definition 5.9 with $m = d$. It satisfies the following master equation, which reads as a system of partial differential equations:

$$
\begin{aligned}
&\partial_t \partial_{x_i} \mathcal{U}(t,x,\mu) + b\Big(t,x,\mu,\hat{\alpha}\big(t,x,\mu,\partial_x \mathcal{U}(t,x,\mu)\big)\Big) \cdot \partial_x \partial_{x_i} \mathcal{U}(t,x,\mu) \\
&\quad + \int_{\mathbb{R}^d} b\Big(t,v,\mu,\hat{\alpha}\big(t,v,\mu,\partial_x \mathcal{U}(t,v,\mu)\big)\Big) \cdot \partial_\mu \partial_{x_i} \mathcal{U}(t,x,\mu)(v) d\mu(v) \\
&\quad + \frac{1}{2} \text{trace}\Big[\big(\sigma\sigma^\dagger + \sigma^0(\sigma^0)^\dagger\big) \partial^2_{xx} \partial_{x_i} \mathcal{U}(t,x,\mu)\Big] \\
&\quad + \frac{1}{2} \int_{\mathbb{R}^d} \text{trace}\Big[\big(\sigma\sigma^\dagger + \sigma^0(\sigma^0)^\dagger\big) \partial_v \partial_\mu \partial_{x_i} \mathcal{U}(t,x,\mu)(v)\Big] d\mu(v) \\
&\quad + \frac{1}{2} \int_{\mathbb{R}^{2d}} \text{trace}\Big[\sigma^0(\sigma^0)^\dagger \partial^2_\mu \partial_{x_i} \mathcal{U}(t,x,\mu)(v,v')\Big] d\mu(v) d\mu(v') \\
&\quad + \int_{\mathbb{R}^d} \text{trace}\Big[\sigma^0\big(\sigma^0\big)^\dagger \partial_x \partial_\mu \partial_{x_i} \mathcal{U}(t,x,\mu)(v)\Big] d\mu(v) \\
&\quad + \partial_{x_i} H\Big(t,x,\mu,\partial_x \mathcal{U}(t,x,\mu),\hat{\alpha}\big(t,x,\mu,\partial_x \mathcal{U}(t,x,\mu)\big)\Big) = 0,
\end{aligned}
\tag{5.117}
$$

for any coordinate $i \in \{1, \cdots, d\}$, with $\partial_x \mathcal{U}(T,x,\mu) = \partial_x g(x,\mu)$ as terminal condition. Above, $\partial_{x_i} \mathcal{U}$ denotes the ith component of the d-dimensional vector valued function $\partial_x \mathcal{U}$. Of course, the vector $\partial_x \mathcal{U}(t,x,\mu)$ may be replaced by $\phi(\partial_x \mathcal{U}(t,x,\mu))$ in the fourth argument of $\partial_{x_i} H$ since Theorem 4.10 ensures that $\partial_x \mathcal{U}(t,x,\mu)$ belongs to the subset of $\mathbb{R}^d$ on which ϕ coincides with the identity.

Importantly, we emphasize that, whenever the function ϕ is the identity, the length c on which Theorem 5.10 applies is determined by the sole parameters L and λ.

Smoothness of the Master Field and Derivation of the Master Equation

We shall deduce from the representation formula (5.116) that for any given $(t,x) \in [0,T] \times \mathbb{R}^d$, $\mathcal{U}(t,x,\cdot)$ is twice differentiable with respect to μ, provided that $T \leq c$, for the same c as above.

We start with the case when the functions $[0,T] \times \mathbb{R}^d \times \mathcal{P}_2(\mathbb{R}^d) \times \mathbb{R}^d \ni (t,x,\mu,y) \mapsto f(t,x,\mu,\hat{\alpha}(t,x,\mu,y))$ and $\mathbb{R}^d \times \mathcal{P}_2(\mathbb{R}^d) \ni (x,\mu) \mapsto g(x,\mu)$ satisfy the same assumption as h in assumption **Smooth Coefficients Order 2** with $q = 2d$ and $w = (x,y)$ when $h = f(\cdot,\cdot,\cdot,\hat{\alpha}(\cdot,\cdot,\cdot,\cdot))$ and with $q = d$ and $w = x$ when $h = g$. Then, there is no difficulty repeating the computations performed in Sections 5.2 and 5.3 and to prove that $\mathcal{U}$ satisfies the same assumption as h in assumption **Smooth Coefficients Order 2** with $q = d$ and $w = x$. Indeed, we already know from the proof of Theorem 5.10 that the processes appearing in the representation formula (5.116) are twice differentiable with respect to the arguments ξ, x, and μ; so, if f and g satisfy assumption **Smooth Coefficients Order 2**, we can easily derive that $\mathcal{U}$ has the aforementioned smoothness property. Appealing to the dynamic programming principle, see Theorem 4.5, we know that, for any $\epsilon > 0$ such that $t + \epsilon \leq T$,

$$\mathcal{U}(t,x,\mu) = \mathbb{E}\bigg[\int_t^{t+\epsilon} f\big(s, X_s^{t,x,\mu}, \mathcal{L}^1(X_s^{t,\xi}), \hat{\alpha}(s, X_s^{t,x,\mu}, \mathcal{L}^1(X_s^{t,\xi}), Y_s^{t,x,\mu})\big)ds$$
$$+\mathcal{U}\big(t+\epsilon, X_{t+\epsilon}^{t,x,\mu}, \mathcal{L}^1(X_{t+\epsilon}^{t,\xi})\big)\bigg].$$

Subtracting $\mathcal{U}(t+\epsilon,x,\mu)$ to both sides and using the smoothness of $\mathcal{U}$ in the directions x and μ, we easily deduce that $\mathcal{U}$ is continuous in time and then jointly continuous in all the arguments. Differentiating twice the above formula with respect to x and μ and subtracting the corresponding derivatives of $\mathcal{U}(t+\epsilon,x,\mu)$, we deduce in the same way that the directional first and second-order derivatives of $\mathcal{U}$ are jointly continuous, including the time variable. Proceeding as in the proof of Theorem 5.10, we deduce that the first and second order derivatives of $\mathcal{U}$ in x and μ satisfy the same continuity properties as in the statement of Theorem 5.10.

Now, expanding $\mathcal{U}$ in the second line of the above dynamic programming principle by means of Itô's formula, see Theorem 4.17, we deduce that $\mathcal{U}$ is right-differentiable in time and that the right-derivative satisfies:

$$\begin{aligned}
&\partial_t\mathcal{U}(t,x,\mu) + b\big(t,x,\mu,\hat{\alpha}(t,x,\mu,\partial_x\mathcal{U}(t,x,\mu))\big)\cdot\partial_x\mathcal{U}(t,x,\mu)\\
&+\int_{\mathbb{R}^d} b\big(t,v,\mu,\hat{\alpha}(t,v,\mu,\partial_x\mathcal{U}(t,v,\mu))\big)\cdot\partial_\mu\mathcal{U}(t,x,\mu)(v)d\mu(v)\\
&+\frac{1}{2}\mathrm{trace}\Big[\big(\sigma\sigma^\dagger+\sigma^0(\sigma^0)^\dagger\big)\partial^2_{xx}\mathcal{U}(t,x,\mu)\Big]\\
&+\frac{1}{2}\int_{\mathbb{R}^d}\mathrm{trace}\Big[\big(\sigma\sigma^\dagger+\sigma^0(\sigma^0)^\dagger\big)\partial_v\partial_\mu\mathcal{U}(t,x,\mu)(v)\Big]d\mu(v)\\
&+\frac{1}{2}\int_{\mathbb{R}^{2d}}\mathrm{trace}\Big[\sigma^0(\sigma^0)^\dagger\partial^2_\mu\mathcal{U}(s,x,\mu)(v,v')\Big]d\mu(v)d\mu(v')\\
&+\int_{\mathbb{R}^d}\mathrm{trace}\Big[\sigma^0\big(\sigma^0\big)^\dagger\partial_x\partial_\mu\mathcal{U}(t,x,\mu)(v)\Big]d\mu(v)\\
&+f\big(t,x,\mu,\hat{\alpha}(t,x,\mu,\partial_x\mathcal{U}(t,x,\mu))\big)=0.
\end{aligned}\tag{5.118}$$

Since all the terms except the first one are known to be jointly continuous in all the arguments, we deduce that $\mathcal{U}$ is time differentiable and that $\partial_t\mathcal{U}$ is jointly continuous. In particular, $\mathcal{U}$ is in the class $\mathfrak{S}_1$ and satisfies the above equation, which is the master equation for mean field games introduced in Chapter 4, see Section 4.4.

However, under assumption **MFG Smooth Coefficients**, the functions $[0,T]\times\mathbb{R}^d\times\mathcal{P}_2(\mathbb{R}^d)\times\mathbb{R}^d \ni (t,x,\mu,y)\mapsto f(t,x,\mu,\hat{\alpha}(t,x,\mu,y))$ and $\mathbb{R}^d\times\mathcal{P}_2(\mathbb{R}^d)\ni(x,\mu)\mapsto g(x,\mu)$ may not satisfy the same assumption as h in assumption **Smooth Coefficients Order 2**. Indeed, under assumption **MFG Smooth Coefficients**, only the coefficients $\partial_x f$ and $\partial_x g$ are assumed to satisfy assumption **Smooth Coefficients Order 2**; in particular, it might happen that neither f nor g are Lipschitz continuous in the variables x or α. However, we still have:

Theorem 5.45 *Under assumptions* **MFG Master Pontryagin** *and* **MFG Smooth Coefficients**, *there exists a constant c, only depending on L, λ, and ϕ, such that the following holds true for $T \leq c$. The master field $\mathcal{U}$ is continuous and differentiable with respect to t, x and μ. The partial derivative $\partial_t \mathcal{U}$ is continuous on $[0,T] \times \mathbb{R}^d \times \mathcal{P}_2(\mathbb{R}^d)$. The partial derivative $\partial_x \mathcal{U}$ belongs to the class $\mathfrak{S}_d$, as defined in Definition* (5.9), *with $m = d$ and with respect to constants L' and Γ' depending on L, λ, Γ, and ϕ. For any $(t,x) \in [0,T] \times \mathbb{R}^d$, the map $\mathcal{P}_2(\mathbb{R}^d) \ni \mu \mapsto \mathcal{U}(t,x,\mu)$ is fully $\mathcal{C}^2$, and the functions:*

$$
\begin{aligned}
&[0,T] \times \mathbb{R}^d \times \mathcal{P}_2(\mathbb{R}^d) \times \mathbb{R}^d \ni (t,x,\mu,v) \\
&\quad \mapsto \frac{1}{1+|x|+M_1(\mu)} \big(\partial_\mu \mathcal{U}(t,x,\mu)(v), \partial_v \partial_\mu \mathcal{U}(t,x,\mu)(v) \big) \in \mathbb{R}^d \times \mathbb{R}^{d \times d}, \\
&[0,T] \times \mathbb{R}^d \times \mathcal{P}_2(\mathbb{R}^d) \times \mathbb{R}^d \times \mathbb{R}^d \ni (t,x,\mu,v,v') \\
&\quad \mapsto \frac{1}{1+|x|+M_1(\mu)} \partial_\mu^2 \mathcal{U}(t,x,\mu)(v,v') \in \mathbb{R}^{d \times d},
\end{aligned}
$$

are bounded by Γ' and jointly continuous in (t,x,μ,v,v') and are Γ'-Lipschitz continuous with respect to (x,μ,v) and to (x,μ,v,v').

Moreover, $\mathcal{U}$ satisfies the master equation for mean field games (5.118) *and $\partial_x \mathcal{U}$ satisfies* (5.117).

Finally, the processes $\boldsymbol{\theta}^{t,\xi} = (\boldsymbol{X}^{t,\xi}, \boldsymbol{Y}^{t,\xi})$ and $\boldsymbol{\theta}^{t,x,\mu} = (\boldsymbol{X}^{t,x,\mu}, \boldsymbol{Y}^{t,x,\mu})$ appearing in the stochastic Pontryagin principle, see (5.114) *and* (5.115), *satisfy the conclusions of Subsections 5.2 and 5.3, see Lemmas 5.24, 5.25, 5.27, 5.35, and 5.38 and Propositions 5.31 and 5.32.*

Whenever ϕ is the identity, the constant c only depends on λ and L.

The statement of Theorem 5.45 gives a new insight into the assumption **MFG Smooth Coefficients**. Indeed, the statement shows that the master field $\mathcal{U}$ inherits the smoothness properties of the coefficients when chosen as in assumption **MFG Smooth Coefficients**, which sounds as a stability property. The fact that assumption **MFG Smooth Coefficients** maps coefficients of the master equation onto a solution satisfying the same properties will play a key role below when we iterate the small time result to construct a global solution of the master equation.

Proof. We first observe that the claim regarding the smoothness of $\partial_x \mathcal{U}$ is a direct consequence of Theorem 5.10. Also, the properties of the processes $\boldsymbol{\theta}^{t,\xi} = (\boldsymbol{X}^{t,\xi}, \boldsymbol{Y}^{t,\xi})$ and $\boldsymbol{\theta}^{t,x,\mu} = (\boldsymbol{X}^{t,x,\mu}, \boldsymbol{Y}^{t,x,\mu})$ follow from the proof of Theorem 5.10. Hence, it suffices to prove the other claims in the statement, the main point being to establish the smoothness of $\mathcal{U}$ in the direction μ.

First Step. In order to recover the same framework as in assumption **Smooth Coefficients Order 2**, we use a truncation argument. For a smooth function $\psi : \mathbb{R}^d \to \mathbb{R}^d$ with compact support included in the d-dimensional ball $B(0,2)$ of center 0 and of radius 2, such that $\psi(x)=x$ for all $x\in\mathbb{R}^d$ with $|x|\le 1$, we let $\psi_n(x)=n\psi(x/n)$ for any integer $n\ge 1$. Then, for $t\in[0,T]$, $x,y\in\mathbb{R}^d$ and $\mu\in\mathcal{P}_2(\mathbb{R}^d)$, we put:

$$F_n(t,(x,y),\mu)=f\big(t,\psi_n(x),\mu\circ\psi_n^{-1},\hat{\alpha}(t,\psi_n(x),\mu\circ\psi_n^{-1},\psi_n(y))\big),$$
$$G_n(x,\mu)=g\big(\psi_n(x),\mu\circ\psi_n^{-1}\big).$$

Following the proof of Lemma 4.15, see also Lemma (Vol I)-5.94, it is easily checked that F_n and G_n satisfy assumption **Smooth Coefficients Order 2** with respect to constants that may depend on λ, L, Γ, and n. We then let, for all $(t,x,\mu)\in[0,T]\times\mathbb{R}^d\times\mathcal{P}_2(\mathbb{R}^d)$,

$$\mathcal{U}_n(t,x,\mu)=\mathbb{E}\bigg[\int_t^T F_n\big(s,\theta_s^{t,x,\mu},\mathcal{L}^1(X_s^{t,\xi})\big)ds+G_n\big(X_T^{t,x,\mu},\mathcal{L}^1(X_T^{t,\xi})\big)\bigg],$$

where $\xi\sim\mu$ and $\boldsymbol{\theta}^{t,x,\mu}=(\boldsymbol{X}^{t,x,\mu},\boldsymbol{Y}^{t,x,\mu})$. Of course, $\mathcal{U}_n$ converges to $\mathcal{U}$ as n tends to ∞, the convergence being uniform on bounded subsets of $[0,T]\times\mathbb{R}^d\times\mathcal{P}_2(\mathbb{R}^d)$.

Moreover, $\mathcal{U}_n$ may be regarded as the decoupling field of the backward SDE:

$$\bar{Y}_s^{n,t,x,\mu}=\mathbb{E}_s\bigg[G_n\big(X_T^{t,x,\mu},\mathcal{L}^1(X_T^{t,\xi})\big)+\int_s^T F_n\big(r,\theta_r^{t,x,\mu},\mathcal{L}^1(X_r^{t,\xi})\big)dr\bigg], \tag{5.119}$$

for $s\in[t,T]$. Writing $(\theta_s^{t,x,\mu}=(X_s^{t,x,\mu},\partial_x\mathcal{U}(s,X_s^{t,x,\mu},\mathcal{L}^1(X_s^{t,\xi}))))_{t\le s\le T}$ and recalling that $\partial_x\mathcal{U}$ is in the class $\mathfrak{S}_d$ with $m=d$, we deduce that (5.119) together with the equation for $\boldsymbol{X}^{t,x,\mu}$ forms a McKean-Vlasov forward-backward system and that its decoupling field is exactly $\mathcal{U}_n$. Then, by Theorem 5.10, $\mathcal{U}_n$ is in the class $\mathfrak{S}_1$ and satisfies the corresponding master equation:

$$\begin{aligned}
&\partial_t\mathcal{U}_n(t,x,\mu)+B\big(t,(x,\partial_x\mathcal{U}(t,x,\mu)),\mu\big)\cdot\partial_x\mathcal{U}_n(t,x,\mu)\\
&+\int_{\mathbb{R}^d}B\big(t,(v,\partial_x\mathcal{U}(t,v,\mu)),\mu\big)\cdot\partial_\mu\mathcal{U}_n(t,x,\mu)(v)d\mu(v)\\
&+\frac{1}{2}\mathrm{trace}\Big[\big(\sigma\sigma^\dagger+\sigma^0(\sigma^0)^\dagger\big)\partial_{xx}^2\mathcal{U}_n(t,x,\mu)\Big]\\
&+\frac{1}{2}\int_{\mathbb{R}^d}\mathrm{trace}\Big[\big(\sigma\sigma^\dagger+\sigma^0(\sigma^0)^\dagger\big)\partial_v\partial_\mu\mathcal{U}_n(t,x,\mu)(v)\Big]d\mu(v)\\
&+\frac{1}{2}\int_{\mathbb{R}^{2d}}\mathrm{trace}\Big[\sigma^0(\sigma^0)^\dagger\partial_\mu^2\mathcal{U}_n(s,x,\mu)(v,v')\Big]d\mu(v)d\mu(v')\\
&+\int_{\mathbb{R}^d}\mathrm{trace}\Big[\sigma^0(\sigma^0)^\dagger\partial_x\partial_\mu\mathcal{U}_n(t,x,\mu)(v)\Big]d\mu(v)\\
&+F_n\big(t,(x,\partial_x\mathcal{U}(t,x,\mu)),\mu\big)=0,
\end{aligned} \tag{5.120}$$

with $\mathcal{U}_n(T,x,\mu)=G_n(x,\mu)$, and where:

$$B(t,(x,y),\mu)=b\big(t,x,\mu,\hat{\alpha}(t,x,\mu,y)\big).$$

Second Step. Now, by combining the smoothness of the coefficients with that of the flows $(x,\mu)\mapsto(\boldsymbol{X}^{t,x,\mu},\boldsymbol{Y}^{t,x,\mu})$ and $\xi\mapsto\boldsymbol{X}^{t,\xi}$, see Subsection 5.2.4, we get that, for all $\xi,\chi\in L^2(\Omega^1,\mathcal{F}_t^1,\mathbb{P}^1;\mathbb{R}^d)$,

$$
\begin{aligned}
\mathbb{E}[\partial_\mu \mathcal{U}_n(t,x,\mathcal{L}^1(\xi)) \cdot \chi] = \mathbb{E}\bigg[\int_t^T \Big(&\partial_w F_n\big(s,\theta_s^{t,x,\mu},\mathcal{L}^1(X_s^{t,\xi})\big) \cdot \partial_\chi \theta_s^{t,x,\xi} \\
&+ \tilde{\mathbb{E}}^1\big[\partial_\mu F_n\big(s,\theta_s^{t,x,\mu},\mathcal{L}^1(X_s^{t,\xi})\big)(\tilde{X}_s^{t,\xi}) \cdot \partial_\chi \tilde{X}_s^{t,\xi}\big]\Big) ds \\
&+ \partial_x G_n\big(X_T^{t,x,\mu},\mathcal{L}^1(X_T^{t,\xi})\big) \cdot \partial_\chi X_T^{t,x,\xi} \\
&+ \tilde{\mathbb{E}}^1\big[\partial_\mu G_n\big(X_T^{t,x,\mu},\mathcal{L}^1(X_T^{t,\xi})\big)(\tilde{X}_T^{t,\xi}) \cdot \partial_\chi \tilde{X}_T^{t,\xi}\big]\bigg],
\end{aligned}
\tag{5.121}
$$

where we used the letter w for the variable (x, y). Above, the exchange of the expectation and derivative symbols may be fully justified by the Lebesgue dominated convergence theorem.

Making use of assumption **MFG Smooth Coefficients** and following the computation in Lemma 4.15, see also Lemma (Vol I)-5.94, for the derivative of the mollified coefficients, it is easy to check that the functions:

$$
\begin{aligned}
&[0,T] \times \mathbb{R}^{2d} \times \mathcal{P}_2(\mathbb{R}^d) \ni (t,w,\mu) \mapsto \frac{1}{1+|w|+M_1(\mu)} \partial_w F_n(t,w,\mu), \\
&[0,T] \times \mathbb{R}^{2d} \times \mathcal{P}_2(\mathbb{R}^d) \times \mathbb{R}^d \ni (t,w,\mu,v) \mapsto \frac{1}{1+|w|+M_1(\mu)} \partial_\mu F_n(t,w,\mu)(v),
\end{aligned}
$$

are bounded, uniformly in $n \geq 1$, and are also Lipschitz continuous in (w, μ), uniformly in $n \geq 1$ and $t \in [0, T]$. Therefore, for any $s \in [t, T]$, the quantities:

$$
\frac{\partial_w F_n\big(s,\theta_s^{t,x,\mu},\mathcal{L}^1(X_s^{t,\xi})\big) \cdot \partial_\chi \theta_s^{t,x,\xi}}{1+|\theta_s^{t,x,\mu}| + \mathbb{E}^1[|X_s^{t,\xi}|]}, \tag{5.122}
$$

and

$$
\frac{\tilde{\mathbb{E}}^1\big[\partial_\mu F_n\big(s,\theta_s^{t,x,\mu},\mathcal{L}^1(X_s^{t,\xi})\big)(\tilde{X}_s^{t,\xi}) \cdot \partial_\chi \tilde{X}_s^{t,\xi}\big]}{1+|\theta_s^{t,x,\mu}| + \mathbb{E}^1[|X_s^{t,\xi}|]}, \tag{5.123}
$$

can be handled as the terms

$$
\partial_w F_n\big(s,\theta_s^{t,x,\mu},\mathcal{L}^1(X_s^{t,\xi})\big) \cdot \partial_\chi \theta_s^{t,x,\xi}
$$

and

$$
\tilde{\mathbb{E}}^1\big[\partial_\mu F_n\big(s,\theta_s^{t,x,\mu},\mathcal{L}^1(X_s^{t,\xi})\big)(\tilde{X}_s^{t,\xi}) \cdot \partial_\chi \tilde{X}_s^{t,\xi}\big]
$$

would have been handled if assumption **Smooth Coefficients Order 2** had been in force. In particular, by Lemmas 5.24, 5.25, and 5.26 and by Proposition 5.13, both (5.122) and (5.123) may be estimated in $L^\infty(\Omega, \mathcal{F}, \mathbb{P}; \mathbb{R}^{2d})$ and $L^\infty(\Omega, \mathcal{F}, \mathbb{P}; \mathbb{R}^d)$ respectively. The bounds are similar to that obtained for $\partial_\chi \boldsymbol{\theta}^{t,x,\xi}$ in the first line of (5.62). Similarly, increments of quantities of the form (5.122) or (5.123), when taken at different triples (x, μ, ξ) and (x', μ', ξ'), satisfy similar bounds to those obtained for $\partial_\chi \boldsymbol{\theta}^{t,x,\xi}$ and $\partial_\chi \boldsymbol{\theta}^{t,x,\xi'}$ in the second line of (5.62) and for $\partial_\chi \boldsymbol{\theta}^{t,x,\xi} - \partial_\chi \boldsymbol{\theta}^{t,x',\xi}$ in the end of the proof of Lemma 5.28.

Based on the above observation, we divide both sides in (5.121) by $1 + |x| + M_1(\mu)$. In lieu of the two first terms in the right-hand side of (5.121), this prompts us to consider:

$$\begin{aligned}&\frac{\partial_w F_n\big(s, \theta_s^{t,x,\mu}, \mathcal{L}^1(X_s^{t,\xi})\big) \cdot \partial_\chi \theta_s^{t,x,\xi}}{1 + |x| + M_1(\mu)}\\&\quad= \frac{1 + |\theta_s^{t,x,\mu}| + \mathbb{E}^1[|X_s^{t,\xi}|]}{1 + |x| + M_1(\mu)} \times \frac{\partial_w F_n\big(s, \theta_s^{t,x,\mu}, \mathcal{L}^1(X_s^{t,\xi})\big) \cdot \partial_\chi \theta_s^{t,x,\xi}}{1 + |\theta_s^{t,x,\mu}| + \mathbb{E}^1[|X_s^{t,\xi}|]},\end{aligned} \tag{5.124}$$

together with:

$$\begin{aligned}&\frac{\tilde{\mathbb{E}}^1\big[\partial_\mu F_n\big(s, \theta_s^{t,x,\mu}, \mathcal{L}^1(X_s^{t,\xi})\big)(\tilde{X}_s^{t,\xi}) \cdot \partial_\chi \tilde{X}_s^{t,\xi}\big]}{1 + |x| + M_1(\mu)}\\&\quad= \frac{1 + |\theta_s^{t,x,\mu}| + \mathbb{E}^1[|X_s^{t,\xi}|]}{1 + |x| + M_1(\mu)} \times \frac{\tilde{\mathbb{E}}^1\big[\partial_\mu F_n\big(s, \theta_s^{t,x,\mu}, \mathcal{L}^1(X_s^{t,\xi})\big)(\tilde{X}_s^{t,\xi}) \cdot \partial_\chi \tilde{X}_s^{t,\xi}\big]}{1 + |\theta_s^{t,x,\mu}| + \mathbb{E}^1[|X_s^{t,\xi}|]}.\end{aligned} \tag{5.125}$$

Of course, we can perform a similar analysis for $\partial_x G_n$ and $\partial_\mu G_n$.

Third Step. We now focus on the first factor in (5.124) and (5.125). By Lemma 5.12, we can find a constant C such that:

$$\mathbb{E}\Big[\frac{1 + |\theta_s^{t,x,\mu}| + \mathbb{E}^1[|X_s^{t,\xi}|]}{1 + |x| + M_1(\mu)}\Big] \le C, \tag{5.126}$$

for $x \in \mathbb{R}^d$, $\mu \in \mathcal{P}_2(\mathbb{R}^d)$ and $\xi \in L^2(\Omega^1, \mathcal{F}_t^1, \mathbb{P}^1; \mathbb{R}^d)$ with $\xi \sim \mu$. Owing to Proposition 5.13, we deduce that:

$$\begin{aligned}&\mathbb{E}\Big[\Big|\frac{1 + |\theta_s^{t,x,\mu}| + \mathbb{E}^1[|X_s^{t,\xi}|]}{1 + |x| + M_1(\mu)} - \frac{1 + |\theta_s^{t,x',\mu'}| + \mathbb{E}^1[|X_s^{t,\xi'}|]}{1 + |x'| + M_1(\mu')}\Big|\Big]\\&\quad\le C\big(|x - x'| + W_1(\mu, \mu')\big),\end{aligned} \tag{5.127}$$

for $x, x' \in \mathbb{R}^d$, $\mu, \mu' \in \mathcal{P}_2(\mathbb{R}^d)$ and $\xi, \xi' \in L^2(\Omega^1, \mathcal{F}_t^1, \mathbb{P}^1; \mathbb{R}^d)$ with $\xi \sim \mu$ and $\xi' \sim \mu'$, and for possibly new value of the constant C.

Returning to (5.121) and recalling the writings (5.124) and (5.125) together with the conclusion of the second step, we conclude that, in comparison to what happens when assumption **Smooth Coefficients Order 2** is in force, similar bounds for $\partial_\mu \mathcal{U}_n(t, x, \mu)(v)$ hold true with additional pre-factor $(1 + |x| + M_1(\mu))^{-1}$. Put it differently, there exists a constant C such that, for all $n \ge 1$, the function:

$$[0, T] \times \mathbb{R}^d \times \mathcal{P}_2(\mathbb{R}^d) \times \mathbb{R}^d \ni (t, x, \mu, v) \mapsto \frac{\partial_\mu \mathcal{U}_n(t, x, \mu)(v)}{1 + |x| + M_1(\mu)}$$

satisfies, for all $t \in [0, T]$, $x, x', v, v' \in \mathbb{R}^d$ and $\mu, \mu' \in \mathcal{P}_2(\mathbb{R}^d)$,

$$\begin{aligned}&\frac{1}{1 + |x| + M_1(\mu)} \big|\partial_\mu \mathcal{U}_n(t, x, \mu)(v)\big| \le C,\\&\Big|\frac{1}{1 + |x| + M_1(\mu)} \partial_\mu \mathcal{U}_n(t, x, \mu)(v) - \frac{1}{1 + |x'| + M_1(\mu')} \partial_\mu \mathcal{U}_n(t, x', \mu')(v')\Big|\\&\quad\le C\big(|x - x'| + |v - v'| + W_1(\mu, \mu')\big).\end{aligned} \tag{5.128}$$

As a consequence, we deduce that, for any $t \in [0,T]$, the functions $(\partial_\mu \mathcal{U}_n(t,\cdot,\cdot)(\cdot))_{n\geq 1}$ extend by continuity to $\mathbb{R}^d \times \mathcal{P}_1(\mathbb{R}^d) \times \mathbb{R}^d$ and that they are equicontinuous on bounded subsets of $\mathbb{R}^d \times \mathcal{P}_2(\mathbb{R}^d) \times \mathbb{R}^d$. Since any bounded subset of $\mathbb{R}^d \times \mathcal{P}_2(\mathbb{R}^d) \times \mathbb{R}^d$ is relatively compact in $\mathbb{R}^d \times \mathcal{P}_1(\mathbb{R}^d) \times \mathbb{R}^d$, we may extract a subsequence that converges uniformly on bounded subsets $\mathbb{R}^d \times \mathcal{P}_2(\mathbb{R}^d) \times \mathbb{R}^d$. Of course, this says that $\mathcal{U}$ is L-differentiable and, for any $(t,x,\mu) \in [0,T] \times \mathbb{R}^d \times \mathcal{P}_2(\mathbb{R}^d)$, the limit of the subsequence of $(\partial_\mu \mathcal{U}_n(t,x,\mu)(\cdot))_{n\geq 1}$ provides a version of $\partial_\mu \mathcal{U}(t,x,\mu)(\cdot)$. In particular, $\partial_\mu \mathcal{U}(t,\cdot,\cdot)(\cdot)$ satisfies (5.128). Passing to the limit in (5.121), we get the following representation formula for $\partial_\mu \mathcal{U}$:

$$
\begin{aligned}
\mathbb{E}\big[\partial_\mu \mathcal{U}(t,x,\mathcal{L}^1(\xi)) \cdot \chi\big] = \mathbb{E}\bigg[\int_t^T & \Big(\partial_w F\big(s,\theta_s^{t,x,\mu},\mathcal{L}^1(X_s^{t,\xi})\big) \cdot \partial_\chi \theta_s^{t,x,\xi} \\
& + \tilde{\mathbb{E}}^1\big[\partial_\mu F\big(s,\theta_s^{t,x,\mu},\mathcal{L}^1(X_s^{t,\xi})\big)(\tilde{X}_s^{t,\xi}) \cdot \partial_\chi \tilde{X}_s^{t,\xi}\big]\Big) ds \\
& + \partial_x G\big(X_T^{t,x,\mu},\mathcal{L}^1(X_T^{t,\xi})\big) \cdot \partial_\chi X_T^{t,x,\xi} \\
& + \tilde{\mathbb{E}}^1\big[\partial_\mu G\big(X_T^{t,x,\mu},\mathcal{L}^1(X_T^{t,\xi})\big)(\tilde{X}_T^{t,\xi}) \cdot \partial_\chi \tilde{X}_T^{t,\xi}\big]\bigg],
\end{aligned}
\tag{5.129}
$$

where

$$
F(t,(x,y),\mu) = f\big(t,x,\mu,\hat{\alpha}(t,x,\mu,y)\big), \quad G(x,\mu) = g(x,\mu).
$$

By writing (5.129) from t to $t+\epsilon$ with G being replaced by $\mathcal{U}(t+\epsilon,\cdot,\cdot)$ and by using (5.128) and the Lipschitz property of $\partial_x \mathcal{U}(t+\epsilon,\cdot,\cdot)$, we can prove that the right-hand side is continuous in time. We prove that $\partial_\mu \mathcal{U}$ is continuous in time by using the same compactness argument as in the proof of Theorem 5.29.

Fourth Step. Fortunately, we may proceed in the same way for the second-order derivatives by differentiating once again (5.121) with respect to ξ in a direction $\zeta \in L^\infty(\Omega^1,\mathcal{F}_t^1,\mathbb{P};\mathbb{R}^d)$. Recall indeed from Propositions 5.32 and 5.33 and from Lemma 5.35 and Proposition 5.39 that the regularity of $\partial_v \partial_\mu \mathcal{U}_n$ and of $\partial_\mu^2 \mathcal{U}_n$ may be investigated through the analysis of $\partial^2_{\zeta,\chi} \boldsymbol{\theta}_t^{t,x,\xi}$ and $\partial^2_{\zeta,\chi} \boldsymbol{\theta}_t^{t,\xi}$ for suitable choices of ζ and χ: $\partial_v \partial_\mu \mathcal{U}_n$ may be represented by choosing $(\varepsilon\chi,\varepsilon\zeta)$ in lieu of (χ,ζ) where ε is independent of (ξ,χ,ζ) and $\zeta = e$ is a constant unit vector, see (5.82), while $\partial_\mu^2 \mathcal{U}_n$ may be represented by choosing $(\varepsilon\chi,(1-\varepsilon)\zeta)$ in lieu of (χ,ζ) where ε is a Bernoulli random variable with parameter $1/2$ independent of (ξ,χ,ζ), see (5.99). And, as above, recall that the estimates we have for $\partial^2_{\zeta,\chi} \boldsymbol{\theta}_t^{t,x,\xi}$ hold true in $L^\infty(\Omega,\mathcal{F},\mathbb{P};\mathbb{R}^d)$ when χ and ζ are chosen as we just explained and $\|\chi\|_\infty$ or $\|\zeta\|_\infty$ is bounded by 1.

Proceeding as above, we complete the analysis of the second-order derivatives.

It remains to prove that $\mathcal{U}$ satisfies the master equation. To do so, it suffices to recall that, for each $n \geq 1$, $\mathcal{U}_n$ is differentiable in time and satisfies the master equation (5.120). Isolating the term $\partial_t \mathcal{U}_n$ and passing to the limit in the remaining ones, we deduce that $\mathcal{U}$ is differentiable in time and satisfies the master equation for mean field games. □

5.4.2 Mean Field Games Over Time Intervals of Arbitrary Lengths

The purpose of this section is to extend the short time solvability result proven above to an interval $[0,T]$ of arbitrary length $T > 0$. As for the stability properties investigated in Section 1.3, see also Chapter (Vol I)-4, the strategy consists in applying iteratively the short time result. As usual with forward-backward systems, the main issue is to bound from below the length of the successive intervals on which the short time result is applied. As already noticed in Section 1.3.1, see also Subsection (Vol I)-4.1.2, the key point is indeed to guarantee that the union of all these small intervals covers the time interval $[0, T]$. Basically, this requires to bound from above the Lipschitz constant of the underlying master field $\partial_x \mathcal{U}$ of (5.114)–(5.115), in both directions x and μ.

The Lipschitz property in the direction x may be proved by means of standard arguments:

1. In case when $\partial_x f$ and $\partial_x g$ are bounded and σ is invertible, it follows somehow from the smoothing properties of the heat kernel, as we already alluded to in Chapter (Vol I)-4, see for instance Theorem (Vol I)-4.12, and in Chapter 1, see Theorem 1.57. See also the proof of Proposition 5.53 below.
2. When σ is not invertible, the Lipschitz property still holds true whenever f and g are convex in the direction x. We refer for instance to Chapter 1, see Theorem 1.60.

It is worth noting that this distinction between the invertible and convex cases is reminiscent of our discussion in Chapters (Vol I)-3 and 1 on the two possible probabilistic approaches for handling stochastic optimization problems.

Actually, the real challenge is to prove that the decoupling field is Lipschitz continuous in the measure argument. In short, we need an *a priori* estimate on the derivative of the master field in the direction μ. Although the idea is quite simple, the possible implementation raises several questions. First, the reader must remember that the derivative in the variable μ reads as a function and, as a result, is of infinite dimension. As a consequence, the choice of the norm used to estimate the derivative really matters: This is a first difficulty. Indeed, the estimate we need on the derivative of the master field must fit the framework used in the short time analysis. Since the small time condition in Theorem 5.45 explicitly depends upon the *supremum norm* of the μ-derivative of the terminal condition $\partial_x g$, the μ-derivative of the decoupling field must be estimated in *supremum norm* as well. This is much demanding: Regarding the tools we have developed so far, see especially Chapter (Vol I)-5, it would be certainly easier to estimate the derivative in L^2 instead of L^∞. Anyhow, as we already mentioned, this would require an analogue of the short time result with just an L^2-bound –instead of an L^∞-bound– on the μ-derivative of $\partial_x g$. We refer to the Notes & Complements at the end of the chapter for references where this program is carried out.

Another difficulty is to identify structural conditions under which we can bound the μ-derivative of the master field. Clearly, it seems hopeless to adapt the first of the two strategies we recalled above for bounding the x-derivative: Except for very few specific cases like those discussed in Section 3.5.2, we are not aware of general cases where the finite dimensional common noise W^0 permits to mollify the master field in the direction μ. In other words, the only conceivable strategy for estimating the μ-derivative of $\partial_x \mu$ is to require suitable monotonicity conditions. Throughout the section, we thus require that the coefficients satisfy the Lasry-Lions monotonicity condition, as recalled in the statement of Proposition 3.34, see also assumption **MFG Master Classical** below.

5.4.3 Main Statement

Here is now the main statement regarding the existence of a classical solution to the master equation for mean field games.

We shall need three types of assumptions: First, we need assumptions **MFG Master Pontryagin** and **MFG Smooth Coefficients**, see Subsection 5.4.1, to be in force in order to apply the short time result; second, we need one of the two assumptions **MFG with a Common Noise HJB** or **MFG with a Common Noise SMP Relaxed**, see Subsections 3.4.1 and 3.4.3, to be in force in order to guarantee the existence of equilibria over time intervals of any length and to bound the derivative of $\partial_x \mathcal{U}$ in the direction x; third, we require assumption **Lasry-Lions Monotonicity**, as stated in Proposition 3.34, see also Subsection (Vol I)-3.4.1, in order to guarantee the uniqueness of the equilibria and to bound the derivative of $\partial_x \mathcal{U}$ in the direction μ.

This prompts us to let:

Assumption (MFG Master Classical). The set of controls A is the entire $\mathbb{R}^k$. Moreover, there exist three constants $L, \Gamma \geq 0$ and $\lambda > 0$ such that:

(A1) The coefficients σ and σ^0 are constant; moreover, the coefficients f and g are continuous with respect to all the variables, the space $\mathcal{P}_2(\mathbb{R}^d)$ being equipped with the 2-Wasserstein distance W_2.

(A2) The drift b is an affine function of α in the sense that it is of the form $b(t, x, \mu, \alpha) = b(t)\alpha$, where the function $[0, T] \ni t \mapsto b(t) \in \mathbb{R}^{d \times q}$ is continuous and bounded by L. Moreover, the running cost f has a separated structure of the form:

$$f(t, x, \mu, \alpha) = f_0(t, x, \mu) + f_1(t, x, \alpha),$$

for $t \in [0, T]$, $x \in \mathbb{R}^d$, $\alpha \in \mathbb{R}^k$ and $\mu \in \mathcal{P}_2(\mathbb{R}^d)$, where f_0 is a continuous function from $[0, T] \times \mathbb{R}^d \times \mathcal{P}_2(\mathbb{R}^d)$ into $\mathbb{R}$ and f_1 is a continuous function from $[0, T] \times \mathbb{R}^d \times A$ into $\mathbb{R}$.

(A3) One of the two assumptions **MFG with a Common Noise HJB** or **MFG with a Common Noise SMP Relaxed** is in force, with respect to the constants Γ and λ.

(A4) Assumption **MFG Smooth Coefficients** holds true.

(A5) The functions $f_0(t, \cdot, \cdot)$ for $t \in [0, T]$, and g are monotone in the following sense:

$$\int_{\mathbb{R}^d} [h(x, \mu) - h(x, \mu')] \, d(\mu - \mu')(x) \geq 0,$$

for all $\mu, \mu' \in \mathcal{P}_2(\mathbb{R}^d)$ and for $h = f_0(t, \cdot, \cdot)$ or $h = g$.

Observe that the Hamiltonian has a separated structure under assumption **MFG Master Classical**. In words, the dependence of the Hamiltonian upon the measure argument can be isolated in a separate function not depending upon the control variable. In particular, the optimizer $\hat{\alpha}$ of the Hamiltonian is a function of $(t, x, y) \in [0, T] \times \mathbb{R}^d \times \mathbb{R}^d$ and does not depend on the measure argument μ.

We let the reader check that assumption **MFG Master Classical** subsumes assumption **MFG Master Pontryagin**. See for instance the proofs of Theorems 4.21 and 4.23, where we show that **(A3)** implies assumptions **FBSDE** and **Decoupling Master** stated in Subsections 4.1.3 and 4.2.2. Importantly, ϕ may be chosen as the identity matrix in assumption **MFG Master Pontryagin**; in that case, the constant c in the statement of Theorem 5.45 only depends on λ and L.

We now claim:

Theorem 5.46 *Let assumption* **MFG Master Classical** *be in force. Then, the following holds true.*

The master field $\mathcal{U}$ is continuous and differentiable with respect to t, x and μ. The partial derivative $\partial_t \mathcal{U}$ is continuous on $[0, T] \times \mathbb{R}^d \times \mathcal{P}_2(\mathbb{R}^d)$. The partial derivative $\partial_x \mathcal{U}$ belongs to the class $\mathfrak{S}_d$, as defined in Definition 5.9, with $m = d$ and with respect to constants L' and Γ' depending on L, λ, T and Γ. For any $(t, x) \in [0, T] \times \mathbb{R}^d$, the map $\mathcal{P}_2(\mathbb{R}^d) \ni \mu \mapsto \mathcal{U}(t, x, \mu)$ is fully C^2, and the functions:

$$
\begin{aligned}
&[0, T] \times \mathbb{R}^d \times \mathcal{P}_2(\mathbb{R}^d) \times \mathbb{R}^d \ni (t, x, \mu, v) \\
&\quad \mapsto \frac{1}{1 + |x| + M_1(\mu)} \big(\partial_\mu \mathcal{U}(t, x, \mu)(v), \partial_v \partial_\mu \mathcal{U}(t, x, \mu)(v) \big) \in \mathbb{R}^d \times \mathbb{R}^{d \times d}, \\
&[0, T] \times \mathbb{R}^d \times \mathcal{P}_2(\mathbb{R}^d) \times \mathbb{R}^d \times \mathbb{R}^d \ni (t, x, \mu, v, v') \\
&\quad \mapsto \frac{1}{1 + |x| + M_1(\mu)} \partial_\mu^2 \mathcal{U}(t, x, \mu)(v, v') \in \mathbb{R}^{d \times d},
\end{aligned}
$$

are bounded by Γ' and jointly continuous in (t, x, μ, v, v') and are Γ'-Lipschitz continuous with respect to (x, μ, v) and to (x, μ, v, v').

Moreover, $\mathcal{U}$ satisfies the master equation for mean field games (5.118) *and $\partial_x \mathcal{U}$ satisfies* (5.117).

Remark 5.47 *Following the statement and the proof of Theorem 5.11, there is one and only function $\mathcal{U}$ such that $\mathcal{U}$ and $\partial_x \mathcal{U}$ satisfy the conclusion of Theorem 5.46.*

Remark 5.48 *The assumption $A = \mathbb{R}^k$ is just used to guarantee that the minimizer $\hat{\alpha}$ is twice continuously differentiable, as stated in Lemma 5.44. Due to boundary effects, this may not be the case when A is strictly included in $\mathbb{R}^k$. Think for instance of the following example:*

$$\hat{\alpha}(y) = \inf_{\alpha \in [-1,1]} \left[\alpha y + \frac{1}{2}\alpha^2 \right] = \begin{cases} -\frac{1}{2}y^2, & \text{if } |y| \leq 1, \\ \frac{1}{2} - y, & \text{if } y > 1, \\ \frac{1}{2} + y, & \text{if } y < -1, \end{cases}$$

in which case the minimizer is not twice differentiable at $y = \pm 1$. *This is what we call a boundary effect.*

Importantly, we observe that the statement of Theorem 5.46 can be strengthened whenever assumption **MFG with a Common Noise HJB** is in force, see (**A3**) in assumption **MFG Master Classical**. To make this clear, we introduce the following stronger version of assumption **MFG Master Classical**:

Assumption (MFG Master Classical HJB). Assumption **MFG Master Classical** is in force with the following requirements:

(**A1**) In condition (**A3**) of assumption **MFG Master Classical**, assumption **MFG with a Common Noise HJB** is in force.

(**A2**) In condition (**A4**) of assumption **MFG Master Classical**, or equivalently in assumption **MFG Smooth Coefficients**, the functions:

$$\mathbb{R}^d \times \mathcal{P}_2(\mathbb{R}^d) \times \mathbb{R}^d \ni (t, x, \mu, v) \mapsto (\partial_\mu f_0(t, x, \mu)(v), \partial_v \partial_\mu f_0(t, x, \mu)(v)),$$
$$\mathbb{R}^d \times \mathcal{P}_2(\mathbb{R}^d) \times \mathbb{R}^d \times \mathbb{R}^d \ni (t, x, \mu, v, v') \mapsto \partial^2_\mu f_0(t, w, \mu)(v, v'),$$

are bounded by Γ and are Γ-Lipschitz continuous with respect to (w, μ, v) and to (w, μ, v, v'), for any $t \in [0, T]$, and similarly with g in lieu of $f_0(t, \cdot, \cdot)$.

We then have the following stronger version of Theorem 5.46:

Theorem 5.49 *Under assumption* **MFG Master Classical HJB**, *the conclusion of Theorem 5.46 holds true. In addition,* $\mathcal{U}$ *is bounded and, for a possibly new value of the constant* Γ' *used in the statement of Theorem 5.46, the functions:*

$$\mathbb{R}^d \times \mathcal{P}_2(\mathbb{R}^d) \times \mathbb{R}^d \ni (t, x, \mu, v) \mapsto \big(\partial_\mu \mathcal{U}(t, x, \mu)(v), \partial_v \partial_\mu \mathcal{U}(t, x, \mu)(v)\big),$$
$$\mathbb{R}^d \times \mathcal{P}_2(\mathbb{R}^d) \times \mathbb{R}^d \times \mathbb{R}^d \ni (t, x, \mu, v, v') \mapsto \partial^2_\mu \mathcal{U}(t, x, \mu)(v, v'),$$

are bounded by Γ' *and are* Γ'*-Lipschitz continuous with respect to* (x, μ, v) *and to* (x, μ, v, v'), *for any* $t \in [0, T]$. *In particular,* $\mathcal{U}$ *belongs to* $\mathfrak{S}_1$, *see Definition 5.9.*

5.4.4 Proof of the Main Statement

The proof of Theorems 5.46 and 5.49 is divided in several steps.

General Strategy

Generally speaking, the construction of a classical solution to the master equation is based upon an induction argument, which consists in applying iteratively the local existence Theorem 5.10. In order to do so, we shall identify general sufficient conditions under which the length of the interval on which local existence holds true is uniformly bounded from below along the induction. This strategy is similar to that explained in Subsection 1.3.3 for solving classical forward-backward systems and in Subsection 1.3.3 for solving forward-backward systems in a random environment.

By (**A3**) in assumption **MFG Master Classical**, we know from Theorems 4.21 and 4.23 that, for any initial condition, the mean field game has a solution. By Proposition 3.34, this solution is unique and is strong. Therefore, by Definition 4.1, the master field $\mathcal{U}$ is well defined. Since assumption **MFG Master Classical** implies assumption **MFG Master Pontryagin** stated in Subsection 5.4.1, the same argument as that explained in Subsection 5.4.1 shows that, for a given initial condition of the mean field game, the unique solution must generate a solution of the Pontryagin adjoint system (5.114). The goal is thus to prove that, under the standing assumption, the two systems (5.114) and (5.115) are uniquely solvable and have a smooth decoupling field. If true, Theorem 4.10 permits to identify the decoupling field with $\partial_x \mathcal{U}$, which allows to restart the analysis from (5.116). So, the crux of the proof is to show by induction that $\partial_x \mathcal{U}$, which is known to exist from Theorem 4.10, is Lipschitz continuous in the variables x and μ, uniformly in time. Indeed, by the same backward induction as in Subsection 1.3.3, this is known to suffice for proving that (5.114) and (5.115) are uniquely solvable.

Throughout the proof, we shall use the following extension of Definition 5.9: For a real $S \in [0, T]$, we denote by $\mathfrak{S}_m([S, T])$ the space of functions $V : [S, T] \times \mathbb{R}^d \times \mathcal{P}_2(\mathbb{R}^d) \ni (t, x, \mu) \mapsto V(t, x, \mu) \in \mathbb{R}^m$ for which we can find a constant $C \geq 0$ such that the function $[0, T] \times \mathbb{R}^d \times \mathcal{P}_2(\mathbb{R}^d) \ni (t, x, \mu) \mapsto V(\max(t, S), x, \mu) \in \mathbb{R}^m$ belongs to $\mathfrak{S}_m$. Equivalently, V belongs to $\mathfrak{S}_m([S, T])$ if it satisfies Definition 5.9 but on $[S, T] \times \mathbb{R}^d \times \mathcal{P}_2(\mathbb{R}^d)$ in lieu of $[0, T] \times \mathbb{R}^d \times \mathcal{P}_2(\mathbb{R}^d)$.

Basically, our goal is to prove by backward induction that $\partial_x \mathcal{U}$ belongs to $\mathfrak{S}_d([S, T])$, for any $S \in [0, T]$. The short time analysis performed in the previous paragraph says that this is indeed true when S is close enough to T. The strategy is thus to decrease the value of S step by step. This prompts us to let:

Hypothesis ($\mathscr{H}(S)$). The conclusion of Theorem 5.46 holds true on $[S, T] \times \mathbb{R}^d \times \mathcal{P}_2(\mathbb{R}^d)$ with respect to some constants L' and Γ'. In particular, the function $\partial_x \mathcal{U}$ belongs to $\mathfrak{S}_d([S, T])$ and satisfies the master equation (5.117) on $[S, T] \times \mathbb{R}^d \times \mathcal{P}_2(\mathbb{R}^d)$; also, the function $\mathcal{U}$ satisfies the master equation for mean field games (5.13).

As already emphasized, we know from Theorem 5.45 that, under assumption **MFG Master Classical**, there exists a real $S \in [0, T)$ such that hypothesis $\mathscr{H}(S)$ holds true.

Within this framework, we have the following lemma:

Lemma 5.50 *In addition to assumption* **MFG Master Classical**, *assume further that hypothesis* $\mathscr{H}(S)$ *holds true for some* $S \in [0, T]$. *Then, for any* $(t, \xi) \in [S, T] \times L^2(\Omega^1, \mathcal{F}_t^1, \mathbb{P}^1; \mathbb{R}^d)$, *the forward-backward system* (5.114) *has a unique solution. Moreover, for any* $(t, x, \mu) \in [S, T] \times \mathbb{R}^d \times \mathcal{P}_2(\mathbb{R}^d)$, *the forward-backward system* (5.115) *also has a unique solution.*

The function $\partial_x \mathcal{U}$ *is the master field of the pair* (5.114)–(5.115). *In particular,* $(X_s^{t,\xi})_{t \le s \le T}$ *solves the McKean-Vlasov SDE:*

$$dX_s^{t,\xi} = b(s)\hat{\alpha}\big(s, X_s^{t,\xi}, \partial_x \mathcal{U}\big(s, X_s^{t,\xi}, \mathcal{L}^1(X_s^{t,\xi})\big)\big)ds + \sigma dW_s + \sigma^0 dW_s^0,$$

for $s \in [t, T]$.

The proof below shows that the solutions to (5.114) and (5.115) are respectively adapted to the completions of the filtrations generated by $(\xi, (W_s^0 - W_t^0, W_s - W_t)_{t \le s \le T})$ and $(W_s^0 - W_t^0, W_s - W_t)_{t \le s \le T}$. We recall that we established a similar property in small time.

Proof. The proof of the first claim on the solvability of (5.114) consists in a mere adaptation of that of Proposition 5.42, with $U = \partial_x \mathcal{U}$, taking advantage of the master equation (5.117) for $\partial_x \mathcal{U}$.

Once the first claim has been proved, the second claim on the solvability of (5.115) follows from the same argument. Indeed, the existence of a solution may be regarded as a consequence of the necessary part in the stochastic Pontryagin principle, but it can be also shown by solving first:

$$dX_s = b(s)\hat{\alpha}\big(s, X_s, \mathcal{L}^1(X_s^{t,\xi})\big)ds + \sigma dW_s + \sigma^0 dW_s^0,$$

for $s \in [t, T]$, with $X_t = x \in \mathbb{R}^d$ as initial condition, and then by letting:

$$\begin{aligned}
Y_s &= \partial_x \mathcal{U}\big(s, X_s, \mathcal{L}^1(X_s^{t,\xi})\big), \\
Z_s &= \partial_x^2 \mathcal{U}\big(s, X_s, \mathcal{L}^1(X_s^{t,\xi})\big)\sigma, \\
Z_s^0 &= \partial_x^2 \mathcal{U}\big(s, X_s, \mathcal{L}^1(X_s^{t,\xi})\big)\sigma^0 + \tilde{\mathbb{E}}^1\big[\partial_\mu \partial_x \mathcal{U}\big(s, X_s, \mathcal{L}^1(X_s^{t,\xi})\big)(\tilde{X}_s^{t,\xi})\big]\sigma^0,
\end{aligned}$$

where we recall once again the convention $\partial_\mu \partial_x \mathcal{U} = ([\partial_\mu [\partial_{x_i} \mathcal{U}]]_j)_{1 \le i,j \le d}$. By Itô's formula, we obtain a solution of (5.115). Now, if $(X_s', Y_s', Z_s', Z_s^{0\prime})$ is another solution of (5.115), we let:

$$\begin{aligned}
\bar{Y}_s' &= \partial_x \mathcal{U}\big(s, X_s', \mathcal{L}^1(X_s^{t,\xi})\big), \\
\bar{Z}_s' &= \partial_x^2 \mathcal{U}\big(s, X_s', \mathcal{L}^1(X_s^{t,\xi})\big)\sigma, \\
\bar{Z}_s^{0\prime} &= \partial_x^2 \mathcal{U}\big(s, X_s', \mathcal{L}^1(X_s^{t,\xi})\big)\sigma^0 + \tilde{\mathbb{E}}^1\big[\partial_\mu \partial_x \mathcal{U}\big(s, X_s', \mathcal{L}^1(X_s^{t,\xi})\big)(\tilde{X}_s^{t,\xi})\big]\sigma^0.
\end{aligned}$$

Expanding $(\bar{Y}'_s)_{t \le s \le T}$ by Itô's formula and proceeding as in the proof of Proposition 5.42, we deduce that $(Y'_s, Z'_s, Z^{0\prime}_s)_{t \le s \le T} = (\bar{Y}'_s, \bar{Z}'_s, \bar{Z}^{0\prime}_s)_{t \le s \le T}$, from which the proof is easily completed. □

On the same model, we have:

Lemma 5.51 *In addition to assumption* **MFG Master Classical**, *assume further that hypothesis $\mathscr{H}(S)$ holds true for some $S \in [0, T]$. Then, for any $(t, \xi) \in [S, T] \times L^2(\Omega^1, \mathcal{F}^1_t, \mathbb{P}^1; \mathbb{R}^d)$, the unique solution to the mean field game with ξ as initial condition at time t is given by $\mathfrak{M} = \mathcal{L}^1((X^{t,\xi}_s, W_s - W_t)_{t \le s \le T})$, where $X^{t,\xi} = (X^{t,\xi}_s)_{t \le s \le T}$ is the forward component of the unique solution to* (5.114) *with ξ as initial condition at time t.*

Proof. As already explained, we know from Theorems 4.21 and 4.23 that, for a given (t, ξ) as in the statement, the mean field game has a solution; by Proposition 3.34, this solution is unique and is strong, namely the equilibrium $\mathfrak{M}$ generated by the solution is adapted to the completion of the filtration generated by $(W^0_s - W^0_t)_{t \le s \le T}$. Hence, the optimal trajectory of the optimal control problem (5.107)–(5.108) in the super-environment $\mathfrak{M}$ is adapted with respect to the completion of the filtration generated by ξ, $(W^0_s - W^0_t)_{t \le s \le T}$ and $(W_s - W_t)_{t \le s \le T}$. Following Proposition 4.7, we may consider the adjoint equation deriving from the necessary condition in the stochastic Pontryagin principle. By the aforementioned measurability properties of the equilibrium and of the optimal trajectory, the extra martingale part therein, which is required to be orthogonal to $\boldsymbol{W}^0$ and $\boldsymbol{W}$, is null. As a consequence, the equilibrium induces a solution to (5.114). We conclude by invoking Lemma 5.50, which ensures that (5.114) is uniquely solvable. □

Also, the monotonicity property is preserved:

Lemma 5.52 *In addition to assumption* **MFG Master Classical**, *assume further that hypothesis $\mathscr{H}(S)$ holds true for some $S \in [0, T]$, then, for every $t \in [S, T]$, the function $\mathcal{U}(t, \cdot, \cdot)$ satisfies the Lasry-Lions monotonicity property.*

The proof of Lemma 5.52 is postponed to the end of the section.

Implementing the Induction Argument

The implementation of the induction principle relies on the next two propositions, whose proofs are also deferred to the end of the section.

Proposition 5.53 *Let assumption* **MFG Master Classical** *be in force. Then, there exists a constant $K_1 \ge 0$ such that for any $S \in [0, T)$ for which $\mathscr{H}(S)$ is satisfied, it holds:*

$$\forall (t, x, \mu) \in [S, T] \times \mathbb{R}^d \times \mathcal{P}_2(\mathbb{R}^d), \quad |\partial^2_x \mathcal{U}(t, x, \mu)| \le K_1.$$

Proposition 5.54 *Let assumption* **MFG Master Classical** *be in force. Then, there exists a constant* $K_2 \geq 0$ *such that for any* $S \in [0, T)$ *for which* $\mathscr{H}(S)$ *is satisfied, it holds:*

$$\forall (t, x, \mu, v) \in [S, T] \times \mathbb{R}^d \times \mathcal{P}_2(\mathbb{R}^d) \times \mathbb{R}^d, \quad |\partial_x \partial_\mu \mathcal{U}(t, x, \mu)(v)| \leq K_2.$$

Provided that Propositions 5.53 and 5.54 hold true, Theorem 5.46 easily follows:

Proof of Theorem 5.46. By Theorem 5.45, we know that there exists a real $S \in [0, T)$ such that hypothesis $\mathscr{H}(S)$ holds true.

If $S = 0$, the proof is over. If not, we proceed by induction. Indeed, Propositions 5.53 and 5.54 apply and say that, for the same two constants K_1 and K_2 as in the statements,

$$\forall (t, x, \mu) \in [S, T] \times \mathbb{R}^d \times \mathcal{P}_2(\mathbb{R}^d), \quad |\partial_x^2 \mathcal{U}(t, x, \mu)| \leq K_1,$$

$$\forall (t, x, \mu, v) \in [S, T] \times \mathbb{R}^d \times \mathcal{P}_2(\mathbb{R}^d) \times \mathbb{R}^d, \quad |\partial_x \partial_\mu \mathcal{U}(t, x, \mu)(v)| \leq K_2.$$

Then, we may consider, on the interval $[0, S]$, the mean field game driven by the same dynamics as in (5.108), but with the new cost functional:

$$\inf_{(\alpha_s)_{0 \leq s \leq T}} \mathbb{E}\left[\int_0^S f(s, X_s, \mu_s, \alpha_s) ds + \mathcal{U}(S, X_S, \mu_S)\right]. \tag{5.130}$$

Obviously, the goal is to apply Theorem 5.45 to the new mean field game involving (5.130) in lieu of (5.107).

In order to do so, we must check that assumptions **MFG Master Pontryagin** and **MFG Smooth Coefficients** from Subsection 5.4.1 are satisfied. Recall indeed that these are the two sets of assumption under which Theorem 5.45 applies. Basically, this requires to check that $\mathcal{U}(S, \cdot, \cdot)$ satisfies the same assumption as g. Thanks to $\mathscr{H}(S)$, it is clear that $\mathcal{U}(S, \cdot, \cdot)$ satisfies the same regularity assumption as g in assumption **MFG Smooth Coefficients**.

We now turn to assumption **MFG Master Pontryagin**. In order to guarantee that assumptions **FBSDE** and **Decoupling Master** from Subsections 4.1.3 and 4.2.2 are satisfied, we shall directly check that, depending upon the framework, one of the two assumptions, **MFG with a Common Noise HJB**, or **MFG with a Common Noise SMP Relaxed**, holds true, see Subsections 3.4.1 and 3.4.3 for a reminder. When assumption **MFG with a Common Noise HJB** is in force, it is pretty clear that $\mathcal{U}(S, \cdot, \cdot)$ satisfies the same assumption as g in **MFG with a Common Noise HJB**. Indeed, by Theorem 1.57, we know that $\partial_x \mathcal{U}$ is bounded; plugging the bound into the representation formula of $\mathcal{U}$ in Definition 4.1, we deduce that $\mathcal{U}$ is also bounded. When assumption **MFG with a Common Noise SMP Relaxed** is in force, it suffices to prove that $\mathcal{U}(S, \cdot, \cdot)$ is convex in the variable x, which follows from a mere variation of (1.64).

Therefore, Theorem 5.45 guarantees that there exists $S' < S$ such that the master field associated with the new cost functional (5.130) is smooth on $[S', S]$ and satisfies the same conclusion as in $\mathscr{H}(S)$. Of course, the main point is that this new master field coincides with the restriction of $\mathcal{U}$ to $[S', S] \times \mathbb{R}^d \times \mathcal{P}_2(\mathbb{R}^d)$. This is a consequence of the dynamic programming principle in Theorem 4.5. This proves that $\partial_x \mathcal{U}$ satisfies the master equation (5.117) on both $[S', S] \times \mathbb{R}^d \times \mathcal{P}_2(\mathbb{R}^d)$ and on $[S, T] \times \mathbb{R}^d \times \mathcal{P}_2(\mathbb{R}^d)$. By continuity of $\partial_x \mathcal{U}$ and of its derivative, it is pretty clear that $\partial_x \mathcal{U}$ is time differentiable and satisfies the master equation (5.117) on the entire $[S', T] \times \mathbb{R}^d \times \mathcal{P}_2(\mathbb{R}^d)$. It belongs to $\mathfrak{S}_d([S', T])$. Proceeding in the same way for $\mathcal{U}$, we deduce that $\mathscr{H}(S')$ holds true.

Importantly, thanks to Propositions 5.53 and 5.54, the length $S - S'$ is only dictated by the various constants in the statement of Theorem 5.46, through the values of K_1 and K_2, and is independent of S. The proof follows by iterating the argument. □

Proof of the Bound for $\partial_x^2 \mathcal{U}$

We now prove Proposition 5.53.

Proof of Proposition 5.53. There are two cases.

First Case. Within the framework of assumption **MFG with a Common Noise SMP Relaxed**, the bound for $\partial_x^2 \mathcal{U}$ is a straightforward consequence of Theorem 1.60.

Second Case. We now focus on the case when assumption **MFG with a Common Noise HJB** holds true. Without any loss of generality, we assume that $\sigma = I_d$. The trick is to define:

$$\bar{X}_s^{t,x} = x + \big(W_s - W_t\big) + \sigma^0\big(W_s^0 - W_t^0\big),$$

and to expand:

$$\Big(\partial_x \mathcal{U}\big(s, \bar{X}_s^{t,x}, \mathcal{L}^1(X_s^{t,\xi})\big)\Big)_{t \le s \le T},$$

by the chain rule and, in the process, to take advantage of the master equation (5.117) satisfied by $\partial_x \mathcal{U}$. We get:

$$\begin{aligned} d\big[\partial_x \mathcal{U}\big(s, \bar{X}_s^{t,x}, \mathcal{L}^1(X_s^{t,\xi})\big)\big] = -\Big[&\partial_x^2 \mathcal{U}\big(s, \bar{X}_s^{t,x}, \mathcal{L}^1(X_s^{t,\xi})\big) b(s) \hat{\alpha}\big(s, \bar{X}_s^{t,x}, \partial_x \mathcal{U}\big(s, \bar{X}_s^{t,x}, \mathcal{L}^1(X_s^{t,\xi})\big)\big) \\ &+ \partial_x H\Big(s, \bar{X}_s^{t,x}, \mathcal{L}^1(X_s^{t,\xi}), \partial_x \mathcal{U}\big(s, \bar{X}_s^{t,x}, \mathcal{L}^1(X_s^{t,\xi})\big), \\ &\quad \hat{\alpha}\big(s, \bar{X}_s^{t,x}, \partial_x \mathcal{U}\big(s, \bar{X}_s^{t,x}, \mathcal{L}^1(X_s^{t,\xi})\big)\big)\Big)\Big] ds + dM_s, \end{aligned}$$

for $s \in [t, T]$, where $(M_s)_{t \le s \le T}$ is a square-integrable martingale. Thanks to the special form of b under the standing assumption, $\partial_x H$ coincides with $\partial_x f$. In particular, by assumption **MFG with a Common Noise HJB**, it is bounded. Letting:

$$\begin{aligned} \bar{F}(s, x, \mu, z) = &\, z\big[b(s) \hat{\alpha}\big(s, x, \partial_x \mathcal{U}(s, x, \mu)\big)\big] \\ &+ \partial_x f\big(s, x, \mu, \partial_x \mathcal{U}(s, x, \mu), \hat{\alpha}\big(s, x, \partial_x \mathcal{U}(s, x, \mu)\big)\big), \end{aligned}$$

for $(s, x, \mu, z) \in [t, T] \times \mathbb{R}^d \times \mathcal{P}_2(\mathbb{R}^d) \times \mathbb{R}^{d \times d}$, we have the shorter writing:

$$\begin{aligned} &d\big[\partial_x \mathcal{U}\big(s, \bar{X}_s^{t,x}, \mathcal{L}^1(X_s^{t,\xi})\big)\big] \\ &= -\bar{F}\Big(s, \bar{X}_s^{t,x}, \mathcal{L}^1(X_s^{t,\xi}), \partial_x^2 \mathcal{U}\big(s, \bar{X}_s^{t,x}, \mathcal{L}^1(X_s^{t,\xi})\big)\Big) ds + dM_s, \quad s \in [t, T]. \end{aligned}$$

Taking expectation, we deduce that:

$$\begin{aligned} \partial_x \mathcal{U}(t, x, \mu) = \mathbb{E}\Big[&\partial_x g\big(\bar{X}_T^{t,x}, \mathcal{L}^1(X_T^{t,\xi})\big) \\ &+ \int_t^T \bar{F}\Big(s, \bar{X}_s^{t,x}, \mathcal{L}^1(X_s^{t,\xi}), \partial_x^2 \mathcal{U}\big(s, \bar{X}_s^{t,x}, \mathcal{L}^1(X_s^{t,\xi})\big)\Big) ds\Big]. \end{aligned}$$

Taking advantage of the specific form of $\bar{X}^{t,x}$ together with the fact that $(\mathcal{L}^1(X_s^{t,\xi}))_{t\le s\le T}$ is independent of W, we deduce that:

$$\begin{aligned}\partial_x \mathcal{U}(t,x,\mu) &= \int_{\mathbb{R}^d} \mathbb{E}\big[\partial_x g\big(y+\sigma^0(W_T^0-W_t^0), \mathcal{L}^1(X_T^{t,\xi})\big)\big] p_{T-t}(x,y)dy \\ &\quad + \int_t^T \int_{\mathbb{R}^d} \mathbb{E}\Big[\bar{F}\Big(s, y+\sigma^0(W_s^0-W_t^0), \mathcal{L}^1(X_s^{t,\xi}), \\ &\qquad\qquad \partial_x^2 \mathcal{U}\big(s, y+\sigma^0(W_s^0-W_t^0), \mathcal{L}^1(X_s^{t,\xi})\big)\Big)\Big] p_{s-t}(x,y)dyds,\end{aligned}$$

where $(p_s(x,y) = s^{-d/2}\varphi_d((x-y)/s))_{s>0,x,y\in\mathbb{R}^d}$ denotes the d-dimensional heat kernel, φ_d standing for the d-dimensional standard Gaussian density.

Now, we recall the standard estimate:

$$\sup_{x\in\mathbb{R}^d}\left|\frac{d}{dx}\int_{\mathbb{R}^d}\psi(y)p_s(x-y)dy\right| \le cs^{-1/2}\sup_{x\in\mathbb{R}^d}|\psi(x)|,$$

for any bounded and measurable function $\psi : \mathbb{R}^d \to \mathbb{R}$ and for a constant c, independent of $s \in (0,T]$ and ψ.

Therefore, recalling that $\partial_x g$ has a bounded derivative in the direction x and that $\bar{F}(s,x,\mu,z)$ is at most of linear growth in z, uniformly in the other variables, we easily deduce that there exists a constant C, only depending on T and the parameters in the assumptions, such that, for all $t \in [S,T]$,

$$\sup_{(x,\mu)\in\mathbb{R}^d\times\mathcal{P}_2(\mathbb{R}^d)}\big|\partial_x^2\mathcal{U}(t,x,\mu)\big| \le C + \int_t^T \frac{C}{\sqrt{s-t}}\Big(1+\sup_{(x,\mu)\in\mathbb{R}^d\times\mathcal{P}_2(\mathbb{R}^d)}\big|\partial_x^2\mathcal{U}(s,x,\mu)\big|\Big)ds.$$

For a new value of C, we get for any $\epsilon \in (0,T]$:

$$\begin{aligned}\sup_{(x,\mu)\in\mathbb{R}^d\times\mathcal{P}_2(\mathbb{R}^d)}\big|\partial_x^2\mathcal{U}(t,x,\mu)\big| &\le C + C\sqrt{\epsilon}\sup_{t\le s\le T}\sup_{(x,\mu)\in\mathbb{R}^d\times\mathcal{P}_2(\mathbb{R}^d)}\big|\partial_x^2\mathcal{U}(s,x,\mu)\big| \\ &\quad + \frac{C}{\sqrt{\epsilon}}\int_t^T \sup_{s\le r\le T}\sup_{(x,\mu)\in\mathbb{R}^d\times\mathcal{P}_2(\mathbb{R}^d)}\big|\partial_x^2\mathcal{U}(r,x,\mu)\big|ds.\end{aligned}$$

Notice that the right-hand side increases as t decreases. Therefore,

$$\begin{aligned}\sup_{t\le s\le T}\sup_{(x,\mu)\in\mathbb{R}^d\times\mathcal{P}_2(\mathbb{R}^d)}\big|\partial_x^2\mathcal{U}(s,x,\mu)\big| &\le C + C\sqrt{\epsilon}\sup_{t\le s\le T}\sup_{(x,\mu)\in\mathbb{R}^d\times\mathcal{P}_2(\mathbb{R}^d)}\big|\partial_x^2\mathcal{U}(s,x,\mu)\big| \\ &\quad + \frac{C}{\sqrt{\epsilon}}\int_t^T \sup_{s\le r\le T}\sup_{(x,\mu)\in\mathbb{R}^d\times\mathcal{P}_2(\mathbb{R}^d)}\big|\partial_x^2\mathcal{U}(r,x,\mu)\big|ds.\end{aligned}$$

Choosing ϵ such that $C\sqrt{\epsilon} = \frac{1}{2}$ and applying Gronwall's lemma, we complete the proof. □

Proof of the Bound for $\partial_x \partial_\mu \mathcal{U}$

We now prove Proposition 5.54. The proof relies on the auxiliary property:

Proposition 5.55 *Let assumption* **MFG Master Classical** *be in force. Then, there exists a function $K_2 : \mathbb{R}_+ \to \mathbb{R}_+$ such that for any $S \in [0, T)$ for which $\mathscr{H}(S)$ is satisfied, the following holds true:*

$$\sup_{t\in[S,T]} \sup_{(x,v)\in\mathbb{R}^d\times\mathbb{R}^d} \sup_{\mu\in\mathcal{P}_2(\mathbb{R}^d)} |\partial_\mu \partial_x \mathcal{U}(t, x, \mu)(v)| \\ \leq K_2\Big(\sup_{t\in[S,T]} \sup_{x\in\mathbb{R}^d} \sup_{\mu\in\mathcal{P}_2(\mathbb{R}^d)} |\partial_x^2 \mathcal{U}(t, x, \mu)| \Big).$$

Proof. The proof starts with a preliminary remark that will be useful in the proof. We observe that, for any $\xi, \chi \in L^2(\Omega, \mathcal{F}, \mathbb{P}; \mathbb{R}^d)$,

$$\begin{aligned} \forall t \in [0, T], \quad & \mathbb{E}\Big[\chi \cdot \Big(\tilde{\mathbb{E}}^1\big[\partial_\mu \partial_x f_0\big(t, \xi, \mathcal{L}^1(\xi)\big)(\tilde{\xi})\tilde{\chi}\big]\Big)\Big] \geq 0, \\ & \mathbb{E}\Big[\chi \cdot \Big(\tilde{\mathbb{E}}^1\big[\partial_\mu \partial_x g\big(\xi, \mathcal{L}^1(\xi)\big)(\tilde{\xi})\tilde{\chi}\big]\Big)\Big] \geq 0, \end{aligned} \tag{5.131}$$

where we recall the useful notation, see (5.78) and (Vol I)-(5.80),

$$\partial_\mu \partial_x f_0(t, x, \mu)(v) y = \bigg(\sum_{j=1}^d \Big(\partial_\mu [\partial_x f_0(t, x, \mu)]_i(v) \Big)_j y_j \bigg)_{1\leq i\leq d} \in \mathbb{R}^d,$$

for $y \in \mathbb{R}^d$, together with $\partial_\mu \partial_x f_0 = [\partial_x \partial_\mu f_0]^\dagger$, and similarly for g, see for instance Subsection 5.1.5.

We only prove the second claim in (5.131), the first one following from the same argument. By continuity of $\partial_x \partial_\mu g$, it suffices to prove the inequality when ξ and χ are bounded. Then, by monotonicity of g, we observe that, for any $s \in \mathbb{R}$,

$$\mathbb{E}\big[g\big(\xi + s\chi, \mathcal{L}^1(\xi + s\chi)\big)\big] - \mathbb{E}\big[g\big(\xi, \mathcal{L}^1(\xi + s\chi)\big)\big] \\ - \Big(\mathbb{E}\big[g\big(\xi + s\chi, \mathcal{L}^1(\xi)\big)\big] - \mathbb{E}\big[g\big(\xi, \mathcal{L}^1(\xi)\big)\big]\Big) \geq 0.$$

Taking advantage of the smoothness of g, we write:

$$\begin{aligned} &\mathbb{E}\big[g\big(\xi + s\chi, \mathcal{L}^1(\xi + s\chi)\big)\big] - \mathbb{E}\big[g\big(\xi, \mathcal{L}^1(\xi + s\chi)\big)\big] \\ &\qquad = \int_0^s \mathbb{E}\big[\partial_x g\big(\xi + r\chi, \mathcal{L}^1(\xi + s\chi)\big) \cdot \chi\big] dr, \\ &\mathbb{E}\big[g\big(\xi + s\chi, \mathcal{L}^1(\xi)\big)\big] - \mathbb{E}\big[g\big(\xi, \mathcal{L}^1(\xi)\big)\big] = \int_0^s \mathbb{E}\big[\partial_x g\big(\xi + r\chi, \mathcal{L}^1(\xi\chi)\big) \cdot \chi\big] dr. \end{aligned}$$

Hence,

$$
\begin{aligned}
&\mathbb{E}\big[g(\xi + s\chi, \mathcal{L}^1(\xi + s\chi))\big] - \mathbb{E}\big[g(\xi, \mathcal{L}^1(\xi + s\chi))\big] \\
&\quad - \Big(\mathbb{E}\big[g(\xi + s\chi, \mathcal{L}^1(\xi))\big] - \mathbb{E}\big[g(\xi, \mathcal{L}^1(\xi))\big]\Big) \\
&= \int_0^s \mathbb{E}\Big[\Big(\partial_x g(\xi + r\chi, \mathcal{L}^1(\xi + s\chi)) - \partial_x g(\xi + r\chi, \mathcal{L}^1(\xi))\Big) \cdot \chi\Big] dr \\
&= \int_0^s \int_0^s \mathbb{E}\Big[\tilde{\mathbb{E}}^1\Big(\partial_\mu \partial_x g(\xi + r\chi, \mathcal{L}^1(\xi + r'\chi))(\xi + r'\chi)\tilde{\chi}\Big) \cdot \chi\Big] dr dr'.
\end{aligned}
$$

Recalling that the left-hand side is nonnegative, dividing by s^2 and taking the limit as s tends to 0, we get:

$$
\mathbb{E}\Big[\chi \cdot \Big(\tilde{\mathbb{E}}^1\big[\partial_\mu \partial_x g(\xi, \mathcal{L}^1(\xi))(\xi)\tilde{\chi}\big]\Big)\Big] \geq 0,
$$

which is precisely the second-line in (5.131).

First Step. Throughout the proof, we shall use the following convention. For $t \in [0, T]$ and $\xi \in L^2(\Omega^1, \mathcal{F}_t^1, \mathbb{P}^1; \mathbb{R}^d)$, we let:

$$
\begin{aligned}
\hat{\boldsymbol{\alpha}}^{t,\xi} &= \big(\hat{\alpha}_s^{t,\xi} = \hat{\alpha}(s, X_s^{t,\xi}, Y_s^{t,\xi})\big)_{t \leq s \leq T}, \\
\boldsymbol{f}_0^{t,\xi} &= \big(f_{0,s}^{t,\xi} = f_0(s, X_s^{t,\xi}, \mathcal{L}^1(X_s^{t,\xi}))\big)_{t \leq s \leq T}, \\
\boldsymbol{f}_1^{t,\xi} &= \big(f_{1,s}^{t,\xi} = f_1(s, X_s^{t,\xi}, \hat{\alpha}(s, X_s^{t,\xi}, Y_s^{t,\xi}))\big)_{t \leq s \leq T}, \\
\boldsymbol{\mathcal{U}}^{t,\xi} &= \big(\mathcal{U}_s^{t,\xi} = \mathcal{U}(s, X_s^{t,\xi}, \mathcal{L}^1(X_s^{t,\xi}))\big)_{t \leq s \leq T}, \quad g^{t,\xi} = g\big(X_T^{t,\xi}, \mathcal{L}^1(X_T^{t,\xi})\big),
\end{aligned}
$$

where $(\boldsymbol{X}^{t,\xi}, \boldsymbol{Y}^{t,\xi})$ is given by (5.114), with similar notations when $\hat{\alpha}$, f_0, f_1, $\mathcal{U}$ and g are replaced by their derivatives. We use a similar convention for any $(t, x, \mu) \in [0, T] \times \mathbb{R}^d \times \mathcal{P}_2(\mathbb{R}^d)$:

$$
\begin{aligned}
\hat{\boldsymbol{\alpha}}^{t,x,\mu} &= \big(\hat{\alpha}_s^{t,x,\mu} = \hat{\alpha}(s, X_s^{t,x,\mu}, Y_s^{t,x,\mu})\big)_{t \leq s \leq T}, \\
\boldsymbol{f}_0^{t,x,\mu} &= \big(f_{0,s}^{t,x,\mu} = f_0(s, X_s^{t,x,\mu}, \mathcal{L}^1(X_s^{t,\xi}))\big)_{t \leq s \leq T}, \\
\boldsymbol{f}_1^{t,x,\mu} &= \big(f_{1,s}^{t,x,\mu} = f_1(s, X_s^{t,x,\mu}, \hat{\alpha}(s, X_s^{t,x,\mu}, Y_s^{t,x,\mu}))\big)_{t \leq s \leq T}, \\
\boldsymbol{\mathcal{U}}^{t,x,\mu} &= \big(\mathcal{U}_s^{t,x,\mu} = \mathcal{U}(s, X_s^{t,x,\mu}, \mathcal{L}^1(X_s^{t,\xi}))\big)_{t \leq s \leq T}, \quad g^{t,x,\mu} = g\big(X_T^{t,x,\mu}, \mathcal{L}^1(X_T^{t,\xi})\big),
\end{aligned}
$$

where $(\boldsymbol{X}^{t,x,\mu}, \boldsymbol{Y}^{t,x,\mu}) = (\boldsymbol{X}^{t,x,\xi}, \boldsymbol{Y}^{t,x,\xi})$ is given by (5.115). With this convention, the dynamics of $(\boldsymbol{X}^{t,\xi}, \boldsymbol{Y}^{t,\xi})$ take the form

$$
\begin{aligned}
dX_s^{t,\xi} &= b(s)\hat{\alpha}_s^{t,\xi} ds + \sigma dW_s + \sigma^0 dW_s^0, \\
dY_s^{t,\xi} &= -\partial_x f_s^{t,\xi} ds + Z_s^{t,\xi} dW_s + Z_s^{0;t,\xi} dW_s^0, \quad s \in [t, T],
\end{aligned}
$$

with terminal condition $Y_T^{t,\xi} = \partial_x g^{t,\xi}$, where, to make it clear, $\partial_x f_s^{t,\xi} = \partial_x f(s, X_s^{t,\xi}, \mathcal{L}^1(X_s^{t,\xi}), \hat{\alpha}(s, X_s^{t,\xi}, Y_s^{t,\xi}))$ and $\partial_x g^{t,\xi} = \partial_x g(X_T^{t,\xi}, \mathcal{L}^1(X_T^{t,\xi}))$. Similarly, for $(t, x, \mu) \in [0, T] \times \mathbb{R}^d \times \mathcal{P}_2(\mathbb{R}^d)$, the equation for $(\boldsymbol{X}^{t,x,\mu}, \boldsymbol{Y}^{t,x,\mu})$ writes:

$$
\begin{aligned}
dX_s^{t,x,\mu} &= b(s)\hat{\alpha}_s^{t,x,\mu} ds + \sigma dW_s + \sigma^0 dW_s^0, \\
dY_s^{t,x,\mu} &= -\partial_x f_s^{t,x,\mu} ds + Z_s^{t,x,\mu} dW_s + Z_s^{0;t,x,\mu} dW_s^0, \quad s \in [t, T],
\end{aligned}
$$

with terminal condition $Y_T^{t,x,\mu} = \partial_x g^{t,x,\mu}$. Above, ξ is distributed according to μ.

We also recall the identities:

$$
Y_s^{t,\xi} = \partial_x \mathcal{U}_s^{t,\xi}, \quad Y_s^{t,x,\mu} = \partial_x \mathcal{U}_s^{t,x,\mu}, \quad s \in [t, T].
$$

Since the restriction of $\partial_x \mathcal{U}$ to $[S, T]$ is known to belong to the class $\mathfrak{S}_d([S, T])$, it is pretty clear that the differentiability properties of the flows

$$
\begin{aligned}
&L^2(\Omega^1, \mathcal{F}_t^1, \mathbb{P}^1; \mathbb{R}^d) \ni \xi \mapsto (X_s^{t,\xi}, Y_s^{t,\xi})_{t \le s \le T}, \\
\text{and} \quad &\mathbb{R}^d \times \mathcal{P}_2(\mathbb{R}^d) \ni (x, \mu) \mapsto (X_s^{t,x,\mu}, Y_s^{t,x,\mu})_{t \le s \le T},
\end{aligned}
$$

hold true as in the conclusions of Subsections 5.2 and 5.3, whatever the length $T - S$ is. Indeed, we may directly regard the forward-backward equations for $\boldsymbol{\theta}^{t,\xi}$ and $\boldsymbol{\theta}^{t,x,\mu}$ as decoupled systems driven by coefficients satisfying assumption **MFG Smooth Coefficients**. Therefore, we may use the notations $\partial_\chi \boldsymbol{\theta}^{t,\xi} = (\partial_\chi \boldsymbol{X}^{t,\xi}, \partial_\chi \boldsymbol{Y}^{t,\xi}) = (\partial_\chi X_s^{t,\xi}, \partial_\chi Y_s^{t,\xi})_{t \le s \le T}$ and $\partial_\chi \boldsymbol{\theta}^{t,x,\xi} = (\partial_\chi \boldsymbol{X}^{t,x,\xi}, \partial_\chi \boldsymbol{Y}^{t,x,\xi}) = (\partial_\chi X_s^{t,x,\xi}, \partial_\chi Y_s^{t,x,\xi})_{t \le s \le T}$ for the respective derivatives of the flows with respect to ξ in the direction χ, for $\chi \in L^2(\Omega^1, \mathcal{F}_t^1, \mathbb{P}^1; \mathbb{R}^d)$. These tangent processes satisfy, for $s \in [t, T]$,

$$
\begin{aligned}
d\partial_\chi X_s^{t,\xi} &= b(s)\big(\partial_x \hat{\alpha}_s^{t,\xi} \partial_\chi X_s^{t,\xi} + \partial_y \hat{\alpha}_s^{t,\xi} \partial_\chi Y_s^{t,\xi}\big) ds, \\
d\partial_\chi Y_s^{t,\xi} &= -\Big(\partial_x^2 f_{0,s}^{t,\xi} \partial_\chi X_s^{t,\xi} + \mathbb{E}^1\big[\partial_\mu \partial_x \tilde{f}_{0,s}^{t,\xi} \partial_\chi \tilde{X}_s^{t,\xi}\big] \\
&\qquad + \partial_x^2 f_{1,s}^{t,\xi} \partial_\chi X_s^{t,\xi} + \partial_{\alpha x}^2 f_{1,s}^{t,\xi} \big[\partial_x \hat{\alpha}_s^{t,\xi} \partial_\chi X_s^{t,\xi} + \partial_y \hat{\alpha}_s^{t,\xi} \partial_\chi Y_s^{t,\xi}\big]\Big) ds \\
&\quad + \partial_\chi Z_s^{t,\xi} dW_s + \partial_\chi Z_s^{0;t,\xi} dW_s^0,
\end{aligned}
\tag{5.132}
$$

where $(\partial_\chi Z_s^{t,\xi})_{t \le s \le T}$ and $(\partial_\chi Z_s^{0;t,\xi})_{t \le s \le T}$ are the $d \times d$-dimensional martingale integrands appearing in the semi-martingale representation of $(\partial_\chi Y_s^{t,\xi})_{t \le s \le T}$, see (5.55)–(5.56). Above, we used the notations:

$$
\begin{aligned}
\partial_{\alpha x} \tilde{f}_{0,s}^{t,\xi} &= \partial_\alpha \partial_x f_0\big(s, X_s^{t,\xi}, \mathcal{L}^1(X_s^{t,\xi})\big)(\tilde{X}_s^{t,\xi}), \\
\partial_\mu \partial_x \tilde{f}_{0,s}^{t,\xi} &= \partial_\mu \partial_x f_0\big(s, X_s^{t,\xi}, \mathcal{L}^1(X_s^{t,\xi})\big)(\tilde{X}_s^{t,\xi}), \quad s \in [t, T],
\end{aligned}
$$

see also (Vol I)-(5.80). In the same way, we consider the dynamics of $(\partial_x X_s^{t,x,\mu})_{t \le s \le T}$ and $(\partial_x Y_s^{t,x,\mu})_{t \le s \le T}$, which take values in $\mathbb{R}^{d \times d}$. We have:

$$
\begin{aligned}
d\partial_x X_s^{t,x,\mu} &= \Big(\partial_x \hat{\alpha}_s^{t,x,\mu} \partial_x X_s^{t,x,\mu} + \partial_y \hat{\alpha}_s^{t,x,\mu} \partial_x Y_s^{t,x,\mu}\Big) ds, \\
d\partial_x Y_s^{t,x,\mu} &= -\Big(\partial_x^2 f_{0,s}^{t,x,\mu} \partial_x X_s^{t,x,\mu} + \partial_x^2 f_{1,s}^{t,x,\mu} \partial_x X_s^{t,x,\mu} \\
&\quad + \partial_{\alpha x}^2 f_{1,s}^{t,x,\mu} \big[\partial_x \hat{\alpha}_s^{t,x,\mu} \partial_x X_s^{t,x,\mu} + \partial_y \hat{\alpha}_s^{t,x,\mu} \partial_x Y_s^{t,x,\mu}\big]\Big) ds \\
&\quad + \partial_x Z_s^{t,x,\mu} dW_s + \partial_x Z_s^{0;t,x,\mu} dW_s^0, \quad s \in [t,T],
\end{aligned} \tag{5.133}
$$

where $(\partial_x Z_s^{t,x,\mu}, \partial_x Z_s^{0;t,x,\mu})_{t \le s \le T}$ are the $(d \times d) \times d$-dimensional martingale integrands appearing in the semi-martingale representation of $(\partial_x Y_s^{t,x,\mu})_{t \le s \le T}$, see (5.65).

Second Step. We now assume that the law of ξ has a finite support, namely $\xi = \sum_{i=1}^N x_i \mathbf{1}_{A_i}$, where $x_1, \cdots, x_N$ are N points in $\mathbb{R}^d$ and $A_1, \cdots, A_N$ are N events in $\mathcal{F}_t^1$. In that case,

$$
(X_s^{t,\xi}, Y_s^{t,\xi}, Z_s^{t,\xi})_{t \le s \le T} = \left(\sum_{i=1}^N X_s^{t,x_i,\mu} \mathbf{1}_{A_i}, \sum_{i=1}^N Y_s^{t,x_i,\mu} \mathbf{1}_{A_i}\right)_{t \le s \le T},
$$

see the proof of Proposition 5.8 for similar arguments. With the current choice of ξ, we let:

$$
\big(\partial_x X_s^{t,\xi,\mu}, \partial_x Y_s^{t,\xi,\mu}\big)_{t \le s \le T} = \left(\sum_{i=1}^N \partial_x X_s^{t,x_i,\mu} \mathbf{1}_{A_i}, \sum_{i=1}^N \partial_x Y_s^{t,x_i,\mu} \mathbf{1}_{A_i}\right)_{t \le s \le T},
$$

and similarly for $(\partial_x Z_s^{t,\xi,\mu}, \partial_x Z_s^{0;t,\xi,\mu})_{t \le s \le T}$. Following (5.133), it satisfies:

$$
\begin{aligned}
d\partial_x X_s^{t,\xi,\mu} &= b(s)\Big(\partial_x \hat{\alpha}_s^{t,\xi} \partial_x X_s^{t,\xi,\mu} + \partial_y \hat{\alpha}_s^{t,\xi} \partial_x Y_s^{t,\xi,\mu}\Big) ds, \\
d\partial_x Y_s^{t,\xi,\mu} &= -\Big(\partial_x^2 f_{0,s}^{t,\xi} \partial_x X_s^{t,\xi,\mu} + \partial_x^2 f_{1,s}^{t,\xi} \partial_x X_s^{t,\xi,\mu} \\
&\quad + \partial_{\alpha x}^2 f_{1,s}^{t,\xi} \big[\partial_x \hat{\alpha}_s^{t,\xi} \partial_x X_s^{t,\xi,\mu} + \partial_y \hat{\alpha}_s^{t,\xi} \partial_x Y_s^{t,\xi,\mu}\big]\Big) ds \\
&\quad + \partial_x Z_s^{t,\xi,\mu} dW_s + \partial_x Z_s^{0;t,\xi,\mu} dW_s^0, \quad s \in [t,T].
\end{aligned} \tag{5.134}
$$

Here the initial condition of the forward equation reads $\partial_x X_t^{t,\xi,\mu} \chi = \chi$ (since $\partial_x X_t^{t,\xi,\mu}$ is the identity matrix) while the terminal condition of the backward equation satisfies $\partial_x Y_T^{t,\xi,\mu} \chi = \partial_{xx}^2 g(X_T^{t,\xi,\mu}, \mathcal{L}^1(X_T^{t,\xi,\mu})) \partial_\chi X_T^{t,\xi,\mu}$, with $\partial_\chi X_T^{t,\xi,\mu} = \partial_x X_T^{t,\xi,\mu} \chi$.

Third Step. We observe that $\partial_\chi \boldsymbol{Y}^{t,\xi}$ can be written:

$$
\partial_\chi Y_s^{t,\xi} = \partial_x^2 \mathcal{U}\big(s, X_s^{t,\xi}, \mathcal{L}^1(X_s^{t,\xi})\big) \partial_\chi X_s^{t,\xi} + \tilde{\mathbb{E}}^1\big[\partial_\mu \partial_x \tilde{\mathcal{U}}_s^{t,\xi} \partial_\chi \tilde{X}_s^{t,\xi}\big],
$$

where, again, we used the notation:

$$
\partial_\mu \partial_x \tilde{\mathcal{U}}_s^{t,\xi} = \partial_\mu \partial_x \mathcal{U}\big(s, X_s^{t,\xi}, \mathcal{L}^1(X_s^{t,\xi})\big)(\tilde{X}_s^{t,\xi}), \quad s \in [t,T].
$$

Letting:

$$\Delta_\chi Y_s^{t,\xi} = \partial_\chi Y_s^{t,\xi} - \partial_x^2 \mathcal{U}(s, X_s^{t,\xi}, \mathcal{L}^1(X_s^{t,\xi}))\partial_\chi X_s^{t,\xi} = \tilde{\mathbb{E}}^1\big[\partial_\mu \partial_x \tilde{\mathcal{U}}_s^{t,\xi} \partial_\chi \tilde{X}_s^{t,\xi}\big], \tag{5.135}$$

we can rewrite (5.132) under the form:

$$\begin{aligned}
d\partial_\chi X_s^{t,\xi} &= b(s)\Big(\partial_x \hat{\alpha}_s^{t,\xi} \partial_\chi X_s^{t,\xi} + \partial_y \hat{\alpha}_s^{t,\xi} \partial_x^2 \mathcal{U}_s^{t,\xi} \partial_\chi X_s^{t,\xi} + \partial_y \hat{\alpha}_s^{t,\xi} \Delta_\chi Y_s^{t,\xi}\Big) ds, \\
d\partial_\chi Y_s^{t,\xi} &= -\Big(\partial_x^2 f_{0,s}^{t,\xi} \partial_\chi X_s^{t,\xi} + \mathbb{E}^1\big[\partial_\mu \partial_x \tilde{f}_{0,s}^{t,\xi} \partial_\chi \tilde{X}_s^{t,\xi}\big] + \partial_x^2 f_{1,s}^{t,\xi} \partial_\chi X_s^{t,\xi} \\
&\quad + \partial_{\alpha x}^2 f_{1,s}^{t,\xi}\big[\partial_x \hat{\alpha}_s^{t,\xi} \partial_\chi X_s^{t,\xi} + \partial_y \hat{\alpha}_s^{t,\xi} \partial_x^2 \mathcal{U}_s^{t,\xi} \partial_\chi X_s^{t,\xi} + \partial_y \hat{\alpha}_s^{t,\xi} \Delta_\chi Y_s^{t,\xi}\big]\Big) ds \\
&\quad + \partial_\chi Z_s^{t,\xi} dW_s + \partial_\chi Z_s^{0;t,\xi} dW_s^0.
\end{aligned} \tag{5.136}$$

A key fact is that the solution of the forward equation in (5.136) can be expressed in terms of the process $(\partial_x X_s^{t,\xi,\mu})_{t \le s \le T}$. Indeed, it is easily checked that $(\partial_x X_s^{t,\xi,\mu})_{t \le s \le T}$ takes values in the set of invertible matrices. Using the fact that $\partial_x Y_s^{t,\xi,\mu} = \partial_x^2 \mathcal{U}^{t,\xi} \partial_x X_s^{t,\xi,\mu}$ together with a standard argument of variation of parameters, we claim:

$$\partial_\chi X_s^{t,\xi} = \partial_x X_s^{t,\xi,\mu}\bigg[\chi + \int_t^s \big(\partial_x X_r^{t,\xi,\mu}\big)^{-1} b(r) \partial_y \hat{\alpha}_r^{t,\xi} \Delta_\chi Y_r^{t,\xi} dr\bigg] = \partial_x X_s^{t,\xi,\mu} \Gamma_s^{t,\xi},$$

where we let:

$$\Gamma_s^{t,\xi} = \chi + \int_t^s \big(\partial_x X_r^{t,\xi,\mu}\big)^{-1} b(r) \partial_y \hat{\alpha}_r^{t,\xi} \Delta_\chi Y_r^{t,\xi} dr, \quad s \in [t, T].$$

Fourth Step. Compute now:

$$\begin{aligned}
d\big[\partial_x Y_s^{t,\xi,\mu} \Gamma_s^{t,\xi}\big] &= -\Big(\partial_x^2 f_{0,s}^{t,\xi} \partial_x X_s^{t,\xi,\mu} \Gamma_s^{t,\xi} + \partial_x^2 f_{1,s}^{t,\xi} \partial_x X_s^{t,\xi,\mu} \Gamma_s^{t,\xi} \\
&\quad + \partial_{\alpha x}^2 f_{1,s}^{t,\xi}\big[\partial_x \hat{\alpha}_s^{t,\xi} \partial_x X_s^{t,\xi,\mu} \Gamma_s^{t,\xi} + \partial_y \hat{\alpha}_s^{t,\xi} \partial_x Y_s^{t,\xi,\mu} \Gamma_s^{t,\xi}\big] \\
&\quad - \partial_x Y_s^{t,\xi,\mu} \big(\partial_x X_s^{t,\xi,\mu}\big)^{-1} b(s) \partial_y \hat{\alpha}_s^{t,\xi} \Delta_\chi Y_s^{t,\xi}\Big) ds \\
&\quad + \big(\partial_x Z_s^{t,\xi,\mu} \Gamma_s^{t,\xi}\big) dW_s + \big(\partial_x Z_s^{0;t,\xi,\mu} \Gamma_s^{t,\xi}\big) dW_s^0,
\end{aligned}$$

where brackets in the last line indicate that $\partial_x Z_s^{t,\xi,\mu}$ and $\partial_x Z_s^{0;t,\xi,\mu}$, which are $\mathbb{R}^{(d \times d) \times d}$-valued, first act on $\Gamma_s^{t,\xi}$, so that $\partial_x Z_s^{t,\xi,\mu} \Gamma_s^{t,\xi}$ and $\partial_x Z_s^{0;t,\xi,\mu} \Gamma_s^{t,\xi}$ take values in $\mathbb{R}^{d \times d}$. We obtain:

$$\begin{aligned}
d\big[\partial_x Y_s^{t,\xi,\mu} \Gamma_s^{t,\xi}\big] &= -\Big(\partial_x^2 f_{0,s}^{t,\xi} \partial_\chi X_s^{t,\xi} + \partial_x^2 f_{1,s}^{t,\xi} \partial_\chi X_s^{t,\xi} \\
&\quad + \partial_{\alpha x}^2 f_{1,s}^{t,\xi}\big[\partial_x \hat{\alpha}_s^{t,\xi} \partial_\chi X_s^{t,\xi} + \partial_y \hat{\alpha}_s^{t,\xi} \partial_x^2 \mathcal{U}_s^{t,\xi} \partial_\chi X_s^{t,\xi}\big] \\
&\quad - \partial_x^2 \mathcal{U}_s^{t,\xi} b(s) \partial_y \hat{\alpha}_s^{t,\xi} \Delta_\chi Y_s^{t,\xi}\Big) ds \\
&\quad + \big(\partial_x Z_s^{t,\xi,\mu} \Gamma_s^{t,\xi}\big) dW_s + \big(\partial_x Z_s^{0;t,\xi,\mu} \Gamma_s^{t,\xi}\big) dW_s^0.
\end{aligned} \tag{5.137}$$

Letting:

$$\Delta_\chi Z_s^{t,\xi} = \partial_\chi Z_s^{t,\xi} - \big(\partial_x Z_s^{t,\xi,\mu} \Gamma_s^{t,\xi}\big), \quad \Delta_\chi Z_s^{0;t,\xi} = \partial_\chi Z_s^{0;t,\xi} - \big(\partial_x Z_s^{0;t,\xi,\mu} \Gamma_s^{t,\xi}\big),$$

and computing the difference between (5.136) and (5.137), we get:

$$\begin{aligned} d\Delta_\chi Y_s^{t,\xi} &= -\Big(\mathbb{E}^1\big[\partial_\mu \partial_x \tilde{f}_{0,s}^{t,\xi} \partial_\chi \tilde{X}_s^{t,\xi}\big] + \partial^2_{\alpha x} f_{1,s}^{t,\xi} \partial_y \hat{\alpha}_s^{t,\xi} \Delta_\chi Y_s^{t,\xi} \\ &\quad + \partial^2_x \mathcal{U}_s^{t,\xi} b(s) \partial_y \hat{\alpha}_s^{t,\xi} \Delta_\chi Y_s^{t,\xi}\Big) ds + \Delta_\chi Z_s^{t,\xi} dW_s + \Delta_\chi Z_s^{0;t,\xi} dW_s^0, \end{aligned} \tag{5.138}$$

for $s \in [t, T]$. With the same convention as above, the terminal condition writes:

$$\Delta_\chi^{t,\xi} Y_T = \tilde{\mathbb{E}}^1\big[\partial_\mu \partial_x \tilde{g}^{t,\xi} \partial_\chi \tilde{X}_T^{t,\xi}\big], \tag{5.139}$$

see (5.135). We now compute:

$$\begin{aligned} &d\big[\partial_\chi X_s^{t,\xi} \cdot \Delta_\chi Y_s^{t,\xi}\big] \\ &= b(s)\Big(\partial_x \hat{\alpha}_s^{t,\xi} \partial_\chi X_s^{t,\xi} + \partial_y \hat{\alpha}_s^{t,\xi} \partial^2_x \mathcal{U}_s^{t,\xi} \partial_\chi X_s^{t,\xi} + \partial_y \hat{\alpha}_s^{t,\xi} \Delta_\chi Y_s^{t,\xi}\Big) \cdot \Delta_\chi Y_s^{t,\xi} \\ &- \partial_\chi X_s^{t,\xi} \cdot \Big(\mathbb{E}^1\big[\partial_\mu \partial_x \tilde{f}_{0,s}^{t,\xi} \partial_\chi \tilde{X}_s^{t,\xi}\big] + \partial^2_{\alpha x} f_{1,s}^{t,\xi} \partial_y \hat{\alpha}_s^{t,\xi} \Delta_\chi Y_s^{t,\xi} + \partial^2_x \mathcal{U}_s^{t,\xi} b(s) \partial_y \hat{\alpha}_s^{t,\xi} \Delta_\chi Y_s^{t,\xi}\Big) ds \\ &+ dM_s, \end{aligned}$$

where $(M_s)_{t \le s \le T}$ is a uniformly integrable martingale.

Fifth Step. Recall the formula:

$$b(t)^\dagger y + \partial_\alpha f_1\big(t, x, \hat{\alpha}(t, x, y)\big) = 0, \tag{5.140}$$

from which get:

$$b(t)^\dagger + \partial^2_\alpha f_1\big(t, x, \hat{\alpha}(t, x, y)\big) \partial_y \hat{\alpha}(t, x, y) = 0,$$

that is,

$$\begin{aligned} \partial_y \hat{\alpha}(t, x, y) &= -\big(\partial^2_\alpha f_1(t, x, \hat{\alpha}(t, x, y))\big)^{-1} b(t)^\dagger, \\ b(t) \partial_y \hat{\alpha}(t, x, y) &= -b(t)\big(\partial^2_\alpha f_1(t, x, \hat{\alpha}(t, x, y))\big)^{-1} b(t)^\dagger. \end{aligned} \tag{5.141}$$

In particular, $b(t)\partial_y \hat{\alpha}(t, x, y)$ is symmetric. Therefore, by symmetry of $b(s)\partial_y \hat{\alpha}_s^{t,\xi}$ and $\partial^2_x \mathcal{U}_s^{t,\xi}$, we obtain:

$$\big(b(s)\partial_y \hat{\alpha}_s^{t,\xi} \partial^2_x \mathcal{U}_s^{t,\xi} \partial_\chi X_s^{t,\xi}\big) \cdot \Delta_\chi Y_s^{t,\xi} = \partial_\chi X_s^{t,\xi} \cdot \big(\partial^2_x \mathcal{U}_s^{t,\xi} b(s) \partial_y \hat{\alpha}_s^{t,\xi} \Delta_\chi Y_s^{t,\xi}\big). \tag{5.142}$$

Moreover, by differentiating (5.140) with respect to x, we get:

$$\partial^2_{x\alpha} f_1\big(t, x, \hat{\alpha}(t, x, y)\big) + \partial^2_\alpha f_1\big(t, x, \hat{\alpha}(t, x, y)\big) \partial_x \hat{\alpha}(t, x, y) = 0,$$

that is,

$$\big(\partial^2_{\alpha x} f_1((t, x, \hat{\alpha}(t, x, y))\big)^\dagger + \partial^2_\alpha f_1\big(t, x, \hat{\alpha}(t, x, y)\big) \partial_x \hat{\alpha}(t, x, y) = 0,$$

so that, multiplying by $b(t)(\partial^2_\alpha f_1(t,x,\hat\alpha(t,x,y)))^{-1}$ and using (5.141), we obtain:

$$b(t)\partial_x\hat\alpha(t,x,y) = \big(\partial_y\hat\alpha(t,x,y)\big)^\dagger\big(\partial^2_{\alpha x} f_1(t,x,\hat\alpha(t,x,y))\big)^\dagger.$$

We deduce:

$$\big(b(s)\partial_x\hat\alpha_s^{t,\xi}\partial_\chi X_s^{t,\xi}\big)\cdot\Delta_\chi Y_s^{t,\xi} = \partial_\chi X_s^{t,\xi}\cdot\big(\partial^2_{\alpha x} f_{1,s}^{t,\xi}\partial_y\hat\alpha_s^{t,\xi}\Delta_\chi Y_s^{t,\xi}\big). \tag{5.143}$$

Plugging (5.142) and (5.143) in the conclusion of the fourth step and using the second line in (5.141), we finally have:

$$\begin{aligned}
&d\big[\partial_\chi X_s^{t,\xi}\cdot\Delta_\chi Y_s^{t,\xi}\big]\\
&= -\Big(\big((\partial^2_\alpha f_{1,s}^{t,\xi})^{-1}b(s)^\dagger\Delta_\chi Y_s^{t,\xi}\big)\cdot\big(b(s)^\dagger\Delta_\chi Y_s^{t,\xi}\big) + \partial_\chi X_s^{t,\xi}\cdot\mathbb{E}^1\big[\partial_\mu\partial_x\tilde f_{0,s}^{t,\xi}\partial_\chi\tilde X_s^{t,\xi}\big]\Big)ds + dM_s.
\end{aligned}$$

Taking expectations and recalling from (5.131) that:

$$\begin{aligned}
&\mathbb{E}\Big[\partial_\chi X_s^{t,\xi}\cdot\tilde{\mathbb{E}}^1\big[\partial_\mu\partial_x\tilde f_{0,s}^{t,\xi}\partial_\chi\tilde X_s^{t,\xi}\big]\Big]\geq 0,\\
&\mathbb{E}\big[\partial_\chi X_T^{t,\xi}\cdot\Delta_\chi Y_T^{t,\xi}\big] = \mathbb{E}\Big[\partial_\chi X_T^{t,\xi}\cdot\tilde{\mathbb{E}}^1\big[\partial_\mu\partial_x\tilde g^{t,\xi}\partial_\chi\tilde X_T^{t,\xi}\big]\Big]\geq 0,
\end{aligned}$$

we deduce:

$$\begin{aligned}
\mathbb{E}\big[\Delta_\chi Y_t^{t,\xi}\cdot\chi\big] &\geq \mathbb{E}\int_t^T \big((\partial^2_\alpha f_{1,s}^{t,\xi})^{-1}b(s)^\dagger\Delta_\chi Y_s^{t,\xi}\big)\cdot\big(b(s)^\dagger\Delta_\chi Y_s\big)^{t,\xi}ds\\
&\geq \Lambda^{-1}\mathbb{E}\int_t^T \big|b(s)^\dagger\Delta_\chi Y_s^{t,\xi}\big|^2 ds.
\end{aligned} \tag{5.144}$$

Sixth Step. We now come back to the dynamics of $(\partial_\chi X_s^{t,\xi})_{t\leq s\leq T}$. From (5.136), we have:

$$d\partial_\chi X_s^{t,\xi} = b(s)\Big(\partial_x\hat\alpha_s^{t,\xi}\partial_\chi X_s^{t,\xi} + \partial_y\hat\alpha_s^{t,\xi}\partial^2_x\mathcal{U}_s^{t,\xi}\partial_\chi X_s^{t,\xi} + \partial_y\hat\alpha_s^{t,\xi}\Delta_\chi Y_s^{t,\xi}\Big)ds.$$

If there exists a constant K_1 such that $|\partial^2_x\mathcal{U}_s^{t,\xi}|\leq K_1$, then, applying Gronwall's lemma, we get:

$$\mathbb{E}\big[\sup_{t\leq s\leq T}|\partial_\chi X_s^{t,\xi}|\big]\leq C\Big(\|\chi\|_1 + \mathbb{E}\int_t^T |b(s)\partial_y\hat\alpha_s^{t,\xi}\Delta_\chi Y_s^{t,\xi}|ds\Big),$$

where C may depend on K_1. Recalling from (5.141) that $b(t)\partial_y\hat\alpha(t,x,y) = -b(t)[\partial^2_\alpha f_1(t,x,\hat\alpha(t,x,y))]^{-1}b(t)^\dagger$ and using the fact that $[\partial^2_\alpha f_1(t,x,\hat\alpha(t,x,y))]^{-1}$ is bounded by λ^{-1}, we deduce that there exists a constant C, depending on Λ, λ, K_1 and T, such that:

$$\begin{aligned}
\mathbb{E}\big[\sup_{t\leq s\leq T}|\partial_\chi X_s^{t,\xi}|\big] &\leq C\Big(\|\chi\|_1 + \mathbb{E}\int_t^T |b(s)^\dagger\Delta_\chi Y_s^{t,\xi}|ds\Big)\\
&\leq C\Big(\|\chi\|_1 + \big|\mathbb{E}^1\big[\chi\cdot\Delta_\chi Y_t^{t,\xi}\big]\big|^{1/2}\Big),
\end{aligned} \tag{5.145}$$

where we used (5.144) to derive the last line. By (5.138) and (5.139), we also have:

$$|\Delta_\chi Y_s^{t,\xi}| \le C\Big(\mathbb{E}_s[\tilde{\mathbb{E}}^1(|\partial_\chi \tilde{X}_T^{t,\xi}|)] + \mathbb{E}_s \int_s^T \Big(\tilde{\mathbb{E}}^1(|\partial_\chi \tilde{X}_u^{t,\xi}|) + |\Delta_\chi Y_u^{t,\xi}|\Big)du\Big).$$

Taking the conditional expectation given $\mathcal{F}_r$, for $r \in [t,s]$, we deduce that, for all $s \in [r,T]$,

$$\mathbb{E}_r[|\Delta_\chi Y_s^{t,\xi}|] \le C\Big(\mathbb{E}_r[\tilde{\mathbb{E}}^1(|\partial_\chi \tilde{X}_T^{t,\xi}|)] + \mathbb{E}_r \int_s^T \Big(\tilde{\mathbb{E}}^1(|\partial_\chi \tilde{X}_u^{t,\xi}|) + |\Delta_\chi Y_u^{t,\xi}|\Big)du\Big).$$

By Gronwall's lemma, we get, for a new value of the constant C,

$$\mathbb{E}_r[|\Delta_\chi Y_s^{t,\xi}|] \le C\Big(\mathbb{E}_r[\tilde{\mathbb{E}}^1(|\partial_\chi \tilde{X}_T^{t,\xi}|)] + \mathbb{E}_r \int_s^T \tilde{\mathbb{E}}^1(|\partial_\chi \tilde{X}_u^{t,\xi}|)du\Big).$$

In particular, taking $r = s$ and multiplying by $|\zeta|$, for $\zeta \in L^2(\Omega^1, \mathcal{F}_t^1, \mathbb{P}^1; \mathbb{R}^d)$, we obtain:

$$|\Delta_\chi Y_s^{t,\xi}|\,|\zeta| \le C|\zeta|\Big(\mathbb{E}_s[\tilde{\mathbb{E}}^1(|\partial_\chi \tilde{X}_T^{t,\xi}|)] + \mathbb{E}_s \int_s^T \tilde{\mathbb{E}}^1(|\partial_\chi \tilde{X}_r^{t,\xi}|)dr\Big).$$

By Lemma 5.16, $\mathbb{E}_s[\tilde{\mathbb{E}}^1(|\partial_\chi \tilde{X}_r^{t,\xi}|)] = \mathbb{E}_s[\mathbb{E}^1(|\partial_\chi X_r^{t,\xi}|)] = \mathbb{E}_s[|\partial_\chi X_r^{t,\xi}|\,|\,\mathcal{F}_s^0]$. In particular, $\mathbb{E}_s[\tilde{\mathbb{E}}^1(|\partial_\chi \tilde{X}_r^{t,\xi}|)]$ is independent of $\mathcal{F}^1$. Therefore, taking expectation in above inequality and making use of (5.145), we get (allowing the constant C to increase from line to line):

$$\sup_{t\le s\le T} \mathbb{E}[|\Delta_\chi Y_s^{t,\xi}|\,|\zeta|] \le C\|\zeta\|_1\Big(\|\chi\|_1 + |\mathbb{E}^1[\chi\cdot \Delta_\chi Y_t^{t,\xi}]|^{1/2}\Big).$$

Specializing the left-hand side at $s = t$, we deduce:

$$\sup_{\|\zeta\|_1\le 1} \mathbb{E}^1[|\Delta_\chi Y_t^{t,\xi}|\,|\zeta|] \le C\Big(\|\chi\|_1 + \|\chi\|_1^{1/2} \sup_{\|\zeta\|_1\le 1} \mathbb{E}^1[|\zeta|\,|\Delta_\chi Y_t^{t,\xi}|]^{1/2}\Big).$$

By a standard convexity argument, we get:

$$\sup_{\|\zeta\|_1\le 1} \mathbb{E}^1[|\Delta_\chi Y_t^{t,\xi}|\,|\zeta|] \le C\|\chi\|_1. \tag{5.146}$$

Last Step. We now complete the proof. We recall from (5.135) that:

$$\Delta_\chi Y_t^{t,\xi} = \tilde{\mathbb{E}}^1\big[\partial_\mu\partial_x\mathcal{U}(t,\xi,\mathcal{L}^1(\xi))(\tilde{\xi})\tilde{\chi}\big],$$

and that (5.146) holds true for random variables ξ taking a finite number of values only.

By regularity of $\partial_x\partial_\mu\mathcal{U}$, (5.146) holds true for any $\xi \in L^2(\Omega^1, \mathcal{F}_t^1, \mathbb{P}^1; \mathbb{R}^d)$. In particular, for any $\xi, \chi \in L^2(\Omega^1, \mathcal{F}_t^1, \mathbb{P}^1; \mathbb{R}^d)$, we have, with probability 1 under $\mathbb{P}^1$,

$$\big|\tilde{\mathbb{E}}^1\big[\partial_\mu\partial_x\mathcal{U}(t,\xi,\mathcal{L}^1(\xi))(\tilde{\xi})\tilde{\chi}\big]\big| \le C\mathbb{E}^1[|\chi|].$$

Since we assumed that $L^2(\Omega^1, \mathcal{F}^1, \mathbb{P}^1; \mathbb{R}^d)$ was separable, see Subsections 5.1.2 and 4.3.1, the above holds true, when ξ is given and with probability 1 under $\mathbb{P}^1$, for all $\chi \in L^2(\Omega^1, \mathcal{F}_t^1, \mathbb{P}^1; \mathbb{R}^d)$. Hence, with probability 1 under $\mathbb{P}^1 \otimes \tilde{\mathbb{P}}^1$,

$$\left|\partial_\mu \partial_x \mathcal{U}\big(t, \xi, \mathcal{L}^1(\xi)\big)(\tilde{\xi})\right| \leq C.$$

We deduce that, whenever ξ has full support,

$$\sup_{t \in [S,T]} \sup_{x, v \in \mathbb{R}^d} \left|\partial_\mu \partial_x \mathcal{U}\big(t, x, \mu\big)(v)\right| \leq C,$$

where $\mu = \mathcal{L}^1(\xi)$. By an approximation argument, the same holds true for any $\mu \in \mathcal{P}_2(\mathbb{R}^d)$. □

Proof of the Lasry-Lions Monotonicity Condition

We now prove Lemma 5.52:

Proof. Given $t \in [S, T]$ and $\xi, \xi' \in L^2(\Omega^1, \mathcal{F}_t^1, \mathbb{P}^1; \mathbb{R}^d)$, we consider the forward components $(X_s^{t,\xi})_{s \in [t,T]}$ and $(X_s^{t,\xi'})_{s \in [t,T]}$ of the FBSDE (5.114), when initialized at time t with ξ and ξ' respectively.

Recalling the representation formulas:

$$Y_s^{t,\xi} = \partial_x \mathcal{U}\big(s, X_s^{t,\xi}, \mathcal{L}^1(X_s^{t,\xi})\big), \quad Y_s^{t,\xi'} = \partial_x \mathcal{U}\big(s, X_s^{t,\xi'}, \mathcal{L}^1(X_s^{t,\xi'})\big), \quad s \in [t, T],$$

we then expand $(\mathcal{U}(s, X_s^{t,\xi}, \mathcal{L}^1(X_s^{t,\xi})) - \mathcal{U}(s, X_s^{t,\xi'}, \mathcal{L}^1(X_s^{t,\xi})))_{s \in [t,T]}$ by Itô's formula. Notice also that, in both terms, the measure argument is driven by ξ.

Notice that the assumption required in the statement of Theorem 4.17 to apply the chain rule is satisfied thanks to hypothesis $\mathscr{H}(S)$. We get, for $s \in [t, T]$,

$$\begin{aligned}
&d\big[\mathcal{U}\big(s, X_s^{t,\xi}, \mathcal{L}^1(X_s^{t,\xi})\big)\big] \\
&= -f\Big(s, X_s^{t,\xi}, \mathcal{L}^1(X_s^{t,\xi}), \hat{\alpha}\big(s, X_s^{t,\xi}, \partial_x \mathcal{U}\big(s, X_s^{t,\xi}, \mathcal{L}^1(X_s^{t,\xi})\big)\big)\Big) ds \\
&\quad + \partial_x \mathcal{U}\big(s, X_s^{t,\xi}, \mathcal{L}^1(X_s^{t,\xi})\big) \cdot \big(\sigma dW_s\big) + \partial_x \mathcal{U}\big(s, X_s^{t,\xi}, \mathcal{L}^1(X_s^{t,\xi})\big) \cdot \big(\sigma^0 dW_s^0\big) \\
&\quad + \tilde{\mathbb{E}}^1\big[\partial_\mu \mathcal{U}\big(s, X_s^{t,\xi}, \mathcal{L}^1(X_s^{t,\xi})\big)(\tilde{X}_s^{t,\xi})\big] \cdot \big(\sigma^0 dW_s^0\big),
\end{aligned} \tag{5.147}$$

and

$$\begin{aligned}
&d\big[\mathcal{U}\big(s, X_s^{t,\xi'}, \mathcal{L}^1(X_s^{t,\xi})\big)\big] \\
&= \Big[-f\Big(s, X_s^{t,\xi'}, \mathcal{L}^1(X_s^{t,\xi}), \hat{\alpha}\big(X_s^{t,\xi'}, \partial_x \mathcal{U}\big(s, X_s^{t,\xi'}, \mathcal{L}^1(X_s^{t,\xi})\big)\big)\Big) \\
&\qquad + \partial_x \mathcal{U}\big(s, X_s^{t,\xi'}, \mathcal{L}^1(X_s^{t,\xi})\big) \cdot \Big(b(s)\hat{\alpha}\big(s, X_s^{t,\xi'}, \partial_x \mathcal{U}\big(s, X_s^{t,\xi'}, \mathcal{L}^1(X_s^{t,\xi'})\big)\big) \\
&\qquad\qquad - b(s)\hat{\alpha}\big(s, X_s^{t,\xi'}, \partial_x \mathcal{U}\big(s, X_s^{t,\xi'}, \mathcal{L}^1(X_s^{t,\xi})\big)\big)\Big)\Big] ds \\
&\quad + \partial_x \mathcal{U}\big(s, X_s^{t,\xi'}, \mathcal{L}^1(X_s^{t,\xi})\big) \cdot \big(\sigma dW_s\big) + \partial_x \mathcal{U}\big(s, X_s^{t,\xi'}, \mathcal{L}^1(X_s^{t,\xi})\big) \cdot \big(\sigma^0 dW_s^0\big) \\
&\quad + \tilde{\mathbb{E}}^1\big[\partial_\mu \mathcal{U}\big(s, X_s^{t,\xi'}, \mathcal{L}^1(X_s^{t,\xi})\big)(\tilde{X}_s^{t,\xi})\big] \cdot \big(\sigma^0 dW_s^0\big).
\end{aligned} \tag{5.148}$$

Subtracting (5.148) from (5.147), we obtain:

$$
\begin{aligned}
& d\big[\mathcal{U}(s, X_s^{t,\xi}, \mathcal{L}^1(X_s^{t,\xi})) - \mathcal{U}(s, X_s^{t,\xi'}, \mathcal{L}^1(X_s^{t,\xi}))\big] \\
&= -\Big[f\Big(s, X_s^{t,\xi}, \mathcal{L}^1(X_s^{t,\xi}), \hat{\alpha}\big(X_s^{t,\xi}, \partial_x \mathcal{U}(s, X_s^{t,\xi}, \mathcal{L}^1(X_s^{t,\xi}))\big)\Big) \\
&\qquad - f\Big(s, X_s^{t,\xi'}, \mathcal{L}^1(X_s^{t,\xi}), \hat{\alpha}\big(s, X_s^{t,\xi'}, \partial_x \mathcal{U}(s, X_s^{t,\xi'}, \mathcal{L}^1(X_s^{t,\xi'}))\big)\Big)\Big] ds \\
&\quad - \Big[H\Big(s, X_s^{t,\xi'}, \mathcal{L}^1(X_s^{t,\xi}), \partial_x \mathcal{U}(s, X_s^{t,\xi'}, \mathcal{L}^1(X_s^{t,\xi})), \\
&\qquad\qquad\qquad \hat{\alpha}\big(s, X_s^{t,\xi'}, \partial_x \mathcal{U}(s, X_s^{t,\xi'}, \mathcal{L}^1(X_s^{t,\xi'}))\big)\Big) \\
&\qquad - H\Big(s, X_s^{t,\xi'}, \mathcal{L}^1(X_s^{t,\xi}), \partial_x \mathcal{U}(s, X_s^{t,\xi'}, \mathcal{L}^1(X_s^{t,\xi})), \\
&\qquad\qquad\qquad \hat{\alpha}\big(s, X_s^{t,\xi'}, \partial_x \mathcal{U}(s, X_s^{t,\xi'}, \mathcal{L}^1(X_s^{t,\xi}))\big)\Big)\Big] ds \\
&\quad + \Big[\partial_x \mathcal{U}(s, X_s^{t,\xi}, \mathcal{L}^1(X_s^{t,\xi})) - \partial_x \mathcal{U}(s, X_s^{t,\xi'}, \mathcal{L}^1(X_s^{t,\xi}))\Big] \cdot (\sigma dW_s) \\
&\quad + \Big[\partial_x \mathcal{U}(s, X_s^{t,\xi}, \mathcal{L}^1(X_s^{t,\xi})) - \partial_x \mathcal{U}(s, X_s^{t,\xi'}, \mathcal{L}^1(X_s^{t,\xi}))\Big] \cdot (\sigma^0 dW_s^0) \\
&\quad + \tilde{\mathbb{E}}^1\Big[\partial_\mu \mathcal{U}(s, X_s^{t,\xi}, \mathcal{L}^1(X_s^{t,\xi}))(\tilde{X}_s^{t,\xi}) - \partial_\mu \mathcal{U}(s, X_s^{t,\xi'}, \mathcal{L}^1(X_s^{t,\xi}))(\tilde{X}_s^{t,\xi})\Big] \cdot (\sigma^0 dW_s^0).
\end{aligned}
$$

Therefore, taking expectation and integrating in s from t to T, we deduce from the fact that H is convex in α and from the fact that $\hat{\alpha}(t, x, y)$ minimizes $H(t, x, \mu, y, \cdot)$ that:

$$
\begin{aligned}
& \mathbb{E}\big[\mathcal{U}(t, \xi, \mathcal{L}^1(\xi)) - \mathcal{U}(t, \xi', \mathcal{L}^1(\xi))\big] \\
&\qquad - \mathbb{E}\int_t^T \Big[f_1\Big(s, X_s^{t,\xi}, \hat{\alpha}\big(s, X_s^{t,\xi}, \partial_x \mathcal{U}(s, X_s^{t,\xi}, \mathcal{L}^1(X_s^{t,\xi}))\big)\Big) \\
&\qquad\qquad - f_1\Big(X_s^{t,\xi'}, \hat{\alpha}\big(s, X_s^{t,\xi'}, \partial_x \mathcal{U}(s, X_s^{t,\xi'}, \mathcal{L}^1(X_s^{t,\xi'}))\big)\Big)\Big] ds \\
&\geq \mathbb{E}\big[g(X_T^{t,\xi}, \mathcal{L}^1(X_T^{t,\xi})) - g(X_T^{t,\xi'}, \mathcal{L}^1(X_T^{t,\xi}))\big] \\
&\qquad + \mathbb{E}\int_t^T \Big(f_0(s, X_s^{t,\xi}, \mathcal{L}^1(X_s^{t,\xi})) - f_0(s, X_s^{t,\xi'}, \mathcal{L}^1(X_s^{t,\xi}))\Big) ds.
\end{aligned}
$$

By exchanging the roles of ξ and ξ' and then summing up, we get:

$$
\begin{aligned}
& \mathbb{E}\Big[\mathcal{U}(t, \xi, \mathcal{L}^1(\xi)) - \mathcal{U}(t, \xi', \mathcal{L}^1(\xi)) - \Big(\mathcal{U}(t, \xi, \mathcal{L}^1(\xi')) - \mathcal{U}(t, \xi', \mathcal{L}^1(\xi'))\Big)\Big] \\
&\geq \mathbb{E}\Big[g(X_T^{t,\xi}, \mathcal{L}^1(X_T^{t,\xi})) - g(X_T^{t,\xi'}, \mathcal{L}^1(X_T^{t,\xi})) \\
&\qquad\qquad - \Big(g(X_T^{t,\xi}, \mathcal{L}^1(X_T^{t,\xi'})) - g(X_T^{t,\xi'}, \mathcal{L}^1(X_T^{t,\xi'}))\Big)\Big] \\
&\quad + \mathbb{E}\int_t^T \Big[f_0(s, X_s^{t,\xi}, \mathcal{L}^1(X_s^{t,\xi})) - f_0(s, X_s^{t,\xi'}, \mathcal{L}^1(X_s^{t,\xi})) \\
&\qquad\qquad - \Big(f_0(s, X_s^{t,\xi}, \mathcal{L}^1(X_s^{t,\xi'})) - f_0(s, X_s^{t,\xi'}, \mathcal{L}^1(X_s^{t,\xi'}))\Big)\Big] ds.
\end{aligned}
$$

Finally, rearranging the terms, we deduce from the Lasry-Lions condition that:

$$\mathbb{E}\Big[\mathcal{U}\big(t,\xi,\mathcal{L}^1(\xi)\big)-\mathcal{U}\big(t,\xi',\mathcal{L}^1(\xi)\big)-\Big(\mathcal{U}\big(t,\xi,\mathcal{L}^1(\xi')\big)-\mathcal{U}\big(t,\xi',\mathcal{L}^1(\xi')\big)\Big)\Big]\geq 0,$$

which completes the proof. □

Proof of Theorem 5.49

Proof of Theorem 5.49. We refer to Subsection 3.4.1 for the details of assumption **MFG with a Common Noise HJB**, which plays a key role in the proof.

First Step. As opposed to assumption **MFG with a Common Noise SMP Relaxed**, see Subsection 3.4.3, assumption **MFG with a Common Noise HJB** forces the coefficients f and g to be bounded in (t,x,μ). In fact, g is bounded and f is of quadratic growth in α, uniformly in the other parameters. Moreover, by Theorem 1.57, the optimal control process $\hat{\boldsymbol{\alpha}}$ is known to be bounded, whatever the initial condition is. Inserting all these bounds into (5.116), we deduce that $\mathcal{U}$ is bounded. This proves the first part of Theorem 5.49.

Second Step. We now turn to the second part of the statement in Theorem 5.49. The strategy is close to that used to prove Theorem 5.45. It is in fact even simpler.

By Theorem 5.46, we know that (5.121) holds true without the subscript n therein. By assumption **MFG with a Common Noise HJB**, $\partial_x f$ is bounded and $\partial_\alpha f$ is at most of linear growth in α. As already mentioned, the optimal control is known to be bounded, so that everything works as if $\partial_\alpha f$ was also bounded. Lastly, (**A2**) in assumption **MFG Master Classical HJB** ensures that $\partial_\mu f$ is bounded. Hence, using the same notation as in (5.121), but without the subscript n, the functions:

$$(t,w,\mu)\mapsto\partial_w F(t,w,\mu),$$
$$(t,w,\mu,v)\mapsto\partial_\mu F(t,w,\mu)(v),$$

can be assumed to be bounded, and similarly with G in lieu of F. Also, by (**A2**) in assumption **MFG Smooth Coefficients**, $\partial_x f$ and $\partial_\alpha f$ are Lipschitz continuous in (x,μ,α), while, by (**A2**) in assumption **MFG Master Classical HJB**, $\partial_\mu f$ is Lipschitz continuous in (x,μ,α,v). As a consequence, the functions:

$$(t,w,\mu)\mapsto\partial_w F(t,w,\mu),$$
$$(t,w,\mu,v)\mapsto\partial_\mu F(t,w,\mu)(v),$$

are Lipschitz continuous, and similarly with G. Duplicating the second step of the proof of Theorem 5.45, we deduce that the function:

$$\mathbb{R}^d\times\mathcal{P}_2(\mathbb{R}^d)\times\mathbb{R}^d\ni(t,x,\mu,v)\mapsto\partial_\mu\mathcal{U}(t,x,\mu)(v)$$

is bounded and Lipschitz.

Third Step. The argument is pretty much the same for the second-order derivatives and consists in a similar adaptation of the fourth step of the proof of Theorem 5.45. □

5.5 Notes & Complements

The master equation for mean field games has the special feature to be posed on the space of probability measures. Because of this, and despite the analysis we provided in this chapter, it remains a quite fascinating and rather mysterious object. However, it is fair to say that it is not the first example of an equation set on the space of measures. For instance Otto [295] gave an interpretation of the porous medium equation as an evolution equation in the space of measures, and Jordan, Kinderlehrer, and Otto [220] showed that the heat equation was also a gradient flow in that framework. Also, the theoretical analysis of Hamilton-Jacobi equations in metric spaces was developed in no small part to treat specific applications for which the underlying metric spaces are spaces of measures. See in particular [20, 152] and the references therein. However, the master equation has the additional feature to combine three challenging features: being nonlocal, nonlinear and of second order.

As we already explained in the Notes & Complements of Chapter 4, the master equation for mean field games has been introduced by Lasry and Lions in Lions' lectures at the *Collège de France* [265], but it is worth emphasizing that it came quite some time after the original description of mean field games by means of the Fokker-Planck / Hamilton-Jacobi-Bellman system presented in Chapter (Vol I)-3. Indeed, it became increasingly clear that the Fokker-Planck / Hamilton-Jacobi-Bellman formulation was not sufficient to capture the entire complexity of mean field games, and Lasry and Lions became hard pressed to come up with a single formulation which could accommodate mean field games with and without common noise.

Beside the analysis provided in [265], the importance of the master equation was acknowledged by several contributions. In the notes he wrote following Lions' lectures, Cardaliaguet [83] already discussed a form of master equation in the particular case of players' states having deterministic dynamics, the solutions to the master equation being understood in the viscosity sense. In the same case of deterministic dynamics, the existence of classical solutions was investigated over short time horizons by Gangbo and Swiech in [168]. Recently, Bensoussan and Yam revisited this result in [54]. Meanwhile, several independent contributions treated stochastic dynamics of various degrees of generality. See for instance [50–52, 97, 182, 183, 237]. These works use different approaches to derive the master equation and compute derivatives on the Wasserstein space. Anyway, their analyses are mostly heuristic in nature.

The road map for a rigorous construction of a classical solution to the master equation over time intervals of arbitrary length may be found in Lions' lectures at the *Collège de France*. Especially, we refer the reader to the video [266], taken from a seminar where Lions gave an outline of a possible proof for investigating the master equation rigorously by means of PDE arguments.

As of the writing of this book, only two preprints appeared with self-contained and rigorous constructions of classical solutions to the master equation for time intervals of arbitrary lengths.

The first one is by Chassagneux, Crisan, and Delarue, [114]. Therein, the authors implement the same kind of strategy as the one developed in this chapter. Namely, the master equation is established in two steps: first, in small time, through the analysis of the flow of the stochastic process solving the corresponding McKean-Vlasov forward-backward system; second, for arbitrary time intervals, through an inductive argument which works under conditions very similar to those used in this chapter. Despite these similarities, there are significant differences between this work and our presentation. First, [114] only refers to the case without common noise, while we here tackle both cases, with and without common noise. Second, the regularity conditions are much weaker in [114]. Indeed, in the present chapter, we strengthened the smoothness assumptions on the coefficients in order to make the analysis more transparent, especially the analysis of the case with common noise. Basically, we require all the derivatives on the Wasserstein space to be bounded in L^∞, which is rather restrictive. In comparison, derivatives are only required to be bounded in L^2 in [114]. The gap between these two sets of conditions is clearly illustrated by the following simple example: the L-derivative of the function $\mathcal{P}_2(\mathbb{R}^d) \ni \mu \mapsto M_2(\mu) = (\int_{\mathbb{R}^d} |x|^2 d\mu(x))^{1/2}$ is $\mathbb{R}^d \ni v \mapsto v/M_2(\mu)$, provided that μ is not the Dirac mass at 0. Obviously, the derivative is uniformly bounded in L^2, but is not in L^∞. Actually, the standing assumptions in this chapter are close to those used in another recent work by Buckdahn, Li, Peng, and Rainer [79]. Therein, the authors implement a similar approach in order to study forward flows, proving that the semigroup of a standard McKean-Vlasov stochastic differential equation with smooth coefficients is the classical solution of a linear PDE defined on $\mathbb{R}^d \times \mathcal{P}_2(\mathbb{R}^d)$, very much in the spirit of Subsection (Vol I)-5.7.4.

The other main result on classical solutions of the master equation is due to Cardaliaguet, Delarue, Lasry, and Lions [86]. The strategy in [86] is somewhat different. Indeed, the starting point is the representation of mean field games with common noise by means of the infinite dimensional forward-backward system (2.37)–(2.38) consisting of a stochastic Fokker-Planck equation and a stochastic Hamilton-Jacobi-Bellman equation. However, once the representation is chosen, the philosophy is pretty much the same as ours. In [86], the goal is also to prove that the flow formed by the solution of the system (2.37)–(2.38) is smooth, even if, therein, smoothness is investigated with respect to the linear functional calculus presented in Subsection (Vol I)-5.4.1 instead of the L-differential calculus used in this chapter. Also, and even if it is only a minor technical difference, the analysis of [86] is done on the torus.

As a final remark, we emphasize once more that the analysis developed in this chapter could be applied to the study of optimal control problems over McKean-Vlasov diffusion processes, like those presented in Chapter (Vol I)-6. We refer to [50–52, 54, 97, 114, 168] for references in that direction.

Convergence and Approximations

6

Abstract

The goal of this chapter is to quantify the relationships between equilibria for finite-player games, as they were defined in Chapter (Vol I)-2, and the solutions of the mean field game problems. We first show that the solution of the limiting mean field game problem can be used to provide approximate Nash equilibria for the corresponding finite-player games, and we quantify the nature of the approximation in terms of the size of the game. Interestingly enough, we prove a similar result for the solution of the optimal control of McKean-Vlasov stochastic dynamics. The very notion of equilibrium used for the finite-player games shed new light on the differences between the two asymptotic problems. Next, we turn to the problem of the convergence of Nash equilibria for finite-player games toward solutions of the mean field game problem. We tackle this challenging problem under more specific assumptions, by means of an analytic approach based on the properties of the master equation when the latter has classical solutions.

Throughout the analysis of finite player games, we shall consider two types of mean field interaction. Indeed, for a given player i interacting with $N-1$ other players, for some $N \geq 2$, we may regard the mean field interaction either as the interaction with the empirical distribution associated with the other players or as the interaction with the empirical distribution associated with all the players, including i itself. In short, the first one reads $\frac{1}{N-1}\sum_{j\neq i}\delta_{x_j}$ and the second one $\frac{1}{N}\sum_{j}\delta_{x_j}$, where $x_1, \cdots, x_N$ denote the states of the N players. Mathematically speaking, there should not

R. Carmona, F. Delarue, *Probabilistic Theory of Mean Field Games with Applications II*, Probability Theory and Stochastic Modelling 84,
DOI 10.1007/978-3-319-56436-4_6

be any major difference between the two cases and we should expect for the same asymptotic behavior. However, for notational convenience, it is sometimes easier to work with either one or the other formulation. So, the reader should not be surprised to see the two conventions appearing in the text.

6.1 Approximate Equilibria for Finite-Player Games

In this section, we show how the solution to a large N limit problem (the solution of a mean field game or the optimal control of McKean-Vlasov dynamics) can be used to construct approximate Nash equilibria for finite-player games. Generally speaking, the accuracy of the approximate equilibria will be given by the rate of convergence of empirical measures in the Wasserstein distance, as stated in Theorem (Vol I)-5.8 proven in Section (Vol I)-5.1.2. In order for this chapter to be as self-contained as possible, we restate this result in the following form.

Lemma 6.1 *If $\mu \in \mathcal{P}_q(\mathbb{R}^d)$ for some $q > 4$, there exists a constant c depending only upon d, q and $M_q(\mu)$ such that:*

$$\mathbb{E}\big[W_2(\bar{\mu}^N, \mu)^2\big] \leq cN^{-2/\max(d,4)}\big(1 + \ln(N)\mathbf{1}_{\{d=4\}}\big),$$

where $\bar{\mu}^N$ denotes the empirical measure of any sample of size N from μ.

Throughout the section, we use the following notation for the rate of convergence appearing in the right-hand side of the above inequality:

$$\epsilon_N = N^{-2/\max(d,4)}\big(1 + \ln(N)\mathbf{1}_{\{d=4\}}\big). \tag{6.1}$$

Estimate in 1-Wasserstein Distance

At some point in this chapter, we shall need the analogue of the result of Lemma 6.1 for the 1-Wasserstein distance W_1. The proof goes along the same line as that given in Chapter (Vol I)-5 and the reader is referred to the Notes & Complements below for references where the argument can be found.

Lemma 6.2 *If $\mu \in \mathcal{P}_q(\mathbb{R}^d)$ for some $q > 2$, there exists a constant c depending only upon d, q and $M_q(\mu)$ such that:*

$$\mathbb{E}\big[W_1(\bar{\mu}^N, \mu)\big] \leq cN^{-1/\max(d,2)}\big(1 + \ln(N)\mathbf{1}_{\{d=2\}}\big),$$

where $\bar{\mu}^N$ denotes the empirical measure of any sample of size N from μ.

Actually, using Remark (Vol I)-5.9, one can prove that the way the above constant c actually depends on μ is through the moment $M_q(\mu)$, the dependence being linear. To be more specific,

$$\mathbb{E}\big[W_1(\bar{\mu}^N,\mu)\big] \le cM_q(\mu)N^{-1/\max(d,2)}\big(1+\ln(N)\mathbf{1}_{\{d=2\}}\big), \tag{6.2}$$

for a constant c depending only on d and q.

The case $q=2$ is of special interest for us since we often work with probability distributions in the Wasserstein space $\mathcal{P}_2(\mathbb{R}^d)$. In the references where the proof of Lemma 6.2 can be found, the bound (6.2) is shown to hold when $q=2$ and $d\ge 3$. When $d\in\{1,2\}$, we can prove the slightly weaker bound given below.

Corollary 6.3 *If $\mu\in\mathcal{P}_2(\mathbb{R}^d)$, then, for any $\eta\in(0,1]$, there exists a constant c_η depending only upon d and η such that:*

$$\mathbb{E}\big[W_1(\bar{\mu}^N,\mu)\big] \le c_\eta M_2(\mu)N^{-1/\max(d,2+\eta)}.$$

When $d\ge 3$, the above bound holds true with $\eta=0$.

Proof. Thanks to (6.2), it suffices to argue the case $d\in\{1,2\}$.

We proceed by a truncation argument. For any given $a>0$, we call π_a the orthogonal projection from $\mathbb{R}^d$ onto the d-dimensional ball of center 0 and of radius $a>0$.

By (6.2), we get that, for any $\eta>0$, there exists a constant $c_\eta>0$ such that

$$\mathbb{E}\big[W_1(\bar{\mu}^N\circ\pi_a^{-1},\mu\circ\pi_a^{-1})\big] \le c_\eta M_{2+\eta}(\mu\circ\pi_a^{-1})N^{-1/2}\big(1+\ln(N)\mathbf{1}_{\{d=2\}}\big).$$

Notice indeed, that, for an N-tuple $(X_1,\cdots,X_N)$ of independent and identically distributed random variables with common distribution μ and with $\bar{\mu}^N$ as empirical distribution, $\bar{\mu}^N\circ\pi_a^{-1}$ is the empirical distribution of $(\pi_a(X_1),\cdots,\pi_a(X_N))$. Then, by the triangle inequality, we get:

$$\begin{aligned}
\mathbb{E}\big[W_1(\bar{\mu}^N,\mu)\big] &\le \mathbb{E}\big[W_1(\bar{\mu}^N,\bar{\mu}^N\circ\pi_a^{-1})\big]+\mathbb{E}\big[W_1(\bar{\mu}^N\circ\pi_a^{-1},\mu\circ\pi_a^{-1})\big]+\mathbb{E}\big[W_1(\mu\circ\pi_a^{-1},\mu)\big]\\
&\le \mathbb{E}\Big[\frac{1}{N}\sum_{i=1}^N|\pi_a(X_i)-X_i|\Big]+\mathbb{E}\big[|\pi_a(X_1)-X_1|\big]\\
&\quad + c_\eta\mathbb{E}\big[|\pi_a(X_1)|^{2+\eta}\big]^{1/(2+\eta)}N^{-1/2}\big(1+\ln(N)\mathbf{1}_{\{d=2\}}\big)\\
&\le 2\mathbb{E}\big[|\pi_a(X_1)-X_1|\big]+c_\eta\mathbb{E}\big[|\pi_a(X_1)|^{2+\eta}\big]^{1/(2+\eta)}N^{-1/2}\big(1+\ln(N)\mathbf{1}_{\{d=2\}}\big).
\end{aligned}$$

Now, we have, for $X\sim\mu$,

$$\begin{aligned}
\mathbb{E}\big[|\pi_a(X_1)-X_1|\big] &= \mathbb{E}\big[|X-\pi_a(X)|\mathbf{1}_{\{|X|\ge a\}}\big]\\
&\le 2\mathbb{E}\big[|X|\mathbf{1}_{\{|X|\ge a\}}\big]\\
&\le \frac{2}{a}\mathbb{E}[|X|^2]=\frac{2}{a}M_2(\mu)^2.
\end{aligned}$$

Therefore,

$$\mathbb{E}\big[W_1(\bar{\mu}^N,\mu)\big] \le \frac{4}{a}M_2(\mu)^2 + c_\eta M_2(\mu)^{2/(2+\eta)} a^{\eta/(2+\eta)} N^{-1/2}\big(1+\ln(N)\mathbf{1}_{\{d=2\}}\big).$$

Now we choose a so that:

$$\frac{1}{a} = M_2(\mu)^{-1}\big[N^{-1/2}\big(1+\ln(N)\mathbf{1}_{\{d=2\}}\big)\big]^{\frac{2+\eta}{2+2\eta}},$$

and deduce that:

$$\mathbb{E}\big[W_1(\bar{\mu}^N,\mu)\big] \le c_\eta M_2(\mu)\big[N^{-1/2}\big(1+\ln(N)\mathbf{1}_{\{d=2\}}\big)\big]^{\frac{2+\eta}{2+2\eta}},$$

where the constant c_η is allowed to increase from line to line. Finally, since $N^{-1/2}\ln(N) \le c_\eta N^{-1/2}N^{\eta/(4+2\eta)} = c_\eta N^{-1/(2+\eta)}$, the proof is complete. □

6.1.1 The Case of the MFGs Without Common Noise

For pedagogical reasons, we start with the case without common noise.

We follow the same plot as in Chapters (Vol I)-3 and (Vol I)-4. We implement the two probabilistic approaches presented in the first volume of the book, using the construction of mean field game equilibria either from the representation of the value function, or from the stochastic maximum principle. We refer to Section (Vol I)-3.3 for a general overview of these two approaches, and to Sections (Vol I)-4.4 and (Vol I)-4.5 for more detailed accounts.

General Strategy

In both cases, we shall use the same trick in order to construct approximate Nash equilibria. This common trick relies on the following key observation. For a mean field game equilibrium $\boldsymbol{\mu} = (\mu_t)_{0\le t\le T} \in \mathcal{C}([0,T];\mathbb{R}^d)$ like those constructed in Chapter (Vol I)-4, there exists a continuous mapping $v : [0,T]\times\mathbb{R}^d \to \mathbb{R}^d$, Lipschitz continuous in x uniformly in time and at most of linear growth in x uniformly in time, v depending on $\boldsymbol{\mu}$, such that in equilibrium (namely under $\boldsymbol{\mu}$), the optimal path of the state has the form:

$$dX_t = b\big(t,X_t,\mu_t,\hat{\alpha}\big(t,X_t,\mu_t,v(t,X_t)\big)\big)dt + \sigma(t,X_t,\mu_t)dW_t, \tag{6.3}$$

for $t\in[0,T]$. Here and in what follows, we use freely the notations of Chapters (Vol I)-3 and (Vol I)-4 for the drift b, the standard deviation or volatility coefficient σ, and the noise $\boldsymbol{W} = (W_t)_{0\le t\le T}$ in the underlying mean field game. The reader may also have a look at the general setting used in Chapter 2, which is basically the same provided that σ^0 is null therein. In this framework, $(\hat{\alpha}(t,X_t,\mu_t,v(t,X_t)))_{0\le t\le T}$ is the optimal control, the function $\hat{\alpha}$ being the unique minimizer of the corresponding reduced Hamiltonian:

$$H^{(r)}(t,x,\mu,y,\alpha) = b(t,x,\mu,\alpha)\cdot y + f(t,x,\mu,\alpha).$$

See for instance Lemma (Vol I)-3.3 or Lemma 1.56. Using also the same notations f and g as in Chapters (Vol I)-3 and (Vol I)-4 for the running and terminal cost functions, the optimal cost of the limiting mean field problem can be written as:

$$J = \mathbb{E}\left[\int_0^T f\big(t, X_t, \mu_t, \hat{\alpha}(t, X_t, \mu_t, v(t, X_t))\big)dt + g(X_T, \mu_T)\right]. \tag{6.4}$$

To rephrase, (6.3) is the optimal path and J is the optimal cost of the optimal control problem:

$$\begin{aligned} &\inf_{\boldsymbol{\alpha}\in\mathbb{A}} J^{\mu}(\boldsymbol{\alpha}), \quad \text{with } J^{\mu}(\boldsymbol{\alpha}) = \mathbb{E}\left[\int_0^T f(t, X_t^{\boldsymbol{\alpha}}, \mu_t, \alpha_t)dt + g(X_T^{\boldsymbol{\alpha}}, \mu_T)\right], \\ &\text{subject to} \\ &dX_t^{\boldsymbol{\alpha}} = b(t, X_t^{\boldsymbol{\alpha}}, \mu_t, \alpha_t)dt + \sigma(t, X_t^{\boldsymbol{\alpha}}, \mu_t)dW_t, \qquad t\in[0,T], \end{aligned} \tag{6.5}$$

where as usual $\mathbb{A}$ is the set of square-integrable and progressively measurable control processes taking values in $A \subset \mathbb{R}^k$.

In our approach based upon the stochastic Pontryagin maximum principle, the existence of the function v is straightforward. Somehow, it is a mere consequence of Theorem (Vol I)-4.53 or of Lemma (Vol I)-4.56, see also Theorem 1.60 in Chapter 1 which is somehow the exact analogue of Lemma (Vol I)-4.56 for cases with a common noise. For instance, the function u in the statement of Theorem (Vol I)-4.53 exactly fits our requirements for v since the process $(Y_t = u(t, X_t))_{0\le t\le T}$ therein solves the adjoint system (Vol I)-(4.69) in Chapter 4 of the first volume. The reader who just has Volume II in hand may easily formulate the same observation by adapting the results obtained for mean field games with a common noise. For instance, Theorem 1.60 says pretty much the same thing, although we used therein the capital letter U instead of u in order to emphasize the fact that U was a random field, which is a specific feature of mean field games with a common noise.

Unfortunately, things are not so clear when constructing the solution from the representation of the value function. Indeed, in this approach, the key element is Theorem (Vol I)-4.45, and therein, the decoupling field u must be interpreted as the value function of the underlying optimal control problem. For that reason, it cannot be the function v we are seeking in (6.3). Once again, the reader may easily reformulate this observation from the sole basis of Theorem 1.57, which provides a similar FBSDE interpretation of the value function but for mean field games with a common noise. Basically, we learnt from Lemma (Vol I)-4.47, see also Theorem 4.10 in the case when the master field exists, that the function v we are looking for should be the derivative of the value function u identified in the statement of Theorem (Vol I)-4.45, or equivalently of the decoupling field identified in the statement of Theorem 1.57. In other words, v should be the gradient of the solution of the Hamilton-Jacobi-Bellman equation in the mean field game system (Vol I)-(3.12) in Chapter 3 of the first volume, the analogue of which in Volume II is (2.37). So, a possible way to establish (6.3) rigorously would be to prove smoothness of the solutions of Hamilton-Jacobi-Bellman equations of the type (Vol I)-(3.12) in Chapter 3 of the first volume. We refer to the Notes & Complements at the end of the chapter for references to these types of results.

For the sake of consistency, we shall adopt another strategy, although it will require more restrictive assumptions. Indeed, in analogy with the proof of Theorem (Vol I)-4.53 or with the strategy used in both Chapters 3 and 5, we can represent the optimal strategy in equilibrium by means of the stochastic Pontryagin maximum principle, and then deduce from standard results for forward-backward SDEs that the adjoint system has a Lipschitz continuous decoupling field. We achieve such an objective in the proof of Theorem 6.4 below. The setting is close to that used in the statement of Theorem (Vol I)-4.45, the analogue of which is Theorem 3.29 in the presence of a common noise. To avoid repeating ourselves too much, we do not restate the definitions of the various assumptions and, to make this volume self-consistent, we shall work under assumption **MFG with a Common Noise HJB**, with $\sigma^0 \equiv 0$, introduced in Subsection 3.4.1 to prove Theorem 3.29. Importantly, assumption **MFG with a Common Noise HJB** subsumes assumption **Necessary SMP in Random Environment** with $\mathcal{X} = \mathcal{P}_2(\mathbb{R}^d)$ as defined in Subsection 1.4.4, the latter assumption providing what is needed to use the necessary part of the stochastic Pontryagin maximum principle. Observe that, although assumption **Necessary SMP in Random Environment** has been stated within the framework of mean field games with common noise, it may be used without any restriction in the current framework of mean field games without common noise.

Theorem 6.4 *Let assumption* **MFG with a Common Noise HJB** *be in force. Assume further that $\partial_x b$, $\partial_x \sigma$, $\partial_x f$, and $\partial_x g$ are Lipschitz continuous with respect to (x,α) and x respectively, uniformly in (t,μ) and μ. Then, for any input $\boldsymbol{\mu} = (\mu_t)_{0\le t\le T}$ with values in $\mathcal{C}([0,T];\mathbb{R}^d)$, the decoupling field u constructed in Theorem (Vol I)-4.45 is differentiable in the space variable, and its gradient $v = \partial_x u : [0,T]\times\mathbb{R}^d \to \mathbb{R}^d$ is bounded and is Lipschitz continuous in x uniformly in $t \in [0,T]$. Moreover, for any initial condition $\xi \in L^2(\Omega,\mathcal{F}_0,\mathbb{P};\mathbb{R}^d)$, the optimal strategy in the optimal control problem (6.5) takes the form $\hat{\boldsymbol{\alpha}} = (\hat\alpha(t,X_t^{0,\xi},\mu_t,v(t,X_t^{0,\xi})))_{0\le t\le T}$, where $\boldsymbol{X}^{0,\xi}$ is the forward component of the unique solution with a bounded martingale integrand of the FBSDE system:*

$$
\begin{cases}
dX_t = b\big(t,X_t,\mu_t,\hat\alpha\big(t,X_t,\mu_t,\sigma(t,X_t,\mu_t)^{-1\dagger}Z_t\big)\big)dt \\
\qquad\quad +\sigma(t,X_t,\mu_t)dW_t, \\
dY_t = -f\big(t,X_t,\mu_t,\hat\alpha\big(t,X_t,\mu_t,\sigma(t,X_t,\mu_t)^{-1\dagger}Z_t\big)\big)dt + Z_t\cdot dW_t,
\end{cases}
\tag{6.6}
$$

for $t\in[0,T]$, with the initial condition $X_0 = \xi \in L^2(\Omega,\mathcal{F}_0,\mathbb{P};\mathbb{R}^d)$ and the terminal condition $Y_T = g(X_T,\mu_T)$.

Remark 6.5 *In the proof of Theorem 6.4 given below, the assumption that $\partial_x b$, $\partial_x \sigma$, $\partial_x f$, and $\partial_x g$ are Lipschitz continuous is only used to prove that v is also Lipschitz continuous in x. In fact, using standard estimates from the theory of uniformly parabolic PDEs, we could prove that the result remains true without assuming that $\partial_x b$, $\partial_x \sigma$, $\partial_x f$ and $\partial_x g$ are Lipschitz. Indeed, the proof below shows that v can be interpreted as the solution of a uniformly parabolic linear equation with bounded coefficients and with a Lipschitz continuous diffusion coefficient in x, and this suffices to get the desired bound on $\partial_x v$. In fact, we used an argument of this type in the proof of Proposition 5.53. We provide more references where these estimates can be found in the Notes & Complements at the end of the chapter.*

Proof. The present proof is inspired by the proofs of Theorem 4.10 and Corollary 4.11.

First Step. Theorem (Vol I)-4.45 says that, for any $(t,x) \in [0,T] \times \mathbb{R}^d$, in the environment $(\mu_s)_{t \le s \le T}$, there is a unique solution to the optimal control problem (6.5) with x as initial condition at time t. The optimal path may be identified to the forward component of the unique solution with a bounded martingale integrand of the FBSDE system (6.6) with $X_t = x$ as initial condition at time t, whose solution we denote by $(X_s^{t,x}, Y_s^{t,x}, Z_s^{t,x})_{t \le s \le T}$.

By the necessary part of the Pontryagin principle, see for instance Theorem 1.59 or Proposition 4.7, we know that $\hat{\boldsymbol{\alpha}}^{t,x} = (\hat{\alpha}(s, X_s^{t,x}, \mu_s, \sigma(s, X_s^{t,x}, \mu_t)^{-1\dagger} Z_s^{t,x}))_{t \le s \le T}$ coincides with $(\hat{\alpha}(s, X_s^{t,x}, \mu_s, v_s^{t,x}))_{t \le s \le T}$, where $(v_s^{t,x})_{t \le s \le T}$ solves the backward equation:

$$dv_s^{t,x} = -\partial_x H\big(s, X_s^{t,x}, \mu_t, v_s^{t,x}, \zeta_s^{t,x}, \hat{\alpha}_s^{t,x}\big) ds + \zeta_s^{t,x} dW_s, \quad s \in [t,T], \tag{6.7}$$

with the terminal condition $v_T^{t,x} = \partial_x g(X_T^{t,x}, \mu_T)$, and where $(\zeta_s^{t,x})_{t \le s \le T}$ is a square-integrable $\mathbb{F}$-progressively measurable process with values in $\mathbb{R}^{d \times d}$. Above H is the full Hamiltonian:

$$H(t,x,\mu,y,z,\alpha) = b(t,x,\mu,\alpha) \cdot y + \sigma(t,x,\mu) \cdot z + f(t,x,\mu,\alpha).$$

Second Step. We can repeat the proof of Theorem 4.10 to prove that, for any $t \in [0,T]$, $u(t,\cdot)$ is continuously differentiable and satisfies for all $x \in \mathbb{R}^d$, $\partial_x u(t,x) = v_t^{t,x}$. In our new framework, the analogue of **(A1)** in assumption **Decoupling Master** in Subsection 4.2.2 follows from the fact that in the proof of Theorem (Vol I)-4.45, we identified solutions of (Vol I)-(4.54) with solutions of a system satisfying the assumption of Lemma (Vol I)-4.9. Also, the bound for v follows from the Lipschitz property of u. The reader may also refer to Theorem 1.57 instead of Theorem (Vol I)-4.45 and to Theorem 1.53 instead of Lemma (Vol I)-4.9.

The argument used in the proof of Theorem 4.10 only shows that $\partial_x u$ is continuous in space. In order to prove that it is Lipschitz continuous, we shall use the stability of the solution $(X_s^{t,x}, Y_s^{t,x}, Z_s^{t,x})_{t \le s \le T}$ with respect to x. We indeed recall from the proof of Theorem (Vol I)-4.45 that $(X_s^{t,x}, Y_s^{t,x}, Z_s^{t,x})_{t \le s \le T}$ may be identified with the solution of a nondegenerate forward-backward system with bounded and Lipschitz continuous coefficients. In particular, Lemma (Vol I)-4.9 implies that:

$$\mathbb{E}\Big[\sup_{t \le s \le T} \big(|X_s^{t,x} - X_s^{t,x'}|^2 + |Y_s^{t,x} - Y_s^{t,x'}|^2\big) + \int_t^T |Z_s^{t,x} - Z_s^{t,x'}|^2 ds\Big] \le C|x - x'|^2, \tag{6.8}$$

for a constant C independent of (t,x). Our goal is to inject the above estimate in (6.7). However, doing so requires a modicum of care. Indeed, in the adjoint system satisfied by $(v_s^{t,x})_{t\le s\le T}$, the driver $\partial_x H$ is only Lipschitz continuous in the variable x for y and z in bounded subsets. While the fact that the Lipschitz constant in x depends on y is not really a problem, the fact that it also depends on z is more difficult to handle. The reason is as follows. The coefficients of (6.7), including the terminal condition, are bounded in x and at most of linear growth in y and z. Therefore, the process $(v_s^{t,x})_{t\le s\le T}$ can be shown to be bounded, uniformly in $(t,x)\in[0,T]\times\mathbb{R}^d$, which shows that the relevant values of the adjoint variable y stay in a bounded subset. However, at this stage of the proof, we do not have a similar a priori bound for $(\zeta_s^{t,x})_{t\le s\le T}$. At this stage, the best we can do is to combine (6.7) and (6.8) and get:

$$
\begin{aligned}
\mathbb{E}\big[\sup_{t\le s\le T}|v_s^{t,x}-v_s^{t,x'}|^2\big] &\le C\Big(\mathbb{E}\big[|X_T^{t,x}-X_T^{t,x'}|^2\big] \\
&\quad + \int_t^T \mathbb{E}\big[|X_s^{t,x}-X_s^{t,x'}|^2(1+|\zeta_s^{t,x}|^2)+|\hat{\alpha}_s^{t,x}-\hat{\alpha}_s^{t,x'}|^2\big]ds\Big) \\
&\le C\Big(|x-x'|^2+\mathbb{E}\int_t^T\big[|X_s^{t,x}-X_s^{t,x'}|^2|\zeta_s^{t,x}|^2\big]ds\Big),
\end{aligned}
$$

which does not permit to conclude. Above, we used the fact that $\hat{\alpha}$ is Lipschitz continuous to pass from the first to the second inequality together with the fact that the process $(Z_s^{t,x})_{t\le s\le T}$ is bounded independently of t and x. The constant C will be allowed to increase from line to line as long as it remains independent of (t,x).

We resolve our quandary as follows. By Cauchy-Schwarz inequality, we get:

$$
\begin{aligned}
&\mathbb{E}\big[\sup_{t\le s\le T}|v_s^{t,x}-v_s^{t,x'}|^2\big] \\
&\quad\le C|x-x'|^2 + C\mathbb{E}\Big[\sup_{t\le s\le T}|X_s^{t,x}-X_s^{t,x'}|^4\Big]^{1/2}\mathbb{E}\Big[\Big(\int_t^T|\zeta_s^{t,x}|^2ds\Big)^2\Big]^{1/2}.
\end{aligned}
$$

In order to complete the proof, we invoke L^4 bounds, instead of L^2 BSDE estimates. By expanding the square of $\boldsymbol{v}^{t,x}$ in (6.7) by Itô's formula and by taking the square again in the resulting expansion, it is pretty standard to derive:

$$
\mathbb{E}\Big[\Big(\int_t^T|\zeta_s^{t,x}|^2ds\Big)^2\Big]\le C.
$$

Similarly, we claim that the following analogue of (6.8) holds:

$$
\mathbb{E}\big[\sup_{t\le s\le T}|X_s^{t,x}-X_s^{t,x'}|^4\big]\le C|x-x'|^4,
$$

the proof of which is similar to that of Lemma (Vol I)-4.9, or equivalently of Theorem 1.57. We refer to the references in the Notes & Complements at the end of the chapter for complete proofs. From these, the Lipschitz property of v easily follows.

Third Step. In order to complete the proof, it remains to show that with probability 1, for all $s \in [t, T]$, $v_s^{t,x} = \partial_x u(s, X_s^{t,x})$. Again, this follows from a suitable adaptation of the proof of Corollary 4.11. This is easier in the present setting since (4.23), with conditioning with respect to the σ-field $\mathcal{F}_s$, is now a straightforward consequence of (6.6). The rest of the proof is similar.

In order to go beyond deterministic initial conditions, it remains to check the result with $(0, \xi)$ in lieu of (t, x) for $\xi \in L^2(\Omega, \mathcal{F}_0, \mathbb{P}; \mathbb{R}^d)$. This can be done by approximating ξ by a sequence of random variables of the form $(\xi_n = \sum_{i=1}^n x_i \mathbf{1}_{A_i})_{n \geq 1}$, with $x_1, \ldots, x_n \in \mathbb{R}^d$ and $A_1, \ldots, A_n \in \mathcal{F}_0$. Observing that, for any $n \geq 1$ and for all $s \in [0, T]$,

$$\begin{aligned}\hat{\alpha}_s^{0,\xi_n} &= \sum_{i=1}^n \mathbf{1}_{A_i} \hat{\alpha}_s^{0,x_i} \\ &= \sum_{i=1}^n \mathbf{1}_{A_i} \hat{\alpha}\big(s, X_s^{0,x_i}, \mu_s, v(s, X_s^{0,x_i})\big) = \hat{\alpha}\big(s, X_s^{0,\xi_n}, \mu_s, v(s, X_s^{0,\xi_n})\big),\end{aligned}$$

where we called $(\hat{\alpha}_s^{0,\xi} = \hat{\alpha}(s, X_s^{0,\xi}, \mu_s, v_s^{0,\xi}))_{0 \leq s \leq T}$ the optimal control under the initial condition $(0, \xi)$, and then letting n to ∞, the result follows by another stability argument. □

In order to proceed with the analysis of approximate Nash equilibria, we shall need more than the assumption of Theorem 6.4. This prompts us to introduce a new assumption.

Assumption (Approximate Nash HJB). On top of assumption **MFG with a Common Noise HJB**, with $\sigma^0 \equiv 0$, assume that $\partial_x b$, $\partial_x \sigma$, $\partial_x f$ and $\partial_x g$ are Lipschitz continuous with respect to (x, α) and x respectively, uniformly in (t, μ) and μ, the functions b and σ are Lipschitz continuous in μ, uniformly with respect to the other parameters, and the functions f and g are locally Lipschitz continuous with respect to μ, the Lipschitz constant being bounded by LR for a constant $L \geq 0$ and for any $R \geq 1$ whenever f and g are restricted to:

$$\{(t, x, \alpha, \mu) \in [0, T] \times \mathbb{R}^d \times A \times \mathcal{P}_2(\mathbb{R}^d) : |x| + |\alpha| + M_2(\mu) \leq R\},$$

and to:

$$\{(x, \mu) \in \mathbb{R}^d \times \mathcal{P}_2(\mathbb{R}^d) : |x| + M_2(\mu) \leq R\}.$$

Approximate Nash Equilibrium Candidates

As a consequence of Theorem 6.4, we know that, under assumption **Approximate Nash HJB**, optimal paths of the control problem (6.5) are of the form (6.3), for a function v as in the statement of Theorem 6.4. Importantly, the same holds true under assumption **MFG Solvability SMP** from Subsection (Vol I)-4.5.1, which we used to prove the existence Theorem (Vol I)-4.53, or equivalently, under assumption

MFG with a Common Noise SMP Relaxed from Subsection 3.4.3, which we used to prove Theorem 3.31, except for the fact that v may no longer be bounded. The latter is the analogue of Theorem (Vol I)-4.53 for mean field games with a common noise, provided we choose $\sigma^0 \equiv 0$ therein. The only slight difference between the two sets of assumptions is that σ is required to be constant under assumption **MFG Solvability SMP**, while we allowed for a more general form of volatility coefficient in assumption **MFG with a Common Noise SMP Relaxed**. Notice also that we could work under assumption **MFG with a Common Noise SMP**, which imposes rather restrictive conditions on the coefficients but allows for any closed convex subset $A \subset \mathbb{R}^k$ as set of controls. In comparison, we recall that A is indeed required to fit the whole $\mathbb{R}^k$ under assumption **MFG with a Common Noise SMP Relaxed**. In any case, it is important to be able to use the formulation (6.3) provided that the function v therein is Lipschitz continuous in x, uniformly in t, and locally bounded in (t, x). In this framework, the systematic strategy described below suggests a candidate for the role of an approximate Nash equilibrium. To make it proper, we need to introduce the analogue of assumption **Approximate Nash HJB** for assumption **MFG with a Common Noise SMP Relaxed**:

Assumption (Approximate Nash SMP). On top of assumption **MFG with a Common Noise SMP Relaxed**, with $\sigma^0 \equiv 0$, functions b and σ are Lipschitz continuous in μ, uniformly with respect to $t \in [0, T]$ and $x \in \mathbb{R}^d$.

Throughout the exposition, we use the same notation $\boldsymbol{\mu} = (\mu_t)_{0 \le t \le T}$ as above for the mean field game equilibrium. As already emphasized, the existence of such an equilibrium can be guaranteed by Theorem (Vol I)-4.44 or Theorem (Vol I)-4.53, or equivalently by Theorem 3.29 or Theorem 3.31. On a probability space $(\Omega, \mathcal{F}, \mathbb{P})$, we also fix a sequence $(\xi^i)_{i \ge 1}$ of independent $\mathbb{R}^d$-valued random variables with $\mu_0 \in \mathcal{P}_2(\mathbb{R}^d)$ as common distribution, and a sequence $(\boldsymbol{W}^i)_{i \ge 1}$ of independent d-dimensional Wiener processes $(\boldsymbol{W}^i = (W^i_t)_{0 \le t \le T})_{i \ge 1}$. The two sequences $(\xi^i)_{i \ge 1}$ and $(\boldsymbol{W}^i)_{i \ge 1}$ are assumed to be independent of each other. For each integer N, we consider the solution $(\boldsymbol{X}^{N,1}, \cdots, \boldsymbol{X}^{N,N}) = (X^{N,1}_t, \cdots, X^{N,N}_t)_{0 \le t \le T}$ of the system of N stochastic differential equations:

$$dX^{N,i}_t = b\big(t, X^{N,i}_t, \bar{\mu}^N_t, \hat{\alpha}\big(t, X^{N,i}_t, \mu_t, v(t, X^{N,i}_t)\big)\big)dt + \sigma(t, X^{N,i}_t, \bar{\mu}^N_t)dW^i_t, \tag{6.9}$$

for $t \in [0, T]$ and $i = 1, \cdots N$, with initial conditions $X^{N,i}_0 = \xi^i$ for $i = 1, \cdots, N$, where as usual, $\bar{\mu}^N_t$ denotes the empirical distribution $\bar{\mu}^N_t = \frac{1}{N}\sum_{j=1}^N \delta_{X^{N,j}_t}$. Equation (6.9) is well posed since v is Lipschitz continuous in x, uniformly in time, and $(v(t, 0))_{0 \le t \le T}$ is bounded. Also, recall that the minimizer $\hat{\alpha}(t, x, \mu, y)$ was proven, in Lemma (Vol I)-3.3 or in Lemma 1.56, to be Lipschitz continuous in the variables

x and y, uniformly in $t \in [0,T]$ and $\mu \in \mathcal{P}_2(\mathbb{R}^d)$, and at most of linear growth in y, uniformly in the other variables. The processes $(\boldsymbol{X}^{N,i})_{1\leq i\leq N}$ give the dynamics of the states of the N players in the stochastic differential game of interest when the players use the control strategies:

$$\hat{\alpha}_t^{N,i} = \hat{\alpha}\big(t, X_t^{N,i}, \mu_t, v(t, X_t^{N,i})\big), \qquad 0 \leq t \leq T, \ \ i \in \{1, \cdots, N\}. \tag{6.10}$$

Notice that these strategies are not only in closed loop form, they are in fact *distributed* since at each time $t \in [0,T]$, a player only needs to know the value of its own private state in order to compute the value of the control to apply at that time; in particular, these strategies are of a very low complexity, which makes them really convenient for a practical use.

The search for Nash equilibria involves the comparison of the players' expected costs for different strategy profiles. In order to do so, we introduce new notations to describe the states and the costs when the players choose different strategy profiles, say $\boldsymbol{\beta}^{(N)} = (\boldsymbol{\beta}^{N,1}, \cdots, \boldsymbol{\beta}^{N,N})$ with $\boldsymbol{\beta}^{N,i} = (\beta_t^{N,i})_{0\leq t\leq T}$ for $i = 1, \cdots, N$. In this case, we denote by $U_t^{N,i}$ the state of player $i \in \{1, \cdots, N\}$ at time $t \in [0,T]$. If the $\boldsymbol{\beta}^{N,i}$s are admissible in the sense that $\boldsymbol{\beta}^{N,i} \in \bar{\mathbb{A}}_{(N)}$, $\bar{\mathbb{A}}_{(N)}$ being defined as the set of square-integrable A-valued processes which are progressively measurable with respect to the complete and right-continuous filtration generated by $(\xi^1, \cdots, \xi^N)$ and $(\boldsymbol{W}^1, \cdots, \boldsymbol{W}^N)$, the dynamics of the state of the i-th player are given by $U_0^{N,i} = \xi^i$ and

$$dU_t^{N,i} = b\big(t, U_t^{N,i}, \bar{\nu}_t^N, \beta_t^{N,i}\big)dt + \sigma\big(t, U_t^{N,i}, \bar{\nu}_t^N\big)dW_t^i, \quad 0 \leq t \leq T, \tag{6.11}$$

with:

$$\bar{\nu}_t^N = \frac{1}{N}\sum_{j=1}^N \delta_{U_t^{N,j}}, \quad 0 \leq t \leq T.$$

For each $1 \leq i \leq N$, we denote by:

$$J^{N,i}\big(\boldsymbol{\beta}^{N,1}, \cdots, \boldsymbol{\beta}^{N,N}\big) = \mathbb{E}\bigg[\int_0^T f\big(t, U_t^{N,i}, \bar{\nu}_t^N, \beta_t^{N,i}\big)dt + g\big(U_T^{N,i}, \bar{\nu}_T^N\big)\bigg], \tag{6.12}$$

the cost to player i.

Our goal is to use the form (6.3) of the optimal path under the limiting mean field game problem to prove that $(\hat{\boldsymbol{\alpha}}^{N,1}, \cdots, \hat{\boldsymbol{\alpha}}^{N,N})$, as constructed in (6.10), is an approximate Nash equilibrium for the N-player game. To do so, we first need a precise definition of the notion of approximate Nash equilibrium.

According to the terminology introduced in Chapter (Vol I)-2, two different notions are conceivable corresponding to the open loop and closed loop versions of the problem. We first follow Definition (Vol I)-2.3 of open loop Nash equilibria.

Definition 6.6 *Given $\epsilon > 0$, an admissible strategy $\boldsymbol{\alpha}^{(N)} = (\boldsymbol{\alpha}^{N,1}, \cdots, \boldsymbol{\alpha}^{N,N})$ is said to be an ϵ-approximate open loop Nash equilibrium if, for each $i \in \{1, \cdots, N\}$,*

$$J^{N,i}\big(\boldsymbol{\alpha}^{(N)}\big) \leq J^{N,i}\big(\boldsymbol{\alpha}^{N,1}, \cdots, \boldsymbol{\alpha}^{N,i-1}, \boldsymbol{\beta}^i, \boldsymbol{\alpha}^{N,i+1}, \cdots, \boldsymbol{\alpha}^{N,N}\big) + \epsilon.$$

for all admissible control $\boldsymbol{\beta}^i \in \bar{\mathbb{A}}_{(N)}$.

In the framework of Definition 6.6, we have the following statement, which provides a first rigorous connection between games with finitely many players and mean field games.

Theorem 6.7 *Under either assumption* **Approximate Nash HJB** *or assumption* **Approximate Nash SMP**, *there exists a sequence $(\varepsilon_N)_{N \geq 1}$, converging to 0 as N tends to ∞, such that the strategies $(\hat{\boldsymbol{\alpha}}^{N,i})_{1 \leq i \leq N}$ defined in* (6.10) *form an ε_N-approximate open loop Nash equilibrium of the N-player game* (6.11)–(6.12). *Precisely, for each $N \geq 1$, for any player $i \in \{1, \cdots, N\}$ and any admissible control strategy $\boldsymbol{\beta}^i \in \bar{\mathbb{A}}_{(N)}$, it holds:*

$$J^{N,i}(\hat{\boldsymbol{\alpha}}^{N,1}, \cdots, \hat{\boldsymbol{\alpha}}^{N,i-1}, \boldsymbol{\beta}^i, \hat{\boldsymbol{\alpha}}^{N,i+1}, \cdots, \hat{\boldsymbol{\alpha}}^{N,N}) \geq J^{N,i}(\hat{\boldsymbol{\alpha}}^{N,1}, \cdots, \hat{\boldsymbol{\alpha}}^{N,N}) - \varepsilon_N. \tag{6.13}$$

If the initial condition μ_0 of the equilibrium is in $\mathcal{P}_q(\mathbb{R}^d)$, for $q > 4$, then we can choose $\varepsilon_N = c\sqrt{\epsilon_N}$, for a constant $c > 0$ independent of N, ϵ_N being defined as in (6.1).

We shall prove a similar result for closed loop Nash equilibria. It will rely on the following definition of an approximate closed loop Nash equilibrium.

Definition 6.8 *Let*

$$\phi^{N,i} : [0,T] \times \mathcal{C}([0,T]; \mathbb{R}^{Nd}) \to A, \qquad i = 1, \cdots, N,$$

be measurable feedback functions, such that the system of stochastic differential equations:

$$\begin{cases} dX_t^{N,i} = b\big(t, X_t^{N,i}, \bar{\mu}_t^N, \phi^{N,i}\big(t, X_{[0,t]}^{(N)}\big)\big)dt + \sigma\big(t, X_t^{N,i}, \bar{\mu}_t^N\big)dW_t^i, & t \in [0,T], \\ X_0^{N,i} = \xi^i, \quad i \in \{1, \cdots, N\}, \end{cases}$$

with

$$\bar{\mu}_t^N = \frac{1}{N} \sum_{j=1}^N \delta_{X_t^{N,j}}, \quad t \in [0,T],$$

is uniquely solvable, where $\boldsymbol{X}^{(N)} = (X_t^{(N)} = (X_t^{N,1}, \cdots, X_t^{N,N}))_{0 \le t \le T}$. *Define the corresponding strategy profile* $\boldsymbol{\alpha}^{(N)} = (\boldsymbol{\alpha}^{N,1}, \cdots, \boldsymbol{\alpha}^{N,N})$ *in closed loop form:*

$$\boldsymbol{\alpha}^{N,i} = \Big(\alpha_t^{N,i} = \phi^{N,i}\big(t, X_{[0,t]}^N\big)\Big)_{0 \le t \le T}, \qquad i = 1, \cdots, N,$$

and assume that it is admissible.

Then, for a given $\epsilon > 0$, $\boldsymbol{\alpha}^{(N)} = (\boldsymbol{\alpha}^{N,1}, \cdots, \boldsymbol{\alpha}^{N,N})$ *is said to be an* ϵ*-approximate closed loop Nash equilibrium if, for each* $i \in \{1, \cdots, N\}$ *and any other measurable feedback function* $\psi^i : [0, T] \times \mathcal{C}([0, T]; \mathbb{R}^{Nd}) \to A$*, it holds:*

$$J^{N,i}\big(\boldsymbol{\alpha}^{(N)}\big) \le J^{N,i}\big(\boldsymbol{\beta}^{(N)}\big) + \epsilon,$$

where $\boldsymbol{\beta}^{N,j} = (\beta_t^{N,j} = \phi^{N,j}(t, U_{[0,t]}^{(N)}))_{0 \le t \le T}$ *for* $j \ne i$ *and* $\boldsymbol{\beta}^{N,i} = (\beta_t^{N,i} = \psi^i(t, U_{[0,t]}^{(N)}))_{0 \le t \le T}$, $(U_t^{(N)} = (U_t^{N,1}, \cdots, U_t^{N,N}))_{0 \le t \le T}$ *satisfying* (6.11) *with* $(U_0^{N,1}, \cdots, U_0^{N,N}) = (\xi^1, \cdots, \xi^N)$ *as initial condition.*

As usual with closed loop Nash equilibria, the definition of $\boldsymbol{\beta}$ in (6.11) is implicit, as it depends upon the solution $\boldsymbol{U}^{(N)} = (\boldsymbol{U}^{N,1}, \cdots, \boldsymbol{U}^{N,N})$ itself. Hence, (6.11) must be understood as a system of stochastic differential equations. It is implicitly required to be well posed and $\boldsymbol{\beta}^{(N)} = (\boldsymbol{\beta}^{N,1}, \cdots, \boldsymbol{\beta}^{N,N})$ is implicitly required to be admissible.

In complete analogy with Theorem 6.7, we have the following result:

Theorem 6.9 *Under either assumption* **Approximate Nash HJB** *or assumption* **Approximate Nash SMP**, *there exists a sequence* $(\varepsilon_N)_{N \ge 1}$, *converging to* 0 *as* N *tends to* ∞, *such that the strategies* $(\hat{\boldsymbol{\alpha}}^{N,i})_{1 \le i \le N}$ *associated with the Markovian feedback functions:*

$$\phi^{N,i} : [0, T] \times \mathcal{C}([0, T]; \mathbb{R}^{Nd}) \ni \big(t, (\boldsymbol{x}^1, \cdots, \boldsymbol{x}^N)\big) \mapsto \hat{\alpha}\big(t, x_t^i, \mu_t, v(t, x_t^i)\big),$$

used in (6.10), *form an* ε_N*-approximate closed loop Nash equilibrium of the* N*-player game* (6.11)–(6.12).

If the initial condition μ_0 *of the equilibrium is in* $\mathcal{P}_q(\mathbb{R}^d)$, *for* $q > 4$, *then we can choose* $\varepsilon_N = c\sqrt{\epsilon_N}$, *for a constant* $c > 0$ *independent of* N, ϵ_N *being defined as in* (6.1).

Proofs of Theorems 6.7 and 6.9

We prove Theorems 6.7 and 6.9 simultaneously, and in so doing, we point out the main differences between the two cases.

Proof. By symmetry, we only need to prove (6.13), or the analog for control strategies in closed loop form, for $i = 1$. Throughout the proof, $\boldsymbol{\beta}^{N,1}, \cdots, \boldsymbol{\beta}^{N,N}$ will denote the alternative tuple of strategies that we aim at comparing with $\hat{\boldsymbol{\alpha}}^{N,1}, \cdots, \hat{\boldsymbol{\alpha}}^{N,N}$. When working with closed loop Nash equilibria, the definition of $\boldsymbol{\beta}^{N,1}, \cdots, \boldsymbol{\beta}^{N,N}$ is already clear from Definition 6.8. When working with open loop Nash equilibria, we just let $\boldsymbol{\beta}^{N,i} = \hat{\alpha}^{N,i}$, for $i \geq 2$.

First Step. Recall that, irrespective of which assumption is in force, there exists a constant C such that $|b(t,x,\mu,\alpha)| \leq C(1 + |x| + |\alpha|)$ and $|\sigma(t,x,\mu)| \leq C(1 + |x|)$. Also, $|\partial_\alpha f(t,x,\mu,\alpha)| \leq C(1 + |\alpha|)$. By Lemma (Vol I)-3.3 or Lemma 1.56, we deduce that $|\hat{\alpha}(t,x,\mu,y)| \leq C(1 + |y|)$. Recall finally that $|v(t,x)| \leq C(1 + |x|)$. Since $\mu_0 \in \mathcal{P}_q(\mathbb{R}^d)$, for $q = 2$ or $q > 4$, we deduce that, in any case,

$$\mathbb{E}\Big[\sup_{0 \leq t \leq T} |X_t^{N,i}|^q\Big] \leq C, \tag{6.14}$$

the constant C being allowed to increase from line to line. In particular,

$$\mathbb{E}\Big[\sup_{0 \leq t \leq T} |\hat{\alpha}_t^{N,i}|^q\Big] \leq C. \tag{6.15}$$

Second Step. Using the same bounds as in the first step, we get from Gronwall's inequality:

$$\mathbb{E}\Big[\sup_{0 \leq t \leq T} |U_t^{N,1}|^2\Big] \leq C\Big(1 + \mathbb{E}\int_0^T |\beta_t^{N,1}|^2 dt\Big). \tag{6.16}$$

Moreover, when working with closed loop Nash equilibria, we have $|\beta_t^{N,i}| \leq C(1 + |U_t^{N,i}|)$ for $i \in \{2, \cdots, N\}$, so that, by Gronwall's lemma again, we get:

$$\mathbb{E}\Big[\sup_{0 \leq t \leq T} |U_s^{N,i}|^q\Big] \leq C, \qquad 2 \leq i \leq N. \tag{6.17}$$

When working with open loop Nash equilibria, (6.17) is also true as it follows from (6.15). After summation, we obtain, in both cases,

$$\frac{1}{N}\sum_{j=1}^N \mathbb{E}\Big[\sup_{0 \leq t \leq T} |U_t^{N,j}|^2\Big] \leq C\Big(1 + \frac{1}{N}\mathbb{E}\int_0^T |\beta_t^{N,1}|^2 dt\Big). \tag{6.18}$$

Third Step. In full analogy with the proof of Theorem 2.12, we now introduce, for the purpose of comparison, the system of decoupled independent and identically distributed states:

$$d\underline{X}_t^i = b\big(t, \underline{X}_t^i, \mu_t, \hat{\alpha}(t, \underline{X}_t^i, \mu_t, v(t, \underline{X}_t^i))\big)dt + \sigma\big(t, \underline{X}_t^i, \mu_t\big)dW_t^i,$$

for $0 \leq t \leq T$ and $i \in \{1, \cdots, N\}$. By construction, the stochastic processes $(\underline{\boldsymbol{X}}^i)_{1 \leq i \leq N}$ are independent copies of $\boldsymbol{X}$ defined in (6.3) and, in particular, $\mathcal{L}(\underline{X}_t^i) = \mu_t$ for any $t \in [0,T]$ and $i \in \{1, \cdots, N\}$. Throughout the rest of the proof, we shall use the notation:

$$\underline{\hat{\alpha}}_t^i = \hat{\alpha}\big(t, \underline{X}_t^i, \mu_t, v(t, \underline{X}_t^i)\big), \quad t \in [0,T], \quad i \in \{1, \cdots, N\}.$$

Following the proof of (6.14), we have:

$$\mathbb{E}\big[\sup_{0\le t\le T}|\underline{X}_t^i|^q\big] \le C. \tag{6.19}$$

Using the regularity properties of the decoupling field v, we deduce from Theorem 2.12:

$$\max_{1\le i\le N}\mathbb{E}\big[\sup_{0\le t\le T}|X_t^{N,i}-\underline{X}_t^i|^2\big] \le \varepsilon_N^2, \tag{6.20}$$

and

$$\sup_{0\le t\le T}\mathbb{E}\big[W_2(\bar{\mu}_t^N,\mu_t)^2\big] \le C\varepsilon_N^2, \tag{6.21}$$

where the sequence $(\varepsilon_N)_{N\ge 1}$ tends to 0 as N tends to ∞ and satisfies the same prescription as in the statement when $q>4$.

Using the local-Lipschitz regularity of the coefficients f and g together with Cauchy-Schwarz' inequality, we get, for each $i\in\{1,\cdots,N\}$:

$$\begin{aligned}
&\big|J-J^{N,i}(\hat{\boldsymbol{\alpha}}^{N,1},\cdots,\hat{\boldsymbol{\alpha}}^{N,N})\big| \\
&= \bigg|\mathbb{E}\bigg[\int_0^T f(t,\underline{X}_t^i,\mu_t,\underline{\hat{\alpha}}_t^i)dt + g(\underline{X}_T^i,\mu_T) - \int_0^T f(t,X_t^{N,i},\bar{\mu}_t^N,\hat{\alpha}_t^{N,i})dt - g(X_T^{N,i},\bar{\mu}_T^N)\bigg]\bigg| \\
&\le C\mathbb{E}\bigg[1+|\underline{X}_T^i|^2+|X_T^{N,i}|^2+\frac{1}{N}\sum_{j=1}^N|X_T^{N,j}|^2\bigg]^{1/2}\times\mathbb{E}\Big[|\underline{X}_T^i-X_T^{N,i}|^2+W_2\big(\mu_T,\bar{\mu}_T^N\big)^2\Big]^{1/2} \\
&\quad + C\int_0^T\bigg\{\mathbb{E}\bigg[1+|\underline{X}_t^i|^2+|X_t^{N,i}|^2+|\underline{\hat{\alpha}}_t^i|^2+|\hat{\alpha}_t^{N,i}|^2+\frac{1}{N}\sum_{j=1}^N|X_t^{N,j}|^2\bigg]^{1/2} \\
&\qquad\qquad \times\mathbb{E}\Big[|\underline{X}_t^i-X_t^{N,i}|^2+|\underline{\hat{\alpha}}_t^i-\hat{\alpha}_t^{N,i}|^2+W_2\big(\mu_t,\bar{\mu}_t^N\big)^2\Big]^{1/2}\bigg\}dt.
\end{aligned}$$

By (6.14), (6.15), and (6.19), we deduce:

$$\begin{aligned}
\big|J-J^{N,i}(\hat{\boldsymbol{\alpha}}^{N,1},\cdots,\hat{\boldsymbol{\alpha}}^{N,N})\big| &\le C\mathbb{E}\Big[|\underline{X}_T^i-X_T^{N,i}|^2+W_2\big(\mu_T,\bar{\mu}_T^N\big)^2\Big]^{1/2} \\
&\quad + C\bigg(\int_0^T\mathbb{E}\Big[|\underline{X}_t^i-X_t^i|^2+|\underline{\hat{\alpha}}_t^i-\hat{\alpha}_t^{N,i}|^2+W_2\big(\mu_t,\bar{\mu}_t^N\big)^2\Big]dt\bigg)^{1/2}.
\end{aligned}$$

Now, by the Lipschitz property of the minimizer $\hat{\alpha}$ proven in Lemma (Vol I)-3.3, see also Lemma 1.56, and by the Lipschitz property of v, we notice that:

$$|\underline{\hat{\alpha}}_t^i-\hat{\alpha}_t^{N,i}| = \big|\hat{\alpha}\big(t,\underline{X}_t^i,\mu_t,v(t,\underline{X}_t^i)\big)-\hat{\alpha}\big(t,X_t^{N,i},\mu_t,v(t,X_t^{N,i})\big)\big| \le c|\underline{X}_t^i-X_t^{N,i}|.$$

Using (6.20) and (6.21), this proves that, for any $1\le i\le N$,

$$J^{N,i}(\hat{\boldsymbol{\alpha}}^{N,1},\cdots,\hat{\boldsymbol{\alpha}}^{N,N}) = J+O(\varepsilon_N). \tag{6.22}$$

Fourth Step. This suggests that, in order to prove inequality (6.13) for $i = 1$, we could restrict ourselves to the comparison of $J^{N,1}(\boldsymbol{\beta}^{N,1}, \boldsymbol{\beta}^{N,2}, \cdots, \boldsymbol{\beta}^{N,N})$ to J. Using the argument which led to (6.16), (6.17) and (6.18), together with the definitions of $\boldsymbol{U}^{N,j}$ and $\boldsymbol{X}^{N,j}$ for $j = 1, \cdots, N$, we get, for any $t \in [0, T]$:

$$\mathbb{E}\Big[\sup_{0\le s\le t} |U_t^{N,i} - X_t^{N,i}|^2\Big] \\ \le \frac{C}{N}\int_0^t \sum_{j=1}^N \mathbb{E}\Big[\sup_{0\le r\le s} |U_r^{N,j} - X_r^{N,j}|^2\Big]ds + C\mathbb{E}\int_0^t |\beta_s^{N,i} - \hat{\alpha}_s^{N,i}|^2 ds,$$

for $i \in \{1, \cdots, N\}$. When working with open loop equilibria and for $i \ge 2$, the second term in the right-hand side is null. When working with closed loop equilibria, we deduce from the form of the strategies that:

$$\mathbb{E}\Big[\sup_{0\le s\le t} |U_t^{N,i} - X_t^{N,i}|^2\Big] \\ \le \frac{C}{N}\int_0^t \sum_{j=1}^N \mathbb{E}\Big[\sup_{0\le r\le s} |U_r^{N,j} - X_r^{N,j}|^2\Big]ds + C\mathbb{E}\int_0^t |X_s^{N,i} - U_s^{N,i}|^2 ds,$$

for $i \ge 2$, which yields, by Gronwall's lemma,

$$\mathbb{E}\Big[\sup_{0\le s\le t} |U_t^{N,i} - X_t^{N,i}|^2\Big] \le \frac{C}{N}\int_0^t \sum_{j=1}^N \mathbb{E}\Big[\sup_{0\le r\le s} |U_r^{N,j} - X_r^{N,j}|^2\Big]ds.$$

Therefore, we have, in any case,

$$\mathbb{E}\Big[\sup_{0\le s\le t} |U_t^{N,1} - X_t^{N,1}|^2\Big] \\ \le \frac{C}{N}\int_0^t \sum_{j=1}^N \mathbb{E}\Big[\sup_{0\le r\le s} |U_r^{N,j} - X_r^{N,j}|^2\Big]ds + C\mathbb{E}\int_0^T |\beta_s^{N,1} - \hat{\alpha}_s^{N,1}|^2 ds,$$

$$\mathbb{E}\Big[\sup_{0\le s\le t} |U_t^{N,i} - X_t^{N,i}|^2\Big] \le \frac{C}{N}\int_0^t \sum_{j=1}^N \mathbb{E}\Big[\sup_{0\le r\le s} |U_r^{N,j} - X_r^{N,j}|^2\Big]ds, \quad i \in \{2, \cdots, N\}.$$

Therefore, using Gronwall's inequality, we get:

$$\frac{1}{N}\sum_{j=1}^N \mathbb{E}\Big[\sup_{0\le t\le T} |U_t^{N,j} - X_t^{N,j}|^2\Big] \le \frac{C}{N}\mathbb{E}\int_0^T |\beta_t^{N,1} - \hat{\alpha}_t^{N,1}|^2 dt, \tag{6.23}$$

so that:

$$\sup_{0\le t\le T} \mathbb{E}[|U_t^{N,i} - X_t^{N,i}|^2] \le \frac{C}{N}\mathbb{E}\int_0^T |\beta_t^{N,1} - \hat{\alpha}_t^{N,1}|^2 dt, \quad 2 \le i \le N. \tag{6.24}$$

Putting together (6.15), (6.20), and (6.24), we see that, for any $\kappa > 0$, there exists a constant C_κ depending on κ such that:

$$\mathbb{E}\int_0^T |\beta_t^{N,1}|^2 dt \leq \kappa \quad \Longrightarrow \quad \max_{2\leq i\leq N} \sup_{0\leq t\leq T} \mathbb{E}\big[|U_t^{N,i} - \underline{X}_t^i|^2\big] \leq C_\kappa \varepsilon_N^2. \tag{6.25}$$

Let us fix $\kappa > 0$ for the moment, and let us assume that $\mathbb{E}\int_0^T |\beta_t^{N,1}|^2 dt \leq \kappa$. Using (6.25) we see that:

$$\frac{1}{N-1}\sum_{j=2}^N \sup_{0\leq t\leq T} \mathbb{E}\big[|U_t^{N,j} - \underline{X}_t^j|^2\big] \leq C_\kappa \varepsilon_N^2, \tag{6.26}$$

for a constant C_κ depending upon κ, and whose value can change from line to line as long as it remains independent of N. Now by the triangle inequality for the Wasserstein distance:

$$\begin{aligned}\mathbb{E}\Big[W_2(\bar{\nu}_t^N, \mu_t)^2\Big] \leq 3\bigg\{\mathbb{E}\bigg[W_2\Big(\frac{1}{N}\sum_{j=1}^N \delta_{U_t^{N,j}}, \frac{1}{N-1}\sum_{j=2}^N \delta_{U_t^{N,j}}\Big)^2\bigg] \\ + \frac{1}{N-1}\sum_{j=2}^N \mathbb{E}\big[|U_t^{N,j} - \underline{X}_t^j|^2\big] + \mathbb{E}\bigg[W_2\Big(\frac{1}{N-1}\sum_{j=2}^N \delta_{\underline{X}_t^j}, \mu_t\Big)^2\bigg]\bigg\}.\end{aligned} \tag{6.27}$$

Noticing that:

$$\mathbb{E}\bigg[W_2\Big(\frac{1}{N}\sum_{j=1}^N \delta_{U_t^{N,j}}, \frac{1}{N-1}\sum_{j=2}^N \delta_{U_t^{N,j}}\Big)^2\bigg] \leq \frac{1}{N(N-1)}\sum_{j=2}^N \mathbb{E}\big[|U_t^{N,1} - U_t^{N,j}|^2\big],$$

which is $O(N^{-1})$ because of (6.16) and (6.18), and plugging this inequality into (6.27), using (6.26) to control the second term, and Theorem 2.12, see also Lemma 6.1, to estimate the third term therein, we conclude that:

$$\sup_{0\leq t\leq T} \mathbb{E}\Big[W_2(\bar{\nu}_t^N, \mu_t)^2\Big] \leq C_\kappa \varepsilon_N^2. \tag{6.28}$$

Last Step. For the final step of the proof, we define $(\bar{U}_t^{N,1})_{0\leq t\leq T}$ as the solution of the SDE:

$$d\bar{U}_t^{N,1} = b(t, \bar{U}_t^{N,1}, \mu_t, \beta_t^{N,1})dt + \sigma(t, \bar{U}_t^{N,1}, \mu_t)dW_t^1, \quad 0 \leq t \leq T\,; \quad \bar{U}_0^{N,1} = \xi^1,$$

so that, from the definition (6.11) of $\boldsymbol{U}^{N,1}$ together with the bound (6.28), we get, by Gronwall's inequality,

$$\sup_{0\leq t\leq T} \mathbb{E}\big[|U_t^{N,1} - \bar{U}_t^{N,1}|^2\big] \leq C_\kappa \varepsilon_N^2. \tag{6.29}$$

Going over the computation leading to (6.22) once more and using (6.28) together with (6.16), (6.17) and (6.18) and the fact that $\mathbb{E}\int_0^T |\beta_t^{N,1}|^2 dt \leq \kappa$, we obtain:

$$J^{N,1}(\boldsymbol{\beta}^{N,1}, \boldsymbol{\beta}^{N,2}, \cdots, \boldsymbol{\beta}^{N,N}) \geq J(\boldsymbol{\beta}^{N,1}) - C_\kappa \varepsilon_N,$$

where $J(\boldsymbol{\beta}^{N,1})$ stands for the mean field cost of $\boldsymbol{\beta}^{N,1}$:

$$J(\boldsymbol{\beta}^{N,1}) = \mathbb{E}\biggl[\int_0^T f(t, \bar{U}_t^{N,1}, \mu_t, \beta_t^{N,1})dt + g(\bar{U}_T^{N,1}, \mu_T)\biggr]. \tag{6.30}$$

Since $J \leq J(\boldsymbol{\beta}^{N,1})$, we get:

$$J^{N,1}(\boldsymbol{\beta}^{N,1}, \boldsymbol{\beta}^{N,2}, \cdots, \boldsymbol{\beta}^{N,N}) \geq J - C_\kappa \varepsilon_N, \tag{6.31}$$

and from (6.22) and (6.31), we easily derive the desired inequality (6.13).

For the proof to be complete, we need to explain how we choose κ, and discuss what happens when $\mathbb{E}\int_0^T |\beta_t^{N,1}|^2 dt > \kappa$. We start with the case when assumption **Approximate Nash SMP** holds. Using the convexity in x of g around $x = 0$, and the convexity of f in (x, α) around $x = 0$ and $\alpha = 0$, we get:

$$\begin{aligned}
&J^{N,1}(\boldsymbol{\beta}^{N,1}, \boldsymbol{\beta}^{N,2}, \cdots, \boldsymbol{\beta}^{N,N}) \\
&\quad \geq \mathbb{E}\biggl[\int_0^T f(t, 0, \bar{\nu}_t^N, 0)dt + g(0, \bar{\nu}_T^N)\biggr] + \lambda \mathbb{E}\int_0^T |\beta_t^{N,1}|^2 dt \\
&\qquad + \mathbb{E}\biggl[\int_0^T \bigl(U_t^{N,1} \cdot \partial_x f(t, 0, \bar{\nu}_t^N, 0) + \beta_t^{N,1} \cdot \partial_\alpha f(t, 0, \bar{\nu}_t^N, 0)\bigr)dt \\
&\qquad\qquad + U_T^{N,1} \cdot \partial_x g(0, \bar{\nu}_T^N)\biggr].
\end{aligned} \tag{6.32}$$

The local-Lipschitz assumption with respect to the Wasserstein distance implies the existence of a constant $C > 0$ such that for any $t \in [0, T]$,

$$\begin{aligned}
\mathbb{E}\bigl[|f(t, 0, \bar{\nu}_t^N, 0) - f(t, 0, \delta_0, 0)|\bigr] &\leq C\mathbb{E}\bigl[1 + M_2(\bar{\nu}_t^N)^2\bigr] \\
&= C\biggl[1 + \frac{1}{N}\sum_{i=1}^N \mathbb{E}\bigl[|U_t^{N,i}|^2\bigr]\biggr],
\end{aligned}$$

with a similar inequality for g. From this, we conclude that:

$$\begin{aligned}
J^{N,1}(\boldsymbol{\beta}^{N,1}, \boldsymbol{\beta}^{N,2}, \cdots, \boldsymbol{\beta}^{N,N}) &\geq \int_0^T f(t, 0, \delta_0, 0)dt + g(0, \delta_0) \\
&+ \mathbb{E}\biggl[\int_0^T \bigl(U_t^{N,1} \cdot \partial_x f(t, 0, \bar{\nu}_t^N, 0) + \beta_t^{N,1} \cdot \partial_\alpha f(t, 0, \bar{\nu}_t^N, 0)\bigr)dt + U_T^{N,1} \cdot \partial_x g(0, \bar{\nu}_T^N)\biggr] \\
&+ \lambda \mathbb{E}\int_0^T |\beta_t^{N,1}|^2 dt - C\biggl[1 + \frac{1}{N}\sum_{i=1}^N \sup_{0 \leq t \leq T} \mathbb{E}\bigl[|U_t^{N,i}|^2\bigr]\biggr].
\end{aligned}$$

Under the standing assumption, we know that $\partial_x g$, $\partial_x f$ and $\partial_\alpha f$ are at most of linear growth in the second moment of the measure argument, so that for any $\delta > 0$, there exists a constant C_δ such that:

$$
\begin{aligned}
&J^{N,1}(\boldsymbol{\beta}^{N,1}, \boldsymbol{\beta}^{N,2}, \cdots, \boldsymbol{\beta}^{N,N}) \\
&\quad \geq \int_0^T f(t, 0, \delta_0, 0)dt + g(0, \delta_0) + \frac{\lambda}{2}\mathbb{E}\int_0^T |\beta_t^{N,1}|^2 dt \\
&\quad\quad - \delta \sup_{0 \leq t \leq T} \mathbb{E}\big[|U_t^{N,1}|^2\big] - C_\delta \bigg(1 + \frac{1}{N}\sum_{i=1}^N \sup_{0 \leq t \leq T} \mathbb{E}\big[|U_t^{N,i}|^2\big]\bigg).
\end{aligned}
\tag{6.33}
$$

Estimates (6.16) and (6.17) show that one can choose δ small enough in (6.33) and find C so that:

$$
J^{N,1}(\boldsymbol{\beta}^{N,1}, \boldsymbol{\beta}^{N,2}, \cdots, \boldsymbol{\beta}^{N,N}) \geq -C + \big(\frac{\lambda}{4} - \frac{C}{N}\big)\mathbb{E}\int_0^T |\beta_t^{N,1}|^2 dt.
$$

This proves that there exists an integer N_0 such that for any integer $N \geq N_0$ and any constant $\bar{\kappa} > 0$, one can choose $\kappa > 0$ such that:

$$
\mathbb{E}\int_0^T |\beta_t^{N,1}|^2 dt \geq \kappa \quad \Longrightarrow \quad J^{N,1}(\boldsymbol{\beta}^{N,1}, \boldsymbol{\beta}^{N,2}, \cdots, \boldsymbol{\beta}^{N,N}) \geq J + \bar{\kappa},
\tag{6.34}
$$

which provides us with the appropriate tool to choose κ and avoid having to consider $(\beta_t^{N,1})_{0 \leq t \leq T}$ whose expected square integral is too large.

We now turn to the case when assumption **Approximate Nash HJB** is in force. In that case, the convexity argument used in (6.32) does not apply anymore. Still, we can obtain a similar result by using the fact that f is convex with respect to α and the fact that, for a given $\alpha_0 \in A$, $f(t, x, \mu, \alpha_0)$, $\partial_\alpha f(t, x, \mu, \alpha_0)$ and $g(x, \mu)$ are uniformly bounded in $(t, x, \mu) \in [0, T] \times \mathbb{R}^d \times \mathcal{P}_2(\mathbb{R}^d)$. Inequality (6.33), with $\inf_{(t,x,\mu)} f(t, x, \mu, \alpha_0)$ in lieu of $f(t, 0, \delta_0, 0)$ and $\inf_{(x,\mu)} g(x, \mu)$ in lieu of $g(0, \delta_0)$, follows quite easily. □

Remark 6.10 *A simple inspection of the last part of the above proof shows that a stronger result actually holds when* $\mathbb{E}\int_0^T |\beta_t^{N,1}|^2 dt \leq \kappa$. *Indeed, the estimates* (6.16), (6.25) *and* (6.28) *can be used as in* (6.22) *to deduce (up to a modification of* C_κ*):*

$$
J^{N,i}(\boldsymbol{\beta}^{N,1}, \boldsymbol{\beta}^{N,2}, \cdots, \boldsymbol{\beta}^{N,N}) \geq J - C_\kappa \varepsilon_N, \quad 2 \leq i \leq N.
\tag{6.35}
$$

6.1.2 The Case of the MFGs with a Common Noise

We now turn to games with a common noise. The objective remains the same: with the same definitions as above for approximate equilibria, we provide a general scheme to construct approximate open and closed loop equilibria given the existence of an MFG equilibrium.

Although the proof goes along the same lines as in the absence of a common noise, the random structure of the equilibrium makes the derivation significantly more intricate. This is especially true when the equilibrium exists in the weak sense only, recall the terminology introduced in Chapter 2. As made clear in the next section, the analysis becomes somehow easier when the equilibrium is understood in the strong sense and uniqueness holds.

To wit, the main difficulty is to provide an analog of (6.3): under the influence of the common noise, the function v appearing in (6.3) becomes a random field. This renders its construction much more intricate than in the case without common noise.

In order to construct v in this challenging environment, we appeal to the general notion of decoupling field introduced in Chapter 1. This step will be the most technical in the proof. To be more specific, we regard v as the decoupling field of some forward-backward system. In order to do so, we work with the adjoint system derived from the stochastic Pontryagin maximum principle for the optimal control problem under the environment formed by the MFG equilibrium. Thanks to Proposition 1.50, this permits to represent the optimal strategy in equilibrium, as a function of the private state of the player, with the subtle provision that this function is random.

Once the existence of a decoupling field has been demonstrated, we can use the same control strategy as in (6.10) in order to build the approximate equilibria. Although it is indeed licit to do so, this does not lead to the same notion of *distributed* strategy as in the case without common noise. Indeed, the resulting strategy is not even in feedback form. In the presence of a common noise, the player needs much more information. Indeed, the general representation formula in Proposition 1.50 says that, in order to implement the control strategy (6.10) in the presence of the common noise, the player needs to observe the realizations of both the common noise and the collective state up to the present time. We make this statement more precise below.

Definition of the Set-Up

Throughout this subsection, the set-up is the same as in the part devoted to mean field games with a common noise, see Chapters 2 and 3. Following the Definition 2.16 of a weak equilibrium together with the presentation in Subsection 3.1.2, we are given:

1. a complete probability space $(\Omega^0, \mathcal{F}^0, \mathbb{P}^0)$, endowed with a complete and right-continuous filtration $\mathbb{F}^0 = (\mathcal{F}_t^0)_{0\le t\le T}$, an $\mathcal{F}_0^0$-measurable initial random probability measure μ_0 on $\mathbb{R}^d$ with $\mathcal{V}_0$ as distribution, and a d-dimensional $\mathbb{F}^0$-Brownian motion $\boldsymbol{W}^0 = (W_t^0)_{0\le t\le T}$,
2. a complete probability space $(\Omega^1, \mathcal{F}^1, \mathbb{P}^1)$ endowed with a complete and right-continuous filtration $\mathbb{F}^1 = (\mathcal{F}_t^1)_{0\le t\le T}$ and a d-dimensional $\mathbb{F}^1$-Brownian motion $\boldsymbol{W} = (W_t)_{0\le t\le T}$.

(continued)

We then denote by $(\Omega, \mathcal{F}, \mathbb{P})$ the completion of the product space $(\Omega^0 \times \Omega^1, \mathcal{F}^0 \otimes \mathcal{F}^1, \mathbb{P}^0 \otimes \mathbb{P}^1)$ and endow it with the filtration $\mathbb{F} = (\mathcal{F}_t)_{0 \le t \le T}$ obtained by augmenting the product filtration $\mathbb{F}^0 \otimes \mathbb{F}^1$ to make it right-continuous, and by completing it.

We recall the useful notation $\mathcal{L}^1(X)(\omega^0) = \mathcal{L}(X(\omega^0, \cdot))$ for $\omega^0 \in \Omega^0$ and a random variable X constructed on Ω, see Subsection 2.1.3.

For a drift function b from $[0,T] \times \mathbb{R}^d \times \mathcal{P}_2(\mathbb{R}^d) \times A$ into $\mathbb{R}^d$, where A is a closed convex subset of $\mathbb{R}^k$, for two (uncontrolled) volatility coefficients σ and σ^0 from $[0,T] \times \mathbb{R}^d \times \mathcal{P}_2(\mathbb{R}^d)$ into $\mathbb{R}^{d \times d}$, for real valued cost functions f and g from $[0,T] \times \mathbb{R}^d \times \mathcal{P}_2(\mathbb{R}^d) \times A$ and $\mathbb{R}^d \times \mathcal{P}_2(\mathbb{R}^d)$, and for a square-integrable $\mathcal{F}_0$-measurable initial condition X_0, we assume that there exists an MFG equilibrium, namely an $\mathcal{F}_T^0$-measurable random variable $\mathfrak{M}$ with values in $\mathcal{P}_2(\mathcal{C}([0,T];\mathbb{R}^{2d}))$ such that: (i) $\mathbb{F}$ is compatible with $(X_0, \boldsymbol{W}^0, \mathfrak{M}, \boldsymbol{W})$; (ii) $\mathfrak{M} \circ (e_0^x)^{-1} = \mathcal{L}^1(X_0)$, where e_t^x is the mapping evaluating the first d coordinates at time t on $\mathcal{C}([0,T];\mathbb{R}^{2d})$; (iii) and

$$\mathfrak{M} = \mathcal{L}^1\big(\boldsymbol{X}, \boldsymbol{W}\big), \tag{6.36}$$

where $\boldsymbol{X}$ is the optimal state process of the random coefficients stochastic control problem:

$$\inf_{\boldsymbol{\alpha}=(\alpha_t)_{0 \le t \le T}} \mathbb{E}\left[\int_0^T f(t, X_t^{\boldsymbol{\alpha}}, \mu_t, \alpha_t)dt + g(X_T^{\boldsymbol{\alpha}}, \mu_T)\right], \tag{6.37}$$

under the dynamic constraint:

$$dX_t^{\boldsymbol{\alpha}} = b\big(t, X_t^{\boldsymbol{\alpha}}, \mu_t, \alpha_t\big)dt + \sigma\big(t, X_t^{\boldsymbol{\alpha}}, \mu_t\big)dW_t + \sigma^0\big(t, X_t^{\boldsymbol{\alpha}}, \mu_t\big)dW_t^0, \tag{6.38}$$

for $t \in [0,T]$, where $\mu_t = \mathfrak{M} \circ (e_t^x)^{-1}$ for all $t \in [0,T]$.

Existence of a weak equilibrium is guaranteed by one of the results obtained in Chapter 3, for instance Theorem 3.29 or Theorem 3.31.

Revisiting the Notion of Decoupling Field

The key point in our analysis is to prove that, under the assumptions of either Theorem 3.29 or Theorem 3.31, the solution of the optimal control problem (6.37)–(6.38) may be represented by means of a decoupled SDE of the form (6.3), namely:

$$\begin{aligned} dX_t &= b\big(t, X_t, \mu_t, \hat{\alpha}\big(t, X_t, \mu_t, V_t(X_t)\big)\big)dt \\ &\quad + \sigma(t, X_t, \mu_t)dW_t + \sigma^0(t, X_t, \mu_t)dW_t^0, \quad t \in [0,T], \end{aligned} \tag{6.39}$$

where $(V_t(x))_{0\le t\le T, x\in\mathbb{R}^d}$ is an $\mathbb{F}^0$-progressively measurable random field from $\Omega^0 \times [0,T] \times \mathbb{R}^d$ into $\mathbb{R}^d$ which is C-Lipschitz in space, for a constant $C \ge 0$. By progressively measurable and C-Lipschitz, we mean that, for any $x \in \mathbb{R}^d$, $(V_t(x))_{0\le t\le T}$ is $\mathbb{F}^0$-progressively measurable and, for any $t \in [0,T]$ and any $\omega^0 \in \Omega^0$, the realization of V_t is a C-Lipschitz function $V_t : \mathbb{R}^d \ni x \mapsto V_t(x) \in \mathbb{R}^d$.

In this regard, we have the following analogue of Theorem 6.4 when σ and σ^0 are independent of x:

Theorem 6.11 *Let assumption* **MFG with a Common Noise HJB** *be in force for some constants $L \ge 0$ and $\lambda > 0$. Assume further that the volatility coefficients σ and σ^0 are independent of x, and that $\partial_x(b,f)$ and $\partial_x g$ are L-Lipschitz continuous with respect to (x,α) and x respectively, uniformly in (t,μ) and μ, for the same constant L as in assumption* **MFG with a Common Noise HJB**. *Then, on the space $(\Omega,\mathcal{F},\mathbb{F},\mathbb{P})$ equipped with $(X_0, \boldsymbol{W}^0, \mathfrak{M}, \boldsymbol{W})$, there exists an $\mathbb{F}^0$-progressively measurable random field $V : \Omega^0 \times [0,T] \times \mathbb{R}^d \to \mathbb{R}^d$, bounded by C and C-Lipschitz continuous in space for a constant C only depending on L and λ, such that the optimal strategy in the optimal control problem* (6.37)–(6.38) *takes the form:*

$$\hat{\alpha}_t = \hat{\alpha}\big(t, X_t, \mu_t, V_t(X_t)\big),$$

for almost every $(t,\omega) \in [0,T] \times \Omega$ under $Leb_1 \otimes \mathbb{P}$.

Remark 6.12 *The assumption that σ and σ^0 are independent of x is mostly for convenience. We strongly believe that the result remains true when σ and σ^0 are smooth functions of x. However, as demonstrated in the proof of Theorem 6.4 for mean field games without common noise, the presence of x in the coefficients σ and σ^0 create additional difficulties, which we prefer to avoid here.*

We refer to Subsection 3.4.1 for the details of assumption **MFG with a Common Noise HJB**. Moreover, we emphasize that, as made clear in the proof below, the statement of Theorem 6.11 still holds even if $\mathfrak{M}$ does not satisfy (6.36).

Proof.

First Step. For any $t \in [0,T]$, we consider the t-initialized set-up $(\Omega, \mathcal{F}, (\mathcal{F}_s)_{t\le s\le T}, \mathbb{P})$ equipped with $(\mathcal{F}_t^{\mathrm{nat},(X_0,\boldsymbol{W}^0,\mathfrak{M},\boldsymbol{W})}, (W_s^0 - W_t^0, \mathfrak{M}_s, W_s - W_t)_{t\le s\le T})$ as input.

Then, for any $x \in \mathbb{R}^d$, we consider the control problem (6.37)–(6.38), initialized at time t, with x as initial condition and $(\mathfrak{M}_s)_{t\le s\le T}$ as environment.

By a straightforward extension of Theorem 1.57 to accommodate the fact that the set-up is here of a generalized form, the optimal control problem has a unique solution. Its state process is given by the forward component of the unique solution, with a bounded martingale integrand, of the FBSDE:

$$\begin{cases} dX_s = b\big(s, X_s, \mu_s, \hat{\alpha}(s, X_s, \mu_s, \sigma(s, \mu_s)^{-1\dagger} Z_s)\big)ds \\ \qquad +\sigma(s, \mu_s)dW_s + \sigma^0(s, \mu_s)dW_s^0, \\ dY_s = -f\big(s, X_s, \mu_s, \hat{\alpha}(s, X_s, \mu_s, \sigma(s, \mu_s)^{-1\dagger} Z_s)\big)ds \\ \qquad +Z_s \cdot dW_s + Z_s^0 \cdot dW_s^0 + dM_s, \end{cases} \tag{6.40}$$

for $s \in [t, T]$, with the terminal condition $Y_T = g(X_T, \mu_T)$, $\boldsymbol{M} = (M_s)_{t \le s \le T}$ being a square-integrable martingale with respect to the filtration $(\mathcal{F}_s)_{t \le s \le T}$, with $M_t = 0$ as initial condition and of zero bracket with $(W_s^0 - W_t^0, W_s - W_t)_{t \le s \le T}$. The unique solution of (6.40) is denoted by:

$$\big(\boldsymbol{X}^{t,x}, \boldsymbol{Y}^{t,x}, \boldsymbol{Z}^{t,x}, \boldsymbol{Z}^{0;t,x}, \boldsymbol{M}^{t,x}\big) = \big(X_s^{t,x}, Y_s^{t,x}, Z_s^{t,x}, Z_s^{0;t,x}, M_s^{t,x}\big)_{t \le s \le T}.$$

Following the convention introduced in the absence of a common noise, we let:

$$\hat{\boldsymbol{\alpha}}^{t,x} = \Big(\hat{\alpha}\big(s, X_s^{t,x}, \mu_s, \sigma(s, \mu_s)^{-1\dagger} Z_s^{t,x}\big)\Big)_{t \le s \le T}.$$

Importantly, the conclusion of Theorem 1.57 and the stability result of Theorem 1.53 imply the existence of a constant C, only depending on the parameters λ and L in assumption **MFG with a Common Noise HJB**, such that, for any $t \in [0, T]$ and $x, x' \in \mathbb{R}^d$,

$$\mathbb{E}\bigg[\sup_{t \le s \le T} \big(|X_s^{t,x} - X_s^{t,x'}|^2 + |Y_s^{t,x} - Y_s^{t,x'}|^2\big) + \int_t^T |Z_s^{t,x} - Z_s^{t,x'}|^2 ds \,\Big|\, \mathcal{F}_t\bigg] \le C|x - x'|^2. \tag{6.41}$$

Second Step. We now make use of the necessary condition in the Pontryagin principle. Since σ and σ^0 are independent of x, we can use the reduced Hamiltonian $H^{(r)}$ in the equation for the adjoint variables. By adapting the statement of Theorem 1.59 in a suitable way to accommodate the fact that the set-up is of a generalized form, we learn that $\hat{\boldsymbol{\alpha}}^{t,x}$ can be rewritten in the form:

$$\hat{\alpha}_s^{t,x} = \hat{\alpha}\big(s, X_s^{t,x}, \mu_s, \upsilon_s^{t,x}\big), \tag{6.42}$$

where $(\upsilon_s^{t,x})_{t \le s \le T}$ is the solution, on the same t-initialized set-up as above, of the uniquely solvable backward equation:

$$\begin{aligned} d\upsilon_s^{t,x} &= -\partial_x H^{(r)}\big(s, X_s^{t,x}, \mu_s, \upsilon_s^{t,x}, \hat{\alpha}_s^{t,x}\big)ds \\ &\quad + \zeta_s^{t,x} dW_s + \zeta_s^{0;t,x}, dW_s^0 + dm_s^{t,x}, \quad s \in [t, T], \end{aligned} \tag{6.43}$$

with the terminal condition $\upsilon_T^{t,x} = \partial_x g(X_T^{t,x}, \mu_T)$, $\boldsymbol{m}^{t,x} = (m_s^{t,x})_{t \le s \le T}$ being a square-integrable martingale with respect to the filtration $(\mathcal{F}_s)_{t \le s \le T}$, with 0 as initial condition and of null bracket with $(W_s^0 - W_t^0, W_s - W_t)_{t \le s \le T}$.

Since $\partial_x f$ and $\partial_x g$ are bounded, there is no difficulty proving that the process $(\upsilon_s^{t,x})_{t \le s \le T}$ is bounded, uniformly in $(t, x) \in [0, T] \times \mathbb{R}^d$. We denote the common bound by the same letter C as in the stability estimate (6.41), so that:

$$\mathbb{P}\big[\sup_{t \le s \le T} |\upsilon_s^{t,x}| \le C\big] = 1. \tag{6.44}$$

Using the bound (6.44) in (6.43) together with the stability estimate (6.41) and proceeding as in the analysis of Example 1.20, we deduce that, for all $x, x' \in \mathbb{R}^d$, with $\mathbb{P}$-probability 1,

$$|v_t^{t,x} - v_t^{t,x'}| \leq C|x - x'|. \tag{6.45}$$

Moreover, using (6.44) and the identification (6.42), on the same t-initialized set-up as above, we can regard the pair $(X_s^{t,x}, v_s^{t,x})_{t\leq s\leq T}$ as the solution of the FBSDE with Lipschitz coefficients:

$$\begin{cases} dX_s = b\big(s, X_s, \mu_s, \hat{\alpha}(s, X_s, \mu_s, v_s)\big)ds \\ \qquad\qquad +\sigma(s, \mu_s)dW_s + \sigma^0(s, \mu_s)dW_s^0, \\ dv_s = -\partial_x H^{(r)}\big(s, X_s, \mu_s, \psi(v_s), \hat{\alpha}(s, X_s, \mu_s, v_s)\big)ds \\ \qquad\qquad +\zeta_s \cdot dW_s + \zeta_s^0 \cdot dW_s^0 + dm_s, \end{cases} \tag{6.46}$$

for $s \in [t, T]$, with $X_t = x$ as initial condition and with $v_T = \partial_x g(X_T, \mu_T)$ as terminal condition. In the derivative of the Hamiltonian, the function ψ is a smooth compactly supported function from $\mathbb{R}^d$ into itself coinciding with the identity on the ball of center 0 and of radius C. Accordingly, (6.45) together with Proposition 1.52 imply that (6.46) is uniquely solvable and admits a decoupling field $(U_t)_{0\leq t\leq T}$. With the same notations as in the statement of Proposition 1.50, it satisfies:

$$v_t^{t,x} = U_t\Big(x, \mathcal{L}\big((W_s^0 - W_t^0, \mathfrak{M}_s)_{t\leq s\leq T} \,|\, \mathcal{F}_t^{\text{nat},(X_0,W^0,\mathfrak{M})}\big), (W_s^0 - W_t^0, \mathfrak{M}_s)_{t\leq s\leq T}\Big)$$

with $\mathbb{P}$-probability 1. Proposition 1.50 asserts that the decoupling field is C-Lipschitz in x. In fact, a careful inspection of the proof of Proposition 1.50, see for instance the third step of the proof of Proposition 1.46, shows that it is also bounded by C. This should not come as a surprise since, in the above identity, $v_t^{t,x}$ itself is bounded by C.

For $t \in [0, T]$, we then let $V_t : \mathbb{R}^d \ni x \mapsto V_t(x) \in \mathbb{R}^d$ be the random field defined by:

$$V_t(x) = U_t\Big(x, \mathcal{L}\big((W_s^0 - W_t^0, \mathfrak{M}_s)_{t\leq s\leq T} \,|\, \mathcal{F}_t^{\text{nat},(X_0,W^0,\mathfrak{M})}\big), (W_s^0 - W_t^0, \mathfrak{M}_s)_{t\leq s\leq T}\Big). \tag{6.47}$$

Third Step. While $(V_t)_{0\leq t\leq T}$ is the quantity we were looking for in the statement, we must still check its measurability in time. In order to do so, we observe that, for $x \in \mathbb{R}^d$ and $0 \leq t \leq s \leq T$, with $\mathbb{P}$-probability 1,

$$V_t(x) - V_s(x) = v_t^{t,x} - v_s^{t,x} + V_s(X_s^{t,x}) - V_s(x),$$

where we used, in the right-hand side, the representation property of the decoupling field. Once more, recall the formula in Proposition 1.50 with s in lieu of t. The right-continuity of the process $(v_s^{t,x})_{t\leq s\leq T}$ and the Lipschitz property of V_s, imply that for any $x \in \mathbb{R}^d$:

$$\lim_{s\searrow t} \mathbb{E}\big[|V_s(x) - V_t(x)|^2\big] = 0, \tag{6.48}$$

where we used the fact that the decoupling field is bounded by C to take the limit in L^2-norm. Now, we let, for any integer $n \geq 1$:

$$V_t^n(x) = V_{\lceil tn/T \rceil/(n/T)}(x), \quad (t,x) \in [0,T] \times \mathbb{R}^d.$$

Then, by (6.48) and by dominated convergence theorem,

$$\lim_{n\to\infty} \sup_{p,q\geq n} \mathbb{E} \int_0^T |V_t^q(x) - V_t^p(x)|^2 dt = 0.$$

Since each process $(V_t^n(x))_{0\leq t\leq T}$, for $n \geq 1$ and $x \in \mathbb{R}^d$, is jointly measurable in (t,ω), we deduce that, for any $x \in \mathbb{Q}^d$ (in other words if x has rational components), we can find a jointly measurable process which we denote by $(\tilde{V}_t(x))_{0\leq t\leq T}$, and for which it holds for almost every $t \in [0,T]$:

$$\mathbb{P}\big[V_t(x) = \tilde{V}_t(x)\big] = 1. \tag{6.49}$$

In particular, for any $x \in \mathbb{Q}^d$, the process $(\tilde{V}_t(x))_{0\leq t\leq T}$ can be assumed to be $\mathbb{F}^0$-adapted and jointly measurable; without any loss of generality, we can assume it to be $\mathbb{F}^0$-progressively measurable. Now, the key point is to observe that we can find a version of $\tilde{V}$ which is Lipschitz continuous in space. To do so, we let for any $t \in [0,T]$:

$$D_t = \bigcap_{x,y\in\mathbb{Q}^d} \big\{|\tilde{V}_t(x) - \tilde{V}_t(y)| \leq C|x-y|\big\}.$$

It is clear that the process $(\mathbf{1}_{D_t})_{0\leq t\leq T}$ is progressively measurable. Moreover, by the Lipschitz property in x of V_t, for almost every $t \in [0,T]$, we have $\mathbb{P}(D_t) = 1$. Therefore, (6.49) remains true if, for any $x \in \mathbb{Q}^d$, we consider $(\mathbf{1}_{D_t}\tilde{V}_t(x))_{0\leq t\leq T}$ in lieu of $(\tilde{V}_t(x))_{0\leq t\leq T}$. Equivalently, we can assume, without any loss of generality, that, for any realization, for all $t \in [0,T]$, for all $x,y \in \mathbb{Q}^d$,

$$|\tilde{V}_t(x) - \tilde{V}_t(y)| \leq C|x-y|,$$

which implies that, for any realization and for all $t \in [0,T]$, $\tilde{V}_t$ extends into a Lipschitz mapping from $\mathbb{R}^d$ into itself. Obviously, for almost every $t \in [0,T]$, for all $x \in \mathbb{R}^d$, (6.49) remains true, which is a crucial fact.

Actually, by continuity, we even have that, for almost every $t \in [0,T]$, with $\mathbb{P}$-probability 1, $V_t(\cdot)$ and $\tilde{V}_t(\cdot)$ coincide.

Fourth Step. In order to complete the proof, we recall that the optimal path of the optimal control problem (6.37)–(6.38), with X_0 as initial condition, is also characterized by the forward component of the adjoint system (6.46), but with X_0 as initial condition at time 0 in lieu of x at time t. Since the latter has been proved to be uniquely solvable on the set-up $(\Omega, \mathcal{F}, (\mathcal{F}_t)_{0\leq t\leq T}, \mathbb{P})$ equipped with $(X_0, W_s^0, \mathfrak{M}_s, W_s)_{0\leq s\leq T}$, we know from Proposition 1.50 that, for any $t \in [0,T]$, $\mathbb{P}[\upsilon_t = V_t(X_t)] = 1$. Recall indeed that Proposition 1.50 permits to represent the backward component in terms of the forward one by means of the decoupling field. Therefore, for almost every $t \in [0,T]$, $\mathbb{P}[\upsilon_t = \tilde{V}_t(X_t)] = 1$. The proof is easily completed by means of Fubini's theorem and Theorem 1.59, the latter providing a representation similar to (6.42). □

In full analogy with assumption **Approximate Nash HJB**, we introduce the following assumption:

Assumption (Approximate Nash with a Common Noise HJB). Let assumption **MFG with a Common Noise HJB** be in force and assume that $\partial_x(b,f)$ and $\partial_x g$ are Lipschitz continuous with respect to (x,α) and x respectively, uniformly in (t,μ) and μ, that the functions σ and σ^0 are independent of x, that the functions b, σ and σ^0 are Lipschitz continuous in μ, uniformly with respect to the other parameters, and that the functions f and g are locally Lipschitz continuous with respect to μ, the Lipschitz constant being bounded, for a constant $L \geq 0$ and for any $R \geq 1$, by LR, when f and g are restricted to:

$$\{(t,x,\alpha,\mu) \in [0,T] \times \mathbb{R}^d \times A \times \mathcal{P}_2(\mathbb{R}^d) : |x| + |\alpha| + M_2(\mu) \leq R\}$$

and

$$\{(x,\mu) \in \mathbb{R}^d \times \mathcal{P}_2(\mathbb{R}^d) : |x| + M_2(\mu) \leq R\}$$

respectively.

As in the absence of common noise, we can work with the optimal trajectories constructed by means of the sufficient condition in the Pontryagin stochastic maximum principle. See for instance Theorem 3.31, which holds true under assumption **MFG with a Common Noise SMP Relaxed** from Subsection 3.4.3 (so that σ and σ^0 may depend on x). The proof is easier in that case since, at each time $t \in [0,T]$, the decoupling field is directly given by Theorem 1.60, though an additional argument is needed to construct a version that is time progressively measurable with respect to the filtration $\mathbb{F}^0$. Generally speaking, this may be achieved by duplicating the third step in the proof of Theorem 6.11. The only difference comes from the fact that the decoupling field is no longer bounded by a constant C. As made clear in the statement of Theorem 1.60, it is at most of linear growth:

$$|V_t(x)| \leq C\Big(1 + |x| + \mathbb{E}^0\big[\sup_{0 \leq s \leq T} M_2(\mu_s)^2 \,|\, \mathcal{F}_t^{\mathrm{nat},X_0,W^0,\mathfrak{M}}\big]^{1/2}\Big), \tag{6.50}$$

which comes from (1.67). We then observe that the process:

$$\Big(\mathbb{E}^0\big[\sup_{0 \leq s \leq T} M_2(\mu_s)^2 \,|\, \mathcal{F}_t^{\mathrm{nat},X_0,W^0,\mathfrak{M}}\big]\Big)_{0 \leq t \leq T}$$

in (6.50) is uniformly integrable, which suffices to replicate the third step in the proof of Theorem 6.11 although V_t is not bounded. Up to this modification regarding the growth of $(V_t(x))_{0\le t\le T, x\in\mathbb{R}^d}$, Theorem 6.11 remains valid. This prompts us to introduce the following assumption, which is the analogue of assumption **Approximate Nash SMP**:

Assumption (Approximate Nash with a Common Noise SMP). On top of assumption **MFG with a Common Noise SMP Relaxed**, the functions b, σ and σ^0 are Lipschitz continuous in μ, uniformly with respect to $t \in [0, T]$ and $x \in \mathbb{R}^d$.

Games with Finitely Many Players

We now consider the analogue of (6.11). Following the construction of the particle system (2.3) in Chapter 2, we assume that the space $(\Omega^1, \mathcal{F}^1, \mathbb{F}^1, \mathbb{P}^1)$ carries a sequence of independent Wiener processes $(W^n)_{n\ge1}$. For the sake of simplicity, we also assume that the common noise reduces to the sole W^0, meaning that the initial states of the players are supported by $(\Omega^1, \mathcal{F}^1, \mathbb{F}^1, \mathbb{P}^1)$ and that $\mu_0 = \mathcal{L}^1(X_0) \in \mathcal{P}_2(\mathbb{R}^d)$ in (6.38) is deterministic. However, we stress that the analysis below can be extended to a more general form of common noise, for example as explained in Remark 2.10 in Chapter 2. We thus assume that $(\Omega^1, \mathcal{F}^1, \mathbb{F}^1, \mathbb{P}^1)$ carries a family of identically distributed and independent $\mathcal{F}_0^1$-measurable random variables $(\xi^n)_{n\ge1}$ with values in $\mathbb{R}^d$, with the same distribution as X_0.

Given the decoupling field constructed in the previous subsection, we consider the analogue of (6.9):

$$
\begin{aligned}
dX_t^{N,i} &= b\big(t, X_t^{N,i}, \bar{\mu}_t^N, \hat{\alpha}\big(t, X_t^{N,i}, \mu_t, V_t(X_t^{N,i})\big)\big)dt \\
&\quad + \sigma(t, X_t^{N,i}, \bar{\mu}_t^N)dW_t^i + \sigma^0(t, X_t^{N,i}, \bar{\mu}_t^N)dW_t^0,
\end{aligned}
\tag{6.51}
$$

for $t \in [0, T]$ and $i \in \{1, \cdots, N\}$, with $(\xi^1, \cdots, \xi^N)$ as initial condition, which means that the players use the control strategies:

$$
\hat{\alpha}_t^{N,i} = \hat{\alpha}\big(t, X_t^{N,i}, \mu_t, V_t(X_t^{N,i})\big), \qquad 0 \le t \le T, \;\; i \in \{1, \cdots, N\}. \tag{6.52}
$$

Up to the fact that V_t is random, the particle system (6.51) is of the same type as the one investigated in Section 2.1. See (2.3). Fortunately, the fact that V_t is random should only be a minor inconvenience in applying the results from Subsection 2.1.4 since the randomness of V_t comes only from $\boldsymbol{W}^0$.

As we already noticed in the introduction of Subsection 6.1.2, the strategies $(\hat{\boldsymbol{\alpha}}^{N,i})_{1\le i\le N}$ are not *distributed*; they are even not in closed feedback form! Owing to the representation formula (6.47) for the decoupling field V_t at time t, the player needs to observe not only the realization of the common noise but also the realization of the *enlarged environment* $\mathfrak{M}$ up to present time. To emphasize this feature, the strategies may be said to be in *semi-feedback form*. What it means exactly may not be clear:

- When the equilibrium is strong, $\mathfrak{M}$ is a function of the environment, see Theorem 2.29. In that case, it suffices for a player to observe the common noise (in addition to its own private state) in order to implement the strategy (6.52), which may be acceptable from a physical or economic point of view.
- When the equilibrium is strictly weak, in the sense that $\mathfrak{M}$ incorporates an additional randomness, independent of $\boldsymbol{W}^0$, there is no way for the player to access the whole realization of $\mathfrak{M}$ from the sole observation of $\boldsymbol{W}^0$. In that case, the construction of a solution to (6.51) requires that all the players use the same realization of the additional randomness entering the definition of V_t, which makes sense since the information carried by $(\boldsymbol{W}^0, \mathfrak{M})$ is understood as the information common to all the players, see Chapter 2. Then, the identity (6.36) together with the conditional propagation of chaos property established in Theorem 2.12 says that the precise value of the realization of $\mathfrak{M}$ should be approximated by the empirical distribution of the system $((X_t^{N,i}, W_t^i)_{0 \le t \le T})_{1 \le i \le N}$; hence it should suffice to observe $((X_t^{N,i}, W_t^i)_{0 \le t \le T})_{1 \le i \le N}$ to provide an approximation of the realization of $\mathfrak{M}$. Although this sounds quite natural, a modicum of care is needed from the practical point of view as the required information is of a high complexity; also, such an empirical estimate would just provide an approximation of the control strategy in (6.51). To make it proper, stability properties of the decoupling field would be needed; however, obtaining such stability properties may be a real issue, which we address at the end of the subsection.

Similar to (6.51), we now define the analogue of (6.11). To do so, we need to adapt the definition of an admissible strategy: We call an admissible tuple of strategies $\boldsymbol{\beta}^{(N)} = (\boldsymbol{\beta}^{N,1}, \cdots, \boldsymbol{\beta}^{N,N})$ with $\boldsymbol{\beta}^{N,i} = (\beta_t^{N,i})_{0 \le t \le T}$ for $i = 1 \le i \le N$, an N-tuple of square-integrable A-valued processes that are progressively measurable with respect to $\mathbb{F}$, while, for games without common noise, we required the strategies to be adapted with respect to the augmentation of the filtration generated by $(\xi^1, \cdots, \xi^N)$ and $(\boldsymbol{W}^0, \boldsymbol{W}^1, \cdots, \boldsymbol{W}^N)$. The enlargement of the filtration is especially useful to handle strategies of the same form as $(\hat{\boldsymbol{\alpha}}^{N,i})_{i=1,\cdots,N}$ in (6.52), which are allowed to depend on $\mathfrak{M}$ and thus which may not be adapted with respect to the filtration generated by $(\xi^1, \cdots, \xi^N)$ and $(\boldsymbol{W}^0, \boldsymbol{W}^1, \cdots, \boldsymbol{W}^N)$. Recalling the standard notation $\mathbb{A}$ for the set of square-integrable A-valued processes that are $\mathbb{F}$-progressively measurable, admissible tuples are just tuples of the form $(\boldsymbol{\beta}^{N,1}, \cdots, \boldsymbol{\beta}^{N,N})$ with $\boldsymbol{\beta}^{N,i} \in \mathbb{A}$ for each $i \in \{1, \cdots, N\}$.

We now provide an analogue to Theorems 6.7 and 6.9. Before we do so, we make the following crucial observation: while we can still expect that the strategies $(\hat{\boldsymbol{\alpha}}^{N,i})_{1 \le i \le N}$ form an approximate Nash equilibrium in open loop, we can no longer expect that they form an approximate equilibrium in closed loop since they are not in closed loop form! Put differently, Definition 6.6 of an approximate Nash equilibrium in open loop is still relevant, but Definition 6.8 must be modified to keep it adapted to the new framework. A natural way to do so is to allow the functions $\phi^{N,1}, \cdots, \phi^{N,N}$ and ψ^i in the definition to be random fields of the form of $\Psi : \Omega^0 \times [0,T] \times \mathcal{C}([0,T];\mathbb{R}^d) \rightarrow A$ such that, for any $S \in [0,T]$, the map $\Psi : \Omega^0 \times [0,S] \times \mathcal{C}([0,T];\mathbb{R}^d) \ni (\omega^0,t,x) \mapsto \Psi(\omega^0,t,x_{[0,t]})$ is $\mathcal{F}^0_S \otimes \mathcal{B}([0,S]) \otimes \mathcal{B}(\mathcal{C}([0,T];\mathbb{R}^d))$-measurable. Such random fields are called $\mathbb{F}^0$-progressively measurable; as already mentioned, corresponding equilibria are said to be in *generalized closed loop form* or in *semi-closed loop form*.

We first claim:

Theorem 6.13 *Under either assumption* **Approximate Nash with a Common Noise HJB** *or assumption* **Approximate Nash with a Common Noise SMP**, *there exists a sequence $(\varepsilon_N)_{N\geq 1}$, converging to 0 as N tends to ∞, such that the strategies $(\hat{\boldsymbol{\alpha}}^{N,i})_{1\leq i\leq N}$ defined in (6.52) form an ε_N-approximate open loop Nash equilibrium of the N-player game (6.37)–(6.38). Precisely, for each $N \geq 1$, for any player $i \in \{1,\cdots,N\}$ and any admissible control strategy $\boldsymbol{\beta}^i \in \mathbb{A}$, it holds:*

$$J^{N,i}(\hat{\boldsymbol{\alpha}}^{N,1},\cdots,\hat{\boldsymbol{\alpha}}^{N,i-1},\boldsymbol{\beta}^i,\hat{\boldsymbol{\alpha}}^{N,i+1},\cdots,\hat{\boldsymbol{\alpha}}^{N,N}) \geq J^{N,i}(\hat{\boldsymbol{\alpha}}^{N,1},\cdots,\hat{\boldsymbol{\alpha}}^{N,N}) - \varepsilon_N, \tag{6.53}$$

with the same definition for $J^{N,i}$ as in (6.12).

If the initial condition X_0 has a finite moment of order q, for some $q > 4$, then we can choose $\varepsilon_N = c\sqrt{\epsilon_N}$, for a constant c independent of N, ϵ_N being defined as in (6.1).

Forewarned by the observations made before the above statement, we solve the problem of approximate closed loop equilibrium in the following way:

Theorem 6.14 *Under either assumption* **Approximate Nash with a Common Noise HJB** *or assumption* **Approximate Nash with a Common Noise SMP**, *there exists a sequence $(\varepsilon_N)_{N\geq 1}$, converging to 0 as N tends to ∞, such that the strategies $(\hat{\boldsymbol{\alpha}}^{N,i})_{1\leq i\leq N}$ associated with the feedback random fields*

$$\phi^{N,i} : [0,T] \times (\mathbb{R}^d)^N \ni \big(t,(x^1,\cdots,x^N)\big) \mapsto \hat{\alpha}\big(t,x^i,\mu_t,V_t(x^i)\big),$$

used in (6.51), *form a generalized closed loop ε_N-approximate Nash equilibrium of the N-player game* (6.37)–(6.38). *Here, the definition of an approximate Nash equilibrium in generalized closed loop form is similar to Definition 6.8, except for the fact that we allow the functions $\phi^{N,1},\cdots,\phi^{N,N}$ and ψ^i in Definition 6.8 to be $\mathbb{F}^0$-progressively measurable random fields from $\Omega^0 \times [0,T] \times \mathcal{C}([0,T];\mathbb{R}^d)$ into A.*

If the initial condition X_0 has a finite moment of order q, for some $q > 4$, then we can choose $\varepsilon_N = c\sqrt{\epsilon_N}$, for a constant c independent of N, ϵ_N being defined as in (6.1).

Once again, we stress the fact that equilibria in generalized closed loop form may not be in closed loop form, since they may depend upon the whole past trajectory of the common noise.

Proof. The proofs of Theorems 6.13 and 6.14 go along the same lines as the proofs of Theorems 6.7 and 6.9.

Here are the main differences. In the first step of the proof, (6.14) and (6.15) are easily proved when assumption **Approximate Nash with a Common Noise HJB** is in force, since, in that case, the decoupling field is bounded. Under assumption **Approximate Nash with a Common Noise SMP**, the decoupling field is no longer bounded but satisfies the bound (6.50). Referring to the proof of Theorem 3.31, see in particular the third step in the proof of Lemma 3.33, there is no difficulty proving that $\mathbb{E}^0[\sup_{0\leq t\leq T} M_2(\mu_t)^q]$ is finite if $\mu_0 \in \mathcal{P}_q(\mathbb{R}^d)$ for $q = 2$ or $q > 4$. Plugging into (6.50), we get (6.14) and (6.15).

Due to the random nature of the decoupling field V_t in (6.39), we need, in the third step of the proof, a generalized version of Theorem 2.12 for particle systems of the same type as (2.3) but with $\mathbb{F}^0$-progressively measurable random coefficients. The proof of Theorem 2.12 can be easily adapted to this more general setting.

The rest of the proof is similar. □

Construction of Approximate Equilibria in True Closed Loop

We now show how to construct true approximated closed loop equilibria when the MFG equilibrium is unique.

The general strategy is as follows. When the equilibrium is unique, we know from Definition 4.1 that we can associate a master field with the mean field game. Provided that the master field is differentiable in space, which is guaranteed by Theorem 4.10 under appropriate conditions, we know from Proposition 4.7 and Corollary 4.11 that the decoupling field V_t used in (6.51) satisfies:

$$V_t(x) = \partial_x \mathcal{U}(t, x, \mu_t).$$

This prompts us to use, instead of the control strategies defined in (6.51)–(6.52), the following ones:

$$\hat{\alpha}_t^{N,i} = \hat{\alpha}\big(t, X_t^{N,i}, \bar{\mu}_t^N, \partial_x \mathcal{U}(t, X_t^{N,i}, \bar{\mu}_t^N)\big), \qquad 0 \leq t \leq T, \;\; i \in \{1, \cdots, N\}, \tag{6.54}$$

in which case the dynamics of the players are given by:

$$\begin{aligned} dX_t^{N,i} &= b\big(t, X_t^{N,i}, \bar{\mu}_t^N, \hat{\alpha}\big(t, X_t^{N,i}, \bar{\mu}_t^N, \partial_x \mathcal{U}(t, X_t^{N,i}, \bar{\mu}_t^N)\big)\big)dt \\ &\quad + \sigma(t, X_t^{N,i}, \bar{\mu}_t^N) dW_t^i + \sigma^0(t, X_t^{N,i}, \bar{\mu}_t^N) dW_t^0, \end{aligned} \tag{6.55}$$

for $t \in [0, T]$ and $i \in \{1, \cdots, N\}$, with $(\xi^1, \cdots, \xi^N)$ as initial condition.

Observe that, in the arguments of the function $\hat{\alpha}$ in the right-hand side of (6.54), we also replaced μ_t by $\bar{\mu}_t^N$. Then, provided that the SDE (6.55) is solvable, the strategies defined by (6.54) are in closed loop form. They are even Markovian. However, they are not distributed.

When the function $\partial_x \mathcal{U}$ is Lipschitz continuous in (x, μ), uniformly in time, and the path $(\partial_x \mathcal{U}(t, 0, \delta_0))_{0\leq t\leq T}$ is bounded, the system (6.55) is uniquely solvable. Moreover, by Theorem 2.12, propagation of chaos holds true and the limit in law is given by the solution of the SDE:

$$dX_t = b\big(t, X_t, \mathcal{L}^1(X_t), \hat{\alpha}\big(t, X_t, \mathcal{L}^1(X_t), \partial_x \mathcal{U}(t, X_t, \mathcal{L}^1(X_t))\big)\big)dt$$
$$+ \sigma(t, X_t, \mathcal{L}^1(X_t))dW_t + \sigma^0(t, X_t, \mathcal{L}^1(X_t))dW_t^0,$$

for $t \in [0, T]$, with $X_0 \sim \mu_0$. Thanks to Proposition 4.7 and Corollary 4.11, the above SDE describes the dynamics of the solution to the mean field game (6.36)–(6.37)–(6.38).

For instance, all these claims are known to be true under assumption **MFG Master Classical** from Subsection 5.4.3, see Theorem 5.46 and Propositions 5.53 and 5.54. We deduce:

Theorem 6.15 *Under assumption* **MFG Master Classical**, *there exists a sequence* $(\varepsilon_N)_{N\geq 1}$, *converging to* 0 *as* N *tends to* ∞, *such that the strategies* $(\hat{\boldsymbol{\alpha}}^{N,i})_{1\leq i\leq N}$ *associated with the feedback functions:*

$$\phi^{N,i} : [0, T] \times (\mathbb{R}^d)^N \ni \big(t, (x^1, \cdots, x^N)\big)$$
$$\mapsto \hat{\alpha}\Big(t, x^i, \frac{1}{N}\sum_{j=1}^N \delta_{x^j}, \partial_x \mathcal{U}\big(t, x^i, \frac{1}{N}\sum_{j=1}^N \delta_{x^j}\big)\Big),$$

used in (6.54) *form an* ε_N*-approximate closed loop Nash equilibrium of the* N*-player game* (6.11)–(6.12) *in the sense of Definition 6.8.*

If μ_0 *belongs to* $\mathcal{P}_q(\mathbb{R}^d)$, *for some* $q > 4$, *then we can choose* $\varepsilon_N = c\sqrt{\epsilon_N}$, *for a constant* c *independent of* N, ϵ_N *being defined as in* (6.1).

Observe that, under assumption **MFG Master Classical**, either assumption **MFG with a Common Noise HJB** or assumption **MFG with a Common Noise SMP Relaxed** is in force. Recall also that σ and σ^0 are constant and that b is independent of (x, μ). In particular, it is pretty obvious that b, σ and σ^0 satisfy the conditions required in assumptions **Approximate Nash with a Common Noise HJB** and **Approximate Nash with a Common Noise SMP**. Regarding the cost functionals f and g, we know that $\partial_\mu f(t, x, \mu, \alpha)(\cdot)$ and $\partial_\mu g(x, \mu)(\cdot)$ are bounded by ΓR if $|x| + |\alpha| + M_1(\mu) \leq R$, see assumption **MFG Smooth Coefficients** in Subsection 5.4.1. Hence, assumption **Approximate Nash with a Common Noise HJB** (resp. assumption **Approximate Nash with a Common Noise SMP**) is satisfied whenever assumption **MFG with a Common Noise HJB** (resp. assumption **MFG with a Common Noise SMP**) holds true in condition **(A3)** of assumption **MFG Master Classical**.

6.1.3 The Case of the Control of McKean-Vlasov SDEs

In this subsection, we address the same question as above but for the optimal control of McKean-Vlasov dynamics investigated in Chapter (Vol I)-6: We show how the solution of the optimal control of McKean-Vlasov diffusion processes can provide equilibria for systems of N players optimizing some common wealth when N tends to ∞. To do so, we explain first what kind of finite-player games we are talking about and we highlight the differences with those arising in mean field games.

As already mentioned several times in the previous chapters, and also in the previous subsections, a McKean-Vlasov SDE describes the asymptotic behavior of a mean field interacting particle system as the number of particles tends to infinity, see in particular Chapter 2. When particles are subject to independent noises, they become asymptotically independent of each other, and each single one satisfies the same McKean-Vlasov SDE in the limit. We called such a phenomenon *propagation of chaos*. When particle evolutions are controlled, each particle attempting to minimize an energy functional, it is natural to investigate equilibria for the finite populations, and study their limits (if any) when the number of particles tends to infinity. This is very similar to the framework of mean field games. However, we are about to show that the solution of the optimal control of McKean-Vlasov SDEs provides strategies leading to approximate equilibria of a nature which is quite different from the Nash equilibria of the theory of mean field games. Our discussion of potential mean field games in Subsection (Vol I)-6.7.2 in Chapter (Vol I)-6 emphasizes the reduction of the mean field game problem to a single optimization problem over an SDE of McKean-Vlasov type. We called this single optimization problem the central planner or representative agent problem. It suggests that the agents in the corresponding finite system should use a *common* control strategy, and seek a different kind of equilibrium. Below, we provide a precise definition for the latter.

Throughout this section, assumption **Control of MKV Dynamics** from Subsection (Vol I)-6.4.1 is assumed to hold. For the sake of completeness, we recall its main features.

First, we recall that σ is allowed to depend on α and that the drift and volatility functions b and σ are linear in μ, x and α, in the sense that:

$$\begin{aligned} b(t,x,\mu,\alpha) &= b_0(t) + b_1(t)x + \bar{b}_1(t)\bar{\mu} + b_2(t)\alpha, \\ \sigma(t,x,\mu,\alpha) &= \sigma_0(t) + \sigma_1(t)x + \bar{\sigma}_1(t)\bar{\mu} + \sigma_2(t)\alpha, \end{aligned}$$

where the various coefficients in the expansions are required to be bounded and where we used the notation $\bar{\mu} = \int x \, d\mu(x)$.

Also, the cost functionals f and g are required to satisfy the following local Lipschitz property:

$$\begin{aligned} &\big|f(t,x',\mu',\alpha') - f(t,x,\mu,\alpha)\big| + \big|g(x',\mu') - g(x,\mu)\big| \\ &\quad \leq L\big[1 + |x'| + |x| + |\alpha'| + |\alpha| + M_2(\mu) + M_2(\mu')\big] \\ &\qquad \times \big[|(x',\alpha') - (x,\alpha)| + W_2(\mu',\mu)\big]. \end{aligned} \tag{6.56}$$

The functions f and g are also required to be differentiable in (x,α,μ) and the derivatives of f and g with respect to (x,α) and x respectively are L-Lipschitz continuous with respect to (x,α,μ) and (x,μ) respectively, the Lipschitz property in the variable μ being understood in the W_2 sense. Moreover,

$$\begin{aligned}
&\mathbb{E}\big[|\partial_\mu f(t,x',\mu',\alpha')(X') - \partial_\mu f(t,x,\mu,\alpha)(X)|^2\big] \\
&\qquad\qquad \le L\big(|(x',\alpha') - (x,\alpha)|^2 + \mathbb{E}\big[|X'-X|^2\big]\big), \\
&\mathbb{E}\big[|\partial_\mu g(x',\mu')(X') - \partial_\mu g(x,\mu)(X)|^2\big] \\
&\qquad\qquad \le L\big(|x'-x|^2 + \mathbb{E}\big[|X'-X|^2\big]\big),
\end{aligned} \tag{6.57}$$

where X and X' have μ and μ' as respective distributions. Importantly, Proposition (Vol I)-5.36 says that the above remains true up to a new value of L if, in the two left-hand sides, the random variable X' is replaced by X while the two measure arguments μ and μ' remain unchanged. Finally, the function f satisfies the convexity property:

$$\begin{aligned}
&f(t,x',\mu',\alpha') - f(t,x,\mu,\alpha) - \partial_{(x,\alpha)} f(t,x,\mu,\alpha)\cdot(x'-x,\alpha'-\alpha) \\
&\qquad - \mathbb{E}\big[\partial_\mu f(t,x,\mu,\alpha)(X)\cdot(X'-X)\big] \ge \lambda|\alpha'-\alpha|^2,
\end{aligned} \tag{6.58}$$

where, as above, X and X' have μ and μ' as respective distributions. Similarly, the function g is also assumed to be convex in (x,μ) in the sense that:

$$\begin{aligned}
&g(x',\mu') - g(x,\mu) - \partial_x g(x,\mu)\cdot(x'-x) \\
&\qquad - \mathbb{E}\big[\partial_\mu g(x,\mu)(X)\cdot(X'-X)\big] \ge 0.
\end{aligned} \tag{6.59}$$

For each integer $N \ge 1$, we consider a stochastic system whose time evolution is given by the system of N coupled stochastic differential equations:

$$dU_t^{N,i} = b\big(t, U_t^{N,i}, \bar{\nu}_t^N, \beta_t^{N,i}\big)dt + \sigma\big(t, U_t^{N,i}, \bar{\nu}_t^N, \beta_t^{N,i}\big)dW_t^i, \quad 1 \le i \le N, \tag{6.60}$$

with:

$$\bar{\nu}_t^N = \frac{1}{N}\sum_{j=1}^N \delta_{U_t^{N,j}},$$

for $t \in [0,T]$ and $U_0^{N,i} = \xi^i$, $1 \le i \le N$, where $(\xi^i)_{i\ge1}$ is a sequence of independent and identically distributed random variables with values in $\mathbb{R}^d$, with $\mu_0 \in \mathcal{P}_2(\mathbb{R}^d)$ as common distribution, and $(\boldsymbol{W}^i = (W_t^i)_{0\le t\le T})_{i\ge1}$ is a sequence of independent d-dimensional Brownian motions on the time interval $[0,T]$. Of course, the families $(\xi^i)_{i\ge1}$ and $(\boldsymbol{W}^i)_{i\ge1}$ are assumed to be independent of each other, and to be constructed on some complete probability space $(\Omega,\mathcal{F},\mathbb{P})$. Here the control strategies $(\boldsymbol{\beta}^{N,1}, \cdots, \boldsymbol{\beta}^{N,N})$ are assumed to be, for $1 \leq i \leq N$, stochastic processes $\boldsymbol{\beta}^{N,i} = (\beta_t^{N,i})_{0 \leq t \leq T}$ that are progressively measurable with respect to the filtration generated by $(\xi^1, \cdots, \xi^N)$ and $(\boldsymbol{W}^1, \cdots, \boldsymbol{W}^N)$, that take values in A, and that have finite L^2 norms over $[0, T] \times \Omega$:

$$\mathbb{E} \int_0^T |\beta_t^{N,i}|^2 dt < +\infty, \qquad i = 1, \cdots, N.$$

One should think of $U_t^{N,i}$ as the (private) state at time t of agent or player $i \in \{1, \cdots, N\}$, $\beta_t^{N,i}$ being the action taken at time t by player i. In this respect, the presentation is pretty similar to that of (6.11). For each $1 \leq i \leq N$, we denote by:

$$J^{N,i}(\boldsymbol{\beta}^{N,1}, \cdots, \boldsymbol{\beta}^{N,N}) = \mathbb{E}\biggl[\int_0^T f\bigl(t, U_t^{N,i}, \bar{\nu}_t^N, \beta_t^{N,i}\bigr)dt + g\bigl(U_T^{N,i}, \bar{\nu}_T^N\bigr)\biggr] \tag{6.61}$$

the cost to the i-th player, which is the analogue of (6.12).

Obviously, we framed the problem in the same set-up as in the case of the mean field game models studied earlier in Subsection 6.1.1; the difference comes from the rule used below for minimizing the cost. Indeed, we now minimize the cost over exchangeable strategies. With a slight abuse of terminology, we shall say that the strategy profile $\boldsymbol{\beta}^N$ is exchangeable if the family $(\xi^i, \boldsymbol{\beta}^{N,i}, \boldsymbol{W}^i)_{1 \leq i \leq N}$ is exchangeable. If that is the case, the costs to all the players are the same, and we can use the notation $\bar{J}^N(\boldsymbol{\beta}^{(N)}) = J^{N,i}(\boldsymbol{\beta}^{(N)})$ for their common value. From a practical point of view, restricting the minimization to exchangeable strategy profiles means that the players agree to use a common policy, which is not the case in the standard mean field game approach. Our first goal is to compute the limit:

$$\lim_{N \to +\infty} \inf_{\boldsymbol{\beta}^{(N)}} \bar{J}^N(\boldsymbol{\beta}^{(N)}),$$

the infimum being taken over exchangeable strategy profiles. Another one is to identify, for each integer N, a specific set of ε-optimal strategies and the corresponding state evolutions.

Limit of the Costs and Non-Markovian Approximate Equilibria

For a square-integrable random variable $\xi \in L^2(\Omega, \mathcal{F}, \mathbb{P}; \mathbb{R}^d)$ and a d-dimensional Brownian motion $\boldsymbol{W} = (W_t)_{0 \leq t \leq T}$, $\boldsymbol{W}$ being independent of ξ, we denote by J the optimal cost:

$$J = \mathbb{E}\biggl[\int_0^T f\bigl(t, X_t, \mu_t, \hat{\alpha}(t, X_t, \mu_t, Y_t, Z_t)\bigr)dt + g(X_T, \mu_T)\biggr], \tag{6.62}$$

where $\hat{\alpha}(t, x, \mu, y, z)$ is the minimizer over $\alpha \in A$ of the Hamiltonian:

$$H(t, x, \mu, y, z, \alpha) = b(t, x, \mu, \alpha) \cdot y + \sigma(t, x, \mu, \alpha) \cdot z + f(t, x, \mu, \alpha), \tag{6.63}$$

for $(t, x, \mu, y, z, \alpha) \in [0, T] \times \mathbb{R}^d \times \mathcal{P}_2(\mathbb{R}^d) \times \mathbb{R}^d \times \mathbb{R}^{d\times d} \times A$ and $(\boldsymbol{X}, \boldsymbol{Y}, \boldsymbol{Z}) = (X_t, Y_t, Z_t)_{0\le t\le T}$ is the solution of the FBSDE:

$$
\left\{
\begin{aligned}
dX_t &= \big[b_0(t) + b_1(t)X_t + \bar{b}_1(t)\mathbb{E}[X_t] \\
&\qquad\qquad + b_2(t)\hat{\alpha}(t, X_t, \mathcal{L}(X_t), Y_t, Z_t)\big]dt \\
&\quad + \big[\sigma_0(t) + \sigma_1(t)X_t + \bar{\sigma}_1(t)\mathbb{E}[X_t] \\
&\qquad\qquad + \sigma_2(t)\hat{\alpha}(t, X_t, \mathcal{L}(X_t), Y_t, Z_t)\big]dW_t, \\
dY_t &= -\big[\partial_x f\big(t, X_t, \mathcal{L}(X_t), \hat{\alpha}(t, X_t, \mathcal{L}(X_t), Y_t, Z_t)\big) \\
&\qquad\qquad + b_1(t)^\dagger Y_t + \sigma_1(t)^\dagger Z_t\big]dt \\
&\quad - \Big[\tilde{\mathbb{E}}\big[\partial_\mu f\big(t, \tilde{X}_t, \mathcal{L}(X_t), \hat{\alpha}(t, \tilde{X}_t, \mathcal{L}(X_t), \tilde{Y}_t, \tilde{Z}_t)\big)(X_t)\big] \\
&\qquad\qquad + \bar{b}_1(t)^\dagger\mathbb{E}[Y_t] + \bar{\sigma}_1(t)^\dagger\mathbb{E}[Z_t]\Big]dt \\
&\quad + Z_t dW_t,
\end{aligned}
\right.
\tag{6.64}
$$

with the initial condition $X_0 = \xi$, and the terminal condition $Y_T = \partial_x g(X_T, \mathcal{L}(X_T)) + \tilde{\mathbb{E}}[\partial_\mu g(\tilde{X}_T, \mathcal{L}(X_T))(X_T)]$. As usual the symbol tilde is used to denote copies of the random variables on a copy $(\tilde{\Omega}, \tilde{\mathcal{F}}, \tilde{\mathbb{P}})$ of $(\Omega, \mathcal{F}, \mathbb{P})$. In (6.62), $(\mu_t)_{0\le t\le T}$ denotes the flow of marginal probability measures $\mu_t = \mathcal{L}(X_t)$, for $0 \le t \le T$.

For the purpose of comparison, for each $i \in \{1, \cdots, N\}$, we introduce the solution $(\underline{\boldsymbol{X}}^i, \underline{\boldsymbol{Y}}^i, \underline{\boldsymbol{Z}}^i)$ of the FBSDE (6.64) when the whole FBSDE is driven by the Wiener process $\boldsymbol{W}^i$ and the initial condition ξ^i. Notice that these triples of processes are independent and identically distributed. In particular, $(\underline{\boldsymbol{X}}^1, \cdots, \underline{\boldsymbol{X}}^N)$ solves the system (6.60) when the empirical distribution $\bar{\nu}_t^N$ providing the interaction is replaced by μ_t and $\beta_t^{N,i}$ is given by $\beta_t^{N,i} = \hat{\underline{\alpha}}_t^i$ with:

$$
\hat{\underline{\alpha}}_t^i = \hat{\alpha}(t, \underline{X}_t^i, \mu_t, \underline{Y}_t^i, \underline{Z}_t^i). \tag{6.65}
$$

Here is the first claim of this subsection:

Theorem 6.16 *Under assumption* **Control of MKV Dynamics**,

$$
\lim_{N\to+\infty} \inf_{\boldsymbol{\beta}^{(N)}} \bar{J}^N(\boldsymbol{\beta}^{(N)}) = J,
$$

the above infimum being taken over all the square integrable strategy profiles $\boldsymbol{\beta}^{(N)} = (\boldsymbol{\beta}^{N,1}, \cdots, \boldsymbol{\beta}^{N,N})$ *such that the family* $(\xi^i, \boldsymbol{\beta}^{N,i}, \boldsymbol{W}^i)_{1\le i\le N}$ *is exchangeable. Moreover, the open loop strategy profile* $\hat{\underline{\boldsymbol{\alpha}}}^{(N)} = (\hat{\underline{\boldsymbol{\alpha}}}^1, \cdots, \hat{\underline{\boldsymbol{\alpha}}}^N)$ *is approximately optimal in the sense that:*

$$
\lim_{N\to+\infty} \bar{J}^N(\hat{\underline{\boldsymbol{\alpha}}}^{(N)}) = J.
$$

Proof. The proof consists in comparing $\bar{J}^N(\boldsymbol{\beta}^{(N)})$ to J for a given exchangeable strategy $\boldsymbol{\beta}^{(N)}$. Once again, we rely on a variant of the Pontryagin stochastic maximum principle proven in Section (Vol I)-6.3, which we spell out below. With the above notation, we have:

$$\bar{J}^N(\boldsymbol{\beta}^{(N)}) - J = \mathbb{E}\left[\int_0^T \big(f(s, U_s^{N,i}, \bar{\nu}_s^N, \beta_s^{N,i}) - f(s, \underline{X}_s^i, \mu_s, \underline{\hat{\alpha}}_s^i)\big)ds\right] + \mathbb{E}\big[g(U_T^{N,i}, \bar{\nu}_T^N) - g(\underline{X}_T^i, \mu_T)\big],$$

for $i \in \{1, \cdots, N\}$. Therefore, we can write:

$$\bar{J}^N(\boldsymbol{\beta}^{(N)}) - J = T_1^i + T_2^i, \tag{6.66}$$

with:

$$\begin{aligned}
T_1^i &= \mathbb{E}\big[(U_T^{N,i} - \underline{X}_T^i) \cdot \underline{Y}_T^i\big] + \mathbb{E}\left[\int_0^T \big(f(s, U_s^{N,i}, \bar{\nu}_s^N, \beta_s^{N,i}) - f(s, \underline{X}_s^i, \mu_s, \underline{\hat{\alpha}}_s^i)\big)ds\right],\\
T_2^i &= \mathbb{E}\big[g(U_T^{N,i}, \bar{\nu}_T^N) - g(\underline{X}_T^i, \mu_T)\big] - \mathbb{E}\big[(U_T^{N,i} - \underline{X}_T^i) \cdot \partial_x g(\underline{X}_T^i, \mu_T)\big]\\
&\quad - \mathbb{E}\tilde{\mathbb{E}}\big[(\tilde{U}_T^{N,i} - \tilde{\underline{X}}_T^i) \cdot \partial_\mu g(\underline{X}_T^i, \mu_T)(\tilde{\underline{X}}_T^i)\big]\\
&= T_{2,1}^i - T_{2,2}^i - T_{2,3}^i,
\end{aligned}$$

where the quantities $T_{2,1}^i$, $T_{2,2}^i$, and $T_{2,3}^i$ have obvious definitions which will be stated explicitly below. Here, the *tilde* corresponds to the independent copies we routinely use when dealing with L-derivatives. Notice that we used Fubini's theorem above in order to handle the terminal condition of the backward equation in (6.64).

First Step. We start with the analysis of T_2^i. Using the diffusive effect of independence, we claim:

$$\begin{aligned}
T_{2,3}^i &= \mathbb{E}\tilde{\mathbb{E}}\big[(\tilde{U}_T^{N,i} - \tilde{\underline{X}}_T^i) \cdot \partial_\mu g(\underline{X}_T^i, \mu_T)(\tilde{\underline{X}}_T^i)\big]\\
&= \frac{1}{N}\sum_{j=1}^N \tilde{\mathbb{E}}\big[(\tilde{U}_T^{N,i} - \tilde{\underline{X}}_T^i) \cdot \partial_\mu g(\tilde{\underline{X}}_T^j, \mu_T)(\tilde{\underline{X}}_T^i)\big]\\
&\quad + O\Bigg(\tilde{\mathbb{E}}\big[|\tilde{U}_T^{N,i} - \tilde{\underline{X}}_T^i|^2\big]^{1/2}\\
&\qquad \times \tilde{\mathbb{E}}\left[\left|\frac{1}{N}\sum_{j=1}^N \partial_\mu g(\tilde{\underline{X}}_T^j, \mu_T)(\tilde{\underline{X}}_T^i) - \mathbb{E}\big[\partial_\mu g(\underline{X}_T^i, \mu_T)(\tilde{\underline{X}}_T^i)\big]\right|^2\right]^{1/2}\Bigg)\\
&= \frac{1}{N}\sum_{j=1}^N \mathbb{E}\big[(U_T^{N,i} - \underline{X}_T^i) \cdot \partial_\mu g(\underline{X}_T^j, \mu_T)(\underline{X}_T^i)\big] + \mathbb{E}\big[|U_T^{N,i} - \underline{X}_T^i|^2\big]^{1/2} O(N^{-1/2}),
\end{aligned}$$

where $O(\cdot)$ stands for the Landau notation in the sense that $|O(x)| \le C|x|$ for a constant C independent of N. Therefore, taking advantage of the exchangeability in order to handle the remainder, we obtain:

$$\frac{1}{N}\sum_{i=1}^{N} T_{2,3}^i = \frac{1}{N^2}\sum_{j=1}^{N}\sum_{i=1}^{N} \mathbb{E}\big[(U_T^{N,i} - \underline{X}_T^i)\cdot \partial_\mu g\big(\underline{X}_T^j, \mu_T)(\underline{X}_T^i)\big]$$
$$+ \mathbb{E}\big[|U_T^{N,1} - \underline{X}_T^1|^2\big]^{1/2} O(N^{-1/2}).$$

Introducing a random variable ϑ from $(\tilde{\Omega}, \tilde{\mathcal{F}}, \tilde{\mathbb{P}})$ into $\mathbb{R}$ with uniform distribution on the set $\{1, \cdots, N\}$ as in the proof of Proposition (Vol I)-5.35, we can write:

$$\frac{1}{N}\sum_{i=1}^{N} T_{2,3}^i = \frac{1}{N}\sum_{j=1}^{N} \mathbb{E}\tilde{\mathbb{E}}\big[(U_T^{N,\vartheta} - \underline{X}_T^\vartheta)\cdot \partial_\mu g\big(\underline{X}_T^j, \mu_T)(\underline{X}_T^\vartheta)\big] + \mathbb{E}\big[|U_T^{N,1} - \underline{X}_T^1|^2\big]^{1/2} O(N^{-1/2}).$$

Finally, defining the flow of empirical measures:

$$\bar{\mu}_t^N = \frac{1}{N}\sum_{j=1}^{N} \delta_{\underline{X}_t^j}, \quad t \in [0, T],$$

and using (**A3**) in assumption **Control of MKV Dynamics**, see also (6.57), the above estimate gives:

$$\frac{1}{N}\sum_{i=1}^{N} T_{2,3}^i = \frac{1}{N}\sum_{j=1}^{N} \mathbb{E}\tilde{\mathbb{E}}\big[(U_T^{N,\vartheta} - \underline{X}_T^\vartheta)\cdot \partial_\mu g\big(\underline{X}_T^j, \bar{\mu}_T^N)(\underline{X}_T^\vartheta)\big] + \mathbb{E}\big[|U_T^{N,1} - \underline{X}_T^1|^2\big]^{1/2} O(\varepsilon_N),$$

where we used the notation ε_N for any function of N which could be used as an upper bound for:

$$\max\left[N^{-1/2}, \mathbb{E}\big[W_2(\bar{\mu}_T^N, \mu_T)^2\big]^{1/2} + \left(\int_0^T \mathbb{E}\big[W_2(\bar{\mu}_t^N, \mu_t)^2\big]dt\right)^{1/2}\right] = O(\varepsilon_N). \tag{6.67}$$

By (Vol I)-(5.19), see also the proof of Theorem 2.12, and by an obvious application of the Lebesgue dominated convergence theorem, the left-hand side tends to 0 as N tends to $+\infty$, since the function $[0, T] \ni t \mapsto \mathbb{E}[W_2(\bar{\mu}_t^N, \mu_t)^2]$ can be bounded independently of N. Therefore, $(\varepsilon_N)_{N\geq 1}$ is always chosen as a sequence that converges to 0 as N tends to $+\infty$. When $\sup_{0\leq t\leq T} |\underline{X}_t^1|$ has a finite moment of order q, for some $q > 4$, Lemma 6.1 says that ε_N can be chosen as $\sqrt{\epsilon_N}$. Going back to (6.66), we get:

$$\frac{1}{N}\sum_{i=1}^{N} T_2^i = \frac{1}{N}\sum_{i=1}^{N} \Big\{\mathbb{E}\big[g(U_T^{N,i}, \bar{\nu}_T^N) - g(\underline{X}_T^i, \bar{\mu}_T^N)\big] - \mathbb{E}\big[(U_T^{N,i} - \underline{X}_T^i)\cdot \partial_x g(\underline{X}_T^i, \bar{\mu}_T^N)\big]$$
$$- \mathbb{E}\tilde{\mathbb{E}}\big[(U_T^{N,\vartheta} - \underline{X}_T^\vartheta)\cdot \partial_\mu g\big(\underline{X}_T^i, \bar{\mu}_T^N)(\underline{X}_T^\vartheta)\big]\Big\}$$
$$+ \big(1 + \mathbb{E}\big[|U_T^{N,1} - \underline{X}_T^1|^2\big]^{1/2}\big) O(\varepsilon_N),$$

where we used the local Lipschitz property of g and (6.67) to replace μ_T by $\bar{\mu}_T^N$.

Noticing that the conditional law of $U_T^{N,\vartheta}$ (respectively $\underline{X}_T^{\vartheta}$) under $\tilde{\mathbb{P}}$ is the empirical distribution $\bar{\nu}_T^N$ (respectively $\bar{\mu}_T^N$), we can use the convexity of g, see (6.59), to get:

$$\frac{1}{N}\sum_{i=1}^N T_2^i \geq \big(1 + \mathbb{E}[|U_T^{N,1} - \underline{X}_T^1|^2]^{1/2}\big)O(\varepsilon_N). \tag{6.68}$$

Second Step. We now turn to the analysis of T_1^i in (6.66). Using Itô's formula and Fubini's theorem, we obtain:

$$\begin{aligned}
T_1^i &= \mathbb{E}\bigg[\int_0^T \big(H(s, U_s^{N,i}, \bar{\nu}_s^N, \underline{Y}_s^i, \underline{Z}_s^i, \beta_s^{N,i}) - H(s, \underline{X}_s^i, \mu_s, \underline{Y}_s^i, \underline{Z}_s^i, \underline{\hat{\alpha}}_s^i)\big)ds\bigg] \\
&\quad - \mathbb{E}\bigg[\int_0^T (U_s^{N,i} - \underline{X}_s^i)\cdot \partial_x H(s, \underline{X}_s^i, \mu_s, \underline{Y}_s^i, \underline{Z}_s^i, \underline{\hat{\alpha}}_s^i)ds\bigg] \\
&\quad - \mathbb{E}\tilde{\mathbb{E}}\bigg[\int_0^T (\tilde{U}_s^{N,i} - \tilde{\underline{X}}_s^i)\cdot \partial_\mu H(s, \underline{X}_s^i, \mu_s, \underline{Y}_s^i, \underline{Z}_s^i, \underline{\hat{\alpha}}_s^i)(\tilde{\underline{X}}_s^i)ds\bigg] \\
&= T_{1,1}^i - T_{1,2}^i - T_{1,3}^i.
\end{aligned} \tag{6.69}$$

Using the regularity properties of the Hamiltonian given by (**A2**) and (**A3**) in assumption **Control of MKV Dynamics**, see also (6.56) and (6.57), together with (6.67), and recalling that the limit processes $(\boldsymbol{X}^i, \boldsymbol{Y}^i, \boldsymbol{Z}^i, \hat{\boldsymbol{\alpha}}^i)_{i\geq 1}$ satisfy the square-integrability property:

$$\sup_{i\geq 1}\mathbb{E}\bigg[\sup_{0\leq t\leq T}\big[|X_t^i|^2 + |Y_t^i|^2\big] + \int_0^T \big[|Z_t^i|^2 + |\hat{\alpha}_t^i|^2\big]dt\bigg] < \infty,$$

we get:

$$\begin{aligned}
T_{1,1}^i &= \mathbb{E}\bigg[\int_0^T \big(H(s, U_s^{N,i}, \bar{\nu}_s^N, \underline{Y}_s^i, \underline{Z}_s^i, \beta_s^{N,i}) - H(s, \underline{X}_s^i, \bar{\mu}_s^N, \underline{Y}_s^i, \underline{Z}_s^i, \underline{\hat{\alpha}}_s^i)\big)ds\bigg] \\
&\quad + O(\varepsilon_N). \\
T_{1,2}^i &= \mathbb{E}\bigg[\int_0^T (U_s^{N,i} - \underline{X}_s^i)\cdot \partial_x H(s, \underline{X}_s^i, \bar{\mu}_s^N, \underline{Y}_s^i, \underline{Z}_s^i, \underline{\hat{\alpha}}_s^i)ds\bigg] \\
&\quad + \bigg(\mathbb{E}\int_0^T |U_s^{N,1} - \underline{X}_s^1|^2 ds\bigg)^{1/2} O(\varepsilon_N).
\end{aligned} \tag{6.70}$$

Finally, using once again the diffusive effect of independence, we have

$$\begin{aligned}
\frac{1}{N}\sum_{i=1}^N T_{1,3}^i &= \frac{1}{N}\sum_{i=1}^N \mathbb{E}\tilde{\mathbb{E}}\bigg[\int_0^T (U_s^{N,i} - \underline{X}_s^i)\cdot \partial_\mu H(s, \tilde{\underline{X}}_s^i, \mu_s, \tilde{\underline{Y}}_s^i, \tilde{\underline{Z}}_s^i, \tilde{\underline{\hat{\alpha}}}_s^i)(\underline{X}_s^i)ds\bigg] \\
&= \frac{1}{N^2}\sum_{j=1}^N\sum_{i=1}^N \mathbb{E}\bigg[\int_0^T (U_s^{N,i} - \underline{X}_s^i)\cdot \partial_\mu H(s, \underline{X}_s^j, \mu_s, \underline{Y}_s^j, \underline{Z}_s^j, \underline{\hat{\alpha}}_s^j)(\underline{X}_s^i)ds\bigg] \\
&\quad + \bigg(\mathbb{E}\int_0^T |U_s^{N,1} - \underline{X}_s^1|^2 ds\bigg)^{1/2} O(N^{-1/2}),
\end{aligned}$$

where we exchanged the copies on the space $(\tilde{\Omega}, \tilde{\mathcal{F}}, \tilde{\mathbb{P}})$ in the first line. By (**A3**) in the assumption, see also (6.57), we deduce:

$$\begin{aligned}
\frac{1}{N}\sum_{i=1}^{N} T_{1,3}^i &= \frac{1}{N}\sum_{i=1}^{N} \mathbb{E}\tilde{\mathbb{E}}\Big[\int_0^T (U_s^{N,\vartheta} - \underline{X}_s^{\vartheta}) \cdot \partial_\mu H(s, \underline{X}_s^i, \mu_s, \underline{Y}_s^i, \underline{Z}_s^i, \underline{\hat{\alpha}}_s^i)(\underline{X}_s^{\vartheta}) ds\Big] \\
&\quad + \Big(\mathbb{E}\int_0^T |U_s^{N,1} - \underline{X}_s^1|^2 ds\Big)^{1/2} O(N^{-1/2}) \\
&= \frac{1}{N}\sum_{i=1}^{N} \mathbb{E}\tilde{\mathbb{E}}\Big[\int_0^T (U_s^{N,\vartheta} - \underline{X}_s^{\vartheta}) \cdot \partial_\mu H(s, \underline{X}_s^i, \bar{\mu}_s^N, \underline{Y}_s^i, \underline{Z}_s^i, \underline{\hat{\alpha}}_s^i)(\underline{X}_s^{\vartheta}) ds\Big] \\
&\quad + \Big(\mathbb{E}\int_0^T |U_s^{N,1} - \underline{X}_s^1|^2 ds\Big)^{1/2} O(\varepsilon_N).
\end{aligned} \tag{6.71}$$

In order to complete the proof, we evaluate the missing term in the Taylor expansion of T_1^i in (6.69), namely:

$$\frac{1}{N}\sum_{i=1}^{N} \mathbb{E}\Big[\int_0^T (\beta_s^{N,i} - \underline{\hat{\alpha}}_s^i) \cdot \partial_\alpha H(s, \underline{X}_s^i, \bar{\mu}_s^N, \underline{Y}_s^i, \underline{Z}_s^i, \underline{\hat{\alpha}}_s^i) ds\Big],$$

in order to benefit from the convexity of H. We use (6.67) once more:

$$\begin{aligned}
&\mathbb{E}\Big[\int_0^T (\beta_s^{N,i} - \underline{\hat{\alpha}}_s^i) \cdot \partial_\alpha H(s, \underline{X}_s^i, \bar{\mu}_s^N, \underline{Y}_s^i, \underline{Z}_s^i, \underline{\hat{\alpha}}_s^i) ds\Big] \\
&\quad = \mathbb{E}\Big[\int_0^T (\beta_s^{N,i} - \underline{\hat{\alpha}}_s^i) \cdot \partial_\alpha H(s, \underline{X}_s^i, \mu_s, \underline{Y}_s^i, \underline{Z}_s^i, \underline{\hat{\alpha}}_s^i) ds\Big] \\
&\qquad + \Big(\mathbb{E}\int_0^T |\beta_s^{N,i} - \underline{\hat{\alpha}}_s^i|^2 ds\Big)^{1/2} O(\varepsilon_N) \\
&\quad \geq \Big(\mathbb{E}\int_0^T |\beta_s^{N,i} - \underline{\hat{\alpha}}_s^i|^2 ds\Big)^{1/2} O(\varepsilon_N),
\end{aligned} \tag{6.72}$$

since $\hat{\alpha}$ is a minimizer for H, see (3.11) in Chapter (Vol I)-3 if needed. Using the convexity of H and taking advantage of the exchangeability, we finally deduce from (6.69), (6.70), (6.71), and (6.72) that there exists a constant $c > 0$ such that:

$$\begin{aligned}
\frac{1}{N}\sum_{i=1}^{N} T_1^i &\geq c\mathbb{E}\int_0^T |\beta_s^{N,1} - \underline{\hat{\alpha}}_s^1|^2 ds \\
&\quad + O(\varepsilon_N)\Big(1 + \sup_{0 \leq t \leq T} \mathbb{E}[|U_t^{N,1} - \underline{X}_t^1|^2] + \mathbb{E}\int_0^T |\beta_s^{N,1} - \underline{\hat{\alpha}}_s^1|^2 ds\Big)^{1/2}.
\end{aligned}$$

Third Step. By (6.66) and (6.68), we deduce that:

$$\bar{J}^N(\boldsymbol{\beta}^{(N)}) \geq J + c\mathbb{E}\int_0^T |\beta_s^{N,1} - \hat{\underline{\alpha}}_s^1|^2 ds$$
$$+ O(\varepsilon_N)\left(1 + \sup_{0\leq t\leq T} \mathbb{E}\big[|U_t^{N,1} - \underline{X}_t^1|^2\big] + \mathbb{E}\int_0^T |\beta_s^{N,1} - \hat{\underline{\alpha}}_s^1|^2 ds\right)^{1/2}.$$

By exchangeability, we have the inequality:

$$\sup_{0\leq t\leq T} \mathbb{E}\big[|U_t^{N,1} - \underline{X}_t^1|^2\big] \leq C\mathbb{E}\int_0^T |\beta_s^{N,1} - \hat{\underline{\alpha}}_s^1|^2 ds,$$

which holds for some constant C independent of N. We deduce that:

$$\bar{J}^N(\boldsymbol{\beta}^{(N)}) \geq J - C\varepsilon_N, \tag{6.73}$$

for a possibly new value of C. This proves that:

$$\liminf_{N\to+\infty} \inf_{\boldsymbol{\beta}^{(N)}} \bar{J}^N(\boldsymbol{\beta}^{(N)}) \geq J.$$

In order to prove Theorem 6.16, it only remains to find a sequence of controls $(\boldsymbol{\beta}^{(N)})_{N\geq 1}$ such that:

$$\limsup_{N\to+\infty} \bar{J}^N(\boldsymbol{\beta}^{(N)}) \leq J.$$

More precisely, we are about to show that:

$$\limsup_{N\to+\infty} \bar{J}^N(\hat{\underline{\boldsymbol{\alpha}}}^{(N)}) \leq J, \tag{6.74}$$

thus proving that $\hat{\underline{\boldsymbol{\alpha}}}^{(N)} = (\underline{\boldsymbol{\alpha}}^1, \cdots, \hat{\underline{\boldsymbol{\alpha}}}^N)$ is an approximate equilibrium, though non-Markovian. Denoting by $(\boldsymbol{X}^{N,1}, \cdots, \boldsymbol{X}^{N,N})$ the solution of (6.60) with $\beta_t^{N,i} = \hat{\underline{\alpha}}_t^i$ and following the proof of Theorem 2.12, we get:

$$\sup_{0\leq t\leq T} \mathbb{E}\big[|X_t^{N,i} - \underline{X}_t^i|^2\big] = \sup_{0\leq t\leq T} \mathbb{E}\big[|X_t^{N,1} - \underline{X}_t^1|^2\big] = O\big((\varepsilon_N)^2\big).$$

It is then plain to derive (6.74), completing the proof. □

Approximate Equilibria with Distributed Closed Loop Controls

When σ doesn't not depend upon the control variable α, it is possible to provide an approximate equilibrium using only distributed controls in closed loop form. Indeed, in this case, the optimizer $\hat{\alpha}$ of the Hamiltonian, as defined in (6.63), does not depend upon the adjoint variable z and reads as $\hat{\alpha}(t, x, \mu, y)$. Also, we learnt from the results of Chapter (Vol I)-6, see Lemma (Vol I)-6.25, that, under the standing assumption, the process $(Y_t)_{0\le t\le T}$ in (6.64) could be represented as $(Y_t = \mathcal{U}(t, X_t, \mathcal{L}(X_t)))_{0\le t\le T}$ for a function $\mathcal{U}$, which we called the master field of the FBSDE (6.64) and which is Lipchitz continuous in the variables x and μ, uniformly in time. Hence, the optimal control $\hat{\boldsymbol{\alpha}} = (\hat{\alpha}_t)_{0\le t\le T}$ in (6.62) gets the *feedback* form:

$$\hat{\alpha}_t = \hat{\alpha}\big(t, X_t, \mu_t, \mathcal{U}(t, X_t, \mu_t)\big), \quad t \in [0, T]. \tag{6.75}$$

In this regard, the function v defined by:

$$v(t, x) = \mathcal{U}(t, x, \mu_t), \quad (t, x) \in [0, T] \times \mathbb{R}^d,$$

plays the same role as the decoupling field v in (6.9).

Of course, it would be desirable to have a similar representation when σ depends on α. However, the result proven in Chapter (Vol I)-6, see Lemma (Vol I)-6.25, does not suffice to do so: If σ depends on α, then $\hat{\alpha}_t$ depends on Z_t, while Lemma (Vol I)-6.25 just provides a representation of Y_t and not of Z_t. Actually, similar to Y_t, we may expect Z_t to be a function of t and X_t. Such a representation is indeed known to hold in the classical *decoupled* forward-backward setting. However, this would require a deeper analysis: First, one would want the feedback function expressing Z_t in terms of X_t to be Lipschitz-continuous at a minimum, as the Lipschitz property ensures that the stochastic differential equation obtained by plugging (6.75) into the forward equation in (6.64) is solvable. In the present context, hoping for such a regularity is mere wishful thinking as it is already very challenging in the standard case, i.e., without any McKean-Vlasov interaction. Second, in any case, the relationship between Z_t and X_t, if it exists, must be rather intricate as Z_t is expected to solve the equation $Z_t = \partial_x v(t, X_t)\sigma(t, X_t, \mathcal{L}(X_t), \hat{\alpha}(t, X_t, \mathcal{L}(X_t), Y_t, Z_t))$, which can be formally derived by identifying the martingale parts when expanding $Y_t = v(t, X_t)$ by a formal application of Itô's formula.

So, assuming the diffusion coefficient σ to be independent of α, we denote by $(\boldsymbol{X}^{N,1}, \cdots, \boldsymbol{X}^{N,N}) = (X_t^{N,1}, \cdots, X_t^{N,N})_{0\le t\le T}$ the solution of the system of N stochastic differential equations:

$$dX_t^{N,i} = b\big(t, X_t^{N,i}, \bar{\mu}_t^N, \hat{\alpha}\big(t, X_t^{N,i}, \mu_t, v(t, X_t^{N,i})\big)\big)dt + \sigma(t, X_t^{N,i}, \bar{\mu}_t^N)dW_t^i, \tag{6.76}$$

with:

$$\bar{\mu}_t^N = \frac{1}{N}\sum_{j=1}^N \delta_{X_t^{N,j}},$$

for $t \in [0, T]$, where we recall that $X_0^{N,1} = \xi^1, \cdots, X_0^{N,N} = \xi^N$ are independent and identically distributed with μ_0 as common distribution. The system (6.76) is well posed because of the Lipschitz and linear growth properties of the decoupling field v, and the fact that the minimizer $\hat{\alpha}(t, x, \mu_t, y)$ is Lipschitz continuous and at most of linear growth in the variables x, μ and y, uniformly in $t \in [0, T]$. The processes

$(\boldsymbol{X}^{N,i})_{1\le i\le N}$ give the dynamics of the private states of the N players in the stochastic control problem of interest when the players use the distributed strategies:

$$\hat{\alpha}_t^{N,i} = \hat{\alpha}\big(t, X_t^{N,i}, \mu_t, v(t, X_t^{N,i})\big), \qquad 0 \le t \le T, \; i = 1, \cdots, N. \tag{6.77}$$

By the linear growth of v and of the minimizer $\hat{\alpha}$, it holds, for any $q \ge 2$ such that $\mu_0 \in \mathcal{P}_q(\mathbb{R}^d)$,

$$\sup_{N\ge 1} \max_{1\le i\le N} \mathbb{E}\big[\sup_{0\le t\le T} |X_t^{N,i}|^q\big] < +\infty, \tag{6.78}$$

the expectation being actually independent of i since the particles are obviously exchangeable. We then have the following approximate equilibrium property:

Theorem 6.17 *In addition to assumption* **Control of MKV Dynamics**, *assume further that σ does not depend upon α and that $\mu_0 \in \mathcal{P}_q(\mathbb{R}^d)$ for some $q > 4$. Then, there exists a constant $c > 0$ such that, for any $N \ge 1$,*

$$\bar{J}^N\big(\boldsymbol{\beta}^{(N)}\big) \ge \bar{J}^N\big(\hat{\boldsymbol{\alpha}}^{(N)}\big) - c\sqrt{\epsilon_N},$$

for any strategy profile $\boldsymbol{\beta}^{(N)} = (\boldsymbol{\beta}^{N,1}, \cdots, \boldsymbol{\beta}^{N,N})$ such that $(\xi^i, \boldsymbol{\beta}^{N,i}, \boldsymbol{W}^i)_{1\le i\le N}$ is exchangeable, where $\bar{J}^N(\boldsymbol{\beta}^{(N)})$ was defined in equation (6.61) *and $\hat{\boldsymbol{\alpha}}^{(N)}$ is given by $(\hat{\boldsymbol{\alpha}}^{N,1}, \cdots, \hat{\boldsymbol{\alpha}}^{N,N})$, as defined in* (6.77).

Proof. We use the same notation as in the proof of Theorem 6.16. In particular, since $\underline{\hat{\alpha}}_t^1 = \hat{\alpha}(t, \underline{X}_t^1, \mu_t, v(t, \underline{X}_t^1))$ for $0 \le t \le T$, see (6.65), the linear growth property of v implies that $\mathbb{E}[\sup_{0\le t\le T} |\underline{X}_t^1|^q] < +\infty$. Because of Lemma 6.1, (6.67) and (6.73) this implies that:

$$\bar{J}^N\big(\boldsymbol{\beta}^{(N)}\big) \ge J - c\sqrt{\epsilon_N}.$$

Moreover, since $v(t, \cdot)$ is Lipschitz continuous uniformly in $t \in [0, T]$, using once again the same classical estimates from Theorem 2.12, we also have:

$$\sup_{0\le t\le T} \mathbb{E}\big[|X_t^{N,i} - \underline{X}_t^i|^2\big] = \sup_{0\le t\le T} \mathbb{E}\big[|X_t^{N,1} - \underline{X}_t^1|^2\big] = O(\epsilon_N),$$

so that:

$$\sup_{0\le t\le T} \mathbb{E}\big[|\hat{\alpha}_t^{N,i} - \underline{\alpha}_t^i|^2\big] = \sup_{0\le t\le T} \mathbb{E}\big[|\alpha_t^{N,1} - \underline{\alpha}_t^1|^2\big] = O(\epsilon_N),$$

for any $1 \le i \le N$. It is then plain to deduce that:

$$\bar{J}^N\big(\hat{\boldsymbol{\alpha}}^{(N)}\big) \le J + c\sqrt{\epsilon_N}.$$

This completes the proof. □

6.2 Limits of Open-Loop N-Player Equilibria

We saw, in the case of some linear quadratic games, that Nash equilibria for N player games converged in some sense to solutions of mean field game problems. See for instance Section (Vol I)-2.4 together with Subsection (Vol I)-3.6.1 and Section (Vol I)-2.5 together with Subsection (Vol I)-3.6.2 in Chapters 2 and 3 of Volume I, and Subsection 4.5.1 in Chapter 4 in Volume II. Moreover, we saw in the previous Section 6.1 that solutions to MFG problems could be used to construct approximate Nash equilibria for finite player games, the larger the game, the better the approximation. The purpose of the next two sections is to investigate this duality in a more systematic way.

6.2.1 Possible Strategies for Passing to the Limit in the N-Player Game

While the statement of the problem is seemingly clear, unfortunately, establishing the convergence of equilibria of N-player games turns out to be a difficult question. It requires more effort than for the construction of approximate Nash equilibria from solutions of the limiting problems. In a way, this should be expected. The construction of approximate Nash equilibria performed in the preceding section is mostly based on the properties of the limiting mean field game. Somehow, it requires little information on the game with finitely many players. Obviously, this cannot be the case when we investigate equilibria of the N-player games and try to control their possible limits.

Generally speaking, a crucial issue when passing to the limit in the N-player game comes from the fact that very few bounds are known to hold uniformly in $N \geq 1$. Basically, the best we can hope for is to prove that equilibria, together with the corresponding strategies, satisfy uniform L^p estimates, for a suitable value of $p \geq 2$. In particular, when working with equilibria in closed loop form, there is little hope to establish uniform smoothness of the feedback functions, except possibly in some specific cases like those addressed in Subsection (Vol I)-7.1.2; recall that, therein, the players only observe their own state. The lack of uniform estimates on the smoothness of the feedback functions prevents the systematic use of any strong compactness method. The situation is slightly better when working with equilibria in open loop forms. The available bounds on the control strategies permit to prove tightness (or compactness) criteria which suffice to pass to the limit in the definition of open loop Nash equilibria. These tightness criteria are similar to those used in Chapter 3 for constructing weak solutions to mean field games with a common noise. We implement this first approach in this section.

While compactness arguments turn out to work quite well with open loop equilibria, they fail, as we just explained, for closed loop equilibria. Another strategy will be needed. We shall present it in the next section. It is based on the fact that closed loop equilibria can be described by means of a system of partial differential equations which we introduced in Chapter (Vol I)-2 and which we called the Nash system. The road map for passing to the limit then relies on the following intuition: as the number N of players tends to ∞, the Nash system should get closer and closer (in a suitable sense) to the master equation introduced in Chapters 4 and 5. Most of the next section is devoted to the rigorous derivation of this claim. In doing so, we restrict ourselves to cases for which the master equation has a classical solution. This allows us to use the smoothness of the limiting solution to establish convergence.

General Setting for the Analysis

In order to simplify the analysis of this section and the next, we assume that the states of the players have simpler dynamics of the form:

$$dX_t^{N,i} = b(t)\alpha_t^{N,i}dt + \sigma dW_t^i + \sigma^0 dW_t^0, \quad t \in [0,T]. \tag{6.79}$$

As a result, we will be in position to directly apply most of the results from the previous chapters that are needed in the analysis, while more general dynamics would have required variations of these results. Also, this simpler form of the dynamics suffices to explain the general philosophy of the strategies used to pass to the limit in the N-player games. If needed, the reader can try to adapt our arguments to handle more complicated models.

In (6.79), $(X_0^{N,1}, \cdots, X_0^{N,N}) = (\xi^1, \cdots, \xi^N)$, where $(\xi^n)_{n\geq 1}$ is a sequence of independent, identically distributed random variables with values in $\mathbb{R}^d$, with $\mu_0 \in \mathcal{P}_2(\mathbb{R}^d)$ as common distribution. Also $(\boldsymbol{W}^n)_{n\geq 0}$ is a sequence of independent d-dimensional Brownian motions, the sequences $(\boldsymbol{W}^n)_{n\geq 0}$ and $(\xi^n)_{n\geq 1}$ being independent. Both are constructed on a probability space $(\Omega, \mathcal{F}, \mathbb{P})$.

In contrast with what we did in the previous section, we assume that the empirical measure appearing in the cost coefficients of the cost functional of player $i \in \{1, \cdots, N\}$ is computed over the states of all the players $j \in \{1, \cdots, N\}$ except player i itself, that is:

$$J^{N,i}(\boldsymbol{\alpha}^{(N)}) = \mathbb{E}\biggl[\int_0^T f\bigl(t, X_t^{N,i}, \bar{\mu}^{N-1}_{X_t^{(N)-i}}, \alpha_t^{N,i}\bigr)dt + g\bigl(X_T^{N,i}, \bar{\mu}^{N-1}_{X_T^{(N)-i}}\bigr)\biggr], \tag{6.80}$$

where $X_t^{(N)} = (X_t^{N,1}, \cdots, X_t^{N,N})$ and, for any N-tuple $\boldsymbol{x} = (x^1, \cdots, x^N) \in (\mathbb{R}^d)^N$, $\bar{\mu}^{N-1}_{\boldsymbol{x}^{-i}}$ stands for:

$$\bar{\mu}^{N-1}_{\boldsymbol{x}^{-i}} = \frac{1}{N-1}\sum_{j=1, j\neq i}^{N} \delta_{x^j}, \tag{6.81}$$

where N is assumed to be larger than or equal to 2. This form is especially convenient for the analysis provided below, but as already explained in Chapter (Vol I)-2, we could also consider cost functionals depending on the full empirical measure $\frac{1}{N}\sum_{j=1}^N \delta_{x^j}$.

6.2.2 Weak Limits of Open Loop N-Player Equilibria

As announced, we first address the limiting behavior of open loop equilibria. Our strategy is based on a weak compactness method close to that used in Chapter 3 to construct weak solutions for mean field games with a common noise.

The analysis relies on the following assumption:

Assumption (Weak Limits of Open Loop N-Player Equilibria).

(A1) The dynamics and the costs are of the form (6.79) and (6.80), the function $b : [0, T] \ni t \mapsto b(t)$ being measurable and bounded.

(A2) Either assumption **MFG with a Common Noise HJB** or assumption **MFG with a Common Noise SMP Relaxed** is in force.

We refer the reader to Subsections 3.4.1 and 3.4.3 for detailed statements of these assumptions. We recall that, under both assumptions, the limiting mean field game has a solution, the proof of the existence of an equilibrium being based upon a weak convergence argument. We shall adapt this argument in order to pass to the limit in the N-player game.

Also, we shall assume the following:

(A3) For each $N \geq 2$, the N-player game has an open loop equilibrium $\hat{\boldsymbol{\alpha}}^{(N)} = (\hat{\boldsymbol{\alpha}}^{N,1}, \cdots, \hat{\boldsymbol{\alpha}}^{N,N})$ satisfying:

$$\forall i \in \{1, \cdots, N\}, \quad \mathbb{E} \int_0^T |\hat{\alpha}_s^{N,i}|^2 ds < \infty.$$

In the definition of a Nash equilibrium, we use implicitly the usual augmentation of the filtration generated by $(\xi^1, \cdots, \xi^N)$ and $(\boldsymbol{W}^0, \cdots, \boldsymbol{W}^N)$ as underlying filtration in the definition of the admissible control strategies. The fact that all the equilibria are defined on the same probability space has no real importance below since our approach relies on weak limit arguments.

For any $i \in \{1, \cdots, N\}$, we let:

$$d\hat{X}_t^{N,i} = b(t)\hat{\alpha}_t^{N,i} dt + \sigma dW_t^i + \sigma^0 dW_t^0, \quad t \in [0, T] \, ; \quad \hat{X}_0^{N,i} = \xi^i,$$

together with:

$$\bar{\mu}_t^{N,i} = \bar{\mu}^{N-1}_{\hat{X}_t^{(N)-i}},$$

where $\hat{\boldsymbol{X}}^{(N)} = (\hat{\boldsymbol{X}}^{N,1}, \cdots, \hat{\boldsymbol{X}}^{N,N})$. See also (6.81).

Stochastic Maximum Principle

For any given $N \geq 2$ and any other A-valued admissible control strategy $\boldsymbol{\beta}^i$, we have:

$$J^{N,i}\big(\boldsymbol{\beta}^i, (\hat{\boldsymbol{\alpha}}^{(N)})^{-i}\big) \geq J^{N,i}\big(\hat{\boldsymbol{\alpha}}^{(N)}\big).$$

The cost in the left-hand side reads:

$$J^{N,i}\big(\boldsymbol{\beta}^i, (\hat{\boldsymbol{\alpha}}^{(N)})^{-i}\big) = \mathbb{E}\bigg[\int_0^T f(t, U_t^{N,i}, \bar{\mu}_t^{N,i}, \beta_t^i)dt + g(U_T^{N,i}, \bar{\mu}_T^{N,i})\bigg],$$

where:

$$dU_t^{N,i} = b(t)\beta_t^i dt + \sigma dW_t^i + \sigma^0 dW_t^0, \quad t \in [0,T]\,; \quad U_0^{N,i} = \xi^i.$$

In particular, the control strategy $\hat{\boldsymbol{\alpha}}^{N,i}$ appears as an optimal control strategy for the minimization of the cost $J^{N,i}(\boldsymbol{\beta}^i, (\hat{\boldsymbol{\alpha}}^{(N)})^{-i})$ over admissible control strategies $\boldsymbol{\beta}^i$.

Although the setting is slightly different since the environment $(\bar{\mu}_t^{N,i})_{0\leq t\leq T}$ is not independent of $\boldsymbol{W}^i$, a mere adaptation of the proof of the Pontryagin principle in Theorem 1.59 shows that, necessarily,

$$\hat{\alpha}_t^{N,i} = \hat{\alpha}(t, \hat{X}_t^{N,i}, \bar{\mu}_t^{N,i}, \hat{Y}_t^{N,i}), \tag{6.82}$$

where:

$$d\hat{Y}_t^{N,i} = -\partial_x H^{(r)}\big(t, \hat{X}_t^{N,i}, \bar{\mu}_t^{N,i}, \hat{Y}_t^{N,i}, \hat{\alpha}_t^{N,i}\big)dt + dM_t^{N,i}, \tag{6.83}$$

where $(M_t^{N,i})_{0\leq t\leq T}$ is a square-integrable continuous martingale. Continuity here follows from the fact that the filtration satisfies the martingale representation property. The function $H^{(r)}$ stands for the reduced Hamiltonian associated with f and b, and $\hat{\alpha}$ is its minimizer, which is known to exist under condition **(A2)** in the standing assumption. Here, we have in fact $\partial_x H^{(r)}(t,x,\mu,y,\alpha) = \partial_x f(t,x,\mu,\alpha)$.

Main Statement

For any $N \geq 2$, we define the empirical measure:

$$\mathfrak{M}^N = \frac{1}{N}\sum_{i=1}^N \delta_{(\hat{X}^{N,i}, W^i)},$$

which reads as a random variable with values in $\mathcal{P}_2(\mathcal{C}([0,T];\mathbb{R}^{2d}))$. The main result of this section appears as the analog of Theorem 3.13.

Theorem 6.18 *Under assumption* **Weak Limits of Open Loop N-Player Equilibria**, *the sequence of probability measures* $(\mathbb{P} \circ (\boldsymbol{W}^0, \mathfrak{M}^N)^{-1})_{N \geq 2}$ *is tight on the space* $\mathcal{C}([0,T];\mathbb{R}^d) \times \mathcal{P}_2(\mathcal{C}([0,T];\mathbb{R}^{2d}))$, *and any weak limit generates a distribution of an equilibrium in the sense of Definition 2.24 for the problem driven by the coefficients* $(b, \sigma, \sigma^0, f, g)$.

Of course, whenever the Lasry-Lions monotonicity condition holds, weak limits are unique and coincide with the distribution of the unique mean field game equilibrium with μ_0 as initial condition.

6.2.3 Proof of the Convergence

Since the proof is similar to that of Theorem 3.13, we only provide a sketch.

Throughout the proof, assumption **Weak Limits of Open Loop N-Player Equilibria** is in force. Also, for any $N \geq 2$, we call ϑ_N a uniform random variable on $\{1, \cdots, N\}$, independent of $((\xi^n)_{n \geq 1}, (\boldsymbol{W}^n)_{n \geq 0})$. We can always consider such a random variable by extending the probability space if necessary.

We then consider the trajectories $\hat{\boldsymbol{X}}^{N,\vartheta_N} = (\hat{X}_t^{N,\vartheta_N})_{0 \leq t \leq T}$ together with the control strategies $\hat{\boldsymbol{\alpha}}^{N,\vartheta_N} = (\hat{\alpha}_t^{N,\vartheta_N})_{0 \leq t \leq T}$, as given by (6.82) and (6.83).

Tightness Properties

We start with the following technical lemma:

Lemma 6.19 *There exists a constant C such that, for all $N \geq 2$,*

$$\sup_{i=1,\cdots,N} \left\{ \mathbb{E}\big[\sup_{0 \leq t \leq T} |\hat{X}_t^{N,i}|^2 \big] + \mathbb{E}\left[\int_0^T |\hat{\alpha}_t^{N,i}|^2 dt \right] \right\} \leq C.$$

Also,

$$\lim_{a \to \infty} \sup_{N \geq 2} \mathbb{E}\big[\sup_{0 \leq t \leq T} |\hat{X}_t^{N,\vartheta_N}|^2 \mathbf{1}_{\{\sup_{0 \leq t \leq T} |\hat{X}_t^{N,\vartheta_N}|^2 \geq a\}} \big] = 0,$$

$$\lim_{a \to \infty} \sup_{N \geq 2} \mathbb{E}\left[\left(\int_0^T |\hat{\alpha}_t^{N,\vartheta_N}|^2 dt \right) \mathbf{1}_{\{\int_0^T |\hat{\alpha}_t^{N,\vartheta_N}|^2 dt \geq a\}} \right] = 0.$$

Proof. There are two cases.

Whenever assumption **MFG with a Common Noise HJB** is in force, the result is pretty straightforward. In that case, both $\partial_x H$ and $\hat{\alpha}$ are at most of linear growth in y uniformly in the other variables. Hence, the processes $(\hat{\boldsymbol{Y}}^{N,i})_{1 \leq i \leq N}$ can be bounded independently of N and, subsequently, the same holds true for the controls $(\hat{\boldsymbol{\alpha}}^{N,i})_{1 \leq i \leq N}$.

The proof is much more involved under assumption **MFG with a Common Noise SMP Relaxed** as we must invoke arguments similar to those used in the proof of Lemma 3.33.

Formally, the computations are the same provided that we replace the symbol $\mathbb{E}_t^{0,n}$ in the original proof by $\mathbb{E}_t\mathbf{E}$, where $\mathbb{E}_t$ denotes the conditional expectation given $(\xi^i, (W_s^i)_{0\le s\le t})_{i\ge 1}$. The notation $\mathbf{E}$ is a shorten notation to indicate that we take the expectation over the sole variable ϑ_N. Observe that it makes sense to do so since ϑ_N is independent of $(\xi^i, \boldsymbol{W}^i)_{i\ge 1}$.

Below, we do not repeat all the arguments of Lemma 3.33. We just point out the differences.

For a given player, say player 1, we construct a control $\boldsymbol{\beta}^{N,1}$ in two different ways. For a given $t \in [0,T]$, $\boldsymbol{\beta}^{N,1}$ is required to coincide with $\hat{\boldsymbol{\alpha}}^{N,1}$ on $[0,t)$. On $[t,T]$, we shall use alternatively one of the following two forms:

$$(i)\ \beta_s^{N,1} = \mathbf{E}\big[\hat{\alpha}_s^{N,\vartheta_N}\big] \quad \text{for } t \le s \le T; \quad (ii)\ \beta_s^{N,1} = 0 \quad \text{for } t \le s \le T. \tag{6.84}$$

Since $A = \mathbb{R}^k$, both choices are admissible. We then let $\boldsymbol{U}^{N,1}$ denote the resulting state:

$$dU_s^{N,1} = b(s)\beta_s^{N,1}ds + \sigma dW_s^1 + \sigma^0 dW_s^0, \quad s \in [0,T].$$

Of course, $(U_s^{N,1})_{0\le s\le t}$ and $(\hat{X}_s^{N,1})_{0\le s\le t}$ are equal. We set $\boldsymbol{\beta}^{(N)} = (\boldsymbol{\beta}^{N,1}, (\hat{\boldsymbol{\alpha}}^{(N)})^{-1})$. Also, the associated controlled trajectory is denoted by $\boldsymbol{U}^{(N)} = (\boldsymbol{U}^{N,1}, \hat{\boldsymbol{X}}^{N,2}, \cdots, \hat{\boldsymbol{X}}^{N,N})$.

As in the proof of Lemma 3.33, we compare the cost $J^{N,1}(\boldsymbol{\beta}^{(N)})$ with the equilibrium cost $J^{N,1}(\hat{\boldsymbol{\alpha}}^{(N)})$ by means of Theorem 1.60. Importantly, the reader can check that (1.63) remains true although the environment $(\bar{\mu}_s^{N,1})_{0\le s\le T}$ is correlated with the noise $\boldsymbol{W}^1$. Throughout the proof, we assume that $N \ge 2$, so that $\frac{1}{N-1} \le \frac{2}{N}$.

First Step. We first consider the alternative (*i*) in (6.84). Following the first step in the proof of Lemma 3.33, we deduce that there exists a constant C such that, for all $N \ge 2$,

$$\sup_{t\le s\le T} \mathbb{E}_t\big[|U_s^{N,1} - \bar{X}_s^N|^2\big] \le C\Big(1 + |\hat{X}_t^{N,1} - \bar{X}_t^N|^2\Big),$$

where $\bar{X}_s^N = \mathbf{E}[\hat{X}_s^{N,\vartheta_N}] = \frac{1}{N}\sum_{i=1}^N \hat{X}_s^{N,i}$. Similarly, we shall use the notation $\bar{\alpha}_s^N = \mathbf{E}[\hat{\alpha}_s^{N,\vartheta_N}] = \frac{1}{N}\sum_{i=1}^N \hat{\alpha}_s^{N,i}$. Then, by the conditional version of Theorem 1.60,

$$\begin{aligned}
&\mathbb{E}_t\Big[g(\hat{X}_T^{N,1}, \bar{\mu}_T^{N,1}) + \int_t^T \Big[\lambda|\hat{\alpha}_s^{N,1} - \bar{\alpha}_s^N|^2 + f(s, \hat{X}_s^{N,1}, \bar{\mu}_s^{N,1}, \hat{\alpha}_s^{N,1})\Big]ds\Big] \\
&\le \mathbb{E}_t\Big[g(U_T^{N,1}, \bar{\mu}_T^{N,1}) + \int_t^T f(s, U_s^{N,1}, \bar{\mu}_s^{N,1}, \bar{\alpha}_s^N)ds\Big] \\
&\le \mathbb{E}_t\Big[g(\bar{X}_T^N, \bar{\mu}_T^{N,1}) + \int_t^T f(s, \bar{X}_s^N, \bar{\mu}_s^{N,1}, \bar{\alpha}_s^N)ds\Big] \\
&\quad + C\Big[1 + \Big(\mathbb{E}_t\big[\sup_{t\le s\le T}|\hat{X}_s^{N,1}|^2 + \mathbf{E}\big[\sup_{t\le s\le T}|\hat{X}_s^{N,\vartheta_N}|^2\big]\big]\Big)^{1/2} \\
&\quad + \Big(\mathbb{E}_t\int_t^T (\mathbf{E}[|\hat{\alpha}_s^{N,\vartheta_N}|^2])ds\Big)^{1/2}\Big]\Big(1 + |\hat{X}_t^{N,1} - \bar{X}_t^N|^2\Big)^{1/2}.
\end{aligned}$$

Observe that:

$$
\begin{aligned}
W_2(\bar\mu_s^{N,1},\bar\mu_s^N)^2 &\le \frac{1}{N(N-1)}\sum_{i\neq 1}|\hat X_s^{N,1}-\hat X_s^{N,i}|^2 \\
&\le \frac{2}{N}|\hat X_s^{N,1}-\bar X_s^N|^2+\frac{4}{N}\mathbf{E}[|\hat X_s^{N,\vartheta_N}-\bar X_s^N|^2],
\end{aligned}
$$

where $\bar\mu_s^N=\frac{1}{N}\sum_{i=1}^N\delta_{\hat X_s^{N,i}}$. We deduce:

$$
\begin{aligned}
&\mathbb{E}_t\Big[g(\hat X_T^{N,1},\bar\mu_T^N)+\int_t^T\Big[\lambda|\hat\alpha_s^{N,1}-\bar\alpha_s^N|^2+f(s,\hat X_s^{N,1},\bar\mu_s^N,\hat\alpha_s^{N,1})\Big]ds\Big] \\
&\le \mathbb{E}_t\Big[g(\bar X_T^N,\bar\mu_T^N)+\int_t^T f(s,\bar X_s^N,\bar\mu_s^N,\bar\alpha_s^N)ds\Big] \\
&\quad + C\Big[1+\Big(\mathbb{E}_t[\sup_{t\le s\le T}|\hat X_s^{N,1}|^2+\mathbf{E}[\sup_{t\le s\le T}|\hat X_s^{N,\vartheta_N}|^2]]\Big)^{1/2} \\
&\quad + \Big(\mathbb{E}_t\int_t^T(|\hat\alpha_s^{N,1}|^2+\mathbf{E}[|\hat\alpha_s^{N,\vartheta_N}|^2])ds\Big)^{1/2}\Big]\Big(1+|\hat X_t^{N,1}-\bar X_t^N|^2 \\
&\quad + \frac{1}{N}\mathbb{E}_t\mathbf{E}[\sup_{t\le s\le T}|\hat X_s^{N,\vartheta_N}-\bar X_s^N|^2]+\frac{1}{N}\mathbb{E}_t[\sup_{t\le s\le T}|\hat X_s^{N,1}-\bar X_s^N|^2]\Big)^{1/2}.
\end{aligned}
$$

By repeating the same argument for any other player $i\neq 1$, by averaging over all the indices in $\{1,\cdots,N\}$, and then by using the same convexity argument as in the proof of Lemma 3.33, the next step is to prove:

$$
\begin{aligned}
&\mathbb{E}_t\mathbf{E}\int_t^T|\hat\alpha_s^{N,\vartheta_N}-\bar\alpha_s^N|^2 ds \\
&\le C\Big[1+\Big(\mathbb{E}_t\mathbf{E}[\sup_{t\le s\le T}|\hat X_s^{N,\vartheta_N}|^2]\Big)^{1/2}+\Big(\mathbb{E}_t\mathbf{E}\int_t^T|\hat\alpha_s^{N,\vartheta_N}|^2ds\Big)^{1/2}\Big] \\
&\quad\times\Big(1+\mathbf{E}[|\hat X_t^{N,\vartheta_N}-\bar X_t^N|^2]+\frac{1}{N}\mathbb{E}_t\mathbf{E}[\sup_{t\le s\le T}|\hat X_s^{N,\vartheta_N}-\bar X_s^N|^2]\Big)^{1/2},
\end{aligned}
$$

and we end up with:

$$
\begin{aligned}
&\mathbb{E}_t\mathbf{E}[\sup_{t\le s\le T}|\hat X_s^{N,\vartheta_N}-\bar X_s^N|^2] \\
&\le C\Big[1+\Big(\mathbb{E}_t\mathbf{E}[\sup_{t\le s\le T}|\hat X_s^{N,\vartheta_N}|^2]\Big)^{1/2}+\Big(\mathbb{E}_t\mathbf{E}\int_t^T|\hat\alpha_s^{N,\vartheta_N}|^2ds\Big)^{1/2}\Big] \\
&\quad\times\Big(1+\mathbf{E}[|\hat X_t^{N,\vartheta_N}-\bar X_t^N|^2]+\frac{1}{N}\mathbb{E}_t\mathbf{E}[\sup_{t\le s\le T}|\hat X_s^{N,\vartheta_N}-\bar X_s^N|^2]\Big)^{1/2}.
\end{aligned}
\tag{6.85}
$$

Second Step. When $\boldsymbol{\beta}^{N,1}$ is null after t, we have $\sup_{N\geq 2}\mathbb{E}_t[\sup_{t\leq s\leq T}|U_s^{N,1}|^2] \leq C(1 + |\hat{X}_t^{N,1}|^2)$. Proceeding as before, we obtain:

$$
\begin{aligned}
&\mathbb{E}_t\Big[g(\hat{X}_T^{N,1},\bar{\mu}_T^N) + \int_t^T \Big[\lambda|\hat{\alpha}_s^{N,1}|^2 + f(s,\hat{X}_s^{N,1},\bar{\mu}_s^N,\hat{\alpha}_s^{N,1})\Big]ds\Big] \\
&\leq \mathbb{E}_t\Big[g(U_T^{N,1},\bar{\mu}_T^N) + \int_t^T f(s,U_s^{N,1},\bar{\mu}_s^N,0)ds\Big] \\
&\quad + C\Big[1 + \Big(\mathbb{E}_t[\sup_{t\leq s\leq T}|\hat{X}_T^{N,1}|^2 + \mathbf{E}[\sup_{t\leq s\leq T}|\hat{X}_s^{N,\vartheta_N}|^2]]\Big)^{1/2} \\
&\qquad + \Big(\mathbb{E}_t\int_t^T (|\hat{\alpha}_s^{N,1}|^2 + \mathbf{E}[|\hat{\alpha}_s^{N,\vartheta_N}|^2])ds\Big)^{1/2}\Big] \\
&\qquad \times \Big(\frac{1}{N}\mathbb{E}_t\mathbf{E}[\sup_{t\leq s\leq T}|\hat{X}_s^{N,\vartheta_N} - \bar{X}_s^N|^2] + \frac{1}{N}\mathbb{E}_t[\sup_{t\leq s\leq T}|\hat{X}_s^{N,1} - \bar{X}_s^N|^2]\Big)^{1/2}.
\end{aligned}
$$

As before, we repeat the same argument for any other player $i \neq 1$. Averaging over all the indices in $\{1,\cdots,N\}$ and using a new convexity argument, we obtain, still by following the steps of the proof of Lemma 3.33:

$$
\begin{aligned}
&\mathbb{E}_t[g(\bar{X}_T^N,\delta_{\bar{X}_T^N})] + \mathbb{E}_t\int_t^T \Big[\lambda\mathbf{E}[|\hat{\alpha}_s^{N,\vartheta_N}|^2] + f(s,\bar{X}_s^N,\delta_{\bar{X}_s^N},0)\Big]ds \\
&\leq \mathbb{E}_t\Big[g(0,\delta_{\bar{X}_T^N}) + \int_t^T f(s,0,\delta_{\bar{X}_s^N},0)ds\Big] \\
&\quad + C\Big(1 + \mathbb{E}_t\mathbf{E}[\sup_{t\leq s\leq T}|\hat{X}_s^{N,\vartheta_N}|^2] + \mathbb{E}_t\mathbf{E}\int_t^T |\hat{\alpha}_s^{N,\vartheta_N}|^2 ds\Big)^{1/2} \\
&\qquad \times \Big(1 + \mathbf{E}[|\hat{X}_t^{N,\vartheta_N}|^2] + \mathbb{E}_t\mathbf{E}[\sup_{t\leq s\leq T}|\hat{X}_s^{N,\vartheta_N} - \bar{X}_s^N|^2]\Big)^{1/2}.
\end{aligned}
$$

Hence, following again the second step in the proof of Lemma 3.33, we get:

$$
\begin{aligned}
&\mathbb{E}_t[\bar{X}_T^N\cdot\partial_x g(0,\delta_{\bar{X}_T^N})] + \mathbb{E}_t\mathbf{E}\int_t^T [\lambda|\hat{\alpha}_s^{N,\vartheta_N}|^2 + \bar{X}_s^N\cdot\partial_x f(s,0,\delta_{\bar{X}_s^N},0)]ds \\
&\leq C\Big(1 + \mathbb{E}_t\mathbf{E}[\sup_{t\leq s\leq T}|\hat{X}_s^{N,\vartheta_N}|^2] + \mathbb{E}_t\mathbf{E}\int_t^T |\hat{\alpha}_s^{N,\vartheta_N}|^2 ds\Big)^{1/2} \\
&\quad \times \Big(1 + \mathbf{E}[|\hat{X}_t^{N,\vartheta_N}|^2] + \mathbb{E}_t\mathbf{E}[\sup_{t\leq s\leq T}|\hat{X}_s^{N,\vartheta_N} - \bar{X}_s^N|^2]\Big)^{1/2}.
\end{aligned}
$$

Using (**A3**) in assumption **MFG with Common Noise SMP Relaxed** together with (6.85), we deduce from Young's inequality that:

$$\mathbb{E}_t\mathbf{E}\Big[\sup_{t\le s\le T}|\hat{X}_s^{N,\vartheta_N}|^2 + \int_t^T |\hat{\alpha}_s^{N,\vartheta_N}|^2 ds\Big]$$
$$\le C_\epsilon\Big(1 + \mathbf{E}[|\hat{X}_t^{N,\vartheta_N}|^2] + \frac{1}{N}\mathbb{E}_t\mathbf{E}[\sup_{t\le s\le T}|\hat{X}_s^{N,\vartheta_N}|^2]\Big)$$
$$+ \epsilon\Big(\mathbb{E}_t\mathbf{E}[\sup_{t\le s\le T}|\hat{X}_s^{N,\vartheta_N}|^2] + \mathbb{E}_t\mathbf{E}\Big[\int_t^T |\hat{\alpha}_s^{N,\vartheta_N}|^2 ds\Big]\Big),$$

for any $\epsilon > 0$, the constant C_ϵ being allowed to depend on ϵ. Choosing first ϵ small enough and then N large enough, we get:

$$\mathbb{E}_t\mathbf{E}\Big[\sup_{t\le s\le T}|\hat{X}_s^{N,\vartheta_N}|^2 + \int_t^T |\hat{\alpha}_s^{N,\vartheta_N}|^2 ds\Big] \le C\Big(1 + \mathbf{E}[|\hat{X}_t^{N,\vartheta_N}|^2]\Big). \tag{6.86}$$

The rest of the second step differs from that of Lemma 3.33. Indeed, we cannot make use the notion of decoupling field because each environment $(\bar{\mu}_s^{N,i})_{0\le s\le T}$, for $i \in \{1,\cdots,N\}$, depends on the noise $\boldsymbol{W}^i$. In order to proceed, we must go back to the beginning of the second step. With the same choice for $\boldsymbol{\beta}^{N,1}$, we have:

$$\mathbb{E}_t\Big[g(\hat{X}_T^{N,1},\bar{\mu}_T^{N,1}) + \int_t^T \Big[\lambda|\hat{\alpha}_s^{N,1}|^2 + f(s,\hat{X}_s^{N,1},\bar{\mu}_s^{N,1},\hat{\alpha}_s^{N,1})\Big]ds\Big]$$
$$\le \mathbb{E}_t\Big[g(U_T^{N,1},\bar{\mu}_T^{N,1}) + \int_t^T f(s,U_s^{N,1},\bar{\mu}_s^{N,1},0)ds\Big]$$
$$\le C\Big(1 + |\hat{X}_t^{N,1}|^2 + \mathbb{E}_t\mathbf{E}[\sup_{t\le s\le T}|\hat{X}_s^{N,\vartheta_N}|^2]\Big)$$
$$\le C(1 + |\hat{X}_t^{N,1}|^2 + \mathbf{E}[|\hat{X}_t^{N,\vartheta_N}|^2]),$$

where we used (6.86). By convexity of the cost functions, this yields:

$$\mathbb{E}_t\Big[g(0,\bar{\mu}_T^{N,1}) + \hat{X}_T^{N,1}\cdot\partial_x g(0,\bar{\mu}_T^{N,1}) + \lambda\int_t^T |\hat{\alpha}_s^{N,1}|^2 ds$$
$$+ \int_t^T \Big[f(s,0,\bar{\mu}_s^{N,1},0) + \hat{X}_s^{N,1}\cdot\partial_x f(s,0,\bar{\mu}_s^{N,1},0) + \hat{\alpha}_s^{N,1}\cdot\partial_\alpha f(s,0,\bar{\mu}_s^{N,1},0)\Big]ds\Big]$$
$$\le C(1 + |\hat{X}_t^{N,1}|^2 + \mathbf{E}[|\hat{X}_t^{N,\vartheta_N}|^2]),$$

from which we get, by using once again (6.86),

$$\mathbb{E}_t\Big[\int_t^T |\hat{\alpha}_s^{N,1}|^2 ds\Big] \le C(1 + |\hat{X}_t^{N,1}|^2 + \mathbf{E}[|\hat{X}_t^{N,\vartheta_N}|^2]),$$

and then,

$$\mathbb{E}_t[\sup_{t\le s\le T}|\hat{X}_s^{N,1}|^2] \le C(1 + |\hat{X}_t^{N,1}|^2 + \mathbf{E}[|\hat{X}_t^{N,\vartheta_N}|^2]).$$

Now, we use the fact that the growths of both $\partial_x H$ and $\hat{\alpha}$ are at most linear in all their arguments. We obtain:

$$\begin{aligned}\mathbb{E}_t\big[\sup_{t\le s\le T}|\hat{Y}_t^{N,1}|^2\big] &\le C\mathbb{E}_t\big[\sup_{t\le s\le T}|\hat{X}_s^{N,1}|^2 + \sup_{t\le s\le T}\mathbf{E}[|\hat{X}_s^{N,\vartheta_N}|^2]\big]\\ &\le C\big(1+|\hat{X}_t^{N,1}|^2+\mathbf{E}[|\hat{X}_t^{N,\vartheta_N}|^2]\big),\end{aligned}$$

and then,

$$|\hat{\alpha}_t^{N,1}| \le C\Big(1+|\hat{X}_t^{N,1}|+\mathbf{E}\big[|\hat{X}_t^{N,\vartheta_N}|^2\big]^{1/2}\Big), \tag{6.87}$$

and similarly for any other player $i \neq 1$.

Third Step. We now proceed as in the proof of Lemma 3.33. Thanks to (6.87), we get, for any $i \in \{1,\cdots,N\}$:

$$\mathbb{E}_0\big[\sup_{0\le s\le t}|\hat{X}_s^{N,i}|^4\big]^{1/2} \le C\Big(1+|\xi^i|^2+\mathbf{E}\big[|\xi^{\vartheta_N}|^2\big]\Big).$$

Hence, for any event D and any $\varepsilon > 0$,

$$\begin{aligned}\frac{1}{N}\sum_{i=1}^N \mathbb{E}\big[\sup_{0\le t\le T}|\hat{X}_t^{N,i}|^2\mathbf{1}_D\big] &\le \frac{1}{N}\sum_{i=1}^N \mathbb{E}\Big[\mathbb{E}_0\big[\sup_{0\le t\le T}|\hat{X}_t^{N,i}|^4\big]^{1/2}\big[\mathbb{P}(D|\mathcal{F}_0)\big]^{1/2}\Big]\\ &\le \frac{C}{N}\sum_{i=1}^N \mathbb{E}\Big[\Big(1+|\xi^i|^2+\mathbf{E}\big[|\xi^{\vartheta_N}|^2\big]\Big)\big[\mathbb{P}(D|\mathcal{F}_0)\big]^{1/2}\Big]\\ &\le C\Big(\varepsilon\mathbb{E}\big[|\xi^1|^2+\frac{1}{\varepsilon}\mathbb{E}\Big[\Big(1+\frac{1}{N}\sum_{i=1}^N|\xi^i|^2\Big)\mathbb{P}(D|\mathcal{F}_0)\Big]\Big)\\ &= C\Big(\varepsilon\mathbb{E}\big[|\xi^1|^2+\frac{1}{\varepsilon}\mathbb{E}\Big[\Big(1+\frac{1}{N}\sum_{i=1}^N|\xi^i|^2\Big)\mathbf{1}_D\Big)\Big]\Big).\end{aligned}$$

Since the sequence $(\frac{1}{N}\sum_{i=1}^N|\xi^i|^2)_{N\ge1}$ is uniformly integrable, this proves that the family $(\sup_{0\le s\le T}|\hat{X}_s^{N,\vartheta_N}|^2)_{N\ge1}$ is uniformly square-integrable. Of course, we also have:

$$\sup_{N\ge1}\mathbb{E}\big[\sup_{0\le s\le T}|\hat{X}_s^{N,i}|^2\big] < \infty,$$

and by (6.87), the proof can be easily completed from there. □

Lemma 6.20 *The sequence $(\mathbb{P} \circ (\hat{X}^{N,\vartheta_N})^{-1})_{N\geq 2}$ of probability measures on the space $\mathcal{C}([0,T];\mathbb{R}^{2d})$ is tight. Moreover, the sequence $(\mathbb{P} \circ (\hat{\alpha}^{N,\vartheta_N})^{-1})_{N\geq 2}$ is tight on $\mathcal{M}([0,T];\mathbb{R}^k)$ equipped with the Meyer-Zheng topology, and any weak limit may be regarded as the law of an A-valued process.*

Proof. The proof is a variation on the proof of Lemma 3.14. By Aldous' criterion and Lemma 6.19, we can prove that the family $((\mathbb{P} \circ (\hat{X}^{N,i})^{-1})_{i=1,\cdots,N})_{N\geq 2}$ is tight on $\mathcal{C}([0,T];\mathbb{R}^d)$. In particular, for a given $\varepsilon > 0$, we can find a compact subset $\mathcal{K} \subset \mathcal{C}([0,T];\mathbb{R}^d)$ such that:

$$\forall N \geq 2,\ \forall i \in \{1,\cdots,N\}, \quad \mathbb{P}\big[\hat{X}^{N,i} \notin \mathcal{K}\big] \leq \varepsilon.$$

Then, to complete the first step of the proof, it suffices to notice that:

$$\mathbb{P}\big(\hat{X}^{N,\vartheta_N} \notin \mathcal{K}\big) = \frac{1}{N}\sum_{i=1}^{N} \mathbb{P}\big[\hat{X}^{N,i} \notin \mathcal{K}\big] \leq \varepsilon.$$

Similarly, we prove the second claim by adapting the arguments used in the proof of Lemma 3.14, taking advantage of the stochastic maximum principle (6.82)–(6.83) to prove the tightness of the sequence $(\hat{Y}^{N,\vartheta_N})_{N\geq 2}$ and of the lemma below to get the tightness of the sequence $((\bar{\mu}_t^{N,\vartheta_N})_{0\leq t\leq T})_{N\geq 2}$. □

As a by-product of the first claim in the statement right above, we obtain the following tightness result:

Lemma 6.21 *The sequence $(\mathbb{P} \circ (\mathfrak{M}^N)^{-1})_{N\geq 2}$ is tight on $\mathcal{P}_2(\mathcal{C}([0,T];\mathbb{R}^{2d}))$.*

Proof. The proof is an adaptation of the proof of Lemma 3.16, noticing that, for any positive function φ on $\mathcal{C}([0,T];\mathbb{R}^{2d})$,

$$\begin{aligned}\mathbb{E}\bigg[\int_{\mathcal{C}([0,T];\mathbb{R}^{2d})} \varphi(\boldsymbol{x},\boldsymbol{w})\, d\mathfrak{M}^N(\boldsymbol{x},\boldsymbol{w})\bigg] &= \frac{1}{N}\sum_{i=1}^{N} \mathbb{E}\big[\varphi(\hat{X}^{N,i}, \boldsymbol{W}^i)\big] \\ &= \mathbb{E}\big[\varphi(\hat{X}^{N,\vartheta_N}, \boldsymbol{W}^{\vartheta_N})\big],\end{aligned}$$

which suffices to duplicate the proof of Lemma 3.16 using in addition Lemma 6.19 in order to prove uniform square integrability. □

Converging Subsequence

We now complete the proof of Theorem 6.18. It suffices to consider a probability space $(\Omega^\infty, \mathcal{F}^\infty, \mathbb{P}^\infty)$ equipped with a process $(\xi^\infty, \boldsymbol{W}^{0,\infty}, \mathfrak{M}^\infty, \boldsymbol{W}^\infty, \hat{X}^\infty, \hat{\alpha}^\infty)$ whose law under $\mathbb{P}^\infty$ is a weak limit of the sequence:

$$\Big(\mathbb{P} \circ \big(\xi^{\vartheta_N}, \boldsymbol{W}^0, \mathfrak{M}^N, \boldsymbol{W}^{\vartheta_N}, \hat{X}^{\vartheta_N}, \hat{\alpha}^{\vartheta_N}\big)^{-1}\Big)_{N\geq 2}$$

on the space:

$$\mathbb{R}^d \times \mathcal{C}([0,T];\mathbb{R}^d) \times \mathcal{P}_2\big(\mathcal{C}([0,T];\mathbb{R}^{2d})\big) \times \mathcal{C}([0,T];\mathbb{R}^d) \\ \times \mathcal{C}([0,T];\mathbb{R}^d) \times \mathscr{M}([0,T];\mathbb{R}^k),$$

the last component being equipped with the Meyer-Zheng topology.

Without any loss of generality, we shall still index by N the subsequence converging to $\mathbb{P}^\infty \circ (\xi^\infty, \boldsymbol{W}^{0,\infty}, \mathfrak{M}^\infty, \boldsymbol{W}^\infty, \hat{\boldsymbol{X}}^\infty, \hat{\boldsymbol{\alpha}}^\infty)^{-1}$.

We start with the obvious preparatory result. Recall that μ_0 is the common distribution of the $(\xi^i)'_{i\geq 1}$s.

Lemma 6.22 *The random variable ξ^∞ has μ_0 as distribution under $\mathcal{P}_2(\mathbb{R}^d)$, while the processes $\boldsymbol{W}^{0,\infty}$ and $\boldsymbol{W}^\infty$ are d-dimensional Brownian motions.*

Proof. We just identify the law of ξ^∞. The law of $\boldsymbol{W}^\infty$ can be identified in the same way, while the fact that $\boldsymbol{W}^{0,\infty}$ is a Brownian motion is obvious. We consider a bounded and continuous function φ from $\mathbb{R}^d$ to $\mathbb{R}$. Then,

$$\begin{aligned}\mathbb{E}^\infty\big[\varphi(\xi^\infty)\big] &= \lim_{N\to\infty} \mathbb{E}\big[\varphi(\xi^{\vartheta_N})\big] \\ &= \lim_{N\to\infty} \frac{1}{N}\sum_{i=1}^N \mathbb{E}\big[\varphi(\xi^i)\big] = \lim_{N\to\infty} \frac{1}{N}\sum_{i=1}^N \mathbb{E}\big[\varphi(\xi^1)\big] = \mathbb{E}\big[\varphi(\xi^1)\big],\end{aligned}$$

which suffices to complete the proof. □

Also, the limit random variables satisfy the required independence property:

Lemma 6.23 *The random variables ξ^∞, $(\boldsymbol{W}^{0,\infty}, \mathfrak{M}^\infty)$ and $\boldsymbol{W}^\infty$ are independent.*

Proof. The proof is a consequence of the law of large numbers, from which we deduce that, for any bounded continuous function φ_1 from $\mathbb{R}^d \times \mathcal{C}([0,T], \mathbb{R}^d)$ into $\mathbb{R}$,

$$\lim_{N\to\infty} \mathbb{E}\bigg[\bigg|\frac{1}{N}\sum_{i=1}^N \varphi_1(\xi^i, \boldsymbol{W}^i) - \mathbb{E}\big[\varphi_1(\xi^1, \boldsymbol{W}^1)\big]\bigg|\bigg] = 0.$$

Now, for any bounded and continuous function φ_0 from $\mathcal{C}([0,T];\mathbb{R}^d) \times \mathcal{P}_2(\mathcal{C}([0,T];\mathbb{R}^{2d}))$ into $\mathbb{R}$, we have:

$$\begin{aligned}\mathbb{E}^\infty\big[\varphi_0(\boldsymbol{W}^{0,\infty}, \mathfrak{M}^\infty)\varphi_1(\xi^\infty, \boldsymbol{W}^\infty)\big] &= \lim_{N\to\infty} \mathbb{E}\big[\varphi_0(\boldsymbol{W}^0, \mathfrak{M}^N)\varphi_1(\xi^{\vartheta_N}, \boldsymbol{W}^{\vartheta_N})\big] \\ &= \lim_{N\to\infty} \mathbb{E}\bigg[\varphi_0(\boldsymbol{W}^0, \mathfrak{M}^N)\bigg(\frac{1}{N}\sum_{i=1}^N \varphi_1(\xi^i, \boldsymbol{W}^i)\bigg)\bigg] \\ &= \mathbb{E}\big[\varphi_1(\xi^1, \boldsymbol{W}^1)\big] \lim_{N\to\infty} \mathbb{E}\big[\varphi_0(\boldsymbol{W}^0, \mathfrak{M}^N)\big],\end{aligned}$$

where we used the law of large numbers to get the last equality. We deduce that:

$$\mathbb{E}^{\infty}\big[\varphi_0(\boldsymbol{W}^{0,\infty}, \mathfrak{M}^{\infty})\varphi_1(\xi^{\infty}, \boldsymbol{W}^{\infty})\big] = \mathbb{E}\big[\varphi_1(\xi^1, \boldsymbol{W}^1)\big]\mathbb{E}\big[\varphi_0(\boldsymbol{W}^{0,\infty}, \mathfrak{M}^{\infty})\big].$$

Recalling that ξ^1 and $\boldsymbol{W}^1$ are independent, we easily complete the proof. □

Next we check the needed compatibility:

Lemma 6.24 *The random variable $\mathfrak{M}^{\infty}$ provides a version of the conditional law of $(\hat{\boldsymbol{X}}^{\infty}, \boldsymbol{W}^{\infty})$ given $(\boldsymbol{W}^{0,\infty}, \mathfrak{M}^{\infty})$. Moreover, the complete and right-continuous filtration $\mathbb{F}^{\infty}$ generated by the limiting process $(\xi^{\infty}, \boldsymbol{W}^{0,\infty}, \mathfrak{M}^{\infty}, \boldsymbol{W}^{\infty}, \hat{\boldsymbol{X}}^{\infty})$ is compatible with $(\xi^{\infty}, \boldsymbol{W}^{0,\infty}, \mathfrak{M}^{\infty}, \boldsymbol{W}^{\infty})$.*

Proof. We start with the first claim. For two real-valued bounded and continuous functions φ on $\mathcal{C}([0,T];\mathbb{R}^{2d})$ and ψ on $\mathcal{C}([0,T];\mathbb{R}^{2d})$, we have:

$$\begin{aligned}
\mathbb{E}\big[\varphi(\hat{\boldsymbol{X}}^{N,\vartheta_N}, \boldsymbol{W}^{\vartheta_N})\psi(\boldsymbol{W}^0, \mathfrak{M}^N)\big] &= \frac{1}{N}\sum_{i=1}^{N}\mathbb{E}\big[\varphi(\hat{\boldsymbol{X}}^{N,i}, \boldsymbol{W}^i)\psi(\boldsymbol{W}^0, \mathfrak{M}^N)\big] \\
&= \mathbb{E}\bigg[\bigg(\int_{\mathcal{C}([0,T];\mathbb{R}^{2d}} \varphi(\boldsymbol{x}, \boldsymbol{w})d\mathfrak{M}^N(\boldsymbol{x}, \boldsymbol{w})\bigg)\psi(\boldsymbol{W}^0, \mathfrak{M}^N)\bigg].
\end{aligned}$$

Passing to the limit, we get:

$$\mathbb{E}\big[\varphi(\boldsymbol{X}^{\infty}, \boldsymbol{W}^{\infty})\psi(\boldsymbol{W}^0, \mathfrak{M}^{\infty})\big] = \mathbb{E}\bigg[\bigg(\int_{\mathcal{C}([0,T];\mathbb{R}^{2d})} \varphi(\boldsymbol{x}, \boldsymbol{w})d\mathfrak{M}^{\infty}(\boldsymbol{x}, \boldsymbol{w})\bigg)\psi(\boldsymbol{W}^{0,\infty}, \mathfrak{M}^{\infty})\bigg],$$

which is enough to complete the proof following the first part of the proof of Proposition 3.12.

The second claim follows from the argument used in the last step of the proof of Proposition 3.12. □

By the previous lemmas, the limiting setting is admissible in the sense that it has the required elements for the optimization problem in the Definition 2.16 of a weak MFG equilibrium. Letting $\mu_t^{\infty} = \mathfrak{M}^{\infty} \circ (e_t^x)^{-1}$ for all $t \in [0,T]$ where e_t^x denotes the mapping evaluating the first d coordinates at time t on $\mathcal{C}([0,T];\mathbb{R}^{2d})$, this optimization problem consists in minimizing:

$$J^{\infty}(\boldsymbol{\beta}^{\infty}) = \mathbb{E}^{\infty}\bigg[\int_0^T f(t, U_t^{\infty}, \mu_t^{\infty}, \beta_t^{\infty})dt + g(U_T^{\infty}, \mu_T^{\infty})\bigg],$$

over the square-integrable $\mathbb{F}^{\infty}$-progressively measurable processes $\boldsymbol{\beta}^{\infty}$ with values in A, where $\boldsymbol{U}^{\infty} = (U_t^{\infty})_{0\le t\le T}$ solves:

$$dU_t^{\infty} = b(t)\beta_t^{\infty}dt + \sigma dW_t^{\infty} + \sigma^0 dW_t^{0,\infty}, \quad t \in [0,T]\,; \quad U_0^{\infty} = \xi^{\infty}.$$

Thanks to (**A2**), this limiting optimization problem has a unique optimal solution, see Theorems 1.57 and 1.60.

What we still need to do in order to complete the proof of Theorem 6.18 is to prove that the optimal solution is in fact $\hat{\boldsymbol{X}}^{\infty}$ itself. Once this is done, the end of the proof of Theorem 6.18 goes along the same lines as the conclusion of the proof of Theorem 3.13, as long as we can transfer the tuple $(\xi^{\infty}, \boldsymbol{W}^{0,\infty}, \mathfrak{M}^{\infty}, \boldsymbol{W}^{\infty})$ onto the corresponding canonical space.

So, here is the last step of the proof of Theorem 6.18.

Lemma 6.25 *There exists a square-integrable and $\mathbb{F}^{\infty}$-progressively measurable process $\hat{\boldsymbol{\alpha}}^{0,\infty} = (\hat{\alpha}_t^{0,\infty})_{0\le t\le T}$ with values in A such that:*

$$d\hat{X}_t^{\infty} = b(t)\hat{\alpha}_t^{0,\infty}dt + \sigma dW_t^{\infty} + \sigma^0 dW_t^{0,\infty}, \quad t \in [0,T]\,; \quad \hat{X}_0^{\infty} = \xi^{\infty},$$

and, for any other square-integrable $\mathbb{F}^{\infty}$-progressively measurable process $\boldsymbol{\beta}^{\infty}$ with values in A, it holds that:

$$J^{\infty}(\hat{\boldsymbol{\alpha}}^{0,\infty}) \le J^{\infty}(\boldsymbol{\beta}^{\infty}).$$

Proof. The proof goes along the same lines as Lemma 3.15. We just provide a sketch.

First Step. We first return to the N-player game. For an admissible control strategy $\boldsymbol{\beta}^{(N)} = (\boldsymbol{\beta}^{N,1}, \cdots, \boldsymbol{\beta}^{N,N})$, we consider:

$$dU_t^{N,i} = b(t)\beta_t^{N,i}dt + \sigma dW_t^i + \sigma^0 dW_t^0, \quad t \in [0,T]\,; \quad U_0^i = \xi^i.$$

By definition of an open loop Nash equilibrium, we have, for any $i \in \{1, \cdots, N\}$,

$$\bar{J}^{N,i}(\boldsymbol{\beta}^{N,i}, (\hat{\boldsymbol{\alpha}}^{(N)})^{-i}) \ge J^{N,i}(\hat{\boldsymbol{\alpha}}^{(N)}),$$

where:

$$J^{N,i}(\boldsymbol{\beta}^{N,i}, (\hat{\boldsymbol{\alpha}}^{(N)})^{-i}) = \mathbb{E}\biggl[\int_0^T f(t, U_t^{N,i}, \bar{\mu}_t^{N,i}, \beta_t^{N,i})dt\biggr].$$

Letting:

$$\bar{\mu}_t^N = \frac{1}{N}\sum_{j=1}^N \delta_{\hat{X}_t^{N,j}}, \quad t \in [0,T],$$

observe that, for all $t \in [0,T]$,

$$W_2(\bar{\mu}_t^{N,i}, \bar{\mu}_t^N)^2 \le \frac{1}{N(N-1)} \sum_{j=1, j\neq i}^N \mathbb{E}[|\hat{X}_t^{N,i} - \hat{X}_t^{N,j}|^2] \le \frac{C}{N},$$

where we used Lemma 6.19 to derive the last inequality. Next, we fix a constant $K \ge 0$, and using the local Lipschitz property of the cost functions, we derive the following bound. If

$$\sup_{1\le i\le N} \mathbb{E}\int_0^T |\beta_t^{N,i}|^2 dt \le K,$$

then,

$$J^{N,i}(\boldsymbol{\beta}^{N,i}, (\hat{\boldsymbol{\alpha}}^{(N)})^{-i}) \leq \mathbb{E}\Big[\int_0^T f(t, U_t^{N,i}, \bar{\mu}_t^N, \beta_t^{N,i})dt + g(U_T^{N,i}, \bar{\mu}_T^N)\Big] + C_K N^{-1/2},$$

where C_K is a constant depending on K but not on N. Performing a similar computation when $\boldsymbol{\beta}^{N,i} = \hat{\boldsymbol{\alpha}}^{N,i}$, we deduce that:

$$\begin{aligned}
&\mathbb{E}\Big[\int_0^T f(t, U_t^{N,i}, \bar{\mu}_t^N, \beta_t^{N,i})dt + g(U_T^{N,i}, \bar{\mu}_T^N)\Big] \\
&\qquad \geq \mathbb{E}\Big[\int_0^T f(t, \hat{X}_t^{N,i}, \bar{\mu}_t^N, \hat{\alpha}_t^{N,i})dt + g(\hat{X}_T^{N,i}, \bar{\mu}_T^N)\Big] - C_K N^{-1/2},
\end{aligned}$$

for a possibly new value of C_K. In particular,

$$\begin{aligned}
&\mathbb{E}\Big[\int_0^T f(t, U_t^{N,\vartheta_N}, \bar{\mu}_t^N, \beta_t^{N,\vartheta_N})dt + g(U_T^{N,\vartheta_N}, \bar{\mu}_T^N)\Big] \\
&\qquad \geq \mathbb{E}\Big[\int_0^T f(t, \hat{X}_t^{N,\vartheta_N}, \bar{\mu}_t^N, \hat{\alpha}_t^{N,\vartheta_N})dt + g(\hat{X}_T^{N,\vartheta_N}, \bar{\mu}_T^N)\Big] - C_K N^{-1/2}.
\end{aligned} \tag{6.88}$$

Second Step. We now make explicit the dynamics of $\hat{\boldsymbol{X}}^{\infty}$. Noticing that:

$$\hat{X}_t^{N,\vartheta_N} = \xi^{\vartheta_N} + \int_0^t b(s)\hat{\alpha}_s^{N,\vartheta_N} ds + \sigma W_t^{\vartheta_N} + \sigma^0 W_t^0, \quad t \in [0,T],$$

we easily get that, under $\mathbb{P}^{\infty}$,

$$\hat{X}_t^{\infty} = \xi^{\infty} + \int_0^t b(s)\hat{\alpha}_s^{\infty} ds + \sigma W_t^{\infty} + \sigma^0 W_t^{0,\infty}, \quad t \in [0,T].$$

Following the proof of Lemma 3.15, we call $(\hat{\alpha}_t^{o,\infty})_{0\leq t\leq T}$ the optional projection of $(\hat{\alpha}_t^{\infty})_{0\leq t\leq T}$ given the filtration $\mathbb{F}^{\infty}$. We then have:

$$\hat{X}_t^{\infty} = \xi^{\infty} + \int_0^t b(s)\hat{\alpha}_s^{o,\infty} ds + \sigma W_t^{\infty} + \sigma^0 W_t^{0,\infty}, \quad t \in [0,T].$$

Third Step. Returning to the first step, we choose, for any $N \geq 2$ and any $i \in \{1, \cdots, N\}$, $\boldsymbol{\beta}^{N,i}$ of the form:

$$\beta_t^{N,i} = \sum_{\ell=0}^{n-1} \mathbf{1}_{[t_\ell^n, t_{\ell+1}^n)}(t)\Phi\Big(t_\ell^n, \xi_0^i, \boldsymbol{W}^0_{\cdot\wedge t_\ell^n}, \mathfrak{M}^N_{\cdot\wedge t_\ell^n}, \boldsymbol{W}^i_{\cdot\wedge t_\ell^n}, \hat{\boldsymbol{X}}^{N,i}_{\cdot\wedge t_\ell^n}\Big), \quad t \in [0,T],$$

where $(\Phi(t_\ell^n, \cdot))_{\ell=0,\cdots,n-1}$ is a collection of bounded and continuous functions, defined on an appropriate domain, and $0 = t_0^n < t_1^n < \cdots < t_n^n = T$ is a mesh whose step size tends to 0 as n tends to ∞. For such a choice,

$$\beta_t^{N,\vartheta_N} = \sum_{\ell=0}^{n-1} \mathbf{1}_{[t_\ell^n, t_{\ell+1}^n)}(t) \Phi\Big(t_\ell^n, \xi_0^{\vartheta_N}, \boldsymbol{W}^0_{\cdot\wedge t_\ell^n}, \mathfrak{M}^N_{\cdot\wedge t_\ell^n}, \boldsymbol{W}^{\vartheta_N}_{\cdot\wedge t_\ell^n}, \hat{\boldsymbol{X}}^{N,\vartheta_N}_{\cdot\wedge t_\ell^n}\Big), \quad t \in [0,T].$$

Duplicating the proof of Lemma 3.15 and using (6.88), we deduce that:

$$J^\infty(\hat{\boldsymbol{\alpha}}^\infty) \le J^\infty(\boldsymbol{\beta}^\infty), \tag{6.89}$$

where:

$$\beta_t^{\infty} = \sum_{\ell=0}^{n-1} \mathbf{1}_{[t_\ell^n, t_{\ell+1}^n)}(t) \Phi\Big(t_\ell^n, X_0^\infty, \boldsymbol{W}^{0,\infty}_{\cdot\wedge t_\ell^n}, \mathfrak{M}^\infty_{\cdot\wedge t_\ell^n}, \boldsymbol{W}^{\infty}_{\cdot\wedge t_\ell^n}, \hat{\boldsymbol{X}}^{\infty}_{\cdot\wedge t_\ell^n}\Big), \quad t \in [0,T]. \tag{6.90}$$

Indeed,

$$\begin{aligned}
&\mathbb{E}\Big[\int_0^T f(t, U_t^\infty, \bar{\mu}_t^\infty, \beta_t^\infty) dt + g(U_T^\infty, \bar{\mu}_T^\infty)\Big] \\
&\quad \ge \mathbb{E}\Big[\int_0^T f(t, X_t^\infty, \bar{\mu}_t^\infty, \alpha_t^\infty) dt + g(X_T^\infty, \bar{\mu}_T^\infty)\Big].
\end{aligned}$$

Observe that the notation in (6.89) is slightly abusive since $\hat{\boldsymbol{\alpha}}^\infty$ may not be $\mathbb{F}^\infty$-measurable. However, by the same convexity argument as in the proof of Lemma 3.15, we get:

$$J^\infty(\hat{\boldsymbol{\alpha}}^{o,\infty}) \le J^\infty(\boldsymbol{\beta}^\infty).$$

By approximating any general admissible control strategy $\boldsymbol{\beta}^\infty$ by processes of the form (6.90), we complete the proof. □

6.3 Limits of Markovian N-Player Equilibria

The purpose of this section is to prove the convergence of Markovian Nash equilibria when the master equation has a smooth solution.

Requiring that the master equation has a smooth solution in order to pass to the limit over Markovian equilibria should not come as a surprise. As explained in Chapter (Vol I)-2, Nash equilibria over Markovian strategies may be characterized by means of the Nash system of nonlinear partial differential equations introduced in (Vol I)-(2.17). We shall restate this system whenever needed. Recall that intuitively, the master equation captures the limiting form of the Nash system when the number of players tends to infinity, as long as the interactions remain of the mean field nature used throughout the book. As emphasized in the introduction of Section 6.2, the main difficulty in handling the asymptotic behavior of the Nash system is the lack of relevant compactness estimates for its solutions, which are defined on spaces of increasingly large dimensions as the number of players grows.

As a result, the goal of the strategy we use in this section is to bypass any use of compactness. In order to do so, we rely on the smoothness properties of the solution of the master equation. Instead of proving that the solution of the Nash system for the N-player game is almost a solution of the master equation, as we could be tempted to do, we shall prove that the solution of the master equation is almost a solution of the Nash system. As one can easily imagine, this approach will require that the solution of the master equation is smooth.

Still, we must keep in mind the sobering counter-example presented in Chapter (Vol I)-7. Indeed, the latter clearly shows that smoothness of the solution of the master equation cannot suffice to guarantee convergence of the Nash equilibria. As exemplified in Chapter (Vol I)-7, there are indeed cases in which the master equation has a smooth solution and still, some Nash equilibria of the N-player games do not converge to the solution of the mean field game. Basically, the counter example constructed in Chapter (Vol I)-7 relies heavily on the fact that, in this specific model, the Nash strategies are of bang-bang type. Obviously, the bang-bang nature of the equilibrium strategies introduces singularities which disrupt the limiting process from finite to infinite games. In the analysis provided below, we completely avoid this kind of phenomenon by making sure that the model is diffusive and the set of actions is continuous.

6.3.1 Main Statement

Throughout the analysis, we work under assumption **MFG Master Classical HJB** introduced in Subsection 5.4.3. As a result, recall Theorem 5.49, the master equation has a unique smooth solution. Importantly, the master field $\mathcal{U}$ and its derivatives of order one and two are bounded and Lipschitz continuous in the space and measure arguments. Notice that this fact would not be true under the conditions of Theorem 5.46. This is part of the rationale for requiring **MFG Master Classical HJB** to be in force, and preferring assumption **MFG with a Common Noise HJB** over assumption **MFG with a Common Noise SMP Relaxed**. See Subsections 3.4.1 and 3.4.3. Another reason for using **MFG Master Classical HJB** is the fact that we shall use consequences of the non-degeneracy assumption in the proof.

Finally, observe that, in contrast with the compactness approach adopted in the previous section on open loop equilibria, we implicitly require the mean field equilibria to be unique since we make use of the master field.

Introducing the N-Nash System

Although the analysis below is possible under the sole assumption that the master equation has a classical solution with bounded derivatives, we found more intuitive to work in the framework of Chapter 5, and in particular, under assumption **MFG Master Classical HJB**. This makes the whole presentation more coherent, and it streamlines the notation. In particular, σ and σ^0 are assumed to be constant, b reduces to a linear function of α, and f has the separated structure:

$$f(t,x,\mu,\alpha) = f_0(t,x,\mu) + f_1(t,x,\alpha),$$

f_0 and g satisfying the Lasry-Lions monotonicity condition. In this framework, the minimizer $\hat{\alpha}(t, x, \mu, y)$ of the Hamiltonian is independent of μ; we shall denote it by $\hat{\alpha}(t, x, y)$.

So the framework is the same as in Subsection 6.1.2 and in particular, the dynamics of the states of the players are subject to a common noise. Accordingly, for any $N \geq 2$, the N-player game has the form:

$$dU_t^{N,i} = b(t)\beta_t^{N,i}dt + \sigma dW_t^i + \sigma^0 dW_t^0, \quad 0 \leq t \leq T, \tag{6.91}$$

for any $i \in \{1, \cdots, N\}$, with $U_0^{N,i} = \xi^i$ as initial condition, where $\xi^1, \cdots, \xi^N$ are independent, identically distributed random variables with common distribution $\mu_0 \in \mathcal{P}_2(\mathbb{R}^d)$, and $\boldsymbol{W}^0, \boldsymbol{W}^1, \cdots, \boldsymbol{W}^N$ are $(N+1)$ independent Wiener processes with values in $\mathbb{R}^d$. Also, the tuples $(\xi^1, \cdots, \xi^N)$ and $(\boldsymbol{W}^0, \cdots, \boldsymbol{W}^N)$ are independent. For each $i \in \{1, \cdots, N\}$, the cost to player i reads:

$$\begin{aligned}
&J^{N,i}(\boldsymbol{\beta}^{N,1}, \cdots, \boldsymbol{\beta}^{N,N}) \\
&= \mathbb{E}\biggl[\int_0^T \bigl(f_0(t, U_t^{N,i}, \bar{\nu}_t^{N,i}) + f_1(t, U_t^{N,i}, \beta_t^{N,i})\bigr)dt + g\bigl(U_T^{N,i}, \bar{\nu}_T^{N,i}\bigr)\biggr],
\end{aligned} \tag{6.92}$$

with the notation:

$$\bar{\nu}_t^{N,i} = \frac{1}{N-1}\sum_{j=1, j\neq i}^{N} \delta_{U_t^{N,j}}.$$

The limiting mean field game can be described according to (6.36)–(6.37)–(6.38).

The thrust of the proof is to test the solution $\mathcal{U}$ of the master equation (4.41) deriving from the mean field game (6.36)–(6.37)–(6.38) as an approximate solution of the aforementioned N-Nash system. Here, the master equation takes the following form. Recall (5.118) from Chapter 5.

$$\begin{aligned}
&\partial_t \mathcal{U}(t, x, \mu) + \bigl(b(t)\hat{\alpha}(t, x, \partial_x \mathcal{U}(t, x, \mu))\bigr) \cdot \partial_x \mathcal{U}(t, x, \mu) \\
&\quad + \int_{\mathbb{R}^d} \bigl(b(t)\hat{\alpha}(t, v, \partial_x \mathcal{U}(t, v, \mu))\bigr) \cdot \partial_\mu \mathcal{U}(t, x, \mu)(v) d\mu(v) \\
&\quad + \frac{1}{2}\mathrm{trace}\Bigl[\bigl(\sigma\sigma^\dagger + \sigma^0(\sigma^0)^\dagger\bigr)\partial_{xx}^2 \mathcal{U}(t, x, \mu)\Bigr] \\
&\quad + \frac{1}{2}\int_{\mathbb{R}^d} \mathrm{trace}\Bigl[\bigl(\sigma\sigma^\dagger + \sigma^0(\sigma^0)^\dagger\bigr)\partial_v \partial_\mu \mathcal{U}(t, x, \mu)(v)\Bigr] d\mu(v) \\
&\quad + \frac{1}{2}\int_{\mathbb{R}^{2d}} \mathrm{trace}\Bigl[\sigma^0(\sigma^0)^\dagger \partial_\mu^2 \mathcal{U}(t, x, \mu)(v, v')\Bigr] d\mu(v) d\mu(v') \\
&\quad + \int_{\mathbb{R}^d} \mathrm{trace}\Bigl[\sigma^0(\sigma^0)^\dagger \partial_x \partial_\mu \mathcal{U}(t, x, \mu)(v)\Bigr] d\mu(v) \\
&\quad + f_0(t, x, \mu) + f_1\bigl(t, x, \hat{\alpha}(t, x, \partial_x \mathcal{U}(t, x, \mu))\bigr) = 0,
\end{aligned} \tag{6.93}$$

for $(t, x, \mu) \in [0, T] \times \mathbb{R}^d \times \mathcal{P}_2(\mathbb{R}^d)$, with $\mathcal{U}(T, \cdot, \cdot) = g$ as terminal condition. Moreover, we know from (Vol I)-(2.17), that the N-player Nash system associated with (6.91)–(6.92) reads:

$$\begin{aligned} &\partial_t v^{N,i}(t, \boldsymbol{x}) + \sum_{j=1}^{N} \big(b(t)\hat{\alpha}\big(t, x^j, D_{x^j} v^{N,j}(t, \boldsymbol{x})\big)\big) \cdot D_{x^j} v^{N,i}(t, \boldsymbol{x}) \\ &\quad + \frac{1}{2}\sum_{j=1}^{N} \mathrm{trace}\big[\sigma\sigma^\dagger D^2_{x^j x^j} v^{N,i}(t, \boldsymbol{x})\big] \\ &\quad + \frac{1}{2}\sum_{j,k=1}^{N} \mathrm{trace}\big[\sigma^0(\sigma^0)^\dagger D^2_{x^j x^k} v^{N,i}(t, \boldsymbol{x})\big] \\ &\quad + f_0(t, x^i, \bar{\mu}^{N-1}_{\boldsymbol{x}^{-i}}) + f_1\big(t, x^i, \hat{\alpha}\big(t, x^i, D_{x^i} v^{N,i}(t, \boldsymbol{x})\big)\big) = 0, \end{aligned} \tag{6.94}$$

for $(t, \boldsymbol{x}) \in [0, T] \times (\mathbb{R}^d)^N$, with the terminal condition:

$$v^{N,i}(T, \boldsymbol{x}) = g(x^i, \bar{\mu}^{N-1}_{\boldsymbol{x}^{-i}}), \quad \boldsymbol{x} \in (\mathbb{R}^d)^N. \tag{6.95}$$

Recall that for $i \in \{1, \cdots, N\}$, $v^{N,i}$ denotes the value function of player i, namely the expected cost to player i when all the players use the equilibrium strategies given by the feedback functions:

$$\Big([0, T] \times (\mathbb{R}^d)^N \ni (t, \boldsymbol{x}) \mapsto \hat{\alpha}^j\Big(t, x^j, D_{x^j} v^{N,j}(t, \boldsymbol{x})\Big)\Big)_{1 \le j \le N}.$$

Here, we used the boldface notation $\boldsymbol{x} = (x^1, \cdots, x^N)$ to denote elements of $(\mathbb{R}^d)^N$. This will allow us to distinguish the generic notation $\partial_{\boldsymbol{x}}$ for partial derivatives in $(\mathbb{R}^d)^N$ from the notation ∂_x for partial derivatives in $\mathbb{R}^d$. For such an $\boldsymbol{x} \in (\mathbb{R}^d)^N$, the empirical measure $\bar{\mu}^{N-1}_{\boldsymbol{x}^{-i}}$ is the uniform distribution over the finite set $\{x^1, \cdots, x^i, x^{i+1}, \cdots, x^N\}$:

$$\bar{\mu}^{N-1}_{\boldsymbol{x}^{-i}} = \frac{1}{N-1} \sum_{j=1, j \neq i}^{N} \delta_{x^j}.$$

In (6.94), D_{x^k} denotes the d-dimensional partial derivative in the direction x^k and $D^2_{x^j x^k}$ denotes the $d \times d$-dimensional partial second order derivative in the directions x^k and x^j. This notation makes it possible to distinguish the d and $d \times d$-dimensional partial derivatives D_{x^k} and $D^2_{x^j x^k}$ from the partial derivatives ∂_{x_k} and $\partial^2_{x_j x_k}$ in the one-dimensional directions x_j and x_k, when $x \in \mathbb{R}^d$.

The rationale for the strategy that consists in comparing (6.93) and (6.94) should be clear. Indeed, we introduced the master field $\mathcal{U}$ in Chapter 4 as the optimal expected cost for a representative player in the mean field game when the population is in equilibrium. This sounds like the asymptotic counterpart of the value functions $(v^{N,i})_{1\le i\le N}$, which appear as we just explained, as the collection of equilibrium costs to the players $1,\cdots,N$ in the N-player game.

Solving the N-Nash System

Before proceeding with the comparison of $\mathcal{U}$ and v^N, we must check that the N-Nash system (6.94)–(6.95) has a solution and that this solution describes the Markovian Nash equilibria of the game. Implicitly, this requires to prove that the N-Nash system is uniquely solvable, that the game has a unique equilibrium for any given initial condition, and that the unique equilibrium can be characterized through the unique solution of the Nash system.

We prove below that all these claims stand under assumption **MFG Master Classical HJB**. As announced earlier, we rewrite the N-Nash system (6.94)–(6.95) in the *canonical* form (Vol I)-(2.17)-(2.18) used in Chapter (Vol I)-2, which we recall under the new labels (6.96)–(6.97):

$$
\begin{aligned}
&\partial_t v^{N,i}(t,\boldsymbol{x}) + \frac{1}{2}\text{trace}\big[\Sigma(t,\boldsymbol{x})\Sigma(t,\boldsymbol{x})^\dagger \partial^2_{\boldsymbol{xx}} v^{N,i}(t,\boldsymbol{x})\big] \\
&\qquad + H^i\big(t,\boldsymbol{x},\partial_{\boldsymbol{x}} v^{N,i}(t,\boldsymbol{x}),\hat{\alpha}^{(N)}(t,\boldsymbol{x},\partial_{\boldsymbol{x}} v^N(t,\boldsymbol{x}))\big) = 0,
\end{aligned} \tag{6.96}
$$

for $(t,\boldsymbol{x}) \in [0,T]\times\mathbb{R}^{Nd}$ and $i\in\{1,\ldots,N\}$, with the terminal condition:

$$
v^{N,i}(T,\boldsymbol{x}) = g^i(\boldsymbol{x}), \quad \boldsymbol{x}\in\mathbb{R}^{Nd}, \quad i\in\{1,\ldots,N\}. \tag{6.97}
$$

In order to go from (6.94)–(6.95) to (6.96)–(6.97), we rewrite (6.91) as a single SDE driven by a generalized drift B with values in $\mathbb{R}^{Nd}$ and a generalized volatility matrix Σ taking values in $\mathbb{R}^{Nd}\times\mathbb{R}^{(N+1)d}$:

$$
\begin{aligned}
&B_i(t,\boldsymbol{x},\boldsymbol{\alpha}) = b(t)\alpha^i, \quad i\in\{1,\cdots,N\}, \\
&\Sigma_{i,j}(t,\boldsymbol{x}) = \begin{cases} \sigma & \text{if } i=j\neq N+1, \\ \sigma^0 & \text{if } j=N+1, \\ 0 & \text{otherwise,} \end{cases}
\end{aligned} \tag{6.98}
$$

where B_i denotes the block of index i in the decomposition of B in N blocks of size d, and $\Sigma_{i,j}$ denotes the block of index (i,j) in the decomposition of Σ as an $N\times(N+1)$ matrix of blocks of size $d\times d$. Accordingly, the noise in (6.91) is regarded as a $d(N+1)$-dimensional Wiener process $\boldsymbol{W}=(W_t)_{0\le t\le T}=(W_t^1,\cdots,W_t^N,W_t^0)_{0\le t\le T}$. Above, $\boldsymbol{\alpha}=(\alpha^1,\cdots,\alpha^N)$ is an element of $A^N=(\mathbb{R}^k)^N$. Below, we shall just write $B(t,\boldsymbol{\alpha})$ for $B(t,\boldsymbol{x},\boldsymbol{\alpha})$ and Σ for $\Sigma(t,\boldsymbol{x})$.

Similarly, we can recover the notations used in Chapter (Vol I)-2 for the cost functionals by letting, for all $i\in\{1,\cdots,N\}$, $\boldsymbol{x}\in(\mathbb{R}^d)^N\cong\mathbb{R}^{Nd}$ and $\alpha\in\mathbb{R}^k$:

$$
f^i(t,\boldsymbol{x},\alpha) = f(t,x^i,\bar{\mu}^{N-1}_{\boldsymbol{x}^{-i}},\alpha), \quad g^i(\boldsymbol{x}) = g(x^i,\bar{\mu}^{N-1}_{\boldsymbol{x}^{-i}}). \tag{6.99}
$$

Using these definitions for B, Σ, $(f^i)_{i=1,\cdots,N}$ and $(g^i)_{i=1,\cdots,N}$, we can easily pass from the formulation (6.94)–(6.95) to (6.96)–(6.97) and vice versa by letting:

$$H^i(t,\boldsymbol{x},\boldsymbol{y},\boldsymbol{\alpha}) = B(t,\boldsymbol{x},\boldsymbol{\alpha}) \cdot \boldsymbol{y} + f^i(t,\boldsymbol{x},\alpha^i), \quad \boldsymbol{x},\boldsymbol{y} \in (\mathbb{R}^d)^N, \ \boldsymbol{\alpha} \in (\mathbb{R}^k)^N,$$

$$\hat{\alpha}^{(N)}(t,\boldsymbol{x},\boldsymbol{y}) = \big(\hat{\alpha}(t,x^i,y^{i,i})\big)_{1\leq i\leq N}, \ \boldsymbol{x} \in (\mathbb{R}^d)^N, \ \boldsymbol{y}=(y^{i,j})_{1\leq i,j\leq N} \in (\mathbb{R}^d)^{N^2} \cong (\mathbb{R}^{Nd})^N.$$

In the definition of H^i, the inner product acts on elements of $(\mathbb{R}^d)^N$.

Unfortunately, Proposition (Vol I)-2.13, as established in Chapter (Vol I)-2 to guarantee the existence and uniqueness of a solution to the Nash system do not apply to (6.96)–(6.97) since A is now assumed to be the entire $\mathbb{R}^k$ while $A^{(N)}$ is required to be bounded in the statement of Proposition (Vol I)-2.13. Still, we can prove a similar result.

Proposition 6.26 *Under assumption* **MFG classical HJB**, *the Nash system* (6.94)–(6.95) *has a unique solution* $v^N = (v^{N,1}, \cdots, v^{N,N})$ *in the space of* $\mathbb{R}^N$*-valued bounded and continuous functions on* $[0,T] \times \mathbb{R}^{Nd}$ *that are differentiable in* $\boldsymbol{x} \in \mathbb{R}^{Nd}$ *with a bounded and continuous gradient on* $[0,T) \times \mathbb{R}^{Nd}$, *and that have generalized first-order derivative in* $t \in [0,T]$ *and second-order derivatives in* $\boldsymbol{x} \in \mathbb{R}^{Nd}$ *belonging to* $L^p_{\text{loc}}([0,T) \times \mathbb{R}^{Nd})$, *for any* $p \geq 1$.

Moreover, the tuple $(\phi^{*1}, \cdots, \phi^{*N})$ *given by:*

$$\phi^{*i}(t,\boldsymbol{x}) = \hat{\alpha}\big(t,x^i,D_{x^i}v^{N,i}(t,\boldsymbol{x})\big), \quad (t,\boldsymbol{x}) \in [0,T] \times (\mathbb{R}^d)^N,$$

is a Markovian Nash equilibrium over bounded Markovian strategy profiles.

As usual, we use the subscript loc to indicate that integrability is only required on compact subsets of the domain.

Proof. When compared to Proposition (Vol I)-2.13, the new difficulty comes from the fact that the set A is not assumed to be bounded. As a result, the running cost $f_1(t,x^i,\hat{\alpha}(t,x^i,D_{x^i}v^{N,i}(t,\boldsymbol{x})))$ in (6.94) is no longer bounded but has quadratic growth in $D_{x^i}v^{N,i}(t,\boldsymbol{x})$. Obviously, this makes the existence of a solution much more challenging to establish. However, if there exists a solution with a bounded gradient, then the cost function f_1 can be assumed to be bounded and, subsequently, we can easily recover the framework of Proposition (Vol I)-2.13. In other words, provided that there is a solution in the same class of functions as in the statement, in which case this peculiar solution has a bounded gradient, the uniqueness of such a solution together with the fact that it induces a Nash equilibrium do follow from the same arguments as those used in Chapter (Vol I)-2 to prove Proposition (Vol I)-2.13. Therefore, the only thing we have to prove is the existence of a solution in the aforementioned class of functions.

In order to proceed, we make use of a truncation argument. We start from the same system as in (6.94), but with a truncated cost function. Namely, in the last line of the equation, we replace $f_1(t,x^i,\hat{\alpha}(t,x^i,D_{x^i}v^{N,i}(t,\boldsymbol{x})))$ by $f_1(t,x^i,\pi(\hat{\alpha}(t,x^i,D_{x^i}v^{N,i}(t,\boldsymbol{x}))))$, where π is a smooth function from $\mathbb{R}^k$ into itself, equal to the identity on the ball of center 0 and radius R and vanishing outside the ball of center 0 and of radius $2R$, for some $R > 0$.

The Lipschitz constant of π is assumed to be less than 2. For such an R, the resulting system (6.94) reads as a system of quasilinear uniformly parabolic equations with bounded coefficients and we can argue as in the proof of Proposition (Vol I)-2.13. See also the Notes & Complements at the end of the chapter. As a result, the system has a solution, still denoted by $v^N = (v^{N,1}, \cdots, v^{N,N})$ within the same class of functions as in the statement. To prove the existence of a solution to (6.94), it suffices to establish a bound for the gradient of v^N independently of π, and thus of R as well. Choosing R large enough, we then deduce that v^N satisfies the original version of (6.94).

The strategy of proof is reminiscent of the approach used in Chapters (Vol I)-4 and 1 to prove Theorems (Vol I)-4.45 and 1.57, as we shall appeal to the theory of quadratic BSDEs. See Chapter (Vol I)-4 for a refresher. However, the present argument is more involved. The reason is that the backward SDEs underpinning the system (6.94) of PDEs are multidimensional, while the backward SDEs used in the proofs of Theorems (Vol I)-4.45 and 1.57 are merely one-dimensional. The multidimensionality is inherent to the fact that we are now dealing with a game, while Theorems (Vol I)-4.45 and 1.57 are concerned with control problems.

The proof is carried out in three steps. The first one is to provide a Hölder estimate of v^N, independently of the value of R and of the details of π. The second step consists in a refined L^2 estimate for the derivatives of v^N. The L^∞ estimate for the derivatives is established in the last step. It is important to keep in mind that the value of N is fixed throughout the proof. In particular, the underlying constants may depend on N. Also, we use the convenient notation $\hat{\alpha}^\pi = \pi \circ \hat{\alpha}$. By Lemma 1.56, $|\hat{\alpha}^\pi(t,x,z)| \leq C(1+|z|)$ for $(t,x,z) \in [0,T] \times \mathbb{R}^d \times \mathbb{R}^d$ and for C independent of π and R.

First Step. As a preliminary remark, we observe that there exists a constant C, independent of π and R, such that:

$$\forall (t,\boldsymbol{x}) \in [0,T] \times (\mathbb{R}^d)^N, \quad |v^{N,1}(t,\boldsymbol{x})| \leq C. \tag{6.100}$$

The proof is as follows. Recall that, under our standing assumption, the gradient in space of each $v^{N,i}$, for $i = 1, \cdots, N$, is bounded (the constant possibly depending upon the details of π and R). As a result, we can consider the collection of diffusion processes:

$$\begin{aligned} dX_t^1 &= \Big[b(t)\hat{\alpha}\big(t, X_t^1, D_{x^1} v^{N,1}(t, X_t)\big) + \delta f^1\big(t, X_t, D_{x^1} v^{N,1}(t, X_t)\big)\Big]dt + \sigma dW_t^1 + \sigma^0 dW_t^0, \\ dX_t^i &= b(t)\hat{\alpha}\big(t, X_t^i, D_{x^i} v^{N,i}(t, X_t)\big)dt + \sigma dW_t^i + \sigma^0 dW_t^0, \quad i \neq 1, \end{aligned} \tag{6.101}$$

for $t \in [t_0, T]$, where $t_0 \in [0,T]$ is some initial time for the process $\boldsymbol{X} = (X_t = (X_t^1, \cdots, X_t^N))_{t_0 \leq t \leq T}$. Above, $\delta f^1 : [0,T] \times (\mathbb{R}^d)^N \times \mathbb{R}^d \to \mathbb{R}^d$ satisfies the following equality:

$$f^1\big(t, \boldsymbol{x}, \hat{\alpha}^\pi(t, x^1, z)\big) = f^1\big(t, \boldsymbol{x}, \hat{\alpha}^\pi(t, x^1, 0)\big) + \delta f^1(t, \boldsymbol{x}, z) \cdot z,$$

for all $(t, \boldsymbol{x} = (x^1, \cdots, x^N), z) \in [0,T] \times (\mathbb{R}^d)^N \times \mathbb{R}^d$. The construction of δf^1 can be achieved along the same lines as in the proofs of Theorem (Vol I)-4.45 and 1.57, namely:

$$\begin{aligned} &\big(\delta f^1(t, \boldsymbol{x}, z)\big)_l \\ &= \frac{f^1\big(t, \boldsymbol{x}, \hat{\alpha}^\pi\big(t, x^1, (0, \cdots, z_l \cdots, z_d)\big)\big) - f^1\big(t, \boldsymbol{x}, \hat{\alpha}^\pi\big(t, x^1, (0, \cdots, z_{l+1}, \cdots, z_d)\big)\big)}{z_l}, \end{aligned}$$

if $z_l \neq 0$ and 0 otherwise, for $l \in \{1, \cdots, d\}$. In particular, allowing the constant C to increase from line to line, as long as it remains independent of π, R and the initial condition of $\boldsymbol{X}$, we have the following bound:

$$\left|\delta f^1(t, \boldsymbol{x}, z)\right| \leq C\big(1 + |z|\big). \tag{6.102}$$

The unique solvability of the SDE (6.101) follows from an argument that we already used in Chapter (Vol I)-2. Indeed, we notice that the drift of the SDE is bounded while the noise is nondegenerate. This suffices to prove the existence and uniqueness of a strong solution, see if needed the references in the Notes & Complements at the end of the chapter. Letting:

$$Y_t^i = v^{N,i}(t, \boldsymbol{X}_t), \quad Z_t^{i,j} = D_{x^j} v^{N,i}(t, \boldsymbol{X}_t), \quad t \in [t_0, T], \quad (i,j) \in \{1, \cdots, N\}^2,$$

applying Itô's formula, and using the system of PDEs satisfied by the tuple of functions $(v^{N,1}, \cdots, v^{N,N})$, we get:

$$dY_t^1 = -f^1\big(t, \boldsymbol{X}_t, \hat{\alpha}^\pi(t, X_t^1, 0)\big)dt + \sum_{j=1}^N Z_t^{1,j}\big(\sigma dW_t^j + \sigma^0 dW_t^0\big), \tag{6.103}$$

for $t \in [t_0, T]$, with the terminal condition $Y_T^1 = g^1(\boldsymbol{X}_T)$. Under the standing assumption, we know that g^1 and $f^1(\cdot, \cdot, \hat{\alpha}^\pi(\cdot, \cdot, 0))$ can be bounded independently of π and R, since $\hat{\alpha}^\pi(\cdot, \cdot, 0)$ itself can be bounded independently of π and R. See assumption **MFG with a Common Noise HJB**. We easily derive (6.100) by initializing the diffusion process $\boldsymbol{X}$ at any time t in $[0, T]$ and at any position $\boldsymbol{x} \in (\mathbb{R}^d)^N$. Of course, a similar bound holds for $v^{N,i}$, for $i \neq 1$.

Actually, we can get more. Taking the square in the backward equation satisfied by $\boldsymbol{Y}^1$, we can find a constant C, independent of R, of the details of π and of the initial condition of the process $\boldsymbol{X}$, such that for any stopping time τ with values in $[t_0, T]$,

$$\mathbb{E}\bigg[\int_\tau^T |Z_s^{1,1}|^2 ds \,|\mathcal{F}_\tau\bigg] \leq C, \tag{6.104}$$

which says that the martingale $(\int_{t_0}^t Z_s^{1,1} dW_s^1)_{t_0 \leq t \leq T}$ is of Bounded Mean Oscillation (BMO), see Definition (Vol I)-4.17 if needed. Importantly, the square of the BMO norm is less than C and is thus bounded independently of R, π and the initial condition of $\boldsymbol{X}$.

For $i \neq 1$, we have:

$$\begin{aligned} dY_t^i &= -f^i\big(t, \boldsymbol{X}_t, \hat{\alpha}^\pi(t, X_t^i, Z_t^{i,i})\big)dt + Z_t^{i,1}\delta f^1\big(t, \boldsymbol{X}_t, \hat{\alpha}^\pi(t, X_t^1, Z_t^{1,1})\big)dt \\ &\quad + \sum_{j=1}^N Z_t^{i,j}\big(\sigma dW_t^j + \sigma^0 dW_t^0\big). \end{aligned}$$

For $\eta \in \mathbb{R}$, we expand $(\exp(\eta Y_t^i))_{t_0 \leq t \leq T}$ by Itô's formula. We get:

$$\begin{aligned} d\big[\exp(\eta Y_t^i)\big] &\geq \exp(\eta Y_t^i)\bigg[\frac{\eta^2}{2}\sum_{j=1}^N |\sigma^\dagger Z_t^{i,j}|^2 - C\eta\Big(1 + |Z_t^{i,i}|^2 + |Z_t^{i,1}|^2 + |Z_t^{1,1}|^2\Big)\bigg]dt \\ &\quad + \eta \exp(\eta Y_t^i)\sum_{j=1}^N Z_t^{i,j}\big(\sigma dW_t^j + \sigma^0 dW_t^0\big). \end{aligned}$$

Recalling that σ is invertible, we can choose η large enough, in terms of C and of the lowest eigenvalue of $\sigma\sigma^\dagger$ only, such that for a new constant c:

$$d\big[\exp(\eta Y_t^i)\big] \geq \exp(\eta Y_t^i)\Big(c\sum_{j=1}^N |Z_t^{i,j}|^2 - C\eta\exp(\eta Y_t^i)|Z_t^{1,1}|^2\Big)dt$$

$$+\,\eta\exp(\eta Y_t^i)\sum_{j=1}^N Z_t^{i,j}\big(\sigma dW_t^j + \sigma^0 dW_t^0\big),$$

where the constant c is independent of R, π and the initial condition of the process $\boldsymbol{X}$. Recalling (6.104) together with the fact that $(Y_t^i = v^{N,i}(t, X_t))_{t_0 \leq t \leq T}$ is bounded, we deduce that, for any stopping time τ with values in $[t_0, T]$:

$$\mathbb{E}\bigg[\int_\tau^T |Z_s^{i,i}|^2 ds\,|\mathcal{F}_\tau\bigg] \leq C.$$

Therefore, all the martingales $(\int_{t_0}^t Z_s^{i,i} dW_s^i)_{t_0 \leq t \leq T}$, for $i = 1, \cdots, N$, are BMO and, most importantly, their BMO norms can be bounded independently of R, π and the initial condition of $\boldsymbol{X}$.

We now return to (6.101). Letting:

$$B_t^1 = b(t)\hat{\alpha}\big(t, X_t^1, D_{x^1}v^{N,1}(t, X_t)\big) + \delta f^1\big(t, X_t, D_{x^1}v^{N,1}(t, X_t)\big),$$
$$B_t^i = b(t)\hat{\alpha}\big(t, X_t^i, D_{x^i}v^{N,i}(t, X_t)\big), \quad i \neq 2,$$

for $t \in [t_0, T]$, we observe from (6.102) that:

$$|B_t^i| \leq C\big(1 + |Z_t^{i,i}|\big),$$

from which we deduce that the martingale $(\sum_{i=1}^N \int_{t_0}^t B_s^i \cdot dW_s^i)_{t_0 \leq t \leq T}$ is also BMO and that its BMO norm is less than C, the constant C being allowed to increase from line to line as long as it remains independent of R, π and the initial condition of $\boldsymbol{X}$. The BMO property implies that the Girsanov density:

$$\mathcal{E}_t = \exp\bigg(-\sum_{i=1}^N \int_{t_0}^t B_s^i \cdot (\sigma^{-1} dW_s^i) - \frac{1}{2}\sum_{i=1}^N \int_{t_0}^t |\sigma^{-1\dagger} B_s^i|^2 ds\bigg), \quad t \in [t_0, T],$$

satisfies:

$$\mathbb{E}\big[(\mathcal{E}_T)^r\big] \leq C,$$

for an exponent $r > 1$ independent of R, π and the initial condition of $\boldsymbol{X}$. Letting $\mathbb{Q} = \mathcal{E}_T \cdot \mathbb{P}$, we deduce that, for any event $E \in \mathcal{F}$,

$$\mathbb{Q}(E) \leq C^{1/r}\mathbb{P}(E)^{r/(r-1)}, \quad \text{that is} \quad \mathbb{P}(E) \geq C^{-(r-1)/r^2}\mathbb{Q}(E)^{(r-1)/r}.$$

The parameters in the above estimate are independent of π, R and the initial conditions. We claim that the above estimate together with the representation formula (6.103) for $v^{N,1}$ and the fact that g^1, f^1 and $\hat{\alpha}(\cdot,\cdot,0)$ are bounded and smooth in x suffice to conclude that $v^{N,1}$ is Hölder continuous on $[0,T] \times (\mathbb{R}^d)^N$, namely:

$$|v^{N,1}(t,\boldsymbol{x}) - v^{N,1}(t',\boldsymbol{x}')| \leq C\big(|t'-t|^{\gamma/2} + |\boldsymbol{x}'-\boldsymbol{x}|^{\gamma}\big), \tag{6.105}$$

for all $(t,\boldsymbol{x}),(t',\boldsymbol{x}') \in [0,T] \times (\mathbb{R}^d)^N$, for some constant C as above and for some exponent $\gamma \in (0,1)$, γ being independent of π and R. Indeed, the above lower bound says that we can control from below the probability that the process $\boldsymbol{X}$ hits a given Borel subset in $\mathbb{R}^{Nd}$ in terms of the probability that a Brownian motion in $\mathbb{R}^{Nd}$ hits the same Borel subset. This turns out to be sufficient to duplicate the so-called Krylov and Safonov estimates for the Hölder regularity of the solutions of second-order parabolic PDEs with measurable coefficients. The proof is not completely trivial, but is in fact by now well known in the literature. In the Notes & Complements at the end of the chapter, we provide several references where the argument is explained in detail, including one in a similar quadratic setting. Of course, a similar argument holds for $v^{N,i}$ with $i \neq 1$.

Second Step. We now change the representation formula of the functions $(v^{N,1},\cdots,v^{N,N})$. Instead of defining $\boldsymbol{X}$ through (6.101), we just let:

$$dX_t^i = \sigma dW_t^i + \sigma^0 dW_t^0,$$

for $t \in [t_0,T]$, where $t_0 \in [0,T]$ is treated as an initial time. As above, we let $X_t = (X_t^1,\cdots,X_t^N)$ together with:

$$Y_t^i = v^{N,i}(t,X_t), \quad Z_t^{i,j} = D_{x^j} v^{N,i}(t,X_t), \quad t \in [t_0,T], \quad (i,j) \in \{1,\cdots,N\}^2.$$

Then,

$$dY_t^i = -F^i(t,X_t,Z_t)dt + \sum_{j=1}^{N} Z_t^{i,j}\big(\sigma dW_t^i + \sigma^0 dW_t^0\big),$$

for $t \in [t_0,T]$, with:

$$F^i(t,\boldsymbol{x},\boldsymbol{z}) = f^i\big(t,\boldsymbol{x},\hat{\alpha}^{\pi}(t,x^i,z^{i,i})\big) + \sum_{j=1}^{N} z^{i,j}\cdot\big(b(t)\hat{\alpha}(t,x_t^j,z^{j,j})\big),$$

for $t \in [0,T]$, $\boldsymbol{x} = (x^i)_{i=1,\cdots,N} \in (\mathbb{R}^d)^N$ and $\boldsymbol{z} = (z^{i,j})_{i,j=1,\cdots,N} \in (\mathbb{R}^d)^{N\times N}$. The only thing that really matters for the following arguments is that:

$$|F^i(t,\boldsymbol{x},\boldsymbol{z})| \leq C\big(1+|\boldsymbol{z}|^2\big),$$

where, as usual, the constant C is independent of π and R.

We claim that there exists a constant $\varrho > 0$, independent of R, π and the initial condition of $\boldsymbol{X}$, such that:

$$\mathbb{E}\biggl[\int_{t_0}^{\tau} \frac{|Z_s|^2}{(s-t_0)^{\beta}} ds\biggr] \le C_{\varrho},$$

where $\beta = \gamma/4$ and τ is the first hitting time:

$$\tau = \inf\bigl\{t \ge t_0 : |X_t - X_{t_0}| \ge \varrho\bigr\} \wedge \bigl(t_0 + \varrho^2\bigr) \wedge T, \tag{6.106}$$

the constant C_{ϱ} being also independent of π, R and the initial condition of the process $\boldsymbol{X}$. The proof works as follows. For a given $\varepsilon > 0$, we consider the process $(|Y_t - Y_{t_0}|^2/(\varepsilon + (t - t_0)^{\beta}))_{t_0 \le t \le \tau}$, where $\boldsymbol{Y} = (Y^1, \cdots, Y^N)$. By Itô's formula, we obtain:

$$\begin{aligned}\mathbb{E}\biggl[\int_{t_0}^{\tau} \frac{|Z_s|^2}{\varepsilon + (s-t_0)^{\beta}} ds\biggr] &\le \mathbb{E}\biggl[\frac{|Y_\tau - Y_{t_0}|^2}{\varepsilon + (\tau - t_0)^{\beta}}\biggr] + 2\mathbb{E}\biggl[\int_{t_0}^{\tau} \frac{(Y_s - Y_{t_0}) \cdot F(s, X_s, Z_s)}{\varepsilon + (s-t_0)^{\beta}} ds\biggr] \\ &\quad + \mathbb{E}\biggl[\int_{t_0}^{\tau} \frac{|Y_s - Y_{t_0}|^2}{(s-t_0)^{1-\beta}(\varepsilon + (s-t_0)^{\beta})^2} ds\biggr].\end{aligned}$$

Recalling the Hölder property (6.105) of $v^N = (v^{N,1}, \cdots, v^{N,N})$, we deduce that for all $t \in [t_0, \tau]$, $|Y_t - Y_{t_0}| \le C\varrho^{\gamma}$. Hence,

$$\begin{aligned}\mathbb{E}\biggl[\int_{t_0}^{\tau} \frac{|Z_s|^2}{\varepsilon + (s-t_0)^{\beta}} ds\biggr] &\le C\biggl(\mathbb{E}\biggl[\frac{(\tau - t_0)^{\gamma} + |X_\tau - X_{t_0}|^{2\gamma}}{\varepsilon + (\tau - t_0)^{\beta}}\biggr] \\ &\quad + \varrho^{\gamma}\mathbb{E}\biggl[\int_{t_0}^{\tau} \frac{1 + |Z_s|^2}{\varepsilon + (s-t_0)^{\beta}} ds\biggr] \\ &\quad + \mathbb{E}\biggl[\int_{t_0}^{\tau} \frac{(s-t_0)^{\gamma} + |X_s - X_{t_0}|^{2\gamma}}{(s-t_0)^{1-\beta}(\varepsilon + (s-t_0)^{\beta})^2} ds\biggr]\biggr).\end{aligned}$$

Since $\beta = \gamma/4$, we finally have:

$$\begin{aligned}\mathbb{E}\biggl[\int_{t_0}^{\tau} \frac{|Z_s|^2}{\varepsilon + (s-t_0)^{\beta}} ds\biggr] &\le C\biggl(1 + \mathbb{E}\biggl[\frac{|X_\tau - X_{t_0}|^2}{\varepsilon^{1/\gamma} + (\tau - t_0)^{1/4}}\biggr]^{\gamma} \\ &\quad + \varrho^{\gamma}\mathbb{E}\biggl[\int_{t_0}^{\tau} \frac{1 + |Z_s|^2}{\varepsilon + (s-t_0)^{\beta}} ds\biggr] \\ &\quad + \mathbb{E}\biggl[\int_{t_0}^{t_0+\varrho^2} \frac{1}{(s-t_0)^{1-\gamma/4}} \mathbb{E}\biggl[\frac{|X_{s\wedge\tau} - X_{t_0}|^2}{\varepsilon^{2/\gamma} + (s \wedge \tau - t_0)^{1/2}}\biggr]^{\gamma} ds\biggr]\biggr).\end{aligned}$$

Choosing ϱ small enough and letting ε tend to 0, we complete the proof by noticing from Itô's formula that:

$$\mathbb{E}\biggl[\frac{|X_{s\wedge\tau} - X_{t_0}|^2}{\varepsilon^{2/\gamma} + (s \wedge \tau - t_0)^{1/2}}\biggr] \le C,$$

with C independent of ε and of the initial condition of $\boldsymbol{X}$.

Third Step. We fix ϱ as above and we consider a smooth cut-off function $\eta : (\mathbb{R}^d)^N \to \mathbb{R}$ equal to 1 on the ball $B(\boldsymbol{x}^0, \varrho/2)$ of center $\boldsymbol{x}^0$ and of radius $\varrho/2$ and vanishing outside the ball $B(\boldsymbol{x}^0, \varrho)$ of center $\boldsymbol{x}^0$ and of radius ϱ, for some $\boldsymbol{x}^0 \in (\mathbb{R}^d)^N$. We then expand $(Y_t^i \eta(X_t))_{t_0 \le t \le \varsigma}$ by Itô's formula, where in analogy with τ in (6.106), we let $\varsigma = \inf\{t \ge t_0 : |X_t - X_{t_0}| \ge \varrho\} \wedge T$. We get:

$$\begin{aligned} Y_{t_0}^i \eta(X_{t_0}) &= \mathbb{E}\Big[g^i(X_\varsigma)\eta(X_\varsigma) + \int_{t_0}^{\varsigma} \Psi^i(s, X_s, Z_s) ds \,|\, \mathcal{F}_{t_0} \Big] \\ &= \mathbb{E}\Big[g^i(X_\varsigma)\eta(X_\varsigma) + \int_{t_0}^{T} \Psi^i(s, X_{s\wedge\varsigma}, Z_{s\wedge\varsigma}) ds \,|\, \mathcal{F}_{t_0} \Big], \end{aligned}$$

where $\Psi^i(t, \boldsymbol{x}, \boldsymbol{z}) = 0$ if $|\boldsymbol{x} - \boldsymbol{x}^0| \ge \varrho$ and $|\Psi^i(t, \boldsymbol{x}, \boldsymbol{z})| \le C(1 + |\boldsymbol{z}|^2)$, for $t \in [0, T]$, $\boldsymbol{x} \in (\mathbb{R}^d)^N$ and $\boldsymbol{z} \in (\mathbb{R}^d)^{N \times N}$. Above, we used the fact that, if $\varsigma < T$, then $|X_\varsigma - X_{t_0}| = \varrho$ and hence $Y_\varsigma^i \eta(X_\varsigma) = g^i(X_\varsigma)\eta(X_\varsigma) = 0$ and $\psi^i(s, X_\varsigma, Z_\varsigma) = 0$. Choosing $X_{t_0} = \boldsymbol{x}$ such that $|\boldsymbol{x} - \boldsymbol{x}^0| \le \varrho/2$, we get:

$$\begin{aligned} v^{N,i}(t_0, \boldsymbol{x}) &= \int_{B(\boldsymbol{x}^0,\varrho)} p_{T-t_0}(\boldsymbol{x}, \boldsymbol{y})(g^i\eta)(\boldsymbol{y}) d\boldsymbol{y} \\ &\quad + \int_{t_0}^{T} \int_{B(\boldsymbol{x}^0,\varrho)} p_{s-t_0}(\boldsymbol{x}, \boldsymbol{y}) \Psi^i\big(s, \boldsymbol{y}, \partial_{\boldsymbol{x}} v^N(s, \boldsymbol{y})\big) d\boldsymbol{y}\, ds, \end{aligned} \tag{6.107}$$

where $(p_t(\boldsymbol{x}, \boldsymbol{y}) = \mathbb{P}[X_{(t_0+t)\wedge\varsigma} \in d\boldsymbol{y} \,|\, X_{t_0} = \boldsymbol{x}])_{t>0, \boldsymbol{x}, \boldsymbol{y} \in B(\boldsymbol{x}^0,\varrho)}$ is the transition kernel of $\boldsymbol{X}_{(t_0+\cdot)\wedge\varsigma}$. Observe by time homogeneity of $\boldsymbol{X}$ that $p_t(\boldsymbol{x}, \boldsymbol{y})$ is independent of t_0. Recalling the standard estimate:

$$\begin{aligned} \Big| \int_{B(\boldsymbol{x}^0,\varrho)} \partial_{\boldsymbol{x}} p_t(\boldsymbol{x}^0, \boldsymbol{y}) \psi(\boldsymbol{y}) d\boldsymbol{y} \Big| &\le \frac{c}{\sqrt{t}} \Big(\int_{B(\boldsymbol{x}^0,\varrho)} p_t(\boldsymbol{x}^0, \boldsymbol{y}) |\psi(\boldsymbol{y})|^2 d\boldsymbol{y} \Big)^{1/2} \\ &= \frac{c}{\sqrt{t}} \mathbb{E}\big[|\psi(X_{(t+t_0)\wedge\varsigma})|^2 \,|\, X_{t_0} = \boldsymbol{x}^0 \big]^{1/2}, \end{aligned}$$

which holds for any bounded and measurable function $\psi : (\mathbb{R}^d)^N \to \mathbb{R}$ and for a constant c independent of t, $\boldsymbol{x}^0$ and ψ (but depending on ϱ). Observing that:

$$\int_{B(\boldsymbol{x}^0,\varrho)} \partial_{\boldsymbol{x}} p_t(\boldsymbol{x}^0, \boldsymbol{y}) d\boldsymbol{y} = 0,$$

we even have, for any constant $\kappa > 0$,

$$\begin{aligned} \Big| \int_{B(\boldsymbol{x}^0,\varrho)} \partial_{\boldsymbol{x}} p_t(\boldsymbol{x}^0, \boldsymbol{y}) \psi(\boldsymbol{y}) d\boldsymbol{y} \Big| &\le \frac{c}{\sqrt{t}} \Big| \int_{B(\boldsymbol{x}^0,\varrho)} p_t(\boldsymbol{x}^0, \boldsymbol{y}) \big(\psi(\boldsymbol{y}) - \kappa\big)^2 d\boldsymbol{y} \Big|^{1/2} \\ &= \frac{c}{\sqrt{t}} \mathbb{E}\big[|\psi(X_{(t+t_0)\wedge\varsigma}) - \kappa|^2 \,|\, X_{t_0} = \boldsymbol{x}^0 \big]^{1/2}. \end{aligned}$$

Therefore, by differentiating (6.107) with respect to $\boldsymbol{x}$ at $\boldsymbol{x} = \boldsymbol{x}^0$ and by using twice the above bound, once with $t = T - t_0$ and $\kappa = g^i(\boldsymbol{x}^0)\eta(\boldsymbol{x}^0)$ and once with $t = s - t_0$ and $\kappa = 0$, we deduce:

$$
\begin{aligned}
|\partial_{\boldsymbol{x}} v^{N,i}(t_0, \boldsymbol{x}^0)| \leq{} & \frac{c}{\sqrt{T - t_0}} \mathbb{E}\Big[\big| g^i(X_\varsigma)\eta(X_\varsigma) - g^i(\boldsymbol{x}^0)\eta(\boldsymbol{x}^0) \big|^2 \,|\, X_{t_0} = \boldsymbol{x}^0 \Big]^{1/2} \\
& + C \int_{t_0}^T \frac{1}{\sqrt{s - t_0}} \Big(1 + \mathbb{E}\Big[\mathbf{1}_{s<\varsigma} |Z_s|^4 \,|\, X_{t_0} = \boldsymbol{x}^0 \Big]^{1/2} \Big) ds,
\end{aligned}
$$

where, in the second line, we used the equality:

$$
\begin{aligned}
\big| \Psi^i(s, X_{s\wedge\varsigma}, Z_{s\wedge\varsigma}) \big| &= \big| \Psi^i\big(s, X_{s\wedge\varsigma}, \partial_{\boldsymbol{x}} v^N(s \wedge \varsigma, X_{s\wedge\varsigma})\big) \big| \\
&= \mathbf{1}_{s<\varsigma} \big| \Psi^i\big(s, X_s, \partial_{\boldsymbol{x}} v^N(s, X_s)\big) \big| \\
&= \mathbf{1}_{s<\varsigma} \big| \Psi^i\big(s, X_s, Z_s\big) \big| \leq C \mathbf{1}_{s<\varsigma} \big(1 + |Z_s|^2 \big).
\end{aligned}
$$

Since $\mathbb{E}[\sup_{t_0 \leq s \leq T} |X_s - X_{t_0}|^2] \leq C(T - t_0)$ and g^i is Lipschitz, we get:

$$
\begin{aligned}
& |\partial_{\boldsymbol{x}} v^{N,i}(t_0, \boldsymbol{x}^0)| \\
& \leq C \Big[1 + \int_{t_0}^T \frac{\sup_{\boldsymbol{y} \in (\mathbb{R}^d)^N} |\partial_{\boldsymbol{x}} v^N(s, \boldsymbol{y})|}{\sqrt{s - t_0}} \mathbb{E}[\mathbf{1}_{s<\varsigma} |Z_s|^2 \,|\, X_{t_0} = \boldsymbol{x}^0]^{1/2} ds \Big] \\
& \leq C \Big[1 + \Big(\int_{t_0}^T \frac{\sup_{\boldsymbol{y} \in (\mathbb{R}^d)^N} |\partial_{\boldsymbol{x}} v^N(s, \boldsymbol{y})|^2}{(s - t_0)^{1-\beta}} ds \Big)^{1/2} \Big(\int_{t_0}^T \frac{\mathbb{E}[\mathbf{1}_{s<\varsigma} |Z_s|^2 \,|\, X_{t_0} = \boldsymbol{x}^0]}{(s - t_0)^\beta} ds \Big)^{1/2} \Big],
\end{aligned}
$$

where we used the Cauchy Schwarz inequality to derive the last line. Assume for a while that $T - t_0 \leq \varrho^2$ for the same ϱ as in the conclusion of the second step. Then, the indicator function $\mathbf{1}_{s<\varsigma}$ in the above bound can be replaced by $\mathbf{1}_{s<\tau}$. Using the conclusion of the second step and then taking the square, we finally have:

$$
|\partial_{\boldsymbol{x}} v^{N,i}(t_0, \boldsymbol{x}^0)|^2 \leq C \Big[1 + \int_{t_0}^T \frac{\sup_{\boldsymbol{y} \in (\mathbb{R}^d)^N} |\partial_{\boldsymbol{x}} v^N(s, \boldsymbol{y})|^2}{(s - t_0)^{1-\beta}} ds \Big].
$$

Obviously, the same holds without the superscript i in the left-hand side. Recalling that the constant C is independent of $\boldsymbol{x}^0$ and t_0, we then conclude as in the proof of Proposition 5.53. Indeed, for a new value of C, we get for any $\epsilon > 0$:

$$
\begin{aligned}
\sup_{\boldsymbol{x} \in (\mathbb{R}^d)^N} |\partial_{\boldsymbol{x}} v^N(t_0, \boldsymbol{x})|^2 \leq{} & C + C\epsilon^\beta \sup_{t_0 \leq t \leq T} \sup_{\boldsymbol{x} \in (\mathbb{R}^d)^N} |\partial_{\boldsymbol{x}} v^N(t, \boldsymbol{x})|^2 \\
& + \frac{C}{\epsilon^{1-\beta}} \int_{t_0}^T \sup_{t \leq s \leq T} \sup_{\boldsymbol{y} \in (\mathbb{R}^d)^N} |\partial_{\boldsymbol{x}} v^N(s, \boldsymbol{y})|^2 dt.
\end{aligned}
$$

Notice that the right-hand side increases as t_0 decreases in $[T - \varrho^2, T]$. Therefore,

$$\sup_{t_0 \le t \le T} \sup_{x \in (\mathbb{R}^d)^N} |\partial_x v^N(t_0, x)|^2 \le C + C\epsilon^\beta \sup_{t_0 \le t \le T} \sup_{x \in (\mathbb{R}^d)^N} |\partial_x v^N(t, x)|^2$$
$$+ \frac{C}{\epsilon^{1-\beta}} \int_{t_0}^T \sup_{t \le s \le T} \sup_{y \in (\mathbb{R}^d)^N} |\partial_x v^N(s, y)|^2 dt.$$

Choosing ϵ such that $C\epsilon^\beta \le \frac{1}{2}$ and applying Gronwall's lemma, we get the required bound for $\partial_x v^N$ on $[T - \varrho^2, T] \times (\mathbb{R}^d)^N$. Recalling that the value of ϱ has been fixed once for all according to the conclusion of the second step, we can duplicate the argument on $[T-2\varrho^2, T-\varrho^2] \times (\mathbb{R}^d)^N$ by letting $v^N(T - \varrho^2, \cdot)$ instead of g play the role of the terminal condition. We complete the proof by iterating the argument a finite number of times. □

As announced, we deduce that the game has a unique Markovian Nash equilibrium with a bounded strategy.

Proposition 6.27 *Under assumption* **MFG Master Classical HJB**, *the Markovian Nash equilibrium provided by the N-Nash system* (6.94)–(6.95) *is the unique equilibrium over bounded Markovian strategies.*

Proof. The proof is rather technical. We refer to Chapter (Vol I)-2 for basic definitions on Markovian Nash equilibria that will be used below. Throughout the proof, we use extensively the notations defined in (6.98) and (6.99).

First Step. For a bounded Markovian Nash equilibrium $(\phi^{*1}, \cdots, \phi^{*N})$ and for a given $i \in \{1, \cdots, N\}$, we consider the optimization problem consisting in minimizing the cost functional:

$$J^{N,i}(\phi^i) = \mathbb{E}\bigg[\int_0^T f\big(t, X_t^{N,i}, \bar{\mu}_t^{N,i}, \phi^i(t, X_t^{(N)})\big)dt + g(X_T^{N,i}, \bar{\mu}_T^{N,i})\bigg],$$

over measurable functions $\phi^i : [0, T] \times \mathbb{R}^{Nd} \to \mathbb{R}^k$ and stochastic processes $X^{(N)} = (X_t^{N,1}, \cdots, X_t^{N,N})_{0 \le t \le T}$ solving systems of the form:

$$dX_t^{N,i} = b(t)\phi^i\big(t, X_t^{(N)}\big)dt + \sigma dW_t^i + \sigma^0 dW_t^0,$$
$$dX_t^{N,j} = b(t)\phi^{*j}\big(t, X_t^{(N)}\big)dt + \sigma dW_t^j + \sigma^0 dW_t^0, \quad j \ne i,$$

for $t \in [0, T]$, with $x = (x^1, \cdots, x^N) \in (\mathbb{R}^d)^N$ as initial condition. As usual, we denote by $\bar{\mu}_t^{N,i}$ the empirical distribution $(N-1)^{-1} \sum_{j \ne i} \delta_{X_t^{N,j}}$. This optimization problem is similar to (Vol I)-(2.14) in Chapter (Vol I)-2.

We recall that, under the standing assumption, the above system of SDEs is uniquely solvable although the feedback functions are not assumed to be Lipschitz continuous. Basically, it follows from the fact that the strategies are required to be bounded and the fact that the matrix $\Sigma\Sigma^\dagger$ is invertible. We refer to the Notes & Complements below for precise references.

Second Step. Following (Vol I)-(2.16), we may associate with the above optimization problem the following HJB equation:

$$\begin{aligned}&\partial_t U^i(t,\boldsymbol{x}) + \frac{1}{2}\mathrm{trace}\big[\Sigma\Sigma^\dagger \partial^2_{\boldsymbol{xx}} U^i(t,\boldsymbol{x})\big]\\ &\qquad + \inf_{\alpha\in\mathbb{R}^k} H^i\big(t,\boldsymbol{x},\partial_{\boldsymbol{x}} U^i(t,\boldsymbol{x}),(\alpha,\phi(t,\boldsymbol{x})^{*-i})\big) = 0,\end{aligned} \tag{6.108}$$

for $(t,\boldsymbol{x}) \in [0,T]\times\mathbb{R}^{Nd}$, with the terminal condition $U^i(T,\boldsymbol{x}) = g^i(\boldsymbol{x})$, where we see $\partial_{\boldsymbol{x}} U^i(t,\boldsymbol{x})$ as a tuple of N vectors of size d, namely:

$$\partial_{\boldsymbol{x}} U^i(t,\boldsymbol{x}) = \big(D_{x^1} U^i(t,\boldsymbol{x}),\cdots,D_{x^N} U^i(t,\boldsymbol{x})\big),$$

and we use the standard notation:

$$(\alpha,\phi(t,\boldsymbol{x})^{*-i}) = \big(\phi^{*1}(t,\boldsymbol{x}),\cdots,\phi^{*(i-1)}(t,\boldsymbol{x}),\alpha,\phi^{*(i+1)}(t,\boldsymbol{x}),\cdots,\phi^{*N}(t,\boldsymbol{x})\big),$$

together with:

$$H^i(t,\boldsymbol{x},\boldsymbol{y},\boldsymbol{\alpha}) = B(t,\boldsymbol{x},\boldsymbol{\alpha})\cdot \boldsymbol{y} + f^i(t,\boldsymbol{x},\alpha^i),$$

where the inner product acts on elements of $(\mathbb{R}^d)^N$. Hence, (6.108) may be rewritten as:

$$\begin{aligned}&\partial_t U^i(t,\boldsymbol{x}) + \frac{1}{2}\mathrm{trace}\big[\Sigma\Sigma^\dagger \partial^2_{\boldsymbol{xx}} U^i(t,\boldsymbol{x})\big] + \sum_{j\neq i} D_{x^j} U^i(t,\boldsymbol{x})\cdot\big(b(t)\phi^{*j}(t,\boldsymbol{x})\big)\\ &\qquad + \inf_{\alpha\in\mathbb{R}^k}\big[D_{x^i} U^i(t,x)\cdot\big(b(t)\alpha\big) + f^i(t,\boldsymbol{x},\alpha)\big] = 0.\end{aligned}$$

The key fact is that, similar to the Nash system, this equation is known to have a strong solution U^i in the space of bounded and continuous functions on $[0,T]\times(\mathbb{R}^d)^N$ that are differentiable in space on $[0,T)\times(\mathbb{R}^d)^N$, with a bounded and continuous gradient on $[0,T)\times(\mathbb{R}^d)^N$, and that have generalized time first-order and space second-order derivatives in $L^p_{\mathrm{loc}}([0,T)\times\mathbb{R}^d)$, for any $p\geq 1$. Although the solution is not classical, it is regular enough to apply, as done in the proof of Proposition (Vol I)-2.13, a generalized version of the chain rule due to Krylov along Itô processes that have a bounded drift and a bounded and uniformly nondegenerate diffusion coefficient. In particular, it permits to duplicate the standard verification argument in stochastic control theory, see Lemma (Vol I)-4.47, whose analogue in Volume II is Proposition 1.55. We deduce that the optimal control problem (6.108) has a unique optimal strategy, given by the feedback function:

$$[0,T]\times\mathbb{R}^{Nd} \ni (t,\boldsymbol{x}) \mapsto \hat{\alpha}\big(t,x^i,D_{x^i} U^i(t,\boldsymbol{x})\big).$$

Third Step. Now, we recall that, by definition, the optimal strategy is already known. It is ϕ^{*i}. We deduce that, for $(X_t^{*(N)} = (X_t^{*N,1},\cdots,X_t^{*N,N}))_{0\leq t\leq T}$ solving the system of SDEs:

$$dX_t^{*N,i} = B\big(t,\phi^{*i}(t,X_t^{*(N)})\big)dt + \Sigma dW_t,\quad t\in[0,T],\quad i\in\{1,\cdots,N\},$$

with $\boldsymbol{x}$ as initial condition, it holds:

$$\mathbb{E}\int_0^T \left|\phi^{*i}(t, X_t^{*(N)}) - \hat{\alpha}\big(t, X_t^{*N,i}, D_{x^i}U^i(t, X_t^{*(N)})\big)\right|^2 dt = 0. \tag{6.109}$$

As a consequence, we deduce that, for any $i \in \{1, \cdots, N\}$,

$$dX_t^{*N,i} = B\big(t, \hat{\alpha}\big(t, X_t^{*N,i}, D_{x^i}U^i(t, X_t^{*(N)})\big)\big)dt + \Sigma dW_t, \quad t \in [0, T],$$

for $i \in \{1, \cdots, N\}$. Recalling that B and Σ are bounded and that $(\Sigma\Sigma^\dagger)^{-1}$ is invertible, we deduce that, for any $t \in (0, T]$, the marginal law of X_t^* is absolutely continuous with respect to the Lebesgue measure and has a positive density. Hence, by (6.109), we also have, for almost every $(t, \boldsymbol{x}) \in [0, T] \times \mathbb{R}^{Nd}$, for all $i \in \{1, \cdots, N\}$,

$$\phi^{*i}(t, \boldsymbol{x}) = \hat{\alpha}\big(t, x^i, D_{x^i}U^i(t, \boldsymbol{x})\big).$$

Therefore, the N-tuple $(U^1, \cdots, U^N)$ is a generalized solution of the system of PDEs:

$$\begin{aligned}
&\partial_t U^i(t, \boldsymbol{x}) + \frac{1}{2}\mathrm{trace}\big[\Sigma\Sigma^\dagger \partial^2_{\boldsymbol{xx}}U^i(t, \boldsymbol{x})\big] + \sum_{j \neq i} D_{x^j}U^i(t, \boldsymbol{x}) \cdot \big(b(t)\hat{\alpha}\big(t, x^j, D_{x^j}U^j(t, \boldsymbol{x})\big)\big) \\
&\qquad + \inf_{\alpha \in \mathbb{R}^k}\big[D_{x^i}U^i(t, \boldsymbol{x}) \cdot \big(b(t)\alpha\big) + f(t, x^i, \bar{\mu}^{N-1}_{\boldsymbol{x}^{-i}}, \alpha)\big] = 0,
\end{aligned}$$

which coincides with the system (6.96)–(6.97).

By the uniqueness property in the statement of Proposition 6.26, we deduce that U^i and $v^{N,i}$ are equal for all $i \in \{1, \cdots, N\}$, where $(v^{N,i})_{i=1,\cdots,N}$ is the solution of the N-Nash system (6.94)–(6.95), as given by Proposition 6.26. Hence, the gradients of U^i and $v^{N,i}$ coincide. Therefore, for almost every $(t, \boldsymbol{x}) \in [0, T] \times \mathbb{R}^{Nd}$, for all $i \in \{1, \cdots, N\}$,

$$\phi^{*i}(t, \boldsymbol{x}) = \hat{\alpha}\big(t, x^i, D_{x^i}v^i(t, \boldsymbol{x})\big),$$

which completes the proof. □

Main Statement

Here is the main result regarding the convergence of the solution of the Nash system.

Theorem 6.28 *Under assumption* **MFG Master Classical HJB**, *there exists a constant C such that, for any $N \geq 1$, any $i \in \{1, \cdots, N\}$ and any $(t_0, \boldsymbol{x}) \in [0, T] \times \mathbb{R}^{Nd}$,*

$$\left|\mathcal{U}(t_0, x^i, \bar{\mu}^{N-1}_{\boldsymbol{x}^{-i}}) - v^{N,i}(t_0, \boldsymbol{x})\right| \leq \frac{C}{N}\Big(1 + |x^i|^2 + \frac{1}{N}\sum_{j=1}^N |x^j|^2\Big)^{1/2}.$$

Moreover, for any $\eta > 0$, there exists a constant $c_\eta > 0$ such that, for any $\mu_0 \in \mathcal{P}_2(\mathbb{R}^d)$ and any N-tuple $(\xi^1, \cdots, \xi^N)$ of independent and identically distributed random variables with μ_0 as distribution:

$$\mathbb{E}\big[|v^{N,i}(t, \boldsymbol{\xi}) - \mathcal{U}(t, \xi^i, \mu_0)|\big] \leq c_\eta M_2(\mu_0)^{1/2} N^{-1/\max(d,2+\eta)},$$

Remark 6.29 *Theorem 6.28 has a striking interpretation. Asymptotically, the value functions of the N-player game may be factorized as a single smooth function of the private state of the current player and of the empirical distribution of the others. In particular, the value functions take the same form as the cost coefficients f_0 and g.*

6.3.2 The Master Equation as an Almost Solution of the N-Nash System

Throughout the subsection, assumption **MFG Master Classical HJB** from Subsection 5.4.3 is in force. In particular, both the master equation (6.93) and the N-Nash system (6.94) are uniquely solvable. As in the previous subsection, the respective solutions are denoted by $\mathcal{U}$ and v^N.

Finite-Dimensional Projection of the Master Field

The main trick in our approach is to regard the tuple of functions

$$u^{N,i}(t, \boldsymbol{x}) = \mathcal{U}\big(t, x_i, \bar{\mu}^{N-1}_{\boldsymbol{x}^{-i}}\big), \quad t \in [0, T], \ \boldsymbol{x} \in (\mathbb{R}^d)^N, \tag{6.110}$$

for $i \in \{1, \cdots, N\}$, as a natural candidate for solving the Nash system (6.94) approximately.

The goal is indeed to prove that the "proxies" $(u^{N,i})_{i=1,\cdots,N}$ almost solve the system (6.94) up to a remainder term that vanishes as N tends to ∞. As a by-product, we shall deduce that the $(u^{N,i})_{i=1,\cdots,N}$ get closer and closer to the "true solutions" $(v^{N,i})_{i=1,\cdots,N}$ when N tends to ∞.

Importantly, we recall that $\mathcal{U}$ satisfies the conclusion of Theorems 5.46 and 5.49. In particular, $\mathcal{U}$ is in the class $\mathfrak{S}_1$ and $\partial_x \mathcal{U}$ is in the class $\mathfrak{S}_d$, see Definition 5.9. Following Proposition 4.13, see also Proposition (Vol I)-5.91, we get:

Proposition 6.30 *For any $N \geq 2$, $i \in \{1, \cdots, N\}$, $u^{N,i}$ is of class $\mathcal{C}^2$ in the space variables and satisfies, for all $\boldsymbol{x} \in \mathbb{R}^{Nd}$,*

$$\begin{aligned}
D_{x^i} u^{N,i}(t, \boldsymbol{x}) &= \partial_x \mathcal{U}(t, x^i, \bar{\mu}^{N-1}_{\boldsymbol{x}^{-i}}), \\
D^2_{x^i x^i} u^{N,i}(t, \boldsymbol{x}) &= \partial^2_x \mathcal{U}(t, x^i, \bar{\mu}^{N-1}_{\boldsymbol{x}^{-i}}), \\
D_{x^j} u^{N,i}(t, \boldsymbol{x}) &= \tfrac{1}{N-1} \partial_\mu \mathcal{U}(t, x^i, \bar{\mu}^{N-1}_{\boldsymbol{x}^{-i}})(x^j) \qquad \text{for } \ j \neq i, \\
D^2_{x^i x^j} u^{N,i}(t, \boldsymbol{x}) &= \tfrac{1}{N-1} \partial_x \partial_\mu \mathcal{U}(t, x^i, \bar{\mu}^{N-1}_{\boldsymbol{x}^{-i}})(x^j) \qquad \text{for } \ j \neq i,
\end{aligned}$$

$$D^2_{x^j x^j} u^{N,i}(t,\boldsymbol{x}) = \tfrac{1}{N-1} \partial_v \partial_\mu \mathcal{U}(t, x^i, \bar{\mu}^{N-1}_{\boldsymbol{x}^{-i}})(x^j) + \tfrac{1}{(N-1)^2} \partial^2_\mu \mathcal{U}(t, x^i, \bar{\mu}^{N-1}_{\boldsymbol{x}^{-i}})(x^j, x^j) \quad \text{for} \quad j \neq i,$$

$$D^2_{x^j x^k} u^{N,i}(t,\boldsymbol{x}) = \tfrac{1}{(N-1)^2} \partial^2_\mu \mathcal{U}(t, x_i, \bar{\mu}^{N-1}_{\boldsymbol{x}^{-i}})(x^k, x^j) \quad \text{for} \quad i, j, k \text{ distinct.}$$

We now show that $(u^{N,i})_{i\in\{1,\cdots,N\}}$ is "almost" a solution of the Nash system (6.94).

Proposition 6.31 *There exist a constant $C \geq 0$ and, for any $N \geq 2$, a collection of functions $(r^{N,i})_{i=1,\cdots,N}$ in $\mathcal{C}([0,T] \times \mathbb{R}^{Nd}; \mathbb{R})$ such that:*

$$\forall (t,\boldsymbol{x}) \in [0,T] \times \mathbb{R}^{Nd}, \quad |r^{N,i}(t,\boldsymbol{x})| \leq \frac{C}{N}\Big(1 + \frac{1}{N}\sum_{j=1}^{N} |x^i - x^j|\Big),$$

and

$$\begin{aligned} &\partial_t u^{N,i}(t,\boldsymbol{x}) + \sum_{j=1}^{N} \big(b(t)\hat{\alpha}\big(t, x^j, D_{x^j} u^{N,j}(t,\boldsymbol{x})\big)\big) \cdot D_{x^j} u^{N,i}(t,\boldsymbol{x}) \\ &+ \frac{1}{2}\sum_{j=1}^{N} \text{trace}\big[\sigma\sigma^\dagger D^2_{x^j x^j} u^{N,i}(t,\boldsymbol{x})\big] \\ &+ \frac{1}{2}\sum_{j,k=1}^{N} \text{trace}\big[\sigma^0(\sigma^0)^\dagger D^2_{x^j x^k} u^{N,i}(t,\boldsymbol{x})\big] \\ &+ f_0(t, x^i, \bar{\mu}^{N-1}_{\boldsymbol{x}^{-i}}) + f_1\big(t, x^i, \hat{\alpha}\big(t, x^i, D_{x^i} u^{N,i}(t,\boldsymbol{x})\big)\big) = r^{N,i}(t,\boldsymbol{x}), \end{aligned} \tag{6.111}$$

for $(t,\boldsymbol{x}) \in [0,T] \times (\mathbb{R}^d)^N$, with the terminal condition:

$$u^{N,i}(T,\boldsymbol{x}) = g(x^i, \bar{\mu}^{N-1}_{\boldsymbol{x}^{-i}}), \quad \boldsymbol{x} \in (\mathbb{R}^d)^N. \tag{6.112}$$

Proof. We emphasize that, throughout the proof, we use the Lipschitz property in the measure argument of $\mathcal{U}$ and its derivatives with respect to the 1-Wasserstein distance W_1. This is in contrast with most of the arguments developed so far, in which we used the 2-Wasserstein distance W_2.

First Step. Making use of the master equation (6.93) with (t,x,μ) therein of the form $(t,x,\mu) = (t, x^i, \bar{\mu}^{N-1}_{\boldsymbol{x}^{-i}})$ for some $\boldsymbol{x} \in \mathbb{R}^{Nd}$ and some $i \in \{1,\cdots,N\}$, we obtain:

$$
\begin{aligned}
&\partial_t \mathcal{U}(t,x^i,\bar{\mu}^{N-1}_{\boldsymbol{x}^{-i}}) + \big(b(t)\hat{\alpha}(t,x^i,\partial_x\mathcal{U}(t,x^i,\bar{\mu}^{N-1}_{\boldsymbol{x}^{-i}}))\big)\cdot \partial_x \mathcal{U}(t,x^i,\bar{\mu}^{N-1}_{\boldsymbol{x}^{-i}})\\
&\quad + \int_{\mathbb{R}^d} \big(b(t)\hat{\alpha}(t,v,\partial_x\mathcal{U}(t,v,\bar{\mu}^{N-1}_{\boldsymbol{x}^{-i}}))\big)\cdot \partial_\mu \mathcal{U}(t,x^i,\bar{\mu}^{N-1}_{\boldsymbol{x}^{-i}})(v)d\bar{\mu}^{N-1}_{\boldsymbol{x}^{-i}}(v)\\
&\quad + \frac{1}{2}\mathrm{trace}\Big[\big(\sigma\sigma^\dagger + \sigma^0(\sigma^0)^\dagger\big)\partial^2_{x}\mathcal{U}(t,x^i,\bar{\mu}^{N-1}_{\boldsymbol{x}^{-i}})\Big]\\
&\quad + \frac{1}{2}\int_{\mathbb{R}^d} \mathrm{trace}\Big[\big(\sigma\sigma^\dagger + \sigma^0(\sigma^0)^\dagger\big)\partial_v\partial_\mu \mathcal{U}(t,x^i,\bar{\mu}^{N-1}_{\boldsymbol{x}^{-i}})(v)\Big]d\bar{\mu}^{N-1}_{\boldsymbol{x}^{-i}}(v)\\
&\quad + \frac{1}{2}\int_{\mathbb{R}^{2d}} \mathrm{trace}\Big[\sigma^0(\sigma^0)^\dagger\partial^2_\mu \mathcal{U}(s,x^i,\bar{\mu}^{N-1}_{\boldsymbol{x}^{-i}})(v,v')\Big]d\bar{\mu}^{N-1}_{\boldsymbol{x}^{-i}}(v)d\bar{\mu}^{N-1}_{\boldsymbol{x}^{-i}}(v')\\
&\quad + \int_{\mathbb{R}^d} \mathrm{trace}\Big[\sigma^0(\sigma^0)^\dagger\partial_x\partial_\mu \mathcal{U}(t,x^i,\bar{\mu}^{N-1}_{\boldsymbol{x}^{-i}})(v)\Big]d\bar{\mu}^{N-1}_{\boldsymbol{x}^{-i}}(v)\\
&\quad + f_0(t,x^i,\bar{\mu}^{N-1}_{\boldsymbol{x}^{-i}}) + f_1\big(t,x^i,\hat{\alpha}\big(t,x^i,\partial_x\mathcal{U}(t,x^i,\bar{\mu}^{N-1}_{\boldsymbol{x}^{-i}})\big)\big) = 0.
\end{aligned}
$$

Recalling the first two lines in the statement of Proposition 6.30, we deduce that $u^{N,i}$ satisfies:

$$
\begin{aligned}
&\partial_t u^{N,i}(t,\boldsymbol{x}) + \big(b(t)\hat{\alpha}(t,x^i,D_{x^i}u^{N,i}(t,\boldsymbol{x}))\big)\cdot D_{x^i}u^{N,i}(t,\boldsymbol{x})\\
&\quad + \int_{\mathbb{R}^d} \big(b(t)\hat{\alpha}\big(t,v,\partial_x\mathcal{U}(t,v,\bar{\mu}^{N-1}_{\boldsymbol{x}^{-i}})\big)\big)\cdot \partial_\mu \mathcal{U}(t,x^i,\bar{\mu}^{N-1}_{\boldsymbol{x}^{-i}})(v)d\bar{\mu}^{N-1}_{\boldsymbol{x}^{-i}}(v)\\
&\quad + \frac{1}{2}\mathrm{trace}\Big[\big(\sigma\sigma^\dagger + \sigma^0(\sigma^0)^\dagger\big)D^2_{x^ix^i}u^{N,i}(t,\boldsymbol{x})\Big]\\
&\quad + \frac{1}{2}\int_{\mathbb{R}^d} \mathrm{trace}\Big[\big(\sigma\sigma^\dagger + \sigma^0(\sigma^0)^\dagger\big)\partial_v\partial_\mu \mathcal{U}(t,x^i,\bar{\mu}^{N-1}_{\boldsymbol{x}^{-i}})(v)\Big]d\bar{\mu}^{N-1}_{\boldsymbol{x}^{-i}}(v)\\
&\quad + \frac{1}{2}\int_{\mathbb{R}^{2d}} \mathrm{trace}\Big[\sigma^0(\sigma^0)^\dagger\partial^2_\mu \mathcal{U}(s,x^i,\bar{\mu}^{N-1}_{\boldsymbol{x}^{-i}})(v,v')\Big]d\bar{\mu}^{N-1}_{\boldsymbol{x}^{-i}}(v)d\bar{\mu}^{N-1}_{\boldsymbol{x}^{-i}}(v')\\
&\quad + \int_{\mathbb{R}^d} \mathrm{trace}\Big[\sigma^0(\sigma^0)^\dagger\partial_x\partial_\mu \mathcal{U}(t,x^i,\bar{\mu}^{N-1}_{\boldsymbol{x}^{-i}})(v)\Big]d\bar{\mu}^{N-1}_{\boldsymbol{x}^{-i}}(v)\\
&\quad + f_0(t,x^i,\bar{\mu}^{N-1}_{\boldsymbol{x}^{-i}}) + f_1\big(t,x^i,\hat{\alpha}\big(t,x^i,D_{x^i}u^{N,i}(t,\boldsymbol{x})\big)\big) = 0.
\end{aligned}
\tag{6.113}
$$

Observe that we replaced $\partial_x\mathcal{U}(t,x^i,\bar{\mu}^{N-1}_{\boldsymbol{x}^{-i}})$ by $D_{x^i}u^{N,i}(t,\boldsymbol{x})$ in the first and last lines, while we replaced $\partial^2_x\mathcal{U}(t,x^i,\bar{\mu}^{N-1}_{\boldsymbol{x}^{-i}})$ by $D^2_{x^ix^i}u^{N,i}(t,\boldsymbol{x})$ in the third line.

Second Step. Now,

$$
\begin{aligned}
&\int_{\mathbb{R}^d} \big(b(t)\hat{\alpha}\big(t,v,\partial_x\mathcal{U}(t,v,\bar{\mu}^{N-1}_{\boldsymbol{x}^{-i}})\big)\big)\cdot \partial_\mu \mathcal{U}(t,x^i,\bar{\mu}^{N-1}_{\boldsymbol{x}^{-i}})(v)d\bar{\mu}^{N-1}_{\boldsymbol{x}^{-i}}(v)\\
&= \frac{1}{N-1}\sum_{j=1, j\neq i}^{N} \big(b(t)\hat{\alpha}\big(t,x^j,\partial_x\mathcal{U}(t,x^j,\bar{\mu}^{N-1}_{\boldsymbol{x}^{-i}})\big)\big)\cdot \partial_\mu \mathcal{U}(t,x^i,\bar{\mu}^{N-1}_{\boldsymbol{x}^{-i}})(x^j).
\end{aligned}
$$

We now make use of the fact that $\partial_x\mathcal{U}$ is Lipschitz continuous with respect to the measure argument with respect to the 1-Wasserstein distance, see the conclusion of Theorem 5.46. Observing that, for $j \neq i$,

$$
W_1(\bar{\mu}^{N-1}_{\boldsymbol{x}^{-i}}, \bar{\mu}^{N-1}_{\boldsymbol{x}^{-j}}) \leq \frac{1}{N-1}|x^i - x^j|,
$$

we deduce that:

$$\begin{aligned}
&\int_{\mathbb{R}^d} \big(b(t)\hat{\alpha}\big(t, v, \partial_x \mathcal{U}(t, v, \bar{\mu}^{N-1}_{x^{-i}})\big)\big) \cdot \partial_\mu \mathcal{U}(t, x^i, \bar{\mu}^{N-1}_{x^{-i}})(v) d\bar{\mu}^{N-1}_{x^{-i}}(v) \\
&= \frac{1}{N-1} \sum_{j=1, j\neq i}^{N} \big(b(t)\hat{\alpha}\big(t, x^j, \partial_x \mathcal{U}(t, x^j, \bar{\mu}^{N-1}_{x^{-j}})\big)\big) \cdot \partial_\mu \mathcal{U}(t, x^i, \bar{\mu}^{N-1}_{x^{-i}})(x^j) \\
&\qquad + O\Big(\frac{1}{(N-1)^2} \sum_{j=1, j\neq i}^{N} |x^i - x^j|\Big) \\
&= \sum_{j=1, j\neq i}^{N} \big(b(t)\hat{\alpha}\big(t, x^j, D_{x^j} u^{N,j}(t, \boldsymbol{x})\big)\big) \cdot D_{x^j} u^{N,i}(t, \boldsymbol{x}) + O\Big(\frac{1}{N^2} \sum_{j=1}^{N} |x^i - x^j|\Big),
\end{aligned}$$

where we used the third equality in the statement of Proposition 6.30 together with the fact that $\partial_x \mathcal{U}$ is Lipschitz with respect to the measure argument to replace $\bar{\mu}^{N-1}_{x^{-i}}$ by $\bar{\mu}^{N-1}_{x^{-j}}$ in the second line. We also used the fact that $\partial_\mu \mathcal{U}$ is bounded, see Theorem 5.49. Here and below, the Landau symbol is considered as uniform in $(t, \boldsymbol{x}) \in [0, T] \times \mathbb{R}^{Nd}$: for a sequence of functions $(y_N)_{N \geq 1}$ of the variables t and $\boldsymbol{x}$, $O(y_N(t, \boldsymbol{x}))$ is a function of $(t, \boldsymbol{x})$ such that $|O(y_N(t, \boldsymbol{x}))| \leq C|y_N(t, \boldsymbol{x})|$ for a constant C independent of N and $(t, \boldsymbol{x})$.

In particular, summing with the second term on the first line of (6.113), we get:

$$\begin{aligned}
&\big(b(t)\hat{\alpha}\big(t, x^i, \bar{\mu}^{N-1}_{x^{-i}}, D_{x^i} u^{N,i}(t, \boldsymbol{x})\big)\big) \cdot D_{x^i} u^{N,i}(t, \boldsymbol{x}) \\
&\quad + \int_{\mathbb{R}^d} \big(b(t)\hat{\alpha}\big(t, v, \partial_x \mathcal{U}(t, v, \bar{\mu}^{N-1}_{x^{-i}})\big)\big) \cdot \partial_\mu \mathcal{U}(t, x^i, \bar{\mu}^{N-1}_{x^{-i}})(v) d\bar{\mu}^{N-1}_{x^{-i}}(v) \\
&= \sum_{j=1}^{N} \big(b(t)\hat{\alpha}\big(t, x^j, D_{x^j} u^{N,j}(t, \boldsymbol{x})\big)\big) \cdot D_{x^j} u^{N,i}(t, \boldsymbol{x}) + O\Big(\frac{1}{N^2} \sum_{j=1}^{N} |x^i - x^j|\Big).
\end{aligned}$$

Third Step. We now return to (6.113) and handle the term on the fourth line therein. It reads:

$$\begin{aligned}
&\frac{1}{2} \int_{\mathbb{R}^d} \text{trace}\Big[\big(\sigma\sigma^\dagger + \sigma^0(\sigma^0)^\dagger\big)\partial_v \partial_\mu \mathcal{U}(t, x^i, \bar{\mu}^{N-1}_{x^{-i}})(v)\Big] d\bar{\mu}^{N-1}_{x^{-i}}(v) \\
&= \frac{1}{2(N-1)} \sum_{j=1, j\neq i}^{N} \text{trace}\Big[\big(\sigma\sigma^\dagger + \sigma^0(\sigma^0)^\dagger\big)\partial_v \partial_\mu \mathcal{U}(t, x^i, \bar{\mu}^{N-1}_{x^{-i}})(x^j)\Big] \\
&= \frac{1}{2} \sum_{j=1, j\neq i}^{N} \text{trace}\Big[\big(\sigma\sigma^\dagger + \sigma^0(\sigma^0)^\dagger\big)\big(D^2_{x^j x^j} u^{N,i}(t, \boldsymbol{x}) \\
&\qquad\qquad - \frac{1}{(N-1)^2} \partial^2_\mu \mathcal{U}(t, x^i, \bar{\mu}^{N-1}_{x^{-i}})(x^j, x^j)\big)\Big] \\
&= \frac{1}{2} \sum_{j=1, j\neq i}^{N} \text{trace}\Big[\big(\sigma\sigma^\dagger + \sigma^0(\sigma^0)^\dagger\big) D^2_{x^j x^j} u^{N,i}(t, \boldsymbol{x})\Big] + O\big(\frac{1}{N}\big),
\end{aligned} \tag{6.114}$$

where we used the fact that $\partial_\mu^2 \mathcal{U}$ is bounded, see Theorem 5.49, together with the fifth line in the statement of Proposition 6.30. We handle the term on the fifth line in (6.113) in the same way:

$$
\begin{aligned}
&\frac{1}{2}\int_{\mathbb{R}^{2d}} \mathrm{trace}\Big[\sigma^0(\sigma^0)^\dagger \partial_\mu^2 \mathcal{U}(t,x^i,\bar{\mu}^{N-1}_{\boldsymbol{x}^{-i}})(v,v')\Big] d\bar{\mu}^{N-1}_{\boldsymbol{x}^{-i}}(v)\bar{\mu}^{N-1}_{\boldsymbol{x}^{-i}}(v') \\
&= \frac{1}{2(N-1)^2}\sum_{j,\ell=1, j,\ell\neq i}^{N} \mathrm{trace}\Big[\sigma^0(\sigma^0)^\dagger \partial_\mu^2 \mathcal{U}(t,x^i,\bar{\mu}^{N-1}_{\boldsymbol{x}^{-i}})(x^j,x^\ell)\Big] \\
&= \frac{1}{2}\sum_{j,\ell=1, j,\ell\neq i, j\neq \ell}^{N} \mathrm{trace}\Big[\sigma^0(\sigma^0)^\dagger D^2_{x^jx^\ell} u^{N,i}(t,\boldsymbol{x})\Big] + O(\frac{1}{N}),
\end{aligned}
\tag{6.115}
$$

where we used the sixth line in the statement of Proposition 6.30 together with the fact that $\sigma^0(\sigma^0)^\dagger$ is symmetric to pass from the second to the third line. Similarly, thanks to the fourth line in the statement of Proposition 6.30, we get:

$$
\begin{aligned}
&\int_{\mathbb{R}^d} \mathrm{trace}\Big[\sigma^0(\sigma^0)^\dagger \partial_x\partial_\mu \mathcal{U}(t,x^i,\bar{\mu}^{N-1}_{\boldsymbol{x}^{-i}})(v)\Big] d\bar{\mu}^{N-1}_{\boldsymbol{x}^{-i}}(v) \\
&= \frac{1}{N-1}\sum_{j=1, j\neq i}^{N} \mathrm{trace}\Big[\sigma^0(\sigma^0)^\dagger \partial_x\partial_\mu \mathcal{U}(t,x^i,\bar{\mu}^{N-1}_{\boldsymbol{x}^{-i}})(x^j)\Big] \\
&= \sum_{j=1, j\neq i}^{N} \mathrm{trace}\Big[\sigma^0(\sigma^0)^\dagger D^2_{x^ix^j} u^{N,i}(t,\boldsymbol{x})\Big],
\end{aligned}
\tag{6.116}
$$

while, as we already explained,

$$
\begin{aligned}
&\frac{1}{2}\mathrm{trace}\Big[\big(\sigma\sigma^\dagger + \sigma^0(\sigma^0)^\dagger\big)\partial^2_{xx}\mathcal{U}(t,x^i,\bar{\mu}^{N-1}_{\boldsymbol{x}^{-i}})\Big] \\
&= \frac{1}{2}\mathrm{trace}\Big[\big(\sigma\sigma^\dagger + \sigma^0(\sigma^0)^\dagger\big)D^2_{x^ix^i} u^{N,i}(t,x^i,\bar{\mu}^{N-1}_{\boldsymbol{x}^{-i}})\Big].
\end{aligned}
\tag{6.117}
$$

Up to remaining terms, the sum of (6.114), (6.115), (6.116), and (6.117) is equal to:

$$
\begin{aligned}
&\frac{1}{2}\sum_{j=1, j\neq i}^{N} \mathrm{trace}\Big[\big(\sigma\sigma^\dagger + \sigma^0(\sigma^0)^\dagger\big)D^2_{x^jx^j} u^{N,i}(t,\boldsymbol{x})\Big] \\
&+ \frac{1}{2}\sum_{j,\ell=1, j,\ell\neq i, j\neq \ell}^{N} \mathrm{trace}\Big[\sigma^0(\sigma^0)^\dagger D^2_{x^jx^\ell} u^{N,i}(t,\boldsymbol{x})\Big] \\
&+ \sum_{j=1, j\neq i}^{N} \mathrm{trace}\Big[\sigma^0(\sigma^0)^\dagger D^2_{x^ix^j} u^{N,i}(t,\boldsymbol{x})\Big] \\
&+ \frac{1}{2}\mathrm{trace}\Big[\big(\sigma\sigma^\dagger + \sigma^0(\sigma^0)^\dagger\big)D^2_{x^ix^i} u^{N,i}(t,x^i,\bar{\mu}^{N-1}_{\boldsymbol{x}^{-i}})\Big],
\end{aligned}
$$

which is equal to:

$$\frac{1}{2}\sum_{j=1}^{N}\mathrm{trace}\Big[(\sigma\sigma^\dagger)D^2_{x^jx^j}u^{N,i}(t,\boldsymbol{x})\Big]+\frac{1}{2}\sum_{j,\ell=1}^{N}\mathrm{trace}\Big[\sigma^0(\sigma^0)^\dagger D^2_{x^jx^\ell}u^{N,i}(t,\boldsymbol{x})\Big].$$

Last Step. Collecting the conclusions of the second and third steps, we easily complete the proof. □

6.3.3 Proving the Convergence of the Nash System

We now turn to the proof of Theorem 6.28. For this, we assume that assumption **MFG Master Classical HJB** is in force and we consider the solution $(v^{N,i})_{i=1,\cdots,N}$ of the Nash system (6.94). By uniqueness of the solution, the $(v^{N,i})_{i=1,\cdots,N}$ must be symmetric. By symmetric, we mean that, for any $\boldsymbol{x}=(x^l)_{l\in\{1,\cdots,N\}}\in\mathbb{R}^{Nd}$ and for any indices $j\neq\ell$, if $\tilde{\boldsymbol{x}}=(\tilde{x}_l)_{l\in\{1,\cdots,N\}}$ is the N-tuple obtained from $\boldsymbol{x}$ by swapping the entries with indices j and ℓ (i.e., $\tilde{x}^l=x^l$ for $l\notin\{j,\ell\}$, $\tilde{x}^j=x^\ell$, $\tilde{x}^\ell=x^j$), then:

$$v^{N,i}(t,\tilde{\boldsymbol{x}})=v^{N,i}(t,\boldsymbol{x})\ \text{if}\ i\notin\{j,\ell\},\quad\text{while}\quad v^{N,i}(t,\tilde{\boldsymbol{x}})=v^{N,\ell}(t,\boldsymbol{x})\ \text{if}\ i=j,$$

which may be reformulated as follows. There exists a function $V^N:[0,T]\times\mathbb{R}^d\times(\mathbb{R}^d)^{N-1}\to\mathbb{R}$ such that, for any $(t,x)\in[0,T]\times\mathbb{R}^d$, the function $(\mathbb{R}^d)^{N-1}\ni(z^1,\cdots,z^{N-1})\mapsto V^N(t,x,(z^1,\cdots,z^{N-1}))$ is invariant under permutation, and for every $i\in\{1,\cdots,N\}$ and $\boldsymbol{x}\in(\mathbb{R}^d)^N$,

$$v^{N,i}(t,\boldsymbol{x})=V^N\big(t,x^i,(x^1,\cdots,x^{i-1},x^{i+1},\cdots,x^N)\big).$$

The above equality should be compared with the statement of Lemma (Vol I)-1.2. The latter says that the function $V^N(t,x,\cdot)$ should factorize, for N large, as a function of a probability measure, provided that it is Lipschitz with respect to the projection of the Wasserstein distance, the Lipschitz constant being uniform in N. As we already emphasized several times, the difficulty is precisely to prove the latter claim as we hardly know how to estimate the regularity of the functions $(v^{N,i})_{1\le i\le N}$ uniformly in N. The best we can prove below is that the functions $(v^{N,i})_{1\le i\le N}$ are bounded, uniformly in N.

Notice also that the functions $(u^{N,i})_{i\in\{1,\cdots,N\}}$ from (6.110) are also symmetric.

The proof of Theorem 6.28 consists in comparing $v^{N,i}$ and $u^{N,i}$ along the equilibrium trajectories of the N-player game, for any $i\in\{1,\cdots,N\}$. For this, let us fix $t_0\in[0,T)$, $\mu_0\in\mathcal{P}_2(\mathbb{R}^d)$ and let $(\xi^i)_{i=1,\cdots,N}$ be a family of N independent and identically distributed random variables of law μ_0. We set $\boldsymbol{\xi}=(\xi^i)_{i=1,\cdots,N}$. Let also $((W^i_t)_{0\le t\le T})_{i=0,\cdots,N}$ be a family of $(N+1)$ independent d-dimensional Brownian motions, independent of $(\xi^i)_{i=1,\cdots,N}$. We consider over the interval $[t_0,T]$, the systems of SDEs with variables $(X_t^{(N)} = (X_t^{N,i})_{i=1,\cdots,N})_{t_0\le t\le T}$ and $(X_t^{*(N)} = (X_t^{*N,i})_{i=1,\cdots,N})_{t_0\le t\le T}$:

$$\begin{aligned} dX_t^{N,i} &= b(t)\hat{\alpha}\big(t, X_t^{N,i}, D_{x^i}u^{N,i}(t, X_t^{(N)})\big)dt + \sigma dW_t^i + \sigma^0 dW_t^0, \\ X_{t_0}^i &= \xi^i, \end{aligned} \tag{6.118}$$

and

$$\begin{aligned} dX_t^{*N,i} &= b(t)\hat{\alpha}\big(t, X_t^{*N,i}, D_{x^i}v^{N,i}(t, X_t^{*(N)})\big)dt + \sigma dW_t^i + \sigma^0 dW_t^0, \\ X_{t_0}^{*N,i} &= \xi^i. \end{aligned} \tag{6.119}$$

By symmetry of the functions $(u^{N,i})_{i=1,\cdots,N}$, the processes $(\boldsymbol{X}^{N,i})_{i=1,\cdots,N}$ are exchangeable. The same holds for the processes $(\boldsymbol{X}^{*N,i})_{i=1,\cdots,N}$ and, actually, the N $\mathbb{R}^{2d}$-valued processes $(\boldsymbol{X}^{N,i}, \boldsymbol{X}^{*N,i})_{i=1,\cdots,N}$ are also exchangeable.

Theorem 6.32 *Under assumption* **MFG Master Classical HJB**, *there exists a constant C such that, for any $t_0 \in [0,T]$, $\mu_0 \in \mathcal{P}_2(\mathbb{R}^d)$ and $N \ge 1$, it holds, for any $i \in \{1,\cdots,N\}$,*

$$\mathbb{E}\big[\sup_{t_0\le t\le T} |X_t^{N,i} - X_t^{*N,i}|^2\big] \le \frac{C(1 + M_2(\mu_0)^2)}{N^2}, \tag{6.120}$$

$$\begin{aligned} \mathbb{E}\bigg[\sup_{t_0\le t\le T} \Big|u^{N,i}(t, X_t^{*(N)}) - v^{N,i}(t, X_t^{*(N)})\Big|^2 \\ + \int_{t_0}^T |D_{x^i}u^{N,i}(t, X_t^{*(N)}) - D_{x^i}v^{N,i}(t, X_t^{*(N)})|^2 dt\bigg] \\ \le \frac{C(1 + M_2(\mu_0)^2)}{N^2}, \end{aligned} \tag{6.121}$$

and, $\mathbb{P}$ almost surely, for all $i = 1,\cdots,N$,

$$\begin{aligned} &|u^{N,i}(t_0, \boldsymbol{\xi}) - v^{N,i}(t_0, \boldsymbol{\xi})| \\ &\le \frac{C}{N}\Big(1 + \frac{1}{N}\sum_{j=1}^N |\xi^i - \xi^j|^2 + \frac{1}{N^2}\sum_{\ell,j=1}^N |\xi^\ell - \xi^j|^2\Big)^{1/2}. \end{aligned} \tag{6.122}$$

Proof. The proof relies on a variant of the Cole-Hopf transformation used to handle the quadratic nature of the cost functional. As a preliminary remark, we observe, by adapting the first step in the proof of Proposition 6.26, that the functions $(v^{N,i})_{1\le i\le N}$ are bounded, uniformly in N.

For simplicity, we shall work with $t_0 = 0$. Throughout the proof, we denote by $\mathbb{F}$ the complete and right-continuous augmentation of the filtration generated by $(\xi^1,\cdots,\xi^N)$ and $(W^0,\cdots,W^N)$. We shall use the shorten notation $\mathbb{E}_0[\cdot] = \mathbb{E}[\cdot|\mathcal{F}_0]$.

First Step. We start with the proof of (6.121). Also, we introduce new notations:

$$U_t^{N,i} = u^{N,i}(t, X_t^{*(N)}), \quad V_t^{N,i} = v^{N,i}(t, X_t^{*(N)}),$$
$$DU_t^{N,i,j} = D_{x^j} u^{N,i}(t, X_t^{*(N)}), \quad DV_t^{N,i,j} = D_{x^j} v^{N,i}(t, X_t^{*(N)}), \quad t \in [0,T].$$

Using the system of equations (6.94) satisfied by the family $(v^{N,i})_{i=1,\cdots,N}$, we deduce from Itô's formula that, for any $i \in \{1, \cdots, N\}$,

$$\begin{aligned} dV_t^{N,i} &= -f\big(t, X_t^{*N,i}, \bar{\mu}_t^{N,i}, \hat{\alpha}(t, X_t^{*N,i}, DV_t^{N,i,i})\big)dt \\ &\quad + \sum_{j=1}^N DV_t^{N,i,j} \cdot \big(\sigma dW_t^j + \sigma^0 dW_t^0\big), \quad t \in [0,T]. \end{aligned} \tag{6.123}$$

Similarly, since $(u^{N,i})_{i=1,\cdots,N}$ satisfies (6.111), we have:

$$\begin{aligned} dU_t^{N,i} &= \Bigg[\sum_{j=1}^N DU_t^{N,i,j} \cdot \big[b(t)\big(\hat{\alpha}(t, X_t^{*N,i}, DV_t^{N,j,j}) - \hat{\alpha}(t, X_t^{*N,i}, DU_t^{N,j,j})\big)\big] \\ &\qquad - f\big(t, X_t^{*N,i}, \bar{\mu}_t^{N,i}, \hat{\alpha}(t, X_t^{*N,i}, DU_t^{N,i,i})\big) + r^{N,i}(t, X_t^{*(N)})\Bigg]dt \\ &\quad + \sum_{j=1}^N DU_t^{N,i,j} \cdot \big(\sigma dW_t^j + \sigma^0 dW_t^0\big), \quad t \in [0,T]. \end{aligned} \tag{6.124}$$

We compute the difference between (6.123) and (6.124), square it and apply Itô's formula again:

$$\begin{aligned} &d\big[U_t^{N,i} - V_t^{N,i}\big]^2 \\ &= \Bigg\{2\sum_{j=1}^N \Big[\big(U_t^{N,i} - V_t^{N,i}\big) \\ &\qquad \times \Big(DU_t^{N,i,j}\big[b(t)\big(\hat{\alpha}(t, X_t^{*N,i}, DV_t^{N,j,j}) - \hat{\alpha}(t, X_t^{*N,i}, DU_t^{N,j,j})\big)\big]\Big)\Big] \\ &\quad + 2\big(U_t^{N,i} - V_t^{N,i}\big)\Big(f\big(t, X_t^{*N,i}, \bar{\mu}_t^{N,i}, \hat{\alpha}(t, X_t^{*N,i}, DV_t^{N,i,i})\big) \\ &\qquad\qquad - f\big(t, X_t^{*N,i}, \bar{\mu}_t^{N,i}, \hat{\alpha}(t, X_t^{*N,i}, DU_t^{N,i,i})\big)\Big) \\ &\quad + 2\big(U_t^{N,i} - V_t^{N,i}\big) r^{N,i}(t, X_t^{*(N)})\Bigg\} dt \\ &\quad + \Bigg[\sum_{j=1}^N \big|\sigma^\dagger\big(DU_t^{N,i,j} - DV_t^{N,i,j}\big)\big|^2 + \Big|\sum_{j=1}^N \Big((\sigma^0)^\dagger\big(DU_t^{N,i,j} - DV_t^{N,i,j}\big)\Big)\Big|^2\Bigg]dt \\ &\quad + 2\sum_{j=1}^N \big(U_t^{N,i} - V_t^{N,i}\big)\Big[\big(DU_t^{N,i,j} - DV_t^{N,i,j}\big) \cdot \big(\sigma dW_t^j + \sigma^0 dW_t^0\big)\Big]. \end{aligned} \tag{6.125}$$

Recall now that $\hat{\alpha}$ is at most of linear growth and Lipschitz continuous in the third argument, uniformly in (t,x), and that f is locally Lipschitz in the last argument. Hence, the term spanning the fourth and fifth lines can be bounded by:

$$
\begin{aligned}
&\Big|\big(U_t^{N,i} - V_t^{N,i}\big)\Big(f\big(t, X_t^{*N,i}, \bar{\mu}_t^{N,i}, \hat{\alpha}(t, X_t^{*N,i}, DV_t^{N,i,i})\big) \\
&\qquad\qquad - f\big(t, X_t^{*N,i}, \bar{\mu}_t^{N,i}, \hat{\alpha}(t, X_t^{*N,i}, DU_t^{N,i,i})\big)\Big)\Big| \\
&\le C\Big(1 + |DU_t^{N,i,i}| + |DV_t^{N,i,i}|\Big)\cdot \big|U_t^{N,i} - V_t^{N,i}\big| \cdot \big|DU_t^{N,i,i} - DV_t^{N,i,i}\big| \\
&\le C\Big(\big|U_t^{N,i} - V_t^{N,i}\big|\big|DU_t^{N,i,i} - DV_t^{N,i,i}\big| + \big|DU_t^{N,i,i} - DV_t^{N,i,i}\big|^2\Big),
\end{aligned}
\tag{6.126}
$$

where we used the fact that $U_t^{N,i}$, $V_t^{N,i}$ and $DU_t^{N,i,i}$ are bounded, independently of i, N and t. Regarding the term in the sixth line of (6.125), we notice from Proposition 6.31 that:

$$
\begin{aligned}
&\mathbb{E}_0\big[|r^{N,i}(t, X_t^{*(N)})|^2\big] \\
&\le \frac{C}{N^2}\Big(1 + \frac{1}{N}\sum_{j=1}^N \mathbb{E}_0\big[|X_t^{*N,i} - X_t^{*N,j}|^2\big]\Big) \\
&\le \frac{C}{N^2}\bigg(1 + \frac{1}{N}\sum_{j=1}^N |\xi^i - \xi^j|^2 + \frac{1}{N}\sum_{j=1}^N \mathbb{E}_0\Big[\int_0^T |DV_t^{N,j,j} - DV_t^{N,i,i}|^2 dt\Big]\bigg) \\
&\le \frac{C}{N^2}\bigg(1 + \frac{1}{N}\sum_{j=1}^N |\xi^i - \xi^j|^2 + \mathbb{E}_0\Big[\int_0^T |DV_t^{N,i,i} - DU_t^{N,i,i}|^2 dt\Big] \\
&\qquad + \frac{1}{N}\sum_{j=1}^N \mathbb{E}_0\Big[\int_0^T |DV_t^{N,j,j} - DU_t^{N,j,j}|^2 dt\Big]\bigg).
\end{aligned}
\tag{6.127}
$$

Returning to the other terms in (6.125), recall also that $DU_t^{N,i,j}$ is bounded by C/N when $i \neq j$, for C independent of i, j, N, and t. Integrating (6.125) from t to T and taking the conditional expectation given $\mathcal{F}_0$, we deduce that:

$$
\begin{aligned}
&\mathbb{E}_0\big[|U_t^{N,i} - V_t^{N,i}|^2\big] \\
&\le \mathbb{E}_0\big[|U_T^{N,i} - V_T^{N,i}|^2\big] + \frac{C}{N^2}\Big(1 + \frac{1}{N}\sum_{j=1}^N |\xi^i - \xi^j|^2\Big) \\
&\quad + C\int_t^T \mathbb{E}_0\big[|U_s^{N,i} - V_s^{N,i}|^2\big]ds + C\int_t^T \mathbb{E}_0\big[|DU_s^{N,i,i} - DV_s^{N,i,i}|^2\big]ds \\
&\quad + \frac{C}{N}\int_t^T \sum_{j=1, j\neq i}^N \mathbb{E}_0\big[|DU_s^{N,j,j} - DV_s^{N,j,j}|^2\big]ds \\
&\quad + \frac{C}{N^2}\int_0^T \mathbb{E}_0\big[|DU_s^{N,i,i} - DV_s^{N,i,i}|^2\big]ds \\
&\quad + \frac{C}{N^3}\int_0^T \sum_{j=1, j\neq i}^N \mathbb{E}_0\big[|DU_s^{N,j,j} - DV_s^{N,j,j}|^2\big]ds.
\end{aligned}
\tag{6.128}
$$

Notice that the terminal condition $U_T^{N,i} - V_T^{N,i}$ is zero. Observe also that the two last integrals run from 0 to T and not from t to T.

Third Step. We now perform a similar computation with $(\cosh[\eta(U_t^{N,i} - V_t^{N,i})] - 1)_{0\le t\le T}$ in lieu of $(|U_t^{N,i} - V_t^{N,i}|^2)_{0\le t\le T}$, where η is a free parameter in $\mathbb{R}$, whose value will be fixed later on.

$$
\begin{aligned}
& d\big[\cosh\big[\eta(U_t^{N,i} - V_t^{N,i})\big]\big] \\
&= \bigg\{\eta \sum_{j=1}^N \bigg[\sinh\big[\eta(U_t^{N,i} - V_t^{N,i})\big] \\
&\qquad \times \Big(DU_t^{N,i,j} \cdot \big[b(t)\big(\hat\alpha(t, X_t^{*N,i}, DV_t^{N,j,j}) - \hat\alpha(t, X_t^{*N,i}, DU_t^{N,j,j})\big)\big]\Big)\bigg] \\
&\quad + \eta \sinh\big[\eta(U_t^{N,i} - V_t^{N,i})\big]\Big(f\big(t, X_t^{*N,i}, \bar\mu_t^{N,i}, \hat\alpha(t, X_t^{*N,i}, DV_t^{N,i,i})\big) \\
&\qquad\qquad - f\big(t, X_t^{*N,i}, \bar\mu_t^{N,i}, \hat\alpha(t, X_t^{*N,i}, DU_t^{N,i,i})\big)\Big) \\
&\quad + \eta \sinh\big[\eta(U_t^{N,i} - V_t^{N,i})\big] r^{N,i}(t, X_t^{*(N)})\bigg\}\, dt \qquad (6.129) \\
&\quad + \frac{\eta^2}{2}\bigg[\cosh\big[\eta(U_t^{N,i} - V_t^{N,i})\big] \sum_{j=1}^N \big|\sigma^\dagger\big(DU_t^{N,i,j} - DV_t^{N,i,j}\big)\big|^2 \\
&\quad + \cosh\big[\eta(U_t^{N,i} - V_t^{N,i})\big]\bigg|\sum_{j=1}^N \Big((\sigma^0)^\dagger\big(DU_t^{N,i,j} - DV_t^{N,i,j}\big)\Big)\bigg|^2\bigg] dt \\
&\quad + \eta \sum_{j=1}^N \sinh\big[\eta(U_t^{N,i} - V_t^{N,i})\big]\Big[\big(DU_t^{N,i,j} - DV_t^{N,i,j}\big) \cdot \big(\sigma dW_t^j + \sigma^0 dW_t^0\big)\Big].
\end{aligned}
$$

Similar to (6.126), we have:

$$
\begin{aligned}
&\Big|\eta \sinh\big[\eta(U_t^{N,i} - V_t^{N,i})\big]\Big(f\big(t, X_t^{*N,i}, \bar\mu_t^{N,i}, \hat\alpha(t, X_t^{*N,i}, DV_t^{N,i,i})\big) \\
&\qquad\qquad - f\big(t, X_t^{*N,i}, \bar\mu_t^{N,i}, \hat\alpha(t, X_t^{*N,i}, DU_t^{N,i,i})\big)\Big)\Big| \\
&\le C\Big(1 + |DU_t^{N,i,i}| + |DV_t^{N,i,i}|\Big)\big|\eta \sinh\big[\eta(U_t^{N,i} - V_t^{N,i})\big]\big| \cdot \big|DU_t^{N,i,i} - DV_t^{N,i,i}\big| \\
&\le C\Big(\big|\eta \sinh\big[\eta(U_t^{N,i} - V_t^{N,i})\big]\big|^2 \\
&\qquad + (1+\eta)\cosh\big[\eta(U_t^{N,i} - V_t^{N,i})\big]\big|DU_t^{N,i,i} - DV_t^{N,i,i}\big|^2\Big).
\end{aligned}
$$

To pass from the second to the third line, we used the standard Young inequality together with the fact that sinh is bounded by cosh.

Following (6.128), but using in addition the fact that σ is uniformly elliptic, we deduce that there exists a new constant C_η, allowed to depend on η, such that:

$$
\begin{aligned}
&\eta^2 \mathbb{E}_0\Big[\int_t^T \cosh\big[\eta(U_t^{N,i} - V_t^{N,i})\big]|DU_s^{N,i,i} - DV_s^{N,i,i}|^2 ds\Big] \\
&\leq \frac{C}{N^2}\Big(1 + \frac{1}{N}\sum_{j=1}^N |\xi^i - \xi^j|^2\Big) \\
&\quad + C_\eta \int_t^T \mathbb{E}_0\big[|U_s^{N,i} - V_s^{N,i}|^2\big] ds \\
&\quad + C(1+\eta)\mathbb{E}_0\Big[\int_t^T \cosh\big[\eta(U_t^{N,i} - V_t^{N,i})\big]|DU_s^{N,i,i} - DV_s^{N,i,i}|^2 ds\Big] \\
&\quad + \frac{C}{N}\mathbb{E}_0 \int_t^T \sum_{j=1, j\neq i}^N \mathbb{E}_0\big[|DU_s^{N,j,j} - DV_s^{N,j,j}|^2\big] ds \\
&\quad + \frac{C}{N^2}\int_0^T \mathbb{E}_0\big[|DU_s^{N,i,i} - DV_s^{N,i,i}|^2\big] ds \\
&\quad + \frac{C}{N^3}\int_0^T \sum_{j=1, j\neq i}^N \mathbb{E}_0\big[|DU_s^{N,j,j} - DV_s^{N,j,j}|^2\big] ds,
\end{aligned}
\tag{6.130}
$$

where, except in the third line, the constant C is independent of η. Also, we used the fact that $(U_s^{N,i})_{0\leq s\leq T}$ and $(V_s^{N,i})_{0\leq s\leq T}$ can be bounded independently of N together with the inequality $|\sinh(r)| \leq \cosh(M)|r|$ for $r \in [-M, M]$ and $M > 0$.

Fourth Step. We add (6.128) and (6.130). Now we choose η large enough so that:

$$
\begin{aligned}
&\mathbb{E}_0\big[|U_t^{N,i} - V_t^{N,i}|^2\big] \\
&\quad + \eta^2 \mathbb{E}_0\Big[\int_t^T \cosh\big[\eta(U_t^{N,i} - V_t^{N,i})\big]|DU_s^{N,i,i} - DV_s^{N,i,i}|^2 ds\Big] \\
&\leq \frac{C}{N^2}\Big(1 + \frac{1}{N}\sum_{j=1}^N |\xi^i - \xi^j|^2\Big) + C_\eta \int_t^T \mathbb{E}_0\big[|U_s^{N,i} - V_s^{N,i}|^2\big] ds \\
&\quad + \frac{C}{N}\mathbb{E}_0 \int_t^T \sum_{j=1, j\neq i}^N \big[|DU_s^{N,j,j} - DV_s^{N,j,j}|^2\big] ds \\
&\quad + \frac{C}{N^2}\int_0^T \mathbb{E}_0\big[|DU_s^{N,i,i} - DV_s^{N,i,i}|^2\big] ds \\
&\quad + \frac{C}{N^3}\int_0^T \sum_{j=1, j\neq i}^N \mathbb{E}_0\big[|DU_s^{N,j,j} - DV_s^{N,j,j}|^2\big] ds,
\end{aligned}
\tag{6.131}
$$

where, as above, the constant C is independent of η, except in the second term on the third line. Taking the mean over the index $i \in \{1, \cdots, N\}$, we get:

$$
\begin{aligned}
&\frac{1}{N}\sum_{i=1}^{N}\mathbb{E}_0\big[|U_t^{N,i}-V_t^{N,i}|^2\big]\\
&\quad+\frac{\eta^2}{N}\sum_{i=1}^{N}\mathbb{E}_0\bigg[\int_t^T\cosh\big[\eta(U_t^{N,i}-V_t^{N,i})\big]|DU_s^{N,i,i}-DV_s^{N,i,i}|^2ds\bigg]\\
&\le\frac{C}{N^2}\bigg(1+\frac{1}{N^2}\sum_{i,j=1}^{N}|\xi^i-\xi^j|^2\bigg)+\frac{C_\eta}{N}\sum_{i=1}^{N}\int_t^T\mathbb{E}_0\big[|U_s^{N,i}-V_s^{N,i}|^2\big]ds\\
&\quad+\frac{C}{N}\mathbb{E}_0\int_t^T\sum_{i=1}^{N}\big[|DU_s^{N,i,i}-DV_s^{N,i,i}|^2\big]ds\\
&\quad+\frac{C}{N^3}\sum_{i=1}^{N}\int_0^T\mathbb{E}_0\big[|DU_s^{N,i,i}-DV_s^{N,i,i}|^2\big]ds,
\end{aligned}
$$

where, once again, the constant C_η in the second term of the third line depends on η. As above, we can get rid of the third term in the right-hand side by choosing η large enough so that $\eta^2\ge 2C$. Then, by applying Gronwall's lemma to the quantity $(\frac{1}{N}\sum_{i=1}^{N}\mathbb{E}_0[|U_t^{N,i}-V_t^{N,i}|^2])_{0\le t\le T}$, we get:

$$
\begin{aligned}
&\sup_{0\le t\le T}\bigg[\frac{1}{N}\sum_{i=1}^{N}\mathbb{E}_0\big[|U_t^{N,i}-V_t^{N,i}|^2\big]\bigg]\\
&\quad+\frac{\eta^2}{N}\sum_{i=1}^{N}\mathbb{E}_0\bigg[\int_0^T\cosh\big[\eta(U_t^{N,i}-V_t^{N,i})\big]|DU_s^{N,i,i}-DV_s^{N,i,i}|^2ds\bigg]\\
&\le\frac{C_\eta}{N^2}\bigg(1+\frac{1}{N^2}\sum_{i,j=1}^{N}|\xi^i-\xi^j|^2\bigg)\\
&\quad+\frac{C_\eta}{N^3}\int_0^T\sum_{i=1}^{N}\mathbb{E}_0\big[|DU_s^{N,i,i}-DV_s^{N,i,i}|^2\big]ds,
\end{aligned}
$$

where, now, the two constants in the right-hand side depend on η. The value of η having been fixed, we can define N_0 as the smallest integer such that $\eta^2\ge 2C_\eta/N^2$. Hence, for $N\ge N_0$, we have:

$$
\begin{aligned}
&\sup_{0\le t\le T}\bigg[\frac{1}{N}\sum_{i=1}^{N}\mathbb{E}_0\big[|U_t^{N,i}-V_t^{N,i}|^2\big]\bigg]\\
&\quad+\frac{\eta^2}{N}\sum_{i=1}^{N}\mathbb{E}_0\bigg[\int_0^T\cosh\big[\eta(U_t^{N,i}-V_t^{N,i})\big]|DU_s^{N,i,i}-DV_s^{N,i,i}|^2ds\bigg] \qquad (6.132)\\
&\le\frac{C_\eta}{N^2}\bigg(1+\frac{1}{N^2}\sum_{i,j=1}^{N}|\xi^i-\xi^j|^2\bigg),
\end{aligned}
$$

for a new value of C_η. Since the value of η has been fixed, we can easily drop it in the above inequality. Namely, we can replace η^2 and cosh in the second line by 1 and C_η in the third line by C. We now insert the resulting form of (6.132) into (6.131). We get:

$$\begin{aligned}
&\mathbb{E}_0\big[|U_t^{N,i} - V_t^{N,i}|^2\big] + \mathbb{E}_0\bigg[\int_t^T |DU_s^{N,i,i} - DV_s^{N,i,i}|^2 ds\bigg] \\
&\leq \frac{C'}{N^2}\Big(1 + \frac{1}{N}\sum_{j=1}^N |\xi^i - \xi^j|^2 + \frac{1}{N^2}\sum_{j,\ell=1}^N |\xi^j - \xi^\ell|^2\Big) + C'\int_t^T \mathbb{E}_0\big[|U_s^{N,i} - V_s^{N,i}|^2\big]ds \\
&\quad + \frac{C'}{N^2}\int_0^T \mathbb{E}_0\big[|DU_s^{N,i,i} - DV_s^{N,i,i}|^2\big]ds,
\end{aligned}$$

which holds true for $N \geq N_0$. We used a new constant C' to distinguish from the preceding constant C, but, similar to C, C' is independent of N and i. Hence, we can play the same game as before. Applying Gronwall's lemma, we can get rid of the last term in the second line, provided that C is allowed to increase. Then, calling N_0' the smallest integer greater than N_0 such that $C'/[(N_0')^2] \leq 1/2$ and then applying Gronwall's lemma, we obtain:

$$\begin{aligned}
&\sup_{0\leq t\leq T} \mathbb{E}_0\big[|U_t^{N,i} - V_t^{N,i}|^2\big] + \mathbb{E}_0\bigg[\int_0^T |DU_s^{N,i,i} - DV_s^{N,i,i}|^2 ds\bigg] \\
&\leq \frac{C'}{N^2}\Big(1 + \frac{1}{N}\sum_{j=1}^N |\xi^i - \xi^j|^2 + \frac{1}{N^2}\sum_{j,\ell=1}^N |\xi^j - \xi^\ell|^2\Big).
\end{aligned} \tag{6.133}$$

This completes the proof of (6.121), except for the fact that the above holds true for $N \geq N_0'$ only. Actually, we can easily bypass this constraint by observing that the left-hand side is bounded, uniformly with respect to the index i, to the initial position of $\boldsymbol{X}^{*(N)}$ and to the integer N, as long as $N \leq N_0'$. Modifying the constant C' accordingly, this permits to conclude.

Last Step. We now derive (6.120) and (6.122). We start with (6.122). Noticing that $U_0^{N,i} - V_0^{N,i} = u^{N,i}(0, \boldsymbol{\xi}) - v^{N,i}(0, \boldsymbol{\xi})$, we deduce from the conclusion of the previous step, see (6.133), that, with probability 1 under $\mathbb{P}$, for all $i \in \{1, \cdots, N\}$,

$$|u^{N,i}(0, \boldsymbol{\xi}) - v^{N,i}(0, \boldsymbol{\xi})| \leq \frac{C}{N}\Big(1 + \frac{1}{N}\sum_{j=1}^N |\xi^i - \xi^j|^2 + \frac{1}{N^2}\sum_{\ell,j=1}^N |\xi^\ell - \xi^j|^2\Big)^{1/2},$$

which is exactly (6.122).

We are now ready to estimate the difference $X_t^{N,i} - X_t^{*N,i}$, for $t \in [0, T]$ and $i \in \{1, \cdots, N\}$. In view of the equation satisfied by the processes $(X_t^{N,i})_{0\leq t\leq T}$ and by $(X_t^{*N,i})_{0\leq t\leq T}$ defined in (6.118) and (6.119), we have:

$$
\begin{aligned}
&|X_t^{N,i} - X_t^{*N,i}| \\
&\leq C \int_0^t \big|\hat{\alpha}\big(X_s^{N,i}, D_{x^i} u^{N,i}(s, X_s^{(N)})\big) - \hat{\alpha}\big(X_s^{*N,i}, D_{x^i} v^{N,i}(s, X_s^{*(N)})\big)\big| ds \\
&\leq C \int_0^t |X_s^{N,i} - X_s^{*N,i}| ds + \frac{C}{N-1} \sum_{j=1, j\neq i}^N \int_0^t |X_s^{N,j} - X_s^{*N,j}| ds \\
&\quad + C \int_0^T \big|DU_s^{N,i,i} - DV_s^{N,i,i}\big| ds,
\end{aligned}
\tag{6.134}
$$

where we used:

$$
\begin{aligned}
&\big|D_{x^i} u^{N,i}(s, X_s^{(N)}) - D_{x^i} v^{N,i}(s, X_s^{*(N)})\big| \\
&\leq \big|D_{x^i} u^{N,i}(s, X_s^{(N)}) - D_{x^i} u^{N,i}(s, X_s^{*(N)})\big| + \big|DU_s^{N,i,i} - DV_s^{N,i,i}\big| \\
&= \big|\partial_x \mathcal{U}\big(s, X_s^{N,i}, \bar{\mu}^{N-1}_{X_s^{(N)-i}}\big) - \partial_x \mathcal{U}\big(s, X_s^{*N,i}, \bar{\mu}^{N-1}_{X_s^{*(N)-i}}\big)\big| + \big|DU_s^{N,i,i} - DV_s^{N,i,i}\big|,
\end{aligned}
$$

which yields:

$$
\begin{aligned}
&\big|D_{x^i} u^{N,i}(s, X_s^{(N)}) - D_{x^i} v^{N,i}(s, X_s^{*(N)})\big| \\
&\leq C\Big(|X_s^{N,i} - X_s^{*N,i}| + \frac{1}{N} \sum_{j=1}^N |X_s^{N,j} - X_s^{*N,j}|\Big) + \big|DU_s^{N,i,i} - DV_s^{N,i,i}\big|.
\end{aligned}
$$

Taking expectations, by exchangeability, by Gronwall's inequality, and by (6.133), we obtain (6.120). □

Remark 6.33 *The reader may observe that, in addition to the existence of a classical solution $\mathcal{U}$ to the master equation satisfying the conclusion of Theorem 5.46, we only used the fact that $\hat{\alpha}$ is bounded and Lipschitz continuous and the fact that f is locally Lipschitz continuous, see* (6.126), (6.128), (6.130), *and* (6.134).

Proof of Theorem 6.28

Proof. We start with the first claim in the statement. To do so, we choose $\mu_0 = \mathcal{N}_d(0, I_d)$ and apply (6.122):

$$
\Big|\mathcal{U}(t_0, \xi^i, \bar{\mu}^{N-1}_{\boldsymbol{\xi}^{-i}}) - v^{N,i}(t_0, \boldsymbol{\xi})\Big| \leq \frac{C}{N}\Big(1 + \frac{1}{N}\sum_{j=1}^N |\xi^i - \xi^j|^2 + \frac{1}{N^2}\sum_{\ell,j=1}^N |\xi^\ell - \xi^j|^2\Big)^{1/2},
$$

for $i \in \{1, \cdots, N\}$. The support of $\boldsymbol{\xi}$ being $(\mathbb{R}^d)^N$, we deduce from the continuity of $\mathcal{U}$ and of the $(v^{N,i})_{i=1,\cdots,N}$ that the above inequality holds for any $\boldsymbol{x} \in (\mathbb{R}^d)^N$:

$$\left|\mathcal{U}(t_0, x^i, \bar{\mu}^{N-1}_{x^{-i}}) - v^{N,i}(t_0, \boldsymbol{x})\right| \le \frac{C}{N}\Big(1 + \frac{1}{N}\sum_{j=1}^{N} |x^i - x^j|^2 + \frac{1}{N^2}\sum_{\ell,j=1}^{N} |x^\ell - x^j|^2\Big)^{1/2},$$

for all $i \in \{1, \cdots, N\}$.

We turn to the second part of the statement. We notice that:

$$\begin{aligned}
\mathbb{E}\big[|u^{N,i}(t, \boldsymbol{\xi}) - \mathcal{U}(t, \xi^i, \mu_0)|\big] &= \mathbb{E}\big[|\mathcal{U}(t, \xi^i, \bar{\mu}^{N-1}_{\boldsymbol{\xi}^{-i}}) - \mathcal{U}(t, \xi^i, \mu_0)|\big] \\
&\le C\mathbb{E}\big[W_1(\bar{\mu}^{N-1}_{\boldsymbol{\xi}^{-i}}, \mu_0)\big],
\end{aligned}$$

where we used the Lipschitz property of $\mathcal{U}$ in the measure argument in the second line. The result follows from Corollary 6.3. □

6.3.4 Propagation of Chaos for the N-Player Game

As a byproduct of our analysis, we now deduce that the optimal trajectories of the N-player game converge to the expected limit. In order to state the result, we use the same set-up as in Chapter 2 for constructing the particle system (2.3), namely the probability space $(\Omega, \mathcal{F}, \mathbb{F}, \mathbb{P})$ reads as the product of two probability spaces $(\Omega^0, \mathcal{F}^0, \mathbb{F}^0, \mathbb{P}^0)$ and $(\Omega^1, \mathcal{F}^1, \mathbb{F}^1, \mathbb{P}^1)$. The space $(\Omega^0, \mathcal{F}^0, \mathbb{F}^0, \mathbb{P}^0)$ carries the d-dimensional Brownian motion $\boldsymbol{W}^0$, which is assumed to be a Brownian motion with respect to $\mathbb{F}^0$ and the space $(\Omega^1, \mathcal{F}^1, \mathbb{F}^1, \mathbb{P}^1)$ carries the N-tuple of independent and identically distributed $\mathcal{F}^1_0$-measurable random variables $(\xi^i)_{i=1,\cdots,N}$, which take values in $\mathbb{R}^d$ and have common distribution $\mu_0 \in \mathcal{P}_2(\mathbb{R}^d)$, together with the tuple of d-dimensional Brownian motions $(\boldsymbol{W}^i)_{i=1,\cdots,N}$, which are assumed to be Brownian motions with respect to the filtration $\mathbb{F}^1$.

Here is the main statement:

Theorem 6.34 *For any $i \in \{1, \cdots, N\}$, the McKean-Vlasov SDE*

$$\begin{aligned}
d\underline{X}^i_t &= b(t)\hat{\alpha}\big(t, \underline{X}^i_t, \partial_x\mathcal{U}(t, \underline{X}^i_t, \mathcal{L}^1(\underline{X}^i_t))\big)dt + \sigma dW^i_t + \sigma^0 dW^0_t, \quad t \in [0, T] \\
\underline{X}^i_0 &= \xi^i.
\end{aligned}$$

is uniquely solvable.

Moreover, the flow of conditional distributions $(\mu_t = \mathcal{L}^1(\underline{X}^i_t))_{0\le t\le T}$ does not depend on i and is the solution to the mean field game with μ_0 as initial condition. Also, for any $\eta > 0$, there exists a constant C_η, independent of N, such that, for all $i \in \{1, \cdots, N\}$,

$$\mathbb{E}\Big[\sup_{0\le t\le T} \big|X^{\star N,i}_t - \underline{X}^i_t\big|\Big] \le C_\eta N^{-1/\max(d,2+\eta)}.$$

Above, $((X_t^{*N,i})_{0\le t\le T})_{i=1,\cdots,N}$ stands for the corresponding "optimal trajectories" of the N-player game, which are associated with the solution $(v^{N,i})_{i=1,\cdots,N}$ of the N-Nash system (6.94). Namely $((X_t^{*N,i})_{0\le t\le T})_{i=1,\cdots,N}$ solve (6.119) with $X_0^{*N,i} = \xi^i$ as initial condition at time 0.

Proof.

First Step. For any $i \in \{1,\cdots,N\}$, the SDE of McKean-Vlasov type:

$$d\underline{X}_t^i = b(t)\hat{\alpha}\big(t,\underline{X}_t^i,\partial_x\mathcal{U}(t,\underline{X}_t^i,\mathcal{L}^1(\underline{X}_t^i))\big)dt + \sigma dW_t^i + \sigma^0 dW_t^0, \quad t\in[t_0,T],$$
$$\underline{X}_0^i = \xi^i,$$

is uniquely solvable since $\partial_x\mathcal{U}$ is Lipschitz continuous in the variables x and μ, see Proposition 2.8. We then call $(\mu_t)_{0\le t\le T}$ the flow of conditional distributions $(\mathcal{L}^1(\underline{X}_t^1))_{0\le t\le T}$. By Proposition 2.11, we also have $(\mu_t = \mathcal{L}^1(\underline{X}_t^i))_{0\le t\le T}$ for any $i \in \{1,\cdots,N\}$. Moreover, by Theorem 5.46, $(\mu_t)_{0\le t\le T}$ is the solution to the mean field game with μ_0 as initial condition, uniqueness following from the fact that the Lasry-Lions monotonicity condition is in force, see Proposition 3.34 and (3.77).

Second Step. We now turn to the second part of the proof. It is a direct application of Theorem 6.32 combined with the following estimate on the distance between $(\underline{X}_t^i)_{0\le t\le T}$ and the solution $(X_t^{N,i})_{0\le t\le T}$ of (6.118):

$$\mathbb{E}\Big[\sup_{0\le t\le T}\big|X_t^{N,i}-\underline{X}_t^i\big|\Big] \le \delta_N, \tag{6.135}$$

for any $i \in \{1,\cdots,N\}$, where the sequence $(\delta_N)_{N\ge 1}$ tends to 0 as N tends to ∞ and can be chosen as $C_\eta N^{-1/\max(d,2+\eta)}$, for any $\eta\in(0,1]$ and for some constant $C_\eta > 0$ depending on η.

Assume for a while that (6.135) is true. Then, by the triangle inequality,

$$\begin{aligned}\mathbb{E}\Big[\sup_{0\le t\le T}\big|X_t^{*N,i}-\underline{X}_t^i\big|\Big] &\le \mathbb{E}\Big[\sup_{0\le t\le T}\big|X_t^{*N,i}-X_t^{N,i}\big|\Big] + \mathbb{E}\Big[\sup_{0\le t\le T}\big|X_t^{N,i}-\underline{X}_t^i\big|\Big]\\ &\le C\big(N^{-1}+\delta_N\big),\end{aligned}$$

where we used (6.120) to pass from the first to the second line.

Third Step. It now remains to check (6.135). For this, we fix $i \in \{1,\cdots,N\}$. Then, for any $t\in[0,T]$, we have:

$$\begin{aligned}\big|X_t^{N,i}-\underline{X}_t^i\big| &\le C\int_0^t \big|\hat{\alpha}\big(s,X_s^{N,i},D_{x^i}u^{N,i}(s,\boldsymbol{X}_s^{(N)})\big)-\hat{\alpha}\big(s,\underline{X}_s^i,\partial_x\mathcal{U}(s,\underline{X}_s^i,\mu_s)\big)\big|ds\\ &\le C\int_0^t \Big|\hat{\alpha}\Big(s,X_s^{N,i},\partial_x\mathcal{U}\big(s,X_s^{N,i},\bar{\mu}^{N-1}_{\boldsymbol{X}_s^{(N)-i}}\big)\Big)-\hat{\alpha}\Big(s,\underline{X}_s^i,\partial_x\mathcal{U}\big(s,\underline{X}_s^i,\bar{\mu}^{N-1}_{\underline{\boldsymbol{X}}_s^{(N)-i}}\big)\Big)\Big|ds\\ &\quad + C\int_0^t \Big|\hat{\alpha}\Big(s,\underline{X}_s^i,\partial_x\mathcal{U}\big(s,\underline{X}_s^i,\bar{\mu}^{N-1}_{\underline{\boldsymbol{X}}_s^{(N)-i}}\big)\Big)-\hat{\alpha}\Big(s,\underline{X}_s^i,\partial_x\mathcal{U}\big(s,\underline{X}_s^i,\mu_s\big)\Big)\Big|ds,\end{aligned}$$

where $\underline{\boldsymbol{X}}^{(N)} = (\underline{X}^1,\cdots,\underline{X}^N)$.

Since $\partial_x \mathcal{U}$ is Lipschitz continuous in the space and measure arguments, with respect to the W_1-distance for the latter one, we get:

$$|X_t^{N,i} - \underline{X}_t^i| \le C \int_0^t \Big[|X_s^{N,i} - \underline{X}_s^i| + W_1\Big(\bar{\mu}^{N-1}_{X_s^{(N)-i}}, \bar{\mu}^{N-1}_{\underline{X}_s^{(N)-i}}\Big) + W_1\Big(\bar{\mu}^{N-1}_{\underline{X}_s^{(N)-i}}, \mu_s\Big)\Big] ds,$$

where:

$$W_1\Big(\bar{\mu}^{N-1}_{X_t^{(N)-i}}, \bar{\mu}^{N-1}_{\underline{X}_t^{(N)-i}}\Big) \le \frac{1}{N-1} \sum_{j=1, j\neq i}^{N} |X_t^{N,j} - \underline{X}_t^j|.$$

Hence,

$$|X_t^{N,i} - \underline{X}_t^i| \le C \int_0^t \Big[|X_s^{N,i} - \underline{X}_s^i| + \frac{1}{N-1} \sum_{j=1, j\neq i}^{N} |X_s^{N,j} - \underline{X}_s^j| + W_1\Big(\bar{\mu}^{N-1}_{\underline{X}_s^{(N)-i}}, \mu_s\Big)\Big] ds, \tag{6.136}$$

where, as usual, the constant C is allowed to increase from line to line. Taking expectations and recalling that the random variables $(X_s^{N,j} - \underline{X}_s^j)_{j\in\{1,\cdots,N\}}$ are exchangeable, we deduce that:

$$\begin{aligned}
\mathbb{E}[|X_t^{N,i} - \underline{X}_t^i|] &\le C \int_0^t \Big(\mathbb{E}[|X_s^{N,i} - \underline{X}_s^i|] + \frac{1}{N-1} \sum_{j=1, j\neq i}^{N} \mathbb{E}[|X_s^{N,j} - \underline{X}_s^j|] \\
&\qquad + \mathbb{E}\big[W_1\big(\bar{\mu}^{N-1}_{\underline{X}_s^{(N)-i}}, \mu_s\big)\big]\Big) ds \\
&\le C \int_0^t \Big(\mathbb{E}[|X_s^{N,i} - \underline{X}_s^i| + \mathbb{E}\Big[W_1\Big(\bar{\mu}^{N-1}_{\underline{X}_s^{(N)-i}}, \mu_s\Big)\Big]\Big) ds.
\end{aligned}$$

By Gronwall's lemma, we get for any $i \in \{1, \cdots, N\}$:

$$\forall t \in [0, T], \quad \mathbb{E}[|X_t^{N,i} - \underline{X}_t^i|] \le C \int_0^t \mathbb{E}\Big[W_1\Big(\bar{\mu}^{N-1}_{\underline{X}_s^{(N)-i}}, \mu_s\Big)\Big] ds.$$

Now, returning to (6.136), taking the supremum over $t \in [0, T]$ and using the above inequality, we deduce that:

$$\forall i \in \{1, \cdots, N\}, \quad \mathbb{E}\Big[\sup_{0\le t\le T} |X_t^{N,i} - \underline{X}_t^i|\Big] \le C \int_0^T \mathbb{E}\Big[W_1\Big(\bar{\mu}^{N-1}_{\underline{X}_s^{(N)-i}}, \mu_s\Big)\Big] ds.$$

In order to complete the proof, it suffices to note that:

$$\sup_{0\le t\le T} \mathbb{E}[|\underline{X}_t^1|^2] < \infty,$$

which permits to apply Corollary 6.3, using the fact that, for almost every $\omega^0 \in \Omega^0$, the processes $(\underline{\boldsymbol{X}}^i(\omega^0, \cdot))_{1\le i\le N}$ are independent under $\mathbb{P}^1$. □

6.4 Notes & Complements

In the absence of common noise, the construction of approximate Nash-equilibria for the N-player game from the solution of the limiting mean field game problem provided in the text is borrowed from the paper [96] of Carmona and Delarue, where the stochastic maximum principle was used for the first time in the solution of mean field games. Actually, the existence of ϵ-approximate Nash equilibria for finite player games already appeared in the original work of Caines, Huang, and Malhamé on Nash certainty equivalence [211], where reliance on the theory of the propagation of chaos is already present. The argument used in the text was also used for simpler models in [53] by Bensoussan, Sung, Yam, and Yung, and in [83] by Cardaliaguet for first order mean field game models. A similar issue was addressed in [236, 238] by Kolokoltsov, Li, and Wang, and by Kolokoltsov, Troeva, and Wang, but with a slightly different twist: there, by means of similar arguments to those used in Subsection (Vol I)-5.7.4 for revisiting the propagation of chaos strategy, the authors show that optimal strategies for mean field games form $O(1/N)$-approximate Nash equilibria when the coefficients of the game are smooth enough. In comparison, in the text we only proved that they form $O(\sqrt{\varepsilon_N})$-approximate Nash equilibria, with ε_N as in (6.1). To the best of our knowledge, the construction of approximate equilibria for mean field games with common noise, as provided in Subsection 6.1.2, had only been addressed by Kolokoltsov and Troeva in [237]. There, the analysis requires the master equation to have a classical solution, in which case approximate Nash equilibria can be constructed, as explained in the text, in closed loop form. The L^4-stability estimate used in the proof of Theorem 6.4 may be found in Delarue [132].

From the practical point of view, numerical methods are needed to compute optimal strategies in mean field games and to plug them into finite player games. We do not address this question in the book and we refer to the following papers for various approximation, discretization, or numerical methods, including finite differences or variational approaches: Achdou and Capuzzo-Dolcetta [4], Achdou, Camili, and Capuzzo-Dolcetta [3], Achdou and Perez [6], Achdou and Porretta [7], Benamou and Carlier [42], Lachapelle, Salomon, and Turinici [252], Cardaliaguet and Hadikhanloo [89], and Guéant [187]. In [2], Achdou, Camilli, and Capuzzo-Dolcetta investigate numerical methods for mean field planning problems in which both the initial and terminal states of the population are prescribed. We refer to the PhD dissertation by Alanko [14] for a probabilistic point of view.

The problem of convergence of Nash equilibria of games with finitely many players towards solutions of mean field games has been known to be more challenging. For closed loop Nash equilibria, convergence was proved for ergodic mean field games by Lasry and Lions in [260], and later on revisited by Bardi and Feleqi in [34], in case when the players only observe the idiosyncratic noise driving their own dynamics. In the latter situation, the Nash system reduces to a coupled system of N partial differential equations in $\mathbb{R}^d$ instead of N equations in $\mathbb{R}^{Nd}$ as in (6.94), and *a priori* estimates for the solutions are available. For *linear-quadratic* models with explicit solutions, convergence was investigated by Bardi in [33]. The approach based on the master equation used in this chapter is due to Cardaliaguet, Delarue, Lasry, and Lions [86]. While its potential for applications seems to be quite large, so far, it has been applied to relatively simple models only. Case in point, the Hamiltonian is assumed to be Lipschitz continuous in the control variable in [86]. The result obtained in this chapter goes one step further since it holds true for quadratic Hamiltonians. Still, the reader should keep in mind the counter-example provided in Subsection (Vol I)-7.2.5 proving that the argument can fail. In contrast with the framework used in this chapter, this latter counter-example addresses mean field games with finite state and control spaces. This makes a subtle difference in the analysis since the minimizer of the Hamiltonian is *de facto* discontinuous with respect to the adjoint variable when the control space is discrete.

The asymptotic analysis of open loop equilibria by means of weak compactness arguments goes back to the works by Fischer [154] and Lacker [255]. Therein, the authors overcome the lack of strong estimates on the solutions to the N-player game by using the notion of *relaxed controls* for which weak compactness criteria are readily available. Here, we bypass the use of *relaxed controls* by relying on the stochastic maximum principle which allows us to prove tightness of the equilibrium control strategies for the Meyer-Zheng topology introduced in Chapter 3. In this regard, our augment is really close to that introduced in Chapter 3 for constructing weak solutions to MFG with a common noise.

Despite the use of a different topology to handle the control strategies, our approach remains quite similar to that developed by Fischer and, especially Lacker. In Lacker's work, compatibility plays a crucial role to prove that weak limits of Nash equilibria are minimizers of the optimal control problem under the environment formed by the limiting empirical distribution, see the Notes and Complements of Chapter 7 for another insight. In our approach, the use of compatibility is not so explicit, but manifests through the preliminary analysis of stochastic control problems in a random environment, as performed in Chapter 1 by means of the stochastic Pontryagin principle. The restriction we imposed on the form of the dynamics of the players, see (6.79), is mostly for convenience and it is likely that it can be dispensed with. After all, such a restriction does not appear in the works of Fischer and Lacker.

As for mean field control problems, the presentation used in Subsection 6.1.3 is mostly taken from the two papers [98] and [99] by Carmona and Delarue and by Carmona, Delarue, and Lachapelle; a similar discussion, but focused on the linear-quadratic setting, may be found in the article [217] by Huang, Caines, and Malhamé. In complete analogy with the analysis performed in Sections 6.2 and 6.3, another interesting question to address is the convergence of optimizers, in systems of finitely many players optimizing a common objective function, towards the solution of a mean field control problem, or equivalently of a control problem of the McKean-Vlasov type. We chose not to discuss the problem in the book, but we refer the interested reader to the contributions [160] by Fornasier and Solombrino and [256] by Lacker.

We conclude with a few words about the results invoked without proof in this chapter. The statement of Lemma 6.2 in the introduction of the chapter is taken from the paper by Fournier and Guillin [161]. Earlier results on the subject may be found in the papers by Barthes and Bordenave [36] and Dereich, Scheutzow, and Schottstedt [135]. We also emphasize the fact that the bound in Corollary 6.3 is certainly not optimal, and that the correction in front of the factor $N^{-1/2}$ should be logarithmic instead of polynomial. For instance, we refer to the earlier paper by Atjai, Komlòs, and Tusnàdy [269].

General results on the smoothness of solutions to semi-linear parabolic PDEs, including Hamilton-Jacobi-Bellman equations, may be found in the monographs by Friedman [162], Ladyzenskaja et al. [258], and Lieberman [264]. We refer to Delarue [133] and Delarue and Guatteri [134] for the corresponding probabilistic approach. However, quadratic systems of the type (6.94)–(6.95) require a specific treatment as they cannot be handled with the results of these references. We refer to the article by Bensoussan and Frehse [49] for the case when the system is set on a bounded domain. In this regard, the proof of Proposition 6.26, which holds for unbounded domains, is essentially new. The idea of combining BMO estimates and Krylov-Safonov theory was inspired by [133]. We refer to the monograph by Bass [38] for a detailed introduction to the results by Krylov and Safonov. The reader may also have a look at the original article [287]. We also refer to Hu and Tang [204] and to Xing and Žitković [340] for related recent results on systems of quadratic backward SDEs. Solvability of the SDE with non-Lipschitz coefficients in the proof of Proposition 6.27 is taken from Veretennikov [336] and Itô-Krylov formula can be found in Chapter II of Krylov's monograph [242].

Extensions for Volume II

7

Abstract
The rationale of this chapter is the same as for the last chapter of the first volume of the book. We leverage the technology developed in the second volume to revisit some of the examples introduced in Chapter (Vol I)-1, and complete their mathematical analysis. We use some of the tools introduced for the analysis of mean field games with a common noise to study important game models which are not amenable to the theory covered by the first volume. These models include extensions to games with minor and major players, games of timing, and some finite state space models. We believe that these mean field game models have a great potential for the quantitative analysis of very important practical applications, and we show how the technology developed in the second volume of the book can be brought to bear on their solutions.

7.1 Mean Field Games with Major and Minor Players

An important requirement of the theory of mean field games is the fact that, when the number of players is large, the influence of one single player on the system becomes asymptotically negligible. This is not the case in many practical applications. For instance, it is in sharp contrast with the reality of the banking system where the actions of a few *Systemically Important Financial Institutions* (SIFI) impact the system no-matter how large the number of small banks is.

7.1.1 Isolating Strong Influential Players

In this section, we study a model with a small number of players which we call *major* and a large number of players with mean field interactions which we call

R. Carmona, F. Delarue, *Probabilistic Theory of Mean Field Games with Applications II*, Probability Theory and Stochastic Modelling 84,
DOI 10.1007/978-3-319-56436-4_7

minor because their influence on the system decreases as their number increases. In the asymptotic regime of large games, the limiting problem is identified as a two-player stochastic differential game, in which the optimization problem faced by the major player is of conditional McKean-Vlasov type, while the optimization problem faced by the representative minor player is a standard control problem. A matching procedure then follows the solution of the two-player game, leading to a characterization of the solution of the limiting problem as an FBSDE of McKean-Vlasov type. In our new setting, the finite-player game is an $(N+1)$-player game including the major player. The construction of approximate Nash equilibria in Subsection 7.1.5 below, involves both minor and major players, justifying our choice for the limiting scheme and the mean field formulation of the problem. It is based on the conditional propagation of chaos results for stochastic differential equations of McKean-Vlasov type developed in Section 2.1 of Chapter 2.

The probabilistic approach advocated throughout the book is especially well suited to the analysis of mean field games with major and minor players. For starters, it is forced on us when one chooses to use, as we do below, open loop controls. Also, and in full analogy with the results obtained for mean field games with a common noise, the persistence of the influence of the major player forces the controls of the minor players to retain a random component. Moreover, while it is clear that the limiting conditional McKean-Vlasov control problem faced by the major player should be amenable to an appropriate version of the Pontryagin stochastic maximum principle, neither the exact form of the stochastic maximum principle nor the conditional propagation of chaos result which we need are covered by the results of Chapter (Vol I)-6, not even the discussion provided in Section 2.1 of Chapter 2. This is the reason why we offer tailor-made statements of the results we need, even if we only hint at their proofs. However, the part of the analysis dealing with the minor players relies on the results on the well posedness of FBSDEs and their associated decoupling fields developed for the solvability of the limiting mean field game problems.

The Finite Player Game Set-Up

The finite player version of the game with major and minor players is as follows. The major player is indexed by 0. It chooses a control strategy $\boldsymbol{\alpha}^{N,0}$ taking values in a convex set $A_0 \subset \mathbb{R}^{k_0}$. The minor players are indexed by $i \in \{1, \cdots, N\}$. Player i chooses a control strategy $\boldsymbol{\alpha}^{N,i}$ taking values in a convex set $A \subset \mathbb{R}^k$. The state of the system at time t is given by a vector $X_t^{(0,N)} = (X_t^{N,0}, X_t^{N,1}, \cdots, X_t^{N,N}) \in \mathbb{R}^{d_0+Nd}$ with dynamics:

$$
\begin{cases}
dX_t^{N,0} = b_0(t, X_t^{N,0}, \bar{\mu}_t^N, \alpha_t^{N,0})dt + \sigma_0(t, X_t^{N,0}, \bar{\mu}_t^N, \alpha_t^{N,0})dW_t^0, \\
dX_t^{N,i} = b(t, X_t^{N,i}, \bar{\mu}_t^N, X_t^{N,0}, \alpha_t^{N,i}, \alpha_t^{N,0})dt \\
\qquad\qquad + \sigma(t, X_t^{N,i}, \bar{\mu}_t^N, X_t^{N,0}, \alpha_t^{N,i}, \alpha_t^{N,0})dW_t^i, \quad i = 1, \cdots, N,
\end{cases}
\tag{7.1}
$$

for $t \in [0, T]$, where $\boldsymbol{W} = (\boldsymbol{W}^i)_{i=0,\cdots,N}$ is a family of independent Wiener processes, and

$$\bar{\mu}_t^N = \frac{1}{N}\sum_{i=1}^N \delta_{X_t^{N,i}} \tag{7.2}$$

is the empirical distribution of the states of the minor players. The Wiener process $\boldsymbol{W}^0$ is assumed to be m_0 dimensional while all the other Wiener processes $\boldsymbol{W}^i$ for $i = 1, \cdots, N$ are assumed to be m-dimensional. The state $X_t^{N,0}$ (and hence b_0) is d_0-dimensional while all the other states $X_t^{N,i}$ (and hence b) are d-dimensional. Here, the set-up is different from what it was in the previous chapters since we allow for pedagogical reasons, the states and the noises to have different dimensions. Finally, for consistency reasons, the matrices σ_0 and σ are $d_0 \times m_0$ and $d \times m$ dimensional respectively. The major player aims at minimizing a cost functional given by:

$$J^{N,0}(\boldsymbol{\alpha}^{N,0}, \boldsymbol{\alpha}^{(N)}) = \mathbb{E}\biggl[\int_0^T f_0(t, X_t^{N,0}, \bar{\mu}_t^N, \alpha_t^{N,0})dt + g_0(X_T^{N,0}, \bar{\mu}_T^N)\biggr], \tag{7.3}$$

and the minor players aim at minimizing the cost functionals defined by:

$$\begin{aligned} &J^{N,i}\bigl(\boldsymbol{\alpha}^{N,0}, \boldsymbol{\alpha}^{(N)}\bigr) \\ &\quad = \mathbb{E}\biggl[\int_0^T f(t, X_t^{N,i}, \bar{\mu}_t^N, X_t^{N,0}, \alpha_t^{N,i}, \alpha_t^{N,0})dt + g(X_T^{N,i}, \bar{\mu}_T^N, X_T^{N,0})\biggr], \end{aligned} \tag{7.4}$$

for $i = 1, \cdots, N$, where we use the notation $\boldsymbol{\alpha}^{(N)}$ for $(\boldsymbol{\alpha}^{N,1}, \cdots, \boldsymbol{\alpha}^{N,N})$. Notice the presence of the state of the major player in the state dynamics and the cost functions of the minor players. Even when the number of minor players is large, the major player can still influence significantly the behavior of the system. This is in sharp contrast with the models considered so far as in all cases, the impact of any given player was becoming negligible as the size of the population increased without bound.

7.1.2 Formulation of the Open Loop MFG Problem

We now formulate the open loop version of the mean field game problem based on the large N behavior of the system. In this limiting regime, the symmetry of the states of the minor players suggests that their empirical distribution $\bar{\mu}_t^N$ should converge toward a probability measure μ_t which should still feel the state of the major player and hence, at least indirectly, the path $W_{[0,t]}^0$ up to time t of the Wiener process W^0. Like in the case of mean field games with a common noise, this limit should act like a random environment depending upon the common noise. This prompts us to use the same setting as in Definition 2.16 of a weak equilibrium of a mean field game with common noise. So, we assume that we are given:

1. a complete probability space $(\Omega^0, \mathcal{F}^0, \mathbb{P}^0)$ endowed with a complete and right-continuous filtration $\mathbb{F}^0 = (\mathcal{F}_t^0)_{0 \leq t \leq T}$ generated by an m_0-dimensional Wiener process $W^0 = (W_t^0)_{0 \leq t \leq T}$,
2. a complete probability space $(\Omega^1, \mathcal{F}^1, \mathbb{P}^1)$ endowed with a complete and right-continuous filtration $\mathbb{F}^1 = (\mathcal{F}_t^1)_{0 \leq t \leq T}$ generated by an m-dimensional Wiener process $W = (W_t)_{0 \leq t \leq T}$.

We then denote by $(\Omega, \mathcal{F}, \mathbb{P})$ the completion of the product space $(\Omega^0 \times \Omega^1, \mathcal{F}^0 \otimes \mathcal{F}^1, \mathbb{P}^0 \otimes \mathbb{P}^1)$, endow it with the filtration $\mathbb{F} = (\mathcal{F}_t)_{0 \leq t \leq T}$ obtained by augmenting the product filtration $\mathbb{F}^0 \otimes \mathbb{F}^1$ to make it right-continuous and by completing it.

These assumptions are different from those of Chapter 2 since we here force the filtration $\mathbb{F}^0$ to be generated by W^0 and similarly for $\mathbb{F}^1$. According to the terminology used for mean field games with common noise, this means that we restrict our analysis to the search of strong solutions. Consequently, there will be no need for compatibility conditions. As in Subsection 2.1.3, we use the notation $\mathcal{L}^1(X)$ to denote the random variable $\mathcal{L}^1(X) : \Omega^0 \ni \omega^0 \mapsto \mathcal{L}(X(\omega^0, \cdot))$ whenever X is a random variable defined on $(\Omega, \mathcal{F}, \mathbb{P})$.

Unfortunately, the parallel with games with a common noise cannot be pushed any further. Indeed, no matter how large N is, the major player's control influences the states of all the minor players, and in particular, their empirical distribution. So when we construct the limiting problem for the major player, it is reasonable to allow the major player to control the stochastic measure flow, instead of fixing it exogenously. So in the limit, the control problem of the major player should be of conditional McKean-Vlasov type with an endogenous measure flow, and the limiting optimization problem of the representative minor player should be standard, with a fixed exogenous measure flow. Because of this asymmetry, we formulate the limiting mean field game problem as a two step procedure as in the classical case, the first step being now a two-player stochastic differential game between the major player and a representative minor player. To be more specific, the solution of a mean field game with major and minor players comprises the two steps:

1. For each fixed $\mathbb{F}^0$-progressively measurable stochastic measure flow $\boldsymbol{\mu} = (\mu_t)_{0 \le t \le T}$, solve the two-player stochastic differential game for open loop Nash equilibria with a state $(X_t^0, X_t, \check{X}_t^0, \check{X}_t)_{0 \le t \le T}$ evolving in time according to:

$$\begin{cases} dX_t^0 = b_0(t, X_t^0, \mathcal{L}^1(X_t), \alpha_t^0)dt + \sigma_0(t, X_t^0, \mathcal{L}^1(X_t), \alpha_t^0)dW_t^0, \\ dX_t = b(t, X_t, \mathcal{L}^1(X_t), X_t^0, \alpha_t, \alpha_t^0)dt + \sigma(t, X_t, \mathcal{L}^1(X_t), X_t^0, \alpha_t, \alpha_t^0)dW_t, \\ d\check{X}_t^0 = b_0(t, \check{X}_t^0, \mu_t, \alpha_t^0)dt + \sigma_0(t, \check{X}_t^0, \mu_t, \alpha_t^0)dW_t^0, \\ d\check{X}_t = b(t, \check{X}_t, \mu_t, \check{X}_t^0, \alpha_t, \alpha_t^0)dt + \sigma(t, \check{X}_t, \mu_t, \check{X}_t^0, \alpha_t, \alpha_t^0)dW_t, \end{cases} \tag{7.5}$$

for $t \in [0, T]$, with initial conditions $X_0^0 = \check{X}_0^0 = x^{00}$ and $X_0 = \check{X}_0 = x^0$, where the control $\boldsymbol{\alpha}^0 = (\alpha_t^0)_{0 \le t \le T}$ of the first player is assumed to be adapted to the filtration $\mathbb{F}^0 = (\mathcal{F}_t^0)_{t \ge 0}$, and the control $\boldsymbol{\alpha} = (\alpha_t)_{0 \le t \le T}$ of the second player is assumed to be adapted to the filtration $\mathbb{F} = (\mathcal{F}_t)_{t \ge 0}$, and where the cost functionals that the two players try to minimize are given by:

$$\begin{aligned} J^0(\boldsymbol{\alpha}^0, \boldsymbol{\alpha}) &= \mathbb{E}\left[\int_0^T f_0\big(t, X_t^0, \mathcal{L}^1(X_t), \alpha_t^0\big)dt + g_0\big(X_T^0, \mathcal{L}^1(X_T)\big)\right], \\ J(\boldsymbol{\alpha}^0, \boldsymbol{\alpha}) &= \mathbb{E}\left[\int_0^T f(t, \check{X}_t, \mu_t, \check{X}_t^0, \alpha_t, \alpha_t^0)dt + g(\check{X}_T, \mu_T, \check{X}_T^0)\right]. \end{aligned} \tag{7.6}$$

2. Enforce the consistency condition:

$$\mu_t = \mathcal{L}^1(X_t), \quad \mathbb{P} - \text{a.s.} \quad t \in [0, T], \tag{7.7}$$

where $(X_t)_{0 \le t \le T}$ is the second component in the state equation (7.5) driven by a Nash equilibrium control $(\boldsymbol{\alpha}^0, \boldsymbol{\alpha})$ found in the first step.

Like in the case of mean field games with a common noise, the above consistency condition requires the solution of a fixed point problem in the space of stochastic measure flows. We present an alternative in the next subsection. In equilibrium, namely once the consistency condition (7.7) is met, $(\boldsymbol{X}^0, \boldsymbol{X})$ and $(\check{\boldsymbol{X}}^0, \check{\boldsymbol{X}})$ coincide, even though they do not ex-ante, since they emerge from different measure flows: $(\boldsymbol{X}^0, \boldsymbol{X})$ is defined with the endogenous measure flow $(\mathcal{L}^1(X_t))_{0 \le t \le T}$, while $(\check{\boldsymbol{X}}^0, \check{\boldsymbol{X}})$ is defined with the exogenous measure flow $\boldsymbol{\mu} = (\mu_t)_{0 \le t \le T}$. We emphasize once more the rationale for this special formulation. When computing its best response to the major player and the other minor players, the representative minor player assumes that the stochastic flow $\boldsymbol{\mu} = (\mu_t)_{0 \le t \le T}$ is fixed, like in the standard approach to mean field games, the random shocks driving the dynamics of the state of the major player justifying the randomness of the flow. The fact that it is responding to a major player who should behave in the environment given by

$\boldsymbol{\mu} = (\mu_t)_{0\le t\le T}$, is the justification for the introduction of $(\check{X}_t^0)_{0\le t\le T}$ in lieu of $(X_t^0)_{0\le t\le T}$ to compute its best response. On the other hand, for reasons already given earlier, the major player computes its best response assuming that the representative minor player uses the endogenous stochastic flow $(\mathcal{L}^1(X_t))_{0\le t\le T}$. So it is responding to the dynamics of $(X_t)_{0\le t\le T}$ instead of the dynamics given by $(\check{X}_t)_{0\le t\le T}$. This explains this apparent doubling of the states which disappears in equilibrium when the consistency condition is satisfied. Accordingly, the cost functional J^0 of the major player is of the McKean-Vlasov type while the cost functional J of the representative minor player is of the standard type. As explained earlier, this is the main feature of our formulation of the problem. We end this subsection with the precise definition of a solution to the mean field game described above.

Definition 7.1 *Given a tuple $(\Omega, \mathcal{F}, \mathbb{P}, \boldsymbol{W}^0, \boldsymbol{W})$ as above, a solution of the mean field game with major and minor players is defined as a couple of controls $(\boldsymbol{\alpha}^0, \boldsymbol{\alpha})$, where $\boldsymbol{\alpha}^0$ is $\mathbb{F}^0$-progressively measurable and $\boldsymbol{\alpha}$ is $\mathbb{F}$-progressively measurable, forming an open loop Nash equilibrium for the two-player game defined in the first step above, and satisfying the consistency condition.*

In the spirit of the results of Section 6.1 of Chapter 6, we shall argue that if we are able to find a fixed point, i.e., a stochastic measure flow $\boldsymbol{\mu} = (\mu_t)_{0\le t\le T}$ satisfying (7.7), then one can construct approximate Nash equilibria for the finite-player game when the number of players is sufficiently large. The precise meaning of this statement will be made clear in Subsection 7.1.5.

Remark 7.2 *The requirement that $\boldsymbol{\alpha}^0$ is $\mathbb{F}^0$-measurable is especially meaningful as it says that the strategy profile of the major player cannot be based upon the private state of the minor player. The typical example for such an $\boldsymbol{\alpha}^0$ is a strategy profile in closed loop form depending on the present private state of the major player and the present conditional distribution of the minor player. By analogy with uniquely solvable mean field games with common noise, we may expect that an optimal $\boldsymbol{\alpha}^0$ is likely to be of this form in equilibrium, although the equilibrium is understood in the open loop sense.*

We close this section with a further emphasis on the importance of the measurability properties of the controls which are permissible in the case of mean field games with major and minor players. We stress that, in contrast with what happened in our previous analysis of standard mean field games, the class of strategy profiles allowed in the limiting formulation really matters now. Indeed, limiting equilibria should differ when computed over strategy profiles in closed loop form. This should be clear when we compute the best response of the major player: if the control α_t of the representative minor player is of the form $\alpha_t = \phi(t, X_t, X_t^0)$, then the dynamics of the state process $\boldsymbol{X} = (X_t)_{0\le t\le T}$ after a perturbation of $\boldsymbol{\alpha}^0$ are not same whether we freeze the sole feedback function ϕ or the whole path $\boldsymbol{\alpha} = (\alpha_t)_{0\le t\le T}$ in the computation of the best response. We come back to this specific question when discussing the passage from the game with finitely many players to the limiting game. In any case, Subsection 7.1.3 below offers an alternative to the open loop framework introduced above, shedding new light on the McKean-Vlasov nature of the optimization problem of the major player.

7.1.3 Aside: Alternative Formulations of the Mean Field Game Problems

The goal of this subsection is to prepare for the extension of the discussion of Subsection (Vol I)-3.1.3 of Chapter (Vol I)-3 to models of mean field games with major and minor players. Meanwhile, we introduce the corresponding notions of equilibria in closed loop and Markovian forms and point out the main differences between the three forms of equilibria.

Since we already went through the procedure many times throughout the book, we do not introduce the finite player game (with major and minor players) from which we usually derive the mean field game formulation by taking the limit ($N \to \infty$) of a large population of minor players. We directly state the mean field game problem. However, in contrast with the presentation of Subsection (Vol I)-3.1.3 where we did not separate the open and closed loop problems, we here treat them separately, even at the risk of annoying the reader with pedantic repetitions. Our reason for this duplication is the following. While it turns out that the solutions to the open and closed loop versions of the standard games often coincide in the mean field limit, this should not be the case for games with major and minor players, as we already alluded to in the previous subsection. This is due to the fact that the finite characteristics of the major player do not disappear in the limit when the number of minor players tends to infinity. We shall illustrate this fact in our discussion of the linear quadratic models below.

We first treat the case of open loop equilibria. This part is a plain rewrite of the previous section introducing the open loop version of the problem. We take advantage of the fact that the filtrations are assumed to be generated by the Wiener processes to write the controls as functions of the paths of these Wiener processes. The benefits of revisiting the open loop set-up are twofold: first it naturally leads to a formulation of the fixed point step on spaces of functions instead of measure flows, and second, it extends in a straightforward manner to closed loop formulations of the problem which we present next.

Open Loop Version of the MFG Problem Revisited

Here, we assume that the controls used by the major player and the representative minor player are of the form:

$$\alpha_t^0 = \phi^0\big(t, W^0_{[0,T]}\big), \quad \text{and} \quad \alpha_t = \phi\big(t, W^0_{[0,T]}, W_{[0,T]}\big), \tag{7.8}$$

for deterministic progressively measurable functions $\phi^0 : [0,T] \times \mathcal{C}([0,T];\mathbb{R}^{m_0}) \to A_0$ and $\phi : [0,T] \times \mathcal{C}([0,T];\mathbb{R}^{m_0}) \times \mathcal{C}([0,T];\mathbb{R}^m) \to A$. Progressive measurability of the function ϕ means in particular that for any $t \in [0,T]$ and $(w^0, w) \in \mathcal{C}([0,T];\mathbb{R}^{m_0}) \times \mathcal{C}([0,T];\mathbb{R}^m)$, the value of $\phi(t, w^0, w)$ depends only upon the restrictions $w^0_{[0,t]}$ and $w_{[0,t]}$ of w^0 and w to the interval $[0,t]$. Similarly for ϕ^0. Our choice for the admissibility of the controls is consistent with our earlier discussion since we assume that the filtrations $\mathbb{F}^0$ and $\mathbb{F}^1$ are generated by the Wiener processes $\boldsymbol{W}^0$ and $\boldsymbol{W}$ respectively. In this framework, the state $(X^0_t)_{0\le t\le T}$ of the major player and the state $(X_t)_{0\le t\le T}$ of the representative minor player in a field of exchangeable minor players evolve according to the dynamic equations:

$$\begin{cases} dX^0_t = b_0(t, X^0_t, \mu_t, \alpha^0_t)dt + \sigma_0(t, X^0_t, \mu_t, \alpha^0_t)dW^0_t, \\ dX_t = b(t, X_t, \mu_t, X^0_t, \alpha_t, \alpha^0_t)dt + \sigma(t, X_t, \mu_t, X^0_t, \alpha_t, \alpha^0_t)dW_t. \end{cases} \tag{7.9}$$

Accordingly, the costs that the players try to minimize are of the form:

$$\begin{cases} J^0(\boldsymbol{\alpha}^0, \boldsymbol{\alpha}) = \mathbb{E}\left[\int_0^T f_0(t, X^0_t, \mu_t, \alpha^0_t)dt + g^0(X^0_T, \mu_T)\right], \\ J(\boldsymbol{\alpha}^0, \boldsymbol{\alpha}) = \mathbb{E}\left[\int_0^T f(t, X_t, \mu_t, X^0_t, \alpha_t, \alpha^0_t)dt + g(X_T, \mu_T, X^0_T)\right]. \end{cases} \tag{7.10}$$

In agreement with (7.5), $(\mu_t)_{0\le t\le T}$ should be understood as a proxy for the flow of empirical measures $(\bar{\mu}^N_t)_{0\le t\le T}$ in the limit $N \to \infty$ when the minor players implement exchangeable strategies. As we explained in our introduction to the mean field game models with a common noise, this limit happens to be $(\mu_t = \mathcal{L}(X_t | W^0_{[0,t]}))_{0\le t\le T}$, the conditional distribution of the state of the representative minor player given the initial path $W^0_{[0,t]}$ of the noise common to all the minor players. Since we are working on a product probability set-up, we can use the notation of the previous subsection and write in that case $\mu_t = \mathcal{L}^1(X_t)$, for $t \in [0,T]$.

We shall see below how to account for the remaining two equations in (7.5).

The Major Player Problem. We assume that the representative minor player uses the open loop control given by the progressively measurable function $\phi : (t, w^0, w) \mapsto \phi(t, w^0, w)$. Hence the problem of the major player is to minimize its expected cost:

$$J^{\phi,0}(\boldsymbol{\alpha}^0) = \mathbb{E}\left[\int_0^T f_0(t, X^0_t, \mu_t, \alpha^0_t)dt + g^0(X^0_T, \mu_T)\right], \tag{7.11}$$

under the dynamical constraints:

$$
\begin{cases}
dX_t^0 = b_0(t, X_t^0, \mu_t, \alpha_t^0)dt + \sigma_0(t, X_t^0, \mu_t, \alpha_t^0)dW_t^0, \\
dX_t = b\big(t, X_t, \mu_t, X_t^0, \phi(t, W_{[0,T]}^0, W_{[0,T]}), \alpha_t^0\big)dt \\
\qquad\qquad + \sigma\big(t, X_t, \mu_t, X_t^0, \phi(t, W_{[0,T]}^0, W_{[0,T]}), \alpha_t^0\big)dW_t^i,
\end{cases}
$$

where each $\mu_t = \mathcal{L}^1(X_t) = \mathcal{L}(X_t | W_{[0,t]}^0)$, for $t \in [0, T]$, denotes the conditional distribution of X_t given $W_{[0,t]}^0$. Since we are considering the open loop version of the problem, we search for minima in the class of controls $\boldsymbol{\alpha}^0$ of the form $(\alpha_t^0 = \phi^0(t, W_{[0,T]}^0))_{0 \le t \le T}$ for a progressively measurable function ϕ^0. So, we frame the major player problem as the search for:

$$
\boldsymbol{\phi}^{0,*}(\phi) = \arg \inf_{\alpha^0 \leftrightarrow \phi^0} J^{\phi,0}(\boldsymbol{\alpha}^0),
$$

where $\boldsymbol{\alpha}^0 \leftrightarrow \phi^0$ means that the infimum is over the set of controls $\boldsymbol{\alpha}^0$ given by progressively measurable functions ϕ^0. For the sake of the present discussion, we assume implicitly that the argument of the minimization is not empty and reduces to a singleton. The important feature of this formulation is that the optimization of the major player appears naturally as an optimal control of the McKean-Vlasov type! In fact, it is an optimal control of the *conditional McKean-Vlasov type* since the distribution appearing in the controlled dynamics is the conditional distribution of the state of the representative minor player.

The Representative Minor Player Problem. To formulate the optimization problem of the representative minor player, as we accounted for through the combination of the last two equations in (7.5) and of the fixed point condition (7.7), we first describe the state of a system comprising a major player and a field of exchangeable minor players. As above, we assume that the major player uses a strategy $\boldsymbol{\alpha}^0$ given by a progressively measurable function ϕ^0 as in $(\alpha_t^0 = \phi^0(t, W_{[0,T]}^0))_{0 \le t \le T}$, and that the representative of the field of minor players uses a strategy $\boldsymbol{\alpha}$ given by a progressively measurable function ϕ in the form $(\alpha_t = \phi(t, W_{[0,T]}^0, W_{[0,T]}))_{0 \le t \le T}$. So, the dynamics of the state of the system are given by:

$$
\begin{cases}
dX_t^0 = b_0\big(t, X_t^0, \mu_t, \phi^0(t, W_{[0,T]}^0)\big)dt \\
\qquad\qquad + \sigma_0\big(t, X_t^0, \mu_t, \phi^0(t, W_{[0,T]}^0)\big)dW_t^0, \\
dX_t = b\big(t, X_t, \mu_t, X_t^0, \phi(t, W_{[0,T]}^0, W_{[0,T]}), \phi^0(t, W_{[0,T]}^0)\big)dt \\
\qquad + \sigma\big(t, X_t, \mu_t, X_t^0, \phi(t, W_{[0,T]}^0, W_{[0,T]}), \phi^0(t, W_{[0,T]}^0)\big)dW_t,
\end{cases}
\tag{7.12}
$$

where, as before, $\mu_t = \mathcal{L}^1(X_t) = \mathcal{L}(X_t | W_{[0,t]}^0)$ is the conditional distribution of X_t given $W_{[0,t]}^0$, for any $t \in [0, T]$. Notice once again that, in the present situation, given the feedback functions ϕ^0 and ϕ, this stochastic differential equation in $\mathbb{R}^{d_0} \times \mathbb{R}^d$ giving the dynamics of the state of the system is of (conditional) McKean-Vlasov type since $(\mu_t)_{0 \le t \le T}$ is the flow of (conditional) distributions of (part of) the state.

As in Subsection (Vol I)-3.1.3, we address the Nash condition for the minor player through the search for the best response that a virtual minor player should implement given the above field of exchangeable minor players, and in the present situation, given the above major player as well. So naturally, we formulate this search for a best response as the result of the optimization problem of an extra minor player which chooses a strategy $\check{\boldsymbol{\alpha}}$ given by a progressively measurable function $\check{\phi}$ in the form $(\check{\alpha}_t = \check{\phi}(t, W^0_{[0,T]}, \check{W}_{[0,T]}))_{0 \le t \le T}$ in order to minimize its expected cost:

$$J^{\phi^0,\phi}(\check{\boldsymbol{\alpha}}) = \mathbb{E}\biggl[\int_0^T f\bigl(t, \check{X}_t, \mu_t, X^0_t, \check{\alpha}_t, \phi^0(t, W^0_{[0,T]})\bigr)dt + g(\check{X}_T, \mu_T, X^0_T)\biggr],$$

where the dynamics of its state $(\check{X}_t)_{0 \le t \le T}$ are given by:

$$\begin{aligned} d\check{X}_t = b\bigl(t, \check{X}_t, \mu_t, X^0_t, \check{\phi}(t, W^0_{[0,T]}, \check{W}_{[0,T]}), \phi^0(t, W^0_{[0,T]})\bigr)dt \\ + \sigma\bigl(t, \check{X}_t, \mu_t, X^0_t, \check{\phi}(t, W^0_{[0,T]}, \check{W}_{[0,T]}), \phi^0(t, W^0_{[0,T]})\bigr)d\check{W}_t, \end{aligned}$$

for a Wiener process $\check{\boldsymbol{W}} = (\check{W}_t)_{0 \le t \le T}$ independent of the other Wiener processes. Notice that this optimization problem *is not* of McKean-Vlasov type. It is merely a classical optimal control problem, though with random coefficients. In particular, $(\mu_t)_{0 \le t \le T}$ is still given by $(\mu_t = \mathcal{L}(X_t | W^0_{[0,t]}))_{0 \le t \le T}$, for the same representative player $(X_t)_{0 \le t \le T}$ as in (7.12), but $\check{X}_t$ may differ from X_t. As stated above, we search for minima in the class of feedback controls $\check{\boldsymbol{\alpha}}$ of the form $(\check{\alpha}_t = \check{\phi}(t, W^0_{[0,T]}, \check{W}_{[0,T]}))_{0 \le t \le T}$. We denote by:

$$\check{\boldsymbol{\phi}}^*(\phi^0, \phi) = \arg \inf_{\check{\alpha} \leftrightarrow \check{\phi}} J^{\phi^0,\phi}(\check{\boldsymbol{\alpha}})$$

the result of the optimization. Again, we assume that the optimal control exists, is given by a progressively measurable function and is unique for the sake of convenience.

We now formulate the existence of a Nash equilibrium for the mean field game with major and minor players as a fixed point of the best response maps identified above. By definition, a couple $(\hat{\boldsymbol{\alpha}}^0, \hat{\boldsymbol{\alpha}})$ of controls given by progressively measurable functions $(\hat{\phi}^0, \hat{\phi})$ as above is a Nash equilibrium for the mean field game with major and minor players if it satisfies the fixed point equation:

$$(\hat{\phi}^0, \hat{\phi}) = \bigl(\boldsymbol{\phi}^{0,*}(\hat{\phi}), \check{\boldsymbol{\phi}}^*(\hat{\phi}^0, \hat{\phi})\bigr). \tag{7.13}$$

Closed Loop Version of the MFG Problem

The way we rewrote the open loop version of the problem may have been rather pompous, but it makes it easy to introduce the closed loop and Markovian versions of the problem. In this subsection, we assume that the controls used by the major player and the representative minor player (taken in a field of exchangeable minor players) are of the form:

$$\alpha_t^0 = \phi^0(t, X^0_{[0,T]}, \mu_t), \quad \text{and} \quad \alpha_t = \phi(t, X_{[0,T]}, \mu_t, X^0_{[0,T]}), \quad i = 1, \cdots, N.$$

for deterministic progressively measurable functions $\phi^0 : [0,T] \times \mathcal{C}([0,T];\mathbb{R}^{d_0}) \times \mathcal{P}_2(\mathbb{R}^d) \to A_0$ and $\phi : [0,T] \times \mathcal{C}([0,T];\mathbb{R}^d) \times \mathcal{P}_2(\mathbb{R}^d) \times \mathcal{C}([0,T];\mathbb{R}^{d_0}) \to A$. The state $(X_t^0)_{0 \le t \le T}$ of the major player and the state $(X_t)_{0 \le t \le T}$ of the representative minor player evolve according to the same dynamic equations (7.9) as before, and the costs are also given by the same formula (7.10), with $\mu_t = \mathcal{L}^1(X_t) = \mathcal{L}(X_t | W^0_{[0,t]})$, for $t \in [0,T]$.

We follow the same procedure as above to describe equilibria in closed loop form.

The Major Player Problem. We assume that the representative minor player uses the progressively measurable feedback function $\phi : (t, \boldsymbol{x}, \mu, \boldsymbol{x}^0) \mapsto \phi(t, \boldsymbol{x}, \mu, \boldsymbol{x}^0)$, so the problem of the major player is to minimize its expected cost (7.11) under the dynamical constraints:

$$\begin{cases} dX_t^0 &= b_0(t, X_t^0, \mu_t, \alpha_t^0)dt + \sigma_0(t, X_t^0, \mu_t, \alpha_t^0)dW_t^0, \\ dX_t &= b\big(t, X_t, \mu_t, X_t^0, \phi(t, X_{[0,T]}, \mu_t, X^0_{[0,T]}), \alpha_t^0\big)dt \\ & \qquad + \sigma\big(t, X_t, \mu_t, X_t^0, \phi(t, X_{[0,T]}, \mu_t, X^0_{[0,T]}), \alpha_t^0\big)dW_t^i, \end{cases}$$

where $\mu_t = \mathcal{L}^1(X_t) = \mathcal{L}(X_t | W^0_{[0,t]})$ denotes the conditional distribution of X_t given $W^0_{[0,t]}$, for any $t \in [0,T]$. As explained earlier, we search for minima in the class of feedback controls $\boldsymbol{\alpha}^0$ of the form $(\alpha_t^0 = \phi^0(t, X^0_{[0,T]}, \mu_t))_{0 \le t \le T}$. So, we frame the major player problem as:

$$\boldsymbol{\phi}^{0,*}(\phi) = \arg \inf_{\alpha^0 \leftrightarrow \phi^0} J^{\phi,0}(\boldsymbol{\alpha}^0),$$

which is an optimal control of the conditional McKean-Vlasov type!

The Representative Minor Player Problem. To formulate the optimization problem of the minor player in the definition of a Nash equilibrium, we first describe the system to which it needs to respond optimally. So, we assume that the major player uses a strategy $\boldsymbol{\alpha}^0$ in feedback form given by a feedback function ϕ^0 so that $(\alpha_t^0 = \phi^0(t, X^0_{[0,T]}, \mu_t))_{0 \le t \le T}$, and that the representative of the field of minor players uses a strategy $\boldsymbol{\alpha}$ given by a progressively measurable feedback function ϕ in the form $(\alpha_t = \phi(t, X_{[0,T]}, \mu_t, X^0_{[0,T]}))_{0\le t\le T}$. Hence, the dynamics of the state of this system are given by:

$$\begin{cases} dX^0_t = b_0\big(t, X^0_t, \mu_t, \phi^0(t, X^0_{[0,T]}, \mu_t)\big)dt + \sigma_0\big(t, X^0_t, \mu_t, \phi^0(t, X^0_{[0,T]}, \mu_t)\big)dW^0_t, \\ dX_t = b\big(t, X_t, \mu_t, X^0_t, \phi(t, X_{[0,T]}, \mu_t, X^0_{[0,T]}), \phi^0(t, X^0_{[0,T]}, \mu_t)\big)dt \\ \qquad\qquad +\sigma\big(t, X_t, \mu_t, X^0_t, \phi(t, X_{[0,T]}, \mu_t, X^0_{[0,T]}), \phi^0(t, X^0_{[0,T]}, \mu_t)\big)dW_t, \end{cases}$$

where as before, $\mu_t = \mathcal{L}^1(X_t) = \mathcal{L}(X_t|W^0_{[0,t]})$ is the conditional distribution of X_t given $W^0_{[0,t]}$, for any $t \in [0,T]$. Again, given the feedback functions ϕ^0 and ϕ, this stochastic differential equation in $\mathbb{R}^{d_0} \times \mathbb{R}^d$ is of (conditional) McKean-Vlasov type.

As before, we address the equilibrium condition for the minor player through the search for the best response of an extra minor player to the major player and to the field of exchangeable minor players. We formulate this best response as the result of the optimization problem of a virtual minor player which chooses a strategy $\check{\boldsymbol{\alpha}}$ given by a progressively measurable feedback function $\check{\phi}$ in the form $(\check{\alpha}_t = \check{\phi}(t, \check{X}_{[0,T]}, \mu_t, X^0_{[0,T]}))_{0\le t\le T}$ in order to minimize its expected cost:

$$J^{\phi^0,\phi}(\check{\boldsymbol{\alpha}}) = \mathbb{E}\bigg[\int_0^T f\big(t, \check{X}_t, \mu_t, X^0_t, \check{\alpha}_t, \phi^0(t, X^0_{[0,T]})\big)dt + g(\check{X}_T, \mu_T, X^0_T)\bigg],$$

where the dynamics of the virtual state $(\check{X}_t)_{0\le t\le T}$ are given by:

$$\begin{aligned} d\check{X}_t &= b\big(t, \check{X}_t, \mu_t, X^0_t, \check{\phi}(t, \check{X}_{[0,T]}, \mu_t, X^0_{[0,T]}), \mu_t, \phi^0(t, X^0_{[0,T]})\big)dt \\ &\qquad + \sigma\big(t, \check{X}_t, \mu_t, X^0_t, \check{\phi}(t, \check{X}_{[0,T]}, \mu_t, X^0_{[0,T]}), \mu_t, \phi^0(t, X^0_{[0,T]})\big)d\check{W}_t, \end{aligned}$$

for a Wiener process $\check{\boldsymbol{W}} = (\check{W}_t)_{0\le t\le T}$ independent of the other Wiener processes. We stress the fact that $(\mu_t)_{0\le t\le T}$ is fixed as it is given by $(\mu_t = \mathcal{L}(X_t|W^0_{[0,t]}))_{0\le t\le T}$. We search for minima in the class of feedback controls $\check{\boldsymbol{\alpha}}$ of the form $(\check{\alpha}_t = \check{\phi}(t, \check{X}_{[0,T]}, \mu_t, X^0_{[0,T]}))_{0\le t\le T}$, and we denote the solution by:

$$\check{\boldsymbol{\phi}}^*(\phi^0, \phi) = \arg\inf_{\check{\alpha}\leftrightarrow\check{\phi}} J^{\phi^0,\phi}(\check{\boldsymbol{\alpha}}).$$

Finally, we define the solution of a Nash equilibrium for the closed loop mean field game with major and minor players as the solution of the same fixed point equation (7.13), except for the fact that the functions $(\hat{\phi}^0, \hat{\phi})$ are now progressively measurable feedback functions of the type considered here.

Markovian Version of the MFG Problem

Here, we assume that the controls used by the major player and the representative minor player are of the form:

$$\alpha_t^0 = \phi^0(t, X_t^0, \mu_t), \quad \text{and} \quad \alpha_t = \phi(t, X_t, \mu_t, X_t^0), \quad i = 1, \cdots, N,$$

for deterministic feedback functions $\phi^0 : [0, T] \times \mathbb{R}^{d_0} \times \mathcal{P}_2(\mathbb{R}^d) \to A_0$ and $\phi : [0, T] \times \mathbb{R}^d \times \mathcal{P}_2(\mathbb{R}^d) \times \mathbb{R}^{d_0} \to A$. The state $(X_t^0)_{0 \le t \le T}$ of the major player and the state $(X_t)_{0 \le t \le T}$ of the representative minor player evolve according to the same dynamic equations (7.9) as before and the costs are also given by the same formula (7.10), with $\mu_t = \mathcal{L}^1(X_t) = \mathcal{L}(X_t | W_{[0,t]}^0)$, for any $t \in [0, T]$.

Remark 7.3 *As we already pointed out in the discussion of the mean field game models with a common noise, the so-called Markovian version of the problem is not really Markovian since the past of the Wiener process driving the dynamics of the state of the major player (which plays the role of the common noise in the current situation) is present in the controls through the proxy μ_t of the conditional distribution of the states of the minor players! Still, the terminology may be fully justified if we consider the whole $\mathbb{R}^{d_0} \times \mathbb{R}^d \times \mathcal{P}_2(\mathbb{R}^d)$ as state space and $(X_t^0, X_t, \mathcal{L}^1(X_t))_{0 \le t \le T}$ as state variable.*

The Major Player Problem. As for equilibria in closed loop form, we assume that the representative minor player uses the feedback function $\phi : (t, x, \mu, x^0) \mapsto \phi(t, x, \mu, x^0)$. Hence the problem of the major player is to minimize its expected cost (7.11) under the dynamical constraints:

$$\begin{cases} dX_t^0 &= b_0(t, X_t^0, \mu_t, \alpha_t^0)dt + \sigma_0(t, X_t^0, \mu_t, \alpha_t^0)dW_t^0, \\ dX_t &= b\big(t, X_t, \mu_t, X_t^0, \phi(t, X_t, \mu_t, X_t^0), \alpha_t^0\big)dt \\ & \qquad + \sigma\big(t, X_t, \mu_t, X_t^0, \phi(t, X_t, \mu_t, X_t^0), \alpha_t^0\big)dW_t^i, \end{cases}$$

where $\mu_t = \mathcal{L}^1(X_t) = \mathcal{L}(X_t | W_{[0,t]}^0)$ denotes the conditional distribution of X_t given $W_{[0,t]}^0$, for any $t \in [0, T]$. We search for minima in the class of feedback controls $\boldsymbol{\alpha}^0$ of the form $(\alpha_t^0 = \phi^0(t, X_t^0, \mu_t))_{0 \le t \le T}$. Accordingly, we frame the major player problem as:

$$\boldsymbol{\phi}^{0,*}(\phi) = \arg \inf_{\alpha^0 \leftrightarrow \phi^0} J^{\phi,0}(\boldsymbol{\alpha}^0).$$

As before, the optimization problem of the major player is of the conditional McKean-Vlasov type.

The Representative Minor Player Problem. As before, in order to formulate the optimization problem of the minor player, we need to describe the system to which it tries to respond optimally. We assume that the major player uses a strategy $\boldsymbol{\alpha}^0$

given by a feedback function ϕ^0 so that $(\alpha_t^0 = \phi^0(t, X_t^0, \mu_t))_{0\le t\le T}$, and that the representative of the field of exchangeable minor players uses a strategy $\boldsymbol{\alpha}$ given by a feedback function ϕ in the form $(\alpha_t = \phi(t, X_t, \mu_t, X_t^0))_{0\le t\le T}$. Hence, the dynamics of the state of this system are given by:

$$\begin{cases} dX_t^0 = b_0\big(t, X_t^0, \mu_t, \phi^0(t, X_t^0, \mu_t)\big)dt + \sigma_0\big(t, X_t^0, \mu_t, \phi^0(t, X_t^0, \mu_t)\big)dW_t^0, \\ dX_t = b\big(t, X_t, \mu_t, X_t^0, \phi(t, X_t, \mu_t, X_t^0), \phi^0(t, X_t^0, \mu_t)\big)dt \\ \qquad\qquad + \sigma\big(t, X_t, \mu_t, X_t^0, \phi(t, X_t, \mu_t, X_t^0), \phi^0(t, X_t^0, \mu_t)\big)dW_t, \end{cases}$$

where as before, $\mu_t = \mathcal{L}^1(X_t) = \mathcal{L}(X_t | W^0_{[0,t]})$ is the conditional distribution of X_t given $W^0_{[0,t]}$. Again, given the feedback functions ϕ^0 and ϕ, this stochastic differential equation in $\mathbb{R}^{d_0} \times \mathbb{R}^d$ is of (conditional) McKean-Vlasov type.

As before, we search for the equilibrium condition for the minor player by solving the optimization problem of an extra virtual minor player which chooses a strategy $\check{\boldsymbol{\alpha}}$ given by a feedback function $\check{\phi}$ in the form $(\check{\alpha}_t = \check{\phi}(t, \check{X}_t, \mu_t, X_t^0))_{0\le t\le T}$ in order to minimize its expected cost:

$$J^{\phi^0,\phi}(\check{\boldsymbol{\alpha}}) = \mathbb{E}\bigg[\int_0^T f\big(t, \check{X}_t, \mu_t, X_t^0, \check{\alpha}_t, \phi^0(t, X_t^0, \mu_t)\big)dt + g(\check{X}_T, \mu_T, X_T^0)\bigg],$$

where the dynamics of the virtual state $(\check{X}_t)_{0\le t\le T}$ are given by:

$$\begin{aligned} d\check{X}_t &= b\big(t, \check{X}_t, \mu_t, X_t^0, \check{\phi}(t, \check{X}_t, \mu_t, X_t^0), \phi^0(t, X_t^0, \mu_t)\big)dt \\ &\quad + \sigma\big(t, \check{X}_t, \mu_t, X_t^0, \check{\phi}(t, \check{X}_t, \mu_t, X_t^0), \phi^0(t, X_t^0, \mu_t)\big)d\check{W}_t, \end{aligned}$$

for a Wiener process $\check{\boldsymbol{W}} = (\check{W}_t)_{0\le t\le T}$ independent of the other Wiener processes. We search for minima in the class of feedback controls $\check{\boldsymbol{\alpha}}$ of the form $(\check{\alpha}_t = \check{\phi}(t, \check{X}_t, \mu_t, X_t^0))_{0\le t\le T}$, and we denote the solution by:

$$\check{\boldsymbol{\phi}}^*(\phi^0, \phi) = \arg \inf_{\check{\alpha}\leftrightarrow\check{\phi}} J^{\phi^0,\phi}(\check{\boldsymbol{\alpha}}).$$

Finally, we define the solution of a Nash equilibrium for the Markovian mean field game with major and minor players as the solution of the same fixed point equation (7.13), except for the fact that the functions $(\hat{\phi}^0, \hat{\phi})$ are now feedback functions of the type considered here.

Next we return to the formulation of the open loop problem given in the previous subsection. We shall use the alternative formulations given in this subsection when we discuss the linear quadratic models in Subsection 7.1.6 where we compare the solutions in the open loop and the closed loop cases.

7.1.4 Back to the General Open Loop Problem

In order to proceed with the analysis of equilibria in open loop form, we introduce the following assumption.

Assumption (Major Minor MFG). The functions b_0, σ_0 and f_0 are defined on $[0,T] \times \mathbb{R}^{d_0} \times \mathcal{P}_2(\mathbb{R}^{d_0}) \times A_0$ with values in $\mathbb{R}^{d_0}$, $\mathbb{R}^{d_0 \times m_0}$ and $\mathbb{R}$ respectively. The functions b, σ and f are defined on $[0,T] \times \mathbb{R}^d \times \mathcal{P}_2(\mathbb{R}^d) \times \mathbb{R}^{d_0} \times A$ with values in $\mathbb{R}^d$, $\mathbb{R}^{d\times m}$ and $\mathbb{R}$ respectively. Also, the real valued functions g_0 and g are defined on $\mathbb{R}^{d_0} \times \mathcal{P}_2(\mathbb{R}^{d_0})$ and $\mathbb{R}^d \times \mathcal{P}_2(\mathbb{R}^d)$ respectively.
Moreover, there exists a constant $L \geq 0$ such that:

(A1) For all $t \in [0,T]$, $x_0, x_0' \in \mathbb{R}^{d_0}$, $x, x' \in \mathbb{R}^d$, $\mu, \mu' \in \mathcal{P}_2(\mathbb{R}^d)$, $\alpha_0, \alpha_0' \in A_0$ and $\alpha, \alpha' \in A$,

$$
\begin{aligned}
&|(b_0,\sigma_0)(t,x_0',\mu',\alpha_0') - (b_0,\sigma_0)(t,x_0,\mu,\alpha_0)| \\
&\quad + |(b,\sigma)(t,x',\mu',x_0',\alpha') - (b,\sigma)(t,x,\mu,x_0,\alpha)| \\
&\leq L\big(|x_0' - x_0| + |x' - x| + |\alpha_0' - \alpha_0| + |\alpha' - \alpha| + W_2(\mu',\mu)\big).
\end{aligned}
$$

(A2) For all $\alpha_0 \in A_0$ and $\alpha \in A$, we have:

$$
\int_0^T \Big(|(b_0,\sigma_0,f_0)(t,0,\delta_0,\alpha_0)|^2 + |(b,\sigma,f)(t,0,\delta_0,0,\alpha)|^2 \Big) dt < \infty.
$$

(A3) For all $x_0, x_0' \in \mathbb{R}^{d_0}$, $\alpha_0, \alpha_0' \in A_0$ and $\mu, \mu' \in \mathcal{P}_2(\mathbb{R}^d)$, we have:

$$
\begin{aligned}
&|(f_0,g_0)(t,x_0',\mu',\alpha_0') - (f_0,g_0)(t,x_0,\mu,\alpha_0)| \\
&\quad \leq L\big(1 + |(x_0',\alpha_0')| + |(x_0,\alpha_0)| + M_2(\mu') + M_2(\mu)\big) \\
&\qquad \times \big(|(x_0',\alpha_0') - (x_0,\alpha_0)| + W_2(\mu',\mu)\big),
\end{aligned}
$$

and for all $x_0 \in \mathbb{R}^{d_0}$, $x, x' \in \mathbb{R}^d$, $\alpha, \alpha' \in A$ and $\mu, \mu' \in \mathcal{P}_2(\mathbb{R}^d)$,

$$
\begin{aligned}
&|(f,g)(t,x',\mu',x_0,\alpha') - (f,g)(t,x,\mu,x_0,\alpha)| \\
&\quad \leq L\big(1 + |(x',\alpha')| + |(x,\alpha)| + |x_0| + M_2(\mu') + M_2(\mu)\big) \\
&\qquad \times \big(|(x',\alpha') - (x,\alpha)| + W_2(\mu,\mu')\big),
\end{aligned}
$$

where as usual $M_2(\mu)^2 = \int_{\mathbb{R}^d} |x|^2 \mu(dx)$.

(continued)

(A4) The functions b_0, b, f, and g are jointly continuously differentiable in x_0, x, α_0, α and μ (in the L-sense) when the time variable is fixed.

Notice that, in order to lighten the notation which is already heavy enough, we removed the dependence of b, σ and f on the control α_0 of the major player. Conditions (**A1**) and (**A2**) guarantee that for all admissible controls, the stochastic differential equations defining the dynamics of the state have unique solutions, while (**A2**) and (**A3**) guarantee that the associated cost functionals are well defined. Condition (**A4**) will be used when we define adjoint processes.

In this subsection, $\mathbb{A}$ (resp. $\mathbb{A}_0$) denotes the space of $\mathbb{F}$ (resp. $\mathbb{F}^0$) progressively measurable processes $\boldsymbol{\alpha}$ (resp. $\boldsymbol{\alpha}^0$) with values in A (resp. A_0) such that:

$$\mathbb{E}\int_0^T |\alpha_t|^2 dt < \infty, \qquad \left(\text{resp. } \mathbb{E}^0\int_0^T |\alpha_t^0|^2 dt < \infty\right).$$

Optimization Problem for the Major Player

In this subsection we consider the limiting two-player game introduced in Subsection 7.1.2, and search for the major player's best response $\boldsymbol{\alpha}^0 \in \mathbb{A}_0$ to a given control strategy $\boldsymbol{\alpha} \in \mathbb{A}$ of the representative minor player. This amounts to solving the optimal control problem based on state controlled dynamics given by:

$$\begin{cases} dX_t^0 = b_0\big(t, X_t^0, \mathcal{L}^1(X_t), \alpha_t^0\big)dt + \sigma_0\big(t, X_t^0, \mathcal{L}^1(X_t), \alpha_t^0\big)dW_t^0, \\ X_0^0 = x^{00}, \\ dX_t = b\big(t, X_t, \mathcal{L}^1(X_t), X_t^0, \alpha_t\big)dt + \sigma\big(t, X_t, \mathcal{L}^1(X_t), X_t^0, \alpha_t\big)dW_t, \\ X_0 = x^0, \end{cases} \tag{7.14}$$

and the cost functional:

$$J^0(\boldsymbol{\alpha}^0, \boldsymbol{\alpha}) = \mathbb{E}\left[\int_0^T f_0\big(t, X_t^0, \mathcal{L}^1(X_t), \alpha_t^0\big)dt + g_0\big(X_T^0, \mathcal{L}^1(X_T)\big)\right], \tag{7.15}$$

where it is assumed that the control strategy $\boldsymbol{\alpha} \in \mathbb{A}$ is given, and where the set of admissible controls is $\mathbb{A}_0$. In what follows, this stochastic control problem will be denoted by ($\mathscr{P}$1). It is of the McKean-Vlasov type. However, it is not directly amenable to the results of Chapter (Vol I)-6 because of the presence of the conditional distributions. For this reason, we take a little detour to develop the tools needed to handle this problem. We do not give detailed proofs because the arguments are very similar to those used in Chapter (Vol I)-6 and Chapter 1.

A Maximum Principle for Conditional McKean-Vlasov Control Problems

We provide a convenient version of the stochastic maximum principle for the optimal control problem that consists in minimizing J^0 given in (7.15) over $\boldsymbol{\alpha}^0 \in \mathbb{A}_0$ and state processes of the form (7.14). To do so, we mostly imitate the arguments introduced in Chapter (Vol I)-6 to handle optimal control problems of the non-conditional McKean-Vlasov type.

Here, the Hamiltonian is defined as:

$$
\begin{aligned}
&H_0(t, x_0, x, \mu, y_0, y, z_{00}, z_{11}, \alpha_0, \alpha) \\
&\quad = b_0(t, x_0, \mu, \alpha_0) \cdot y_0 + b(t, x, \mu, x_0, \alpha) \cdot y \\
&\qquad + \sigma_0(t, x_0, \mu, \alpha_0) \cdot z_{00} + \sigma(t, x, \mu, x_0, \alpha) \cdot z_{11} + f_0(t, x_0, \mu, \alpha_0),
\end{aligned} \tag{7.16}
$$

for $t \in [0, T]$, $x_0 \in \mathbb{R}^{d_0}$, $x \in \mathbb{R}^d$, $\mu \in \mathcal{P}_2(\mathbb{R}^d)$, $y_0 \in \mathbb{R}^{d_0}$, $y \in \mathbb{R}^d$, $z_{00} \in \mathbb{R}^{d_0 \times m_0}$, $z_{11} \in \mathbb{R}^{d \times m}$, $\alpha_0 \in A_0$ and $\alpha \in A$.

Adjoint Equations. The adjoint processes comprise two tuples of stochastic processes, $(\boldsymbol{P}^0, \boldsymbol{P})$ and $(\boldsymbol{Q}^{00}, \boldsymbol{Q}^{01}, \boldsymbol{Q}^{10}, \boldsymbol{Q}^{11})$: $\boldsymbol{P}^0$ denotes the adjoint process associated with $\boldsymbol{X}^0$ and $\boldsymbol{Q}^{00}$ and $\boldsymbol{Q}^{01}$ its martingale representation terms with respect to $\boldsymbol{W}^0$ and $\boldsymbol{W}$; similarly, $\boldsymbol{P}$ denotes the adjoint process associated with $\boldsymbol{X}$ and $\boldsymbol{Q}^{10}$ and $\boldsymbol{Q}^{11}$ its martingale representation terms. To avoid any possible confusion with the notations used below in the maximum principle for the minor player, we here use the letters $(\boldsymbol{P}, \boldsymbol{Q})$ for the adjoint processes in lieu of $(\boldsymbol{Y}, \boldsymbol{Z})$.

The dynamics of $\boldsymbol{P}^0$ and $\boldsymbol{P}$ are given as follows: $\boldsymbol{P}^0$ is intended to account for the sensitivity of the Hamiltonian under variations of $\boldsymbol{X}^0$, while $\boldsymbol{P}$ is intended to account for the sensitivity of H^0 under variations of $\boldsymbol{X}$ and the conditional law of $\boldsymbol{X}$ given $\boldsymbol{W}^0$.

For a tuple $(\boldsymbol{X}^0, \boldsymbol{X}, \boldsymbol{\alpha}^0, \boldsymbol{\alpha})$, we get as backward equation for $\boldsymbol{P}^0$:

$$
dP_t^0 = -\partial_{x_0} H_0\big(t, \underset{\smile}{X}_t, \mathcal{L}^1(X_t), \underset{\smile}{P}_t, \underset{\smile}{Q}_t, \alpha_t^0, \alpha_t\big)dt + Q_t^{00} dW_t^0 + Q_t^{01} dW_t, \tag{7.17}
$$

for $t \in [0, T]$, with the terminal boundary condition $Y_T^0 = \partial_{x_0} g_0(X_T^0, \mathcal{L}^1(X_T))$, where, to lighten the notations, we write $\underset{\smile}{X} = (\boldsymbol{X}^0, \boldsymbol{X})$, $\underset{\smile}{P} = (\boldsymbol{P}^0, \boldsymbol{P})$ and $\underset{\smile}{Q} = (\boldsymbol{Q}^{00}, \boldsymbol{Q}^{01}, \boldsymbol{Q}^{10}, \boldsymbol{Q}^{11})$.

The derivation of the backward equation for $\boldsymbol{P}$ obeys the same principle except that, in addition, it must incorporate the derivatives of H^0 and g_0 with respect to the measure argument. The form of the derivative terms is similar to that in the backward equation (Vol I)-(6.31) in the maximum principle for optimal control problems of the unconditional McKean-Vlasov type, except that, due to the conditioning, the expectation on the copy $\tilde{\Omega}$ appearing in (Vol I)-(6.31) now becomes an expectation on some copy $\tilde{\Omega}^1$ of the sole Ω^1 in lieu of the whole Ω. We get as backward equation for $\boldsymbol{P}$:

$$
\begin{aligned}
dP_t &= -\partial_x H_0(t, \underset{\smile}{X}_t, \mathcal{L}^1(X_t), \underset{\smile}{P}_t, \underset{\smile}{Q}_t, \alpha_t^0, \alpha_t)dt + Q_t^{10} dW_t^0 + Q_t^{11} dW_t \\
&\quad - \tilde{\mathbb{E}}^1\big[\partial_\mu H_0\big(t, \tilde{\underset{\smile}{X}}_t, \mathcal{L}^1(X_t), \tilde{\underset{\smile}{P}}_t, \tilde{\underset{\smile}{Q}}_t, \alpha_t^0, \tilde{\alpha}_t\big)(X_t)\big]dt,
\end{aligned} \tag{7.18}
$$

for $t \in [0,T]$, with terminal condition $P_T = \mathbb{E}^1\big[\partial_\mu g_0\big(X_T^0, \mathcal{L}^1(X_T)\big)(X_T)\big]$. We used the same notation as in Subsection 4.3.3: $(\tilde{\Omega}^1, \tilde{\mathcal{F}}^1, \tilde{\mathbb{P}}^1)$ is a copy of the probability space $(\Omega^1, \mathcal{F}^1, \mathbb{P}^1)$, the expectation under $\tilde{\mathbb{P}}^1$ being denoted by $\tilde{\mathbb{E}}^1$. Given such a copy, together with a random variable X defined on $\Omega = \Omega^0 \times \Omega^1$, we denote by $\tilde{X}$ the random variable defined as a copy of X on the space $\tilde{\Omega} = \Omega^0 \times \tilde{\Omega}^1$. Notice that, in the second line of (7.18), there is no need to write $\tilde{\alpha}_t^0$ in $\tilde{\mathbb{E}}^1$ since α_t^0 is $\mathcal{F}_t^0$-measurable. Similarly for X_T^0 in the terminal condition.

Therefore, we conclude that, for each tuple $(\boldsymbol{X}^0, \boldsymbol{X}, \boldsymbol{\alpha}^0, \boldsymbol{\alpha})$, the associated adjoint process $(\boldsymbol{P}^0, \boldsymbol{P}, \boldsymbol{Q}^{00}, \boldsymbol{Q}^{01}, \boldsymbol{Q}^{10}, \boldsymbol{Q}^{11})$ is defined as the solution of the backward stochastic differential equation:

$$\begin{cases} dP_t^0 = -\,\partial_{x_0} H_0\big(t, \underline{X}_t, \mathcal{L}^1(X_t), \underline{P}_t, \underline{Q}_t, \alpha_t^0, \alpha_t\big)dt + Q_t^{00} dW_t^0 + Q_t^{01} dW_t, \\ dP_t = -\,\partial_x H_0\big(t, \underline{X}_t, \mathcal{L}^1(X_t), \underline{P}_t, \underline{Q}_t, \alpha_t^0, \alpha_t\big)dt + Q_t^{10} dW_t^0 + Q_t^{11} dW_t \\ \qquad - \tilde{\mathbb{E}}^1\big[\partial_\mu H_0\big(t, \tilde{\underline{X}}_t, \mathcal{L}^1(X_t), \tilde{\underline{P}}_t, \tilde{\underline{Q}}_t, \alpha_t^0, \tilde{\alpha}_t\big)(X_t)\big]dt, \\ P_T^0 = \partial_{x_0} g_0(X_T^0, \mathcal{L}^1(X_T)), \quad P_T = \mathbb{E}^1\big[\partial_\mu g_0(X_T^0, \mathcal{L}^1(X_T))(X_T)\big]. \end{cases} \tag{7.19}$$

Despite the presence of the conditional distributions in the coefficients, the standard proofs of existence and uniqueness of solutions of BSDEs with Lipschitz coefficients still apply to (7.19), in full analogy with the discussion following Definition (Vol I)-6.5. In order to avoid too many detours away from the route to an equilibrium, we do not state these existence and uniqueness results. The reader can easily formulate them if he or she feels compelled to do so. We merely note that their assumptions are satisfied here. Indeed, under assumption **Major Minor MFG**, the derivatives of b, b_0, σ and σ_0 in (x_0, x) are bounded, and the derivatives in μ are bounded in L^2.

In the following we systematically add a bar to denote the expectation under $\mathbb{P}^1$. For example, $\bar{P}_t^0$ stands for $\mathbb{E}^1[P_t^0]$.

Necessary Form of the Pontryagin Principle. The necessary form of the maximum principle can be derived by duplicating the proof of Theorem (Vol I)-6.14. Whenever A_0 is a convex set and H_0 is convex in the parameter α_0, any control strategy $\boldsymbol{\alpha}^0 = (\alpha_t^0)_{0 \le t \le T}$ which minimizes $J^0(\boldsymbol{\alpha}^0, \boldsymbol{\alpha})$ for a given $\boldsymbol{\alpha} \in \mathbb{A}_0$ should satisfy:

$$\mathbb{E}^1\big[H_0\big(t, \underline{X}_t, \mathcal{L}^1(X_t), \underline{P}_t, \underline{Q}_t, \alpha_t^0, \alpha_t\big)\big] \le \mathbb{E}^1\big[H_0\big(t, \underline{X}_t, \mathcal{L}^1(X_t), \underline{P}_t, \underline{Q}_t, \beta, \alpha_t\big)\big],$$

$dt \otimes d\mathbb{P}$ almost-everywhere for every $\beta \in A_0$. Observe that, in contrast with the statement of Theorem (Vol I)-6.14, we now take the expectation under $\mathbb{P}^1$ in the minimization of the Hamiltonian. We must do so to account for the fact that $\boldsymbol{\alpha}^0$ is merely $\mathbb{F}^0$-progressively measurable. Equivalently, with the same notation as in the proof of Theorem (Vol I)-6.14, the process $\boldsymbol{\beta}$ therein is just $\mathbb{F}^0$-progressively measurable.

Actually, using the special form of H_0, we can get rid of the expectation $\mathbb{E}^1$ in the minimization of the Hamiltonian. Indeed, the necessary condition may be rewritten (observe the bars over the process $\bar{\boldsymbol{P}}^0$ and $\bar{\boldsymbol{Q}}^{00}$):

$$\forall \beta \in A_0, \quad H_{00}\big(t, X_t, \mathcal{L}^1(X_t), \bar{P}_t^0, \bar{Q}_t^{00}, \alpha_t^0\big) \leq H_{00}\big(t, X_t, \mathcal{L}^1(X_t), \bar{P}_t^0, \bar{Q}_t^{00}, \beta\big),$$

where H_{00} stands for the *reduced* Hamiltonian:

$$\begin{aligned} H_{00}(t, x_0, \mu, y_0, z_{00}, \alpha_0) &= b_0(t, x_0, \mu, \alpha_0) \cdot y_0 \\ &\quad + \sigma_0(t, x_0, \mu, \alpha_0) \cdot z_{00} + f_0(t, x_0, \mu, \alpha_0). \end{aligned}$$

This prompts us to formulate the following assumption:

Assumption (Major Hamiltonian). The set A_0 is convex and H_{00} is convex in α_0. Moreover, for each fixed $(t, x_0, \mu, y_0, z_{00})$, there exists a unique minimizer of the reduced Hamiltonian H_{00} as a function of α_0. It is denoted by $\hat{\alpha}^0(t, x_0, \mu, y_0, z_{00})$.

Then, the necessary part of the Pontryagin stochastic maximum, used as a principle to guide our intuition, suggests that as long as the control strategy $\boldsymbol{\alpha}^0 = (\alpha_t^0)_{0 \leq t \leq T}$ is optimal, it must be of the form $(\alpha_t^0 = \hat{\alpha}^0(X_t^0, X_t, \bar{P}_t^0, \bar{Q}_t^{00}))_{0 \leq t \leq T}$.

Sufficient Form of the Pontryagin Principle. Under suitable convexity conditions on the coefficients and for a given $\boldsymbol{\alpha} \in \mathbb{A}$, $(\hat{\alpha}_t^0 = \hat{\alpha}^0(t, X_t^0, \mathcal{L}^1(X_t), \bar{P}_t^0, \bar{Q}_t^{00}))_{0 \leq t \leq T}$ is an optimizer of the problem $\inf_{\boldsymbol{\alpha}^0 \in \mathbb{A}_0} J^0(\boldsymbol{\alpha}^0, \boldsymbol{\alpha})$ given in (7.15) if we can solve the forward-backward stochastic differential equation:

$$\begin{cases} dX_t^0 = b_0\big(t, X_t^0, \mathcal{L}^1(X_t), \hat{\alpha}_t^0\big)dt + \sigma_0\big(t, X_t^0, \mathcal{L}^1(X_t), \hat{\alpha}_t^0\big)dW_t^0, \\ dX_t = b\big(t, X_t, \mathcal{L}^1(X_t), X_t^0, \alpha_t\big)dt + \sigma\big(t, X_t, \mathcal{L}^1(X_t), X_t^0, \alpha_t\big)dW_t, \\ dP_t^0 = -\partial_{x_0} H_0\big(t, \underline{X}_t, \mathcal{L}^1(X_t), \underline{P}_t, \underline{Q}_t, \hat{\alpha}_t^0, \alpha_t\big)dt + Q_t^{00} dW_t^0 + Q_t^{01} dW_t, \\ dP_t = -\partial_x H_0\big(t, \underline{X}_t, \mathcal{L}^1(X_t), \underline{P}_t, \underline{Q}_t, \hat{\alpha}_t^0, \alpha_t\big)dt + Q_t^{10} dW_t^0 + Q_t^{11} dW_t \\ \qquad - \tilde{\mathbb{E}}^1\big[\partial_\mu H_0\big(t, \tilde{\underline{X}}_t, \mathcal{L}^1(X_t), \tilde{\underline{P}}_t, \tilde{\underline{Q}}_t, \hat{\alpha}_t^0, \tilde{\alpha}_t\big)(X_t)\big]dt, \end{cases} \tag{7.20}$$

with the prescription that $\hat{\alpha}_t^0 = \hat{\alpha}^0(t, X_t^0, \mathcal{L}^1(X_t), \bar{P}_t^0, \bar{Q}_t^{00})$ for $t \in [0, T]$ and with the initial and terminal conditions given by:

$$\begin{cases} X_0^0 = x^{00}, \quad X_0 = x^0, \\ P_T^0 = \partial_{x_0} g_0\big(X_T^0, \mathcal{L}^1(X_T)\big), \\ P_T = \mathbb{E}^1\big[\partial_\mu g_0\big(X_T^0, \mathcal{L}^1(X_T)\big)(X_T)\big]. \end{cases}$$

We shall solve this general FBSDE only in a few particular cases. For the sake of convenience, we give a name to the convexity assumption needed for the sufficient part of the stochastic maximum principle proved above:

Assumption (Major Convexity). The function $\mathbb{R}^{d_0} \times \mathcal{P}_2(\mathbb{R}^d) \ni (x, \mu) \mapsto g(x, \mu)$ is convex and for each fixed $(t, y_0, y, z_{00}, z_{11}, \alpha)$, the function:

$$\mathbb{R}^{d_0} \times \mathbb{R}^d \times \mathcal{P}_2(\mathbb{R}^d) \times A_0 \ni (x_0, x, \mu, \alpha_0) \mapsto H_0(t, x_0, x, \mu, y_0, y, z_{00}, z_{11}, \alpha_0, \alpha)$$

is also convex.

Observe that it does not suffice to require that the reduced Hamiltonian H_{00} is convex. We really need the whole Hamiltonian H_0 to be jointly convex in (x_0, x, μ).

We then have the following result:

Proposition 7.4 *Let assumptions* **Major Minor MFG**, **Major Hamiltonian** *and* **Major Convexity** *be in force. If, for a given* $\boldsymbol{\alpha} \in \mathbb{A}$, *the process:*

$$(\boldsymbol{X}^0, \boldsymbol{X}, \boldsymbol{P}^0, \boldsymbol{P}, \boldsymbol{Q}^{00}, \boldsymbol{Q}^{01}, \boldsymbol{Q}^{10}, \boldsymbol{Q}^{11})$$

is a solution of the FBSDE (7.20) with the constraint:

$$\hat{\alpha}_t^0 = \hat{\alpha}^0\big(t, X_t^0, \mathcal{L}^1(X_t), \bar{P}_t^0, \bar{Q}_t^{00}\big), \quad t \in [0, T],$$

then $\hat{\boldsymbol{\alpha}}^0$ *is optimal for the problem* $(\mathscr{P}1)$ *and* $(\boldsymbol{X}^0, \boldsymbol{X})$ *is the associated optimally controlled state process.*

Optimization Problem for the Representative Minor Player

For the representative minor player's best response optimization problem, for each fixed admissible control $\boldsymbol{\alpha}^0 = (\alpha_t^0)_{0 \le t \le T}$ of the major player, we fix a stochastic measure flow $\boldsymbol{\mu}$, that is a continuous $\mathbb{F}^0$-adapted process $\boldsymbol{\mu} = (\mu_t)_{0 \le t \le T}$ with values in $\mathcal{P}_2(\mathbb{R}^d)$ such that $\mathbb{E}^0[\sup_{0\le t\le T} M_2(\mu_t)^2] < \infty$ and we solve the control problem based on the controlled dynamics:

$$\begin{cases} d\check{X}_t^0 = b_0(t, \check{X}_t^0, \mu_t, \alpha_t^0)dt + \sigma_0(t, \check{X}_t^0, \mu_t, \alpha_t^0)dW_t^0, & \check{X}_0^0 = x^{00}, \\ d\check{X}_t = b(t, \check{X}_t, \mu_t, \check{X}_t^0, \alpha_t)dt + \sigma(t, \check{X}_t, \mu_t, \check{X}_t^0, \alpha_t)dW_t, & \check{X}_0 = x^0, \end{cases} \tag{7.21}$$

and the cost functional:

$$J(\boldsymbol{\alpha}^0, \boldsymbol{\alpha}) = \mathbb{E}\left[\int_0^T f(t, \check{X}_t, \mu_t, \check{X}_t^0, \alpha_t) + g(\check{X}_T, \mu_T, \check{X}_T^0)\right]. \tag{7.22}$$

Note that since $\boldsymbol{\alpha}^0$ and $\boldsymbol{\mu}$ are fixed, the first SDE in (7.21) can be solved separately, and its solution will appear in the second SDE of (7.21) and the cost functional only as an exogenous source of randomness. If we choose the set of admissible controls for the representative minor player to be $\mathbb{A}$, this problem is a standard non-Markovian stochastic control problem which we denote by ($\mathscr{P}2$). Because of this special structure, namely the fact that the process $\check{\boldsymbol{X}}^0$ can be determined *off line*, we only introduce adjoint processes for $\check{\boldsymbol{X}} = (\check{X}_t)_{0\le t\le T}$. We denote by $\underset{\sim}{\boldsymbol{Z}} = (\boldsymbol{Z}^{10}, \boldsymbol{Z}^{11})$ its martingale representation terms with respect to $\boldsymbol{W}^0$ and $\boldsymbol{W}$. Also, we use the reduced Hamiltonian:

$$\begin{aligned} H(t, x_0, x, \mu, y, z_{11}, \alpha) = b(t, x, \mu, x_0, \alpha) \cdot y \\ + \sigma(t, x, \mu, x_0, \alpha) \cdot z_{11} + f(t, x, \mu, x_0, \alpha). \end{aligned} \tag{7.23}$$

From Chapter 1 we know that, for each admissible control strategy $\boldsymbol{\alpha}$, the adjoint process $(\boldsymbol{Y}, \underset{\sim}{\boldsymbol{Z}}) = (\boldsymbol{Y}, \boldsymbol{Z}^{10}, \boldsymbol{Z}^{11})$ associated with $\boldsymbol{\alpha}$ is given as the solution of the BSDE:

$$\begin{cases} dY_t = -\partial_x H(t, \underset{\sim}{\check{X}}_t, \mu_t, Y_t, Z_t^{11}, \alpha_t)dt + Z_t^{10}dW_t^0 + Z_t^{11}dW_t, \\ Y_T = \partial_x g(\check{X}_T, \mu_T, \check{X}_T^0), \end{cases} \tag{7.24}$$

where $\underset{\sim}{\check{\boldsymbol{X}}} = (\check{\boldsymbol{X}}^0, \check{\boldsymbol{X}})$. Observe that, in comparison with Chapter 1, there is no need to discuss any compatibility condition since the filtrations $\mathbb{F}^0$ and $\mathbb{F}$ are assumed to be generated by $\boldsymbol{W}^0$ and $(\boldsymbol{W}^0, \boldsymbol{W})$.

The existence of the adjoint processes associated with a given admissible control strategy $\boldsymbol{\alpha}$ is a consequence of the standard existence result of solutions of BSDEs since the partial derivatives of b and σ with respect to x_0 and x are bounded.

Necessary Form of the Pontryagin Principle. Again, whenever A is convex and H is convex in α, the necessary part of the Pontryagin stochastic maximum principle suggests that, if the admissible control $\boldsymbol{\alpha} = (\alpha_t)_{0\le t\le T}$ is optimal, the Hamiltonian (7.23) should be minimized along the trajectory of $(\boldsymbol{X}^0, \boldsymbol{X}, \boldsymbol{Y}, \underset{\sim}{\boldsymbol{Z}})$.

In analogy with our analysis of the major player optimization problem, we introduce a tailor-made hypothesis for the minimization of this Hamiltonian.

Assumption (Minor Hamiltonian). The set A is convex and H is convex in α. Moreover, for each fixed $(t, x_0, x, \mu, y, z_{11})$, there exists a unique minimizer of the above reduced Hamiltonian H as a function of α. This minimizer is denoted by $\hat{\alpha}(t, x_0, x, \mu, y, z_{11})$.

So given assumption **Minor Hamiltonian**, any optimal admissible control $\boldsymbol{\alpha} = (\alpha_t)_{0\le t\le T} \in \mathbb{A}$ should be of the form $(\alpha_t = \hat{\alpha}(t, \check{X}_t^0, \check{X}_t, \mu_t, Y_t, Z_t^{11}))_{0\le t\le T}$.

Sufficient Form of the Pontryagin principle. The sufficient part of the stochastic maximum principle may be easily derived. Under suitable convexity conditions on the coefficients and for a given $\boldsymbol{\alpha}^0 \in \mathbb{A}_0$, $(\hat{\alpha}_t = \hat{\alpha}(t, \check{X}_t^0, \check{X}_t, \mu_t, Y_t, Z_t^{11}))_{0\le t\le T}$ is an optimizer of the problem $\inf_{\boldsymbol{\alpha}\in\mathbb{A}} J(\boldsymbol{\alpha}^0, \boldsymbol{\alpha})$ if we can solve the forward-backward stochastic differential equation obtained by plugging this minimizer into the controlled dynamics and the adjoint BSDE (7.24):

$$\begin{cases} d\check{X}_t^0 = b(t, \check{X}_t^0, \mu_t, \alpha_t^0)dt + \sigma(t, \check{X}_t^0, \mu_t, \alpha_t^0)dW_t^0, \\ d\check{X}_t = b(t, \check{X}_t, \mu_t, \check{X}_t^0, \hat{\alpha}_t)dt + \sigma(t, \check{X}_t, \mu_t, \check{X}_t^0, \hat{\alpha}_t)dW_t, \\ dY_t = -\partial_x H(t, \check{X}_t, \mu_t, Y_t, Z_t^{11}, \hat{\alpha}_t)dt + Z_t^{10}dW_t^0 + Z_t^{11}dW_t, \end{cases} \tag{7.25}$$

for $t \in [0, T]$, with the prescription $\hat{\alpha}_t = \hat{\alpha}(t, \check{X}_t^0, \check{X}_t, \mu_t, Y_t, Z_t^{11})$ for $t \in [0, T]$ and with the initial and terminal conditions given by:

$$\check{X}_0^0 = x^{00}, \quad \check{X}_0 = x^0, \quad Y_T = \partial_x g(\check{X}_T, \mu_T, \check{X}_T^0).$$

As before, we shall make use of the following convexity assumption:

Assumption (Minor Convexity). The function $\mathbb{R}^d \ni x \mapsto g(x, \mu, x_0)$ is convex for each fixed (x_0, μ). Moreover, the function:

$$\mathbb{R}^d \times A \ni (x, \alpha) \mapsto H(t, x_0, x, \mu, y, z_{11}, \alpha)$$

is also convex for each fixed (t, x_0, μ, y, z_{11}).

We have the following proposition:

Proposition 7.5 *Let assumptions* **Major Minor MFG**, **Minor Hamiltonian** *and* **Minor Convexity** *be in force. If, for a given* $\boldsymbol{\alpha}^0 \in \mathbb{A}_0$ *with* $\check{\boldsymbol{X}}^0$ *as associated controlled process, the stochastic process* $(\check{\boldsymbol{X}}, \boldsymbol{Y}, \boldsymbol{Z}^{10}, \boldsymbol{Z}^{11})$ *solves the FBSDE (7.25), then the control:*

$$\hat{\alpha}_t = \hat{\alpha}(t, \check{X}_t^0, \check{X}_t, \mu_t, Y_t, Z_t^{11}), \quad t \in [0, T],$$

is an optimal control for the problem ($\mathscr{P}2$), *and* $\check{\boldsymbol{X}}$ *is the associated optimally controlled state process.*

Nash Equilibrium for the Limiting Two-Player Game

In order to construct a Nash equilibrium for the two-player game described in the first step of our formulation of the mean field game with major and minor players, we assume that assumptions **Major Minor MFG**, **Major Hamiltonian**, **Major Convexity**, **Minor Hamiltonian**, and **Minor Convexity** hold and, for a continuous $\mathbb{F}^0$-adapted process $\boldsymbol{\mu}$ with values in $\mathcal{P}_2(\mathbb{R}^d)$ such that $\mathbb{E}^0[\sup_{0 \le t \le T} M_2(\mu_t)^2] < \infty$, we consider the FBSDE:

$$\begin{cases} dX_t^0 = b_0(t, X_t^0, \mathcal{L}^1(X_t), \hat{\alpha}_t^0)dt + \sigma_0(t, X_t^0, \mathcal{L}^1(X_t), \hat{\alpha}_t^0)dW_t^0, \\ dX_t = b(t, X_t, \mathcal{L}^1(X_t), X_t^0, \hat{\alpha}_t)dt + \sigma(t, X_t, \mathcal{L}^1(X_t), X_t^0, \hat{\alpha}_t)dW_t, \\ d\check{X}_t^0 = b_0(t, \check{X}_t^0, \mu_t, \hat{\alpha}_t^0)dt + \sigma_0(t, \check{X}_t^0, \mathcal{L}^1(X_t), \hat{\alpha}_t^0)dW_t^0, \\ d\check{X}_t = b(t, \check{X}_t, \mu_t, \check{X}_t^0, \hat{\alpha}_t)dt + \sigma(t, \check{X}_t, \mathcal{L}^1(X_t), \check{X}_t^0, \hat{\alpha}_t)dW_t, \\ dP_t^0 = -\partial_{x_0} H_0(t, \underset{\sim}{X}_t, \mathcal{L}^1(X_t), \underset{\sim}{P}_t, \underset{\sim}{Q}_t, \hat{\alpha}_t^0, \hat{\alpha}_t)dt + Q_t^{00}dW_t^0 + Q_t^{01}dW_t, \\ dP_t = -\partial_x H_0(t, \underset{\sim}{X}_t, \mathcal{L}^1(X_t), \underset{\sim}{P}_t, \underset{\sim}{Q}_t, \hat{\alpha}_t^0, \hat{\alpha}_t)dt + Q_t^{10}dW_t^0 + Q_t^{11}dW_t \\ \qquad - \tilde{\mathbb{E}}^1\big[\partial_\mu H_0(t, \tilde{\underset{\sim}{X}}_t, \mathcal{L}^1(X_t), \tilde{\underset{\sim}{P}}_t, \tilde{\underset{\sim}{Q}}_t, \hat{\alpha}_t^0, \tilde{\hat{\alpha}}_t)(X_t)\big]dt, \\ dY_t = -\partial_x H(t, \underset{\sim}{\check{X}}_t, \mu_t, Y_t, Z_t^{11}, \hat{\alpha}_t)dt + Z_t^{10}dW_t^0 + Z_t^{11}dW_t, \end{cases} \tag{7.26}$$

with the initial and terminal conditions given by:

$$\begin{cases} X_0^0 = x^{00}, \quad X_0 = x^0, \quad \check{X}_0^0 = x^{00}, \quad \check{X}_0 = x^0, \\ P_T^0 = \partial_{x_0} g_0(X_T^0, \mathcal{L}^1(X_T)), \quad P_T = \mathbb{E}^1\big[\partial_\mu g_0(X_T^0, \mathcal{L}(X_T))(X_T)\big], \\ Y_T = \partial_x g(\check{X}_T, \mu_T, \check{X}_T^0), \end{cases}$$

with:

$$\hat{\alpha}_t^0 = \hat{\alpha}^0(t, X_t^0, \mathcal{L}^1(X_t), \bar{P}_t^0, \bar{Q}_t^{00}), \quad \hat{\alpha}_t = \hat{\alpha}(t, \check{X}_t^0, \check{X}_t, \mu_t, Y_t, Z_t^{11}), \quad t \in [0, T].$$

If this FBSDE has a solution, then by definition of a Nash equilibrium, $(\hat{\boldsymbol{\alpha}}^0, \hat{\boldsymbol{\alpha}})$ is a Nash equilibrium for the two-player stochastic differential game (7.5)–(7.6).

The Consistency Condition

Now, in the limiting formulation of the major minor problem, the consistency condition (7.7) imposes the additional mean field constraint:

$$\mu_t = \mathcal{L}^1(X_t), \quad \forall t \in [0, T].$$

Plugging it into FBSDE (7.26) gives the following ultimate FBSDE:

$$\begin{cases} dX_t^0 = b_0(t, X_t^0, \mathcal{L}^1(X_t), \hat{\alpha}_t^0)dt + \sigma_0(t, X_t^0, \mathcal{L}^1(X_t), \hat{\alpha}_t^0)dW_t^0, \\ dX_t = b(t, X_t, \mathcal{L}^1(X_t), X_t^0, \hat{\alpha}_t)dt + \sigma(t, X_t, \mathcal{L}^1(X_t), X_t^0, \hat{\alpha}_t)dW_t, \\ dP_t^0 = -\partial_{x_0} H_0(t, \underset{\sim}{X}_t, \mathcal{L}^1(X_t), \underset{\sim}{P}_t, \underset{\sim}{Q}_t, \hat{\alpha}_t^0, \hat{\alpha}_t)dt + Q_t^{00} dW_t^0 + Q_t^{01} dW_t, \\ dP_t = -\partial_x H_0(t, \underset{\sim}{X}_t, \mathcal{L}^1(X_t), \underset{\sim}{P}_t, \underset{\sim}{Q}_t, \hat{\alpha}_t^0, \hat{\alpha}_t)dt + Q_t^{10} dW_t^0 + Q_t^{11} dW_t \\ \qquad - \tilde{\mathbb{E}}^1\big[\partial_\mu H_0(t, \tilde{\underset{\sim}{X}}_t, \mathcal{L}^1(X_t), \tilde{\underset{\sim}{P}}_t, \tilde{\underset{\sim}{Q}}_t, \hat{\alpha}_t^0, \tilde{\hat{\alpha}}_t)(X_t)\big]dt, \\ dY_t = -\partial_x H(t, \underset{\sim}{X}_t, \mathcal{L}^1(X_t), Y_t, Z_t^{11}, \hat{\alpha}_t)dt + Z_t^{10} dW_t^0 + Z_t^{11} dW_t, \end{cases} \tag{7.27}$$

with initial and terminal conditions given by:

$$\begin{cases} X_0^0 = x^{00}, \quad X_0 = x^0, \\ P_T^0 = \partial_{x_0} g_0(X_T^0, \mathcal{L}^1(X_T)), \quad P_T = \mathbb{E}^1\big[\partial_\mu g_0(X_T^0, \mathcal{L}^1(X_T))(X_T)\big], \\ Y_T = \partial_x g(X_T, \mathcal{L}^1(X_T), X_T^0), \end{cases} \tag{7.28}$$

where for $0 \leq t \leq T$, we define:

$$\hat{\alpha}_t^0 = \hat{\alpha}^0\big(t, X_t^0, \mathcal{L}^1(X_t), \bar{P}_t^0, \bar{Q}_t^{00}\big), \quad \hat{\alpha}_t = \hat{\alpha}\big(t, X_t^0, X_t, \mathcal{L}^1(X_t), Y_t, Z_t^{11}\big),$$

Remark 7.6 *In equilibrium, $(\boldsymbol{X}^0, \boldsymbol{X})$ and $(\check{\boldsymbol{X}}^0, \check{\boldsymbol{X}})$ are the same and we could replace the consistency condition by $\mu_t = \mathcal{L}^1(\check{X}_t) = \mathcal{L}^1(X_t)$, for $0 \leq t \leq T$.*

Proving existence and uniqueness for conditional McKean-Vlasov FBSDEs of the form of (7.27) is as hard as solving mean field games with common noise and we shall not attempt to address this question here. We shall only consider it in the case of linear quadratic models.

7.1.5 Conditional Propagation of Chaos and ϵ-Nash Equilibria

In this subsection we prove that solutions of the limiting mean field game problem with major and minor players induce approximate Nash equilibria for the finite player game (7.1)–(7.3)–(7.4). This is very similar to what is done in Section 6.1 of Chapter 6. Our interest in this result is that it justifies the formulation we chose for the major minor mean field game problem.

Throughout the analysis, assumptions **Major Minor MFG**, **Major Hamiltonian**, **Minor Hamiltonian**, **Major Convexity** and **Minor Convexity** are in force. In addition, we also assume for the sake of simplicity that the volatility coefficients are constant.

Assumption (Major Minor Convergence).

(A1) The diffusion coefficients σ_0 and σ are constant matrices.

A convenient consequence of assumption **Major Minor Convergence** is that the minimizer $\hat{\alpha}(t, x_0, x, \mu, y, z_{11})$ of the Hamiltonian $A \ni \alpha \mapsto H(t, x_0, x, \mu, y, z_{11}, \alpha)$ in assumption **Minor Hamiltonian** is independent of z_{11}.

In order to proceed, we use the same set-up as in Chapter 2 for constructing the particle system (2.3) and in Chapter 6 for constructing ϵ-Nash equilibria to finite player games associated with standard mean field games. As above, the probability space $(\Omega, \mathcal{F}, \mathbb{F}, \mathbb{P})$ is the product of two probability spaces $(\Omega^0, \mathcal{F}^0, \mathbb{F}^0, \mathbb{P}^0)$ and $(\Omega^1, \mathcal{F}^1, \mathbb{F}^1, \mathbb{P}^1)$. The space $(\Omega^0, \mathcal{F}^0, \mathbb{F}^0, \mathbb{P}^0)$ carries the m_0-dimensional Wiener process $\boldsymbol{W}^0$ and $\mathbb{F}^0$ is the complete filtration generated by $\boldsymbol{W}^0$. The probability space $(\Omega^1, \mathcal{F}^1, \mathbb{F}^1, \mathbb{P}^1)$ carries a sequence of m-dimensional independent Wiener processes $(\boldsymbol{W}^i)_{i \geq 1}$, which are assumed to be Brownian motions with respect to the filtration $\mathbb{F}^1$. We then denote by $(\Omega, \mathcal{F}, \mathbb{P})$ the completion of the product space $(\Omega, \mathcal{F}^0 \otimes \mathcal{F}^1, \mathbb{P}^0 \otimes \mathbb{P}^1)$ and endow it with the filtration $\mathbb{F} = (\mathcal{F}_t)_{0 \leq t \leq T}$ obtained by augmenting the product filtration $\mathbb{F}^0 \otimes \mathbb{F}^1$ to make it right-continuous and by completing it. For any integer $N \geq 1$, we then call $\mathbb{F}^{(0,N)}$ the completed right continuous filtration generated by $(\boldsymbol{W}^0, \cdots, \boldsymbol{W}^N)$.

Below, we use freely the notation introduced at the beginning of this section for games with one major player and N minor players. We shall say that a strategy $\boldsymbol{\alpha}^{N,0}$ is admissible for the major player if $\boldsymbol{\alpha}^{N,0} \in \mathbb{A}_0$. Motivated by the statement of Theorem 2.12, we say that $\boldsymbol{\alpha}^{N,0}$ is (κ, q)-admissible for some $\kappa \geq 0$ and $q > 4$, and we write $\boldsymbol{\alpha}^{N,0} \in \mathbb{A}_0^{\kappa,q}$ if:

$$\mathbb{E}\left[\int_0^T |\alpha_t^{N,0}|^q dt\right] \leq \kappa.$$

A strategy $\boldsymbol{\alpha}^{N,i}$ is admissible for the i-th minor player if it is $\mathbb{F}^{(0,N)}$-progressively measurable, takes its values in $A \subset \mathbb{R}^k$, and if its energy:

$$\mathbb{E}\left[\int_0^T |\alpha_t^{N,i}|^2 dt\right]$$

is finite. Then, we write $\boldsymbol{\alpha}^{N,i} \in \bar{\mathbb{A}}_{(0,N)}$. The strategy $\boldsymbol{\alpha}^{N,i}$ is said to be κ-admissible if this energy is not greater than κ, in which case we write $\boldsymbol{\alpha}^{N,i} \in \bar{\mathbb{A}}^{\kappa}_{(0,N)}$.

In the present context, the definition of an ϵ-Nash equilibrium introduced in Chapter 6 takes the following form:

Definition 7.7 *A set of admissible strategies* $\boldsymbol{\alpha}^{(0,N)} = (\boldsymbol{\alpha}^{N,0}, \boldsymbol{\alpha}^{N,1}, \cdots, \boldsymbol{\alpha}^{N,N})$ *in* $\mathbb{A}_0^{\kappa,q} \times (\bar{\mathbb{A}}^{\kappa}_{(0,N)})^N$*, for some* $\kappa \geq 0$ *and* $q > 4$*, is called an* ϵ*-Nash equilibrium in* $\mathbb{A}_0^{\kappa,q} \times (\bar{\mathbb{A}}^{\kappa}_{(0,N)})^N$ *for the stochastic differential game with major and minor players* (7.1)–(7.3)–(7.4) *if, for each* $\boldsymbol{\beta}^0 \in \mathbb{A}_0^{\kappa,q}$,

$$J^{N,0}\big(\boldsymbol{\alpha}^{N,0}, \boldsymbol{\alpha}^{N,1}, \cdots, \boldsymbol{\alpha}^{N,N}\big) \leq J^{N,0}\big(\boldsymbol{\beta}^0, \boldsymbol{\alpha}^{N,1}, \cdots, \boldsymbol{\alpha}^{N,N}\big) + \epsilon,$$

and, for each $i \in \{1, \cdots, N\}$ *and* $\boldsymbol{\beta}^i \in \bar{\mathbb{A}}^{\kappa}_{(0,N)}$,

$$\begin{aligned} &J^{N,i}\big(\boldsymbol{\alpha}^{N,0}, \boldsymbol{\alpha}^{N,1}, \cdots, \boldsymbol{\alpha}^{N,N}\big) \\ &\quad \leq J^{N,i}\big(\boldsymbol{\alpha}^{N,0}, \boldsymbol{\alpha}^{N,1}, \cdots, \boldsymbol{\alpha}^{N,i-1}, \boldsymbol{\beta}^i, \boldsymbol{\alpha}^{N,i+1}, \cdots, \boldsymbol{\alpha}^{N,N}\big) + \epsilon. \end{aligned}$$

Before stating and proving the main result of this section, we introduce, in addition to **(A1)** in assumption **Major Minor Convergence**, the following assumptions:

There exist constants $q > 4$ and κ such that:

(A2) The FBSDE (7.27) with $\boldsymbol{W} = \boldsymbol{W}^1$ and $\mathbb{F} = \mathbb{F}^{(0,1)}$ has a unique solution. Moreover, there exists a collection of random variables $(V_t(x))_{0 \leq t \leq T, x \in \mathbb{R}^d}$ from Ω^0 into $\mathbb{R}^d$ such that, for any $x \in \mathbb{R}^d$, the process $(V_t(x))_{0 \leq t \leq T}$ is $\mathbb{F}^0$-progressively measurable, for any $t \in [0, T]$ and any $\omega^0 \in \Omega^0$, the realization of $V_t : \mathbb{R}^d \ni x \mapsto V_t(x)$ is a Lipschitz function whose Lipschitz constant can be bounded uniformly in (t, ω^0), and for almost every (t, ω^0) under $\text{Leb}_1 \otimes \mathbb{P}^0$,

$$Y_t = V_t(X_t).$$

(A3) The process $\hat{\boldsymbol{\alpha}}^0$ satisfies:

$$\mathbb{E}^0 \int_0^T |\hat{\alpha}_t^0|^q dt \leq \kappa.$$

(continued)

(A4) The minimizer $\hat{\alpha}$ in assumption **Minor Hamiltonian** is Lipschitz in (x, y), uniformly in the other variables. Moreover, the solution to (7.27) satisfies:

$$\mathbb{E}^0 \int_0^T \left|\hat{\alpha}\big(t, X_t^0, 0, \mathcal{L}^1(X_t), V_t(0)\big)\right|^q dt \leq \kappa.$$

In general, assumption **(A2)** is merely *wishful thinking* as it may be difficult to check in practice. A concrete sufficient condition of well posedness and the existence of a decoupling field $(V_t(x))_{0 \leq t \leq T, x \in \mathbb{R}^d}$ will be given for the linear quadratic Gaussian (LQG) models which we study next. Of course, the assumption that $\boldsymbol{W} = \boldsymbol{W}^1$ does not play any role in the analysis of the equation. It is here to guarantee that the notations are consistent with our new description of $(\Omega, \mathcal{F}, \mathbb{F}, \mathbb{P})$. In particular, it is implicitly assumed that the same results hold with $\boldsymbol{W}^i$, for any $i \geq 2$, in lieu of $\boldsymbol{W}^1$, meaning that the FBSDE would be uniquely solvable as well with the same decoupling field $(V_t(x))_{0 \leq t \leq T}$. We refer to Subsection 6.1.2 for a detailed review of these facts for mean field games with a common noise.

Below, for $t \in [0, T]$ and $x \in \mathbb{R}^d$, we use the convenient notation $\hat{\alpha}(t, x)$ for the $\mathcal{F}_t^0$-measurable random variable $\hat{\alpha}(t, X_t^0, x, \mathcal{L}^1(X_t), V_t(x))$, where $(\boldsymbol{X}^0, \boldsymbol{X})$ is given by the FBSDE (7.27).

The following theorem shows that the construction of approximate Nash equilibria from the solution of a mean field game problem is still possible in the presence of major and minor players.

Theorem 7.8 *Let assumptions* **Major Minor MFG**, **Major Hamiltonian**, **Minor Hamiltonian**, **Major Convexity**, **Minor Convexity**, *and* **Major Minor Convergence** *be in force. Then, for any $N \geq 1$, there exists a constant $C_{\kappa,q}$, depending on κ and q, but not on N, such that the strategy in partial feedback form given by $(\hat{\alpha}_t^0, (\hat{\alpha}(t, X_t^{N,i}))_{1 \leq i \leq N})_{0 \leq t \leq T}$ is a $(C_{\kappa,q}\sqrt{\epsilon_N})$-Nash equilibrium for the $(N+1)$-player game* (7.1)–(7.3)–(7.4) *in $\mathbb{A}_0^{\kappa,q} \times (\bar{\mathbb{A}}_{(0,N)}^{\kappa})^N$, where:*

$$\epsilon_N = N^{-2/\max(d,4)}\big(1 + \ln(N)\mathbf{1}_{\{d=4\}}\big).$$

Observe that the definition of the strategy used in the statement is implicit since it depends on the optimal path itself. Namely, $(\boldsymbol{X}^{N,1}, \ldots, \boldsymbol{X}^{N,N})$ is given as the solution of a stochastic differential equation with random coefficients obtained by plugging the strategy into (7.1). This equation is uniquely solvable under the standing assumption. Notice also that the strategy profile is not really in feedback form because of the dependence upon $\boldsymbol{W}^0$. Obviously, the strategy played by the major player, namely $\hat{\boldsymbol{\alpha}}^0$, is in open loop. On the other hand, the strategy proposed for each minor player, namely $(\hat{\alpha}(t, X_t^{N,i}))_{0 \leq t \leq T}$ for $i \in \{1, \cdots, N\}$, has a mixed structure. The sole information appearing in feedback form is the private state of the minor player i itself while, in analogy with the construction of approximate Nash equilibria for mean field games with a common noise, the information related to the major player is encapsulated into the open loop structure of the strategy profile. According to the terminology introduced in Chapter 6, see Subsection 6.1.2, the strategy played by the minor player i may be said to be in *semi-closed feedback form*.

As for mean field games with a common noise, the construction of a control strategy in complete feedback form would require a more detailed analysis of the limiting game, say, for instance, of a relevant version of the master equation.

Proof.

First Step. For a fixed N, we investigate the fluctuations in the system obtained by letting the players use the strategies defined in the statement. When all the players apply the prescribed controls, the resulting controlled states, which we denote by $(\hat{X}^{N,i})_{0\le i\le N}$, satisfy:

$$\begin{cases} d\hat{X}_t^{N,0} = b_0\big(t,\hat{X}_t^{N,0},\hat{\mu}_t^N,\hat{\alpha}_t^0\big)dt + \sigma_0 dW_t^0, \quad \hat{X}_0^{N,0} = x^{00}, \\ d\hat{X}_t^{N,i} = b\big(t,\hat{X}_t^{N,i},\hat{\mu}_t^N,\hat{X}_t^{N,0},\hat{\alpha}(t,\hat{X}_t^{N,i})\big)dt + \sigma dW_t^i, \quad \hat{X}_0^{N,i} = x^0, \\ \qquad i = 1,\cdots,N, \end{cases} \tag{7.29}$$

where we define:

$$\hat{\mu}_t^N = \frac{1}{N}\sum_{i=1}^N \delta_{\hat{X}_t^{N,i}}.$$

Following the approach presented in Chapter 6, we define the limiting nonlinear processes as the solution of:

$$\begin{cases} dX_t^0 = b_0\big(t,X_t^0,\mathcal{L}^1(\underline{X}_t^1),\hat{\underline{\alpha}}_t^0\big)dt + \sigma_0 dW_t^0, \quad X_0^0 = x^{00}, \\ d\underline{X}_t^i = b\big(t,\underline{X}_t^i,\mathcal{L}^1(\underline{X}_t^i),X_t^0,\hat{\alpha}(t,\underline{X}_t^i)\big)dt + \sigma dW_t^i, \quad \underline{X}_0^i = x^0, \quad i \ge 1. \end{cases} \tag{7.30}$$

Observe from Proposition 2.11 together with the assumption that the decoupling field is independent of the Wiener process $\boldsymbol{W}$ driving the dynamics of the state of the minor player, that $\boldsymbol{X}^0$ in (7.30) coincides with $\boldsymbol{X}^0$ in (7.26), which makes licit the use of the same notation. By the same argument, we have $\mathcal{L}^1(\underline{X}_t^i) = \mathcal{L}^1(\underline{X}_t^1)$, for all $t \in [0,T]$. Also, in order to have the same notation for the major and the minor players in the limiting game, we shall sometimes write $\underline{\boldsymbol{X}}^0$ for $\boldsymbol{X}^0$.

Recall also that $(\boldsymbol{W}^i)_{i\ge 0}$ is an infinite sequence of independent standard Wiener processes. The stochastic measure flow $(\mathcal{L}^1(\underline{X}_t^1))_{0\le t\le T}$ will be sometimes denoted by $\boldsymbol{\mu} = (\mu_t)_{0\le t\le T}$ in the following. Standard estimates on the solutions of forward McKean-Vlasov equations extended to the conditional case as in Theorem 2.12 in Chapter 2 give the existence of a constant C such that:

$$\max_{0\le i\le N} \mathbb{E}\Big[\sup_{0\le t\le T} |\hat{X}_t^{N,i} - \underline{X}_t^i|^2\Big] \le C\epsilon_N, \tag{7.31}$$

and by applying the usual upper bound for the 2-Wasserstein distance, we also have:

$$\mathbb{E}\Big[\sup_{0\le t\le T} W_2\Big(\hat{\mu}_t^N, \frac{1}{N}\sum_{i=1}^N \delta_{\underline{X}_t^i}\Big)^2\Big] \le C\epsilon_N, \tag{7.32}$$

where C is independent of N, but depends upon κ, q, and the Lipschitz constants of b_0 and b in the variables x_0, μ, and x, and of $\hat{\alpha}$ and $(V_t(\cdot))_{0\le t\le T}$ in the variable x. The dependence upon κ and q comes through the following bound. Thanks to **(A3)** in assumption **Major Minor Convergence**,

$$\mathbb{E}^0\int_0^T |\hat{\alpha}_t^0|^q dt \le \kappa, \tag{7.33}$$

so that, together with the bound $\mathbb{E}^0[\sup_{0\le t\le T} M_2(\mathcal{L}^1(X_t))^2] < \infty$, we obtain $\mathbb{E}^0[\sup_{0\le t\le T}|X_t^0|^q] \le C_{\kappa,q}$ for a constant $C_{\kappa,q}$ depending on κ and q. Combining with **(A4)**, we deduce that $\mathbb{E}^0[\sup_{0\le t\le T}|\underline{X}_t^1|^q] \le C_{\kappa,q}$, which is the required bound to let the machinery of Theorem 2.12 work.

Next, we turn our attention to the cost functionals. We define:

$$\hat{J}^{N,0} = \mathbb{E}\Big[\int_0^T f_0(t, \hat{X}_t^{N,0}, \hat{\mu}_t^N, \hat{\alpha}_t^0)dt + g_0(\hat{X}_T^{N,0}, \hat{\mu}_T^N)\Big],$$

$$J^0 = \mathbb{E}\Big[\int_0^T f_0(t, X_t^0, \mu_t, \hat{\alpha}_t^0)dt + g_0(X_T^0, \mu_T)\Big],$$

and we have, by **(A3)** in assumption **Major Minor MFG** that:

$$\begin{aligned}
|\hat{J}^{N,0} - J^0| &= \Big|\mathbb{E}\Big[\int_0^T f_0(t, \hat{X}_t^{N,0}, \hat{\mu}_t^N, \hat{\alpha}_t^0)dt + g_0(\hat{X}_T^{0,N}, \hat{\mu}_T^N)\Big] \\
&\qquad - \mathbb{E}\Big[\int_0^T f_0(t, X_t^0, \mu_t, \hat{\alpha}_t^0)dt + g_0(\hat{X}_T^0, \mu_T)\Big]\Big| \\
&\le C\Big[\mathbb{E}\int_0^T\Big[\Big(1 + |\hat{X}_t^{N,0}| + |X_t^0| + |\hat{\alpha}_t^0| + M_2(\hat{\mu}_t^N) + M_2(\mu_t)\Big) \\
&\qquad \times \Big(|\hat{X}_t^{N,0} - X_t^0| + W_2(\hat{\mu}_t^N, \mu_t)\Big)\Big]dt \\
&\quad + \mathbb{E}\Big[\Big(1 + |\hat{X}_T^{N,0}| + |X_T^0| + M_2(\hat{\mu}_T^N) + M_2(\mu_T)\Big) \\
&\qquad \times \Big(|\hat{X}_T^{0,N} - X_T^0| + W_2(\hat{\mu}_T^N, \mu_T)\Big)\Big]\Big].
\end{aligned}$$

Hence, by Cauchy Schwarz inequality,

$$
\begin{aligned}
&|\hat{J}^{N,0} - J^0| \\
&\leq C\mathbb{E}\Bigl[\int_0^T \Bigl(1 + |\hat{X}_t^{N,0}|^2 + |X_t^0|^2 + \frac{1}{N}\sum_{i=1}^N [|\hat{X}_t^{N,i}|^2 + |\underline{X}_t^1|^2]\Bigr)dt\Bigr]^{1/2} \\
&\qquad \times \mathbb{E}\Bigl[\int_0^T \Bigl(|\hat{X}_t^{N,0} - X_t^0|^2 + W_2(\hat{\mu}_t^N, \mu_t)^2\Bigr)dt\Bigr]^{1/2} \\
&\quad + C\mathbb{E}\Bigl[\Bigl(1 + |\hat{X}_T^{N,0}| + |X_T^0| + \frac{1}{N}\sum_{i=1}^N [|\hat{X}_T^{N,i}|^2 + |\underline{X}_T^1|^2]\Bigr)^2\Bigr]^{1/2} \\
&\qquad \times \mathbb{E}\Bigl[\Bigl(|\hat{X}_T^{N,0} - X_T^0| + W_2(\hat{\mu}_T^N, \mu_T)\Bigr)^2\Bigr]^{1/2}.
\end{aligned}
\tag{7.34}
$$

By (7.31), we have:

$$
\max_{0 \leq i \leq N} \mathbb{E}\bigl[\sup_{0 \leq t \leq T} |\hat{X}_t^{N,i}|^2\bigr] \leq C_{\kappa,q}.
$$

By applying (7.31) and (7.32), we deduce that:

$$
\hat{J}^{N,0} = J^0 + O(\epsilon_N^{1/2}). \tag{7.35}
$$

Second Step. Assume now that the major player uses a different admissible control $\boldsymbol{\beta}^0 \in \mathbb{A}_0^{\kappa,q}$, and that the minor players still use the strategies in open loop form $(\hat{\alpha}(t, \hat{X}_t^{N,i}))_{0 \leq t \leq T, 1 \leq i \leq N}$. The resulting perturbed state processes will be denoted by $(\boldsymbol{U}^{N,i})_{0 \leq i \leq N}$. They solve the system:

$$
\begin{cases}
dU_t^{N,0} = b_0(t, U_t^{N,0}, \bar{\nu}_t^N, \beta_t^0)dt + \sigma_0 dW_t^0, \quad U_0^{N,0} = x^{00}, \\
dU_t^{N,i} = b\bigl(t, U_t^{N,i}, \bar{\nu}_t^N, U_t^{N,0}, \hat{\alpha}(t, \hat{X}_t^{N,i})\bigr)dt + \sigma dW_t^i, \quad U_0^{N,i} = x^0, \\
\qquad\qquad\qquad\qquad\qquad\qquad\qquad\qquad\qquad i = 1, \cdots, N,
\end{cases}
\tag{7.36}
$$

where as usual,

$$
\bar{\nu}_t^N = \frac{1}{N}\sum_{i=1}^N \delta_{U_t^{N,i}}.
$$

Note that $(\boldsymbol{U}^{N,i})_{1 \leq i \leq N}$, and thus $(\bar{\nu}_t^N)_{0 \leq t \leq T}$, are not $\mathbb{F}^0$-progressively measurable in general. In order to apply Theorem 2.12 of Subsection 2.1.4 in Chapter 2, we combine (7.30) and (7.36) and consider the limiting nonlinear processes defined as the solution of:

$$
\begin{cases}
dX_t^0 = b_0(t, X_t^0, \mathcal{L}^1(\underline{X}_t^1), \hat{\alpha}_t^0)dt + \sigma_0 dW_t^0, \quad X_0^0 = x^{00}, \\
d\underline{X}_t^i = b\bigl(t, \underline{X}_t^i, \mathcal{L}^1(\underline{X}_t^i), X_t^0, \hat{\alpha}(t, \underline{X}_t^i)\bigr)dt + \sigma dW_t^i, \quad \underline{X}_0^i = x^0, \quad i \geq 1, \\
dU_t^0 = b_0(t, U_t^0, \mathcal{L}^1(\underline{U}_t^1), \beta_t^0)dt + \sigma_0 dW_t^0, \quad U_0^0 = x^{00}, \\
d\underline{U}_t^i = b\bigl(t, \underline{U}_t^i, \mathcal{L}^1(\underline{U}_t^i), U_t^0, \hat{\alpha}(t, \underline{X}_t^i)\bigr)dt + \sigma dW_t^i, \quad \underline{U}_0^i = x^0, \quad i \geq 1.
\end{cases}
\tag{7.37}
$$

Adapting the argument used in Theorem 2.12 to the present situation and using (7.31), we deduce that there exists a constant C' such that:

$$\mathbb{E}\Big[\sup_{0\le t\le T}|U_t^{N,i}-\underline{U}_t^i|^2\Big]\le C'\epsilon_N,$$

where C' depends upon T, the Lipschitz constants of b_0 and b in the variables x_0, μ, and x, and of $\hat\alpha$ and $(V_t(\cdot))_{0\le t\le T}$ in the variable x, and upon:

$$\mathbb{E}\int_0^T|\beta_t^0|^q dt,$$

for the same q as above. Since the latter is assumed to be less than κ, we can replace C' by the constant $C_{\kappa,q}$ which we used before, provided that it is allowed to increase from line to line.

Making use of (**A2**) and (**A4**) in assumption **Major Minor Convergence**, it also holds:

$$\mathbb{E}\int_0^T|\beta_t^0|^q dt\le\kappa\Rightarrow\mathbb{E}\Big[\sup_{0\le t\le T}\big(|U_t^0|^q+|\underline{U}_t^1|^q\big)\Big]\le C_{\kappa,q}.$$

Using the same estimates as in (7.34), we deduce that, for all $\boldsymbol{\beta}^0\in\mathbb{A}_0^{\kappa,q}$,

$$\Big|J^{N,0}\big(\boldsymbol{\beta}^0,\big((\hat\alpha(t,\hat X_t^{N,i}))_{0\le t\le T}\big)_{i=1,\cdots,N}\big)-J^0\big(\boldsymbol{\beta}^0,(\hat\alpha(t,\underline{X}_t^1))_{0\le t\le T}\big)\Big|\le C_{\kappa,q}\sqrt{\epsilon_N},\tag{7.38}$$

where $J^0(\boldsymbol{\beta}^0,(\hat\alpha(t,\underline{X}_t^1))_{0\le t\le T})$ is defined as in (7.6). Finally, since $(\hat\alpha_t^0,\hat\alpha(t,\underline{X}_t^1))_{0\le t\le T}$ solves the limiting two-player game problem driven by $\boldsymbol{W}^0$ and $\boldsymbol{W}^1$, it is clear that:

$$J^0\le J^0\big(\boldsymbol{\beta}^0,(\hat\alpha(t,\underline{X}_t^1))_{0\le t\le T}\big),\tag{7.39}$$

and combining (7.35), (7.38) and (7.39), we get the desired result for the major player.

Third Step. We now consider the case when a minor player changes its strategy unilaterally, and without loss of generality we consider the case when the minor player with index 1 changes its strategy to $\boldsymbol{\beta}\in\bar{\mathbb{A}}^{\kappa}_{(0,N)}$. This part of the proof mimics very closely the proofs of Theorems 6.7 and 6.13 of Sections 6.1.1 and 6.1.2 in Chapter 6, and we will refer to these proofs for the details which we skip in the present argumentation. The resulting perturbed controlled dynamics are now given by:

$$\begin{cases}dU_t^{N,0}=b_0(t,U_t^{N,0},\bar\nu_t^N,\hat\alpha_t^0)dt+\sigma_0 dW_t^0, & U_0^{N,0}=x^{00},\\ dU_t^{N,1}=b(t,U_t^{N,1},\bar\nu_t^N,U_t^{N,0},\beta_t)dt+\sigma dW_t^1, & U_0^{N,1}=x^0,\\ dU_t^{N,i}=b\big(t,U_t^{N,i},\bar\nu_t^N,U_t^{N,0},\hat\alpha(t,\hat X_t^{N,i})\big)dt+\sigma dW_t^i, & U_0^{N,i}=x^0,\\ & i=2,\cdots,N.\end{cases}\tag{7.40}$$

Using the usual estimates on the difference between $\boldsymbol{U}^{N,i}$ and $\hat{\boldsymbol{X}}^{N,i}$, and applying Gronwall's inequality, we show that there exists a constant C, independent of N, such that:

$$\mathbb{E}\Big[\sup_{0\le t\le T}|U_t^{N,0}-\hat{X}_t^{N,0}|^2\Big]+\frac{1}{N}\sum_{i=1}^{N}\mathbb{E}\Big[\sup_{0\le t\le T}|U_t^{N,i}-\hat{X}_t^{N,i}|^2\Big] \le \frac{C}{N}\int_0^T|\beta_t-\hat{\alpha}(t,\hat{X}_t^{N,1})|^2dt. \tag{7.41}$$

Combining the above bound, the growth properties of $\hat{\alpha}$, and (7.31), we see that:

$$\mathbb{E}\int_0^T|\beta_t|^2dt\le\kappa \Rightarrow \mathbb{E}\Big[\sup_{0\le t\le T}|U_t^{N,0}-X_t^0|^2\Big]+\mathbb{E}\Big[\sup_{0\le t\le T}W_2(\bar{\nu}_t^N,\mu_t)^2\Big]\le C_{\kappa,q}\epsilon_N.$$

We hence conclude that, when $\mathbb{E}\int_0^T|\beta_t|^2dt\le\kappa$, we have:

$$\mathbb{E}\Big[\sup_{0\le t\le T}|U_t^{N,1}-\underline{U}_t^1|^2\Big]\le C_{\kappa,q}\epsilon_N,$$

where $\underline{\boldsymbol{U}}^1$ is the solution of the stochastic differential equation:

$$d\underline{U}_t^1=b(t,\underline{U}_t^1,\mu_t,X_t^0,\beta_t)dt+\sigma dW_t^1,\quad \underline{U}_0^1=x^0, \tag{7.42}$$

where $\boldsymbol{\mu}$ and $\boldsymbol{X}^0$ are from the solution of the FBSDE (7.27). We then conclude in the same way as for the major player. □

Remark 7.9 *In the analysis of the best response of player* 1*, we can allow the other minor players, namely players with an index $i\in\{2,\cdots,N\}$, to implement a strategy in semi-closed feedback form, namely to use $(\hat{\alpha}(t,U_t^{N,i}))_{0\le t\le T}$ instead of $(\hat{\alpha}(t,\hat{X}_t^{N,i}))_{0\le t\le T}$ in* (7.40)*. The analysis would be the same. This is consistent with the fact that, for mean field games without common noise, limiting equilibria provide approximate equilibria of the N-player game in both open loop and Markovian form.*

However, in the analysis of the best response of the major player, we cannot do so. If one modifies the strategy profiles of the minor players in (7.36)*, one changes the corresponding limiting equation in* (7.37)*. As a result, the argument used in the second step of the above proof fails. This is still another instance of the simple fact that, in the limiting mean field game with major and minor players, open loop and closed loop equilibria may differ when computing the best response of the major player.*

7.1.6 The Linear Quadratic Case

In this section, we consider the mean field game with major and minor players issued from a finite player game in which the dynamics of the states of the players are given by the following linear stochastic differential equations:

$$\begin{cases} dX_t^{N,0} = (L_0 X_t^{N,0} + B_0 \alpha_t^{N,0} + F_0 \bar{X}_t^N)dt + D_0 dW_t^0, \\ dX_t^{N,i} = (LX_t^{N,i} + B\alpha_t^{N,i} + F\bar{X}_t^N + GX_t^0)dt + DdW_t^i, \qquad 1 \leq i \leq N, \end{cases}$$

for $t \in [0, T]$, and we choose $A_0 = \mathbb{R}^{k_0}$ and $A = \mathbb{R}^k$ for the sets of possible values of the controls. Note that the coefficients are deterministic constant matrices independent of time. The real matrices L_0, B_0, F_0, and D_0 are of dimensions $d_0 \times d_0$, $d_0 \times k_0$, $d_0 \times d$, and $d_0 \times m_0$ respectively. Similarly, the real matrices L, B, F, G, and D are of dimensions $d \times d$, $d \times k$, $d \times d$, $d \times d_0$, and $d \times m_0$ respectively. The cost functionals for the major and minor players are given by:

$$\begin{aligned} & J^{N,0}(\boldsymbol{\alpha}^{N,0}, \cdots, \boldsymbol{\alpha}^{N,N}) \\ & \quad = \mathbb{E}\biggl[\int_0^T \Bigl[\bigl(X_t^{N,0} - \psi_0(\bar{X}_t^N)\bigr)^\dagger \Gamma_0 \bigl(X_t^{N,0} - \psi_0(\bar{X}_t^N)\bigr) + (\alpha_t^{N,0})^\dagger R_0 \alpha_t^{N,0}\Bigr] dt\biggr], \\ & J^{N,i}(\boldsymbol{\alpha}^{N,0}, \cdots, \boldsymbol{\alpha}^{N,N}) \\ & \quad = \mathbb{E}\biggl[\int_0^T \Bigl[\bigl(X_t^{N,i} - \psi(X_t^{N,0}, \bar{X}_t^N)\bigr)^\dagger \Gamma \bigl(X_t^{N,i} - \psi(X_t^{N,0}, \bar{X}_t^N)\bigr) \\ & \qquad\qquad\qquad + (\alpha_t^{N,i})^\dagger R \alpha_t^{N,i}\Bigr] dt\biggr], \end{aligned}$$

in which Γ_0, Γ are nonnegative semi-definite symmetric matrices of dimensions $d_0 \times d_0$, $d \times d$ and R_0 and R are (strictly) positive definite symmetric matrices of dimensions $k_0 \times k_0$ and $k \times k$, and where the functions ψ_0 and ψ are defined by:

$$\psi_0(x) = K_0 x + \eta_0, \quad \psi(x^0, x) = Kx^0 + K_1 x + \eta, \quad x^0 \in \mathbb{R}^{d_0},\ x \in \mathbb{R}^d,$$

for some fixed $d_0 \times d$, $d \times d_0$ and $d \times d$ matrices K_0, K and K_1, and some fixed $\eta_0 \in \mathbb{R}^{d_0}$ and $\eta \in \mathbb{R}^d$. Here, $\bar{X}_t^N$ stands for the empirical mean $(X_t^{N,1} + \cdots + X_t^{N,N})/N$.

We propose to study this linear quadratic model with the different approaches introduced earlier, starting with the open loop formulation of Subsection 7.1.2 based on a limiting two-player game and a fixed point on measure flows.

Solution in the Original Open-Loop Formulation

Formulating the limiting game accordingly, we observe that all the aforementioned assumptions **Major Minor MFG**, **Major Hamiltonian**, **Minor Hamiltonian**, **Major Convexity**, and **Minor Convexity** hold in the present linear quadratic setting. Also, the non-Markovian conditional McKean-Vlasov FBSDE (7.27), with:

$$\hat{\alpha}_t^0 = -\tfrac{1}{2} R_0^{-1} B_0^\dagger \mathbb{E}^1[P_t^0], \quad \hat{\alpha}_t = -\tfrac{1}{2} R^{-1} B^\dagger Y_t, \quad t \in [0, T],$$

can be rewritten as:

$$\begin{cases} dX_t^0 = \big(L_0 X_t^0 - \frac{1}{2} B_0 R_0^{-1} B_0^\dagger \mathbb{E}^1[P_t^0] + F_0 \mathbb{E}^1[X_t]\big) dt + D_0 dW_t^0, \\ dX_t = \big(L X_t - \frac{1}{2} B R^{-1} B^\dagger Y_t + F \mathbb{E}^1[X_t] + G X_t^0\big) dt + D dW_t, \\ dP_t^0 = \big(-L_0^\dagger P_t^0 - G^\dagger P_t - 2\Gamma_0 (X_t^0 - \psi_0(\mathbb{E}^1[X_t]))\big) dt \\ \qquad + Q_t^{00} dW_t^0 + Q_t^{01} dW_t, \\ dP_t = - L^\dagger P_t dt + Q_t^{10} dW_t^0 + Q_t^{11} dW_t \\ \qquad - \Big(F_0^\dagger \mathbb{E}^1[P_t^0] + F^\dagger \mathbb{E}^1[P_t] - 2K_0^\dagger \Gamma_0 (X_t^0 - \psi_0(\mathbb{E}^1[X_t]))\Big) dt, \\ dY_t = \big(-L^\dagger Y_t - 2\Gamma (X_t - \psi(X_t^0, \mathbb{E}^1[X_t]))\big) dt + Z_t^0 dW_t^0 + Z_t dW_t, \end{cases}$$

for $t \in [0, T]$, with the initial and terminal conditions given by:

$$X_0^0 = x^{00}, \ X_0 = x^0, \ P_T^0 = P_T = Y_T = 0.$$

As before, we use a bar to denote the conditional expectation in order to simplify the notation. Doing so, we arrive at the following FBSDE:

$$\begin{cases} dX_t^0 = \big(L_0 X_t^0 - \frac{1}{2} B_0 R_0^{-1} B_0^\dagger \bar{P}_t^0 + F_0 \bar{X}_t\big) dt + D_0 dW_t^0, \\ dX_t = \big(L X_t - \frac{1}{2} B R^{-1} B^\dagger Y_t + F \bar{X}_t + G X_t^0\big) dt + D dW_t, \\ dP_t^0 = \Big(-L_0^\dagger P_t^0 - G^\dagger P_t - 2\Gamma_0 \big(X_t^0 - K_0 \bar{X}_t - \eta_0\big)\Big) dt \\ \qquad + Q_t^{00} dW_t^0 + Q_t^{01} dW_t, \\ dP_t = - L^\dagger P_t dt + Q_t^{10} dW_t^0 + Q_t^{11} dW_t \\ \qquad - \Big(F_0^\dagger \bar{P}_t^0 + F^\dagger \bar{P}_t - 2K_0^\dagger \Gamma_0 \big(X_t^0 - K_0 \bar{X}_t - \eta_0\big)\Big) dt, \\ dY_t = \big(-L^\dagger Y_t - 2\Gamma X_t + 2\Gamma K X_t^0 + 2\Gamma K_1 \bar{X}_t + 2\Gamma \eta\big) dt \\ \qquad + Z_t^0 dW_t^0 + Z_t dW_t. \end{cases} \tag{7.43}$$

We rewrite this FBSDE one more time by taking the expectation under $\mathbb{P}^1$. Notice that since $\boldsymbol{X}_t^0$ is already $\mathbb{F}^0$-progressively measurable, it will not get a bar, and its notation will remain unchanged.

$$
\begin{cases}
dX_t^0 = \left(L_0 X_t^0 - \frac{1}{2} B_0 R_0^{-1} B_0^\dagger \bar{P}_t^0 + F_0 \bar{X}_t\right)dt + D_0 dW_t^0, \\
d\bar{X}_t = \left(L\bar{X}_t - \frac{1}{2} B R^{-1} B^\dagger \bar{Y}_t + F\bar{X}_t + G X_t^0\right)dt, \\
d\bar{P}_t^0 = \Big(-L_0^\dagger \bar{P}_t^0 - G^\dagger \bar{P}_t - 2\Gamma_0\big(X_t^0 - K_0\bar{X}_t - \eta_0\big)\Big)dt \\
\qquad\qquad + \bar{Q}_t^{00} dW_t^0, \\
d\bar{P}_t = -L^\dagger \bar{P}_t dt + \bar{Q}_t^{10} dW_t^0 \\
\qquad\qquad - \Big(F_0^\dagger \bar{P}_t^0 + F^\dagger \bar{P}_t - 2K_0^\dagger \Gamma_0\big(X_t^0 - K_0\bar{X}_t - \eta_0\big)\Big)dt \\
d\bar{Y}_t = \big(-L^\dagger \bar{Y}_t - 2\Gamma \bar{X}_t + 2\Gamma K X_t^0 + 2\Gamma K_1 \bar{X}_t + 2\Gamma\eta\big)dt \\
\qquad\qquad + \bar{Z}_t^0 dW_t^0,
\end{cases}
\tag{7.44}
$$

for $t \in [0, T]$. If we use $\boldsymbol{\mathcal{X}}$ and $\boldsymbol{\mathcal{Y}}$ to denote $(\boldsymbol{X}^0, \bar{\boldsymbol{X}})$ and $(\bar{\boldsymbol{P}}^0, \bar{\boldsymbol{P}}, \bar{\boldsymbol{Y}})$, we can write the above FBSDE in the following standard form:

$$
\begin{cases}
d\mathcal{X}_t = (\mathfrak{L}\mathcal{X}_t + \mathfrak{B}\mathcal{Y}_t)dt + \mathfrak{D}dW_t^0, \\
d\mathcal{Y}_t = -(\hat{\mathfrak{L}}\mathcal{X}_t + \hat{\mathfrak{B}}\mathcal{Y}_t + \hat{\mathfrak{C}})dt + \mathcal{Z}_t dW_t^0,
\end{cases}
\tag{7.45}
$$

for $t \in [0, T]$, with initial and terminal conditions given by:

$$
\mathcal{X}_0 = \begin{bmatrix} x^{00} \\ x^0 \end{bmatrix}, \quad \mathcal{Y}_T = \begin{bmatrix} 0 \\ 0 \\ 0 \end{bmatrix},
$$

and the coefficients given by:

$$
\mathfrak{L} = \begin{bmatrix} L_0 & F_0 \\ G & L+F \end{bmatrix}, \quad \mathfrak{B} = \begin{bmatrix} -\frac{1}{2}B_0 R_0^{-1} B_0^\dagger & 0 & 0 \\ 0 & 0 & -\frac{1}{2} B R^{-1} B^\dagger \end{bmatrix}, \quad \mathfrak{D} = \begin{bmatrix} D_0 \\ 0 \end{bmatrix},
$$

$$
\hat{\mathfrak{L}} = \begin{bmatrix} 2\Gamma_0 & -2\Gamma_0 K_0 \\ -2K_0^\dagger \Gamma_0 & 2K_0^\dagger \Gamma_0 K_0 \\ -2\Gamma K & 2\Gamma - 2\Gamma K_1 \end{bmatrix}, \quad \hat{\mathfrak{B}} = \begin{bmatrix} L_0^\dagger & G^\dagger & 0 \\ F_0^\dagger & L^\dagger + F^\dagger & 0 \\ 0 & 0 & L^\dagger \end{bmatrix}, \quad \hat{\mathfrak{C}} = \begin{bmatrix} -2\Gamma_0\eta_0 \\ 2K_0^\dagger \Gamma_0 \eta_0 \\ -2\Gamma\eta \end{bmatrix}.
$$

Riccati Equation

We look for solutions with an affine decoupling field, namely for solutions in the form $\mathcal{Y}_t = S_t \mathcal{X}_t + s_t$, for $t \in [0, T]$, where $(S_t)_{0 \le t \le T}$ and $(s_t)_{0 \le t \le T}$ are two deterministic functions defined on $[0, T]$ with values in $\mathbb{R}^{(d_0+2d)\times(d_0+d)}$ and $\mathbb{R}^{d_0+2d}$ respectively.

In order to proceed, we thus consider the following matrix Riccati equation on $[0, T]$ with terminal condition at time T:

$$\dot{S}_t + S_t\mathfrak{L} + \hat{\mathfrak{B}}S_t + S_t\mathfrak{B}S_t + \hat{\mathfrak{L}} = 0, \quad t \in [0,T]\ ; \quad S_T = 0, \tag{7.46}$$

and the linear ordinary differential equation:

$$\dot{s}_t = -(\hat{\mathfrak{B}} + S_t\mathfrak{B})s_t - \hat{\mathfrak{C}}, \quad t \in [0,T]\ ; \quad s_T = 0. \tag{7.47}$$

We observe that, when $(S_t)_{0\le t\le T}$ is well defined, the backward equation (7.47) is always uniquely solvable. We state the following proposition without proof since we went through the same argument several times already in the book.

Proposition 7.10 *If the matrix Riccati equation (7.46) and the backward ODE (7.47) are well posed, i.e., admit unique solutions denoted by:*

$$S_t = \begin{bmatrix} S_t^{1,1} & S_t^{1,2} \\ S_t^{2,1} & S_t^{2,2} \\ S_t^{3,1} & S_t^{3,2} \end{bmatrix}, \qquad \textit{and} \qquad s_t = \begin{bmatrix} s_t^1 \\ s_t^2 \\ s_t^3 \end{bmatrix}, \quad t \in [0,T],$$

then the FBSDE (7.44) is uniquely solvable. The first two components in the solution, namely $\boldsymbol{\mathcal{X}} = (\boldsymbol{X}^0, \bar{\boldsymbol{X}})$, *are given by the solution of the linear SDE:*

$$\begin{cases} dX_t^0 = \big[L_0X_t^0 - \frac{1}{2}B_0R_0^{-1}B_0^\dagger\big(S_t^{1,1}X_t^0 + S_t^{1,2}\bar{X}_t + s_t^1\big) + F_0\bar{X}_t\big]dt + D_0dW_t^0, \\ d\bar{X}_t = \big[L\bar{X}_t - \frac{1}{2}BR^{-1}B^\dagger\big(S_t^{3,1}X_t^0 + S_t^{3,2}\bar{X}_t + s_t^3\big) + F\bar{X}_t + GX_t^0\big]dt, \end{cases}$$

with initial conditions given by:

$$X_0^0 = x^{00}, \quad \bar{X}_0 = x^0,$$

and the process $\boldsymbol{\mathcal{Y}} = (\bar{\boldsymbol{P}}^0, \bar{\boldsymbol{P}}, \bar{\boldsymbol{Y}})$ *is given by:*

$$\begin{cases} \bar{P}_t^0 = S_t^{1,1}X_t^0 + S_t^{1,2}\bar{X}_t + s_t^1, \\ \bar{P}_t = S_t^{2,1}X_t^0 + S_t^{2,2}\bar{X}_t + s_t^2, \\ \bar{Y}_t = S_t^{3,1}X_t^0 + S_t^{3,2}\bar{X}_t + s_t^3. \end{cases}$$

We now turn to the original conditional FBSDE (7.43). Now that $\boldsymbol{X}^0$, $\bar{\boldsymbol{X}}$, $\bar{\boldsymbol{P}}^0$ and $\bar{\boldsymbol{P}}$ are identified, we plug their values into the FBSDE which in turn, becomes a standard linear FBSDE with random coefficients. Using the fact that $\boldsymbol{X}^0$, $\bar{\boldsymbol{X}}$, $\bar{\boldsymbol{P}}^0$ and $\bar{\boldsymbol{P}}$ are actually solutions of linear SDEs with deterministic coefficients, we have the following result.

Proposition 7.11 *Under the assumption of Proposition 7.10, the FBSDE (7.43) has a unique solution. Moreover, there exist a deterministic $\mathbb{R}^{(d_0+2d)\times(d_0+d)}$-matrix valued function $(\gamma_t)_{0\le t\le T}$ and an $\mathbb{F}^0$-progressively measurable process $(\kappa_t)_{0\le t\le T}$ with values in $\mathbb{R}^d$ such that:*

$$Y_t = \gamma_t X_t + \kappa_t, \qquad t \in [0,T], \tag{7.48}$$

with $\mathbb{E}^0[\sup_{0\le t\le T}|\kappa_t|^q] < \infty$, for all $q \ge 1$.

Proof. We plug $\boldsymbol{X}^0$, $\bar{\boldsymbol{X}}$, $\bar{\boldsymbol{Y}}$, $\bar{\boldsymbol{P}}^0$ and $\bar{\boldsymbol{P}}$ into (7.43). We readily observe that the second and last equations form a standard FBSDE with random coefficients. The structure of this FBSDE is standard in the sense that it can be derived from the stochastic optimal control problem based on state controlled dynamics given by:

$$d\check{X}_t = \big(L\check{X}_t + B\check{\alpha}_t + \check{\mu}_t\big)dt + DdW_t, \quad t \in [0,T]\,; \quad \check{X}_0 = x^0,$$

and the cost functional:

$$\check{J}(\check{\boldsymbol{\alpha}}) = \mathbb{E}\bigg[\int_0^T \Big[\big(\check{X}_t - (KX_t^0 + K_1\bar{X}_t + \eta)\big)^\dagger \Gamma\big(\check{X}_t - (KX_t^0 + K_1\bar{X}_t + \eta)\big) + \check{\alpha}_t^\dagger R\check{\alpha}_t\Big]dt\bigg],$$

where $(\check{\mu}_t = FX_t + GX_t^0)_{0\le t\le T}$. Taking advantage of the fact that Γ is nonnegative semi-definite and that R is positive definite, we may tackle this problem by standard arguments. This yields (7.48). Meanwhile, we observe that the third and fourth equations in (7.43) form a standard BSDE whose well posedness is well known. The identification of the processes $\boldsymbol{P}^0$ and $\boldsymbol{P}$ then follows. □

In order to solve the Riccati equation (7.46) we introduce the $(2d_0+3d)\times(2d_0+3d)$-matrix $\mathfrak{T}$ defined as:

$$\mathfrak{T} = \begin{bmatrix} \mathfrak{L} & \mathfrak{B} \\ -\hat{\mathfrak{L}} & -\hat{\mathfrak{B}} \end{bmatrix},$$

where we recall that $\mathfrak{L}$ is of size $(d_0+d)\times(d_0+d)$, $\mathfrak{B}$ is of size $(d_0+d)\times(d_0+2d)$, $\hat{\mathfrak{L}}$ is of size $(d_0+2d)\times(d_0+d)$ and $\hat{\mathfrak{B}}$ is of size $(d_0+2d)\times(d_0+2d)$. We then denote by $(\Psi(t))_{t\in\mathbb{R}}$ the $(2d_0+3d)\times(d_0+d)$-matrix valued solution to the ODE:

$$\begin{cases} \dfrac{d}{dt}\Psi(t) = \mathfrak{T}\Psi(t), \quad t \in [0,T], \\ \Psi(T) = \begin{bmatrix} I_{d_0+d} \\ 0_{(d_0+2d)\times(d_0+d)} \end{bmatrix}, \end{cases}$$

where I_{d_0+d} is the identity matrix of size d_0+d and $0_{(d_0+2d)\times(d_0+d)}$ is the zero matrix of size $(d_0+2d)\times(d_0+d)$. Obviously, $\Psi(t)$ can be decomposed in block form:

$$\Psi(t) = \begin{bmatrix} \Gamma_t^1 \\ \Gamma_t^2 \end{bmatrix}, \quad t \in [0,T],$$

where Γ_t^1 is of size $(d_0+d)\times(d_0+d)$ and Γ_t^2 is of size $(d_0+2d)\times(d_0+d)$.

We state without proof a standard sufficient condition for the unique solvability of (7.46), known as Radon's lemma. See the Notes & Complements at the end of the chapter for references.

Lemma 7.12 *If the $(d_0+d)\times(d_0+d)$-matrix Γ_t^1 is invertible for each $t \in [0,T]$, then,*

$$S_t = \Gamma_t^2 \left(\Gamma_t^1\right)^{-1}, \quad t \in [0,T],$$

is the unique solution of the Riccati equation (7.46).

Under the assumption of Lemma 7.12, the above two propositions say that assumption **Major Minor Convergence** is satisfied with:

$$V_t(x) = \gamma_t x + \kappa_t, \quad (t,x) \in [0,T]\times\mathbb{R}^d,$$

and

$$\begin{aligned}
\hat{\alpha}_t^0 &= -\tfrac{1}{2} R_0^{-1} B_0^\dagger \bar{P}_t^0, \quad t \in [0,T], \\
\hat{\alpha}(t,x) &= -\tfrac{1}{2} R B^\dagger (\gamma_t x + \kappa_t), \quad (t,x) \in [0,T]\times\mathbb{R}^d.
\end{aligned}$$

Therefore, Theorem 7.8 applies.

7.1.7 An Enlightening Example

As explained in the Notes & Complements at the end of the chapter, our formulation of the mean field game with major and minor players is more involved than what one usually finds in the existing literature. There, the control problem faced by the major player is not of conditional McKean-Vlasov type, and in particular, the measure flow is not endogenous to the controller. The goal of this section is to provide another justification of the choice for our earlier major minor game formulation. We argue that it is necessary for the limiting problem to be a *bona fide* two-player game instead of two consecutive control problems, even if it is at the cost of adding a second fixed point problem, coming from the Nash equilibrium for the two-player game, on top of the mean field fixed point problem.

Our argument is based on the analysis of a concrete model of a $(N+1)$-player game with one major and N minor players for which we can show that, in the limit $N \to \infty$, the equilibrium problem converges toward our formulation of the mean field game.

For the sake of simplicity, we set the model in one dimension, so we assume $d = d_0 = m = m_0 = k = k_0 = 1$. Also, we choose $A_0 = A = \mathbb{R}$. We then consider the $(N+1)$-player game whose state dynamics are given by:

$$\begin{cases} dX_t^{N,0} = \Big(\frac{a}{N}\sum_{i=1}^{N} X_t^{N,i} + b\alpha_t^{N,0}\Big)dt + D_0 dW_t^0, \quad X_0^{N,0} = x^{00}, \\ dX_t^{N,i} = cX_t^{N,0}dt + D dW_t^i, \quad X_0^{N,i} = x^0, \quad i = 1, 2, \cdots, N, \end{cases}$$

D and D_0 now standing for scalars, and the objective function of the major player being given by:

$$J^{N,0}\big(\boldsymbol{\alpha}^{N,0}, \cdots, \boldsymbol{\alpha}^{N,N}\big) = \mathbb{E}\bigg[\int_0^T \Big(q|X_t^{N,0}|^2 + |\alpha_t^{N,0}|^2\Big)dt\bigg],$$

with $q \geq 0$, and the objective functions of the minor players by:

$$J^{N,i}\big(\boldsymbol{\alpha}^{N,0}, \cdots, \boldsymbol{\alpha}^{N,N}\big) = \mathbb{E}\bigg[\int_0^T |\alpha_t^{N,i}|^2 dt\bigg],$$

and we search for an open loop Nash equilibrium. Obviously, the minor players' best responses are always identically 0, regardless of other players' control processes. Therefore, the only remaining problem is to determine the major player's best response, which amounts to solving a stochastic control problem, before taking care of the fixed point step.

Finite-Player Game Nash Equilibrium

We use the same framework as in Subsection 7.1.5, except for the following fact. We allow the admissible controls for the major player to be general square-integrable $\mathbb{F}^{(0,N)}$-progressively measurable processes with values in $A_0 = \mathbb{R}$. In particular, they may not be adapted with respect to the sole noise $\boldsymbol{W}^0$. Although it does not seem consistent with the analysis provided in Subsection 7.1.5, we emphasize the fact that the equilibria constructed right below fit asymptotically the setting used in Subsection 7.1.5, in the sense that the controls for the major player are asymptotically independent of the private noises $(\boldsymbol{W}^i)_{i\geq 1}$. This suffices for our illustration.

As we shall apply the stochastic maximum principle, we introduce the Hamiltonian of the major player. It is given by:

$$H^0\big(t,(x_0,\cdots,x_N),(y_0,\cdots,y_N),\alpha\big)$$
$$= \Big(\frac{a}{N}\sum_{i=1}^N x_i + b\alpha_0\Big)y_0 + cx_0\sum_{i=1}^N y_i + qx_0^2 + \alpha_0^2,$$

for $(x_0,\cdots,x_N),(y_0,\cdots,y_N) \in \mathbb{R}^{N+1}$ and $\alpha_0 \in \mathbb{R}$.

Its minimization as a function of α_0 is straightforward. The minimizer is $\hat{\alpha}_0 = -by_0/2$. Applying the *game version* of the Pontryagin stochastic maximum principle leads to the FBSDE:

$$\begin{cases} dX_t^{N,0} = \Big(\frac{a}{N}\sum_{i=1}^N X_t^{N,i} - \frac{1}{2}b^2 Y_t^{N,0}\Big)dt + D_0 dW_t^0, \\ dX_t^{N,i} = cX_t^{N,0}dt + DdW_t^i, \quad 1 \le i \le N, \\ dY_t^{N,0} = -\Big(c\sum_{i=1}^N Y_t^{N,i} + 2qX_t^{N,0}\Big)dt + \sum_{j=0}^N Z_t^{N,0,j}dW_t^j, \\ dY_t^{N,i} = -\frac{a}{N}Y_t^{N,0}dt + \sum_{j=0}^N Z_t^{N,i,j}dW_t^j, \quad 1 \le i \le N, \end{cases}$$

for $t \in [0,T]$, with $X_0^{N,0} = x^{00}$ and $X_0^{N,i} = x^0$, for $i = 1,\cdots,N$, as initial conditions, and $Y_T^{N,0} = Y_T^{N,1} = \cdots = Y_T^{N,N} = 0$ as terminal conditions. The fact that the optimal control identified by the Pontryagin stochastic maximum principle is $\hat{\alpha}_t^{N,0} = -bY_t^{N,0}/2$ implies that the state variables and the adjoint variables of the minor players only enter the above equations through their aggregate empirical averages. So we rewrite the problem in terms of the processes:

$$\bar{X}_t^N = \frac{1}{N}\sum_{i=1}^N X_t^{N,i}, \quad \bar{W}_t^N = \frac{1}{N}\sum_{i=1}^N W_t^{N,i}, \quad Y_t^N = \sum_{i=1}^N Y_t^{N,i}, \quad t \in [0,T],$$

where we put a bar on both $\bar{X}_t^N$ and $\bar{W}_t^N$ since they are averages, while we do not do the same for the last term, since it is not an average. With these notations, the above FBSDE can be rewritten as:

$$\begin{cases} dX_t^{N,0} = \big(a\bar{X}_t^N - \frac{1}{2}b^2 Y_t^{N,0}\big)dt + D_0 dW_t^0, \\ d\bar{X}_t^N = cX_t^{N,0}dt + Dd\bar{W}_t^N, \\ dY_t^{N,0} = -\big(cY_t^N + 2qX_t^{N,0}\big)dt + \sum_{j=1}^N Z_t^{N,0,j}dW_t^j, \\ dY_t^N = -aY_t^{N,0}dt + \sum_{i=1}^N\sum_{j=0}^N Z_t^{N,i,j}dW_t^j. \end{cases}$$

According to the scheme used so far for solving affine FBSDEs, we make the ansatz that the decoupling field is affine, namely that it is of the form:

$$\begin{bmatrix} Y_t^{N,0} \\ Y_t^N \end{bmatrix} = S_t \begin{bmatrix} X_t^{N,0} \\ \bar{X}_t^N \end{bmatrix}, \quad t \in [0, T],$$

for a suitable deterministic 2×2-matrix valued function $(S_t)_{0 \le t \le T}$. We deduce that the solvability of the above FBSDE reduces to the solvability of:

$$\dot{S}_t + S_t \mathfrak{L} + \hat{\mathfrak{B}} S_t + S_t \mathfrak{B} S_t + \hat{\mathfrak{L}} = 0, \quad t \in [0, T] \, ; \quad S_T = 0, \tag{7.49}$$

with:

$$\mathfrak{L} = \begin{bmatrix} 0 & a \\ c & 0 \end{bmatrix}, \quad \mathfrak{B} = \begin{bmatrix} -\frac{1}{2}b^2 & 0 \\ 0 & 0 \end{bmatrix}, \quad \hat{\mathfrak{L}} = \begin{bmatrix} 2q & 0 \\ 0 & 0 \end{bmatrix}, \quad \hat{\mathfrak{B}} = \begin{bmatrix} 0 & c \\ a & 0 \end{bmatrix}.$$

For the same reasons as before, we write the 2×2 matrix S_t in the form:

$$S_t = \begin{bmatrix} S_t^{0,0} & S_t^{0,1} \\ S_t^{1,0} & S_t^{1,1} \end{bmatrix}, \quad t \in [0, T].$$

If the Riccati equation (7.49) is well posed, we derive the equilibrium dynamics of the state of the major player (as well as the dynamics of the average state of the minor players) by solving the following linear forward stochastic differential system:

$$\begin{cases} dX_t^{N,0} = \big[a\bar{X}_t^N - \frac{1}{2}b^2\big(S_t^{0,0}X_t^{N,0} + S_t^{0,1}\bar{X}_t^N\big)\big]dt + D_0 dW_t^0, \\ d\bar{X}_t^N = cX_t^{N,0}dt + D d\bar{W}_t^N, \end{cases} \tag{7.50}$$

for $t \in [0, T]$ with x^{00} and x^0 as respective initial conditions.

Once this is done, the optimal control strategy of the major player is given by $(\hat{\alpha}_t^{N,0} = -\frac{1}{2}bY_t^{N,0})_{0 \le t \le T}$ where $(Y_t^{N,0})_{0 \le t \le T}$ is obtained by using the affine decoupling field.

The Limiting Mean Field Game

Using the same notation as before for the limiting game, see Subsection 7.1.2, our formulation of the mean field game problem with major and minor players here requires the solution of the McKean-Vlasov control problem consisting of the controlled dynamics:

$$\begin{cases} dX_t^0 = \big(a\mathbb{E}^1[X_t] + b\alpha_t^0\big)dt + D_0 dW_t^0, \\ dX_t = cX_t^0 dt + D dW_t, \end{cases}$$

for $t \in [0, T]$, with x^{00} and x^0 as respective initial conditions, the objective function being given by:

$$J^0(\boldsymbol{\alpha}^0) = \mathbb{E}\left[\int_0^T \left(q(X_t^0)^2 + (\alpha_t^0)^2\right) dt\right],$$

where $\boldsymbol{\alpha}^0$ is asked to be $\mathbb{F}^0$-progressively measurable.

The minimizer $\hat{\alpha}^0$ of the Hamiltonian being the same as before, the stochastic maximum principle (7.27) leads to the FBSDE:

$$\begin{cases} dX_t^0 = \left(a\bar{X}_t - \frac{1}{2}b^2\bar{P}_t^0\right)dt + D_0 dW_t^0, \\ dX_t = cX_t^0 dt + D dW_t, \\ dP_t^0 = -\left(2qX_t^0 + cP_t\right)dt + Q_t^{00} dW_t^0 + Q_t^{01} dW_t, \\ dP_t = -a\bar{P}_t^0 dt + Q_t^{10} dW_t^0 + Q_t^{11} dW_t, \end{cases}$$

for $t \in [0, T]$, with $P_T^0 = P_T = 0$, where we added a bar to denote the expectation under $\mathbb{P}^1$. Taking expectations under $\mathbb{P}^1$ in the third line, we get:

$$\begin{cases} dX_t^0 = \left(a\bar{X}_t - \frac{1}{2}b^2\bar{P}_t^0\right)dt + D_0 dW_t^0, \\ d\bar{X}_t = cX_t^0 dt, \\ d\bar{P}_t^0 = -\left(2qX_t^0 + c\bar{P}_t\right)dt + \bar{Q}_t^{00} dW_t^0, \\ d\bar{P}_t = -a\bar{P}_t^0 dt + \bar{Q}_t^{10} dW_t^0, \end{cases} \tag{7.51}$$

for $t \in [0, T]$. Searching for an affine decoupling field, we see that the associated Riccati equation is again (7.49). Once it is solved, we then solve the forward SDE:

$$\begin{cases} dX_t^0 = \left[a\bar{X}_t - \frac{1}{2}b^2\left(S_t^{0,0}X_t^0 + S_t^{0,1}\bar{X}_t\right)\right]dt + D_0 dW_t^0, \\ d\bar{X}_t = cX_t^0 dt, \end{cases} \tag{7.52}$$

for $t \in [0, T]$, in order to obtain the solution of our problem. For the same reasons as before, the optimal control $\hat{\boldsymbol{\alpha}}^0$ is given by $(\hat{\alpha}_t^0 = -\frac{1}{2}b\bar{P}_t^0)_{0 \le t \le T}$.

The following proposition supports our formulation of the mean field game problem in the presence of major and minor players, at least under the assumptions of this section:

Proposition 7.13 *If the games with finitely many players and the mean field game with major and minor players are based on the same* $\mathbf{W}^0$, *and* $(\Omega^1, \mathcal{F}^1, \mathbb{P}^1)$ *carries both* $\mathbf{W}$ *and an infinite sequence of independent Brownian motions* $(\mathbf{W}^n)_{n \ge 1}$ *accounting for the noises driving the minor players' states, we have:*

$$\lim_{N\to\infty} \mathbb{E}\Big[\sup_{0\le t\le T}\Big(|X_t^{N,0}-X_t^0|^2+|\bar{X}_t^N-\bar{X}_t|^2+|Y_t^{N,0}-\bar{P}_t^0|^2+|Y_t^N-\bar{P}_t|^2\Big)\Big]=0.$$

In particular,

$$\lim_{N\to\infty} \mathbb{E}\Big[\sup_{0\le t\le T}|\hat{\alpha}_t^{N,0}-\hat{\alpha}_t^0|^2\Big]=0.$$

Proof. We use the expressions of $\boldsymbol{X}^{N,0}$, $\boldsymbol{X}^0$, $\bar{\boldsymbol{X}}^N$ and $\bar{\boldsymbol{X}}$ given by solving the SDEs (7.50) and (7.52) together with the fact:

$$\lim_{N\to\infty} \mathbb{E}\Big[\sup_{0\le t\le T}|\bar{W}_t^N|^2\Big]=0.$$

Forming the difference between $\boldsymbol{X}^{N,0}$ and $\boldsymbol{X}^0$ on the one hand, and $\bar{\boldsymbol{X}}^N$ and $\bar{\boldsymbol{X}}$ on the other hand, we easily deduce that:

$$\lim_{N\to\infty} \mathbb{E}\Big[\sup_{0\le t\le T}\Big(|X_t^{N,0}-X_t^0|^2+|\bar{X}_t^N-\bar{X}_t|^2\Big)\Big]=0.$$

The other claims follow from the representation of the involved processes by means of the solution $(S_t)_{0\le t\le T}$ to the Riccati equation. □

Comparison with Still a Different Formulation

Simpler formulations of the limiting mean field game problems are possible. One of them could be to first fix an $\mathbb{F}^0$-progressively measurable process $\boldsymbol{m}=(m_t)_{0\le t\le T}$ as a proxy for the conditional expectation of the states of the minor players, and then minimize the objective functional:

$$J^0(\boldsymbol{\alpha}^0)=\mathbb{E}\Big[\int_0^T[q(X_t^0)^2+(\alpha_t^0)^2]dt\Big],$$

under the dynamical constraint:

$$dX_t^0=(am_t+b\alpha_t^0)dt+D_0dW_t^0,\quad t\in[0,T]\ ;\quad X_0^0=x_0^0.$$

This is a standard control problem, though non-Markovian because $\boldsymbol{m}$ may depend upon the past. It can be solved using standard methods without worrying about possible McKean-Vlasov features. By applying the classical Pontryagin maximum principle and by noticing that the minimizer of the Hamiltonian is still the same, we arrive at the following FBSDE characterizing the optimally controlled system:

$$\begin{cases} dX_t^0=(am_t-\frac{1}{2}b^2Y_t^0)dt+D_0dW_t^0,\\ dY_t^0=-2qX_t^0dt+Z_t^0dW_t^0,\end{cases}$$

with the initial condition $X_0^0 = x^{00}$ and the terminal condition $Y_T^0 = 0$. We then impose the consistency condition:

$$m_t = \mathbb{E}^1[X_t], \quad t \in [0, T],$$

where $X = (X_t)_{0 \le t \le T}$ is the state of the representative minor player:

$$dX_t = cX_t^0 dt + DdW_t, \quad t \in [0, T] \; ; \quad X_0 = x^0.$$

Letting $(\bar{X}_t = \mathbb{E}^1[X_t])_{0 \le t \le T}$ we end up with the following FBSDE:

$$\begin{cases} dX_t^0 = \big(a\bar{X}_t - \frac{1}{2}b^2 Y_t^0\big)dt + D_0 dW_t^0, \\ d\bar{X}_t = cX_t^0 dt, \\ dY_t^0 = -2qX_t^0 dt + Z_t^0 dW_t^0, \end{cases} \tag{7.53}$$

for $t \in [0, T]$, with $X_0^0 = x^{00}$, $\bar{X}_0 = x^0$ and $Y_T^0 = 0$.

The comparison between (7.53) and (7.51) will be based on the following proposition.

Proposition 7.14 *Assume that $a, c, q, x^{00} \neq 0$. Then, the set of time instants $t \in [0, T]$ for which $\mathbb{P}[\bar{P}_t^0 \neq Y_t^0] > 0$ has a positive Lebesgue measure.*

Proof. We prove the result by contradiction. Assume that, for almost every $t \in [0, T]$, $\mathbb{P}[\bar{P}_t^0 = Y_t^0] = 1$. This implies that the first two (forward) equations in the systems (7.51) and (7.53) are identical. By uniqueness of solutions of SDEs, we conclude that each of the two processes $\boldsymbol{X}^0$ and $\bar{\boldsymbol{X}}$ is the same whatever the formulation used to describe the limiting mean field game. Computing the difference between the third equations of (7.51) and (7.53) and using the fact that $c \neq 0$, we deduce that $\bar{\boldsymbol{P}}$ is 0 by uniqueness of the Itô decomposition of $\bar{\boldsymbol{P}}^0 - \boldsymbol{Y}^0$. Since $a \neq 0$, we deduce in the same way that $\bar{\boldsymbol{P}}^0$ is 0 by uniqueness of the Itô decomposition in the fourth equation in (7.51). By the same argument, we see from the third equation in (7.51) that $\boldsymbol{X}^0$ is also 0. This is a contradiction since $x^{00} \neq 0$. □

Since the optimal control of the major player identified by the present scheme is still given by $(\hat{\alpha}_t^0 = -bY_t^0/2)_{0 \le t \le T}$, Propositions 7.13 and 7.14 imply that the two formulations lead to different optimal control strategies for the major player. Still, notice the Nash equilibria for the finite-player games converge towards the strategy produced by the original formulation touted earlier in this section, and not toward the strategy of the alternative formulation considered in this subsection.

7.1.8 Alternative Approaches to the Linear Quadratic Models

We now implement the alternative approaches introduced in Subsection 7.1.3 in the particular case of the Linear Quadratic (LQ) Mean Field Games (MFGs) with major and minor players.

Existence of Open-Loop Equilibria

We first consider the alternative approach to the open loop problem. Recall that the dynamics of the state $(X_t^0)_{0\le t\le T}$ of the major player and the state $(X_t)_{0\le t\le T}$ of the representative player of the field of exchangeable minor players are given by:

$$\begin{cases} dX_t^0 &= (L_0X_t^0 + B_0\alpha_t^0 + F_0\bar{X}_t)dt + D_0dW_t^0, \\ dX_t &= (LX_t + B\alpha_t + F\bar{X}_t + GX_t^0)dt + DdW_t, \end{cases} \tag{7.54}$$

where $(\bar{X}_t = \mathbb{E}^1[X_t])_{0\le t\le T}$ is the conditional expectation of $(X_t)_{0\le t\le T}$ with respect to the history of the Wiener process $\boldsymbol{W}^0$ up to time t. The cost functionals for the major and minor players are given by:

$$J^0(\boldsymbol{\alpha}^0,\boldsymbol{\alpha}) = \mathbb{E}\left[\int_0^T \left[(X_t^0 - K_0\bar{X}_t - \eta_0)^\dagger \Gamma_0(X_t^0 - K_0\bar{X}_t - \eta_0) + \alpha_t^{0\dagger}R_0\alpha_t^0\right]dt\right],$$

$$J(\boldsymbol{\alpha}^0,\boldsymbol{\alpha}) = \mathbb{E}\left[\int_0^T \left[(X_t - KX_t^0 - K_1\bar{X}_t - \eta)^\dagger \Gamma(X_t - KX_t^0 - K_1\bar{X}_t - \eta) \right.\right.$$
$$\left.\left. + \alpha_t^\dagger R\alpha_t\right]dt\right],$$

where we recall that Γ, Γ_0, R, and R_0 are symmetric matrices, Γ and Γ_0 being assumed to be nonnegative semi-definite, and R and R_0 being assumed to be positive definite. Taking conditional expectations in the equation for the state of the representative minor player we get:

$$d\bar{X}_t = \left[(L+F)\bar{X}_t + B\bar{\alpha}_t + GX_t^0\right]dt, \tag{7.55}$$

with $\bar{\alpha}_t = \mathbb{E}^1[\alpha_t]$, for any $t \in [0,T]$. The idea is now to express the optimization problem of the major player over the dynamics of the couple $(\bar{X}_t, X_t^0)_{0\le t\le T}$ which, in contrast with the dynamics of the original couple $(X_t, X_t^0)_{0\le t\le T}$, are not of the McKean-Vlasov type. In order to do so, we introduce the following notation:

$$\mathcal{X}_t = \begin{bmatrix} X_t^0 \\ \bar{X}_t \end{bmatrix}, \quad \mathfrak{L} = \begin{bmatrix} L_0 & F_0 \\ G & L+F \end{bmatrix}, \quad \mathfrak{B}_{00} = \begin{bmatrix} B_0 \\ 0 \end{bmatrix}, \quad \mathfrak{B}_0 = \begin{bmatrix} 0 \\ B \end{bmatrix}, \quad \mathfrak{D} = \begin{bmatrix} D_0 \\ 0 \end{bmatrix},$$

$$\hat{\mathfrak{L}}_0 = \begin{bmatrix} \Gamma_0 & -\Gamma_0K_0 \\ -K_0^\dagger\Gamma_0 & K_0^\dagger\Gamma_0K_0 \end{bmatrix}, \quad \hat{\mathfrak{C}}_0 = \begin{bmatrix} -\Gamma_0\eta_0 \\ K_0^\dagger\Gamma_0\eta_0 \end{bmatrix}.$$

Notice also that, the fact that the matrix Γ_0 is symmetric nonnegative definite implies that $\hat{\mathfrak{L}}_0$ is also symmetric nonnegative definite. This will play a crucial role when we address the solvability of certain matrix Riccati equations. The optimization problem of the major player becomes:

$$\inf_{\alpha^0 \in \mathbb{A}_0} \mathbb{E}\left[\int_0^T [\mathcal{X}_t^\dagger \hat{\mathfrak{L}}_0 \mathcal{X}_t + 2\mathcal{X}_t^\dagger \hat{\mathfrak{C}}_0 + \eta_0^\dagger \Gamma_0 \eta_0 + \alpha_t^{0\dagger} R_0 \alpha_t^0] dt\right],$$

where the controlled dynamics are given by:

$$d\mathcal{X}_t = (\mathfrak{L}\mathcal{X}_t + \mathfrak{B}_{00}\alpha_t^0 + \mathfrak{B}_0\bar{\alpha}_t)dt + \mathfrak{D}dW_t^0. \tag{7.56}$$

The reduced Hamiltonian is given by:

$$\begin{aligned} &H^{(r),\boldsymbol{\alpha}}(t,x,y,\alpha^0) \\ &\quad = y^\dagger(\mathfrak{L}x + \mathfrak{B}_{00}\alpha^0 + \mathfrak{B}_0\bar{\alpha}_t) + x^\dagger \hat{\mathfrak{L}}_0 x + 2x^\dagger \hat{\mathfrak{C}}_0 + \eta_0^\dagger \Gamma_0 \eta_0 + \alpha^{0\dagger} R_0 \alpha^0, \end{aligned}$$

where $x, y \in \mathbb{R}^{d_0+d}$. Here we added the superscript $\boldsymbol{\alpha}$ for the Hamiltonian in order to emphasize the fact that the optimization of the major player is performed assuming that the representative minor player is using the strategy $\boldsymbol{\alpha} \in \mathbb{A}$. As usual, $(\bar{\alpha}_t = \mathbb{E}^1[\alpha_t])_{0 \le t \le T}$ and, obviously, $H^{(r),\boldsymbol{\alpha}}$ is a random function, in which the randomness comes from the realization of the minor player's control. However we see that almost surely $\mathbb{R}^{d_0+d} \times A_0 \ni (x, \alpha^0) \to H^{(r),\boldsymbol{\alpha}}(t,x,y,\alpha^0)$ is jointly convex, and we can use the sufficient condition of the stochastic maximum principle, see for instance Subsection 1.4.4. Therefore the minimizer of the reduced Hamiltonian and the optimal control are given by:

$$\hat{\alpha}^0(t,x,y) = -\tfrac{1}{2} R_0^{-1} \mathfrak{B}_{00}^\dagger y, \qquad \text{and} \qquad \hat{\alpha}_t^0 = -\tfrac{1}{2} R_0^{-1} \mathfrak{B}_{00}^\dagger \mathcal{Y}_t^0, \qquad t \in [0,T],$$

respectively, where $(\mathcal{X}_t, \mathcal{Y}_t^0)_{0 \le t \le T}$ solves the forward-backward stochastic differential equation:

$$\begin{cases} d\mathcal{X}_t &= (\mathfrak{L}\mathcal{X}_t - \tfrac{1}{2}\mathfrak{B}_{00} R_0^{-1} \mathfrak{B}_{00}^\dagger \mathcal{Y}_t^0 + \mathfrak{B}_0 \bar{\alpha}_t)dt + \mathfrak{D}dW_t^0, \\ d\mathcal{Y}_t^0 &= -(\mathfrak{L}^\dagger \mathcal{Y}_t^0 + 2\hat{\mathfrak{L}}_0 \mathcal{X}_t + 2\hat{\mathfrak{C}}_0)dt + Z_t^0 dW_t^0, \quad \mathcal{Y}_T^0 = 0. \end{cases} \tag{7.57}$$

We now address the equilibrium condition for the minor player through the search for the best response of an extra minor player to the major player and to the field of exchangeable minor players. We fix an admissible strategy $\boldsymbol{\alpha}^0 \in \mathbb{A}_0$ for the major player and an admissible strategy $\boldsymbol{\alpha} \in \mathbb{A}$ for the representative of the field of exchangeable minor players. We let $(\bar{\alpha}_t = \mathbb{E}^1[\alpha_t])_{0 \le t \le T}$. This prescription leads to the time evolution of the state of a system given by (7.54), equation (7.55) after taking conditional expectations, and finally the dynamic equation (7.56). Given this background state evolution, the extra virtual minor player needs to solve the optimization problem:

$$\inf_{\check{\alpha} \in \mathbb{A}} \mathbb{E}\left[\int_0^T [(\check{X}_t - \psi(\mathcal{X}_t))^\dagger \Gamma (\check{X}_t - \psi(\mathcal{X}_t)) + \check{\alpha}_t^\dagger R \check{\alpha}_t] dt\right],$$

where the dynamics of the controlled state $(\check{X}_t)_{0\le t\le T}$ are given by:

$$d\check{X}_t = \big(L\check{X}_t + B\check{\alpha}_t + F\bar{X}_t + GX_t^0\big)dt + DdW_t.$$

Note that the process $(\mathcal{X}_t)_{0\le t\le T}$ is merely part of the random coefficients of the optimization problem. We introduce the reduced Hamiltonian:

$$\begin{aligned} H^{(r),\alpha^0,\alpha}(t,\check{x},\check{y},\check{\alpha}) &= \check{y}^\dagger\big(L\check{x} + B\check{\alpha} + F\bar{X}_t + GX_t^0\big) \\ &\quad + \big(\check{x} - \psi(\mathcal{X}_t)\big)^\dagger \Gamma\big(\check{x} - \psi(\mathcal{X}_t)\big) + \check{\alpha}^\dagger R\check{\alpha}, \end{aligned}$$

where $\check{x},\check{y} \in \mathbb{R}^d$ and $\check{\alpha} \in \mathbb{R}^k$. Once again, we use the superscript $(\boldsymbol{\alpha}^0,\boldsymbol{\alpha})$ to emphasize the fact that the optimization is performed under the environment created by the major player using strategy $\boldsymbol{\alpha}^0$ and the population of exchangeable minor players using $\boldsymbol{\alpha}$, leading to the use of its conditional mean $\bar{\boldsymbol{\alpha}}$. The Hamiltonian $H^{(r),\alpha^0,\alpha}$ depends on the random realization of the environment, and is almost surely jointly convex in $(\check{x},\check{\alpha})$. Applying the stochastic maximum principle, the optimal control exists and is given by $(\check{\alpha}_t = -\frac{1}{2}R^{-1}B^\dagger\check{Y}_t)_{0\le t\le T}$, where $(\check{\boldsymbol{X}},\check{\boldsymbol{Y}})$ solves the following FBSDE:

$$\begin{cases} d\check{X}_t &= \big(L\check{X}_t - \frac{1}{2}BR^{-1}B^\dagger\check{Y}_t + F\bar{X}_t + GX_t^0\big)dt + DdW_t, \\ d\check{Y}_t &= -\big[L^\dagger\check{Y}_t + 2\Gamma\big(\check{X}_t - \psi(\mathcal{X}_t)\big)\big]dt + Z_t dW_t + Z_t^0 dW_t^0, \end{cases} \tag{7.58}$$

with terminal condition $\check{Y}_T = 0$. Recall that in this FBSDE, the process $(\mathcal{X}_t)_{0\le t\le T}$ only plays the part of a random coefficient. It is determined *off line* by solving the standard stochastic differential equation (7.56) for the given values of $\boldsymbol{\alpha}^0$ and $\boldsymbol{\alpha}$, which at this stage of the proof (i.e., before considering the fixed point step) may differ from the controls used in the forward equation of (7.57).

Now that we are done characterizing the solutions of both optimization problems, we identify the fixed point constraint in the framework given by the characterizations of the two optimization problems. The fixed point condition (7.13) characterizing Nash equilibria in the current set-up says that:

$$\hat{\alpha}_t^0 = -\tfrac{1}{2}R_0^{-1}\mathfrak{B}_{00}^\dagger\mathcal{Y}_t^0, \quad t \in [0,T],$$

where $(\mathcal{Y}_t^0)_{0\le t\le T}$ is the backward component of the solution of (7.57) with $(\bar{\alpha}_t = \mathbb{E}^1[\check{\alpha}_t])_{0\le t\le T}$, where:

$$\hat{\alpha}_t = -\tfrac{1}{2}R^{-1}B^\dagger\check{Y}_t, \quad t \in [0,T],$$

where $(\check{Y}_t)_{0\le t\le T}$ is the backward component of the solution of (7.58) in which the random coefficient $(\mathcal{X}_t)_{0\le t\le T}$ solves (7.56) with the processes $\boldsymbol{\alpha}^0 = (\hat{\alpha}_t^0)_{0\le t\le T}$ and $\bar{\boldsymbol{\alpha}} = (\mathbb{E}^1[\hat{\alpha}_t])_{0\le t\le T}$ just defined.

The optimal controls for the major and representative minor players are functions of the solution of the following FBSDE which we obtain by putting together the FBSDEs (7.57) and (7.58) characterizing the major and virtual minor players' optimization problems:

$$\begin{cases} d\mathcal{X}_t = \big(\mathfrak{L}\mathcal{X}_t - \frac{1}{2}\mathfrak{B}_{00}R_0^{-1}\mathfrak{B}_{00}^\dagger\mathcal{Y}_t^0 - \frac{1}{2}\mathfrak{B}_0R^{-1}B^\dagger\mathbb{E}^1[\check{Y}_t]\big)dt + \mathfrak{D}dW_t^0, \\ d\check{X}_t = \big(L\check{X}_t - \frac{1}{2}BR^{-1}B^\dagger\check{Y}_t + F\mathcal{X}_t^2 + G\mathcal{X}_t^1\big)dt + DdW_t, \\ d\mathcal{Y}_t^0 = -\big(\mathfrak{L}^\dagger\mathcal{Y}_t^0 + 2\hat{\mathfrak{L}}_0\mathcal{X}_t + 2\hat{\mathfrak{C}}_0\big)dt + Z_t^0dW_t^0, \\ d\check{Y}_t = -\big[L^\dagger\check{Y}_t + 2\Gamma\big(\check{X}_t - \psi(\mathcal{X}_t)\big)\big]dt + Z_tdW_t + Z_t^0dW_t^0, \\ \mathcal{X}_0 = \begin{bmatrix} x^{00} \\ x^0 \end{bmatrix}, \quad \check{X}_0 = x^0, \quad \mathcal{Y}_T^0 = 0, \quad \check{Y}_T = 0, \end{cases} \tag{7.59}$$

where we denoted by $\mathcal{X}_t^1$ and $\mathcal{X}_t^2$ the two blocks of dimension d_0 and d of $\mathcal{X}_t$.

We summarize the above discussion in the form of a verification theorem for open loop Nash equilibria.

Proposition 7.15 *If the system* (7.59) *admits a solution, then the linear quadratic mean field game problem with major and minor players admits an open loop Nash equilibrium.*

The equilibrium strategy $(\hat{\boldsymbol{\alpha}}^0, \hat{\boldsymbol{\alpha}})$ *is given by* $(\hat{\alpha}_t^0 = -\frac{1}{2}R_0^{-1}\mathfrak{B}_{00}^\dagger\mathcal{Y}_t^0)_{0\le t\le T}$ *for the major player and* $(\hat{\alpha}_t = -\frac{1}{2}R^{-1}B\check{Y}_t)_{0\le t\le T}$ *for the minor player.*

The way the system (7.59) is stated is a natural conclusion of the search for equilibrium as formulated by the fixed point step following the two optimization problems. However, as convenient as can be, simple remarks can help the solution of this system. First we notice that one could solve for $(\mathcal{X}_t, \mathcal{Y}_t^0)_{0\le t\le T}$ by solving the FBSDE formed by the first and the third equations if we knew the values of $(\bar{Y}_t = \mathbb{E}^1[\check{Y}_t])_{0\le t\le T}$. By taking expectation with respect to $\mathbb{P}^1$ in the second equation, and subtracting the result from the equation satisfied by the second component of the first equation, we identify $(\mathbb{E}^1[\check{X}_t])_{0\le t\le T}$ with $(\mathcal{X}_t^2)_{0\le t\le T}$ because they have the same initial conditions. Next, by taking expectation with respect to $\mathbb{P}^1$ in the fourth equation, we see that $(\bar{Y}_t)_{0\le t\le T}$ should satisfy:

$$d\bar{Y}_t = \; -\big[L^\dagger\bar{Y}_t + 2\Gamma\big(\mathcal{X}_t^2 - \psi(\mathcal{X}_t)\big)\big]dt + \bar{Z}_t^0dW_t^0, \; \bar{Y}_T = 0.$$

Consequently, the solution of (7.59) also satisfies:

$$\begin{cases} d\mathcal{X}_t = \big(\mathfrak{L}\mathcal{X}_t - \frac{1}{2}\mathfrak{B}_{00}R_0^{-1}\mathfrak{B}_{00}^\dagger\mathcal{Y}_t^0 - \frac{1}{2}\mathfrak{B}_0R^{-1}B^\dagger\bar{Y}_t\big)dt + \mathfrak{D}dW_t^0, \\ d\mathcal{Y}_t^0 = -\big(\mathfrak{L}^\dagger\mathcal{Y}_t^0 + 2\hat{\mathfrak{L}}_0\mathcal{X}_t + 2\hat{\mathfrak{C}}_0\big)dt + Z_t^0dW_t^0, \\ d\bar{Y}_t = -\big[L^\dagger\bar{Y}_t + 2\Gamma\big(\mathcal{X}_t^2 - \psi(\mathcal{X}_t)\big)\big]dt + \bar{Z}_t^0dW_t^0, \\ \mathcal{Y}_T^0 = 0, \quad \bar{Y}_T = 0. \end{cases} \tag{7.60}$$

Our final remark is that the solution of system (7.60) is not only necessary, but also sufficient. Indeed, once it is solved, one can solve for $(\check{X}_t, \check{Y}_t)_{0\le t\le T}$ by solving the affine FBSDE with random coefficients formed by the second and fourth equations of (7.59) and check that $(\mathbb{E}^1[\check{Y}_t])_{0\le t\le T}$ is indeed the solution of the third equation of (7.60).

Identifying $(\mathcal{X}_t)_{0\le t\le T}$ with $\left(\begin{bmatrix}\bar{X}^0_t\\ \bar{X}_t\end{bmatrix}\right)_{0\le t\le T}$ and $(\mathcal{Y}^0_t)_{0\le t\le T}$ with $\left(\begin{bmatrix}\bar{P}^0_t\\ \bar{P}_t\end{bmatrix}\right)_{0\le t\le T}$, we recognize the FBSDE (7.60) as the original (7.44), showing that the alternative approach to the search of a Nash equilibrium leads to the solution found in Subsection 7.1.6.

Existence of Closed Loop Equilibria

In this section we implement the closed loop alternative formulation of the equilibrium problem. Since we expect that the optimal controls will be in feedback form, we search directly for Markovian controls. In other words, we assume that the controls used by the major player and the representative minor players are respectively of the form:

$$\alpha^0_t = \phi^0(t, X^0_t, \bar{X}_t), \qquad \text{and} \qquad \alpha_t = \phi(t, X_t, X^0_t, \bar{X}_t), \quad t \in [0,T],$$

for some $\mathbb{R}^{k_0}$ and $\mathbb{R}^k$ valued deterministic functions ϕ^0 and ϕ defined on $[0,T]\times\mathbb{R}^{d_0}\times\mathbb{R}^d$ and $[0,T]\times\mathbb{R}^d\times\mathbb{R}^{d_0}\times\mathbb{R}^d$ respectively. As before, we assume that $A_0 = \mathbb{R}^{k_0}$ and $A = \mathbb{R}^k$ for the sake of simplicity. In this way, the major player can only observe its own state and the mean of the states in the field of minor players, while the representative minor player can observe its own state, the state of the major player, as well as the mean of the exchangeable minor players' states. Several times already, we hinted at the fact that this version of the equilibrium problem is more difficult than in the open loop case. As a result, we shall not try to compute the best response map everywhere before we search for some of its fixed points. Instead, we identify a special set of controls $\boldsymbol{\alpha}^0$ and $\boldsymbol{\alpha}$ which is left stable by the best response map, and we search for fixed points in this set only. To be more specific, we compute the best responses to controls $\boldsymbol{\alpha}^0$ and $\boldsymbol{\alpha}$ given by feedback functions ϕ^0 and ϕ assumed to be affine in their arguments. In other words, we assume that the controls $\boldsymbol{\alpha}^0$ and $\boldsymbol{\alpha}$ are of the form:

$$\alpha^0_t = \phi^0(t, X^0_t, \bar{X}_t) = \phi^0_0(t) + \phi^0_1(t)X^0_t + \phi^0_2(t)\bar{X}_t, \tag{7.61}$$

$$\alpha_t = \phi(t, X_t, X^0_t, \bar{X}_t) = \phi_0(t) + \phi_1(t)X_t + \phi_2(t)X^0_t + \phi_3(t)\bar{X}_t, \tag{7.62}$$

for $t \in [0,T]$, where the functions $[0,T] \ni t \mapsto \phi^0_i(t)$ for $i = 0,1,2$ and $[0,T] \ni t \mapsto \phi_i(t)$ for $i = 0,1,2,3$ are matrix-valued deterministic continuous functions with the appropriate dimensions. To be more specific, $\phi^0_0(t) \in \mathbb{R}^{k_0}$, $\phi^0_1(t) \in \mathbb{R}^{k_0\times d_0}$, $\phi^0_2(t) \in \mathbb{R}^{k_0\times d}$, $\phi_0(t) \in \mathbb{R}^k$, $\phi_1(t) \in \mathbb{R}^{k\times d}$, $\phi_2(t) \in \mathbb{R}^{k\times d_0}$ and $\phi_3(t) \in \mathbb{R}^{k\times d}$.

We first consider the major player's optimization problem. We assume that the representative minor player uses strategy $(\alpha_t = \phi(t, X_t, X^0_t, \bar{X}_t))_{0\le t\le T}$ as specified in

(7.62). Next we look for the controls $\boldsymbol{\alpha}^0$ which could be used by the major player to minimize its expected cost. Note that for this optimization problem, we do not assume that the controls $\boldsymbol{\alpha}^0$ are of the form (7.61). Indeed, as we shall make it clear below, the optimal control will be automatically of this form, which will suffice to render our approach consistent. The dynamics of the system are then given by:

$$\begin{cases} dX_t^0 = \big(L_0 X_t^0 + B_0\alpha_t^0 + F_0\bar{X}_t\big)dt + D_0 dW_t^0, \\ dX_t = \big[B\phi_0(t) + \big(L + B\phi_1(t)\big)X_t \\ \qquad + \big(B\phi_2(t) + G\big)X_t^0 + \big(B\phi_3(t) + F\big)\bar{X}_t\big)\big]dt + D dW_t, \end{cases} \tag{7.63}$$

where as before, for each $t \in [0,T]$, $\bar{X}_t = \mathbb{E}^1[X_t]$ is the conditional expectation of X_t with respect to the filtration generated by the history of the Wiener process $\boldsymbol{W}^0$ up to time t. Like in the case of the search for open loop equilibria, we replace the optimization over the above dynamics of a conditional McKean-Vlasov type by an optimization over standard dynamics by taking conditional expectations in the equation for the state of the representative minor player. Doing so, we get:

$$d\bar{X}_t = \big[B\phi_0(t) + \big(L + F + B[\phi_1(t) + \phi_3(t)]\big)\bar{X}_t + \big(G + B\phi_2(t)\big)X_t^0\big]dt. \tag{7.64}$$

As in the case of the open loop version of the equilibrium problem, we express the optimization problem of the major player over the dynamics of the couple $(\mathcal{X}_t = (X_t^0, \bar{X}_t))_{0\le t\le T}$. In order to do so, we use the same notations $\mathfrak{L}$, $\hat{\mathfrak{L}}_0$, $\mathfrak{B}_0$, $\mathfrak{B}_{00}$, $\hat{\mathfrak{C}}_0$ and $\mathfrak{D}$ as in the case of our analysis of the open loop problem, and we introduce the following new ones:

$$\mathfrak{L}_0^{\boldsymbol{\phi}}(t) = \begin{bmatrix} L_0 & F_0 \\ G + B\phi_2(t) & L + F + B[\phi_1(t) + \phi_3(t)] \end{bmatrix}, \quad \mathfrak{A}_0^{\boldsymbol{\phi}}(t) = \begin{bmatrix} 0 \\ B\phi_0(t) \end{bmatrix},$$

where we wrote $\boldsymbol{\phi}$ for the tuple of functions $(\phi_i)_{i=0,\cdots,3}$. Then, the optimization problem of the major player can be formulated exactly as in the open loop case as the minimization:

$$\inf_{\boldsymbol{\alpha}^0\in\mathbb{A}_0} \mathbb{E}\bigg[\int_0^T [\mathcal{X}_t^\dagger \hat{\mathfrak{L}}_0 \mathcal{X}_t + 2\mathcal{X}_t^\dagger \hat{\mathfrak{C}}_0 + \eta_0^\dagger \Gamma_0 \eta_0 + \alpha_t^{0\dagger} R_0 \alpha_t^0]dt\bigg],$$

where the controlled dynamics are given by:

$$d\mathcal{X}_t = \big[\mathfrak{L}_0^{\boldsymbol{\phi}}(t)\mathcal{X}_t + \mathfrak{B}_{00}\alpha_t^0 + \mathfrak{A}_0^{\boldsymbol{\phi}}(t)\big]dt + \mathfrak{D}dW_t^0. \tag{7.65}$$

The reduced Hamiltonian (minus the term $\eta_0^\dagger \Gamma_0 \eta_0$ which is irrelevant) is given by:

$$\begin{aligned} &H^{(r),\boldsymbol{\phi}}(t,x,y,\alpha^0) \\ &= y^\dagger\big[\mathfrak{L}_0^{\boldsymbol{\phi}}(t)x + \mathfrak{B}_{00}\alpha^0 + \mathfrak{A}_0^{\boldsymbol{\phi}}(t)\big] + x^\dagger \hat{\mathfrak{L}}_0 x + 2x^\dagger \hat{\mathfrak{C}}_0 + \alpha^{0\dagger} R_0 \alpha^0, \end{aligned}$$

where $x, y \in \mathbb{R}^{d_0+d}$ and $\alpha^0 \in \mathbb{R}^{k_0}$. Applying the stochastic maximum principle, we find that the optimal control is given as before by $(\hat{\alpha}^0_t = -\frac{1}{2}R_0^{-1}\mathfrak{B}_{00}^\dagger \mathcal{Y}^0_t)_{0\le t\le T}$, where $(\mathcal{X}_t, \mathcal{Y}^0_t, \mathcal{Z}^0_t)_{0\le t\le T}$ solves the linear FBSDE:

$$\begin{cases} d\mathcal{X}_t = \big[\mathfrak{L}_0^\phi(t)\mathcal{X}_t - \frac{1}{2}\mathfrak{B}_{00}R_0^{-1}\mathfrak{B}_{00}^\dagger \mathcal{Y}^0_t + \mathfrak{A}_0^\phi(t)\big]dt + \mathfrak{D}dW^0_t \\ d\mathcal{Y}^0_t = -\big[\mathfrak{L}_0^\phi(t)^\dagger \mathcal{Y}^0_t + 2\hat{\mathfrak{L}}_0\mathcal{X}_t + 2\hat{\mathfrak{C}}_0\big]dt + \mathcal{Z}^0_t dW^0_t, \quad \mathcal{Y}^0_T = 0. \end{cases} \tag{7.66}$$

This FBSDE being affine, we expect the decoupling field to be affine as well, so we search for a solution of the form $(\mathcal{Y}^0_t = \gamma_t\mathcal{X}_t + \kappa_t)_{0\le t\le T}$ for two deterministic functions $[0,T] \ni t \mapsto \gamma_t \in \mathbb{R}^{(d_0+d)\times(d_0+d)}$ and $[0,T] \ni t \mapsto \kappa_t \in \mathbb{R}^{d_0+d}$. We compute $d\mathcal{Y}^0_t$ applying Itô's formula to this ansatz, and using the expression for $d\mathcal{X}_t$ given by the forward equation. Identifying term by term the result with the right-hand side of the backward component of the above FBSDE, we obtain the following system of ordinary differential equations:

$$\begin{cases} \dot{\gamma}_t - \frac{1}{2}\gamma_t\mathfrak{B}_{00}R_0^{-1}\mathfrak{B}_{00}^\dagger\gamma_t + \gamma_t\mathfrak{L}_0^\phi(t) + \big[\mathfrak{L}_0^\phi(t)\big]^\dagger\gamma_t + 2\hat{\mathfrak{L}}_0 = 0, \\ \gamma_T = 0, \\ \dot{\kappa}_t + \big(\big[\mathfrak{L}_0^\phi(t)\big]^\dagger - \frac{1}{2}\gamma_t\mathfrak{B}_{00}R_0^{-1}\mathfrak{B}_{00}^\dagger\big)\kappa_t + \gamma_t\mathfrak{A}_0^\phi(t) + 2\hat{\mathfrak{C}}_0 = 0, \\ \kappa_T = 0. \end{cases} \tag{7.67}$$

For any choice of a continuous strategy $t \mapsto (\phi_0(t), \phi_1(t), \phi_2(t), \phi_3(t))$, the first equation is a standard matrix Riccati differential equation. Since the coefficients are continuous and $\hat{\mathfrak{L}}_0$ is nonnegative definite, the equation admits a unique global solution over $[0,T]$ for any $T > 0$. Recall that R_0 is symmetric and positive definite. Injecting the solution $[0,T] \ni t \mapsto \gamma_t$ into the second equation yields a linear ordinary differential equation with continuous coefficients for which the global unique solvability also holds. Therefore the FBSDE (7.66) is uniquely solvable and the optimal control exists and is given by:

$$\alpha^{0*}_t = -\frac{1}{2}R_0^{-1}\mathfrak{B}_{00}^\dagger\gamma_t\mathcal{X}_t - \frac{1}{2}R_0^{-1}\mathfrak{B}_{00}^\dagger\kappa_t, \quad t \in [0,T], \tag{7.68}$$

which is an affine function of X^0_t and $\bar{X}_t$.

As before, we address the equilibrium condition for the minor player through the search for the best response of an extra minor player to the major player and to the field of exchangeable minor players. According to the strategy outlined earlier, we compute its best response to controls of a specific form. So we assume that the major player uses the feedback strategy $(\alpha^0_t = \phi^0(t, X^0_t, \bar{X}_t))_{0\le t\le T}$ and the representative of the other minor players uses the feedback strategy $(\alpha_t = \phi(t, X_t, X^0_t, \bar{X}_t))_{0\le t\le T}$ of the forms (7.61) and (7.62) respectively. These choices lead to the dynamics of the state $\left(\mathcal{X}_t = \begin{bmatrix} X^0_t \\ \bar{X}_t \end{bmatrix}\right)_{0\le t\le T}$ given by:

$$d\mathcal{X}_t = \big[\mathfrak{L}^{(\phi,\phi^0)}(t)\mathcal{X}_t + \mathfrak{A}^{(\phi,\phi^0)}(t)\big]dt + \mathfrak{D}dW^0_t,$$

with:

$$\mathfrak{L}^{(\phi,\phi^0)}(t) = \begin{bmatrix} L_0 + B_0\phi_1^0(t) & F_0 + B_0\phi_2^0(t) \\ G + B\phi_2(t) & L + F + B(\phi_1(t) + \phi_3(t)) \end{bmatrix},$$

$$\mathfrak{A}^{(\phi,\phi^0)}(t) = \begin{bmatrix} B_0\phi_0^0(t) \\ B\phi_0(t) \end{bmatrix},$$

where $\boldsymbol{\phi}^0 = (\phi_0^0, \phi_1^0, \phi_2^0)$. In this environment, we search for the best response of a virtual minor player trying to minimize as earlier,

$$\inf_{\check{\alpha} \in \mathbb{A}} \mathbb{E}\bigg[\int_0^T [(\check{X}_t - \psi(\mathcal{X}_t))^\dagger \Gamma (\check{X}_t - \psi(\mathcal{X}_t)) + \check{\alpha}_t^\dagger R \check{\alpha}_t] dt\bigg],$$

where the dynamics of the controlled state $(\check{X}_t)_{0 \le t \le T}$ are given as before by:

$$d\check{X}_t = \big(L\check{X}_t + B\check{\alpha}_t + F\bar{X}_t + GX_t^0\big)dt + DdW_t.$$

Again the process $(\mathcal{X}_t)_{0 \le t \le T}$ is merely part of the random coefficients of the optimization problem. We introduce the reduced Hamiltonian:

$$\begin{aligned} H^{(r),\phi^0,\phi}(t, \check{x}, \check{y}, \check{\alpha}) &= \check{y}^\dagger \big(L\check{x} + B\check{\alpha} + F\bar{X}_t + GX_t^0\big) \\ &\quad + \big(\check{x} - \psi(\mathcal{X}_t)\big)^\dagger \Gamma \big(\check{x} - \psi(\mathcal{X}_t)\big) + \check{\alpha}^\dagger R \check{\alpha}, \end{aligned}$$

for $(\check{x}, \check{y}, \check{\alpha}) \in \mathbb{R}^d \times \mathbb{R}^d \times \mathbb{R}^k$. We find that the optimal control is given by $(\check{\alpha}_t^* = -\frac{1}{2}R^{-1}B^\dagger \check{Y}_t)_{0 \le t \le T}$, where $(\check{X}_t, \mathcal{X}_t, \check{Y}_t, \check{Z}_t, \check{Z}_t^0)_{0 \le t \le T}$ solves the linear FBSDE:

$$\begin{cases} d\check{X}_t = \big(L\check{X}_t - \frac{1}{2}BR^{-1}B^\dagger \check{Y}_t + F\bar{X}_t + GX_t^0\big)dt + DdW_t, \\ d\mathcal{X}_t = \big(\mathfrak{L}^{(\phi,\phi^0)}(t)\mathcal{X}_t + \mathfrak{A}^{(\phi,\phi^0)}(t)\big)dt + \mathfrak{D}dW_t^0, \\ d\check{Y}_t = -\big(L^\dagger \check{Y}_t + 2\Gamma\check{X}_t - 2\Gamma\psi(\mathcal{X}_t)\big)dt + \check{Z}_t dW_t + \check{Z}_t^0 dW_t^0, \\ \check{Y}_T = 0. \end{cases}$$

Again we search for a solution of the form $(\check{Y}_t = \mathcal{S}_t\mathcal{X}_t + S_t\check{X}_t + s_t)_{0 \le t \le T}$ for continuous deterministic functions $[0, T] \ni t \mapsto \mathcal{S}_t \in \mathbb{R}^{d \times (d_0 + d)}$, $[0, T] \ni t \mapsto S_t \in \mathbb{R}^{d \times d}$ and $[0, T] \ni t \mapsto s_t \in \mathbb{R}^d$. Proceeding as before, we see that these functions provide a solution to the above FBSDE if and only if they solve the system of ordinary differential equations:

$$\begin{cases} \dot{S}_t + S_t L + L^\dagger S_t - \frac{1}{2}S_t BR^{-1}B^\dagger S_t + 2\Gamma = 0, \quad S_T = 0, \\ \dot{\mathcal{S}}_t + \mathcal{S}_t \mathfrak{L}^{(\phi,\phi^0)}(t) + L^\dagger \mathcal{S}_t - \frac{1}{2}S_t BR^{-1}B^\dagger \mathcal{S}_t \\ \qquad\qquad + S_t[G, F] - 2\Gamma[K, K_1] = 0, \quad \mathcal{S}_T = 0, \\ \dot{s}_t + \big(L^\dagger - \frac{1}{2}S_t BR^{-1}B^\dagger\big)s_t + \mathcal{S}_t \mathfrak{A}^{(\phi,\phi^0)}(t) - 2\Gamma\eta = 0, \quad s_T = 0. \end{cases} \tag{7.69}$$

The first equation is a standard symmetric matrix Riccati equation. As before, the fact that Γ is symmetric and nonnegative definite and R is symmetric and positive definite implies that this Riccati equation has a unique solution on $[0,T]$. Note that its solution $(S_t)_{0\le t\le T}$ is symmetric and independent of the input feedback functions $\boldsymbol{\phi}^0$ and $\boldsymbol{\phi}$ giving the controls chosen by the major player and the exchangeable minor players. Injecting the solution $(S_t)_{0\le t\le T}$ into the second and third equations, leads to a linear system of ordinary differential equations which can be readily solved. Given such a solution, we find that the optimal control can be expressed as:

$$\check{\alpha}_t^* = -\tfrac{1}{2}R^{-1}B^\dagger\big[\mathcal{S}_t\mathcal{X}_t + S_t\check{X}_t + s_t\big], \quad t\in[0,T], \tag{7.70}$$

which is indeed an affine function of $\check{X}_t$, X_t^0 and $\bar{X}_t$.

Now that the two optimization problems are solved and that we showed that the family of affine feedback controls $(\alpha_t^0 = \phi^0(t,X_t^0,\bar{X}_t))_{0\le t\le T}$ and $(\alpha_t = \phi(t,X_t,X_t^0,\bar{X}_t))_{0\le t\le T}$ of the forms (7.61) and (7.62) is left invariant by the best response maps, we tackle the fixed point step in this subset of feedback controls. For such a fixed point, we must have:

$$\check{X}_t = X_t, \quad t\in[0,T],$$

together with:

$$\alpha_t^{0,*} = \phi^0(t,X_t^0,\bar{X}_t) = \phi_0^0(t) + \phi_1^0(t)X_t^0 + \phi_2^0(t)\bar{X}_t, \quad t\in[0,T],$$

and

$$\check{\alpha}_t^* = \phi(t,X_t,X_t^0,\bar{X}_t) = \phi_0(t) + \phi_1(t)X_t + \phi_2(t)X_t^0 + \phi_3(t)\bar{X}_t, \quad t\in[0,T],$$

which translates into the following equations:

$$\begin{aligned}
&\big[\phi_1^0(t),\phi_2^0(t)\big] = -\tfrac{1}{2}R_0^{-1}\mathfrak{B}_{00}^\dagger\gamma_t, \;\; \phi_0^0(t) = -\tfrac{1}{2}R_0^{-1}\mathfrak{B}_{00}^\dagger\kappa_t,\\
&\big[\phi_2(t),\phi_3(t)\big] = -\tfrac{1}{2}R^{-1}B^\dagger\mathcal{S}_t, \;\; \phi_1(t) = -\tfrac{1}{2}R^{-1}B^\dagger S_t, \;\; \phi_0(t) = -\tfrac{1}{2}R^{-1}B^\dagger s_t.
\end{aligned}$$

To complete the construction of the equilibrium, it thus remains to determine the quantities $(\gamma_t)_{0\le t\le T}$, $(\kappa_t)_{0\le t\le T}$, $(\mathcal{S}_t)_{0\le t\le T}$, $(S_t)_{0\le t\le T}$ and $(s_t)_{0\le t\le T}$ from the systems (7.67) and (7.69). Notice that $(S_t)_{0\le t\le T}$ can be obtained by solving the first equation of (7.69) on its own. As for $(\gamma_t)_{0\le t\le T}$, $(\kappa_t)_{0\le t\le T}$, $(\mathcal{S}_t)_{0\le t\le T}$ and $(s_t)_{0\le t\le T}$, they solve a coupled system obtained by replacing ϕ_0^0, ϕ_1^0, ϕ_2^0, ϕ_0, ϕ_1, ϕ_2 and ϕ_3 by their explicit expressions in (7.67) and (7.69). In other words, we can solve for $(S_t)_{0\le t\le T}$ by solving first the first equation of (7.69), and then group the remaining four equations into two systems of ordinary differential equations as follows:

$$
\begin{cases}
\dot{\gamma}_t + \gamma_t\big[\mathfrak{L}(t) - \frac{1}{2}\mathfrak{B}_0 R^{-1} B^\dagger \mathcal{S}_t\big] + \big[\mathfrak{L}(t) - \frac{1}{2}\mathfrak{B}_0 R^{-1} B^\dagger \mathcal{S}_t\big]^\dagger \gamma_t \\
\qquad\qquad - \frac{1}{2}\gamma_t \mathfrak{B}_{00} R_0^{-1} \mathfrak{B}_{00}^\dagger \gamma_t + 2\hat{\mathfrak{L}}_0 = 0, \\
\dot{\mathcal{S}}(t) + \mathcal{S}_t \mathfrak{L}(t) + \big[L^\dagger - \frac{1}{2} S_t B R^{-1} B^\dagger\big]\mathcal{S}_t - \frac{1}{2}\mathcal{S}_t \mathfrak{B}_0 R^{-1} B^\dagger \mathcal{S}_t \\
\qquad - \frac{1}{2}\mathcal{S}_t \mathfrak{B}_{00} R_0^{-1} \mathfrak{B}_{00}^\dagger \gamma_t + \big[S_t G - 2\Gamma K, S_t F - 2\Gamma K_1\big] = 0,
\end{cases}
\tag{7.71}
$$

and

$$
\begin{cases}
\dot{\kappa}_t + \big[\mathfrak{L}(t) - \frac{1}{2}\mathfrak{B}_0 R^{-1} B^\dagger \mathcal{S}_t\big]^\dagger \kappa_t - \frac{1}{2}\gamma_t \mathfrak{B}_{00} R_0^{-1} \mathfrak{B}_{00}^\dagger \kappa_t \\
\qquad\qquad - \frac{1}{2}\gamma_t \mathfrak{B}_0 R^{-1} B^\dagger s_t + 2\hat{\mathfrak{C}}_0 = 0, \\
\dot{s}_t + \big[L^\dagger - \frac{1}{2} S_t B R^{-1} B^\dagger\big] s_t - \frac{1}{2}\mathcal{S}_t \mathfrak{B}_{00} R_0^{-1} \mathfrak{B}_{00}^\dagger \kappa_t \\
\qquad\qquad - \frac{1}{2}\mathcal{S}_t \mathfrak{B}_0 R^{-1} B^\dagger s_t - 2\Gamma\eta = 0,
\end{cases}
\tag{7.72}
$$

with 0 as terminal condition, where we used the notation:

$$
\mathfrak{L}(t) = \mathfrak{L} - \frac{1}{2}\begin{bmatrix} 0 & 0 \\ 0 & BR^{-1}B^\dagger S_t \end{bmatrix}.
$$

The first system (7.71) comprises two mildly coupled matrix Riccati equations, while the system (7.72), once the solutions of the first system are identified and substituted for, is a plain linear system whose solution is standard. In other words, the functions $[0,T] \ni t \mapsto \kappa_t$ and $[0,T] \ni t \mapsto s_t$ can easily be determined once a solution $[0,T \ni t \mapsto (\gamma_t, \mathcal{S}_t)$ of system (7.71) is found. In essence, we proved the following verification theorem.

Proposition 7.16 *If the system* (7.71) *of matrix Riccati equations is well posed, then there exists a Nash equilibrium in the family of linear closed loop feedback controls, the optimal controls for the major and minor players being given by the strategies* (7.68) *and* (7.70).

Numerical Application

In this subsection, we present the results of the implementation of the above results to a simple linear quadratic model inspired by the Cucker-Smale flocking model described in Chapter 1 of Volume I. For the purpose of graphical illustrations, we formulate the model in terms of the velocity only. This would correspond to the case $\beta = 0$ in the model of Chapter (Vol I)-1.

We propose a simplistic model for a bee swarm relocating to a new nest site. According to the prevalent theory of the *streaker bee hypothesis* , only a very small number of bees, typically less than 5% of the swarm, know the new location and their role is to lead the swarm by flying at high speed through the swarm to lead by inviting the other bees to adopt the same velocity. In our model, the (small number of) streaker bees will be modeled as the major player. The other bees in the swarm will be modeled as the minor player.

We denote by $V_t^{N,0}$ the velocity of the streaker bee at time t, and by $V_t^{N,i}$ for $i = 1, \cdots, N$ the velocity of the ith bee in the swarm. The *leader bee* and the *follower bees* control their velocities through their drifts $(\alpha_t^{N,0})_{0 \le t \le T}$ and $((\alpha_t^{N,i})_{0 \le t \le T})_{i=1,\cdots,N}$. In other words, we assume that their velocities $(V_t^{N,0})_{0 \le t \le T}$ and $((V_t^{N,i})_{0 \le t \le T})_{i=1,\cdots,N}$ satisfy:

$$\begin{cases} dV_t^{N,0} = \alpha_t^{N,0} dt + \Sigma_0 dW_t^0, \\ dV_t^{N,i} = \alpha_t^{N,i} dt + \Sigma dW_t^i, \qquad i = 1, \cdots, N, \end{cases} \tag{7.73}$$

where $\boldsymbol{W}^0$ and $(\boldsymbol{W}^i)_{i=1,\cdots,N}$ are independent standard Wiener processes of dimension $m \ge 1$, and where Σ_0 and Σ are constant volatility matrices of dimension $d \times m$, for some $d \ge 1$ (we should think of $d_0 = d = 3$). We also consider a deterministic function $[0, T] \ni t \mapsto \nu_t \in \mathbb{R}^d$ representing the ideal velocity which the streaker bee would like to have in order to get to the location of the new nest. The objective of the major player is to make sure that its velocity is as close as possible to the target velocity $(\nu_t)_{0 \le t \le T}$, while at the same time keeping a reasonable distance from the bulk of the other bees not to lose its influence on them. We denote by $(\bar{V}_t^N = \frac{1}{N} \sum_{i=1}^N V_t^{N,i})_{0 \le t \le T}$ the average velocity of the bees in the swarm. To capture the main stylized facts of the streaker bee theory, we assume that the streaker bee tries to minimize its cost over the time horizon T as given by:

$$\begin{aligned} &J^{N,0}\bigl(\boldsymbol{\alpha}^{N,0}, (\boldsymbol{\alpha}^{N,i})_{i=1,\cdots,N}\bigr) \\ &\quad = \mathbb{E}\biggl[\int_0^T \Bigl(k_0 |V_t^{N,0} - \nu_t|^2 + k_1 |V_t^{N,0} - \bar{V}_t^N|^2 + (1 - k_0 - k_1)|\alpha_t^0|^2\Bigr) dt\biggr], \end{aligned}$$

where k_0 and k_1 are positive real numbers satisfying $k_0 + k_1 \le 1$. Similarly, each bee in the swarm faces a tradeoff between keeping up with the streaker and staying close to its peers. We capture this dilemma in the minimization of the objective function defined as:

$$\begin{aligned} &J^{N,i}\bigl(\boldsymbol{\alpha}^{N,0}, (\boldsymbol{\alpha}^{N,i})_{i=1,\cdots,N}\bigr) \\ &\quad = \mathbb{E}\biggl[\int_0^T \Bigl(l_0 |V_t^{i,N} - V_t^{0,N}|^2 + l_1 |V_t^{i,N} - \bar{V}_t^N|^2 + (1 - l_0 - l_1)|\alpha_t^i|^2\Bigr) dt\biggr], \end{aligned}$$

where l_0 and l_1 are positive real numbers satisfying $l_0 + l_1 \le 1$. It is plain to check that in the mean field game limit, we can fit this swarming model into the framework of linear quadratic models studied earlier, by simply doubling the state variable. More precisely, we define:

$$X_t^0 = \begin{bmatrix} V_t^0 \\ V_t^0 \end{bmatrix}, \quad X_t = \begin{bmatrix} V_t \\ V_t \end{bmatrix}, \quad \bar{X}_t = \begin{bmatrix} \bar{V}_t \\ \bar{V}_t \end{bmatrix}, \qquad t \in [0, T],$$

and we can use the results of the mean field games with major and minor players studied in this section as long as we set:

$$L_0 = L = F_0 = F = G = \begin{bmatrix} 0 & 0 \\ 0 & 0 \end{bmatrix}, \quad R_0 = (1 - k_0 - k_1)I, \quad R = (1 - l_0 - l_1)I,$$

$$K = \begin{bmatrix} I & 0 \\ 0 & 0 \end{bmatrix}, \quad K_0 = K_1 = \begin{bmatrix} 0 & 0 \\ 0 & I \end{bmatrix}, \quad \Gamma_0 = \begin{bmatrix} k_0 I & 0 \\ 0 & k_1 I \end{bmatrix}, \quad \Gamma = \begin{bmatrix} l_0 I & 0 \\ 0 & l_1 I \end{bmatrix},$$

$$\eta_0(t) = \begin{bmatrix} \nu_t \\ 0 \end{bmatrix}, \quad \eta = \begin{bmatrix} 0 \\ 0 \end{bmatrix}, \quad B_0 = B = \begin{bmatrix} I \\ I \end{bmatrix}, \quad D_0 = \begin{bmatrix} \Sigma_0 \\ \Sigma_0 \end{bmatrix}, \quad D = \begin{bmatrix} \Sigma \\ \Sigma \end{bmatrix}.$$

For the purpose of the numerical simulations which we demonstrate below, we consider a 2-dimensional swarm, and use two different target velocities $[0, T] \ni t \mapsto \nu_t \in \mathbb{R}^2$ to illustrate clearly the impact of the relative values of the constants $(k_j)_{j=0,1}$ and $(l_j)_{j=0,1}$ on the behavior of the different bees. Figures 7.1 and 7.2 confirm the

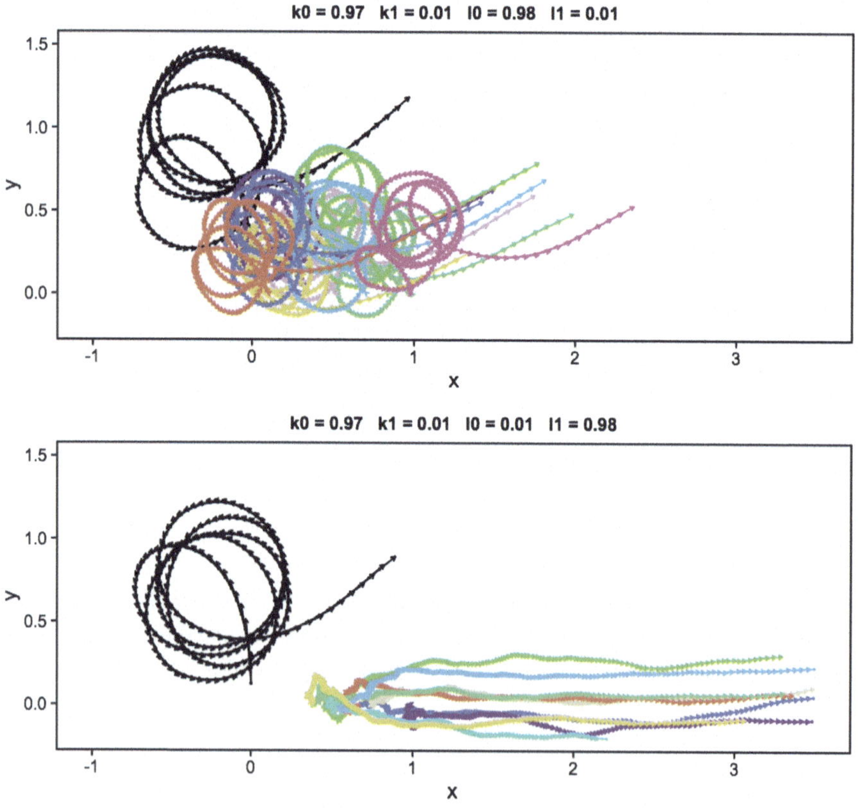

Fig. 7.1 Equilibrium velocities and trajectories of the major player (streaker bee) and the minor players (swarm bees) for two different sets of parameters and a circular target velocity.

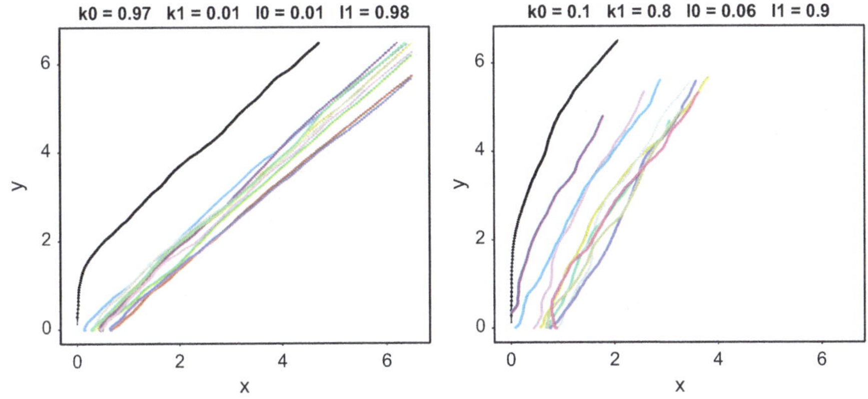

Fig. 7.2 Equilibrium velocities and trajectories of the major player (streaker bee) and the minor players (swarm bees) for two sets of parameters when the target velocity is constant and equal to $[1, 1]^\dagger$.

natural intuition: the larger the constant k_0 the closer the equilibrium velocity of the streaker to the target velocity, and the bees in the swarm will follow suit if the constant l_0 is relatively large. In both cases, we choose $\Sigma_0 = \Sigma = 0.5I_2$, and for each choice of coefficients k_0, k_1, l_0, l_1, we use a simple Euler scheme to solve numerically the system of matrix Riccati equations (7.71) over the horizon $[0, T] = [0, 5]$. Then we simulate the dynamics of the leader and the N followers defined in (7.73), where we assign the equilibrium strategy of the mean field game to the major and the minor players. In both figures, the velocity and the trajectory of the streaker bee are given in black and the velocities and the trajectories of a small sample of bees in the swarm are plotted in color. Their initial positions were chosen uniformly over the interval $[0, 1] \times \{0\}$.

For the experiment reported in Figure 7.1, we use the target velocity $(\nu_t = [-2\pi \sin(2\pi t), 2\pi \cos(2\pi t)]^\dagger)_{0 \le t \le T}$. We clearly see that the major player tries very hard to mimic the target velocity because the coefficient k_0 is relatively large compared to k_1 and $1 - k_0 - k_1$. In the top plot, the bees in the swarm try to have the same velocities because the coefficient l_0 is dominant. However, in the bottom plot, the bees in the swarm pay more attention to the remaining ones because the coefficient l_1 is now dominant.

Figure 7.2 was produced with a constant target velocity $(\nu_t = [1, 1]^\dagger)_{0 \le t \le T}$ and similar values of the coefficients $(k_j)_{j=0,1}$ and $(l_j)_{j=0,1}$ and the rationale for what they demonstrate is exactly the same as in the case of Figure 7.1.

We conclude with a numerical experiment to illustrate the conditional propagation of chaos discussed in Subsection 7.1.5. Our results are reproduced in Figure 7.3. We expose the phenomenon in the following way. For a given number of worker bees, say N, we fix a realization of the Wiener process driving the dynamics of the streaker. Given the fixed realization of the common Wiener process, we simulate

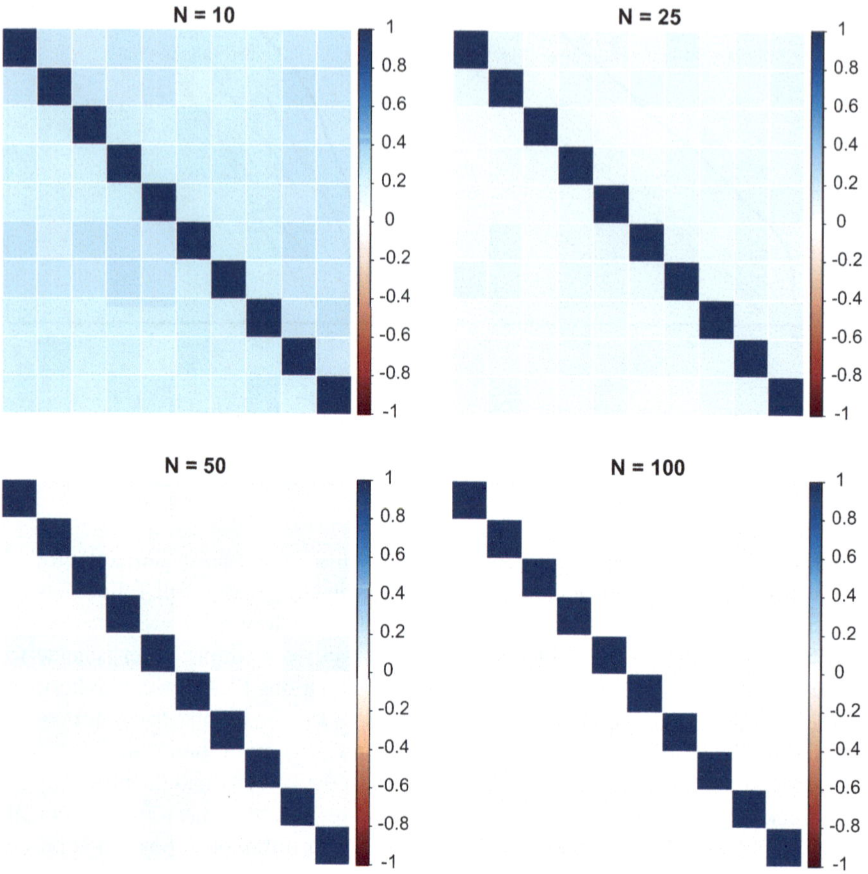

Fig. 7.3 Conditional correlation of the states of 10 minor players given the common noise, as the total number of minor players increases from $N = 10$ to $N = 100$.

M copies of the optimal path of the velocity of the streaker, say $(V_t^{N,0})_{0 \le t \le T}$, and of the optimal paths of the velocities of the N worker bees, say $(V_t^{N,i})_{0 \le t \le T}$, for $i = 1, \dots, N$, using the same fixed Wiener process for the streaker, but independent copies of Wiener processes for each of the N worker bees. Then, for a given time t, we compute the sample correlation matrix of the first components $V_t^{i,N,(1)}$, for the first 10 worker bees, namely for $i = 1, \dots, 10$. Obviously, the result would be exactly the same if we chose the second component instead. Finally, we compute the average of the correlation matrix across time $t \le T$. Figure 7.3 displays the average correlation matrices of the first component of the velocities of the first 10 worker bees in a hive of size $N = 10, 25, 50, 100$. Conditional propagation of chaos says that, conditional on the trajectory of the common Wiener process, the velocities of the first 10 worker bees should become independent when the total number N of worker bees goes to ∞. So we expect that the conditional correlation matrix of the velocities of the first 10 worker bees converge toward the 10×10 identity matrix. This is confirmed by the plots produced in Figure 7.3.

7.1.9 An Example with Finitely Many States

We end our analysis of mean field games with major and minor players with a discussion of game models with finite state spaces. Even though a detailed analysis of the finite player games would provide a clear justification for the state dynamics chosen for the limiting mean field game problem, in the interest of time and space, we skip the description of the finite player games and formulate directly the mean field game problem.

We assume that the state of the system is given by a generic point (x_0, x) in the product $E_0 \times E$ of two finite sets. Here x_0 (resp. x) represents the state of the major (resp. representative minor) player. Also, we assume that the time evolution of the state of the major (resp. representative minor) player is given by a continuous time stochastic process $\boldsymbol{X}^0 = (X_t^0)_{0 \le t \le T}$ (resp. $\boldsymbol{X} = (X_t)_{0 \le t \le T}$) in E^0 (resp. E) whose jump rates are given by a function of the form:

$$[0, T] \times E_0 \times E_0 \times \mathcal{P}(E) \times A_0 \ni (t, x_0, x_0', \mu, \alpha_0) \mapsto \lambda_t^0(x_0, x_0', \mu, \alpha_0)$$

$$\Big(\text{resp.} \quad [0, T] \times E \times E \times \mathcal{P}(E) \times E_0 \times A_0 \times A \ni (t, x, x', \mu, x_0, \alpha_0, \alpha) \\ \mapsto \lambda_t(x, x', \mu, x_0, \alpha_0, \alpha)\Big),$$

which emphasizes the fact that the dynamics of the state of the major player depend upon the statistical distribution μ of the states of the minor players while the dynamics of the states of the minor players depend upon the same distribution μ and the state x_0 of the major player, which is consistent with the stochastic differential game model stated in (7.1). Notice that we also allow the dynamics of the state of the representative minor player to depend upon the control α_0 used by the major player. We want to keep the present discussion at a rather informal level, and for this reason, we do not state precise definitions for the spaces of admissible control strategies $\mathbb{A}_0$ and $\mathbb{A}$. However, we shall limit ourselves to control strategies in feedback form. As usual, the spaces of controls A_0 and A are closed subsets of Euclidean spaces $\mathbb{R}^{k_0}$ and $\mathbb{R}^k$.

In full analogy with the previous discussions in this section, the running and terminal cost functions of the major and the minor players are denoted by f_0, g_0, f and g, so that when the major player and the representative minor player choose the strategy profiles $\boldsymbol{\alpha}^0$ and $\boldsymbol{\alpha}$ respectively, the expected costs to the major and minor players are given by:

$$J^0(\boldsymbol{\alpha}^0, \boldsymbol{\alpha}) = \mathbb{E}\bigg[\int_0^T f_0(t, X_t^0, \mu_t, \alpha_t^0)dt + g_0(X_T^0, \mu_T)\bigg],$$

and

$$J(\alpha^0, \alpha) = \mathbb{E}\left[\int_0^T f(t, X_t, \mu_t, X_t^0, \alpha_t^0, \alpha_t)dt + g(X_T, \mu_T, X_T^0)\right].$$

7.1.10 The Search for Nash Equilibria

Without going into the gory details of the analysis, we describe the steps of the alternative formulation outlined in Subsection 7.1.3 to construct the best response map and identify Nash equilibria in Markovian feedback form as its fixed points.

Major Player's Problem. We fix an admissible strategy $\boldsymbol{\alpha} \in \mathbb{A}$ in feedback form $(\alpha_t = \phi(t, X_t, \mu_t, X_t^0))_{0 \le t \le T}$ for the representative minor player, and solve the optimal control problem of the major player given that all the minor players use the feedback function ϕ. Assuming that it exists and is unique, we denote by $\hat{\phi}^0 = \boldsymbol{\phi}^{0,*}(\phi)$ the feedback function giving the optimal strategy of this optimization problem. So, given that all the minor players use control strategies based on the same feedback function ϕ, the best response of the major player is to use the strategy $\boldsymbol{\alpha}^{0,*}$ given by the feedback function $\hat{\phi}^0$ solving the optimal control problem:

$$\inf_{\alpha^0 \leftrightarrow \phi^0 \in \mathbb{A}^0} \mathbb{E}\left[\int_0^T f_0(t, X_t^0, \mu_t, \phi^0(t, X_t^0, \mu_t))dt + g^0(X_T^0, \mu_T)\right],$$

where $(X_t^0, \mu_t)_{0 \le t \le T}$ is the continuous time Markov process which we now describe. In the above formula, the (random) measure μ_t is the distribution at time t of the state of a representative minor player. We rely on the fact that the process $(X_t^0, \mu_t)_{0 \le t \le T}$ is sufficient (in the sense of sufficiency in mathematical statistics) for the dynamics and the expected costs of the major player. Also, it is Markovian. The infinitesimal generator $\mathcal{G}^0_{\phi^0,\phi}$ of this Markov process is given by:

$$\begin{aligned}
&\big[\mathcal{G}^0_{\phi^0,\phi} F\big](t, x^0, \mu) \\
&= \frac{\partial F}{\partial t}(t, x^0, \mu) + \sum_{y^0 \in E_0} F(t, y^0, \mu)\lambda_t^0\big(x^0, y^0, \mu, \phi^0(t, x^0, \mu)\big) \\
&\quad + \sum_{x' \in E} \frac{\partial F}{\partial \mu(\{x'\})}(t, x^0, \mu) \sum_{x \in E} \mu(\{x\})\lambda_t\big(x, x', \mu, x^0, \phi^0(t, x^0, \mu), \phi(t, x, \mu, x^0)\big),
\end{aligned}$$

for a smooth function $F : [0, T] \times E_0 \times \mathcal{P}(E) \to \mathbb{R}$, which we regard rather abusively as a smooth function of t, x^0 and $(\mu(\{x\}))_{x \in E}$. Recalling that the families $(\lambda_t^0(x^0, y^0, \mu, \phi^0(t, x^0, \mu)))_{x^0, y^0 \in E}$ and $(\lambda_t(x', x'', \mu, x^0, \phi^0(t, x^0, \mu), \phi(t, x', x^0, \mu)))_{x', x'' \in E}$ are Q-matrices, we get:

$$
\begin{aligned}
&\big[\mathcal{G}^0_{\phi^0,\phi} F\big](t,x^0,\mu) \\
&= \frac{\partial F}{\partial t}(t,x^0,\mu) + \sum_{y^0 \neq x^0} \big[F(t,y^0,\mu) - F(t,x^0,\mu)\big]\lambda^0_t\big(x^0,y^0,\mu,\phi^0(t,x^0,\mu)\big) \\
&\quad + \sum_{x' \in E} \frac{\partial F}{\partial \mu(\{x'\})}(t,x^0,\mu) \sum_{x \neq x'} \mu(\{x\})\lambda_t\big(x,x',\mu,x^0,\phi^0(t,x^0,\mu),\phi(t,x,\mu,x^0)\big) \\
&\quad - \sum_{x' \in E} \Big[\frac{\partial F}{\partial \mu(\{x'\})}(t,x^0,\mu)\mu(\{x'\}) \\
&\qquad\qquad \times \sum_{x \neq x'} \lambda_t\big(x',x,\mu,x^0,\phi^0(t,x^0,\mu),\phi(t,x',\mu,x^0)\big)\Big] \\
&= \frac{\partial F}{\partial t}(t,x^0,\mu) + \sum_{y^0 \neq x^0} \big[F(t,y^0,\mu) - F(t,x^0,\mu)\big]\lambda^0_t\big(x^0,y^0,\mu,\phi^0(t,x^0,\mu)\big) \\
&\quad + \sum_{x' \in E} \sum_{x \neq x'} \Big[\Big(\frac{\partial F}{\partial \mu(\{x'\})}(t,x^0,\mu) - \frac{\partial F}{\partial \mu(\{x\})}(t,x^0,\mu)\Big) \\
&\qquad\qquad \times \mu(\{x\})\lambda_t\big(x,x',\mu,x^0,\phi^0(t,x^0,\mu),\phi(t,x,\mu,x^0)\big)\Big],
\end{aligned}
$$

where we exchanged x and x' on the fourth and fifth lines to derive the final expression. The definition of this infinitesimal generator can be understood in the following way. The partial derivative with respect to the time variable is present because we are using space time Markov processes as the transition rates depend upon time. The next term corresponds to transitions from state x^0 to y^0 in the state X^0_t of the major player when the feedback function at time t is $\phi^0(t,\cdot,\cdot)$. The last two summations correspond to transitions in the probability μ_t, which may be derived from the Fokker-Planck equation satisfied by the marginal law of the representative player, see for instance Subsection (Vol I)-7.2.2.

We refer to Subsection (Vol I)-5.4.4 for a complete overview on the differentiability properties of a real-valued function defined on $\mathcal{P}(E)$. If we denote by e_1, e_2, $\cdots$, e_d the elements of E, we may identify probability measures $\mu \in \mathcal{P}(E)$ with elements of the simplex $\mathcal{S}_d = \{(p_1,\cdots,p_{d-1},p_d) \in [0,1]^d : \sum_{i=1}^d p_i = 1\}$. Since $\mathcal{S}_d$ is in one-to-one correspondence with the set $\mathcal{S}_{d-1,\leq} = \{(p_1,\cdots,p_{d-1}) \in [0,1]^{d-1} : \sum_{i=1}^{d-1} p_i \leq 1\}$, smooth functions on $\mathcal{P}(E)$ can be viewed as smooth functions on $\mathcal{S}_{d-1,\leq}$, the latter having a nonempty interior in $\mathbb{R}^{d-1}$. In this regard, Proposition (Vol I)-5.66 and Corollary (Vol I)-5.67 make the connection between the derivatives of a function defined on the simplex and the linear functional derivative of a function defined on $\mathcal{P}(E)$. Importantly, it permits to rewrite the above formula in terms of both the functional linear derivative of F and the derivatives on the $(d-1)$-dimensional domain $\mathcal{S}_{d-1,\leq}$, as $([\partial F/\partial \mu(\{x'\})](t,x^0,\mu) - [\partial F/\partial \mu(\{x\})](t,x^0,\mu))_{x' \in E \setminus \{x\}}$ may be regarded as the derivatives of F with respect to $(\mu(\{x'\}))_{x' \in E \setminus \{x\}}$ seen as an element of $\mathcal{S}_{d-1,\leq}$.

Representative Minor Player's Problem. We first single out a minor player, which we also called before a virtual player, and we search for its best response to the rest of the other players when all the remaining minor players are exchangeable. So we fix an admissible strategy $\boldsymbol{\alpha}^0 \in \mathbb{A}^0$ of the form $(\alpha_t^0 = \phi^0(t, X_t^0, \mu_t))_{0\le t\le T}$ for the major player, and an admissible strategy $\boldsymbol{\alpha} \in \mathbb{A}$ of the form $(\alpha_t = \phi(t, X_t, \mu_t, X_t^0))_{0\le t\le T}$ for the representative of the exchangeable minor players. We then assume that the minor player which we singled out responds to the other players by choosing an admissible strategy $\check{\boldsymbol{\alpha}} \in \mathbb{A}$ of the form $(\check{\alpha}_t = \check{\phi}(t, \check{X}_t, \mu_t, X_t^0))_{0\le t\le T}$ minimizing the expected cost:

$$\inf_{\check{\alpha}\leftrightarrow\check{\phi}\in\mathbb{A}} \mathbb{E}\biggl[\int_0^T f\bigl(t, \check{X}_t, \mu_t, X_t^0, \check{\phi}(t, \check{X}_t, \mu_t, X_t^0), \phi^0(t, X_t^0, \mu_t)\bigr)dt \\ + g(\check{X}_T, \mu_T, X_T^0)\biggr],$$

where the process $(\check{X}_t, \mu_t, X_t^0)_{0\le t\le T}$ is a Markov process with infinitesimal generator $\mathcal{G}_{\phi^0,\phi,\check{\phi}}$ given by:

$$\begin{aligned}
&\bigl[\mathcal{G}_{\phi^0,\phi,\check{\phi}} F\bigr](t,x,\mu,x^0) \\
&= \frac{\partial F}{\partial t}(t,x,\mu,x^0) + \sum_{y^0\in E_0} F(t,x,\mu,y^0)\lambda_t^0\bigl(t,x^0,y^0,\mu,\phi^0(t,x^0,\mu)\bigr) \\
&\quad + \sum_{x'\in E} F(t,x',\mu,x^0)\lambda_t\bigl(x,x',\mu,x^0,\phi^0(t,x^0,\mu),\check{\phi}(t,x,\mu,x^0)\bigr) \\
&\quad + \sum_{x'\in E}\sum_{x''\in E}\Bigl[\frac{\partial F}{\partial \mu(\{x'\})}(t,x,\mu,x^0)\mu(\{x''\}) \\
&\qquad\qquad \times \lambda_t\bigl(x'',x',\mu,x^0,\phi^0(t,x^0,\mu),\phi(t,x'',\mu,x^0)\bigr)\Bigr].
\end{aligned}$$

The various terms appearing in the above definition of the infinitesimal generator can be understood as before. The only new term is the summation on the third line, which corresponds to transitions in which the state of the representative minor player jumps from x'' to x'.

We shall assume that the minimizer of the above optimal control problem exists and that it is unique and we shall denote by $\hat{\phi} = \boldsymbol{\phi}^*(\phi^0, \phi)$ the optimal feedback function providing the solution of this optimal control problem.

Search for a Fixed Point of the Best Response Map. A Nash equilibrium for the mean field game with major and minor players is a fixed point $[\hat{\phi}^0, \hat{\phi}] = [\boldsymbol{\phi}^{0,*}(\hat{\phi}), \boldsymbol{\phi}^*(\hat{\phi}^0, \hat{\phi})]$.

We shall not pursue the construction of Nash equilibria for the system at this level of generality. See the Notes & Complements at the end of the chapter for a reference to the construction of equilibria in small time. Instead, we concentrate on the extension to the major/minor set-up of the cyber security model introduced in Chapter (Vol I)-1, and studied in Chapter (Vol I)-7.

A Form of the Cyber Security Model with Major and Minor Players

We now revisit the cyber-security mean field game model studied in Subsection (Vol I)-7.2.3 of Chapter (Vol I)-7, with the firm intention to extend it to a full-fledged model with major and minor players. This form of the game was already touted in the introduction to the model we gave in Subsection (Vol I)-1.6.2 of Chapter (Vol I)-1. The major player is a botnet herder or a hacker, who put machines into his or her control by installing malicious software. Minor players are computer owners who are susceptible to the hacker's attacks. The dynamics of the states of the computers of the minor players are the same as the dynamics already described in Subsection (Vol I)-7.2.3 where we discuss a simplified version of the model in which the attacker did not control the intensity of the attacks. In order to avoid referring too often to Volume I, and for the current presentation to be self-contained, we repeat some of the definitions already given in Chapter (Vol I)-7. The vulnerable network computers can be in one of $d = 4$ states:

- DI: defended infected;
- DS: defended and susceptible to infection;
- UI: unprotected infected;
- US: unprotected and susceptible to infection;

so the state space of the minor players is $E = \{\mathrm{DI}, \mathrm{DS}, \mathrm{UI}, \mathrm{US}\}$. In this simplistic model, the rate λ_t is independent of t and each network computer owner can choose one of two actions, that is $A = \{0, 1\}$. Action $\alpha = 0$ means that the computer owner is happy with its level of protection (Defended or Unprotected) and does not try to change its own state. On the other hand, action $\alpha = 1$ means that the computer owner is willing to update the level of protection of its computer and switch to the other state (Unprotected or Defended). In the latter case, updating occurs after an exponential time with parameter $\lambda > 0$, which accounts for the speed of the response of the defense system.

When infected, a computer may recover at a rate depending on its protection level: the recovery rate is denoted by $q_{\mathrm{rec}}^{\mathrm{D}}$ for a protected computer and by $q_{\mathrm{rec}}^{\mathrm{U}}$ for an unprotected one.

Conversely, a computer may become infected in two ways, either directly from the attacks of the hacker or indirectly from infected computers that spread out the infection. The rate of direct infection depends upon the intensity of the attacks, as fixed by the botnet herder. This intensity is denoted by $\alpha^0 \in A_0$ with $A_0 = [0, \infty)$, and the rate of direct infection of a protected computer is $\alpha^0 q_{\mathrm{inf}}^{\mathrm{D}}$ while the rate of direct infection of an unprotected computer is $\alpha^0 q_{\mathrm{inf}}^{\mathrm{U}}$. Also, the rates of infection spreading from infected to susceptible computers depend upon the distribution μ of the states within the population of computers. The rate of infection of an unprotected susceptible computer by other unprotected infected computers is $\beta_{\mathrm{UU}}\mu(\{\mathrm{UI}\})$, the rate of infection of a protected susceptible computer by other unprotected infected computers is $\beta_{\mathrm{UD}}\mu(\{\mathrm{UI}\})$, the rate of infection of an unprotected susceptible computer by other protected infected computers is $\beta_{\mathrm{DU}}\mu(\{\mathrm{DI}\})$, and the rate of infection of a protected susceptible computer by other protected infected computers is $\beta_{\mathrm{DD}}\mu\{\mathrm{DI}\}$.

As a result of these assumptions, the infinitesimal rates of transition of the state of a typical minor player are given by the Q-matrix $\lambda_t(x, x', \mu, \alpha^0, \alpha)$ (recall that $\alpha = 0$ or $\alpha = 1$):

$$\lambda_t(\cdot, \cdot, \mu, \alpha^0, \alpha) =$$

$$\begin{array}{c} \\ \mathrm{DI} \\ \mathrm{DS} \\ \mathrm{UI} \\ \mathrm{US} \end{array}
\begin{array}{cccc}
\mathrm{DI} & \mathrm{DS} & \mathrm{UI} & \mathrm{US} \\
\left[\begin{array}{c} \cdots \\ \alpha^0 q^{\mathrm{D}}_{\mathrm{inf}} + \beta_{\mathrm{DD}}\mu(\{\mathrm{DI}\}) + \beta_{\mathrm{UD}}\mu(\{\mathrm{UI}\}) \\ \alpha\lambda \\ 0 \end{array}\right. &
\begin{array}{c} q^{\mathrm{D}}_{\mathrm{rec}} \\ \cdots \\ 0 \\ \alpha\lambda \end{array} &
\begin{array}{c} \alpha\lambda \\ 0 \\ \cdots \\ \alpha^0 q^{\mathrm{U}}_{\mathrm{inf}} + \beta_{\mathrm{UU}}\mu(\{\mathrm{UI}\}) + \beta_{\mathrm{DU}}\mu(\{\mathrm{DI}\}) \end{array} &
\left.\begin{array}{c} 0 \\ \alpha\lambda \\ q^{\mathrm{U}}_{\mathrm{rec}} \\ \cdots \end{array}\right]
\end{array}$$

where all the instances of $\cdots$ should be replaced by the negative of the sum of the entries of the row in which the dots $\cdots$ appear on the diagonal.

In contrast with what we did for the minor players, we do not specify the dynamics nor the state space of the major player. In order to proceed, it suffices to know the value of its control, which is here given by α^0. Notice in particular that, in this model, the rate λ_t depends not only on the action of the minor player, but also on the control of the major player. The first models of stochastic differential games with major and minor players we introduced in Subsection 7.1.1, where we described the finite player games, included this type of interaction. We dropped this dependence upon the control of the major player for the sake of simplicity, but in the particular example at hand, one may view α^0 as the state as well as the control of the major player.

We now specify the costs incurred by each of the players. Each computer owner (minor player), with X_t as state at time t, pays a fee k_{D} per unit of time for the defense of its system, and incurs a loss k_{I} per unit of time for losses resulting from infection. Recalling the standard notations $\boldsymbol{\alpha}^0$ and $\boldsymbol{\alpha}$ for the strategies of the major and minor players, the expected cost to the minor player is given by:

$$J(\boldsymbol{\alpha}^0, \boldsymbol{\alpha}) = \mathbb{E}\left[\int_0^T \big(k_{\mathrm{D}}\mathbf{1}_{\mathrm{D}} + k_{\mathrm{I}}\mathbf{1}_{\mathrm{I}}\big)(X_t)dt\right],$$

where we use the notation $\mathrm{D} = \{\mathrm{DI}, \mathrm{DS}\}$ and $\mathrm{I} = \{\mathrm{DI}, \mathrm{UI}\}$. If the attacker (major player) chooses at time t the attack strategy α^0_t as given by a deterministic function ϕ^0 of time and the statistical distribution $\mu_t = \mathcal{L}(X_t)$ of the states of the computers, its instantaneous cost is $k_{\mathrm{H}}\phi^0(t, \mu_t)^2/2$. Notice that in the more general examples considered earlier, μ_t was random since it was the conditional distribution of the state X_t given the noise driving the dynamics of the state of the major player. However, since the major player only enters the model through $\boldsymbol{\alpha}^0$, which is deterministic, the measure μ_t is deterministic here. We also assume that if the computer network is in state μ_t, its reward is given by a function f_0 defined on $\mathcal{P}(E)$ so that the total expected cost the attacker tries to minimize is given by:

$$
\begin{aligned}
&J^0(\boldsymbol{\alpha}^0, \boldsymbol{\alpha}) \\
&= \mathbb{E}\Big[\int_0^T \Big(-f_0(\mu_t) + \frac{k_{\mathrm{H}}}{2}(\alpha_t^0)^2\Big)dt\Big] = \int_0^T \Big(-f_0(\mu_t) + \frac{k_{\mathrm{H}}}{2}(\alpha_t^0)^2\Big)dt,
\end{aligned}
\tag{7.74}
$$

because the quantity appearing in the expectation is deterministic.

Search for Nash Equilibria

We now describe what the three steps of the search for Nash equilibria reduce to in the particular case of the model at hand. We emphasize once more the fact that the state of the major player does not enter the model.

Optimization Problem of the Major Player. This optimization problem is rather straightforward. Indeed, the cost (7.74) to the major player is a deterministic function of the deterministic flow $\boldsymbol{\mu} = (\mu_t)_{0\le t\le T}$ of probability measures and we can capture the dynamics of this flow $\boldsymbol{\mu}$ by an ODE; recall that at each time t, the measure μ_t can be viewed as an element of $\mathcal{P}(E)$, which can be identified to the probability simplex in $\mathbb{R}^4$.

Following the prescriptions for the optimization problem of the major player which we outlined earlier, we fix a distributed Markovian strategy $\boldsymbol{\alpha} = (\alpha_t)_{0\le t\le T}$ given by a feedback function ϕ in the sense that $\alpha_t = \phi(t, X_t)$, and we try to identify the best response of the hacker when all the computer owners use the same feedback function ϕ. Observe that, the flow $\boldsymbol{\mu} = (\mu_t)_{0\le t\le T}$ being deterministic, it is implicitly hidden in ϕ through the time argument t. As explained earlier, since the model does not involve the state X_t^0 of the hacker, we look for strategies in the form $\alpha_t^0 = \phi^0(t, \mu_t)$ to minimize the cost to the hacker given in (7.74). As emphasized several times already, this optimization problem should be of the McKean-Vlasov type. However, since the dynamics of the state of the major player are not present in the current model, the dynamics over which the optimization is performed can be expressed directly in terms of the time evolution of the probability distribution μ_t. To be more specific, the optimization problem of the major player becomes the minimization:

$$
\inf_{\alpha^0} \int_0^T \Big(\frac{k_{\mathrm{H}}}{2}(\alpha_t^0)^2 - f_0(\mu_t)\Big)dt,
$$

under the dynamic constraint:

$$
\partial_t \mu_t(\{x\}) = \sum_{x' \in E} q_t(x', x)\mu_t(\{x'\}) \tag{7.75}
$$

where the Q-matrix q_t is given by:

$$q_t(x, x') = \lambda_t(x, x', \mu_t, \alpha_t^0, \phi(t, x)), \qquad 0 \le t \le T, \ \ x, x' \in E.$$

Equation (7.75) is an ordinary (vector) differential equation in 4 dimensions. Even though the form (7.75) suggests that this equation is linear, the fact that the measure μ_t appears in the definition of the Q-matrix makes it a nonlinear equation. To be specific, we can use the two possible values of the matrix λ_t recalled above to write completely the system giving the time evolution of the measure μ_t.

$$\left\{\begin{aligned}
\partial_t \mu_t(\mathrm{DI}) &= q^{\mathrm{D}}_{\mathrm{inf}} \alpha_t^0 \mu_t(\mathrm{DS}) - q^{\mathrm{D}}_{\mathrm{rec}} \mu_t(\mathrm{DI}) + \beta_{\mathrm{DD}} \mu_t(\mathrm{DI}) \mu_t(\mathrm{DS}) \\
&\qquad + \beta_{\mathrm{UD}} \mu_t(\mathrm{UI}) \mu_t(\mathrm{DS}) + \lambda[\phi(t, \mathrm{UI}) \mu_t(\mathrm{UI}) - \phi(t, \mathrm{DI}) \mu_t(\mathrm{DI})] \\
\partial_t \mu_t(\mathrm{DS}) &= -q^{\mathrm{UD}}_{\mathrm{inf}} \alpha_t^0 \mu_t(\mathrm{DS}) + q^{\mathrm{D}}_{\mathrm{rec}} \mu_t(\mathrm{DI}) - \beta_{\mathrm{DD}} \mu_t(\mathrm{DI}) \mu_t(\mathrm{DS}) \\
&\qquad - \beta_{\mathrm{UD}} \mu_t(\mathrm{UI}) \mu_t(\mathrm{DS}) + \lambda[\phi(t, \mathrm{US}) \mu_t(\mathrm{US}) - \phi(t, \mathrm{DS}) \mu_t(\mathrm{DS})] \\
\partial_t \mu_t(\mathrm{UI}) &= q^{\mathrm{U}}_{\mathrm{inf}} \alpha_t^0 \mu_t(\mathrm{US}) - q^{\mathrm{U}}_{\mathrm{rec}} \mu_t(\mathrm{UI}) + \beta_{\mathrm{DU}} \mu_t(\mathrm{DI}) \mu_t(\mathrm{US}) \\
&\qquad + \beta_{\mathrm{UU}} \mu_t(\mathrm{UI}) \mu_t(\mathrm{US}) + \lambda[\phi(t, \mathrm{DI}) \mu_t(\mathrm{DI}) - \phi(t, \mathrm{UI}) \mu_t(\mathrm{UI})] \\
\partial_t \mu_t(\mathrm{US}) &= -q^{\mathrm{U}}_{\mathrm{inf}} \alpha_t^0 \mu_t(\mathrm{US}) + q^{\mathrm{U}}_{\mathrm{rec}} \mu_t(\mathrm{UI}) - \beta_{\mathrm{DU}} \mu_t(\mathrm{DI}) \mu_t(\mathrm{US}) \\
&\qquad - \beta_{\mathrm{UU}} \mu_t(\mathrm{UI}) \mu_t(\mathrm{US}) + \lambda[\phi(t, \mathrm{DS}) \mu_t(\mathrm{DS}) - \phi(t, \mathrm{US}) \mu_t(\mathrm{US})].
\end{aligned}\right. \tag{7.76}$$

For the sake of notation, we write $\mu_t(\mathrm{XX})$ for $\mu_t(\{\mathrm{XX}\})$ for $\mathrm{XX} = \mathrm{DI}, \mathrm{DS}, \mathrm{UI}, \mathrm{US}$. For further convenience, we rewrite the system (7.76) in shorter form:

$$\frac{d\mu_t}{dt} = \alpha_t^0 B^0(\mu_t) + B(t, \mu_t),$$

where μ_t and $B^0(\mu)$ are the 4-dimensional vectors:

$$\mu_t = \begin{bmatrix} \mu_t(\mathrm{DI}) \\ \mu_t(\mathrm{DS}) \\ \mu_t(\mathrm{UI}) \\ \mu_t(\mathrm{US}) \end{bmatrix}, \qquad B^0(\mu) = \begin{bmatrix} q^{\mathrm{D}}_{\mathrm{inf}} \mu(\mathrm{DS}) \\ -q^{\mathrm{D}}_{\mathrm{inf}} \mu(\mathrm{DS}) \\ q^{\mathrm{U}}_{\mathrm{inf}} \mu(\mathrm{US}) \\ -q^{\mathrm{U}}_{\mathrm{inf}} \mu(\mathrm{US}) \end{bmatrix},$$

and $B(t, \mu)$ is given by:

$$B(t, \mu) = \begin{bmatrix}
-q^{\mathrm{D}}_{\mathrm{rec}} \mu(\mathrm{DI}) + \beta_{\mathrm{DD}} \mu(\mathrm{DI}) \mu(\mathrm{DS}) + \beta_{\mathrm{UD}} \mu(\mathrm{UI}) \mu(\mathrm{DS}) \\
\qquad + \lambda[\phi(t, \mathrm{UI}) \mu(\mathrm{UI}) - \phi(t, \mathrm{DI}) \mu(\mathrm{DI})] \\
q^{\mathrm{D}}_{\mathrm{rec}} \mu(\mathrm{DI}) - \beta_{\mathrm{DD}} \mu(\mathrm{DI}) \mu(\mathrm{DS}) - \beta_{\mathrm{UD}} \mu(\mathrm{UI}) \mu(\mathrm{DS}) \\
\qquad + \lambda[\phi(t, \mathrm{US}) \mu(\mathrm{US}) - \phi(t, \mathrm{DS}) \mu(\mathrm{DS})] \\
-q^{\mathrm{U}}_{\mathrm{rec}} \mu(\mathrm{UI}) + \beta_{\mathrm{DU}} \mu(\mathrm{DI}) \mu(\mathrm{US}) + \beta_{\mathrm{UU}} \mu(\mathrm{UI}) \mu(\mathrm{US}) \\
\qquad + \lambda[\phi(t, \mathrm{DI}) \mu(\mathrm{DI}) - \phi(t, \mathrm{UI}) \mu(\mathrm{UI})] \\
q^{\mathrm{U}}_{\mathrm{rec}} \mu(\mathrm{UI}) - \beta_{\mathrm{DU}} \mu(\mathrm{DI}) \mu(\mathrm{US}) - \beta_{\mathrm{UU}} \mu(\mathrm{UI}) \mu(\mathrm{US}) \\
\qquad + \lambda[\phi(t, \mathrm{DS}) \mu(\mathrm{DS}) - \phi(t, \mathrm{US}) \mu(\mathrm{US})]
\end{bmatrix},$$

and for consistency, we label the components of these vectors by DI, DS, UI, and US respectively. The dynamic equation (7.76) is linear in α_t^0 and $B(\mu)$ is Lipschitz in μ as the coordinates of μ live in $[0, 1]$. Hence, existence of solutions is pretty straightforward.

This being said, the optimization problem of the attacker appears as a standard (deterministic) optimal control problem. Its value function can be obtained by solving the HJB equation:

$$\partial_t V(t,\mu) + \mathcal{H}^*\Big(t,\mu,\frac{\delta}{\delta\mu}V(t,\mu)\Big) = 0, \tag{7.77}$$

with terminal condition $V(T,\mu) = 0$, where the function $\mathcal{H}^*$ is defined as:

$$\mathcal{H}^*(t,\mu,h) = \begin{cases} B(t,\mu)\cdot h - \frac{1}{2k_H}[B^0(\mu)\cdot h]^2 - f_0(\mu) & \text{if } B^0(\mu)\cdot h \le 0 \\ B(t,\mu)\cdot h - f_0(\mu) & \text{if } B^0(\mu)\cdot h > 0. \end{cases} \tag{7.78}$$

We need to consider two cases to define the Hamiltonian appearing in the HJB equation (7.77) because of the constraint on the control of the major player: the control α^0 needs to be nonnegative (i.e., $A_0 = [0,\infty)$) because of its interpretation as an intensity. Under these conditions, the value of the optimal control $\hat{\boldsymbol{\alpha}}^0$ is given by the feedback function $\hat{\phi}^0(t,\mu)$ defined by:

$$\hat{\phi}^0(t,\mu) = \begin{cases} -\frac{1}{k_H}B^0(\mu)\cdot\frac{\delta}{\delta\mu}V(t,\mu) & \text{if } B^0(\mu)\cdot\frac{\delta}{\delta\mu}V(t,\mu) \le 0 \\ 0 & \text{if } B^0(\mu)\cdot\frac{\delta}{\delta\mu}V(t,\mu) > 0. \end{cases} \tag{7.79}$$

As usual, $[\delta/\delta\mu]V$ is regarded as the gradient of V with respect to the four inputs $\mu(\{\mathrm{DI}\})$, $\mu(\{\mathrm{DS}\})$, $\mu(\{\mathrm{UI}\})$, $\mu(\{\mathrm{US}\})$, see Proposition (Vol I)-5.66.

Assuming that this deterministic control problem is well posed, and restoring the dependence of the optimization on the feedback function ϕ used by the minor players, we denote by $\phi^{0,*}(\phi)$ the optimal feedback function $\hat{\phi}^0(t,\mu) = \frac{1}{k_H}[B^0(\mu)\cdot\partial_\mu V(t,\mu)]_-$ where as usual, we denote by x_- the negative part of the real number x.

Optimization Problem of the Minor Players. We now turn to the optimization problem of the individual computer owners. We single out a computer owner (minor player), and according to the strategy introduced earlier, we assume that the major player (hacker) and the other minor players (all the other computer owners) chose their respective strategies, and we search for the best response of the singled out minor player. So we assume that the hacker chose an attack intensity $\boldsymbol{\alpha}^0 = (\alpha_t^0)_{0\le t\le T}$ given by a deterministic function $\alpha_t^0 = \phi^0(t,\mu_t)$, that all the other minor players are using the same distributed feedback function ϕ, and we look for the feedback function $\check{\phi}$ which produces the best response of the representative computer owner. In other words, we look for $\check{\phi}$ to minimize the expected cost:

$$J^{\phi^0,\phi}(\check{\phi}) = \mathbb{E}\Big[\int_0^T (k_{\mathrm{D}}\mathbf{1}_{\mathrm{D}} + k_{\mathrm{I}}\mathbf{1}_{\mathrm{I}})(\check{X}_t)dt\Big], \tag{7.80}$$

under the dynamic constraint that the state $\check{X} = (\check{X}_t)_{0\le t\le T}$ evolves as a Markov process having the Q-matrix:

$$\check{q}_t(x,x') = \lambda_t\big(x,x',\mu_t,\phi^0(t,\mu_t),\check{\phi}(t,x)\big), \qquad 0 \le t \le T, \quad x,x' \in E, \tag{7.81}$$

where μ_t is the distribution of the (other) minor players when they all use the feedback function ϕ and the major player uses the strategy given by the feedback function ϕ^0. Recall that the state of the hacker is not involved in the model. Put in a different way, once ϕ^0 and ϕ are chosen, the flow $\boldsymbol{\mu} = (\mu_t)_{0\le t\le T}$ of probability measures on E is nothing but the flow of marginal distributions of the Markov process $\boldsymbol{X} = (X_t)_{0\le t\le T}$ having Q-matrix:

$$q_t(x,x') = \lambda_t\big(x,x',\mu_t,\phi^0(t,\mu_t),\phi(t,x)\big), \qquad 0 \le t \le T, \quad x,x' \in E.$$

Since the measure μ_t appears in the Q-matrix, this flow of measures solves a nonlinear Kolmogorov equation of the McKean-Vlasov type. This equation is exactly equation (7.76) with α_t^0 replaced by $\phi^0(t,\mu_t)$.

So given the choices of ϕ^0 and ϕ, the minor player looks for the probability flow $\boldsymbol{\mu} = (\mu_t)_{0\le t\le T}$ solving the nonlinear Kolmogorov equation (7.76), and uses this flow to compute and minimize the expected cost (7.80) over the controls $\check{\phi}$ under the dynamic constraint that the controlled process $\check{X} = (\check{X}_t)_{0\le t\le T}$ should have the Q-matrix (7.81). We denote the optimal control by $\hat{\phi} = \boldsymbol{\phi}^*(\phi^0,\phi)$.

Nash Equilibrium as a Fixed Point of the Best Response Map. According to what we learned in this chapter, we have a Nash equilibrium for the system when we can solve the two optimization problems described above in such a way that the solutions give a fixed point of the best response map. In other words, a couple (ϕ^0,ϕ) representing the attack intensity of the hacker and the defense strategy of a typical computer owner form a Nash equilibrium for the system if:

$$\phi^0 = \boldsymbol{\phi}^{0,*}(\phi), \qquad \text{and} \qquad \phi = \boldsymbol{\phi}^*(\phi^0,\phi).$$

Clearly, this fixed point can be rewritten in terms of ϕ only:

$$\phi = \boldsymbol{\phi}^*\big(\boldsymbol{\phi}^{0,*}(\phi),\phi\big). \tag{7.82}$$

This last formula can be used as the basis for an iterative procedure to find a Nash equilibrium:

1. start with a feedback function ϕ;
2. compute $\hat{\phi}^0 = \boldsymbol{\phi}^{0,*}(\phi)$ by solving the major player optimization problem as described above, namely by solving the HJB equation (7.77) and setting $\hat{\phi}^0(t,\mu) = \frac{1}{k_H}[B^0(\mu)\cdot \frac{\delta}{\delta\mu}V(t,\mu)]_-$;

3. still with the starting ϕ and the ϕ^0 given by $\hat{\phi}^0$ just found, compute $\hat{\phi} = \boldsymbol{\phi}^*(\boldsymbol{\phi}^{0,*}(\phi), \phi)$ by solving the minor player optimization problem as described above;
4. if $\hat{\phi}$ coincides with the feedback function ϕ we started from, i.e., if $\hat{\phi} = \phi$, we are done, otherwise we substitute $\hat{\phi}$ for ϕ and we iterate.

This iteration may not converge, even if a fixed point (and hence a Nash equilibrium) exists. We just mention it as a natural first attempt to find numerically Nash equilibria for the system.

In full analogy with the numerical implementations reported in Subsection (Vol I)-7.2.3, we implemented numerically the search strategy described in the above bullet points. For the purpose of illustration, we used the following parameters: $\beta\{\mathrm{UU}\} = \beta\{\mathrm{DU}\} = 0.3$, $\beta\{\mathrm{UD}\} = \beta\{\mathrm{DD}\} = 0.4$, for the rates of contamination; $q^{\mathrm{D}}_{\mathrm{rec}} = q^{\mathrm{U}}_{\mathrm{rec}} = 0.4$, for the rates of recovery; $q^{\mathrm{D}}_{\mathrm{inf}} = 0.3$ and $q^{\mathrm{U}}_{\mathrm{inf}} = 0.4$ for the rates of infection; $T = 10$ for the time horizon; $k_{\mathrm{I}} = 0.5$ for the cost of being infected, $k_{\mathrm{D}} = 0.3$ for the cost of being defended and $\lambda = 0.2$ for the speed of response. As for the major player, we first chose $k_{\mathrm{H}} = 1$ and $f_0(\mu) \equiv 0$. In this case, the major player does not get rewarded for his attacks, while still paying a quadratic penalty for the intensity of its attacks. As a result, its optimal strategy is to choose $\hat{\alpha}^0 \equiv 0$. The left plot on the top row of Figure 7.4 shows the time evolution of the distribution $(\mu_t)_{0 \le t \le T}$ of the states of the minor players (target computers) as computed from the implementation of the above numerical scheme. For comparison purposes, the right plot on the top row shows the result of the computation of the distribution $(\mu_t)_{0 \le t \le T}$ of the states of the minor players as computed from the algorithm described in Chapter (Vol I)-7 in the absence of a major player by setting the parameter v_{H} therein to 0 (or equivalently by setting $q^{\mathrm{U}}_{\mathrm{inf}}$ and $q^{\mathrm{D}}_{\mathrm{inf}}$ to 0) and solving the master equation. Despite the oscillations visible in the left plot, the results are consistent, demonstrating the potential of the above iteration scheme. However, numerical results are not always as nice. For the purpose of illustration, we added in the bottom pane, a plot of the time evolution of the distribution $(\mu_t)_{0 \le t \le T}$ as computed from the implementation of the above numerical scheme with $k_H = 0.02$ and $f_0(\mu) = k_0(\mu(\{UI\}) + \mu(\{DI\}))$ with $k_0 = 0.05$ rewarding the attacker when the proportion of infected computers is high. Clearly the resulting dynamic equilibrium picked up by the numerical computations is different. The proportion of unprotected computers susceptible to infection does not grow as much, while at the same time, the proportion of infected computers which were unprotected is much higher. Both facts should be expected given the fact that the attacker is now rewarded for his attacks.

Despite the annoying oscillations due to insufficient resolution in the computations, one can see a significant change in the relative proportion of defended computers (in anticipation of the aggressive behavior of the attacker), even if the proportion of infected while protected computers decreases toward the end of the period. The same anticipation of the aggressive behavior of the attacker can also explain the smaller proportion of unprotected computers susceptible to be victims of the attacks.

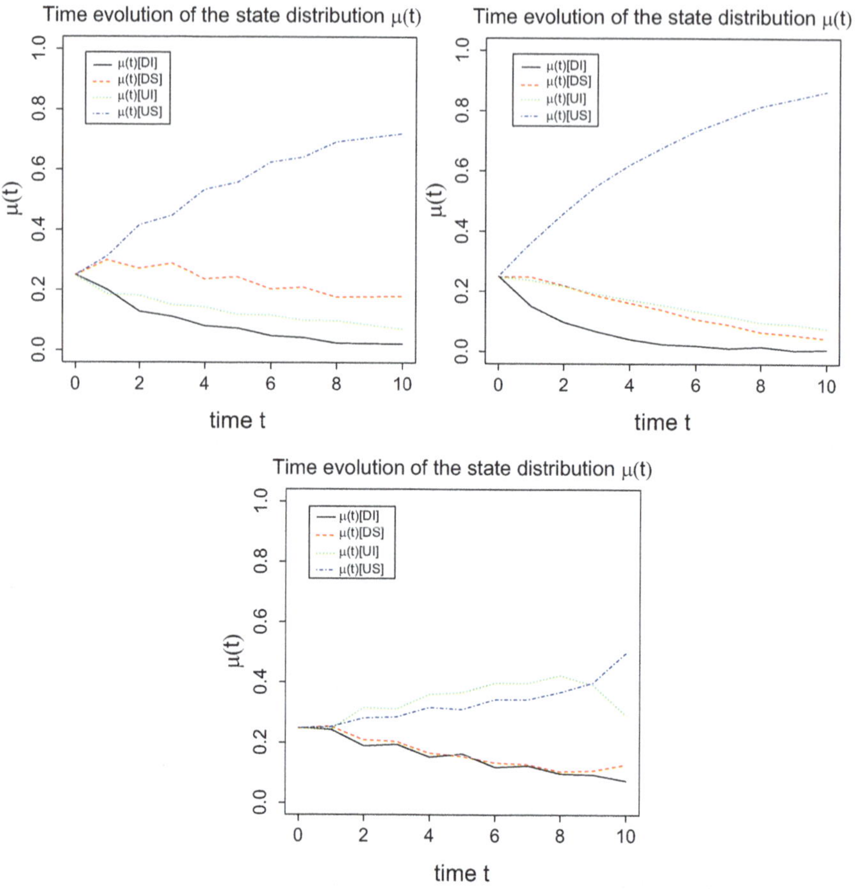

Fig. 7.4 Time evolution in equilibrium, of the distribution μ_t of the states of the computers in the network for the initial condition μ_0: $\mu_0 = (0.25, 0.25, 0.25, 0.25)$ when the major player is not rewarded for its attacks, i.e., when $f_0(\mu) \equiv 0$ (left pane on the top row), in the absence of major player and $v = 0$ (right pane on the top row), and with $f_0(\mu) = k_0(\mu(\{UI\}) + \mu(\{DI\}))$ with $k_0 = 0.05$ (bottom pane).

7.2 Mean Field Games of Timing

This section builds on the discussion of economic models of bank runs introduced in Section (Vol I)-1.2 of Chapter (Vol I)-1, and especially the diffusion version presented there. As a follow up, we introduce a more general continuous time set-up for mean field games of timing, and we propose solutions in the spirit of the games models studied in the book. We do not aim at the greatest generality. The subject is relatively young (and presumably immature) and as a result, we err on the side of a proof of concept rather than to attempt an exposé of a general theory.

7.2.1 Revisiting the Bank Run Diffusion Model of Chapter 1 (First Volume)

We use the diffusion model of bank run introduced in Chapter (Vol I)-1 as a motivation for the abstract set-up we propose for mean field games of timing. In that particular model, the states (observations / private signals) $(X_t^i)_{i=1,\cdots,N}$ at time t of the N players are assumed to satisfy:

$$dX_t^i = dY_t + \sigma dW_t^i, \quad 0 \le t \le T, \tag{7.83}$$

for a fixed time horizon T and for a common unobserved signal $\boldsymbol{Y} = (Y_t)_{0\le t\le T}$ given by an Itô process:

$$dY_t = b_t dt + \sigma_t dW_t^0, \quad 0 \le t \le T, \tag{7.84}$$

where $(\boldsymbol{W}^i = (W_t^i)_{0\le t\le T})_{i=0,\cdots,N}$ are $N+1$ independent Wiener processes with values in $\mathbb{R}$ and $(b_t)_{0\le t\le T}$ and $(\sigma_t)_{0\le t\le T}$ are square integrable processes that are progressively measurable with respect to the filtration generated by $\boldsymbol{W}^0$. All these processes are constructed on a complete probability space $(\Omega, \mathcal{F}, \mathbb{P})$. Obviously, and as we did throughout the book, we use the notation $\boldsymbol{W}^0$ for the common source of noise, and the notation $\boldsymbol{W}^i$ for the idiosyncratic source of noise to player i.

For each $i \in \{1, \cdots, N\}$, player i chooses a time $\tau^i \in [0, T]$ on the basis of its available information at that time, trying to maximize a quantity of the form:

$$J^i(\tau^1, \cdots, \tau^N) = \mathbb{E}[F(\boldsymbol{W}^0, \boldsymbol{W}^i, \bar{\mu}_\tau^N, \tau^i)], \tag{7.85}$$

where as usual, we denote by:

$$\bar{\mu}_\tau^N = \frac{1}{N}\sum_{i=1}^N \delta_{\tau^i} \tag{7.86}$$

the empirical distribution of $\tau = (\tau^1, \cdots, \tau^N)$. In the model introduced in Section (Vol I)-1.2, the functional $F(\boldsymbol{W}^0, \boldsymbol{W}, \mu, t)$ was given by:

$$F(\boldsymbol{W}^0, \boldsymbol{W}, \mu, t) = \exp\bigl((\bar{r} - r)t\bigr)\Bigl[1 \wedge \Bigl(L(Y_t) - \mu\bigl([0, t)\bigr)\Bigr)^+\Bigr], \tag{7.87}$$

as long as $L(Y_s) - \mu([0, s))$ is positive for all $s < t$, and by 0 otherwise, for any $t \in [0, T]$ and any probability distribution $\mu \in \mathcal{P}([0, T])$, where $\bar{r}$ and r are two rates such that $\bar{r} > r$ and $L : y \mapsto L(y)$ is some deterministic function.

Each τ^i should be a stopping time with respect to the right-continuous filtration $\mathbb{F}^{X^i}$ generated by the process $\boldsymbol{X}^i$ and augmented with $\mathbb{P}$-null sets of $\mathcal{F}$. Indeed, this filtration represents the information available to player i, and the choice of the stopping time τ^i is the actual decision made by that player.

General Formulation

We generalize the set-up of Chapter (Vol I)-1 by assuming that the reward function F may take a more general form than (7.87) and that, in lieu of (7.83)–(7.84), the private states $(\boldsymbol{X}^i = (X^i_t)_{0\le t\le T})_{i=1,\cdots,N}$ of the players are given by general measurable functions of the sample paths of the common noise $\boldsymbol{W}^0$ and of the idiosyncratic ones $(\boldsymbol{W}^i)_{i=1,\cdots,N}$. In order to do so, we shall regard the function F as a general measurable map on the canonical space $\mathcal{C}([0,T])\times\mathcal{C}([0,T])\times\mathcal{P}([0,T])\times[0,T]$, where we use the short notation $\mathcal{C}([0,T])$ for $\mathcal{C}([0,T];\mathbb{R})$. The canonical process on the path space $\mathcal{C}([0,T])\times\mathcal{C}([0,T])$ will be denoted by $(\boldsymbol{w}^0,\boldsymbol{w})$, while the canonical random variable on $\mathcal{P}([0,T])$ will be denoted by ν. We recall from Proposition (Vol I)-5.7 that the Borel σ-field on $\mathcal{P}([0,T])$ coincides with the σ-field generated by $(\nu([0,t]))_{0\le t\le T}$.

Given these notations, our basic assumption reads as follows.

Assumption (MFG of Timing Set-Up)

(A1) For $i\in\{1,\cdots,N\}$ and $t\in[0,T]$, the private state X^i_t of player i at time t is given in the form $X^i_t=[X(\boldsymbol{W}^0,\boldsymbol{W}^i)]_t$ for a single measurable function:

$$X:\mathcal{C}([0,T])\times\mathcal{C}([0,T])\to\mathcal{C}([0,T];\mathbb{R}^d),$$

which is adapted, in the sense that for each fixed $t\in[0,T]$, $[X(\boldsymbol{w}^0,\boldsymbol{w})]_t$ is measurable with respect to the product σ-field $\sigma\{\boldsymbol{w}^0_{\cdot\wedge t}\}\otimes\sigma\{\boldsymbol{w}_{\cdot\wedge t}\}$ generated by the stopped processes $\boldsymbol{w}^0_{\cdot\wedge t}$ and $\boldsymbol{w}_{\cdot\wedge t}$.

(A2) The reward function F is a real valued bounded measurable function on the product space $\mathcal{C}([0,T])\times\mathcal{C}([0,T])\times\mathcal{P}([0,T])\times[0,T]$ which is progressively measurable in the sense that for each fixed $t\in[0,T]$, the restriction of F to $\mathcal{C}([0,T])\times\mathcal{C}([0,T])\times\mathcal{P}([0,T])\times[0,t]$ is measurable with respect to the σ-field $\sigma\{\boldsymbol{w}^0_{\cdot\wedge t}\}\otimes\sigma\{\boldsymbol{w}_{\cdot\wedge t}\}\otimes\sigma\{\nu([0,s]);\,0\le s\le t\}\otimes\mathcal{B}([0,t])$.

In the bank run model, the quantity $F(\boldsymbol{w}^0,\boldsymbol{w},\nu,t)$ represents the reward to a player for exercising its timing option at time t in a scenario where the sample trajectories of the common and idiosyncratic noises are given by the realizations of $\boldsymbol{w}^0$ and $\boldsymbol{w}$, and the distribution of the times of withdrawal in the population is given by the realization of ν.

It is important to keep in mind that $\mathbb{F}^{X^i}$ may be strictly smaller than the complete filtration $\mathbb{F}^{(W^0,W^i)}$ generated by $(\boldsymbol{W}^0,\boldsymbol{W}^i)$, which is automatically right-continuous. Denoting by $\mathcal{S}_{X^i}$ the set of $[0,T]$-valued stopping times for the filtration $\mathbb{F}^{X^i}$, we shall write $\tau^i\in\mathcal{S}_{X^i}$. More generally, for a real-valued process $\boldsymbol{X}$, we shall denote by $\mathcal{S}_X$ the set of $[0,T]$-valued stopping times for the complete and right-continuous filtration $\mathbb{F}^X$ generated by the process $\boldsymbol{X}$.

Our goal is to identify the right notion of Nash equilibrium for these models, find reasonable sufficient conditions for their existence, and study their structure when they do exist. We start with the following definition:

Definition 7.17 *If $\epsilon > 0$, a tuple of stopping times $(\tau^{1,*}, \cdots, \tau^{N,*})$, with $\tau^{i,*} \in \mathcal{S}_{X^i}$ for each $i \in \{1, \cdots, N\}$, is said to be an ϵ-Nash equilibrium if, for every $i \in \{1, \cdots, N\}$ and $\tau \in \mathcal{S}_{X^i}$, we have:*

$$\mathbb{E}[F(\boldsymbol{W}^0, \boldsymbol{W}^i, \bar{\mu}^{N,-i}, \tau^{i,*})] \geq \mathbb{E}[F(\boldsymbol{W}^0, \boldsymbol{W}^i, \bar{\mu}^{N,-i}, \tau)] - \epsilon,$$

$\bar{\mu}^{N,-i}$ denoting the empirical distribution of $(\tau^{1,*}, \cdots, \tau^{i-1,*}, \tau^{i+1,*}, \cdots, \tau^{N,*})$.

This definition is reminiscent of the definition of an approximate Nash equilibrium over open loop strategies which we used for stochastic differential games. Recall for example Definition 6.6. However, there are major differences between these definitions. Indeed, not only the players cannot observe the states of the other players, which is a typical feature of open loop equilibria, but also they cannot even observe their private noises, since the stopping time chosen by player i, for each $i \in \{1, \cdots, N\}$, is required to be in $\mathcal{S}_{X^i}$. This makes a subtle, though significant, difference in the analysis. Anyhow, since the players interact through the empirical distributions of the stopping times they choose, it is reasonable to expect that an asymptotic description of the game will take the form of a mean field game, which we shall call a mean field game of timing. In full analogy with the results of Section 6.2, we shall prove that any weak limit point as $N \to \infty$ of $(\epsilon_N)_{N \geq 1}$-Nash equilibria, for any sequence $(\epsilon_N)_{N \geq 1}$ converging to 0, is a solution to the asymptotic mean field game, provided this notion of solution is defined in a weak sense which we spell out in Definition 7.37 below.

7.2.2 Formulation of the MFG of Timing Problem

In this subsection, we only deal with two independent Wiener processes, $\boldsymbol{W}^0$ and $\boldsymbol{W}$, constructed on the same probability space $(\Omega, \mathcal{F}, \mathbb{P})$ as before, and we consider the corresponding state process $\boldsymbol{X} = ([X(\boldsymbol{W}^0, \boldsymbol{W})]_t)_{0 \leq t \leq T}$ given by the mapping X introduced in the statement of assumption **MFG of Timing Set-Up**. As in the case of stochastic differential games, the presence of the common noise $\boldsymbol{W}^0$ will force the introduction of strong and weak notions of equilibrium. We postpone the technicalities of the analysis of weak solutions to the next subsection and we first concentrate on the notion of strong solution. For each random measure μ on $[0, T]$ and each $[0, T]$-valued random variable τ, we denote by:

$$J(\mu, \tau) = \mathbb{E}[F(\boldsymbol{W}^0, \boldsymbol{W}, \mu, \tau)] \tag{7.88}$$

the corresponding expected reward.

In full analogy with the definition of a strong solution to a mean field stochastic differential game given in Subsection 2.2.1, we introduce the following definition of a strong equilibrium for a mean field game of timing:

Definition 7.18 *A stopping time $\tau^* \in \mathcal{S}_X$ is said to be a strong MFG equilibrium if, for every $\tau \in \mathcal{S}_X$, we have:*

$$J(\mu, \tau^*) \geq J(\mu, \tau),$$

with $\mu = \mathcal{L}(\tau^* | \boldsymbol{W}^0)$.

It is worth mentioning that, in contrast with mean field stochastic differential games, the state process $\boldsymbol{X}$ is here insensitive to the mean field interaction.

Coming back to the motivating example of the bank run model for inspiration, we realize that the problem of the depositor is to choose a stopping time in $\mathcal{S}_X$ which is as large as possible to maximize the interests earned by the deposit, but not too large to miss its chance to get back as much of its investment as possible. Clearly, its reward depends upon the sample distribution of the times at which the other depositors withdraw their deposits. So in the mean field game formulation, a typical depositor will choose a withdrawal time to maximize his/her response to the distribution of the times of withdrawal of the other depositors (this is a rather standard optimal stopping problem), and in equilibrium, one expects that the distribution of the optimal time of withdrawal chosen by the specific depositor coincides with the statistical distribution of the times of withdrawals of all the depositors (this is the usual fixed point step). This formulation of the fixed point step would be appropriate in the absence of the common noise $\boldsymbol{W}^0$, namely if the value of the assets of the bank was deterministic. In the presence of $\boldsymbol{W}^0$, the fixed point step needs to be taken with *conditional distributions*, and this makes the problem much more difficult as we saw in the chapters devoted to mean field games with a common noise.

In preparation for the solution of the problem in the presence of $\boldsymbol{W}^0$, we emphasize the relevance of the notion of random environment introduced in Section 1.1.1 of Chapter 1 in the present context of games of timing. As in Chapter 1, we may assume that the probability space $(\Omega, \mathcal{F}, \mathbb{P})$ is the completion of the product $(\Omega^0 \times \Omega^1, \mathcal{F}^0 \otimes \mathcal{F}^1, \mathbb{P}^0 \otimes \mathbb{P}^1)$ of two complete probability spaces $(\Omega^0, \mathcal{F}^0, \mathbb{P}^0)$ and $(\Omega^1, \mathcal{F}^1, \mathbb{P}^1)$, and that $\boldsymbol{W}^0$ and $\boldsymbol{W}$ are independent Wiener processes originally defined on Ω^0 and Ω^1 respectively, and naturally extended to Ω. Of course, we could assume that $\boldsymbol{W}^0$ and $\boldsymbol{W}$ are more general processes, but we shall not consider such a higher level of generality.

We then say that μ is a random environment if it is a random measure on $[0, T]$ defined on the probability space $(\Omega^0, \mathcal{F}^0, \mathbb{P}^0)$. In other words, it is an $\mathcal{F}^0$-measurable function $\Omega^0 \ni \omega^0 \mapsto \mu(\omega^0, \cdot) \in \mathcal{P}([0, T])$ which can be used as a proxy for the conditional distribution of the time at which the generic player decides to stop the game if the trajectory of the common noise is $\boldsymbol{W}^0(\omega^0)$. In this context, we can define an MFG of timing problem as the set of the following two steps:

(i) For each random environment μ, solve:

$$\hat{\theta} \in \arg\sup_{\theta \in \mathcal{S}_X} J(\mu, \theta).$$

(ii) Find μ so that, for $\mathbb{P}^0$-almost every $\omega^0 \in \Omega^0$,

$$\forall t \in [0, T], \ \mu(\omega^0, [0, t]) = \mathbb{P}[\hat{\theta} \leq t \,|\, \boldsymbol{W}^0](\omega^0).$$

The above formulation is pretty similar to that used for mean field stochastic differential games with a common noise, see for instance (2.18)–(2.19)–(2.20). Step (i) is to find the best response under the random environment μ, while step (ii) is the fixed point step. Also notice that the fixed point condition in this latter step (ii) forces the solution of the mean field game to be adapted with respect to the common noise $\boldsymbol{W}^0$, which is precisely what we called a strong solution in Chapter 2. Alternatively, we could reformulate step (ii) by choosing Ω^0 as the canonical space $\mathcal{C}([0, T])$ and $\boldsymbol{W}^0$ as the canonical process $\boldsymbol{w}^0$ and then, by requiring that, with $\mathbb{P}^0$-probability 1,

$$\forall t \in [0, T], \quad \mu([0, t]) = \mathbb{P}^1[\hat{\theta} \leq t].$$

Actually, step (ii) says more regarding the measurability properties of μ. Indeed, since the event $\{\hat{\theta} \leq t\}$ belongs to $\mathcal{F}_t^X \subset \mathcal{F}_t^{(\boldsymbol{W}^0, \boldsymbol{W})}$, $\mu([0, t])$ must coincide with $\mathbb{P}[\hat{\theta} \leq t \,|\, \mathcal{F}_t^{\boldsymbol{W}^0}]$ if μ is a solution of the mean field game. This proves that, in order for μ to be a solution, it must be adapted in the sense that for each $t \in [0, T]$, $\mu([0, t])$ must be $\mathcal{F}_t^{\boldsymbol{W}^0}$-measurable. Equivalently, the process $(\mu([0, t]))_{0 \leq t \leq T}$, which is càd-làg, must be $\mathbb{F}^{\boldsymbol{W}^0}$-progressively measurable.

Order Structures and Supermodularity

In this section, we study strong equilibria by taking advantage of the order structure of the set of controls, the set of stopping times in the present situation. This is in sharp contrast with most of the analyses performed in the book, which are most often based on fixed point theorems for continuous maps on topological spaces. The efficiency of these order-based arguments was demonstrated in the original works on unimodular games, or games with strategic complementarities, in the static case with no timing decision. We refer the reader to the Notes & Complements of Chapter (Vol I)-1 for references.

Obviously, we use the standard order structure on $[0, T]$. On $\mathcal{P}([0, T])$ however, we use the *dominance order*, also called *stochastic order*, according to which $\mu \preceq \mu'$ if $\mu([0, s]) \geq \mu'([0, s])$ for all $s \in [0, T]$. With this notion in hand, we articulate the main assumptions under which we shall prove existence of strong equilibria:

Assumption (MFG of Timing Regularity).

(A1) For each fixed $(w^0, w) \in \mathcal{C}([0,T]) \times \mathcal{C}([0,T])$, the function $(\mu, t) \mapsto F(w^0, w, \mu, t)$ is continuous.

(A2) For each fixed $(w^0, w, \mu) \in \mathcal{C}([0,T]) \times \mathcal{C}([0,T]) \times \mathcal{P}([0,T])$, the function $t \mapsto F(w^0, w, \mu, t)$ is upper semicontinuous.

(A3) For each pair of measurable and $\mathcal{W}_1$-almost surely adapted functions $\mu, \mu' : \mathcal{C}([0,T]) \to \mathcal{P}([0,T])$ satisfying $\mathcal{W}_1[\boldsymbol{w}^0 : \mu(\boldsymbol{w}^0) \preceq \mu'(\boldsymbol{w}^0)] = 1$, the process $(M_t)_{0 \le t \le T}$ defined by:

$$M_t = F\big(\boldsymbol{w}^0, \boldsymbol{w}, \mu'(\boldsymbol{w}^0), t\big) - F\big(\boldsymbol{w}^0, \boldsymbol{w}, \mu(\boldsymbol{w}^0), t\big)$$

is a right-continuous sub-martingale on $\mathcal{C}([0,T]) \times \mathcal{C}([0,T])$ equipped with $\mathcal{W}_1 \otimes \mathcal{W}_1$, where $\mathcal{W}_1$ is the one-dimensional Wiener measure on $\mathcal{C}([0,T])$.

Obviously, assumption (**A1**) implies (**A2**), so that, when using assumption **MFG of Timing Regularity**, we shall specify if we mean it with either (**A1**) or (**A2**).

In (**A3**), we say that μ is $\mathcal{W}_1$-almost surely adapted if, for all $t \in [0,T]$, $\mu([0,t])$ is measurable with respect to the completion of the σ-field $\sigma\{\boldsymbol{w}^0_{\cdot \wedge t}\}$, where $\boldsymbol{w}^0$ is the canonical process on $\mathcal{C}([0,T])$. Condition (**A3**) is obviously satisfied when the function F has itself increasing differences in the sense that, for each fixed $(w^0, w) \in \mathcal{C}([0,T])^2$, the inequalities $t \le t'$ and $\mu \preceq \mu'$, with $t, t' \in [0,T]$ and $\mu, \mu' \in \mathcal{P}([0,T])$, imply:

$$F(w^0, w, \mu', t') - F(w^0, w, \mu', t) \ge F(w^0, w, \mu, t') - F(w^0, w, \mu, t).$$

Condition (**A3**) is rather natural in bank run models. It means that for larger μ, F increases more in time than for smaller μ, which may be justified as follows. If $\mu \preceq \mu'$, at any given time, more people have already withdrawn under μ, so the reward for an agent waiting from t to $t' > t$ should not exceed the reward under μ'. Put in other words, as people run to the bank earlier under μ, the cost of waiting from t to $t' > t$ should be greater under μ since $\mu \preceq \mu'$. An interesting consequence of condition (**A3**) as stated is the fact that if it holds, the expected reward J, recall Definition (7.88), has also increasing differences in the sense that if $\mu \preceq \mu'$ almost surely in the sense of stochastic order (i.e., if $\mu'([0,t]) \le \mu([0,t])$ a.s. for each $t \in [0,T]$), and if $\tau \le \tau'$ are stopping times, taking expectations in the submartingale property of $(M_t)_{0 \le t \le T}$ posited in condition (**A3**) yields:

$$J(\mu', \tau') - J(\mu', \tau) \ge J(\mu, \tau') - J(\mu, \tau). \tag{7.89}$$

In this form, condition (**A3**) says that for *larger* μ, the expected reward increases more rapidly in time than it does for *smaller* μ. These hypotheses introduce strategic complementarities in the game and recast the game of timing model as a supermodular game.

It is worth mentioning that unfortunately, condition (**A3**) does not hold for the specific form of reward functional (7.87), even if we close the interval in the quantity $\mu([0,t])$ to make the whole right-continuous. Indeed, for given $s,t \in [0,T]$, with $s < t$, $a \in (0,1)$ and $\varepsilon > 0$ such that $a - \varepsilon \geq 0$, it is pretty straightforward to construct two probability measures μ and μ' in $\mathcal{P}([0,T])$ such that $\mu \preceq \mu'$, $\mu([0,s)) = \mu([0,t)) = \mu'([0,t)) = a$ and $\mu'([0,s)) = a - \varepsilon$. Therefore, under (**A3**), we would get by comparing the two values at time t and at time s:

$$0 \geq \mathbb{E}\Big[e^{(\bar{r}-r)(s\wedge\varrho)}\Big(1 \wedge \big[L(Y_{s\wedge\varrho}) + \varepsilon - a\big]^+\Big)\Big] - \mathbb{E}\Big[e^{(\bar{r}-r)(s\wedge\varrho)}\Big(1 \wedge \big[L(Y_{s\wedge\varrho}) - a\big]^+\Big)\Big],$$

where $\varrho = \inf\{r \geq 0 : L(Y_r) \leq a\}$. Obviously, the above is not possible in full generality.

However, we can provide a general class of reward functionals F for which (**A3**) in assumption **MFG of Timing Regularity** holds.

Proposition 7.19 *Assume that the reward functional F is of the form:*

$$F(w^0, w, \mu, t) = f\big(w_t^0, w_t, \rho * \mu(t), t\big), \quad t \in [0,T],$$

for $w^0, w \in \mathcal{C}([0,T])$ and $\mu \in \mathcal{P}([0,T])$, some real valued functions f and ρ on $\mathbb{R}\times\mathbb{R}\times\mathbb{R}\times[0,T]$ and $\mathbb{R}$ respectively, both functions being bounded and continuous, the support of ρ being included in $[0,+\infty)$, and with the notation:

$$\rho * \mu(t) = \int_{[0,T]} \rho(t-s)d\mu(s), \quad t \in [0,T].$$

*If the function $(x^0,x,y,t) \mapsto f(x^0,x,y,t)$ is twice differentiable in $(x^0,x) \in \mathbb{R}\times\mathbb{R}$, with bounded and continuous derivatives on the whole space, and is once continuously differentiable in y and t, with bounded and continuous derivatives on the whole space, and if ρ is nondecreasing, continuously differentiable and convex and f, $\partial_y f$ and $\frac{1}{2}(\partial^2_{x^0} + \partial^2_x)f + \partial_t f$ are nonincreasing in y, then the reward functional F satisfies (**A1**) and (**A3**) in assumption **MFG of timing regularity.***

Proof. Continuity of F in the last two arguments is easily checked since f itself is continuous. This proves (**A1**).

In order to prove (**A3**), we consider two measurable and almost surely adapted functions μ and μ', as in the statement of assumption **MFG of Timing Regularity**, with $\mu \preceq \mu'$. Recalling that $(\boldsymbol{w}^0, \boldsymbol{w}) = (w_t^0, w_t)_{0\leq t\leq T}$ denotes the canonical process on the space $\mathcal{C}([0,T])^2$ equipped with the Wiener measure $\mathcal{W}_1 \otimes \mathcal{W}_1$, Itô's formula yields:

$$
\begin{aligned}
& df\big(w_t^0, w_t, \rho * \mu(t), t\big) \\
&= \Big[\Big(\frac{1}{2}(\partial_{x^0}^2 + \partial_x^2)f + \partial_t f\Big)\big(w_t^0, w_t, \rho * \mu(t), t\big) + \partial_y f\big(w_t^0, w_t, \varrho * \mu(t), t\big)\varrho' * \mu(t)\Big]dt \\
&\quad + \partial_{x^0} f\big(w_t^0, w_t, \rho * \mu(t), t\big)dw_t^0 + \partial_x f\big(w_t^0, w_t, \rho * \mu(t), t\big)dw_t,
\end{aligned}
$$

where we used the fact that the process $(\rho * \mu(t))_{0 \le t \le T}$ is adapted to the completion of the filtration generated by $\boldsymbol{w}^0$ since ρ is supported by $[0, +\infty)$. In order to prove (**A3**), it then suffices to compare the dt-terms corresponding to μ and μ'. Their difference writes:

$$
\begin{aligned}
& \Big(\frac{1}{2}(\partial_{x^0}^2 + \partial_x^2)f + \partial_t f\Big)\big(w_t^0, w_t, \rho * \mu'(t), t\big) \\
&\qquad - \Big(\frac{1}{2}(\partial_{x^0}^2 + \partial_x^2)f + \partial_t f\Big)\big(w_t^0, w_t, \rho * \mu(t), t\big) \\
&\qquad + \partial_y f\big(w_t^0, w_t, \rho * \mu'(t), t\big)\rho' * \mu'(t) - \partial_y f\big(w_t^0, w_t, \rho * \mu(t), t\big)\rho' * \mu(t).
\end{aligned}
$$

Now note that, if $m \preceq m'$, with $m, m' \in \mathcal{P}([0, T])$, then $\int_0^T g dm \ge \int_0^T g dm'$ for every nonincreasing function g on $[0, T]$. In particular, since ρ and ρ' are nondecreasing, we have:

$$
\rho * \mu(t) \ge \rho * \mu'(t), \quad \rho' * \mu(t) \ge \rho' * \mu'(t),
$$

where we used the fact that $\rho(t - \cdot)$ and $\rho'(t - \cdot)$ are nonincreasing. We complete the proof by noticing that $\partial_y f$ is non-positive and ρ' is nonnegative. □

7.2.3 Existence of Strong Equilibria for MFGs of Timing

In this section, we prove existence of strong mean field game of timing equilibria. For the sake of completeness, we first recall several definitions and classical results from lattice theory which we use throughout. See the Notes & Complements at the end of the chapter for precise references to papers and textbooks where proofs and further material can be found.

Definition 7.20 *A partially ordered set $(\mathcal{S}, \le)$ is said to be a lattice if, for any $x, y \in \mathcal{S}$, the set $\{x, y\}$ has a least upper bound $x \vee y$ and a greatest lower bound $x \wedge y$:*

$$
\begin{aligned}
& x \vee y = \min\{z \in \mathcal{S} : \ z \ge x, z \ge y\} \in \mathcal{S}, \\
& \textit{and} \qquad\qquad\qquad\qquad\qquad\qquad\qquad\qquad (7.90) \\
& x \wedge y = \max\{z \in \mathcal{S} : \ z \le x, z \le y\} \in \mathcal{S}.
\end{aligned}
$$

A lattice $(\mathcal{S}, \le)$ is said to be complete if every subset $S \subset \mathcal{S}$ has a greatest lower bound $\inf S$ *and a least upper bound* $\sup S$, *with the convention that* $\inf \emptyset = \sup \mathcal{S}$ *and* $\sup \emptyset = \inf \mathcal{S}$.

Example 7.21. *Let $\mathbb{F} = (\mathcal{F}_t)_{t\geq 0}$ be a filtration on a probability space $(\Omega, \mathcal{F}, \mathbb{P})$ and let us denote by $\mathcal{S}$ the set of $\mathbb{F}$-stopping times. We equip $\mathcal{S}$ with the partial order $\tau \leq \theta$ if $\tau(\omega) \leq \theta(\omega)$ for $\mathbb{P}$ - almost every $\omega \in \Omega$. In this example, unless we make an explicit mention to the contrary, all the stopping times are with respect to the same filtration $\mathbb{F}$. The set $\mathcal{S}$ is a lattice because if $\tau, \theta \in \mathcal{S}$ then $\tau \vee \theta$ and $\tau \wedge \theta$ are also stopping times, i.e., elements of $\mathcal{S}$. If $(\tau_\iota)_{\iota\in\mathcal{I}} \subset \mathcal{S}$ is a (possibly uncountable) family of stopping times, we define:*

$$\tau^* = \text{ess}\sup_{\iota\in\mathcal{I}}\tau_\iota, \qquad \textit{and} \qquad \tau_* = \text{ess}\inf_{\iota\in\mathcal{I}}\tau_\iota,$$

and notice that τ^* appears as the $\mathbb{P}$-almost sure limit of an increasing sequence of stopping times (recall that we already know that $\mathcal{S}$ is a lattice). Consequently, τ^* is itself a stopping time for the filtration $\mathbb{F}$. The situation is not as clean for the infimum. Indeed, because of the lattice property, the essential infimum τ_* appears as the almost sure limit of a decreasing sequence of stopping times, but this only guarantees the fact that τ_* is a stopping time for the filtration $\mathbb{F}_+ = (\mathcal{F}_{t+})_{t\geq 0}$. The conclusion is that the partially ordered set $\mathcal{S}$ of $\mathbb{F}$-stopping times is a complete lattice whenever the filtration $\mathbb{F}$ is right continuous.

The above definitions and the preceding example have been chosen for the sole purpose of using Tarski's fixed point theorem, which we now state without proof.

Theorem 7.22 *If $\mathcal{S}$ is a complete lattice and $\Phi : \mathcal{S} \ni x \mapsto \Phi(x) \in \mathcal{S}$ is order preserving in the sense that $\Phi(x) \leq \Phi(y)$ whenever $x, y \in \mathcal{S}$ are such that $x \leq y$, then the set of fixed points of Φ is a nonempty complete lattice.*

We shall also use the form of Topkis monotonicity theorem stated in Theorem 7.24 below for a special class of supermodular functions whose definition we recall first.

Definition 7.23 *A real valued function f on a lattice $(\mathcal{S}, \leq)$ is said to be supermodular if, for all $x, y \in \mathcal{S}$,*

$$f(x \vee y) + f(x \wedge y) \geq f(x) + f(y). \tag{7.91}$$

Theorem 7.24 *Let $(\mathcal{S}_1, \leq_1)$ be a lattice and $(\mathcal{S}_2, \leq_2)$ be a partially ordered set. Suppose that $f : \mathcal{S}_1 \times \mathcal{S}_2 \ni (x, y) \mapsto f(x, y) \in \mathbb{R}$ is super-modular in x for any given $y \in \mathcal{S}_2$ and satisfies:*

$$\begin{aligned}
&\forall x, x' \in \mathcal{S}_1,\ y, y' \in \mathcal{S}_2, \\
&\qquad (x \leq_1 x', \quad y \leq_2 y') \Rightarrow f(x', y') - f(x, y') \geq f(x', y) - f(x, y),
\end{aligned}$$

in which case f is said to have increasing differences in x and y. Then, for $x, x' \in \mathcal{S}_1$ and $y, y' \in \mathcal{S}_2$ such that $y \leq_2 y'$, $x \in \arg\max(f(\cdot, y))$ and $x' \in \arg\max(f(\cdot, y'))$, *it holds that* $x \wedge x' \in \arg\max(f(\cdot, y))$ *and* $x \vee x' \in \arg\max(f(\cdot, y'))$. *In particular, for any* $y \in \mathcal{S}_2$, *the set of maximizers of* $f(\cdot, y)$ *is a sub-lattice, in the sense that it is closed under* $\wedge$ *and* $\vee$.

Finally, we recall a useful optimization result for order upper semi-continuous function over a complete lattice $\mathcal{S}$.

Definition 7.25 *Let* $(\mathcal{S}, \le)$ *be a complete lattice. A function* $f : \mathcal{S} \to \mathbb{R} \cup \{-\infty\}$ *is said to be order upper semi-continuous if, for any totally ordered subset* $\mathcal{T} \subset \mathcal{S}$,

$$\inf_{x \in \mathcal{T}} \sup_{y \in \mathcal{T}: y \ge x} f(y) \le f\bigl(\sup \mathcal{T}\bigr), \quad \textit{and} \quad \inf_{x \in \mathcal{T}} \sup_{y \in \mathcal{T}: y \le x} f(y) \le f\bigl(\inf \mathcal{T}\bigr).$$

Theorem 7.26 *If* $(\mathcal{S}, \le)$ *is a complete lattice and* $f : \mathcal{S} \to \mathbb{R} \cup \{-\infty\}$ *is supermodular and order upper semi-continuous, the set* $\operatorname{argmax}(f)$ *of maximizers of* f *is a (non-empty) complete sub-lattice of* $\mathcal{S}$. *In particular, the supremum and infimum of every subset of* $\arg\max(f)$ *are in* $\arg\max(f)$.

We can now state and prove the main result of this section.

Theorem 7.27 *Let* $(\Omega^0, \mathcal{F}^0, \mathbb{P}^0)$ *and* $(\Omega^1, \mathcal{F}^1, \mathbb{P}^1)$ *both denote the completion of the Wiener space* $\mathcal{C}([0, T])$ *equipped with the Wiener measure and call* $\boldsymbol{w}^0$ *and* $\boldsymbol{w}^1$ *the canonical processes on* Ω^0 *and on* Ω^1 *respectively. As usual, call* $(\Omega, \mathcal{F}, \mathbb{P})$ *the completion of the product of the two probability spaces and set* $\boldsymbol{X} = X(\boldsymbol{w}^0, \boldsymbol{w})$.

Under assumption **MFG of Timing Set-Up** *and under conditions* (**A2**) *and* (**A3**) *of* **MFG of Timing Regularity**, *there exists a strong equilibrium on* $(\Omega, \mathcal{F}, \mathbb{P})$. *Moreover, if the continuity condition* (**A1**) *of* **MFG of Timing Regularity** *is assumed instead of the semicontinuity condition* (**A2**), *then there exist strong equilibria* τ^* *and* θ^* *such that, for any other strong equilibrium* τ, *we have* $\theta_* \le \tau \le \tau^*$ *almost-surely.*

Proof. As above, the complete (and necessarily right-continuous) augmentation of the filtration generated by $\boldsymbol{w}^0$ on $(\Omega^0, \mathcal{F}^0, \mathbb{P}^0)$ is denoted by $\mathbb{F}^0$ and the complete and right-continuous augmentation of the filtration generated by $\boldsymbol{X}$ on $(\Omega, \mathcal{F}, \mathbb{P})$ is denoted by $\mathbb{F}^{\boldsymbol{X}}$.

Let us denote by $\mathcal{M}_T^0$ the space of admissible random environments, namely the set of random measures $\mu : \Omega^0 \ni w^0 \mapsto \mu(w^0) \in \mathcal{P}([0, T])$ which are adapted in the sense that, for each $t \in [0, T]$, the random variable $\mu([0, t])$ is $\mathcal{F}_t^0$-measurable. We then view the expected reward J defined in (7.88) as a real valued function on $\mathcal{M}_T^0 \times \mathcal{S}_X$. Notice that, for any $\mu \in \mathcal{M}_T^0$, $J(\mu, \cdot)$ is supermodular in the sense of Definition 7.23 when $\mathcal{S}_X$ is equipped with the order defined in Example 7.21. Indeed, taking expectations on both sides of the equality:

$$F(\boldsymbol{w}^0, \boldsymbol{w}, \mu, \tau \vee \tau') + F(\boldsymbol{w}^0, \boldsymbol{w}, \mu, \tau \wedge \tau') = F(\boldsymbol{w}^0, \boldsymbol{w}, \mu, \tau) + F(\boldsymbol{w}^0, \boldsymbol{w}, \mu, \tau'),$$

for any $\tau, \tau' \in \mathcal{S}_X$, we find that:

$$J(\mu, \tau \vee \tau') + J(\mu, \tau \wedge \tau') = J(\mu, \tau) + J(\mu, \tau').$$

Equipping $\mathcal{M}_T^0$ with the partial order $\mu \leq \mu'$ if $\mathbb{P}^0[w^0 \in \Omega^0; \mu(w^0) \preceq \mu'(w^0)] = 1$, where $\preceq$ is the stochastic order on $\mathcal{P}([0,T])$, we observe from (7.89) that J has increasing differences in μ and τ. Hence, Topkis Theorem 7.24 implies that the set-valued map:

$$\mathcal{M}_T^0 \ni \mu \mapsto \Phi(\mu) = \arg\max_{\tau \in \mathcal{S}_X} J(\mu, \tau)$$

is nondecreasing in the strong set order in the sense that, whenever $\mu, \mu' \in \mathcal{M}_T^0$ satisfy $\mu \leq \mu'$, and whenever $\tau \in \Phi(\mu)$ and $\tau' \in \Phi(\mu')$, we have $\tau \vee \tau' \in \Phi(\mu')$ and $\tau \wedge \tau' \in \Phi(\mu)$. Also, by Fatou's lemma and by Example 7.21, and since F is bounded and upper semicontinuous, J is order upper semicontinuous in τ, as defined in Definition 7.25. By Theorem 7.26, this implies that for every μ, $\Phi(\mu)$ is a nonempty complete sub-lattice of $\mathcal{S}_X$. Recall indeed that $\mathcal{S}_X$ is a complete lattice since we work with the right-continuous completion of the natural filtration generated by $\boldsymbol{X}$. In particular, $\Phi(\mu)$ has a maximum, which we denote by $\phi^*(\mu)$ and a minimum which we denote by $\phi_*(\mu)$. Note that $\phi^* : \mathcal{M}_T^0 \to \mathcal{S}_X$ is nondecreasing in the sense that $\mu \leq \mu'$ implies $\phi^*(\mu) \leq \phi^*(\mu')$. Moreover, it is plain to check that the function $\psi : \mathcal{S}_X \to \mathcal{M}_T^0$ defined by $\psi(\tau) = \mathcal{L}(\tau|\boldsymbol{w}^0) = \mathcal{L}^1(\tau)$ is monotone. Thus $\phi^* \circ \psi$ is a monotone map from $\mathcal{S}_X$ to itself, and since $\mathcal{S}_X$ is a complete lattice, we conclude from Tarski's fixed point Theorem 7.22 that there exists τ such that $\tau = \phi^*(\psi(\tau))$. It is readily verified that any such fixed point τ is a strong equilibrium for the mean field game of timing in the sense of Definition 7.18.

We now assume that F is not only upper semicontinuous, but also lower semicontinuous, and we complete the proof. Starting from $\tau_0 \equiv T$, we define $\tau_i = \phi^* \circ \psi(\tau_{i-1})$ for $i \geq 1$ by induction. Clearly, $\tau_1 \leq \tau_0$, and if we assume $\tau_i \leq \tau_{i-1}$, the monotonicity of $\phi^* \circ \psi$ proved earlier implies $\tau_{i+1} = \phi^* \circ \psi(\tau_i) \leq \phi^* \circ \psi(\tau_{i-1}) = \tau_i$. If we define τ^* as the a.s. limit of the nonincreasing sequence of stopping times $(\tau_i)_{i\geq 1}$, then $\tau^* \in \mathcal{S}_X$ since we are working with a right continuous filtration. Under these conditions, $(\psi(\tau_i))_{i\geq 1}$ converges almost surely to $\psi(\tau^*)$. Recalling that, for any $\sigma \in \mathcal{S}_X$ and any $i \geq 1$, $J(\psi(\tau_i), \tau_{i+1}) \geq J(\psi(\tau_i), \sigma)$, we deduce from dominated convergence and by continuity of F that $J(\psi(\tau^*), \tau^*) \geq J(\psi(\tau^*), \sigma)$. That is, τ^* is a mean field game of timing equilibrium in the strong sense.

Similarly, define $\theta_0 \equiv 0$, and by induction $\theta_i = \phi_* \circ \psi(\theta_{i-1})$ for $i \geq 1$. Clearly, $\theta_0 \leq \theta_1$, and as above, we prove by induction that $\theta_{i-1} \leq \theta_i$. Next, we define θ_* as the a.s. limit of the nondecreasing sequence of stopping times $(\theta_i)_{i\geq 1}$. Then we argue as before that $\theta_* \in \mathcal{S}_X$ is a fixed point of the map $\phi_* \circ \psi$ and thus a strong equilibrium.

Finally, it is plain to check that if τ is any equilibrium in the strong sense, it is a fixed point of the set-valued map $\Phi \circ \psi$, in the sense that $\tau \in \Phi(\psi(\tau))$. Trivially, $\theta_0 = 0 \leq \tau \leq T = \tau_0$. Applying $\phi_* \circ \psi$ and $\phi^* \circ \psi$ repeatedly to the left and right sides, respectively, we conclude that $\theta_n \leq \tau \leq \tau_n$ for each n, and thus $\theta_* \leq \tau \leq \tau^*$. □

The above proof shows that, under the full continuity assumption, there is no need to use Tarski's theorem to prove existence of a solution to the mean field game since τ^* and θ^* are constructed inductively.

7.2.4 Randomized Measures and Stopping Times

As an alternative to the notion of strong solutions defined in Subsection 7.2.2, and as we did for stochastic differential mean field games, we introduce a notion of weak solution for mean field games of timing. The rationale for doing so is pretty much the same as for stochastic differential mean field games. The construction of strong solutions successfully achieved in the previous subsection was based upon rather constraining order preserving conditions, engineered in no small part to apply Tarski's fixed point theorem for mappings on a complete lattice. However, based on our experience with our first successes with the construction of equilibria for mean field games, we may think of an alternative approach based upon a fixed point theorem on a topological space. This is precisely the strategy we used in Chapter (Vol I)-4, and Chapter 3 of this volume for solving stochastic differential games by means of Schauder's fixed point theorem.

However, we learnt from the first part of this volume that, in the presence of a common noise, the space carrying the equilibria, or equivalently the space carrying the fixed points, becomes much too large to identify compact subsets for topologies which could be used for the search of such fixed points. This observation was the basis for the procedure implemented in Chapter 3, which consisted in discretizing first the common noise in order to reduce the size of the ambient space for the fixed points and then in passing to the limit in the weak sense along the discretization. As we saw in the analysis, the use of weak convergence arguments inherently carries losses of measurability and the limiting objects appear as *externally randomized* in the sense that they involve an external signal that comes in addition to the original noises $\boldsymbol{W}^0$ and $\boldsymbol{W}$. This shortcoming is clearly illustrated by the following classical fact about weak convergence and measurability. If $(X_n)_{n\geq 1}$ is a sequence of random variables converging in distribution toward X and if, for each $n \geq 1$, X_n is measurable with respect to the σ-field $\sigma\{Y\}$ generated by another random element Y (in other words, if X_n is a function of Y), then there is no guarantee that the weak limit X will be a function of Y and be $\sigma\{Y\}$-measurable. For this reason, we shall reformulate the problem in a way which is analogous to the weak formulation of stochastic differential mean field games introduced in Chapter 2. Namely, we shall allow the random measure μ to be measurable with respect to a larger σ-field than the one generated by the sole common noise $\boldsymbol{W}^0$. Put it differently, we shall extend the space Ω^0 and use a larger space instead of the Wiener space $\mathcal{C}([0,T])$, as we did in the statement of Theorem 7.27. Also, similar to the strategy used throughout the first part of this volume, we shall regard the random measure μ as the conditional law of both the stopping time and the idiosyncratic noise $\boldsymbol{W}$. Actually, we shall even ask for a relaxed version of the weak formulation. In words, we shall consider solutions that are *doubly weak*. In contrast with the solutions to stochastic differential mean field games constructed in Chapters 2 and 3, we shall allow the control itself to carry its own *external randomization*, which leads to something analog to the notion of relaxed control introduced in Chapter (Vol I)-6 for handling stochastic mean field control problems. This directs us to the notion of randomized stopping times, which is the counterpart of the notion of relaxed control. In the present discussion, we emphasize two very important properties which make the use of randomized stopping times quite attractive: *i*) their space is compact for natural topologies, and *ii*) in many instances, randomized stopping times can be approximated by regular stopping times.

For the sake of simplicity and tractability, we limit our analysis to the case of *full observation* when the completion of the σ-field generated by $\boldsymbol{X}$ coincides with the completion of the σ-field generated by $(\boldsymbol{W}^0, \boldsymbol{W})$, the completions being taken under the law of $(\boldsymbol{W}^0, \boldsymbol{W})$. For instance, this is the case if X is the identity on $\mathcal{C}([0,T]) \times \mathcal{C}([0,T])$.

In order to construct our externally randomized solutions, we need to extend the canonical space $\mathcal{C}([0,T]) \times \mathcal{C}([0,T])$ used for the construction of strong equilibria in the proof of Theorem 7.27. To be specific, we let:

$$\Omega_{\text{canon}} = \mathcal{C}([0,T])^2 \times \mathcal{P}_1\big([0,T] \times \mathcal{C}([0,T])\big) \times [0,T].$$

In other words, we choose $\Omega^0 = \mathcal{C}([0,T]) \times \mathcal{P}_1([0,T] \times \mathcal{C}([0,T]))$ and we shall regard the environment as the canonical variable on $\mathcal{P}_1([0,T] \times \mathcal{C}([0,T]))$, by which we mean its identity map. The rationale for regarding the environment as a probability measure on the enlarged $[0,T] \times \mathcal{C}([0,T])$, and not merely a probability measure on $[0,T]$, is exactly the same as for stochastic differential mean field games. Basically, we shall regard the environment as the joint law of the stopping time and of the idiosyncratic noise. This will guarantee for free that all the external randomizations underpinning our notion of weak solutions are compatible.

Below, we always equip $\mathcal{P}_1([0,T] \times \mathcal{C}([0,T]))$ with the 1-Wasserstein distance. We then endow Ω_{canon} with its Borel σ-field $\mathcal{B}(\Omega_{\text{canon}})$ and we denote the canonical or identity mapping by $(\boldsymbol{w}^0, \boldsymbol{w}, \nu, \varrho)$. At some point, it will be useful to regard each of the functions $\boldsymbol{w}^0$, $\boldsymbol{w}$, ν and ϱ as the identity on the corresponding factor of the product $\mathcal{C}([0,T])^2 \times \mathcal{P}_1([0,T] \times \mathcal{C}([0,T])) \times [0,T]$. For instance, ϱ will be regarded as the identity on $[0,T]$.

For any given probability measure $\mathbb{P}$ on $(\Omega_{\text{canon}}, \mathcal{B}(\Omega_{\text{canon}}))$, we denote (with a slight abuse of notation as we keep the same symbol for the probability $\mathbb{P}$) by $(\Omega_{\text{canon}}, \mathcal{F}, \mathbb{P})$ the corresponding complete probability space. Obviously, we shall only consider probability measures $\mathbb{P}$ under which $(\boldsymbol{w}^0, \boldsymbol{w})$ is a 2-dimensional Brownian motion with respect to the filtration generated by $(\boldsymbol{w}^0, \boldsymbol{w}, \nu, \varrho)$, namely the filtration $(\sigma\{\boldsymbol{w}^0_{\cdot\wedge t}, \boldsymbol{w}_{\cdot\wedge t}, \nu \circ \mathcal{E}_t^{-1}, \varrho \wedge t\})_{0 \le t \le T}$, where:

$$\mathcal{E}_t : [0,T] \times \mathcal{C}([0,T]) \ni (s, w) \mapsto (s \wedge t, w_{\cdot\wedge t}).$$

The marginal law of $\mathbb{P}$ on $\mathcal{C}([0,T])^2$ will be denoted by $\mathcal{W}$. It will be fixed throughout the subsequent analysis.

Throughout the rest of this subsection, we shall automatically include the following condition in assumption **MFG of Timing Set-Up**.

Assumption (MFG of Timing Set-Up).

(A3) Under the probability measure $\mathcal{W}$ on $\mathcal{C}([0,T])^2$, $\boldsymbol{w}^0$ and $\boldsymbol{w}$ are two independent Brownian motions. Moreover, the completion of the filtration generated by $\boldsymbol{X} = X(\boldsymbol{w}^0, \boldsymbol{w})$ under $\mathcal{W}$ coincides with the completion of the filtration generated by $(\boldsymbol{w}^0, \boldsymbol{w})$.

Undoubtedly, the most typical instance of such a probability measure $\mathcal{W}$ is $\mathcal{W} = \mathcal{W}_1 \otimes \mathcal{W}_1 = \mathcal{W}_1^{\otimes 2}$ where $\mathcal{W}_1$ is the standard Wiener measure. However, the use of this notation makes it possible to cover cases when the initial condition of $(\boldsymbol{w}^0, \boldsymbol{w})$ is random and different from $(0, 0)$.

This subsection is a fact gathering intended to introduce the necessary measure theoretic notions and the topological properties of the spaces of randomized stopping times.

Randomized Measures

We shall define randomized measures and randomized stopping times in two consecutive steps, the first one being to define a randomized measure properly, independently of what a randomized stopping time should be. To do so, we shall restrict ourselves to the space:

$$\Omega_{\mathrm{input}} = \mathcal{C}([0,T])^2 \times \mathcal{P}_1\big([0,T] \times \mathcal{C}([0,T])\big),$$

which is obviously smaller than Ω_{canon}. As above, the canonical random variable on Ω_{input} is denoted by $(\boldsymbol{w}^0, \boldsymbol{w}, \nu)$.

Definition 7.28 *A probability measure* $\mathbb{Q}$ *on* $(\Omega_{\mathrm{input}}, \mathcal{B}(\Omega_{\mathrm{input}}))$ *is said to induce a randomized measure if the first marginal of* $\mathbb{Q}$ *on* $\mathcal{C}([0,T])^2$ *matches* $\mathcal{W}$ *and if* $(\boldsymbol{w}^0, \boldsymbol{w})$ *is a Brownian motion with respect to the filtration generated by* $(\boldsymbol{w}^0, \boldsymbol{w}, \nu)$ *under* $\mathbb{Q}$*, and thus with respect to its right-filtration as well, or equivalently if, for every* $t \in [0,T]$*, the* σ*-field* $\mathcal{F}_t^{\mathrm{nat},\nu}$ *is conditionally independent of* $\mathcal{F}_T^{\mathrm{nat},(\boldsymbol{w}^0,\boldsymbol{w})}$ *given* $\mathcal{F}_t^{\mathrm{nat},(\boldsymbol{w}^0,\boldsymbol{w})}$*, where we set:*

$$\mathcal{F}_t^{\mathrm{nat},\nu} = \sigma\big\{\nu\big(C\big);\ C \in \mathcal{F}_t^{\mathrm{nat},(\varrho,\boldsymbol{w})}\big\},$$

$\mathbb{F}^{\mathrm{nat},(\varrho,\boldsymbol{w})} = (\sigma\{(\varrho \wedge t, \boldsymbol{w}_{\cdot\wedge t})\})_{0 \le t \le T}$ *denoting the canonical filtration on* $[0,T] \times \mathcal{C}([0,T])$*. We call* $\mathcal{M}$ *the set of such randomized measures.*

Using the fact that the σ-field generated by $\varrho \wedge t$ on $[0,T]$ or equivalently by the function $[0,T] \ni s \mapsto s \wedge t$ is the same as the sub-σ-field of $[0,T]$ generated by $\mathcal{B}([0,t))$, we can rewrite $\mathcal{F}_t^{\mathrm{nat},\nu}$ as $\sigma\{\nu\big(C\big);\ C \in \mathcal{B}([0,t)) \otimes \mathcal{F}_t^{\mathrm{nat},\boldsymbol{w}}\}$.

Definition 7.28 is obviously satisfied whenever ν is progressively measurable with respect to the $\mathbb{Q}$-completion of the filtration generated by $\boldsymbol{w}^0$, in the sense that $\mathcal{F}_t^{\mathrm{nat},\nu}$ is included in the completion of $\sigma\{\boldsymbol{w}^0_{\cdot\wedge t}\}$, for all $t \in [0,T]$. As we shall see next, so is the case when dealing with strong solutions of mean field games of timing. Whenever ν is no longer adapted with respect to the common noise, we require some compatibility condition to make the solution admissible. Such a compatibility condition is consistent with that used in the first part of this volume for solving stochastic differential mean field games. In short, we require that the additional external randomization used in the definition of ν does not induce any bias in the future realizations of the two noises $\boldsymbol{w}^0$ and $\boldsymbol{w}$.

Throughout the analysis below, we shall use the following result:

Theorem 7.29 *The set $\mathcal{M}$ is convex and closed for the weak topology on $\mathcal{P}(\Omega_{\mathrm{input}})$.*

Proof.

First Step. We first check that $\mathcal{M}$ is convex. To do so, we observe that the constraint requiring the marginal on $\mathcal{C}([0,T])^2$ of any $\mathbb{Q} \in \mathcal{M}$ to match $\mathcal{W}$ is obviously convex. Regarding the compatibility condition, we proceed as follows. For any $t \in [0,T]$, the σ-field $\mathcal{F}_t^{\mathrm{nat},\nu}$ is conditionally independent of $\mathcal{F}_T^{\mathrm{nat},(\boldsymbol{w}^0,\boldsymbol{w})}$ given $\mathcal{F}_t^{\mathrm{nat},(\boldsymbol{w}^0,\boldsymbol{w})}$ if and only if:

$$\mathbb{E}^{\mathbb{Q}}\big[\phi_t(\nu)\psi(\boldsymbol{w}^0,\boldsymbol{w})\psi_t(\boldsymbol{w}^0,\boldsymbol{w})\big] = \mathbb{E}^{\mathbb{Q}}\big[\mathbb{E}^{\mathbb{Q}}[\phi_t(\nu)\,|\,\mathcal{F}_t^{\mathrm{nat},(\boldsymbol{w}^0,\boldsymbol{w})}]\psi(\boldsymbol{w}^0,\boldsymbol{w})\psi_t(\boldsymbol{w}^0,\boldsymbol{w})\big],$$

for every triple of bounded functions ϕ_t, ψ and ψ_t that are measurable with respect to $\mathcal{F}_t^{\mathrm{nat},\nu}$, $\mathcal{F}_T^{\mathrm{nat},(\boldsymbol{w}^0,\boldsymbol{w})}$ and $\mathcal{F}_t^{\mathrm{nat},(\boldsymbol{w}^0,\boldsymbol{w})}$ respectively. Above, we put a superscript in the expectation in order to emphasize the dependence of the expectation upon the probability measure $\mathbb{Q}$. Regarding $\boldsymbol{w}^0$ and $\boldsymbol{w}$ as random variables on $\mathcal{C}([0,T])^2$, we then notice that the right-hand side above can be rewritten in the form:

$$\mathbb{E}^{\mathcal{W}}\big[\mathbb{E}^{\mathbb{Q}}[\phi_t(\nu)\,|\,\mathcal{F}_t^{\mathrm{nat},(\boldsymbol{w}^0,\boldsymbol{w})}]\psi(\boldsymbol{w}^0,\boldsymbol{w})\psi_t(\boldsymbol{w}^0,\boldsymbol{w})\big],$$

since $\mathbb{E}^{\mathbb{Q}}[\phi_t(\nu)\,|\,\mathcal{F}_t^{\mathrm{nat},(\boldsymbol{w}^0,\boldsymbol{w})}]$ is $\sigma\{(\boldsymbol{w}^0,\boldsymbol{w})\}$-measurable. Convexity easily follows.

Second Step. We consider a sequence $(\mathbb{Q}_n)_{n\geq 1}$ of elements of $\mathcal{M}$ converging to some $\mathbb{Q}$. We will show that $\mathbb{Q}$ also belongs to $\mathcal{M}$. Obviously, it holds that $\mathbb{Q}\circ(\boldsymbol{w}^0,\boldsymbol{w})^{-1} = \mathcal{W}$.

We now prove that $(\boldsymbol{w}^0,\boldsymbol{w})$ is a Brownian motion with respect to the filtration generated by $(\boldsymbol{w}^0,\boldsymbol{w},\nu)$ under $\mathbb{Q}$. To do so, we define the process $(\nu_t = \nu\circ\mathcal{E}_t^{-1})_{0\leq t\leq T}$, where we recall that $\mathcal{E}_t : [0,T]\times\mathcal{C}([0,T]) \ni (s,w) \mapsto (s\wedge t, w_{\cdot\wedge t})$. Hence, we can write $\mathbb{F}^{\mathrm{nat},(\boldsymbol{w}^0,\boldsymbol{w},\nu)}$ as the natural filtration generated by the process $(w_t^0, w_t, \nu_t)_{0\leq t\leq T}$, which takes values in the space Ω_{input} and is continuous. Since $(\boldsymbol{w}^0,\boldsymbol{w})$ is a Brownian motion with respect to $\mathbb{F}^{\mathrm{nat},(\boldsymbol{w}^0,\boldsymbol{w},\nu)}$ under $\mathbb{Q}_n$ for each $n\geq 1$, it is pretty standard to prove that the same holds true under the limiting probability $\mathbb{Q}$. □

Randomized Stopping Times

As explained earlier, we need to randomize not only externally the environment ν modeling the distribution of the stopping time chosen by the agent, but also the stopping time itself. This requires to work on the extended canonical space Ω_{canon}, which we defined as:

$$\Omega_{\text{canon}} = \mathcal{C}([0,T])^2 \times \mathcal{P}_1\big([0,T] \times \mathcal{C}([0,T])\big) \times [0,T] = \Omega_{\text{input}} \times [0,T].$$

In order to define properly a randomized stopping time, we shall need to freeze the marginal law of any element of $\mathcal{P}(\Omega_{\text{canon}})$ on Ω_{input}.

Definition 7.30 *Let $\mathbb{F}^{\text{nat},\varrho} = (\mathcal{F}_t^{\text{nat},\varrho})_{0 \le t \le T}$ be the filtration generated by ρ. For a given $\mathbb{Q}$ in $\mathcal{M}$, a probability measure $\mathbb{P}$ on $(\Omega_{\text{canon}}, \mathcal{B}(\Omega_{\text{canon}}))$, admitting $\mathbb{Q}$ as marginal law on Ω_{input}, is said to generate a randomized stopping time if, for every $t \in [0,T]$, the σ-field $\mathcal{F}_t^{\text{nat},\varrho}$ is conditionally independent of the σ-field $\mathcal{F}_T^{\text{nat},(\boldsymbol{w}^0,\boldsymbol{w},\nu)}$ given $\mathcal{F}_t^{\text{nat},(\boldsymbol{w}^0,\boldsymbol{w},\nu)}$. We denote by $\mathcal{R}(\mathbb{Q})$ the set of such probability measures.*

Definition 7.30 has the following interpretation. We shall regard a weak equilibrium as a probability measure $\mathbb{P}$ on Ω_{canon}, under which the random variables ν and ϱ form an externally randomized pair of environment and stopping time. As we already explained, we shall work with the completion of $(\Omega_{\text{canon}}, \mathcal{B}(\Omega_{\text{canon}}), \mathbb{P})$ and, to ease notations, we shall still denote by $\mathbb{P}$ the completed measure. Observe that the external randomization is trivially admissible when ν is adapted with respect to the completion $\mathbb{F}^{(\boldsymbol{w}^0,\boldsymbol{w})}$ of the filtration generated by $(\boldsymbol{w}^0, \boldsymbol{w})$, and ϱ is a stopping time with respect to $\mathbb{F}^{(\boldsymbol{w}^0,\boldsymbol{w})}$, recall (**A3**) in Assumption **MFG of Timing Set-Up**. This is the framework used for handling strong equilibria. In the general case, the external randomization is required to satisfy a compatibility condition, which is reminiscent of that introduced in Chapter 2 to handle stochastic differential mean field games with a common noise. In analogy with Definition 7.28, this compatibility condition takes the form of a conditional independence property. It says that, conditional on the information supplied by the observation of the signal $\boldsymbol{X}$ and of the environment ν up until time t, the additional information used by the representative player to decide whether it stops strictly before t is independent of the future of the two noises $\boldsymbol{w}^0$ and $\boldsymbol{w}$ and of the environment ν. Observe that we say *strictly before t* since all the events $\{\varrho \le s\}$, for $s < t$, belong to $\mathcal{F}_t^{\text{nat},\varrho}$, while the event $\{\varrho \le t\}$ belongs to $\mathcal{F}_{t+}^{\text{nat},\varrho}$ but not to $\mathcal{F}_t^{\text{nat},\varrho}$. All these requirements are obviously true when ϱ is a stopping time with respect to $\mathbb{F}^{(\boldsymbol{w}^0,\boldsymbol{w},\nu)}$. In that case, conditional on $\mathcal{F}_t^{\text{nat},(\boldsymbol{w}^0,\boldsymbol{w},\nu)}$, $\mathcal{F}_{t+}^{\text{nat},\varrho}$ is $\mathbb{P}$-almost surely trivial. Actually, the key point is that this situation is somehow typical under the compatibility condition.

Theorem 7.31 *For a given probability measure $\mathbb{Q} \in \mathcal{M}$, the set $\mathcal{R}(\mathbb{Q})$ is a convex subset of $\mathcal{P}(\Omega_{\text{canon}})$. Moreover, for any $\mathbb{P} \in \mathcal{R}(\mathbb{Q})$, there exists a sequence of continuous functions $(\tilde{\tau}_n)_{n \ge 1}$ from Ω_{input} to $[0,T]$ such that each $\tilde{\tau}_n(\boldsymbol{w}^0, \boldsymbol{w}, \nu)$ is a stopping time with respect to the filtration $\mathbb{F}^{\text{nat},(\boldsymbol{w}^0,\boldsymbol{w},\nu)}$ and the sequence $(\mathbb{Q} \circ (\boldsymbol{w}^0, \boldsymbol{w}, \nu, \tilde{\tau}_n(\boldsymbol{w}^0, \boldsymbol{w}, \nu))^{-1})_{n \ge 1}$ converges to $\mathbb{P}$. If furthermore the completion of the filtration generated by $(\boldsymbol{w}^0, \boldsymbol{w}, \nu)$ is right-continuous, then $\mathcal{R}(\mathbb{Q})$ is closed.*

We shall use the notation $\mathcal{R}_0(\mathbb{Q})$ for the set of elements $\mathbb{P}$ of $\mathcal{R}(\mathbb{Q})$ under which the random variable ϱ is a stopping time with respect to the completion of the filtration $\mathbb{F}^{\mathrm{nat},(w^0,w,\nu)}$ under $\mathbb{P}$. Among other things, the above result implies that $\mathcal{R}_0(\mathbb{Q})$ is dense in $\mathcal{R}(\mathbb{Q})$.

The proof of Theorem 7.31 relies on several technical results, so we postpone it to a later subsection in order to state and prove (most of) these technical results.

7.2.5 Approximation of Adapted Processes Under Compatibility

The first lemma concerns the approximation of $\sigma\{\varrho\}$-measurable functionals by continuous functionals.

Lemma 7.32 *For any $\mu \in \mathcal{P}([0,T])$, $t \in [0,T]$, and any bounded $\mathcal{F}_{t+}^{\mathrm{nat},\varrho}$-measurable function $g : [0,T] \to \mathbb{R}$, there exists a sequence of uniformly bounded $\mathcal{F}_{t+}^{\mathrm{nat},\varrho}$-measurable functions $(g_n)_{n\geq 1}$ such that $g_n \to g$ in $L^1(\mu)$ and each g_n is continuous at every point but t.*

Proof. First notice that, being $\mathcal{F}_{t+}^{\mathrm{nat},\varrho}$-measurable, g is necessarily of the form:

$$g(s) = h(s)\mathbf{1}_{[0,t]}(s) + c\mathbf{1}_{(t,T]}(s),$$

for some bounded measurable function $h : [0,t] \to \mathbb{R}$ and some constant $c \in \mathbb{R}$. Now let $(h_n)_{n\geq 1}$ be a sequence of real valued continuous functions on $[0,t]$ approximating h in $L^1(\mu|_{[0,t]})$. For each $n \geq 1$, define:

$$g_n(s) = h_n(s)\mathbf{1}_{[0,t]}(s) + c\mathbf{1}_{(t,T]}(s), \quad s \in [0,T].$$

This sequence of functions has the desired properties. □

The next lemma will be quite useful to investigate stopping times in terms of càd-làg processes.

Lemma 7.33 *Let $\widetilde{T} : \mathcal{D}([0,T];[0,1]) \to [0,1]$ be defined by:*

$$\widetilde{T}(h) = \inf\{t \in [0,T] : h(t) \geq \tfrac{1}{2}\}, \quad h \in \mathcal{D}([0,T];[0,1]),$$

with $\inf \emptyset = T$. Then, $\widetilde{T}$ is continuous at each point $h \in \mathcal{D}([0,T];[0,1])$ that is càd-làg and nondecreasing and satisfies $h(0) = 0$, $h(T) = 1$ together with the following property for any $t \in [0,T]$:

$$\Bigl(h(t) \geq \tfrac{1}{2} \text{ and } h(t-) \leq \tfrac{1}{2}\Bigr) \Rightarrow t = \widetilde{T}(h). \tag{7.92}$$

Above, $\mathcal{D}([0,T];[0,1])$ is equipped with the $J1$-Skorohod topology.

Before we prove Lemma 7.33, we make the following observations. Notice that if h is càd-làg and nondecreasing and satisfies $h(0) = 0$ and $h(T) = 1$, then it always satisfies $h(\widetilde{T}(h)) \geq 1/2$ and $h(\widetilde{T}(h)-) \leq 1/2$, but $\widetilde{T}(h)$ may not be the unique point satisfying such a property. It is the unique point in the following two cases. First, if h is of the form $\mathbf{1}_{[s,T]}$, for some $s \in [0, T]$, then $\widetilde{T}(h) = s$ and for any other point $t \in [0, T] \setminus \{s\}$, $\{h(t), h(t-)\}$ is either equal to $\{0\}$ or $\{1\}$. Second, if h is strictly increasing, then for any $t > \widetilde{T}(h)$, it holds $h(t-) > h(\widetilde{T}(h)) \geq 1/2$, and for any $t < \widetilde{T}(h)$, it holds $h(t) < 1/2$.

Proof of Lemma 7.33. To prove the claimed continuity, we let $(h_n)_{n\geq 1}$ be a sequence of functions in the space $\mathcal{D}([0, T]; [0, 1])$ converging to h, where h is as in the statement. Since $h(0) = 0$ and $h(T) = 1$, we know that $h_n(0) < 1/2$ and $h_n(T) > 1/2$, for n large enough; to simplify, we can assume it to be true for all $n \geq 1$. It is straightforward to check that the sequence $(\widetilde{T}(h_n))_{n\geq 1}$ is bounded. Suppose that, along a subsequence $(n(p))_{p\geq 1}$, $\widetilde{T}(h_{n(p)}) \to t$, for some $t \in [0, T]$. We recall from the properties of the $J1$ topology, see the Notes & Complements below, that the sole fact that $(h_n)_{n\geq 1}$ converges to h and that $(\widetilde{T}(h_{n(p)}))_{p\geq 1}$ converges to t suffices to show that $(h_{n(p)}(\widetilde{T}(h_{n(p)})))_{p\geq 1}$ has at least one limit point and at most two, which are $h(t-)$ and $h(t)$. As $h(t) \geq h(t-)$ and $h_{n(p)}(\widetilde{T}(h_{n(p)})) \geq 1/2$ for all $p \geq 1$, we must have $h(t) \geq 1/2$. In particular, $t > 0$. Similarly, for $\epsilon > 0$ small enough, the sequence $(h_{n(p)}(\widetilde{T}(h_{n(p)}) - \epsilon))_{p\geq 1}$ has at least one limit point and at most two which are $h((t-\epsilon)-)$ and $h(t-\epsilon)$. Since $h_{n(p)}(\widetilde{T}(h_{n(p)}) - \epsilon) < 1/2$ for all $p \geq 1$, we conclude that $h((t-\epsilon)-) \leq 1/2$ and then $h(t-) \leq 1/2$. Hence, by assumption, $t = \widetilde{T}(h)$ and $(\widetilde{T}(h_n))_{n\geq 1}$ has a unique limit, which is $\widetilde{T}(h)$. □

The proof of Theorem 7.31 relies on still another technical result, which we shall use rather intensively below. This result is a general lemma on *stable convergence*. We already appealed to it in a very particular setting in Chapter (Vol I)-6. We refer to the Notes & Complements below for references where the proof of the statement can be found.

Lemma 7.34 *Let E and E' be Polish spaces. Suppose that a sequence $(\mathbf{P}_n)_{n\geq 1}$ of probability measures on $E\times E'$ converges weakly toward $\mathbf{P} \in \mathcal{P}(E\times E')$, and suppose that all the $(\mathbf{P}_n)_{n\geq 1}$'s have the same E-marginal, say $m(\cdot) = \mathbf{P}_n(\cdot \times E')$ for all $n \geq 1$. Then, $\mathbf{P}$ has also m as E-marginal. Moreover, for every bounded measurable function $\phi : E \times E' \to \mathbb{R}$ such that $\phi(x, \cdot)$ is continuous on E' for m-almost every $x \in E$, we have:*

$$\int_{E\times E'} \phi \, d\mathbf{P}_n \to \int_{E\times E'} \phi \, d\mathbf{P}, \qquad \textit{as } n \to \infty.$$

The reader will easily convince herself / himself of the interest of Lemma 7.34 for our purpose. We shall make use of it with $E = \mathcal{C}([0, T])^2$ and $m = \mathcal{W}$. In this respect, our use of Lemma 7.34 for proving Theorem 7.31 is reminiscent of the definition of the Baxter-Chacon topology on the set of randomized stopping times, see again the Notes & Complements below.

The last ingredient we need to prove Theorem 7.31 is the following special approximation result for processes on Polish spaces.

Theorem 7.35 *Let $\mathcal{Z}$ be a Polish space and $\mathcal{Y}$ be a convex subset of a normed vector space. On a common probability space, consider an n-tuple $(Y_1, \cdots, Y_n)$ of $\mathcal{Y}$-valued random variables and a continuous $\mathcal{Z}$-valued process $\mathbf{Z} = (Z_t)_{0 \le t \le T}$. Denoting by $\mathbb{F}^{\mathrm{nat},\mathbf{Z}} = (\mathcal{F}_t^{\mathrm{nat},\mathbf{Z}})_{0 \le t \le T}$ the filtration generated by the process $\mathbf{Z}$, assume that for any $i \in \{1, \cdots, n\}$, $\sigma\{Y_1, \cdots, Y_i\}$ is conditionally independent of $\mathcal{F}_T^{\mathrm{nat},\mathbf{Z}}$ given $\mathcal{F}_{t_i}^{\mathrm{nat},\mathbf{Z}}$, where $0 \le t_1 < \cdots < t_n \le T$ is a fixed subdivision of $[0, T]$.*

If the law of $\mathbf{Z}_{\cdot \wedge t_1}$ is atomless (that is singletons have zero probability measure under the law of $\mathbf{Z}_{\cdot \wedge t_1}$), then there exists a sequence of functions $(h_1^N, \cdots, h_n^N)_{N \ge 1}$, each h_i^N being a continuous and $\mathcal{F}_{t_i}^{\mathrm{nat},\mathcal{Z}}$-measurable function from $\mathcal{C}([0, T]; \mathcal{Z})$ into $\mathcal{Y}$, where $\mathbb{F}^{\mathrm{nat},\mathcal{Z}}$ is the natural filtration on $\mathcal{C}([0, T]; \mathcal{Z})$, such that the sequence $(\mathbf{Z}, h_1^N(\mathbf{Z}), \cdots, h_n^N(\mathbf{Z}))_{N \ge 1}$ converges in law to $(\mathbf{Z}, Y_1, \cdots, Y_n)$ as $N \to \infty$.

The proof relies on the following lemma. We refer the reader to the Notes & Complements at the end of the chapter for a reference to its proof.

Lemma 7.36 *Let E be a Polish space and E' be a convex subset of a normed vector space. For a given $\mu \in \mathcal{P}(E)$, call $\mathfrak{S}_\mu = \{\mathbb{P} \in \mathcal{P}(E \times E');\ \mathbb{P}(\cdot \times E') = \mu\}$ the set of probability measures on $E \times E'$ with first marginal μ. If μ is atomless, then the set:*

$$\big\{\mu(dx)\delta_{\phi(x)}(dx') \in \mathcal{P}(E \times E');\ \phi : E \to E' \text{ is continuous}\big\}$$

is dense in $\mathfrak{S}_\mu$ for the weak convergence.

Proof of Theorem 7.35. The proof relies on repeated applications of Lemma 7.36. We shall use the result of this lemma with E successively given by $E = \mathcal{C}([0, T]; \mathcal{Z})$, $E = \mathcal{C}([0, T]; \mathcal{Z}) \times \mathcal{Y}$, $E = \mathcal{C}([0, T]; \mathcal{Z}) \times \mathcal{Y}^2, \ldots$, the first factor being always equipped with the uniform topology, and E' being given by $\mathcal{Y}$.

First Step. Since $\mathcal{L}(\mathbf{Z}_{\cdot \wedge t_1})$ is atomless, we deduce from Lemma 7.36 with $\mu = \mathcal{L}(\mathbf{Z}_{\cdot \wedge t_1})$ that there exists a sequence of continuous functions $(h_1^N)_{N \ge 1}$ from $\mathcal{C}([0, T]; \mathcal{Z})$ into $\mathcal{Y}$ such that $(\mathbf{Z}_{\cdot \wedge t_1}, h_1^N(\mathbf{Z}_{\cdot \wedge t_1}))_{N \ge 1}$ converges in law to $(\mathbf{Z}_{\cdot \wedge t_1}, Y_1)$ as $N \to \infty$. Using the compatibility assumption, we now show that in fact, $(\mathbf{Z}, h_1^N(\mathbf{Z}_{\cdot \wedge t_1}))_{N \ge 1}$ converges in law toward $(\mathbf{Z}, Y_1)$. To do so, we let $\phi : \mathcal{C}([0, T]; \mathcal{Z}) \to \mathbb{R}$ and $\psi : \mathcal{Y} \to \mathbb{R}$ be two bounded continuous functions. Denoting by $\mathbf{P}$ the underlying probability measure and by $\mathbf{E}$ the corresponding expectation, we have:

$$\begin{aligned}
\lim_{N \to \infty} \mathbf{E}\big[\phi(\mathbf{Z})\psi\big(h_1^N(\mathbf{Z}_{\cdot \wedge t_1})\big)\big] &= \lim_{N \to \infty} \mathbf{E}\Big[\mathbf{E}\big[\phi(\mathbf{Z}) \,|\, \mathcal{F}_{t_1}^{\mathrm{nat},\mathbf{Z}}\big]\psi\big(h_1^N(\mathbf{Z}_{\cdot \wedge t_1})\big)\Big] \\
&= \mathbf{E}\Big[\mathbf{E}\big[\phi(\mathbf{Z}) \,|\, \mathcal{F}_{t_1}^{\mathrm{nat},\mathbf{Z}}\big]\psi(Y_1)\Big],
\end{aligned}$$

where we used Lemma 7.34 and the fact that $\mathbf{E}[\phi(\mathbf{Z}) \,|\, \mathcal{F}_{t_1}^{\mathrm{nat},\mathbf{Z}}]$ could be written as a measurable function of $\mathbf{Z}_{\cdot \wedge t_1}$. We now use the fact that $\mathbf{Z}$ and Y_1 are conditionally independent given $\mathcal{F}_{t_1}^{\mathrm{nat},\mathbf{Z}}$.

We get:

$$\begin{aligned}\lim_{N\to\infty} \mathbf{E}\big[\phi(\mathbf{Z})\psi\big(h_1^N(\mathbf{Z}_{\cdot\wedge t_1})\big)\big] &= \mathbf{E}\Big[\mathbf{E}\big[\phi(\mathbf{Z})\,|\,\mathcal{F}_{t_1}^{\mathrm{nat},\mathbf{Z}}\big]\mathbf{E}\big[\psi(Y_1)\,|\,\mathcal{F}_{t_1}^{\mathrm{nat},\mathbf{Z}}\big]\Big]\\ &= \mathbf{E}\Big[\mathbf{E}\big[\phi(\mathbf{Z})\psi(Y_1)\,|\,\mathcal{F}_{t_1}^{\mathrm{nat},\mathbf{Z}}\big]\Big]\\ &= \mathbf{E}\big[\phi(\mathbf{Z})\psi(Y_1)\big].\end{aligned}$$

Since the class of functions of the form $\mathcal{C}([0,T];\mathcal{Z})\times\mathcal{Y}\ni(z,y)\mapsto\phi(z)\psi(y)$, with ϕ and ψ as above, is convergence determining, see the references in the Notes & Complements below, we conclude that $(\mathbf{Z}, h_1^N(\mathbf{Z}_{\cdot\wedge t_1}))_{N\geq 1}$ converges in law to $(\mathbf{Z}, Y_1)$. Replacing h_1^N by $h_1^N\circ\mathcal{E}_{t_1}$, where $\mathcal{E}_{t_1}:\mathcal{C}([0,T];\mathcal{Z})\ni z\to z_{\cdot\wedge t_1}\in\mathcal{C}([0,T];\mathcal{Z})$, we have that $(\mathbf{Z}, h_1^N(\mathbf{Z}))_{N\geq 1}$ converges to $(\mathbf{Z}, Y_1)$, where h_1^N is $\mathcal{F}_{t_1}^{\mathrm{nat},\mathcal{Z}}$ measurable.

Second Step. We proceed inductively as follows. We assume that, for a given $1\leq i<n$, we constructed functions $h_1^N,\cdots,h_i^N$ as in the statement in such a way that:

$$\lim_{N\to\infty}\mathbf{P}\circ\big(\mathbf{Z}, h_1^N(\mathbf{Z}),\cdots,h_i^N(\mathbf{Z})\big)^{-1} = \mathbf{P}\circ(\mathbf{Z}, Y_1,\cdots,Y_i)^{-1}, \tag{7.93}$$

where $\mathbf{P}$ is the underlying probability measure. Next, we construct h_{i+1}^N. To do so, we observe that $(\mathcal{L}(\mathbf{Z}_{\cdot\wedge t_{i+1}}), Y_1,\cdots,Y_i)$ is necessarily atomless as otherwise $\mathcal{L}(\mathbf{Z}_{\cdot\wedge t_1})$ itself would have an atom. Hence, using again Lemma 7.36, we can find a sequence of continuous functions $(\hat{h}^N)_{N\geq 1}$ from $\mathcal{C}([0,T];\mathcal{Z})\times\mathcal{Y}^i$ into $\mathcal{Y}$ such that:

$$\begin{aligned}&\lim_{N\to\infty}\mathbf{P}\circ\Big(\mathbf{Z}_{\cdot\wedge t_{i+1}}, Y_1,\cdots,Y_i,\hat{h}^N\big(\mathbf{Z}_{\cdot\wedge t_{i+1}}, Y_1,\cdots,Y_i\big)\Big)^{-1}\\ &\quad= \mathbf{P}\circ\big(\mathbf{Z}_{\cdot\wedge t_{i+1}}, Y_1,\cdots,Y_i,Y_{i+1}\big)^{-1}.\end{aligned}$$

Note that $\mathbf{Z}$ and $(Y_1,\cdots,Y_{i+1})$ are conditionally independent given $\mathbf{Z}_{\cdot\wedge t_{i+1}}$. Using the same argument as above, it follows that in fact:

$$\begin{aligned}&\lim_{N\to\infty}\mathbf{P}\circ\Big(\mathbf{Z}, Y_1,\cdots,Y_i,\hat{h}^N\big(\mathbf{Z}_{\cdot\wedge t_{i+1}}, Y_1,\cdots,Y_i\big)\Big)^{-1}\\ &\quad= \mathbf{P}\circ\big(\mathbf{Z}, Y_1,\cdots,Y_i,Y_{i+1}\big)^{-1}.\end{aligned} \tag{7.94}$$

By continuity of $\hat{h}^N$, the limit (7.93) implies that, for each N,

$$\begin{aligned}&\lim_{k\to\infty}\mathbf{P}\circ\Big(\mathbf{Z}, h_1^k(\mathbf{Z}),\cdots,h_i^k(\mathbf{Z}),\hat{h}^N\big(\mathbf{Z}_{\cdot\wedge t_{i+1}}, h_1^k(\mathbf{Z}),\cdots,h_i^k(\mathbf{Z})\big)\Big)^{-1}\\ &\quad= \mathbf{P}\circ\Big(\mathbf{Z}, Y_1,\cdots,Y_i,\hat{h}^N\big(\mathbf{Z}_{\cdot\wedge t_{i+1}}, Y_1,\cdots,Y_i\big)\Big)^{-1}.\end{aligned} \tag{7.95}$$

Combining the two limits (7.94) and (7.95), we can find a subsequence $(k_N)_{N\geq 1}$ such that:

$$\lim_{N\to\infty} \mathbf{P} \circ \Big(\mathbf{Z}, h_1^{k_N}(\mathbf{Z}), \cdots, h_i^{k_N}(\mathbf{Z}), \hat{h}^N\big(\mathbf{Z}_{\cdot\wedge t_{i+1}}, h_1^{k_N}(\mathbf{Z}), \cdots, h_i^{k_N}(\mathbf{Z})\big)\Big)^{-1}$$
$$= \mathbf{P} \circ \big(\mathbf{Z}, Y_1, \cdots, Y_i, Y_{i+1}\big)^{-1}.$$

Relabeling $h_1^{k_N}, \cdots, h_i^{k_N}$ by $h_1^N, \cdots, h_i^N$ and letting $h_{i+1}^N(\cdot) = \hat{h}^N(\mathcal{E}_{t_{i+1}}(\cdot), h_1^N(\cdot), \cdots, h_i^N(\cdot))$, this proves that we can iterate the construction. □

We now have all the required ingredients to prove Theorem 7.31.

Proof of Theorem 7.31

Proof of Theorem 7.31. Throughout the proof, we use the convenient notation $\boldsymbol{\theta} = (\boldsymbol{w}^0, \boldsymbol{w}, \nu)$.

First Step. We first check that $\mathcal{R}(\mathbb{Q})$ is convex. The proof is analogous to that of the first step of the proof of Theorem 7.29. Indeed, the constraint requiring the marginal law on Ω_{input} to match $\mathbb{Q}$ is clearly convex. In order to handle the compatibility condition, we notice that, for any $t \in [0, T]$, the σ-field $\mathcal{F}_t^{\text{nat},\varrho}$ is conditionally independent of $\boldsymbol{\theta}$ given the σ-field $\mathcal{F}_t^{\text{nat},\boldsymbol{\theta}}$ if and only if:

$$\mathbb{E}^{\mathbb{P}}\big[\zeta_t(\varrho)\psi(\boldsymbol{\theta})\psi_t(\boldsymbol{\theta})\big] = \mathbb{E}^{\mathbb{P}}\big[\mathbb{E}^{\mathbb{P}}[\zeta_t(\varrho) \,|\, \mathcal{F}_t^{\text{nat},\boldsymbol{\theta}}]\psi(\boldsymbol{\theta})\psi_t(\boldsymbol{\theta})\big], \tag{7.96}$$

for every triple of bounded functions ζ_t, ψ and ψ_t that are measurable with respect to $\mathcal{F}_t^{\text{nat},\varrho}$, $\sigma\{\boldsymbol{\theta}\}$ and $\mathcal{F}_t^{\text{nat},\boldsymbol{\theta}}$ respectively. Above, we put a superscript in the expectation in order to emphasize the dependence of the expectation upon the probability measure $\mathbb{P}$.

Recalling that $\mathbb{Q}$ in Definition 7.30 is fixed throughout the proof, we notice that the right-hand side in (7.96) can be rewritten in the form:

$$\mathbb{E}^{\mathbb{Q}}\big[\mathbb{E}^{\mathbb{P}}[\zeta_t(\varrho) \,|\, \mathcal{F}_t^{\text{nat},\boldsymbol{\theta}}]\psi(\boldsymbol{\theta})\psi_t(\boldsymbol{\theta})\big],$$

since $\mathbb{E}^{\mathbb{P}}[\zeta_t(\varrho) \,|\, \mathcal{F}_t^{\text{nat},\boldsymbol{\theta}}]$ is $\sigma(\{\boldsymbol{\theta}\})$-measurable. Of course, to do so, we must regard $\boldsymbol{\theta}$ as a random variable on Ω_{input}. Convexity follows as in the proof of Theorem 7.29.

Second Step. We now prove that $\mathcal{R}(\mathbb{Q})$ is included in the closure of $\mathcal{R}^0(\mathbb{Q})$. For any given $n \geq 1$, we let:

$$\varrho^n = \begin{cases} \frac{T}{2^n}\lceil 2^n \frac{\varrho}{T}\rceil + \frac{T}{2^n} & \text{if} \quad \varrho \leq T - \frac{T}{2^n}, \\ T & \text{if} \quad \varrho \in \big(T - \frac{T}{2^n}, T\big]. \end{cases}$$

Obviously, $(\varrho^n)_{n\geq 1}$ converges pointwise to ϱ. It thus suffices to approximate each $\mathbb{P} \circ (\boldsymbol{w}^0, \boldsymbol{w}, \nu, \varrho^n)^{-1}$ by a sequence in $\mathcal{R}^0(\mathbb{Q})$. To do so, we observe that, for any $i \in \{1, \cdots, 2^n - 1\}$, the event $\{\varrho^n \leq t_i\}$ coincides with $\{\varrho \leq t_{i-1}\}$ and thus belongs to $\mathcal{F}_{t_{i-1}+}^{\text{nat},\varrho} \subset \mathcal{F}_{t_i}^{\text{nat},\varrho}$, where $t_i = iT/2^n$. We also notice that the process $(H_t^n)_{0\leq t\leq T}$ defined by $H_t^n = \mathbf{1}_{\{\varrho^n\leq t\}}$ may be rewritten as:

$$H_t^n = \sum_{i=1}^{2^n-1} \mathbf{1}_{\{\varrho^n\leq t_i\}}\mathbf{1}_{[t_i,t_{i+1})}(t) + \mathbf{1}_{\{T\}}(t), \quad t \in [0, T], \tag{7.97}$$

where we used the fact that $\{\varrho^n = 0\} = \emptyset$. Next, we use Theorem 7.35 on the probability space Ω_{canon} with $\mathcal{Y} = [0, 1]$ and $Y_i = \mathbf{1}_{\{\varrho^n\leq t_i\}}$ for $i = 1, \cdots, 2^n - 1$, $\mathcal{Z} = \mathbb{R} \times \mathbb{R} \times \mathcal{P}_1$

$([0,T]\times\mathcal{C}([0,T]))$ and $\mathbf{Z} = (w_t^0, w_t, \nu_t)_{0\le t\le T}$ where $\nu_t = \nu\circ\mathcal{E}_t^{-1}$ for $\mathcal{E}_t$ defined by $\mathcal{E}_t : [0,T]\times\mathcal{C}([0,T]) \ni (s,w) \mapsto (s\wedge t, w_{\cdot\wedge t})$. Notice that as required, $\mathcal{L}(\mathbf{Z}_{\cdot\wedge t_1})$ is atomless since the law of $(w_{t_1}^0, w_{t_1})$ is diffuse as $t_1 > 0$. Theorem 7.35 says that there exists, for each $N \ge 1$, a family of continuous functions $(h_i^N)_{1\le i\le 2^n-1}$ from $\mathcal{C}([0,T];\mathcal{Z})$ into $\mathcal{Y} = [0,1]$, each h_i^N being $\mathcal{F}_{t_i}^{\mathrm{nat},\mathcal{Z}}$-measurable, such that:

$$\lim_{N\to\infty} \mathbb{P}\circ\Big(\mathbf{Z}, \big(h_i^N(\mathbf{Z})\big)_{1\le i\le 2^n-1}\Big)^{-1} = \mathbb{P}\circ\Big(\mathbf{Z}, \big(\mathbf{1}_{\{\varrho^n\le t_i\}}\big)_{1\le i\le 2^n-1}\Big)^{-1}.$$

For each $i\in\{1,\cdots,2^n-1\}$, we introduce the function $\tilde h_i^N$ defined on $\mathcal{C}([0,T])^2\times\mathcal{P}_1([0,T]\times\mathcal{C}([0,T]))$ by:

$$\tilde h_i^N(w^0,w,m) = h_i^N\big((w_t^0, w_t, m\circ\mathcal{E}_t^{-1})_{0\le t\le T}\big),$$

for $(w^0,w,m)\in\mathcal{C}([0,T])^2\times\mathcal{P}_1([0,T]\times\mathcal{C}([0,T]))$. The function $\tilde h_i^N$ will be easier to manipulate than h_i^N mostly because h_i^N is defined on the larger space $\mathcal{C}([0,T];\mathcal{Z})$. Observe that, for a sequence $(w^{0,p}, w^p, m^p)_{p\ge1}$ converging to (w^0,w,m) in the space $\mathcal{C}([0,T])^2\times\mathcal{P}_1([0,T]\times\mathcal{C}([0,T]))$, we have for any $p\ge 1$:

$$\sup_{0\le t\le T} W_1\big(m^p\circ\mathcal{E}_t^{-1}, m\circ\mathcal{E}_t^{-1}\big) \le W_1(m^p, m),$$

and thus $((w_t^{0,p}, w_t^p, m^p\circ\mathcal{E}_t^{-1})_{0\le t\le T})_{p\ge1}$ converges to $(w_t^0, w_t, m\circ\mathcal{E}_t^{-1})_{0\le t\le T}$ in $\mathcal{C}([0,T];\mathcal{Z})$. This shows that $\tilde h_i^N$ is continuous. Similarly, we can check that $\tilde h_i^N$ is $\mathcal{F}_{t_i}^{\mathrm{nat},\boldsymbol\theta}$-measurable. Indeed, by definition, we have $\mathcal{F}_{t_i}^{\mathrm{nat},\boldsymbol\theta} = \sigma\{(\boldsymbol{w}_{\cdot\wedge t_i}^0, \boldsymbol{w}_{\cdot\wedge t_i}, \nu\circ\mathcal{E}_{t_i}^{-1})\}$. So, the random variable $\mathcal{C}([0,T])^2\times\mathcal{P}_1([0,T]\times\mathcal{C}([0,T])) \ni (\boldsymbol{w}^0,\boldsymbol{w},\nu)\mapsto (w_t^0, w_t, \nu\circ\mathcal{E}_t^{-1})_{0\le t\le T}\in\mathcal{C}([0,T];\mathcal{Z})$ is $\mathcal{F}_{t_i}^{\mathrm{nat},\boldsymbol\theta}/\mathcal{F}_{t_i}^{\mathrm{nat},\mathcal{Z}}$ measurable.

Now for $t\in[0,T]$, we define:

$$H_t^{n,N}(w^0,w,m) = \sum_{i=1}^{2^n-1} \tilde h_i^N(w^0,w,m)\mathbf{1}_{[t_i,t_{i+1})}(t) + \mathbf{1}_{\{T\}}(t), \tag{7.98}$$

for all $w^0, w\in\mathcal{C}([0,T])$ and $m\in\mathcal{P}_1([0,T]\times\mathcal{C}([0,T]))$. We deduce that:

$$\lim_{N\to\infty}\mathbb{P}\circ\big(\mathbf{Z}, H_\cdot^{n,N}(\boldsymbol\theta)\big)^{-1} = \mathbb{P}\circ\big(\mathbf{Z}, \boldsymbol{H}_\cdot^n\big)^{-1},$$

where we used the formula (7.97) for $\boldsymbol{H}^n = (H_t^n)_{0\le t\le T}$ together with the fact that $\tilde h_i^N(\boldsymbol\theta) = h_i^N(\mathbf{Z})$. Here the second component is regarded as an element of the space $\mathcal{D}([0,T];[0,1])$. As usual it denotes the space of càd-làg functions from $[0,T]$ into $[0,1]$ equipped with the Skorohod topology $J1$. Observing that the mapping $\mathcal{C}([0,T];\mathcal{Z}) \ni (w_s^0, w_s, m_s)_{0\le s\le T} \mapsto ((w_s^0)_{0\le s\le T}, (w_s)_{0\le s\le T}, m_T) \in \mathcal{C}([0,T])^2\times\mathcal{P}_1([0,T]\times\mathcal{C}([0,T]))$ is continuous and maps the random variable $\mathbf{Z}$ onto $\boldsymbol\theta$, we also have:

$$\lim_{N\to\infty}\mathbb{P}\circ\big(\boldsymbol\theta, H_\cdot^{n,N}(\boldsymbol\theta)\big)^{-1} = \mathbb{P}\circ\big(\boldsymbol\theta, \boldsymbol{H}_\cdot^n\big)^{-1},$$

the convergence taking place in the space of probability measures on $\mathcal{C}([0,T])^2\times\mathcal{P}_1([0,T]\times\mathcal{C}([0,T]))\times\mathcal{D}([0,T];[0,1])$.

Using the function $\tilde{T}$ identified in Lemma 7.33, we deduce that:

$$\lim_{N\to\infty} \mathbb{P} \circ \Big(\boldsymbol{\theta}, \tilde{T}\big(H^{n,N}_{\cdot}(\boldsymbol{\theta})\big)\Big)^{-1} = \mathbb{P} \circ \big(\boldsymbol{\theta}, \tilde{T}(\boldsymbol{H}^{n}_{\cdot})\big)^{-1} \\ = \mathbb{P} \circ \big(\boldsymbol{\theta}, \varrho^{n}\big)^{-1}. \tag{7.99}$$

A crucial fact to complete the proof is to show that, without any loss of generality, the function $\mathcal{C}([0,T])^2 \times \mathcal{P}_1([0,T] \times \mathcal{C}([0,T])) \ni (w^0, w, m) \mapsto \tilde{T}(H^{n,N}_{\cdot}(w^0, w, m))$ can be assumed to be continuous. To do so, notice that replacing $\tilde{h}^N_i(w^0, w, m)$ by $\max_{1\le j\le i} \tilde{h}^N_j(w^0, w, m)$ in the definition (7.98) of $H^{n,N}_t$ does not change the value of $\tilde{T}(H^{n,N}_{\cdot}(\boldsymbol{\theta}))$ while making $H^{n,N}_{\cdot}(\boldsymbol{\theta})$ nondecreasing in time. Still, condition (7.92) of Lemma 7.33 may not be satisfied. So we redefine each $H^{n,N}_t$ for $t \in [0,T]$ by:

$$H^{n,N}_t(w^0, w, m) \\ = \frac{N-1}{N}\bigg(\sum_{i=1}^{2^n-1}\big(\max_{1\le j\le i} \tilde{h}^N_j(w^0, w, m)\big)\mathbf{1}_{[t_i, t_{i+1})}(t) + \mathbf{1}_{\{T\}}(t)\bigg) + \frac{t}{TN},$$

to preserve the limit (7.99), to force strict monotonicity (notice that the above form of $H^{n,N}_{\cdot}$ is now strictly increasing in time) and thus to satisfy (7.92). Since $H^{n,N}_0(w^0, w, m) = 0$ and $H^{n,N}_T(w^0, w, m) = 1$, we can now apply Lemma 7.33 for fixed values of n and N, as (w^0, w, m) varies. In particular, with this choice, the function $\mathcal{C}([0,T])^2 \times \mathcal{P}_1([0,T] \times \mathcal{C}([0,T])) \ni (w^0, w, m) \mapsto \widetilde{T}(H^{n,N}_{\cdot}(w^0, w, m))$ is continuous.

It is easily checked that, for any $t \in [0,T]$, $\widetilde{T}(H^{n,N}_{\cdot}(\boldsymbol{w}^0, \boldsymbol{w}, \nu)) \le t$ if and only if $H^{n,N}_t(\boldsymbol{w}^0, \boldsymbol{w}, \nu) \ge 1/2$. Since $H^{n,N}_t(\boldsymbol{w}^0, \boldsymbol{w}, \nu)$ is $\mathcal{F}^{\mathrm{nat},(\boldsymbol{w}^0,\boldsymbol{w},\nu)}_t$-measurable, this shows that $\widetilde{T}(H^{n,N}_{\cdot}(\boldsymbol{w}^0, \boldsymbol{w}, \nu))$ is a stopping time with respect to $\mathbb{F}^{\mathrm{nat},(\boldsymbol{w}^0,\boldsymbol{w},\nu)}$.

Third Step. We conclude the proof by showing that $\mathcal{R}(\mathbb{Q})$ is closed if the completion of $\mathbb{F}^{\mathrm{nat},\boldsymbol{\theta}}$ is right-continuous. Then, in order to check the compatibility constraint in Definition 7.30, we need to prove that, for any $t \in [0,T]$, $\mathcal{F}^{\mathrm{nat},\varrho}_{t+}$ and $\mathcal{F}^{\mathrm{nat},\boldsymbol{\theta}}_T$ are conditionally independent given $\mathcal{F}^{\mathrm{nat},\boldsymbol{\theta}}_t$. Let $(\mathbb{P}_n)_{n\ge1}$ be a sequence in $\mathcal{R}(\mathbb{Q})$ weakly converging toward $\mathbb{P} \in \mathcal{P}(\Omega_{\mathrm{canon}})$. Clearly, $\mathbb{P} \circ \boldsymbol{\theta}^{-1} = \lim_{n\to\infty} \mathbb{P}_n \circ \boldsymbol{\theta}^{-1} = \mathbb{Q}$. Now, we consider $t \in [0,T]$ such that $\mathbb{P}[\varrho = t] = 0$, and we use Lemma 7.32 to identify a function $g : [0,T] \to \mathbb{R}$ which is $\mathcal{F}^{\mathrm{nat},\varrho}_{t+}$-measurable and continuous $\mathbb{P} \circ \varrho^{-1}$-almost everywhere. We then let $\psi_t, \psi : \Omega_{\mathrm{input}} \to \mathbb{R}$ be bounded and measurable functions with respect to $\mathcal{F}^{\mathrm{nat},\boldsymbol{\theta}}_t$ and $\mathcal{F}^{\mathrm{nat},\boldsymbol{\theta}}_T$ respectively, and ϕ_t be a $\mathcal{F}^{\mathrm{nat},\boldsymbol{\theta}}_t$-measurable function from $\mathcal{C}(\Omega_{\mathrm{input}})$ into $\mathbb{R}$ such that:

$$\phi_t(\boldsymbol{\theta}) = \mathbb{E}^{\mathbb{P}}\Big[\psi(\boldsymbol{\theta}) \,|\, \mathcal{F}^{\mathrm{nat},\boldsymbol{\theta}}_t\Big].$$

Notice that $\phi_t(\boldsymbol{\theta}) = \mathbb{E}^{\mathbb{P}_n}[\psi(\boldsymbol{\theta}) \,|\, \mathcal{F}^{\mathrm{nat},\boldsymbol{\theta}}_t]$ for each $n \ge 1$ since $\mathbb{P}_n \circ \boldsymbol{\theta}^{-1} = \mathbb{P} \circ \boldsymbol{\theta}^{-1} = \mathbb{Q}$. Also, owing to the fact that $\mathbb{P}_n$ belongs to $\mathcal{R}(\mathbb{Q})$ for each $n \ge 1$, we know that:

$$\mathbb{E}^{\mathbb{P}_n}\Big[\psi(\boldsymbol{\theta})g(\varrho)\psi_t(\boldsymbol{\theta})\Big] = \mathbb{E}^{\mathbb{P}_n}\Big[\mathbb{E}^{\mathbb{P}_n}\big[\psi(\boldsymbol{\theta})g(\varrho) \,|\, \mathcal{F}^{\mathrm{nat},\boldsymbol{\theta}}_t\big]\psi_t(\boldsymbol{\theta})\Big] \\ = \mathbb{E}^{\mathbb{P}_n}\Big[\phi_t(\boldsymbol{\theta})\mathbb{E}^{\mathbb{P}_n}\big[g(\varrho) \,|\, \mathcal{F}^{\mathrm{nat},\boldsymbol{\theta}}_t\big]\psi_t(\boldsymbol{\theta})\Big] \\ = \mathbb{E}^{\mathbb{P}_n}\Big[\phi_t(\boldsymbol{\theta})g(\varrho)\psi_t(\boldsymbol{\theta})\Big],$$

where we used the fact that $\phi_t(\boldsymbol{\theta})$ is $\mathcal{F}_t^{\mathrm{nat},\boldsymbol{\theta}}$-measurable. Thus, by Lemma 7.34,

$$\begin{aligned}\mathbb{E}^{\mathbb{P}}\Big[\psi(\boldsymbol{\theta})g(\varrho)\psi_t(\boldsymbol{\theta})\Big] &= \lim_{n\to\infty}\mathbb{E}^{\mathbb{P}_n}\Big[\psi(\boldsymbol{\theta})g(\varrho)\psi_t(\boldsymbol{\theta})\Big]\\ &= \lim_{n\to\infty}\mathbb{E}^{\mathbb{P}_n}\Big[\phi_t(\boldsymbol{\theta})g(\varrho)\psi_t(\boldsymbol{\theta})\Big]\\ &= \mathbb{E}^{\mathbb{P}}\big[\phi_t(\boldsymbol{\theta})g(\varrho)\psi_t(\boldsymbol{\theta})\big]\\ &= \mathbb{E}^{\mathbb{P}}\Big[\mathbb{E}^{\mathbb{P}}\big[\psi(\boldsymbol{\theta})\,|\,\mathcal{F}_t^{\mathrm{nat},\boldsymbol{\theta}}\big]g(\varrho)\psi_t(\boldsymbol{\theta})\Big].\end{aligned}$$

Consequently, we can use Lemma 7.32 to conclude that:

$$\mathbb{E}^{\mathbb{P}}\Big[\psi(\boldsymbol{\theta})g(\varrho)\psi_t(\boldsymbol{\theta})\Big] = \mathbb{E}^{\mathbb{P}}\Big[\mathbb{E}^{\mathbb{P}}\big[\psi(\boldsymbol{\theta})\,|\,\mathcal{F}_t^{\mathrm{nat},\boldsymbol{\theta}}\big]g(\varrho)\psi_t(\boldsymbol{\theta})\Big],$$

for every bounded $\mathcal{F}_{t+}^{\mathrm{nat},\varrho}$-measurable g, not just those which are almost everywhere continuous. This shows that $\mathcal{F}_{t+}^{\mathrm{nat},\varrho}$ is conditionally independent of $\mathcal{F}_T^{\mathrm{nat},\boldsymbol{\theta}}$ given $\mathcal{F}_t^{\mathrm{nat},\boldsymbol{\theta}}$ for every $t \in [0,T]$ satisfying $\mathbb{P}[\varrho = t] = 0$.

Last Step. In order to complete the proof, it remains to show that this conditional independence property is valid, not only for the time indices $t \in [0,T]$ for which $\mathbb{P}[\varrho = t] = 0$, but in fact for all the time indices $t \in [0,T]$. To do so, we fix $t \in [0,T]$ and we pick a decreasing sequence $(t_n)_{n\geq 1}$ converging to t, such that $\mathbb{P}[\varrho = t_n] = 0$ for all $n \geq 1$. As above, we let $g : [0,T] \to \mathbb{R}$ and $\psi_t, \psi : \Omega_{\mathrm{input}} \to \mathbb{R}$ be three bounded measurable functions with respect to $\mathcal{F}_{t+}^{\mathrm{nat},\varrho}$, $\mathcal{F}_t^{\mathrm{nat},\boldsymbol{\theta}}$ and $\mathcal{F}_T^{\mathrm{nat},\boldsymbol{\theta}}$ respectively. Since $\mathcal{F}_{t+}^{\mathrm{nat},\varrho} \subset \mathcal{F}_{t_n}^{\mathrm{nat},\varrho}$ and $\mathcal{F}_t^{\mathrm{nat},\boldsymbol{\theta}} \subset \mathcal{F}_{t_n}^{\mathrm{nat},\boldsymbol{\theta}}$ for all $n \geq 1$, we deduce from the third step that:

$$\mathbb{E}^{\mathbb{P}}\Big[\psi(\boldsymbol{\theta})g(\varrho)\psi_t(\boldsymbol{\theta})\Big] = \mathbb{E}^{\mathbb{P}}\Big[\mathbb{E}^{\mathbb{P}}\big[\psi(\boldsymbol{\theta})\,|\,\mathcal{F}_{t_n}^{\mathrm{nat},\boldsymbol{\theta}}\big]g(\varrho)\psi_t(\boldsymbol{\theta})\Big].$$

We then conclude by letting n tend to ∞, and by using the fact that $\mathcal{F}_{t+}^{\mathrm{nat},\boldsymbol{\theta}}$ is included in the completion of $\mathcal{F}_t^{\mathrm{nat},\boldsymbol{\theta}}$. □

7.2.6 Equilibria in the Weak Sense for MFGs of Timing

Building upon the analogy with the classical treatment of strong and weak solutions of stochastic differential equations, but also of strong and weak solutions of stochastic differential mean field games as investigated in the first part of this volume, we are prompted to associate with each equilibrium in the strong sense a canonical candidate for being an equilibrium in the weak sense. Namely, if τ^* is an equilibrium in the strong sense on $\mathcal{C}([0,T])^2$, it is natural to consider the probability measure:

$$\mathbb{P}^* = \mathcal{W} \circ \big(w^0, w, \mathcal{L}\big((\tau^*, w)|w^0\big), \tau^*\big)^{-1} \tag{7.100}$$

on the canonical space Ω_{canon} and expect this law to be the epitome of a solution to the problem in the weak sense. In fact, with the definition we are about to state, this guesstimate becomes a theorem. See Theorem 7.39 below. Obviously, the conditional law $\mathcal{L}((\tau^*, \boldsymbol{w})|\boldsymbol{w}^0)$ appearing in the right-hand side of (7.100) is the conditional law under $\mathcal{W}$.

To make this intuition clear, we need first a proper definition of what a weak solution is. Recalling that $\boldsymbol{w}^0$, $\boldsymbol{w}$, ν and ϱ denotes the canonical variables on Ω_{canon} and denoting by e be the canonical projection from $[0, T] \times \mathcal{C}([0, T])$ onto $[0, T]$, we introduce the following definition:

Definition 7.37 *A probability measure $\mathbb{P}$ on the canonical space Ω_{canon} is said to be a weak equilibrium for the MFG of timing if*

1. *Under $\mathbb{P}$, the process $(\boldsymbol{w}^0, \boldsymbol{w})$ has distribution $\mathcal{W}$ and is a two-dimensional Brownian motion process with respect to the natural filtration of $(\boldsymbol{w}^0, \boldsymbol{w}, \nu, \varrho)$;*
2. *The pair $(\boldsymbol{w}^0, \nu)$ is independent of $\boldsymbol{w}$ under $\mathbb{P}$;*
3. *The process $(\boldsymbol{w}^0, \boldsymbol{w}, \nu)$ is compatible with the filtration generated by the process $(\boldsymbol{w}^0, \boldsymbol{w}, \nu, \varrho)$ in the sense that, under $\mathbb{P}$, $\mathcal{F}_t^{\text{nat},\varrho}$ is conditionally independent of $(\boldsymbol{w}^0, \boldsymbol{w}, \nu)$ given $\mathcal{F}_t^{\text{nat},(\boldsymbol{w}^0,\boldsymbol{w},\nu)}$, for every $t \in [0, T]$.*
4. *The probability measure $\mathbb{P}$ belongs to* $\arg\sup_{\mathbb{P}'} \mathbb{E}^{\mathbb{P}'}[F(\boldsymbol{w}^0, \boldsymbol{w}, \nu \circ e^{-1}, \varrho)]$, *where the supremum is taken over all the probability measures $\mathbb{P}'$ on Ω_{input} satisfying 1–3 as well as $\mathbb{P}' \circ (\boldsymbol{w}^0, \boldsymbol{w}, \nu)^{-1} = \mathbb{P} \circ (\boldsymbol{w}^0, \boldsymbol{w}, \nu)^{-1}$;*
5. *The weak fixed point condition holds:*

$$\mathbb{P}[(\varrho, \boldsymbol{w}) \in \cdot \,|\, \boldsymbol{w}^0, \nu] = \nu(\cdot), \quad \mathbb{P} - a.s..$$

Of course, the notion of compatibility used in condition 3 is reminiscent of that introduced in Definition 2.16. As explained earlier, the rationale for such a compatibility constraint is that we cannot expect ν to remain $\sigma\{\boldsymbol{w}^0\}$-measurable or ϱ to be $\sigma\{\boldsymbol{w}^0, \boldsymbol{w}\}$-measurable after taking weak limits. Somehow, condition 3 is here to capture an important structure we do retain in these limits. Also, the rationale for enlarging the support of the probability measure in the fixed point condition is the same as in stochastic differential mean field games. Somehow, it renders the compatibility condition 3 stable under weak convergence of weak equilibria. This is illustrated by the following lemma.

Lemma 7.38 *A probability measure $\mathbb{P} \in \mathcal{P}(\Omega_{\text{canon}})$ satisfying condition 5 in Definition 7.37 also satisfies condition 3 if the σ-fields:*

$$\sigma\{w_s - w_t;\, t \le s \le T\} \quad \textit{and} \quad \mathcal{F}_T^{\text{nat},(\boldsymbol{w}^0,\nu)} \vee \mathcal{F}_t^{\text{nat},(\boldsymbol{w}^0,\boldsymbol{w},\nu,\varrho)}$$

are independent under $\mathbb{P}$, for every $t \in [0, T]$.

In short, Lemma 7.38 says that it suffices to check an independence instead of a conditional independence property in order to prove compatibility. Of course, this sounds much easier since independence is, generally speaking, stable under weak convergence.

Proof of Lemma 7.38. Let $\mathbb{P} \in \mathcal{P}(\Omega_{\text{canon}})$ satisfy condition 5. For a given $t \in [0, T]$, we then consider five bounded test functions g, ψ_t, ψ, ζ_t and ζ, $g : [0, T] \to \mathbb{R}$ being $\mathcal{F}_t^{\text{nat},\varrho}$-measurable, $\psi_t, \psi : \mathcal{C}([0, T]) \times \mathcal{P}_1([0, T] \times \mathcal{C}([0, T])) \to \mathbb{R}$ being measurable with respect to $\mathcal{F}_t^{\text{nat},(w^0,\nu)}$ and $\mathcal{F}_T^{\text{nat},(w^0,\nu)}$, and $\zeta_t, \zeta : \mathcal{C}([0, T]) \to \mathbb{R}$ being measurable with respect to $\mathcal{F}_t^{\text{nat},w}$ and $\sigma\{w_s - w_t;\ t \le s \le T\}$ respectively.

Then, if $\sigma\{w_s - w_t;\ t \le s \le T\}$ and $\mathcal{F}_T^{\text{nat},(w^0,\nu)} \vee \mathcal{F}_t^{\text{nat},(w^0,w,\nu,\varrho)}$ are independent, condition 5 yields:

$$
\begin{aligned}
&\mathbb{E}^{\mathbb{P}}\Big[g(\varrho)\psi_t(w^0, \nu)\psi(w^0, \nu)\zeta_t(w)\zeta(w)\Big] \\
&= \mathbb{E}^{\mathbb{P}}\Big[g(\varrho)\psi_t(w^0, \nu)\psi(w^0, \nu)\zeta_t(w)\Big]\mathbb{E}^{\mathbb{P}}\big[\zeta(w)\big] \\
&= \mathbb{E}^{\mathbb{P}}\Big[\Big(\int_{[0,T]\times\mathcal{C}([0,T])} g(s)\zeta_t(w)d\nu(s, w)\Big)\psi_t(w^0, \nu)\psi(w^0, \nu)\Big]\mathbb{E}^{\mathbb{P}}\big[\zeta(w)\big].
\end{aligned}
$$

Since $\int_{[0,T]\times\mathcal{C}([0,T])} g(s)\zeta_t(w)d\nu(s, w)$ is $\mathcal{F}_t^{\text{nat},\nu}$-measurable, we can take the conditional expectation given $\mathcal{F}_t^{\text{nat},(w^0,\nu)}$ in the first expectation on the right-hand side and then proceed backward. We deduce:

$$
\begin{aligned}
&\mathbb{E}^{\mathbb{P}}\Big[g(\varrho)\psi_t(w^0, \nu)\psi(w^0, \nu)\zeta_t(w)\zeta(w)\Big] \\
&= \mathbb{E}^{\mathbb{P}}\Big[\Big(\int_{[0,T]\times\mathcal{C}([0,T])} g(s)\zeta_t(w)d\nu(s, w)\Big)\psi_t(w^0, \nu)\mathbb{E}^{\mathbb{P}}\big[\psi(w^0, \nu)\,|\,\mathcal{F}_t^{\text{nat},(w^0,\nu)}\big]\Big]\mathbb{E}^{\mathbb{P}}\big[\zeta(w)\big] \\
&= \mathbb{E}^{\mathbb{P}}\Big[g(\varrho)\zeta_t(w)\psi_t(w^0, \nu)\mathbb{E}^{\mathbb{P}}\big[\psi(w^0, \nu)\,|\,\mathcal{F}_t^{\text{nat},(w^0,\nu)}\big]\Big]\mathbb{E}^{\mathbb{P}}\big[\zeta(w)\big] \\
&= \mathbb{E}^{\mathbb{P}}\Big[g(\varrho)\zeta_t(w)\psi_t(w^0, \nu)\mathbb{E}^{\mathbb{P}}\big[\psi(w^0, \nu)\zeta(w)\,|\,\mathcal{F}_t^{\text{nat},(w^0,\nu)}\big]\Big],
\end{aligned}
$$

which suffices to complete the proof. □

The following result shows that our definition of a weak equilibrium is consistent with that of a strong equilibrium.

Theorem 7.39 *Assume that assumption* **MFG of Timing Set-Up** *(including* **(A3)***) and condition* **(A1)** *in assumption* **MFG of Timing Regularity** *are in force. If τ^* is a strong equilibrium in the sense of Definition 7.18 on the canonical space $\mathcal{C}([0, T])^2$, then the probability measure $\mathbb{P}^*$ defined by* (7.100) *is a weak equilibrium in the sense of Definition 7.37.*

Proof. We first observe that $\mathbb{P}^*$ satisfies property 2 of Definition 7.37 of a weak equilibrium.

First Step. Next, we prove conditions 1 and 3. To do so, we prove that, for every $t \in [0,T]$, $\mathcal{F}_t^{\mathrm{nat},\varrho}$ is included in $\mathcal{F}_t^{(w^0,w)}$ and $\mathcal{F}_t^{\mathrm{nat},\nu}$ is included in $\mathcal{F}_t^{w^0}$, where $\mathbb{F}^{(w^0,w)}$ and $\mathbb{F}^{w^0}$ are the $\mathbb{P}^*$-completions of the filtrations generated by (w^0,w) and w^0 respectively.

The first claim is pretty clear. Indeed, since τ^* is an $\mathbb{F}^{(w^0,w)}$-stopping time, we deduce that $\mathcal{F}_t^{\mathrm{nat},\varrho}$ is contained in $\mathcal{F}_t^{(w^0,w)}$ for any $t \in [0,T]$.

To prove the second claim, we start with the following observation. For any given $t \in [0,T]$ and any bounded $\mathcal{F}_t^{\mathrm{nat},(w^0,w)}$-measurable function $g_t : \mathcal{C}([0,T])^2 \to \mathbb{R}$, we have:

$$\mathbb{E}^{\mathcal{W}}\big[g_t(w^0,w)\,|\,w^0\big] = \mathbb{E}^{\mathcal{W}}\big[g_t(w^0,w)\,|\,\mathcal{F}_t^{\mathrm{nat},w^0}\big],$$

almost surely, where $\mathbb{E}^{\mathcal{W}}$ denotes the expectation on $\mathcal{C}([0,T])^2$ under $\mathcal{W}$. Therefore, for any $s \in [0,t)$ and any $C \in \mathcal{F}_t^{\mathrm{nat},w}$, $\mathcal{W}[\tau^* \le s, w \in C\,|\,w^0] = \mathcal{W}[\tau^* \le s, w \in C\,|\,\mathcal{F}_t^{\mathrm{nat},w^0}]$ with probability 1 under $\mathcal{W}$. Therefore, letting $\mu(\cdot) = \mathcal{W}[(\tau^*,w) \in \cdot\,|\,w^0]$, we deduce that $\mu([0,s]\times C) \in \mathcal{F}_t^{w^0}$. This shows that $\mathcal{F}_t^{\mathrm{nat},\nu}$ is included in $\mathcal{F}_t^{w^0}$.

Properties 1 and 3 easily follow. Also, observe that the weak fixed point condition 5 holds because μ is $\sigma\{w^0\}$-measurable.

Second Step. It remains to check the optimality condition 4. First, we observe from the equilibrium property of τ^* that:

$$\begin{aligned}\mathbb{E}^{\mathbb{P}^*}\big[F\big(w^0,w,\nu\circ e^{-1},\varrho\big)\big] &= \mathbb{E}^{\mathbb{P}^*}\big[F\big(w^0,w,\mu\circ e^{-1},\tau^*\big)\big]\\ &\ge \mathbb{E}^{\mathbb{P}^*}\big[F\big(w^0,w,\mu\circ e^{-1},\sigma\big)\big],\end{aligned} \tag{7.101}$$

for every $\mathbb{F}^{(w^0,w)}$-stopping time σ defined on the canonical probability space $\mathcal{C}([0,T])^2$. Also, since μ is $\sigma\{w^0\}$-measurable, there exists a measurable map $\tilde{\mu} : \mathcal{C}([0,T]) \to \mathcal{P}_1([0,T]\times\mathcal{C}([0,T]))$ such that $\mu = \tilde{\mu}(w^0)$ and the function $\mathcal{C}([0,T]) \ni w^0 \mapsto [\tilde{\mu}(w^0)](C)$ is measurable with respect to the completion of $\mathcal{F}_t^{\mathrm{nat},w^0}$ under the Wiener measure for any $t \in [0,T]$ and any $C \in \mathcal{F}_t^{\mathrm{nat},\nu}$. Hence, for $\mathbb{P}' \in \mathcal{P}(\Omega_{\mathrm{canon}})$ satisfying properties 1–3 of Definition 7.37 as well as $\mathbb{P}'\circ(w^0,w,\nu)^{-1} = \mathbb{P}^*\circ(w^0,w,\nu)^{-1} = \mathcal{W}\circ(w^0,w,\tilde{\mu}(w^0))^{-1}$, Theorem 7.31 with $\mathbb{Q} = \mathcal{W}\circ(w^0,w,\tilde{\mu}(w^0))^{-1}$ implies that there exists a sequence of $\mathbb{F}^{(w^0,w)}$-stopping times $(\sigma_n)_{n\ge1}$, of the form $\sigma_n = \tilde{\tau}_n(w^0,w,\tilde{\mu}(w^0))$ for each $n \ge 1$, such that:

$$\mathbb{P}' = \lim_{n\to\infty} \mathbb{P}'\circ\big(w^0,w,\tilde{\mu}(w^0),\sigma_n\big)^{-1}.$$

The above identity may be rewritten:

$$\mathbb{P}' = \lim_{n\to\infty} \mathcal{W}\circ\big(w^0,w,\tilde{\mu}(w^0),\tilde{\tau}_n(w^0,w,\tilde{\mu}(w^0))\big)^{-1}.$$

Using (7.101) we get:

$$\begin{aligned}\mathbb{E}^{\mathbb{P}^*}\big[F\big(w^0,w,\nu\circ e^{-1},\varrho\big)\big] &\ge \lim_{n\to\infty} \mathbb{E}^{\mathcal{W}}\big[F\big(w^0,w,\tilde{\mu}(w^0)\circ e^{-1},\tilde{\tau}_n(w^0,w,\tilde{\mu}(w^0))\big)\big]\\ &= \mathbb{E}^{\mathbb{P}'}\big[F\big(w^0,w,\nu\circ e^{-1},\varrho\big)\big],\end{aligned}$$

where the last equality follows from the continuity of F in the last argument and Lemma 7.34.

□

7.2.7 Weak Equilibria as Limits of Finite Player Games Equilibria

In this subsection, we show that Cesaro limits of approximate equilibria of finite player games are weak equilibria. This is the analogue of Theorem 6.18 for mean field games of timing.

Theorem 7.40 *Let assumption* **MFG of Timing Set-Up** *(including* **(A3)***) and condition (***A1***) in assumption* **MFG of Timing Regularity** *be in force, and consider a complete probability space* $(\Xi, \mathcal{G}, \mathbf{P})$ *equipped with a sequence* $(\mathbf{W}^i)_{i\geq 0}$ *of independent Brownian motions, each pair* $(\mathbf{W}^0, \mathbf{W}^i)$, *for* $i \geq 1$, *being distributed according to* $\mathcal{W}$. *Assume further that* $(\epsilon_N)_{N\geq 1}$ *is a sequence of positive numbers converging to* 0, *and that on* $(\Xi, \mathcal{G}, \mathbf{P})$, *for each* $N \geq 1$, $(\tau^{1,N,*}, \cdots, \tau^{N,N,*})$ *is an* ϵ_N *- Nash equilibrium of the* N*-player game associated with* (7.85) *and* (7.86). *If, for each* $N \geq 1$, *we define the measure* $\mathbb{P}^N$ *by the Cesaro mean:*

$$\mathbb{P}^N = \frac{1}{N}\sum_{i=1}^{N} \mathbf{P} \circ \left(\mathbf{W}^0, \mathbf{W}^i, \bar{\mathfrak{m}}^N_{(\tau^{\cdot,N,*},\mathbf{W}^{\cdot})}, \tau^{i,N,*}\right)^{-1}, \tag{7.102}$$

with:

$$\bar{\mathfrak{m}}^N_{(\tau^{\cdot,N,*},\mathbf{W}^{\cdot})} = \frac{1}{N}\sum_{i=1}^{N} \delta_{(\tau^{i,N,*},\mathbf{W}^i)}, \tag{7.103}$$

then the sequence $(\mathbb{P}^N)_{N\geq 1}$ *is tight on* Ω_{canon} *and any limit point of this sequence is a weak MFG of timing equilibrium.*

In the proof of Theorem 7.40, one of the important consequences of the compatibility assumption will come into play by allowing to approximate randomized stopping times with nonrandomized stopping times.

Proof. Throughout the proof, we denote by $\mathbf{E}^{\mathbf{P}}$. the expectation associated with $\mathbf{P}$. Also, for any $N \geq 1$, we let $\bar{\mathfrak{m}}^N$ be the empirical distribution (7.103), i.e., we drop the subscript $(\tau^{\cdot,N,*}, \mathbf{W}^{\cdot})$. We notice that, for each $N \geq 1$ and any $i \in \{1, \cdots, N\}$, $\mathbf{P} \circ (\mathbf{W}^0, \mathbf{W}^i, \bar{\mathfrak{m}}^N)^{-1}$ has $\mathcal{W}$ as marginal law on $\mathcal{C}([0,T])^2$. Proceeding as in the proof of Theorem 7.31, this constraint is convex. Hence, $\mathbb{Q}^N = \mathbb{P}^N \circ (w^0, w, \nu)^{-1}$ also has $\mathcal{W}$ as marginal law on $\mathcal{C}([0,T])^2$.

Tightness of $(\mathbb{P}^N \circ (w^0, w)^{-1})_{N\geq 1}$ is straightforward. Tightness of $(\mathbb{P}^N \circ \varrho^{-1})_{N\geq 1}$ is an immediate consequence of the fact that $[0,T]$ is compact. Tightness of $(\mathbb{P}^N \circ \nu^{-1})_{N\geq 1}$ may be proved as follows. For any compact subset $K \subset \mathcal{C}([0,T])$ and for every $N \geq 1$, $(\mathbb{P}^N \circ \nu^{-1})([0,T] \times K^{\complement})$ is equal to:

$$\bar{\mathfrak{m}}^N([0,T] \times K^{\complement}) = \frac{1}{N}\sum_{i=1}^{N} \mathbf{1}_{\{W^i \in K^{\complement}\}},$$

from which we deduce that:

$$\mathbb{E}^{\mathbb{P}^N}\big[\nu([0,T]\times K^{\complement})\big] = \mathbf{E}^{\mathbf{P}}\big[\bar{\mathfrak{m}}^N([0,T]\times K^{\complement})\big] = \mathbf{P}\big[\mathbf{W}^1 \in \mathcal{K}^{\complement}\big].$$

Proceeding as in the proof of Lemma 3.16, we deduce that the sequence $(\mathbb{P}^N \circ \nu^{-1})_{N\geq 1}$ is tight on $\mathcal{P}([0,T]\times \mathcal{C}([0,T]))$ equipped with the topology of weak convergence. To prove tightness on $\mathcal{P}_1([0,T]\times \mathcal{C}([0,T]))$, we compute:

$$\begin{aligned}
&\mathbb{E}^{\mathbb{P}^N}\bigg[\int_{[0,T]\times\mathcal{C}([0,T])} \sup_{0\leq t\leq T} |w_t|^2 d\nu(t,w)\bigg] \\
&= \mathbf{E}^{\mathbf{P}}\bigg[\int_{[0,T]\times\mathcal{C}([0,T])} \sup_{0\leq t\leq T} |w_t|^2 d\bar{\mathfrak{m}}^N(t,w)\bigg] = \mathbf{E}^{\mathbf{P}}\big[\sup_{0\leq t\leq T} |W_t^1|^2\big],
\end{aligned}$$

and we conclude as in the proof of Lemma 3.16.

We deduce that the sequence $(\mathbb{P}^N)_{N\geq 1}$ is tight on Ω_{canon}. We denote by $\mathbb{P}$ any limit point of $(\mathbb{P}^N)_{N\geq 1}$. Working with a subsequence if necessary, we shall assume without any loss of generality that $(\mathbb{P}^N)_{N\geq 1}$ converges to $\mathbb{P}$. Clearly, $\mathbb{Q} = \mathbb{P}\circ(\boldsymbol{w}^0,\boldsymbol{w},\nu)^{-1}$ has $\mathcal{W}$ as marginal law on $\mathcal{C}([0,T])^2$. We now check step by step that $\mathbb{P}$ is weak solution of the mean field game.

First Step. To prove the fixed point condition 5 of the definition of a weak equilibrium, take a bounded and continuous real-valued function f on $\mathcal{C}([0,T])\times\mathcal{P}_1([0,T]\times\mathcal{C}([0,T]))$ and a bounded and continuous real-valued function g on $[0,T]\times\mathcal{C}([0,T])$. Notice that:

$$\begin{aligned}
\mathbb{E}^{\mathbb{P}}\big[f(\boldsymbol{w}^0,\nu)g(\varrho,\boldsymbol{w})\big] &= \lim_{N\to\infty}\frac{1}{N}\sum_{i=1}^N \mathbf{E}^{\mathbf{P}}\big[f(\mathbf{W}^0,\bar{\mathfrak{m}}^N)g(\tau^{i,N,*},\mathbf{W}^i)\big] \\
&= \lim_{N\to\infty}\mathbf{E}^{\mathbf{P}}\bigg[f(\mathbf{W}^0,\bar{\mathfrak{m}}^N)\int g\,d\bar{\mathfrak{m}}^N\bigg] = \mathbb{E}^{\mathbb{P}}\bigg[f(\boldsymbol{w}^0,\nu)\int g\,d\nu\bigg],
\end{aligned}$$

which is sufficient to conclude.

Second Step. We now check condition 1. We first notice that, for each $N \geq 1$ and any $i\in\{1,\cdots,N\}$, the stopping time $\tau^{i,N,*}$ is a stopping time with respect to the completion $\mathbb{F}^{(\mathbf{W}^0,\mathbf{W}^i)}$ of the filtration generated by $(\mathbf{W}^0,\mathbf{W}^i)$ under $\mathbf{P}$. Hence, the filtration generated by the process $(t\wedge\tau^{i,N,*})_{0\leq t\leq T}$ is included in $\mathbb{F}^{(\mathbf{W}^0,\mathbf{W}^i)}$. Also, we deduce that, for any $t\in[0,T]$, any $C\in\mathcal{F}_t^{\text{nat},(\varrho,\boldsymbol{w})}$, $\bar{\mathfrak{m}}^N(C)$ belongs to $\mathcal{F}_t^{(\mathbf{W}^0,\cdots,\mathbf{W}^N)}$, where $\mathbb{F}^{(\mathbf{W}^0,\cdots,\mathbf{W}^N)}$ is the completion of the filtration generated by $(\mathbf{W}^0,\mathbf{W}^1,\cdots,\mathbf{W}^N)$.

As a result, for each $N\geq 1$ and any $i\in\{1,\cdots,N\}$, $(\boldsymbol{w}^0,\boldsymbol{w})$ is a Brownian motion with respect to the natural filtration generated by $(\boldsymbol{w}^0,\boldsymbol{w},\nu,\varrho)$ under the probability measure $\mathbb{P}^{i,N} = \mathbf{P}\circ(\mathbf{W}^0,\mathbf{W}^i,\bar{\mathfrak{m}}^N,\tau^{i,N,*})^{-1}$. By convexity, $(\boldsymbol{w}^0,\boldsymbol{w})$ is also a Brownian motion with respect to $(\boldsymbol{w}^0,\boldsymbol{w},\nu,\varrho)$ under each $\mathbb{P}^N$, $N\geq 1$. Proceeding as in the proof of Theorem 7.29, it is plain to deduce that the same holds under the limiting probability $\mathbb{P}$. This proves condition 1.

Third Step. We now prove conditions 2 and 3. Since condition 5 has already been proved, it suffices to verify the criterion established in Lemma 7.38.

We start with the following observation. For each $N\geq 1$ and any $i\in\{1,\cdots,N\}$, and for all $t\in[0,T]$, the σ-field $\sigma\{\mathbf{W}^i\}$ is independent of $\sigma\{\mathbf{W}^0,\cdots,\mathbf{W}^{i-1},\mathbf{W}^{i+1},\cdots,\mathbf{W}^N\}$ and the σ-field $\sigma\{W_s^i - W_t^i;\, t\leq s\leq T\}$ is also independent of the σ-field $\mathcal{F}_t^{(\mathbf{W}^0,\cdots,\mathbf{W}^N)}\vee\mathcal{F}_T^{(\mathbf{W}^0,\cdots,\mathbf{W}^{i-1},\mathbf{W}^{i+1},\mathbf{W}^N)}$.

Letting:

$$\bar{\mathfrak{m}}^{-i,N} = \frac{1}{N-1}\sum_{j=1,j\neq i}^{N} \delta_{(\tau^{j,N,*},\boldsymbol{W}^j)},$$

we deduce that, under $\mathbf{P}$, the σ-fields $\sigma\{\boldsymbol{W}^i\}$ and $\sigma\{\boldsymbol{W}^0, \bar{\mathfrak{m}}^{-i,N}\}$ are independent, and similarly, the σ-fields $\sigma\{W_s^i - W_t^i;\ t \leq s \leq T\}$ and $\sigma\{\boldsymbol{W}^i_{\cdot\wedge t}, \tau^{i,N,*} \wedge t\} \vee \sigma\{\boldsymbol{W}^0, \bar{\mathfrak{m}}^{-i,N}\}$ are also independent. Hence, on the canonical space, the σ-fields $\sigma\{\boldsymbol{w}\}$ and $\sigma\{\boldsymbol{w}^0, \nu\}$ on the one hand, and the σ-fields $\sigma\{w_s - w_t;\ t \leq s \leq T\}$ and $\mathcal{F}_T^{\mathrm{nat},(\boldsymbol{w}^0,\nu)} \vee \mathcal{F}_t^{\mathrm{nat},(\boldsymbol{w}^0,\boldsymbol{w},\nu,\varrho)}$ on the other hand, are independent under the probability measure $\mathbb{P}^{-i,N}$ given by:

$$\mathbb{P}^{-i,N} = \mathbf{P}\circ\big(\boldsymbol{W}^0, \boldsymbol{W}^i, \bar{\mathfrak{m}}^{-i,N}, \tau^{i,N,*}\big)^{-1}.$$

Following the convexity argument used above, we deduce that the σ-fields $\sigma\{\boldsymbol{w}\}$ and $\sigma\{\boldsymbol{w}^0, \nu\}$ on the one hand, and the σ-fields $\sigma\{w_s - w_t;\ t \leq s \leq T\}$ and $\mathcal{F}_T^{\mathrm{nat},(\boldsymbol{w}^0,\nu)} \vee \mathcal{F}_t^{\mathrm{nat},(\boldsymbol{w}^0,\boldsymbol{w},\nu,\varrho)}$ on the other hand, are independent under the probability measure $\bar{\mathbb{P}}^N$ defined by:

$$\bar{\mathbb{P}}^N = \frac{1}{N}\sum_{i=1}^{N}\mathbb{P}^{-i,N}.$$

The next step in the proof of condition 3 is to observe that:

$$W_1\big(\bar{\mathbb{P}}^N, \mathbb{P}^N\big) \leq \frac{1}{N}\sum_{i=1}^{N} W_1\big(\mathbb{P}^{-i,N}, \mathbb{P}^{i,N}\big),$$

for all $N \geq 1$, where W_1 here denotes the 1-Wasserstein distance on Ω_{canon}, Ω_{canon} itself being equipped with the ℓ_1-distance induced by the distances of each of its components. Notice now that, for each $i \in \{1, \cdots, N\}$,

$$W_1\big(\mathbb{P}^{-i,N}, \mathbb{P}^{i,N}\big) \leq \mathbb{E}\big[W_1\big(\bar{\mathfrak{m}}^{-i,N}, \bar{\mathfrak{m}}^N\big)\big],$$

where we denoted by the same symbol W_1 the Wasserstein distance on $\mathcal{P}_1([0,T]\times\mathcal{C}([0,T]))$ in the right-hand side. It is then pretty straightforward to bound the last term in the above inequality by c/N, for a constant c independent of N. We end up with:

$$W_1\big(\mathbb{P}^{-i,N}, \mathbb{P}^{i,N}\big) \leq \frac{c}{N}. \quad \text{which implies} \quad W_1\big(\bar{\mathbb{P}}^N, \mathbb{P}^N\big) \leq \frac{c}{N}.$$

Therefore, $(\bar{\mathbb{P}}^N)_{N\geq 1}$ weakly converges to $\mathbb{P}$.

Since the σ-fields $\sigma\{\boldsymbol{w}\}$ and $\sigma\{\boldsymbol{w}^0, \nu\}$ on the one hand, and the σ-fields $\sigma\{w_s - w_t;\ t \leq s \leq T\}$ and $\mathcal{F}_T^{\mathrm{nat},(\boldsymbol{w}^0,\nu)} \vee \mathcal{F}_t^{\mathrm{nat},(\boldsymbol{w}^0,\boldsymbol{w},\nu,\varrho)}$ on the other hand, are independent under the probability measure $\bar{\mathbb{P}}^N$, for each $N \geq 1$, the same holds under the limiting measure $\mathbb{P}$. This proves condition 2 and, by Lemma 7.38, this proves condition 3 as well.

Fourth Step. Finally, we show that $\mathbb{P}$ satisfies the optimality condition 4 in Definition 7.37. We let $\mathbb{Q} = \mathbb{P}\circ(\boldsymbol{w}^0, \boldsymbol{w}, \nu)^{-1}$. By the conclusion of the third step, we know that $\mathbb{P} \in \mathcal{R}(\mathbb{Q})$. By Theorem 7.31, we can approximate any $\mathbb{P}' \in \mathcal{R}(\mathbb{Q})$ by a sequence of the form

$(\mathbb{Q} \circ (\boldsymbol{w}^0, \boldsymbol{w}, \nu, \tilde{\sigma}_n(\boldsymbol{w}^0, \boldsymbol{w}, \nu))^{-1})_{n\geq 1}$ where, for each $n \geq 1$, $\tilde{\sigma}_n$ is a continuous function from Ω_{input} to $[0, T]$, and $\tilde{\sigma}_n(\boldsymbol{w}^0, \boldsymbol{w}, \nu)$ is a stopping time with respect to the completion of the filtration $\mathbb{F}^{\text{nat},(\boldsymbol{w}^0,\boldsymbol{w},\nu)}$ under $\mathbb{Q}$. Therefore, by continuity of F in the last argument and by Lemma 7.34, see for instance the end of the proof of Theorem 7.39, it suffices to show:

$$\mathbb{E}^{\mathbb{P}}\big[F\big(\boldsymbol{w}^0, \boldsymbol{w}, \nu \circ e^{-1}, \varrho\big)\big] \geq \mathbb{E}^{\mathbb{Q}}\big[F\big(\boldsymbol{w}^0, \boldsymbol{w}, \nu \circ e^{-1}, \tilde{\sigma}(\boldsymbol{w}^0, \boldsymbol{w}, \nu)\big)\big]$$

for every continuous function $\tilde{\sigma}$ from Ω_{input} to $[0, T]$ such that $\tilde{\sigma}(\boldsymbol{w}^0, \boldsymbol{w}, \nu)$ is a stopping time respect to the $\mathbb{Q}$-completion of the filtration $\mathbb{F}^{\text{nat},(\boldsymbol{w}^0,\boldsymbol{w},\nu)}$.

Now fix such a mapping $\tilde{\sigma}$. The difficulty we face below is that, under $\mathbb{P}^N$, $\tilde{\sigma}(\boldsymbol{w}^0, \boldsymbol{w}, \nu)$ is not a stopping time with respect to $\mathbb{F}^{(\boldsymbol{w}^0,\boldsymbol{w})}$. So, in order to proceed, we use the following fact, which we prove in the fifth step below. Since $[0, T] \times \mathcal{C}([0, T])$ is a Polish space, its Borel σ-field is generated by a countable field; hence, by Carathéodory's theorem, we can define $\mathbf{E}^{\mathbf{P}}[\bar{\mathfrak{m}}^N \mid \boldsymbol{W}^0]$ as the random probability measure determined by $\mathbf{E}^{\mathbf{P}}[\bar{\mathfrak{m}}^N \mid \boldsymbol{W}^0](C) = \mathbf{E}^{\mathbf{P}}[\bar{\mathfrak{m}}^N(C) \mid \boldsymbol{W}^0]$ for $C \in [0, T] \times \mathcal{C}([0, T])$. We claim that, with $\mathbf{P}$-probability 1, it holds:

$$\lim_{N\to\infty} W_1\big(\bar{\mathfrak{m}}^N, \mathbf{E}^{\mathbf{P}}[\bar{\mathfrak{m}}^N \mid \boldsymbol{W}^0]\big) = 0, \tag{7.104}$$

the proof of which is deferred to the fifth step below. Importantly, for any $t \in [0, T]$ and any $C \in \mathcal{F}_t^{\text{nat},(\varrho,\boldsymbol{w})}$, $\mathbf{E}^{\mathbf{P}}[\bar{\mathfrak{m}}^N \mid \boldsymbol{W}^0](C)$ is $\mathcal{F}_t^{\boldsymbol{W}^0}$-measurable. We deduce that, for each $N \geq 1$, there exists a measurable function $\tilde{\mathfrak{m}}^N$ from $\mathcal{C}([0, T])$ into $\mathcal{P}_1([0, T] \times \mathcal{C}([0, T]))$ such that, for all $C \in \mathcal{F}_t^{\text{nat},(\varrho,\boldsymbol{w})}$, $[\tilde{\mathfrak{m}}^N(\boldsymbol{W}^0)](C)$ is $\mathcal{F}_t^{\boldsymbol{W}^0}$-measurable and, with $\mathbf{P}$-probability 1,

$$\lim_{N\to\infty} W_1\big(\bar{\mathfrak{m}}^N, \tilde{\mathfrak{m}}^N(\boldsymbol{W}^0)\big) = 0. \tag{7.105}$$

For each $N \geq 1$ and any $i \in \{1, \cdots, N\}$, we then let $\sigma^{i,N} = \tilde{\sigma}(\boldsymbol{W}^0, \boldsymbol{W}^i, \tilde{\mathfrak{m}}^N(\boldsymbol{W}^0))$ for any $i \in \{1, \cdots, N\}$. By construction, $\sigma^{i,N}$ is an $\mathbb{F}^{(\boldsymbol{W}^0,\boldsymbol{W}^i)}$-stopping time. In particular, the Nash property implies:

$$\begin{aligned}
\mathbb{E}^{\mathbb{P}}\big[F\big(\boldsymbol{w}^0, \boldsymbol{w}, \nu \circ e^{-1}, \varrho\big)\big] &= \lim_{N\to\infty} \frac{1}{N} \sum_{i=1}^N \mathbf{E}^{\mathbf{P}}\big[F\big(\boldsymbol{W}^0, \boldsymbol{W}^i, \bar{\mu}^N, \tau^{i,N,*}\big)\big] \\
&\geq \limsup_{N\to\infty} \frac{1}{N} \sum_{i=1}^N \mathbf{E}^{\mathbf{P}}\big[F\big(\boldsymbol{W}^0, \boldsymbol{W}^i, \bar{\mu}^N_{(\sigma^{i,N},\tau^{-i,N,*})}, \sigma^{i,N}\big)\big],
\end{aligned}$$

where we used the notations:

$$\bar{\mu}^N = \frac{1}{N} \sum_{j=1}^N \delta_{\tau^{j,N,*}}, \quad \bar{\mu}^N_{(\sigma^{i,N},\tau^{-i,N,*})} = \frac{1}{N} \sum_{j=1, j\neq i}^N \delta_{\tau^{j,N,*}} + \frac{1}{N} \delta_{\sigma^{i,N}}.$$

Now, we notice that:

$$W_1\big(\bar{\mu}^N_{(\sigma^{i,N},\tau^{-i,N,*})}, \bar{\mu}^N\big) \leq \frac{T}{N},$$

where, here, W_1 is the Wasserstein distance on $\mathcal{P}([0, T])$. Combining the above inequality and (7.105), and recalling that F is continuous in its last two arguments and that the random variables $(\bar{\mathfrak{m}}^N)_{N\geq 1}$ are tight on $\mathcal{P}_1([0, T] \times \mathcal{C}([0, T]))$, we deduce that

$$\lim_{N\to 0}\frac{1}{N}\sum_{i=1}^{N}\Big|\mathbf{E}^{\mathbf{P}}\big[F\big(\mathbf{W}^0,\mathbf{W}^i,\bar{\mu}^N_{(\sigma^{i,N},\tau^{-i,N,*})},\sigma^{i,N}\big)$$
$$-F\big(\mathbf{W}^0,\mathbf{W}^i,\bar{\mu}^N,\tilde{\sigma}(\mathbf{W}^0,\mathbf{W}^i,\bar{\mathfrak{m}}^N)\big)\big]\Big|=0.$$

Thus,

$$\mathbb{E}^{\mathbb{P}}\big[F\big(\mathbf{w}^0,\mathbf{w},\nu\circ e^{-1},\varrho\big)\big]\geq\limsup_{N\to\infty}\frac{1}{N}\sum_{i=1}^{N}\mathbf{E}^{\mathbf{P}}\big[F\big(\mathbf{W}^0,\mathbf{W}^i,\bar{\mu}^N,\tilde{\sigma}(\mathbf{W}^0,\mathbf{W}^i,\bar{\mathfrak{m}}^N)\big)\big]$$
$$=\mathbb{E}^{\mathbb{P}}\big[F\big(\mathbf{w}^0,\mathbf{w},\nu\circ e^{-1},\tilde{\sigma}(\mathbf{w}^0,\mathbf{w},\nu)\big)\big],$$

which gives the desired bound.

Fifth Step. We now prove (7.104). First, we notice that for any bounded and measurable function f from $[0,T]\times\mathcal{C}([0,T])$ into $\mathbb{R}$:

$$\mathbf{E}^{\mathbf{P}}\bigg[\bigg(\int f(t,w)d\bar{\mathfrak{m}}^N(t,w)-\mathbf{E}^{\mathbf{P}}\bigg[\int f(t,w)d\bar{\mathfrak{m}}^N(t,w)\,|\,\mathbf{W}^0\bigg]\bigg)^4\bigg]$$
$$=\mathbf{E}^{\mathbf{P}}\bigg[\bigg(\frac{1}{N}\sum_{j=1}^{N}\Big[f(\tau^{i,N,*},\mathbf{W}^i)-\mathbf{E}^{\mathbf{P}}\big[f(\tau^{i,N,*},\mathbf{W}^i)\,|\,\mathbf{W}^0\big]\Big]\bigg)^4\bigg].$$

Since the variables $(\tau^{i,N,*},\mathbf{W}^i)_{i\geq 1}$ are conditionally independent given $\mathbf{W}^0$, we deduce, by duplicating the proof of the strong law of large numbers for independent and identically distributed variables with a finite fourth moment, that:

$$\mathbf{E}^{\mathbf{P}}\bigg[\bigg(\int f(t,w)d\bar{\mathfrak{m}}^N(t,w)-\mathbf{E}^{\mathbf{P}}\bigg[\int f(t,w)d\bar{\mathfrak{m}}^N(t,w)\,|\,\mathbf{W}^0\bigg]\bigg)^4\bigg]\leq\frac{C_f}{N^2},$$

for a constant C_f only depending on f. Hence, for any bounded measurable function f from $[0,T]\times\mathcal{C}([0,T])$ into $\mathbb{R}$, with probability 1 under $\mathbf{P}$,

$$\lim_{N\to\infty}\bigg(\int_{[0,T]\times\mathcal{C}([0,T])}f(t,w)d\big(\bar{\mathfrak{m}}^N-\mathbf{E}^{\mathbf{P}}[\bar{\mathfrak{m}}^N\,|\,\mathbf{W}^0]\big)(t,w)\bigg)=0. \tag{7.106}$$

For any integer $n\geq 1$, call π_n the projection from $[0,T]\times\mathcal{C}([0,T])$ onto $[0,T]\times\mathbb{R}^{n+1}$ defined by $\pi_n(t,w)=(t,(w_{kT/n})_{0\leq k\leq n})$, for any $(t,w)\in[0,T]\times\mathcal{C}([0,T])$. Using the fact that the set of continuous functions with compact support from $[0,T]\times\mathbb{R}^{n+1}$ into $\mathbb{R}$ is separable, we deduce from (7.106) that, with $\mathbf{P}$-probability 1, for every continuous function with compact support from $[0,T]\times\mathbb{R}^{n+1}$ into $\mathbb{R}$,

$$\lim_{N\to 0}\bigg(\int_{[0,T]\times\mathbb{R}^{n+1}}f(t,x)d\big(\bar{\mathfrak{m}}^N\circ\pi_n^{-1}\big)(t,x)$$
$$-\int_{[0,T]\times\mathbb{R}^{n+1}}f(t,x)d\big(\mathbf{E}^{\mathbf{P}}[\bar{\mathfrak{m}}^N\,|\,\mathbf{W}^0]\circ\pi_n^{-1}\big)(t,x)\bigg)=0. \tag{7.107}$$

We then observe that there exists a compact subset $\mathcal{K}$ of $\mathcal{P}([0,T]\times\mathcal{C}([0,T]))$ such that, with $\mathbf{P}$-probability 1, the sequence $(\mathbf{E}^{\mathbf{P}}[\bar{\mathfrak{m}}^N \mid \boldsymbol{W}^0])_{N\geq 1}$ is included in $\mathcal{K}$. Indeed, with probability 1 under $\mathbf{P}$, for any compact subset $K\subset\mathcal{C}([0,T])$,

$$\mathbf{E}^{\mathbf{P}}[\bar{\mathfrak{m}}^N \mid \boldsymbol{W}^0]\big([0,T]\times K^{\complement}\big) = \mathbf{P}\big[\boldsymbol{W}^1 \in K^{\complement}\big].$$

Hence, with probability 1, for any $n\geq 1$, the sequence $(\mathbf{E}^{\mathbf{P}}[\bar{\mathfrak{m}}^N \mid \boldsymbol{W}^0]\circ\pi_n^{-1})_{N\geq 1}$ is tight on $[0,T]\times\mathbb{R}^{n+1}$. By (7.107), so is $(\bar{\mathfrak{m}}^N\circ\pi_n^{-1})_{N\geq 1}$. As a result, (7.107) says that, for any $n\geq 1$ and almost surely, the Lévy-Prokhorov distance between $\bar{\mathfrak{m}}^N\circ\pi_n^{-1}$ and $\mathbf{E}^{\mathbf{P}}[\bar{\mathfrak{m}}^N \mid \boldsymbol{W}^0]\circ\pi_n^{-1}$ tends to 0 as N tends to ∞.

For any integer $n\geq 1$, we call i_n the injection from $[0,T]\times\mathbb{R}^{n+1}$ into $[0,T]\times\mathcal{C}([0,T])$ that maps a tuple $(t,(x_0,x_1,\cdots,x_n))$ to the pair (t,w) where w is the linearly interpolated path satisfying $w_{iT/n}=x_i$, for all $i\in\{0,\dots,n\}$. Obviously, the Lévy-Prokhorov distance between $\bar{\mathfrak{m}}^N\circ\pi_n^{-1}\circ i_n^{-1}=\bar{\mathfrak{m}}^N\circ\mathfrak{E}_n^{-1}$ and $\mathbf{E}^{\mathbf{P}}[\bar{\mathfrak{m}}^N \mid \boldsymbol{W}^0]\circ\pi_n^{-1}\circ i_n^{-1}=\mathbf{E}^{\mathbf{P}}[\bar{\mathfrak{m}}^N \mid \boldsymbol{W}^0]\circ\mathfrak{E}_n^{-1}$ tends to 0 as N tends to ∞, where $\mathfrak{E}_n=i_n\circ\pi_n:[0,T]\times\mathcal{C}([0,T])\rightarrow[0,T]\times\mathcal{C}([0,T])$ may be written in the form $\mathfrak{E}_n=(\mathfrak{E}_n^{(1)},\mathfrak{E}_n^{(2)})$, $\mathfrak{E}_n^{(1)}$ matching the identity on $[0,T]$ and $\mathfrak{E}_n^{(2)}$ mapping a path w to its linear interpolation at times $(iT/n)_{0\leq i\leq n}$. We claim that the convergence also holds in 1-Wasserstein distance. Indeed, we have the following uniform integrability property:

$$\int_{[0,T]\times\mathcal{C}([0,T])}\sup_{0\leq t\leq T}|w_t|^2 d\big(\bar{\mathfrak{m}}^N\circ\mathfrak{E}_n^{-1}\big)(t,w)\leq\frac{1}{N}\sum_{i=1}^N\sup_{0\leq t\leq T}|\boldsymbol{W}_t^i|^2,$$

which is uniformly bounded in N, with $\mathbf{P}$-probability 1, by the law of large numbers. Hence, with $\mathbf{P}$-probability 1, the sequence $(\bar{\mathfrak{m}}^N\circ\mathfrak{E}_n^{-1})_{N\geq 1}$ is uniformly square integrable. We have a similar argument for $(\mathbf{E}^{\mathbf{P}}[\bar{\mathfrak{m}}^N \mid \boldsymbol{W}^0]\circ\mathfrak{E}_n^{-1})_{N\geq 1}$. We finally obtain, for every $n\geq 1$,

$$\lim_{N\rightarrow\infty}W_1\big(\bar{\mathfrak{m}}^N\circ\mathfrak{E}_n^{-1},\mathbf{E}^{\mathbf{P}}[\bar{\mathfrak{m}}^N \mid \boldsymbol{W}^0]\circ\mathfrak{E}_n^{-1}\big)=0.$$

Hence, for every $n\geq 1$,

$$\begin{aligned}\limsup_{N\rightarrow\infty}W_1\big(\bar{\mathfrak{m}}^N,\mathbf{E}^{\mathbf{P}}[\bar{\mathfrak{m}}^N \mid \boldsymbol{W}^0]\big)&\leq\limsup_{N\rightarrow\infty}W_1\big(\bar{\mathfrak{m}}^N,\bar{\mathfrak{m}}^N\circ\mathfrak{E}_n^{-1}\big)\\&\quad+\limsup_{N\rightarrow\infty}W_1\Big(\mathbf{E}^{\mathbf{P}}[\bar{\mathfrak{m}}^N \mid \boldsymbol{W}^0],\mathbf{E}^{\mathbf{P}}[\bar{\mathfrak{m}}^N \mid \boldsymbol{W}^0]\circ\mathfrak{E}_n^{-1}\Big).\end{aligned}$$

It remains to prove that the two terms in the right-hand side tend to 0 as n tends to ∞.

We proceed as follows. For any 1-Lipschitz continuous function f from $[0,T]\times\mathcal{C}([0,T])$, we have:

$$\begin{aligned}\left|\int f(t,w)d\big(\bar{\mathfrak{m}}^N-\bar{\mathfrak{m}}^N\circ\mathfrak{E}_n^{-1}\big)(t,w)\right|&=\left|\int\Big(f(t,w)-f\big(t,\mathfrak{E}_n^{(2)}(w)\big)\Big)d\bar{\mathfrak{m}}^N(t,w)\right|\\&\leq\int\sup_{0\leq t\leq T}\big|w_t-\big(\mathfrak{E}_n^{(2)}(w)\big)_t\big|d\bar{\mathfrak{m}}^N(t,w)\\&=\frac{1}{N}\sum_{i=1}^N\sup_{0\leq t\leq T}\big|W_t^i-\big(\mathfrak{E}_n^{(2)}(\boldsymbol{W}^i)\big)_t\big|.\end{aligned}$$

Taking the supremum over f, we deduce from the Kantorovich-Rubinstein duality theorem, see for instance Corollary (Vol I)-5.4, and from the law of large numbers that, with $\mathbf{P}$-probability 1,

$$\lim_{n\to\infty} \limsup_{N\to\infty} W_1\big(\bar{\mathfrak{m}}^N \circ \mathfrak{E}_n^{-1}, \bar{\mathfrak{m}}^N\big) = 0.$$

Similarly, with $\mathbf{P}$-probability 1,

$$\lim_{n\to\infty} \sup_{N\geq 1} W_1\big(\mathbf{E}^{\mathbf{P}}[\bar{\mathfrak{m}}^N \mid \boldsymbol{W}^0] \circ \mathfrak{E}_n^{-1}, \mathbf{E}^{\mathbf{P}}[\bar{\mathfrak{m}}^N \mid \boldsymbol{W}^0]\big) = 0,$$

which completes the proof of (7.104). □

7.2.8 Existence of Weak Equilibria Under Continuity Assumptions

This subsection is devoted to the proof of the following general existence result for equilibria in the weak sense.

Theorem 7.41 *Assume that assumption* **MFG of Timing Set-Up** *(including* **(A3)***) and condition* **(A1)** *in assumption* **MFG of Timing Regularity** *hold. Then there exists a weak equilibrium to the mean field game of timing.*

Discretization of the Conditioning

In order to prove Theorem 7.41, we shall use the strategy which proved to be successful in the analysis of mean field games with a common noise. See for example the introduction of Chapter 3 for a description of this approach based on the construction of strong equilibria when the *common random shocks* are restricted to a finite number of possible values. As a first step, we prove existence of an equilibrium when the conditioning in the matching problem is restricted to a finitely discretized version of the common noise. To make the framework and the notations consistent with the material from Chapter 3, we shall assume that $\boldsymbol{W}_0^0 = 0$, or equivalently that $\boldsymbol{w}_0^0 = 0$ under $\mathcal{W}^1$; we let the reader adapt the arguments to the general case.

In order to construct the discrete conditioning, we proceed as in Subsection 3.3.1. We choose two integers $\ell, n \geq 1$, ℓ referring to the step size of the space grid and n to the step size of the time grid. For $\Lambda = 2^\ell$, we then let Π_Λ be the mapping from $\mathbb{R}$ into itself defined by:

$$\Pi_\Lambda : \mathbb{R} \ni x \mapsto \begin{cases} \Lambda^{-1}\lfloor \Lambda x \rfloor & \text{if } |x| \leq \Lambda, \\ \Lambda \,\text{sign}(x) & \text{if } |x| > \Lambda. \end{cases}$$

For any integer $j \geq 1$, we also consider the projection $\Pi_{\Lambda,j}$ from $\mathbb{R}^j$ into itself defined iteratively by:

$$\Pi_{\Lambda,1} \equiv \Pi_\Lambda,$$
$$\Pi_{\Lambda,j+1}(x^1,\cdots,x^{j+1}) = (y^1,\cdots,y^j,y^{j+1}), \quad (x^1,\cdots,x^{j+1}) \in \mathbb{R}^{j+1},$$

where:

$$(y^1,\cdots,y^j) = \Pi_{\Lambda,j}(x^1,\cdots,x^j) \in \mathbb{R}^j, \quad y^{j+1} = \Pi_\Lambda\big(y^j + x^{j+1} - x^j\big) \in \mathbb{R}.$$

For the sake of convenience, we restate the already proven result of Lemma 3.17.

Lemma 7.42 *With the above notation, for* $(x^1,\cdots,x^j) \in \mathbb{R}^j$ *such that, for any* $i \in \{1,\cdots,j\}$, $|x^i| \leq \Lambda - 1$, *let:*

$$(y^1,\cdots,y^j) = \Pi_{\Lambda,j}(x^1,\cdots,x^j).$$

If $j \leq \Lambda$, *then, for each* $i \in \{1,\cdots,j\}$, $|x^i - y^i| \leq i/\Lambda$.

Given an integer n, we let $N = 2^n$, and we consider the dyadic time mesh:

$$t_i^N = \frac{iT}{N}, \quad i \in \{0,1,\cdots,N\}, \tag{7.108}$$

and, given the canonical process $\boldsymbol{w}^0 = (w_t^0)_{0\leq t\leq T}$, we define the random variables $V_i(\boldsymbol{w}^0)$, for $i = 1,\cdots,N$, by:

$$\big(V_1(\boldsymbol{w}^0),\cdots,V_N(\boldsymbol{w}^0)\big) = \Pi_{\Lambda,N}\big(w^0_{t_1^N},\cdots,w^0_{t_N^N}\big), \tag{7.109}$$

and we often rewrite the left-hand side as $\pi_{\Lambda,N}(\boldsymbol{w}^0)$. Similarly, we restate the result of the already proven Lemma 3.18, where we use the same notation as before:

$$\mathbb{J} = \{-\Lambda, -\Lambda + 1/\Lambda, -\Lambda + 2/\Lambda, \cdots, \Lambda - 1/\Lambda, \Lambda\}.$$

Lemma 7.43 *When viewed as random variables on* $\mathcal{C}([0,T])$, *the* V_j*'s have the following property: given* $i = 1,\cdots,N$, *the random vector* $(V_1,\cdots,V_i)$ *has the whole* $\mathbb{J}^i$ *as support.*

The random variables $V_1,\cdots,V_N$ must be understood as a discretization of the common noise $\boldsymbol{w}^0$. Following (3.34), we call a *discretized environment* a mapping $\vartheta : \mathbb{J}^N \ni (v_1,\cdots,v_N) \mapsto \vartheta(v_1,\cdots,v_N,\cdot) \in \mathcal{P}(\mathbb{T}_N)$, where $\mathbb{T}_N = \{t_0^N, t_1^N, \cdots, t_N^N\}$, such that, for all $i \in \{0,\cdots,N-1\}$, the function:

$$\mathbb{J}^N \ni (v_1,\cdots,v_N) \mapsto \vartheta\big(v_1,\cdots,v_N,\{t_0^N,\cdots,t_i^N\}\big)$$

is just a function of $(v_1,\cdots,v_i)$, or equivalently the function:

$$\mathbb{J}^N \ni (v_1, \cdots, v_N) \mapsto \vartheta\big(v_1, \cdots, v_N, \{t_i^N\}\big)$$

is just a function of $(v_1, \cdots, v_i)$. In particular, the function $\mathbb{J}^N \ni (v_1, \cdots, v_N) \mapsto \vartheta(v_1, \cdots, v_N, \{t_0^N\})$ is required to be constant. We denote by $\mathfrak{F}^{N,\text{adapt}}$ the set of such discretized environments. For each $\vartheta \in \mathfrak{F}^{N,\text{adapt}}$ and for all $(v_1, \cdots, v_N) \in \mathbb{J}^N$, we may regard $\vartheta(v_1, \cdots, v_N, \cdot)$ as an element of $\mathcal{P}([0, T])$. We then notice that for $\vartheta \in \mathfrak{F}^{N,\text{adapt}}$, the process $(\vartheta(\pi_{\Lambda,N}(\boldsymbol{w}^0), [0, t]))_{0 \le t \le T}$ is adapted with respect to the filtration $\mathbb{F}^{\text{nat},\boldsymbol{w}^0}$.

Throughout the analysis, we equip $\mathfrak{F}^{N,\text{adapt}}$ with the topology of pointwise convergence. In other words, a sequence $(\vartheta_n)_{n\ge1}$ converges to ϑ if, for all $(v_1, \cdots, v_N) \in \mathbb{J}^N$ and $i \in \{0, \cdots, N\}$,

$$\lim_{n\to\infty} \vartheta_n\big(v_1, \cdots, v_N, \{t_i^N\}\big) = \vartheta\big(v_1, \cdots, v_N, \{t_i^N\}\big).$$

It is easily checked that $\mathfrak{F}^{N,\text{adapt}}$ is then a compact metric space. Moreover, it is also clear that $(\vartheta_n)_{n\ge1}$ converges to ϑ if and only if for all $(v_1, \cdots, v_N) \in \mathbb{J}^N$, the sequence of probability measures $(\vartheta_n(v_1, \cdots, v_N, \cdot))_{n\ge1}$ converges to $\vartheta(v_1, \cdots, v_N, \cdot)$ in $\mathcal{P}([0, T])$.

On the space $\mathcal{C}([0, T])^2$ equipped with the measure $\mathcal{W}$, see **(A3)** in assumption **MFG of Timing Set-Up**, we define the following discretized mean field game of timing.

(i) For each discretized environment $\vartheta \in \mathfrak{F}^{N,\text{adapt}}$, solve:

$$\hat{\theta} \in \arg \sup_{\theta \in \mathcal{S}_{(\boldsymbol{w}^0,\boldsymbol{w})}} J\big(\vartheta(\pi_{\Lambda,N}(\boldsymbol{w}^0), \cdot), \theta\big).$$

(ii) Find $\vartheta \in \mathfrak{F}^{N,\text{adapt}}$ so that, for all $v_1, \cdots, v_N \in \mathbb{J}$, for all $i \in \{0, \cdots, N\}$,

$$\vartheta\big(v_1, \cdots, v_N, \{t_i^N\}\big) = \mathbb{P}\big[t_{i-1}^N < \hat{\theta} \le t_i^N \mid \pi_{\Lambda,N}(\boldsymbol{w}^0) = (v_1, \cdots, v_N)\big],$$

with the convention that $t_{-1} = -\infty$.

Except for the fact that the stopping times are from the set $\mathcal{S}_{(\boldsymbol{w}^0,\boldsymbol{w})}$ instead of $\mathcal{S}_X$ and that the environment is discretized, the above definition could be understood as the definition of a strong equilibrium. However, even though the environment is discretized, we need to appeal to the notion of randomized stopping time and compactify the set of stopping times. This leaves us with the following notion of semi-strong equilibrium.

Definition 7.44 *A discretized environment $\vartheta \in \mathfrak{F}^{N,\text{adapt}}$ is said to be a semi-strong discretized equilibrium if there exists a probability measure $\mathbb{P}$ on the canonical space Ω_{canon} such that $\mathbb{P} \circ (\boldsymbol{w}^0, \boldsymbol{w}) = \mathcal{W}$, $\mathbb{P}[\nu \circ e^{-1} = \vartheta(\pi_{\Lambda,N}(\boldsymbol{w}^0), \cdot)] = 1$, where e is the first coordinate projection from $[0, T] \times \mathcal{C}([0, T])$ onto $[0, T]$, and:*

1. *The process* $(\boldsymbol{w}^0, \boldsymbol{w})$ *is a Brownian motion with respect to the natural filtration generated by* $(\boldsymbol{w}^0, \boldsymbol{w}, \varrho)$ *under* $\mathbb{P}$*;*
2. *The process* ν *is adapted with respect to the completion of the filtration generated by* $\boldsymbol{w}^0$ *under* $\mathbb{P}$ *in the sense that, for all* $t \in [0, T]$, $\mathcal{F}_t^{\mathrm{nat},\nu} \subset \mathcal{F}_{t+}^{\mathrm{nat},\nu} \subset \mathcal{F}_t^{\boldsymbol{w}^0}$*;*
3. *The probability measure* $\mathbb{P}$ *belongs to* $\arg\sup_{\mathbb{P}'} \mathbb{E}^{\mathbb{P}'}[F(\boldsymbol{w}^0, \boldsymbol{w}, \nu \circ e^{-1}, \varrho)]$*, where the supremum is taken over all the probability measures* $\mathbb{P}'$ *on* Ω_{canon} *satisfying 1,* $\mathbb{P}' \circ (\boldsymbol{w}^0, \boldsymbol{w})^{-1} = \mathcal{W}$ *and* $\mathbb{P}'[\nu \circ e^{-1} = \vartheta(\pi_{\Lambda,N}(\boldsymbol{w}^0), \cdot)] = 1$*;*
4. *The weak fixed point condition holds:*

$$\mathbb{P}\big[\varrho \in (t_{i-1}, t_i] \,|\, \pi_{\Lambda,N}(\boldsymbol{w}^0)\big] = \nu \circ e^{-1}(\{t_i\}), \quad i \in \{0, \cdots, N\}, \quad \mathbb{P}-a.s.$$

Of course, the reader will observe that, except for the fixed point condition, the other conditions in Definition 7.44 are consistent with that ones required in Definition 7.37. The fact that the fixed point condition is formulated in terms of the conditional law of the sole ϱ should not come as a surprise. We do so because the environment is required to be $\mathbb{F}^{\boldsymbol{w}^0}$-adapted. Basically, only the marginal law of ν on $[0, T]$ matters. Given the conditional law of the randomized stopping time ϱ, the form of the full-fledged lift ν in $\mathcal{P}_1([0, T] \times \mathcal{C}([0, T]))$ does not really matter in the construction of a semi-strong equilibrium. If needed, we can choose it to make it consistent with the fixed point condition used in Definition 7.37, but, in fact, we can also choose it in a purely arbitrary way, as long as condition 2 is satisfied. For instance, if ϑ is a semi-strong discretized equilibrium under some probability $\mathbb{P}$, then it is also a semi-strong discretized equilibrium under $\tilde{\mathbb{P}} = \mathbb{P} \circ (\boldsymbol{w}^0, \boldsymbol{w}, \vartheta(\pi_{\Lambda,N}(\boldsymbol{w}^0), \cdot) \otimes \mathcal{W}_1, \varrho)^{-1}$, where $\mathcal{W}_1$ is the one-dimensional Wiener measure on $[0, T]$. Then, if we let $\mathbb{Q} = \mathcal{W} \circ (\boldsymbol{w}^0, \boldsymbol{w}, \vartheta(\pi_{\Lambda,N}(\boldsymbol{w}^0), \cdot) \otimes \mathcal{W}_1)^{-1}$, then $\mathbb{Q} \in \mathcal{M}$ and $\tilde{\mathbb{P}} \in \mathcal{R}(\mathbb{Q})$, with $\mathcal{M}$ and $\mathcal{R}(\mathbb{Q})$ as in Definition 7.28 and Definition 7.30.

The following lemma is the analogue of the results proven in Subsection 3.3.1.

Lemma 7.45 *For any* $\ell, n \geq 1$*, there exists a semi-strong discretized equilibrium.*

In preparation for the proof of Lemma 7.45, we state without proof two classical results from the theory of set-valued functions. See the Notes & Complements at the end of the chapter for references. We shall need the following definition.

Definition 7.46 *If* $\mathcal{X}$ *and* $\mathcal{Y}$ *are metric spaces, a set-valued map* ψ *from* $\mathcal{X}$ *to* $\mathcal{Y}$*, or alternatively a function* ψ *from* $\mathcal{X}$ *to the power set* $2^{\mathcal{Y}}$ *of* $\mathcal{Y}$*, is said to be upper hemicontinuous if, for any* $x \in \mathcal{X}$ *and any sequence* $(x_k, y_k)_{k\geq 1}$ *in* $\mathcal{X} \times \mathcal{Y}$ *such that* $(x_k)_{k\geq 1}$ *converges to* x *and* $y_k \in \psi(x_k)$ *for all* $k \geq 1$*, the sequence* $(y_k)_{k\geq 1}$ *has a limit point in* $\psi(x)$*.*

The set-valued map ψ *is said to be lower hemicontinuous if, for any* $x \in \mathcal{X}$*, any sequence* $(x_k)_{k\geq 1}$ *in* $\mathcal{X}$ *converging to* x *and any* $y \in \psi(x)$*, there exist an increasing sequence of positive integers* $(k(p))_{p\geq 1}$ *and a sequence* $(y_p)_{p\geq 1}$ *in* $\mathcal{Y}$ *such that* $y_p \in \psi(x_{k(p)})$ *for* $p \geq 1$ *and* $(y_p)_{p\geq 1}$ *converges to* y*.*

The set-valued map ψ is said to be continuous if it is both lower and upper hemicontinuous.

The advised reader will notice that our sequential definition of upper hemicontinuity is in fact stronger than the usual one since we implicitly require $\psi(x)$ to be compact for any $x \in \mathcal{X}$.

The following statement is known as Berge's maximum theorem.

Theorem 7.47 *If $\mathcal{X}$ and $\mathcal{Y}$ are metric spaces, and $\psi : \mathcal{X} \to 2^{\mathcal{Y}}$ and $f : \mathcal{X} \times \mathcal{Y} \to \mathbb{R}$ are continuous functions such that $\psi(x)$ is a nonempty compact subset of $\mathcal{Y}$ for any $x \in \mathcal{X}$, then the set-valued map:*

$$\mathcal{X} \ni x \mapsto \{y \in \mathcal{Y} : f(x,y) = \max_{z \in \psi(x)} f(x,z)\} = \{y \in \mathcal{Y} : y \in \arg\max_{z \in \psi(x)} f(x,z)\}$$

has nonempty compact values and is upper hemicontinuous.

We now recall a very useful fixed point existence result known as Kakutani-Fan-Glickbserg's theorem. It plays in the subsequent analysis, the same role as Schauder's theorem in the analysis of stochastic differential mean field games.

Theorem 7.48 *Let $\mathcal{K}$ be a nonempty compact convex subset of a normed vector space $\mathcal{X}$. If $\psi : \mathcal{K} \to 2^{\mathcal{K}}$ is a set-valued map such that the graph $\{(x,y) \in \mathcal{K} \times \mathcal{K} : y \in \psi(x)\}$ is closed and $\psi(x)$ is a convex subset of $\mathcal{K}$ for all $x \in \mathcal{K}$, then ψ admits a fixed point x^* in the sense that $x^* \in \psi(x^*)$.*

Proof of Lemma 7.45. Throughout the proof, the values of ℓ and n are fixed.

First Step. For a discretized environment ϑ, we let $\mathbb{Q}_\vartheta = \mathcal{W} \circ (\boldsymbol{w}^0, \boldsymbol{w}, \vartheta(\pi_{\Lambda,N}(\boldsymbol{w}^0),\cdot) \otimes \mathcal{W}_1)^{-1}$. Recall that from Theorem 7.31, we know that $\mathcal{R}(\mathbb{Q}_\vartheta)$ is a convex closed subset of $\mathcal{P}(\Omega_{\text{canon}})$. Also, it is compact. Indeed, the first two marginal projections of any probability measure in $\mathcal{R}(\mathbb{Q}_\vartheta)$ are fixed, and the third marginal is a random measure which belongs to a fixed compact subset of $\mathcal{P}([0,T] \times \mathcal{C}([0,T];\mathbb{R}))$ since the first component is bounded while the second one is fixed. We also notice that any $\mathbb{P} \in \mathcal{R}(\mathbb{Q}_\vartheta)$ automatically satisfies condition 1 in Definition 7.44. We recast the present setting in the framework of Theorem 7.47. We set $\mathcal{X} = \mathfrak{F}^{N,\text{adapt}}$ and we choose $\mathcal{Y}$ as the subset of $\mathcal{P}(\Omega_{\text{canon}})$ formed by all the probability measures $\mathbb{P}$ having $\mathcal{W}$ as marginal on $\mathcal{C}([0,T])^2$. Next we define:

$$\psi(\vartheta) = \mathcal{R}\big(\mathcal{W} \circ (\boldsymbol{w}^0, \boldsymbol{w}, \vartheta(\pi_{\Lambda,N}(\boldsymbol{w}^0),\cdot) \otimes \mathcal{W}_1)^{-1}\big) = \mathcal{R}(\mathbb{Q}_\vartheta),$$

for $\vartheta \in \mathcal{X} = \mathfrak{F}^{N,\text{adapt}}$, and:

$$f(\vartheta, \mathbb{P}) = \mathbb{E}^{\mathbb{P}}\big[F(\boldsymbol{w}^0, \boldsymbol{w}, \nu \circ e^{-1}, \varrho)\big],$$

for $(\vartheta, \mathbb{P}) \in \mathcal{X} \times \mathcal{Y}$. The function f is continuous because of Lemma 7.34. We now check that ψ is continuous in the sense of Definition 7.46 so we can use Berge's theorem and conclude that the map:

$$\vartheta \mapsto \Phi(\vartheta) = \arg \max_{\mathbb{P}\in\mathcal{R}(\mathbb{Q}_\vartheta)} \mathbb{E}^{\mathbb{P}}\big[F\big(w^0, w, \nu \circ e^{-1}, \varrho\big)\big].$$

has nonempty compact values and is upper hemicontinuous.

We start with the proof of the upper hemicontinuity. To do so, we consider a sequence $(\vartheta_k, \mathbb{P}_k)_{k\geq 1}$ in $\mathcal{X}\times\mathcal{Y}$, such that $\mathbb{P}_k \in \psi(\vartheta_k)$, for all $k \geq 1$, and $(\vartheta_k)_{k\geq 1}$ converges to some ϑ. It is easily checked that the 1-Wasserstein distance between $\mathbb{P}_k$ and $\mathbb{P}_k \circ (w^0, w, \vartheta(\pi_{\Lambda,N}(w^0), \cdot)\otimes \mathcal{W}_1, \varrho)^{-1}$ tends to 0 as k tends to ∞. Moreover, for all $k \geq 1$, $\mathbb{P}_k \circ (w^0, w, \vartheta(\pi_{\Lambda,N}(w^0), \cdot) \otimes \mathcal{W}_1, \varrho)^{-1}$ belongs to $\mathcal{R}(\mathbb{Q}) = \psi(\vartheta)$ where $\mathbb{Q} = \mathcal{W} \circ (w^0, w, \vartheta(\pi_{\Lambda,N}(w^0), \cdot) \otimes \mathcal{W}_1)^{-1}$. Since $\mathcal{R}(\mathbb{Q})$ is closed, see Theorem 7.31, we deduce that $\mathbb{P} \in \mathcal{R}(\mathbb{Q})$. Lower hemicontinuity is proved in the same way. Namely, if $(\vartheta_k)_{k\geq 1}$ converges to some ϑ in $\mathcal{X}$, then, for any $\mathbb{P} \in \psi(\vartheta)$, we let $\mathbb{P}_k = \mathbb{P}\circ(w^0, w, \vartheta_k(\pi_{\Lambda,N}(w^0), \cdot)\otimes\mathcal{W}_1, \varrho)^{-1}$. Once again, the 1-Wasserstein distance between $\mathbb{P}_k$ and $\mathbb{P}$ tends to 0 as k tends to ∞ and, for all $k \geq 1$, $\mathbb{P}_k \in \psi(\vartheta_k)$.

Hence, Berge's theorem implies that $\Phi(\vartheta)$ is nonempty compact convex for each ϑ and, most importantly, the set-valued function Φ is upper hemicontinuous on $\mathfrak{F}^{N,\text{adapt}}$.

Second Step. Finally, we define a mapping Ψ from $\mathfrak{F}^{N,\text{adapt}}$ into $2^{\mathfrak{F}^{N,\text{adapt}}}$ as follows. For $\vartheta \in \mathfrak{F}^{N,\text{adapt}}$ and $\mathbb{P} \in \Phi(\vartheta)$, we define the map $\psi(\vartheta, \mathbb{P})$ from $\mathbb{J}^N$ into $\mathcal{P}(\mathbb{T}_N)$ by:

$$\psi(\vartheta, \mathbb{P})\big(v_1, \cdots, v_N, \{t_i\}\big) = \mathbb{P}\big[\varrho \in (t_{i-1}, t_i] \,|\, \pi_{\Lambda,N}(w^0) = (v_1, \cdots, v_N)\big],$$

for $i \in \{0, \cdots, N\}$, and $v_1, \cdots, v_N \in \mathbb{J}^N$. Since $\mathbb{P} \in \mathcal{R}(\mathbb{Q}_\vartheta)$, we know that, for any $i \in \{0, \cdots, N\}$, $\mathcal{F}^{\text{nat},\varrho}_{t_i+}$ is independent of $\sigma\{(w^0_s - w^0_{t_i}, w_s - w_{t_i}); t_i \leq s \leq T\}$ under $\mathbb{P}$. Observing that the event $\{\pi_{\Lambda,N}(w^0) = (v_1, \cdots, v_N)\}$ may be rewritten as $\{(V_1(w^0), \cdots, V_i(w^0)) = (v_1, \cdots, v_i)\} \cap C$, for an event $C \in \sigma\{w^0_s - w^0_{t_i}; t_i \leq s \leq T\}$, we deduce that:

$$\psi(\vartheta, \mathbb{P})\big(v_1, \cdots, v_N, \{t_i\}\big) = \mathbb{P}\big[\varrho \in (t_{i-1}, t_i] \,|\, (V_1(w^0), \cdots, V_i(w^0)) = (v_1, \cdots, v_i)\big],$$

which proves that $\psi(\vartheta, \mathbb{P})$ belongs to $\mathfrak{F}^{N,\text{adapt}}$. Hence, we define the set valued map Ψ by $\Psi(\vartheta) = \{\psi(\vartheta, \mathbb{P});\ \mathbb{P} \in \Phi(\vartheta)\}$ as a subset of $\mathfrak{F}^{N,\text{adapt}}$. Writing:

$$\begin{aligned}
&\psi(\vartheta, \mathbb{P})\big(v_1, \cdots, v_N, \{t_i\}\big) \\
&\quad = \frac{\mathbb{P}\big[\varrho \in (t_{i-1}, t_i],\, (V_1(w^0), \cdots, V_i(w^0)) = (v_1, \cdots, v_i)\big]}{\mathcal{W}\big[(V_1(w^0), \cdots, V_i(w^0)) = (v_1, \cdots, v_i)\big]},
\end{aligned} \tag{7.110}$$

we see that $\psi(\vartheta, \cdot)$ is a convex function, since $\Phi(\vartheta)$ is a convex subset of $\mathcal{Y}$. Hence, $\Psi(\vartheta)$ is a convex subset of $\mathfrak{F}^{N,\text{adapt}}$.

We finally check that the graph of Ψ is closed. For a sequence $(\vartheta_k, \theta_k)_{n\geq 1}$ converging to some (ϑ, θ) such that $\theta_k \in \Psi(\vartheta_k)$ for all $k \geq 1$, we know, that for each $k \geq 1$, there exists $\mathbb{P}_k \in \Phi(\vartheta_k)$ such that $\theta_k = \psi(\vartheta_k, \mathbb{P}_k)$. From the upper hemicontinuity of Φ, we get that $(\mathbb{P}_k)_{k\geq 1}$ has a limit point $\mathbb{P}$ in $\Phi(\vartheta)$. Passing to the limit in (7.110), we get that $\theta = \psi(\vartheta, \mathbb{P})$, which proves that $\theta \in \Psi(\vartheta)$.

By Kakutani-Fan-Glicksberg's theorem, Ψ admits a fixed point, which produces the desired equilibrium. □

Proof of Existence in the General Setting

Proof of Theorem 7.41. We follow step by step the proof of Theorem 7.40.

For any integer $n \geq 1$, we consider a semi-strong discretized equilibrium, say ϑ^n, for the mean field game of timing, as constructed in the proof of Lemma 7.45 with $N = 2^n$ and $\Lambda = \Lambda_N$, with $\Lambda_N = 4^n = N^2$. Following Definition 7.44, we associate with ϑ^n a probability measure $\mathbb{P}^n$ on Ω_{canon}. We let $\pi_{(N)} \equiv \pi_{\Lambda_N,N}$ and then:

$$\bar{\mathbb{P}}^n = \mathbb{P}^n \circ \big(\boldsymbol{w}^0, \boldsymbol{w}, \mathcal{L}(\varrho, \boldsymbol{w} \,|\, \pi_{(N)}(\boldsymbol{w}^0)), \varrho\big)^{-1},$$

where $\mathcal{L}$ is used to denote the conditional law under $\mathbb{P}^n$. So, in contrast with the proof of Lemma 7.45, we now force the environment to satisfy the fixed point condition. Following the second step in the proof of Lemma 7.45, we can prove that, under $\bar{\mathbb{P}}^n$, ν_{t_i} is $\mathcal{F}^{\boldsymbol{w}^0}_{t_i}$-measurable. Also, with $\bar{\mathbb{P}}^n$-probability 1, for all $i \in \{0, \cdots, N\}$,

$$\big(\nu \circ e^{-1}\big)\big((t^N_{i-1}, t^N_i]\big) = \mathbb{P}^n\big[\varrho \in (t^N_{i-1}, t^N_i] \,|\, \pi_{(N)}(\boldsymbol{w}^0)\big] = \vartheta^n\big(\pi_{(N)}(\boldsymbol{w}^0), \{t^N_i\}\big),$$

Consequently, for any 1-Lipschitz function f from $[0, T]$ to $\mathbb{R}$,

$$\begin{aligned}
&\left|\int_0^T f(t) d\big[\big(\nu \circ e^{-1}\big) - \vartheta^n\big(\pi_{(N)}(\boldsymbol{w}^0), \cdot\big)\big]\right| \\
&= \left|\sum_{i=0}^N \int_{(t^N_{i-1}, t^N_i]} f(t) d\big[\big(\nu \circ e^{-1}\big) - \vartheta^n\big(\pi_{(N)}(\boldsymbol{w}^0), \cdot\big)\big](t)\right| \\
&\leq \left|\sum_{i=0}^N f(t^N_i)\big[\big(\nu \circ e^{-1}\big)\big((t^N_{i-1}, t^N_i]\big) - \vartheta^n\big(\pi_{(N)}(\boldsymbol{w}^0), \{t^N_i\}\big)\big]\right| + \frac{T}{N} = \frac{T}{N},
\end{aligned}$$

from which we deduce that $W_1(\nu \circ e^{-1}, \vartheta^n(\pi_{(N)}(\boldsymbol{w}^0), \cdot)) \leq T/N$ with probability 1 under $\bar{\mathbb{P}}^n$.

In preparation for the proof, we also make the following observation in order to get rid of $\pi_{(N)}$ in the conditioning. By Lemma 7.42, we have $\sup_{t \in \mathbb{T}_N} |w^0_t - [i^{(2)}_N(\pi_{(N)}(\boldsymbol{w}^0))]_t| \leq 1/2^n$ on the event $\{\sup_{0 \leq t \leq T} |w^0_t| \leq 4^n - 1\}$, where $i^{(2)}_N : \mathbb{R}^N \rightarrow \mathcal{C}([0, T])$ maps a tuple $(x_1, \cdots, x_N)$ to the piecewise linear path interpolating the points $0, x_1, \cdots, x_N$ at times $(t^i_N)_{i=0,\cdots,N}$. As a byproduct, we easily deduce that:

$$\forall \varepsilon > 0, \quad \lim_{n \to \infty} \bar{\mathbb{P}}^n\big[\sup_{t \in [0,T]} |w^0_t - \big[i^{(2)}_N(\pi_{(N)}(\boldsymbol{w}^0))\big]_t| > \varepsilon\big] = 0.$$

As another preliminary step, we notice that the sequence $(\bar{\mathbb{P}}^n)_{n \geq 1}$ is tight on Ω_{canon}. The argument is similar to that used in the proof of Theorem 7.40. We call $\mathbb{P}$ a limit point. Working with a subsequence if necessary, we shall assume without any loss of generality that $(\bar{\mathbb{P}}^n)_{n \geq 1}$ converges to $\mathbb{P}$.

First Step. We first prove condition 5 in Definition 7.37. To do so, we consider two real valued bounded and continuous functions f and g on $\mathcal{C}([0, T]) \times \mathcal{P}_1([0, T] \times \mathcal{C}([0, T]))$ and $[0, T] \times \mathcal{C}([0, T])$ respectively, and we notice that the preliminary steps taken above imply:

$$\lim_{n \to \infty} \mathbb{E}^{\bar{\mathbb{P}}^n}\Big[\big|f(\boldsymbol{w}^0, \nu) - f\big(i^{(2)}_N(\pi_{(N)}(\boldsymbol{w}^0)), \nu\big)\big|\Big] = 0.$$

Therefore,

$$\begin{aligned}\mathbb{E}^{\mathbb{P}}\big[f(\boldsymbol{w}^0,\nu)g(\varrho,\boldsymbol{w})\big] &= \lim_{n\to\infty}\mathbb{E}^{\bar{\mathbb{P}}^n}\big[f(\boldsymbol{w}^0,\nu)g(\varrho,\boldsymbol{w})\big]\\ &= \lim_{n\to\infty}\mathbb{E}^{\bar{\mathbb{P}}^n}\Big[f\big(i_N^{(2)}(\pi_{(N)}(\boldsymbol{w}^0)),\nu\big)g(\varrho,\boldsymbol{w})\Big]\\ &= \lim_{n\to\infty}\mathbb{E}^{\bar{\mathbb{P}}^n}\Big[f\big(i_N^{(2)}(\pi_{(N)}(\boldsymbol{w}^0)),\nu\big)\int_{[0,T]\times\mathcal{C}([0,T])} g(t,w)d\nu(t,w)\Big],\end{aligned}$$

where we used the fact that, under $\bar{\mathbb{P}}^n$, ν is measurable with respect to the completion of $\sigma\{\pi_{(N)}(\boldsymbol{w}^0)\}$, which implies that, with $\bar{\mathbb{P}}^n$-probability 1, $\nu = \mathcal{L}((\varrho,\boldsymbol{w})\mid\pi_{(N)}(\boldsymbol{w}^0),\nu)$. Hence,

$$\begin{aligned}\mathbb{E}^{\mathbb{P}}\big[f(\boldsymbol{w}^0,\nu)g(\varrho,\boldsymbol{w})\big] &= \lim_{n\to\infty}\mathbb{E}^{\bar{\mathbb{P}}^n}\Big[f(\boldsymbol{w}^0,\nu)\int_{[0,T]\times\mathcal{C}([0,T])} g(t,w)d\nu(t,w)\Big],\\ &= \mathbb{E}^{\mathbb{P}}\Big[f(\boldsymbol{w}^0,\nu)\int_{[0,T]\times\mathcal{C}([0,T])} g(t,w)d\nu(t,w)\Big],\end{aligned}$$

which proves condition 5.

Second Step. We now prove conditions 1 and 2 in Definition 7.37. Condition 2 is easily checked. Indeed, for each $n \geq 1$, the processes $(\boldsymbol{w}^0,\nu)$ and $\boldsymbol{w}$ are independent under $\bar{\mathbb{P}}^n$. This remains true under the limiting probability $\mathbb{P}$.

In order to check condition 1, we recall that $(\boldsymbol{w}^0,\boldsymbol{w})$ is a Brownian motion with respect to the natural filtration generated by $(\boldsymbol{w}^0,\boldsymbol{w},\varrho)$ under $\bar{\mathbb{P}}^n$, for each $n \geq 1$. We also recall that, for any $n \geq 1$, any $i \in \{1,\cdots,N\}$ and any $t \in (t_{i-1}^N, t_i^N]$, ν_t is $\mathcal{F}_{t_i^N}^{\boldsymbol{w}^0}$-measurable. Hence, $\sigma\{(w_s^0-w_{t_i^N}^0, w_s-w_{t_i^N});\ s\in[t_i^N,T]\}$ is independent of $(w_s^0,w_s,\nu_s,\varrho_s)_{0\leq s\leq t_i^N}$ under $\bar{\mathbb{P}}^n$. Passing to the limit, we deduce that for any dyadic $t \in [0,T]$, $\sigma\{(w_s^0-w_t^0, w_s-w_t);\ s\in[t,T]\}$ is independent of $(w_s^0,w_s,\nu_s,\varrho_s)_{0\leq s\leq t}$ under the limiting probability $\mathbb{P}$. By a density argument, the same is true for any $t \in [0,T]$.

Third Step. We now check that condition 3 holds. Since condition 5 has been verified, we may invoke Lemma 7.38. Namely, it suffices to prove that the σ-fields $\sigma\{w_s-w_t,\ t\leq s\leq T\}$ and $\mathcal{F}_T^{\mathrm{nat},(\boldsymbol{w}^0,\nu)}\vee\mathcal{F}_t^{\mathrm{nat},(\boldsymbol{w}^0,\boldsymbol{w},\nu,\varrho)}$ are independent under $\mathbb{P}$, for every $t\in[0,T]$. Again, it is enough to notice that this holds true under $\bar{\mathbb{P}}^n$, for each $n\geq 1$ and each $t\in\{t_0^N,\cdots,t_N^N\}$, which is quite obvious since, for each $n\geq 1$ and each $i\in\{0,\cdots,N\}$, $\nu_{t_i^N}$ is $\mathcal{F}_{t_i^N}^{\boldsymbol{w}^0}$-measurable under $\bar{\mathbb{P}}^n$ and $(\boldsymbol{w}^0,\boldsymbol{w})$ is a Brownian motion with respect to the filtration generated by $(\boldsymbol{w}^0,\boldsymbol{w},\varrho)$.

Fourth Step. The last step is to prove the optimality condition 4 under $\mathbb{P}$. The strategy is similar to that used in the fourth step of the proof of Theorem 7.40. We let $\mathbb{Q} = \mathbb{P}\circ(\boldsymbol{w}^0,\boldsymbol{w},\nu)^{-1}$. By the conclusion of the third step, $\mathbb{P}\in\mathcal{R}(\mathbb{Q})$ and, by Theorem 7.31, we can approximate any $\mathbb{P}'\in\mathcal{R}(\mathbb{Q})$ by a sequence of the form $(\mathbb{Q}\circ(\boldsymbol{w}^0,\boldsymbol{w},\nu,\tilde{\sigma}_n(\boldsymbol{w}^0,\boldsymbol{w},\nu))^{-1})_{n\geq 1}$ where, for each $n\geq 1$, $\tilde{\sigma}_n$ is a continuous function from Ω_{input} to $[0,T]$ and $\tilde{\sigma}_n(\boldsymbol{w}^0,\boldsymbol{w},\nu)$ is a stopping time with respect to the completion of the filtration $\mathbb{F}^{\mathrm{nat},(\boldsymbol{w}^0,\boldsymbol{w},\nu)}$ under $\mathbb{Q}$. Therefore, by continuity of F in the last argument and by Lemma 7.34, see for instance the end of the proof of Theorem 7.39, it suffices to show:

$$\mathbb{E}^{\mathbb{P}}\big[F\big(w^0, w, \nu \circ e^{-1}, \varrho\big)\big] \geq \mathbb{E}^{\mathbb{Q}}\big[F\big(w^0, w, \nu \circ e^{-1}, \tilde{\sigma}(w^0, w, \nu)\big)\big], \tag{7.111}$$

for every continuous function $\tilde{\sigma}$ from Ω_{input} to $[0, T]$ such that $\tilde{\sigma}(w^0, w, \nu)$ is a stopping time with respect to the $\mathbb{Q}$-completion of the filtration $\mathbb{F}^{\text{nat},(w^0,w,\nu)}$.

Now fix such a mapping $\tilde{\sigma}$. Since it is continuous,

$$\mathbb{E}^{\mathbb{Q}}\big[F\big(w^0, w, \nu \circ e^{-1}, \tilde{\sigma}(w^0, w, \nu)\big)\big] = \lim_{n \to \infty} \mathbb{E}^{\bar{\mathbb{P}}^n}\big[F\big(w^0, w, \nu \circ e^{-1}, \tilde{\sigma}(w^0, w, \nu)\big)\big].$$

Recalling from the preliminary step that $W_1(\nu \circ e^{-1}, \vartheta^n((\pi_{(N)}(w^0), \cdot))) \leq 1/N$ under each $\bar{\mathbb{P}}^n$, $n \geq 1$, we deduce that:

$$\begin{aligned} &\mathbb{E}^{\mathbb{P}}\big[F\big(w^0, w, \nu \circ e^{-1}, \tilde{\sigma}(w^0, w, \nu)\big)\big] \\ &\quad = \lim_{n \to \infty} \mathbb{E}^{\bar{\mathbb{P}}^n}\big[F\big(w^0, w, \vartheta^n((\pi_{(N)}(w^0), \cdot), \tilde{\sigma}(w^0, w, \nu)\big)\big]. \end{aligned} \tag{7.112}$$

Under $\bar{\mathbb{P}}^n$, $\nu = \mathcal{L}((\varrho, w) \,|\, \pi_{(N)}(w^0))$, where $\mathcal{L}$ is used to denote the conditional law under $\mathbb{P}^n$ or under $\bar{\mathbb{P}}^n$, the two of them being the same in that case. Hence, there exists a measurable function $\tilde{\mu}^n : \mathcal{C}([0, T]) \to \mathcal{P}_1([0, T] \times \mathcal{C}([0, T]))$ such that $\nu = \tilde{\mu}^n(w^0)$ and the random variable $[\tilde{\mu}^n(w^0)](C)$ is measurable with respect to the completion of $\mathcal{F}_t^{\text{nat},w^0}$ under the Wiener measure, for any $t \in [0, T]$ and any $C \in \mathcal{F}_t^{\text{nat},\nu}$. Therefore, under each $\bar{\mathbb{P}}^n$, $\tilde{\sigma}(w^0, w, \nu)$ may be rewritten as $\tilde{\sigma}(w^0, w, \tilde{\mu}^n(w^0))$, which is a stopping time with respect to the filtration $\mathbb{F}^{(w^0,w)}$. Hence, for each $n \geq 1$,

$$\begin{aligned} &\mathbb{E}^{\bar{\mathbb{P}}^n}\big[F\big(w^0, w, \vartheta^n((\pi_N(w^0), \cdot), \tilde{\sigma}(w^0, w, \nu)\big)\big] \\ &= \mathbb{E}^{\mathbb{P}^n}\big[F\big(w^0, w, \nu \circ e^{-1}, \tilde{\sigma}(w^0, w, \tilde{\mu}^n(w^0))\big)\big], \end{aligned}$$

where $\mathbb{P}^n$ is the probability measure we associated with the semi-strong equilibrium ϑ^n by means of Definition 7.44. Hence, by condition 3 in Definition 7.44, we have:

$$\begin{aligned} &\mathbb{E}^{\bar{\mathbb{P}}^n}\big[F\big(w^0, w, \vartheta^n(\pi_N(w^0), \cdot), \tilde{\sigma}(w^0, w, \nu)\big)\big] \\ &= \mathbb{E}^{\mathbb{P}^n}\big[F\big(w^0, w, \nu \circ e^{-1}, \tilde{\sigma}(w^0, w, \tilde{\mu}^n(w^0))\big)\big] \\ &\leq \mathbb{E}^{\mathbb{P}^n}\big[F\big(w^0, w, \nu \circ e^{-1}, \varrho\big)\big] \\ &= \mathbb{E}^{\bar{\mathbb{P}}^n}\big[F\big(w^0, w, \vartheta^n((\pi_N(w^0), \cdot), \varrho\big)\big]. \end{aligned}$$

The limit of the first term in the left-hand side is given by (7.112). We proceed in a similar way to get the limit of the right-hand side. We obtain:

$$\begin{aligned} \lim_{n \to \infty} \mathbb{E}^{\bar{\mathbb{P}}^n}\big[F\big(w^0, w, \vartheta^n(\pi_N(w^0), \cdot), \varrho\big)\big] &= \lim_{n \to \infty} \mathbb{E}^{\bar{\mathbb{P}}^n}\big[F\big(w^0, w, \nu \circ e^{-1}, \varrho\big)\big] \\ &= \mathbb{E}^{\mathbb{P}}\big[F\big(w^0, w, \nu \circ e^{-1}, \varrho\big)\big], \end{aligned}$$

from which we get (7.111). □

7.2.9 Mean Field Games of Timing with Major and Minor Players

We conclude this second chapter of extensions with the description of a practical application of importance in the financial industry. We frame the model as a mean field game of timing with major and minor players. Given the technical difficulties already existing in the solution of mean field games of timing, we shall not attempt to solve it mathematically. However, the blow-by-blow account of the features of the situation fit the framework of this section like a glove and we could not resist the temptation to present it.

Callable Convertible Bonds

Convertible bonds are debts issued by companies looking for external funds to finance long-term investments. As such, they are subject to default of the issuer. For this reason they should offer a unique set of incentives in order to be attractive and competitive with the Treasury (essentially default free) bonds with comparable lengths.

To make them attractive to investors, even with interest rates lower than the default free interest rates, issuing companies often embed in the indenture of the bond an option to exchange the security for a given number of shares. This number is determined by a conversion ratio whose value at time t we denote by c_t. So convertible bonds are hybrid derivatives with a fixed income component (the interest coupon payments) and an equity component (stock shares). They are very attractive to investors interested in the *upside* (in case the stock price appreciates significantly) with little or no *downside* (except possibly for the occurrence of default) due to the bond protection. Convertible bonds were extremely popular and their market volume increased very fast until the credit crunch of May 2005 due to the credit downgrade of GM and Ford and, later on, the financial crisis of 2007.

Recall that in a typical corporate bond scenario, the seller 1) collects the nominal (i.e., the loan amount) at inception, 2) pays coupons (interest) at regular time intervals, 3) returns the nominal at maturity of the bond, while the buyer (bond holder) 1) pays the nominal upfront, 2) receives coupon payments (interest) at regular time intervals, 3) retrieves the nominal amount at maturity if no default occurred, 4) gets the recovery pay-out (proportion of nominal) in case of default before maturity.

When the bond is *callable*, the seller can at a time of his/her *choosing* (which we shall model mathematically by a stopping time) 1) return the nominal (loan amount) to the investors, 2) stop paying interest coupons. To be specific, a call provision allows the issuer to force the holders to redeem their certificates for an agreed upon cash amount $\overline{C}$. If recall takes place at time t, the actual amount received by the holder of the bond is $\overline{C} \vee (c_t S_t) + A_t$ where A_t represents the *accrued interests* defined by:

$$A_t = r \frac{t - T_i}{T_{i+1} - T_i}, \qquad T_i \le t < T_{i+1}, \tag{7.113}$$

where the T_i's are the dates of the planned coupon payments.

When the bond is *convertible*, the buyer can at a time of his/her *choosing* (which we shall model mathematically by a stopping time) 1) request his/her original investment and walk away (game over), 2) or convert the amount originally invested into company shares. The number of shares per dollar invested is specified in the indenture of the bond as the conversion ratio c_t. The contract ends the first time one of the two counterparties exercises its right.

To shed some light on the definition given above, a convertible bond should be viewed as a corporate (hence *defaultable*) issue for which the investors have the option to exchange each of their certificates for the given number c_t of shares in the company stock if they choose to exercise their exchange option at time t. This simple description of the conversion option justifies the terminology *convertible bond*, but it does not do justice to the high complexity of most issue indentures. For the sake of completeness, we list a few of the commonly clauses included in convertible bond prospectuses.

- A *Put* or *Redemption* provision allows the holder to redeem the bond for an agreed upon amount of cash $\overline{P}$.
- Special clauses specify what is to be expected at maturity, e.g., the holder is often allowed to convert at maturity T if default or call did not occur before.
- A *Put / Redemption* protection specifies a period of time (most often in the form $[0, T_P]$) during which the bond cannot be put or redeemed.
- Similarly, a *Call* protection specifies a period of time (most often in the form $[0, T_C]$) during which the bond cannot be called.
- To cite another example of the many *perls* we can find in corporate bond indentures, we mention that if a call notice of length δ is included in the indenture, upon call of the bond at time τ_c by the issuer, each holder can choose to convert the bond into c_t shares of stock on any day of the interval $[\tau_c, (\tau_c+\delta)\wedge T]$ as long as the investor gives one day notice to the issuer.
- The definition of *default* varies from one bond issue to the next, and the credit events accepted as triggers for the special liquidation of a convertible bond are not limited to bond issuers seeking bankruptcy protection such as Chapter 7 or Chapter 11. Obviously, these are not references to earlier chapters of this book. These chapters are from the lawyers' parlance and they refer to the US bankruptcy code. Indeed, missed payments, credit downgrades, mergers, and acquisitions or other forms of restructuring are often accepted as triggers.
- The procedure used to determine the recovery payment in case of default is clearly articulated in the prospectus of a convertible bond, and it does not always involve auctions or lawyers. For the purpose of this introductory discussion, we assume that it is given by a random variable R_t if default occurs at time t, in such a way that the process $R = \{R_t\}_{t\geq 0}$ is adapted.

Most of these intricate clauses are rarely discussed in the academic literature, and since our goal is to give an example of mean field game of timing with major and minor players, we shall also ignore most of them, and keep only the features relevant to the mean field models we want to motivate.

Set-Up of the Model

The first thing we need is a stochastic process $\boldsymbol{S} = (S_t)_{0 \le t \le T}$ giving the price of the stock of the company. We shall specify the dynamics of this process later on in the discussion. The next element of the model is the time of default τ which we shall take as the time of the first arrival of a Cox process with intensity $\boldsymbol{\gamma} = (\gamma_t)_{0 \le t \le T}$ which we shall also specify later on.

We now introduce for each investor, a couple of processes which play an important role in the analysis of the bond.

- For each investor i, the process $\boldsymbol{U}^i = (U^i_t)_{0 \le t \le T}$ gives the present value at time $t = 0$ of the cumulative cash flows to investor i from the issuer before and including time $t \le T$, should the issuer decide to call the bond at time t while investor i has not exercised any option yet.
- For each investor i, the process $\boldsymbol{L}^i = (L^i_t)_{0 \le t \le T}$ gives the present value at time $t = 0$ of the cumulative cash flows to investor i, before-and-including time t, should he/she decide to convert at time t, while the issuer has not exercised any option yet.

With these notation in hand, a typical convertible bond scenario can be described as follows.

- Bond holder i chooses a *strategy* in the form of a stopping time τ^i with respect to the filtration $\mathbb{F}^i = (\mathcal{F}^i_t)_{t \ge 0}$ comprising his/her information.
- In parallel, the bond seller chooses a strategy τ^0, a stopping time with respect to his/her filtration $\mathbb{F}^0 = (\mathcal{F}^0_t)_{t \ge 0}$

With these choices, the *present value* of all the payments to the bond holder i from the seller is given by the quantity:

$$R^i(\tau^0, \boldsymbol{\tau}) = \begin{cases} L_{\tau^i}, & \text{whenever } \tau^i \le \tau^0 \text{ or } \tau^i = \tau^0 < T \\ \xi, & \text{whenever } \tau^i = \tau^0 = T \\ U_{\tau^0}, & \text{whenever } \tau^0 < \tau^i \end{cases}$$

where $\boldsymbol{\tau} = (\tau^i)_{1 \le i \le N}$ is the set and strategies of the bond investors. The random variable ξ represents the payment to the bond holder at maturity when neither party exercises its right before maturity.

So the mathematical problem is for the bond holders to maximize:

$$J^i(\tau^0, \boldsymbol{\tau}) = \mathbb{E}[R^i(\tau^0, \boldsymbol{\tau})]$$

while the issuer of the bond tries to minimize:

$$J^0(\tau^0, \boldsymbol{\tau}) = \mathbb{E}\Big[\frac{1}{N}\sum_{i=1}^{N} R^i(\tau^0, \boldsymbol{\tau})\Big]$$

in the form of a zero sum game.

The above description does not highlight the interaction between the investors, and does not explain how their individual decisions affect the system. The answer is captured by the word *dilution*. Since holding stocks from a company is a form of partial ownership of the business, if the value of the assets of the company do not change and new shares are issued, the value of each share decreases. There are many reason for a company to issue new shares. We describe a simple example to explain why we claim that the interaction between the bond investors is of a mean field nature. Without any loss of generality, we may assume that all the bond holders invested the same nominal amount in the bond, and we normalize this amount to 1 for the sake of the present discussion. The conversion option states that, if bond investor i decides to convert at time t , namely if $\tau^i = t$, he/she is entitled to receive c_r shares of the company to redeem his/her bond certificate. As a result, the company needs to issue new shares in the amount $c_r \Delta N_t$ where:

$$\Delta N_t = \sum_{i=1}^{N} \mathbf{1}_{\tau^i = t}$$

is the number of investors converting at time t. So if the value of one share of stock is S_{t-} at time t just before the issuance of new shares, the new (theoretical) value of one share after the new issuance becomes:

$$S_{t+} = \frac{N_{t-} S_{t-} + c_r \Delta N_t I_S}{N_{t-} + \Delta N_t},$$

where N_{t-} is the number of outstanding shares just before the new issuance, S_{t-} is the value of one share price, again, just before the new issuance, and I_S is the price at which the new shares are issued (which is usually stipulated in the indenture of the convertible bond). Our point here is to emphasize that the proportion of bond holders converting their bond holding is affecting the value held by the shareholders.

7.2.10 An Explicitly Solvable Toy Model

This last subsection is devoted to the discussion of a simple toy model of a mean field game of timing which can be solved as explicitly as one can hope for. However, it is fair to say that some of the assumptions we need to make for the solution to be possible are highly unrealistic. They preclude many practical applications, including those to bank runs.

We start with two real-valued processes $\boldsymbol{X} = (X_t)_{t \geq 0}$ and $\boldsymbol{Y} = (Y_t)_{t \geq 0}$ with right-continuous and left-limited paths. Both are defined on a common probability space $(\Omega, \mathcal{F}, \mathbb{P})$, and both are assumed to be adapted to a filtration $\mathbb{F} = (\mathcal{F}_t)_{t \geq 0}$. In what follows, the filtration $\mathbb{F}$ is understood as the information available to a representative agent. The information contained in the process $\boldsymbol{Y}$ is assumed to be public. In this regard, $\boldsymbol{Y}$ will be seen as an observable common noise. This model is thus a model of full information as opposed to the diffusion model of bank run discussed in the introduction of this section, for which the common signal $\boldsymbol{Y}$ was not observed. The process $\boldsymbol{X}$ should be interpreted as a private signal only available to the representative agent.

In order to distinguish the common information from the private one, we introduce another filtration, denoted by $\mathbb{G}$. It represents the common knowledge of the players involved in the game. Subsequently, it is assumed to be a sub-filtration of $\mathbb{F}$.

We now add two extra ingredients needed to finalize the set-up of this game of timing. We assume that:

- $\boldsymbol{r} = (r_t)_{t \geq 0}$ is a real valued right-continuous locally integrable and $\mathbb{G}$-progressively measurable process;
- $\boldsymbol{\gamma} = (\gamma_t)_{t \geq 0}$ is a nonnegative right-continuous locally integrable and $\mathbb{F}$-progressively measurable process.

The quantity r_t plays the role of an interest rate at time t. As a $\mathcal{G}_t$-measurable random variable, it is observed by all the players. On the other hand, the value γ_t of $\boldsymbol{\gamma}$ at time t appears as the instantaneous intensity of occurrence of a failure (say a default on some financial market for example), as perceived by the representative player. To be more precise, we define the stopping time:

$$\theta = \inf\Big\{t \geq 0 : \int_0^t \gamma_s ds = \xi\Big\}, \tag{7.114}$$

where ξ is an exponentially distributed random variable with mean 1 which is independent of $\mathbb{F}$. The variable θ is the time of the first jump of a Cox process with intensity $\boldsymbol{\gamma}$. It will be understood as a failure or default time whose perception is subjective to the representative agent.

The game of timing which we propose is based on an idea very similar to the model resulting from our discussion of bank runs. Each player chooses a stopping time for the filtration of its own available information with the goal to maximize an expected reward. The reward comprises two components. An interest component driven by the process $\boldsymbol{r} = (r_t)_{t \geq 0}$ suggests that the player should stop the game as late as possible in order to accumulate interest payments. However, the longer the player stays in the game, the more likely default becomes. This default, as perceived by the player, occurs with intensity $\boldsymbol{\gamma} = (\gamma_t)_{t \geq 0}$, and if default occurs before the decision to stop, all is lost. Under some technical conditions, this optimization problem can be solved using standard techniques. See Lemma 7.50 below.

Because of possible interactions between the various players, the solutions of the individual optimization problems can only lead to an equilibrium if some form of consistency condition is satisfied. Since we are interested in mean field games, we assume that the players interact through the empirical proportion of players which are still in the game or, equivalently, through the theoretical conditional probability that the representative player is still in the game. We shall denote by ρ_t

the cumulative distribution function of this conditional law at time t. In other words, we shall assume that the default intensity γ_t at time t depends upon ρ_t, creating the need for a consistency condition.

In this context, a solution to the mean field game of timing problem can be defined in the following way.

1. *Individual optimal stopping problem.* For each $\mathbb{G}$-progressively measurable $[0,1]$-valued process $\boldsymbol{\rho} = (\rho_t)_{t\geq 0}$ with nondecreasing sample paths, solve the optimal stopping problem:

$$\tau^* \in \arg\sup_{\tau\in\mathcal{S}} \mathbb{E}\Big[e^{\int_0^\tau r_s ds}\mathbf{1}_{\{\theta>\tau\}\cup\{\theta=\infty\}}\Big], \tag{7.115}$$

where $\mathcal{S}$ denotes the set of stopping times with respect to the filtration $\mathbb{F}$.

2. *Fixed point condition.* Find a process $\boldsymbol{\rho} = (\rho_t)_{t\geq 0}$ and a solution τ^* of the above optimal stopping problem such that:

$$\mathbb{P}\big[\tau^* \leq t \,|\, \mathcal{G}_t\big] = \rho_t, \qquad \mathbb{P} - a.s., \quad \text{for all } t \geq 0. \tag{7.116}$$

Definition 7.49 *A process $\boldsymbol{\rho} = (\rho_t)_{t\geq 0}$ and a a stopping time τ^* are said to form an equilibrium for the mean field game of timing if they satisfy the two properties 1. and 2. above.*

We first consider the optimal stopping problem for each individual player.

Lemma 7.50 *Let us assume that the function $[0,\infty) \ni t \mapsto (r_t)_+ = \max(r_t, 0)$ is $\mathbb{P}$-almost surely integrable on $[0,\infty)$, or that $\inf\{t \geq 0 : \gamma_t - r_t \geq 0\} < \infty$ $\mathbb{P}$-almost surely. Furthermore, we assume that the function $[0,\infty) \ni t \mapsto \gamma_t - r_t$ is nondecreasing $\mathbb{P}$-almost surely. Then, the $\mathbb{F}$-stopping time:*

$$\tau^* = \inf\{t \geq 0 : \gamma_t - r_t \geq 0\} \tag{7.117}$$

solves the optimal stopping problem (7.115). *In fact, τ^* is the minimal solution if the supremum in* (7.115) *is finite, and it is the unique solution if $[0,\infty) \ni t \mapsto \gamma_t - r_t$ has strictly increasing paths.*

Proof. Let $\tau \in \mathcal{S}$ be such that $\int_0^\tau (r_s)_+ ds < \infty$ $\mathbb{P}$-a.s., and compute:

$$\begin{aligned}\mathbb{E}\Big[e^{\int_0^\tau r_s ds}\mathbf{1}_{\{\theta>\tau\}\cup\{\theta=\infty\}}\Big] &= \mathbb{E}\Big[\mathbb{E}\Big[e^{\int_0^\tau r_s ds}\mathbf{1}_{\{\theta>\tau\}\cup\{\theta=\infty\}} \,\big|\, \mathcal{F}_\tau\Big]\Big] \\ &= \mathbb{E}\Big[\mathbb{E}\Big[e^{\int_0^\tau (r_s-\gamma_s)ds} \,\big|\, \mathcal{F}_\tau\Big]\Big],\end{aligned}$$

where we used the fact that:

$$\mathbb{P}\Big[\{\theta > \tau\} \cup \{\theta = \infty\} \,|\, \mathcal{F}_\tau\Big] = \exp\Big(-\int_0^\tau \gamma_s ds\Big).$$

Hence,

$$\mathbb{E}\Big[e^{\int_0^\tau r_s ds}\mathbf{1}_{\{\theta>\tau\}\cup\{\theta=\infty\}}\Big] = \mathbb{E}\Big[e^{\int_0^\tau (r_s-\gamma_s)ds}\Big]$$
$$= \mathbb{E}\Big[e^{\int_0^{\tau\wedge\tau^*}(r_s-\gamma_s)ds}e^{\int_{\tau\wedge\tau^*}^{\tau}(r_s-\gamma_s)ds}\Big].$$

This last equality allows to conclude when $(\boldsymbol{r})_+$ is integrable because $(r_t - \gamma_t)_{t\geq 0}$ is nonincreasing. If $(\boldsymbol{r})_+$ is not integrable but $\inf\{t \geq 0 : \gamma_t - r_t \geq 0\} < \infty$ $\mathbb{P}$-as., we can go over the same argument with τ replaced by $\tau \wedge N$ and control the limit $N \to \infty$ with Fatou's lemma. The remainder of the proof (minimality and uniqueness) follows easily. □

Existence of equilibria will be proven in the following framework. We assume that there exists a deterministic continuous function $r : [0,\infty) \times \mathbb{R} \ni (t,y) \mapsto r(t,y) \in \mathbb{R}$ such that the interest rate process $\boldsymbol{r}$ given by $\boldsymbol{r} = (r_t = r(t,Y_t))_{t\geq 0}$ has nonincreasing sample paths and the positive part of $\boldsymbol{r}$ is $\mathbb{P}$-almost surely integrable on $[0,\infty)$, as in the assumption of Lemma 7.50. Also, the intensity process $\boldsymbol{\gamma} = (\gamma_t)_{t\geq 0}$ is given by a formula of the form $\gamma_t = g(t, X_t, Y_t, \rho_t)$ for $t \geq 0$ where $g : [0,\infty)\times\mathbb{R}\times\mathbb{R}\times[0,1] \ni (t,x,y,\rho) \mapsto g(t,x,y,\rho) \in [0,\infty)$ is a continuous function such that, for each fixed (t,y,ρ), the function $\mathbb{R} \ni x \mapsto g_{t,y,\rho}(x) = g(t,x,y,\rho)$ maps $\mathbb{R}$ onto itself, is strictly increasing and thus invertible, and its inverse, which we denote by $g^{-1}_{t,y,\rho}$, is a continuous function.

Also, we assume that $\mathbb{G}$ coincides with the filtration generated by $\boldsymbol{Y}$ and, for each $t \geq 0$ and each $\boldsymbol{y} \in \mathcal{D}([0,T];\mathbb{R})$, we let $F_{t,\boldsymbol{y}}$ be the cumulative distribution function of the regular conditional distribution of X_t given $\boldsymbol{Y}_{\cdot\wedge t}$, namely:

$$\forall x \in \mathbb{R}, \quad F_{t,\boldsymbol{y}}(x) = \mathbb{P}\big[X_t \leq x \,|\, \boldsymbol{Y}_{\cdot\wedge t} = \boldsymbol{y}\big].$$

We assume that $F_{t,\boldsymbol{y}}$ is a continuous function of its variable x.

Proposition 7.51 *Assume that for each $t \in [0,\infty)$, there exists a measurable function $\rho(t,\cdot,\cdot) : \mathcal{D}([0,t];\mathbb{R}) \times \mathbb{R} \ni (\boldsymbol{y}, r) \mapsto \rho(t,\boldsymbol{y},r) \in [0,1]$ such that $u = \rho(t,\boldsymbol{y},r)$ solves the equation:*

$$1 - u = F_{t,\boldsymbol{y}}\Big(g^{-1}_{t,y_t,u}(r)\Big),$$

and the processes $\boldsymbol{\rho} = (\rho_t)_{t\geq 0}$ and $\boldsymbol{\gamma} = (\gamma_t)_{t\geq 0}$ defined by:

$$\rho_t = \rho\big(t, \boldsymbol{Y}_{\cdot\wedge t}, r(t,Y_t)\big), \quad \text{and} \quad \gamma_t = g(t, X_t, Y_t, \rho_t), \quad t \geq 0, \tag{7.118}$$

are nondecreasing and right-continuous.

Call τ^* the stopping time $\tau^* = \inf\{t \geq 0 : \gamma_t - r_t = 0\}$ as in (7.117). Then $\boldsymbol{\rho} = (\rho_t)_{t\geq 0}$ and τ^* form an equilibrium for the mean field game of timing.

Proof. We define the processes $\boldsymbol{\rho} = (\rho_t)_{t\geq 0}$ and $\boldsymbol{\gamma} = (\gamma_t)_{t\geq 0}$ as in (7.118). Observe that $\boldsymbol{\rho}$ is progressively measurable with respect to the filtration generated by $\boldsymbol{Y}$ and that $\boldsymbol{\gamma}$ is progressively measurable with respect to $\mathbb{F}$. Since $\boldsymbol{r}$ is nonincreasing, it is clear that $\boldsymbol{\gamma} - \boldsymbol{r}$ is nondecreasing. Hence, the assumptions of Lemma 7.50 are satisfied and τ^* is an optimal stopping time under the environment $\boldsymbol{\rho}$.

It then remains to check the consistency condition. For any $t \geq 0$,

$$\begin{aligned}\mathbb{P}\big[\tau^* \leq t \,|\, \mathcal{G}_t\big] &= \mathbb{P}\big[\gamma_t \geq r_t \,|\, \mathcal{G}_t\big] \\ &= \mathbb{P}\big[g(t, X_t, Y_t, \rho_t) \geq r(t, Y_t) \,|\, \mathcal{G}_t\big].\end{aligned}$$

Since $\mathcal{G}_t = \sigma\{\boldsymbol{Y}_{\cdot\wedge t}\}$ and ρ_t is $\sigma\{\boldsymbol{Y}_{\cdot\wedge t}\}$-measurable, we get:

$$\begin{aligned}\mathbb{P}\big[\tau^* \leq t \,|\, \mathcal{G}_t\big] &= \mathbb{P}\big[g(t, X_t, Y_t, \rho_t) \geq r(t, Y_t) \,|\, \sigma\{\boldsymbol{Y}_{\cdot\wedge t}\}\big] \\ &= \mathbb{P}\big[X_t \geq g^{-1}_{t,Y_t,\rho_t}\big(r(t, Y_t)\big) \,|\, \sigma\{\boldsymbol{Y}_{\cdot\wedge t}\}\big] \\ &= 1 - F_{t,\boldsymbol{Y}_{\cdot\wedge t}}\big(g^{-1}_{t,Y_t,\rho_t}(r(t, Y_t)\big) = \rho_t,\end{aligned}$$

which completes the proof, where we used the fact that $F_{t,\boldsymbol{y}}$ is continuous in x. □

Example. A simple example of intensity process satisfying the above conditions is given by the additive model:

$$\gamma_t = X_t + Y_t + c\rho_t, \quad t \geq 0,$$

where $c > 0$ is a deterministic constant.

Let us assume for instance that $\boldsymbol{r} = (r_t = r)_{t\geq 0}$ for a fixed constant $r \geq 1$, that $(X_t = X)_{t\geq 0}$ where X is a random variable uniformly distributed on $[r-1, r]$, and that $\boldsymbol{Y} = (Y_t)_{t\geq 0}$ is a process with $Y_0 = 0$ and right-continuous strictly increasing sample paths independent of X. Then,

$$F_{t,\boldsymbol{y}}(x) = \mathbb{P}\big[X \leq x\big] = 1 \wedge (x + 1 - r)_+.$$

Also,

$$g(t, x, y, \rho) = x + y + c\rho,$$

and, for r as above,

$$g^{-1}_{t,y,\rho}(r) = r - y - c\rho.$$

Since $\boldsymbol{Y}$ takes nonnegative values, the equation defining $\rho(t, \boldsymbol{y}, r)$ reads:

$$\begin{aligned}1 - u = F_{t,\boldsymbol{y}}\big(g^{-1}_{t,y_t,u}(r)\big) &\Leftrightarrow 1 - u = 1 \wedge (1 - y_t - cu)_+ \\ &\Leftrightarrow 1 - u = (1 - y_t - cu)_+ \Leftrightarrow u = 1 \wedge \big(\frac{y_t}{1-c}\big).\end{aligned}$$

We find $\rho_t = 1 \wedge [(1-c)^{-1} Y_t]$.

7.3 Notes & Complements

In [207], Huang introduced a linear-quadratic infinite-horizon model with a major player, whose influence does not fade away when the number of players tends to infinity. In a joint work with Nguyen [291], he then considered the finite-horizon counterpart of the model, and Nourian and Caines generalized this model to the nonlinear case in [293]. These models are usually called *"mean field games with major and minor players"*. Unfortunately, the scheme proposed in [291, 293] fails to accommodate the case where the state of the major player enters the dynamics of the minor players. To be more specific, in [291,293], the major player influences the minor players solely via their cost functionals. Subsequently, Nguyen and Huang proposed in [292] a new scheme to solve the general case for linear-quadratic-Gaussian (LQG for short) games in which the major player's state enters the dynamics of the minor players. The limiting control problem for the major player is solved by what the authors call "anticipative variational calculation". In [44], Bensoussan, Chau, and Yam take, as in [293], a stochastic Hamilton-Jacobi-Bellman approach to tackle a type of general mean field games with major and minor players. The limiting problem is characterized by a set of stochastic PDEs and the asymmetry between the major and minor players is pointed out. However, the formulation of the limiting mean field game problem used in [44] is different from ours. Indeed, instead of looking for a global Nash equilibrium of the whole system, including the major and minor players, the authors propose to tackle the problem as a Stackelberg game where the major player goes first. Then, the minor players solve a mean field game problem in the random environment created by the random process of the state of the major player. In this set-up, the major player chooses its own control to minimize its expected cost, assuming that the response of the minor players to the choice of its control will be to put themselves in the unique mean field equilibrium in the random environment induced by the control of the major player.

The contents of Section 7.1 are borrowed from the paper [110] by Carmona and Zhu, and the recent technical report [106] by Carmona and Wang. Differently from [44], we use a probabilistic approach based on an appropriate version of the Pontryagin stochastic maximum principle for conditional McKean-Vlasov dynamics in order to solve the embedded stochastic control problems. Also, we define the limiting problem as a two-player game as opposed to the sequential optimization problems of the Stackelberg formulation of [44]. We believe that this is the right formulation of Nash equilibrium for mean field games with major and minor players. To wit, we stress that the finite-player game in [44] is a N-player game including only the minor players, the major player being treated as exogenous, with no active participation in the game. The associated propagation of chaos is then just a randomized version of the usual propagation of chaos associated with the usual mean field games, and the limiting scheme cannot be completely justified as a Nash equilibrium for the whole system. In contrast, we define the finite-player game as an $(N + 1)$-player game including the major player. The construction of approximate Nash equilibria is proved for the minor players and, most importantly, for the major player as well.

The linear-quadratic-Gaussian (LQG) stochastic control problems are among the best-understood models in stochastic control theory. It is thus natural to look for more explicit results of the major-minor mean field games under the LQG setting. As explained above, this special class of models was first considered by Huang, Nguyen, Caines, and Nourian. The extended mean field game model of optimal execution introduced in Chapter (Vol I)-1 and solved in Subsection (Vol I)-4.7.1 of Chapter (Vol I)-4 was extended by Huang, Jaimungal, and Nourian in [218] to include a major trader. In the absence of idiosyncratic noise, and when the initial conditions of the minor player states are independent and identically distributed, these two authors formulate a fixed point equilibrium problem when the rate of trading of the major player is restricted to be a linear function of the average rate of trading of the minor players; they solve this fixed point problem in the infinite horizon stationary case.

Our discussion of the cyber security model as a mean field game model with major and minor players is borrowed from the paper [106] of Carmona and Wang. It is a straightforward extension of the four state model for the behavior of computer owners facing cyber attacks by hackers proposed by Kolokoltsov and Bensoussan in [235] and discussed in Chapter (Vol I)-7.

Games of timing similar to those considered in this chapter were considered in an abstract setting by Huang and Li in [206]. The presentation of Section 7.2 is mostly inspired by the recent paper [104] by Carmona, Delarue, and Lacker, The toy model presented in Subsection 7.2.10 is borrowed from Nutz' recent technical report [294].

Regarding the construction of a strong equilibrium, the interested reader can find a proof of Tarski's fixed point theorem in [330] and of Topkis' theorem in the fundamental paper [284] of Milgrom and Roberts, both theorems being used in the proof of Theorem 7.27. The compatibility conditions used in our analysis of weak equilibria were introduced and used in Chapter 1. However, the way we use them for constructing solutions of mean field games, and for identifying the asymptotic behavior of a sequence of finite player game approximate Nash equilibria, is somewhat different from what is done in the first part of this volume. This should shed a new light on the meaning of compatibility. In Chapters 1, 2 and 3, compatibility provides a framework that guarantees the fairness of the underlying stochastic control problem, in the sense that the information used by the player to choose its own control strategy must not introduce any bias in the future realizations of the environment. From a mathematical point of view, we made an intensive use of this property when we investigated the FBSDE formulation of the optimal stochastic control problem. Here, our approach is somewhat different and the crux of the proof is the density property stated in Theorem 7.31. It is entirely based upon the notion of compatibility. This density argument is the cornerstone of both the proof of the existence of a solution and the asymptotic analysis of finite player games. For a similar use of compatibility in the analysis of stochastic differential mean field games, we refer to the papers of Lacker [255] and of Carmona, Delarue, and Lacker [100].

Our notion of randomized stopping times is inspired by the work of Baxter and Chacon in [40] where the authors prove compactness of the space of randomized stopping times. While they assume that the filtration is right-continuous, we do not require this assumption in our analysis. The reason is as follows. As made clear by the statement of Theorem 7.31, right-continuity of the filtration is a sufficient condition for the set of randomized stopping times to be closed under weak convergence, and this is all we need in the proof of Lemma 7.45. However, it is worth mentioning that we do not invoke any closeness property of the randomized stopping times for taking limits in the proofs of Theorems 7.40 and 7.41. The key point in our approach is to make use of the fixed point condition in the definition of a weak equilibrium in order to guarantee that the limiting set-up is admissible, in the sense that it satisfies the required compatibility condition. Notice also that, in contrast with [40], we no longer pay any attention to the definition of a specific topology on the set of randomized stopping times. Instead, we make an intensive use of Lemma 7.34, the proof of which may be found in the original article of Jacod and Mémin [216] or in the monograph [196] by Häusler and Luschgy.

Other tools or materials used in the proof may be found in various textbooks. For instance, we refer to the monograph by Ethier and Kurtz [149] for details on the Skorohod topology $J1$ and on convergence in law on Polish spaces. See for instance Proposition 3.6.5 therein, which we applied in the proof of Lemma 7.33, and Proposition 3.4.6 for classes of convergence determining functions, as used in the proof of Theorem 7.35. We also refer to Aliprantis and Border's monograph [17] for general results on set valued functions, including the statement and proof of Kakutani-Fan-Glicksberg's theorem. The proof of Lemma 7.36 stated in the last Subsection 7.2.5 may be found in [100].

We believe that our discussion of callable-convertible bonds is original in the sense that we do not know of any instance of published model framed as a game between the issuer and a large population of investors. In the mid 2000's, stylized forms of zero-sum two-player games, typically Dynkin games of timing, were proposed for pricing purposes. This was initiated by Kifer in [229], who introduced game options, and this culminated in the works of Kallsen and Kühn [222,223], who proved that a no-arbitrage price was given by the value function of a Dynkin game. These results set the stage for further theoretical developments. See for example the works by Bielecki, Crepey, Jeanblanc, and Rutkowski [326,327].

References

1. Y. Achdou, F. Buera, J.M. Lasry, P.L. Lions, and B. Moll. Partial differential equation models in macroeconomics. *Philosophical Transactions of the Royal Society, A*, 372, Oct. 2014.
2. Y. Achdou, F. Camilli, and I. Capuzzo-Dolcetta. Mean field games: numerical methods for the planning problem. *SIAM Journal on Control and Optimization*, 50:77–109, 2010.
3. Y. Achdou, F. Camilli, and I. Capuzzo-Dolcetta. Mean field games: convergence of a finite difference method. *SIAM Journal on Numerical Analysis*, 51:2585–2612, 2013.
4. Y. Achdou and I. Capuzzo-Dolcetta. Mean field games: numerical methods. *SIAM Journal on Numerical Analysis*, 48:1136–1162, 2010.
5. Y. Achdou and M. Laurière. On the system of partial differential equations arising in mean field type control. *Discrete and Continuous Dynamical Systems, A*, 35:3879–3900, 2015.
6. Y. Achdou and V. Perez. Iterative strategies for solving linearized discrete mean field games systems. *Networks and Heterogeneous Media*, 7:197–217, 2012.
7. Y. Achdou and A. Porretta. Convergence of a finite difference scheme to weak solutions of the system of partial differential equations arising in mean field games. *SIAM Journal on Numerical Analysis*, 54:161–186.
8. S. Adlakha and R. Johari. Mean field equilibrium in dynamic games with strategic complementarities. *Operations Research*, 61:971–989, 2013.
9. N. Aghbal and R. Carmona. A solvable mean field game with interactions through the control. Technical report, Princeton University, 2014.
10. P. Aghion and P. Howitt. A model of growth through creative destruction. *Econometrica*, 60:323–352, 1992.
11. S. Ahuja. Wellposedness of mean field games with common noise under a weak monotonicity condition. *SIAM Journal on Control and Optimization*, 54:30–48, 2016.
12. S.R. Aiyagari. Uninsured idiosyncratic risk and aggregate saving. *The Quarterly Journal of Economics*, 109:659–684, 1994.
13. M. Aizenman and B. Simon. Brownian motion and Harnack inequality for Schrödinger operators. *Communications in Pure and Applied Mathematics*, 35:209–273, 1982.
14. S. Alanko. *Regression-based Monte Carlo methods for solving nonlinear PDEs*. PhD thesis, New York University, 2015.
15. D. Aldous. Weak convergence and the general theory of processes. Unpublished notes. http://www.stat.berkeley.edu/~aldous/Papers/weak-gtp.pdf, 1983.
16. D. Aldous. Exchangeability and related topics. In *Ecole d'Eté de Probabilités de Saint Flour 1983*. Volume 1117 of Lecture Notes in Mathematics, pages 1–198. Springer-Verlag Berlin Heidelberg, 1985.
17. C. D. Aliprantis and K. Border. *Infinite Dimensional Analysis. Third Edition*. Springer-Verlag Berlin Heidelberg, 2006.
18. R. Almgren and N. Chriss. Optimal execution of portfolio transactions. *Journal of Risk*, 3:5–39, 2001.

R. Carmona, F. Delarue, *Probabilistic Theory of Mean Field Games with Applications II*, Probability Theory and Stochastic Modelling 84,
DOI 10.1007/978-3-319-56436-4

19. N. Almulla, R. Ferreira, and D. Gomes. Two numerical approaches to stationary mean-field games. *Dynamic Games and Applications*, 7:657–682, 2016.
20. L. Ambrosio and J. Feng. On a class of first order Hamilton-Jacobi equations in metric spaces. *Journal of Differential Equations*, 256:2194–2245, 2014.
21. L. Ambrosio, N. Gigli, and G. Savaré. *Gradient flows in metric spaces and in the Wasserstein space of probability measures*. Birkhäuser Basel, 2004.
22. T.T.K. An and B. Oksendal. Maximum principle for stochastic differential games with partial information. *Journal of Optimization Theory and Applications*, 139:463–483, 2008.
23. T.T.K. An and B. Oksendal. A maximum principle for stochastic differential games with g-expectations and partial information. *Stochastics*, 84:137–155, 2012.
24. D. Andersson and B. Djehiche. A maximum principle for SDEs of mean-field type. *Applied Mathematics & Optimization*, 63:341–356, 2010.
25. F. Antonelli. Backward-forward stochastic differential equations. *Annals of Applied Probability*, 3:777–793, 1993.
26. F. Antonelli and J. Ma. Weak solutions of forward-backward SDE's. *Stochastic Analysis and Applications*, 21(3):493–514, 2003.
27. R. Aumann. Markets with a continuum of traders. *Econometrica*, 32:39–50, 1964.
28. R. J. Aumann. Existence of competitive equilibrium in markets with continuum of traders. *Econometrica*, 34:1–17, 1966.
29. X. Tan C. Zhou B. Bouchard, D. Possamaï. A unified approach to a priori estimates for supersolutions of BSDEs in general filtrations. *Annales de l'institut Henri Poincaré, Probabilités et Statistiques*, to appear.
30. R. Bafico and P. Baldi. Small random perturbations of Peano phenomena. *Stochastics*, 6:279–292, 1982.
31. F. Baghery and B. Oksendal. A maximum principle for stochastic control with partial information. *Stochastic Analysis and Applications*, 25:705–717, 2007.
32. K. Bahlali, B. Mezerdi, M. N'zi, and Y. Ouknine. Weak solutions and a Yamada-Watanabe theorem for FBSDEs. *Random Operators and Stochastic Equations*, 15:271–285, 2007.
33. M. Bardi. Explicit solutions of some linear quadratic mean field games. *Networks and Heterogeneous Media*, 7:243–261, 2012.
34. M. Bardi and E. Feleqi. Nonlinear elliptic systems and mean field games. *Nonlinear Differential Equations and Applications NoDEA*, 23:44, 2016.
35. M. Bardi and F. Priuli. Linear-quadratic N-person and mean-field games with ergodic cost. *SIAM Journal on Control and Optimization*, 52:3022–3052, 2014.
36. F. Barthe and C. Bordenave. Combinatorial optimization over two random point sets. In C. Donati-Martin et al., editors, *Séminaire de Probabilités XLV*. Volume 2046 of Lecture Notes in Mathematics, pages 483–536. Springer International Publilshing, 2013.
37. R. Basna, A. Hilbert, and V.N. Kolokolstov. An approximate Nash equilibrium for pure jump Markov games of mean-field-type on continuous state space. *Stochastics*, 89:967–993, 2017.
38. R.F. Bass. *Diffusions and Elliptic Operators*. Springer-Verlag New York, 1998.
39. R.F. Bass and P. Hsu. Some potential theory for reflecting Brownian motion in Hölder and Lipschitz domains. *Annals of Probability*, 19:486–508, 1991.
40. J.R. Baxter and R.V. Chacon. Compactness of stopping times. *Zeitschrift für Wahrscheinlichkeitstheorie und verwandte Gebiete*, 40:169–181, 1977.
41. J.-D. Benamou and Y. Brenier. A computational fluid mechanics solution to the Monge-Kantorovich mass transfer problem. *Numerische Mathematik*, 84:375–393, 2000.
42. J.D. Benamou and G. Carlier. Augmented Lagrangian methods for transport optimization, mean field games and degenerate elliptic equations. *Journal of Optimization Theory and Applications*, 167:1–26, 2015.
43. J.D. Benamou, G. Carlier, and N. Bonne. An augmented Lagrangian numerical approach to solving mean field games. Technical report, INRIA, 2013. https://hal.inria.fr/hal-00922349/
44. A. Bensoussan, M.H.M. Chau, and S.C.P. Yam. Mean field games with a dominating player. *Applied Mathematics & Optimization*, 74:91–128, 2016.

45. A. Bensoussan and J. Frehse. Nonlinear elliptic systems in stochastic game theory. *Journal für die reine und angewandte Mathematik*, 350:23–67, 1984.
46. A. Bensoussan and J. Frehse. On Bellman equations of ergodic control in $\mathbf{R}^n$. *Journal für die reine und angewandte Mathematik*, 492:125–160, 1992.
47. A. Bensoussan and J. Frehse. Ergodic Bellman systems for stochastic games in arbitrary dimension. *Proceedings of the Royal Society of Edinburgh, A*, 449:65–77, 1995.
48. A. Bensoussan and J. Frehse. Stochastic games for N players. *Journal of Optimization Theory and Applications*, 105:543–565, 2000.
49. A. Bensoussan and J. Frehse. Smooth solutions of systems of quasilinear parabolic equations. *ESAIM: Control, Optimisation and Calculus of Variations*, 8:169–193, 2010.
50. A. Bensoussan, J. Frehse, and P. Yam. *Mean Field Games and Mean Field Type Control Theory*. SpringerBriefs in Mathematics. Springer-Verlag New York, 2013.
51. A. Bensoussan, J. Frehse, and S. C. P. Yam. The master equation in mean field theory. *Journal de Mathématiques Pures et Appliquées*, 2014.
52. A. Bensoussan, J. Frehse, and S. C. P. Yam. On the interpretation of the master equation. *Stochastic Processes and their Applications*, 127:2093–2137, 2016.
53. A. Bensoussan, K.C.J. Sung, S.C.P. Yam, and S.P. Yung. Linear quadratic mean field games. *Journal of Optimization Theory and Applications*, 169:469–529, 2016.
54. A. Bensoussan and S. C. P. Yam. Control problem on space of random variables and master equation. Technical report, http://arxiv.org/abs/1508.00713, 2015.
55. D.P. Bertsekas. *Nonlinear Programming*. Athena Scientific, 1995.
56. D.P. Bertsekas and S.E. Shreve. *Stochastic Optimal Control: The Discrete Time Case*. Academic Press, 1978.
57. P. Billingsley. *Convergence of Probability Measures. Third edition.* John Wiley & Sons, Inc., 1995.
58. P. Billingsley. *Probability and Measure. Second edition.* John Wiley & Sons, Inc., 1999.
59. A. Bisin, U. Horst, and O. Özgür. Rational expectations equilibria of economies with local interactions. *Journal of Economic Theory*, 127:74–116, 2006.
60. J.-M. Bismut. Théorie probabiliste du contrôle des diffusions, *Memoirs of the American Mathematical Society*, 167(4), 1976.
61. J.-M. Bismut. Conjugate convex functions in optimal stochastic control. *Journal of Mathematical Analysis and Applications*, 44:384–404, 1973.
62. J.M. Bismut. An introductory approach to duality in optimal stochastic control. *SIAM Review*, 20:62–78, 1978.
63. D. Blackwell and L.E. Dubins. An extension of Skorohod's almost sure convergence theorem. *Proceedings of the American Mathematical Society*, 89:691–692, 1983.
64. V.I. Bogachev. *Measure Theory, Volume 2*. Springer-Verlag Berlin Heidelberg, 2007.
65. V.S. Borkar. Controlled diffusion processes. *Probability Surveys*, 2:213–244, 2005.
66. V.S. Borkar. Ergodic control of diffusion processes. In Marta Sanz-Solé et al., editors, *Proceedings of the International Congress of Mathematics, Madrid, Spain*, pages 1299–1309. European Mathematical Society, 2006.
67. P. Brémaud. *Point Processes and Queues: Martingale Dynamics*. Springer Series in Statistics. Springer-Verlag New York, 1981.
68. P. Brémaud and M. Yor. Changes of filtrations and of probability measures. *Zeitschrift für Wahrscheinlichkeitstheorie und verwandte Gebiete*, 45:269–295, 1978.
69. Y. Brenier. Polar factorization and monotone rearrangement of vector-valued functions. *Communications on Pure and Applied Mathematics*, 44:375–417, 1991.
70. P. Briand and Y. Hu. Stability of BSDEs with random terminal time and homogenization of semilinear elliptic PDEs. *Journal of Functional Analysis*, 155:455–494, 1998.
71. P. Briand and Y. Hu. BSDE with quadratic growth and unbounded terminal value. *Probability Theory and Related Fields*, 136:604–618, 2006.
72. G. Brunick and S. Shreve. Mimicking an Itô process by a solution of a stochastic differential equation. *Annals of Applied Probability*, 23:1584–1628, 2013.

73. J. Bryant. A model of reserves, bank runs and deposit insurance. *Journal of Banking and Finance*, 4:335–344, 1980.
74. R. Buckdahn, B. Djehiche, and J. Li. Mean field backward stochastic differential equations and related partial differential equations. *Stochastic Processes and their Applications*, 119:3133–3154, 2007.
75. R. Buckdahn, B. Djehiche, J. Li, and S. Peng. Mean field backward stochastic differential equations: A limit approach. *Annals of Probability*, 37:1524–1565, 2009.
76. R. Buckdahn and H.-J. Engelbert. A backward stochastic differential equation without strong solution. *Theory of Probability and its Applications*, 50:284–289, 2006.
77. R. Buckdahn and H.-J. Engelbert. On the continuity of weak solutions of backward stochastic differential equations. *Theory of Probability and its Applications*, 52:152–160, 2008.
78. R. Buckdahn, H. J. Engelbert, and A. Răşcanu. On weak solutions of backward stochastic differential equations. *Theory of Probability and its Applications*, 49:16–50, 2005.
79. R. Buckdahn, J. Li, S. Peng, and C. Rainer. Mean-field stochastic differential equations and associated PDEs. *Annals of Probability*, 45:824–878, 2017.
80. R. Buckdahn and S. Peng. Stationary backward stochastic differential equations and associated partial differential equations. *Probability Theory and Related Fields*, 115:383–399, 1999.
81. K. Burdzy, W. Kang, and K. Ramanan. The Skorokhod problem in a time-dependent interval. *Stochastic Processes and their Applications*, 119:428–452, 2009.
82. F. Camilli and C. Marchi. Stationary mean field games systems defined on networks. *SIAM Journal on Control and Optimization*, 54:1085–1103, 2016.
83. P. Cardaliaguet. Notes from P.L. Lions' lectures at the Collège de France. Technical report, https://www.ceremade.dauphine.fr/$\sim$cardalia/MFG100629.pdf, 2012.
84. P. Cardaliaguet. Long time average of first order mean field games and weak KAM theory. *Dynamic Games and Applications*, 3:473–488, 2013.
85. P. Cardaliaguet. Weak solutions for first order mean field games with local coupling. In P. Bettiol et al., editors, *Analysis and Geometry in Control Theory and its Applications*. Springer INdAM Series, pages 111–158. Springer International Publishing, 2015.
86. P. Cardaliaguet, F. Delarue, J.-M. Lasry, and P.-L. Lions. The master equation and the convergence problem in mean field games. Technical report, http://arxiv.org/abs/1509.02505, 2015.
87. P. Cardaliaguet and J. Graber. Mean field games systems of first order. *ESAIM: Control, Optimisation and Calculus of Variations*, 21:690–722, 2015.
88. P. Cardaliaguet, J. Graber, A. Porretta, and D. Tonon. Second order mean field games with degenerate diffusion and local coupling. *Nonlinear Differential Equations and Applications NoDEA*, 22:1287–1317, 2015.
89. P. Cardaliaguet and S. Hadikhanloo. Learning in mean field games: the fictitious play. *ESAIM: Control, Optimisation and Calculus of Variations*, 23:569–591, 2017.
90. P. Cardaliaguet, J.M. Lasry, P.L. Lions, and A. Porretta. Long time average of mean field games. *Networks and Heterogeneous Media*, 7:279–301, 2012.
91. P. Cardaliaguet, J.M. Lasry, P.L. Lions, and A. Porretta. Long time average of mean field games with a nonlocal coupling. *SIAM Journal on Control and Optimization*, 51:3558–3591, 2013.
92. P. Cardaliaguet, A. R. Mészáros, and F. Santambrogio. First order mean field games with density constraints: Pressure equals price. *SIAM Journal on Control and Optimization*, 54:2672–2709, 2016.
93. E. A. Carlen. Conservative diffusions. *Communication in Mathematical Physics*, 94:293–315, 1984.
94. R. Carmona. *Lectures on BSDEs, Stochastic Control and Stochastic Differential Games*. SIAM, 2015.
95. R. Carmona and F. Delarue. Mean field forward-backward stochastic differential equations. *Electronic Communications in Probability*, 2013.

96. R. Carmona and F. Delarue. Probabilistic analysis of mean field games. *SIAM Journal on Control and Optimization*, 51:2705–2734, 2013.
97. R. Carmona and F. Delarue. The master equation for large population equilibriums. In D. Crisan, B. Hambly, T. Zariphopoulou, editors, *Stochastic Analysis and Applications 2014: In Honour of Terry Lyons*, pages 77–128. Springer Cham, 2014.
98. R. Carmona and F. Delarue. Forward-backward stochastic differential equations and controlled Mckean Vlasov dynamics. *Annals of Probability*, 43:2647–2700, 2015.
99. R. Carmona, F. Delarue, and A. Lachapelle. Control of McKean-Vlasov versus mean field games. *Mathematics and Financial Economics*, 7:131–166, 2013.
100. R. Carmona, F. Delarue, and D. Lacker. Mean field games with common noise. *Annals of Probability*, 44:3740–3803, 2016.
101. R. Carmona, J.P. Fouque, M. Moussavi, and L.H. Sun. Systemic risk and stochastic games with delay. Technical report, 2016. https://arxiv.org/abs/1607.06373
102. R. Carmona, J.P. Fouque, and L.H. Sun. Mean field games and systemic risk: a toy model. *Communications in Mathematical Sciences*, 13:911–933, 2015.
103. R. Carmona and D. Lacker. A probabilistic weak formulation of mean field games and applications. *Annals of Applied Probability*, 25:1189–1231, 2015.
104. R. Carmona, F. Delarue, and D. Lacker. Mean field games of timing and models for bank runs. *Applied Mathematics & Optimization*, 76:217–260, 2017.
105. R. Carmona and D. Nualart. *Nonlinear Stochastic Integrators, Equations and Flows*. Gordon & Breach, 1990.
106. R. Carmona and P. Wang. An alternative approach to mean field game with major and minor players, and applications to herders impacts. *Applied Mathematics & Optimization*, 76:5–27, 2017.
107. R. Carmona and K. Webster. The self financing condition in high frequency markets. *Finance Stochastics*, to appear.
108. R. Carmona and K. Webster. A Stackelberg equilibrium for the Limit Order Book. *Mathematical Finance*, to appear.
109. R. Carmona and W.I. Zheng. Reflecting Brownian motions and comparison theorems for Neumann heat kernels. *Journal of Functional Analysis*, 123:109–128, 1994.
110. R. Carmona and G. Zhu. A probabilistic approach to mean field games with major and minor players. *Annals of Applied Probability*, 26:1535–1580, 2016.
111. C. Ceci, A. Cretarola, and F. Russo. BSDEs under partial information and financial applications. *Stochastic Processes and their Applications*, 124:2628–2653, 2014.
112. U. Cetin, H.M. Soner, and N. Touzi. Options hedging under liquidity costs. *Finance Stochastics*, 14:317–341, 2010.
113. P. Chan and R. Sircar. Bertrand and Cournot mean field games. *Applied Mathematics & Optimization*, 71:533–569, 2015.
114. J.F. Chassagneux, D. Crisan, and F. Delarue. McKean-vlasov FBSDEs and related master equation. Technical report, http://arxiv.org/abs/1411.3009, 2015.
115. P. G. Ciarlet. *Introduction to Numerical Linear Algebra and Optimisation*. Cambridge Texts in Applied Mathematics. Cambridge University Press, 1989.
116. E. Çinlar. *Probability and Stochastics*. Graduate Texts in Mathematics. Springer-Verlag New York, 2011.
117. M. Cirant. Multi-population mean field games systems with Neumann boundary conditions. *Journal de Mathématiques Pures et Appliquées*, 103:1294–1315, 2015.
118. M. Cirant. Stationary focusing mean-field games. *Communications in Partial Differential Equations*, 41:1324–1346, 2016.
119. M. Cirant and G. Verzini. Bifurcation and segregation in quadratic two-populations mean field games systems. *ESAIM: Control, Optimisation and Calculus of Variations*, 23:1145–1177, 2017.
120. M. Coghi and F. Flandoli. Propagation of chaos for interacting particles subject to environmental noise. *Annals of Applied Probability*, 26:1407–1442, 2016.

121. M.G Crandall and P.L. Lions. Viscosity solutions of Hamilton-Jacobi equations in infinite dimensions. i. Uniqueness of viscosity solutions. *Journal of Functional Analysis*, 62:379–396, 1985.
122. M.G Crandall and P.L. Lions. Viscosity solutions of Hamilton-Jacobi equations in infinite dimensions. ii. Existence of viscosity solutions. *Journal of Functional Analysis*, 65:368–405, 1986.
123. M.G Crandall and P.L. Lions. Viscosity solutions of Hamilton-Jacobi equations in infinite dimensions. iii. *Journal of Functional Analysis*, 68:214–247, 1986.
124. M.G Crandall and P.L. Lions. Viscosity solutions of Hamilton-Jacobi equations in infinite dimensions. iv. Hamiltonians with unbounded linear terms. *Journal of Functional Analysis*, 90:3237–283, 1990.
125. F. Cucker and E. Mordecki. Flocking in noisy environments. *Journal de Mathématiques Pures et Appliquées*, 89:278–296, 2008.
126. F. Cucker and S. Smale. Emergent behavior in flocks. *IEEE Transactions on Automatic Control*, 52:852–862, 2007.
127. V. Voskanyana D.A. Gomes, S. Patrizi. On the existence of classical solutions for stationary extended mean field games. *Nonlinear Analysis: Theory, Methods & Applications*, 99:49–79, 2014.
128. G. Da Prato and J. Zabczyk. *Ergodicity for Infinite Dimensional Systems*. Cambridge University Press, 1996.
129. E.B. Davies. Spectral properties of compact manifolds and changes of metric. *American Journal of Mathematics*, 112:15–39, 1990.
130. D. Dawson and J. Vaillancourt. Stochastic McKean-Vlasov equations. *NoDEA. Nonlinear Differential Equations and Applications*, 2(2):199–229, 1995.
131. A. Debussche, Y. Hu, and G. Tessitore. Ergodic BSDEs under weak dissipative assumptions. *Stochastic Processes and their Applications*, 121:407–426, 2011.
132. F. Delarue. On the existence and uniqueness of solutions to FBSDEs in a non-degenerate case. *Stochastic Processes and their Applications*, 99:209–286, 2002.
133. F. Delarue. Estimates of the solutions of a system of quasi-linear PDEs. a probabilistic scheme. In J. Azéma et al., editors, *Séminaire de Probabilités XXXVII*, pages 290–332. Springer-Verlag Berlin Heidelberg, 2003.
134. F. Delarue and G. Guatteri. Weak existence and uniqueness for FBSDEs. *Stochastic Processes and their Applications*, 116:1712–1742, 2006.
135. S. Dereich, M. Scheutzow, and R. Schottstedt. Constructive quantization: approximation by empirical measures. *Annales Institut Henri Poincaré, Probabilités Statistiques*, 49:1183–1203, 2013.
136. D.W. Diamond and P.H. Dybvig. Bank runs, deposit insurance, and liquidity. *The Journal of Political Economy*, (91):401–419, 1983.
137. B. Djehiche, A. Tcheukam Siwe, and H. Tembine. Mean field-type games in engineering. Technical report, https://arxiv.org/abs/1605.03281, 2016.
138. B. Djehiche, H. Tembine, and R. Tempone. A stochastic maximum principle for risk-sensitive mean-field type control. *IEEE Transactions on Automatic Control*, 60:2640–2649, 2015.
139. J. Doncel, N. Gast, and B. Gaujal. Are mean-field games the limits of finite stochastic games? *SIGMETRICS Performance Evaluation Review*, 44:18–20, 2016.
140. J. Doncel, N. Gast, and B. Gaujal. Mean-field games with explicit interactions. Technical report, https://hal.inria.fr/hal-01277098/file/main.pdf, 2016.
141. G. Dos Reis. *Some advances on quadratic BSDE: Theory - Numerics - Applications*. LAP LAMBERT Academic Publishing, 2011.
142. R. Duboscq and A. Réveillac. Stochastic regularization effects of semi-martingales on random functions. *Journal de Mathématiques Pures et Appliquées*, 106:1141–1173, 2016.
143. R.M. Dudley. *Real Analysis and Probability*. Wadsworth & Brooks/Cole, 1989.
144. D. Duffie, G. Giroux, and G. Manso. Information percolation. *American Economics Journal: Microeconomic Theory*, 2:1, 2010.

145. D. Duffie, S. Malamud, and G. Manso. Information percolation with equilibrium search dynamics. *Econometrica*, 77:1513–1574, 2009.
146. D. Duffie, S. Malamud, and G. Manso. Information percolation in segmented markets. *Journal of Economic Theory*, 153:1–32, 2014.
147. D. Duffie and G. Manso. Information percolation in large markets. *American Economic Review, Papers and Proceedings*, 97:203–209, 2007.
148. D. Duffie and Y. Sun. Existence of independent random matching. *Annals of Applied Probability*, 17:385–419, 2007.
149. S. Ethier and T. Kurtz. *Markov Processes: Characterization and Convergence*. John Wiley & Sons, Inc., 2005.
150. G. Fabbri, F. Gozzi, and A. Swiech. *Stochastic Optimal Control in Infinite Dimensions: Dynamic Programming and HJB Equations*. Probability Theory and Stochastic Modelling. Springer International Publishing, 2017.
151. E. Feleqi. The derivation of ergodic mean field game equations for several population of players. *Dynamic Games and Applications*, 3:523–536, 2013.
152. J. Feng and M. Katsoulakis. A comparison principle for Hamilton-Jacobi equations related to controlled gradient flows in infinite dimensions. *Archive for Rational Mechanics and Analysis*, 192:275–310, 2009.
153. R. Ferreira and D. Gomes. Existence of weak solutions to stationary mean field games through variational inequalities. Technical report, http://arxiv.org/abs/1512.05828, 2015.
154. M. Fischer. On the connection between symmetric N-player games and mean field games. *Annals of Applied Probability*, 27:757–810, 2017.
155. M. Fischer and G. Livieri. Continuous time mean-variance portfolio optimization through the mean field approach. *ESAIM: Probability and Statistics*, 20:30–44, 2016.
156. F. Flandoli. *Random Perturbation of PDEs and Fluid Dynamics: Ecole d'été de probabilités de Saint-Flour XL*. Volume 2015 of Lecture Notes in Mathematics. Springer-Verlag Berlin Heidelberg, 2011.
157. W.H. Fleming and M. Soner. *Controlled Markov Processes and Viscosity Solutions*. Stochastic Modelling and Applied Probability. Springer-Verlag, New York, 2010.
158. W.H. Fleming. Generalized solutions in optimal stochastic control. In *Proceedings of the Second Kingston Conference on Differential Games*, pages 147–165. Marcel Dekker, 1977.
159. R. Foguen Tchuendom. Restoration of uniqueness of Nash equilibria for a class of linear-quadratic mean field games with common noise. *Dynamic Games and Applications*, to appear.
160. M. Fornasier and F. Solombrino. Mean-field optimal control. *ESAIM: Control, Optimisation and Calculus of Variations*, 20:1123–1152, 2014.
161. N. Fournier and A. Guillin. On the rate of convergence in the Wasserstein distance of the empirical measure. *Probability Theory and Related Fields*, 162:707–738, 2015.
162. A. Friedman. *Partial differential equations of parabolic type*. Prentice-Hall, Englewood Cliffs, N.J., first edition, 1964.
163. A. Friedman. Stochastic differential games. *Journal of Differential Equations*, 11:79–108, 1972.
164. D. Fudenberg and D. Levine. Open-loop and closed-loop equilibria in dynamic games with many players. *Journal of Economic Theory*, 44:1–18, 1988.
165. D. Fudenberg and J. Tirole. *Game Theory*. MIT Press, 1991.
166. M. Fuhrman and G. Tessitore. Infinite horizon backward stochastic differential equations and elliptic equations in Hilbert spaces. *Annals of Probability*, 30:607–660, 2004.
167. W. Gangbo, T. Nguyen, and A. Tudorascu. Hamilton-Jacobi equations in the Wasserstein space. *Methods and Applications of Analysis*, 15:155–184, 2008.
168. W. Gangbo and A. Swiech. Existence of a solution to an equation arising from the theory of mean field games. *Journal of Differential Equations*, 259:6573–6643, 2015.
169. N. Gast and B. Gaujal. A mean field approach for optimization in discrete time. *Journal of Discrete Event Dynamic Systems*, 21:63–101, 2011.

170. N. Gast, B. Gaujal, and J.-Y. Le Boudec. Mean field for Markov decision processes: from discrete to continuous optimization. *IEEE Transactions on Automatic Control*, 57:2266–2280, 2012.
171. J. Gatheral, A. Schied, and A. Slynko. Transient linear price impact and Fredholm integral equations. *Mathematical Finance*, 22:445–474, 2012.
172. R. Gayduk and S. Nadtochiy. Liquidity effects of trading frequency. *Mathematical Finance*, to appear.
173. I.I. Gihman and A.V. Skorohod. *Controlled Stochastic Processes*. Springer-Verlag Berlin Heidelberg New York, 1979.
174. D.A. Gomes, J. Mohr, and R.R. Souza. Discrete time, finite state space mean field games. *Journal de Mathématiques Pures et Appliquées*, 93:308–328, 2010.
175. D.A. Gomes, J. Mohr, and R.R. Souza. Continuous time finite state mean field games. *Applied Mathematics & Optimization*, 68:99–143, 2013.
176. D.A. Gomes, L. Nurbekyan, and E. Pimentel. *Economic Models and Mean-field Games Theory*. Publicaões Matemáticas, IMPA, Rio, Brazil, 2015.
177. D.A. Gomes and E. Pimentel. Time-dependent mean-field games with logarithmic nonlinearities. *SIAM Journal of Mathematical Analysis*, 47:3798–3812, 2015.
178. D.A. Gomes and E. Pimentel. Local regularity for mean-field games in the whole space. *Minimax Theory and its Applications*, 1:65–82, 2016.
179. D.A. Gomes, E. Pimentel, and H. Sánchez-Morgado. Time-dependent mean-field games in the sub- quadratic case. *Communications in Partial Differential Equations*, 40:40–76, 2015.
180. D.A. Gomes, E. Pimentel, and H. Sánchez-Morgado. Time-dependent mean-field games in the superquadratic case. *ESAIM: Control, Optimisation and Calculus of Variations*, 22:562–580, 2016.
181. D.A. Gomes, E. Pimentel, and V. Voskanyan. *Regularity Theory for Mean-Field Game Systems*. SpringerBriefs in Mathematics Springer International Publishing, 2016.
182. D.A. Gomes and J. Saude. Mean field games models - a brief survey. *Dynamic Games and Applications*, 4:110–154, 2014.
183. D.A. Gomes and V. Voskanyan. Extended mean field games. *SIAM Journal on Control and Optimization*, 54:1030–1055, 2016.
184. A. Granas and J. Dugundji. *Fixed point theory*. Springer Monographs in Mathematics. Springer-Verlag New York, 2003.
185. O. Guéant. From infinity to one: The reduction of some mean field games to a global control problem. *Cahier de la Chaire Finance et Développement Durable*, 42, 2011.
186. O. Guéant. Mean field games equations with quadratic Hamiltonian: A specific approach. *Mathematical Models and Methods in Applied Sciences*, 22:291–303, 2012.
187. O. Guéant. New numerical methods for mean field games with quadratic costs. *Networks and Heterogeneous Media*, 2:315–336, 2012.
188. O. Guéant. Existence and uniqueness result for mean field games with congestion effect on graphs. *Applied Mathematics & Optimization*, 72:291–303, 2015.
189. O. Guéant, J.M. Lasry, and P.L. Lions. Mean field games and applications. In R. Carmona et al., editors, *Paris Princeton Lectures on Mathematical Finance 2010*. Volume 2003 of Lecture Notes in Mathematics. Springer-Verlag Berlin Heidelberg, 2010.
190. X. Guo and O. Hernández-Lerma. *Continuous-Time Markov Decision Processes*. Stochastic Modelling and Applied Probability. Springer-Verlag Berlin Heidelberg, 2009.
191. I. Gyöngy. Mimicking the one-dimensional marginal distributions of processes having an Itô differential. *Probability Theory and Related Fields*, 71:501–516, 1986.
192. S. Haadem, B. Øksendal, and F. Proske. Maximum principles for jump diffusion processes with infinite horizon. *Automatica*, 49:2267–2275, 2013.
193. S. Hamadène. Backward-forward SDE's and stochastic differential games. *Stochastic Processes and their Applications*, 77:1–15, 1998.
194. S. Hamadène. Nonzero-sum linear quadratic stochastic differential games and backward forward equations. *Stochastic Analysis and Applications*, 17:117–130, 1999.

195. S. Hamadène and J.P. Lepeltier. Backward equations, stochastic control and zero-sum stochastic differential games. *Stochastics and Stochastic Reports*, 54:221–231, 1995.
196. E. Häusler and H. Luschgy. *Stable Convergence and Stable Limit Theorems*. Probability Theory and Stochastic Modelling. Springer International Publishing, 1995.
197. Z. He and W. Xiong. Dynamic debt runs. *Review of Financial Studies*, 25:1799–1843, 2012.
198. J. Horowitz and R.L. Karandikar. Mean rates of convergence of empirical measures in the Wasserstein metric. *Journal of Computational and Applied Mathematics*, 55:261–273, 1994.
199. U. Horst. Ergodic fluctuations in a stock market model with interacting agents: the mean field case. *Discussion paper No. 106, Sonderforschungbereich 373, Humboldt Universität, Berlin*, 1999.
200. U. Horst. Stationary equilibria in discounted stochastic games with weakly interacting players. *Games and Economic Behavior*, 51:83–108, 2005.
201. J.A. Hosking. A stochastic maximum principle for a stochastic differential game of a mean-field type. *Applied Mathematics & Optimization*, 66:415–454, 2012.
202. Y. Hu. Stochastic maximum principle. In John Baillieul, Tariq Samad, editors, *Encyclopedia of Systems and Control*, pages 1347–1350. Springer-Verlag London, 2015.
203. Y. Hu and S. Peng. Maximum principle for semilinear stochastic evolution control systems. *Stochastics and Stochastic Reports*, 33:159–180, 1990.
204. Y. Hu and S. Tang. Multi-dimensional backward stochastic differential equations of diagonally quadratic generators. *Stochastic Processes and their Applications*, 126:1066–1086, 2016.
205. Y. Hu and G. Tessitore. BSDE on an infinite horizon and elliptic PDEs in infinite dimension. *Nonlinear Differential Equations and Applications NoDEA*, 14:825–846, 2007.
206. C-F. Huang and L. Li. Continuous time stopping games with monotone reward structures. *Mathematics of Operations Research*, 15:496–507.
207. M. Huang. Large-population LQG games involving a major player: the Nash equivalence principle. *SIAM Journal on Control and Optimization*, 48:3318–3353, 2010.
208. M. Huang. A mean field accumulation game with HARA utility. *Dynamics Games and Applications*, 3:446–472, 2013.
209. M. Huang. Mean field capital accumulation games: the long time behavior. In *Proceedings of the 52nd IEEE Conference on Decision and Control*, pages 2499–2504. 2013.
210. M. Huang, P.E. Caines, and R.P. Malhamé. Individual and mass behavior in large population stochastic wireless power control problems: centralized and Nash equilibrium solutions. In *Proceedings of the 42nd IEEE International Conference on Decision and Control*, pages 98–103. 2003.
211. M. Huang, P.E. Caines, and R.P. Malhamé. Large population stochastic dynamic games: closed-loop McKean-Vlasov systems and the Nash certainty equivalence principle. *Communications in Information and Systems*, 6:221–252, 2006.
212. M. Huang, P.E. Caines, and R.P. Malhamé. Large population cost coupled LQG problems with nonuniform agents: individual mass behavior and decentralized ϵ-Nash equilibria. *IEEE Transactions on Automatic Control*, 52:1560–1571, 2007.
213. M. Huang, R.P. Malhamé, and P.E. Caines. Nash equilibria for large population linear stochastic systems with weakly coupled agents. In R.P. Malhamé, E.K. Boukas, editors, *Analysis, Control and Optimization of Complex Dynamic Systems*, pages 215–252. Springer-US, 2005.
214. J. Jacod. Multivariate point processes: predictable projections, Radon-Nykodym derivatives, representation of martingales. *Zeitschrift für Wahrscheinlichkeitstheorie und verwandte. Gebiete*, 31:235–253, 1975.
215. J. Jacod. Weak and strong solutions of stochastic differential equations. *Stochastics*, 3:171–191, 1980.
216. J. Jacod and J. Mémin. Weak and strong solutions of stochastic differential equations: Existence and stability. In D. Williams, editor, *Stochastic Integrals*. Volume 851 of Lecture Notes in Mathematics, pages 169–212. Springer-Verlag Berlin Heidelberg, 1981.

217. M. Huang, P.E. Caines, and R.P. Malhamé. Social optima in mean field LQG control: centralized and decentralized strategies. *IEEE Transactions on Automatic Control*, 57(7):1736–1751, 2012.
218. X. Huang, S. Jaimungal and M. Nourian. Mean-field game strategies for optimal execution. Technical report, University of Toronto, https://papers.ssrn.com/sol3/papers.cfm?abstract_id=2578733, 2017.
219. M. Jeanblanc and Y. Le Cam. Immersion property and credit risk modelling. In F. Delbaen, M. Rásonyi, C. Stricker, editors, *Optimality and Risk-Modern Trends in Mathematical Finance*, pages 99–132. Springer-Verlag Berlin Heidelberg, 2010.
220. R. Jordan, D. Kinderlehrer, and F. Otto. The variational formulation of the Fokker-Planck equation. *SIAM Journal on Mathematical Analysis*, 29:1–17, 1998.
221. B. Jourdain, S. Meleard, and W. Woyczynski. Nonlinear SDEs driven by Lévy processes and related PDEs. *ALEA, Latin American Journal of Probability*, 4:1–29, 2008.
222. J. Kallsen and C. Kühn. Pricing derivatives of american and game type in incomplete markets. *Finance and Stochastics*, 8:261–284, 2004.
223. J. Kallsen and C. Kühn. Convertible bonds: Financial derivatives of game type. In A. Kyprianou, W. Schoutens, P. Wilmott, editors, *Exotic Option Pricing and Advanced Lévy Models*, pages 277–288. John Wiley & Sons, Inc., 2005.
224. N. El Karoui and S.J. Huang. A general result of existence and uniqueness of backward stochastic differential equations. In N. El Karoui, L. Mazliak, editors, *Backward stochastic differential equations*, Research Notes in Mathematics, pages 27–36. Pitman, Longman, Harlow, 1997.
225. N. El Karoui, D.H. Nguyen, and M. Jeanblanc-Picqué. Compactification methods in the control of degenerate diffusions: existence of an optimal control. *Stochastics*, 20:169–219, 1987.
226. N. El Karoui, S. Peng, and M.C. Quenez. Backward stochastic differential equations in finance. *Mathematical Finance*, 7:1–71, 1997.
227. N. Kazamaki. *Continuous Exponential Martingales and BMO*. Volume 1579 of Lecture Notes in Mathematics. Springer-Verlag Berlin Heidelberg, 1994.
228. R.Z. Khasminskii. *Stochastic Stability of Differential Equations*. Sijthoff & Noordhoff, 1980.
229. Y. Kifer. Game options. *Finance and Stochastics*, 4:443–463, 2000.
230. J.F.C. Kingman. Uses of exchangeability. *Annals of Probability*, 6:183–197, 1978.
231. M.Y. Kitaev and V. Rykov. *Controlled Queuing Systems*. CRC Press, 1995.
232. M. Kobylanski. Backward stochastic differential equations and partial differential equations with quadratic growth. *Annals of Probability*, 28:558–602, 2000.
233. V.N. Kolokolstov. Nonlinear Markov semigroups and interacting Lévy processes. *Journal of Statistical Physics*, 126:585–642, 2007.
234. V.N. Kolokolstov. *Nonlinear Markov processes and kinetic equations*. Cambridge University Press, Cambridge, 2010.
235. V.N. Kolokolstov and A. Bensoussan. Mean-field-game model for botnet defense in cyber-security. *Applied Mathematics & Optimization*, 74:669–692, 2016.
236. V.N. Kolokolstov, J. Li, and W. Yang. Mean field games and nonlinear Markov processes. Technical report, http://arxiv.org/abs/1112.3744, 2011.
237. V.N. Kolokolstov and M. Troeva. On the mean field games with common noise and the McKean-Vlasov SPDEs. Technical report, http://arxiv.org/abs/1506.04594, 2015.
238. V.N. Kolokolstov, M. Troeva, and W. Yang. On the rate of convergence for the mean field approximation of controlled diffusions with large number of players. *Dynamic Games and Applications*, 4:208–230, 2013.
239. V.N. Kolokoltsov. Nonlinear Markov games on a finite state space (mean-field and binary interactions). *International Journal of Statistics and Probability*, 1:77–91, 2012.
240. T. Kruse and A. Popier. BSDEs with monotone generator driven by Brownian and Poisson noises in a general filtration. *Stochastics*, 88:491–539, 2016.
241. P. Krusell and Jr. A. Smith. Income and wealth heterogeneity in the macroeconomy. *Journal of Political Economy*, 106:867–896, 1998.

242. N. Krylov. *Controlled Diffusion Processes*. Stochastic Modelling and Applied Probability. Springer-Verlag Berlin Heidelberg, 1980.
243. H. Kunita. *Stochastic Flows and Stochastic Differential Equations*. Cambridge Studies in Advanced Mathematics. Cambridge University Press, 1990.
244. H. Kunita and S. Watanabe. On square integrable martingales. *Nagoya Mathematical Journal*, 30:209–245, 1967.
245. T.G. Kurtz. Random time changes and convergence in distribution under the Meyer-Zheng conditions. *Annals of Applied Probability*, 19:1010–1034, 1991.
246. T.G. Kurtz. The Yamada-Watanabe-Engelbert theorem for general stochastic equations and inequalities. *Electronic Journal of Probability*, 12:951–965, 2007.
247. T.G. Kurtz. Weak and strong solutions of general stochastic models. *Electronic Journal of Probability*, 19:1–16, 2014.
248. T.G. Kurtz and P. Protter. Weak limit theorems for stochastic integrals and stochastic differential equations. *Annals of Probability*, 19:1035–1070, 1991.
249. T.G. Kurtz and J. Xiong. Particle representations for a class of nonlinear SPDEs. *Stochastic Processes and their Applications*, 83(1):103–126, 1999.
250. T.G. Kurtz and J. Xiong. A stochastic evolution equation arising from the fluctuations of a class of interacting particle systems. *Communications in Mathematical Sciences*, 2(3):325–358, 2004.
251. A. Lachapelle, J.M. Lasry, C.A. Lehalle, and P.L. Lions. Efficiency of the price formation process in the presence of high frequency participants: a mean field games analysis. *Mathematics and Financial Economics*, 10:223–262, 2016.
252. A. Lachapelle, J. Salomon, and G. Turinici. Computation of mean field equilibria in economics. *Mathematical Models and Methods in Applied Sciences*, 20:567–588, 2010.
253. A. Lachapelle and M.T. Wolfram. On a mean field game approach modeling congestion and aversion in pedestrian crowds. *Transportation Research Part B: Methodological*, 45:1572–1589, 2011.
254. D. Lacker. Mean field games via controlled martingale problems: Existence of markovian equilibria. *Stochastic Processes and their Applications*, 125:2856–2894, 2015.
255. D. Lacker. A general characterization of the mean field limit for stochastic differential games. *Probability Theory and Related Fields*, 165:581–648, 2016.
256. D. Lacker. Limit theory for controlled McKean-Vlasov dynamics. http://arxiv.org/1609.08064, 2016.
257. D. Lacker and K. Webster. Translation invariant mean field games with common noise. *Electronic Communications in Probability*, 20, 2015.
258. O.A. Ladyzenskaja, V.A. Solonnikov, and N. N. Ural'ceva. *Linear and Quasi-linear Equations of Parabolic Type*. Translations of Mathematical Monographs. American Mathematical Society, 1968.
259. J.M. Lasry and P.L. Lions. A remark on regularization in Hilbert spaces. *Israël Journal of Mathematics*, 55, 1986.
260. J.M. Lasry and P.L. Lions. Jeux à champ moyen I. Le cas stationnaire. *Comptes Rendus de l'Académie des Sciences de Paris, ser. I*, 343:619–625, 2006.
261. J.M. Lasry and P.L. Lions. Jeux à champ moyen II. Horizon fini et contrôle optimal. *Comptes Rendus de l'Académie des Sciences de Paris, ser. I*, 343:679–684, 2006.
262. J.M. Lasry and P.L. Lions. Mean field games. *Japanese Journal of Mathematics*, 2:229–260, 2007.
263. M. Laurière and O. Pironneau. Dynamic programming for mean field type control. *Comptes Rendus Mathematique, ser. I*, 352:707–713, 2014.
264. G M Lieberman. *Second Order Parabolic Differential Equations*. World Scientific, 1996.
265. P.L. Lions. Théorie des jeux à champs moyen et applications. Lectures at the Collège de France. http://www.college-de-france.fr/default/EN/all/equ_der/cours_et_seminaires.htm, 2007–2008.

266. P.L. Lions. Estimées nouvelles pour les équations quasilinéaires. Seminar in Applied Mathematics at the Collège de France. http://www.college-de-france.fr/site/pierre-louis-lions/seminar-2014-11-14-11h15.htm, 2014.
267. P.L. Lions and A.S. Sznitman. Stochastic differential equations with reflecting boundary conditions. *Communications on Pure and Applied Mathematics*, 37:511–537, 1984.
268. K. Lye and J. Wing. Game strategies in network security. *International Journal on Information Security*, 4:71–86, 2005.
269. J. Komlós M. Ajtai and G. Tusnàdy. On optimal matchings. *Combinatorica*, 4:259–264, 1983.
270. Y. Hu M. Fuhrman and G. Tessitore. Ergodic BSDEs and optimal ergodic control in Banach spaces. *SIAM Journal on Control and Optimization*, 48:1542–1566, 2009.
271. J. Ma, P. Protter, and J. Yong. Solving forward-backward stochastic differential equations explicitly – a four step scheme. *Probability Theory and Related Fields*, 98:339–359, 1994.
272. J. Ma, Z. Wu, D. Zhang, and J. Zhang. On well-posedness of forward-backward SDEs - a unified approach. *Annals of Applied Probability*, 25:2168–2214, 2015.
273. J. Ma, H. Yin, and J. Zhang. On non-Markovian forward backward SDEs and backward stochastic PDEs. *Stochastic Processes and their Applications*, 122:3980–4004, 2012.
274. J. Ma and J. Yong. *Forward-Backward Stochastic Differential Equations and their Applications*. Volume 1702 of Lecture Notes in Mathematics. Springer-Verlag Berlin Heidelberg, 2007.
275. J. Ma and J. Zhang. Path regularity for solutions of backward stochastic differential equations. *Probability Theory and Related Fields*, 122:163–190, 2002.
276. B. Maslowski and P. Veverka. Sufficient stochastic maximum principle for discounted control problem. *Applied Mathematics & Optimization*, 70:225–252, 2014.
277. H.P. McKean. A class of Markov processes associated with nonlinear parabolic equations. *Proceedings of the National Academy of Science*, 56:1907–1911, 1966.
278. H.P. McKean. Propagation of chaos for a class of nonlinear parabolic equations. *Lecture Series in Differential Equations*, 7:41–57, 1967.
279. S. Méléard. Asymptotic behaviour of some interacting particle systems; McKean-Vlasov and Boltzmann models. In D. Talay, L. Denis, L. Tubaro, editors, *Probabilistic Models for Nonlinear Partial Differential Equations*. Volume 1627 of Lecture Notes in Mathematics, pages 42–95. Springer-Verlag Berlin Heidelberg, 1996.
280. P.-A. Meyer and Zheng. Tightness criteria for laws of semimartingales. *Annales de l'Institut Henri Poincaré, Probabilités et Statistiques*, 20:353–372, 1984.
281. T. Meyer-Brandis, B. Oksendal, and X.Y. Zhou. A mean field stochastic maximum principle via Malliavin calculus. *Stochastics*, 84:643–666, 2012.
282. T. Mikami. Markov marginal problems and their applications to Markov optimal control. In W.M. McEneaney, G.G. Yin, Q., Zhang, editors, *Stochastic Analysis, Control, Optimization and Applications, A Volume in Honor of W.H. Fleming*, pages 457–476. Boston, Birkhäuser, 1999.
283. T. Mikami. Monge's problem with a quadratic cost by the zero-noise limit of h-path processes. *Probability Theory and Related Fields*, 29:245–260, 2004.
284. P. Milgrom and J. Roberts. Rationalizability, learning, and equilibrium in games with strategic complementarities. *Econometrica*, 58:1255–1277, 1990.
285. M. Nourian, P.E. Caines, and R.P. Malhamé. Mean field analysis of controlled Cucker-Smale type flocking: Linear analysis and perturbation equations. In S. Bittanti, editor, *Proceedings of the 18th IFAC World Congress, Milan, August 2011*, pages 4471–4476. Curran Associates, Inc., 2011.
286. S. Morris and H.S. Shin. Unique equilibrium in a model of self-fulfilling currency attacks. *American Economic Review*, 88:587–597, 1998.
287. M. Safonov N. Krylov. An estimate for the probability of a diffusion process hitting a set of positive measure. *Doklady Akademii nauk SSSR*, 245:18–20, 1979.
288. J. Nash. Equilibrium points in n-person games. *Proceedings of the National Academy of Sciences of the USA*, 36:48–49, 1950.
289. J. Nash. Non-cooperative games. *Annals of Mathematics*, 54:286–295, 1951.

290. K. Nguyen, T. Alpcan, and T. Basar. Stochastic games for security in networks with interdependent nodes. In *Proceedings of the 2009 International Conference on Game Theory for Networks, 13–15 May, 2009, Istanbul* pages 697–703, 2009.
291. S. Nguyen and M. Huang. Linear-quadratic-Gaussian mixed games with continuum-parametrized minor players. *SIAM Journal on Control and Optimization*, 50:2907–2937, 2012.
292. S. Nguyen and M. Huang. Mean field LQG games with mass behavior responsive to a major player. In *Proceedings of the 51st IEEE Conference on Decision and Control*, pages 5792–5797, 2012.
293. M. Nourian and P. Caines. ϵ-Nash mean field game theory for nonlinear stochastic dynamical systems with major and minor agents. *SIAM Journal on Control and Optimization*, 51:3302–3331, 2013.
294. M. Nutz. A mean field game of optimal stopping. Technical report, Columbia University, https://arxiv.org/abs/1605.09112, 2016.
295. F. Otto. The geometry of dissipative evolution equations: the porous medium equation. *Communications in Partial Differential Equations*, 26:101–174, 2001.
296. E. Pardoux. Homogenization of linear and semilinear second order parabolic PDEs with periodic coefficients: A probabilistic approach. *Journal of Functional Analysis*, 167:469–520, 1999.
297. E. Pardoux and S. Peng. Adapted solution of a backward stochastic differential equation. *Systems & Control Letters*, 14:55–61, 1990.
298. E. Pardoux and S. Peng. Backward SDEs and quasilinear PDEs. In B. L. Rozovskii and R. B. Sowers, editors, *Stochastic Partial Differential Equations and Their Applications*. Volume 176 of Lecture Notes in Control and Information Sciences. Springer-Verlag Berlin Heidelberg, 1992.
299. E. Pardoux and A. Răşcanu. *Stochastic Differential Equations, Backward SDEs, Partial Differential Equations*. Stochastic Modelling and Applied Probability. Springer International Publishing, 2014.
300. E. Pardoux and S. Tang. Forward-backward stochastic differential equations and quasilinear parabolic PDEs. *Probability Theory and Related Fields*, 114:123–150, 1999.
301. K.R. Parthasarathy. *Probability on Metric Spaces*. Chelsea AMS Publishing, 1967.
302. S. Peng. A general stochastic maximum principle for optimal control problems. *SIAM Journal on Control and Optimization*, 2:966–979, 1990.
303. S. Peng. Probabilistic interpretation for systems of quasilinear parabolic partial differential equations. *Stochastics and Stochastics Reports*, 37:61–74, 1991.
304. S. Peng. A generalized dynamic programming principle and Hamilton-Jacobi-Bellman equation. *Stochastics and Stochastics Reports*, 38:119–134, 1992.
305. S. Peng. Stochastic Hamilton Jacobi Bellman equations. *SIAM Journal on Control and Optimization*, 30:284–304, 1992.
306. S. Peng and Y. Shi. Infinite horizon forward-backward stochastic differential equations. *Stochastic Processes and their Applications*, 85:75–92, 2000.
307. S. Peng and Z. Wu. Fully coupled forward-backward stochastic differential equations and applications to optimal control. *SIAM Journal on Control and Optimization*, 37:825–843, 1999.
308. J.P. Penot. *Calculus Without Derivatives*. Graduate Texts in Mathematics. Springer-Verlag New York, 2012.
309. H. Pham. On some recent aspects of stochastic control and their applications. *Probability Surveys*, 2:506–549, 2005.
310. H. Pham. *Continuous-time Stochastic Control and Optimization with Financial Applications*. Stochastic Modelling and Applied Probability. Springer-Verlag Berlin Heidelberg, 2009.
311. H. Pham and X. Wei. Bellman equation and viscosity solutions for mean field stochastic control problem. *ESAIM: Control, Optimisation and Calculus of Variations*, to appear.
312. H. Pham and X. Wei. Dynamic programming for optimal control of stochastic McKean-Vlasov dynamics. *SIAM Journal on Control and Optimization*, 55:1069–1101, 2017.

313. E. Pimentel and V. Voskanyan. Regularity theory for second order stationary mean-field games. *Indiana University Mathematics Journal*, 66:1–22, 2017.
314. F. Priuli. Linear-quadratic N-person and mean-field games: Infinite horizon games with discounted cost and singular limits. *Dynamic Games and Applications*, 5:397–419, 2015.
315. P. Protter. *Stochastic Integration and Differential Equations. A New Approach.* Stochastic Modelling and Applied Probability. Springer-Verlag Berlin Heidelberg, 1990.
316. J. Quastel and S.R.S. Varadhan. Diffusion semigroups and diffusion processes corresponding to degenerate divergence form operators. *Communications on Pure and Applied Mathematics*, 50:667–706, 1997.
317. S.T. Rachev and L. Ruschendorf. *Mass Transportation Problems I: Theory*. Probability and Its Applications. Springer-Verlag New York, 1998.
318. M. Reed and B. Simon. *Methods of Modern Mathematical Physics. I. Functional analysis.* Academic Press, Inc. San Diego, 1980.
319. J.C. Rochet and X. Vives. Coordination failures and the lender of last resort. *Journal of the European Economic Associateion*, 2:1116–1148, 2004.
320. R. T. Rockafellar. *Convex Analysis*. Princeton University Press, 1970.
321. M. Royer. BSDEs with a random terminal time driven by a monotone generator and their links with PDEs. *Stochastics and Stochastic Reports*, 76:281–307, 2004.
322. W. Rudin. *Real and Complex Analysis*. McGraw-Hill, New York, 1966.
323. J. Schauder. Der fixpunktsatz in funktionalräumen. *Studia Mathematica*, 2:171–180, 1930.
324. Y. Sun. The exact law of large numbers via Fubini extension and characterization of insurable risks. *Journal of Economic Theory*, 126:31–69, 2006.
325. A.S. Sznitman. Topics in propagation of chaos. In P-L Hennequin, editor, *Ecole de Probabilités de Saint Flour, XIX-1989*. Volume 1464 of Lecture Notes in Mathematics, pages 165–251. Springer-Verlag Berlin Heidelberg, 1989.
326. M. Jeanblanc T. Bielecki and M. Rutkowski. Hedging of defaultable claims. In R. Carmona et al., editors, *Paris Princeton Lectures on Mathematical Finance 2003*. Volume 1847 of Lecture Notes in Mathematics, pages 1–132. Springer-Verlag Berlin Heidelberg, 2004.
327. M. Jeanblanc T. Bielecki, S. Crepey and M. Rutkowski. Arbitrage pricing of defaultable game options with applications to convertible bonds. *Quantitative Finance*, 8:795–810, 2008.
328. X. Tan and N. Touzi. Optimal transportation under controlled stochastic dynamics. *Annals of Probability*, 41:3201–3240, 2013.
329. H. Tanaka. Stochastic differential equations with reflecting boundary condition in convex regions. *Hiroshima Mathematical Journal*, 9:163–177, 1979.
330. A. Tarski. A lattice theoretic fixed point theorem. *Pacific Journal of Mathematics*, 5:285–309, 1955.
331. R. Temam. *Navier-Stokes Equations*. AMS Chelsea, 1984.
332. H. Tembine, Q. Zhu, and T. Basar. Risk-sensitive mean-field stochastic differential games. *IEEE Transactions on Automatic Control*, 59:835–850, 2014.
333. R. Tevzadze. Solvability of backward stochastic differential equations with quadratic growth. *Stochastic Processes and their Applications*, 118:503–515, 2008.
334. N. Touzi. *Optimal Stochastic Control, Stochastic Target Problems, and Backward SDE*. Fields Institute Monographs. Springer-Verlag New York, 2012.
335. J. Vaillancourt. On the existence of random McKean-Vlasov limits for triangular arrays of exchangeable diffusions. *Stochastic Analysis and Applications*, 6(4):431–446, 1988.
336. A. Y. Veretennikov. Strong solutions and explicit formulas for solutions of stochastic integral equations. *Matematicheskii Sbornik*, 111:434–452, 1980.
337. C. Villani. *Topics in Optimal Transportation*. Graduate Studies in Mathematics. American Mathematical Society, 2003.
338. C. Villani. *Optimal Transport, Old and New*. Grundlehren der mathematischen Wissenschaften. Springer-Verlag Berlin Heidelberg, 2009.
339. X. Vives. Nash equilibrium with strategic complementarities. *Journal of Mathematical Economics*, 19:305–321, 1990.

340. H. Xing and G. Žitković. A class of globally solvable Markovian quadratic BSDE systems and applications. *The Annals of Probability*, to appear.
341. J. Yong. Linear forward backward stochastic differential equations. *Applied Mathematics & Optimization*, 39:93–119, 1999.
342. J. Yong. Linear forward backward stochastic differential equations with random coefficients. *Probability Theory and Related Fields*, 135:53–83, 2006.
343. J. Yong and X. Zhou. *Stochastic Controls: Hamiltonian Systems and HJB Equations*. Stochastic Modelling and Applied Probability. Springer-Verlag New York, 1999.
344. L.C. Young. *Calculus of variations and control theory*. W.B. Saunders, Philadelphia, 1969.
345. E. Zeidler. *Nonlinear Functional Analysis and its Applications I: Fixed-Point Theorems*. Springer-Verlag New York, 1986.
346. A. K. Zvonkin. A transformation of the phase space of a diffusion process that will remove the drift. *Matematicheskii Sbornik*, 93:129–149, 1974.

Assumption Index

Note: Page numbers in *italics* refer to *Volume I* and those in Roman refer to Volume II

R. Carmona, F. Delarue, *Probabilistic Theory of Mean Field Games with Applications II*, Probability Theory and Stochastic Modelling 84,
DOI 10.1007/978-3-319-56436-4

Notation Index

Note: Page numbers in *italics* refer to *Volume I* and those in Roman refer to Volume II

Symbols

R. Carmona, F. Delarue, *Probabilistic Theory of Mean Field Games with Applications II*, Probability Theory and Stochastic Modelling 84,
DOI 10.1007/978-3-319-56436-4

Author Index

Note: Page numbers in *italics* refer to *Volume I* and those in Roman refer to Volume II

R. Carmona, F. Delarue, *Probabilistic Theory of Mean Field Games with Applications II*, Probability Theory and Stochastic Modelling 84,
DOI 10.1007/978-3-319-56436-4

Subject Index

Note: Page numbers in *italics* refer to *Volume I* and those in Roman refer to Volume II

R. Carmona, F. Delarue, *Probabilistic Theory of Mean Field Games with Applications II*, Probability Theory and Stochastic Modelling 84,
DOI 10.1007/978-3-319-56436-4

The manufacturer's authorised representative in the EU is Springer Nature Customer Service Centre GmbH, Europaplatz 3, 69115 Heidelberg, Germany. If you have any concerns regarding our products, please contact ProductSafety@springernature.com

Printed and bound by CPI Group (UK) Ltd, Croydon, CR0 4YY
06/10/2026
02228868-0008